Elemental Geosystems

Combination of satellite *Terra* MODIS sensor image and *GOES* satellite image produces a true-color view of North and South America from 35,000 km (22,000 mi) in space. [Courtesy of MODIS Science Team, Goddard Space Flight Center, NASA and NOAA.]

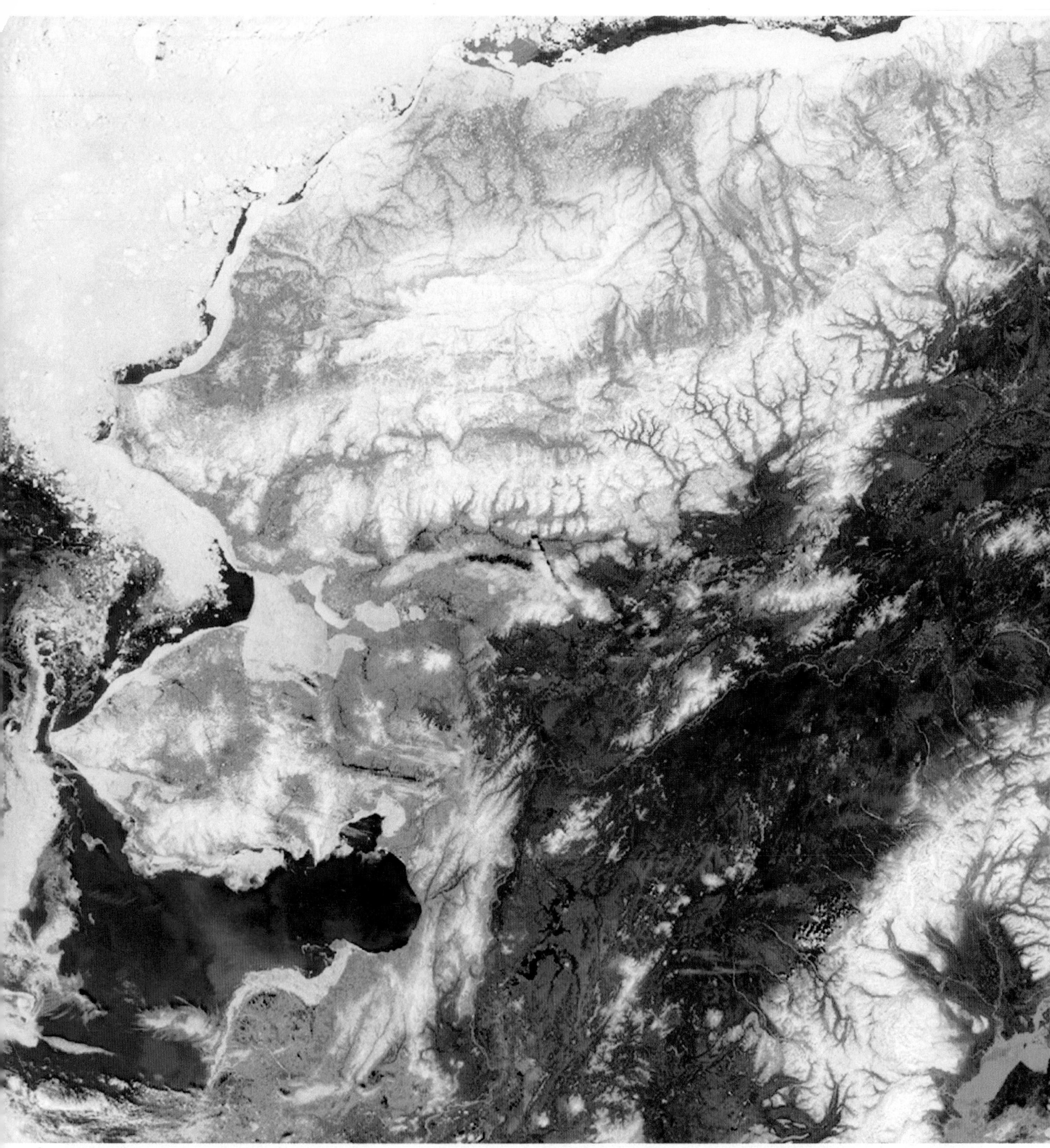

Spring 2002 arrives in Alaska, from the Alaska Range in the south to the Brooks Range in the north. Mount McKinley and Denali National Park are slightly right of center. South of the Alaska Range, Cook Inlet is loaded with sediment from runoff. Glacial ice in Alaska is undergoing a major meltdown as temperatures increase. The Columbia Glacier (east of Cook Inlet) retreated 12 km and lost 400 m in thickness between 1982 and 2000. A study of 67 glaciers estimated that Alaska's melt was contributing more than twice the volume of water to sea-level rise than Greenland's glacial melt water. News Report 14.1 discusses these profound changes in Alaska. [*Terra* MODIS image, May 21, 2002, courtesy of MODIS Land Rapid Response Team NASA/GSFC.]

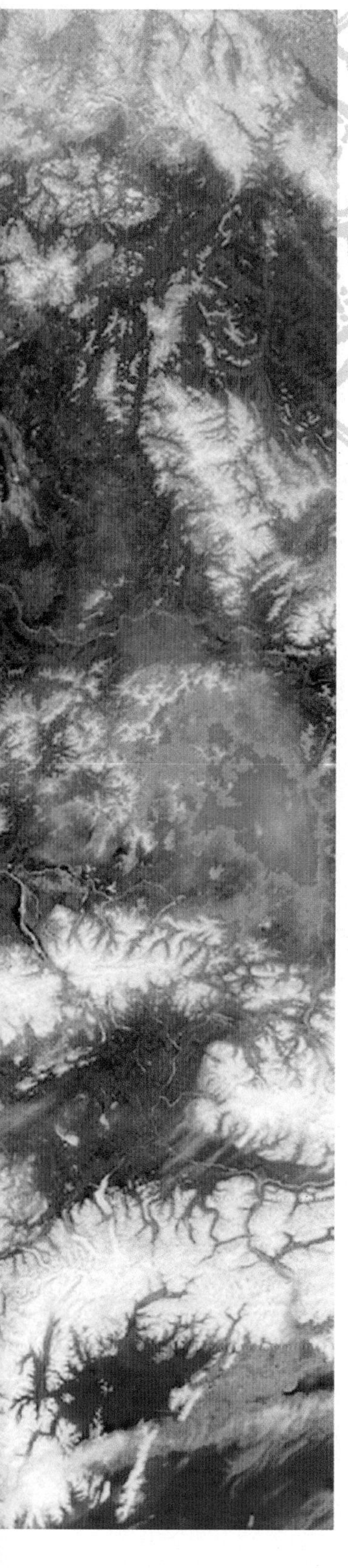

Elemental Geosystems

Fourth Edition

Robert W. Christopherson

Pearson Education, Inc.
Upper Saddle River, New Jersey 07458

Library of Congress Cataloging-in-Publication Data

Christopherson, Robert W.
Elemental Geosystems / Robert W. Christopherson.—4th ed.
p. cm.
Includes bibliographical references (p.).
ISBN 0-13-101553-2 (pbk.)
1. Physical geography. I. Title.

GB54.5.C47 2004
910'.02—dc21 2003042993

Geoscience Executive Editor: *Daniel E. Kaveney*
Associate Editor: *Amanda Grifffith*
Editor in Chief, Life and Geosciences: *Sheri L. Snavely*
Vice President of Production and Manufacturing: *David W. Riccardi*
Executive Managing Editor: *Kathleen Schiaparelli*
Assistant Managing Editor: *Beth Sweeten*
Production Editor: *Kim Dellas*
Copy Editor: *Marcia Youngman*
Manager of Formatting: *Jim Sullivan*
Electronic Production Specialist/ Electronic Page Makeup: *Joanne Del Ben*
Production Assistant: *Nancy Bauer*
Production Assistant to the Author: *Bobbé Christopherson*
Manufacturing Buyer: *Alan Fischer*
Manufacturing Manager: *Trudy Pisciotti*
Senior Marketing Manager: *Christine Henry*
Media Editor: *Chris Rapp*
Assistant Managing Editor, Science Media: *Nicole Bush*
Editorial Assistant: *Margaret Ziegler*
Art Director: *Kenny Beck*
Interior Designer: *Joseph Sengotta*
Cover Designer: *Bruce Kenselaar*
AV Editor: *Adam Velthaus*
Art Studio: *Precision Graphics*
Director of Creative Services: *Paul Belfanti*
Director of Design: *Carole Anson*
Front Cover Photo: Tibetan landscape, Mount Kangbochen (7295 m, 23,933 ft). [Galen Rowell/Mountain Light Photography, Inc.]
Inside Front-Cover Image: [MODIS Science Team, Goddard Space Flight Center, NASA and NOAA.]
Title Page Image: *Terra* MODIS image, May 21, 2002, courtesy of MODIS Land Rapid Response Team NASA/GSFC; see http://terra.nasa.gov/.
Dedication page quote: B. Kingsolver, *Small Wonder* (New York: Harper Collins Publishers, 2002), p. 39.

Pearson Education, Inc.
Upper Saddle River, New Jersey 07458

Printed in the United States of America
10 9 8 7 6 5 4 3 2

ISBN 0-13-101553-2

Pearson Education LTD., *London*
Pearson Education Australia PTY, Limited, *Sydney*
Pearson Education Singapore, Pte. Ltd
Pearson Education North Asia Ltd, *Hong Kong*
Pearson Education Canada, Ltd., *Toronto*
Pearson Educación de Mexico, S.A. de C.V.
Pearson Education—Japan, *Tokyo*
Pearson Education Malaysia, Pte. Ltd

To all the students and teachers of Earth,
our home planet, and a sustainable future.

And to the late Galen and Barbara Cushman Rowell
for their vision of the world in photographs and words.

The land still provides our genesis, however we
might like to forget that our food comes from dank,
muddy earth, that the oxygen in our lungs was
recently inside a leaf, and that every newspaper or
book we may pick up is made from the hearts of
trees that died for the sake of our imagined lives.
What you hold in your hands right now,
beneath these words,
is consecrated air and time and sunlight. . . .

—Barbara Kingsolver

Brief Contents

Contents

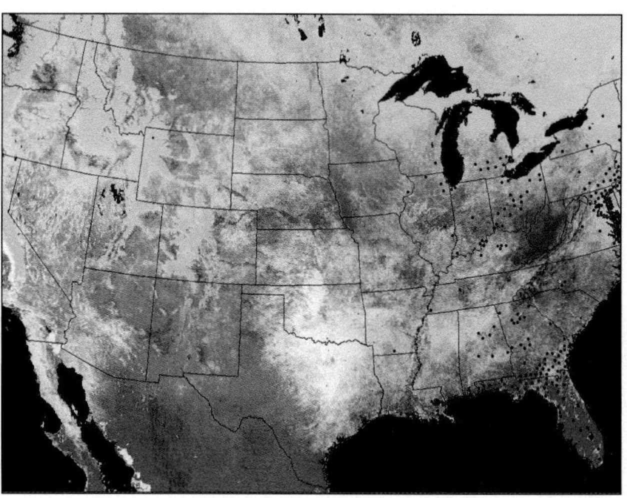

7 Water Resources 221

PART 3 Earth's Changing Landscape Systems 247

8 The Dynamic Planet 249

9 Earthquakes and Volcanoes 277

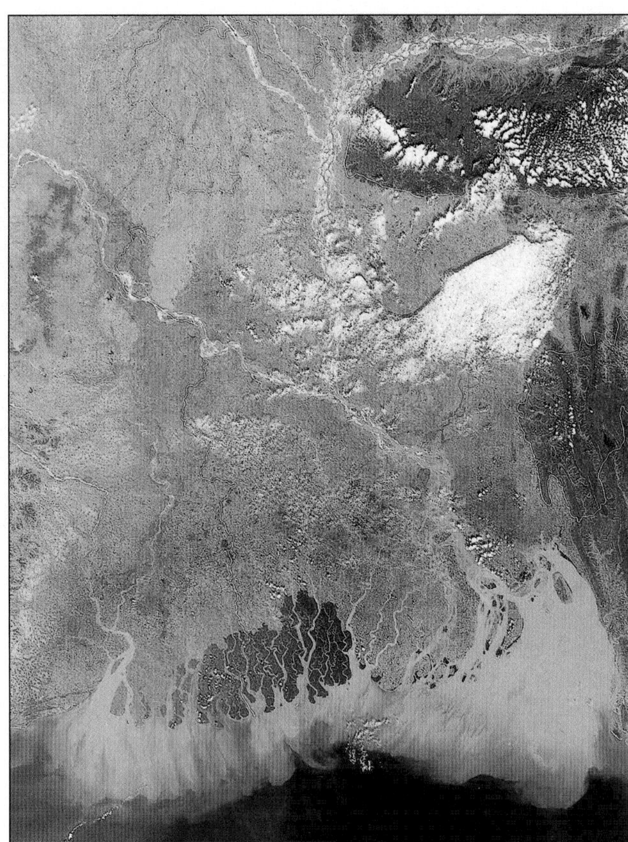

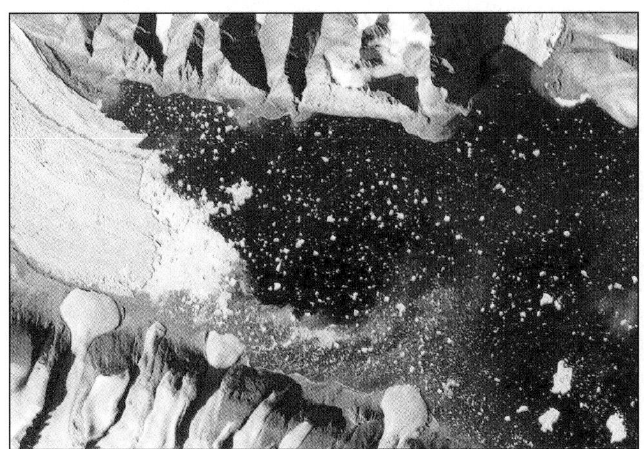

Preface

Welcome to physical geography and the fourth edition of *Elemental Geosystems*. This book builds on the previous three editions and on the five editions of its companion text *Geosystems, An Introduction to Physical Geography*. Students and teachers alike continue to express appreciation for the systems organization, readability, scientific accuracy, up-to-date coverage and relevancy, clarity of the summary and review sections, the functional beauty of the photographs, art, cartography, and the many integrated figures in the text that combine media.

The world community is responding to global concerns over the condition of Earth's physical, biological, and chemical systems. The globalization of the world economies seems paralleled by a global scientific inquiry into the state of the environment.

Armed with the spatial analysis tools of geographic science, physical geographers are well equipped to participate in a planetary understanding of environmental conditions. U.N. Secretary General Kofi Annan, recipient of the 2001 Nobel Peace Prize, spoke to the Association of American Geographers annual meeting, stating,

> As you know only too well the signs of severe environmental distress are all around us.... The idea of interdependence is old hat to geographers, but for most people it is a new garment they are only now trying on for size.... I look forward to working with you in that all-important journey.

Elemental Geosystems Communicates the Science of Physical Geography

The goal of physical geography is to explain the spatial dimension of Earth's dynamic systems—its energy, air, water, weather, climate, tectonics, landforms, rocks, soils, plants, ecosystems, and biomes. Understanding human-Earth relations is part of the challenge of physical geography—to create a holistic (or complete) view of the planet and its inhabitants.

Elemental Geosystems analyzes the worldwide impact of environmental events, bringing together many physical factors to create a complete picture of Earth system operations. A good example is the 1991 eruption of Mount Pinatubo in the Philippines. The global implications of this major event (one of the largest eruptions in the 20th century) are woven through several chapters of the book (see Figure 1.5 for a summary). Global climate change and its related potential effects are part of the fabric in six chapters. These content threads, among many, weave together diverse topics crucial to an understanding of physical geography.

This edition of *Elemental Geosystems* features more than 450 photographs from across the globe and 90 remote-sensing images from a wide variety of orbital platforms. To assist with spatial analysis and location, more than 100 maps are utilized, and more than 250 illustrations explain concepts. New compound arts help you see the concepts you are studying. Here are two examples: Focus Study 13.1, art and 7 photos look at coastal planning; Figure 16.19, satellite image, map, and 10 photos examine succession in the area blasted by Mount St. Helens.

Systems Organization Makes *Elemental Geosystems* Flow

Each section of this book is organized around the flow of energy, materials, and information. *Elemental Geosystems* presents subjects in the same sequence in which they occur in nature. In this way you and your teacher logically progress through topics that unfold according to the flow of individual systems, or in accord with time and the flow of events. See Figure 1.6 for an illustration of this systems organization.

For flexibility, *Elemental Geosystems* is divided into four parts, each containing chapters that link content in logical groupings. The diagram below, from Figure 1.7 illustrates our part structure. A quick check of the Table of Contents and this illustration shows you the order of chapters within these four parts.

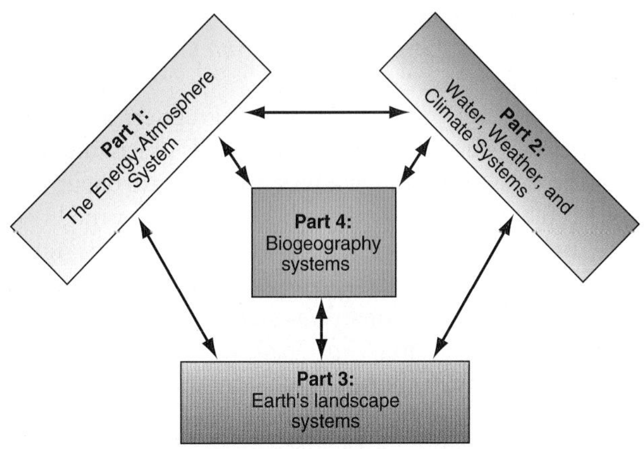

The text culminates with Chapter 17, "Earth and the Human Denominator," a unique capstone chapter that summarizes physical geography as an important discipline to help us understand Earth's present status and possible future. Think of the world's population and the totality of our impact as the *human denominator*. Just as the denominator in a fraction tells how many parts a whole is divided into, so the growing human population and the increasing demand for resources and rising planetary impact suggest how much the whole Earth system must adjust. This chapter is sure to stimulate further thought and discussion, dealing as it does with the most profound issue of our time, Earth's stewardship.

Elemental Geosystems Is a Text That Teaches

Elemental Geosystems is written to help you in the learning process. **Boldface** words are defined where they first appear in the text. These terms are collected in the Glossary alphabetically, with a chapter-number reference. *Italics* are used in the text to emphasize other words and phrases of importance. Also, in the introduction to each chapter, a new feature called "In this chapter" gives you an overview.

An important continuing feature is a list of Key Learning Concepts that opens each chapter, stating what you should be able to do upon completing the chapter. These objectives are keyed to the main headings in the chapter. At the end of each chapter is a unique Summary and Review section that corresponds to the Key Learning Concepts. Grouped under each learning concept are a narrative review that defines the boldfaced terms, a key term list with page numbers, and specific review questions for that concept. You can conveniently review each concept, test your understanding with review questions, check key terms in the glossary, and then return to the chapter and the next learning concept.

A Critical Thinking section ends each chapter, challenging you to take the next step with information from the chapter. *Key learning concepts* help you determine what you want to learn, the *text* helps you develop information and more questions, *summary and review* helps you assess what you have learned and what more you might want to know about the subject, and *critical thinking* provokes action and application.

"Career Link" essays feature geographers and other scientists in a variety of professional fields practicing their spatial analysis craft. You will read about an astronaut with over 1200 hours in orbit, a weather forecaster at the Forecast Systems Lab, a park ranger who works at the active lava flows in Hawai'i, an environmental scientist, a hydrologist with the National Weather Service, a snow avalanche specialist, and an expert on global scale ecosystems, among others.

Continued coverage of Canadian physical geography includes text, figures, and maps of periglacial landscapes and Canadian soils (see Appendix B and color soils map). Canadian data on a variety of subjects are portrayed on dozens of different maps in combination with the data for the United States—physical geography does not stop at the United States-Canadian border!

"Focus Study" essays, completely revised and several new to this edition, provide additional explanation of diverse key topics: the stratospheric ozone predicament, solar energy collection and wind power, forecasting near-record hurricane seasons, the El Niño phenomenon, the status of the High Plains Aquifer, the record low flows of the Colorado River, floodplain strategies, an environmental approach to shoreline planning, the Mount St. Helens eruption, the present status of the Colorado River, and the loss of biodiversity.

"News Reports" relate topics of special interest. These are some examples: GPS, careers in GIS, a 34-kilometer sky dive to study the atmosphere, jet streams and airline flight times, how one culture harvests fog, the UV Index, coordination of global climate change research with many URLs presented, the new height measurements of Mount Everest, artificial scouring of the Grand Canyon in an attempt to restore beaches and habitats, the record glacial meltdown in Alaska, and the Gaia hypothesis.

We now live on a planet served by the Internet and its World Wide Web. The fact that we have Internet access into almost all the compartments aboard Spaceship Earth is clearly evident in *Elemental Geosystems*. You will find more than 180 URLs (Internet addresses) in the body of the text (printed in blue color and boldface). Given the fluid nature of the Internet, URLs were rechecked at press time for accuracy. If some URLs changed since publication, you can most likely find the new location using elements of the old address. Our Internet link begins with a new Table 1.1 presenting the URLs for major geography organizations.

The Geosystems Learning/Teaching Package

The fourth edition provides a complete physical geography program for you and your teacher.

For You the Student:

- ***Elemental Geosystems Student Animations CD***, a CD-ROM by Robert Christopherson is packaged with each copy of the text. This exciting CD contains 30 new animations illustrating key concepts in the text, along with support material from the text for every animation. The CD also contains numerous satellite loops of various phenomena. Instantly graded self-tests follow the animations, with pop-up details to reinforce correct answers and your learning. See the text's Walkthrough preview for more details.
- ***Student Study Guide***, Fourth Edition (0-13-101554-0), by Robert Christopherson and Charlie Thomsen. The study guide includes additional

learning objectives, a complete chapter outline, critical thinking exercises, problems and short essay work using actual figures from the text, and a self-test with answer key in the back.

- *Geosystems WWW Site*: This site gives you the opportunity to further explore topics presented in the book using the Internet. The site contains numerous review exercises (from which you get immediate feedback), exercises to expand your understanding of physical geography, and resources for further exploration. This Web site provides an excellent opportunity from which to start using the Internet for the study of geography. Please visit the site at **www.prenhall.com/christopherson**.
- ***Science on the Internet: A Student's Guide*** (0-13-028253-7) by Andrew T. Stull and Harry Nickla is a guide to the Internet and World Wide Web specific to geography.
- ***Prentice Hall–New York Times Themes of the Times*** supplements, especially Physical Geography (0-13-142636-2), and also Environmental Science (0-13-142638-9), reprint significant recent articles on related topics. These are available at no charge from your local Prentice Hall representative; ask your teacher about a copy.

For You the Teacher:

Elemental Geosystems is designed to give you flexibility in presenting your course. The text is true to each scientific discipline from which it draws subject matter. This diversity is a strength of physical geography, yet makes it difficult to cover the entire book in a school term. You should feel free to customize use of the text based on your specialty or emphasis. The four-part structure of chapters, systems organization within each chapter, and focus study and news report features will all assist you in sampling some chapters while covering others to greater depth. The following materials are available to assist you—have a great class!

- ***Instructor's Resource Manual***, Fourth Edition (0-13-101566-4), by Robert Christopherson and Charlie Thomsen: The Instructor's Resource Manual, intended as a resource for both new and experienced teachers, includes lecture outlines and key terms, additional source materials, teaching tips, complete annotation of chapter review questions, and a list of overhead transparencies. This is also available on the IRCD.
- ***Instructor's Resource CD-ROM***, Fourth Edition (0-13-101558-3): This edition contains all of the figures and some of the photographs from the text. Images are high-resolution, low compression in high quality digital form. The software makes customizing your multimedia presentations easy. You can organize figures in any order you want, add labels, lines, and your own artwork to them using an overlay tool, integrate materials from other sources, and edit and annotate lecture notes. New to this edition are the Geoscience Animations and customizable ***PowerPoint*** lecture presentations, ready for classroom.
- ***Test Item File*** (0-13-101559-1): The Test Item File contains many test questions drawn from the book, available in printed format. This is also available on the IRCD.
- ***Test Manager Geosystems Test Bank*** (0-13-101557-5), by Robert Christopherson and Charlie Thomsen: This collaboration has produced the most extensive and fully revised test item file available in physical geography. This test bank employs **TestGen-EQ** software. **TestGen-EQ** is a computerized test generator that lets you view and edit test bank questions, transfer questions to tests, and print customized formats. Included is the **QuizMaster-EQ** program that lets you administer tests on a computer network, record student scores, and print diagnostic reports. Mac and IBM/DOS computer formats are served.
- ***Overhead Transparencies*** (0-13-101556-7) includes more than 300 illustrations from the text on 300 transparencies, all enlarged for excellent classroom visibility.
- ***Applied Physical Geography—Geosystems in the Laboratory***, Fifth Edition (0-13-034823-6), by Robert Christopherson and Gail Hobbs of Pierce College; Reviewer comments and the feedback from users were very positive for the third edition. The new fifth edition is the result of a careful revision. Twenty lab exercises, divided into logical sections, allow flexibility in presentation. Each exercise comes with a list of learning concepts. Our manual is the only one that comes with its own complete glossary and stereolenses for viewing photo stereopairs and stereomaps in the manual. A complete ***Solutions and Answers Manual*** is available to teachers (0-13-034815-5).

Acknowledgments

As in all past editions, I recognize my family, for they continue to grant us support in reaching our *Geosystems* goals—both our moms, my sister Lynne, brothers Randy and Marty, and our children Keri, Matt, Reneé, and Steve. And the next generation: Chavon, Bryce, Payton, Brock, Trevor, Blake, and our newest Chase. When I look into our grandchildren's faces, it tells me why we need to work toward a sustainable future, one that works for the children.

I give special gratitude to all the students and colleagues during my 29 years at American River College for defining the importance of Earth's future, for their questions, and their enthusiasm. To all students and teachers this text remains dedicated. The tragic loss of Galen and Barbara Rowell further signals the importance of life.

They contributed so much to our world view and to our appreciation and awe at the foot of Earth's wonders—I add them to our dedication. One of Galen's photos is on our cover.

My thanks go to the many authors and scientists who published research, articles, and books that enriched my work. And, although unnamed here, to all the correspondence received from students and teachers from across the globe who shared with me over the Internet, e-mail, FAX, and phone—a continuing appreciated dialogue. Thanks also to all the colleagues who served as reviewers on one or more editions, who participated in our focus groups, or who offered helpful suggestions at our national and regional geography meetings. I am grateful to all of them for their generosity of ideas and sacrifice of time. Here is a master list of all our reviewers.

Ted J. Alsop, *Utah State University*
Ward Barrett, *University of Minnesota*
David Berner, *Normandale Community College*
Franco Biondi, *University of Nevada, Reno*
Peter D. Blanken, *University of Colorado–Boulder*
David R. Butler, *Southwest Texas State University*
Mary-Louise Byrne, *Wilfrid Laurier University*
Ian A. Campbell, *University of Alberta–Edmonton*
Randall S. Cerveny, *Arizona State University*
Fred Chambers, *University of Colorado–Boulder*
Muncel Chang, *Butte College and California State University–Chico*
Andrew Comrie, *University of Arizona*
C. Mark Cowell, *Indiana State University*
Richard A. Crooker, *Kutztown University*
Armando M. da Silva, *Towson State University*
Dirk H. de Boer, *University of Saskatchewan*
Mario P. Delisio, *Boise State University*
Joseph R. Desloges, *University of Toronto*
Lee R. Dexter, *Northern Arizona University*
Don W. Duckson, Jr., *Frostburg State University*
Christopher H. Exline, *University of Nevada–Reno*
Michael M. Folsom, *Eastern Washington University*
Mark Francek, *Central Michigan University*
Glen Fredlund, *University of Wisconsin–Milwaukee*
David E. Greenland, *University of North Carolina–Chapel Hill*
Duane Griffin, *Bucknell University*
Barry N. Haack, *George Mason University*
Roy Haggerty, *Oregon State University*
John W. Hall, *Louisiana State University–Shreveport*
Vern Harnapp, *University of Akron*
Jason "Jake" Haugland, *University of Colorado–Boulder*
Gail Hobbs, *Pierce College*
David A. Howarth, *University of Louisville*
Patricia G. Humbertson, *Youngstown State University*
David W. Icenogle, *Auburn University*
Philip L. Jackson, *Oregon State University*
J. Peter Johnson, Jr., *Carleton University*
Guy King, *California State University–Chico*
Ronald G. Knapp, *SUNY–The College at New Paltz*
Peter W. Knightes, *Central Texas College*
Thomas Krabacher, *California State University–Sacramento*
Richard Kurzhals, *Grand Rapids Junior College*
Steve Ladochy, *California State University, Los Angeles*
Robert D. Larson, *Southwest Texas State University*
Paul R. Larson, *Southern Utah University*
Elena Lioubimtseva, *Grand Valley State University*
Joyce Lundberg, *Carleton University*
W. Andrew Marcus, *Montana State University*
Elliot G. McIntire, *California State University, Northridge*
Norman Meek, *California State University, San Bernardino*
Sherry Morea-Oaks, *Boulder, CO*
Patrick Moss, *University of Wisconsin–Madison*
Lawrence C. Nkemdirim, *University of Calgary*
Andrew Oliphant, *San Francisco State University*
John E. Oliver, *Indiana State University*
Bradley M. Opdyke, *Michigan State University*
Richard L. Orndorff, *University of Nevada, Las Vegas*
Patrick Pease, *East Carolina University*
James Penn, *Southeastern Louisiana University*
Greg Pope, *Montclair State University*
Robin J. Rapai, *University of North Dakota*
Philip D. Renner, *American River College*
William C. Rense, *Shippensburg University*
Dar Roberts, *University of California–Santa Barbara*
Wolf Roder, *University of Cincinnati*
Robert Rohli, *Louisiana State University*
Bill Russell, *L.A. Pierce College*
Dorothy Sack, *Ohio University*
Glenn R. Sebastian, *University of South Alabama*
Daniel A. Selwa, *U.S.C. Coastal Carolina College*
Thomas W. Small, *Frostburg State University*
Daniel J. Smith, *University of Victoria*
Stephen J. Stadler, *Oklahoma State University*
Michael Talbot, *Pima Community College*
Susanna T.Y. Tong, *University of Cincinnati*
Suzanne Traub-Metlay, *Front Range Community College*
David Weide, *University of Nevada–Las Vegas*
Thomas B. Williams, *Western Illinois University*
Brenton M. Yarnal, *Pennsylvania State University*
Stephen R. Yool, *University of Arizona*

I extend my continuing gratitude to the editorial, production, and sales staff of Prentice Hall. Thanks to ESM President Paul Corey for his leadership and friendship from the beginnings of the *Geosystems* books. Thanks to Dan Kaveney, Geosciences Executive Editor, who is dedicated, innovative, and energetic as a geographer and friend. I feel these two executives are like my brothers in this effort, for they are positive forces in these texts and in geographic education. My compliments to a talented and wise Amanda Griffith who managed this project and all my ancillaries, and was always there with answers. My appreciation to Chris Rapp for expertise on all the CD projects and his help and guidance on the dramatic new animations CD. And thanks to Ginger Birkeland for work on the *Geosystems* Web site. Thanks to Margaret Ziegler's organizational skills in the Geosciences office. And to all the staff for allowing me to participate in the entire publishing process.

My thanks to Kim Dellas, project Production Editor, for such expertise and care in converting all my materials into this textbook and to copy editor Marcia Youngman for such detail. Thanks to the art and design team at Prentice Hall for this powerful cover and beautiful text design and for letting me in on many decisions. The many sales representatives that spend months in the field communicating the *Elemental Geosystems* approach are a tremendous asset to the book, thanks and safe travels to them, always.

As on every book I mention my partnership with a special collaborator, photographer, and production assistant, Bobbé Christopherson. She works tirelessly on all the textbook projects. She catalogues photographs, prepares our extensive figure and photo logs, processes and prints satellite imagery and photos, copy edits, obtains permissions, and assists me in proofing art and editing final pages. Bobbé is an activist in Earth matters. Her eclectic interests embrace birds, invertebrate marine biology, plants and wildflowers, the Hawaiian volcanoes, and her nature photography. And she is my best friend, wife, and colleague.

Physical geography teaches us a holistic view of the intricate supporting web that is Earth's environment and our place in it. Dramatic changes that demand our understanding are occurring in many human-Earth relations, as we alter physical, chemical, and biological systems. All things considered, this is a critical time for you to be enrolled in a physical geography course! The best to you in your studies—and, as always, ***carpe diem!***

Robert W. Christopherson
P. O. Box 128
Lincoln, California 95648-0128
E-mail: bobobbe@aol.com
Web site: **http://www.prenhall.com/christopherson**

HALLMARK FEATURES

Elemental Geosystems, 4e features a strong emphasis on the *science* of physical geography, coupled with a focus on the student's learning process. To facilitate this approach, the book focuses on four main areas: organization, currency, writing, and art.

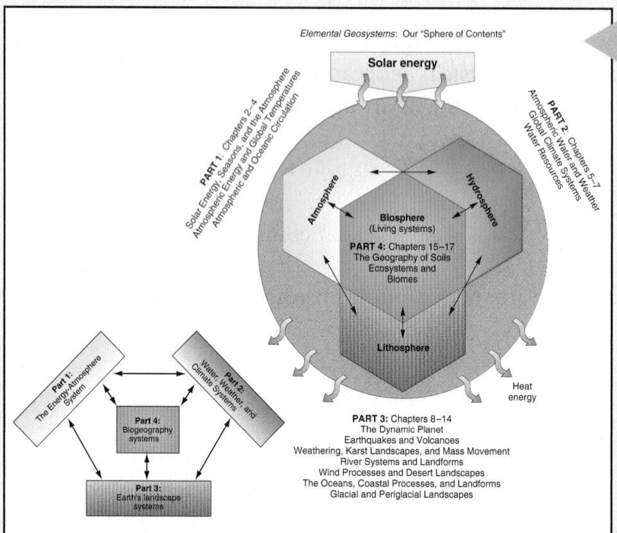

UNIQUE ORGANIZATION AND FLOW OF TOPICS

Elemental Geosystems four-part organization and chapter sequence treats subjects as process systems, essentially nonmathematical, with text material organized in the same direction as the flow of energy and matter in the environment. This edition, as in its previous editions, has served to update the field and bring physical geography into play as an important Earth systems science.

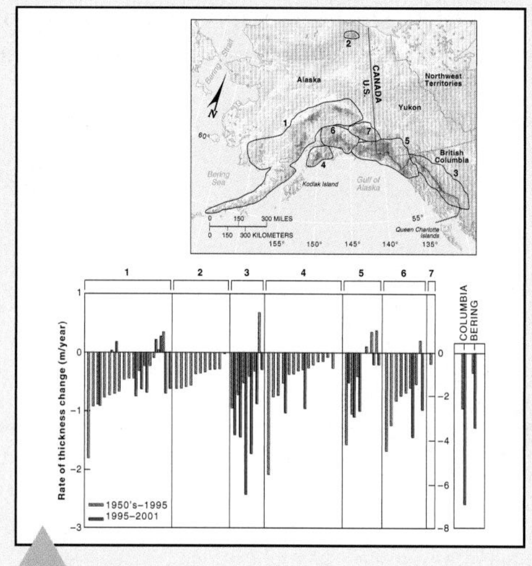

WRITING AND PEDAGOGY

The book features a strong writing and pedagogical program. The writing program is characterized by numerous interesting examples for students, by a strong focus on physical geography's applicability to understanding current events, and vignettes that highlight some interesting careers in physical geography. The pedagogical program includes extensive study materials at the beginning and end of every chapter, a study guide written by Robert W. Christopherson, and a free online study guide available to all students.

CURRENCY

This text is the most up-to-date physical geography text available. Currency has always been a hallmark of ***Elemental Geosystems*** and this edition is no exception. A few examples of the author's use of currency include: a new study about the negative net mass balance in Alaskan glaciers, the latest weather technology (including ASOS and AWIPS), cirrus cloud studies following the airline 3-day shutdown after 9/11/01 events, and agenda of the 2002 Earth Summit as it relates to physical geography.

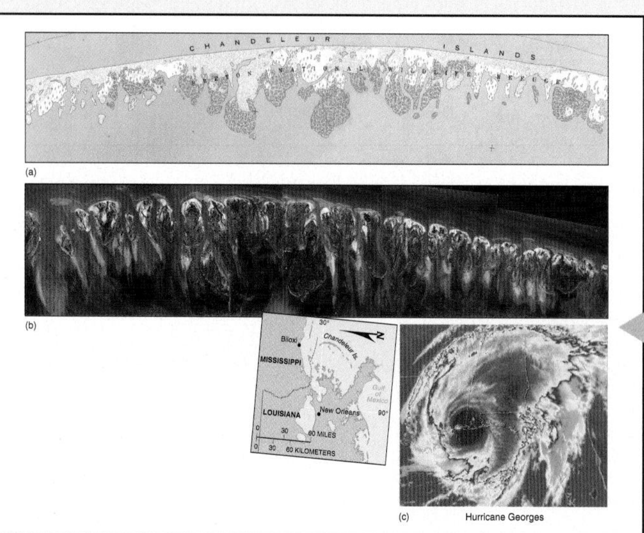

ART AND CARTOGRAPHY

Elemental Geosystems highlights the field of physical geography through the extensive use of maps, photographs, satellite images, and line drawings. The book's signature technique is to combine maps with other forms of visual media to give a clearer and more complete picture of the concept illustrated.

NEW TO THIS EDITION

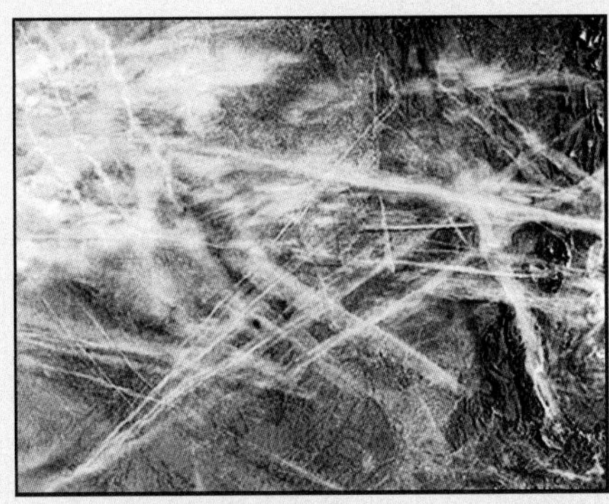

Satellite image of jet contrails triggering cloud development

THOROUGHLY REVISED AND UPDATED THROUGHOUT

Currency has always been a hallmark of ***Elemental Geosystems,*** and this edition has been thoroughly updated. Much of the updated material draws from events and studies from as recent as 2002. A few examples include a new study about the negative net mass balance in Alaskan glaciers, the latest weather technology (including ASOS and AWIPS), cirrus cloud studies following the airline 3-day shutdown after 9/11/01 events, and especially low Colorado River flows, 2000 - 2002 as it relates to physical geography.

NEW STUDENT ANIMATION CD-ROM

The new student CD-ROM is packaged automatically with each copy of the text and features 30 animations of key concepts in physical geography. These animations are interactive and will greatly augment classroom presentation and student studies. The author has taken the Prentice Hall Geoscience animations and added pedagogy to help students understand and learn more efficiently about physical processes. This CD-ROM is a unique product to the market, and referenced throughout the text by CD, Satellite Loop, and Notebook icons.

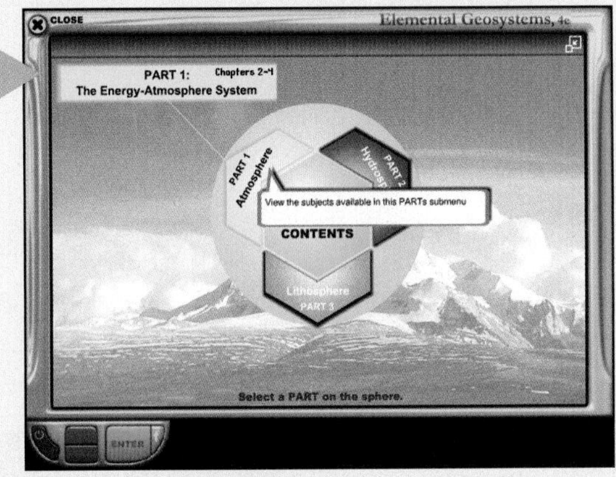

REVISED COVERAGE OF GLOBAL CLIMATE SYSTEMS

Chapter 6 Global Climate

The causes of climate are labeled on each climograph and the maps are altered to reflect cause. In this way students can see the causative factors that produce the conditions on which the station's classification is based. The previous use of the Köppen classification systems is diminished, yet retained in an Appendix for those still using the systems explicit classification criteria.

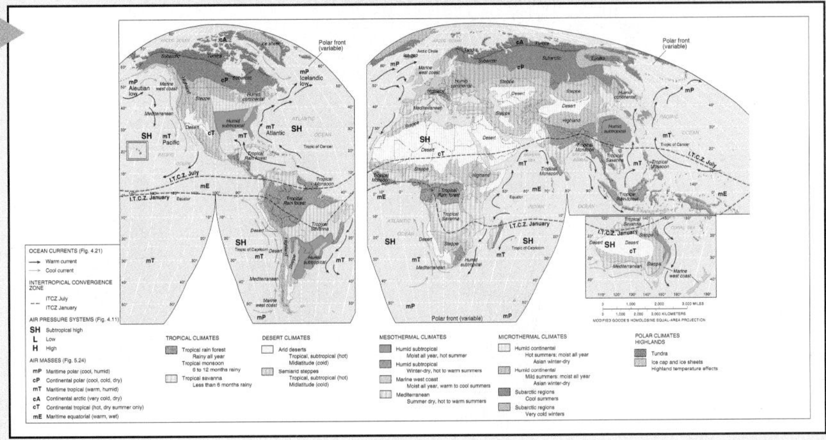

ons of Geography

Association of American Ge- 2001, offered this assessment:

l, the signs of severe envi- around us. Unsustainable ly into the fabric of mod- n threatens food security. tens biodiversity. Water c health, and fierce com- ay well become a source the future . . . .the over- ientific experts have con-

In this chapter: Our study of *Elemental Geosystems*—Earth systems—begins in this chapter with a look at the science of physical geography and the geographic tools we use. Physical geography is key to studying entire Earth systems because of its integrative approach.

Physical geographers utilize systems to study the environment. Therefore, we discuss systems and the feedback mechanisms that influence system operations. We then consider location, a key theme of geographic inquiry—the latitude, longitude, and time coordinates that inscribe Earth's surface, and the global positioning systems (GPS) technologies to measure them. The study of

"IN THIS CHAPTER"

After a brief introduction in each chapter, there is a new heading called "In This Chapter." This feature provides solid pedagogic support and provides the students a way to help summarize chapters when reviewing for exams.

SUPERIOR GRAPHICS PROGRAM

Elemental Geosystems employs a sophisticated, yet extremely accessible, system of maps, photographs, line drawings, and satellite images. The fourth edition features a refinement of this basic philosophy as well as a dramatic expansion of the remotely-sensed images employed.

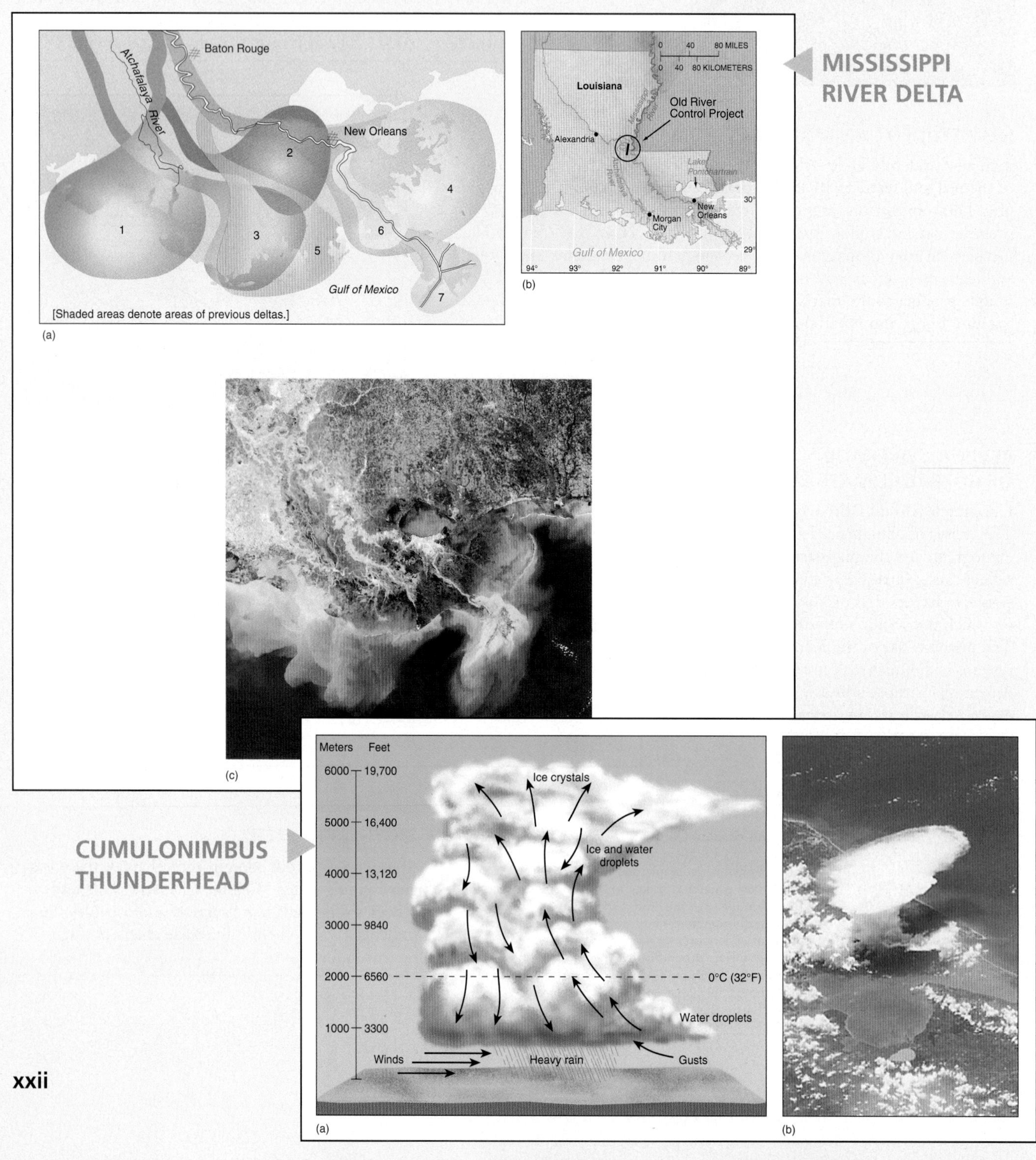

THE MOUNT ST. HELENS ERUPTION SEQUENCE AND CORRESPONDING SCHEMATICS

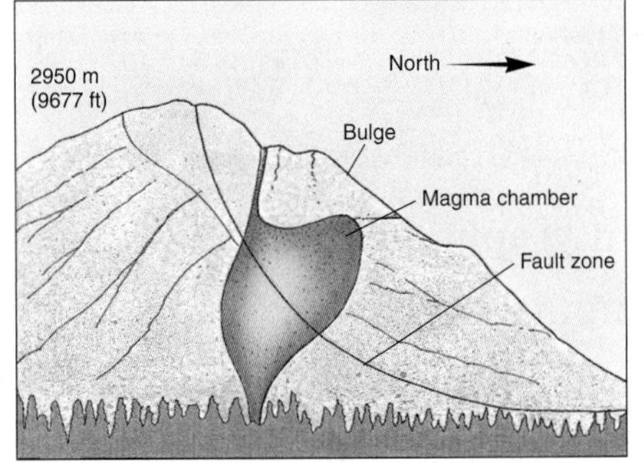

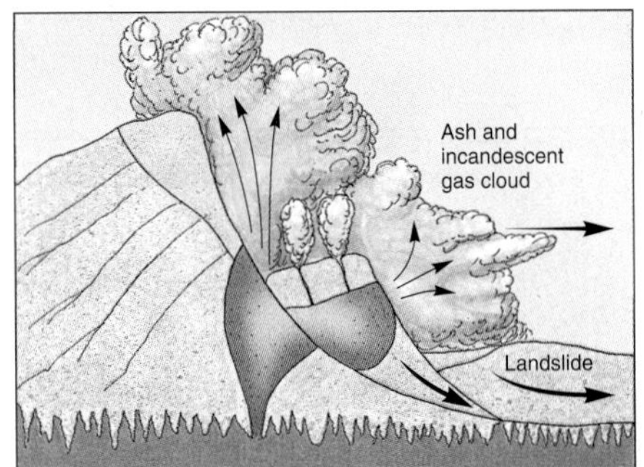

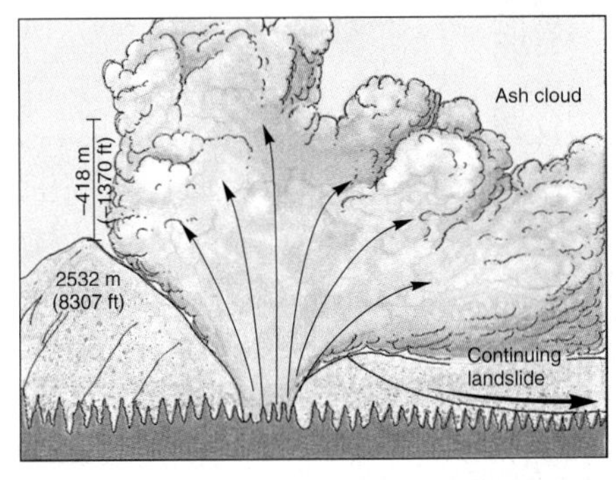

MASS BALANCE AND RETREATING ALPINE GLACIER

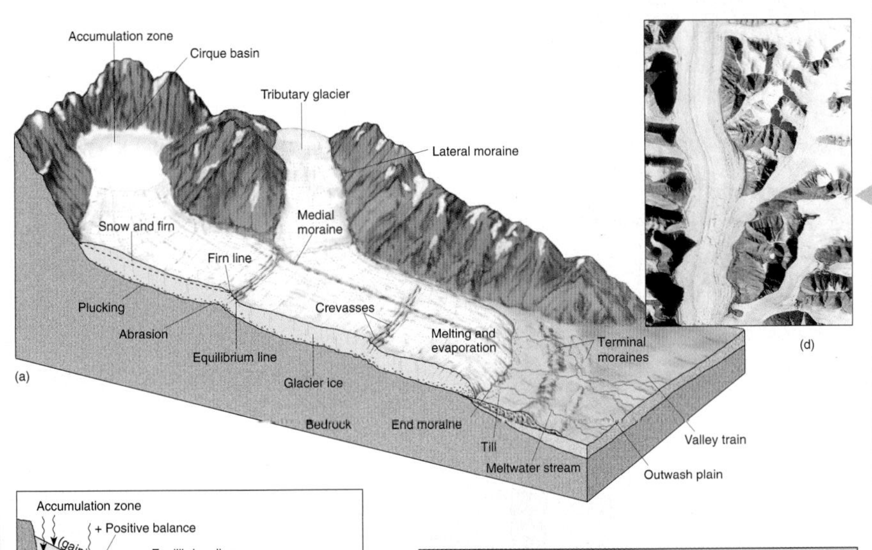

(c)

STUDENT ANIMATION CD-ROM

In keeping with the text's emphasis on graphics, the book now features animations of key concepts. The topics and concepts animated were chosen by a panel of physical geographers as the most difficult concepts and processes for students to visualize. These animations have been specifically crafted to help students learn these concepts and processes more quickly and effectively.

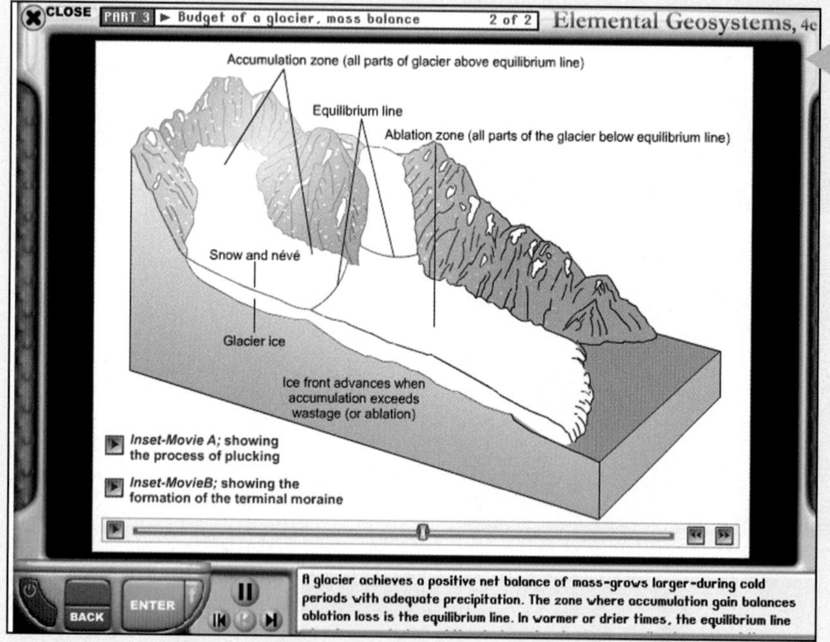

GLACIAL PROCESSES AND BUDGET

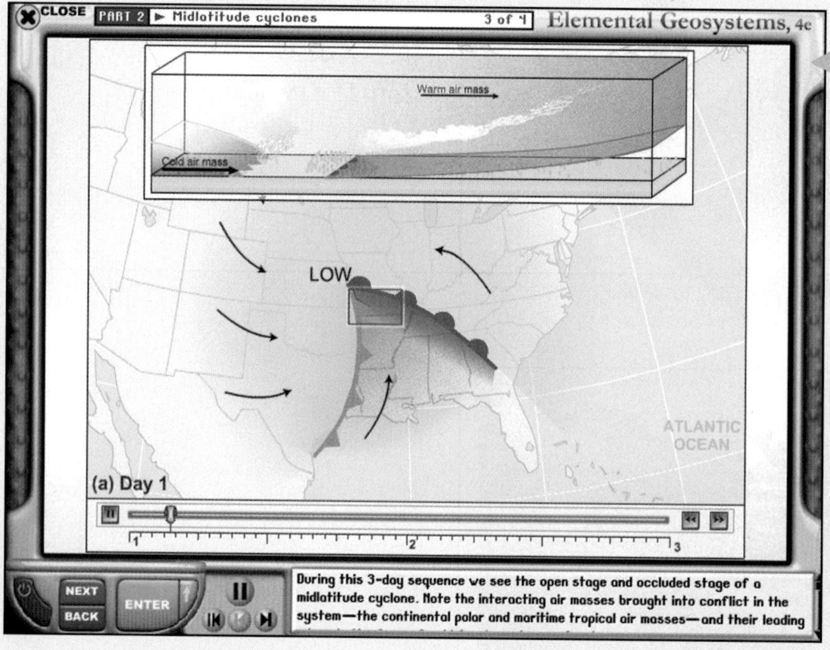

MIDLATITUDE CYCLONES

Each animation allows the student to control the action. The user can replay the animation, control the pace of the animation, and stop and start the animation anywhere in its sequence. In order to facilitate effective independent study, Robert W. Christopherson has written accompanying narrations to each animation.

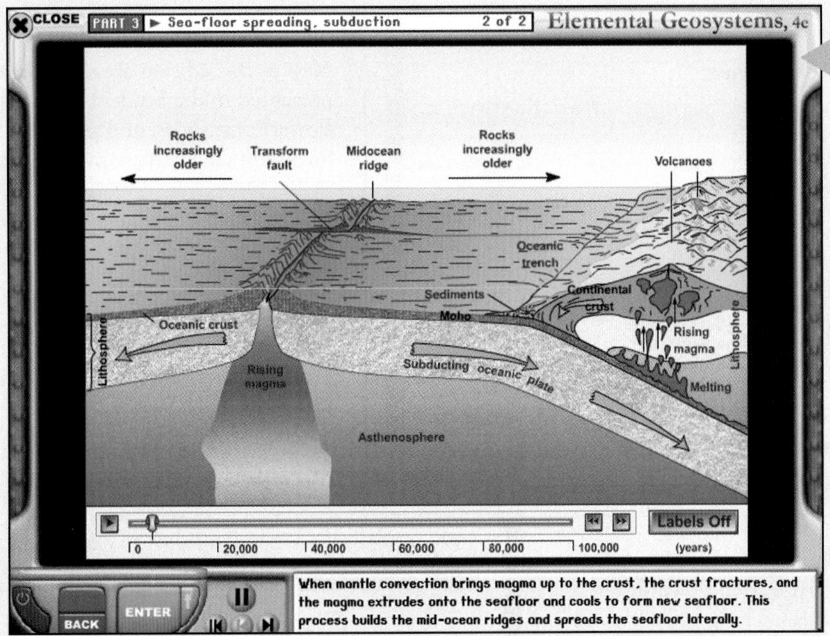

SEAFLOOR SPREADING AND PLATE BOUNDARIES

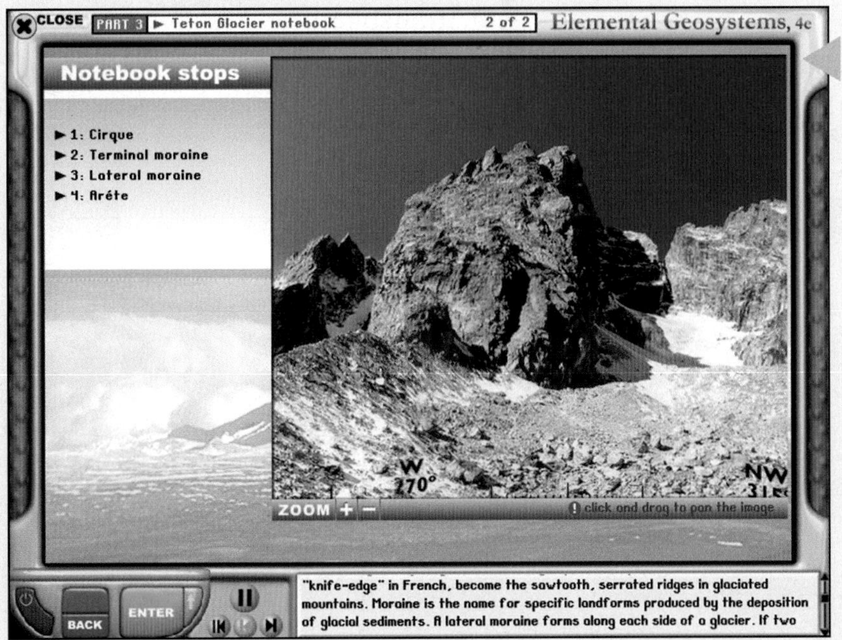

TETON GLACIER NOTEBOOK

LECTURE PRESENTATION TOOLS

The Instructor's Resource CD-ROM (0-13-101558-3) provides high-quality electronic versions of photos and illustrations from the book, as well as additional photographs and customizable PowerPoint lecture presentations.

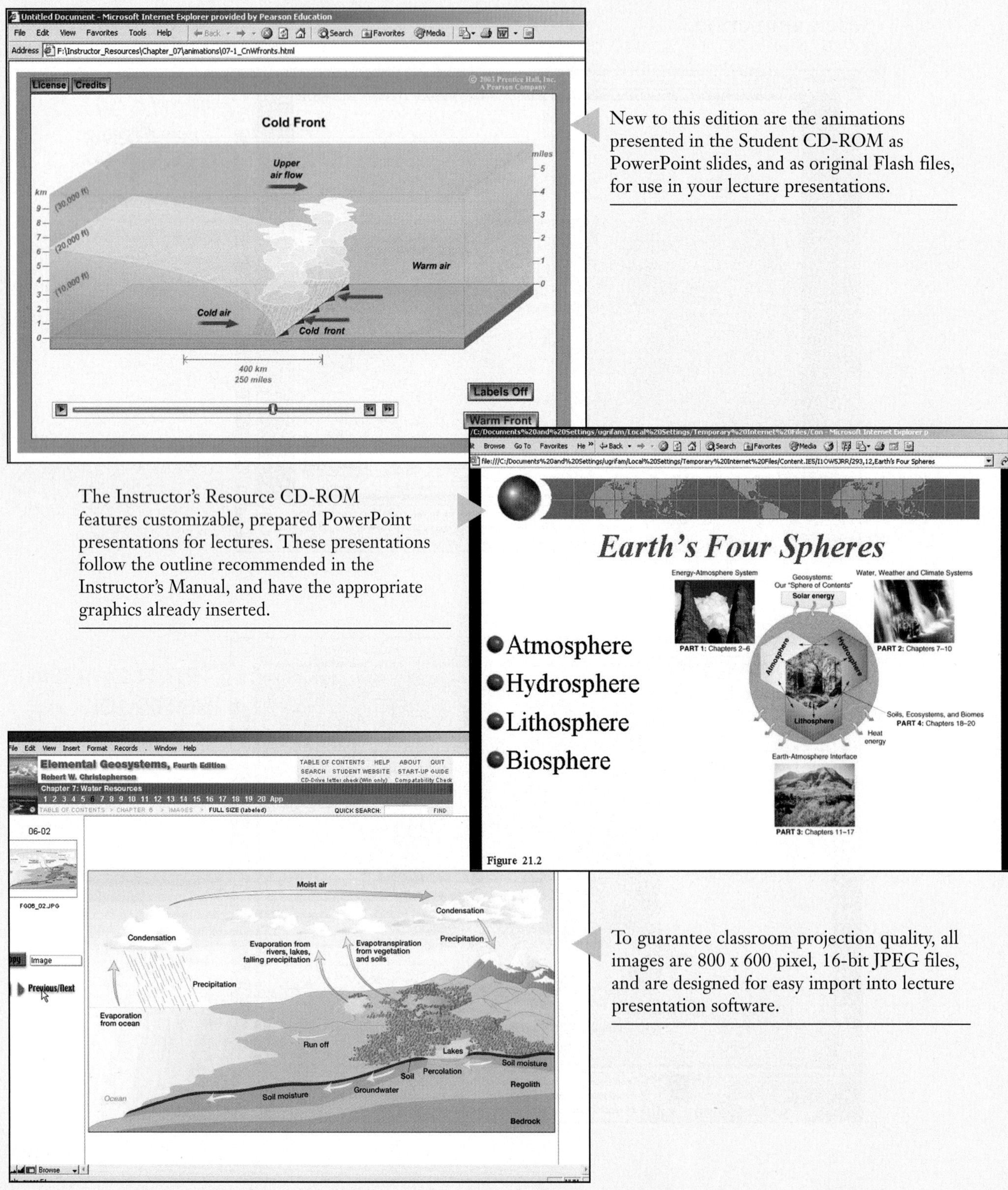

New to this edition are the animations presented in the Student CD-ROM as PowerPoint slides, and as original Flash files, for use in your lecture presentations.

The Instructor's Resource CD-ROM features customizable, prepared PowerPoint presentations for lectures. These presentations follow the outline recommended in the Instructor's Manual, and have the appropriate graphics already inserted.

To guarantee classroom projection quality, all images are 800 x 600 pixel, 16-bit JPEG files, and are designed for easy import into lecture presentation software.

INSTRUCTOR'S TEACHING PACKAGE

INSTRUCTOR'S MANUAL (0-13-101566-4)

Intended as a resource for both new and experienced instructors, the Instructor's Manual includes a variety of lecture outlines, additional source materials, teaching tips, advice about how to integrate visual supplements (including Web-based resources), and various other ideas for the classroom.

TEST GEN EQ (0-13-101557-5)

Computerized test generator that lets you view and edit test bank questions, transfer questions to tests, and print the test in a variety of customized formats. Included in each package is the QuizMaster EQ program that lets you administer tests on a computer network, record student scores, and print diagnostic reports.

TEST ITEM FILE (0-13-101559-1)

The test item file is a printed bank of test item questions, a valuable complement to an instructor's own test/quiz files.

TRANSPARENCIES (0-13-101556-7)

Dramatically expanded transparency set includes over 300 pieces of art from the text, all enlarged for excellent classroom visibility.

STUDENT LEARNING PACKAGE

NEW STUDENT CD-ROM

The new Student CD-ROM is packaged automatically with each copy of the text. Robert has taken the Prentice Hall Geoscience animations and added pedagogy to help students understand and learn more efficiently about physical processes.

STUDENT STUDY GUIDE (0-13-101554-0)

The study guide includes additional learning objectives, a complete chapter outline, critical thinking exercises, problems and short essay work using actual figures from the text, and a self-test with answer key in the back.

APPLIED PHYSICAL GEOGRAPHY – GEOSYSTEMS IN THE LABORATORY, 5E (0-13-034823-6)

The new edition is a result of careful revision and contains 20 lab exercises, divided into logical sections, allowing for flexibility in their presentation. Each exercise comes with a list of learning concepts. This is the only manual that comes with its own complete glossary and stereo lenses and stereo maps for viewing photo stereo pairs presented in the manual. A complete **Solutions Manual** is available for the instructor (0-13-034815-5).

FREE ONLINE STUDY GUIDE www.prenhall.com/christopherson

This innovative online resource center is keyed by chapter to the text. It provides Key Terms, Chapter Exercises, Additional Internet Resources, and Additional Readings to support and enhance students' study of physical geography. The CW also offers instructors access to Syllabus Manager, Prentice Hall's online syllabus creation and management tool.

OPTIONAL RESOURCES

RAND MCNALLY ATLAS OF WORLD GEOGRAPHY

The *Rand McNally Atlas of World Geography* includes more than 65 physical and political maps of the world, continents, regions, countries and vicinities. Free when packaged with the text.

To order this package use ISBN: 0-13-183751-6

THE GEOGRAPHY COLORING BOOK, 3/E BY WYNN KAPIT

The *Geography Coloring Book* is a unique educational tool that introduces students to the countries of the world and the states of the United States. Each section begins with a plate containing a political map, a physical map, and/or a regional map. Through active participation and coloring the maps, students gain a broader understanding of material and retain more information. Free when packaged with the text.

To order this package use ISBN: 0-13-183752-4

BUILDING GEOGRAPHIC LITERACY: AN INTERACTIVE APPROACH, 4/E BY CHARLES A. STANSFIELD

Innovative in its approach, this interactive workbook enables students to learn place geography and, as they learn it, to reinforce and apply their knowledge by constructing thematic maps that convey physical, economic, cultural, or political characteristics of places. Free when packaged with the text.

To order this package use ISBN: 0-13-183749-4

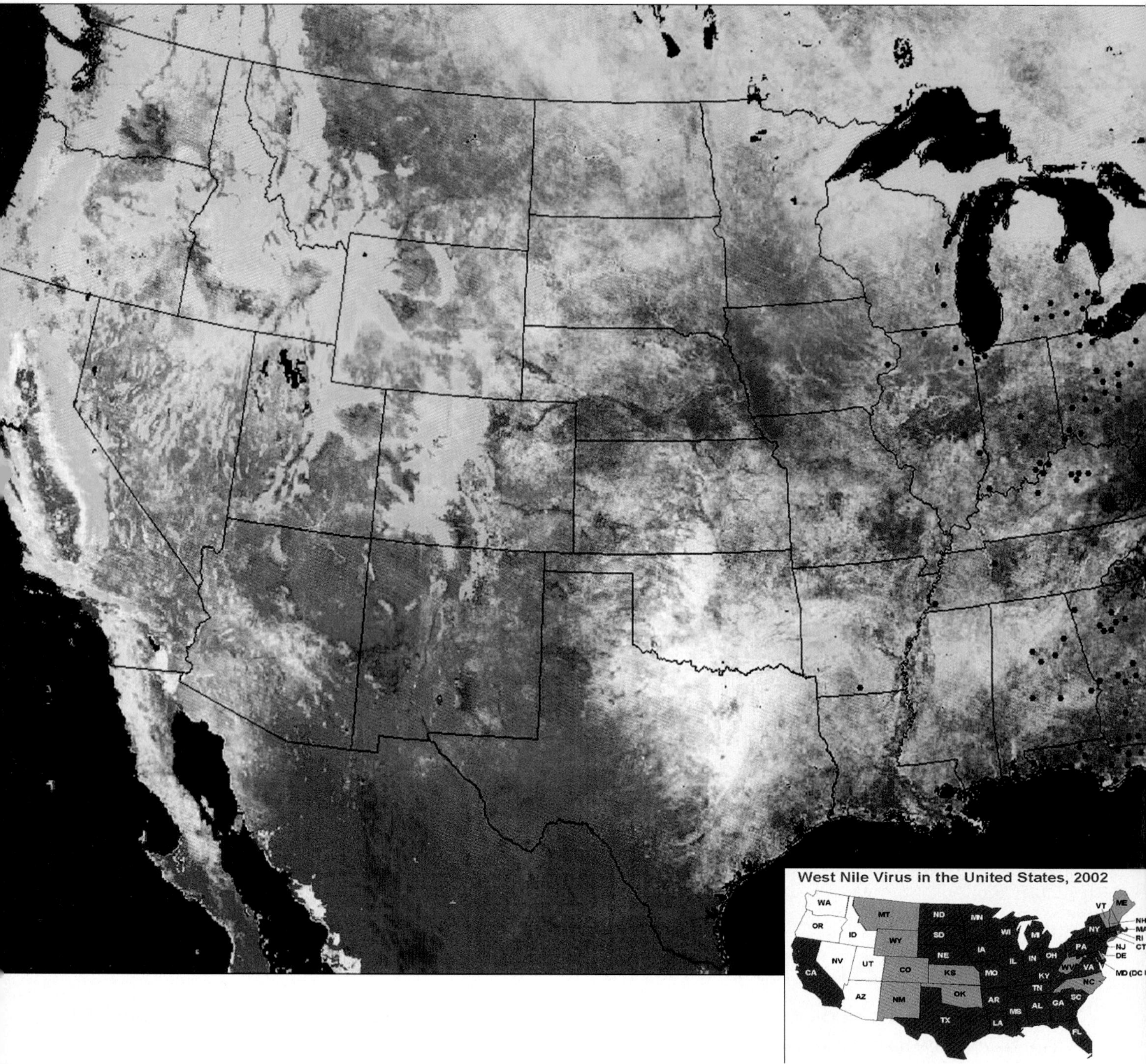

Temperature, humidity, and vegetation variables (1997–2000) help predict spread of disease. A NASA geographic information systems map uses satellite-derived temperature and vegetation (a humidity indicator) data to track and predict the spread of the West Nile virus. The map shows average surface temperatures (reds), where annual temperature range is a factor (blues), and where annual peak temperatures are key (greens)—more intense colors equal higher values. The higher temperatures, warmer winters, and higher humidity, and the earlier peak temperatures are areas where mosquito populations thrive. The unusually mild winter of 1998–99 provided the mosquito carriers favorable conditions to spread the disease. Black dots show where infected birds were found through October 2001. Thousands of wild birds have died in the Midwest, East, and South. The inset map portrays states reporting infected birds and humans through September 2002. For more information go to http://www.gsfc.nasa.gov/topstory/20020204westnile.html and http://www.cdc.gov/ncidod/dvbid/westnile/index.htm. [Composite image from *NOAA-15* AVHRR, NASA/GSFC/International Research Partnership for Infectious Disease; Inset map courtesy of Centers for Disease Control, Atlanta, GA.]

1 Foundations of Geography

Key Learning Concepts

By knowing and understanding the key learning concepts in this chapter, you should be able to:

- *Define* geography and physical geography in particular.
- *Describe* systems analysis and open and closed systems, and *relate* these concepts to Earth systems.
- *Explain* Earth's reference grid: latitude and longitude, and latitudinal geographic zones and time.
- *Define* cartography and mapping basics: map scale and map projections.
- *Describe* remote sensing and *explain* geographic information systems (GIS) as tools used in geographic analysis.

Welcome to physical geography and *Elemental Geosystems*, fourth edition. Today is a scientifically exciting and challenging time to be enrolled in a physical geography course. Physical geography deals with our environment and the powerful forces that influence our lives and the many ways we are altering Earth's systems. *Elemental Geosystems* is an assessment of the operation of Earth's physical systems.

Earth's physical and natural diversity continues to fascinate us: lush rain forests straddle the equator, stark deserts bake in the subtropics, cool moist climates dominate northwestern coastlines, and perpetual cold hugs the polar regions. Human societies should consider services received from these diverse Earth systems. This natural "work" of Earth's physical, chemical, and biological systems has an estimated worth of some US$35 trillion annually. Ironically, the investment costs to sustain these life-support systems are comparatively low.

Physical geography provides elemental information needed to understand and sustain our planetary journey in this new century. United Nations Secretary General

Kofi Annan, speaking to the Association of American Geographers annual meeting in 2001, offered this assessment:

> As you know only too well, the signs of severe environmental distress are all around us. Unsustainable practices are woven deeply into the fabric of modern life. Land degradation threatens food security. Forest destruction threatens biodiversity. Water pollution threatens public health, and fierce competition for freshwater may well become a source of conflict and wars in the future … the overwhelming majority of scientific experts have concluded that climate change is occurring, that humans are contributing, and that we cannot wait any longer to take action … environmental problems build up over time, and take an equally long time to remedy.

Scientists and governments scramble to understand the worldwide impact of global changes in weather, climate, water resources, and the landscape. Over the past several decades civilization experienced unprecedented environmental conditions: increasing air and ocean temperatures; stratospheric ozone losses; increased rates of sea-level rise with record glacial ice melt from the mountains of Alaska, Asia, Europe, and South America; earlier springs and later falls (for longer summers); heightened thunderstorm intensity; increasing plant and animal species extinctions; massive floods in Europe and elsewhere, contrasted with severe droughts in the United States, Canada, and China with related wildfires. Such unexpected change is costly. As an example, in 1998 weather-related damage topped US$90 billion, nine times greater than the previous annual average.

Why do all of these conditions occur? How are these events different from past experience? Why does the environment and rates of global change vary from equator to midlatitudes, between deserts and polar regions? How does solar energy influence the distribution of trees, soils, climates, and lifestyles? How does solar energy produce the patterns of wind, weather, and ocean currents? How do natural systems affect human populations, and, in turn, what impact are humans having on these same natural systems? In this book, we explore those questions, and more, through geography's unique perspective. Once again, welcome to an exploration of physical geography!

We live in an extraordinary era of **Earth systems science**, a science that contributes to our emerging view of Earth as a complete entity—an interacting set of physical, chemical, and biological systems that produce a whole Earth. Physical geography greatly adds to this quest with its critical *spatial* questions concerning Earth's physical systems and their interaction with living things. Physical geographers analyze interactions and changes that are occurring in natural systems to better quantify Earth systems over diverse *temporal* (time) and *spatial* (space or area) scales, and address the question of how these changes might affect life on Earth.

In this chapter: Our study of *Elemental Geosystems*—Earth systems—begins in this chapter with a look at the science of physical geography and the geographic tools we use. Physical geography is key to studying entire Earth systems because of its integrative approach.

Physical geographers utilize systems to study the environment. Therefore, we discuss systems and the feedback mechanisms that influence system operations. We then consider location, a key theme of geographic inquiry—the latitude, longitude, and time coordinates that inscribe Earth's surface, and the global positioning systems (GPS) technologies to measure them. The study of longitude and a universal time system provide us with interesting insights into our lives. Next we examine maps as critical tools that geographers use to portray physical and cultural information. This chapter concludes with an overview of the technology that is adding exciting new dimensions to geography: remote sensing from space and computer-based geographic information systems (GIS).

The Science of Geography

Geography (from *geo*, "Earth," and *graphein*, "to write,") is the science that studies the relationships among natural systems, geographic areas, society, cultural activities, and the interdependence of all these *over space*. The term **spatial** refers to the nature and character of physical space, its measurement, and the distribution of things within it. Humans are spatial actors. For example, think of your own route to the classroom or library today and how you used your knowledge of street patterns, traffic trouble spots, parking spaces, or bike rack locations to minimize walking distance.

To guide geographic education, the "National Geography Standards" for learning were prepared by the Association of American Geographers (AAG) and the National Council for Geographic Education (NCGE) in response to *Goals 2000*, the Educate America Act of 1994 (see *Geography for Life*, prepared by the Geography Education Project for the AAG, NCGE, American Geographical Society, and National Geographic Society, 1994). For more information see the listing of some important geography organizations and their URLs in Table 1.1.

We simplify these national standards of geographic science using five important spatial themes: **location**, **movement**, **place**, **region**, and **human-Earth relationships**, illustrated and defined in Figure 1.1. *Elemental Geosystems* draws on each theme.

Geographic Analysis

Within these five geographic themes, geography is governed by a *method* rather than a specific body of knowledge, and the method is **spatial analysis**. Using this method, geography synthesizes (brings together) knowledge from many fields, integrating information to form a whole Earth concept. Geographers view phenomena as occurring in spaces and areas having distinctive characteristics. The language of geography reflects

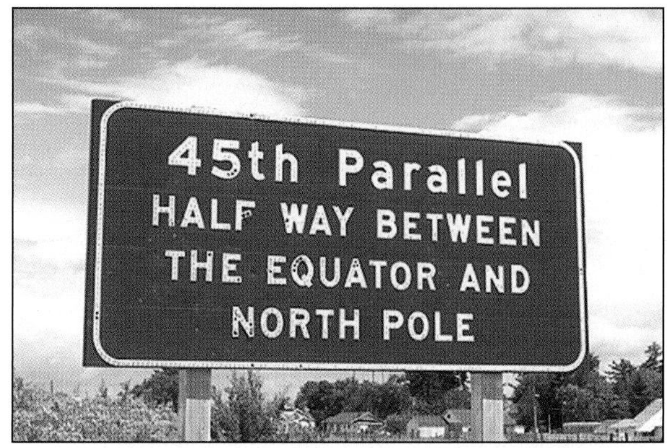

Location

Absolute and relative location on Earth. Location answers the question "*Where?*" The specific planetary address of a location. This freeway sign is posted on Interstate 5 in Oregon telling drivers their position of Earth.

Region

Areas having uniform characteristics; how they form and change; their relation to other regions. The American "South" is a distinct region in the U.S., bounded by the Gulf Coast on the south and Atlantic Ocean on the east.

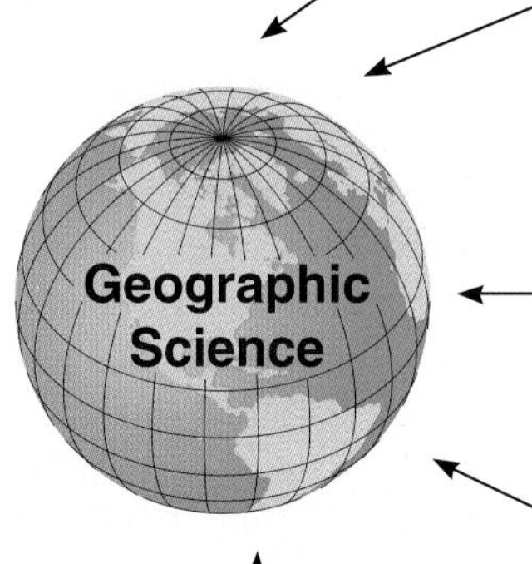

Human–Earth Relationships

Humans and the environment: resource exploitation, hazard perception, and environmental pollution and modification. The 2002 floods that struck Europe were unprecedented in history, causing many governments and people to rethink hazard assessment and planning.

Place

Tangible and intangible living and nonliving characteristics that make each place unique. No two places on Earth are exactly alike. Central Park in New York City is unique in its answer to the place of nature in the human city.

Movement

Communication, movement, circulation, and diffusion across Earth's surface. Global interdependence links all regions and places. Tropical storm Allison is a product of atmospheric circulation. On June 11, 2001, Allison assaults Alabama and Mississippi after drenching Houston with record rains—

FIGURE 1.1 Five themes of geographic science.
Five fundamental themes in geography defined—location, place, movement, regions, and Human-Earth relationships. [Place photo by Bobbé Christopherson; Region and Movement *Terra* images from MODIS Land Rapid Response Team, NASA; Human-Earth Relationships photo of Soane River in France by © Stephane Ruet/Corbis Sygma; and Location photo by author.]]

Table 1.1

A Few Geography Organizations	URL Addresses
American Geographical Society	http://www.amergeog.org/
Association of American Geographers*	http://www.aag.org/
National Council for Geographic Education	http://www.ncge.org/
National Geographic Society	http://www.nationalgeographic.com/
Canadian Association of Geographers	http://venus.uwindsor.ca/cag/cagindex.html
Royal Canadian Geographical Society	http://www.rcgs.org/
Institute of Australian Geographers	http://www.iag.org.au/
Australian Geography Teachers Association	http://www.agta.asn.au/
European Geography Association	http://egea.geog.uu.nl/
Royal Geographical Society, Institute of British Geographers	http://www.rgs.org/
Complete global listing of geography organizations	http://www.geoggeol.fau.edu/prof_org/GeographyPO.html

*Includes nine regional divisions: East Lakes, Great Plains/Rocky Mountain, Middle Atlantic, Middle States, New England—St. Lawrence Valley, Pacific Coast Regional, Southeastern, Southwestern, West Lakes.

this spatial view: *space, territory, zone, pattern, distribution, place, location, region, sphere, province,* and *distance*. Geographers analyze the differences and similarities among places and locations.

Process, a set of actions or mechanisms that operate in some special order, is central to geographic analysis. As examples in *Elemental Geosystems*, numerous processes are involved in Earth's vast water-atmosphere-weather system, in continental crust movements and earthquake occurrences, or in ecosystem functions. Geographers use spatial analysis to examine how Earth's processes interact over space or area.

Therefore, **physical geography** is the *spatial analysis of all the physical elements and processes that make up the environment: energy, air, water, weather, climate, landforms, soils, animals, plants, microorganisms, and Earth itself.* We add to this the oldest theme in the geographic tradition, that of human activity. As a science, geographers employ the **scientific method**. Focus Study 1.1 (pp. 8–9) explains this essential process of science.

The Geographic Continuum

Geography is eclectic, integrating a wide range of subject matter from diverse fields; virtually any subject can be examined geographically. Figure 1.2 shows a continuous distribution—a continuum—along which the content of geography is arranged. To the continuum's left are disciplines in the physical and life sciences; those in the human/cultural sciences are on the right. Lists at either end of the continuum feature specialties within geography that draw from these disciplines.

This continuum reflects a basic division within geography—*physical* geography versus *human/cultural* geography. This duality is paralleled in society by the tendency of those who live in more developed countries (MDCs) to distance themselves from their life-sustaining environment and to think of themselves as separate from the physical functions of Earth. In contrast, many people in less developed countries (LDCs) live closer to nature and are aware of its importance in their lives.

Regardless of our philosophies toward Earth, we all depend on Earth's systems to provide oxygen, water, nutrients, energy, and materials. Our modern world requires that we shift our study of geographic processes, and perhaps our philosophies, toward the center of the continuum to attain a more *holistic*, or balanced, perspective. Some past civilizations adapted to crises, whereas others failed. Perhaps this ability to adapt is the key. If so, understanding our relationship to Earth's physical geography and innumerable physical processes is of great importance to human survival.

> ... during the last few decades, humans have emerged as a new force of nature. We are modifying physical, chemical, and biological systems in new ways, at faster rates, and over larger *spatial* scales than ever recorded on Earth. Humans have unwittingly embarked upon a grand experiment with our planet. The outcome of this experiment is unknown, but has profound implications for all of life on Earth.*

Earth Systems Concepts

The word *system* pervades our lives daily: "Check the car's cooling system"; "How does the grading system work?"; "There is a weather system approaching." Systems of many kinds surround us. *Systems analysis* began with studies of energy and temperature (thermodynamics) in the nineteenth century and was fostered in engineering during World War II. Today, geographers use systems methodology as an analytical tool. In this textbook's 4 parts and 17 chapters, content is organized along logical flow paths consistent with nature and systems thinking.

*Jane Lubchenco, Presidential Address, American Association for the Advancement of Science, February 15, 1997.

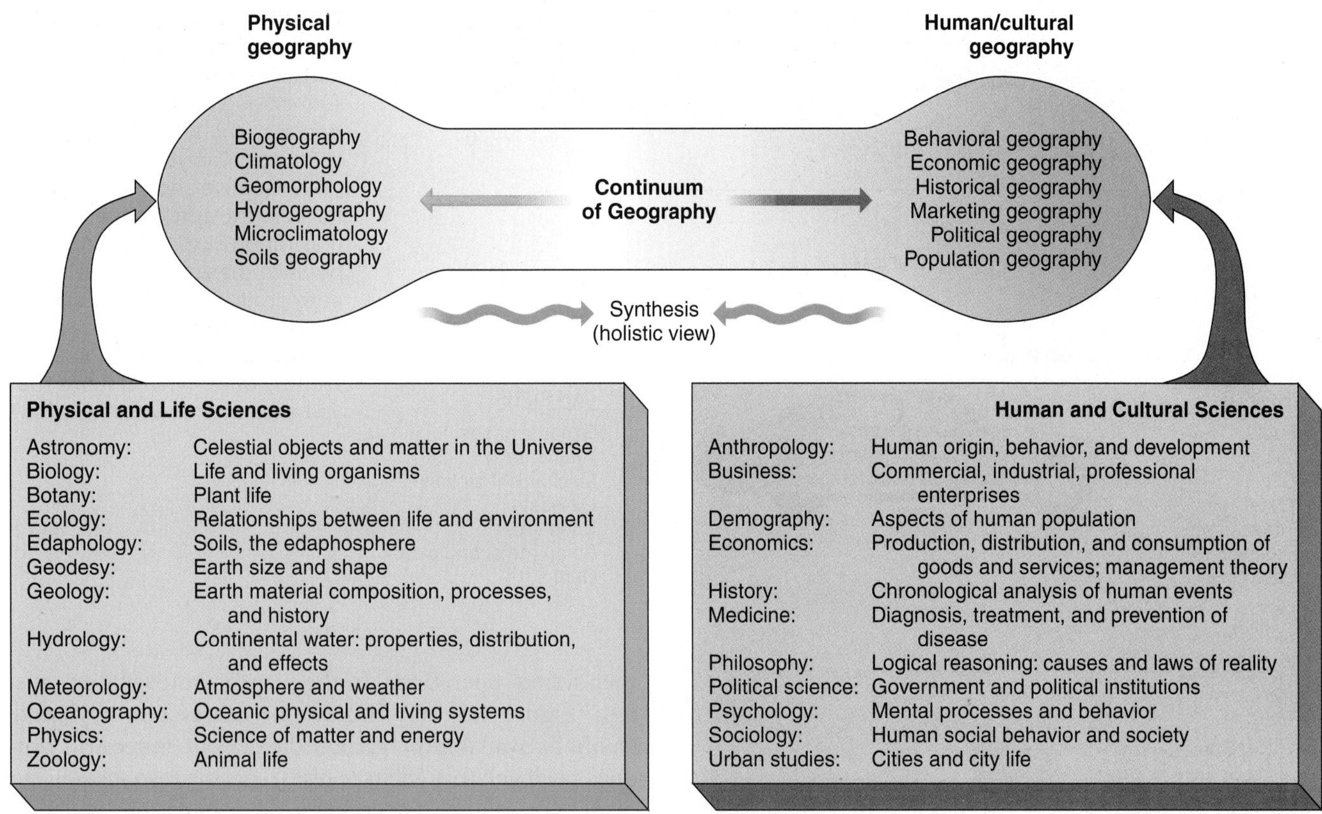

FIGURE 1.2 The content of geography.
Distribution of geographic content along a continuum (a continuous distribution). Our focus is physical geography, but with an integration of the human component. Movement toward the middle of the continuum suggests synthesis of Earth topics and human topics. Examine the subjects listed in the two boxes. Do you find any subjects you have studied? Think for a moment and recall any spatial aspects you remember from those courses.

Systems Theory

Simply stated, a **system** is any ordered, interrelated set of things and their attributes, linked by flows of energy and matter, as distinct from the surrounding environment outside the system. The elements within a system may be arranged in a series or interwoven with one another. A system comprises any number of subsystems. Most Earth systems are dynamic (energetic, in motion) because of the tremendous infusion of radiant energy from the Sun. Within Earth's systems, both matter and energy are stored and retrieved, and energy is transformed from one type to another. (Remember: *matter* is mass that assumes a physical shape and occupies space; *energy* is a capacity to change the motion of, or to do work on, matter.)

Open System. Systems in nature are generally not self-contained: Inputs of energy and matter flow into the system, and outputs of energy and matter flow from the system. Such a system is called an **open system**. Earth is an open system *in terms of energy*, for solar energy enters freely and heat energy leaves freely back into space. Most natural systems are open in terms of energy. Figure 1.3 illustrates an open system.

Closed System. A system that is shut off from the surrounding environment so that it is self-contained is a **closed system**. Although such closed systems are rarely found in nature, Earth is essentially a closed system *in terms of physical matter and resources*—air, water, and material resources. The only exceptions are the slow escape of lightweight gases (such as hydrogen) from the atmosphere into space and the input of frequent but tiny meteors and cosmic and meteoric dust. Earth's physical materials are finite (limited). No matter how numerous and daring the technological reorganizations of matter become, our physical base is, for all practical purposes, fixed. This is it!

System Example. Figure 1.4 illustrates a simple open-flow system, using plant photosynthesis as an example. In *photosynthesis*, plants use an energy input (certain wavelengths of sunlight) and material inputs of water, nutrients, and carbon dioxide. The photosynthetic process converts these inputs to stored chemical energy in the form of plant sugars (carbohydrates). The process also releases an output from the plant system: the oxygen we breathe.

Reversing the process shown in Figure 1.4, plants derive energy for their operations from respiration. In *respiration*, the plant consumes inputs of chemical energy

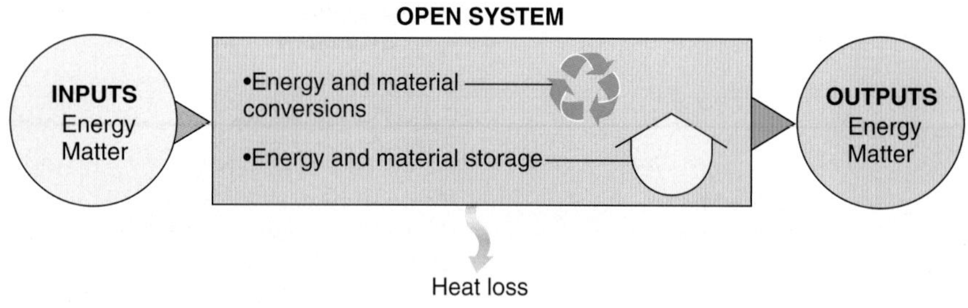

FIGURE 1.3 An open system. In an open system, inputs of matter and energy are transformed and stored as the system operates and system outputs are produced. Can you describe possible negative and positive feedbacks in the automobile illustration? Expand your viewpoint to the entire system of auto production, from raw materials, to assembly, to sales, car accidents, to junk yards. Can you identify other open systems that you encounter in your daily life?

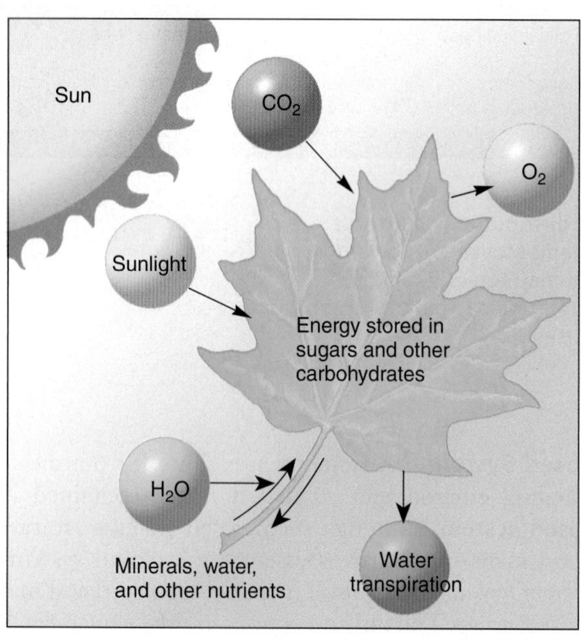

FIGURE 1.4 A leaf is a natural open system. Simple open system of a plant leaf. Plants take light, carbon dioxide (CO_2), water (H_2O), and nutrients as system inputs to produce outputs of oxygen (O_2), and carbohydrates (sugars) through the photosynthesis process. Plant respiration is simply a reverse of this process.

(carbohydrates) and oxygen, and releases outputs of carbon dioxide, water, and heat energy into the environment. Thus, a plant acts as an open system. (Photosynthesis and respiration processes are discussed further in Chapter 16.)

System Feedback. As a system operates, it generates outputs that influence its own operations. These outputs function as "information" that is returned to various points in the system via pathways called **feedback loops**. Feedback information can control (or at least guide) further system operations. In the plant's photosynthetic system (Figure 1.4), any increase or decrease in daylength (sunlight availability), carbon dioxide, or water will produce feedback that elicits (causes) specific responses in the plant. For example, decreasing the water input will slow the growth process; increasing daylength will increase the growth process within limits.

If the feedback *information* discourages response in the system, it is called **negative feedback**—like bad reviews affecting ticket sales for a film or play. *Further production in the system decreases the growth of the system.* Such negative feedback causes self-regulation in a natural system, stabilizing and maintaining the system. Think of a weight-loss diet—the human body is an open system. When you stand on the scales, the good or bad news reported about your weight acts as negative feedback to you.

If feedback *information* encourages increased response in the system, it is called **positive feedback**—like good reviews affecting ticket sales for a film. *Further production in the system stimulates the growth of the system.* Unchecked positive feedback in a system can create a runaway ("snowballing") condition. In natural systems, such unchecked growth will reach a critical limit, leading to instability, disruption, or death.

Think of the numerous devastating wildfires that occurred in Africa, Mexico, and the western United States in 2002 as examples of positive feedback. As the fires burned, they dried surrounding wet shrubs and green wood, thus providing more fuel for combustion. The greater the fire, the greater becomes the availability of fuel, and thus more fire is possible—a positive feedback for the fire. Control of the input of flammable fuel and oxygen is the key to extinguishing such fires.

Can you describe possible negative and positive feedback in the automobile illustration in Figure 1.3? For instance, how would the inputs and outputs be affected if tailpipe exhaust emission standards are weakened or strengthened? What if the automobile is operated at high elevation? Or if the car is a sports utility vehicle (SUV) that weighs 2300 kg (5070 lb)? Or if the price of gasoline increases or decreases? Or if consumers continue buying more efficient hybrid-electric cars? Take a moment to assess the effect of such changes on automobile system operations.

System Equilibrium. Most systems maintain structure and character over time. An energy and material system that remains balanced over time, where conditions are constant or recur, is in a *steady-state condition*. When the rates of inputs and outputs in the system are equal and the amounts of energy and matter in storage within the system are constant (or more realistically, as they fluctuate around a stable average), the system is in **steady-state equilibrium**.

However, a steady-state system fluctuating around an average value may demonstrate a changing trend over time, a condition described as *dynamic equilibrium*. These changing trends of either increasing or decreasing system operations may appear gradual. Examples of dynamic equilibrium include long-term climatic changes and the present pattern of increasing temperatures in the atmosphere and ocean, or the erosion and loss of many beaches and barrier islands in the late 1990s along some coastlines.

Note that given the nature of systems to maintain their operations, they tend to resist abrupt change. However, a system may reach a *threshold* at which it can no longer maintain its character, so it lurches to a new operational level. The relatively sudden collapse of ice shelves surrounding a portion of Antarctica serves as an example of systems reaching a threshold and changing to a new status, that of disintegration. The bleaching (death) of living coral reefs worldwide accelerated dramatically since 1997. Warming conditions and some pollution in the ocean led to such a *threshold* and coral system collapse. Science is hurrying to identify any such thresholds in natural system operations.

Mount Pinatubo—Global System Impact. A particularly dramatic example of interactions between volcanic eruptions and Earth systems illustrates the strength of spatial analysis in physical geography and the systems organization of this textbook. Mount Pinatubo in the Philippines erupted violently in 1991, injecting 15 to 20 million tons of ash and sulfuric acid mist into the upper atmosphere (Figure 1.5). This was the second greatest eruption during the twentieth century; Mount Katmai in Alaska (1912) is the only one greater. The eruption materials from Mount Pinatubo affected Earth systems in several ways noted on the map.

As you progress through this book, you see the story of Mount Pinatubo and its implications woven through seven chapters: Chapter 1 (systems theory), Chapter 3 (effects on energy budgets in the atmosphere), Chapter 4 (satellite images of the spread of debris by atmospheric winds), Chapter 6 (temporary effect

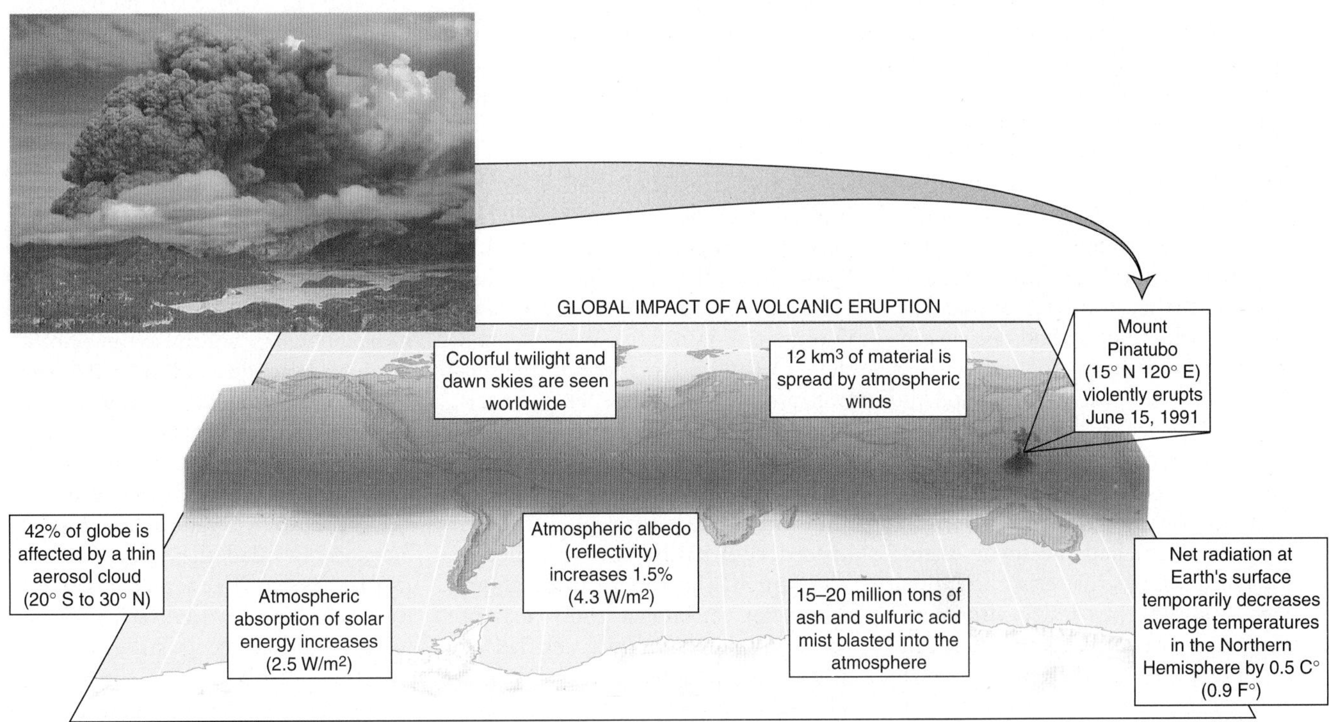

FIGURE 1.5 The eruption of Mount Pinatubo.
The 1991 Mount Pinatubo eruption widely affected the Earth-atmosphere system. Geographers and other scientists use the latest technology to study how such eruptions affect the atmosphere's dynamic equilibrium. For a summary of the impacts from this eruption refer to the discussion of volcanoes in Chapter 9. [Photo by Van Cappellen/REA/SABA.]

Focus Study 1.1

The Scientific Method

The term *scientific method* may have an aura of complexity that it should not. The scientific method is simply the application of common sense in an organized and objective way. A scientist observes, makes a general statement to summarize the observations, formulates a hypothesis, conducts experiments to test the hypothesis, and develops a theory and governing scientific laws. Sir Isaac Newton (1642–1727) developed this method of discovering the patterns of nature, although the term scientific method was applied later.

Scientists are curious about nature and appreciate the challenge of problem solving. Society depends on the discoveries of science to solve mysteries, find solutions, and promote progress. *Complexity* dominates nature, making several outcomes possible as a system operates. Science serves an important function to reduce such uncertainty. Yet the more knowledge we have the more the uncertainty and awareness of other possible scenarios (outcomes and events) increase. This in turn demands more precise and aggressive science. A danger in society occurs when scientific uncertainty fuels an antiscience viewpoint. As we realize that scientific principles of complexity and chaos dominate natural and human-made systems, the need for critical thinking and the scientific method deepens in all aspects of life.

Follow the scientific method illustration in Figure 1 as you read. The scientific method begins with our perception of the real world and a determination of what we know, what we want to know, and the many unanswered questions that exist. Scientists who study the physical environment turn to nature for clues that they can observe and measure. They see what data are needed and begin to collect those data. Then these observations and data are analyzed to identify consistent patterns that may be present. This search for patterns requires *inductive reasoning*, or the process of drawing generalizations from specific facts. This step is important in modern Earth systems sciences, in which the goal is to understand *a whole functioning Earth*, rather than isolated, small compartments of information. Such understanding allows the scientist to construct models that simulate general operations of Earth systems.

If patterns are discovered, the researcher may formulate a *hypothesis*—a formal generalization of a principle. Examples include the planetesimal hypothesis, nuclear-winter hypothesis, and moisture-benefits-from-hurricanes hypothesis. Further observations are related to the general principles established by the hypothesis. Further data gathered may support or disprove the hypothesis, or predictions made according to the hypothesis may prove accurate or inaccurate. All these findings provide feedback to adjust data collection and model building and to refine the hypothesis statement. Verification of the hypothesis after exhaustive testing may lead to its elevation to the status of a *theory*.

A theory is constructed on the basis of several extensively tested hypotheses. Theories represent truly broad general principles—unifying concepts that tie together the laws that govern nature (for example, the theory of relativity, theory of evolution, atomic theory, Big Bang theory, stratospheric ozone depletion theory, or plate tectonics theory). A theory is a powerful device with which to understand both the order and chaos (disorder) in nature. Using a theory allows predictions to be made about things not yet known, the effects of which can be tested and verified or disproved through real evidence. The value of a theory is the continued observation, testing, understanding, and pursuit of knowledge that the theory stimulates. A general theory reinforces our perception of the real world, acting as positive feedback.

Pure science does not make value judgments. Instead, pure science provides people and their institutions with objective information on which to base their own value judgments. Social and political judgments about the applications of science are increasingly critical as Earth's natural systems respond to the impact of modern civilization. Quoting again from Lubchenco's 1997 AAAS Presidential Address:

> Science alone does not hold the power to achieve the goal of greater sustainability, but scientific knowledge and wisdom are needed to help inform decisions that will enable society to move toward that end.

The growing awareness that human activity is producing global change places increasing pressure on scientists to participate in decision making. Numerous editorials in scientific journals have called for such *applied science* involvement. An example, discussed in Chapter 2, is provided by F. Sherwood Rowland and Mario Molina, who first proposed a hypothesis that certain human-made chemicals caused damaging reactions in the stratosphere. Subsequently, surface, atmosphere, and satellite measurements confirmed the photochemical reactions and provided data to map the real losses in stratospheric ozone that were occurring. International treaties and agreements to ban the chemical culprits followed. In 1995, the Royal Swedish Academy of Sciences awarded these three scientists the Nobel Prize for chemistry for their pioneering work. Such successful applied science is strengthening society's resolve and provoking treaties to ban problem chemicals and practices and introduce sustainable substitutes.

(continued)

Focus Study 1.1 *(continued)*

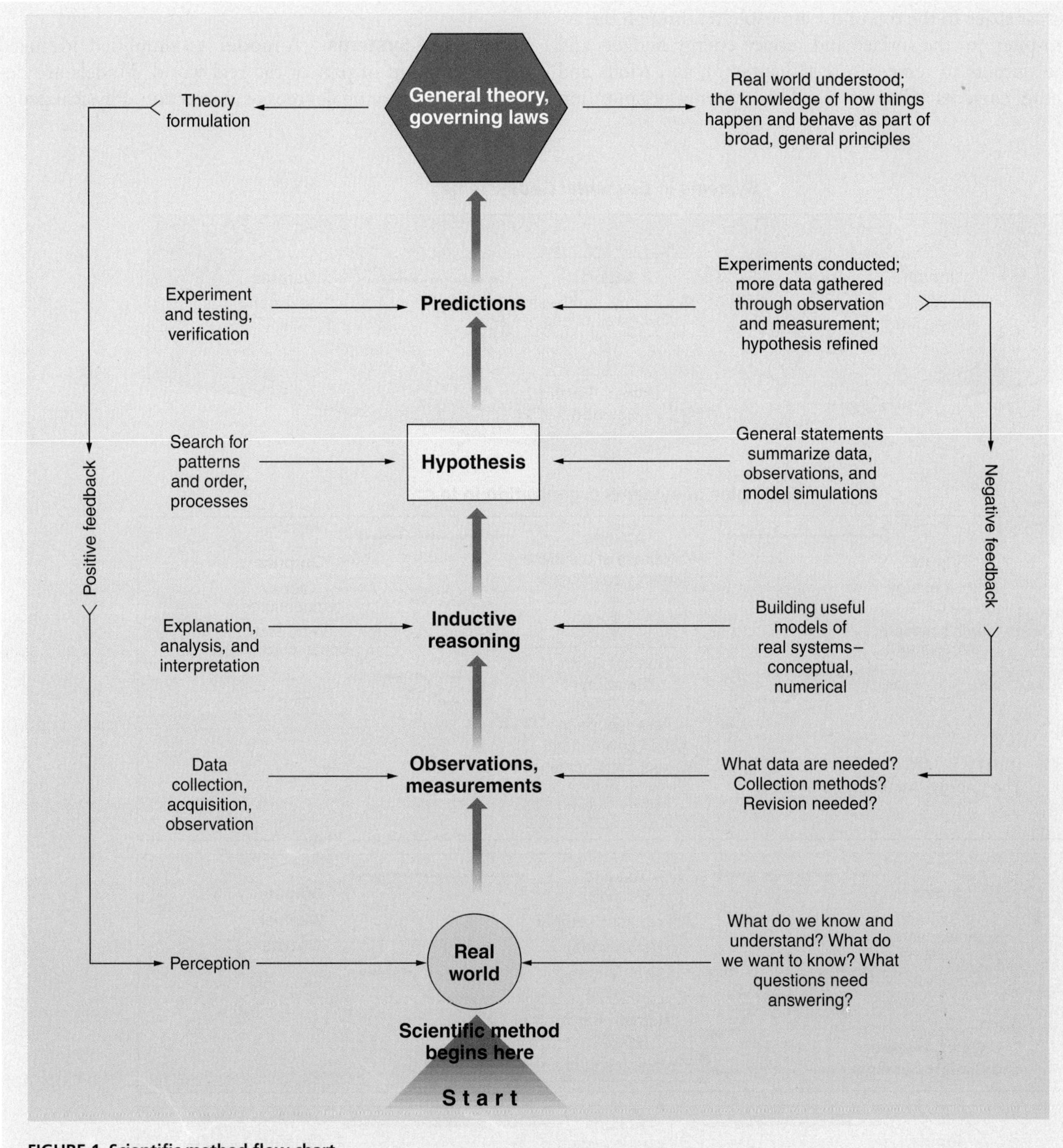

FIGURE 1 Scientific method flow chart.
The scientific method process: from perceptions to observations, reasoning, hypothesis, predictions, and possibly to general theory and natural laws.

on global atmospheric temperatures), Chapters 8 and 9 (volcanic processes), and Chapter 14 (past climatic effects of volcanoes).

Systems in *Elemental Geosystems*. To further illustrate the conceptual systems organization used in this textbook, notice that chapters and portions of chapters are presented in simple sequences organized around the flow of energy, materials, and information. This book presents subjects in the same sequence in which they occur in nature. In this way a logical progression is followed through topics that unfold according to the flow within individual systems, or in accord with time and the flow of events.

The sequence—*input* (components and driving force), *actions* (movements and processes), *outputs* (results and consequences), and *human impacts/impact on humans* (measure of relevance)—is seen in Part 1 (Figure 1.6). We start at the Sun in Chapter 2 and follow the energy flow across space to the top of the atmosphere, through the atmosphere to the surface and surface energy budgets and the outputs of temperature (Chapter 3), and winds and ocean currents (Chapter 4). This systems organization carries throughout the text. The systems organization of Part 2 and Chapter 13 also is illustrated.

Models of Systems. A **model** is a simplified, idealized representation of part of the real world. Models are designed with varying degrees of abstraction. Physical geog-

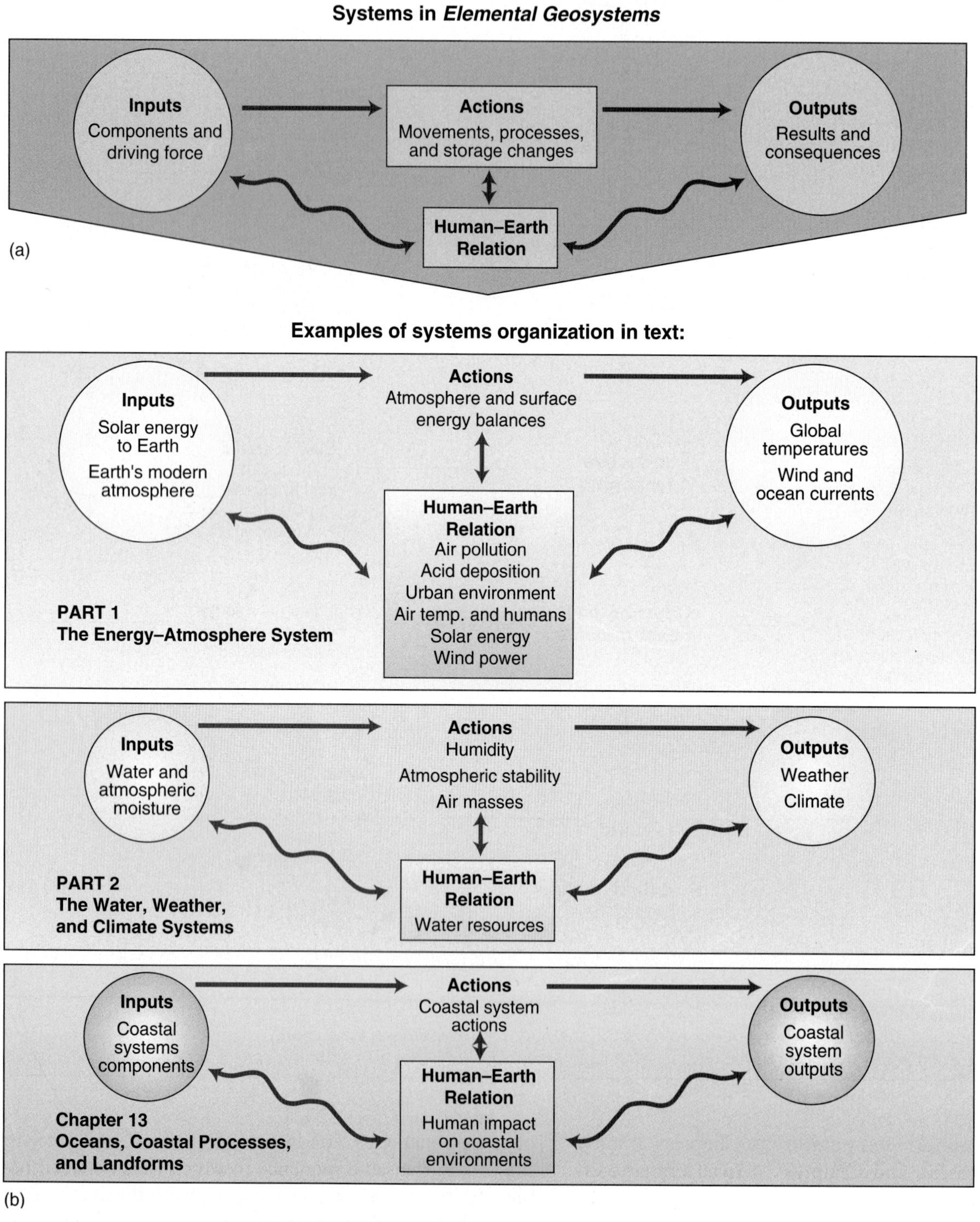

Systems in *Elemental Geosystems*

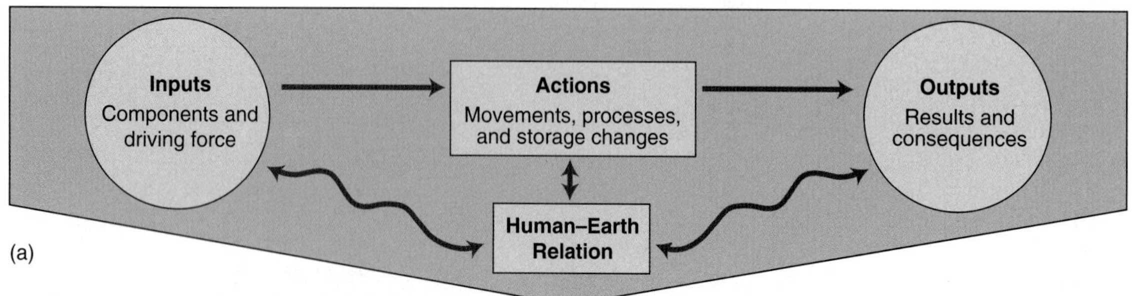

Examples of systems organization in text:

Inputs
Solar energy to Earth
Earth's modern atmosphere
Actions
Atmosphere and surface energy balances
Outputs
Global temperatures
Wind and ocean currents
Human–Earth Relation
Air pollution
Acid deposition
Urban environment
Air temp. and humans
Solar energy
Wind power
PART 1
The Energy–Atmosphere System

Inputs
Water and atmospheric moisture
Actions
Humidity
Atmospheric stability
Air masses
Outputs
Weather
Climate
Human–Earth Relation
Water resources
PART 2
The Water, Weather, and Climate Systems

Inputs
Coastal systems components
Actions
Coastal system actions
Outputs
Coastal system outputs
Human–Earth Relation
Human impact on coastal environments
Chapter 13
Oceans, Coastal Processes, and Landforms

(b)

FIGURE 1.6 The systems in *Elemental Geosystems*.
(a) Chapters, sections, and topics are organized around simple flow systems, or around time and the flow of events, at various scales. (b) As an example here are Parts 1 and 2. Chapter 13, which examines coastal processes, provides a chapter example. Note the outline headings used: Coastal System Components, Coastal System Actions, Coastal System Outputs, and Human Impact on Coastal Environments—along the systems flow shown in (a).

raphers construct simple system models to demonstrate complex associations in the environment, such as the hydrologic cycle that models Earth's entire water system. The simplicity of a model makes a system easier to comprehend (example: Figures 1.3 and 1.4). A model also allows predictions, but such predictions are only as good as the assumptions and accuracy built into the model. Imposing a model too rigidly on a natural setting can lead to mistakes, so it is best to view a model for what it is—a simplification. Have you ever built a model of something?

We discuss many system models in this text, including the hydrologic cycle, water balance, surface energy budgets, earthquakes and faulting, glacier mass budgets, soil profiles, and various ecosystems. Computer-based models are in use to study most natural systems—from climate change to Earth's interior. Let us look at a general model of Earth's major systems.

Earth's Four Spheres

Earth's surface is where four immense open systems interact. Figure 1.7 shows three **abiotic** (nonliving) systems overlapping to form the realm of the **biotic** (living) system. The abiotic spheres are the *atmosphere*, *hydrosphere*, and *lithosphere*. The biotic sphere is called the *biosphere*. Because these four models are not independent units in nature, their boundaries must be understood as transition zones rather than sharp edges.

Atmosphere (Part 1, Chapters 2–4). The **atmosphere** is a thin, gaseous veil surrounding Earth, held to the planet by the force of gravity. Formed by gases arising from within Earth's crust and interior, and the exhalations of all life over time, the lower atmosphere is unique in the Solar System. It is a combination of

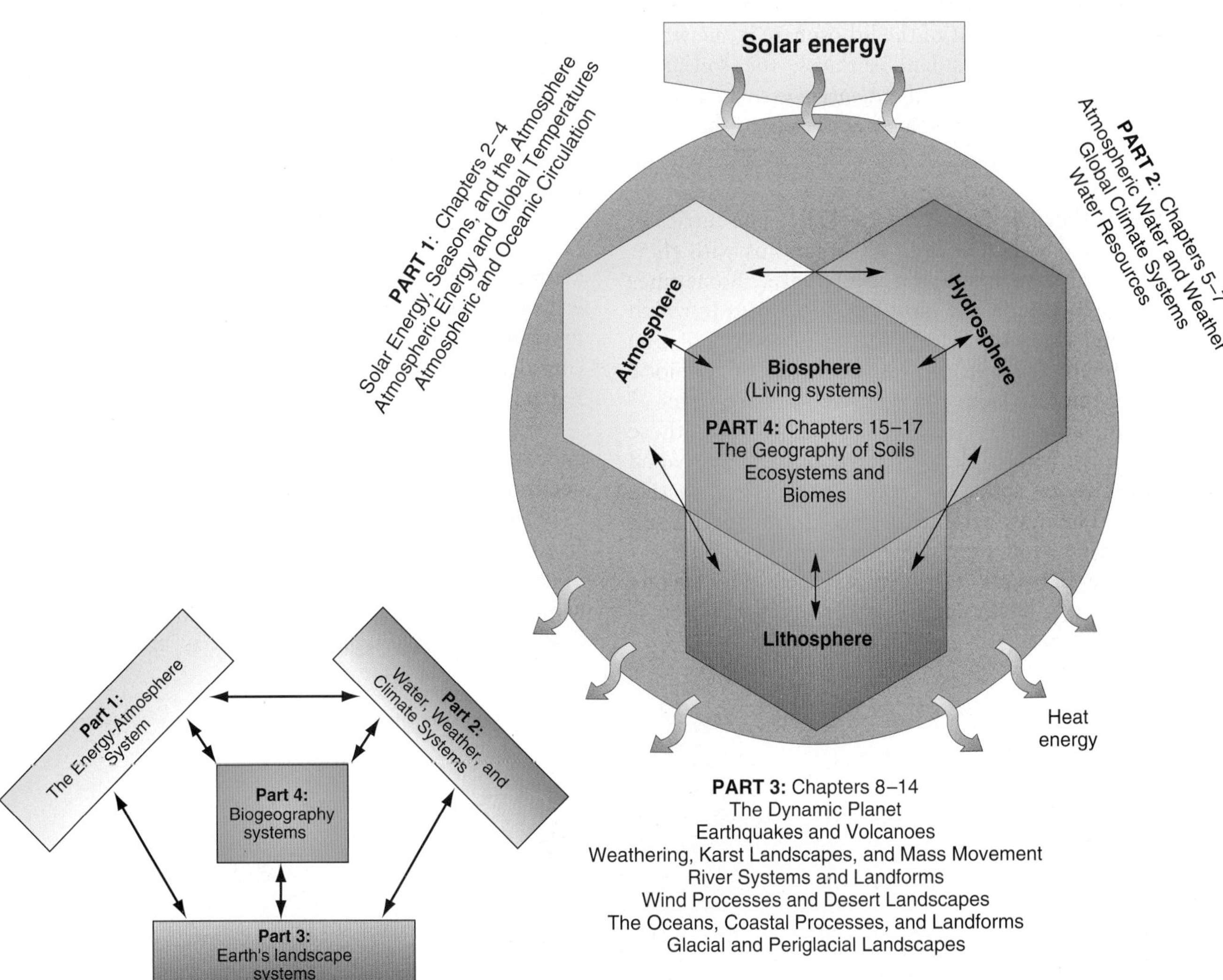

FIGURE 1.7 Earth's four spheres.
Each sphere is a model for the four "parts" used to organize *Elemental Geosystems*: our "sphere of contents."
Part 1—Atmosphere: The Energy-Atmosphere System
Part 2—Hydrosphere: The Water, Weather, and Climate Systems
Part 3—Lithosphere: Earth's Changing Landscape Systems
Part 4—Biosphere: Biogeography Systems

nitrogen, oxygen, argon, carbon dioxide, water vapor, and small amounts of trace gases.

Hydrosphere (Part 2, Chapters 5–7). Earth's waters exist in the atmosphere, on the surface, and in the crust near the surface. Collectively, these areas form the **hydrosphere**. Water occurs in two forms, fresh and saline (salty), and has important heat properties as well as playing its extraordinary role as a solvent. Among the planets in the Solar System, only Earth possesses surface water in any quantity. As an exception, Jupiter's moon Europa appears to have frozen water oceans. Mars seems to have significant subsurface water.

Lithosphere (Part 3, Chapters 8–14). Earth's crust and a portion of the upper mantle directly below the crust form the **lithosphere**. The crust is quite brittle compared to the layers beneath it, which are in motion in response to an uneven distribution of heat and pressure. An important component of the lithosphere is *soil*, which generally covers Earth's land surfaces; the soil layer sometimes is referred to as the *edaphosphere*. (In a broad sense, the term *lithosphere* sometimes refers to the entire solid planet.)

Biosphere (Part 4, Chapters 15–17). The intricate, interconnected web that links all organisms with their physical environment is the **biosphere**. Sometimes called the **ecosphere**, the biosphere is the area in which physical and chemical factors form the context of life. The biosphere exists in the overlap among the abiotic spheres, extending from the seafloor to about 8 km (5 mi) into the atmosphere. Life is sustainable within these natural limits. In turn, life processes powerfully shaped the other three spheres through various interactive processes. The biosphere has evolved, reorganized itself at times, faced extinction, gained new vitality, and managed to flourish overall. Earth's biosphere is the only one known in the Solar System; thus, life as we know it is unique to Earth.

A Spherical Planet

We have all heard that some people in the past believed that Earth was flat. Yet the roundness or *sphericity* of Earth is not as modern a concept as many believe. For instance, the Greek mathematician and philosopher Pythagoras (ca. 580–500 B.C.) determined through observation that Earth is round. We do not know what Pythagoras observed to conclude that Earth is a sphere, but we can guess.

He might have noticed ships sailing beyond the horizon and apparently sinking below the surface, only to arrive back at port with dry decks. Perhaps he noticed Earth's curved shadow cast on the lunar surface during an eclipse of the Moon. He might have deduced that the Sun and Moon are not just disks but are spherical, and that Earth must be a sphere as well.

Earth's sphericity was generally accepted by the educated populace as early as the first century A.D. Christopher Columbus, for example, knew he was sailing around a sphere in 1492; that is why he expected to reach the East Indies. In 1687 Sir Isaac Newton postulated that the round Earth, along with the other planets, could not be perfectly spherical. Newton reasoned that Earth is slightly misshapen by its spinning, making it bulge through the equator and flatten at the poles.

Earth's equatorial bulge and its polar oblateness (slight flattening) are universally accepted and confirmed by satellite observations. **Geodesy** is the science that attempts to determine Earth's shape and size by surveys and mathematical calculations. The modern era of precise Earth measurement is called the "geoidal epoch" because Earth is considered a **geoid**, meaning literally that "the shape of Earth is uniquely Earth-shaped," not a perfect sphere.

Figure 1.8 gives Earth's measurements. Its circumference was first measured over 2200 years ago by the Greek geographer, astronomer, and librarian Eratosthenes (ca. 276–195 B.C.). The ingenious reasoning by which he arrived at his calculation deserves a quick look.

Measuring Earth in 247 B.C. Eratosthenes, foremost among early geographers, served as the librarian of

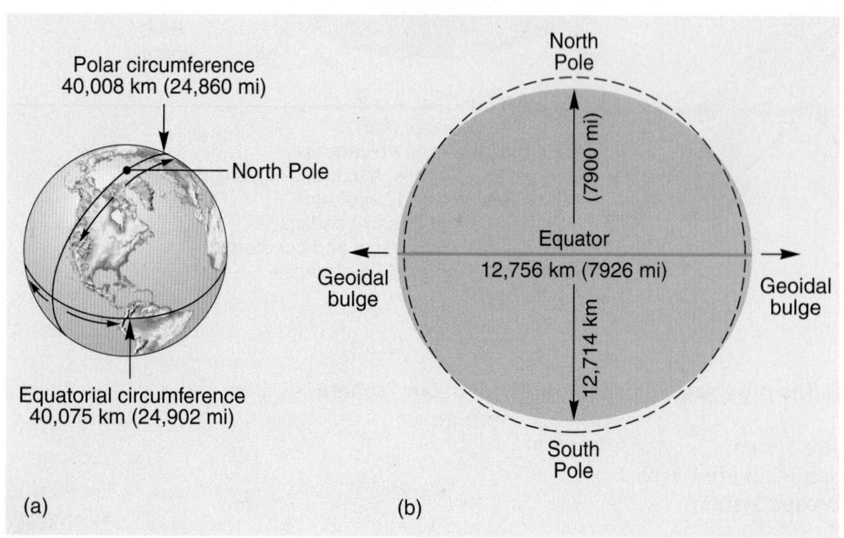

FIGURE 1.8 Earth's dimensions. Earth's circumference (a) and diameter (b)—equatorial and polar—are shown. The dashed line is a perfect circle for reference to Earth's geoid.

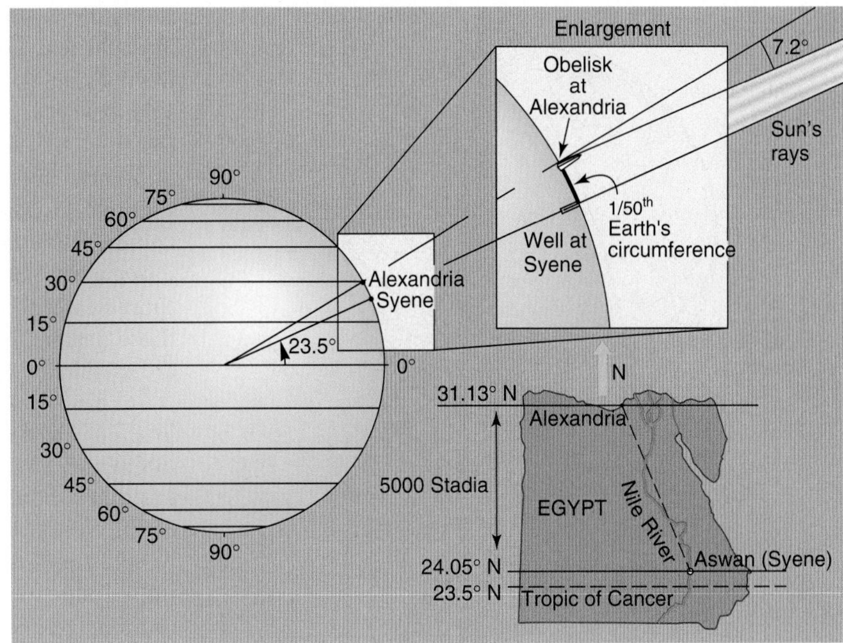

FIGURE 1.9 Eratosthenes' calculation. Eratosthenes' work teaches the value of observing carefully and integrating all observations with previous learning. Calculating Earth's circumference required application of his knowledge of Earth-Sun relationships, geometry, and geography to his keen observations to prove his perception about Earth's size. (Several values can be used for the distance represented by a Greek stadium—the 185 m used here represents an average value.)

Alexandria in Egypt during the third century B.C. He was in a position of scientific leadership, for Alexandria's library was the finest in the ancient world. Among his achievements was calculation of Earth's polar circumference to a high level of accuracy, quite a feat for 247 B.C. Here's how he did it.

Travelers told Eratosthenes that on June 21 they had seen the Sun's rays shine directly to the bottom of a well at Syene, the location of present-day Aswan, Egypt (Figure 1.9). This meant that the Sun had to be directly overhead. North of Syene in Alexandria, Eratosthenes knew from his own observations that the Sun's rays never were directly overhead, even at noon on June 21, the longest day of the year and the day on which the Sun is at its northernmost position in the sky. Unlike objects in Syene, objects in Alexandria always cast a noontime shadow. Using the considerable geometric knowledge of the era, Eratosthenes conducted an experiment.

In Alexandria at noon on June 21, he measured the angle of a shadow cast by an obelisk, a column used for telling time by the Sun. Knowing the height of the obelisk and measuring the length of the shadow from its base, he solved the triangle for the angle of the Sun's rays, which he determined to be 7.2°. However, at Syene on the same day, the angle of the Sun's rays was 0° from a perpendicular—that is, the Sun was directly overhead.

Geometric principles told Eratosthenes that the distance on the ground between Alexandria and Syene formed an arc of Earth's circumference equal to the angle of the Sun's rays at Alexandria. Since 7.2° is roughly 1/50 of the 360° in Earth's total circumference (360° ÷ 7.2° = 50), the distance between Alexandria and Syene must represent approximately 1/50 of Earth's total circumference.

Eratosthenes measured the surface distance between the two cities as 5000 stadia. He then multiplied 5000 stadia by 50 to determine that Earth's polar circumference is about 250,000 stadia. Eratosthenes' calculations convert to roughly 46,250 km (28,738 mi), which is remarkably close to the correct value of 40,008 km (24,860 mi) for Earth's polar circumference. Not bad for 247 B.C.!

Location and Time on Earth

An essential for geographic science is a coordinated grid system to determine location on Earth, a system of coordinates agreed to by all peoples. The terms *latitude* and *longitude* were in use on maps as early as the first century A.D., with the concepts themselves dating to Eratosthenes and others.

The geographer, astronomer, and mathematician Ptolemy (ca. A.D. 90–168) contributed greatly to modern maps, and many of his terms are still used today. Ptolemy divided the circle into 360 *degrees* (360°), with each degree comprising 60 *minutes* (60′), and each minute including 60 *seconds* (60″), in a manner adapted from the ancient Babylonians. He located places using these degrees, minutes, and seconds, although the precise length of a degree of latitude and a degree of longitude on Earth's surface remained unresolved for the next 17 centuries.

Latitude

Latitude *is an angular distance north or south of the equator*, measured from the center of Earth (Figure 1.10a). Because Earth's equator divides the distance between the North Pole and the South Pole exactly in half, it is assigned the value of 0° latitude. Angles of latitude increase toward the North Pole at 90° north latitude, and increase toward the South Pole at 90° south latitude.

A line connecting all points along the same latitudinal angle is called a **parallel**. On a map or globe, the lines designating these angles of latitude run east and west, parallel to the equator. Thus, in Figure 1.10b, *latitude* is the name of the angle (49° north latitude), *parallel* names

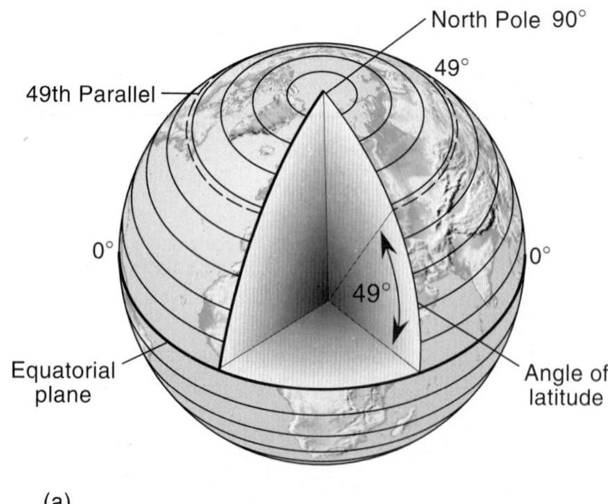

(a)

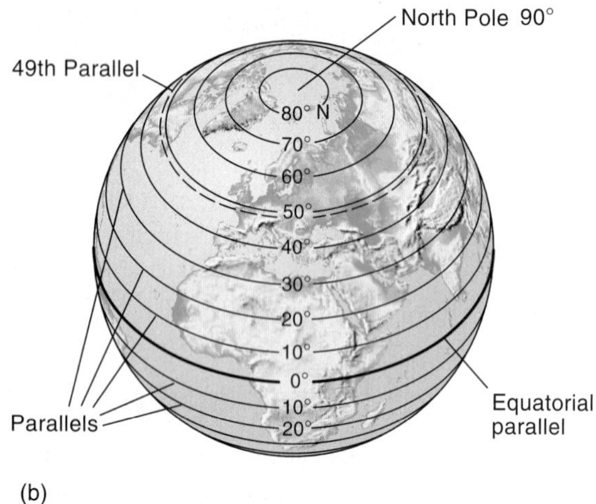

(b)

FIGURE 1.10 Parallels of latitude.
(a) Latitude is measured in degrees north or south of the equator, which is 0°. Each pole is at 90°. Note the measurement of 49° latitude. (b) These angles of latitude determine parallels along Earth's surface. Do you know your present latitude?

the line (49th parallel), and both indicate distance north of the equator. The 49th parallel is a significant one in the Western Hemisphere, for it forms the boundary between Canada and the United States from Minnesota to the Pacific Ocean.

Latitude is readily determined by reference to *fixed celestial objects* such as the Sun or the stars, a method dating to ancient times. During daylight hours the angle of the Sun above the horizon indicates the observer's latitude, after adjustment is made for the seasonal tilt of Earth and time of day. Because Polaris, the North Star, is almost directly overhead at the North Pole, persons anywhere in the Northern Hemisphere can determine their latitude at night simply by sighting Polaris and measuring its angle above the local horizon. In the Southern Hemisphere, Polaris cannot be seen (it is below the horizon), so observers sight on the Southern Cross (*Crux Australis*) constellation that points to a celestial location above the South Pole.

Latitudinal Geographic Zones. Natural environments differ in appearance from the warm and humid equator to the frigid poles. These differences result from the amount of solar energy received, which varies by latitude and season of the year. As a convenience, geographers identify *latitudinal geographic zones* as regions with fairly consistent qualities. Figure 1.11 portrays these zones, their locations, and their names: *equatorial*, *tropical*, *subtropical*, *midlatitude*, *subarctic* or *subantarctic*, and *arctic* or *antarctic*. "Lower latitudes" are those nearer the equator, whereas "higher latitudes" refer to those nearer the poles. These generalized latitudinal zones are useful concepts but are not rigid boundaries.

The *Tropic of Cancer* (23.5° north parallel) and the *Tropic of Capricorn* (23.5° south parallel), discussed further in Chapter 2, are the most extreme north and south parallels that experience perpendicular (directly overhead) rays of the Sun at local noon. The Arctic Circle (66.5° north parallel) and the Antarctic Circle (66.5° south parallel) are the parallels farthest from the poles that experience 24 uninterrupted hours of night during local winter, or day during local summer.

Longitude

Longitude *is an angular distance east or west of a point on Earth's surface*, measured from the center of Earth (Figure 1.12a). On a map or globe, the lines designating these angles of longitude run north and south at right angles (90°) to the equator and to all parallels. A line connecting all points along the same longitude is a **meridian.** In the figure a longitudinal angle of 60° E is measured. Thus, *longitude* is the name of the angle, *meridian* names the line, and both indicate distance east or west of an arbitrary prime meridian (Figure 1.12b). Earth's **prime meridian** (0°) passes through the old Royal Observatory at Greenwich, England, as set by treaty—the *Greenwich prime meridian*.

We have noted that latitude is easily determined by sighting the Sun, North Star, or Southern Cross, but a method of accurately determining longitude, especially at sea, remained a major difficulty until the mid-1700s. In addition, a specific determination of longitude was needed before world standard time could be established. The interesting geographic quest for longitude and time is the topic of Focus Study 1.2.

Today, latitude, longitude, elevation, and Earth measurements are accurately calibrated using a handheld instrument that reads radio signals from satellites, a technology known as the **global positioning system (GPS)**. News Report 1.1 discusses the dramatic spatial applications of GPS.

Great Circles and Small Circles

Great circles and small circles are important concepts that help summarize latitude and longitude (Figure 1.13). A **great circle** is any circle of Earth's circumference whose center coincides with the center of Earth. An infinite number of great circles can be drawn on Earth. Every meridian is half of a great circle that passes through the poles. On flat

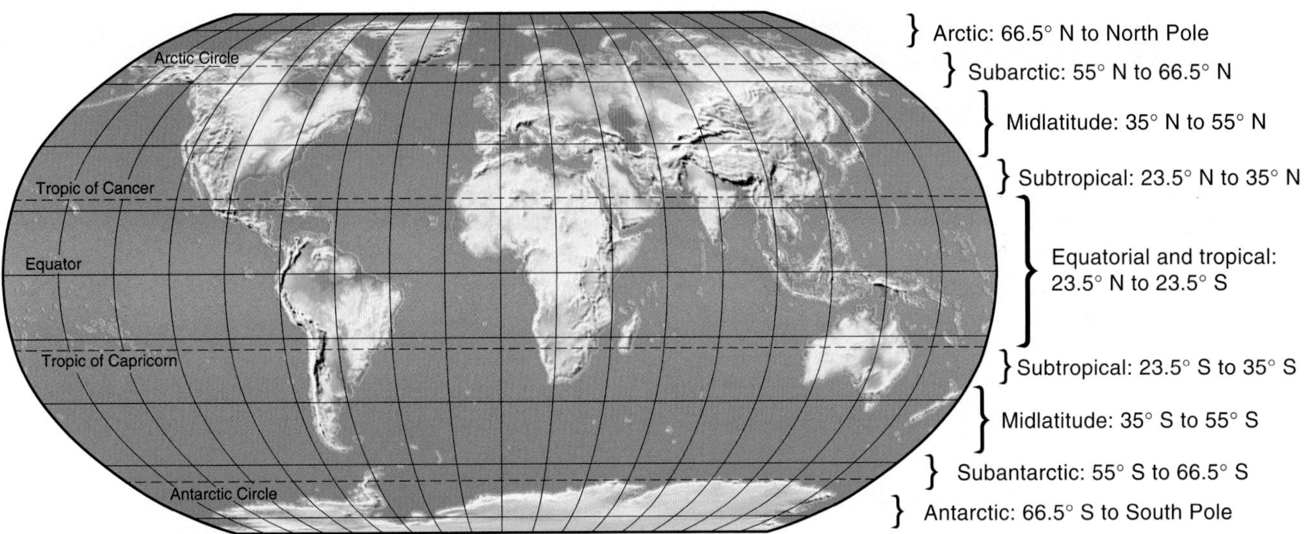

FIGURE 1.11 Latitudinal geographic zones.
Geographic zones are generalizations that characterize various regions by latitude. Think of these as transitioning into one another over broad areas.

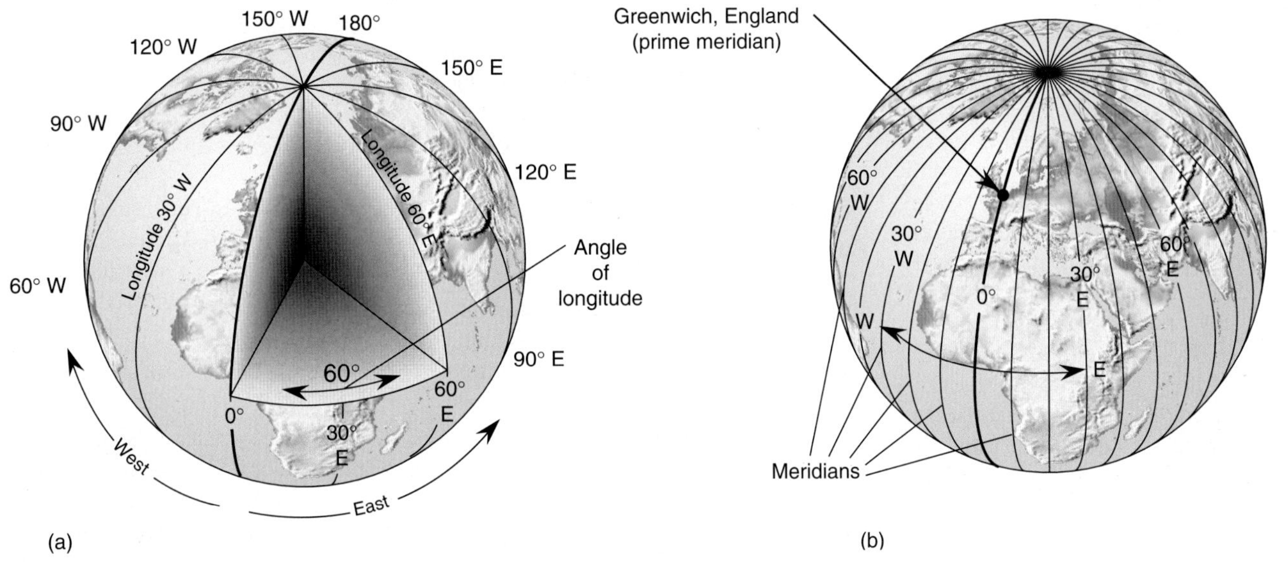

FIGURE 1.12 Meridians of longitude.
(a) Longitude is measured in degrees east or west of a 0° starting line, the prime meridian. Note the measurement of 60° E longitude. (b) Angles of longitude measured from this prime meridian determine other meridians. The prime meridian is drawn from the North Pole through the Royal Observatory in Greenwich, England, to the South Pole. North America is west of Greenwich, therefore it is in the Western Hemisphere. Do you know your present longitude?

maps, airline and shipping routes appear to arch their way across oceans and landmasses. These are *great circle routes*, the shortest distance between two points on Earth.

Only one parallel is a great circle—the equatorial parallel. All other parallels diminish in length toward the poles and, along with any other non-great circle that one might draw, constitute **small circles**—circles whose centers do not coincide with Earth's center.

Prime Meridian and Standard Time

Coordination of international trade, airline schedules, business activities, and daily living depends on a common time system. Today we take for granted standard time zones and an agreed-upon prime meridian. Most nations in the early 1800s used their own national prime meridians, creating confusion in global mapping and clock time. As a result, no standard time existed between or even within countries.

Setting time was not so great a problem in small European countries, most of which are less than 15° wide. In North America, which covers more than 90° of longitude (the equivalent of six 15° time zones), the problem was serious. In 1870, travelers going from Maine to San Francisco made 22 adjustments to their watches to stay consistent

Focus Study 1.2

The Timely Search for Longitude

Unlike latitude, longitude cannot be determined readily from fixed celestial bodies. The problem is Earth's rotation, which constantly changes the apparent position of the Sun and stars. Determining longitude is particularly critical at sea, when no landmarks are in sight.

In his historical novel *Shogun*, author James Clavell expressed the frustration of the longitude problem through his pilot, Blackthorn:

> Find how to fix longitude and you're the richest man in the world. ... The Queen, God bless her, 'll give you ten thousand pound and dukedom for answer to the riddle. ... Out of sight of land you're always lost, lad.*

In the early 1600s, Galileo explained that longitude could be measured by using two clocks. Any point on Earth takes 24 hours to travel around the full 360° of one rotation (one day), regardless of its latitude. If you divide 360° by 24 hours, you find that any point on Earth travels through 15° of longitude every hour. Thus, if there were a way to measure time accurately at sea, a comparison of two clocks could give a value for longitude. One clock would indicate the time back at home port (Figure 1). The other clock would be reset at local noon each day, as determined by the highest Sun position in the sky (solar zenith). The time difference then would indicate the longitudinal difference traveled: 1 hour for each 15° of longitude. The principle was right; all that was needed were accurate clocks. Unfortunately, the pendulum clock invented by Christian Huygens in 1656 did not work on the rolling deck of a ship at sea!

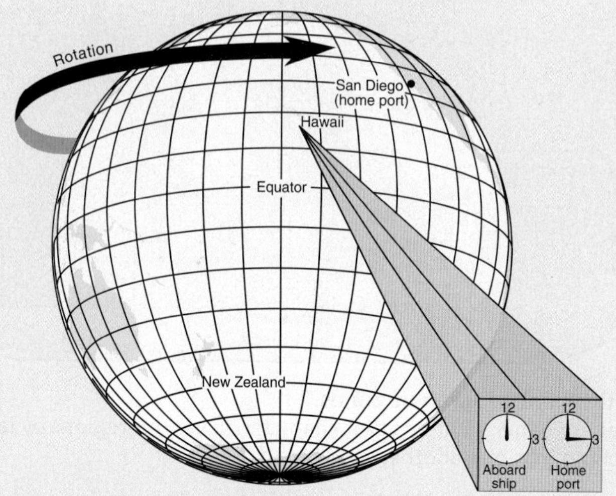

FIGURE 1 Clock times determine longitude.
Using two clocks on a ship to determine longitude. For example, if the shipboard clock reads local noon and the clock set for home port reads 3:00 P.M., ship time is 3 hours earlier than home time. Therefore, calculating 3 hours at 15° per hour puts the ship at 45° west longitude from home port.

In 1707, the British lost four ships and 2000 men in a sea tragedy that was blamed specifically on the longitude problem. In response, Parliament demanded action and authorized a prize worth over $2 million in today's dollars to the first successful inventor of an accurate seafaring clock. The Board of Longitude was established to judge any devices submitted.

John Harrison, a self-taught country clockmaker, began work on the problem in 1728 and finally produced his brilliant marine chronometer, known as Number 4, in 1760. The clock was tested on a voyage to Jamaica in 1761. When taken ashore and compared to land-based longitude, Harrison's ingenious Number 4 was only 5 seconds slow, an error that translates to only 1.25′ or 2.3 km (1.4 mi), well within Parliament's standard. After many delays, Harrison finally received most of the prize money in his last years of life.

> With his marine clocks, John Harrison tested the waters of space–time. He succeeded, against all odds, in using the fourth—temporal—dimension to link points on the three-dimensional globe. He wrested the world's whereabouts from the stars, and locked the secret in a pocket watch.†

From that time it was possible to determine longitude accurately on land and sea, as long as everyone agreed upon a prime meridian to use as a reference for time comparisons. In this modern era of atomic clocks and satellites in mathematically precise orbits, we have far greater accuracy available for the determination of longitude, precise navigation, and location on Earth's surface.

*J. Clavell, *Shogun* (New York: Delacorte Press, a division of Dell Publishing Group, Inc., 1975), p.10.

†Dava Sobel, *Longitude, The Story of a Lone Genius Who Solved the Greatest Scientific Problem of His Time* (New York: Walker and Co., 1995), p. 175.

News Report 1.1

GPS: A Personal Locator

The Global Positioning System (GPS) comprises 24 orbiting satellites, in six orbital planes, that transmit navigational signals for Earth-bound use (backup GPS satellites are in orbital storage as replacements). Originally devised in the 1970s by the Department of Defense for military purposes, the present system is commercially available worldwide.

A small receiver, some about the size of a pocket radio, receives signals from four or more satellites at the same time, calculates latitude and longitude within 10-m accuracy (33 ft) and elevation within 15 m (49 ft), and displays the results. With the shutdown in 2000 of the Pentagon Selective Availability, commercial resolution is the same as for military applications and its Precise Positioning Service (PPS). *Differential GPS (DGPS)* increases accuracy by comparing readings with another base station (reference receiver) for differential correction (Figure 1).

GPS is useful for diverse applications, such as ocean navigation, land surveying, tracking small changes in Earth's crust, managing the movement of fleets of trucks, mining and resource mapping, tracking wildlife (birds and land and marine animals) migration and behavior, and environmental planning. Relative to earthquakes in southern California, JPL operates the GPS Observation Office that monitors a network of 250 seismic stations. Detecting and mapping an ominous bulge on an Oregon volcano is another present-day application. Also, GPS is useful to the backpacker and sportsperson. Commercial airlines use GPS to improve accuracy of routes flown and thus increase fuel efficiency. Scientists used GPS to accurately determine the height of Mount Everest in the Himalayan Mountains—now 8850 m compared to the former 8848 m (29,035 ft, 29,028 ft). In contrast GPS measurements of Mount Kilimanjaro lowered its summit from 5895 m to a lower 5892 m (19,340 ft, 19,330 ft).

FIGURE 1 GPS in action. A GPS unit in operation for surveying and location analysis in an environmental study. [Photo courtesy of Trimble Navigation Ltd., Sunnyvale, California.]

FIGURE 2 GPS on Mount Everest. A Trimble 4800 global positioning system (GPS) installation establishes Earth's highest benchmark. The installation is only 18 m (60 ft) from the 8850 m (29,035 ft) summit of Mount Everest. Scientists calculated this new measurement of the mountain's height and announced the results in 1999. [Photo and GPS installation by explorer, climber Wally Berg, May 20, 1998.]

Farmers use GPS to determine crop yields on specific parts of their farms. A detailed plot map is made to guide the farmer as to where more fertilizer, proper seed distribution, irrigation applications, or other work is needed. A computer and GPS unit on board the farm equipment guides the work. This is the precision science of *variable-rate technology*, made possible by GPS. Hawkeye Community College in Waterloo, Iowa, offers a certificate program in Precision Agriculture, (see http://www.ag.hawkeye.cc.ia.us/).

The importance of GPS to geography is obvious because this technology reduces the need to maintain ground control points for location, mapping, and spatial analysis. Instead, geographers working in the field can determine their position accurately as they work. Boundaries and data points in a study area are easily determined and entered into a database, reducing the need for traditional surveys. For this and myriad other applications, GPS sales are exceeding $10 billion a year. As additional frequencies are added in 2003 and 2006, accuracy will increase significantly. Also, the European Union plans to launch its own GPS system of 20 satellites beginning in 2005. (For a GPS overview see http://www.colorado.Edu/geography/gcraft/notes/gps/gps_f.html.)

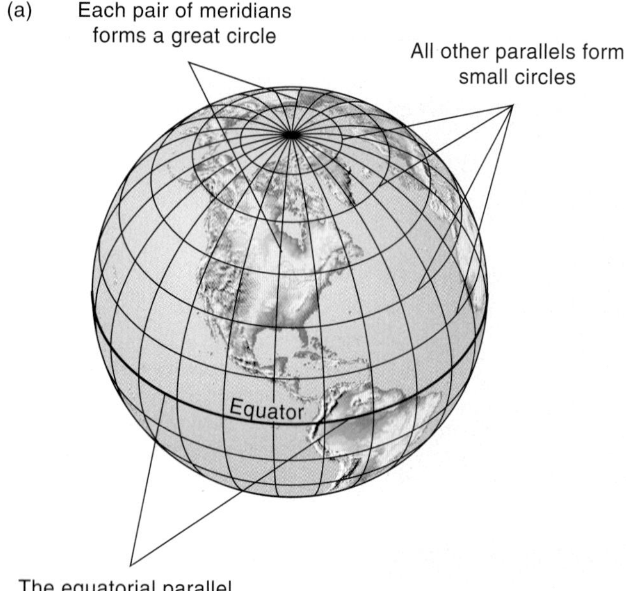

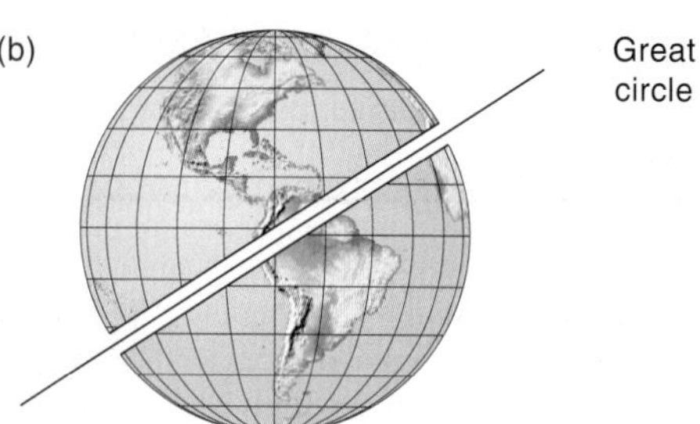

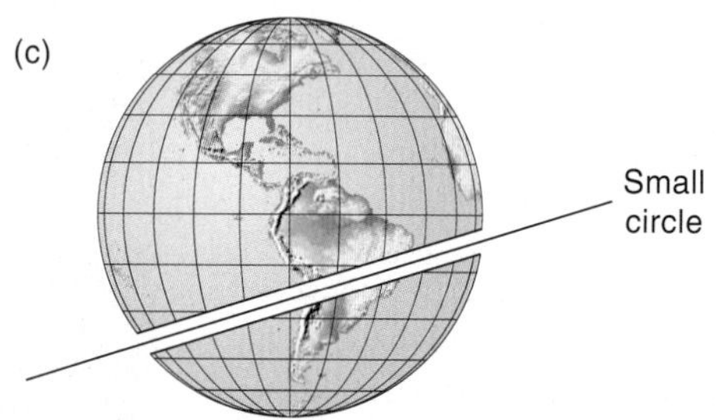

FIGURE 1.13 Great circles and small circles.
(a) Examples of great circles and small circles on Earth. (b) Any plane that divides Earth into equal halves will intersect the globe along a great circle; this great circle is a full circumference of the globe and is the shortest distance between any two surface points. (c) Any plane that splits the globe into unequal portions will intersect the globe along a small circle.

with local railroad time. In Canada, Sir Sanford Fleming led the fight for standard time and for an international agreement on a prime meridian.

Representatives from 27 countries attended an the International Meridian Conference in Washington, D.C., in 1884. After lengthy debate, most participants chose the Royal Observatory at Greenwich in England as the place for the prime meridian of 0° longitude. The Observatory was highly respected, and more than 70% of the world's merchant ships already used the London prime meridian. Thus, a world standard was established—**Greenwich Mean Time (GMT)**—and the first international time system was set (see **http://www.gmt2000.co.uk/meridian/place/plco0a1.htm**).

Because Earth revolves 360° every 24 hours, or 15° per hour (360° ÷ 24 = 15°), a time zone of 1 hour spans each 15° of longitude (7.5° on either side of a *controlling meridian*). Assuming it is 9:00 P.M. in Greenwich, then it's 4:00 P.M. in Baltimore (+5 hrs), 3:00 P.M. in Oklahoma City (+6 hrs), 2:00 P.M. in Salt Lake City (+7 hrs), 1:00 P.M. in Seattle and Los Angeles (+8 hrs), noon in Anchorage (+9 hrs), and 11:00 A.M. in Honolulu (+10 hrs). To the east, it is midnight in Ar Riyāḍ, Saudi Arabia (−3 hrs). (The designation A.M. is for *ante meridiem*, "before noon," whereas P.M. is for *post meridiem*, "after noon." A 24-hour clock avoids the use of these designations.)

As you can see from the modern international time zone map in Figure 1.14, national borders and political considerations distort time zone boundaries. For example, China spans four time zones, but its government decided to keep the entire country operating at the same time. Thus, in some parts of China clocks are several hours off from what the Sun is doing. In the United States, parts of Florida and west Texas are in the same time zone.

International Date Line. An important corollary of the prime meridian is the 180° meridian on the opposite side of the planet. This meridian is called the **International Date Line (IDL)** and marks the place where each day officially begins (12:01 A.M.). From this "line" the new day sweeps westward. This *westward* movement of time is created by the planet's turning *eastward* on its axis. At the IDL, the west side of the line is always one day ahead of the east side. No matter what time of day it is when the line is crossed, the calendar changes a day (Figure 1.15).

The choice of Greenwich, which located the date line in the sparsely populated Pacific Ocean, minimizes most local confusion. However, the consternation of early explorers before the date-line concept was understood is interesting. For example, Magellan's crew returned from the first circumnavigation of Earth in 1522, confident from the ship's log that the day of their arrival was a Wednesday, September 7. They were shocked when informed by insistent local residents that it was actually a Thursday, September 8. Of course, without an International Date Line they had no idea that they must advance a day some-

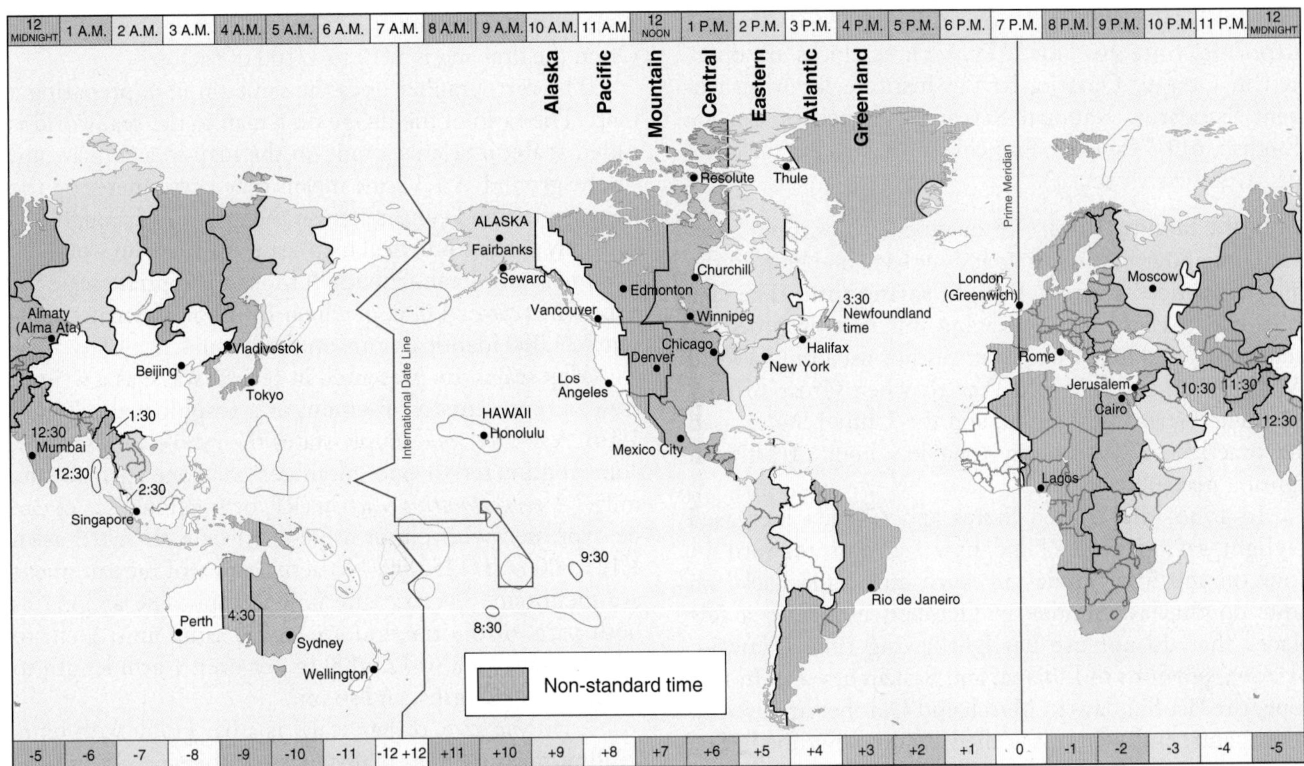

FIGURE 1.14 Modern international standard time zones.
Numbers along the bottom of the map indicate how many hours each zone is earlier (plus sign) or later (minus sign) than the Coordinated Universal Time (UTC) standard at the prime meridian. The United States has five time zones; Canada is divided into six. The island country of Kiribati moved the International Date Line (IDL) to its eastern margin (150° west longitude) to be the first to experience each new day. These distortions of the IDL only apply to the countries and their territorial waters and not to international waters between them and the 180th meridian. [Adapted from Standard Time Zone Chart of the World, Defense Mapping Agency, Bethesda, Maryland.]

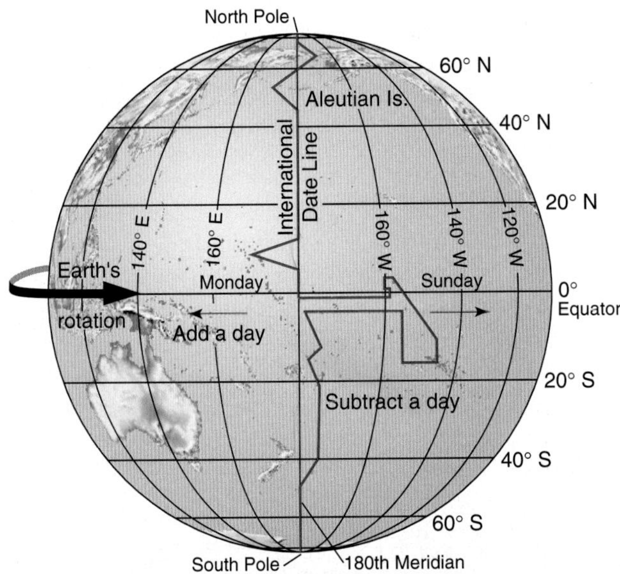

FIGURE 1.15 International Date Line.
International Date Line location, approximately along the 180th meridian (see also Figure 1.14). Note that it is officially one day later west of the IDL.

where when sailing around the world in a westward direction. Imagine the confusion as the crew accounted for each day in their log!

Coordinated Universal Time. Greenwich Mean Time was replaced by a Universal Time (UT) system in 1928. Since 1972, **Coordinated Universal Time (UTC)** is the legal reference for official time in all countries.* Although the prime meridian still runs through Greenwich, UTC is based on average time calculations collected by the International Bureau of Weights and Measures (BIPM) near Paris and broadcast worldwide. Progress in accurately measuring time advanced rapidly with the invention of a quartz clock in 1939, atomic clocks in the early 1950s, and fountaining cesium atomic clocks in 2000.

Time and Frequency Services of the National Institute for Standards and Technology (NIST), U.S. Department of Commerce in Boulder, Colorado, operates the newest clock *NIST-F1*. *NIST-F1* replaced *NIST-7*, which is still operational and participates in UTC. (For more on

*UTC is in use because agreement was not reached on whether to use the English word order, CUT, or the French order, TUC. UTC was the compromise and is recommended for all timekeeping applications; use of the term GMT is discouraged.

time call 303-499-7111 or 808-335-4363; or see http://nist.time.gov/ for UTC.) Three clocks are operated in Ottawa, Ontario, by the Institute for Measurement Standards, National Research Council of Canada (English, 613-745-1578; French, 613-745-9426; for more see http://www.nrc.ca/).

Daylight Saving Time. In many countries, time is set ahead one hour in the spring and set back one hour in the fall—a practice known as **daylight saving time**. The idea to extend daylight in the evening was first proposed by Benjamin Franklin, although it was not used until World War I and again in World War II, when Great Britain, Australia, Germany, Canada, and the United States used the practice to save energy (one less hour of artificial lighting needed).

In 1986, the United States and Canada increased daylight saving time. Time now "springs forward" 1 hour on the first Sunday in April and "falls back" an hour on the last Sunday in October, except in a few places that do not use daylight saving time (Hawai'i, Arizona, portions of Indiana, and Saskatchewan). In Europe, the last Sundays in March and October are used to begin and end what is called "summer time." See http://webexhibits.org/daylightsaving/.

Maps, Scales, and Projections

The earliest known graphic map presentations date to 2300 B.C., when the Babylonians used clay tablets to record information about the region of the Tigris and Euphrates Rivers (the area of modern-day Iraq). Today the making of maps and charts is a specialized science as well as an art, blending aspects of geography, engineering, mathematics, graphics, computer science, and artistic specialties. It is similar in ways to architecture, in which aesthetics and utility are combined to produce an end product.

A **map** is a generalized view of an area, usually some portion of Earth's surface, as seen from above and greatly reduced in size. The part of geography that embodies mapmaking is called **cartography**. Maps are tools with which geographers show spatial information and analyze spatial relationships. We all use maps at some time to visualize our location and our relation to other places, to plan a trip, or to coordinate commercial and economic activities. Have you found yourself looking at a map, planning real and imagined adventures to far-distant places? Maps are wonderful tools! Learning a few basics about maps is essential to our study of physical geography.

Map Scales

Architects, toy designers, and mapmakers have something in common: they all create *scale models*. They reduce real things and places to the more convenient scale of a model car, or plane, a diagram, or a map. An architect renders a blueprint of a building to guide the contractors, selecting a scale so that one centimeter (or inch) on the drawing represents so many meters (or feet) on the proposed building. Often the drawing is 1/50 to 1/100 of real size.

The cartographer does the same thing in preparing a map. The ratio of the image on a map to the real world is called **scale**; it relates a unit on the map to a similar unit on the ground. A 1:1 scale means that a centimeter on the map represents a centimeter on the ground (although this certainly is an impractical map scale, for the map would be as large as the area mapped). A more appropriate scale for a local map is 1:24,000, in which 1 unit on the map represents 24,000 identical units on the ground.

Map scales are presented in several ways: as a written scale, a representative fraction, or a graphic scale (Figure 1.16). A *written scale* simply states the ratio—for example, "one centimeter to one kilometer" or "one inch to one mile." A *representative fraction* (RF, or fractional scale) can be expressed with either a colon (:) or a slash (/), as in 1:125,000 or 1/125,000. No actual units of measurement are mentioned because any unit is applicable as long as both parts of the fraction are in the same unit: 1 cm to 125,000 cm, 1 in. to 125,000 in., or even 1 arm length to 125,000 arm lengths, and so on.

A *graphic scale*, or bar scale, is a bar graph with units to allow measurement of distances on the map. An important advantage of a graphic scale is that it changes along with the map if the map is enlarged or reduced. In contrast, written and fractional scales become incorrect with enlargement or reduction: you can shrink a map from 1:24,000 to 1:63,360, but the scale will still say "1 in. = 2000 ft," instead of the new correct scale of 1 in. to 5280 ft (1 in. = 1 mi).

Scales are called *small*, *medium*, and *large*, depending on the ratio described. In relative terms, a scale of 1:24,000 is a *large scale*, whereas a scale of 1:50,000,000 is a *small scale*. The greater the denominator in a fractional scale (or the number on the right in a ratio expression), the smaller the scale and the more abstract the map must be in relation to what is being mapped. Examples of selected representative fractions and written scales are listed in Table 1.2 for small-, medium-, and large-scale maps.

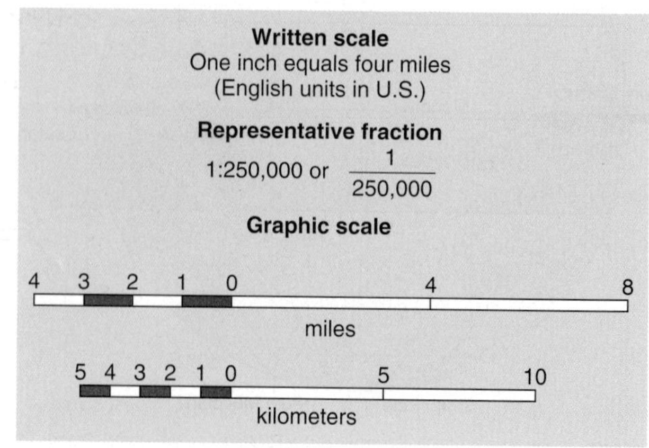

FIGURE 1.16 Map scale.
Three common expressions of map scale—written scale, representative fraction, and graphic scale.

Table 1.2 Sample Representative Fractions and Written Scales for Small-, Medium-, and Large-Scale Maps

System	Scale Size	Representative Fraction	Written Scale
English	Small	1:3,168,000	1 in. = 50 mi
		1:2,500,000	1 in. = 40 mi
		1:1,000,000	1 in. = 16 mi
		1:500,000	1 in. = 8 mi
		1:250,000	1 in. = 4 mi
	Medium	1:125,000	1 in. = 2 mi
		1:63,360 (or 1:62,500)	1 in. = 1 mi
		1:31,680	1 in. = 0.5 mi
		1:30,000	1 in. = 2500 ft
	Large	1:24,000	1 in. = 2000 ft
System		**Representative Fraction**	**Written Scale**
Metric		1:1,000,000	1 cm = 10.0 km
		1:50,000	1 cm = 0.50 km
		1:25,000	1 cm = 0.25 km
		1:20,000	1 cm = 0.20 km

If globes and maps are available in your library or classroom, check to see the scale at which they are drawn. See if you can find examples of written, representative, and graphic scales on wall maps, on highway maps, and in atlases.

Map Projections

A globe is not always a helpful representation of Earth. Travelers, for example, need more detailed information than a globe can provide, and large globes are hard to carry on trips. Consequently, to provide localized detail, cartographers prepare large-scale flat maps, which are two-dimensional representations (scale models) of our three-dimensional Earth.

A globe is the only true representation of *distance*, *direction*, *area*, *shape*, and *proximity*. A flat map distorts these properties. Therefore, in preparing a flat map, the cartographer must decide which characteristic to preserve, which to distort, and how much distortion is acceptable.

To understand this problem, consider these important properties of a *globe*.

- Parallels always are parallel to each other, always are evenly spaced along meridians, and always decrease in length toward the poles.
- Meridians converge at both poles and are evenly spaced along any individual parallel.
- The distance between meridians decreases toward poles, with the spacing between meridians at the 60th parallel equal to one-half the equatorial spacing.
- Parallels and meridians always cross each other at right angles.

The problem is that all these qualities cannot be reproduced on the same flat surface. Simply taking a globe apart and laying it flat on a table illustrates the difficulty faced by cartographers in constructing a flat map (Figure 1.17). You can see the empty spaces that open up between the sections, or *gores*, of the globe. This reduction of the spherical Earth to a flat surface is called a **map projection**. Thus, no flat map projection of Earth can ever have all the features of a globe. Flat maps always possess some degree of distortion—much less for large-scale maps representing a few kilometers, much more for small-scale maps covering individual countries, continents, or the entire world.

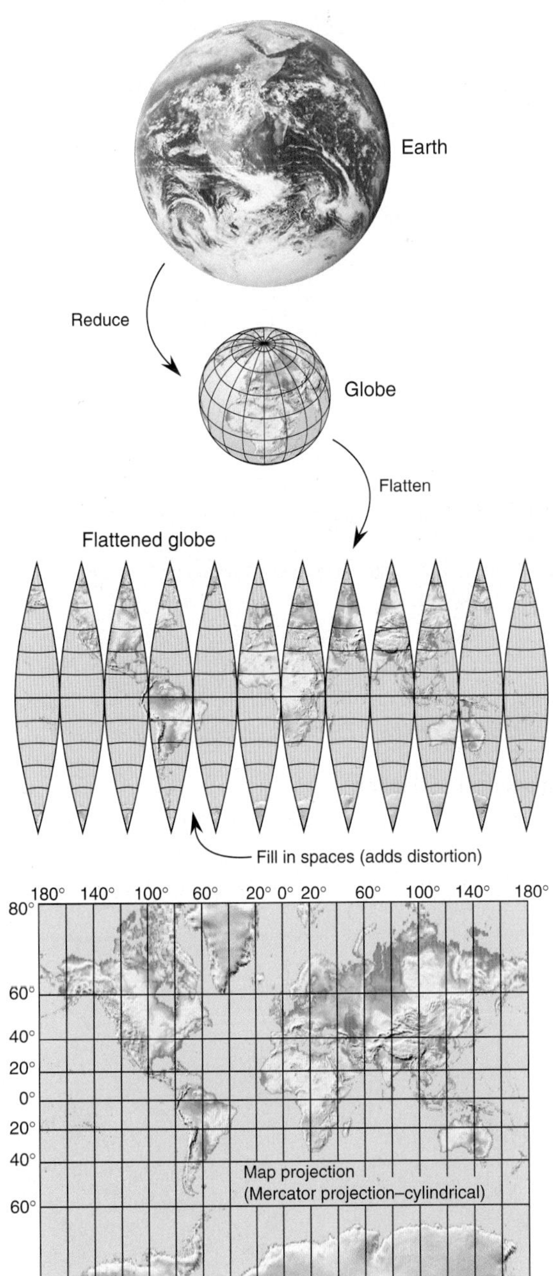

FIGURE 1.17 From globe to flat map.
Conversion of the globe to a flat map projection requires decisions about which properties to preserve and what amount of distortion is acceptable. [Astronaut photo, NASA.]

Properties of Projections. *The best projection is always determined by its intended use.* There are many possibilities, four of which are shown in Figure 1.18. The major decisions in selecting a map projection involve the properties of **equal area** (equivalence) and **true shape** (conformality). If a cartographer selects equal area as the desired trait, as for a map showing the distribution of world climates, then shape must be sacrificed by *stretching* and *shearing* (allowing parallels and meridians to cross at other than right angles). On an equal-area map, a coin covers the same amount of surface area, no matter where you place it on the map.

Conversely, if a cartographer selects the property of true shape, as for a map used for navigational purposes, then equal area must be sacrificed, and the scale will actually change from one region of the map to another.

The Nature and Classes of Projections. Despite modern cartographic technology utilizing mathematical constructions and computer-assisted graphics, the word *projection* still is used. It comes from times past, when a globe literally was projected onto a surface. The globe was constructed of wire parallels and meridians. A light source then cast a pattern of latitude and longitude lines from the globe onto the desired geometric surfaces, such as a *cylinder*, *plane*, or *cone*.

Figure 1.18 illustrates the general classes of map projections and the perspectives from which they are generated. The classes shown include the *cylindrical*, *planar* (or azimuthal), and *conic*. Another class of projections that cannot be derived from this physical-perspective approach is the nonperspective *oval* class. Still others are derived from purely mathematical calculations.

With all projections, the contact line or contact point between the wire globe and the projection surface—called a *standard line (or point)—is the only place where all globe properties are preserved.* Thus, a *standard parallel* or *standard*

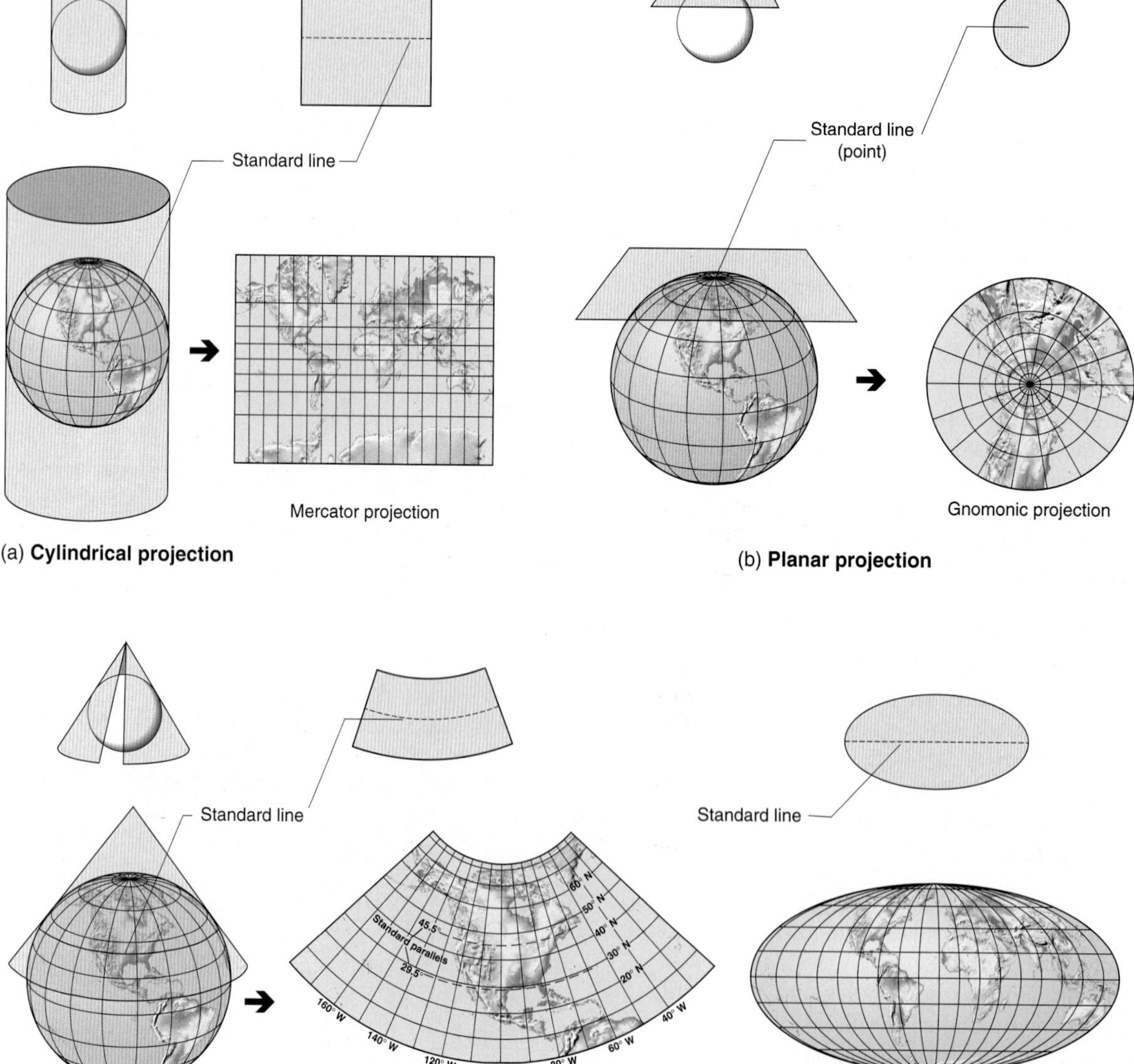

FIGURE 1.18 Map projection classes.
Four general classes and perspectives of map projections—cylindrical, planar, conic, and oval projections.

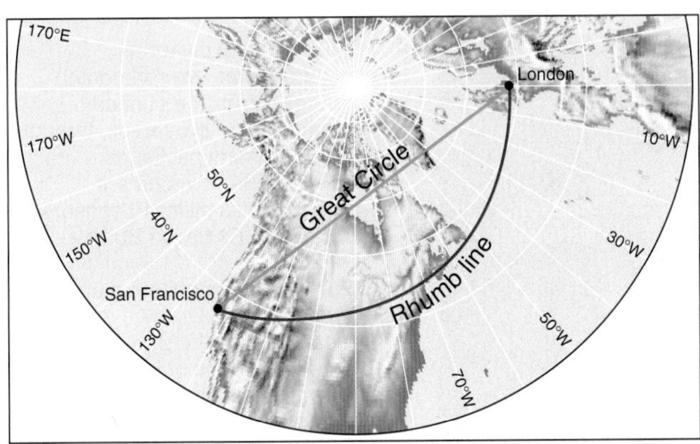

(a) Gnomonic Projection

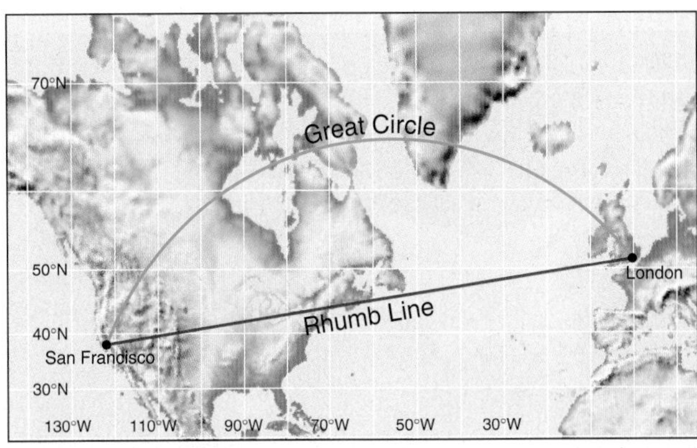

(b) Mercator Projection (conformal, true shape)

FIGURE 1.19 Determining great-circle routes.
Comparison of rhumb lines and great-circle routes from San Francisco to London on gnomonic (a) and Mercator (b) projections. All great-circle routes are the shortest distance between any two points because on the gnomonic map an arc of a great circle is a straight line. Note that straight lines of constant direction (or bearing) on a Mercator projection—rhumb lines—are not the shortest route in terms of distance.

meridian is a standard line true to scale along its entire length without any distortion. Areas away from this critical tangent line or point become increasingly distorted. Consequently, this area of optimum spatial properties should be centered on the region of greatest interest so that greatest accuracy is preserved.

The commonly used **Mercator projection** (from Gerardus Mercator, A.D. 1569) is a cylindrical projection (Figure 1.18a). I chose a Mercator projection to show cylindrical projections in Figure 1.18. The Mercator is a true-shape projection, with meridians appearing as equally spaced straight lines and parallels appearing as straight lines that come closer together near the equator. The poles are infinitely stretched, with the 84th north parallel and 84th south parallel fixed at the same length as that of the equator. Locally, the shape is accurate and recognizable for navigation; however, the scale varies with latitude.

Unfortunately, Mercator classroom maps present false notions of the size (area) of midlatitude and poleward landmasses. A dramatic example on the cylindrical projection in Figure 1.18a is Greenland, which looks bigger than all of South America. In reality, Greenland is only one-eighth the size of South America and is actually 20% smaller than Argentina alone!

The advantage of the Mercator projection is that lines of constant direction, called **rhumb lines**, are straight and thus facilitate plotting directions between two points (Figure 1.19b). Thus, the Mercator projection is useful in navigation and has been the standard for nautical charts prepared by the National Ocean Service since 1910 (formerly U.S. Coast and Geodetic Survey).

The *gnomonic* or *planar projection* in Figure 1.18b (and 1.19a) is generated with a light source at the center of a globe projecting onto a plane touching the globe's surface. The resulting severe distortion prevents showing a full hemisphere on one projection. However, a valuable feature is derived: All great-circle routes, which are the shortest distance between two points on Earth's surface, are projected as straight lines. The great-circle routes plotted on a gnomonic projection then can be transferred to a true-direction projection, such as the Mercator, for determination of precise compass headings (Figure 1.19b). For more information on maps used in this text and standard map symbols, turn to Appendix A, "Maps Used in This Text and Topographic Maps."

Remote Sensing and GIS

Geographers now probe, analyze, and map our home planet through remote sensing and geographic information systems (GIS). These technologies are enhancing our understanding of Earth. Geographers use remote-sensing data to study weather, water vapor, humid and arid lands, vegetation, snow and ice, the seasonal variation of atmospheric and oceanic circulation and temperatures, sea level, atmospheric chemistry, geologic features, soil losses and sedimentation, events, and the human activities that are producing global change.

Remote Sensing

In this era of observations from orbit outside the atmosphere and from aircraft within it, scientists are obtaining a wide array of remotely sensed data (Figure 1.20). Remote sensing is an important tool for the spatial analysis of Earth's environment. You will see more than 100 remote sensing images in this textbook.

When we observe the environment with our eyes, we are sensing the shape, size, and color of objects from a distance, utilizing the visible-wavelength portion of the electromagnetic spectrum. Similarly, the film in a camera senses the wavelengths for which it was designed (visible light or infrared) and is exposed by the energy that is reflected and emitted from a scene. Our eyes and cameras both represent familiar means of **remote sensing** of information about a distant subject, without physical contact.

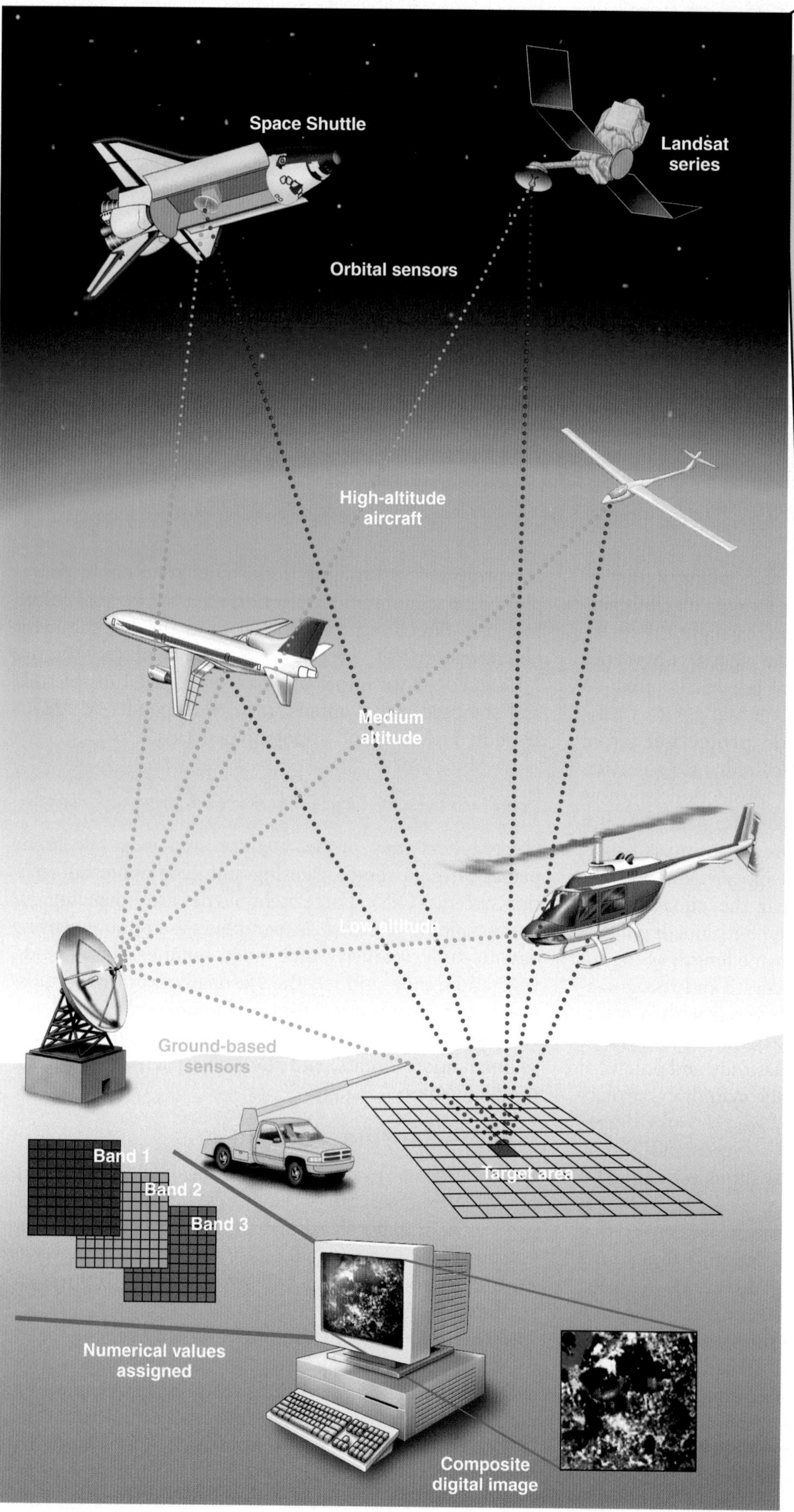

A sample of orbital platforms

- ***Terra*** **and *Aqua:*** environmental change, error-free surface images, cloud properties, through five instrument packages
- ***ENVISAT:*** ESA environment-monitoring satellite; 10 sensors
- ***ERBS:*** Earth Radiation Budget Satellite
- ***GOES:*** weather monitoring and forecasting
- ***Landsat:*** First in 1972 to *Landsat-7* in 1999, millions of images for Earth systems science and global change
- **NOAA*:*** First in 1978 through *NOAA-15* and *NOAA-16*, global data, short- and long-term weather forecasts
- **TRMM*:*** Tropical Rainfall Measuring Mission, includes lightning detection, and global energy budget measurements
- ***UARS:*** since 1991 measuring atmospheric chemistry and ozone layer changes
- ***SeaStar:*** carries the SeaWiFS (Sea-viewing Wide Field-of-View instrument) to observe Earth's oceans and microscopic marine plants
- **TOMS-EP*:*** Total Ozone Mapping Spectrometer, monitoring stratospheric ozone, similar instruments on *NIMBUS-7* and *Meteor-3*

For more info see:
http://www.gsfc.nasa.gov/indepth/earth_esm.html

FIGURE 1.20 Remote-sensing technologies.
Remote sensing from orbiting spacecraft, aircraft in the atmosphere, and ground-based sensors measure and monitor Earth's systems. Computers process data to produce digital images for analysis. Many of the physical systems discussed in this text are studied using these technologies. A sample of remote sensing platforms is in the margin. A Space Shuttle is shown in its inverted orbital flight mode (illustration not to scale).

Remote sensors on satellites and other craft sense a broader range of wavelengths than do our eyes and photographic films. This is because various surface materials absorb and reflect the Sun's radiation in different and characteristic ways. These differences can be remotely sensed by their reflected wavelengths: ultraviolet, visible light, reflected infrared (near- and shortwave), thermal infrared, and microwave radar (see Figure 2.6 for an illustration of these different wavelengths).

Satellites do not take conventional film photographs. Rather, they record electronic *digital images* that are transmitted to Earth-based receivers in a manner similar to television broadcasts. A scene is scanned and broken into *pixels* (picture elements) identified by coordinates known as *lines* (horizontal rows) and *samples* (vertical columns). You can see that the pixel count for an image runs well into the millions.

Digital data are processed in many ways to enhance their utility: false and simulated natural color, enhanced contrast, signal filtering, and different levels of sampling and resolution. Active and passive are two types of remote-sensing systems.

Active Remote Sensing. Active systems direct a beam of energy at a surface and analyze the energy that is reflected back. An example is *radar* (*ra*dio *d*etection *a*nd *r*anging). A radar transmitter emits short bursts of energy that have relatively long wavelengths (0.3 to 10 m) toward land or water, penetrating clouds and darkness. Energy reflected back, known as *backscatter*, is received by a radar receiver and analyzed. An example is the computer image of wind and sea-surface patterns over the Pacific in Figure 4.6a, developed from 150,000 radar-derived measurements made on a single day by the *Seasat* satellite.

In addition, the National Aeronautics and Space Administration (NASA) sent imaging radar systems into orbit on several Space Shuttle missions. The subjects of study included oceanography, landforms and geology, and biogeography. Shuttle missions in 1994 by *Endeavour* and *Atlantis* marked dramatic contributions to Earth observations using radar and other sensors to study stratospheric ozone, weather, volcanic activity, earthquakes, and water resources, among many subjects.

One Space Shuttle mission in September 1994 was appropriately loaded with radar sensors to study volcanoes. Only eight hours after launch, the Kliuchevskoi Volcano on the Kamchatka Peninsula of Russia erupted unexpectedly. The shuttle radar was able to see through ash and smoke and expose lava flows and the volcanic eruption in dramatic images (Figure 1.21). Previously this volcano had erupted in 1737 and 1945. Astronaut Mission Specialist Dr. Thomas Jones operated the radar and camera to make the image and photo in the figure. He is profiled in a Career Link at the end of this chapter. Chapter 9 discusses volcanic processes.

The European Space Agency (ESA) now operates three Earth resource satellites. *ERS 1* and *2* work in tandem, producing a spectacular 10-cm (3.9-in.) resolution, imaging the same area at different times, producing a three-dimensional data set. ESA's newest environment-monitoring satellite, *Envisat*, went into service in 2002. In addition to a more sophisticated radar than *ERS*, *Envisat* carries nine passive sensor packages, making it quite a multitalented craft. It monitors ocean temperatures, sea level, wave patterns, polar ice, forests, biological activity in the oceans and on land, cloud heights, atmospheric ozone and pollution, and carbon dioxide concentrations (see http://envisat.esa.int/). Along with these ESA satellites, the Canadian satellite, *Radarsat*, and the Japanese, *JERS-1*, image in radar wavelengths.

Passive Remote Sensing. Passive remote-sensing systems record energy radiated from a surface, particularly visible light and infrared. Our own eyes are passive remote sensors, as was the *Apollo 17* astronaut camera that took the picture of Earth on the back cover of this book, or the *Terra* satellite image of Alaska on this book's title page.

Passive remote sensors on five *Landsat* satellites launched by the United States provide a variety of data, as shown in images of the Appalachian Mountains in Chapter 9, river deltas in Chapter 11, and the Malaspina and other glaciers in Alaska in Chapter 14. Three *Landsats* are operational—*Landsats 4, 5*, and the newest *Landsat 7*—although *Landsat 4* no longer gathers images and is used for orbital tests. The Worldwide Reference System (WRS) makes it easy to sample *Landsat* satellite imagery worldwide by specifying exact image coordinates. See http://ltpwww.gsfc.nasa.gov/LANDSAT/CAMPAIGN_DOCS/MAIN/Outline.html for an image gallery and other links.

National Oceanic and Atmospheric Administration (NOAA) polar-orbiting satellites carry the *advanced very high resolution radiometer* (*AVHRR*) sensors aboard *NOAA-14*, *NOAA-15*, and *NOAA-16*, which are sensitive in visible and infrared wavelengths (NOAA, see http://www.noaa.gov/). The incredible image of Hurricane Andrew (Chapter 5), among others in this text, were produced by an AVHRR system. AVHRR is primarily used for sensing day or night clouds, snow, and ice; for monitoring forest fires and surface temperature; and for determining natural and planted vegetation patterns.

Key to NASA's Earth Observing System (EOS) is satellite *Terra*, which began beaming back data and images in 2000 (see http://terra.nasa.gov/), followed in 2002 by another satellite in the series called *Aqua*. Five instrument packages observe Earth systems in detail, exploring the atmosphere, landscapes, oceans, environmental change, climate, and wildfires, among other abilities. For example, the Clouds and the Earth's Radiant Energy System (CERES) instruments aboard *Terra* monitor the Earth's energy balance, giving new insights into climate change (see Chapter 3). These monitors offer the most accurate global radiation and energy measurements ever available. Another instrument set, the Moderate-resolution Imaging Spectroradiometer (MODIS), sees all of

FIGURE 1.21 A volcanic eruption seen from orbit. (a) Photograph (passive, visible light) and (b) image (active, radar) of the eruption of the Kliuchevskoi Volcano on the Kamchatka Peninsula of Siberia, Russia, as captured by the Space Shuttle *Endeavour*, September 1994. (c) *Terra* image of the Kamchatka Peninsula made June 26, 2002. Snow cover highlights the many active volcanoes. [(a) and (b) JPL photo/NASA; (c) *Terra* image, June 26, 2002, from MODIS Land Rapid Response Team, NASA/GSFC.]

Earth's surface every 1–2 days in 36 spectral bands, thereby expanding on AVHRR capabilities.

GOES-10, on line in 1998, operates above 135° W longitude to monitor the West Coast and the eastern Pacific Ocean. *GOES-8* is positioned above 75° W longitude to monitor central and eastern North America and the western Atlantic. *GOES-11* and *GOES-12*, launched in 2000 and 2001 respectively, are stored in orbit ready to replace the older satellites in the *GOES* series, as needed (Figure 1.22). See the Geostationary Satellite Server at http://www.goes.noaa.gov/, *GOES* Project Science at http://rsd.gsfc.nasa.gov/goes/, or http://www.ghcc.msfc.nasa.gov/GOES/. The image of Earth on the inside front cover of this book includes a cloud snapshot from *GOES* combined with *Terra* MODIS images over a 16-day period in 2000.

Other satellites used for weather include Japan's *GMS-5* weather satellite and China's *Feng Yun-2* covering the Far East, and *METEOSAT-7* for Europe and Africa, operated by ESA. (See the Remote Sensing Virtual Library at http://www.vtt.fi/tte/research/tte1/tte14/virtual for remote sensing links; click on "Satellite Data" for specific coverage.)

Geographic Information Systems (GIS)

Remote sensing is an important tool for acquiring large volumes of spatial data for powerful information handling systems. The next step is storing, processing, and retrieving those data in useful ways. Computers allow integration of geographic information from direct surveys (on-the-ground mapping) and remote sensing in complex ways never before possible.

A **geographic information system (GIS)** is a computer-based, data-processing tool for gathering, manipulating, and analyzing geographic information. Through a GIS, Earth and human phenomena are analyzed over time. GIS is a rapidly expanding career field in

FIGURE 1.22 ***GOES-12*** **first image.** New environmental satellite in 2001, the first image by *GOES-12* demonstrates excellent image quality from its 37,500 km (23,300 mi) orbital post. This satellite, along with *GOES-11*, is stored in orbit to replace either of the existing satellites as needed. [Image courtesy of NOAA.]

many sectors of the economy. Regardless of your academic major, the ability to analyze data spatially is important. Be sure to check some of the URLs listed for information on professional career directions in this exciting field (News Report 1.2).

The beginning component for any GIS is a coordinate system such as latitude-longitude, which establishes reference points against which to position data. The coordinate system is digitized, along with all areas, points, and lines. Remotely sensed imagery and data are then added on the coordinate system.

News Report 1.2

Careers in GIS

Geographic information system (GIS) methodology offers great career opportunities in industry, government, business, criminal justice, marketing, teaching, sales, military, and other fields. Right now, geographers trained in GIS analyze all aspects of Earth's physical systems and human activities. They map ecosystems and monitor the declining diversity of plant and animal species. In the chapter-opening computer map you can see how the spread of the West Nile virus is predicted through GIS overlays of temperature, vegetation, and humidity, to determine areas favorable to the mosquito carrier. Geographers plan, design, and survey urban developments, follow the trends of global warming, study the impact of human population, and analyze air and water pollution.

We are in the midst of a real GIS revolution! Potent new careers are emerging in almost every academic area as GIS training programs are implemented. For a list of current trends in GIS simply enter this topic in your search engine; for an alphabetized GIS resources list check out http://www.geo.ed.ac.uk/home/giswww.html.

GIS degree programs are available at many colleges and universities. GIS curriculum and certificate programs are now available at many community colleges. A consortium of three universities forms the National Center for Geographic Information and Analysis (NCGIA) for GIS education, research, outreach, and model generation. These GIS centers are Department of Geography, University of California Santa Barbara, Santa Barbara, CA 93106; Department of Surveying and Engineering, University of Maine, Orono, ME 04669; and State University of New York–Buffalo, Buffalo, NY 14260. (See http://www.ncgia.ucsb.edu/ or http://www.ncgia.maine.edu/ or http://www.geog.buffalo.edu/ncgia.)

A GIS is capable of analyzing patterns and relationships within a single data plane, such as the floodplain or soil layer in Figure 1.23. The GIS also can generate an *overlay analysis* in which two or more data planes interact. Various assumptions, comparisons, and policies can be tested. When the layers are combined, the resulting synthesis—a *composite overlay*—is a valuable product, ready for use in analyzing complex problems. Such a composite is shown for the state of Washington. A research study may follow specific points or areas through the complex of overlay planes. The utility of a GIS compared with that of a fixed static map is the ability of the GIS to manipulate the variables for analysis—a GIS is a dynamic and changeable map.

Before the advent of computers, an environmental-impact analysis required someone to gather data and painstakingly hand produce overlays of information to determine positive and negative impacts of a project or event. Today, a computer-driven GIS handles this layered information, which assesses the complex interconnections among different components. In this way, subtle changes in one element of a landscape may be identified as having a powerful impact elsewhere. (For GIS resources and information, see http://www.geo.ed.ac.uk/home/giswww.html and http://www.usgs.gov/research/gis/title.html.)

One of the most extensive and longest-operating systems is the Canada Geographic Information System (CGIS). Environmental data about natural features, resources, and land use were taken from maps, aerial photographs, and orbital sources, reduced to map segments, and entered into the CGIS. The development of this system has progressed with the ongoing Canada land inventory project. As an example, see the use of GIS in Canada's national park system, http://www2.parkscanada.gc.ca/natress/inf_pa1/GIS/GIS_E.HTM

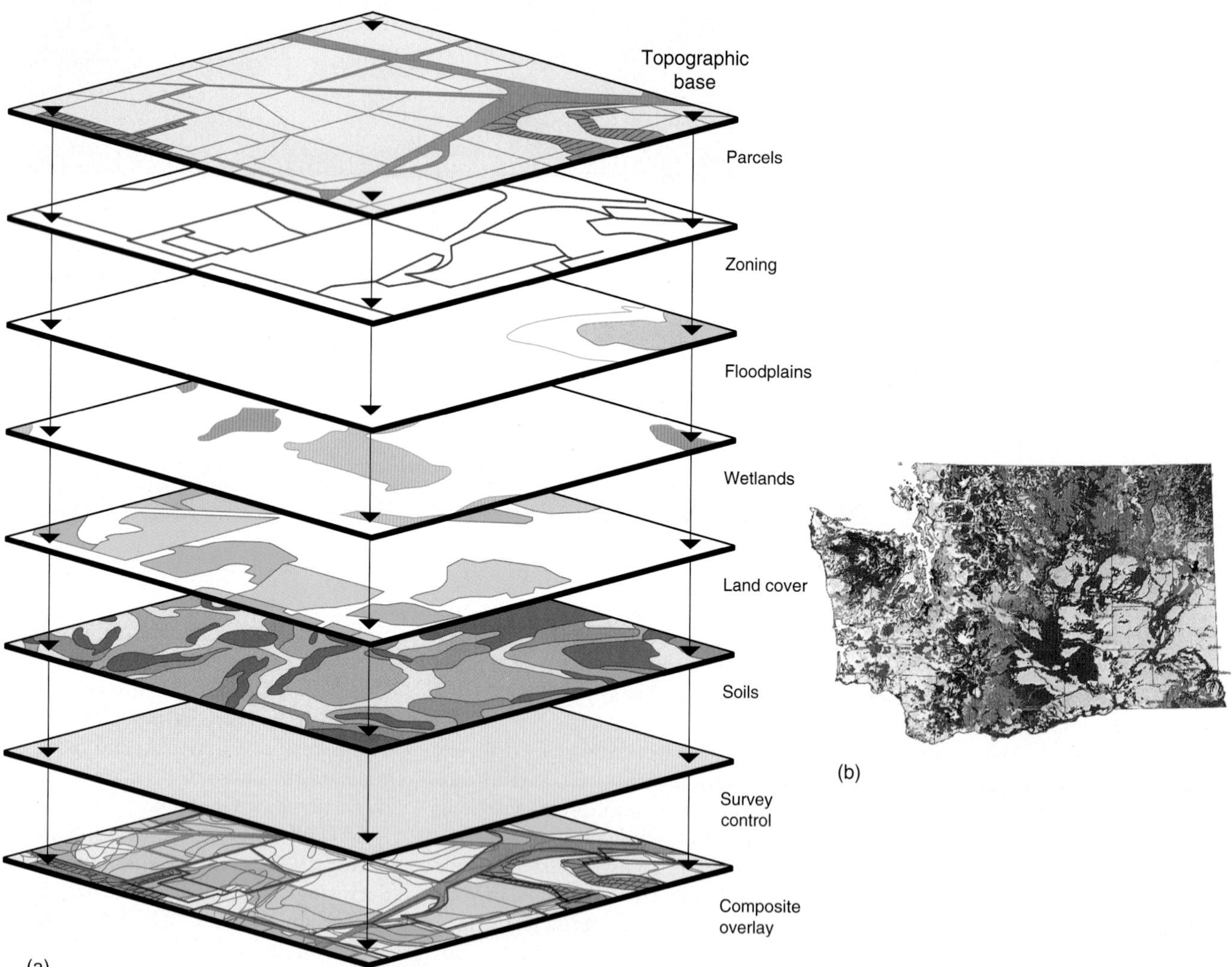

FIGURE 1.23 A geographic information system (GIS) model and map output.
(a) Layered spatial data in a GIS format. (b) Comprehensive land-cover GIS map of Washington state using 100-hectare polygons for compositing data from satellite, aerial photos, and ground checking. Colors denote 35 types of cover, divided between eastern and western regions. The goal is to evaluate the protection of species and biodiversity. [(a) After USGS. (b) Courtesy of Kelly M. Cassidy, *GAP Analysis of Washington State, An Evaluation of the Protection of Biodiversity*, Vol. 5, Map 3, p. 143, Washington Cooperative Fish and Wildlife Research Unit, Seattle, Washington, 1997.]

Summary and Review—Foundations of Geography

To assist you, here is a review of the Key Learning Concepts listed on this chapter's title page, in handy summary form. Each concept review concludes with a list of the key terms from the chapter, their page numbers, and review questions. Such summary and review sections follow each chapter in this text.

● *Define* geography and physical geography in particular.

Geography brings together disciplines from the physical and life sciences with the cultural and human sciences to attain a holistic view of Earth—an essential aspect of the emerging **Earth systems sciences. Geography** is a science of method, a special way of analyzing phenomena over space; **spatial** refers to the nature and character of physical space. Geography integrates a wide range of subject matter. Geographic education recognizes five major themes: **location, place, movement, region**, and **human-Earth relationships** (including environmental concerns). Geography's **spatial analysis** method is used to study the interdependence among geographic areas, natural systems, society, and cultural activities over space. **Process**—that is, analyzing a set of actions or mechanisms that operate in some special order—is central to geographic synthesis.

Physical geography applies spatial analysis to all the physical elements and processes that make up the environment: energy, air, water, weather, climate, landforms, soils, animals, plants, and Earth itself. Understanding the complex relations among these elements is important to human survival because Earth's physical systems and human society are so intertwined. The development of hypotheses and theories about the Universe, Earth, and life involves the **scientific method**.

Earth systems science (p. 2)
geography (p. 2)
spatial (p. 2)
location (p. 2)
place (p. 2)
movement (p. 2)
region (p. 2)
human-Earth relationships (p. 2)
spatial analysis (p. 2)
process (p. 4)
physical geography (p. 4)
scientific method (p. 4)

1. What is unique about the science of geography? On the basis of information in this chapter, define physical geography and review the geographic approach.
2. Assess your geographic literacy by examining atlases and maps. What types of maps have you used—political? physical? topographic? Do you know what projections they employed? Do you know the names and locations of the four oceans, seven continents, and individual countries? Can you identify the new countries that have emerged since 1990?
3. Suggest a representative example for each of the five geographic themes and use that theme in a sentence.
4. Have you made decisions today that involve geographic concepts discussed within the five themes presented? Explain briefly.

● *Describe* systems analysis and open and closed systems, and *relate* these concepts to Earth systems.

A **system** is any ordered, related set of things and their attributes, as distinct from their surrounding environment. Systems analysis is an important organizational and analytical tool used by geographers. Earth is an **open system** in terms of energy, receiving energy from the Sun, but is essentially a **closed system** in terms of matter and physical resources.

As a system operates, "information" is returned to various points in the system via pathways called **feedback loops**. If the feedback information discourages response in the system, it is called **negative feedback**. (Further production in the system decreases the growth of the system.) If feedback information encourages response in the system, it is called **positive feedback**. (Further production in the system stimulates the growth of the system.) When the rates of inputs and outputs in the system are equal and the amounts of energy and matter in storage within the system are constant (or as they fluctuate around a stable average), the system is in **steady-state equilibrium**. Geographers often construct a simplified **model** of natural systems to better understand them.

Four immense open systems powerfully interact at Earth's surface: three nonliving **abiotic** systems (**atmosphere, hydrosphere**, and **lithosphere**) and a living **biotic** system (**biosphere**, or **ecosphere**).

system (p. 5)
open system (p. 5)
closed system (p. 5)
feedback loops (p. 6)
negative feedback (p. 6)
positive feedback (p. 6)
steady-state equilibrium (p. 7)
model (p. 10)
abiotic (p. 11)
biotic (p. 11)
atmosphere (p. 11)
hydrosphere (p. 12)
lithosphere (p. 12)
biosphere (p. 12)
ecosphere (p. 12)

5. What is a system? What are open systems, closed systems, and negative feedback? When is a system in a steady-state equilibrium condition? What type of system (open or closed) is a human body? A lake? A wheat plant?
6. Describe Earth as a system in terms of both energy and matter.
7. What are the three abiotic spheres (nonliving) that make up Earth's environment? Relate these to the biotic (living) sphere: the biosphere.

● *Explain* Earth's reference grid: latitude, longitude, and latitudinal geographic zones and time.

The science that studies Earth's shape and size is **geodesy**. Earth bulges slightly through the equator and is oblate (flattened) at the poles, producing a misshapen spheroid called a **geoid**. Absolute location on Earth is described with a specific reference grid of **parallels** of **latitude** (measuring distances north and south of the equator) and **meridians** of **longitude** (measuring distances east and west of a prime meridian). A historic breakthrough in navigation and timekeeping occurred with the establishment of an international **prime meridian** (0° through Greenwich, England,) and the invention of precise chronometers that enabled accurate measurement of longitude. Latitude, longitude, and elevation are accurately calibrated using a handheld **global positioning system (GPS)** instrument that reads radio signals from satellites. A **great circle** is any circle of Earth's circumference whose center coincides with the center of Earth. Great circle routes are the shortest distance between two points on Earth. **Small circles** are those whose centers do not coincide with Earth's center.

The prime meridian provided the basis for **Greenwich Mean Time (GMT)**, the world's first universal time system. A corollary of the prime meridian is the 180° meridian, the **International Date Line**, which marks the place where each day officially begins. Today, **Coordinated Universal Time (UTC)** is the worldwide official time standard and the basis for international time zones. **Daylight saving time** is a seasonal change of clocks by one hour in summer months.

geodesy (p. 12)
geoid (p. 12)
latitude (p. 13)
parallel (p. 13)
longitude (p. 14)
meridian (p. 14)
prime meridian (p. 14)
global positioning system (GPS) (p. 14)
great circle (p. 14)
small circles (p. 15)
Greenwich Mean Time (GMT) (p. 18)
International Date Line (p. 18)
Coordinated Universal Time (UTC) (p. 19)
daylight saving time (p. 20)

8. Draw a simple sketch describing Earth's shape and size.
9. What are the latitude and longitude coordinates (in degrees, minutes, and seconds) of your present location? Where can you find this information?
10. Define latitude and parallel and define longitude and meridian using a simple sketch with labels.
11. Identify the various latitudinal geographic zones that roughly subdivide Earth's surface. In which zone do you live?
12. What does timekeeping have to do with longitude? Explain this relationship. How is Coordinated Universal Time (UTC) determined on Earth?
13. What and where is the prime meridian? How was the location originally selected? Describe the meridian that is opposite the prime meridian on Earth's surface.
14. Define a great circle, great circle routes, and a small circle. In terms of these concepts, describe the equator, other parallels, and meridians.

● *Define* cartography and mapping basics: map scale and map projections.

A **map** is a generalized view of an area, usually some portion of Earth's surface, as seen from above, and greatly reduced in size. The science and art of mapmaking is called **cartography**. Maps are used by geographers for the spatial portrayal of Earth's physical systems. **Scale** is the ratio of the image on a map to the real world; it relates a unit on the map to an identical unit on the ground. Cartographers create **map projections** for specific purposes, selecting the best compromise of projection for each application. Compromise always is necessary because Earth's round, three-dimensional surface cannot be exactly duplicated on a flat, two-dimensional map. **Equal area** (equivalence), **true shape** (conformality), true direction, and true distance all are considerations in selecting a projection. Straight lines on a **Mercator projection** are called **rhumb lines** of constant direction.

map (p. 20)
cartography (p. 20)
scale (p. 20)
map projections (p. 21)
equal area (p. 22)
true shape (p. 22)
Mercator projection (p. 23)
rhumb lines (p. 23)

15. Define cartography. Explain why it is an integrative discipline.
16. What is map scale? In what three ways is it expressed on a map?
17. State whether each of the following ratios is a large scale, medium scale, or small scale: 1:3,168,000, 1:24,000, 1:250,000.
18. Describe the differences between the characteristics of a globe and those that result when a flat map is prepared.
19. What type of map projection is used in Figure 1.11? In Figure 1.14? (See Appendix A for a discussion of maps used in this text.)

● *Describe* remote sensing and *explain* geographic information systems (GIS) as tools used in geographic analysis.

The operation of Earth's systems is disclosed through orbital and aerial **remote sensing**. Satellites do not take photographs, but record images that are transmitted to Earth-based receivers. Satellite images are recorded in digital form for later processing, enhancement, and generation. Aerial photographs have been used for years to improve the accuracy of surface maps.

The mountain of data already collected has led to the development of **geographic information system (GIS)** technology. Computers process geographic information from direct surveys and remote sensing in complex ways never before possible. GIS methodology is an important step in better understanding Earth's systems and is a vital career opportunity for geographers. The science of physical geography is in a unique position to synthesize the spatial, environmental, and human aspects of our increasingly complex relationship with our home planet—Earth.

remote sensing (p. 23)
geographic information system (GIS) (p. 26)

20. What is remote sensing? What are you viewing when you observe a weather satellite image on TV or in the newspaper? Explain.

21. If you were in charge of planning for development of a large tract of land, how would GIS methodologies assist you? How might planning and zoning be affected if a portion of the tract in the GIS is a floodplain or prime agricultural land?

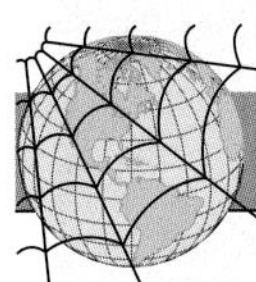

NetWork

The *Elemental Geosystems Home Page* provides on-line resources for this chapter on the World Wide Web. Once on the Home Page, click on this textbook, scroll the Table of Contents menu, select this chapter, and click "Begin." You will find self-tests that are graded, review exercises, specific updates for items in the chapter, and in "Destinations" many links to interesting related pathways on the Internet. *Elemental Geosystems* is found at **http://www.prenhall.com/christopherson.**

Critical Thinking

A. Select a location (your campus, home, workplace, a public place, or a city) and determine the following: latitude, longitude, and elevation. Describe the resources you used to gather this geographic information. Have you ever used a GPS unit to determine these aspects of your location?

B. The various geographic information technologies discussed in your text (GIS, GPS, remote sensing) promise to revolutionize many aspects of modern life. How are they used now, and what sorts of changes might we expect in the near future? Use the Netsearch and Destinations sections of the *Elemental Geosystems* Web Site to assist you.

C. Reexamine the chapter-opening map image of temperature factors, vegetation, and humidity that are favorable to the spread of West Nile virus through its mosquito vector (carrier). Imagine you are a policy maker armed with this geographic information and GIS capability. What kinds of action would you take? How would you keep the public up to speed?

D. Let's say there is a world globe in the library or geography department that is 61 cm (24 in.) in diameter. We know that Earth has an equatorial diameter of 12,756 km (7926 mi), so the scale of the globe is the ratio of 61 cm to 12,756 km. We divide Earth's actual diameter by the globe's diameter (12,756 km ÷ 61 cm) and determine that 1 cm of the globe's diameter equals about _____ cm of Earth's actual diameter. Thus, the representative fraction for the globe is expressed in centimeters as 1/_____. (Hint: 1 km = 1000 m, 1 m = 100 cm, therefore Earth's diameter, 12,756 km, represents 1,275,600,000 cm.)

Career Link 1.1

Thomas D. Jones, Ph.D., Astronaut, Earth Observer, and Geographer

I first met Dr. Thomas Jones at an annual meeting of the National Council for Geographic Education (NCGE) in Indianapolis. Earth was the feature as the audience orbited the planet through NASA photos and imagery piloted by his enthusiastic geographic analysis. At one point the photo and radar image of the Kliuchevskoi volcanic eruption, on the Kamchatka Peninsula, appeared on the screen. The same photo and image are in Figure 1.21 in *Elemental Geosystems.* Afterward, I asked astronaut Tom Jones to sign the figure in a copy of my other book, *Geosystems*—a real thrill.

Here was the man that made the photo and operated the Spaceborne Imaging Radar (SIR–C/X-Band Synthetic Apature Radar) in the Space Radar Laboratory during flight STS-68. His enthusiasm for Earth observation and his spatial analysis of natural and human phenomena—from cities, to rock formations, to hydrology, and weather—makes him a real friend of geographic education. I interviewed Tom in June 2001 at the Johnson Space Center (JSC) in Building 9, where the simulators for both the Shuttle and components of the International Space Station (ISS) are housed for training (Figure 1).

Tom was born and raised in Maryland, living there through high school graduation and becoming a National Merit Scholar. I asked him if he had an early interest in geography. "Yes, elementary school geography was fascinating. Even as early as first grade, I can remember looking at maps with my older second-grade friend next door. We were fascinated by landforms, straits of water and isthmus formations of land, and shapes of continents, as we turned through each

(continued)

Career Link 1.1 *(continued)*

page of an old atlas. In Boy Scouts, I loved compass work and map reading, and all the outdoor activities. These early interests feed right into piloting, where you have a map right on your knee and must observe geography."

He received his B.S. degree as a Distinguished Graduate from the U.S. Air Force Academy and served for 6 years on active duty as pilot and commander on B-52D strategic bombers. Tom completed more than 2000 hours of flying time, achieving the rank of Captain. He stated, "Spending thousands of hours in aircraft is a good way to learn how to look carefully at landscapes from above. You look for landmarks and geographic features. I have always thought it valuable in my life to have so many hours viewing Earth from above, first in planes, then from orbit."

Tom Jones continued his education at the University of Arizona, earning a Ph.D. in planetary science. He said, "I used remote sensing to study the water and mineral content of asteroids—those chunks of rock between Mars and Jupiter that are the source of many meteors that Earth encounters. We did telescopic surveys of dark asteroids, searching for water. I tried to link these spectroscopic fingerprints to meteorites (meteors that hit Earth) that we had in the lab. Any water and minerals found on asteroids are possible future resources for space travel."

Tom joined NASA in 1990 and became an astronaut in 1991. He flew on Space Shuttle *Endeavour* as a Mission Specialist on STS-59 (April 1994) and STS-68 (October 1994). This later flight is when he captured the volcanic eruption image and photo in Figure 1.21. Tom next went into space aboard *Columbia* in STS-80 (1996), where he operated the robotic arm to launch a satellite and, incidentally, made the photo of Mount Everest at dawn seen in News Report 9.1. On his fourth flight aboard *Atlantis* in STS-98 (February 2001), he worked installing the Destiny Laboratory Module for the International Space Station (ISS).

I asked him to compare his four Shuttle launches. He explained, "They were different because of launch conditions. The main engines ignite and build up thrust, then the solid-fuel booster rockets fire, clamps holding the Shuttle upright explosively break, and you begin to lift. Whatever winds might be blowing at launch are sensed by accelerometers. They also sense any variations in thrust from the engines and begin adjusting the engine nozzles to compensate. If it is smooth outside, the ride is easy. However, a little wind sends the craft into these slight compensating moves and your seat shakes left and right, forward and back. Overall, the Shuttle is a great ride for you only feel 3.0 Gs on assent and 1.7 Gs on reentry" (1G = Earth's surface gravity).

FIGURE 1 Astronaut Thomas Jones. Astronaut Mission Specialist Thomas Jones stands in front of a Space Shuttle simulator at the Johnson Space Center, Houston, Texas. Dr. Jones logged 1272 hours in orbit aboard four Shuttle flights. [Photo by Bobbé Christopherson.]

I asked what it felt like as he fled gravity. "After the main engine cuts off some 8.5 minutes out, you go to 0 Gs and weightlessness. With the disappearance of acceleration, you are instantly in free-fall. We are strapped in so tight we don't notice much. So the first thing I did on my initial flight was unzip my glove and let it float in front of me in the cabin. 'Yep! I am really here in orbit,' I remember thinking."

I inquired about the *Endeavour* flight in October 1994. Tom said, "We were going to study volcanoes with the SIR-C/X-SAR, including the volcanic complex on Kamchatka. So it was on our charts and flight plan. We never expected that the biggest volcano in Asia would erupt right after launch and present us with such a wonderful opportunity. We saw it on our first orbit! This huge smudge on the horizon looked like the strangest thunderstorm. Then we realized that Kliuchevskoi had blown.

"Engineers on Earth quickly reprogrammed the radar between orbits and we got right over it on the second pass. We tended the radar and used our cameras to photograph. On subsequent days a storm covered the area, and when it cleared you could see the lava and hot mud streaking the fresh snow—quite a sight." Tom added, "Earth gave us a gift. We were supposed to fly 6 weeks earlier and would have missed this chance to see an active eruption."

When working outside the Shuttle above the protective layers of the atmosphere, an astronaut must wear protection from the Sun's radiation, micrometeor impacts, and the intense cold. The spacesuit must regulate the temperature differences experienced, from nearly 120°C (156°F) where the Sun strikes the spacesuit, to −250°C (−156°F) when in shadows. Oxygen and water must be provided and carbon dioxide buildup managed. Here on Earth's surface the atmosphere does all this for us. Imagine designing a spacesuit that does everything portrayed in Figure 2.18 in this book.

Tom Jones completed three extravehicular activities (EVAs) on his last flight, totaling 19 hours (Figure 2a). He and his partner Bob Curbeam installed the U.S. Destiny Laboratory Module on the ISS. Relative to his spacesuit (extravehicular mobility unit, or EMU), Tom said, "The limiting factor in a spacesuit is the carbon dioxide scrubber because you can't replenish that during a space walk. You can replenish your water supply, oxygen supply, and electrical battery charge by simply plugging in your umbilical for

(continued)

a short time. But the CO_2-scrubber needs to be replaced, and that requires going back in after 8 hours."

I asked him about any feeling of vulnerability during his EVA while he was an Earth satellite drifting in orbit at more than 28,100 kmph (17,500 mph). "Intellectually you are aware of that, and you can stop and think about your independence from things as you look over the brilliant Earth, but most of the time you are focused on the work, and the spacesuit is almost invisible to you. It works so well that you get quite comfortable, you forget that you are next to a vacuum travelling 10 times faster than a bullet, or in such a harsh environment. And since you are floating within the suit, with few pressure points on your body, you are quite comfortable."

He continued, "Despite the thermal protection of the EMU spacesuit, when the Sun rises, which it does every 90 minutes in orbit, I could feel the warmth when the light hit me—the heat energy is conducted through the layers of the suit. This is not uncomfortable, just a warming sensation. The thermal tubing in the suit that regulates temperature goes throughout the suit—ankles, to wrists, to neckline. If it does get warm in the suit, you can adjust your thermostat. With sunset, I felt that warmth drop away and when my feet were on a work platform, I felt the coldness of the metal plate and the heat energy conduct out through the feet of the suit, despite the bulky socks I wore. If you feel your feet chilling, you adjust the suit temperature up."

I asked about Earth observation time. "Everyone gets training for Earth observing, including classroom work in physical and human geography and geology. For my last mission we attended lectures and got experience during actual camera practice. We have NASA's *Space Shuttle Physiographic Atlas* (1:10,000,000 scale), from the Earth Observation Project, organized along flight paths (west to east). Clear, bold labels used in the atlas help us orient the camera to specific photographic targets. Decal black circles are placed on charts so you know what needs to be photographed.

"The Earth Sciences Team chooses three dozen or more candidates for intense observation, depending on research needs, or studies of a particular ecological phenomena, trend, or situation. We have electronic maps on our laptops, so we can click on a site and it will tell us the time when the target is within range, and it will suggest camera and film to use. Houston sends a daily Earth Observation Bulletin that lists the times for certain scenes, mission elapsed time to the target, correlating with map notations."

I asked, "After 1272 hours in Earth orbit, more than 52 days total, what are some of your thoughts about Earth?" Tom answered, "Earth never fails to amaze me and captivate me with the beauty it presents, the ever-changing aspects of light, the changing vision, for you are always seeing Earth in a new way. Every time I looked out the window I saw a different angle or lighting that changes something I might have seen many times before. I always found something new to be amazed at! Looking at Earth refreshes you instantly.

"My days in orbit are a privilege, to see Earth from that distance is a lifelong memory and one that will never leave me. Yet it comes with a sense of regret because on this last mission I know there was so much I could have seen, but we had an important mission to complete and there wasn't time to linger by the window. This is the mixed blessing of working in space."

As to the future, Tom wants to focus on further research related to his dissertation topic and nonfiction writing for the general public on space travel and possible missions to asteroids. He said, "The Moon is a close-by testing place for equipment, so we will no doubt be returning. We need to build support for future space exploration."

In closing, I asked Tom about the insignia patch for STS-98 (shown here in Figure 2b). He answered, "The crew thought it was important to show the Earth observing window in the Destiny Laboratory Module, with Earth reflected in it. Uncovering the protective coating to begin operations of this crystal-clear viewing portal was a real thrill for Bob and me on our EVA." This says it all about Tom and the enrichment he gives to geographic education—he has opened a portal for us to better see Earth, our Home Planet.

(a)

(b)

FIGURE 2 Space walk February 2001. (a) Astronaut Thomas Jones completed three space walks, totaling 19 hours, working on the International Space Station *Destiny* module installation. He is waving to crew members inside *Atlantis*. His partner in the space walks was Robert L. Curbeam. (b) Insignia for the Atlantic mission; note the reflection of Earth in the large window in the Destiny module. See http://www.spaceflight.nasa.gov/gallery/images/shuttle/sts-98/ndxpage1.html. [Space Shuttle photograph STS98-E-5195 and mission insignia courtesy of NASA.]

PART ONE

The Energy-Atmosphere System

Sunset, clouds, ocean waves, and coastal rocks near Pacific Grove, California. [Photo by Bobbé Christopherson.]

Our planet and our lives are powered by radiant energy from the star that is closest to Earth—the Sun. For more than 4.6 billion years, solar energy has traveled across interplanetary space to Earth, where a small portion of the solar output is intercepted. Because of Earth's curvature, the energy at the top of the atmosphere is unevenly distributed, creating imbalances from the equator to each pole—the equatorial region experiences energy surpluses; the polar regions experience energy deficits. This unevenness of energy receipt empowers circulations in the atmosphere, in the ocean, and on land. The pulse of seasonal change varies the distribution of energy during the year.

Earth's atmosphere acts as an efficient filter, absorbing most harmful radiation, charged particles, and space debris so that they do not reach Earth's surface. Surface energy balances are established, giving rise to global patterns of temperature, winds, and ocean currents. Each of us depends on many systems that are set into motion by energy from the Sun. These systems are the subject of Part One.

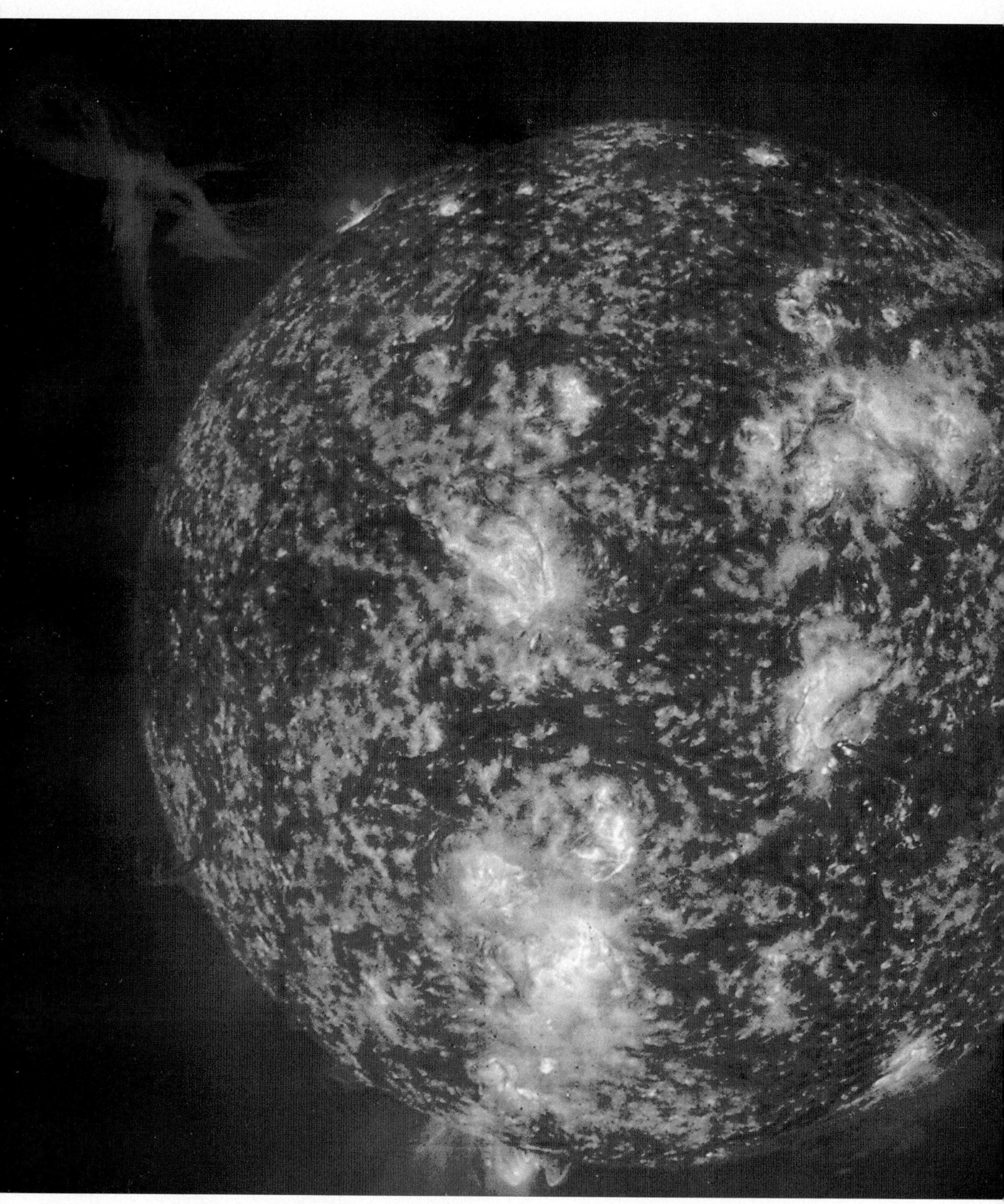

A dramatic Sun captured by instruments aboard the *SOHO* satellite, February 12, 2001. A twirling prominence rises into the Sun's corona, and other prominences are also visible. [Image courtesy of *SOHO/EIT* (Solar and Heliospheric Observatory/Extreme Ultraviolet Imaging Telescope) Consortium. *SOHO* is an international project of cooperation between the European Space Agency and NASA. See http://sohowww.nascom.nasa.gov/.]

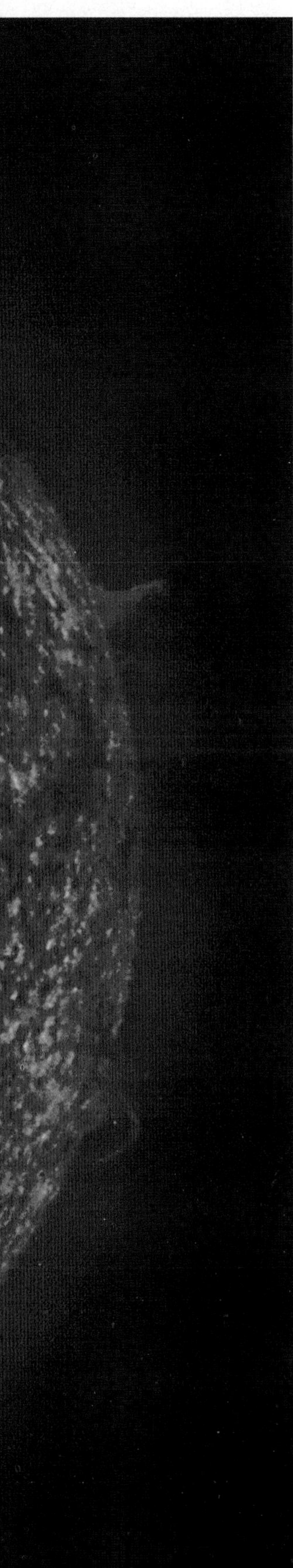

2

Solar Energy, Seasons, and the Atmosphere

Key Learning Concepts

By knowing and understanding the key learning concepts in this chapter, you should be able to:

- *Distinguish* among galaxies, stars, and planets and *locate* Earth.
- *Describe* the Sun's operation and *explain* the characteristics of the solar wind and the electromagnetic spectrum of radiant energy.
- *Define* solar altitude, solar declination, and daylength and *describe* the annual variability of each—Earth's seasonality.
- *Construct* a general model of the atmosphere based on composition, temperature, and function and *diagram* this model in a simple sketch.
- *Describe* conditions within the stratosphere; specifically, *review* the function and status of the ozonosphere (ozone layer).
- *Distinguish* between natural and anthropogenic variable gases and materials in the lower atmosphere and *describe* the sources and effects of air pollution and acid deposition.

The Universe is populated with millions of galaxies. One of these is our own Milky Way Galaxy, consisting of billions of stars. Among these stars is an average yellow star we call the Sun, although the dramatic *SOHO* satellite image that opens this chapter seems anything but average! Our Sun radiates energy in all directions and upon its family of orbiting planets. Of special interest to us is the solar energy that falls on the third planet, our immediate home.

In this chapter: Solar energy at the top of the atmosphere sets into motion the winds, weather systems, and ocean currents that greatly influence our lives. This solar energy input to the atmosphere plus Earth's tilt, orientation, revolution, and rotation produce daily, seasonal, and annual patterns of changing daylength and Sun angles.

Earth's atmosphere is a unique reservoir of gases—the product of nearly 5 billion years of development. We all participate in the atmosphere with each breath we take. We examine the atmosphere's structure, composition, and function. Our study also must include the spatial aspects of human-induced problems that affect the atmosphere, such as the stratospheric ozone problem, air pollution, and acid deposition.

The Solar System, Sun, and Earth

ANIMATION **Nebular Hypothesis**

Our Solar System is located on a remote, trailing edge of the **Milky Way Galaxy**, a flattened, disk-shaped mass estimated to contain nearly 400 billion stars. From our Earth-bound perspective in the Milky Way, the galaxy appears to stretch across the night sky like a narrow band of hazy light. On a clear night the unaided eye can see only a few thousand of these billions of stars (Figure 2.1a, b).

According to prevailing theory, our Solar System condensed from a large, slowly rotating, collapsing cloud of dust and gas called a *nebula.* **Gravity**, the mutual attracting force exerted by the mass of an object upon all other objects, was the key organizing force in this condensing solar nebula. The beginnings of the formation of the Sun and its Solar System are estimated to have occurred more than 4.6 billion years ago.

The concept that suns condense from nebular clouds and planetesimals form in orbits around their central masses is called the **planetesimal hypothesis**, or *dust-cloud hypothesis*. The planets appear to accrete (growth by accumulation) from dust, gases, and icy comets that are drawn by gravity into collision and coalescence. Astronomers are observing this formation process under way in other parts of the galaxy. Through 2002 they have tracked more than 100 planets orbiting around other stars.

Dimensions, Distances, and Earth's Orbit

The **speed of light** is 299,792 kmps (186,282 mps), which is about 9.5 trillion km or nearly 6 trillion mi per year. This tremendous distance that light travels in a year is known as a *light-year* and is used as a unit of measurement for the vast universe.

For spatial comparison, our Moon is an average distance of 384,400 km (238,866 mi) from Earth, or about 1.28 seconds in terms of light speed. Our entire Solar System is approximately 11 hours in diameter, measured by light speed. In contrast, the Milky Way is about 100,000 light-years from side to side, and the known Universe that is observable from Earth stretches approximately 12 billion light-years in all directions.

Earth's orbit around the Sun is presently elliptical—an oval-shaped path. Earth's average distance from the Sun is approximately 150 million km (93 million mi), which means that light reaches Earth from the Sun in an average of 8 minutes and 20 seconds. Earth is at **perihelion** (its closest position to the Sun) on January 3 and at **aphelion** (its farthest position from the Sun) on July 4 (Figure 2.1d).

Solar Energy: From Sun to Earth

Our Sun is unique to us and yet commonplace in our galaxy. It is the ultimate energy source for most life processes in our biosphere. The Sun is the only object in the entire Solar System that produces thermonuclear energy. Clearly the Sun is the dominant object in our region of space.

The solar mass produces tremendous pressure and high temperatures deep in its dense interior. Under these conditions, the Sun's abundant atoms of hydrogen are forced together, and pairs of hydrogen nuclei are fused (joined). This process, called **fusion**, liberates enormous quantities of energy. A sunny day can seem so peaceful, certainly belying the violence proceeding on the Sun. The Sun's principal outputs consist of the solar wind and radiant energy in portions of the electromagnetic spectrum. Let us trace each of these emissions across space to Earth.

Solar Wind

The Sun constantly emits ionized (electrically charged) particles (principally electrons and protons) that surge outward in all directions from the Sun's surface. The term **solar wind** was first applied to this phenomenon in 1958. The solar wind grows stronger during periods of increased sunspot activity. **Sunspots** are large magnetic storms that reveal solar activity (Figure 2.2). These surface disturbances produce flares and prominences. In addition, outbursts of charged material referred to as *coronal mass ejections* contribute to the solar wind flow of material to space.

A regular cycle exists for sunspot occurrences, averaging 11 years from maximum to maximum; however, the cycle may vary from 7 to 17 years. The 1990–1991 maximum was the most intense ever observed. A sunspot minimum occurred in 1997 and a maximum in 2001 maintains the average. (For more on the sunspot cycle see **http://wwwssl.msfc.nasa.gov/ssL/pad/solar/sunspots.htm**.)

Earth's outer defense against the charged particles of the solar wind is the **magnetosphere**, which is a magnetic field surrounding Earth, generated by dynamo-like motions within our planet. The magnetosphere deflects the solar wind toward both poles, so that only a small portion of it enters the atmosphere.

Because the solar wind does not reach Earth's surface, research on this phenomenon must be conducted in space. On July 20, 1969, the *Apollo XI* astronauts deployed a solar wind experiment on the lunar surface that exhibited particle impacts that confirmed the presence and character of the solar wind (Figure 2.3).

The solar wind creates **auroras** in the upper atmosphere when absorbed energy is reradiated as light energy of varying colors. These lighting effects are the *aurora borealis* (northern lights) and *aurora australis* (southern lights). The auroras generally are visible poleward of 65° latitude when the solar wind is active. More intense

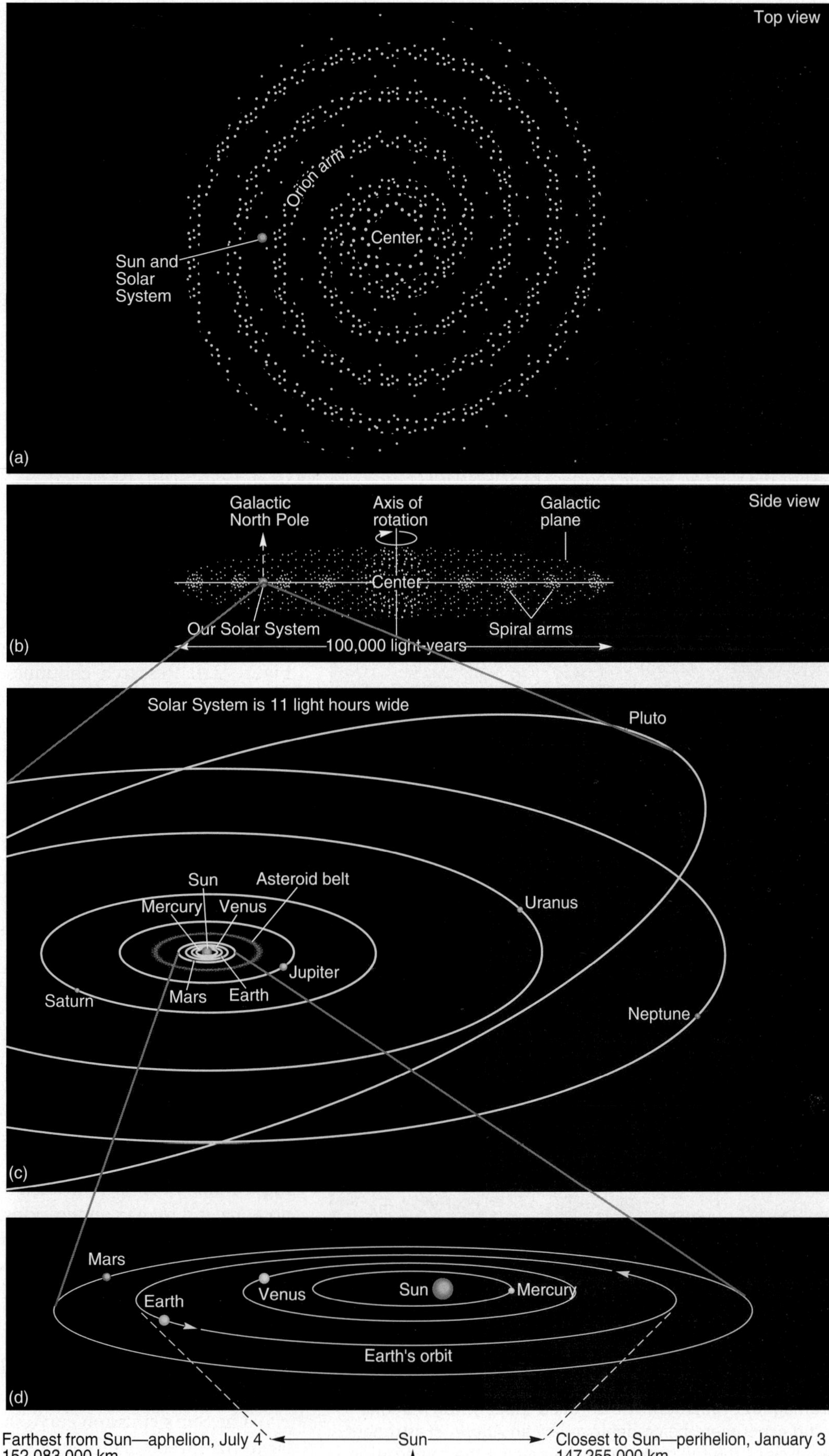

FIGURE 2.1 Milky Way Galaxy, Solar System, and Earth's orbit. The Milky Way Galaxy viewed from above (a) and cross-section side view (b). (c) Our Solar System of nine planets and asteroids is some 30,000 light-years from the center of the Galaxy. All of the planets except Pluto have orbits closely aligned to the plane of the ecliptic. (d) The four inner terrestrial planets and the structure of Earth's elliptical orbit, illustrating perihelion (closest) and aphelion (farthest) positions during the year. Have you ever observed the Milky Way Galaxy in the night sky?

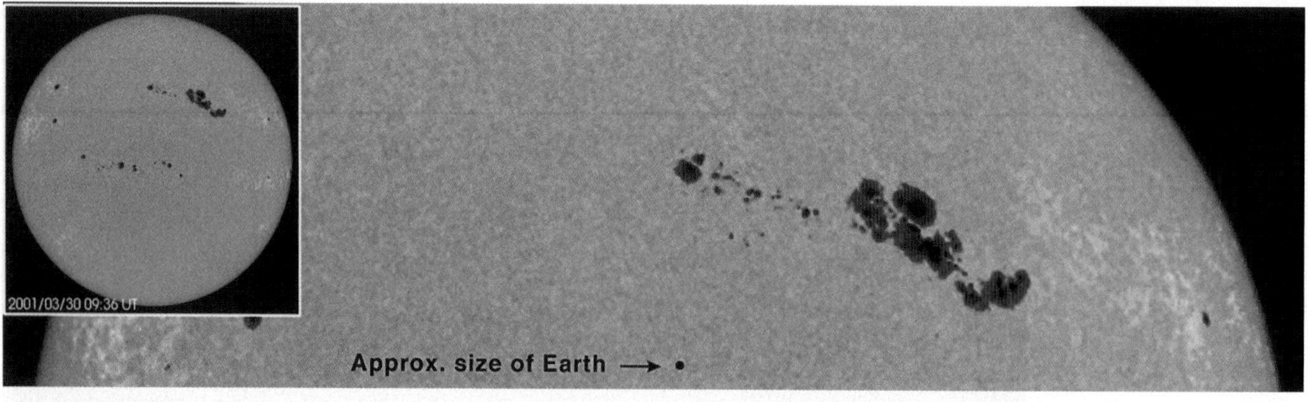

FIGURE 2.2 Images of the Sun and sunspots.
The Sun and large sunspot group in a recent active cycle, imaged March 30, 2001, by the MDI (Michelson Doppler Imager) instrument aboard satellite *SOHO*. This group was the source of numerous flares and coronal mass ejections, including the largest flare in 25 years on April 2. The area within the sunspot group is more than 13 times the entire surface area of Earth—Earth is shown for scale. Sunspots appear as visible dark patches because of their lower temperature relative to the rest of the surface. [Image courtesy of *SOHO/MDI* Consortium, NASA and ESA. *SOHO* (Solar and Heliospheric Observatory). MDI is from Stanford-Lockheed/Martin Institute for Space Research.]

FIGURE 2.3 Astronaut and solar wind experiment.
Without a protective atmosphere, the lunar surface receives charged particles of the solar wind and all of the Sun's electromagnetic radiation. The unrolled sheet of foil is a solar wind experiment being deployed by an *Apollo XI* astronaut. Analysis of the foil back on Earth revealed the composition of the solar wind. Why wouldn't this experiment work if deployed on Earth's surface? [NASA photo.]

outbursts produce auroras at lower latitudes. They appear as folded sheets of green, yellow, blue, and red light that undulate across the skies of high latitudes, as shown in Figure 2.4. Research continues as to possible links between the sunspot cycle and patterns of drought and periods of wetness—the sunspot-weather connection. (For auroral activity see http://www.sec.noaa.gov/pmap/ and http://www.gi.alaska.edu.)

Electromagnetic Spectrum of Radiant Energy

The key solar input to life is electromagnetic energy. Solar radiation occupies a portion of the **electromagnetic spectrum** of radiant energy. This radiant energy travels at the speed of light to Earth. The total spectrum of this radiant energy is made up of different wavelengths. A **wavelength** is the distance between corresponding points on any two successive waves (Figure 2.5). The number of waves passing a fixed point in one second is the *frequency*.

The Sun emits radiant energy composed of 8% ultraviolet, X-ray, and gamma ray wavelengths; 47% visible light wavelengths; and 45% infrared wavelengths. A portion of the electromagnetic spectrum is illustrated in Figure 2.6; note the wavelengths at which various phenomena and human applications of energy occur.

An important physical law states that all objects radiate energy in wavelengths related to their individual surface temperatures: the hotter the object, the shorter the wavelengths emitted. This law holds true for the Sun and Earth. Figure 2.7 shows that the *hot-body* Sun, with a surface temperature of about 6000°C (11,000°F), radiates shorter wavelength energy.

The Sun's emission curve shown in the figure is similar to that predicted for an idealized *blackbody radiator*. An ideal blackbody emits as much radiant energy as it absorbs—the hotter the blackbody, the more radiation it

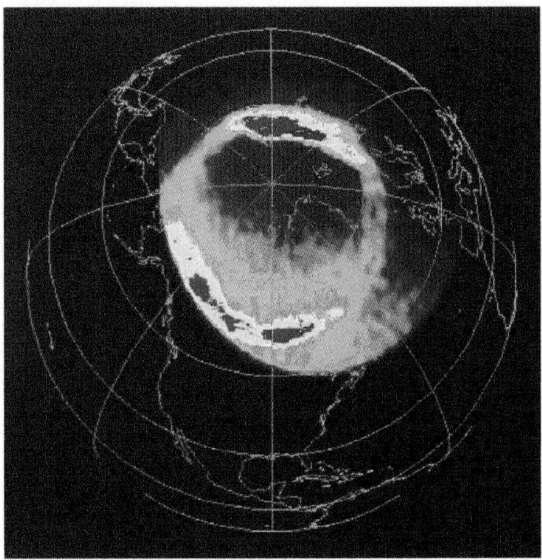

(a)

(b)

FIGURE 2.4 Auroras from space and from the ground in Alaska.
(a) *Polar* satellite false-color image of auroral halo over Earth's North Pole region from April 1996—the UVI sensor is able to capture the aurora on the day and night sides of Earth. (b) Surface view of the aurora borealis in the night sky over Alaska. [(a) Image from the Ultraviolet Imager (UVI) aboard *Polar* satellite courtesy of NASA; (b) aurora photo by Johnny Johnson/Tony Stone Images, Inc.]

emits at all wavelengths, with shorter wavelengths dominant at higher temperatures.

Earth is a cooler radiating body, so longer wavelengths are emitted. Lower temperatures at Earth's surface produce radiation mostly in the infrared portion of the spectrum.

Figure 2.8 illustrates the flows of energy into and out of Earth systems. To summarize, the solar spectrum is *shortwave radiation* that peaks in the visible wavelengths, and Earth's spectrum is *longwave radiation* concentrated in infrared wavelengths. In Chapter 3, we see that Earth, clouds, sky, ground, and all things that are terrestrial are *cool-body* radiators.

Energy at the Top of the Atmosphere

The region at the top of the atmosphere, approximately 480 km (300 mi) above Earth's surface, is the **thermopause**. It is the outer boundary of Earth's energy system and provides a useful point at which to assess the arriving solar radiation before it is diminished by passage through the atmosphere.

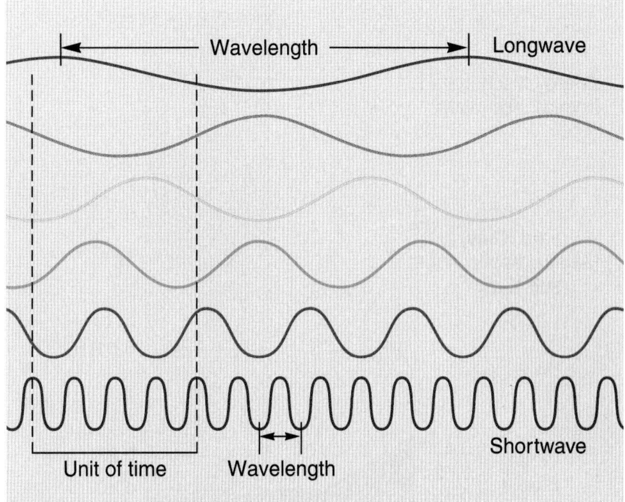

FIGURE 2.5 Wavelength and frequency.
Wavelength and frequency describe the electromagnetic spectrum. These are two ways of defining the same phenomenon—electromagnetic wave motion. Shorter wavelengths are higher in frequency, whereas longer wavelengths are lower in frequency.

Intercepted Energy and the Solar Constant. Earth's distance from the Sun results in its interception of only one two-billionth of the Sun's total energy output. Nevertheless, this tiny fraction of the Sun's overall output represents an enormous amount of energy input into Earth's systems. **Insolation** refers to *in*tercepted *sol*ar radi*ation*.

Knowing the amount of insolation intercepted by Earth is important to climatologists and other scientists. The **solar constant** is the average value of insolation received at the thermopause when Earth is at its average distance from the Sun. That value of the solar constant is 1372 watts per square meter (W/m^2).* As we follow insolation through the atmosphere to Earth's surface in Chapter 3, we see the value of the solar constant reduced by half or more through reflection, scattering, and absorption of shortwave radiation.

Uneven Distribution of Insolation. Earth's curved surface presents a continuous varying angle to the incoming parallel rays of insolation (Figure 2.9). The latitudinal variation in the angle of solar rays results in an uneven global distribution of insolation. The only point receiving

*A *watt* is equal to one joule (a unit of energy) per second and is the standard unit of power in the International System of Units (SI). (See the conversion tables in Appendix C of this text for more information on measurement conversions.) In nonmetric *calorie* heat units, the solar constant is expressed as approximately 2 calories per square centimeter per minute, or 2 *langleys* per minute (a langley being 1 cal/cm^2). A calorie is the amount of energy required to raise the temperature of one gram of water (at 15°C) one degree Celsius and is equal to 4.184 joules.

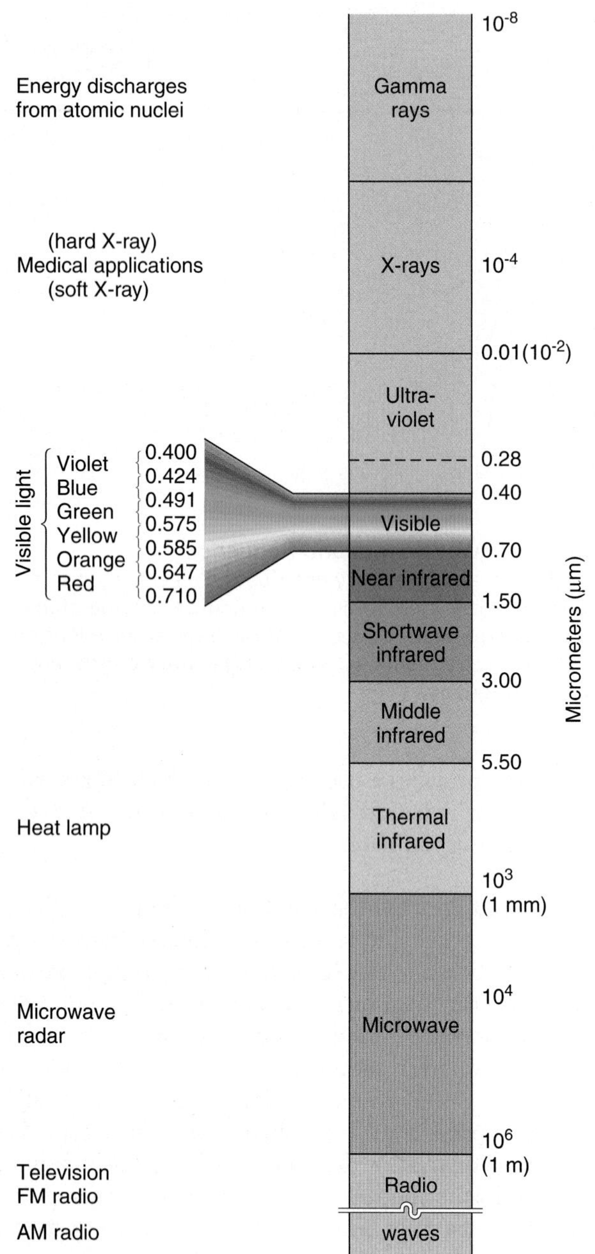

FIGURE 2.6 A portion of the electromagnetic spectrum of radiant energy.
Shorter wavelengths are toward the top of the spectrum; longer wavelengths are toward the bottom.

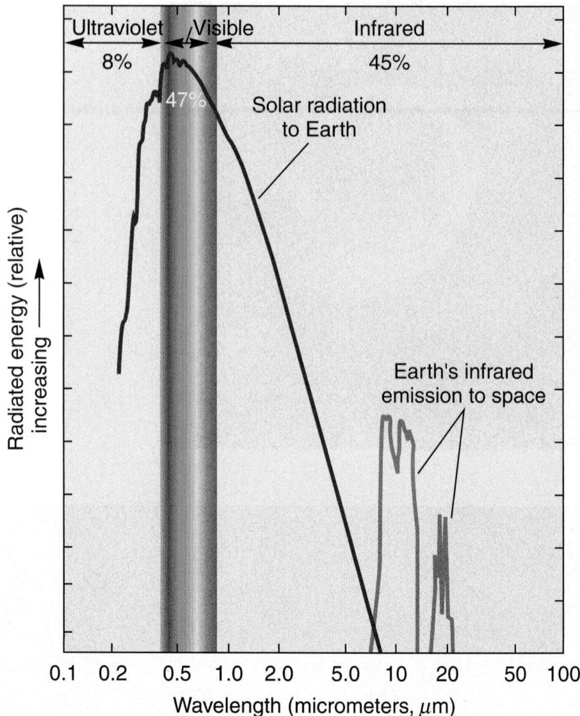

FIGURE 2.7 Solar and terrestrial energy distribution by wavelength.
The solar output peaks in shorter wavelengths of visible light in relation to its higher surface temperature, whereas Earth's emissions are concentrated in the infrared portion of the spectrum in relation to its lower surface temperature.

insolation perpendicular to the surface (from directly overhead) is the **subsolar point**. During the year this point occurs at lower latitudes between the tropics (23.5° north and 23.5° south latitudes), where the energy received is more concentrated. All other places away from the subsolar point receive insolation at an angle less than 90° and thus experience more diffuse energy; this effect is pronounced at higher latitudes.

The thermopause above the equatorial region receives 2.5 times more insolation annually than the thermopause above the poles. Of lesser importance is the fact that the lower-angle solar rays toward the poles must pass through a greater thickness of atmosphere, resulting in greater losses of energy due to scattering, absorption, and reflection.

Global Net Radiation. The *Earth Radiation Budget* (*ERB*) instrument aboard several satellites measures shortwave and longwave flows of energy at the top of the atmosphere. ERB sensors collected the data used to develop the map in Figure 2.10. This map shows *net radiation*, or the balance between incoming shortwave and outgoing longwave radiation. Note the latitudinal energy imbalance in net radiation—positive values in lower latitudes and negative values toward the poles.

In middle and high latitudes, approximately poleward of 36° north and south latitudes, net radiation is negative. This happens because Earth's climate system loses more energy to space than it gains from the Sun as measured at the top of the atmosphere. In the lower atmosphere, these polar energy deficits are offset by flows of heat energy from tropical energy surpluses, as we shall see in Chapters 3 and 4. The atmosphere and ocean form a giant heat engine, driven by differences in energy from place to place and causing major circulations within the lower atmosphere and in the ocean. The largest net radiation values are above the tropical oceans along a narrow equatorial zone, averaging 80 W/m^2. Net radiation minimums are lowest over Antarctica.

Of interest is the –20 W/m^2 area over the Sahara desert region, where usually clear skies—which permit

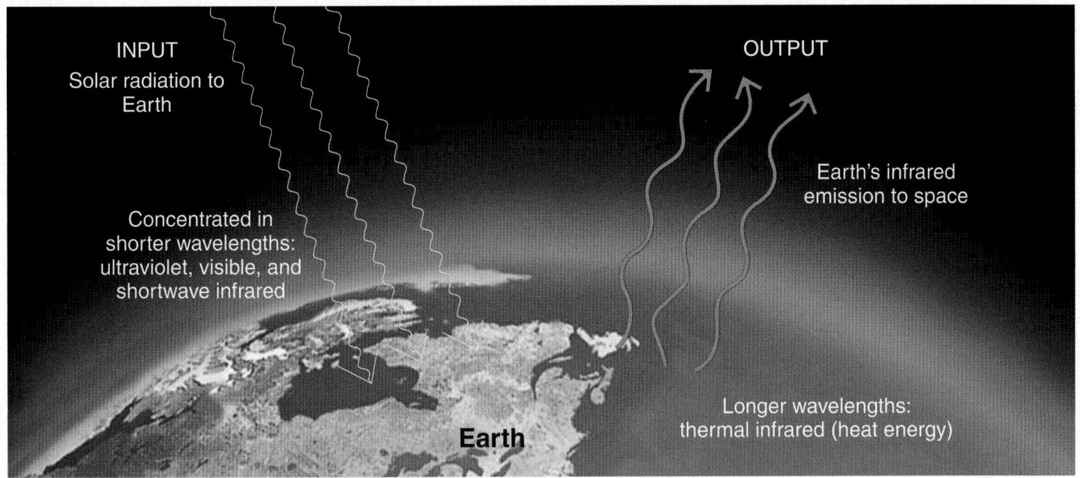

FIGURE 2.8 Earth's energy budget simplified.
Solar radiation is concentrated in shorter wavelengths. Earth emits longer wavelengths of infrared to the atmosphere and eventually to space.

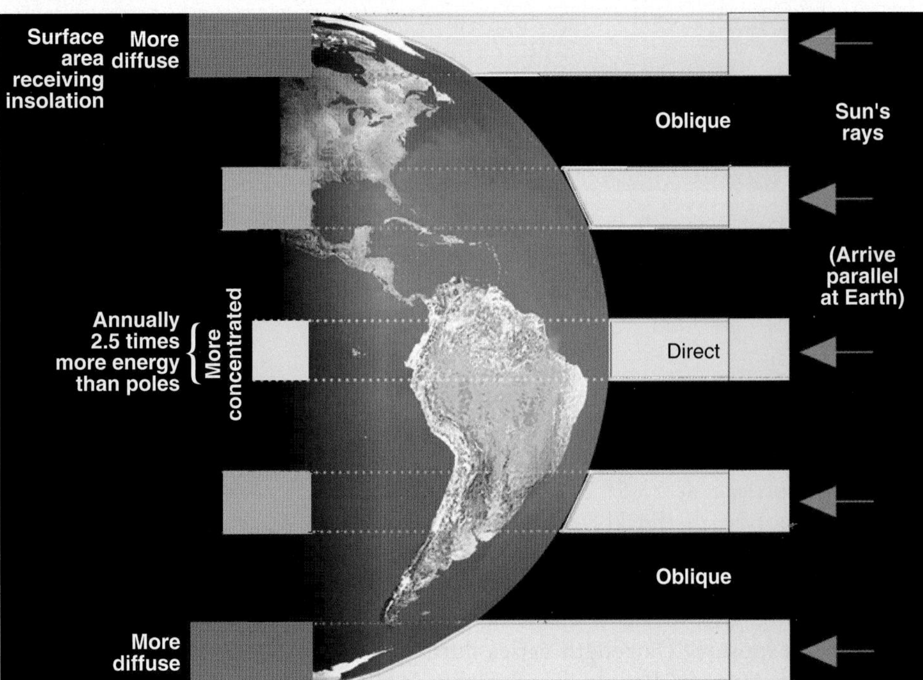

FIGURE 2.9 Insolation receipts and Earth's curved surface.
Solar insolation angles determine the concentration of energy received at each latitude. Lower latitudes receive more concentrated energy from a more direct solar beam. Higher latitudes receive slanting (oblique) rays and more diffuse energy. Note the area covered by identical columns of solar energy arriving at Earth's surface at higher latitudes (more diffuse, larger area covered) and at lower latitudes (more concentrated, smaller area covered).

large longwave radiation losses from Earth's surface—and reflective (light-colored) surfaces work together to reduce net radiation values measured at the thermopause. Clouds in the lower atmosphere also affect net radiation patterns at the top of the atmosphere by reflecting greater amounts of shortwave energy to space.

Having examined the flow of solar energy to Earth, let us now look at the nature of seasonal change as it affects these energy receipts.

The Seasons

ANIMATION **Earth-Sun Relations, Seasons**

The periodic rhythm of warmth and cold, dawn and daylight, twilight and night has fascinated humans for centuries. In fact, many ancient societies demonstrated a greater awareness of seasonal change than modern peoples and formally commemorated natural energy rhythms with festivals and monuments. Presently, warm-season length is on the increase worldwide in response to global temperature change.

Seasonality

Seasonality refers to both the seasonal variation of the Sun's position above the horizon and the changing daylengths during the year. Seasonal variations are a response to changes in the Sun's **altitude**, or the angle between the horizon and the Sun. At sunrise or sunset, the Sun is at the horizon, so its altitude is 0°. If during the day the Sun reaches halfway between the horizon and directly overhead, it is at 45°, and if directly overhead it is at 90° altitude (called the *zenith*). The Sun is directly overhead only at the subsolar point.

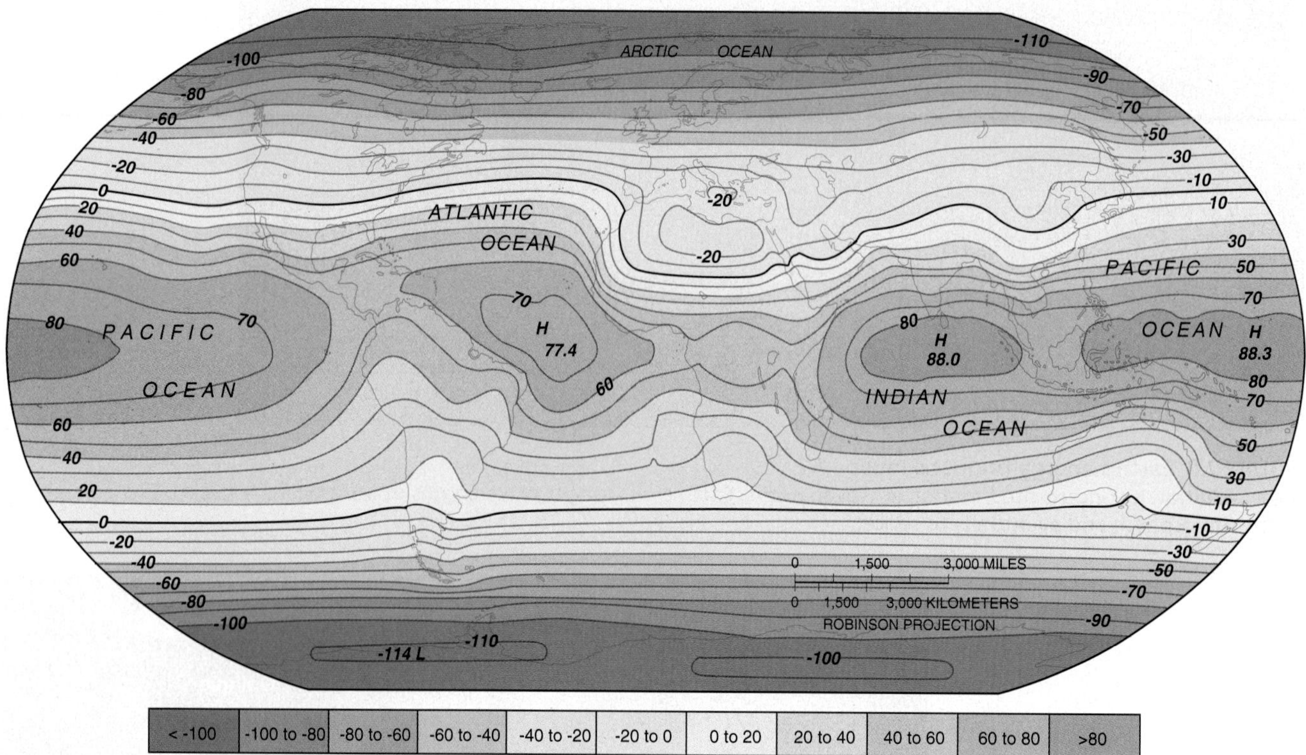

FIGURE 2.10 Daily net radiation patterns at top of atmosphere. Average daily net radiation flows for a 9-year period (1979 to 1987) measured at the top of the atmosphere by the Earth Radiation Budget (ERB) instrument in orbit. Units are watts per square meter (W/m^2). [Map courtesy of Dr. H. Lee Kyle, Goddard Space Flight Center, NASA.]

The Sun's **declination** is the latitude of the subsolar point. Declination annually migrates through 47° of latitude between the *Tropic of Cancer* at 23.5° N and the *Tropic of Capricorn* at 23.5° S latitude. The subsolar point does not reach the U.S. or Canadian mainland. Other than Hawai'i, all other states and provinces are too far north.

Seasonality also means a changing **daylength**, or duration of exposure. Daylength varies during the year depending on latitude. People living at the equator always receive equal hours of day and night, whereas those living along 40° N or S latitude experience about 6 hours' difference in daylight hours between winter and summer. Those at 50° N or S latitude experience almost 8 hours of annual daylength variation. At the polar extremes, the range extends from a 6-month period of no insolation and darkness to a 6-month period of continuous 24-hour days.

Factors That Influence Seasonal Change

Seasons result from variations in the Sun's altitude above the horizon, declination latitude, and daylength. These in turn are created by several physical factors that operate together: Earth's revolution in orbit around the Sun, its daily rotation on its axis, its tilted axis, the unchanging orientation of its axis, and its sphericity. We now briefly describe these factors, which are summarized in Table 2.1. Of course, the essential ingredient is having a single source of radiant energy—the Sun.

Revolution. The structure of Earth's orbit and **revolution** about the Sun are shown in Figure 2.1 and Figure 2.11. Note the distinction between orbital *revolution* and *rotation*—the spinning of Earth on its axis. At an average distance from the Sun of 150 million km

Table 2.1 Five Reasons for Seasons

Factor	Description
Revolution	Orbit around the Sun; requires 365.24 days to complete at 107,280 kmph (66,660 mph)
Rotation	Earth turning on its axis; takes approximately 24 hours to complete at 1675 kmph (1041 mph) at the equator
Tilt	Axis is aligned at a 23.5° angle from a perpendicular to the plane of the ecliptic (the plane of Earth's orbit)
Axial parallelism	Remains in a fixed alignment, with Polaris directly overhead at the North Pole throughout the year
Sphericity	Appears as an oblate spheroid to the Sun's parallel rays; the geoid

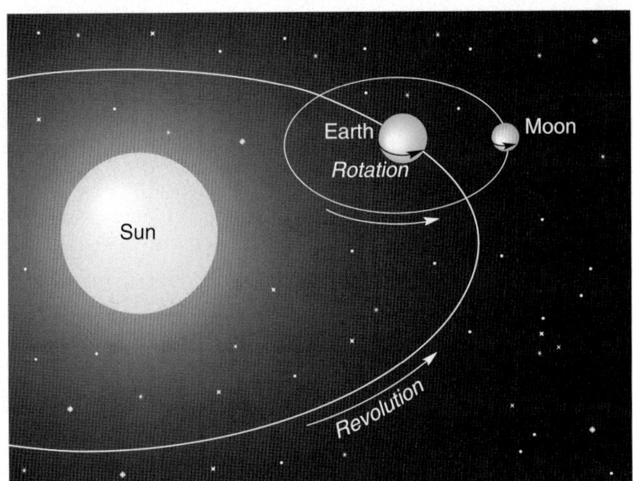

FIGURE 2.11 Earth's revolution and rotation.
Earth's *revolution* about the Sun and *rotation* on its axis, as viewed from above Earth's orbit. Note the Moon's rotation on its axis and revolution about Earth are counterclockwise as well.

(93 million mi), Earth completes its annual orbit in 365.24 days at speeds averaging 107,280 kmph (66,660 mph) in a counterclockwise direction when viewed from above Earth's North Pole. Earth's revolution determines the *length of the year* and therefore the *duration of the seasons.*

Rotation. Earth's **rotation**, or turning, is a complex motion that averages 24 hours in duration. A true day varies slightly from 24 hours, but by international agreement a day is considered to be exactly 24 hours (86,400 seconds) in length. Rotation determines daylength, causes the apparent deflection of winds and ocean currents, and produces the twice-daily rise and fall of the ocean tides in relation to the gravitational pull of the Sun and Moon.

Earth rotates about its *axis*, an imaginary line extending through the planet from the geographic North Pole to the South Pole. When viewed from above the North Pole, Earth rotates counterclockwise around this axis; viewed from above the equator it moves west to east, or eastward. This west-to-east rotation creates the Sun's *apparent* daily journey from east to west, even though the Sun remains in a fixed position in the center of the Solar System. Similarly, the Moon revolves around Earth and rotates counterclockwise on its axis (Figure 2.11).

Earth's rotation produces the diurnal (daily) pattern of day and night. The dividing line between day and night is called the **circle of illumination** (as illustrated in Figure 2.13). Because this day-night dividing circle of illumination intersects the equator, *daylength at the equator is always evenly divided*—12 hours of day and 12 hours of night. All other latitudes experience uneven daylength through the seasons, except for 2 days a year, on the equinoxes.

Tilt of Earth's Axis. To understand Earth's *axial tilt*, imagine Earth's elliptical orbit about the Sun as a plane, with half of the Sun and Earth above the plane and half below. (It may help to envision two spheres floating in water, where the water's surface forms a plane.) This flat surface is termed the **plane of the ecliptic**. Now imagine a perpendicular line passing through the plane. From this perpendicular, Earth's axis is tilted 23.5°. It forms a 66.5° angle from the plane itself (Figure 2.12). The axis through Earth's two poles points just slightly off Polaris, which is appropriately called the North Star.

Axial Parallelism. Throughout our annual journey around the Sun, Earth's axis *maintains the same alignment* relative to the plane of the ecliptic and to Polaris and the other stars. You can see this alignment in Figure 2.13. Note that in each position of Earth shown revolving about the Sun, the axis is oriented identically, or parallel to itself. This is known as **axial parallelism**.

Annual March of the Seasons

The combined effect of all these physical factors is the annual march of the seasons on Earth. Daylength is the most evident way of sensing changes in season at latitudes away from the equator. The extremes of daylength occur in December and June. The times around December 21 and June 21 are termed the *solstices.* They mark the times of the year when the Sun's declination places it directly over one of the two *tropics*, parallels of latitude that represent the Sun's farthest northerly or southerly position. ("Tropic" is from the Latin *tropicus*, meaning a turn or change, so a tropic latitude is where the Sun's declination appears to stand still briefly—Sun stance, or *sol stice*; then it turns and heads toward the other tropic.) Table 2.2 presents the key seasonal anniversary dates, their names, and the subsolar point location (declination).

Figure 2.13 demonstrates the annual march of the seasons and illustrates Earth's relationship to the Sun during each month of the year. Let us begin with December

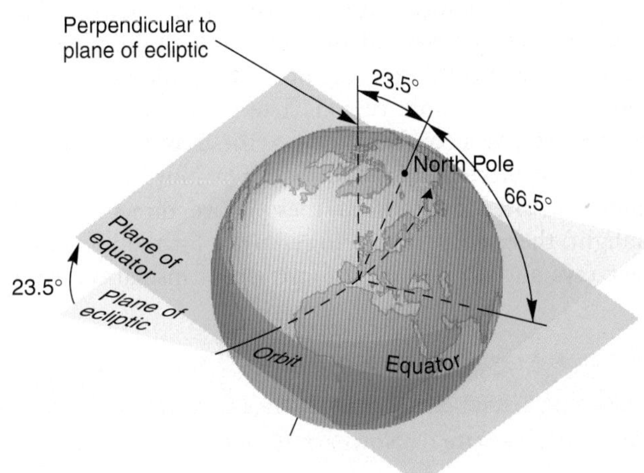

FIGURE 2.12 The ecliptic—plane of Earth's orbit—and Earth's tilt.
The plane of Earth's equator is 23.5° offset from the plane of the ecliptic. This is because Earth is tilted 23.5° from the plane of the ecliptic.

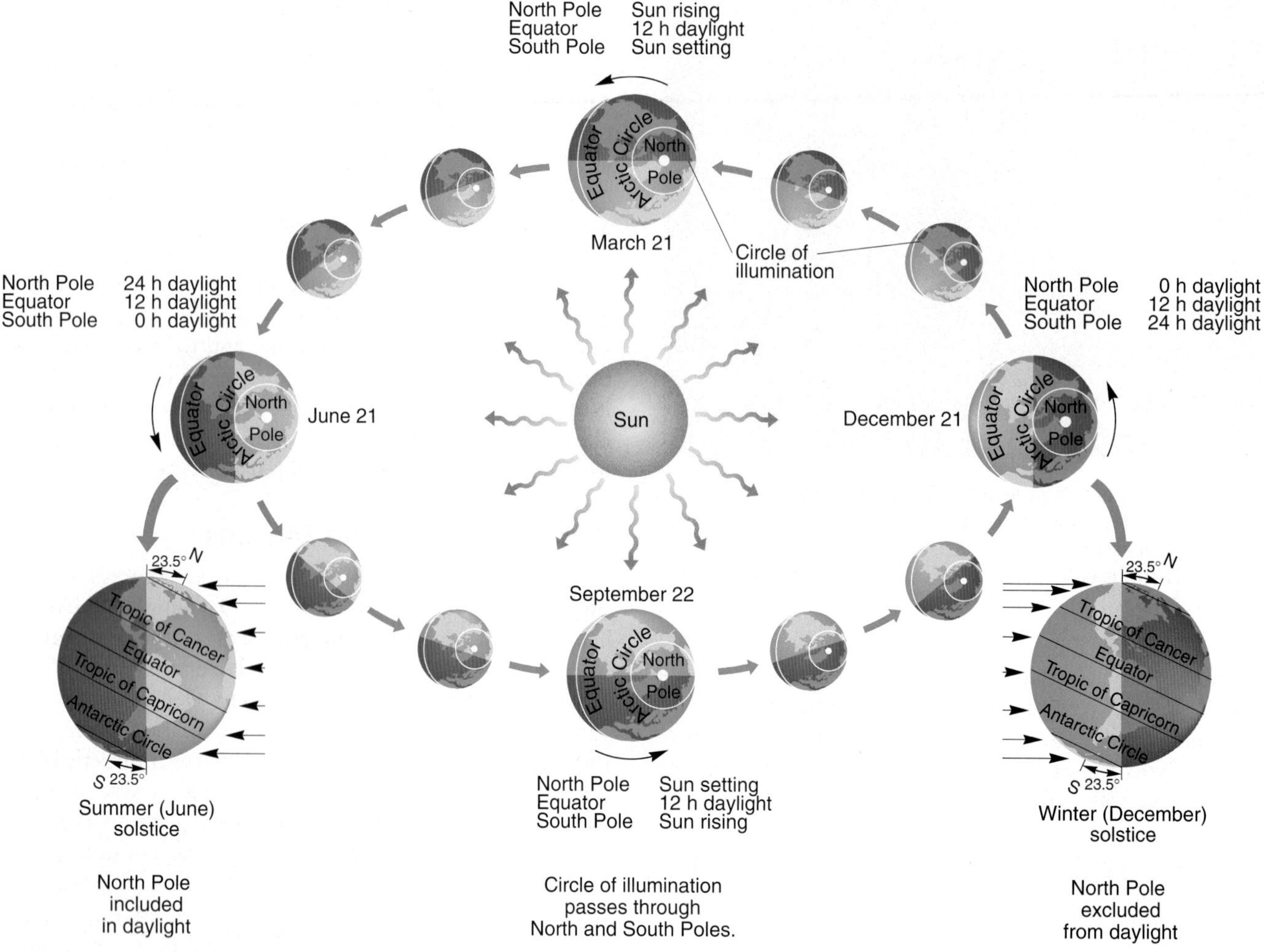

FIGURE 2.13 Annual march of the seasons.
Annual march of the seasons as Earth revolves about the Sun. Shading indicates the changing position of the circle of illumination in the Northern Hemisphere (polar views) and in both Northern and Southern Hemispheres (profile views). Note the hours of daylight for the equator and the poles. To follow the text, begin on the right side at December 21 and move counterclockwise.

(the globe to the far right in the figure). On December 21 or 22, at the **winter solstice** (literally, "winter Sun stance") or **December solstice**, the circle of illumination excludes the North Pole region from sunlight and includes the South Pole region. The subsolar point is at 23.5° S latitude, a parallel known as the **Tropic of Capricorn**, at the moment of the solstice. The Northern Hemisphere is tilted away from these more direct rays of sunlight, thereby creating a lower angle for the incoming solar rays and a more diffuse pattern of insolation, thus causing our northern winter. Examine the photo of Earth on the back cover of this book. During what month do you think it was taken?

From 66.5° N latitude to 90° N (the North Pole), the Sun remains below the horizon the entire day. This latitude (66.5° N) marks the *Arctic Circle*, the southernmost parallel (in the Northern Hemisphere) that experiences a 24-hour period of darkness. During the next 3 months, daylength and solar angles gradually increase in the Northern Hemisphere as Earth completes one-fourth of its orbit.

Table 2.2 Annual March of the Seasons

Approximate Date	Northern Hemisphere Name	Location of the Subsolar Point
December 21–22	Winter solstice (December solstice)	23.5° S latitude (Tropic of Capricorn)
March 20–21	Vernal equinox (March equinox)	0° (equator)
June 20–21	Summer solstice (June solstice)	23.5° N latitude (Tropic of Cancer)
September 22–23	Autumnal equinox (September equinox)	0° (equator)

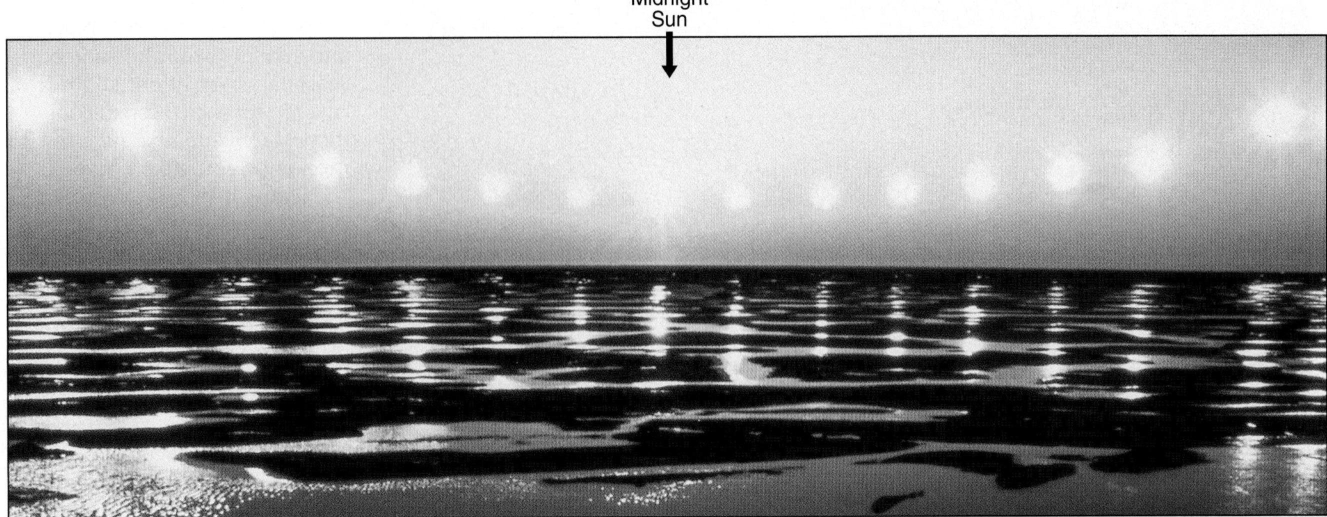

FIGURE 2.14 The midnight Sun.
The midnight Sun north of the Arctic Circle captured in a series of 18 exposures on the same piece of film. The camera is facing due north. Midnight is the exposure showing the Sun closest to the horizon. The photographer removed the lens cap at regular intervals to make the multiple exposures. [Photos © by Gary Braasch/Tony Stone Images.]

The **vernal equinox**, or **March equinox**, occurs on March 20 or 21. At that time, the circle of illumination passes through both poles, so that all locations on Earth experience a 12-hour day and a 12-hour night. Those living around 40° N latitude (New York, Denver) have gained 3 hours of daylight since the December solstice. At the North Pole, the Sun peeks above the horizon for the first time since the previous September; at the South Pole the Sun is setting, marking the beginning of a long, dark winter.

From March, the seasons move on to June 20 or 21 and the **summer solstice**, or **June solstice**. The subsolar point has now shifted from the equator to 23.5° N latitude, the **Tropic of Cancer**. Because the circle of illumination now includes the North Pole region, everything north of the Arctic Circle receives 24 hours of daylight—the "Midnight Sun" (Figure 2.14). In contrast, the region from the *Antarctic Circle* to the South Pole (66.5°–90° S latitude) is in darkness the entire day.

Note that the orientation of Earth's axis has remained fixed relative to the heavens and parallel to its previous position 6 months earlier. The Northern Hemisphere in June is tilted toward the Sun, receives higher Sun angles, and experiences longer days—and therefore more insolation—than the Southern Hemisphere. Those living at 40° N latitude now receive more than 15 hours of sunlight a day, which is 6 hours more than in December.

September 22 or 23 is the time of the **autumnal equinox**, or **September equinox**, when Earth's orientation is such that the circle of illumination again passes through both poles so that all parts of the globe receive a 12-hour day and a 12-hour night. The subsolar point has returned to the equator, with days growing shorter to the north and longer to the south. Researchers stationed at the South Pole see the disk of the Sun just rising, ending their 6 months of night. In the Northern Hemisphere, fall arrives, a time of many colorful changes in the landscape.

Seasonal Observations. For most of you reading this text in the Northern Hemisphere, the point of sunrise migrates along the horizon from the southeast in December to the northeast in June. The point of sunset migrates from the southwest to the northwest during those same times. The Sun's altitude at local noon at 40° N latitude migrates from a 26° angle above the horizon at the winter (December) solstice to a 73° angle above the horizon at the summer (June) solstice—a range of 47° (Figure 2.15). Think back over the past year. What seasonal changes have you observed in Sun angles, vegetation, temperatures, and weather?

Atmospheric Composition, Temperature, and Function

ANIMATION Ozone Breakdown, Ozone Hole

Insolation cascades through the atmosphere toward Earth's surface, powering the physical systems in the atmosphere, in the oceans, and on land—varying through the seasonal changes just discussed. Along the way the atmosphere works as an efficient filter, removing harmful radiation from sunlight. Let's now examine this unique atmosphere—its composition, temperature, and function.

The modern atmosphere probably is the fourth general atmosphere in Earth's history. This modern atmosphere is a gaseous mixture of ancient origin, the sum of all the exhalations and inhalations of life on Earth throughout time. The principal substance of this atmosphere is

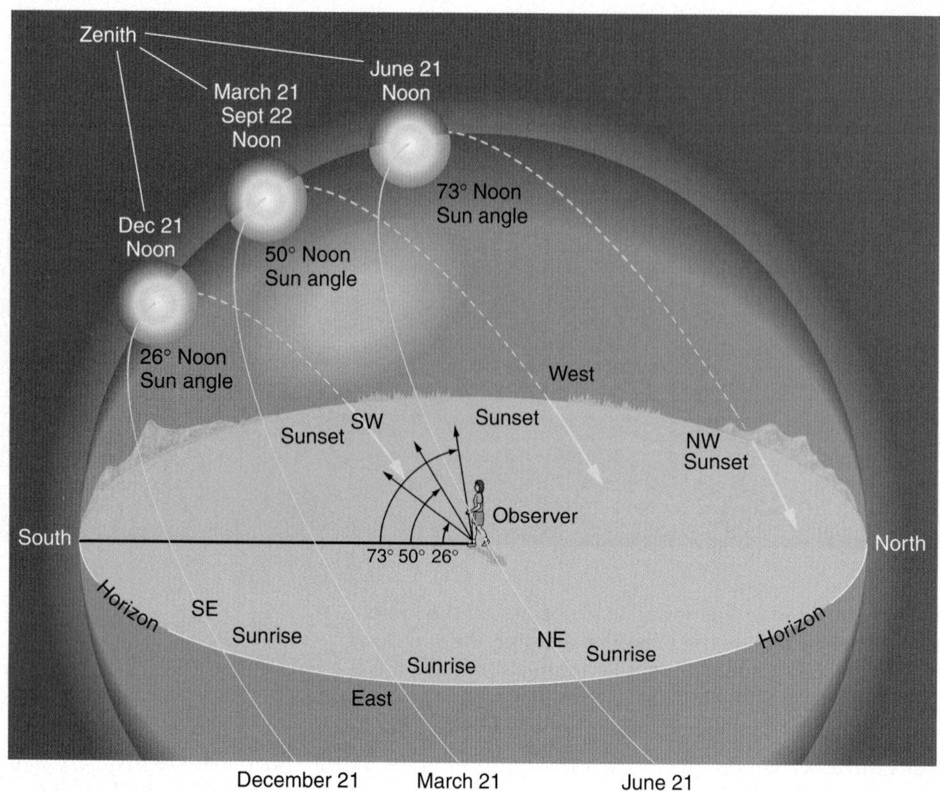

FIGURE 2.15 Seasonal observations—sunrise, noon, and sunset through the year. Seasonal observations at 40° N latitude for the December solstice, March equinox, June solstice, and September equinox. The Sun's altitude increases from 26° in December to 73° above the horizon in June—a difference of 47°.

air, the medium of life as well as a major industrial and chemical raw material. *Air* is a simple additive mixture of gases that is naturally odorless, colorless, tasteless, and formless, blended so thoroughly that it behaves as if it were a single gas.

In his book *The Lives of a Cell*, physician and self-styled "biology-watcher" Lewis Thomas compared the atmosphere of Earth to an enormous cell membrane. The membrane around a cell regulates the interactions between the cell's delicate inner workings and the potentially disruptive outer environment. Each cell membrane is very selective as to what it will and will not allow to pass through. The modern atmosphere acts as Earth's protective membrane, as Thomas described so vividly (Figure 2.16).

As a practical matter, we consider the top of our atmosphere to be around 480 km (300 mi) above Earth's surface, the same altitude we use for measuring the solar constant and initial insolation receipt. Beyond that altitude, the atmosphere is rarefied (nearly a vacuum) and is called the **exosphere**, which means "outer sphere." It contains scarce, lightweight hydrogen and helium atoms, weakly bound by gravity as far as 32,000 km (20,000 mi) from Earth.

Atmospheric Profile

Earth's modern atmosphere is arranged in a series of imperfectly shaped concentric "shells" or "spheres" that grade into one another, all bound to the planet by gravity. As critical as the atmosphere is to us, it represents only the thinnest envelope, amounting to less than one-millionth of Earth's total mass. Important to the following discussion is Figure 2.17, a vertical cross section, or side view, of Earth's atmosphere. We simplify our discussion of the atmosphere by looking at three aspects: its *composition*, *temperature*, and *function*. These categories are shown along the left side of Figure 2.17a.

This 480-km atmospheric profile exerts its weight, pressing downward under the pull of gravity. Air molecules create air pressure through their motion, size, and number. Pressure is exerted on all surfaces in contact with the air. The weight (force over a unit area) of the atmosphere, or **air pressure**, pushes in on all of us. Fortunately, that same pressure also exists inside us, pushing outward; otherwise we would be crushed by the mass of air around us.

The atmosphere exerts an average force of approximately 1 kg/cm^2 (14.7 lb/in.^{2}) at sea level. Under the influence of gravity, air is compressed and therefore denser near Earth's surface; it thins rapidly with increasing altitude (Figure 2.18a). Consequently, over half the total mass of the atmosphere is compressed below 5500 m (18,000 ft), 75% below 10,700 m (35,100 ft), and 90% is compressed below 16,000 m (52,500 ft). All but 0.1% of the atmosphere exists within an altitude of 50 km (31 mi), as shown in the pressure profile in Figure 2.18b (percentage column is farthest to the right).

Few people are aware that in routine air travel they are sitting above 80% of the total atmospheric volume! To better understand this pressure profile, imagine sky diving from a high-altitude balloon 33 km (20 mi) above Earth. What would you experience? How fast would you fall? What sounds would you hear? See News Report 2.1 to read about someone who did this very thing.

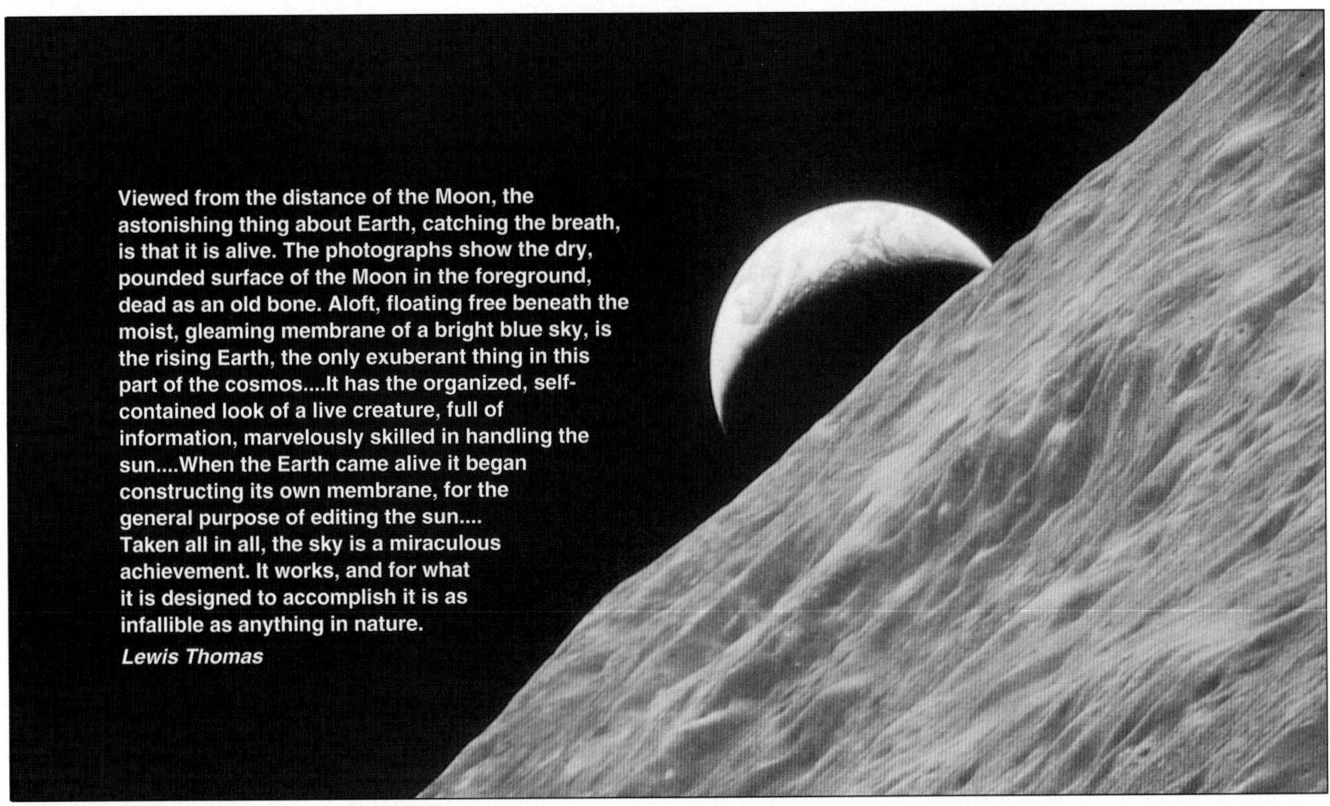

FIGURE 2.16 Earthrise.
Earthrise over the "...dead as an old bone..." lunar surface. [NASA photo. Quotation from "The World's Biggest Membrane" from *The Lives of a Cell* by Lewis Thomas. Copyright © 1973 by the Massachusetts Medical Society. Originally published in the *New England Journal of Medicine.*]

At sea level, the atmosphere exerts a pressure of 1013.2 mb (*millibar*, *mb*, force per square meter of surface area) or 29.92 in. of mercury, as measured by a *barometer*. In Canada and other countries, normal air pressure is expressed as 101.32 kPa (*kilopascal*; 1 kPa = 10 mb). For more on air pressure, the instruments that measure it, and the role air pressure plays in generating winds, see Chapter 4.

Atmospheric Composition Criterion

Using chemical *composition* as a criterion, we find that the atmosphere divides into two broad regions, the *heterosphere* (80 km to 480 km altitude) and the *homosphere* (Earth's surface to 80 km altitude). As you read, note that we follow the same path that incoming solar radiation travels through the atmosphere to Earth's surface.

Heterosphere. The **heterosphere** is defined as the outer atmosphere in terms of composition. It begins at about 80 km (50 mi) altitude and extends outward to the transition to the exosphere and interplanetary space (Figure 2.17). The International Space Station and most Space Shuttle missions orbit in the upper heterosphere.

As the prefix *hetero-* implies, this region is not uniform—its gases are *not evenly mixed*. This distribution is quite different from the nicely blended gases we breathe near Earth's surface, in the homosphere. Gases in the heterosphere are distributed in distinct layers that are sorted by gravity according to their atomic weight, with the lightest elements (hydrogen and helium) at the margins of outer space and the heavier elements (oxygen and nitrogen) dominant in the lower heterosphere. Less than 0.001% of the atmosphere's mass is in the heterosphere.

Homosphere. Between the heterosphere and Earth's surface is the other compositional shell of the atmosphere, the **homosphere**. This region extends from the surface to an altitude of 80 km (50 mi). Even though the atmosphere rapidly changes density in the homosphere, decreasing with increasing altitude, the blend (proportion) of gases is nearly uniform throughout the homosphere. The only exceptions are the concentration of ozone (O_3) in the "ozone layer," from 19 to 50 km (12 to 31 mi), and the variations in water vapor, carbon dioxide, pollutants, and some trace chemicals in the lowest portion of the atmosphere.

The stable mixture of gases throughout the homosphere evolved slowly. The present proportion, which includes oxygen, was attained approximately 500 million years ago. Figure 2.19 presents by volume the stable ingredients that constitute dry, clean air in the homosphere.

The homosphere is a vast reservoir of relatively inert *nitrogen*, originating principally from volcanic sources.

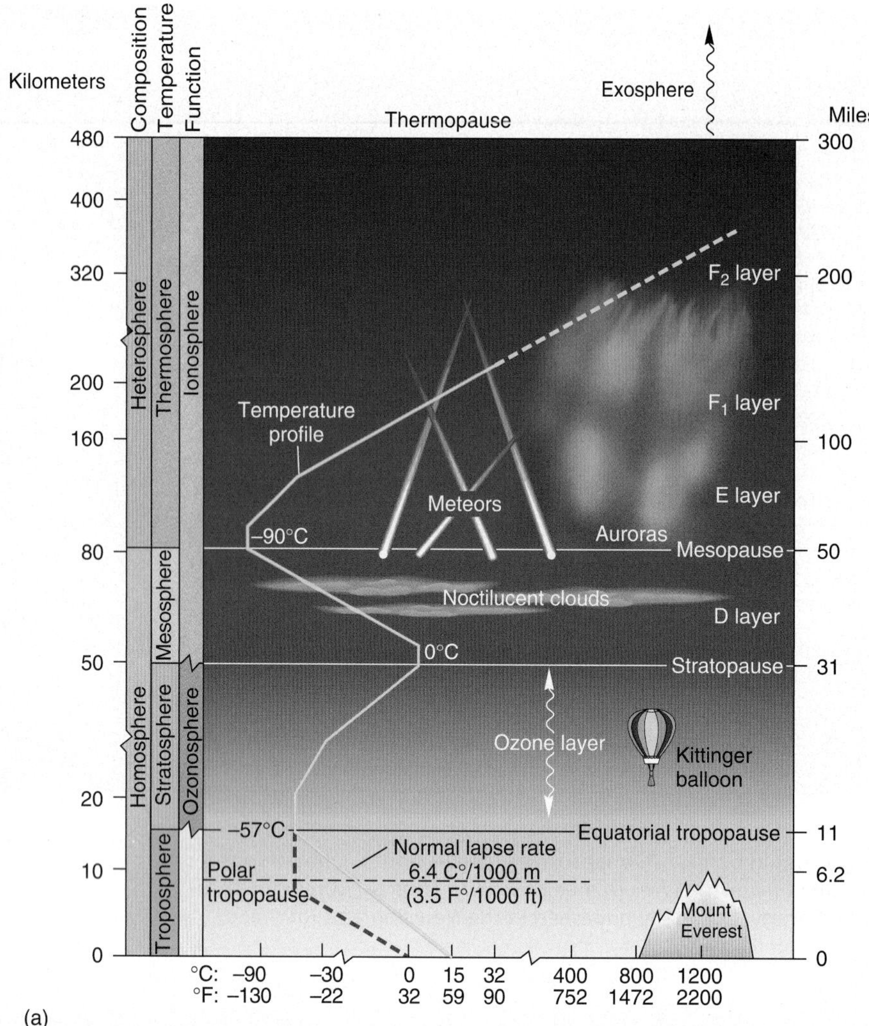

FIGURE 2.17 Modern atmosphere profile.
(a) An integrated chart of our modern atmosphere. Columns show division of the atmosphere by composition, temperature, and function. The chart spans from Earth's surface to the thermopause at 480 km (300 mi). (The small balloon shows the height achieved by Kittinger, discussed in News Report 2.1.) (b) Space Shuttle astronauts captured a dramatic sunset through various atmospheric layers across the "edge" of our planet—called Earth's limb. A silhouetted cumulonimbus thunderhead cloud is seen rising to the tropopause. [Space Shuttle photo from NASA.]

Nitrogen is a key element of life, yet we exhale all the nitrogen that we inhale. This apparent contradiction is explained by the fact that nitrogen is absorbed into our bodies not from the air we breathe but through compounds in food. In the soil, nitrogen is bound to these compounds by nitrogen-fixing bacteria, and it is returned to the atmosphere by denitrifying bacteria that remove nitrogen from organic materials. The nitrogen cycle discussion is in Chapter 16.

Oxygen, a by-product of photosynthesis, also is essential for life processes. Slight spatial variations occur in the percentage of oxygen in the atmosphere because of variations in photosynthetic rates with latitude, seasonal changes, and the lag time as atmospheric circulation slowly mixes the air.

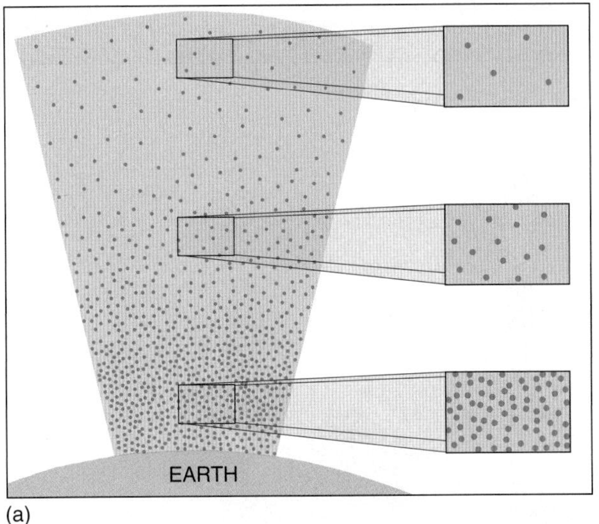

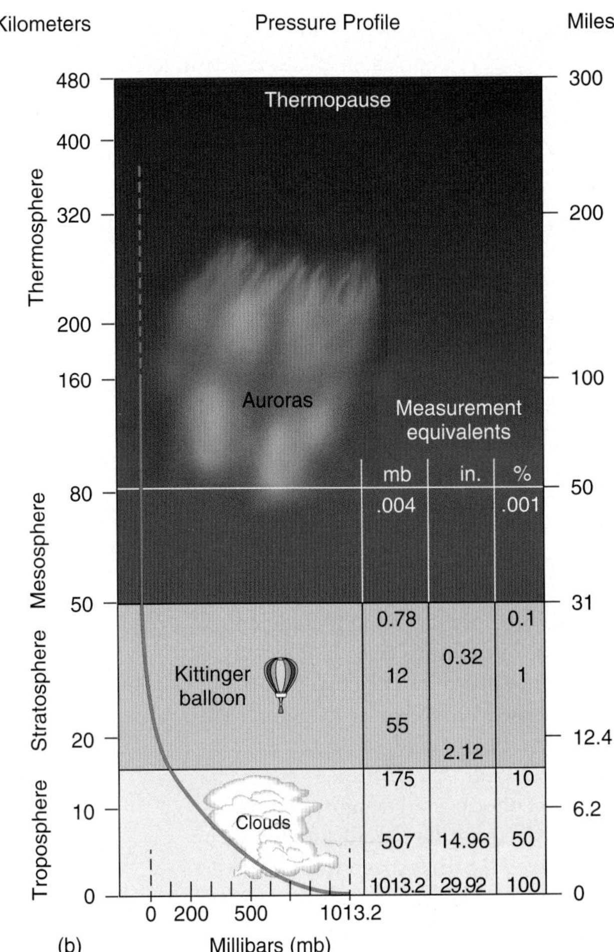

Kilometers	mb	in.	%	Miles
80	.004		.001	50
50	0.78		0.1	31
	12	0.32	1	
20	55	2.12		12.4
	175		10	
10	507	14.96	50	6.2
0	1013.2	29.92	100	0

FIGURE 2.18 Density decreases with altitude.
(a) The atmosphere, denser near Earth's surface, rapidly decreases in density with altitude. The difference in density is easily measured because air exerts its weight as pressure. Have you experienced pressure changes that you could feel on your eardrums? How high above sea level were you at the time? (b) Pressure profile of the atmosphere. Note the rapid decrease in atmospheric pressure with altitude and that about 90% of the atmospheric mass resides in the troposphere.

Although it forms about one-fifth of the atmosphere, oxygen forms compounds that compose about half of Earth's crust. Oxygen readily reacts with many elements to form these materials. Both nitrogen and oxygen reserves in the atmosphere are so extensive that, at present, they far exceed human capabilities to disrupt or deplete them.

The gas *argon*, constituting about 1% of the homosphere, is completely inert (an unreactive "noble" gas) and therefore is unusable in life processes. Argon is a residue from the radioactive decay of an isotope (form) of potassium called potassium-40 (symbolized ^{40}K). Because industry has found uses for inert argon (in light bulbs, welding, and some lasers), it is extracted or "mined" from the atmosphere, in addition to nitrogen and oxygen.

Carbon dioxide is a natural by-product of life processes. It is essentially a stable atmospheric component, qualifying it for inclusion in Figure 2.19. Although its present percentage in the atmosphere is small at over 0.03793%, it is important in maintaining global temperatures. Its percentage has been increasing over the past 200 years as a result of human activities. The implications of this increase to global warming and climate change are discussed in Chapter 6.

Atmospheric Temperature Criterion

Shifting to temperature as a criterion, the atmosphere has four distinct zones—the *thermosphere*, *mesosphere*,

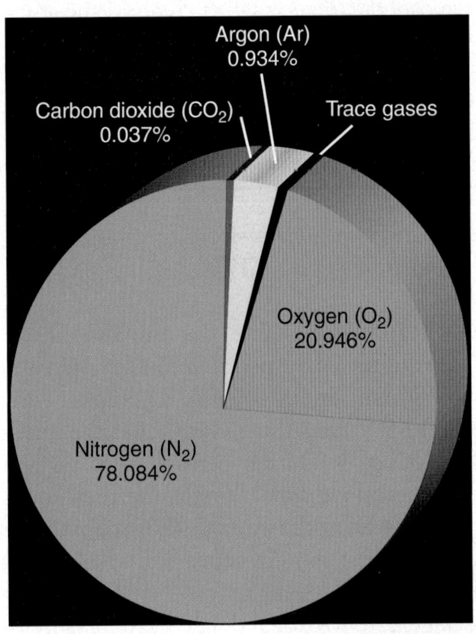

FIGURE 2.19 Composition of the homosphere.
Stable components of the modern atmosphere (percentage concentration by volume). Average carbon dioxide concentration measured in 2002 at Mauna Loa, Hawai'i (go to http://cdiac.esd.ornl.gov/ftp/maunaloa-co2/maunaloa.co2).

News Report 2.1

Falling through the Atmosphere—The Highest Sky Dive

Imagine a small, unpressurized compartment large enough for only one person, floating at 31.3 km (19.5 mi) altitude, carried to such height by a helium-filled balloon. Even though it is daytime, the sky is almost black, for this is above all but 1.0% of the atmospheric mass. The air pressure at 31,300 m (102,800 ft) is barely measurable—this altitude is used as the beginning of space in aircraft testing. The year is 1961. (See the position of this balloon on Figures 2.17a and 2.18b.)

Captain Joseph Kittinger, Jr., an Air Force officer, stood looking out of the opening in his capsule. He paused for the view of Earth's curved horizon, the New Mexico landscape more than 31 km below, and the dark heavens above, and then leaped into the stratospheric void (Figure 1). What do you think he would initially hear and feel? At what speed would he fall? What would the temperature be? How long would he fall before opening his parachute?

He heard nothing, no rushing sound. The fabric of his pressure suit did not flutter, for there was not enough air to create friction against the fibers. Because of these factors he had no sensation of movement until he glanced back at the balloon. It retreated rapidly from his motionless perception.

His speed was remarkable. In the lower atmosphere, objects fall at "terminal velocity," reaching about 200 kmph (125 mph) in a few seconds. However, in the rarefied stratosphere the lack of air resistance permits incredible velocity. Captain Kittinger quickly accelerated to 988 kmph (614 mph), slightly less than the speed of sound at sea level.

His free fall took him through the stratosphere and its ozone layer. As he encountered denser layers of atmospheric gases, his free fall was slowed by frictional drag. As he flew past 15,240 m (50,000 ft), he had slowed to 400 kmph (250 mph). He dropped into the lower atmosphere, finally falling below regular airplane altitudes. His free fall lasted 4 minutes and 25 seconds, slowing to terminal velocity and the opening of his main chute at 5500 m (18,000 ft). He safely drifted to Earth's surface. This remarkable 13-minute 35-second voyage through 99% of the atmospheric mass remains a record to this day. (In 2003, two sky divers plan to jump from balloon gondolas at 40 km (25 mi) altitude, in individual projects called *Space Jump* and *Stratoquest*.)

FIGURE 1 Stratospheric leap into history.
Moments into Captain Joseph Kittinger's historic exploration of the atmosphere, captured by a remotely triggered camera. The clouds are more than 26,000 m (85,000 ft) below him. He carries an instrument pack on his seat, his main chute, and pure oxygen for his breathing mask. [Volkmar Wentzel/NGS Image Collection, used by permission from the National Geographic Society, *National Geographic Magazine*, Dec. 1960, p. 855. All rights reserved.]

stratosphere, and *troposphere*—from thermopause to Earth's surface (labeled in Figure 2.17).

Thermosphere. We define the **thermosphere** ("heat sphere") as roughly corresponding to the heterosphere (80 km out to 480 km, or 50–300 mi). The upper limit of the thermosphere is called the *thermopause* (the suffix *-pause* means "to change"). During periods of a less active Sun (fewer sunspots and coronal bursts), the thermopause may lower in altitude from the average 480 km (300 mi) to only 250 km altitude (155 mi). An active Sun will cause the outer atmosphere to swell to an altitude of 550 km (340 mi), where it can create frictional drag on satellites in low orbit.

The temperature profile in Figure 2.17 (yellow curve) shows that temperatures rise sharply in the thermosphere, to 1200°C (2200°F) and higher. Despite such high temperatures, the thermosphere is not "hot" in the way you might expect. Temperature and heat are different concepts. The intense solar radiation in this portion of the atmosphere excites individual molecules (principally nitrogen and oxygen) to high levels of vibration. This **kinetic energy**, the energy of motion, is the vibrational energy that we measure as *temperature*.

However, the actual heat involved is very small. The reason is that the density of the molecules is so low that little actual *heat*—the flow of kinetic energy from one body to another because of a temperature difference between them—is produced. Heating in the atmosphere near Earth's surface is different because the greater number of molecules in the denser atmosphere transmit their kinetic energy as **sensible heat**, meaning that we can measure its temperature. (Density, temperature, and heat capacity determine the sensible heat of a substance.)

Mesosphere. The **mesosphere** is the area from 50 to 80 km (30 to 50 mi) above Earth and is the highest in altitude of the three temperature regions within the homosphere. As Figure 2.17 shows, the mesosphere's outer boundary, the *mesopause*, is the coldest portion of the atmosphere, averaging −90°C (−130°F), although that temperature may vary considerably (25 to 30 C°, or 45 to 54 F°). Very low pressures (low density of molecules) exist in the mesosphere.

Stratosphere. The **stratosphere** extends from 18 to 50 km (11 to 31 mi) from Earth's surface. Temperatures increase with altitude throughout the stratosphere, from −57°C (−70°F) at 18 km (tropopause), warming to 0°C (32°F) at 50 km, the altitude of the stratosphere's outer boundary, the *stratopause*. This is the location of the ozone layer.

Troposphere. The **troposphere** is the final layer encountered by incoming solar radiation as it surges through the atmosphere to the surface. It is the home of the biosphere, the atmospheric layer that supports life, and the region of principal weather activity.

The troposphere contains approximately 90% of the total mass of the atmosphere and the bulk of all water vapor, clouds, weather, air pollution, and life forms. The *tropopause*, its upper limit, is defined by an average temperature of −57°C (−70°F), but its exact elevation varies with the season, latitude, and surface temperatures and pressures. Near the equator, because of intense heating from the surface, the tropopause occurs at 18 km (11 mi); in the middle latitudes, it occurs at 12 km (8 mi); and at the North and South Poles it is only 8 km (5 mi) or less above Earth's surface.

Figure 2.20 illustrates the normal temperature profile within the troposphere during daytime. As the graph shows, temperatures decrease rapidly with increasing altitude at an average of 6.4 C° per kilometer (3.5 F° per 1000 feet), a rate known as the **normal lapse rate**. This temperature plot also appears in Figure 2.17.

The normal lapse rate is an average. The **environmental lapse rate** is the actual lapse rate at any particular time and place, which may deviate considerably because of local weather conditions. This variation in temperature gradient in the lower troposphere is central to our discussion of weather processes (Chapter 5).

In the stratosphere, the marked warming with increasing altitude causes the tropopause to act like a lid, essentially preventing whatever is in the cooler (denser) air below from mixing into the warmer (less dense) stratosphere. However, the tropopause may be disrupted above the midlatitudes wherever jet streams produce vertical turbulence and an interchange between the troposphere and the stratosphere (Chapter 4).

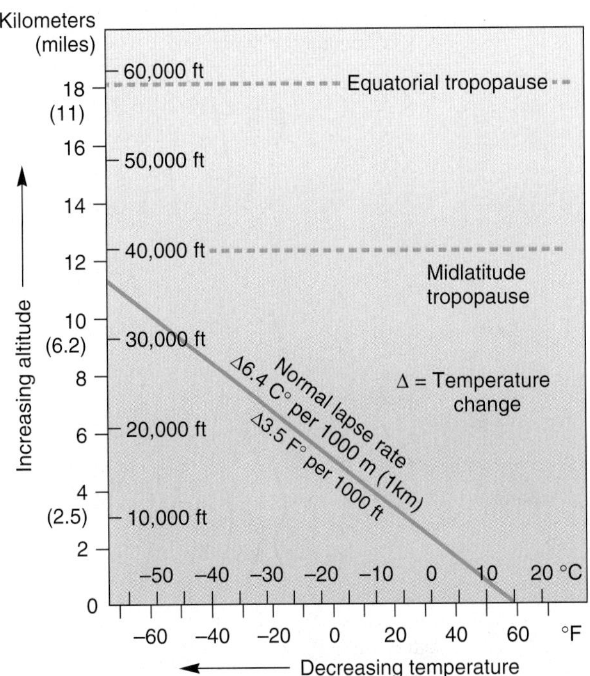

FIGURE 2.20 The temperature profile of the troposphere. During daytime, temperature decreases at a rate known as the *normal lapse rate*. Scientists use a concept called the *standard atmosphere* as an accepted description of air temperature and pressure changes with altitude. The values in this profile are part of the standard atmosphere. Note the approximate locations of the equatorial and midlatitude tropopauses.

Atmospheric Function Criterion

Looking at our final atmospheric criterion of *function*, we find that the atmosphere has two specific zones that remove most of the harmful wavelengths of incoming solar radiation and charged particles: the *ionosphere* and the *ozonosphere* (*ozone layer*). Figure 2.21 depicts in a general way the absorption of radiation by the various functional layers of the atmosphere.

Ionosphere. The outer functional layer, the **ionosphere**, extends throughout the thermosphere and into the mesosphere below (Figure 2.17). The ionosphere absorbs cosmic rays, gamma rays, X-rays, and shorter wavelengths of ultraviolet radiation, changing atoms to positively charged ions and giving the ionosphere its name. The glowing auroral lights occur principally within the ionosphere.

Ozonosphere. That portion of the stratosphere that contains an increased level of ozone is the **ozonosphere**, or **ozone layer**. Ozone is a highly reactive oxygen molecule made up of three oxygen atoms (O_3) instead of the usual two atoms (O_2) that make up oxygen gas. Ozone

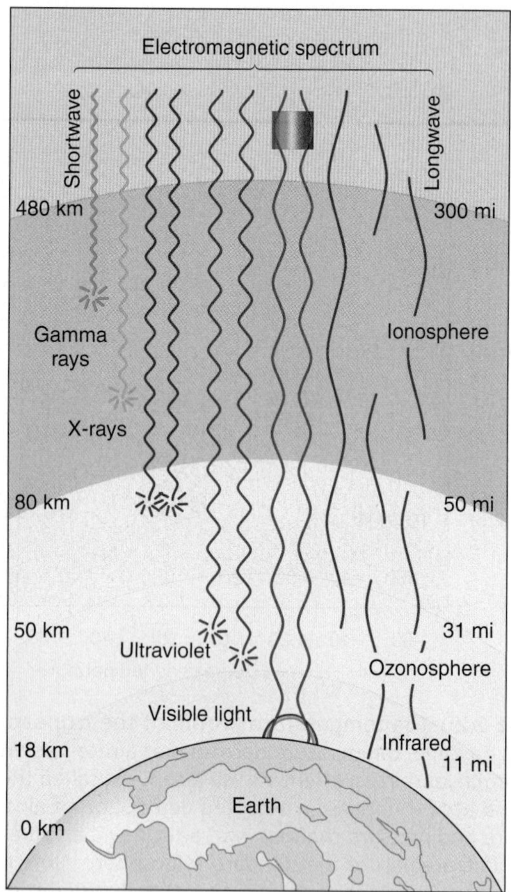

FIGURE 2.21 The atmosphere protects Earth's surface. As solar energy passes through the atmosphere, the shortest wavelengths are absorbed. Only a fraction of the ultraviolet radiation, and most of the visible light and infrared, reaches Earth's surface. When they are above these protective layers, astronauts must wear a spacesuit for them to survive; the suit duplicates this filtering process.

absorbs wavelengths of ultraviolet light (0.1–0.3 μm) and subsequently reradiates this energy at longer wavelengths, as infrared radiation. This process converts most harmful ultraviolet radiation, effectively "filtering" it and safeguarding life at Earth's surface.

The ozone layer is presumed to have been relatively stable over the past several hundred million years (allowing for daily and seasonal fluctuations). Today, however, it is in a state of continuous change. Focus Study 2.1 presents an analysis of the crisis in this critical portion of our atmosphere. Fortunately, international treaties to prevent further losses appear to be working.

Variable Atmospheric Components

The troposphere contains natural and human-caused variable gases, particles, and other chemicals. The spatial aspects of these variables are important applied topics in physical geography and the study of Earth's atmosphere.

Air pollution is not a new problem. Romans complained over 2000 years ago about the foul air of their cities. Filling Roman air was the stench of open sewers, smoke from fires, and fumes from ceramic-making kilns and smelters (furnaces) that converted ores into metals. In human experience, cities are always the place where the environment's natural ability to process and recycle waste is most taxed. Regulations to curb human-caused air pollution have met with great success, although much remains to be done. Before we discuss these topics, let's examine some natural sources of air pollution.

Natural Sources

Natural air pollution sources produce a greater quantity of pollutants—nitrogen oxides, carbon monoxide, hydrocarbons from plants and trees, and carbon dioxide—than do human-made sources. Table 2.3 lists some of these natural sources and the substances they contribute to the air. However, any attempt to diminish the impact of human-made air pollution through a comparison with natural sources is irrelevant, for we have evolved with and adapted to the natural ingredients in the air and not to what we have introduced. We have not evolved in relation to the comparatively recent concentrations of *anthropogenic* (human-caused) contaminants in our metropolitan regions.

A dramatic natural source of pollution was the 1991 eruption of Mount Pinatubo in the Philippines (15° N 120° E), perhaps the last century's largest eruption. This event injected between 15 and 20 million tons of sulfur dioxide (SO_2) into the stratosphere. The spread of these emissions is shown in a sequence of satellite images that begins Chapter 4.

The devastating wildfires on several continents, including widespread fires across the western United States, produced natural air pollution. Soot, ash, and gases darkened skies and damaged health in affected regions. Wind patterns spread the pollution from the fires to nearby cities, closing airports and forcing evacuations to avoid the health-related dangers. During August 2002 in Alaska, record heat and dryness, coupled with more than 12,000

Table 2.3 Sources of Natural Variable Gases and Materials

Sources	Contribution
Volcanoes	Sulfur oxides, particulates
Forest fires	Carbon monoxide and dioxide, nitrogen oxides, particulates
Plants	Hydrocarbons, pollens
Decaying plants	Methane, hydrogen sulfides
Soil	Dust and viruses
Ocean	Salt spray and particulates

Focus Study 2.1

Stratospheric Ozone Losses: A Worldwide Health Hazard

Consider:

- The stratospheric ozone above Antarctica during the months of Antarctic spring (September to November) reached record losses in 2000, 2001, and 2002, covering an area three times larger than the United States. Each year, the "ozone hole" (actually a thinning) widens and deepens. The protective ozone layer is being depleted over southern South America, southern Africa, Australia, and New Zealand.
- An international scientific consensus confirmed previous assessments of the anthropogenic (human-caused) disruption of the ozone layer—chlorine atoms and chlorine monoxide molecules in the stratosphere are of human origin. (See *Scientific Assessment of Ozone Depletion*, by NASA, NOAA, United Nations Environment Programme, and World Meteorological Organization.)
- At Earth's opposite pole, a similar ozone depletion over the Arctic annually exceeds 30% to 45% lower than average. Canadian and U.S. governments report an "ultraviolet index" to help the public protect themselves.
- Environment Canada operates a scientific observatory to monitor ozone losses over Canada. This high-Arctic facility is at a remote weather station near Eureka on Ellesmere Island, Nunavut, Canada, about 1000 km (620 mi) from the North Pole (see http://www.ec.gc.ca/ozone/indexe.htm).
- Overall ozone losses in the midlatitudes are continuing at 6% to 8% per decade. In North America, related skin cancers are increasing, totaling more than 1 million cases a year, of which some 41,000 are malignant melanomas leading to an average of 10,000 deaths annually. Skin cancer is increasing at 4.0% per year. Most affected are light-skinned persons who live at higher elevations and those who work principally outdoors.

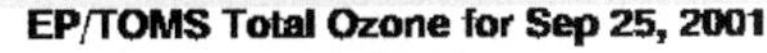

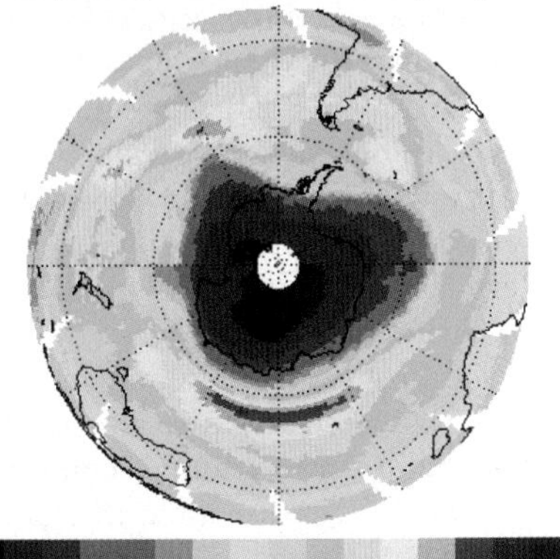

FIGURE 1 The Antarctic ozone hole. TOMS image for September 2001, uses a color scale to represent ozone concentrations in Dobson units, with purples for amounts less than 180, black below 125. (One Dobson unit equals 2.69 × 1016 molecules of O^3/m^3.) Measurements have dropped below 100 Dobson units. The ozone "hole" has grown larger since 1979, covering a record 25 million km^2 (9 million mi^2) or more than double the surface area of Antarctica—an area larger than Canada, the United States, and Mexico combined. [Image from NASA.]

- Increased ultraviolet is affecting atmospheric chemistry, biological systems, oceanic phytoplankton (small photosynthetic organisms that form the basis of the ocean's primary food production) and fisheries, crop yields, and human skin, eye tissues, and immunity.

More ultraviolet radiation than ever before is breaking through Earth's protective ozone layer. What is happening in the stratosphere everywhere and in the polar regions specifically? Why is this happening, and how are people and their governments responding? What effects does this have on you personally?

Monitoring Earth's Fragile Safety Screen

A sample from the ozone layer's densest part (at 29 km, or 18 mi altitude) contains only 1 part ozone per 4 million parts of air—compressed to surface pressure, it would be only 3 mm thick. Yet, this rarefied layer was in steady-state equilibrium for several hundred million years, absorbing intense ultraviolet radiation and permitting life to proceed safely on Earth.

The ozone layer has been monitored since the 1920s. Ground stations with instrumented balloons, aircraft, orbiting satellites, and a 30-station ozone-monitoring network (mostly in North America) observe stratospheric ozone. The total ozone mapping spectrometer (TOMS) began operations in 1978 aboard *Nimbus-7*, later on the *Upper Atmosphere Research Satellite* (*UARS*), *Adeos*, and *Meteor-3*, and, at the time of this writing, aboard the *Earth Probe* satellite. The September 2001 image is in Figure 1. (See http://jwocky.gsfc.nasa.gov/.)

Ozone Losses Explained

What is causing the decline in stratospheric ozone? In 1974, two atmospheric chemists, F. Sherwood Rowland and Mario Molina, hypothesized that some synthetic chemicals were releasing chlorine atoms that decompose ozone. These **chlorofluorocarbons**, or **CFCs**, are synthetic molecules of chlorine, fluorine, and carbon. (See Rowland and Molina's report: "Stratospheric sink for chlorofluoromethanes: Chlorine atom catalyzed destruction of ozone," *Nature* 249 (1974): 810.)

CFCs are stable (inert) under conditions at Earth's surface and they possess remarkable heat properties. Both qualities made them valuable as propellants in aerosol sprays and as refrigerants. Also, some 45% of CFCs were solvents in the electronics industry and used as foaming agents. Being inert, CFC molecules do not dissolve in water and do not break down in biological processes. (In contrast, chlorine compounds derived from

(continued)

Focus Study 2.1 *(continued)*

volcanic eruptions and the ocean are water-soluble and rarely reach the stratosphere.)

Researchers Rowland and Molina hypothesized that stable CFC molecules slowly migrate into the stratosphere, where intense ultraviolet radiation splits them, freeing chlorine (Cl) atoms. These Cl atoms produce a complex set of reactions that break up ozone molecules (O_3) and leaves oxygen gas molecules (O_2) in their place. Oxygen gas molecules are transparent to ultraviolet radiation. The effect is severe, for a single chlorine atom decomposes more than 100,000 ozone molecules. The long residence time of chlorine atoms in the ozone layer (40 to 100 years) is likely to produce long-term consequences through this century from the chlorine already in place. More than 22 million metric tons (24 million tons) of CFCs were sold worldwide and subsequently released into the atmosphere by 1998. (See **http://www.epa.gov/ozone/index.html**.)

Political Realities—an International Response and the Future

Between 1976 and 1979, Canada, Sweden, Norway, and the state of Oregon banned CFC propellants in aerosols, and a U.S. federal ban began in 1978. However, more than half of U.S. production was exempted, including CFCs used as air-conditioning refrigerants and to make polyurethane foam. CFC sales initially dropped in the late 1970s, but they rose again under a 1981 presidential order that permitted the export and sale of these banned products. Sales increased and hit a new peak in 1987 at 1.2 million metric tons (1.32 million tons).

Chemical manufacturers once claimed that no hard evidence existed to prove the ozone-depletion model, and they successfully delayed remedial action for 15 years. Today, with extensive scientific evidence and verification of losses, even the CFC manufacturers admit that the problem is serious. A few remaining critics are outside of the scientific community.

The *Montreal Protocol on Substances That Deplete the Ozone Layer*, as amended three times in 1990, 1992, and 1997, aims to reduce and eliminate CFC damage to the ozone layer. CFC sales are declining as many countries reduce demand and as worldwide industry phases in alternative chemicals. All production of harmful CFCs will cease by 2010. If the Protocol is fully implemented, it is estimated that the stratosphere will return to nominal conditions during this century. (See the United Nations Ozone Secretariat at **http://www.unep.org/index-en.shtml** and the Montreal Protocol at **http://www.ec.gc.ca/ozone/**.)

Mario Molina stated, "It was frustrating for many years, but it really paid off with the Protocol, which was a marvelous example of what the international community can do working together. We can see from atmospheric measurements that it is already working" (*Nature* 389, September 18, 1997, p. 219). For their work, Rowland, Molina, and another colleague, Paul Crutzen, received the 1995 Nobel Prize for Chemistry. In making the award, the Royal Swedish Academy of Sciences said, "By explaining the chemical mechanism that affects the thickness of the ozone layer, the three researchers have contributed to our salvation from a global environmental problem that could have catastrophic consequences. It has been possible to make far-reaching decisions on prohibiting the release of the gases that destroy ozone."

Ozone Losses over the Poles

How do Northern Hemisphere CFCs become concentrated over the South Pole? Evidently, chlorine freed in the Northern Hemisphere midlatitudes concentrates over Antarctica through the work of atmospheric winds. Persistent cold temperatures over the South Pole and the presence of thin, icy clouds in the stratosphere promote development of ozone losses. Over the North Pole, conditions are more changeable, so the depletion area is smaller, although growing each year.

Polar stratospheric clouds (*PSCs*) are thin clouds that are important catalysts in the release of chlorine for ozone-depleting reactions. During the long, cold winter months, a tight circulation pattern forms over the Antarctic continent—the polar vortex. Figure 2a graphs the negative correlation between ClO and O_3 over the Antarctic continent within the polar vortex. You can see that concentrations of ClO increase toward the pole as O_3 levels decline (Figure 2b). The ozone hole usually fills by early December with stratospheric ozone from lower latitudes, thus thinning ozone in middle and high latitudes.

UV Index Helps Save Your Skin

The television weather report and the newspaper weather page regularly include the *UV Index* in forecasts. This item is reported widely by the National Weather Service (NWS) and the Environmental Protection Agency (EPA). Table 1 presents a sampling of the ultraviolet index numbers and their application to two skin types.

As stratospheric ozone levels continue to thin, and eventually stabilize, surface exposure to cancer-causing radiation climbs. The public now is alerted to take extra precautions in the form of sunscreens, hats, and sunglasses. Remember, damage is cumulative and it may be decades before you experience the ill effects triggered by this summer's sunburn. (For more information, contact the American Cancer Society, 800-227-2345, **http://www.cancer.org/**; or write the American Academy of Dermatology, P.O. Box 681069, Schaumburg, IL 60168-1069.) The scientific community hopes that, with the international actions taken and the hazard to life reduced, science will have scored a significant victory.

(continued)

Focus Study 2.1 *(continued)*

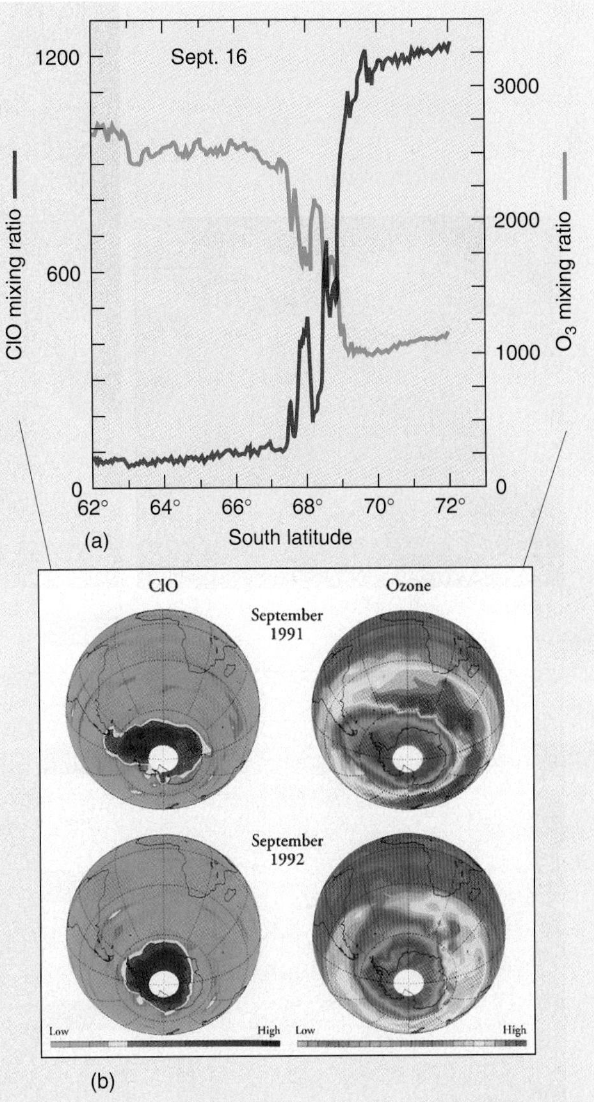

FIGURE 2 Chemical evidence of ozone damage by humans. (a) The negative correlation between ClO and O_3 over the Antarctic continent poleward of 68° S latitude. (b) The data were collected from flights during September 1987 at stratospheric altitudes. Chlorine monoxide (ClO) and ozone (O_3) concentrations above 20 km (12.5 mi) measured during September 1991 and 1992—the beginning of the Antarctic spring. [(a) Data from NASA; (b) Jet Propulsion Laboratory and Goddard Space Flight Center.]

Table 1 UV Index (EPA)

Exposure Category/ Index Value	Minutes to Burn for "Never Tans" (Most Susceptible)	Minutes to Burn for "Rarely Burns" (Least Susceptible)
Minimal 0–2	30 minutes	>120 minutes
Low 4	15 minutes	75 minutes
Moderate 6	10 minutes	50 minutes
High 8	7.5 minutes	35 minutes
Very high 10–15	6 minutes	30 minutes
	<4 minutes	20 minutes

lightning strikes, combined to spark numerous wildfires (Figure 2.22).

Natural Factors That Affect Air Pollution

The problems resulting from both natural and human-made atmospheric contaminants are made worse by several important natural factors. Among

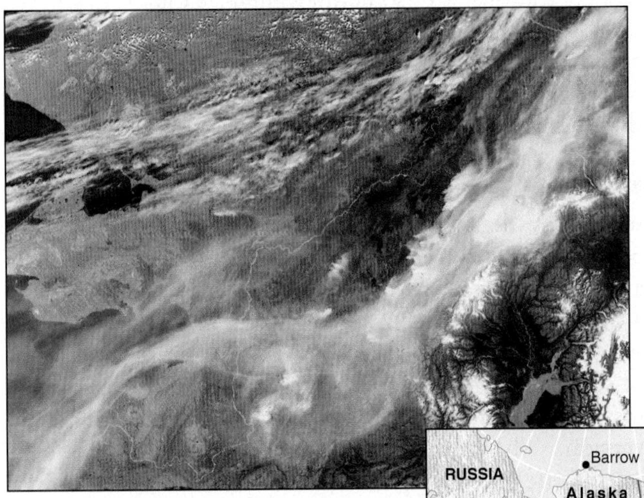

FIGURE 2.22 Alaskan wildfires fill the atmosphere with smoke.
Terra image of wildfires in drought- and high-temperature-plagued portions of Alaska, August 4, 2002. More than 117,000 hectare (289,000 acres) were estimated to be burning in the image; stretching northeastward to the Brooks Range. Nearly 607,000 hectare (1.5 million acres) burned in wildfires in Alaska during 2002. The smoke contains particulate matter (dust, smoke, soot, ash), nitrogen oxides, carbon monoxide, and volatile organic compounds). [*Terra* MODIS sensor image courtesy of the MODIS Land Rapid Response Team, NASA/GSFC.]

(a)

(b)

FIGURE 2.23 Natural variable dust in the atmosphere. (a) A dust storm in central Nevada. (b) A *Terra* orbital view of a sinuous windblown dust plume flowing northward from Africa, across the Mediterranean Sea, toward Turkey. The Nile Delta appears obscured by the dust storm. (c) Alkali dust, a serious air pollutant, rises from the exposed shorelines of Mono Lake, California. [(a) Photo by author; (b) *Terra* MODIS sensor image from 5/13/2001 courtesy of the MODIS Land Rapid Response Team NASA/GSFC; (c) photo by Bobbé Christopherson.]

(c)

these are wind, local and regional landscape characteristics, and temperature inversions in the troposphere.

Wind. Winds gather and move pollutants from one area to another, sometimes reducing the concentration of pollution in one location while increasing it in another. Wind can produce dramatic episodes of dust movement. (Dust is defined as particles less than 62 μm, or 0.0025 in.) Traveling on prevailing winds, dust from Africa contributes to the soils of South America and Europe. Such movement is confirmed by chemical analysis, frequently employed by scientists to track dust to its source area (Figure 2.23).

Such air movements make the atmosphere's condition an international issue. Indeed, prevailing winds transport air pollution from the United States to Canada, causing much complaint and negotiation between the two governments. Pollution in North America is tracked to Europe, adding to European air pollution problems. In Europe, the cross-boundary drift of pollution is a major issue because of the proximity of countries. This issue led in part to Europe's unification and the European Union (EU).

Local and Regional Landscapes. Local and regional landscapes are another important factor in air pollution. Surrounding mountains and hills can form barriers to air movement or can direct pollutants from one area to another. Some of the worst incidents have resulted when local landscapes have trapped and concentrated air pollution.

Places such as Iceland and Hawai'i have their own natural pollution with which to deal. During periods of sustained volcanic activity at Kīlauea, some 2000 metric tons (2200 tons) of sulfur dioxide are produced a day. Concentrations are sometimes high enough to merit broadcast warnings about health concerns, losses to agriculture, and other economic impacts from volcanic smog and acid rain. Hawaiians coined the word *vog* to describe their *vo*lcanic sm*og*.

Temperature Inversion. Vertical temperature and atmospheric density distribution in the troposphere also can worsen pollution conditions. A **temperature inversion** occurs when the normal temperature decrease with altitude (normal lapse rate) begins to *increase* at some altitude. This can happen at any point from ground level to several thousand meters. Figure 2.24 compares a normal temperature profile with that of a temperature inversion. The normal profile (Figure 2.24a) permits warmer (less dense) air at the surface to rise, ventilating the valley and moderating surface pollution. But a warm air inversion (Figure 2.24b) prevents the rise of cooler (denser) air beneath, halting the vertical mixing of pollutants with other atmospheric gases. Thus, instead of being carried away, pollutants are trapped under the *inversion layer*.

FIGURE 2.24 Normal and inverted temperature profiles. (a) A comparison of a normal temperature profile in the atmosphere with (b) a temperature inversion in the lower atmosphere. Note how the warmer air layer prevents mixing of the denser (cooler) air below the inversion, thereby trapping pollution. (c) An inversion layer is visible in the morning hours over a valley. [Photo by Bobbé Christopherson.]

Anthropogenic Pollution

Anthropogenic, or human-caused, air pollution remains most prevalent in urbanized regions. Approximately 2% of annual deaths in the United States are attributable to air pollution—some 50,000 people. Comparable risks are identified in Canada, Europe, Mexico, Asia, and elsewhere.

The human population is moving to cities and is therefore coming into increasing contact with air pollution. By the year 2010, approximately 3.3 billion people (48% of the world population) will live in metropolitan regions, some one-third with unhealthful levels of air pollution. This represents a potentially massive public health issue in the new century.

Table 2.4 lists the names, chemical symbols, principal sources, and impacts of variable anthropogenic components in the air. The first eight pollutants in the table result from combustion of fossil fuels in transportation (specifically automobiles, including light trucks) and at stationary sources such as power plants and factories. Overall, automobiles contribute about 60% of United States and 50% of Canadian human-caused air pollution. (See http://www.epa.gov/.)

In the United States in 2000, transportation produced 77% of the carbon monoxide, 47% of volatile organic compounds (VOC), 56% of the nitrogen oxides, and 25% of particulates. Environment Canada, an agency similar to the U.S. Environmental Protection Agency (EPA), reported that with Canada's smaller transportation fleet, Canadian automobile emissions contribute 40% of the carbon monoxide, 21% of the VOCs, and 60% of the nitrogen oxides in Canada (see http://www.ec.gc.ca/).

Strangely, the U.S. fleet of 2003 model cars and trucks worsened in gas mileage from 2002—decreasing from 23.9 mpg to 20.8 mpg (17.6 mpg for SUVs and light trucks). This poor showing is principally due to gas-guzzling sport utility vehicles and pickups that account for more than half of new sales. Efficiency standards for this class of vehicles, lower than for cars, have not changed since 1975.

The apparent manageability of this transportation-pollution problem is interesting. In one California study, only 7% of the vehicles contributed half of the carbon monoxide and only 10% contributed half of the VOC pollution. These "gross polluting" vehicles are not old cars—a common misconception—but include new cars. In random highway checks, 41% of vehicles had pollution equipment

Table 2.4 Anthropogenic Gases and Materials in the Lower Atmosphere

Name	Symbol	Source	Description and Effects of Criteria Pollutants
Carbon monoxide	CO	Incomplete combustion of fuels	Odorless, colorless, tasteless gas Toxicity: affinity for hemoglobin Displaces O_2 in bloodstream 50 to 100 ppm causes headaches, vision and judgment losses
Nitrogen oxides	NO_x (NO, NO_2)	High temperature/pressure combustion	Reddish-brown choking gas Inflames respiratory system, destroys lung tissue Damages plants 3 to 5 ppm is dangerous
Volatile organic compounds	VOC	Incomplete combustion of fossil fuels such as gasoline; cleaning and paint solvents	Prime agents of ozone formation
Ozone	O_3	Photochemical reactions	Highly reactive, unstable gas Oxidizes surfaces, dries rubber and elastic Damages plants at 0.01 to 0.09 ppm Agricultural loses at 0.1 ppm 0.3 to 1.0 ppm irritates eyes, nose, throat
Peroxyacetyl nitrates	PAN	Photochemical reactions	Produced by NO + VOC photochemistry No human health effects Major damage to plants, forests, crops
Sulfur oxides	SO_x (SO_2, SO_3)	Combustion of sulfur-containing fuels	Colorless; irritating smell 0.1 to 1 ppm impairs breathing; taste threshold Human asthma, bronchitis, emphysema Leads to acid deposition
Particulate matter	PM	Dust, dirt, soot, salt, metals, organics; fugitive dust from agriculture, construction, roads, and wind erosion	Complex mixture of solid and aerosol particles Dust, smoke, and haze affect visibility Various health effects: bronchitis, pulmonary function PM_{10} negative health effects established by researchers
Carbon dioxide	CO_2	Complete combustion, mainly from fossil fuel consumption	Principal greenhouse gas Atmospheric concentration increasing 60% of greenhouse warming effect
Methane	CH_4	Organic processes	Secondary greenhouse gas Atmospheric concentration increasing 12% of greenhouse warming effect
Water vapor	H_2O	Combustion processes, steam	See Chapter 5 for more on the role of water vapor in the atmosphere

that was deliberately tampered with and 25% had defective or missing emission controls. Reducing air pollution from the transportation sector does not pose many mysteries.

Stationary sources, such as electric power plants and industrial plants that use fossil fuels, contribute the most sulfur oxides and particulates. For this reason, concentrations of these substances are focused in the Northern Hemisphere and the industrial, developed countries.

The last three gases shown in Table 2.4 are discussed elsewhere in this text: Water vapor is examined with water and weather (Chapter 5); carbon dioxide and methane are covered with greenhouse gases and climate (Chapters 3 and 6).

Photochemical Smog Pollution. Photochemical smog was not generally experienced in the past, but developed with the advent of the automobile. Today it is the major component of anthropogenic air pollution. **Photochemical smog** results from the interaction of sunlight and the combustion products in automobile and light truck exhaust (nitrogen oxides and VOCs). Although the term *smog*—a combination of the words *sm*oke and f*og*—is a misnomer, it is generally used to describe this phenomenon. Smog is responsible for the hazy sky and reduced sunlight in many of our cities, as shown in China and the U.S. Mid-Atlantic states in Figure 2.25—both regions under an air pollution seige.

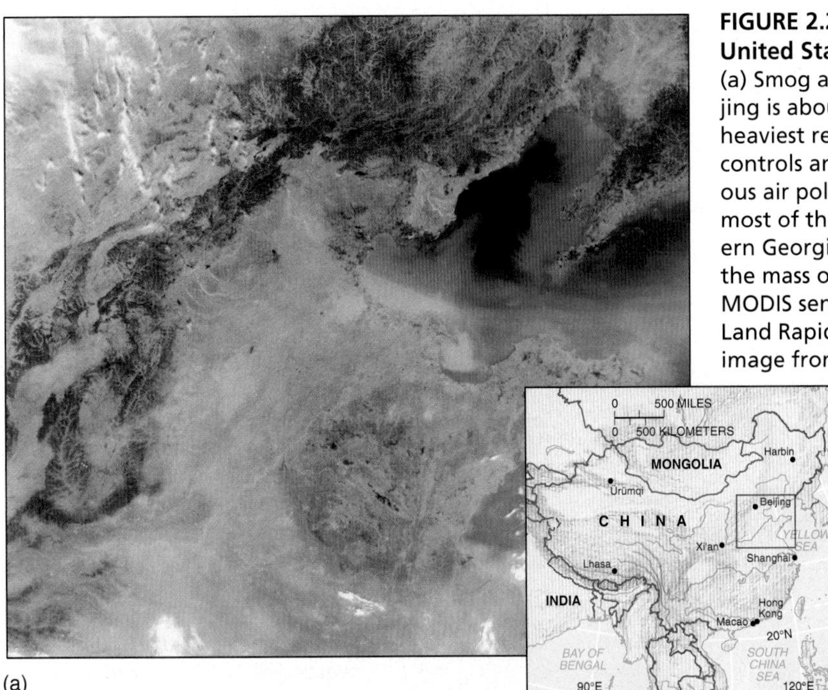

(a)

(b)

FIGURE 2.25 Serious air pollution in China and the United States.
(a) Smog and haze blanket most of eastern China. Beijing is about 150 km (93 mi) west of the coast where the heaviest region of pollution hovers. Poor air pollution controls and the burning of coal contributes. (b) A serious air pollution alert with unhealthy conditions masks most of the Mid-Atlantic region of the U.S.—from northern Georgia to New York. Prevailing winds were blowing the mass offshore over the Atlantic Ocean. [(a) *Terra* MODIS sensor image from 3/12/2002 courtesy of MODIS Land Rapid Response Team NASA/GSFC; (b) *Terra* MODIS image from 6/11/2002 courtesy of Liam Gumley, Space Science and Engineering Center, University of Wisconsin–Madison and NASA.]

Figure 2.26 summarizes how car exhaust is converted into major air pollutants—*ozone*, **peroxyacetyl nitrates (PAN)**, and *nitric acid*. PAN produces no known health effect in humans, but it is particularly damaging to plants, including both agricultural crops and forests. Damage in California is estimated to exceed $1 billion a year and several billion dollars nationwide in the farming and forestry sectors. For several reasons children are at greatest risk from ozone pollution—one in four children in U.S. cities is at risk of developing health problems from ozone pollution. This ratio is significant; it means that more than 12 million children are vulnerable in those cities with the worst polluted air (Los Angeles, New York City, Atlanta, Houston, and Detroit).

Industrial Smog and Sulfur Oxides. Over the past 300 years, except in some developing countries, coal has slowly replaced wood as the basic fuel used by society. The Industrial Revolution required high-grade energy to run machines. The air pollution associated with coal-burning industries is known as **industrial smog** (Figure 2.27). The term *smog* was coined by a London physician at the beginning of the twentieth century to describe the combination of fog and smoke containing sulfur gases (sulfur is an impurity in fossil fuels).

Once in the atmosphere, **sulfur dioxide** (SO_2) reacts with oxygen (O) to form sulfur trioxide (SO_3), which is highly reactive and, in the presence of water or water vapor, forms **sulfate aerosols**, tiny particles about 0.1 to 1 mm in diameter. Sulfuric acid (H_2SO_4) can form even in moderately polluted air, at normal temperatures. Sulfur dioxide-laden air is dangerous to health, corrodes metals, and deteriorates stone building materials at accelerated rates. Sulfuric acid deposition, added to nitric acid deposition, has increased in severity since it was first described in the 1970s. Focus Study 2.2 discusses this vital atmospheric issue.

In the United States, coal-burning electric utilities and steel manufacturing are the main sources of sulfur dioxide, principally in the East and Midwest. And, because of the prevailing movement of air masses, they are the main sources of sulfur dioxide in adjacent Canadian regions. As much as 70% of Canadian sulfur dioxide is initiated within the United States.

Particulates. **Particulate matter (PM)** is a diverse mixture of fine particles, both solid and aerosol, that impact human health. Haze, smoke,

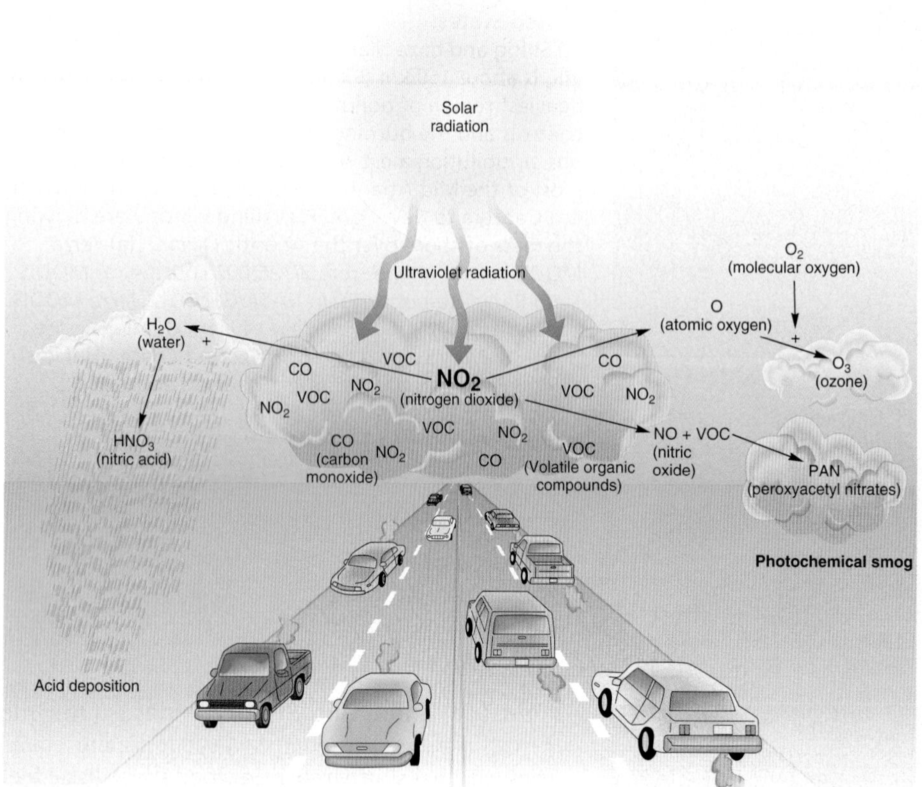

FIGURE 2.26 Photochemical reactions.
Photochemical reactions are produced through the interaction of automobile exhaust (NO_2, VOCs, CO) and ultraviolet radiation in sunlight. The high temperatures in modern automobile engines cause reactions that produce nitrogen dioxide (NO_2). This nitrogen dioxide, derived from automobiles and to a lesser extent from power plants, is highly reactive with ultraviolet light. The reaction liberates atomic oxygen (O) and a nitric oxide (NO) molecule from the NO_2. The free oxygen atom combines with an oxygen molecule (O_2) to form the oxidant ozone (O_3); this same gas that is so beneficial in the stratosphere is an air-pollution hazard at Earth's surface. In addition, the nitric oxide (NO) molecule reacts with VOCs to produce a family of chemicals called peroxyacetyl nitrates (PAN). To the left, note the formation of nitric acid, a contributor to acid deposition.

FIGURE 2.27 Typical industrial smog.
Pollution generated by industry differs from that produced by transportation. Industrial pollution has high concentrations of sulfur oxides, particulates, and carbon dioxide. [Photo by author.]

and dust are visible reminders of particulate material in the air we breathe. PM_{10}, particulate smaller than 10 microns (10 μm) in diameter, was designated a matter for concern in 1987. $PM_{2.5}$ is currently being debated as an appropriate standard for human health. Studies in Provo and Orem, Utah (1989), Philadelphia (1992), and other cities, and by the American Cancer Society (1995), established links between PM pollution and health. Nationally, a study of major cities disclosed a 26% greater risk of premature death due to respirable particulate pollution as compared with nonpolluted air—further driving up medical costs and related expenses.

In Utah County, Utah, researchers correlated concentrations with increased rates of hospitalization for bronchitis, asthma, pneumonia, and pleurisy (especially in children), and greatly increased medical costs. The major source for the PM in this region shut down in 2001, leading to significant improvement in air quality and health-related issues. Similar studies of children affected by related illnesses in seven other cities revealed sickness rates twice as high for the city having the dirtiest air as for the city having the cleanest air. Asthma prevalence has nearly doubled since 1980 in the U.S. Representatives of pollution sources dispute these findings and the concern generated by such studies.

We are now contributing significantly to the creation of the **anthropogenic atmosphere**, a tentative label for Earth's next (fifth) atmosphere. The urban air we breathe today may be just a preview. What is the air quality like where you live, work, and go to college? How might you find out its status?

Benefits of the Clean Air Act

Concentration of many air pollutants declined over the past several decades because of Clean Air Act (CAA) legislation (1970, 1977, 1990), resulting in the saving of trillions of dollars in avoided health, economic, and environmental losses. Despite this success, air pollution controls are subject to a continuing political debate and an actual relaxation of some standards. According to a report prepared by the EPA, the 2001 emissions of five primary

Focus Study 2.2

Acid Deposition: A Continuing Blight on the Landscape

Acid deposition is a major environmental problem in some areas of the United States, Canada, Europe, and Asia. Such deposition is most familiar as "acid rain," but it also occurs as "acid snow" and in dry form as dust or aerosols. (Aerosols are tiny liquid droplets or solid particles.) In addition, winds can carry the acid-producing chemicals many kilometers from their sources before they settle on the landscape, where they enter streams and lakes as runoff and groundwater flows.

Acid deposition is causally linked to serious problems: declining fish populations and fish kills in the northeastern United States, southeastern Canada, Sweden, and Norway; widespread forest damage in these same places and Germany; widespread changes in soil chemistry; and damage to buildings, sculptures, and historic artifacts. It is a meteorological, chemical, and biological phenomena. In New Hampshire's Hubbard Brook Experimental Forest, a study covering 1960 to the present found half the nutrient calcium and magnesium base cations were leached from the soil (more on cations in Chapter 15). Excess acids are the cause of the decline.

The acidity of precipitation is measured on the pH scale, which expresses the relative abundance of free hydrogen ions in a solution. Free hydrogen ions in a solution are what make an acid corrosive, for they easily combine with other ions. The pH scale is logarithmic: each whole number represents a 10-fold change. A pH of 7.0 is neutral (neither acidic nor basic). Values less than 7.0 are increasingly *acidic*, and values greater than 7.0 are increasingly *basic*, or *alkaline*. (A pH scale for soil acidity and alkalinity is portrayed graphically in Chapter 15.)

Natural precipitation dissolves carbon dioxide from the atmosphere to form carbonic acid. This process releases hydrogen ions and produces an average pH reading for precipitation of 5.65. The normal range for precipitation is 5.3–6.0. Thus, normal precipitation is always slightly acidic.

Some anthropogenic gases are converted to acids in the atmosphere and then are removed by wet and dry deposition processes. Specifically, nitrogen and sulfur oxides released in the combustion of fossil fuels can produce nitric acid and sulfuric acid in the atmosphere.

Acid Precipitation Damage

Precipitation as acidic as pH 2.0 has fallen in the eastern United States, Scandinavia, and Europe. By comparison, vinegar and lemon juice register slightly less than pH 3.0. Aquatic plant and animal life perishes when lakes drop below pH 4.8.

More than 50,000 lakes and some 100,000 km (62,000 mi) of streams in the United States and Canada are at a pH level below normal (i.e., below pH 5.3), with several hundred lakes incapable of supporting any aquatic life—15% of the lakes in New England and 41% in the Adirondack Mountains. Acid deposition causes the release of aluminum and magnesium from clay minerals in

(continued)

(a)

(b)

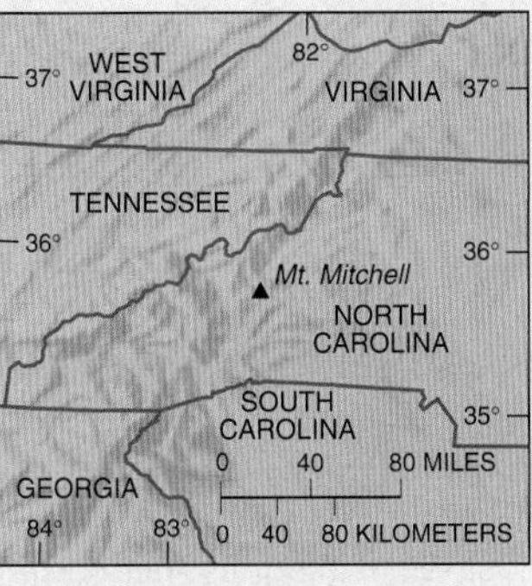

FIGURE 1 The blight of acid deposition. The harm done to forests and crops by acid deposition is well established, especially in Europe, here in the Czech Republic (a) and in the forests of the Appalachian Mountains in the United States, here in the forests on Mount Mitchell (b). [(a) Photo by Simon Fraser/Science Photo Library/Photo Researchers, Inc.; (b) Will and Deni McIntyre/Photo Researchers, Inc.]

Focus Study 2.2 *(continued)*

the soil, and both of these are harmful to fish and plant communities.

Also, relatively harmless mercury deposits in lake-bottom sediments convert in acidified lake waters into highly toxic *methylmercury*, which is deadly to aquatic life. Local health advisories in two provinces and 22 U.S. states are regularly issued to warn those who fish of the methylmercury problem. Mercury atoms rapidly bond with carbon and move through biological systems as an *organometallic compound*.

Damage to forests results from the rearrangement of soil nutrients, the death of soil microorganisms, and an aluminum-induced calcium deficiency that is currently under investigation. The most advanced impact is seen in forests in Europe, especially in eastern Europe, principally because of its long history of burning coal and the density of industrial activity. In Germany and Poland up to 50% of the forests are dead or damaged; in Switzerland 30% are afflicted (Figure 1a).

In the United States, regional-scale decline in forest cover is significant, especially red spruce and sugar maples. In some maples, aluminum is collecting around rootlets; in spruce, acid fogs and rains leach calcium from needles directly. Affected trees are susceptible to winter cold, insects, and droughts. In New England, some stands of spruce are as much as 75% affected, as evidenced through analysis of tree-growth rings which become narrower in adverse growing years. An indicator of forest damage is the reduction by almost half of the annual production of U.S. and Canadian maple sugar. Trees at higher elevations in the Appalachians are injured by acid-laden cloud cover (Figure 1b).

Government estimates of damage in the United States, Canada, and Europe exceed $50 billion annually. Because wind and weather patterns are international, efforts at reducing acidic deposition also must be international in scope. The decline of sulfur dioxide by more than 40% between 1973 and the present is a result of the U.S. Clean Air Act. Researchers found a correlation between these reductions and a reduction in the geographic area affected by wet deposition of sulfur. Figure 2 maps the reduction in sulfate deposition between the 1983–1988 and 1995–1999 periods. However, this progress is only a beginning. According to a study in *BioScience*, power plants and other sources must cut emissions 80% beyond the Clean Air Act mandate.

Ten leading acid deposition researchers reported in an extensive study in *BioScience*, March 2001,

> Model calculations suggest that the greater the reduction in atmospheric sulfur deposition, the greater the magnitude and rate of chemical recovery. Less aggressive proposals for controls of sulfur emissions will result in slower chemical and biological recovery and in delays in regaining the services of a fully functional ecosystem.... North America and Europe are in the midst of a large-scale experiment. Sulfuric

(continued)

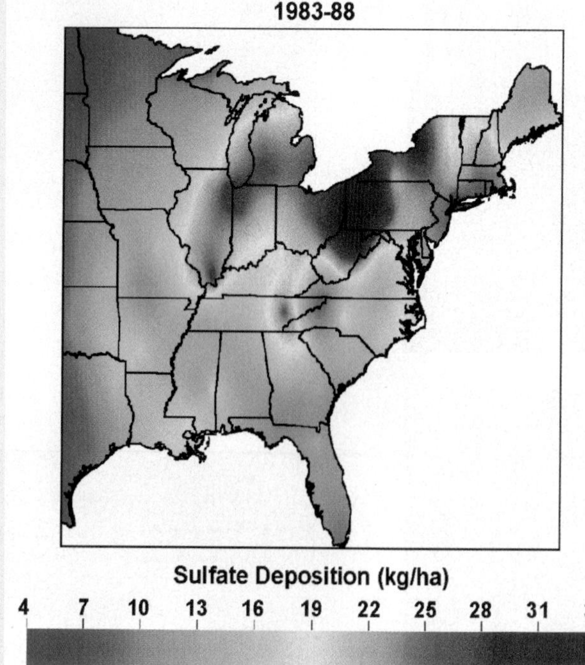

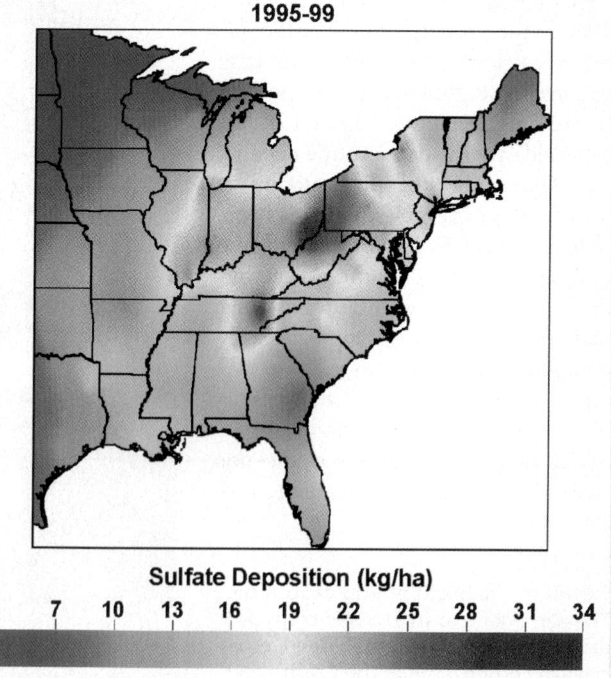

FIGURE 2 Improvement in sulfate wet deposition rate.
Spatial portrayal of annual sulfate (principally SO_4) wet deposition on the landscape, 1983–88 and 1995–99 in kilograms per hectare. Clean Air Act regulations have lowered levels of emissions that add acid to the environment. According to a new study more improvement is necessary to help ecosystems recover. [Maps courtesy of J.W. Lynch and J.A. Grimm, U.S. Forest Service, Northeast Forest Experimental Station, Northern Global Change Research Program.]

Focus Study 2.2 *(continued)*

and nitric acids have acidified soils, lakes, and streams, thereby stressing or killing terrestrial and aquatic biota.*

At best, acid deposition is an issue of global spatial significance for which science is providing strong incentives for action. Reductions in troublesome emissions are closely tied to energy conservation and therefore directly related to production of greenhouse gases and global warming concerns—thus, linking these environmental issues to political actions and the fossil fuel transnational corporations.

* C. T. Driscoll, *et al.*, "Acidic Deposition in the Northeastern United States: Sources and Inputs, Ecosystems Effects, and Management Strategies," *BioScience* 51 (March 2001): 195.

lation (1970, 1977, 1990), resulting in the saving of trillions of dollars in avoided health, economic, and environmental losses. Despite this success, air pollution controls are subject to a continuing political debate and an actual relaxation of some standards. According to a report prepared by the EPA, the 2001 emissions of five primary pollutants totaled 177 million metric tons (195 million tons), compared to 224 million metric tons (246 million tons) in 1970, the first year of the CAA. Figure 2.28 illustrates the trends in CO, NO_x, VOCs, SO_2, PM_{10}, and lead (Pb). Only nitrogen oxides increased between 1970 and 2001. In Canada, between 1980 and 1997, sulfur dioxide emissions decreased 44%.

To be justified, abatement (mitigation and prevention) costs must not exceed the financial benefits derived from reducing pollution damage. Compliance with the CAA affected patterns of industrial production, employment, and capital investment. Although these expenditures must be viewed as investments that generated benefits and opportunities, the dislocation of workers in some regions was severe: reductions in high-sulfur coal mining and cutbacks in polluting industries such as steel, for example.

In 1990, Congress requested the EPA to answer the following question: How do the overall health, welfare, ecological, and economic benefits of CAA programs compare with the costs of these programs? In response, the EPA performed an exhaustive cost-benefit analysis and published a draft report in 1996 and a final report in 1997. *The Benefits of the Clean Air Act, 1970 to 1990* (Office of Policy, Planning, and Evaluation, U.S. EPA) reported the following findings:

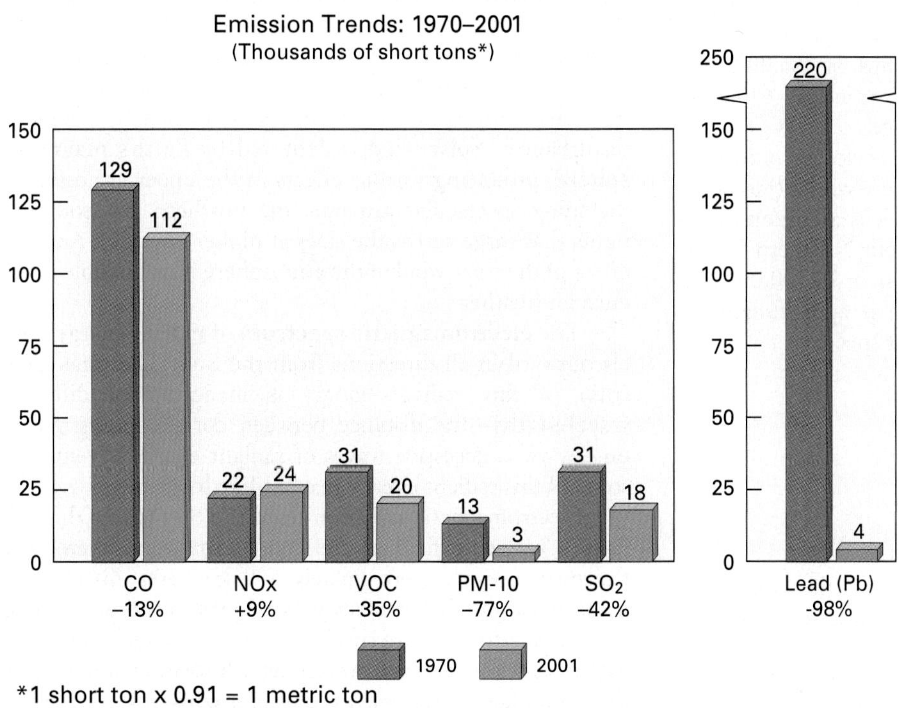

FIGURE 2.28 Trends in air pollutants—1970–2001.
Sulfur dioxide lowered due to scrubbers on smokestacks and emission controls. Nitrogen oxides increased, but volatile organic compounds and carbon monoxide decreased, all three involve exhaust emission controls. Particulate material declined. A huge lead reduction of −98%, or 234,000 tons resulted from the use of unleaded gas and reduced industrial emissions. [Adapted from Office of Air Quality, *National Air Quality and Emission Trends Report and Latest Findings on National Air Quality,* U.S. EPA.]

$49.4 trillion, with a central mean of *$22.2 trillion.* (The uncertainty of the assessment is indicated by the range of benefit estimates.)

- Therefore, the *net financial benefit* of the CAA is *$21.7 trillion.* "The finding is overwhelming. The benefits far exceed the costs of the CAA in the first 20 years," said Richard Morgenstern, associate administrator for policy planning and evaluation at the EPA.

The benefits to society, directly and indirectly, have been widespread across the entire population: improved health and environment, less lead to harm children, lowered cancer rates, less acid deposition, and an estimated 206,000 fewer deaths related to air pollution in 1990 alone, among many benefits described. Further, these benefits continue during a period (1970–2001) in which the U.S. population grew by 22% and the economy expanded by 161%.

As you reflect on this chapter and our modern atmosphere, the treaties to protect stratospheric ozone, and the EPA study of benefits from the CAA, these positive outcomes should be refreshing. Society knew what to do, took action, and reaped enormous economic and health benefits. An important role for physical geographers is to explain these global impacts through spatial analysis and to guide an informed citizenry toward enhanced understanding.

Summary and Review—Solar Energy, Seasons, and the Atmosphere

● *Distinguish* among galaxies, stars, and planets and locate Earth.

Our Solar System—Sun and nine planets—is located on a remote, trailing edge of the **Milky Way Galaxy**, a flattened, disk-shaped mass estimated to contain up to 400 billion stars. **Gravity**, the mutual attracting force exerted by the mass of an object upon all other objects, is an organizing force in the Universe. The process of suns (stars) condensing from nebular clouds with planetesimals (protoplanets) forming in orbits around their central masses is the **planetesimal hypothesis**.

The Solar System, planets, and Earth began to condense from a nebular cloud of dust, gas, debris, and icy comets approximately 4.6 billion years ago. Distances in space are so vast that the **speed of light** (300,000 kmps, or 186,000 mps, which is about 9.5 trillion kilometers, or nearly 6 trillion miles, per year) is used to express distance.

In its orbit, Earth is at **perihelion** (its closest position to the Sun) during our Northern Hemisphere winter (January 3 at 147,255,000 km, or 91,500,000 mi). It is at **aphelion** (its farthest position from the Sun) during our Northern Hemisphere summer (July 4 at 152,083,000 km, or 94,500,000 mi). Earth's average distance from the Sun is approximately 8 minutes and 20 seconds in terms of light speed.

Milky Way Galaxy (p. 38)
gravity (p. 38)
planetesimal hypothesis (p. 38)
speed of light (p. 38)
perihelion (p. 38)
aphelion (p. 38)

1. Describe the Sun's status among stars in the Milky Way Galaxy. Describe the Sun's location, size, and relationship to its planets.
2. If you have seen the Milky Way at night, briefly describe it. Use specifics from the text in your description.
3. Briefly describe Earth's origin as part of the Solar System.
4. Compare the nine planets of the Solar System and their distances from the Sun.
5. How far is Earth from the Sun in terms of light speed? In terms of kilometers and miles? Relate this distance to the shape of Earth's orbit during the year.
6. Briefly describe the relationship among these concepts: Universe, Milky Way Galaxy, Solar System, Sun, and Planet Earth.
7. Within which of Earth's atmospheres, past or present, did photosynthesis begin? When photosynthesis began, what atmospheric gas began to accumulate?

● *Describe* the Sun's operation and *explain* the characteristics of the solar wind and the electromagnetic spectrum of radiant energy.

The **fusion** process—hydrogen atoms forced together under tremendous temperature and pressure in the Sun's interior—generates incredible quantities of energy. Solar energy in the form of charged particles of **solar wind** travels out in all directions from disturbances on the Sun. The Sun's most conspicuous features are large **sunspots**, caused by magnetic disturbances. Solar wind is deflected by Earth's **magnetosphere**, producing various effects in the upper atmosphere, including spectacular **auroras**, the northern and southern lights that surge across the skies at higher latitudes. Another effect of the solar wind in the atmosphere is its possible influence on weather.

The **electromagnetic spectrum** of radiant energy travels outward in all directions from the Sun. The total spectrum of this radiant energy is made up of different **wavelengths**—the distance between corresponding points on any two successive waves of radiant energy. Eventually, some of this radiant energy reaches Earth's surface.

Electromagnetic radiation from the Sun passes through Earth's magnetic field to the top of the atmosphere—the **thermopause**, at approximately 500 km (300 mi) altitude. Solar radiation that reaches a horizontal plane at Earth is called **insolation**, a term specifically applied to radiation arriving at Earth's surface and atmosphere. Insolation at the top of the atmosphere is expressed as the **solar constant**: the average insolation received at the thermopause when Earth is at its average distance from the Sun. The solar constant is measured as 1372 W/m^2 (2.0 $cal/cm^2/min$; 2 langleys/min. The place receiving maximum insolation is the **subsolar point**, where solar rays are perpendicular to the surface (radiating from directly overhead). All other locations away from the subsolar point receive slanting rays and more diffuse energy.

fusion (p. 38)
solar wind (p. 38)
sunspots (p. 38)
magnetosphere (p. 38)
auroras (p. 38)
electromagnetic spectrum (p. 40)
wavelength (p. 40)
thermopause (p. 41)
insolation (p. 41)
solar constant (p. 41)
subsolar point (p. 42)

8. How does the Sun produce such tremendous quantities of energy?
9. What is the sunspot cycle? At what stage was the cycle in the year 2001?
10. Describe Earth's magnetosphere and its effects on the solar wind and the electromagnetic spectrum.
11. Compare the filtering aspects of the atmosphere to the astronaut's spacesuit shown in Figure 2.3, specifically the solar wind.
12. Describe the various segments of the electromagnetic spectrum, from shortest to longest wavelength. What wavelengths are mainly produced by the Sun? Which are principally radiated by Earth to space?
13. What is the solar constant? Why is it important to know?
14. Select 0° or 90° latitude on Figure 2.10 and compare approximately the amount of energy received. What differences do you observe? On the map where do you find the largest net radiation? The lowest net radiation?
15. If Earth were flat and oriented perpendicularly to incoming solar radiation (insolation), what would be the latitudinal distribution of solar energy at the top of the atmosphere?

● ***Define*** **solar altitude, solar declination, and daylength and** ***describe*** **the annual variability of each—Earth's seasonality.**

The angle between the Sun and the horizon is the Sun's **altitude**. The Sun's **declination** is the latitude of the subsolar point. Declination annually migrates through 47° of latitude, moving between the *Tropic of Cancer* at 23.5° N (June) and the *Tropic of Capricorn* at 23.5° S latitude (December). Seasonality means an annual change in the Sun's altitude and changing **daylength**, or duration of exposure.

Earth's distinct seasons are produced by interactions of **revolution** (annual orbit about the Sun), **rotation** (turning on the axis), *axial tilt* (23.5° from a perpendicular to the **plane of the ecliptic**—an imaginary plane touching all points of Earth's orbit), **axial parallelism** (the parallel alignment of the axis throughout the year), and sphericity. Earth rotates about its *axis*, an imaginary line extending through the planet from the geographic North Pole to the South Pole. As it rotates, the traveling boundary that divides daylight and darkness is called the **circle of illumination**.

On December 21 or 22, at the moment of the **winter solstice** ("winter Sun stance"), or **December solstice**, the circle of illumination excludes the North Pole but includes the South Pole. The subsolar point is at 23.5° S latitude, the parallel called the **Tropic of Capricorn**. The moment of the **vernal equinox**, or **March equinox**, occurs on March 20 or 21. At that time, the circle of illumination passes through both poles so that all locations on Earth experience a 12-hour day and a 12-hour night.

June 20 or 21 is the moment of the **summer solstice**, or **June solstice**. The subsolar point now has shifted from the equator to 23.5° N latitude, the **Tropic of Cancer**. Because the circle of illumination now includes the North Polar region, everything north of the Arctic Circle receives 24 hours of daylight—the "midnight Sun." September 22 or 23 is the time of the **autumnal equinox**, or **September equinox**, when Earth's orientation is such that the circle of illumination again passes through both poles so that all parts of the globe experience a 12-hour day and a 12-hour night.

altitude (p. 43)
declination (p. 44)
daylength (p. 44)
revolution (p. 44)
rotation (p. 45)
circle of illumination (p. 45)
plane of the ecliptic (p. 45)
axial parallelism (p. 45)
winter solstice, December solstice (p. 46)
Tropic of Capricorn (p. 46)
vernal equinox, March equinox (p. 47)
summer solstice, June solstice (p. 47)
Tropic of Cancer (p. 47)
autumnal equinox, September equinox (p. 47)

16. Contrast the concept of seasonality between the equator and the polar regions.
17. The concept of seasonality refers to two specific observations. How do these two aspects of seasonality change during a year at 0° latitude? At 40°? At 90°?
18. Differentiate between the Sun's altitude and its declination at Earth's surface.
19. For the latitude at which you live, how does daylength vary during the year? How does the Sun's altitude vary? Does your local newspaper publish a weather calendar containing such information?
20. List the five physical factors that operate together to produce seasons.
21. Describe Earth's revolution and rotation, and differentiate between them.
22. Define Earth's present tilt relative to its orbit about the Sun.

● ***Construct*** **a general model of the atmosphere based on composition, temperature, and function and** ***diagram*** **this model in a simple sketch.**

Our modern atmosphere is a gaseous mixture so evenly mixed it behaves as if it were a single gas. It is naturally odorless, colorless, tasteless, and formless. The principal substance of this atmosphere is air—the medium of life.

Above 480 km (300 mi) altitude, the atmosphere is rarefied (nearly a vacuum) and is called the **exosphere**, which means "outer sphere." The weight (force over a unit area) of the atmosphere, exerted on all surfaces, is termed **air pressure**. It decreases rapidly with altitude.

By *composition*, we divide the atmosphere into the **heterosphere**, extending from 480 km to 80 km, and the **homosphere**, extending from 80 km to Earth's surface. Within the heterosphere and using *temperature* as a criterion, we identify the **thermosphere**. Its upper limit, called the *thermopause*, is at approximately 480 km altitude. **Kinetic energy**, the energy of motion, is the vibrational energy that we measure and call temperature. However, the actual heat produced in the thermosphere is very small. The density of the molecules is so low that little actual *heat*, the flow of kinetic energy from one body to another because of a temperature difference between them, is produced. Nearer Earth's surface the greater number of molecules in the denser atmosphere transmit their kinetic energy as **sensible heat**, meaning that we can feel it.

The homosphere includes the **mesosphere**, **stratosphere**, and **troposphere**, as defined by temperature criteria. The normal temperature profile within the troposphere during the daytime decreases rapidly with increasing altitude at an average of 6.4 C° per km (3.5 F° per 1000 ft), a rate known as the **normal lapse rate**. The top of the troposphere is wherever a temperature of −57°C (−70°F) is recorded, a transition known as the *tropopause*. The actual lapse rate at any particular time and place may deviate considerably because of local weather conditions and is called the **environmental lapse rate**.

We distinguish a region in the heterosphere by its *function*. The **ionosphere** absorbs cosmic rays, gamma rays, X-rays, and shorter wavelengths of ultraviolet radiation and converts them into kinetic energy. A functional region within the stratosphere is the **ozonosphere**, or **ozone layer**, which absorbs life-threatening ultraviolet radiation, subsequently raising the temperature of the stratosphere.

exosphere (p. 48)
air pressure (p. 48)
heterosphere (p. 49)
homosphere (p. 49)
thermosphere (p. 52)
kinetic energy (p. 52)
sensible heat (p. 53)
mesosphere (p. 53)
stratosphere (p. 53)
troposphere (p. 53)
normal lapse rate (p. 53)
environmental lapse rate (p. 53)
ionosphere (p. 53)
ozonosphere, ozone layer (p. 53)

23. What is air? Where did the components in Earth's present atmosphere originate?
24. In view of the analogy by Lewis Thomas, characterize the various functions the atmosphere performs that protect the surface environment.
25. What three distinct criteria are employed in dividing the atmosphere for study?
26. Describe the overall temperature profile of the atmosphere, and list the four layers defined by temperature.
27. Describe the two divisions of the atmosphere on the basis of composition.
28. What are the two primary functional layers of the atmosphere and what does each do?

- ***Describe* conditions within the stratosphere; specifically, *review* the function and status of the ozonosphere (ozone layer).**

The overall reduction of the stratospheric ozonosphere, or ozone layer, during the past several decades represents a hazard for society and many natural systems and is caused by chemicals introduced into the atmosphere by humans. Since World War II, quantities of human-made **chlorofluorocarbons (CFCs)** and bromine-containing compounds have made their way into the stratosphere. The increased ultraviolet light at those altitudes breaks down these stable chemical compounds, thus freeing chlorine and bromine atoms. These atoms act as catalysts in reactions that destroy ozone molecules.

chlorofluorocarbons (CFCs) (p. 55)

29. Why is stratospheric ozone (O_3) so important? Describe the effects created by increases in ultraviolet light reaching the surface.
30. Summarize the ozone predicament and present trends and any treaties that intend to protect the ozone layer.
31. Evaluate Crutzen, Rowland, and Molina's use of the scientific method in investigating stratospheric ozone depletion.

- ***Distinguish* between natural and anthropogenic variable gases and materials in the lower atmosphere and *describe* the sources and effects of air pollution and acid deposition.**

Within the troposphere, both natural and human-caused variable gases, particles, and other chemicals are part of the atmosphere. We coevolved with natural "pollution" and thus are adapted to it. But we are not adapted to cope with our own anthropogenic pollution. It constitutes a major health threat, particularly where people are concentrated in cities.

Vertical temperature and atmospheric density distribution in the troposphere also can worsen pollution conditions. A **temperature inversion** occurs when the normal temperature decrease with altitude (normal lapse rate) reverses. In other words, temperature begins to increase at some altitude.

Photochemical smog results from the interaction of sunlight and the products of automobile exhaust, the single largest contributor of pollution that produces smog. Car exhaust, containing *nitrogen dioxide* and *volatile organic compounds (VOCs)*, in the presence of ultraviolet light in sunlight converts into major air pollutants—*ozone*, **peroxyacetyl nitrates (PAN)**, and *nitric acid*. The principal photochemical by-products include *ozone* (O_3), which causes negative health effects, oxidizes surfaces, and kills or damages plants; and peroxyacetyl nitrates (PAN), which produce no known health effects in humans but are particularly damaging to plants, including both agricultural crops and forests. **Particulate matter (PM)** is a diverse mixture of fine particles, both solid and aerosol, that impact human health.

The distribution of human-produced **industrial smog** over North America, Europe, and Asia is related to transportation and electrical production. Such characteristic pollution contains **sulfur dioxide**. Sulfur dioxide in the atmosphere reacts to produce **sulfate aerosols**, which produce sulfuric acid (H_2SO_4) deposition and affect the Earth energy budget by scattering and reflecting solar energy.

Energy conservation and efficiency and reducing emissions are essential strategies for abating air pollution. Earth's next atmosphere most accurately may be described as the **anthropogenic atmosphere** (human-influenced atmosphere).

temperature inversion (p. 58)
photochemical smog (p. 60)
peroxyacetyl nitrates (PAN) (p. 61)
particulate matter (PM) (p. 61)
industrial smog (p. 61)
sulfur dioxide (p. 61)
sulfate aerosols (p. 61)
anthropogenic atmosphere (p. 62)

32. Why are anthropogenic gases more significant to human health than are those produced from natural sources?
33. In what ways does a temperature inversion worsen an air pollution episode? Why?
34. What is the difference between industrial smog and photochemical smog?
35. Describe the relationship between automobiles and the production of ozone and PAN in city air. What are the principal negative impacts of these gases?
36. How are sulfur impurities in fossil fuels related to the formation of acid in the atmosphere and acid deposition on the land?
37. In summary, what are the results from the first 20 years under Clean Air Act regulations? In your opinion, do you see viable arguments for its repeal?

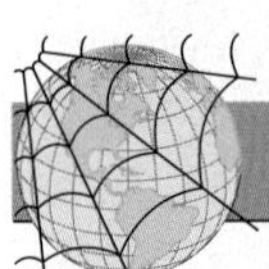

Network

The *Elemental Geosystems Home Page* provides on-line resources for this chapter on the World Wide Web. Once on the Home Page, click on this textbook, scroll the Table of Contents menu, select this chapter, and click "Begin." You will find self-tests that are graded, review exercises, specific updates for items in the chapter, and in "Destinations" many links to interesting related pathways on the Internet. *Elemental Geosystems* is found at **http://www.prenhall.com/christopherson**.

Critical Thinking

A. Using the concepts in Figure 2.15, use a protractor and stick or ruler to measure the angle of the Sun's altitude at noon (or 1 P.M. if in daylight saving time). Do not look at the Sun; rather, with your back to the Sun, align the stick so that it casts no shadow as you measure its rays against the protractor. Write this measurement in your notebook and affix a Post-It™ in the textbook to remind you to repeat the measurement near the end of the semester or quarter. Compare and analyze the different measurements you determine.

B. To determine the total ozone column at your present location, go to the TOMS Home Page at **http://toms.gsfc.nasa.gov/teacher/ozone_overhead.html**, "What was the total ozone column at your house?" Select a point on the map or enter your latitude and longitude, and the date you want to check. The ozone column refers principally to stratospheric ozone and not to the photochemical pollutant in the lower troposphere. Note the instrument and satellite platform used. (Note also the limitations listed on the extent of data availability.) For several different dates, when do the lowest values occur? The highest values? Briefly explain what your results mean. How do you interpret the values found?

C. Science has determined the root causes of stratospheric ozone depletion, photochemical smog, industrial smog, and acid deposition in streams and soils. In several cases treaties and legislation have successfully halted worsening conditions. The success of the Clean Air Act is an excellent example. In your opinion, why is the public generally unaware of the details? What are the difficulties in instructing the public? Why is such media attention given to antiscience and nonscientific opinions? Take a moment and brainstorm recommendations for action, education, and public awareness on these issues.

Jet contrails foster cloud development. New studies found that contrail-stimulated clouds affect daily temperature ranges and Earth's atmospheric energy budget. [Photo by Bobbé Christopherson.]

3 Atmospheric Energy and Global Temperatures

Key Learning Concepts

By knowing and understanding the key learning concepts in this chapter, you should be able to:

- *Identify* the pathways of solar energy through the troposphere to Earth's surface: transmission, refraction, albedo (reflectivity), scattering, diffuse radiation, conduction, convection, and advection.
- *Describe* the greenhouse effect and the patterns of global net radiation and surface energy balances.
- *Review* the temperature concepts and temperature controls that produce global temperature patterns.
- *Interpret* the pattern of Earth's temperatures for January and July and annual temperature ranges.
- *Contrast* wind chill and heat index and *determine* human response to these apparent temperature effects.
- *Portray* typical urban heat island conditions and *contrast* the microclimatology of urban areas with that of surrounding rural environments.

Earth's biosphere pulses with flows of energy. Think for a moment of your own life and activities—all reflect atmosphere and surface energy patterns, such as the seasons, climate, and the daily weather we experience. Find a nice spot outdoors to read this chapter so you can actually observe the concepts happening as you learn—depending on local weather and seasons, of course.

In this chapter: This chapter examines the cascade of energy through the atmosphere to Earth's surface as insolation is absorbed and redirected along various pathways in the troposphere. An important output produced by this atmosphere and surface energy system is the present pattern of global temperatures. Global temperature patterns presently are changing in a warming trend that is affecting us all. Also, you will find out what apparent temperature means as wind and humidity alter the temperatures we sense. Finally, the cities in which we live alter surface energy characteristics, so the temperatures and climates of our urban areas differ from those of surrounding rural areas.

Energy Essentials

Global albedo values

SATELLITE LOOP

In a photograph of Earth taken from space, the pattern of surface response to incoming insolation is clearly visible (see the back cover of this text). Land and water surfaces, clouds, and atmospheric gases and dust intercept solar energy. The flows of energy are manifest in swirling weather patterns, powerful oceanic currents, and the varied distribution of vegetation. Specific energy patterns differ for deserts, oceans, mountain tops, rain forests, and ice-covered landscapes. In addition, clear or cloudy weather may mean a 75% difference in the amount of energy reaching the surface, because clouds reflect incoming energy.

Energy Pathways and Principles

Earth's atmosphere and surface are heated by solar energy, which is unevenly distributed by latitude and which fluctuates seasonally. Figure 3.1 is a simplified flow diagram of shortwave and longwave radiation in the Earth-atmosphere system. You will find it helpful to refer to this figure, and the more-detailed energy balance illustration in Figure 3.10, as you read through the following section. We first look at some important pathways and principles for insolation as it passes through the atmosphere to Earth's surface.

Transmission refers to the passage of shortwave and longwave energy through either the atmosphere or water. The atmosphere and surface eventually reradiate heat energy back to space, and this energy, together with reflected energy, equals the initial solar input. Thus, a *balance* of energy input and output exists in our atmosphere.

Insolation Input. Insolation is the single energy input driving the Earth-atmosphere system. The world map in Figure 3.2 shows the distribution of average annual solar energy at Earth's surface. It includes all radiation arriving at Earth's surface, both direct and diffuse (or downward scattered).

Several patterns are notable on the map. Insolation decreases poleward from about 25° latitude in both the Northern and Southern Hemispheres. Consistent daylength and high Sun altitude produce average annual values of 180–200 W/m^2 throughout the equatorial and tropical latitudes. In general, greater insolation of 240–280 W/m^2 occurs in low-latitude deserts worldwide because of frequently cloudless skies. Note the energy pattern in the cloudless subtropical deserts in both hemispheres (for example, the Sonoran, Saharan, Arabian, Gobi, Atacama, Namib, Kalahari, and Australian Deserts).

Scattering (Diffuse Radiation). Insolation encounters an increasing density of atmospheric gases as it travels toward the surface. Gas molecules redirect radiation, changing the direction of the light's movement *without altering its wavelengths.* This phenomenon is known as **scattering** and represents 7% of Earth's reflectivity (see Figure 3.10). Dust particles, pollutants, ice, cloud droplets, and water vapor produce further scattering. Air pollution—smog and haze—produces scattering and an almost white sky because the larger particles associated with air pollution act to scatter all wavelengths of the visible light.

The angle of the Sun's rays determines the thickness of atmosphere they must pass through to reach the

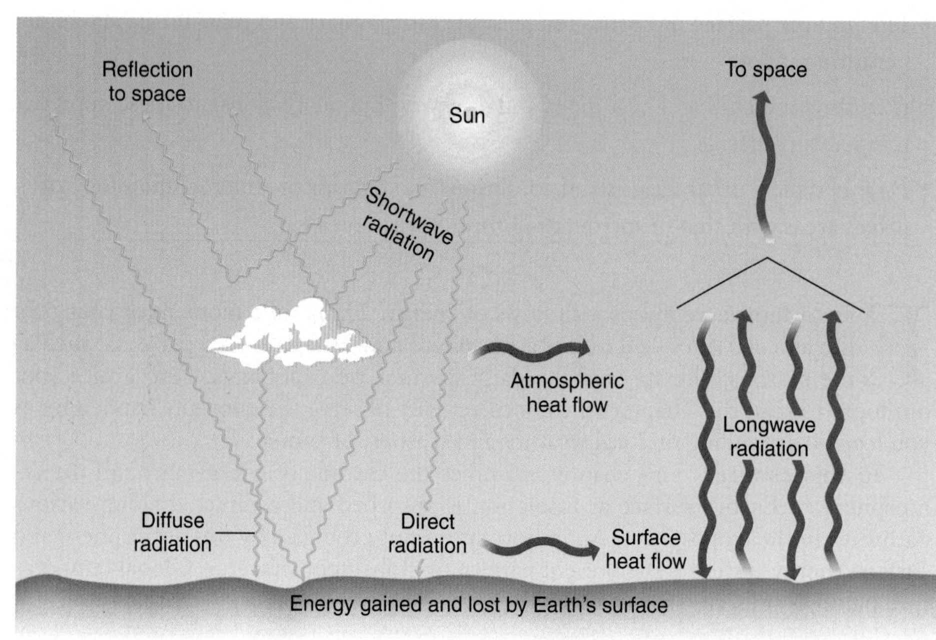

FIGURE 3.1 Energy gained and lost by Earth's surface and atmosphere.
Simplified view of the Earth-atmosphere energy system—circuits include incoming shortwave insolation, reflected shortwave radiation, and outgoing longwave radiation.

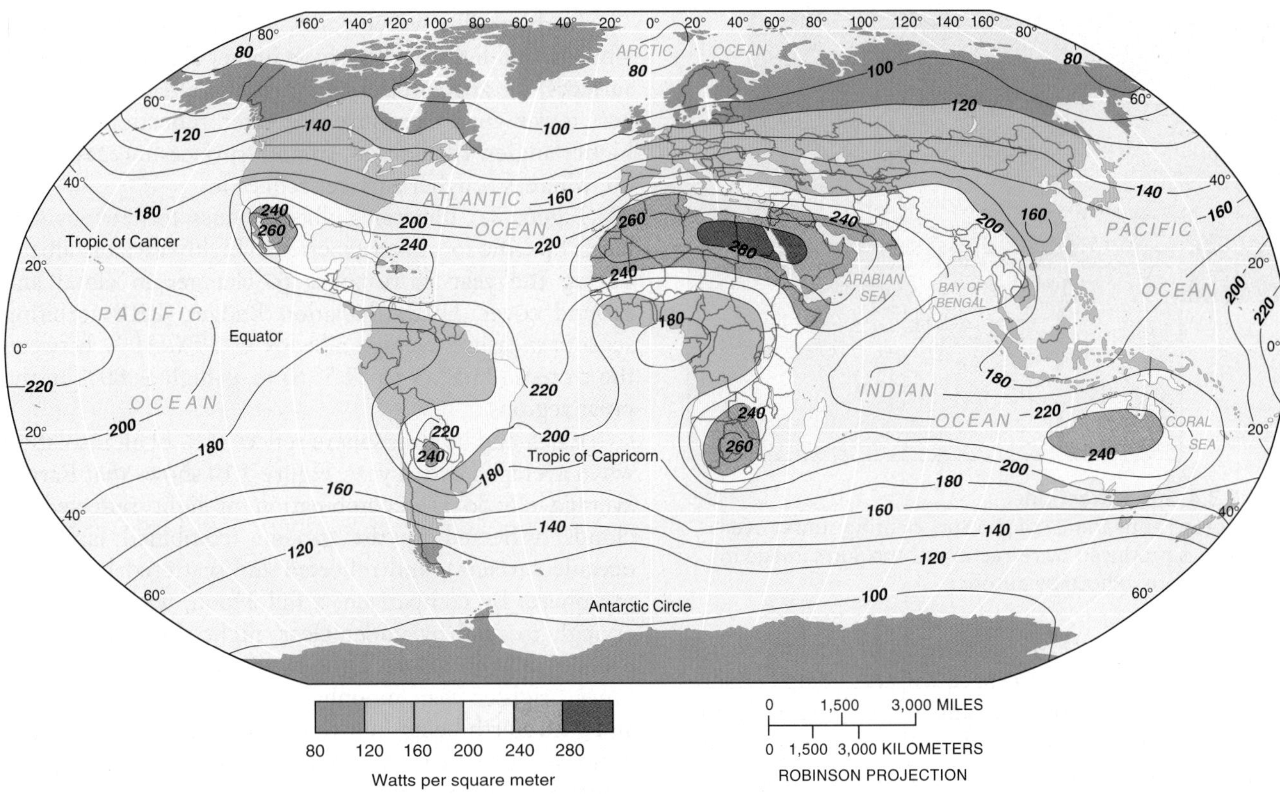

FIGURE 3.2 Insolation at Earth's surface.
Average annual solar radiation receipt on a horizontal surface at ground level in watts per square meter (100 W/m^2 = 75 kcal/cm^2/yr). [After M. I. Budyko, *The Heat Balance of the Earth's Surface* (Washington, D.C.: U.S. Department of Commerce, 1958), p. 99.]

surface. Direct rays experience less scattering and absorption than do low, oblique-angle rays that must travel farther through the atmosphere.

Figure 3.10 shows that some incoming insolation is diffused by clouds and atmosphere and is transmitted to Earth as **diffuse radiation**, the downward component of scattered light. For instance, on a cloudy day this light is multidirectional and thus casts shadowless light on the ground.

Refraction. When insolation enters the atmosphere, it passes from one medium to another (from virtually empty space to atmospheric gases) and is subject to a change of speed, which also shifts its direction—a bending action called **refraction**. In the same way, a crystal or prism refracts light passing through it, bending different wavelengths to different degrees, separating the light into its component colors to display the spectrum. A rainbow is created when visible light passes through myriad raindrops and is refracted and reflected toward the observer at a precise angle (Figure 3.3). Another example of refraction is a *mirage*, an image that appears near the horizon where light waves are refracted by layers of air of differing temperatures (and resulting differing densities) on a hot day.

The Sun's image is refracted in its passage from space through the atmosphere, and so, at sunrise, we see the Sun about 4 minutes before it actually peeks over the horizon. Similarly, the Sun actually sets at sunset, but its image is refracted over the horizon for about 4 minutes afterward. The distortion of the setting Sun in Figure 3.4 is a product of refraction. (The inset diagram explains what is happening in the photo.)

FIGURE 3.3 A rainbow.
A rainbow is produced when raindrops refract and reflect light. Note that the color order in the rainbow is distributed with the shortest wavelengths on the inside of the bow and the longest wavelengths on the outside of the bow. [Photo by author.]

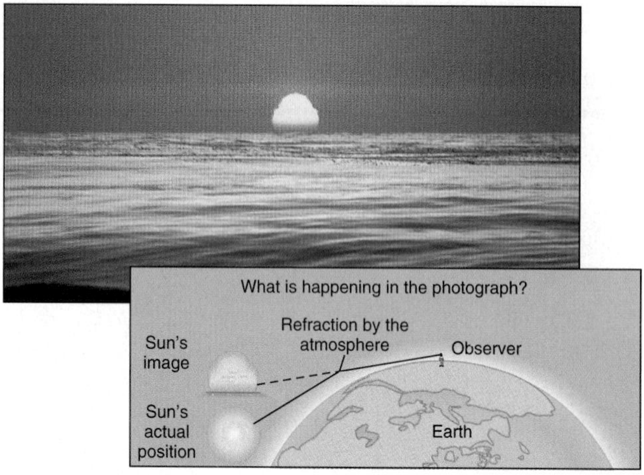

FIGURE 3.4 Sun refraction.
The distorted appearance of the Sun, nearing sunset over the ocean, is produced by refraction of the Sun's image in the atmosphere. [Photo by author.]

Albedo and Reflection. A portion of arriving energy bounces directly back into space without being converted into heat or performing any work. The reflective quality of a surface is its **albedo**. Albedo is the percentage of insolation that is reflected. This returned energy is called **reflection**, a term that applies to both visible and ultraviolet light. Albedo is an important control over the amount of insolation that is available to heat a surface.

In the visible wavelengths, darker colors have lower albedos, and lighter colors have higher albedos. On water surfaces, the angle of the solar rays also affects albedo values; lower angles produce a greater reflection than do higher angles. In addition, smooth surfaces increase albedo, whereas rougher surfaces reduce it.

Figure 3.5 illustrates albedo values for various surfaces. Specific locations experience highly variable albedo during the year in response to changes in cloud and ground cover. Earth Radiation Budget (ERB) orbiting sensors measure average albedos of 19%–38% between the tropics (23.5° N to 23.5° S) to as high as 80% in the polar regions.

Earth and its atmosphere reflect 31% of all insolation when averaged over a year. Figure 3.10 shows that Earth's average albedo is a combination of light reflected by clouds, reflected by the ground (combined land and oceanic surfaces), and reflected and scattered by the atmosphere. By comparison, a full Moon, which is bright enough to read by under clear, night skies, has only a 6%–8% albedo value. Thus, with *earthshine* being four times brighter than moonlight (four times the albedo), and with Earth being four times larger than the Moon, it is no surprise that astronauts report how startling our planet looks from space.

Clouds and the Atmosphere's Albedo. An unpredictable factor in the tropospheric energy budget, and therefore in refining climatic models, is the role of clouds.

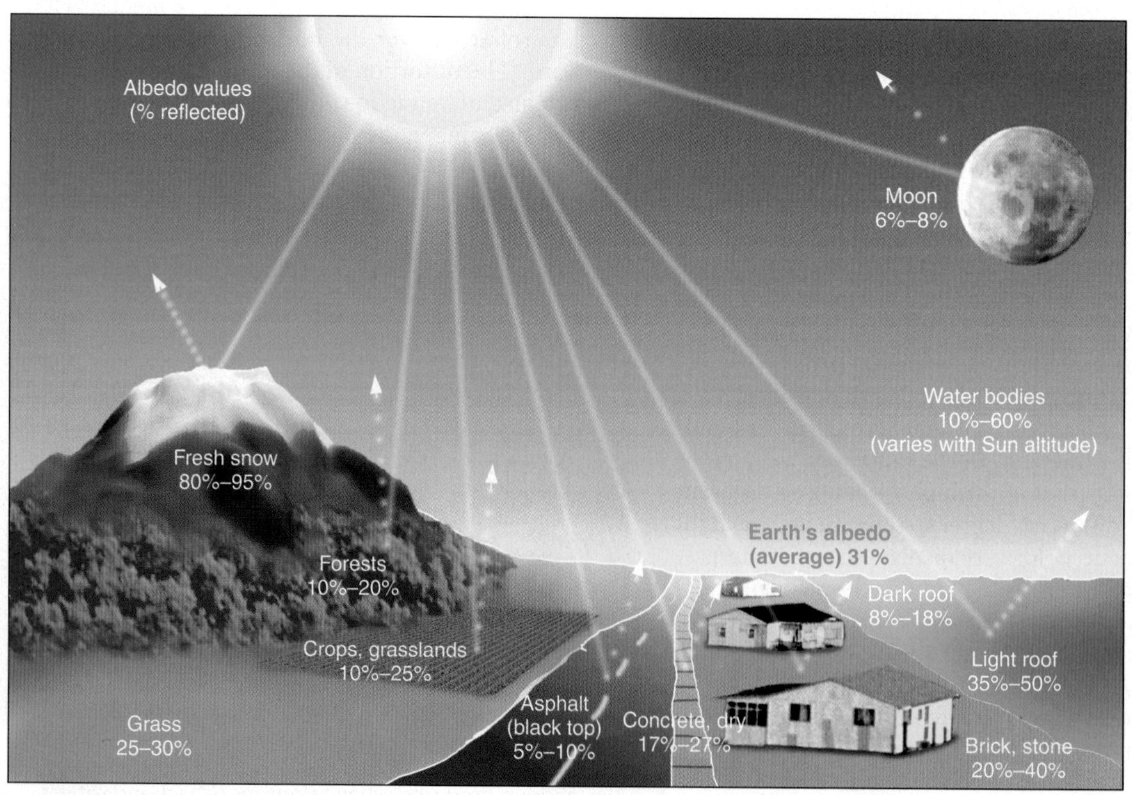

FIGURE 3.5 Various albedo values.
Selected albedos for different surfaces. Generally, light surfaces are more reflective and thus have higher albedo values. Darker surfaces are less reflective. [Data from M. I. Budyko, 1958, *The Heat Balance of the Earth's Surface*, Washington, D.C., U.S. Department of Commerce, p. 36.]

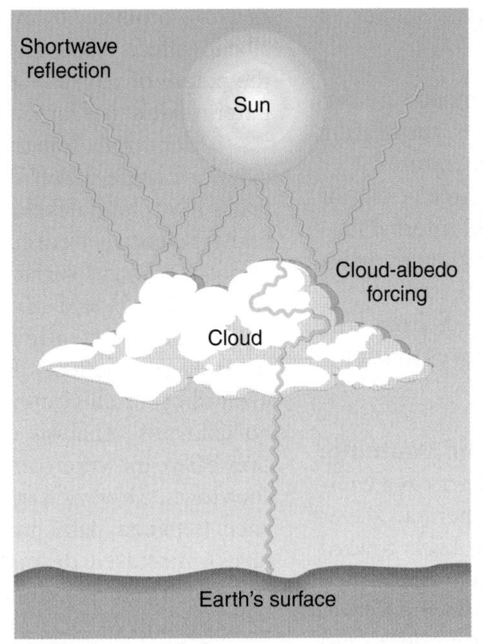

(a) Shortwave radiation

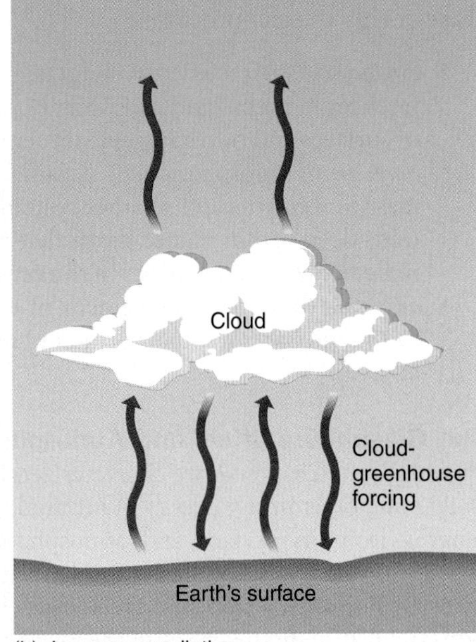

(b) Longwave radiation

FIGURE 3.6 The effects of clouds on shortwave and longwave radiation.
(a) Shortwave radiation is reflected and scattered by clouds; a high percentage is returned to space. (b) Longwave radiation emitted by Earth is absorbed and reradiated by clouds; some infrared energy is radiated to space and some back toward the surface.

Clouds reflect insolation and thus cool Earth's surface. An *increase in albedo* caused by clouds is described by the term **cloud-albedo forcing**. Yet clouds act as insulation, trapping longwave radiation from Earth and raising minimum temperatures. An *increase in greenhouse warming* caused by clouds is described as **cloud-greenhouse forcing**. Figure 3.6 illustrates the general effects of clouds on shortwave radiation and longwave radiation.

Absorption. **Absorption** is the *assimilation* of radiation and its *conversion* from one form to another. Insolation (both direct and diffuse) that is not part of the 31% reflected from Earth's surfaces is absorbed and converted into infrared radiation or is used by plants in photosynthesis. The temperature of the absorbing surface is raised in the process, causing that surface to radiate more total energy at shorter wavelengths. In addition to absorption by land and water surfaces, absorption also occurs in atmospheric gases, dust, clouds, and stratospheric ozone.

Conduction, Convection, and Advection. Several means transfer heat energy in a system. **Conduction** is the molecule-to-molecule transfer of heat energy as it diffuses through a substance. As molecules warm, their vibration increases, causing collisions that produce motion in neighboring molecules, thus transferring heat from warmer to cooler materials.

Different materials (gases, liquids, and solids) conduct sensible heat directionally from areas of higher temperature to those of lower temperature. This heat flow transfers energy through matter at varying rates, depending on the conductivity of the material. Earth's land surface is a better conductor than air; moist air is a slightly better conductor than dry air.

Energy also is transferred through gases and liquids by movements called **convection**, when the physical mixing involves a strong vertical motion. In the atmosphere or bodies of water, warmer (less dense) masses tend to rise and cooler (denser) masses tend to sink, establishing patterns of convection. When a lateral (horizontal) motion is dominant, the term **advection** applies. Sensible heat is transported physically through the medium in these ways.

You commonly experience such energy flows in the kitchen: Energy is conducted through the handle of a pan, or boiling water bubbles in the saucepan in convective motions (Figure 3.7). Also, a kitchen may be equipped with a convection oven that uses a fan to circulate heated air to uniformly cook food.

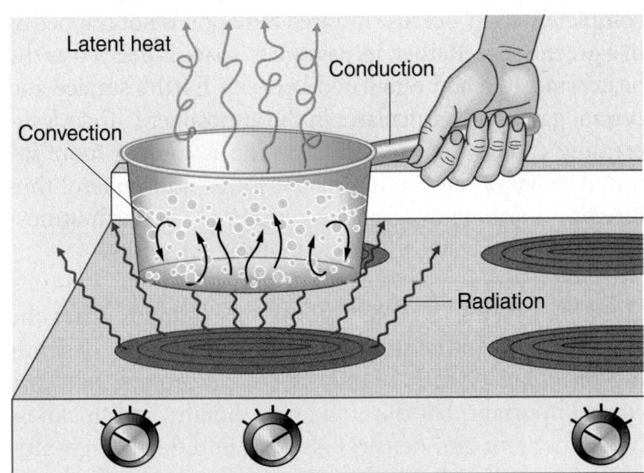

FIGURE 3.7 Heat energy transfer processes.
A pan of water heating on the stove illustrates heat transfer. Infrared energy *radiates* from the burner to the saucepan and the air. Energy *conducts* through the molecules of the pan and the handle. The water physically mixes, carrying heat energy by *convection*. The energy in the water and handle is measurable as *sensible heat*. The vapor leaving the surface of the water contains the *latent heat* absorbed in the change of water to a vapor.

In physical geography, we find many examples of these energy transfer processes:

- *conduction* (surface energy budgets, temperature differences between land and water bodies, the heating of surfaces and overlying air, soil temperatures);
- *convection* (atmospheric and oceanic circulation, air mass movements and weather systems, internal motions deep within Planet Earth that produce a magnetic field and movements in the crust); and
- *advection* (horizontal movement of winds from land to sea and back, fog that forms and moves to another area).

The Greenhouse Effect and Atmospheric Warming. Previously, in Chapter 2, we characterized Earth as a cool-body radiator, emitting energy in infrared (thermal) wavelengths from its surface and atmosphere back toward space. However, some of this infrared radiation is absorbed by carbon dioxide, water vapor, methane, chlorofluorocarbons (CFCs), and other gases in the lower atmosphere and is then reradiated to Earth, thus delaying heat loss to space. This counterradiation process is an important factor in warming the lower atmosphere. The approximate similarity between this process and the way a greenhouse operates gives the process its name—the **greenhouse effect**.

In a greenhouse, the glass is transparent to shortwave insolation, allowing light to pass through to the soil, plants, and materials inside. The absorbed energy is then radiated as infrared energy back toward the glass, but the glass effectively traps both the infrared wavelengths and the warmed air inside the greenhouse. Thus, the glass acts as a one-way filter, allowing the light in but not allowing the heat out. The same process also can be observed in a car parked in direct sunlight.

In the atmosphere, the greenhouse analogy does not completely apply because infrared radiation is not trapped as in a greenhouse. Rather, its passage to space is delayed as the heat is radiated and reradiated between Earth's surface and certain gases and particulates in the atmosphere. Today's increasing carbon dioxide concentration is causing more infrared radiation absorption in the lower atmosphere, thus forcing a warming trend and disruption of the Earth-atmosphere energy system, according to many scientists.

Clouds and Earth's "Greenhouse." Clouds affect the heating of the lower atmosphere in several ways, depending on cloud type. Not only is the percentage of cloud cover important, but the cloud type, height, and thickness (water content and density) also has an effect. High-altitude, ice-crystal clouds reflect insolation with albedos of about 50%, whereas thick, lower cloud cover reflects about 90% of incoming insolation.

To understand the actual effects on the atmosphere's energy budget, however, we must consider both transmission of shortwave and longwave radiation and cloud type. Figure 3.8a portrays the *cloud-greenhouse forcing* caused by high clouds (warming, because their greenhouse effects exceed their albedo effects); and Figure 3.8b portrays the *cloud-albedo forcing* produced by lower, thicker clouds (cooling, because albedo effects exceed greenhouse effects). Understanding the nature of global cloud cover is crucial in refining computer models that forecast global climate change.

Jet contrails (*condensation trails*) produce high cirrus clouds (Figure 3.8c). The chapter-opening photo illustrates how cloud development is stimulated by aircraft exhaust. Scientists were suspicious that these clouds affected atmospheric and surface temperatures. The tragedy that struck the World Trade Center, and humanity, on 9/11/01, inadvertantly provided researchers with a chance to study these effects. Following 9/11, there was a 3-day grounding of all commercial airline traffic and therefore no contrails. Analysis of weather data for 4000 stations over 30 years were compared with data during the 3-day shutdown. *Diurnal temperature range* (DTR)—the difference between daily maximum and minimum temperatures—increased during the 3 days. The elimination of contrail-stimulated clouds allowed more incoming insolation through to the surface and more outgoing longwave energy to leave the surface, causing slightly higher afternoon and slightly lower nighttime temperature readings across the United States.* This study may prove helpful in the spatial analysis of possible impacts of future aircraft design on the Earth-atmosphere energy budget.

To better understand the role of clouds, NASA is operating the CERES program, an acronym for Clouds and Earth's Radiant Energy System. Collection of data by CERES sensors cover both reflected and emitted radiation. Figure 3.9 shows the first global monthly images, in flux, meaning energy "flow," from satellite *Terra* for March 2000. In Figure 3.9a lighter regions indicate where more sunlight is reflected into space than is absorbed—for example, by light land surfaces such as deserts, or by cloud cover such as over tropical lands. Green and blue areas illustrate where less light is reflected.

In Figure 3.9b orange and red indicate regions where more heat was absorbed and emitted to space, whereas less heat energy is escaping in the blue and purple areas. Those blue regions of lower longwave emissions over tropical lands are due to tall, thick clouds along the equatorial convergence (Amazon, equatorial Africa, and Indonesia); these clouds also caused higher shortwave reflection. Subtropical desert regions exhibit greater longwave radiation emissions owing to the presence of little cloud cover and greater radiative energy losses from surfaces that have absorbed a lot of energy, as we saw in Figure 3.2.

Earth-Atmosphere Radiation Balance

The Earth-atmosphere energy system naturally balances in a steady-state equilibrium. The natural energy balance occurs through energy transfers that are *radiative* and *nonradiative*. Radiative transfer is by infrared radiation between the surface and the atmosphere. Nonradiative

*See David J. Travis, et. al., "Contrails reduce daily temperature range," *Nature* 418 (August 8, 2002): 601, from the Department of Geography and Geology, University of Wisconsin–Whitewater.

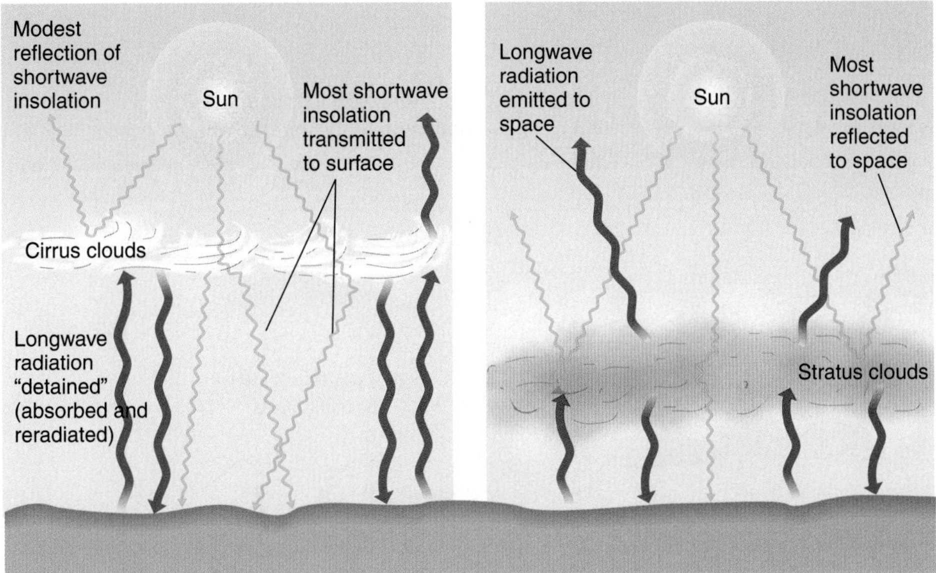

(a) High clouds: net greenhouse forcing and atmospheric warming

(b) Low clouds: net albedo forcing and atmospheric cooling

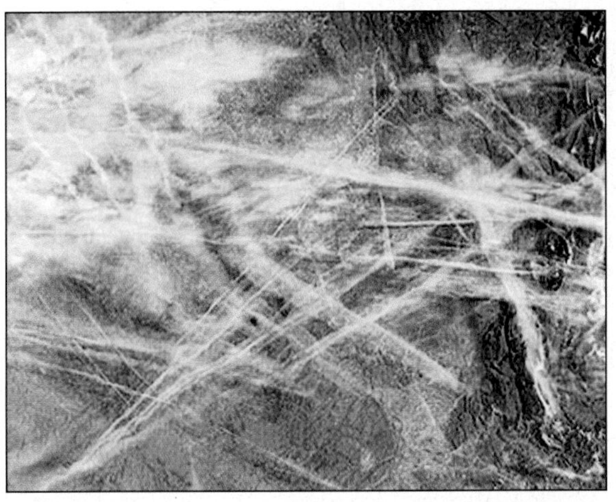

(c)

FIGURE 3.8 Energy effects of two cloud types.
(a) High, ice-crystal clouds (called *cirrus*) transmit most of the insolation. However, they absorb and delay losses of outgoing longwave infrared, producing a greater greenhouse forcing and a net warming of Earth. (b) Low, thick clouds (stratus) reflect most of the incoming insolation and radiate longwave infrared to space, producing a greater albedo forcing and a net cooling of Earth. (c) Jet-airliner exhaust triggers cirrus cloud development. Newer, thinner contrails spread to form cloud cover. [(c) Astronaut photo May 15, 2002, courtesy of NASA-JSC.]

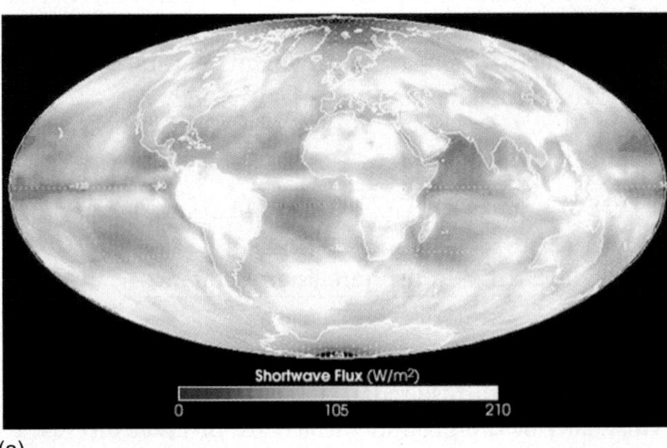

(a)

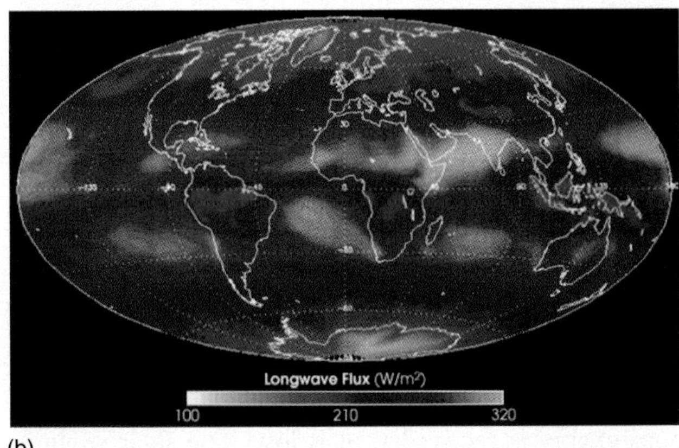

(b)

FIGURE 3.9 Shortwave and longwave images show Earth's radiation budget components.
The CERES sensors aboard *Terra* made these portraits in March 2000, capturing (a) outgoing shortwave energy flux reflected from clouds, land, and water—Earth's albedo and (b) longwave energy flux emitted by surfaces back to space. The scale beneath each image displays value gradations in watts per square meter. [Images courtesy of CERES Instrument Team, Langley Research Center, NASA.]

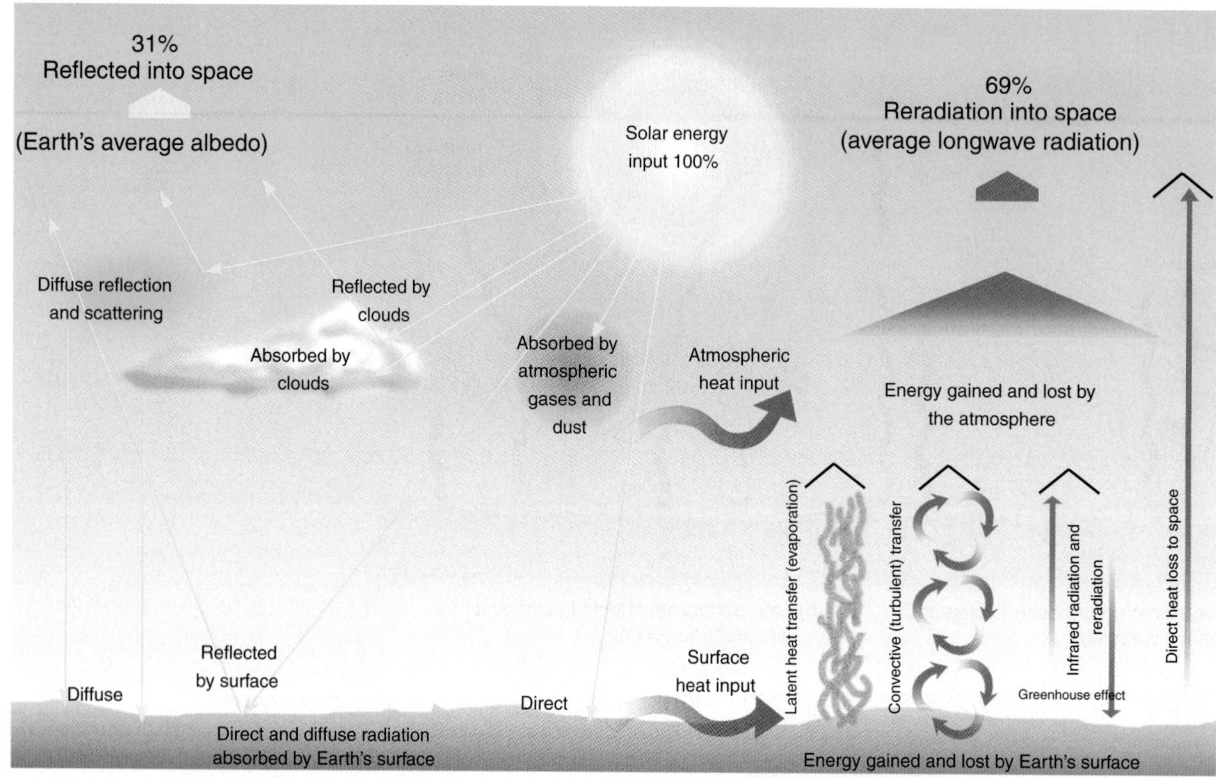

FIGURE 3.10 Earth-atmosphere energy budget.
Solar energy cascades through the lower atmosphere, where it is absorbed, reflected, and scattered. Clouds, atmosphere, and the surface reflect 31% of this insolation back to space. Atmospheric gases and dust and Earth's surface absorb energy and radiate infrared radiation. Earth and atmosphere exchange energy through latent heat transfer in water vapor, convective transfer (moving air), and infrared radiation (right-hand portion of the illustration). Eventually, Earth reradiates absorbed energy in infrared wavelengths (−69%) that, when added to Earth's average albedo (reflected energy, −31%), yields a total energy output equal to the insolation input of 100%.

(physical motion) transfers include convection, conduction, and the latent heat of evaporation (heat absorbed by water vapor when water evaporates).

Figure 3.10 summarizes the Earth-atmosphere radiation balance. It brings together all the elements discussed to this point in the chapter by following 100% of energy through the system. To summarize, of 100% of the solar energy arriving, Earth's average albedo reflects 31% into space. The atmospheric heat input comprises absorption by atmospheric clouds, dust, gases, and stratospheric ozone. This leaves less than half of the incoming insolation to actually reach Earth's surface as direct and diffuse radiation.

Figure 3.11 summarizes this radiation balance for all shortwave and longwave energy by latitude. In the equatorial zone, energy surpluses dominate, for in those areas more energy is received than is lost. The solar angle is high, with consistent daylength. However, deficits exist in the polar regions, where more energy is lost than gained. At the poles, the Sun is extremely low in the sky, snow and ice surfaces are light and reflective, and for 6 months during the year no insolation is received.

This overall imbalance of insolation and heating from equator to the poles is the basis for major circulations within the lower atmosphere and in the ocean. The transfer agents are global wind circulation, ocean currents, weather systems, and related phenomena. Tropical cyclones (hurricanes and typhoons) represent dramatic, concentrated examples of such energy and mass transfers. As you go about your daily activities, use these dynamic natural systems as reminders of the constant flow of solar energy.

Energy at Earth's Surface

Earth-Atmosphere Energy Balance

Solar energy is the principal heat source at Earth's surface. The direct and diffuse radiation and infrared radiation arriving daily at the ground surface are of great interest to geographers. Examine the terms along the ground surface in Figure 3.10.

Daily Radiation Patterns

The fluctuating daily pattern of incoming shortwave energy and outgoing longwave energy at the surface, and resultant air temperature, are shown in Figure 3.12. This

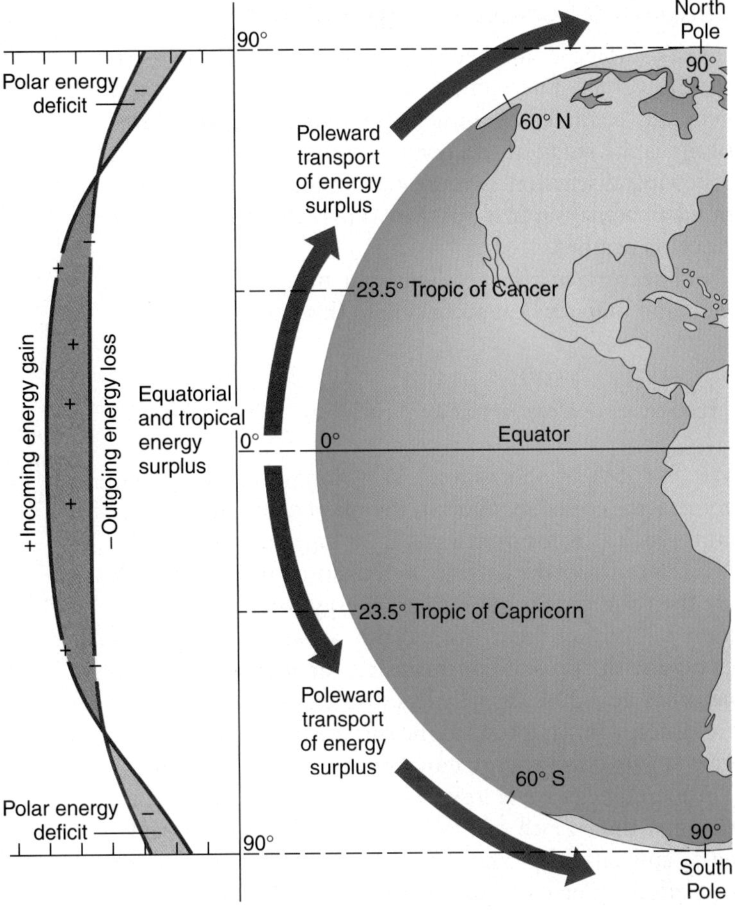

FIGURE 3.11 Energy budget by latitude.
Earth's energy surpluses and deficits by latitude produce poleward transport of energy and mass in each hemisphere—atmospheric circulation and ocean currents.

graph represents conditions for bare soil on a cloudless day in the middle latitudes. Incoming energy arrives during daylight, beginning at sunrise, peaking at noon, and ending at sunset. The shape and height of this insolation curve varies with season and latitude. The highest trend for such a curve occurs at the time of the summer solstice (around June 21 in the Northern Hemisphere and December 21 in the Southern Hemisphere). The temperature plot also responds to seasons and variations in input.

The relationship between the insolation curve and the temperature curve on the graph is interesting. They do not align; there is a lag. As long as the incoming energy exceeds the outgoing energy, temperature continues to increase during the day, not peaking until the incoming energy begins to diminish in the afternoon as the Sun loses altitude.

The warmest time of day occurs not at the moment of maximum insolation but at that moment *when a maximum of insolation is absorbed*. Thus, this temperature lag places the warmest time of day 3 to 4 hours after solar noon as absorbed heat is supplied to the atmosphere from the ground. Then, as the insolation input decreases toward sunset, the amount of heat lost exceeds the input; temperatures begin to drop until the surface has radiated away the maximum amount of energy, just at dawn.

The annual pattern of insolation and temperature exhibits a similar lag. For the Northern Hemisphere, January is usually the coldest month, occurring after the winter solstice, the shortest day in December. Similarly, the warmest months of July and August in the Northern Hemisphere occur after the summer solstice, the longest day in June.

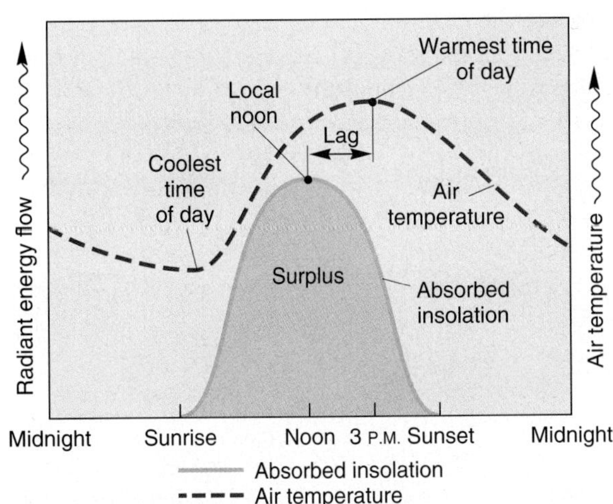

FIGURE 3.12 Daily radiation curves.
Sample radiation curves for a typical day show the changes in insolation (orange), and air temperature (dashed). Comparing the curves demonstrates a lag between local noon (the insolation peak for the day) and the warmest time of day.

Simplified Surface Energy Balance

Earth's surface is supplied with energy that daily and seasonally varies. The climate at or near Earth's surface is generally termed the *boundary layer climate*. **Microclimatology** is the study of this portion of the atmosphere. The following discussion is more meaningful if you keep in mind an actual surface—perhaps a park, a front yard, or a place on campus.

The surface receives light and heat, and it reflects light and radiates heat according to this basic scheme:

$$\underset{\text{(Insolation)}}{+SW\downarrow}\quad \underset{\text{(Reflection)}}{-SW\uparrow}\quad \underset{\text{(Infrared)}}{+LW\downarrow}\quad \underset{\text{(Infrared)}}{-LW\uparrow} = \underset{\text{(Net Radiation)}}{\text{NET R}}$$

We use SW = shortwave, LW = longwave (you may come across other symbols in the microclimatology literature, such as K for shortwave, L for longwave, and Q^* for NET R). Along the surface, as illustrated in Figure 3.13, are the components of a surface energy balance. The column of soil continues to a depth at which energy exchange with surrounding materials, or with the surface, becomes negligible, usually less than a meter.

Energy moving toward the surface is regarded as positive (a gain), and energy moving away from the surface is considered negative (a loss). The amount of insolation arriving at the surface ($+SW\downarrow$) varies with season, cloudiness, and latitude. The albedo dictates the amount of insolation reflected at the surface ($-SW\uparrow$). Darker and rougher surfaces *reflect* less energy, *absorb* more energy, and therefore produce more net radiation. Lighter and smoother surfaces reflect more shortwave energy, leading to less net radiation.

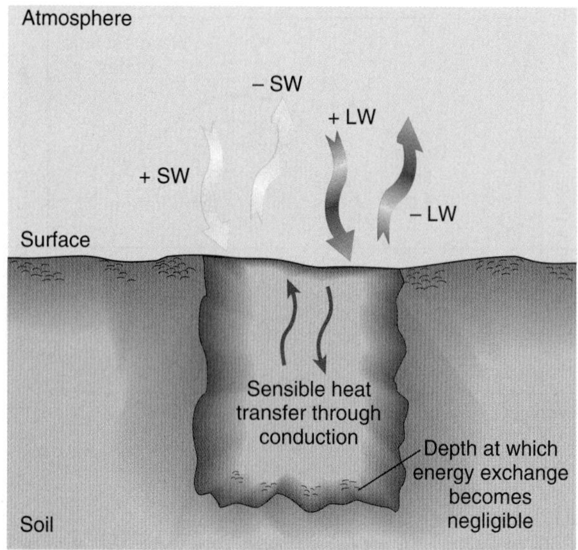

FIGURE 3.13 Surface energy budget.
Input and output components of a surface energy balance for a typical soil column.

Surface albedo values ($-SW\uparrow$) dictate the amount of insolation reflected, and therefore not absorbed, at the surface. As an example, imagine a snow-covered landscape as a surface energy system. Snow has high reflectivity, sending sunlight back to space and reducing the amount of insolation absorbed. Consequently, surfaces and air temperatures are lower, and thus the snow cover does not melt. This is a system with *positive feedback*. (Remember from Chapter 1 that positive feedback increases response in the system: more cooling—more snow—increasing albedo—more cooling—then colder, drier air—then less snow—and so on.)

Adding and subtracting the heat flow at the surface completes the calculation of net radiation (NET R), or the balance of all radiation at Earth's surface—shortwave and longwave.

Net Radiation. **Net radiation (NET R)**, the net radiation (of all wavelengths) available at Earth's surface, is the final outcome of the entire radiation-balance process discussed in this chapter. NET R is expended from a nonvegetated surface through three pathways:

- *Turbulent sensible heat transfer*, the quantity of sensible heat energy physically transferred from air to surface, and surface to air, through convection and conduction;
- *Latent heat of evaporation*, the heat energy that is stored in water vapor as water evaporates. Large quantities of latent heat are absorbed into water vapor during its change of state from liquid to gas;
- *Ground heating and cooling*, energy that flows into and out of the ground surface by conduction.

Understanding the NET R balance is essential to solar energy technologies that concentrate shortwave and longwave energy for use. Solar energy offers great potential worldwide and will see expanded use in the near future. Focus Study 3.1 briefly reviews solar energy strategies.

Variation in the expenditure of NET R among *sensible heat* (heat we can feel), *latent heat*, and ground heating and cooling produces the variety of environments we experience in nature. Figure 3.14a (p. 83) illustrates a desert environment, in which few clouds, sparse plants, and little water result in high sensible heat production and hot ground. Figure 3.14b shows the area near Pitt Meadows, British Columbia, featuring a moist, vegetated environment of irrigated mixed orchard and rye grass. Here, high amounts of energy are expended for latent heat of evaporation, with more moderate amounts spent as sensible heat.

Now that we have examined the flow of solar energy across space, through the atmosphere, and to Earth's surface, we turn our attention to the pattern of global temperatures and, in the next chapter, to the flow of atmospheric and oceanic circulation.

Focus Study 3.1

Solar Energy Collection and Concentration

Consider the following:

- Photographs from the early 1900s of residences in southern California and Florida show solar (flat-plate) water heaters on many rooftops. Early twentieth-century newspaper ads and merchandise catalogues featured the Climax solar water heater (1905) and the Day and Night water heater (1909). Applied solar energy principles and technologies are old. However, low-priced natural gas and oil displaced many of these early applications.
- The insolation receipt in just 35 minutes at the surface of the United States exceeds the amount of energy derived from the burning of fossil fuels (coal, oil, natural gas) in a year.
- An average building in the United States receives 6 to 10 times more energy from the Sun hitting its exterior than is required to heat the inside!
- Photovoltaic (photo-electric cell) prices dropped from $75 per watt in 1975 to $2 per watt in 2001. New installed solar-electric capacity exceeded 390 megawatts in 2001, an increase of 36% over 2000. In 1980 only 6.5 megawatts total were installed worldwide.

Not only does insolation warm Earth's surface, but it also provides an inexhaustible supply of energy far into the future for humanity. Sunlight is direct and renewable and has been collected for centuries through various technologies. Yet it is underutilized.

Rural villages in developing countries could benefit greatly from the simplest, most cost-effective solar application—the *solar-panel cooker*. For example, women in Kenya walk many kilometers collecting fuelwood for cooking fires (Figure 1a). Each village and refugee camp is surrounded by impoverished land, stripped bare of wood. Using simple cardboard solar cookers, villagers are able to cook meals and sanitize their drinking water right at their home without scavenging for wood (Figure 1b). (See Solar Cookers International at **http://solarcooking.org/**.) In such countries, the pressing need is for *decentralized* energy sources, appropriate in scale to everyday needs, such as cooking, heating water, and pasteurization. *Net per capita* (per person) *cost for solar cookers is far less than for centralized electrical production*, regardless of fuel source.

Collecting and Concentrating Solar Energy

Any surface that receives light from the Sun is a *solar collector*. But the diffuse nature of solar energy received at the surface requires that it be collected, concentrated, transformed, and stored to be useful. Space heating is the simplest application. Windows that are carefully designed and placed allow sunlight to shine into a building, where it is absorbed and converted into sensible heat. Here we have an everyday application of the greenhouse effect.

A *passive solar system* captures heat energy and stores it in a "thermal mass," such as water-filled tanks, adobe, tile, or concrete. An *active solar system* involves heating water or air in a collector and then pumping it through a plumbing system to a tank where it can provide hot water for direct use or for space heating.

Solar energy systems can generate heat energy of an appropriate scale for approximately half the present

(continued)

(a)

(b)

FIGURE 1 The solar-cooking solution.
(a) Five women haul firewood many miles to the Dadaab refugee camp in northeastern Kenya.
(b) Kenyan women in training to use their solar panel cookers, which do not require scavenging the countryside for scarce fuelwood. These simple cookers collect and focus direct and diffuse insolation using a shiny surface and an enclosed box or cooking bag that traps infrared radiation. (This is a small-scale, efficient application of the greenhouse effect.) Construction is easy, using cardboard components. Temperatures easily exceed 105°C (220°F) for baking, boiling, purifying water, and sterilizing instruments. [Photos by Solar Cookers International, Sacramento, California.]

Focus Study 3.1 *(continued)*

domestic applications in the United States (space heating and water heating). In marginal climates, solar-assisted water and space heating is feasible as a backup; even in New England, Northern Plains states, and southern Canada, solar collection systems prove effective. (See the National Solar Radiation Data Base at http://rredc.nrel.gov/solar/pubs/NSRDB/.)

Focusing (concentrating) mirrors, such as Fresnel lenses, or parabolic (curved surface) troughs and dishes can be used to attain very high temperatures to heat water or other heat-storing fluids. Kramer Junction, California, about 225 km (140 mi) northeast of Los Angeles, has the world's largest operating solar electric-generating facility, with a capacity of 150 MW (megawatts; 150 million watts). Long troughs of computer-guided curved mirrors concentrate sunlight to create temperatures of 390°C (735°F) in vacuum-sealed tubes filled with synthetic oil. The heated oil is then used to heat water; the heated water produces steam that rotates turbines to generate cost-effective electricity. The facility converts 23% of the sunlight it receives into electricity during peak hours (Figure 2a).

The National Renewable Energy Laboratory (NREL, http://www.nrel.gov and the National Center for Photovoltaics at http://www.nrel.gov/ncpv/) was established in 1974 to coordinate solar energy research, development, and testing in partnership with private industry. NREL's headquarters building in Golden, Colorado, features the latest in passive and active solar design.

Electricity Directly from Sunlight

Producing electricity by *photovoltaic cells* (PVs) is a technology that has been used in spacecraft since 1958. Familiar to us all are the solar cells in pocket calculators (over 100 million units now in use). When sunlight shines upon a semiconductor material in these cells, it stimulates a flow of electrons (an electrical current) in the cell. PV cells are arranged in modules that can be assembled in large arrays. At NREL's Outdoor Test Facility, successful tests continue on arrays of prototype solar cells. Some of these panels have been in operation for more than 20 years (Figure 2b).

The efficiency of these cells has improved to the level that they are generally cost-competitive. NREL developed a copper-indium-gallium solar cell that achieves an astonishing 18% conversion rate of sunlight to electricity. New cells that are 50% more efficient than conventional crystalline-silicon cells are now available for satellites.

As of 1998, some 250,000 homes in Mexico, Indonesia, South Africa, India, and elsewhere have PV roof systems. Norway, at the same high latitudes as Alaska, has 60,000 units operating and is adding about 8,000 new PV systems a year. Some 200 photovoltaic power systems are operating in the Navajo Nation of northeastern Arizona. Innovative uses abound, as shown in Figure 2c. (See the Department of Energy's "Photovoltaic Home Page," at http://www.eren.doe.gov/pv/.)

In 2001 worldwide photovoltaic production increased to 391 megawatts, a 35% increase over the previous year and the fourth year in a row that growth exceeded 30%. PVs represent more than half of all solar-energy installations. Production leader Japan expanded its production to 45% of the global production. U.S. production reached 100 MW in 2001, but this was mostly for export to Japan and Europe. Europe produced 86 MW, led by Germany. Despite this international boom in photovoltaic electric power generation, the United States' administration proposed sweeping budget cuts in its photovoltaic and solar programs for the 2002 and 2003 budgets.

Obvious drawbacks of both solar-heating and solar-electric systems are periods of cloudiness and night,

(continued)

(a)

(b)

(c)

FIGURE 2 Solar thermal energy production.
(a) Kramer Junction (http://www.kjcsolar.com) 150 MW solar-thermal energy installation in southern California. (b) NREL Outdoor Test Facility in Golden, Colorado, where a variety of photovoltaic cell arrays successfully convert sunlight directly into electricity. (c) The Sacramento Municipal Utility District installed a PV array that doubles as parking lot cover. [Photos by Bobbé Christopherson.]

Focus Study 3.1 *(continued)*

which inhibit operations. Research is under way to enhance energy-storage technologies, such as hydrogen fuel production (using energy to extract hydrogen from water for later use in producing more energy), and to improve battery technology.

The Promise of Solar Energy

Solar energy is a wise choice for the future. It is directly available to the consumer; it is based on a renewable energy source of an appropriate scale for end-use needs (it matches them well); and most solar strategies are labor-intensive (rather than capital-intensive centralized power production). Solar is preferable to further development of our decreasing fossil fuel reserves, further increases in oil imports and tanker spills, investment in foreign military operations, or more troubled nuclear power and its many problems.

Whether or not the alternative path of solar energy is followed comes down to political control. The technology is ready for installation and is cost-effective when all direct and indirect costs are considered. NREL asserts in its mission statement that the laboratory wants to "... lead the nation toward a sustainable energy future by developing renewable energy technologies, improving energy efficiency, advancing related science, and engineering commercialization." Unfortunately, during the early 1980s and again in the late 1990s and 2000s, progress was slowed by political decisions and intense lobbying by traditional fuel interests.

(a) El Mirage—hot, dry

(b) Pitt Meadows—cool, moist

FIGURE 3.14 Net radiation comparison and the landscape.
(a) In the California desert east of Los Angeles (at about 35° N), the daily expenditures of net radiation produce a desert landscape of high temperature and sparse vegetation. (b) Irrigated blueberry orchards are characteristic of the agricultural activity near Pitt Meadows, British Columbia (at about 49° N), which has a moist environment and moderate temperatures. [Photos by author.]

Temperature Concepts and Controls

Air temperature has a remarkable influence on our lives. What temperature is it now (both indoors and outdoors) as you read these words? How is temperature measured and what does the value mean? How is today's air temperature controlling your plans for the day? Our bodies sense temperature and subjectively judge comfort, reacting to changing temperatures with predictable responses. A variety of temperature regimes worldwide affect cultures, decision making, and resources consumed. We read of heat waves and cold spells affecting people, crops, events, and energy consumption.

Heat and temperature are not the same. *Heat* is a form of energy that flows from one system or object to another because the two are at different temperatures. **Temperature** is a measure of the average kinetic energy (motion) of individual molecules in matter. We feel the effect of temperature as the *sensible heat* transfer from warmer objects to cooler objects. For instance, when you jump into a cool lake you can sense the heat transfer from your skin to the water as kinetic energy leaves your body and flows to the water; a chill develops.

Temperature and heat are related because changes in temperature are caused by the absorption or emission (gain or loss) of heat energy. The term *heat energy* is frequently used to describe energy that is added to or removed from a system or substance.

Temperature Scales

The temperature at which all atomic and molecular motion in matter completely stops is called 0° *absolute temperature* (commonly, "*absolute zero*"). Its value on the different temperature-measuring scales is −273° Celsius (C), −459.4° Fahrenheit (F), and 0 Kelvin (K). Figure 3.15 compares these three scales. Formulas for converting between Celsius, SI (Système International), and English units are in Appendix D of this text.

The Fahrenheit scale places the melting point of ice at 32°F and the boiling point of water at 212°F, with 180 subdivisions between. The scale is named for Daniel G. Fahrenheit, a German physicist (1686–1736), who used these odd values based on the coldest temperature he could achieve in his laboratory. About a year after the adoption of the Fahrenheit scale, the Celsius scale (formerly called centigrade) was developed by Swedish astronomer Anders Celsius (1701–1744). He placed the melting point of ice at 0° and the boiling point of water at 100° at sea level, dividing his system using a decimal scale. British physicist Lord Kelvin (born William Thomson, 1824–1907), proposed his scale in 1848. Kelvin is used in science because temperature readings start at absolute zero and thus are proportional to the actual kinetic energy in a material; melting point of ice is 273 K, and its boiling point of water is 373 K.

Most countries use the Celsius scale to express temperature. The United States remains the only major country still using the Fahrenheit scale. This textbook presents Celsius (with Fahrenheit equivalents in parentheses) throughout, to help bridge this transitional era in the United States. The continuing pressure of the scientific community, not to mention the rest of the world, makes adoption of Celsius and SI units inevitable for the United States.

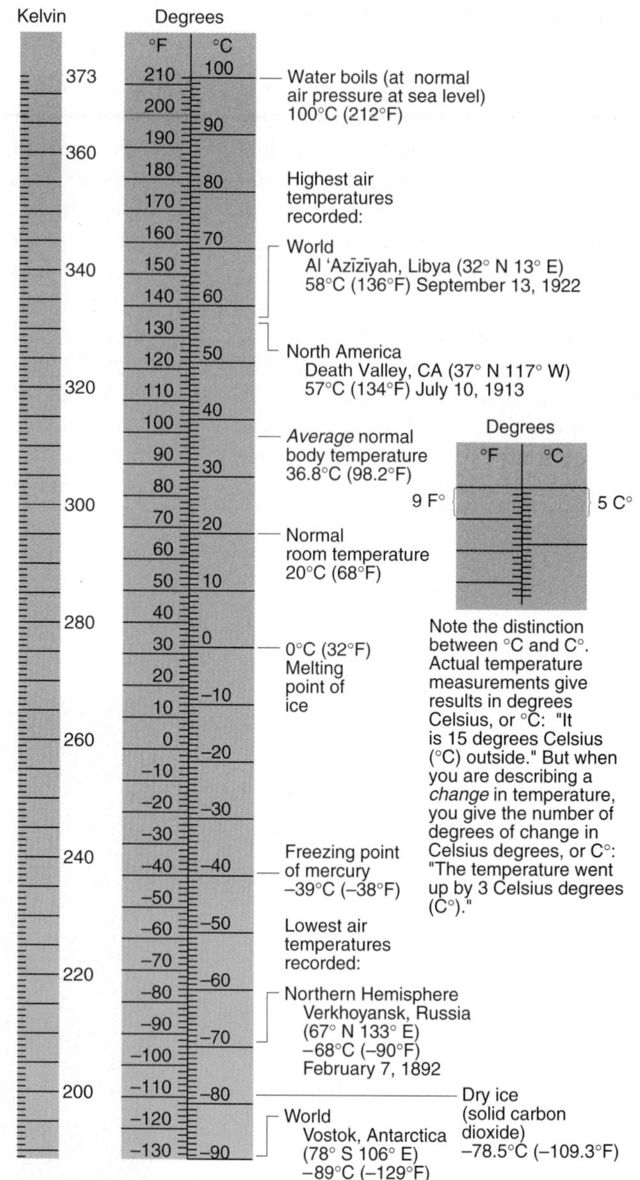

FIGURE 3.15 Temperature scales.
Scales for expressing temperature in Kelvin (K) and degrees Celsius (°C) and Fahrenheit (°F). (Note that there is only one melting point for ice, but there are many freezing points for water, ranging from 32°F down to −40°F, depending on its purity and volume and certain conditions in the atmosphere.)

Measuring Temperature

Outdoor temperature usually is measured with a mercury thermometer or alcohol thermometer in a sealed glass tube. (Fahrenheit invented the alcohol and mercury thermometers.) Alcohol is preferred in cold climates because it freezes at −112°C (−170°F), whereas mercury freezes at

−39°C (−38.2°F). The principle of these thermometers is simple: When these fluids are heated, they expand; upon cooling, they contract. A thermometer stores fluid in a small reservoir at one end and is marked with calibrations to measure the expansion or contraction of the fluid, which reflects the temperature of the thermometer's environment.

Thermometers for standardized official readings are placed outdoors, in small shelters that are white (for high albedo) and louvered (for ventilation) to avoid overheating of the instruments. In increasing use are *thermisters* that measure temperature by sensing the electrical resistance of a semi-conducting material. A thermister is in the shelter in Figure 3.16.

Temperature readings are taken daily, sometimes hourly, at more than 15,400 weather stations worldwide. One of the goals of the Global Climate Observing System (GCOS) is to establish a reference network of one station per 250,000 km^2. Some stations with recording equipment also report the duration of temperatures, rates of rise or fall, and variation over time throughout the day and night. (See the World Meteorological Organization at http://www.wmo.ch/.)

Daily minimum-maximum readings are averaged to get the *daily mean temperature*. The *monthly mean temperature* is the total of daily mean temperatures for the month divided by the number of days in the month. An *annual temperature range* expresses the difference between the lowest and highest monthly mean temperatures for a given year. If you install a thermometer for outdoor temperature reading, be sure to avoid direct sunlight on the instrument and place it in an area of good ventilation.

FIGURE 3.16 Instrument shelter.
This standard thermometer instrument shelter is white (for high albedo) and louvered (for ventilation), replacing the traditional cotton-region shelter. [Photo by Bobbé Christopherson.]

Principal Temperature Controls

The interaction of complex control mechanisms produces Earth's temperature patterns. These principal influences upon temperatures include latitude, altitude, cloud cover, and land-water heating differences.

Latitude. Insolation is the single most important influence on temperature variations. Figure 2.9 showed how insolation intensity decreases as we move away from the subsolar point, a point that migrates between the Tropic of Cancer and the Tropic of Capricorn—that is, between 23.5° N and 23.5° S—during the year. In addition, daylength and Sun angle change throughout the year, increasing in seasonal effect with increasing latitude. The five cities graphed in Figure 3.17 demonstrate the effects of latitudinal position. From equator to poles, Earth ranges from continually warm, to seasonally variable, to continually cold.

Altitude. Within the troposphere, temperatures decrease with increasing altitude above Earth's surface (recall that the normal lapse rate of temperature change with altitude is 6.4 C° /1000 m or 3.5 F° /1000 ft—see Figure 2.20). Thus, worldwide, mountainous areas experience lower temperatures than do regions nearer sea level, even at similar latitudes. The density of the atmosphere also diminishes with increasing altitude. In fact, the density of the atmosphere at an elevation of 5500 m (18,000 ft) is about half of that at sea level. As the atmosphere thins, its ability to absorb and radiate heat is reduced.

The consequences are that, at higher elevations, average air temperatures are lower, nighttime cooling increases, and the temperature range between day and night and between areas of sunlight and shadow also is greater. Temperatures may decrease noticeably in the shadows and shortly after sunset. Surfaces both gain heat rapidly and lose heat rapidly to the thinner atmosphere. Also, at higher elevations, the insolation received is more intense because of the reduced mass of atmospheric gases. As a result of this intensity, the ultraviolet energy component makes sunburn a distinct hazard.

The snowline seen in mountain areas indicates where winter snowfall exceeds the amount of snow lost through summer melting and evaporation. The snowline's location is a function both of elevation and latitude, which means that glaciers can exist even at equatorial latitudes if the elevation is great enough. In equatorial mountains, the snowline occurs at approximately 5000 m (16,400 ft), and permanent ice fields and glaciers exist on equatorial mountain summits in the Andes and East Africa. With increasing latitude toward the poles, snowlines gradually descend in elevation from 2700 m (8850 ft) in the midlatitudes to lower than 900 m (2950 ft) in southern Greenland.

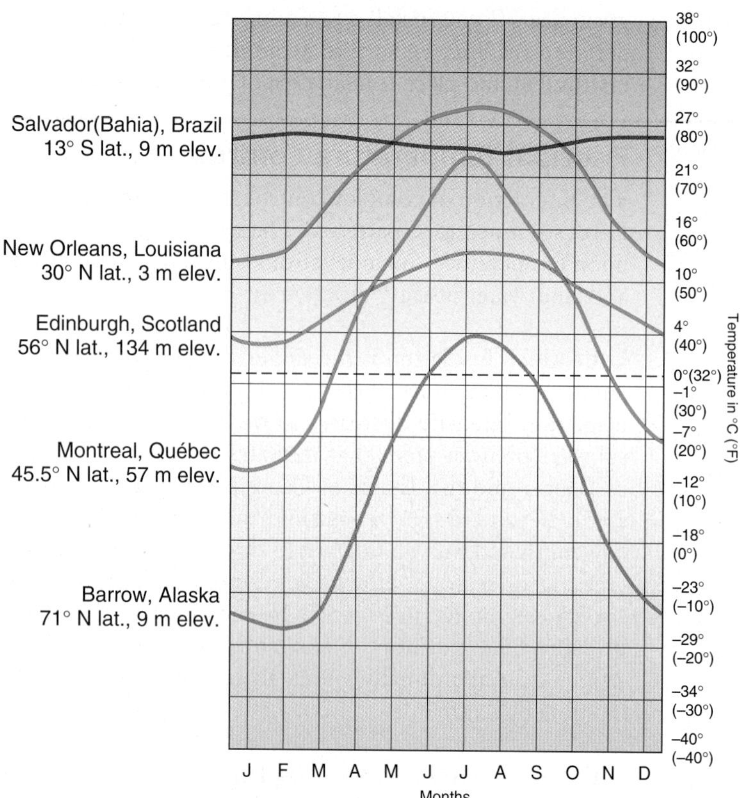

FIGURE 3.17 Latitude affects temperatures. A comparison of five cities from near the equator to above the Arctic Circle demonstrates changing seasonality and increasing differences between average minimum and maximum temperatures.

Two cities in Bolivia illustrate the interaction of the two temperature controls, latitude and altitude. Figure 3.18 displays temperature data for the cities of Concepción and La Paz, which are near the same latitude (about 16° S). Note the elevation, average annual temperature, and precipitation for each location. The hot, humid climate of Concepción at its much lower elevation stands in marked contrast to the cool, dry climate of highland La Paz.

The moderate temperature and moisture conditions characteristic of La Paz lead to the formation of more fertile soils than those found in the warmer, wetter climate of Concepción. People living around La Paz actually grow wheat, barley, and potatoes—crops characteristically grown in the cooler midlatitudes—despite the fact that La Paz is 4103 m (13,461 ft) above sea level (Figure 3.19). (For comparison, the summit of Pikes Peak in Colorado is at 4301 m or 14,111 ft, and Mount Rainier, Washington, is at 4392 m or 14,410 ft.)

Cloud Cover. Orbiting satellites reveal that approximately 50% of Earth is cloud covered at any given moment. Clouds moderate temperature, and their effect varies with cloud type, height, and density discussed earlier in this chapter. In general, they lower daily maximum temperatures by reflecting insolation as a result of their high albedo values; and they raise nighttime minimum temperatures by acting as insulation and radiating longwave energy, preventing rapid energy loss. Clouds also reduce latitudinal and seasonal temperature differences.

Clouds are the most variable factor influencing Earth's radiation budget, making them the subject of much investigation and simulation in computer models of atmospheric behavior. The International Satellite Cloud Climatology Project, part of the World Climate Research Programme, is presently in the midst of such research. The Clouds and the Earth Radiant Energy System (CERES) sensors aboard *TRMM* and *Terra* satellites are assessing cloud effects on longwave, shortwave, and net radiation patterns as never before possible. (For more about clouds, see the ISCCP and WCRP Home Pages at **http://isccp.giss.nasa.gov/** and **http://www.wmo.ch/web/wcrp/wcrp-home.html**.)

Land-Water Heating Differences

The irregular arrangement of landmasses and water bodies on Earth contributes to the overall pattern of temperature. The physical nature of the substances themselves—rock and soil vs. water—is the reason for these **land-water heating differences**. More moderate temperature patterns are associated with water bodies, compared with more extreme temperatures inland. Figure 3.20 visually summarizes the following discussion.

Evaporation. More of the energy arriving at the ocean's surface is expended for *evaporation* than is expended over a comparable area of land. An estimated 84% of all evaporation on Earth is from the oceans. When water evaporates and thus changes to water vapor, *heat energy is absorbed in the process and is stored in the water vapor*. This stored heat energy is called *latent heat* and is discussed in Chapter 5. You can experience this evaporative heat loss (cooling) by wetting the back of your hand and then blowing on the moist skin. Sensible heat energy is drawn from your skin to supply some of the energy for the water's

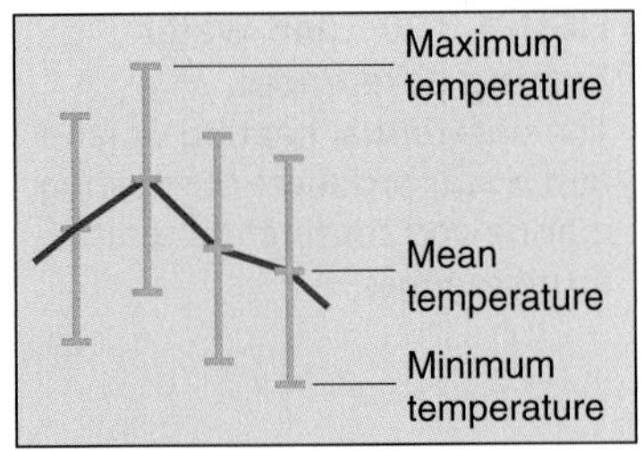

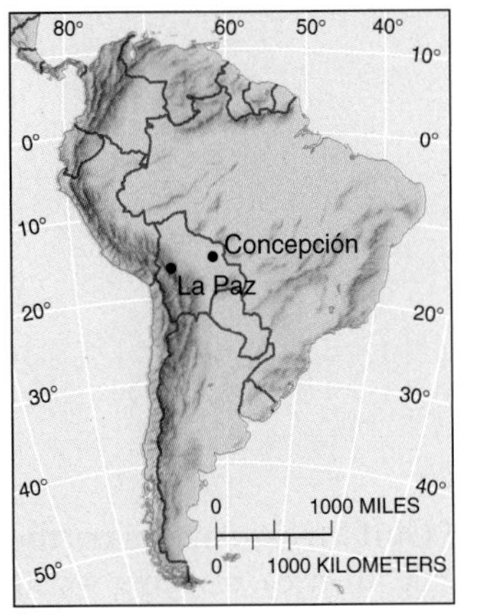

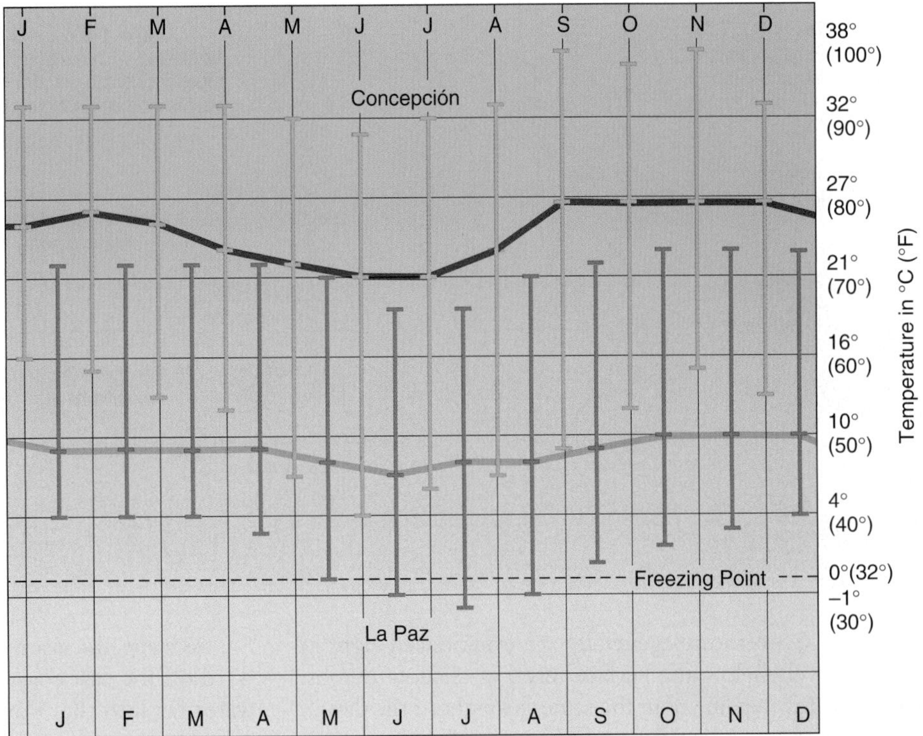

	Station	
	Concepción, Bolivia	**La Paz, Bolivia**
Latitude/longitude	16° 15′ S 62° 03′ W	16° 30′ S 68° 10′ W
Elevation	**490 m (1608 ft)**	**4103 m (13,461 ft)**
Avg. ann. temperature	24°C (75.2°F)	9°C (48.2°F)
Ann. temperature range	5 C° (9 F°)	3 C° (5.4 F°)
Ann. precipitation	121.2 cm (47.7 in.)	55.5 cm (21.9 in.)
Population	10,000	810,300 (Administrative division 1.6 million)

FIGURE 3.18 Combined affects of latitude and altitude.
Comparison of temperature patterns in La Paz and Concepción, Bolivia.

evaporation, and you feel the cooling. Similarly, as surface water evaporates, substantial energy is absorbed, resulting in a lowering of nearby air temperatures. The land, containing far less water, experiences less evaporation and therefore is moderated less by evaporative cooling.

Transparency. The transmission of light obviously differs between soil and water; solid ground is opaque, whereas water is *transparent*. Consequently, light striking a soil surface does not penetrate but is absorbed, heating the ground surface. That heat is accumulated during times of exposure and is rapidly lost at night or in shadows. Maximum and

FIGURE 3.19 High-elevation farming.
People in these high-elevation villages grow potatoes and wheat in view of the permanent ice-covered peaks of the Andes Mountains. The combination of low latitude and high elevation creates consistent but moderate temperatures throughout the year, averaging about 9°C (48°F). [Photo by Mireille Vautier/Woodfin Camp & Associates.]

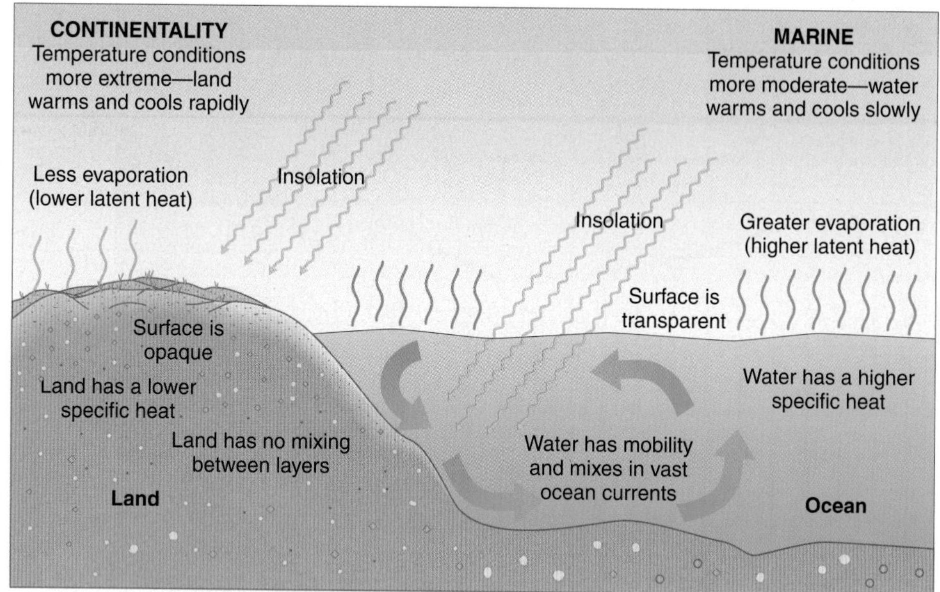

FIGURE 3.20 Land-water heating differences. The differential heating of land and water produces contrasting marine and continental temperature regimes.

minimum temperatures generally are experienced right at ground level. Below the surface, even at shallow depths, temperatures remain about the same throughout the day.

In contrast, when light reaches a body of water, it penetrates the surface because of water's **transparency**, transmitting light to an average depth of 60 m (200 ft) in the ocean. This illuminated zone is known as the *photic layer* and reaches a depth in some ocean waters of 300 m (1000 ft). This characteristic of water results in the distribution of available heat energy over a much greater depth and volume, forming a larger heat reservoir than that made up of the surface layers of the land.

Specific Heat. When equal volumes of water and land are compared, water requires far more heat to raise its temperature than does land. In other words, *water can hold more heat than can soil or rock*, and therefore water is said to have a higher **specific heat**. A given volume of water represents a more substantial heat reservoir than an equal volume of land, so that changing the temperature of the oceanic heat reservoir is a slower process than changing the temperature of land.

Movement. Land is a rigid, solid material, whereas water is a fluid and is capable of movement. Differing temperatures and currents result in a mixing of cooler and warmer waters, and that *mixing spreads the available heat over an even greater volume* than if the water were still. Surface water and deeper waters mix, redistributing heat energy. Both ocean and land surfaces radiate heat at night, but land loses its heat energy more rapidly than does the moving mass of the oceanic heat reservoir.

Ocean Currents and Sea-Surface Temperatures. Although ocean currents are discussed in greater detail in the next chapter, the influence of currents on temperature and weather requires mention here. In an air mass, water vapor content is affected by ocean temperatures, because warm water tends to energize overlying air through high evaporation rates and transfers of latent heat.

As a specific example, the **Gulf Stream** moves northward off the east coast of North America, carrying warm water far into the North Atlantic (Figure 3.21). As a result, the southern third of Iceland experiences much milder temperatures than would be expected for a latitude of 65° N, just below the Arctic Circle (66.5°). In Reykjavík, on the southwestern coast of Iceland, monthly temperatures average above freezing during all months of the year. The Gulf Stream affects Scandinavia and northwestern Europe in the same manner.

In the western Pacific Ocean, the *Kuroshio* or Japan Current, similar to the Gulf Stream, functions much the same in its warming effect on Japan and the Aleutians. In contrast, along midlatitude west coasts, cool ocean currents moderate air temperatures.

Remote sensing from satellites is providing sea-surface temperature (SST) data that correlate well with actual sea-surface measurements. This correlation permits a thorough global assessment of SSTs in programs such as Tropical Ocean Global Atmosphere (TOGA) and Coupled Ocean-Atmosphere Response Experiment (COARE). (See **http://lwf.ncdc.noaa.gov/oa/coare/index.html**.)

Figure 3.22 displays SST data for February and July 1999. The *Western Pacific Warm Pool* is the red and maroon area in the southwestern Pacific Ocean (north of New Guinea) with temperatures above 30°C (86°F). This region has the highest average ocean temperatures in the world. Note the seasonal change in ocean temperatures; for example compare the waters around Australia on the two images. Worldwide average SSTs increased from 1982 through 2000 to record levels for the past several hundred years.

Summary of Marine Effects vs. Continental Effects. Figure 3.20 summarizes the operation of all the land-water temperature controls: evaporation, transparency, specific heat, movement, and ocean currents and sea-surface temperatures. The term **marine effect**, or *maritime*, is used to describe locations that exhibit the moderating influences

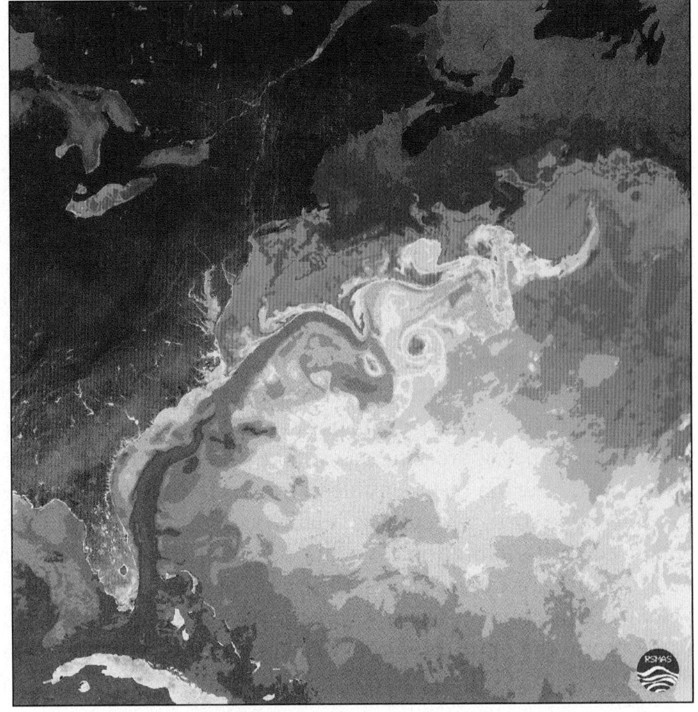

FIGURE 3.21 The Gulf Stream.
Satellite image of the warm Gulf Stream, the streamlike flow in red, orange, and yellow colors, as it flows northward along the North American East coast. Instruments sensitive to infrared wavelengths produced this remote-sensing image. Temperature differences are distinguished by computer-enhanced coloration: reds/oranges = 25°–29°C (76°–84°F); yellows/greens = 17°–24°C (63°–75°F); blues = 10°–16°C (50°–61°F); and purples = 2°–9°C (36°–48°F). [Imagery by Rosenstiel School of Marine and Atmospheric Science, University of Miami.]

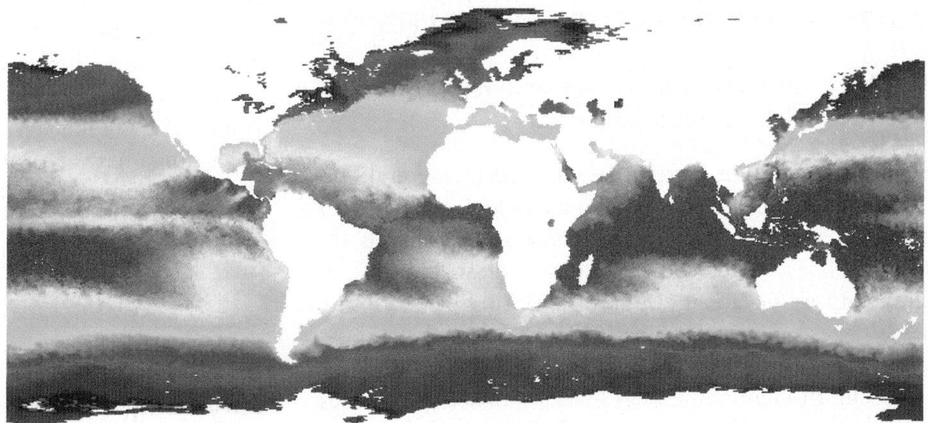

(a) February

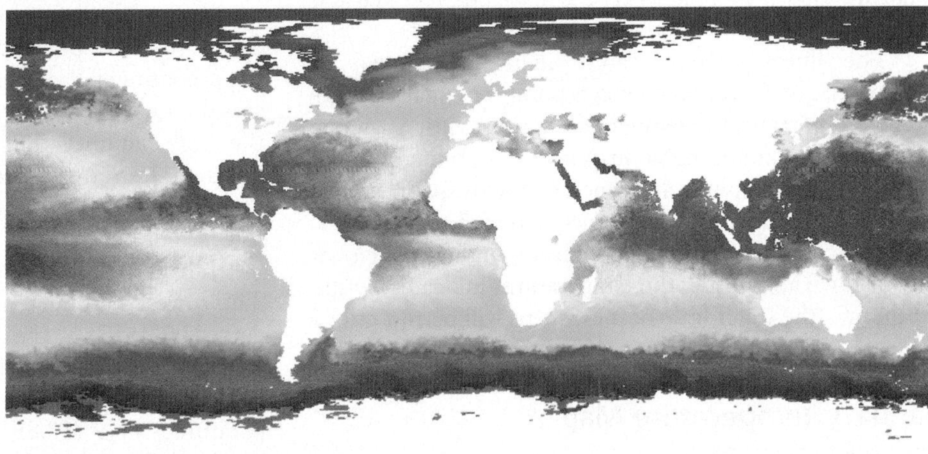

(b) July

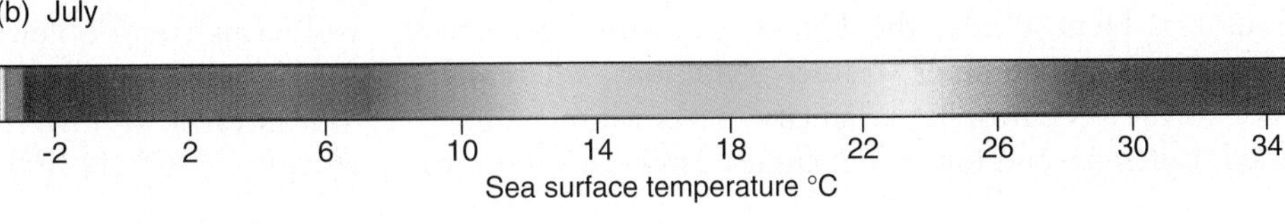

FIGURE 3.22 Sea-surface temperatures.
Average annual sea-surface temperatures for (a) February and (b) July 1999 from satellites in the NOAA/NASA Pathfinder AVHRR data set, aboard *NOAA-7, -9, -11*, and *-14*. The Western Pacific Warm Pool, the warmest area of all oceans, is well defined. These remotely sensed data are closely correlated with actual measurements of the ocean's surface temperature. [Satellite image data were obtained from the NASA Physical Oceanography Distributed Active Archive Center at the Jet Propulsion Laboratory, California Institute of Technology.]

of the ocean, usually along coastlines or on islands. **Continental effect**, or a condition of *continentality*, refers to the condition of areas that are less affected by the sea and therefore have a greater range between maximum and minimum temperatures diurnally and seasonally.

The cities of San Francisco, California, and Wichita, Kansas, provide us with a comparison of marine and continental conditions (Figure 3.23). Both cities are at approximately 37° 40′ N latitude and are at 5 m (16.4 ft) and 403 m (1321 ft) in elevation, respectively. Summer fog plays a role in delaying until September the warmest summer month in San Francisco. In 90 years of weather records there, summer maximum temperatures have risen above 32.2°C (90°F) an average of only once per year and have dropped below freezing an average of once per year. In contrast, Wichita is susceptible to freezes from late October to mid-April, with diurnal (daily) variations slightly enhanced by its elevation. Wichita has temperatures of 32.2°C (90°F) or higher at least 65 days each year. West of Wichita, winters increase in severity with increasing distance from the moderating influences of invading air masses from the Gulf of Mexico.

Earth's Temperature Patterns

SATELLITE LOOP

Global sea-surface temperatures
Global surface temperatures

Earth's temperature patterns result from the several controlling factors just discussed. Now let's look at their combined effect, portrayed by the temperature maps in this section. These maps show worldwide mean monthly air temperatures for January (Figure 3.24) and July (see Figure 3.26). These maps, along with Figure 3.28, which shows annual temperature ranges (January compared to July), also are useful in identifying areas that experience the greatest annual extremes. We use maps for January and July instead of the solstice months (December and June) because, as explained earlier, a lag occurs between insolation received and maximum or minimum temperatures experienced.

The lines on temperature maps are known as **isotherms**. An isotherm connects points of equal temperature and portrays the temperature pattern, just as a contour line on a topographic map portrays points of equal elevation. Geographers are concerned with spatial analysis of temperatures, and isotherms help with this analysis. With each temperature map, begin by finding your own city or town and noting the temperatures indicated by the isotherms (the small scale of these maps will permit only a general determination).

January Temperature Map

Figure 3.24 is a map of mean January temperatures. In the Southern Hemisphere, the higher Sun altitude means longer days and summer weather conditions, whereas in the Northern Hemisphere, January marks winter's shortened daylength and lower Sun angles. Isotherms indicate the general decrease in insolation and net radiation with distance from the equator.

The **thermal equator**, a line connecting all points of highest mean temperature (the dashed-red line on the map, roughly 27°C, 80°F), trends southward into the interior of South America and Africa, indicating higher temperatures over landmasses. In the Northern Hemisphere, isotherms trend equatorward as cold air chills the continental interiors. The oceans, on the other hand, appear more moderate, with warmer conditions extending farther north than over land at comparable latitudes.

The coldest area on the map is in northeastern Siberia in Russia. The cold experienced there relates to consistent clear, dry, calm air; small insolation input; and an inland location far from any moderating maritime effects. Both Verkhoyansk and Oymyakon, Russia, have experienced a minimum temperature of −68°C (−90°F) and a daily average of −50.5°C (−58.9°F) for January (Figure 3.25b). Verkhoyansk experiences at least 7 months of temperatures below freezing, including at least 4 months below −34°C (−30°F)! People do live and work in Verkhoyansk, which has a population of 1400. In contrast, these same towns experience maximum temperatures of +37°C (+98°F) in July.

Trondheim, Norway, is near the latitude of Verkhoyansk and at a similar elevation. However, Trondheim's coastal location moderates its annual temperature regime (Figure 3.25a): January minimum and maximum temperatures range between −17° and +8°C (+1.4° and +46°F), and the minimum/maximum range for July is from +5° to +27°C (+41° to +81°F). The most extreme minimum and maximum temperatures ever recorded in Trondheim are −30° and +35°C (−22° and +95°F)—quite a difference from the extremes at Verkhoyansk.

July Temperature Map

Average July temperatures are presented in Figure 3.26. The longer days of summer and higher Sun altitude now are in the Northern Hemisphere. Winter occupies the Southern Hemisphere, although it is milder there because large continental landmasses are absent and oceans and seas dominate. The thermal equator shifts northward with the high summer Sun and reaches the Persian Gulf–Pakistan–Iran area. The Persian Gulf is the site of the highest recorded SST of 36°C (96°F), difficult to imagine for a sea or ocean body.

July is a time of 24-hour-long nights in Antarctica. The lowest natural temperature reported on Earth occurred in 1983, at the Russian research base at Vostok, Antarctica (78° 27′ S, elevation 3420 m or 11,220 ft)—a frigid −89.2°C (−128.6°F). For comparison, such a temperature is 11C° (19.8F°) colder than dry ice (solid carbon dioxide)! During July in the Northern Hemisphere, isotherms trend poleward over land as higher temperatures dominate continental interiors. July temperatures in Verkhoyansk average more than 13°C (56°F), which represents a 63C° (113F°) seasonal variation between winter

FIGURE 3.23 Marine and continental cities—United States.
Comparison of temperatures in coastal San Francisco, California, and continental Wichita, Kansas. [(a) San Francisco photo by author; (b) Wichita region photo by © Philip Gould/Corbis.]

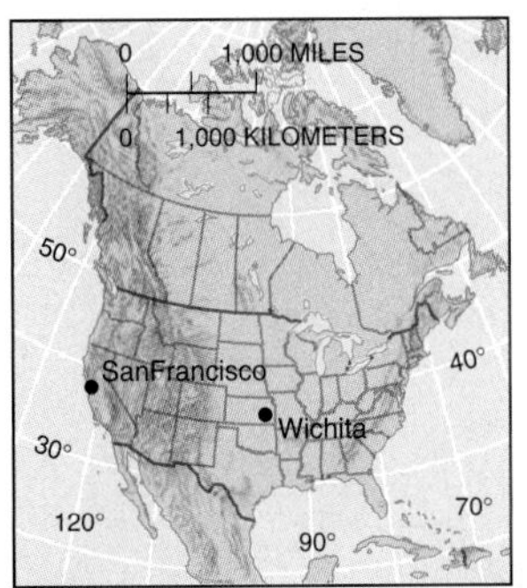

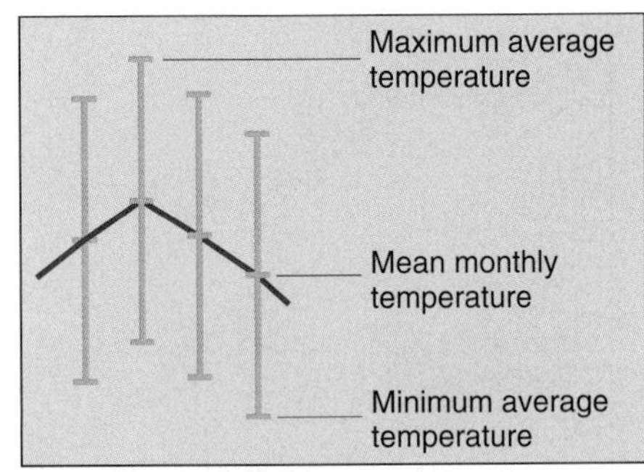

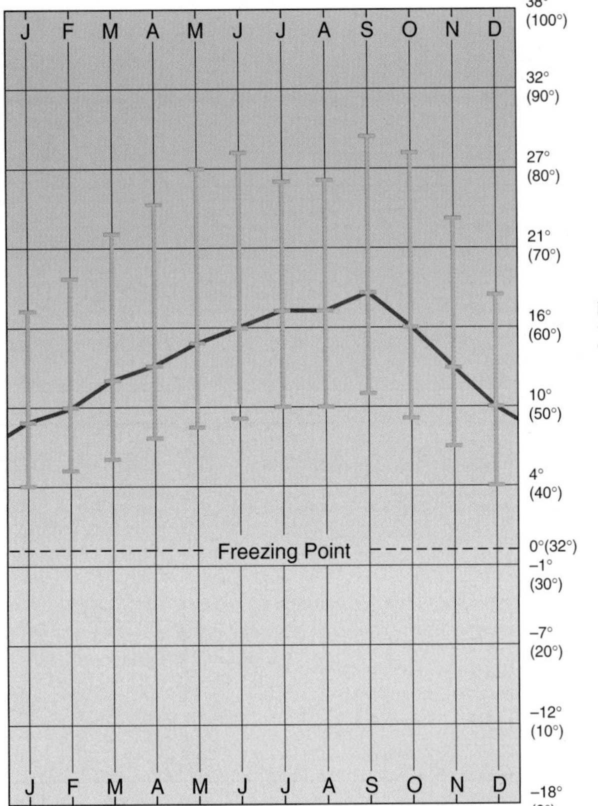

Station: San Francisco, California
Lat/long: 37° 37′ N 122° 23′ W
Avg. ann. temp.: 14°C (57.2°F)
Total ann. precip.: 47.5 cm (18.7 in.)
Elevation: 5 m (16.4 ft)
Population: 750,000
Ann. temp. range: 9 C° (16.2 F°)

(a)

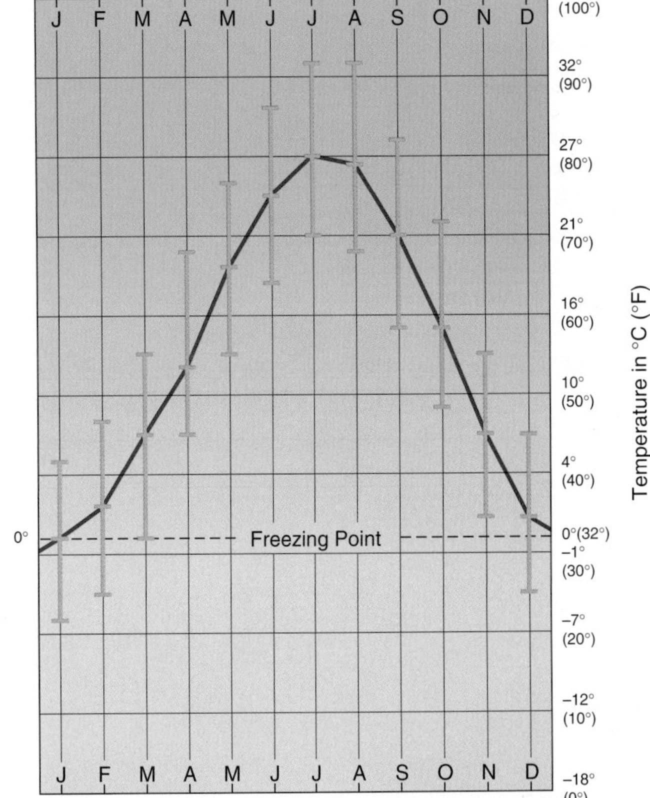

Station: Wichita, Kansas
Lat/long: 37° 39′ N 97° 25′ W
Avg. ann. temp.: 13.7°C (56.6°F)
Total ann. precip.: 72.2 cm (28.4 in.)
Elevation: 402.6 m (1321 ft)
Population: 280,000
Ann. temp. range: 27 C° (48.6 F°)

(b)

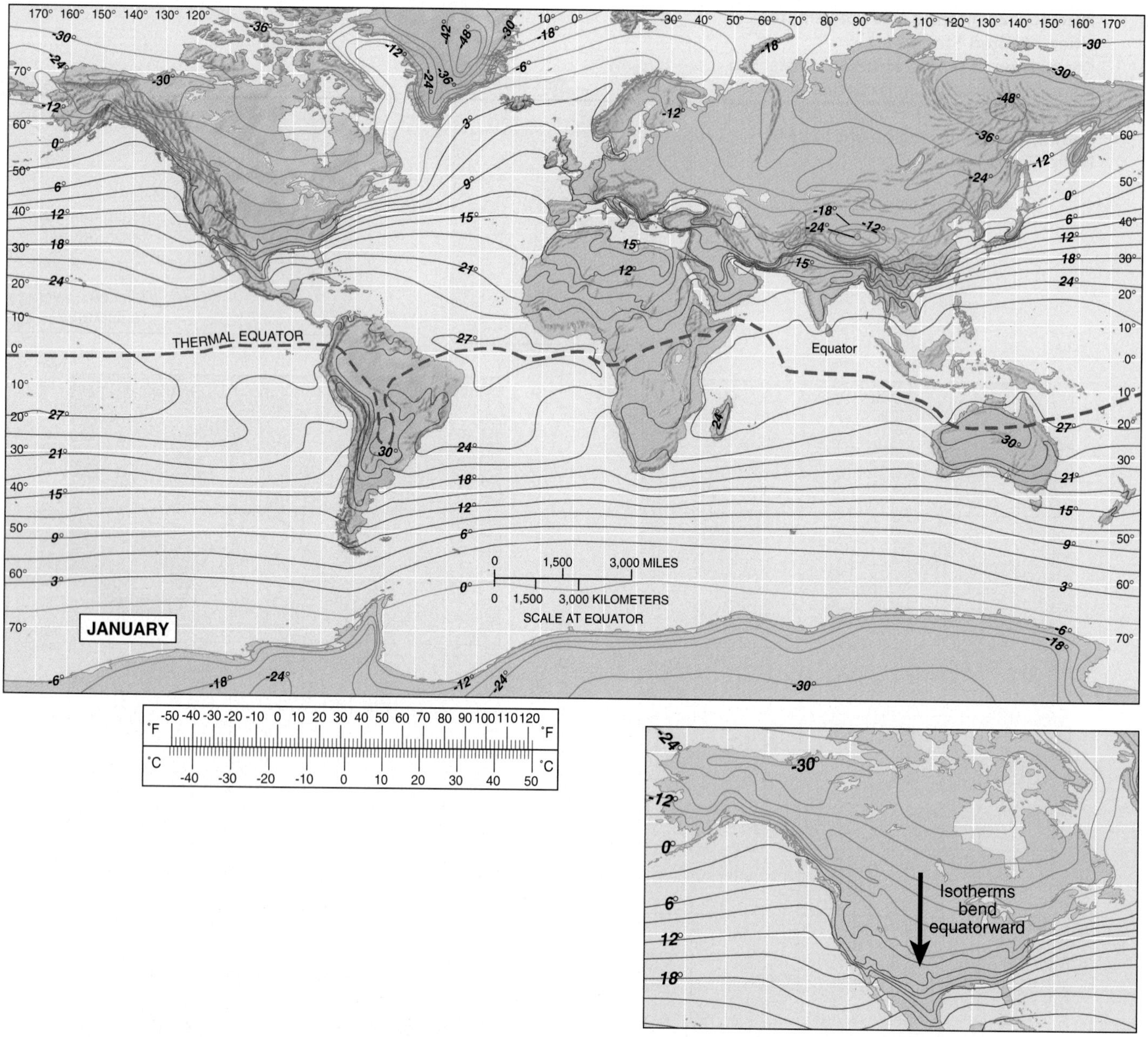

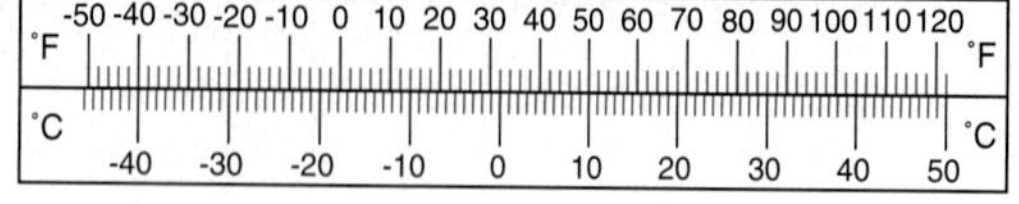

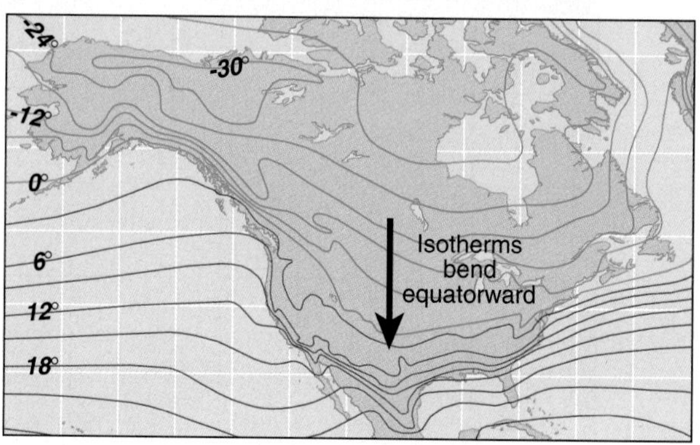

FIGURE 3.24 Global temperature for January.
Temperatures are in degrees Celsius (converted to Fahrenheit by means of the scale) and are equated to sea level to compensate for the effects of landforms. Note the inset map of North America and the equatorward trending isotherms in the interior. (Compare to Figure 3.26.) [Adapted from National Climatic Data Center, *Monthly Climatic Data for the World*, 47, no. 1, January 1994. Prepared in cooperation with the WMO and NOAA.]

and summer averages. The Verkhoyansk region of Siberia is probably the best example of continentality on Earth.

The hottest places on Earth occur in Northern Hemisphere deserts during July. These deserts are areas of clear skies and strong surface heating, with virtually no surface water and few plants. Locations such as the Sonoran Desert of North America and the Sahara of Africa are prime examples. Africa has recorded shade temperatures in excess of 58°C (136°F), a record set on September 13, 1922 at Al 'Azīzīyah, Libya (32° 32′ N; 112 m or 367 ft elevation).

The highest maximum and annual average temperatures in North America occurred in Death Valley, California, where the Greenland Ranch Station (37° N; −54.3 m or −178 ft below sea level) reached 57°C (134°F) in 1913. During June 1994 temperatures hit a high of 53.3°C (128°F) in Death Valley (Figure 3.27). Such arid and hot lands are discussed further in Chapter 12.

Annual Range of Temperatures

The pattern of marine effects versus continental effects emerges dramatically when the range in averages is mapped for a full year (Figure 3.28). As you might expect, the largest temperature ranges occur in subpolar

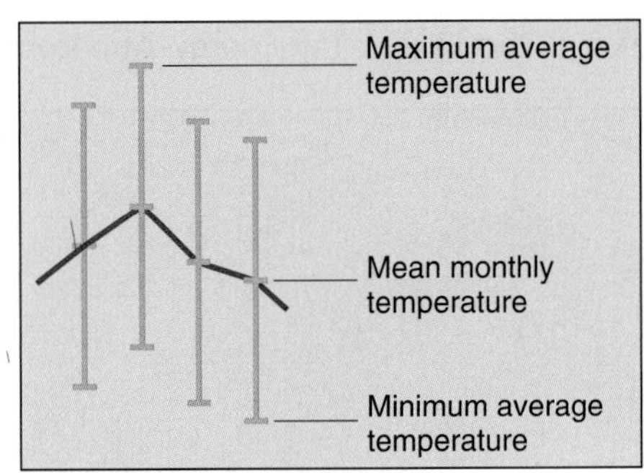

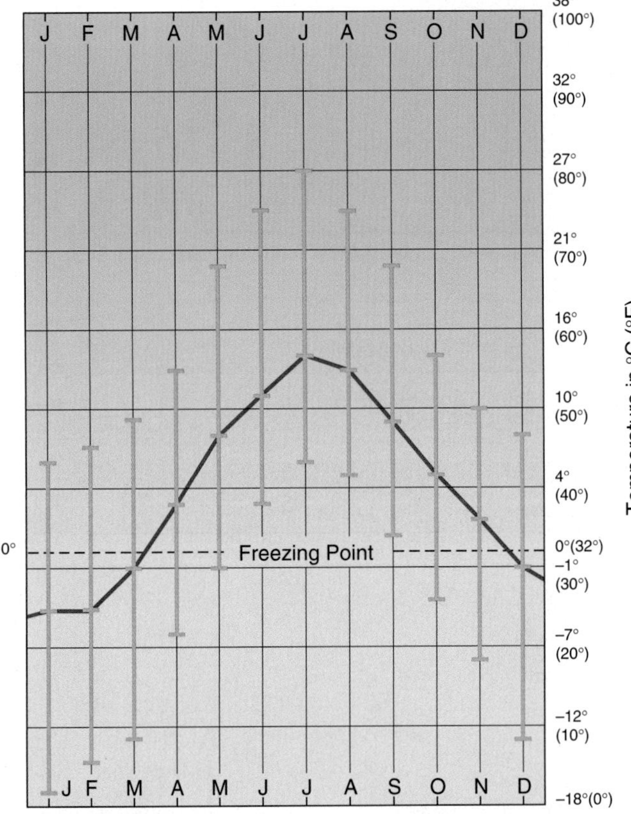

Station: Trondheim, Norway
Lat/long: 63° 25′ N 10° 27′ E
Avg. ann. temp.: 5°C (41°F)
Total ann. precip.: 85.7 cm (33.7 in.)
Elevation: 115 m (377.3 ft)
Population: 135,000
Ann. temp. range: 17 C° (30.6 F°)

Station: Verkhoyansk, Russia
Lat/long: 67° 35′ N 135° 23′ E
Avg. ann. temp.: −15°C (5°F)
Total ann. precip.: 15.5 cm (6.1 in.)
Elevation: 137 m (449.5 ft)
Population: 1400
Ann. temp. range: 63 C° (113.4 F°)

(a)

(b)

FIGURE 3.25 Marine and continental cities—Eurasia.
Comparison of temperatures in coastal Trondheim, Norway (a), and continental Siberian Russia (b). Note that the freezing levels on the two graphs are positioned differently to accommodate the contrasting data. [Photos by (a) Norman Benton/Peter Arnold, Inc., (b) TASS/Sovfoto/Eastfoto.]

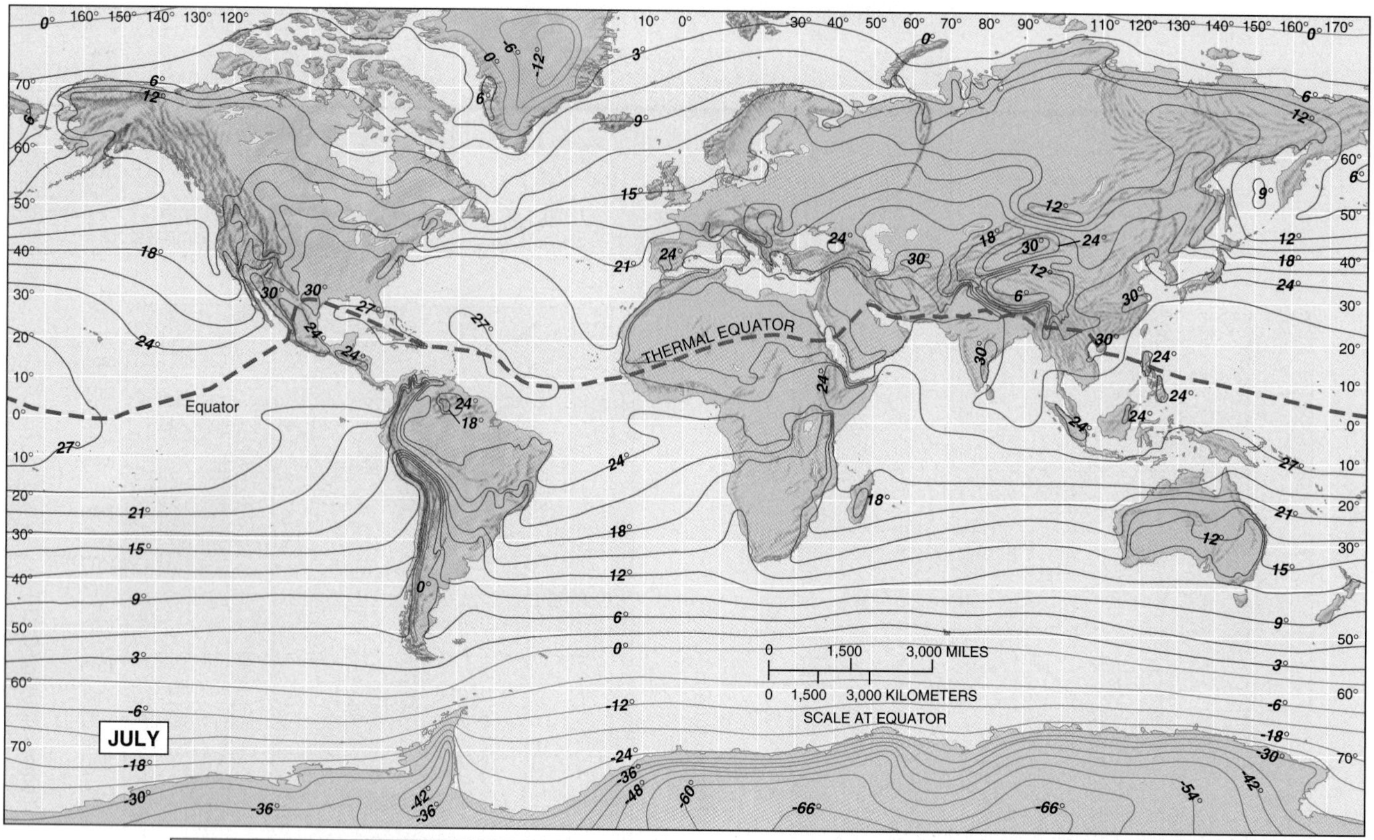

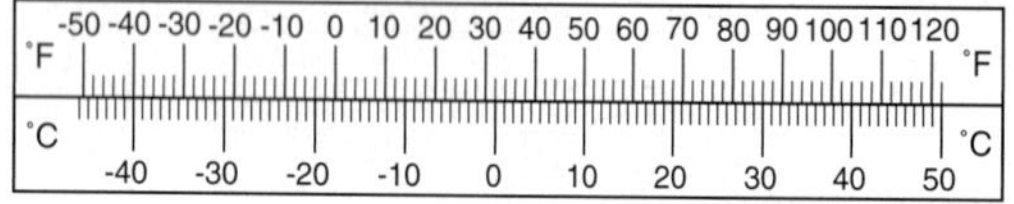

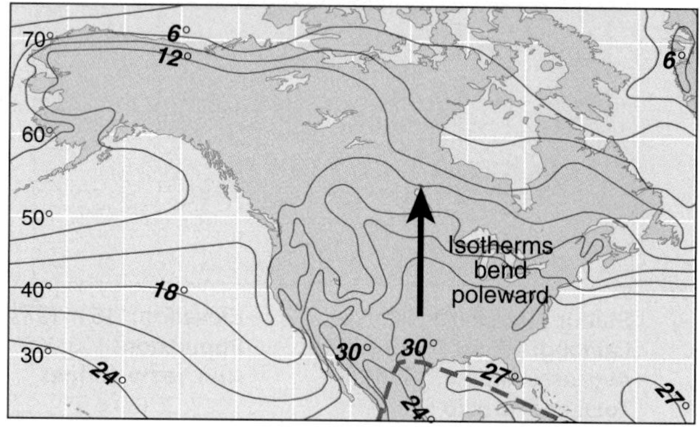

FIGURE 3.26 Global temperatures for July.
Temperatures are in degrees Celsius (converted to Fahrenheit by means of the scale) and are equated to sea level to compensate for the effects of landforms. Note the inset map of North America and the poleward trending isotherms in the interior. (Compare to Figure 3.24.) [Adapted from National Climatic Data Center, *Monthly Climatic Data for the World*, 47, no. 7, July 1994. Prepared in cooperation with the WMO and NOAA.]

locations in North America and Asia, where average ranges of 64 C° (115 F°) are recorded. The Southern Hemisphere, on the other hand, has little seasonal variation in mean temperatures, although the landmasses present there do produce some increase in temperature range.

For example, in January (see Figure 3.24) Australia is dominated by isotherms of 20° to 30°C (68° to 86°F), whereas in July (see Figure 3.26) Australia experiences isotherms of 10° to 20°C (50° to 68°F). Generally speaking, Southern Hemisphere patterns are more marine, whereas Northern Hemisphere patterns are more continental.

FIGURE 3.27 Weather instrumentation in Death Valley, California.
Weather instruments placed near Badwater, the lowest elevation in the Western Hemisphere (−86 m, −282 ft below sea level), record the harsh conditions on the salt-encrusted playa (intermittent lake bed). The USGS is studying temperature and wind to determine the evaporation-transpiration rates as part of a regional study of groundwater resources. [Photo by Bobbé Christopherson.]

The Northern Hemisphere, with greater land area overall, registers a slightly higher average surface temperature than does the Southern Hemisphere.

Imagine living in some of these regions and the degree to which your personal comfort adaptations would need to adjust. See Focus Study 3.2 for more on air temperature and the human body. The summer heat-wave deaths in 1995, and several summers since then in the United States and Canada, remain powerful reminders of the impact of air temperatures on our lives.

The Urban Environment

For most of you reading this book, an urban landscape affects the temperatures you feel each day. Urban microclimates generally differ from those of nearby nonurban areas. In fact, the surface energy characteristics of urban areas are similar to desert locations. Because almost 50% of the world's population will be living in cities by the year 2010, urban microclimatology and other specific environmental effects related to cities are important topics for physical geographers.

The physical characteristics of urbanized regions produce the **urban heat island**; that is, on average both maximum and minimum temperatures are higher in urban areas than in nearby rural settings. Table 3.1 lists five urban characteristics and the resulting temperature and moisture effects produced; these traits are illustrated in Figure 3.29. Every major city produces its own **dust dome** of airborne pollution, which can be blown from the city in elongated plumes; as noted in the table such domes affect urban energy budgets. Table 3.2 compares climatic factors of rural and urban environments. The worldwide trend toward greater urbanization is placing more and more people on urban heat islands.

Figure 3.30 illustrates a generalized cross section of a typical urban heat island, showing increasing temperatures toward the downtown central business district. Note temperatures drop over areas of trees and parks. Sensible heat is lessened because of latent heat of evaporation and plant effects. Urban forests are important factors in cooling cities. NASA measurements in a mall parking lot found temperatures of 48°C (118°F); however, a small planter with trees in the same lot was significantly cooler at 32°C (90°F)—16 C° lower in temperature!

NASA launched its Urban Heat Island Pilot Project (UHIPP) in 1997, through its Global Hydrology and Climate Center, to better understand the role of cities in climate. Thermal infrared measurements were made from NASA's research jet over Atlanta, Georgia; Sacramento, California; Baton Rouge, Louisiana; and Salt Lake City, Utah. Images were produced and urban heat island profiles mapped. Teachers and students on the ground assisted efforts by making temperature measurements at the same time as the flights. Instruments carried by balloons tracked vertical temperature, relative humidity, and air pressure profiles. (See **http://www.ghcc.msfc. nasa.gov/ghcc_research.html.**)

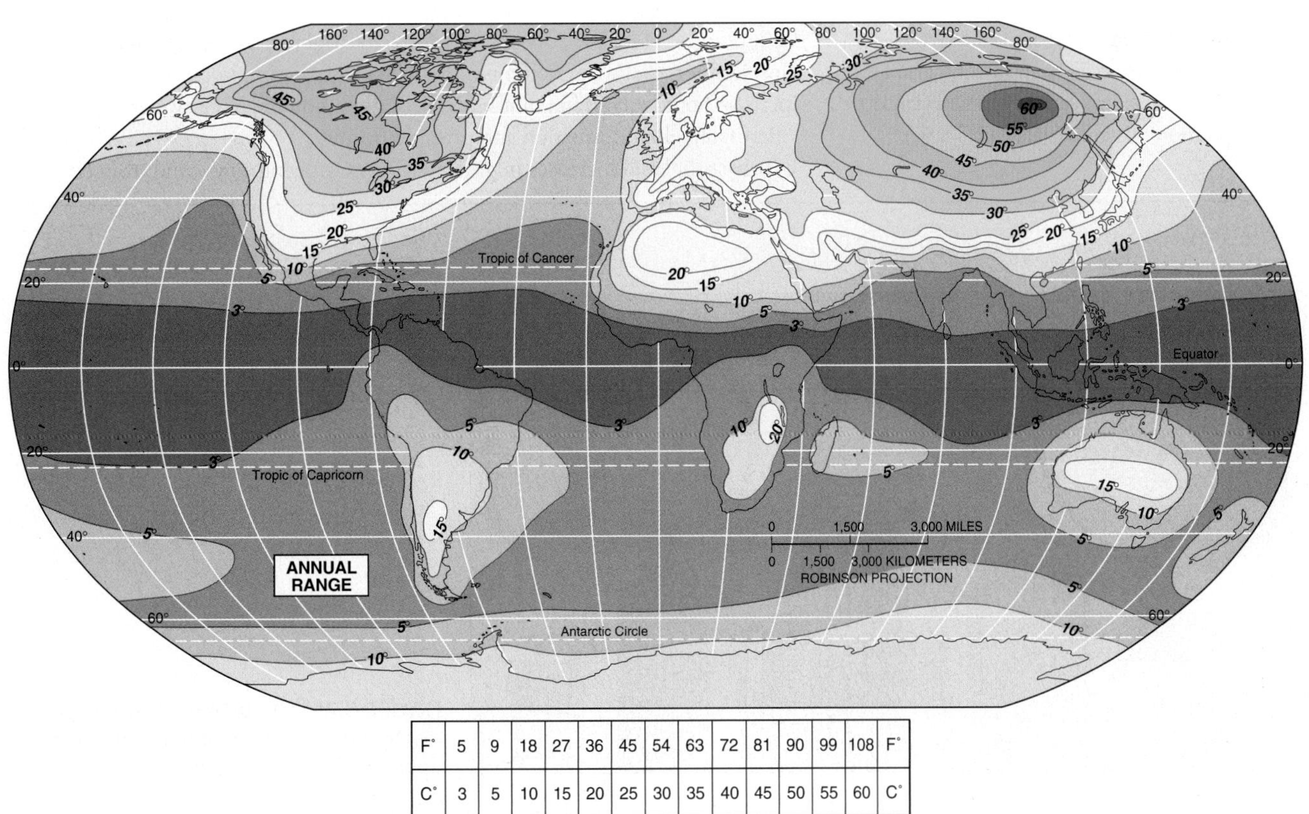

FIGURE 3.28 Global annual temperature ranges.
Annual range of global temperatures between January and July in Celsius (Fahrenheit) degrees.

Figure 3.31 is a thermal infrared image of a portion of Sacramento, California (taken midday). As you look across a small portion of this Sacramento landscape, what strategies do you think would reduce the urban heat island effect? Possibilities include lighter surfaces for buildings, streets, and parking lots; more reflective roofs; more trees, parks, and open space. (The images from all the test cities are available at http://wwwghcc.msfc.nasa.gov/uhipp/urban_uhipp.html.)

Table 3.1 Urban Physical Characteristics and Conditions

Urban characteristics	Results and conditions
Urban surfaces typically are metal, glass, asphalt, concrete, or stone, and their energy characteristics respond differently from natural surfaces	Albedos of urban surfaces are lower, leading to higher net radiation values Urban surfaces expend more energy as sensible heat than do nonurban areas (70% of the net radiation to sensible heat) Surfaces conduct up to three times more energy than wet sandy soil and thus are warmer During the day and evening, temperatures above urban surfaces are higher than those above natural areas
Irregular geometric shapes in a city affect radiation patterns and winds	Incoming insolation is caught in mazelike reflection and radiation "canyons" Delayed energy is conducted into surface materials, thus increasing temperatures Buildings interrupt wind flows, diminishing heat loss through advective (horizontal) movement Maximum heat island effects occur on calm, clear days and nights
Human activity alters the heat characteristics of cities	In summer, urban electricity production and use of fossil fuels release energy equivalent to 25%–50% of insolation In winter, urban-generated sensible heat averages 250% greater than arriving insolation, reducing winter heating requirements
Many urban surfaces are sealed (built on and paved), so water cannot reach the soil	Central business district surfaces average 50% sealed, suburbs average 20% sealed, producing more water runoff Urban areas respond as a desert landscape: A storm may cause a flash flood over the hard, sparsely vegetated surfaces, to be followed by dry conditions a few hours later
Air pollution, including gases and aerosols, greater in urban areas than in comparable natural settings; increased convection and precipitation possible	Pollution increases the atmosphere's reflectivity above a city, reducing insolation and absorbing infrared radiation, reradiating infrared downward Increased particulates in pollution are condensation nuclei for water vapor, increasing cloud formation and precipitation Urban-stimulated increases in precipitation may occur downwind from cities

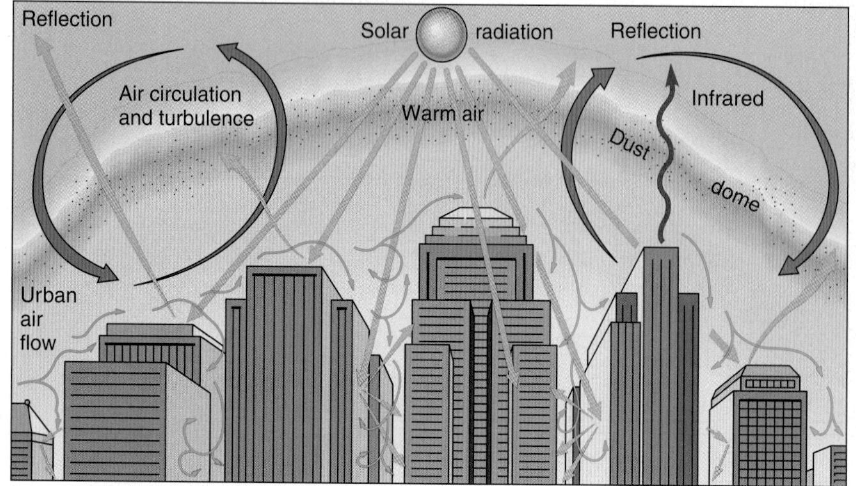

FIGURE 3.29 The urban environment. Insolation, wind movements, and dust dome in city environments.

Table 3.2 Average Differences in Climatic Elements Between Urban and Rural Environments

Element	Urban Compared with Rural Environs
Contaminants	
Condensation nuclei	10 times more
Particulates	10 times more
Gaseous admixtures	5–25 times more
Radiation	
Total on horizontal surface	0%–20% less
Ultraviolet, winter	30% less
Ultraviolet, summer	5% less
Sunshine duration	5%–15% less
Cloudiness	
Clouds	5%–10% more
Fog, winter	100% more
Fog, summer	30% more
Precipitation	
Amounts	5%–15% more
Days with <5 mm (0.2 in.)	10% more
Snowfall, inner city	5%–10% less
Precipitation (cont.)	
Snowfall, downwind (lee) of city	10% more
Thunderstorms	10%–15% more
Temperature	
Annual mean	0.5–3.0 C° (0.9–5.4 F°) more
Winter minima (average)	1.0–2 C° (1.8–3.6 F°) more
Summer maxima	1.0–3 C° (1.8–3.0 F°) more
Heating degree days	10% less
Relative Humidity	
Annual mean	6% less
Winter	2% less
Summer	8% less
Wind Speed	
Annual mean	20%–30% less
Extreme gusts	10%–20% less
Calm	5%–20% more

Source: H. E. Landsberg, *The Urban Climate*, International Geophysics Series, vol. 28 (1981), p. 258. Reprinted by permission from Academic Press.

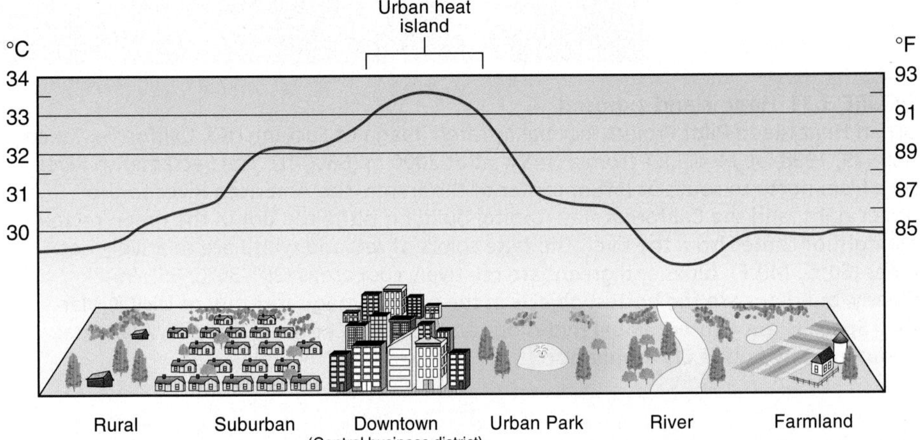

FIGURE 3.30 Urban heat island profile.
Generalized cross section of a typical urban heat island. The trend of the temperature gradient is described by the terms *cliff* at the edge of the city, where temperatures steeply rise; *plateau* over the suburban built-up area, and *peak* where temperature is highest at the urban core. Note the slight cooling over the park and river.

FIGURE 3.31 Heat island exposed.
Urban Heat Island Pilot Project thermal infrared image of Sacramento, California, taken June 29, 1998, at 1 P.M. PDT from a NASA jet at 2000 m (6600 ft). The Sacramento River runs from north to south at left of center of the frame, the American River to the upper right, and the California state capitol building is the red dot in the green rectangle right of center from the river. The false colors of red and white are relatively hot areas (60°C, 140°F); blues and greens are relatively cool areas (29°–36°C, 85°–96°F). Clearly, buildings are the hottest objects in the scene. [Image courtesy of UHIPP, Marshall Space Flight Center, Global Hydrology and Climate Center, Huntsville, Alabama; photo inset by Bobbé Christopherson.]

Focus Study 3.2

Air Temperature and the Human Body

We humans are capable of sensing small changes in temperature in the environment around us. Our perception of temperature is described by the terms *apparent temperature*, or *sensible temperature*. This perception of temperature varies among individuals and cultures. Through complex mechanisms, our bodies maintain an average internal temperature ranging within a degree of 36.8°C (98.2°F), slightly lower in the morning or in cold weather and slightly higher at emotional times or during exercise and work.

The water vapor content of air, wind speed, and air temperature taken together affect each individual's sense of comfort. High temperatures, high humidity, and low winds produce the most heat discomfort, whereas low humidity and strong winds enhance cooling rates. Although modern heating and cooling systems, if available, can reduce the impact of uncomfortable temperatures indoors, the danger to human life from excessive heat or cold persists outdoors. When changes occur in the surrounding air, the human body reacts in various ways to maintain its core temperature and to protect the brain at all cost.

The *wind chill index* is important to those who experience winters with freezing temperatures. The wind-chill factor indicates the enhanced rate at which body heat is lost to the air. As wind speeds increase, heat loss

(continued)

Focus Study 3.2 *(continued)*

from the skin increases. A formula for calculating these relationships was developed in 1945. This older index tended to overestimate heat loss from skin.

The NWS and the Meteorological Services of Canada (MSC, http://www.msc-smc.ec.gc.ca/) revised the wind chill formula. The new Wind Chill Temperature (WCT) Index improves the accuracy of heat loss calculations. Figure 1 is the new version that went into service November 2001.

The *heat index* indicates the human body's reaction to air temperature and water vapor. The water vapor in air is expressed as relative humidity, a concept presented in Chapter 5. For now, we simply can assume that the amount of water vapor in the air affects the evaporation rate of perspiration on the skin because the more water vapor in the air (the higher the humidity), the less water from perspiration the air can absorb through evaporation, thus reducing natural evaporative cooling. The heat index indicates how the air feels to an average person—in other words, its *apparent temperature*.

Figure 2 is an abbreviated version of the heat index used by the National Weather Service (NWS) and now included in its daily weather summaries during appropriate months. The table beneath the graph describes the effects of heat-index categories on higher-risk groups. A combination of high temperature and high humidity can severely reduce the body's natural ability to regulate internal temperature (see http://weather.noaa.gov/weather/hwave.html).There is a distinct possibility that future humans may experience greater temperature-related challenges.

Summer 1995 Heat-Index Deaths

For nearly a week during July 1995, Chicago's heat-index values went to category I in dwellings that lacked air conditioning. Chicago had never experienced a 48-hour period during which temperatures did not go below 31.5°C (89°F)!

With high pressure (hot, stable air) dominating from the Midwest to the Atlantic and moist air from the Gulf of Mexico, all the ingredients for this disaster were present. Afternoon temperatures were above 32°C (90°F) for a week in Chicago, 38°C (100°F) in South Dakota, and 40°C (104°F) in Toledo, Ohio, and New York City; the temperature hit 43°C (109°F) in Omaha, Nebraska. Cities in New England that had reached 37.7°C (100°F) only twice in their history, exceeded that for 4 or 5 days during July 1995!

(continued)

Actual Air Temperature in °C (°F)

Wind speed, kmph (mph) / Calm	4° (40°)	−1° (30°)	−7° (20°)	−12° (10°)	−18° (0°)	−23° (−10°)	−29° (−20°)	−34° (−30°)	−40° (−40°)
8 (5)	2° (36°)	−4° (25°)	−11° (13°)	−17° (1°)	−24° (−11°)	−30° (−22°)	−37° (−34°)	−43° (−46°)	−49° (−57°)
16 (10)	1° (34°)	−6° (21°)	−13° (9°)	−20° (−4°)	−27° (−16°)	−33° (−28°)	−41° (−41°)	−47° (−53°)	−54° (−66°)
24 (15)	0° (32°)	−7° (19°)	−14° (6°)	−22° (−7°)	−28° (−19°)	−36° (−32°)	−43° (−45°)	−50° (−58°)	−57° (−71°)
32 (20)	−1° (30°)	−8° (17°)	−16° (4°)	−23° (−9°)	−30° (−22°)	−37° (−35°)	−44° (−48°)	−52° (−61°)	−59° (−74°)
40 (25)	−2° (29°)	−9° (16°)	−16° (3°)	−24° (−11°)	−31° (−24°)	−38° (−37°)	−46° (−51°)	−53° (−64°)	−61° (−78°)
48 (30)	−2° (28°)	−9° (15°)	−17° (−1°)	−24° (−12°)	−32° (−26°)	−39° (−39°)	−47° (−53°)	−55° (−67°)	−62° (−80°)
56 (35)	−2° (28°)	−10° (14°)	−18° (0°)	−26° (−14°)	−33° (−27°)	−41° (−41°)	−48° (−55°)	−56° (−69°)	−63° (−82°)
64 (40)	−3° (27°)	−11° (13°)	−18° (−1°)	−26° (−15°)	−34° (−29°)	−42° (−43°)	−49° (−57°)	−57° (−71°)	−64° (−84°)
72 (45)	−3° (26°)	−11° (12°)	−19° (−2°)	−27° (−16°)	−34° (−30°)	−42° (−44°)	−50° (−58°)	−58° (−72°)	−66° (−86°)
80 (50)	−3° (26°)	−11° (12°)	−19° (−3°)	−27° (−17°)	−35° (−31°)	−43° (−45°)	−51° (−60°)	−59° (−74°)	−67° (−88°)

Frostbite times: 30 min. 10 min. 5 min.

FIGURE 1 Wind Chill Temperature Index.
WCT Index factor for various temperatures and wind speeds. For quick calculations see http://205.156.54.206/om/windchill/index.shtml#calculator. (In Canada, heat loss is expressed in watts per square meter.) [Adapted from the National Weather Service and Meteorological Services of Canada, version 11/01/01.]

Focus Study 3.2 *(continued)*

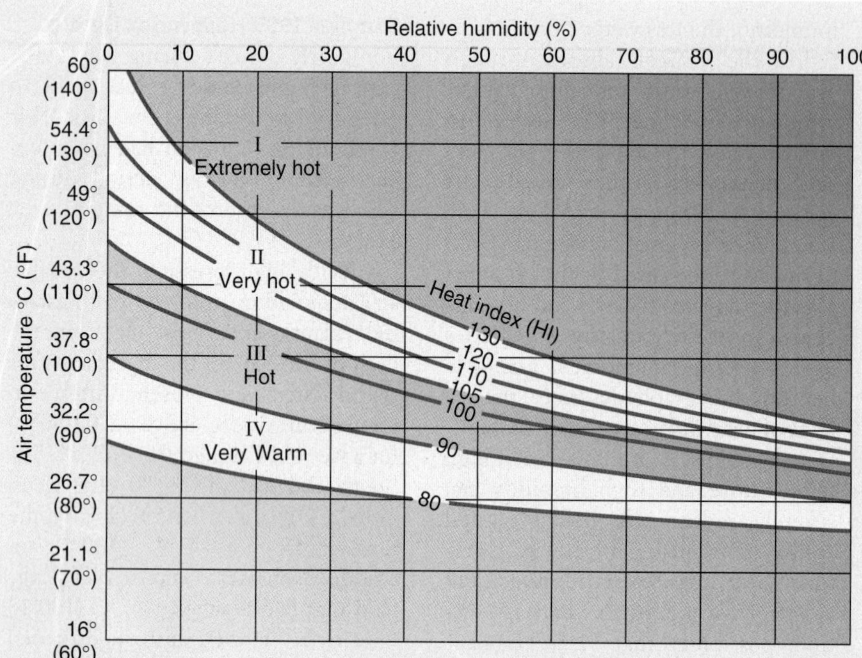

Level of concern	Category	Heat Index Apparent Temperature	General Effect of Heat Index on People in High-Risk Groups
Extreme danger	I	54°C (130°F) or higher	Heat/sunstroke highly likely with continued exposure
Danger	II	41° – 54°C (105° – 130°F)	Sunstroke, heat cramps, or heat exhaustion likely and heatstroke possible with prolonged exposure and/or physical activity
Extreme caution	III	32° – 41°C (90° – 105°F)	Sunstroke, heat cramps, and heat exhaustion possible with prolonged exposure and/or physical activity
Caution	IV	27° – 32°C (80° – 90°F)	Fatigue possible with prolonged exposure and/or physical activity

FIGURE 2 Heat index.
Heat-index graph for various temperatures and relative humidity levels. [Courtesy of National Weather Service.]

This combination of high temperatures and water vapor produced stifling conditions. The sick and elderly, particularly, were affected. On July 13 in Chicago, some apartments without air conditioning exceeded indoor heat-index temperatures of 54°C (130°F) when the official temperature at Midway Airport reached a record 41°C (106°F). The death toll in Chicago was especially severe; 700 people died. Nearly 1000 people died overall in the Midwest and East from these conditions.

The rush for medical help quickly swamped hospitals, which had to turn away hundreds of patients. The rate at which the heat claimed its victims was far greater than the coroner could handle, so the dead were stored in many refrigerated trucks parked outside the morgue (Figure 3). We can only wonder how many lives would have been saved if refrigeration for the living was available.

FIGURE 3 Chicago heat wave.
The Midwest and eastern United States were hit with a devastating heat wave during July 1995. City emergency, medical, and coroner services were overwhelmed by the tragedy. Refrigerated trucks served as a temporary morgue in Chicago. [Photo by Dave Weaver/Gamma Liaison, Inc.]

Summary and Review—Atmospheric Energy and Global Temperatures

● *Identify* the pathways of solar energy through the troposphere to Earth's surface: transmission, refraction, albedo (reflectivity), scattering, diffuse radiation, conduction, convection, and advection.

Earth's biosphere is powered by radiant energy from the Sun that cascades through complex circuits to the surface. Our budget of atmospheric energy comprises shortwave radiation *inputs* (ultraviolet light, visible light, and near-infrared wavelengths) and longwave radiation *outputs* (thermal infrared).

Transmission refers to the passage of shortwave and longwave energy through either the atmosphere or water. Insolation encounters an increasing density of atmospheric gases that absorb and reemit it *without altering its wavelengths*. This **scattering** represents 7% of Earth's reflectivity, or albedo. Some incoming insolation is diffused by clouds and atmosphere and is transmitted to Earth as **diffuse radiation**, the downward component of scattered light. The speed of insolation entering the atmosphere changes as it passes from one medium to another; the change of speed causes a bending action called **refraction**.

A portion of arriving energy bounces directly back into space without being converted into heat or performing any work. **Albedo** is the reflective quality (intrinsic brightness) of a surface. This returned energy is called **reflection**. The amount reflected is an important control over the amount of insolation that is available for absorption by a surface. We state albedo as the percentage of insolation that is reflected. Earth and its atmosphere reflect 31% of all insolation when averaged over a year.

An increase in albedo and reflection of shortwave radiation caused by clouds is described by the term **cloud-albedo forcing**. Also, clouds can act as insulation, thus trapping longwave radiation and raising minimum temperatures. An increase in greenhouse warming caused by clouds is described by the term **cloud-greenhouse forcing**.

Absorption is the assimilation of radiation by molecules of a substance and its conversion from one form to another—for example, visible light to infrared radiation. **Conduction** is the molecule-to-molecule transfer of energy as it diffuses through a substance. Energy also is transferred by gases and liquids through **convection** (when the physical mixing involves a strong vertical motion) or **advection** (when the dominant motion is horizontal). In the atmosphere or bodies of water, warmer portions tend to rise (they are less dense) and cooler portions tend to sink (they are more dense), establishing patterns of convection.

transmission (p. 72)
scattering (p. 72)
diffuse radiation (p. 73)
refraction (p. 73)
albedo (p. 74)
reflection (p. 74)
cloud-albedo forcing (p. 75)
cloud-greenhouse forcing (p. 75)
absorption (p. 75)
conduction (p. 75)
convection (p. 75)
advection (p. 75)

1. Diagram a simple energy balance for the troposphere. Label each shortwave and longwave component and the directional aspects of related flows.
2. Define refraction. How is it related to daylength? To a rainbow? To the beautiful colors of a sunset?
3. List several types of surfaces and their albedo values. Explain the differences among these surfaces. What determines the reflectivity of a surface?
4. Define the concepts transmission, absorption, diffuse radiation, conduction, and convection.

● *Describe* the greenhouse effect and the patterns of global net radiation and surface energy balances.

The Earth-atmosphere energy system naturally balances itself in a steady-state equilibrium. It does so through energy transfers that are *nonradiative* (convection, conduction, and the latent heat of evaporation) and *radiative* (by infrared radiation between the surface, the atmosphere, and space).

Some infrared radiation is absorbed by carbon dioxide, water vapor, methane, CFCs (chlorofluorcarbons), and other gases in the lower atmosphere and is then reradiated to Earth, thus delaying energy loss to space. This process is the **greenhouse effect**. In the atmosphere, infrared radiation is not actually trapped, as it would be in a greenhouse, but its passage to space is delayed (heat energy is detained in the atmosphere) as it is absorbed and reradiated.

Between the tropics, high insolation angle and consistent daylength cause more energy to be gained than lost (there are energy surpluses). In the polar regions, an extremely low insolation angle, highly reflective surfaces, and up to 6 months of no insolation annually cause more energy to be lost (there are energy deficits). This imbalance of net radiation from tropical surpluses to the polar deficits drives a vast global circulation of both energy and mass.

Surface energy balances are used to summarize the energy expenditure for any location. Surface energy measurements are used as an analytical tool of **microclimatology**. Adding and subtracting the energy flow at the surface lets us calculate **net radiation (NET R)**, or the balance of all radiation at Earth's surface—shortwave (SW) and longwave (LW).

The greatest insolation input occurs at the time of the summer solstice in each hemisphere. Air temperature responds to seasons and variations in insolation input. Within a 24-hour day, air temperature peaks between 3:00 and 4:00 P.M. and dips to its lowest point at or slightly after sunrise.

Air temperature lags behind each day's peak insolation. The warmest time of day occurs not at the moment of maximum insolation but at that moment when a maximum of insolation is absorbed.

greenhouse effect (p. 76)
microclimatology (p. 80)
net radiation (NET R) (p. 80)

5. What are the similarities and differences between an actual greenhouse and the gaseous atmospheric greenhouse? Why is Earth's greenhouse changing?
6. Generalize the pattern of global net radiation. How might this pattern drive the atmospheric weather machine? (See Figure 3.11.)
7. In terms of surface energy balance, explain the term net radiation (NET R).
8. What are the expenditure pathways for surface NET R? What kind of work is accomplished?
9. What is the role played by latent heat of evaporation in surface energy budgets?
10. Compare the daily surface energy patterns of El Mirage, California, and Pitt Meadows, British Columbia. Explain the differences.
11. Why is there a temperature lag between the highest Sun altitude and the warmest time of day? Relate your answer to insolation and temperature patterns during the day.

- ***Review* the temperature concepts and temperature controls that produce global temperature patterns.**

Temperature is a measure of the average kinetic energy (motion) of individual molecules in matter. Principal controls and influences upon temperature patterns include latitude (the distance north or south of the equator), altitude (location above sea level), cloud cover (reflect, absorb, and reradiate energy), and land-water heating differences. The physical nature of land (rock and soil) and water (oceans, seas, and lakes) is the reason for **land-water heating differences**, the fact that land heats and cools faster than water. Moderate temperature patterns are associated with water bodies, and extreme temperatures occur inland. The five controls that differ between land and water surfaces are: evaporation, transparency, specific heat, movement, and ocean currents and sea-surface temperatures.

Light penetrates water because of its **transparency**. Water is clear, and light transmits to an average depth of 60 m (200 ft) in the ocean. This penetration distributes available heat energy over a much greater volume than could occur on opaque land, thus a larger energy reservoir is formed. When equal volumes of water and land are compared, water requires far more energy to increase its temperature than does land. In other words, water can hold more energy than can soil or rock, so water has a higher **specific heat**, the heat capacity of a substance, averaging about four times that of soil.

Ocean currents affect temperature. An example of the effect of ocean currents is the **Gulf Stream**, which moves northward off the east coast of North America, carrying warm water far into the North Atlantic. As a result, the southern third of Iceland experiences much milder temperatures than would be expected for a latitude of 65° N, just below the Arctic Circle (66.5°).

Marine effect, or *maritime*, describes locations that exhibit the moderating influences of the ocean, usually along coastlines or on islands. **Continental effect** refers to the condition of areas that are less affected by the sea and therefore have a greater range between maximum and minimum temperatures diurnally and yearly.

temperature (p. 84)
land-water heating differences (p. 86)
transparency (p. 88)
specific heat (p. 88)
Gulf Stream (p. 88)
marine effect (p. 88)
continental effect (p. 90)

12. Explain the effect of altitude on air temperature. Why is air at higher altitudes lower in temperature? Why does it feel cooler standing in shadows at higher altitude than at lower altitude?
13. What noticeable effect does air density have on the absorption and radiation of energy? What role does altitude play in that process?
14. How is it possible to grow moderate-climate type crops such as wheat, barley, and potatoes at an elevation of 4103 m (13,460 ft) near La Paz, Bolivia, so near the equator?
15. Describe the effect of cloud cover with regard to Earth's temperature patterns—review cloud-albedo forcing and cloud-greenhouse forcing.

- ***Interpret* the pattern of Earth's temperatures for January and July and annual temperature ranges.**

Maps for January and July instead of the solstice months of December and June are used for temperature comparison because of the natural lag that occurs between insolation received and maximum or minimum temperatures experienced. Each line on these temperature maps is an **isotherm**, an isoline that connects points of equal temperature. Isotherms portray temperature patterns.

Isotherms generally are zonal, trending east-west, parallel to the equator. They mark the general decrease in insolation and net radiation with distance from the equator. The **thermal equator** (isoline connecting all points of highest mean temperature) trends southward in January and shifts northward with the high summer Sun in July. In January it extends farther south into the interior of South America and Africa, indicating higher temperatures over landmasses.

In the Northern Hemisphere in January, isotherms shift equatorward as cold air chills the continental interiors. The coldest area on the map is in Russia, specifically northeastern Siberia. The intense cold experienced there results from winter conditions of consistent clear, dry, calm air; small insolation input; and an inland location far from any moderating maritime effects.

isotherm (p. 90)
thermal equator (p. 90)

16. What is the thermal equator? Describe its location in January and in July. Explain why it shifts position annually.
17. Observe trends in the pattern of isolines over North America and compare the January average temperature map with the July map. Why do the patterns shift locations?

18. Describe and explain the extreme temperature range experienced in north-central Siberia between January and July.
19. Where are the hottest places on Earth? Are they near the equator or elsewhere? Explain. Where is the coldest place on Earth?

- ***Contrast*** **wind chill and heat index and** ***determine*** **human response to these apparent temperature effects.**

The wind-chill factor indicates the enhanced rate at which body heat is lost to the air. As wind speeds increase, heat loss from the skin increases. The *heat index* indicates the human body's reaction to air temperature and water vapor. The level of humidity in the air affects our natural ability to cool through evaporation from skin.

20. What is the wind chill temperature on a day with an air temperature of −12°C (10°F) and a wind speed of 32 kmph (20 mph)?
21. On a day when temperature reaches 37.8°C (100°F), how does a relative humidity reading of 50% affect apparent temperature?

- ***Portray*** **typical urban heat island conditions and** ***contrast*** **the micro-climatology of urban areas with that of surrounding rural environments.**

A growing percentage of Earth's people live in cities and experience their unique set of altered microclimatic effects: increased conduction, lower albedos, higher NET R values, increased water runoff, complex radiation and reflection patterns, anthropogenic heating, and the gases, dusts, and aerosols of urban pollution. Urban surfaces of metal, glass, asphalt, concrete, and stone conduct up to three times more energy than wet sandy soil and thus are warmed as described by the term **urban heat island**. Air pollution, including gases and aerosols, is greater in urban areas than in rural ones. Every major city produces its own **dust dome** of airborne pollution.

urban heat island (p. 95)
dust dome (p. 95)

22. What is the basis for the urban heat island concept? Describe the climatic effects attributable to urban as compared with nonurban environments. What did NASA determine from the UHIPP overflight of Sacramento (Figure 3.30)?
23. Have you experienced any condition graphed in Focus Study 3.2, Figures 1 or 2? Explain.
24. Assess the potential for solar energy applications in our society. What are some negatives? What are some positives?

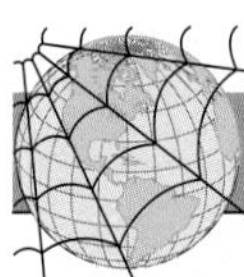

NetWork

The *Elemental Geosystems Home Page* provides on-line resources for this chapter on the World Wide Web. Once on the Home Page, click on this textbook, scroll the Table of Contents menu, select this chapter, and click "Begin." You will find self-tests that are graded, review exercises, specific updates for items in the chapter, and in "Destinations" many links to interesting related pathways on the Internet. *Elemental Geosystems* is found at **http://www.prenhall.com/christopherson**.

Critical Thinking

A. Given what you now know about reflection, albedo, absorption, and net radiation expenditures, assess your wardrobe (fabrics and colors), house or apartment (colors of walls and roof, orientation relative to the Sun), automobile (color, use of sun shades), bicycle seat (color), and other aspects of your environment to determine a personal energy quiz. What grade do you give yourself? Next summer, be cool!

B. Several URLs relating to solar energy applications are listed in the text and in the Destinations section on the Web page for Chapter 3. Take an hour and explore the Internet for a personal assessment of these technologies (solar thermal, solar-electric, photovoltaic cells, solar-box cookers, and the like). As we near the end of the fossil fuel era, these available technologies will become part of the fabric of our lives. Briefly describe your results. Given these findings, determine if there is local availability of solar technology in your area.

C. With each temperature map (Figures 3.24 and 3.26), begin by finding your own city or town and noting the temperatures indicated by the isotherms for January, July, and the annual temperature range. Record the information from these maps in your notebook. As you work through the different maps throughout this text, note atmospheric pressure and winds, annual precipitation, climate type, landforms, soil orders, vegetation, and terrestrial biomes. By the end of the course you will have a complete physical geography profile for your regional environment.

The winds over Lake Michigan fill the sails of this four-masted sailing schooner, a class "B" tall ship, just offshore from Chicago. Serious questions arise about the feasibility of again harvesting the winds to assist in powering the world's merchant fleet and oil tankers. Estimates of fuel savings range from 15% to 50%. [Photo by Bobbé Christopherson.]

4 Atmospheric and Oceanic Circulation

Key Learning Concepts

By knowing and understanding the key learning concepts in this chapter, you should be able to:

- *Define* the concept of air pressure and *portray* the pattern of global pressure systems on isobaric maps.
- *Define* wind and *describe* how wind is measured, how wind direction is determined, and how winds are named.
- *Explain* the four driving forces within the atmosphere—gravity, pressure gradient force, Coriolis force, and friction force—and *describe* the primary high- and low-pressure areas and principal winds.
- *Describe* upper-air circulation and its support role for surface systems and *define* the jet streams.
- *Explain* several types of local winds: land-sea breezes, mountain-valley breezes, katabatic winds, and the regional monsoons.
- *Describe* the basic pattern of Earth's major surface and deep ocean currents.

Early in April 1815, on the island Sumbawa in present-day Indonesia, the volcano Tambora erupted violently, releasing an estimated 150 km^3 (36 mi^3) of material, 25 times the volume produced by the 1980 Mount St. Helens eruption in Washington State. Some materials from Tambora—the ash and acid mists—were carried worldwide by global atmospheric circulation, creating a stratospheric dust veil. The result was both a higher atmospheric albedo and absorption of energy by the debris injected into the stratosphere. Beautiful, colorful sunrises and sunsets resulting from the spreading dust were seen worldwide for several years after the eruption. Scientists in 1815 lacked the remote-sensing capability of satellite technology and had no way of knowing the global impact of Tambora's eruption.

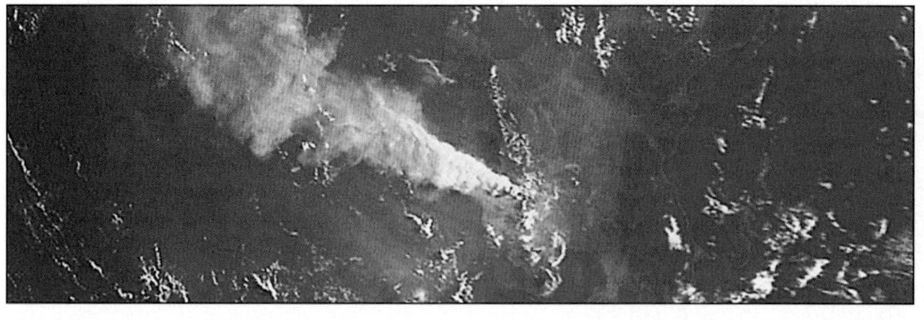

(a)

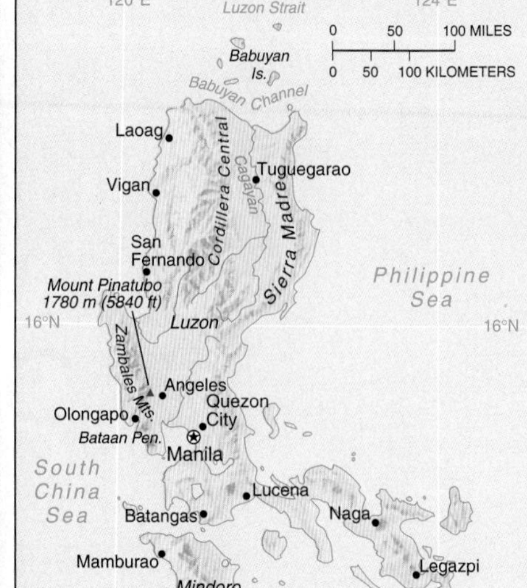

Luzon Island, Philippines

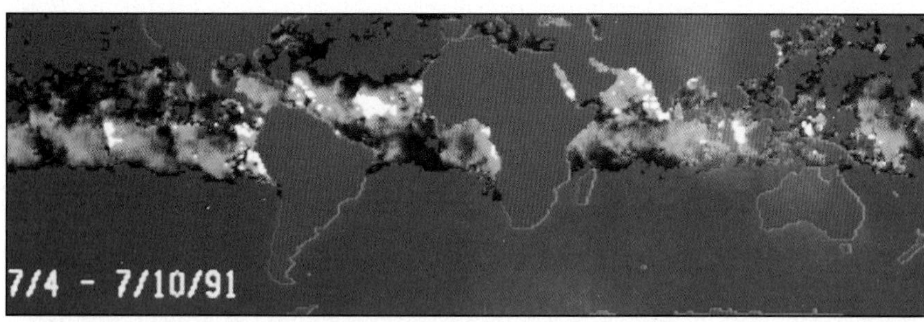

(b)

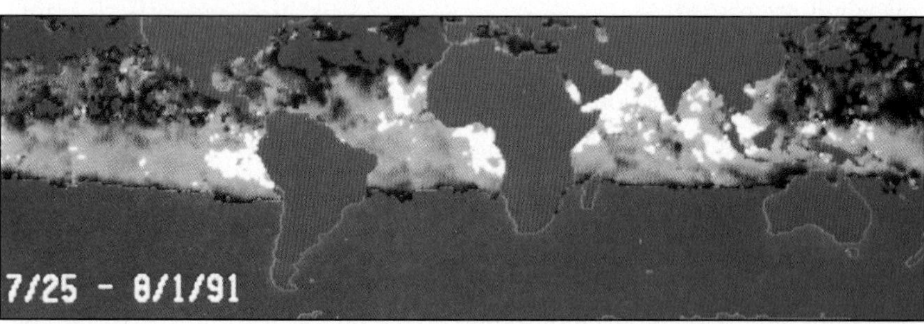

(c)

(d)

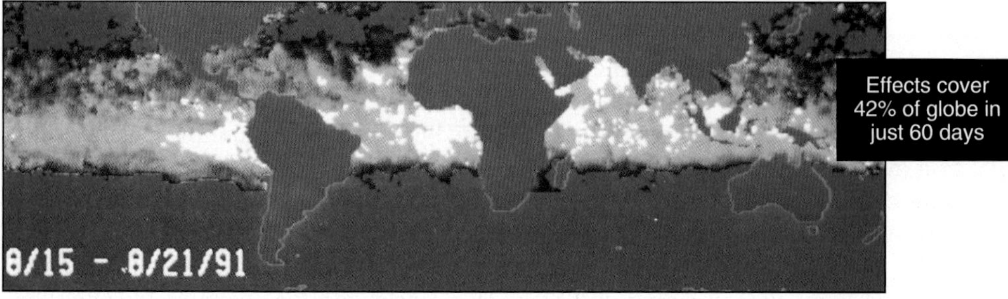

(e)

FIGURE 4.1 Volcanic debris spreads worldwide.
These false-color images show aerosols from Mount Pinatubo, smoke from fires, and dust storms, all swept about the globe by the general atmospheric circulation. Image (a) is a satellite image of the volcano as it erupted. The dramatic increase in aerosols from mid-June 1991 (b) to mid-August (e) spanned the area from 20° S to 30° N, covering about 42% of the globe. (The false-color denotes the atmosphere's aerosol optical thickness, or AOT. Lowest AOT values are dull yellow, medium values are bright yellow, and highest values are white.) [(a) AVHRR satellite eruption image courtesy of EROS Data Center; (b–e) AVHRR satellite images of aerosols courtesy of Dr. Larry L. Stowe, NESDIS, NOAA. Used by permission.]

After 635 years of dormancy, Mount Pinatubo in the Philippines erupted on June 15, 1991, an event of tremendous atmospheric impact. This volcanic explosion lofted millions of tons of ash, dust, and SO_2 (aerosols) into the atmosphere. (The volume of material blasted from Tambora was about eight times greater than that from Pinatubo.) Materials rose into the stratosphere. The increase in atmospheric albedo (about 1.3%) produced by these aerosols affects the *aerosol optical thickness* (AOT) of the atmosphere. Thus, albedo measurements by satellite gave scientists an estimate of the overall aerosol load generated by Mount Pinatubo and a unique insight into the dynamics of atmospheric circulation.

Figure 4.1 shows combined *NOAA-11* images made at 2-week intervals that clearly track the reflected solar radiation from these aerosols, carried by the atmospheric circulation. The earliest image (b), near the time of the eruption, shows winds blowing dust westward from Africa, smoke from the Kuwaiti oil well fires of the Persian Gulf War, and some haze off the East Coast of the United States. Also visible is the Pinatubo aerosol layer beginning to emerge north of Indonesia. Atmospheric winds spread the eruption debris worldwide in just 60 days. (Figure 1.5 shows this spread and its global spatial effects.)

In this chapter: we examine the dynamic circulation of Earth's atmosphere that carried Tambora's debris and Mount Pinatubo's acid aerosols worldwide and that carries the atmosphere's common ingredients, such as oxygen, carbon dioxide, and water vapor, around the globe. The driving forces that produce surface winds are pressure gradient, Coriolis, and friction. We examine the circulation of Earth's atmosphere and the patterns of global winds, including principal pressure systems and winds. We also consider Earth's wind-driven oceanic currents. The energy driving all this movement comes from one source: the Sun.

Wind Essentials

Earth's atmospheric circulation is an important transfer mechanism for both energy and mass. In the process, the imbalance between equatorial energy surpluses and polar energy deficits is partly resolved, Earth's weather patterns generated, and ocean currents produced. Human-caused pollution also is spread by this circulation, far from its points of origin. The fluid movement of the atmosphere socializes humanity more than any other natural or cultural factor. Our atmosphere makes all the world a spatially linked society—one person's or nation's exhalation is another's breath intake.

Atmospheric circulation is generally categorized at three levels: *primary circulation* (general), *secondary circulation* of migratory high-pressure and low-pressure systems, and *tertiary circulation*, which includes local winds and temporal weather patterns. Winds that move principally north or south along meridians are known as *meridional flows*. Winds moving east or west along parallels of latitude are called *zonal flows*.

Air Pressure and Its Measurement

Important to an understanding of wind is the concept of air pressure and its measurement and expression. The gases that constitute air create pressure through their motion, size, and number. This pressure is exerted on all surfaces in contact with the air. The weight of the atmosphere, or **air pressure**, crushes in on all of us. Fortunately, that same pressure also exists inside us, pushing outward; otherwise we would be smashed by the mass of air around us.

In A.D. 1643, a pupil of Galileo's named Evangelista Torricelli was working on a mine drainage problem. This led him to discover a method for measuring air pressure (Figure 4.2a). He knew that pumps in the mine were able to draw water upward only about 10 m (33 ft) but did not know why. Careful observation led him to discover that this problem was not caused by the pumps but by the atmosphere itself. Torricelli noted that the water level in the vertical pipe fluctuated from day to day. He thought correctly that air pressure varied over time and with changing weather conditions. A pump of the type shown works by creating a vacuum above the water, which allows the pressure of the atmosphere that is pushing down on water in the mine to force water up the pipe.

To simulate the mine-pump situation, Torricelli devised an instrument at Galileo's suggestion. Instead of water, he used a much denser fluid, mercury (Hg), so that he could use a glass laboratory tube only 1 m high. He sealed the glass tube at one end, filled it with mercury, and then inverted it into a dish of mercury (Figure 4.2b). Torricelli established that the average height of the column of mercury in the tube was 760 mm (29.92 in.). He concluded that the column of mercury was counterbalanced by the mass of surrounding air exerting an equivalent pressure on the mercury in the dish. Thus, the 10 m (33 ft) limit to which the mine pumps could draw water was controlled by the mass of the atmosphere. Using similar instruments today, *normal sea level pressure* is expressed as

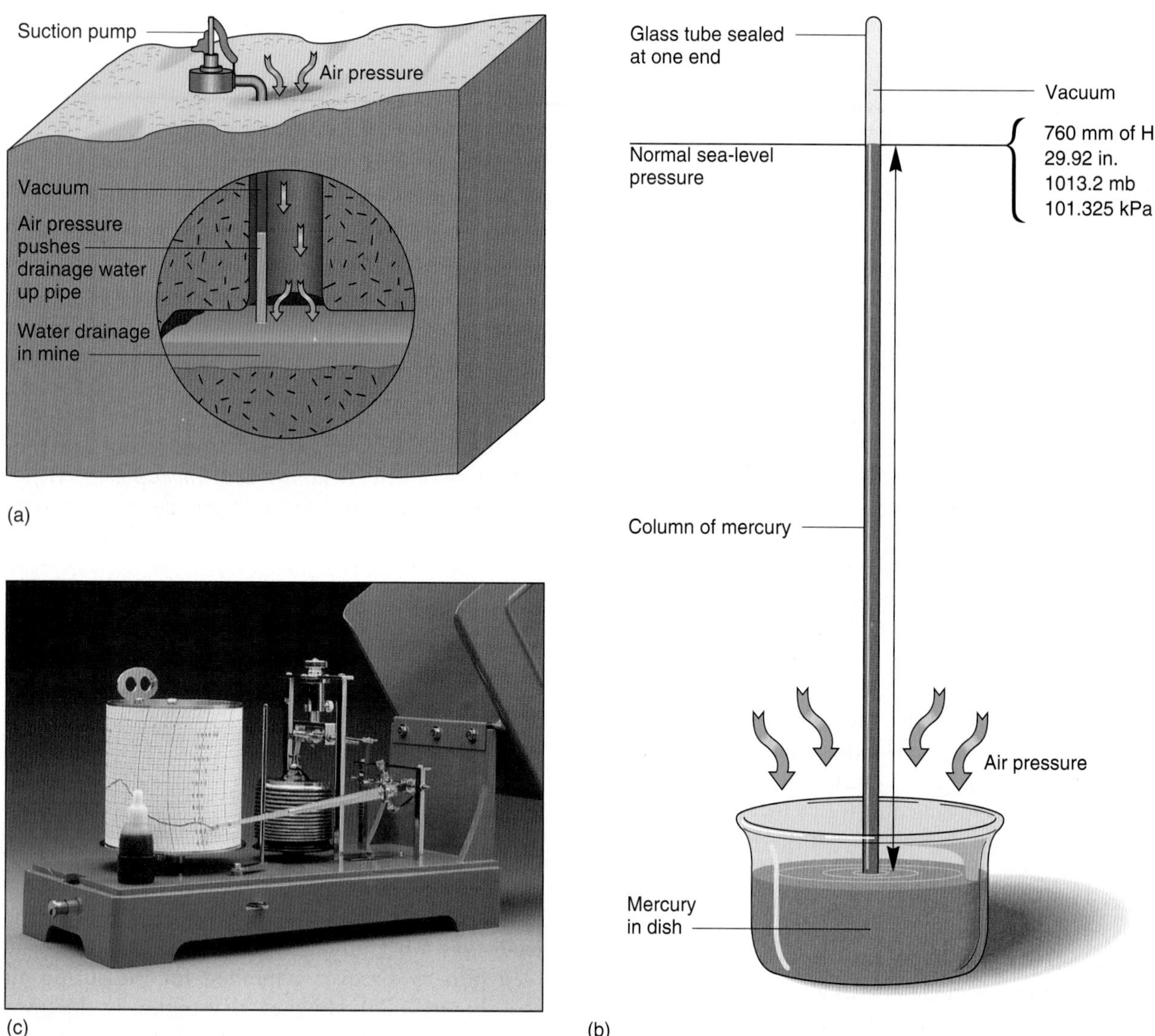

FIGURE 4.2 Developing a barometer.
Evangelista Torricelli developed the barometer to measure air pressure as a by-product of trying to solve a mine drainage problem (a). Instruments used to measure atmospheric pressure include an idealized model of a mercury barometer (b) and an aneroid barometer (c). Have you used a barometer? If so, what type was it? [(c) Photo courtesy of Qualimetrics, Inc., Sacramento, Calif.]

1013.2 millibars of mercury (a way of expressing force per square meter of surface area), 29.92 in. of mercury, or 101.32 kPa (kilopascal; 1 kilopascal = 10 millibars).

Any instrument that measures air pressure is a *barometer*. Torricelli developed a **mercury barometer**. A more compact design that works without a meter-long tube of mercury is called an **aneroid barometer** (Figure 4.2c). Inside this type of barometer, a small chamber is partially emptied of air, sealed, and connected to a mechanism that is sensitive to changes in the chamber. As air pressure varies, the mechanism responds to the difference and moves the needle on the dial.

Figure 4.3 illustrates comparative scales in millibars and inches used to express air pressure and its relative force. Strong high pressure to deep low pressure can range from 1050 to 980 mb. The figure also indicates the extreme highest and lowest pressures ever recorded in the United States, Canada, and on Earth.

Wind: Description and Measurement

Simply stated, **wind** is the horizontal motion of air across Earth's surface. Turbulence adds wind updrafts and downdrafts and a vertical component to this definition. *Differences in air pressure (density) between one location and another produce wind.* Wind's two principal properties are speed and direction, and instruments are used to measure each. Wind speed is measured with an **anemometer** and is expressed in kilometers per hour (kmph), miles per hour (mph), meters per second (mps), or knots. (A *knot* is

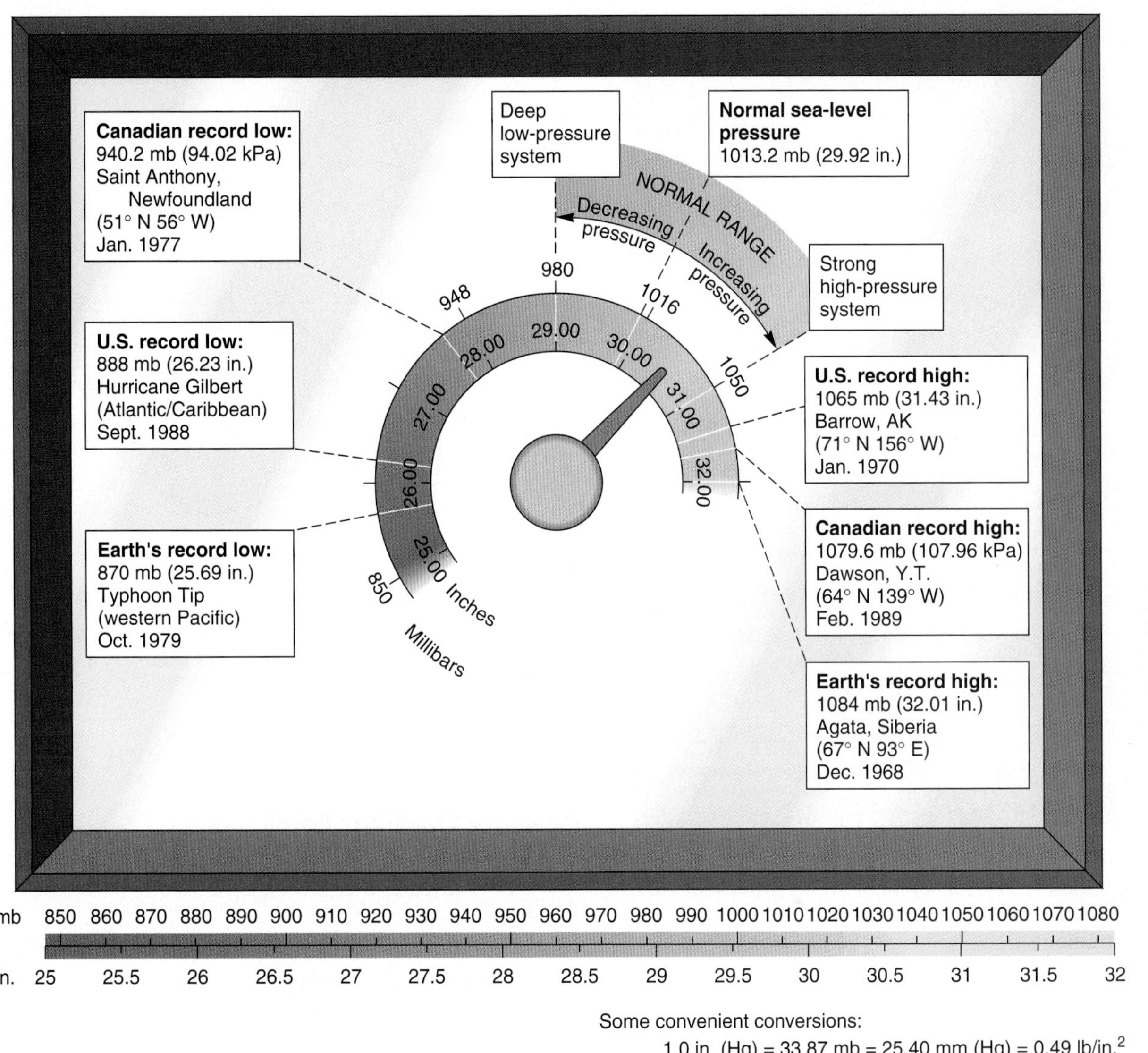

FIGURE 4.3 Air pressure readings and conversions.
Scales for expressing barometric air pressure in millibars and inches, with average air pressure values and recorded pressure extremes. Canadian values include kilopascal equivalents (10 mb = 1 kPa).

a nautical mile per hour, equivalent to 1.85 kmph, or 1.15 mph.) Wind direction is determined with a **wind vane**; the standard measurement is taken 10 m (33 ft) above the ground to avoid local effects of topography on wind direction (Figure 4.4).

Winds are named for the direction *from which they originate*. For example, a wind from the west is a *westerly* wind (it blows eastward); a wind out of the south is a *southerly* wind (it blows northward). Figure 4.5 illustrates a simple wind compass, naming the 16 principal wind directions.

Global Winds

The primary circulation of winds across Earth has fascinated travelers, sailors, and scientists for centuries, although only in the modern era is a true picture emerging of the pattern of global winds. A remarkable NASA *Seasat* satellite image combining 150,000 measurements made during a single day portrays surface winds (Figure 4.6). Wind drives waves on the ocean surface that combine to form patterns of currents. *Seasat* used radar to measure the motion and direction of waves from their backscatter to the satellite, producing this portrait of surface wind fields over the Pacific Ocean.

The patterns in the figure are the result of specific forces at work in the atmosphere: *pressure gradient force, Coriolis force, friction force*, and *gravity*. These forces are discussed next. As we progress through this chapter, you may want to refer to this *Seasat* image to identify the winds, eddies, and vortices it portrays.

FIGURE 4.4 Wind vane and anemometer.
Instruments used to measure wind direction (wind vane, flat blade) and wind speed (anemometer, three cups). [Photo by Belfort Instruments.]

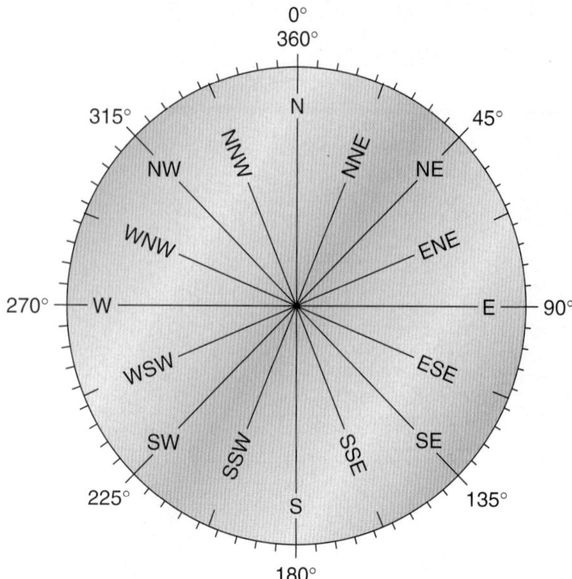

FIGURE 4.5 A wind compass.
Sixteen wind directions identified. Winds are named for the direction from which they originate. For example, a wind from the west is a westerly wind.

Driving Forces within the Atmosphere

Several forces determine both speed and direction of winds:

- Earth's *gravitational force* on the atmosphere is practically uniform, equally compressing the atmosphere near the ground worldwide. Density decreases as altitude increases.
- The **pressure gradient force** drives air from areas of higher barometric pressure to areas of lower barometric pressure, causing winds.
- The **Coriolis force**, a deflective force, appears to deflect wind from a straight path in relation to Earth's rotating surface—to the right in the Northern Hemisphere and to the left in the Southern Hemisphere.
- The **friction force** drags on the wind as it moves across surfaces; it decreases with height above the surface.

All four of these forces operate on moving air and water and affect the circulation patterns of global winds. The following sections describe the actions of the pressure gradient, and Coriolis and friction forces.

Pressure Gradient Force

Cyclone and anticyclones

High- and low-pressure areas that exist in the atmosphere result principally from unequal heating at Earth's surface and from certain dynamic forces in the atmosphere. A pressure gradient is the difference in atmospheric pressure between areas of higher pressure and lower pressure. An **isobar** is an *isoline* (a line along which there is a constant value) plotted on a weather map to connect points of equal pressure. The distance between isobars indicates the degree of pressure difference. Isobars on a weather map are used to make a spatial analysis of pressure patterns (Figure 4.7).

Just as closer contour lines on a topographic map mark a steeper slope, so do closer isobars denote steepness in the pressure gradient. In Figure 4.7a, note the spacing of the isobars (green lines). A steep gradient causes faster air movement from a high-pressure area to a low-pressure area. Isobars spaced at greater distances from one another mark a more gradual pressure gradient, one that creates a slower air flow. Along a horizontal surface, the pressure gradient force acts at right angles to the isobars. Note the location of steep and gradual pressure gradients and their relationship to wind intensity on the map in Figure 4.7b.

Figure 4.8 illustrates the combined effect of the forces that direct wind. The pressure gradient force acting alone is shown in Figure 4.8a from two perspectives. As air descends in the high-pressure area, a field of subsiding, or sinking, air develops. Air moves out of the high-pressure area in a flow described as diverging. High-pressure areas feature *descending, diverging* air flows. Conversely, air moving into a

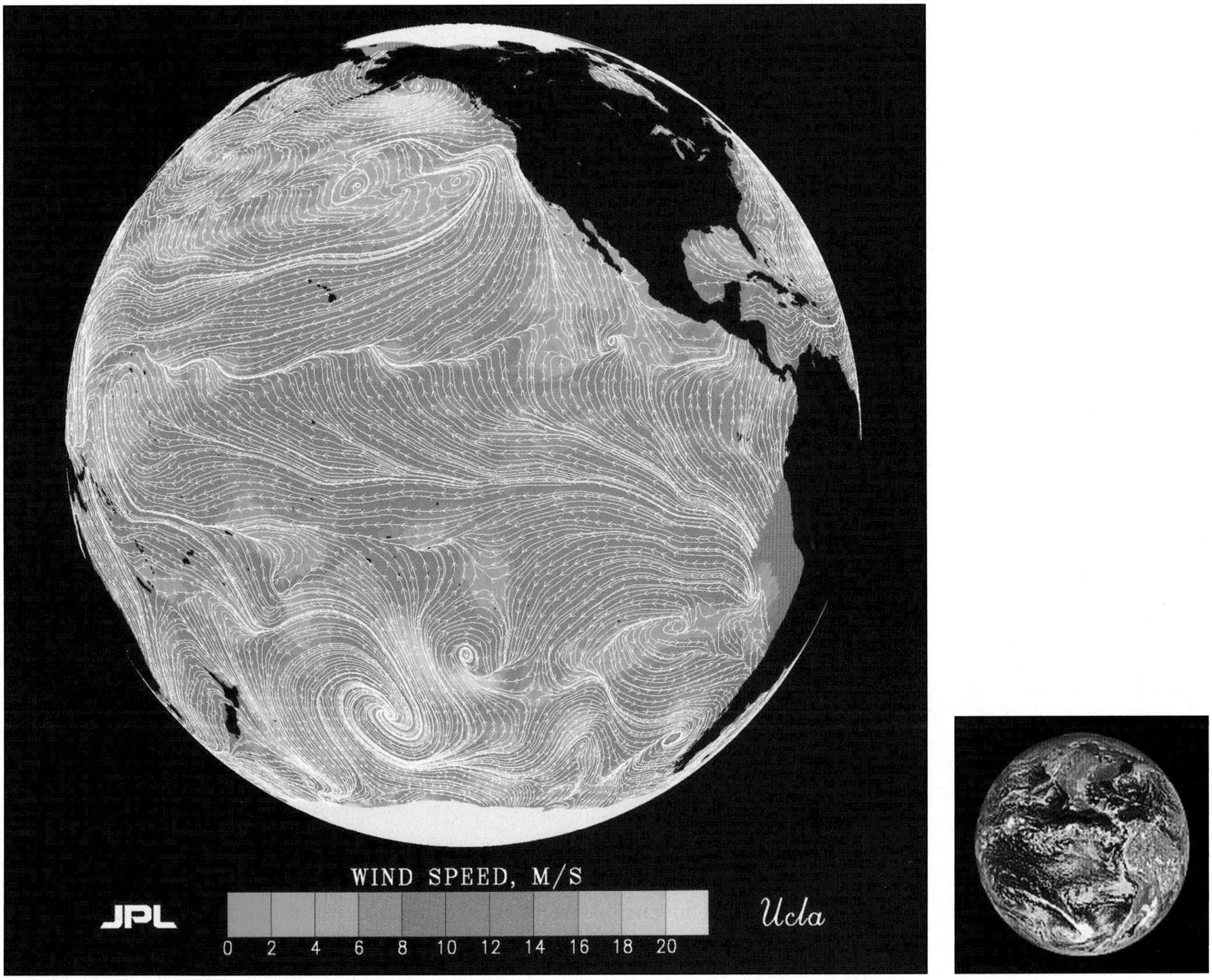

(a)

(b)

FIGURE 4.6 Wind portrait of the Pacific Ocean.
(a) Surface wind measured by radar scatterometer aboard the *Seasat* satellite on September 14, 1978. Scientists analyzed 150,000 measurements to produce this image. Colors correlate to wind speeds (see scale), and the white arrows note wind direction. Can you identify the pattern of trade winds, westerlies, high-pressure cells, and low-pressure cells from the cloud patterns on the image (b)? [*Seasat* Image courtesy of Dr. Peter Woiceshyn, Jet Propulsion Laboratory, Pasadena. Satellite image inset from Cornell University.]

low-pressure area does so with a converging flow. Thus, low-pressure areas feature *converging, ascending* air flows.

Coriolis Force

Coriolis Force

You might expect surface winds to move in a straight line from areas of higher pressure to areas of lower pressure, but they do not. The reason for this is the Coriolis force. The *Coriolis force* deflects from a straight path any object that flies or flows across Earth's surface, be it wind, an airplane, or ocean currents. The deflection produced by the Coriolis force is a product of Earth's rotation. Because Earth rotates eastward, objects that move in an absolute straight line over a distance (such as winds and ocean currents) appear to curve to the right in the Northern Hemisphere and to the left in the Southern Hemisphere.*

Earth's rotational speed varies with latitude, increasing from 0 kmph at the poles to 1675 kmph (1041 mph) at the equator (noted on Figure 4.9b). This means that materials at different latitudes possess different amounts of angular momentum. The Coriolis force is zero along the equator, increasing to half the maximum deflection at

*Note that the text calls this a force. The label *force* is appropriate because, as the physicist Sir Isaac Newton (1643–1727) stated, when something is accelerating over a space, a force is in operation. This apparent force (in classical mechanics, an inertial force) acts as an effect on moving objects. It is named for Gaspard Coriolis, a French mathematician and researcher of applied mechanics, who first described the phenomenon in 1831.

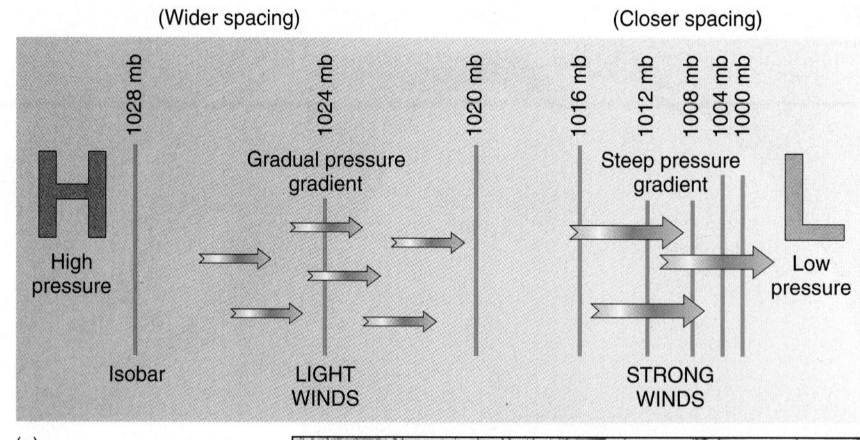

(a)

(b)

FIGURE 4.7 Pressure gradient determines wind speed. (a) Pressure gradient portrait. (b) On a weather map the closer spacing of isobars represents a steeper pressure gradient, which produces strong winds; wider spacing of isobars denotes a gradual pressure gradient, which leads to light winds.

30° N and S latitudes, and to maximum deflection flowing away from the poles (Figure 4.9a).

Another example of the effect of this force: A pilot leaves the North Pole and flies due south toward Quito, Ecuador, traveling a direct route (Figure 4.9b). If Earth were not rotating, the aircraft would simply travel along a meridian of longitude and arrive at Quito. However, Earth is rotating beneath the aircraft's flight path—slower where the plane started and increasingly faster toward Quito. If the pilot does not allow for this rotation, the plane will reach the equator over the ocean far to the west of the intended destination (Quito has a greater angular momentum than the plane). Pilots must correct for this Coriolis deflection in their navigation calculations. The deflection occurs regardless of the direction in which the object is moving (Figure 4.9c).

How does the Coriolis force affect wind aloft? As air rises from the surface through the lowest levels of the atmosphere, leaving the drag of surface friction behind, its speed increases, thus increasing the Coriolis force, spiraling the winds to the right in the Northern Hemisphere or to the left in the Southern Hemisphere. In the upper troposphere, the Coriolis force just balances the pressure gradient force. Consequently, the winds between high-pressure and low-pressure areas aloft flow parallel to the isobars.

Figure 4.8b illustrates the combined effect of the pressure gradient force and the Coriolis force on air currents aloft, producing **geostrophic winds**. Geostrophic winds are characteristic of upper tropospheric circulation. The air does not flow directly from high to low but around the pressure areas instead, remaining parallel to the isobars and producing the characteristic pattern shown on the upper-air weather chart in Figure 4.10. Note the inset illustration showing the effects of the pressure gradient and Coriolis forces that produce a geostrophic flow of air.

Friction Force

The effect of surface friction extends to a height of about 500 m (1650 ft) and varies with surface texture, wind speed, time of day and year, and atmospheric conditions. Generally, rougher surfaces produce more friction. Figure 4.8c adds the effect of friction to the Coriolis and pressure gradient forces on wind movements.

Near the surface, friction disrupts the equilibrium established in geostrophic wind flows between the pressure gradient and Coriolis forces—note the inset illustration in Figure 4.8c. Because surface friction decreases wind speed, it reduces the effect of the Coriolis force and causes winds in the Northern Hemisphere to move across isobars at an angle, spiraling out from a high-pressure area clockwise to form an **anticyclone** and spiraling into a low-pressure area counterclockwise to form a **cyclone**. In the Southern Hemisphere these circulation patterns are reversed, flowing out of high-pressure cells counterclockwise and into low-pressure cells clockwise.

FIGURE 4.8 Three physical forces that produce winds. (a) The pressure gradient force. (b) The Coriolis force added to the effects of the pressure gradient force produces a geostrophic wind flow in the upper atmosphere. And (c) the friction force considered together with the other forces produces characteristic surface winds. Note the reverse circulation pattern in the Southern Hemisphere. The three inset diagrams show the interaction of forces that form prevailing wind flow, geostrophic, and surface winds. (The gravitational force is assumed.)

Top view and side view of air movement in an idealized high-pressure area and low-pressure area on a nonrotating Earth.

High
Low
Top view
Isobar Isobar
Pressure gradient force
H
L
High pressure
Low pressure
Side view
High
Low
Earth's surface
Descending, diverging
Ascending, converging
(a) Pressure gradient force alone

Earth's rotation adds the Coriolis force and a "twist" to air movements. High-pressure and low-pressure areas develop a rotary motion, and wind flowing between highs and lows flows parallel to isobars.

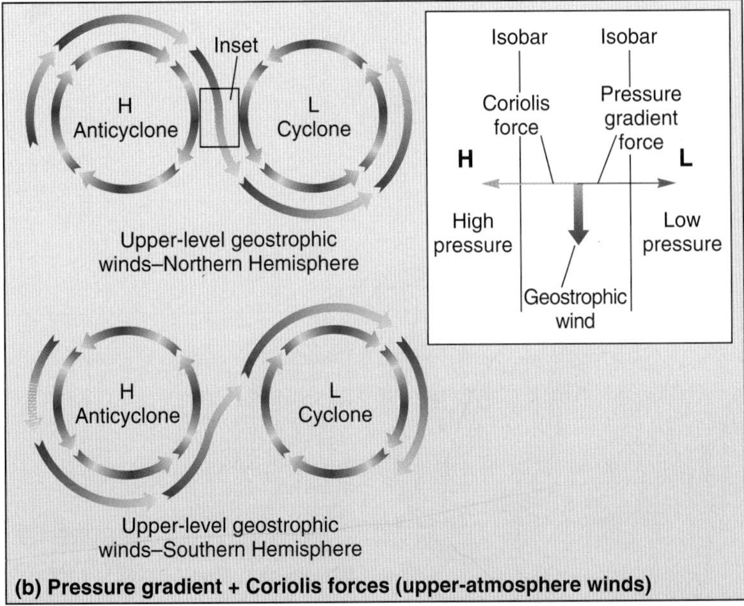

(b) Pressure gradient + Coriolis forces (upper-atmosphere winds)

Surface friction adds a countering force to Coriolis, producing winds that spiral out of a high-pressure area and into a low-pressure area. Surface winds cross isobars at an angle.

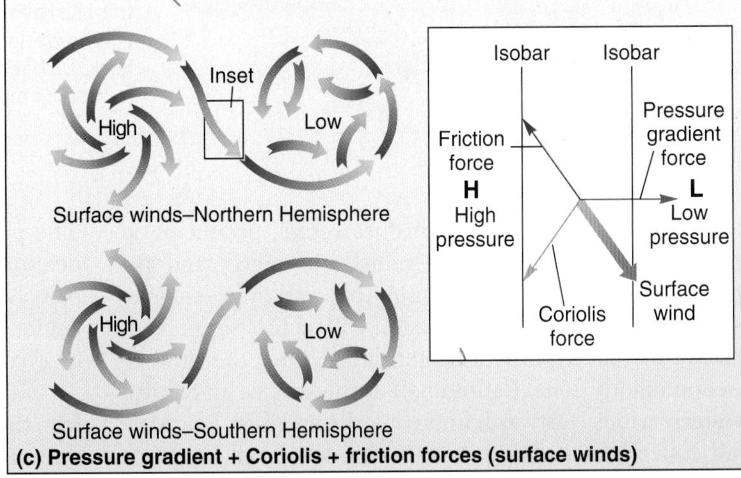

(c) Pressure gradient + Coriolis + friction forces (surface winds)

Atmospheric Patterns of Motion

SATELLITE LOOP
Global infrared

ANIMATION
Global Wind Circulation, Hadley Cells

With these forces and motions in mind, we are ready to build a general model of total atmospheric circulation and better understand the earlier *Seasat* image (see Figure 4.6). If Earth did not rotate, then the warmer, less dense air of the equatorial regions would rise, creating low pressure at the surface, and the colder and denser air of the poles would sink, creating high pressure at the surface. The net effect would be a meridional flow of winds established by this simple pressure gradient.

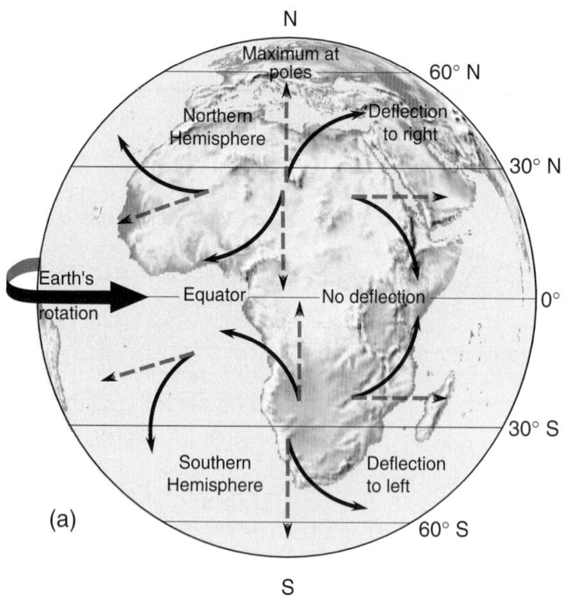

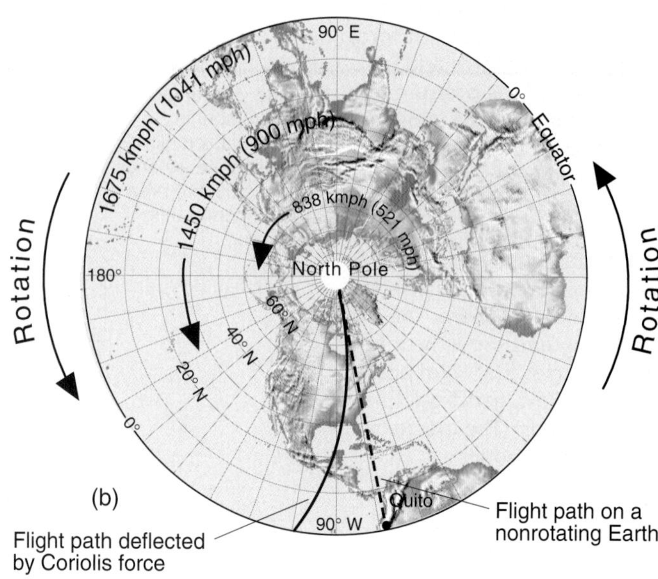

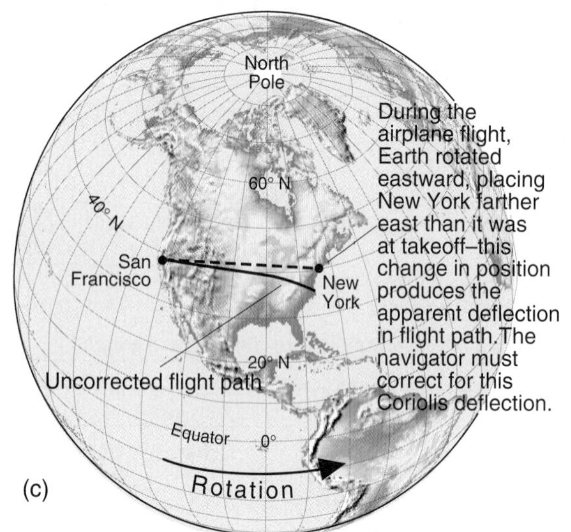

FIGURE 4.9 The Coriolis force—an apparent deflection. (a) Apparent deflection to the right of a straight line in the Northern Hemisphere and apparent deflection to the left in the Southern Hemisphere. (b) Coriolis deflection of a flight path between the North Pole and Quito, Ecuador, which is on the equator. (c) The deflection of a flight path between San Francisco and New York. Deflection from a straight path occurs regardless of the direction of movement.

However, in reality, Earth does rotate, producing a more complex system for the transfer of energy and air masses from equatorial surpluses to polar deficits—one with waves, streams, and eddies on a planetary scale. And instead of this hypothetical meridional flow, the flow is predominantly zonal (latitudinal) at the surface and aloft: westerly winds (eastward) in the middle and high latitudes and easterly winds (westward) in the equatorial latitudes of both hemispheres.

Primary High-Pressure and Low-Pressure Areas

ANIMATION **Global Patterns of Pressure**

The following discussion of Earth's pressure and wind patterns refers often to Figure 4.11, which shows January and July isobaric maps of average surface barometric pressure. The primary high- and low-pressure areas on Earth appear on these isobaric maps as interrupted cells or uneven belts of similar pressure that stretch across the face of the planet. Between these areas flow the primary winds, noted in adventure and myth throughout human experience.

Secondary highs and lows, from a few hundred to a few thousand kilometers in diameter and hundreds to thousands of meters high, are formed within the primary pressure areas. These pressure systems seasonally migrate to produce changing weather and climate patterns in the regions over which they pass.

Four identifiable pressure areas cover the Northern Hemisphere; a similar set is found in the Southern Hemisphere. In each hemisphere, two of the pressure areas are specifically stimulated by thermal (temperature) factors: the **equatorial low-pressure trough** and the weak **polar high-pressure cells**, both north and south. The other two areas are formed by dynamic factors: the **subtropical high-pressure cells** and the **subpolar low-pressure**

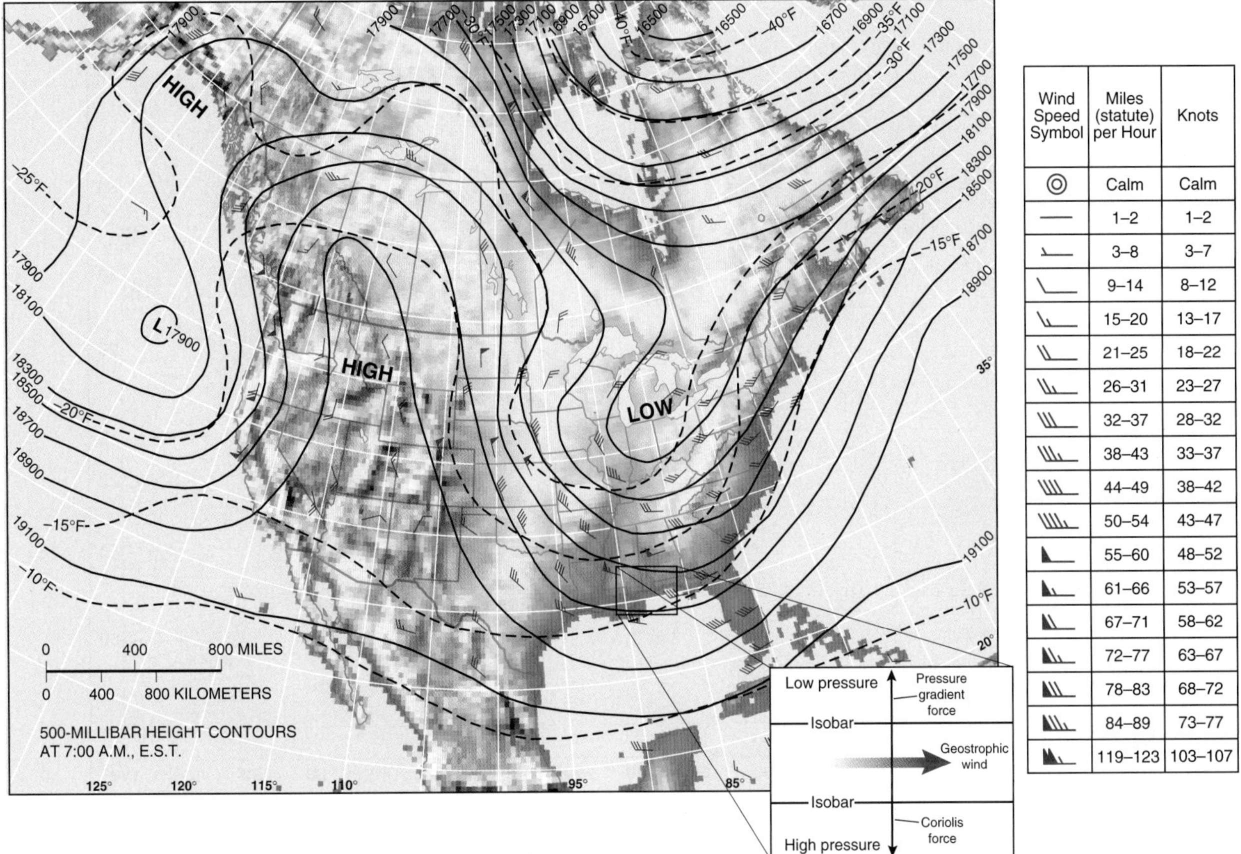

Wind Speed Symbol	Miles (statute) per Hour	Knots
◎	Calm	Calm
	1–2	1–2
	3–8	3–7
	9–14	8–12
	15–20	13–17
	21–25	18–22
	26–31	23–27
	32–37	28–32
	38–43	33–37
	44–49	38–42
	50–54	43–47
	55–60	48–52
	61–66	53–57
	67–71	58–62
	72–77	63–67
	78–83	68–72
	84–89	73–77
	119–123	103–107

FIGURE 4.10 Geostrophic winds on a 500-mb map.
Isobaric chart for an April day. Contours show height at which 500 mb pressure occurs (in feet). The pattern of contours reveals geostrophic wind patterns in the troposphere at approximately 5500 m (18,000 ft) altitude. See the "ridge" of high pressure over the Intermountain West and the "trough" of low pressure over the Great Lakes region. Note the inset diagram showing the interaction of forces that form prevailing geostrophic winds. [Data provided by National Weather Service, NOAA.]

cells. Table 4.1 summarizes the characteristics of these pressure areas, and we now examine each.

Equatorial Low-Pressure Trough: Clouds and Rain. The equatorial low-pressure trough is an elongated, narrow band of low pressure that nearly girdles Earth, following an undulating linear axis. Constant high Sun altitude and consistent daylength make large amounts of energy available in this region throughout the year. The warming creates lighter, less dense, ascending air, with winds converging all along the extent of the trough.

This converging air is extremely moist and full of latent heat energy. Vertical cloud columns frequently reach the tropopause, and precipitation is heavy throughout this zone. The combination of heating and convergence, forces air aloft and forms the **intertropical convergence zone (ITCZ)**. Figure 4.12 is a satellite image of Earth showing the equatorial low-pressure trough. The ITCZ is identified by bands of clouds associated with the convergence of winds along the equator—compare this with the ITCZ plot on the December global pressure map in Figure 4.11.

During summer, a marked wet season accompanies the shifting ITCZ over various regions. The maps in Figure 4.11 show the location of the ITCZ in January and July. In January, the zone crosses northern Australia and dips southward in eastern Africa and South America, whereas by July the zone has shifted northward in the Americas and as far north as Pakistan and southern Asia.

Figure 4.13 shows two views of this circulation and resulting *Hadley cells*, named for the eighteenth-century English scientist who described the *trade winds*. This equatorial low-pressure system of converging, ascending air generates such cells. These winds converging on the equatorial low-pressure trough are known as the *northeast trade winds* (in the Northern Hemisphere) and *southeast trade winds* (in the Southern Hemisphere), or generally as **trade winds**.

The trade winds pick up large quantities of moisture as they return through the Hadley circulation cell for another cycle of uplift and condensation. Within the ITCZ, winds are calm or mildly variable because of the even pressure gradient and the strong vertical ascent of air. Sixteenth-century sailors complained about getting caught here as being in the *doldrums*.

The rising air from the equatorial low-pressure area spirals upward into a geostrophic flow to the north and south. These upper-air winds turn eastward, flowing

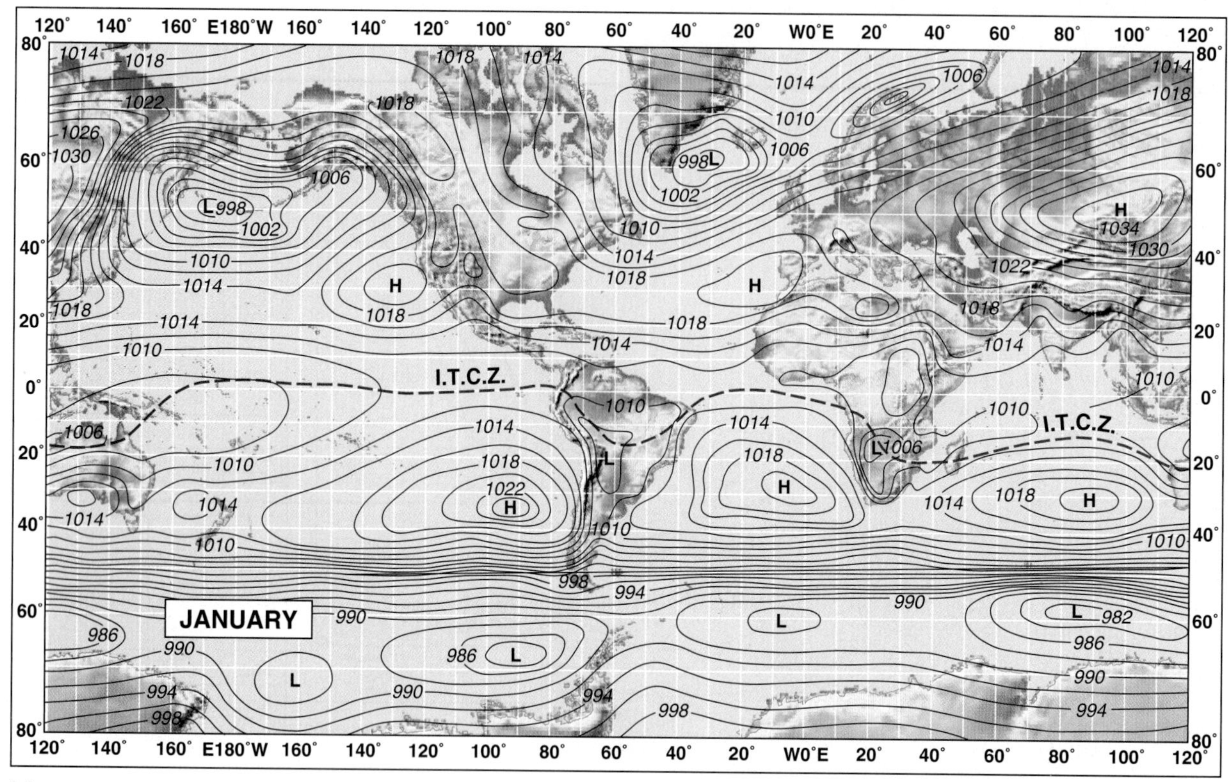

(a)

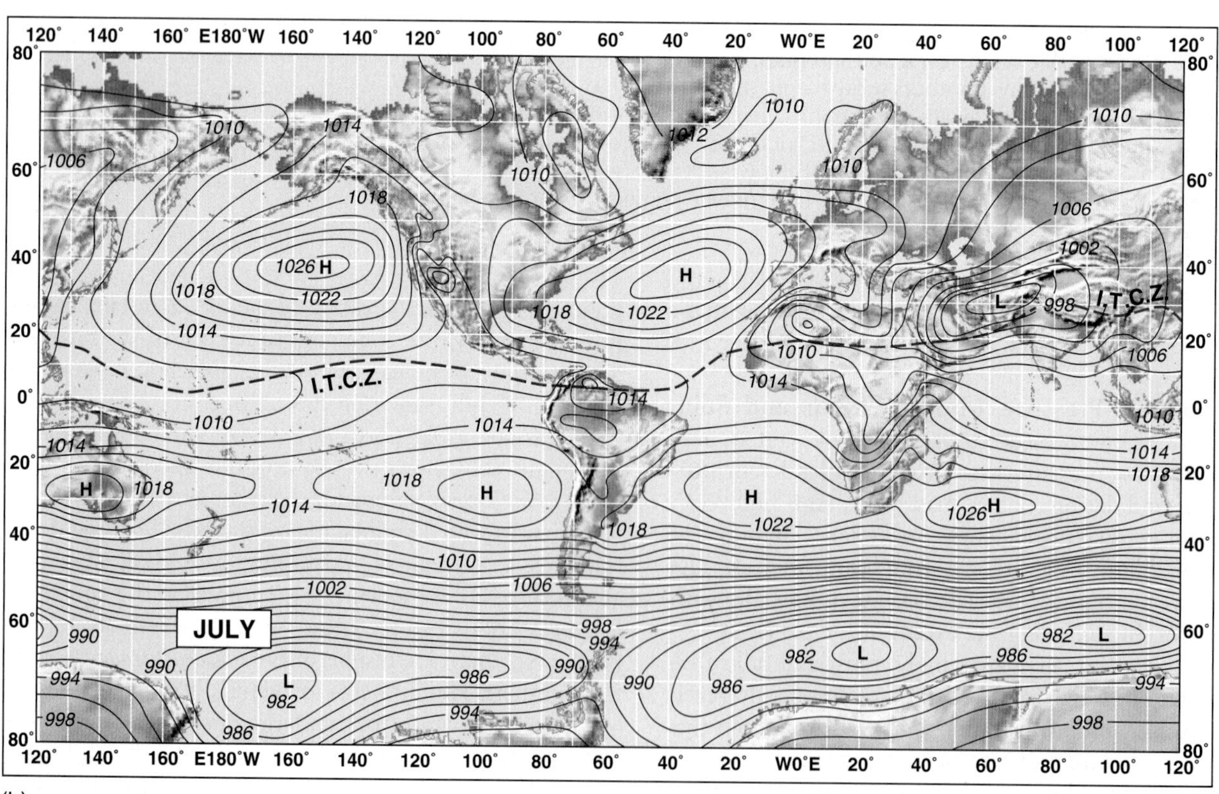

(b)

FIGURE 4.11 Global barometric pressures.
Average surface barometric pressure (millibars) for January and July. Dashed red line marks the intertropical convergence zone (ITCZ). [Adapted from National Climatic Data Center, *Monthly Climatic Data for the World*, 46, no. 1, January and July 1993. Prepared in cooperation with the WMO and NOAA.]

Table 4.1 Four Hemispheric Pressure Areas

Name	Cause	Location	Air Temperature/Moisture
Polar high-pressure cells	Thermal	90° N 90° S	Cold/dry
Subpolar low-pressure cells	Dynamic	60° N 60° S	Cool/wet
Subtropical high-pressure cells	Dynamic	20° to 35° N and S	Hot/dry
Equatorial low-pressure trough	Thermal	10° N to 10° S	Warm/wet

FIGURE 4.12 Clouds portray equatorial and subtropical circulation patterns. An interrupted band of clouds denotes the intertropical convergence zone (ITCZ), flanked by several regions dominated by subtropical high-pressure systems. This natural-color image was made by the *Galileo* spacecraft during its December 1990 flyby of Earth on its voyage to Jupiter. [Image from the late Dr. W. Reid Thompson. Used by permission.]

from west to east, beginning at about 20° N and S, forming descending anticyclonic flows above the subtropics. In each hemisphere, this circulation pattern appears most vertically symmetrical near the equinoxes of March and September.

Subtropical High-Pressure Cells: Hot, Dry Desert Air. Between 20° and 35° latitude in both hemispheres, a broad high-pressure zone of hot, dry air is evident across the globe (see Figures 4.11 and 4.13). It is indicated by the clear, frequently cloudless skies of the satellite imagery in Figure 4.12. The dynamic cause of these subtropical anticyclones is too complex to detail here, but generally they form as air in the Hadley cell descends in these latitudes, constantly supported by a dynamic mechanism in the upper-air circulation.

Air above the subtropics is mechanically pushed downward and is heated as it compresses on its descent to the surface, as illustrated in Figure 4.13. Warmer air has a greater capacity to hold water vapor than does cooler air, making this descending warm air relatively dry. The air is also dry because heavy precipitation along the equatorial circulation removed moisture.

Surface air diverging from the subtropical high-pressure cells generates Earth's principal surface winds: the westerlies and the trade winds. The **westerlies** are the dominant surface winds from the subtropics to high latitudes. They diminish somewhat in summer and are stronger in winter.

If you examine Figure 4.11, you will find several high-pressure areas. In the Northern Hemisphere, the Atlantic subtropical high-pressure cell is called the *Bermuda high* (in the western Atlantic) or the *Azores high* (when it migrates to the eastern Atlantic). The area under this subtropical high features clear, warm waters and large quantities of *Sargassum* (a seaweed), which gives the area its name—the Sargasso Sea. The *Pacific high*, or *Hawaiian high*, dominates the Pacific in July, retreating southward in January. In the Southern Hemisphere, three large high-pressure centers dominate the Pacific, Atlantic, and Indian Oceans and tend to move along parallels in shifting zonal positions.

The entire high-pressure system migrates with the summer high Sun, fluctuating about 5°–10° in latitude. The eastern (right-hand) sides of these anticyclonic systems are drier and more stable and feature cooler ocean currents than do the western sides. Thus, subtropical and midlatitude west coasts are influenced by the drier eastern side of these systems and therefore experience dry-summer conditions. In fact, Earth's major deserts generally occur within the subtropical belt and extend to the west coast of each continent except Antarctica.

The western (left-hand) sides of subtropical high-pressure cells tend to be moist and unstable, overlying warm ocean currents, as characterized in Figure 4.14. This condition causes warm, moist conditions in Hawai'i, Japan, southeastern China, and the southeastern United States. Crews lamented the hot, dry, stable air and calms in these high-pressure cells in the era of sailing ships, calling them the *horse latitudes*. The *Oxford English Dictionary* is uncertain of this negative nickname's actual meaning or origin.

Subpolar Low-Pressure Cells: Cool and Moist Air. The January map in Figure 4.11 shows two low-pressure cyclonic cells over the oceans around 60° N latitude. The

North Pacific *Aleutian low* is near the Aleutian Islands; the North Atlantic *Icelandic low* is near Iceland. Both cells are dominant in winter and weaken significantly or disappear altogether in the summer with the strengthening of high-pressure systems in the subtropics. The area of contrast between cold and warm air masses forms a contact zone known as the **polar front**, which encircles Earth at this latitude, focused in these low-pressure areas.

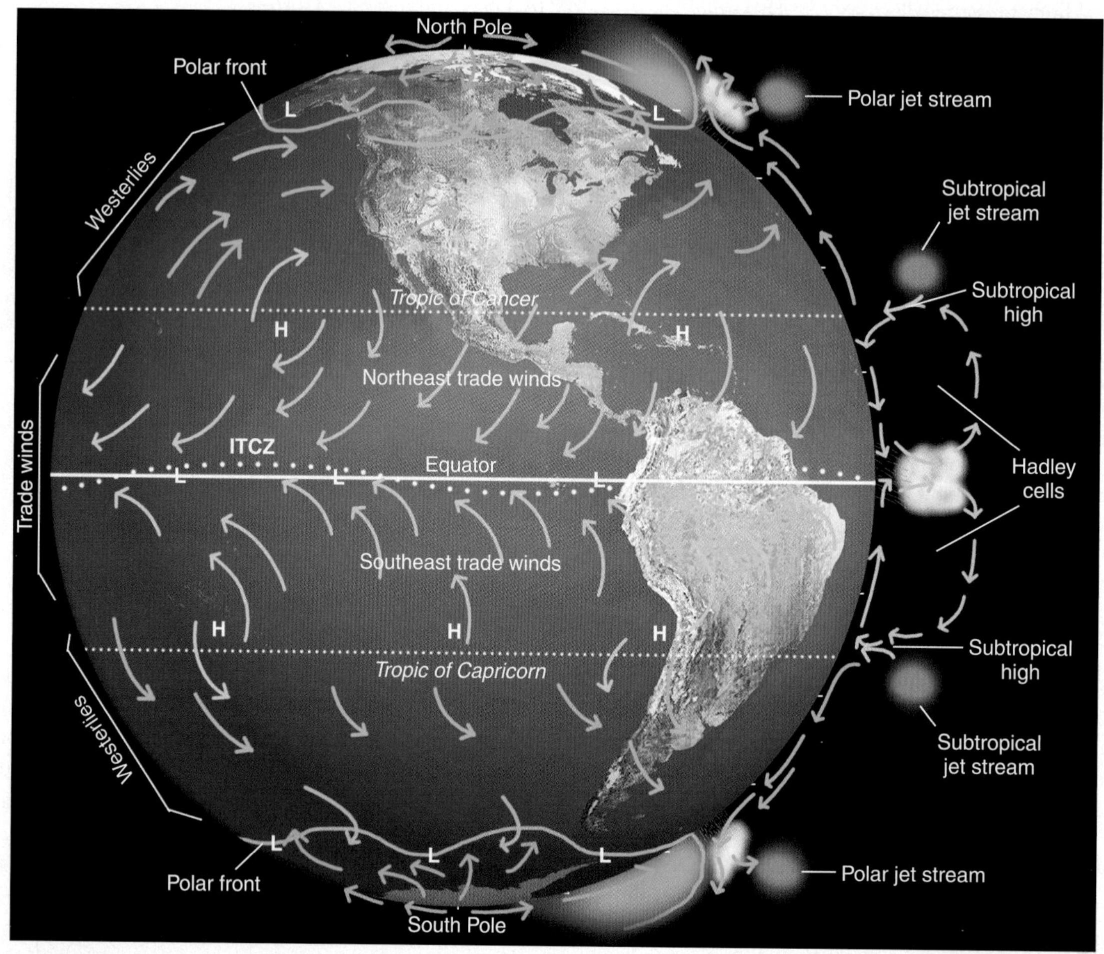

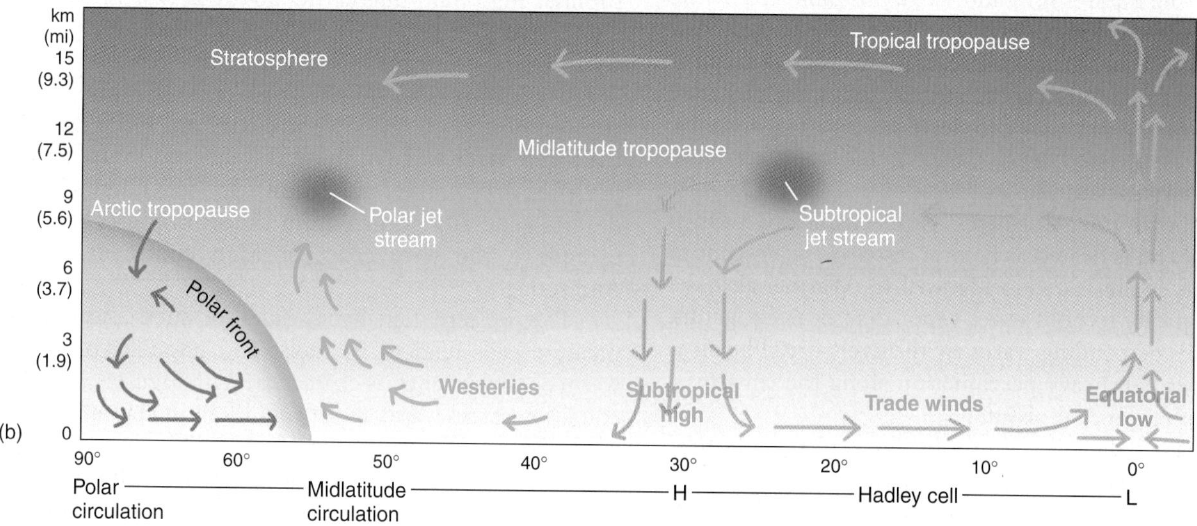

FIGURE 4.13 General atmospheric circulation model.
Two views of the general atmospheric circulation: (a) general circulation schematic; (b) equator-to-pole cross-section for the Northern Hemisphere. Both views show Hadley cells, subtropical highs, polar front, the subpolar low-pressure cells, and approximate locations of the subtropical and polar jet streams.

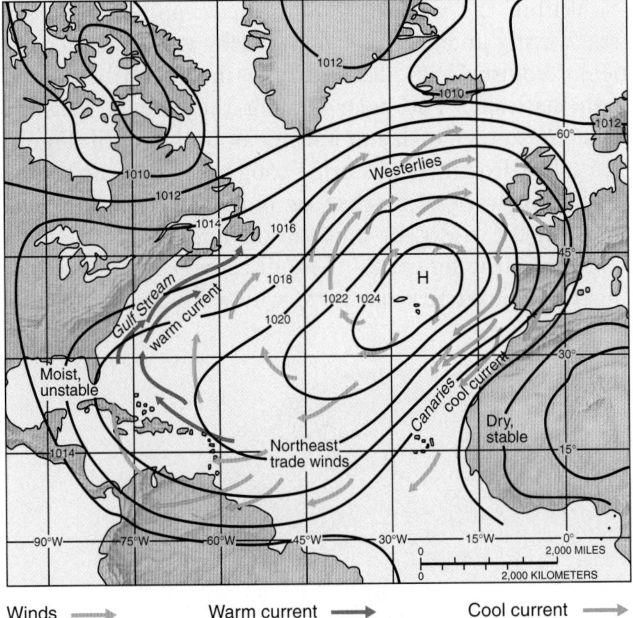

FIGURE 4.14 Subtropical high pressure system in the Atlantic.
Characteristic circulation and climate conditions related to Atlantic subtropical high-pressure anticyclone in the Northern Hemisphere. Note the clockwise wind pattern and different ocean temperatures off each coast—cool currents and deserts in Africa, warm currents and humid conditions in the southeastern U.S.

Figures 4.13a and b illustrate this confrontation between warmer, moist air from the westerlies and colder, drier air from the polar and Arctic regions. The heavier cold air displaces warmer air upward, forcing cooling and condensation in the lifted air. Low-pressure cyclonic storms migrate out of the Aleutian and Icelandic frontal areas and may produce precipitation in North America and Europe, respectively.

In the Southern Hemisphere, a noncontinuous belt of subpolar cyclonic pressure systems surrounds Antarctica. The spiraling cloud patterns produced by these cyclonic systems are visible on the satellite image in Figure 4.15. Severe cyclonic storms can cross Antarctica, producing strong winds and new snowfall, though small in amount.

Polar High-Pressure Cells: Frigid, Dry Deserts. The weakness of the polar high-pressure cell at each pole may argue against their inclusion among pressure areas on Earth. The polar atmospheric mass is small, receiving little energy to put it into motion. The variable cold, dry winds moving away from the polar region are anticyclonic, descending and diverging clockwise in the Northern Hemisphere (counterclockwise in the Southern Hemisphere) and forming the weak, variable winds of the **polar easterlies**.

Of the two polar regions, the **Antarctic high** is significant in terms of both strength and persistence. Less pronounced is the polar high-pressure cell in the Arctic. When it does form, it tends to locate over the colder northern continental areas in winter (Canadian and Siberian highs) rather than directly over the relatively warmer Arctic Ocean.

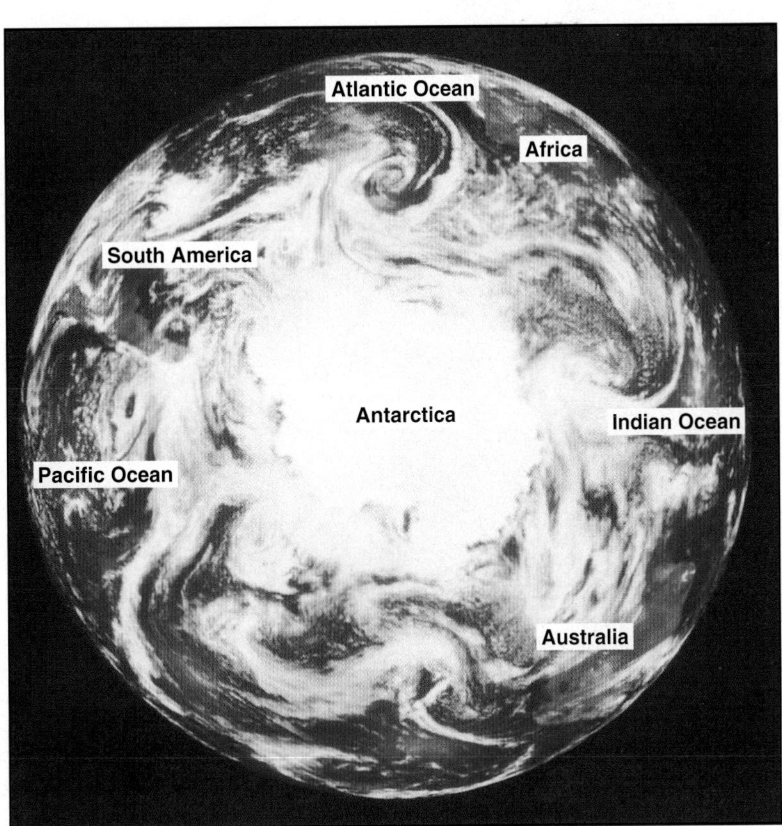

FIGURE 4.15 Clouds portray subpolar and polar circulation patterns.
Centered on Antarctica, this satellite image shows a series of subpolar low-pressure cyclones in the Southern Hemisphere. Antarctica is fully illuminated by a midsummer, December Sun. This natural color image was made by the *Galileo* spacecraft during its 1990 flyby of Earth. [Image from the late Dr. W. Reid Thompson. Used by permission.]

Upper Atmospheric Circulation

Jet Stream, Rossby waves

Middle and upper tropospheric circulation is an important component of the atmosphere's general circulation. As described previously, these upper atmosphere winds tend to flow west to east from the subtropics to the poles.

Within the westerly flow of geostrophic winds are great waving undulations called **Rossby waves**, named for meteorologist C. G. Rossby, who first described them mathematically. The polar front is the contact between colder air to the north and warmer air to the south (Figure 4.16). The Rossby waves bring tongues of cold air southward, with warmer tropical air moving northward. The development of Rossby waves in the upper-air circulation is shown in the figure. As these disturbances mature, distinct cyclonic circulations form, with warmer air and cold-

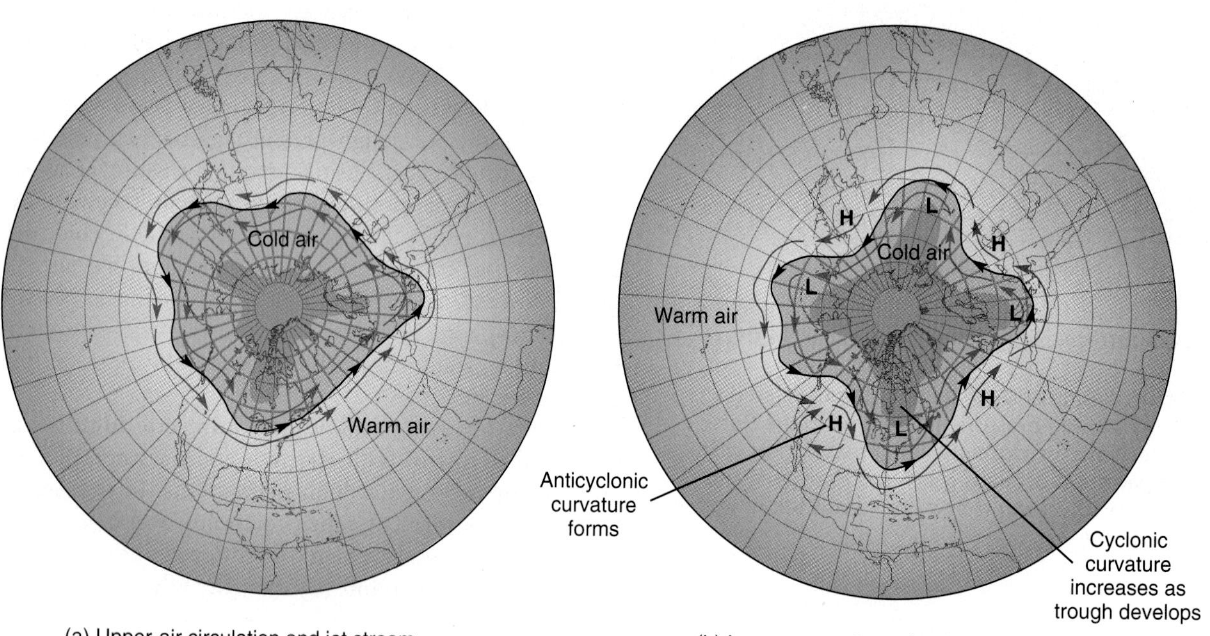

(a) Upper-air circulation and jet stream begin to gently undulate.

(b) Longwave patterns begin to form Rossby waves.

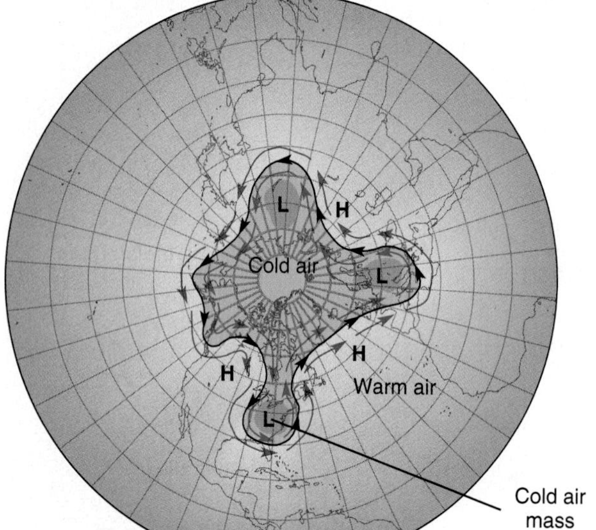

(c) Strong development of waves produces cells of cold and warm air—high-pressure ridges and low-pressure troughs.

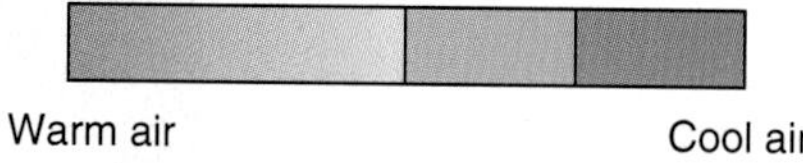

Polar stereographic projections azimuthal and conformal

FIGURE 4.16 Rossby upper-atmosphere waves. Development of longwaves in the upper-air circulation first described by C. G. Rossby in 1938 and detailed by J. Namias in 1952. [Adapted from J. Namias, NOAA.]

er air mixing along distinct fronts. These wave-and-eddy formations and other upper-air flows support the development of cyclonic storm systems at the surface. These Rossby waves develop in relation to jet stream flows.

The pattern of ridges and troughs in the upper-air wind flow is important in sustaining surface cyclonic (low) and anticyclonic (high) circulation (refer to Figure 4.10). Note the wind-speed indicators in Figure 4.10 near the ridge (over Alberta, Saskatchewan, Montana, and Wyoming); now compare these with the wind-speed indicators around the trough (over Kentucky, West Virginia, the New England states, and Maritimes). Also, note the wind relationships off the Pacific Coast.

Frequently, the upper-air wind flow generates and sustains surface pressure systems. Near ridges in the isobaric surface, winds slow and converge (pile up), supporting descending air and surface high pressure. In contrast, winds near the area of maximum wind speeds in the isobaric surface accelerate and diverge (spread out), supporting ascending air and surface lows.

Jet Streams. The most prominent movement in these upper-level westerly wind flows is the **jet stream**, an irregular, concentrated band of wind occurring at several different locations. Rather flattened in vertical cross section, the jet streams normally are 160–480 km (100–300 mi) wide by 900–2150 m (3000–7000 ft) thick, with core speeds that can exceed 300 kmph (190 mph).

The *polar jet stream* is located at the tropopause along the polar front, at altitudes between 7600 and 10,700 m (24,900–35,100 ft), meandering between 30° and 70° N latitude. Shifting of daily and seasonal locations supports surface weather patterns. The polar jet stream can migrate as far south as Texas, steering colder air masses into North America and influencing surface storm paths traveling eastward. In the summer, the polar jet stream exerts less influence on storms by staying far poleward. Figure 4.17 shows a stylized view of a polar jet stream and two jet streams over North America (see News Report 4.1 about the jet stream and airline schedules).

In subtropical latitudes, near the boundary between tropical and midlatitude air, another jet stream flows near the tropopause. This *subtropical jet stream* ranges from 9100 to 13,700 m (29,850–45,000 ft) in altitude; and although it is generally weaker, it can reach greater speeds than those of the polar jet stream. The subtropical jet stream meanders from 20° to 50° latitude and may occur over North America simultaneously with the polar jet stream. The cross section of the troposphere in Figure 4.13b depicts the general location of these two jet streams.

Local Winds

Several winds that form in response to local terrain deserve mention in our discussion of atmospheric circulation.

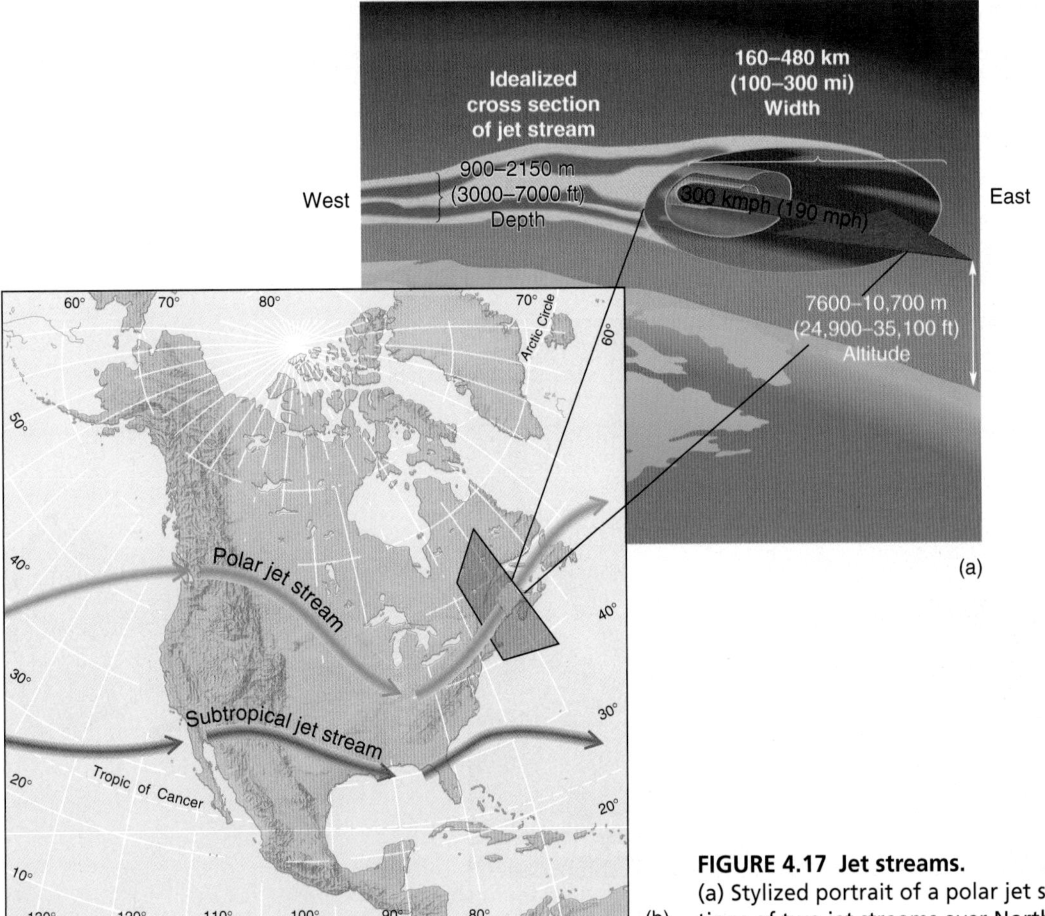

FIGURE 4.17 Jet streams.
(a) Stylized portrait of a polar jet stream. (b) Average locations of two jet streams over North America.

These local effects can, of course, be overwhelmed by weather systems passing through an area.

Land-sea breezes occur on most coastlines (Figure 4.18). The different heating characteristics of land and water surfaces create these winds (Chapter 3). During the day, land heats faster and becomes warmer than the water offshore. Because warm air is less dense, it rises and triggers an onshore flow of cooler marine air, usually stronger in the afternoon. At night, inland areas cool (radiate heat) faster than offshore waters. As a result, the cooler air over the land subsides and flows offshore over the warmer water, where the air lifts. This night pattern reverses the process that developed during the day.

Well inland from the Pacific Ocean—160 km (100 mi)—Sacramento, California, demonstrates the sea-breeze effect. The city is at 39° N and 5 m (17 ft) elevation. The average July maximum and minimum temperatures are 34°C and 14°C (93°F and 57°F). The evening cooling by the natural flow of marine air establishes a monthly mean of only 24°C (75°F), despite high daytime temperatures.

A somewhat similar exchange creates **mountain-valley breezes**. Mountain air cools rapidly at night, and valley air heats rapidly during the day (Figure 4.19). Thus, warm air rises upslope during the day, particularly in the afternoon; at night, cooler air subsides downslope into the valleys. Chapter 5 discusses another type of downslope wind.

Katabatic winds, or gravity drainage winds, are significant on a larger regional scale than mountain-valley breezes, under certain conditions. These winds usually are stronger than mountain-valley winds. An elevated plateau or highland is essential for katabatic wind, where layers of air at the surface cool, become denser, and flow downslope. The ferocious winds that can blow off the ice sheets of Antarctica and Greenland are katabatic in nature.

Worldwide, a variety of terrains produce such winds and bear many local names. The *mistral* of the Rhône

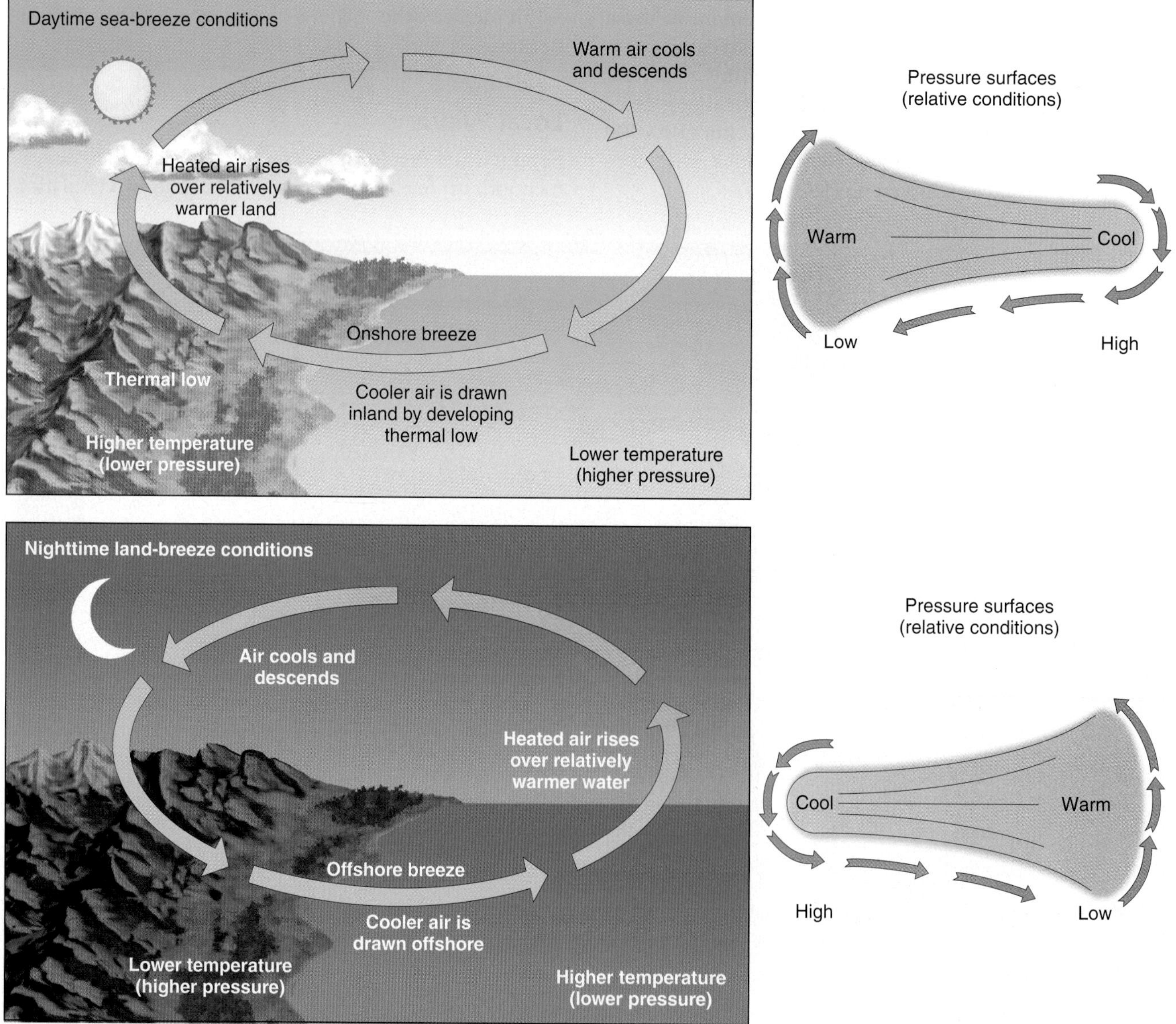

FIGURE 4.18 Land-sea breeze characteristics for day and night.

Valley in southern France can cause frost damage to vineyards as the cold north winds move over the region to the Gulf of Lions and the Mediterranean Sea. The frequently stronger *bora*, driven by the cold air of winter high-pressure systems inland, flows across the Adriatic Coast to the west and south. In Alaska such winds are called the *taku*.

Santa Ana winds are another local wind type, generated by high pressure over the Great Basin of the western

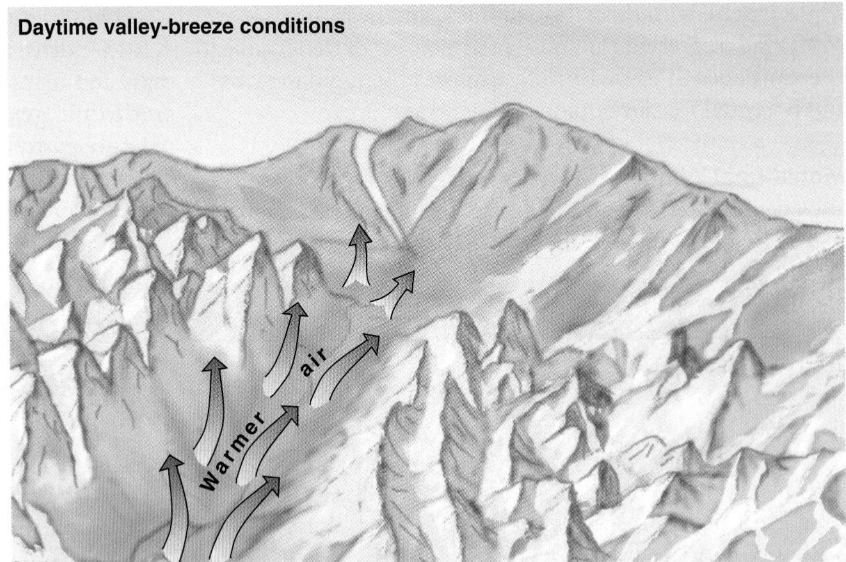

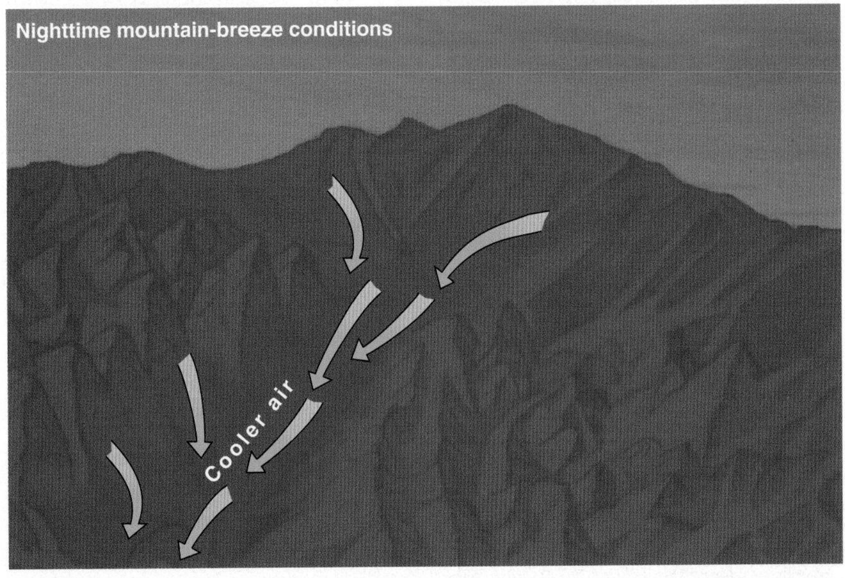

FIGURE 4.19 Patterns of mountain-valley breeze for day and night.

News Report 4.1

Jet Streams Affect Flight Times

As airplane travel increased during World War II, flight crews reported strong headwinds on routes to the west. In some cases, airplanes leaving San Francisco for the Pacific were turned back by opposing wind in the upper troposphere. These were the jet streams.

Next time you plan a flight, note that airline schedules reflect the presence of these upper-level westerly winds, for they allot shorter flight times from west to east and longer flight times from east to west. I recently flew round-trip between Sacramento, California, and Philadelphia, Pennsylvania, and experienced a difference of 1 hour and 5 minutes between shorter eastbound (tail wind) and longer westbound (head wind) flights—a difference caused by a particularly strong jet stream.

Also important to both military and civilian aircraft is the effect the jet streams have on fuel consumption and the existence of air turbulence. Seasonal adjustments are necessary because the jet streams in both hemispheres tend to weaken during each hemisphere's summer and strengthen during winter as the streams shift closer to the equator.

United States. A strong, dry wind is produced that flows out across the desert to southern California coastal areas. The air is heated by compression as it flows from higher to lower elevations, and with increasing speed it moves through constricting valleys to the southwest. These winds irritate with their dust, dryness, and heat.

Wind represents a substantial source of renewable energy. Focus Study 4.1 briefly explores the rapid growth and potential for development of wind resources.

Monsoonal Winds. Regional wind systems that seasonally change direction are important in some areas. Examples occur in the tropics over Southeast Asia, Indonesia, India, northern Australia, equatorial Africa, and southern Arizona. These winds involve an annual cycle of returning precipitation with the summer Sun and are named after the Arabic word for season, *mausim*, or **monsoon**.

The monsoons of southern and eastern Asia (Figure 4.20) are driven by the location and size of the Asian landmass and its proximity to the Indian Ocean. Also important to the generation of monsoonal flows are wind and pressure patterns in the upper-air circulation.

The extreme temperature range from summer to winter over the Asian landmass is due to its continentality,

Focus Study 4.1

Wind Power: an Energy Resource for the Present and Future

The principles of wind power are ancient, but the technology is modern and the benefits are worth pursuing. In more-developed countries, energy sources are dominated by the use of nonrenewable fuels—coal, gas, oil, nuclear—and centralized energy production. In less-developed countries, however, many rely on renewable energy—small hydroelectric plants, wind-power systems, diminishing wood supplies, and solar energy—for cooking, heating, and pumping water. These resources are considered renewable because they are not depleted in the span of a human lifetime.

Rough estimates place the global wind resource at 500% more than present global energy use. The Financial Times World Renewable Energy Conference held in Brussels, Belgium, in 1999, proclaimed that wind power could supply 10% of world energy needs and create 1.7 million new jobs by the year 2020.

A Brief Assessment

Wind-generated energy resources are the fastest-growing energy technology, in terms of new capacity. There was more than a 1000% increase in capacity between 1990 and 2002, reaching 24,800 MW in 2001, 30,000 MW by the end of 2002 (up 40 percent over 2000!). More than 70 percent of the wind installations are in Europe, with Germany leading the way with 10,650 MW, or 3.7 percent of their electricity supply (Figure 1). The European Wind Energy Association announced an installed capacity goal of 60,000 MW by 2010, more two times the worldwide total in 2002.

The U.S. market tends to swing widely in response to political decisions concerning tax credits and incentives, as compared to favorable policies in Europe. This is the case despite the fact that the U.S. wind-energy potential exceeds present electricity consumption by 300%. Following costly delays (layoffs to industry personnel and stilled production), in renewal of the federal Wind Energy Production Tax Credit, a 2002 extension through 2003 will make 2003 a record year for new installations.

(continued)

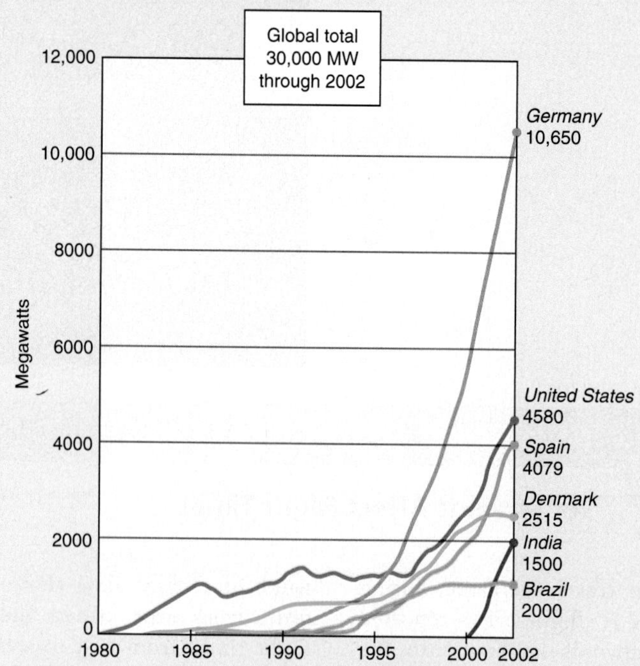

FIGURE 1 Installed wind-generating capacity. In 1980 only 10 MW of wind-generated capacity was in operation in the world. By 2002 it surged to 30,000 MW. New capacity is growing at record rates in Germany, Denmark, Spain, and India. [Sources: *Vital Signs 2002*, Worldwatch Institute and the United Nations Environment Programme, New York, NY, © 2002; the American Wind Energy Association; and the European Wind Energy Association.]

Focus Study 4.1 *(continued)*

To put these numbers in meaningful perspective, every 10,000 MW of wind-generation capacity reduces carbon dioxide emissions by 33 million metric tons if it replaces coal, or 21 million metric tons if it replaces mixed fossil fuels. As an example, if countries rally and create a proposed US$400-billion-dollar industry by installing 500,000 MW of worldwide wind capacity by 2020, as much as a fourth of carbon dioxide emissions would be avoided. For the United States, 225,000 wind turbines would reduce coal emissions of carbon dioxide by 59%. For comparison, in the 1997 Kyoto Protocol the United States proposed reducing greenhouse gas emissions by just 7% below 1990 levels.

Economics Update

Natural gas prices in California reached 15 to 20 cents per kilowatt-hour (kWh) in the early 2000 price spike (electricity on the spot [short-term, daily] market hit 33 cents/kWh), whereas wind-power contracts came in at 3 cents/kWh in the same year. In Europe, smaller turbines produced at 4 cents per kWh. Two Stanford University engineers calculated that the contracted price of 3 cents/kWh in the United States is cheaper than coal (estimated at 5.5 to 8.3 cents per kWh), when considering all costs.* Lower industry estimates for coal costs exclude consideration of coal-mine dust health impacts, acid deposition, smog, and global warming—significant when you consider that the black-lung disease program to date has cost some $70 billion alone.

The Nature of Wind Energy

Power generation from wind is site-specific, because conditions that produce adequate winds are limited to certain areas. Wind resources are greatest in three basic settings: (1) along coastlines that are influenced by trade winds and westerly winds; (2) where mountain passes constrict air flow and interior valleys develop thermal low-pressure areas, thus drawing air across the landscape; or (3) where localized winds occur, such as katabatic or monsoonal flows. Cash-short developing countries are generally located in areas blessed by such steady winds. Where wind is reliable less than 25%–30% of the time, only small-scale uses are economically feasible. Improvement in energy storage strategies, such as the production of hydrogen from water, could cover windless days.

The winds of North and South Dakota and Texas could meet all the U.S. electrical needs. One study done in Wyoming found that land selling for $100 per hectare (2.47 acres) would yield about $25,000 worth of wind-generated electricity annually.

Farmers in Iowa and Minnesota receive $2000 in annual income from a leased turbine and $20,000 a year from an owned turbine that takes one-quarter acre to site. The corn harvested annually from that same site is worth about $100. Farmers are exploring this income source with assistance from the American Corn Growers Association (ACGA) and are forming wind-energy cooperatives. The limitation seems to be a shortage of transmission capacity from rural sites to cities. (See the ACGA report

(continued)

(a)

(b)

FIGURE 2
(a) Wind farm near the Cajon Pass in southern California, along Interstate 15. (b) Wind farms near Harle, East Friesland, Germany. [Photos by (a) Bobbé Christopherson; (b) © Uwe Walz/CORBIS.]

*See M. Z. Jacobsen and G. M. Masters, "Exploiting Wind vs. Coal," *Science* 293 (August 24, 2001): 1438; also see American Wind Energy Association, "Global Wind Energy Market Report," May 2001.

Focus Study 4.1 *(continued)*

at http://www.acga.org/programs/2001WindCCS/.)

The Benefits of the Wind Resource

Economic reality will perhaps override further delaying actions, especially where peak winds are in concert with peak electrical demand for air cooling, space heating, or agricultural water pumping. The two Stanford engineers, in a study cited previously in this focus study, stated,

> Much of the recent energy debate in the United States has focused on increasing coal use. However, the cost of wind energy is now less than coal. Shifting from coal to wind would address health, environmental, and energy problems (p. 1438).

Needless to say, whether or not governments and transnational energy corporations begin full-scale planning and implementation of renewable resources now, energy resources and international realities of the near future will necessitate expansion. By the middle of this century, wind-generated electricity will be routine, along with other renewable energy sources, conservation, and energy efficiency. Germany, Spain, and Denmark in the European Union are leading the way for now.

---

reflecting its isolation from the modifying effects of the ocean. An intense high-pressure anticyclone dominates the continental landmass in winter (Figure 4.11a), whereas the equatorial low-pressure trough dominates the central area of the Indian Ocean. Resultant cold, dry winds blow from the Asian interior over the Himalayas, downslope, and across India, producing average temperatures between 15° and 20°C (60° to 68°F) at lower elevations. These dry winds desiccate (dehydrate) the landscape and then give way to hot weather from March through May.

During the June–September wet period, the subsolar point shifts northward to the Tropic of Cancer, near the mouths of the Indus and Ganges Rivers. The intertropical convergence zone (ITCZ) shifts northward over southern Asia, and the Asian continental interior develops a thermal low pressure, associated with high average temperatures. Meanwhile, the Indian Ocean, with a surface temperature of 30°C (86°F), is under the influence of subtropical high pressure. As a result, hot, dry subtropical air sweeps over the warm ocean, producing extremely high evaporation rates (Figure 4.20b).

By the time this mass of air reaches India and the ITCZ, it is laden with moisture in thunderous, dark clouds. The warmth of the land lends additional lifting to

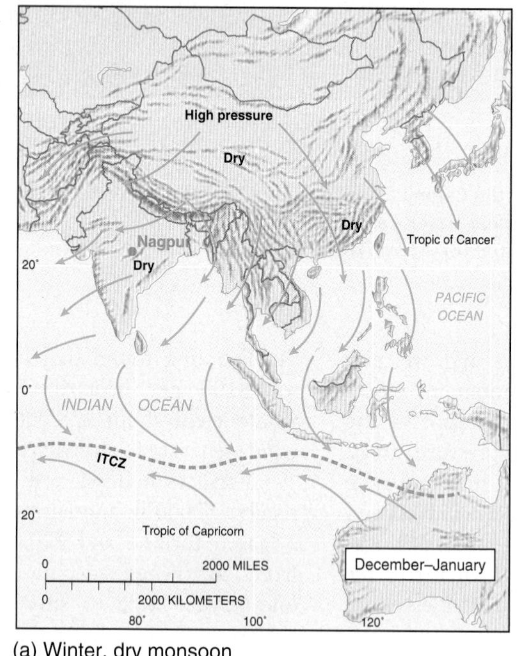

(a) Winter, dry monsoon

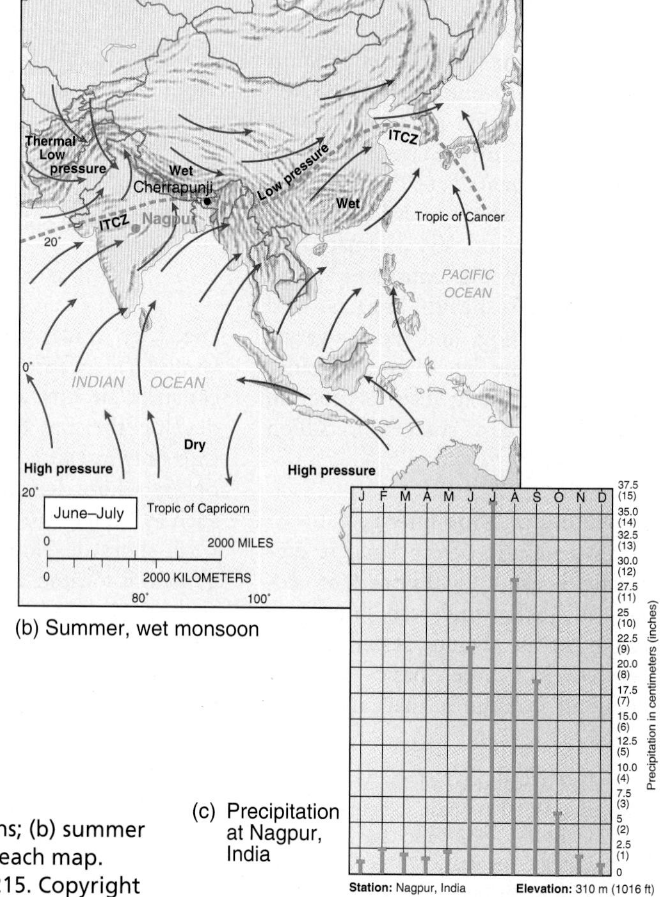

(b) Summer, wet monsoon

(c) Precipitation at Nagpur, India

FIGURE 4.20 The Asian monsoon.
Asian monsoon patterns: (a) winter pressure and wind patterns; (b) summer pressure and wind patterns. Note the location of the ITCZ on each map. [Adapted from Joseph E. Van Riper, *Man's Physical World*, p. 215. Copyright 1971 by McGraw-Hill. Adapted by permission.]

the incoming air, as do the Himalayas, forcing the air mass to higher altitudes. These conditions produce the wet monsoon of India, where world-record rainfalls have occurred. When the monsoonal rains arrive in June, they are welcome relief from the dust, heat, and parched land of Asia's springtime. Note the precipitation distribution on the inset climograph for Nagpur, India, in Figure 4.20.

Oceanic Currents

Earth's atmospheric and oceanic circulations are closely interrelated. The driving force for ocean currents is the frictional drag of the winds. Also important in shaping these currents is the interplay of the Coriolis force, density differences caused by temperature and salinity of the water, the configuration of the continents and ocean floor, and astronomical forces (the tides).

Surface Currents

The general patterns of major ocean currents are shown in Figure 4.21. Coriolis force deflects ocean currents because they flow over distance and through time. However, their pattern of deflection is not as tightly circular as that of the atmosphere. Compare this map with that of Earth's pressure and wind systems (Figure 4.11), and you can see that ocean currents are driven by the circulation around subtropical high-pressure cells in both hemispheres. (Remember: In the Northern Hemisphere, winds and ocean currents move *clockwise* about high pressure cells—note the currents in the North Pacific and North Atlantic on the map. In the Southern Hemisphere, the *counterclockwise* circulation about a high is evident on the map.) News Report 4.2 dramatically portrays this clockwise flow around the Pacific Ocean basin in the Northern Hemisphere. These circulation systems are known as *gyres*.

Along the full extent of ocean areas adjoining the equator, trade winds drive the oceans westward in a concentrated channel. These currents are kept near the equator by a Coriolis force influence that weakens near the equator. As the surface current approaches the western margins of the oceans, the water actually piles up an average of 15 cm (6 in.). From this western edge, ocean water then spills northward and southward in strong currents.

In the Northern Hemisphere, the *Gulf Stream* and the *Kuroshio* move forcefully northward with their speed and depth increased by the constriction of the area they occupy. The warm, deep blue water of the ribbonlike Gulf Stream usually is 50–80 km (30–50 mi) wide and 1.5–2.0 km (0.9–1.2 mi) deep, moving at 3–10 kmph (1.8–6.2 mph). In 24 hours, ocean water can move 70–240 km (40–150 mi) in the Gulf Stream, although a complete circuit around an entire gyre may take a year.

Deep Currents

Where surface water is swept away from a coast, an **upwelling current** occurs. This cool water generally is nutrient-rich and rises from great depths to replace the vacating water. Such cold upwelling currents occur off the Pacific coasts of North and South America and the subtropical and midlatitude west coast of Africa, and are some of the world's prime commercial fishing areas. In other portions of the sea where there is an accumulation of water—as at the western end of an equatorial current, the Labrador Sea, or along the margins of Antarctica—excess

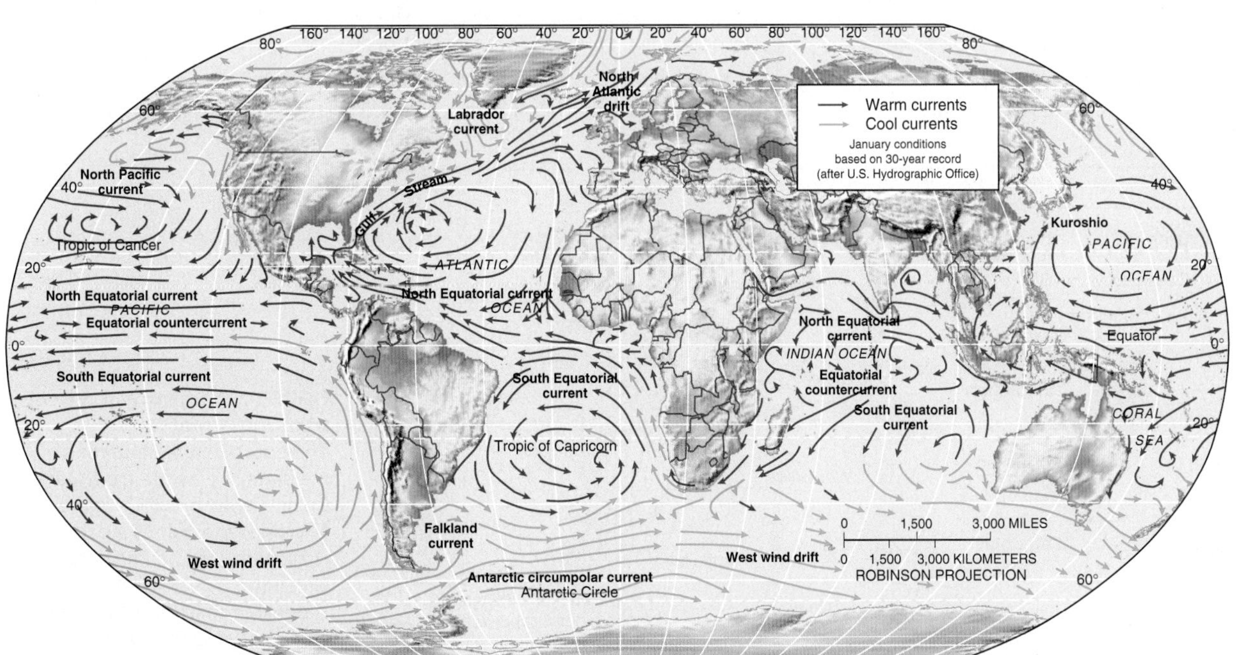

FIGURE 4.21 Major ocean currents.
[After the U.S. Navy Hydrographic Office.]

water gravitates downward in a **downwelling current.** Such downwelling zones generate important mixing currents along the ocean floor and travel the full extent of the ocean basins, carrying heat energy and salinity.

Imagine a continuous channel of water beginning with cold water downwelling in the North Atlantic, flowing deep and strong to upwellings in the Indian Ocean and North Pacific (Figure 4.22). Here it warms and then is carried in surface currents back to the North Atlantic. A complete circuit of the current may require 1000 years from downwelling in the Labrador Sea off Greenland to its reemergence in the southern Indian Ocean and return. Even deeper Antarctic bottom water flows northward in the Atlantic Basin beneath these currents. Scientists are assessing the impact of global warming on the functioning of these essential ocean circulations.

News Report 4.2

A Message in a Bottle and Rubber Duckies

To give you an idea of the dynamic circulation of the ocean, consider the following examples. A 9-year-old child at Dana Point, California (33.5° N), a small seaside community south of Los Angeles, placed a letter in a glass juice bottle in July 1992 and tossed it into the waves. Thoughts of distant lands and fabled characters filled the child's imagination as the bottle disappeared. The vast circulation around the Pacific high, clockwise-circulating gyre, took command (Figure 1).

Three years passed before ocean currents carried the bottle to the coral reefs and white sands of Mogmog, a small island in Micronesia (7° N). A 7-year-old child there found a pen pal from afar and immediately mailed a photo and card to the sender of the message. Imagine the journey of that note from California—traveling through storms and calms, clear moonlit nights and typhoons—as it floated on ancient currents as the Spanish Galleons once had.

In January 1994, a powerful storm ravaged a large container ship from Hong Kong loaded with toys and other goods. One of the containers on board split apart in the wind off the coast of Japan, dumping nearly 30,000 rubber ducks, turtles, and frogs into the North Pacific. Westerly winds and the North Pacific current swept this floating cargo across the ocean to the coasts of Alaska, Canada, Oregon, and California. Other toys, still adrift, went through the Bering Sea and into the Arctic Ocean (this route is indicated by the dashed line in Figure 1). These toys will eventually end up in the Atlantic Ocean as they drift around the Arctic Ocean frozen in pack ice.

The high-floating message bottle and rubber ducks offered a much better opportunity to study winds than did an earlier spill of 60,000 athletic shoes near Japan. The low-floating shoes were tracked across the Pacific Ocean until they landed in the Pacific Northwest. By the way, the shoes that did not make landfall headed (or footed) back around the Pacific gyre into the tropics and westward, back toward Japan! Scientists are using incidents such as these to learn more about wind systems and ocean currents.

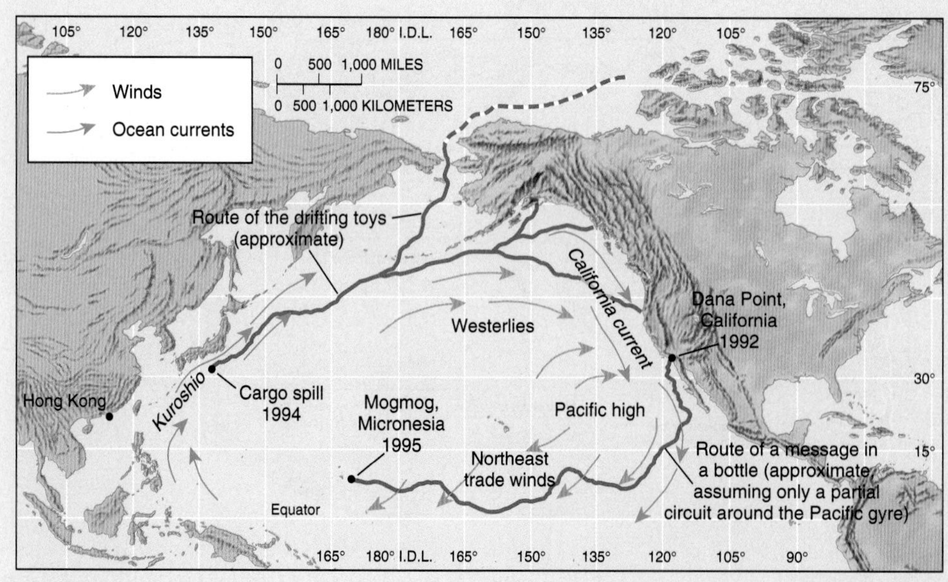

FIGURE 1 Pacific Ocean currents transport artifacts. The approximate route of a message in a bottle from Dana Point, California, to Mogmog, Micronesia, assumes a partial circuit around the Pacific gyre. Given the 3-year travel time, it is possible that the message circumnavigated the Pacific Ocean more than once. The rubber duckies voyaged eastward across the Pacific and beyond.

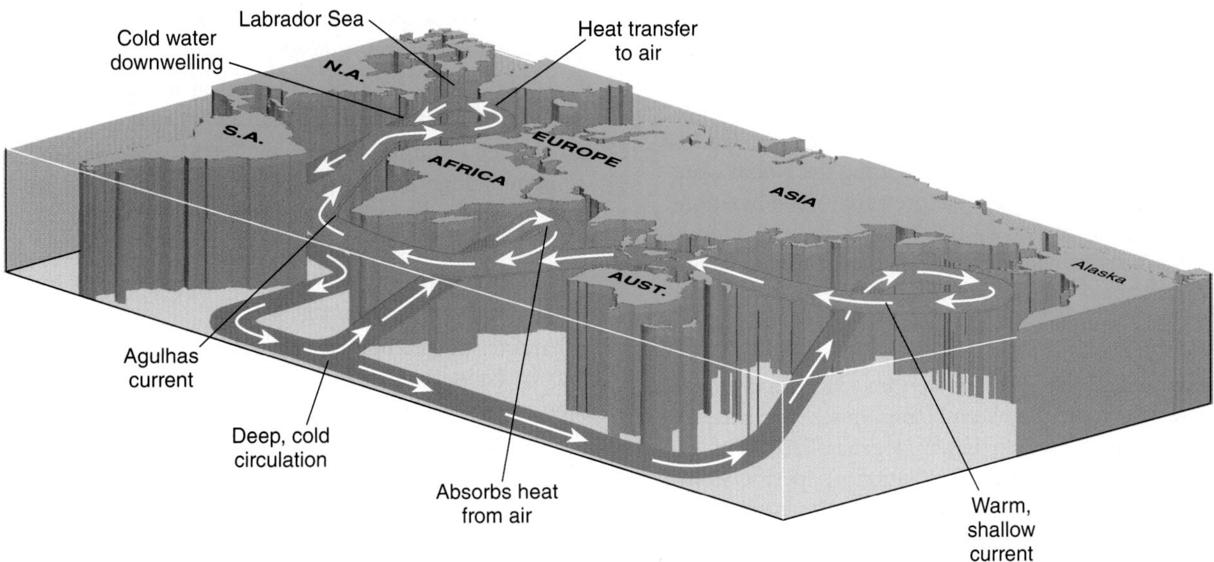

FIGURE 4.22 Deep-ocean circulation.
Scientists are deciphering centuries-long deep circulation in the oceans. This global circulation mimics a vast conveyor belt of water drawing heat from some regions and transporting it for release in others.

Summary and Review—Atmospheric and Oceanic Circulation

● *Define* the concept of air pressure and *portray* the pattern of global pressure systems on isobaric maps.

Volcanic eruptions such as those of Tambora in 1815 and Mount Pinatubo in 1991 dramatically demonstrate the power of global winds to disperse aerosols and pollution worldwide in a matter of weeks. Atmospheric circulation facilitates important transfers of energy and mass on Earth, thus maintaining Earth's natural balances. Earth's atmospheric and oceanic circulations represent a vast heat engine powered by the Sun.

The weight (created by motion, size, and number of molecules) of the atmosphere is **air pressure**, which exerts an average force of approximately 1 kg/cm^2 (14.7 $lb/in.^2$). Air pressure is measured with a **mercury barometer** at the surface (mercury in a tube—closed at one end and open at the other, with the open end placed in a vessel of mercury—that changes level in response to pressure changes) or an **aneroid barometer** (a closed cell, partially evacuated of air, that detects changes in pressure).

air pressure (p. 107)
mercury barometer (p. 108)
aneroid barometer (p. 108)

1. How does air exert pressure? Describe the basic instrument used to measure air pressure. Compare the operation of two different types of instruments discussed.
2. What is normal sea-level pressure in millimeters? Millibars? Inches? Kilopascals?
3. What is a possible explanation for the beautiful sunrises and sunsets during the summer of 1992 in North America? Relate your answer to global circulation.
4. Explain this statement: "The atmosphere socializes humanity, making all the world a spatially linked society." Illustrate your answer with some examples.

● *Define* wind and *describe* how wind is measured, how wind direction is determined, and how winds are named.

Wind is the horizontal movement of air across Earth's surface. Its speed is measured with an **anemometer** (a device with cups that are pushed by the wind) and its direction with a **wind vane** (a flat blade or surface that is directed by the wind).

wind (p. 108)
anemometer (p. 108)
wind vane (p. 109)

5. Define wind. How is it measured? How is its direction determined?
6. Distinguish among primary, secondary, and tertiary general classifications of global atmospheric circulation.
7. How was the image in Figure 4.6a produced? What does it demonstrate about winds in the Pacific?

● *Explain* the four driving forces within the atmosphere—gravity, pressure gradient force, Coriolis force, and friction force—and *describe* the primary high- and low-pressure areas and principal winds.

Earth's gravitational force on the atmosphere operates uniformly worldwide. Winds are directed and driven by the **pressure gradient force** (air moves from areas of high pressure to areas of low pressure); the **Coriolis force** (an apparent deflection in the path of winds or ocean currents caused by the rotation of Earth, deflecting objects to the right in the Northern Hemisphere and to the left in the Southern Hemisphere); and **friction force** (Earth's varied surfaces exert a drag on wind movements in opposition to the pressure gradient). Maps portray air pressure patterns using the **isobar**—an isoline that connects points of equal pressure. A combination of the pressure gradient and Coriolis forces alone produces **geostrophic winds**, which move parallel to isobars, characteristic of winds above the surface frictional layer.

Winds descend and diverge, spiraling outward to form an **anticyclone** (clockwise in the Northern Hemisphere), and they converge and ascend, spiraling upward to form a **cyclone** (counterclockwise in the Northern Hemisphere). The pattern of high and low pressures on Earth in generalized belts in each hemisphere produces the distribution of specific wind systems. These primary pressure regions are the **equatorial low-pressure trough**, the weak **polar high-pressure cells** (at both the North and South Poles), and the **subtropical high-pressure cells** and **subpolar low-pressure cells.**

All along the equator winds converge into the equatorial low, creating the **intertropical convergence zone (ITCZ).** Air rises along the equator and descends in the subtropics in each hemisphere. The winds returning to the ITCZ from the northeast in the Northern Hemisphere and from the southeast in the Southern Hemisphere produce the **trade winds.**

Winds flowing out of the subtropics to higher latitudes produce the **westerlies** in either hemisphere. The subtropical high-pressure cells on Earth, generally between 20° and 35° in either hemisphere, are variously named the *Bermuda high*, *Azores high*, and *Pacific high*.

Along the polar front and a series of low-pressure cells, the *Aleutian low* and *Icelandic low* dominate the North Pacific and Atlantic, respectively. This region of contrast between colder air toward the poles and warmer air equatorward is called the **polar front.** The weak and variable **polar easterlies** diverge from the polar high-pressure cells, particularly the **Antarctic high.**

pressure gradient force (p. 110)
Coriolis force (p. 110)
friction force (p. 110)
isobar (p. 110)
geostrophic winds (p. 112)
anticyclone (p. 112)
cyclone (p. 113)
equatorial low-pressure trough (p. 114)
polar high-pressure cells (p. 114)
subtropical high-pressure cells (p. 114)
subpolar low-pressure cells (p. 114)
intertropical convergence zone (ITCZ) (p. 115)
trade winds (p. 115)
westerlies (p. 117)
polar front (p. 118)
polar easterlies (p. 119)
Antarctic high (p. 119)

8. What does an isobaric map of surface air pressure portray? Contrast pressures over North America for January and July.
9. Describe the effect of the Coriolis force. Explain how it apparently deflects atmospheric and oceanic circulations.
10. What are geostrophic winds, and where are they encountered in the atmosphere?
11. Describe the horizontal and vertical air motions in a high-pressure anticyclone and in a low-pressure cyclone.
12. Construct a simple diagram of Earth's general circulation, including the four principal pressure belts or zones and the three principal wind systems.
13. How is the intertropical convergence zone (ITCZ) related to the equatorial low-pressure trough? How might it appear on a satellite image?
14. Characterize the belt of subtropical high pressure on Earth: Name the specific cells. Describe the generation of westerlies and trade winds. Discuss sailing conditions.
15. What is the relation among the Aleutian low, the Icelandic low, and migratory low-pressure cyclonic storms in North America? In Europe?

- ***Describe* upper-air circulation and its support role for surface systems and *define* the jet streams.**

Vast, flowing longwave undulations in upper-air westerlies form wave motions called **Rossby waves.** The prominent movements in upper level, westerly winds are streams of high speed winds called the **jet streams.** Depending on their latitudinal position in either hemisphere, they are termed the *polar jet stream* or the *subtropical jet stream.*

Rossby waves (p. 120)
jet streams (p. 121)

16. What is the relation between wind speed and the spacing of isobars?
17. How are undulations in the upper-air circulation (ridges and troughs) related to surface pressure systems? To divergence aloft and surface lows? To convergence aloft and surface highs? (See Figure 4.10 and text.)
18. Relate the jet-stream phenomenon to general upper-air circulation. How is the presence of this circulation related to airline schedules from New York to San Francisco and the return trip to New York?

- ***Explain* several types of local winds: land-sea breezes, mountain-valley breezes, katabatic winds, and the regional monsoons.**

Different heating characteristics of land and water surfaces create **land-sea breezes. Mountain-valley breezes** are caused by temperature differences during the day and evening between valleys and mountain summits. **Katabatic winds**, or gravity drainage winds, are of larger regional scale and are usually stronger than mountain-valley breezes, under certain conditions. An elevated plateau or highland is essential for katabatic wind, where layers of air at the surface cool, become denser, and flow downslope.

Intense, seasonally shifting wind systems occur in the tropics over Southeast Asia, Indonesia, India, northern Australia, equatorial Africa, and southern Arizona. These winds involve an annual cycle of returning precipitation with the summer Sun and are named after the Arabic word for season, *mausim*, or **monsoon.** The monsoons of southern and eastern Asia are driven by the location and size of the Asian landmass and its proximity to the Indian Ocean.

land-sea breezes (p. 122)
mountain-valley breezes (p. 122)
katabatic winds (p. 122)
monsoon (p. 124)

19. People living along coastlines generally experience variations in winds from day to night. Explain the factors that produce these changing wind patterns.
20. The arrangement of mountains and nearby valleys produces local wind patterns. Explain the day and night winds that might develop.
21. Describe the seasonal pressure patterns that produce the Asian monsoonal wind and precipitation patterns. Contrast January and July conditions.

● *Describe* the basic pattern of Earth's major surface and deep ocean currents.

Ocean currents are primarily caused by the frictional drag of wind and occur worldwide at varying intensities, temperatures, and speeds, both along the surface and at great depths in the oceanic basins. The circulation around subtropical high-pressure cells in both hemispheres is notable on the ocean circulation map—these *gyres* usually are offset toward the western side of each ocean basin.

The trade winds converge along the ITCZ and push enormous quantities of water westward, generating the Gulf Stream and Kuroshio currents. Where surface water is swept away from a coast, either by surface divergence (induced by the Coriolis force) or by offshore winds, an **upwelling current** occurs. This cool water generally is nutrient-rich and rises from great depths to replace the vacating water. In other portions of the sea where there is an accumulation of water, the excess water gravitates downward in a **downwelling current**. These currents generate important mixing actions that flow along the ocean floor and travel the full extent of the ocean basins, carrying heat energy and salinity.

upwelling current (p. 127)
downwelling current (p. 128)

22. What is the relationship between global atmospheric circulation and ocean currents? Relate oceanic gyres to patterns of subtropical high pressure.
23. Define the western intensification. How is it related to the Gulf Stream and Kuroshio currents?
24. Where on Earth are upwelling currents experienced? What is the nature of these currents?
25. What is meant by deep-ocean circulation? At what rates do these currents flow? How might this circulation be related to the Gulf Stream in the western Atlantic Ocean?

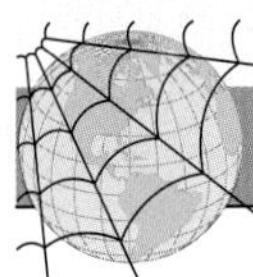

NetWork

The *Elemental Geosystems Home Page* provides on-line resources for this chapter on the World Wide Web. Once on the Home Page, click on this textbook, scroll the Table of Contents menu, select this chapter, and click "Begin." You will find self-tests that are graded, review exercises, specific updates for items in the chapter, and in "Destinations" many links to interesting related pathways on the Internet. *Elemental Geosystems* is found at http://www.prenhall.com/christopherson.

Critical Thinking

A. Locate an anemometer on your campus. Where is it located? Describe the installation after a visual inspection. Is it part of a complete weather station? If so, who is operating the station? What were the level of record winds that have occurred on your campus?

B. Go to http://www.awea.org/ and the American Wind Energy Association. Sample the materials presented as you assess Focus Study 4.1 and the potential for wind-generated electricity. What are your thoughts concerning this resource: its potential, reasons for delays, and any necessities you perceive? What countries seem to be leading the way? Locate the nearest wind farm to your present location.

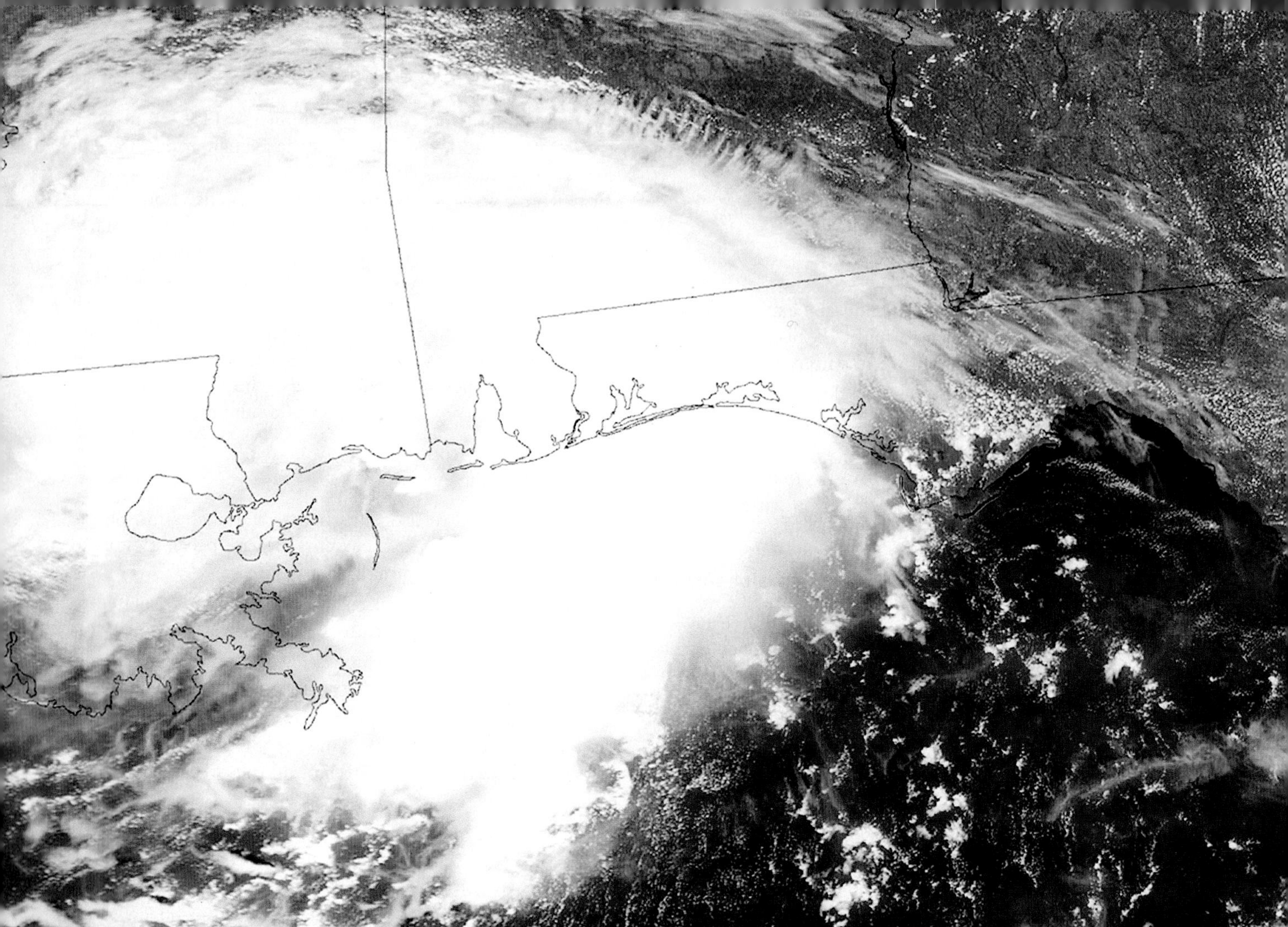

PART TWO

Water, Weather, and Climate Systems

Remnants of Tropical Storm Bertha deluge Mississippi and Louisiana with heavy precipitation, approaching 18 cm (7 in.) in the region. Tropical Storm Cristobol is taking shape off South Carolina. [*Terra* Image from August 5, 2002, courtesy of the Land Rapid Response Team, NASA/GSFC.]

Earth is the water planet among planets in the Solar System. In Part 2, we analyze the origin, quantity, distribution, and remarkable qualities of water. Water has a leading role in the vast drama played out daily on Earth's stage. In this ongoing play, the actors are large air masses that come into conflict, moving and shifting to dominate different regions. Migrating cyclonic systems that cross the middle latitudes and tropical storms, fueled by water and its ability to absorb and release vast quantities of heat energy, bring surplus energy poleward in both hemispheres.

We see the spatial implications over time of the energy-atmosphere and water-weather systems that generate Earth's climate patterns. And we analyze the human impact on the climate system. As an ultimate output of these systems, we examine water resources—the specifics of the hydrologic cycle and how water circulates over Earth. We look at the water-balance concept as a useful way to understand water-resource relationships, whether global, regional, or local. Important water resources include rivers, lakes, groundwater, and oceans. This "pale blue dot" in space is indeed the water planet.

SOLAR ENERGY
Atmosphere
Hydrosphere
Biosphere
Lithosphere

A supercell tornado descends from the cloud base near Spearman, Texas. Strong hail is falling to the left of the tornado. [Photo by Howard Bluestein, all rights reserved.]

5

Atmospheric Water and Weather

Key Learning Concepts

By knowing and understanding the key learning concepts in this chapter, you should be able to:

- *Describe* the origin of Earth's waters, the present distribution, and water's unique heat properties.
- *Define* humidity, relative humidity, and dew-point temperature and *illustrate* three atmospheric conditions—unstable, conditionally unstable, and stable.
- *Explain* cloud formation and *identify* major cloud classes and types, including fog.
- *Describe* air masses that affect North America and *identify* four types of atmospheric lifting mechanisms.
- *List* the measurable elements of weather and *describe* the life cycle of a midlatitude cyclone.
- *Analyze* various types of violent weather—thunderstorms, tornadoes, and tropical cyclones.

> Walden is blue at one time and green at another, even from the same point of view. Lying between the earth and the heavens, it partakes of the color of both.... A lake is the landscape's most beautiful and expressive feature. It is earth's eye; looking into which the beholder measures the depth of his own nature.... Sky water. It needs no fence. Nations come and go without defiling it. It is a mirror which no stone can crack ... Nature continually repairs ... a mirror in which all impurity presented to it sinks, swept and dusted by the sun's hazy brush. A field of water ... is continually receiving new life and motion from above. It is intermediate in its nature between land and sky.*

*Henry David Thoreau, *Walden*, (© 1969 by Merrill Publishing, originally published 1854), pp. 192, 202, 204.

Thus did Thoreau speak of the water so dear to him—Walden Pond in Massachusetts, where he lived along the shore.

Water is critical to our daily lives and a principal compound in nature. It covers 71% of Earth (by area) and in the Solar System occurs in such significant quantities only on our planet. Pure water is naturally colorless, odorless, and tasteless and weighs 1 gram per cubic centimeter or 1 kg per liter (62.3 lb/ft^3 or 8.337 lb/gal).

Water constitutes nearly 70% of our bodies by weight and is the major ingredient in plants, animals, and our food. A human can survive 50 to 60 days without food, but only 2 or 3 days without water. The water we use must be adequate in quantity as well as quality for its many tasks—everything from personal hygiene to vast national water projects. Water indeed occupies that place between land and sky, mediating energy and shaping both the lithosphere and atmosphere, as Thoreau revealed.

In this chapter: We examine water on Earth and the dynamics of atmospheric moisture and stability—the essentials of weather. The key questions answered include: What were the origins of Earth's water? How much is there? Where is it located? What are its unique heat properties and how is it critical in powering Earth's weather systems? Condensation of water vapor and atmospheric conditions of stability and instability are key to cloud formation. We follow huge air masses across North America, observe powerful lifting mechanisms in the atmosphere, examine migrating cyclonic systems with attendant cold and warm fronts, and conclude with a tour of violent and dramatic weather. The spatial implications of weather and its relationship to human activities strongly link meteorology to physical geography and this chapter.

Water on Earth: Location and Properties

Water on Earth formed within the planet, reaching Earth's surface in an ongoing process called **outgassing**, by which water and water vapor emerge from deep within the crust—25 km (15.5 mi) or deeper below Earth's surface (Figure 5.1). In the early atmosphere, massive quantities of outgassed water vapor condensed and fell back to Earth in torrential rains. The lowest places across the face of Earth began to fill with water: first ponds, then lakes and seas, and eventually ocean-sized bodies of water. Massive flows of water washed over the landscape, carrying both dissolved and undissolved elements to these early seas and oceans.

Quantity Equilibrium

Today water is the most common compound on the surface of Earth, having achieved the present volume of 1.36 billion km^3 (326 million mi^3) approximately 2 billion years ago. This quantity has remained relatively constant, even though water is continuously being lost from the system, escaping to space or breaking down and forming new compounds with other elements. Lost water is replaced by pristine water not previously at the surface, which emerges from within Earth. The net result of these inputs and outputs to water quantity is a steady-state equilibrium in Earth's hydrosphere.

Despite this overall net balance in quantity, worldwide changes in sea level do occur; this change is called **eustasy**. Eustatic sea-level changes are specifically related to changes in volume of water in the oceans. Some of these changes are explained by the amount of water stored on Earth as ice. As more water is bound up in glaciers (in mountains worldwide) and in ice sheets (Greenland and Antarctica), sea level lowers, whereas a warmer era reduces the quantity of water stored as ice and thus raises sea level. Some 18,000 years ago, during the most recent ice age, sea level was more than 100 m (330 ft) less than today; 40,000 years ago it was 150 m (490 ft) lower. Over the past 100 years glacial ice has retreated, causing mean sea level to rise, and it still is rising worldwide at a rate not previously seen as higher temperatures melt more ice. *Glacio-eustatic* effects on sea level are discussed further in Chapter 13.

Distribution of Earth's Water

If you briefly examine a globe, it becomes obvious that most of Earth's continental land is in the Northern Hemisphere, whereas the Southern Hemisphere is dominated

FIGURE 5.1 Water outgassing from the crust. Outgassing of water from Earth's crust in the geothermal area near Wairakei on the North Island of New Zealand. [Photo by Bill Bachman/Photo Researchers, Inc.]

by water. In fact, from certain perspectives Earth appears to have a distinct oceanic hemisphere and a distinct land hemisphere (Figure 5.2).

An illustration of the present location of all of Earth's liquid and frozen water—whether fresh or saline, surface or underground—is in Figure 5.3. The oceans contain 97.22% of all water. Only 2.78% of all of Earth's water is freshwater, or nonoceanic (Figure 5.3a).

Of all the freshwater on Earth, about 22% exists as groundwater and soil moisture (Figure 5.3b). On the surface, the greatest single repository of freshwater is ice. Ice sheets and glaciers account for the majority of all freshwater on Earth. The remaining freshwater, although very familiar to us, in lakes, rivers, and streams, represents but a small quantity—less than 1% (Figure 5.3c).

Think for a moment of the weather occurring in the atmosphere worldwide and the many flowing rivers and streams. Combined, they amount to only 0.033% of all freshwater! This small amount is very dynamic, however. A water molecule traveling along atmospheric and surface-water paths moves through the entire ocean-atmosphere-precipitation-runoff cycle in less than 2 weeks, whereas a water molecule located in deep-ocean circulation, groundwater, or a glacier moves slowly, taking thousands of years to migrate through the system.

Unique Properties of Water

Earth's distance from the Sun places it within a remarkable temperate zone, compared with the other planets. This temperate location allows all states of water—ice, water, and water vapor—to occur naturally on Earth. Water is composed of two atoms of hydrogen and one of

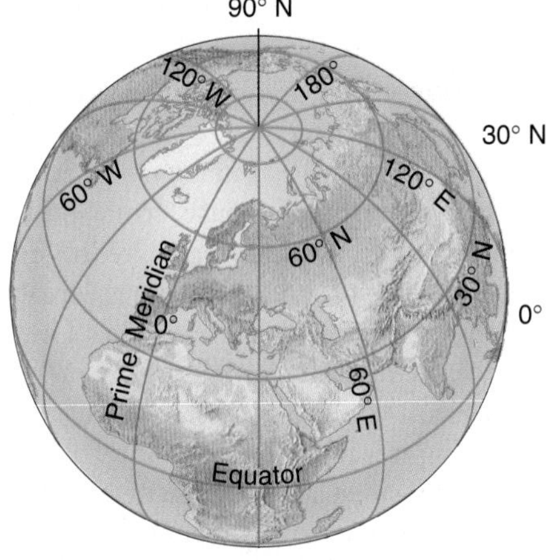

FIGURE 5.2 Land and water hemispheres.
Two perspectives that divide Earth's surface into an ocean hemisphere and a land hemisphere.

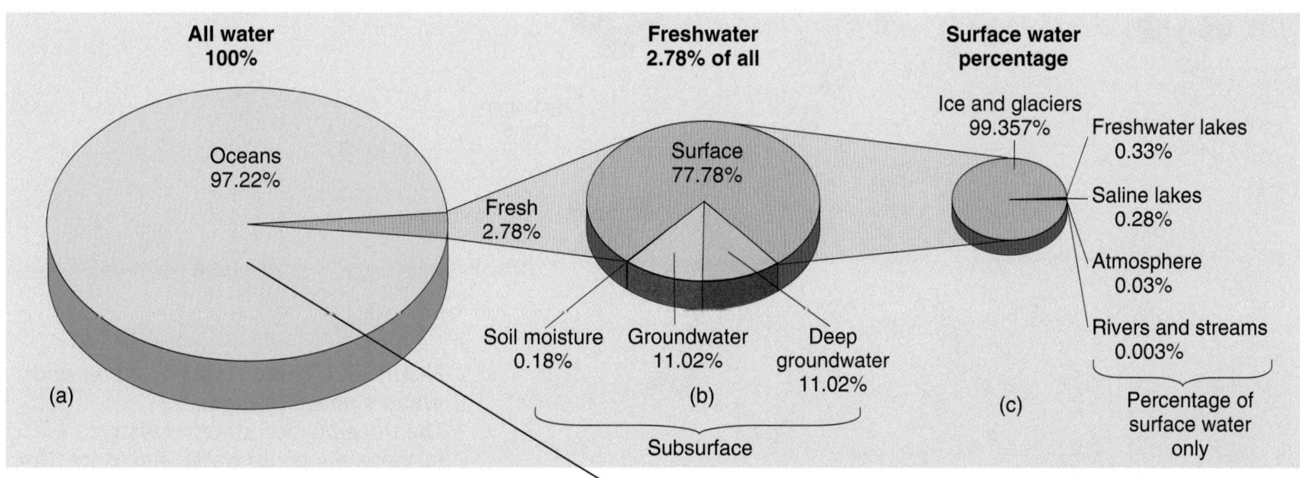

Ocean	Earth's Ocean Area (%)	*Area (km² [mi²])	*Volume (km³ [mi³])	Mean Depth of Main Basin (m [ft])
Pacific	48	179,670 (69,370)	724,330 (173,700)	4280 (14,040)
Atlantic	28	106,450 (41,100)	355,280 (85,200)	3930 (12,890)
Indian	20	74,930 (28,930)	292,310 (70,100)	3960 (12,900)
Arctic	4	14,090 (5440)	17,100 (4100)	1205 (3950)

*Data in thousands (000): includes all marginal seas.

FIGURE 5.3 Ocean and freshwater distribution on Earth.
(a) All of Earth's water; (b) freshwater, surface and subsurface; (c) distribution of surface freshwater alone.

oxygen, which readily bond. The resulting water molecule exhibits a unique *stability*, is a *versatile solvent*, and possesses *extraordinary heat characteristics*. As the most common compound on the surface of Earth, water exhibits uncommon properties.

Because of the nature of the hydrogen-oxygen bond, the hydrogen side of the water molecule has a positive charge and the oxygen side a negative charge (see Figure 5.4). This *polarity* of the water molecule explains why water "acts wet" and dissolves so many other molecules and elements. Thus, because of its solvent activity, pure water is rare in nature.

Water molecules are attracted to each other because of their polarity; the positive (hydrogen) side of one water molecule is attracted to the negative (oxygen) side of another. (Note this bonding pattern in the molecular illustrations in Figure 5.4b and c.) The bonding between water molecules is called *hydrogen bonding*. The effects of hydrogen bonding in water are observable in everyday life. Hydrogen bonding creates *surface tension* and is the cause of capillarity, which you observe when you use a paper towel. The towel draws water upward through its fibers because each molecule is pulling on its neighbor. Capillary action is an important component of soil moisture processes, discussed in Chapters 7 and 15. It is important to note that, without hydrogen bonding, water would be a gas at normal surface temperatures.

Heat Properties

For water to change from one state to another (solid, liquid, gas), heat energy must be added to it or released from it. To cause a change of state, the amount of heat energy must be sufficient to affect the hydrogen bonds between the molecules. This relationship between water and heat energy is an important driving force in producing weather. In fact, the heat exchanged in the phase changes of water provides over 30% of the energy powering the general circulation of the atmosphere. Figure 5.4 presents the terms used to describe each **phase change**.

Melting and *freezing* describe the phase change between solid and liquid. The terms *condensation* and *evaporation*, or *vaporization* at boiling temperature, apply to the change between liquid and vapor. The term **sublimation** refers to the direct change of water vapor to ice or ice to water vapor; sometimes the term *deposition* is used if water vapor attaches itself directly to an ice crystal. The deposition of water vapor as ice may form frost on surfaces.

Ice, the Solid Phase. As water cools, it behaves like most compounds and contracts in volume, reaching its greatest density at 4°C (39°F). But below that temperature, water behaves very differently from other compounds and begins to expand as more hydrogen bonds form among the slower-moving molecules, creating the hexagonal structures shown in Figure 5.4c. This expansion continues to a temperature of −29°C (−20°F), with up to a 9% increase in volume possible.

As shown in Figure 5.5a, the rigid internal structure of ice dictates the six-sided appearance of all ice crystals, which can loosely combine to form snowflakes. This six-sided preference applies to ice crystals of all shapes: plates, columns, needles, and dendrites (branching or treelike

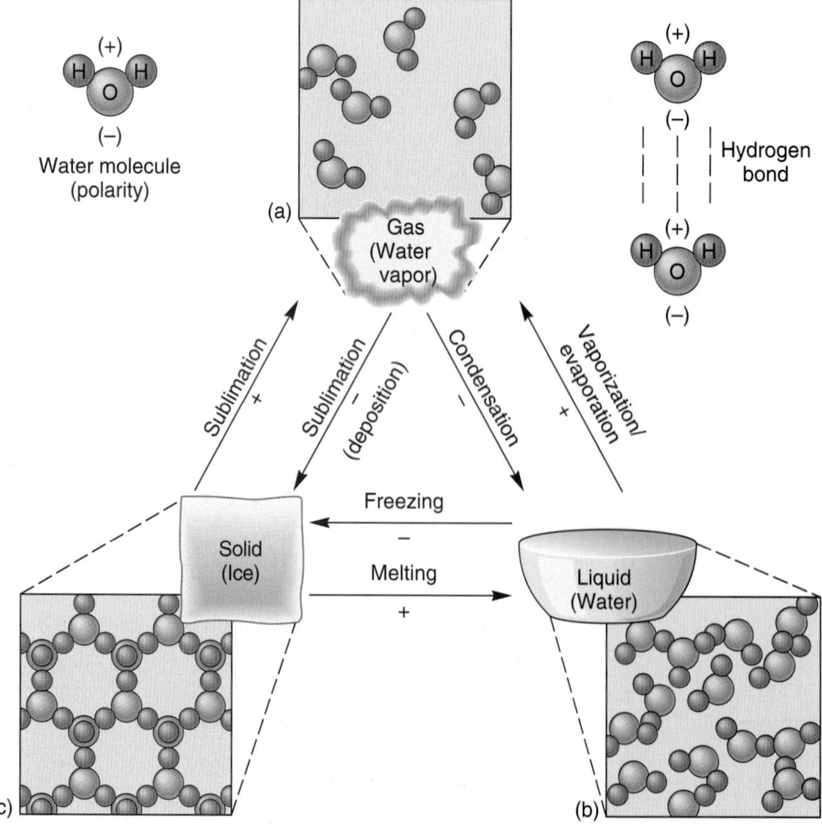

FIGURE 5.4 Three states of water and water's phase changes.
The three physical states of water: (a) water vapor, (b) water, and (c) ice. Note the molecular arrangement in each state and the terms that describe the changes from one phase to another as energy is either absorbed or released. Also note how the polarity of water molecules bonds them to one another, loosely in the liquid state and firmly in the solid state. The plus and minus symbols on the phase changes denote whether heat energy is absorbed (+) or liberated (−) during the phase change.

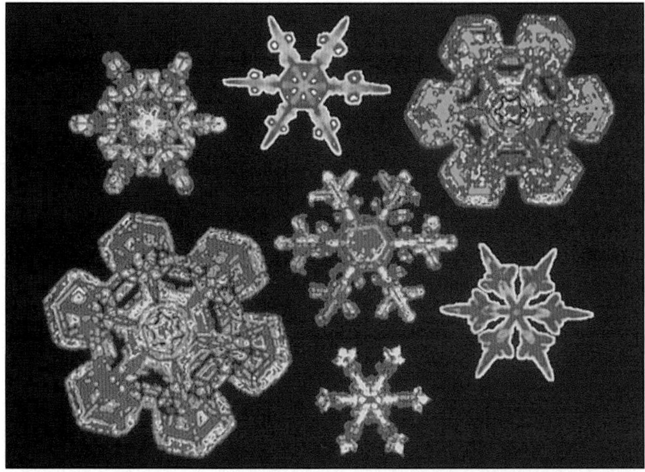
(a)

(b)

(c)

FIGURE 5.5 The uniqueness of ice forms.
(a) Ice-crystal patterns are dictated by the internal structure between water molecules. (b) This structure also explains why ice is less dense than water and why ice floats, as demonstrated by an iceberg adrift in Disko Bay, Greenland. (c) Frost—delicate ice crystals formed by the deposit of water vapor directly as ice. [(a) Photo enhancement © Scott Camazine/Photo Researchers, Inc. after W. A. Bentley; (b) photo by George Hunter/Tony Stone Images; (c) photo by author.]

forms)—a unique interaction of randomness and the determinism of physical principles. (For more on ice crystals and snowflakes see http://www.its.caltech.edu/~atomic/snowcrystals/physics/physics.htm or see http://www.lpsi.barc.usda.gov/emusnow/.)

The expansion in volume that accompanies the freezing process results in a decrease in density (the same number of molecules occupy greater space). Specifically, ice is 0.91 times the density of water, and so it floats. Without this change in density, much of Earth's freshwater would be bound in masses of ice on the ocean floor. Instead, we have floating icebergs, with approximately 1/11 (9%) of their mass exposed and 10/11 (91%) hidden beneath the ocean's surface (Figure 5.5b). News Report 5.1 discusses the power exerted by this volume increase.

Water, the Liquid Phase. As a liquid, water assumes the shape of its container and is a noncompressible fluid, quite different from solid, rigid ice. For ice to melt, heat energy must increase the motion of the water molecules to break some of the hydrogen bonds. Despite the fact that there is no change in sensible temperature between ice at 0°C and water at 0°C, 80 calories of heat must be added for the phase change of 1 gram of ice to 1 gram of water (Figure 5.6, upper left). This heat is called **latent heat** because it is stored within the water and is liberated with a phase reversal when a gram of water freezes. These 80 calories are the *latent heat of melting* and *of freezing*.

To raise the temperature of 1 gram of water from freezing at 0°C (32°F) to boiling at 100°C (212°F), we must add 100 additional calories, gaining an increase of 1 C° (1.8 F°) for each calorie added (Figure 5.6, top center).

Water Vapor, the Gas Phase. Water vapor is an invisible and compressible gas in which each molecule moves independently (Figure 5.4a). To accomplish the phase change from liquid to vapor at boiling temperature, under normal sea-level pressure, 540 calories must be added to 1 gram of boiling water (Figure 5.6, upper right). Those calories are the **latent heat of vaporization.** When water vapor condenses to a liquid, each gram gives up its hidden 540 calories as the **latent heat of condensation.**

To summarize, taking 1 gram of ice at 0°C and changing it to water vapor at 100°C—changing it from a solid, to a liquid, to a gas—*absorbs* 720 calories (80 cal + 100 cal + 540 cal). Or reversing the process, changing 1 gram of water vapor at 100°C to ice at 0°C *liberates* 720 calories.

Heat Properties of Water in Nature. In a lake, stream, or in soil water, at 20°C (68°F), every gram of water that

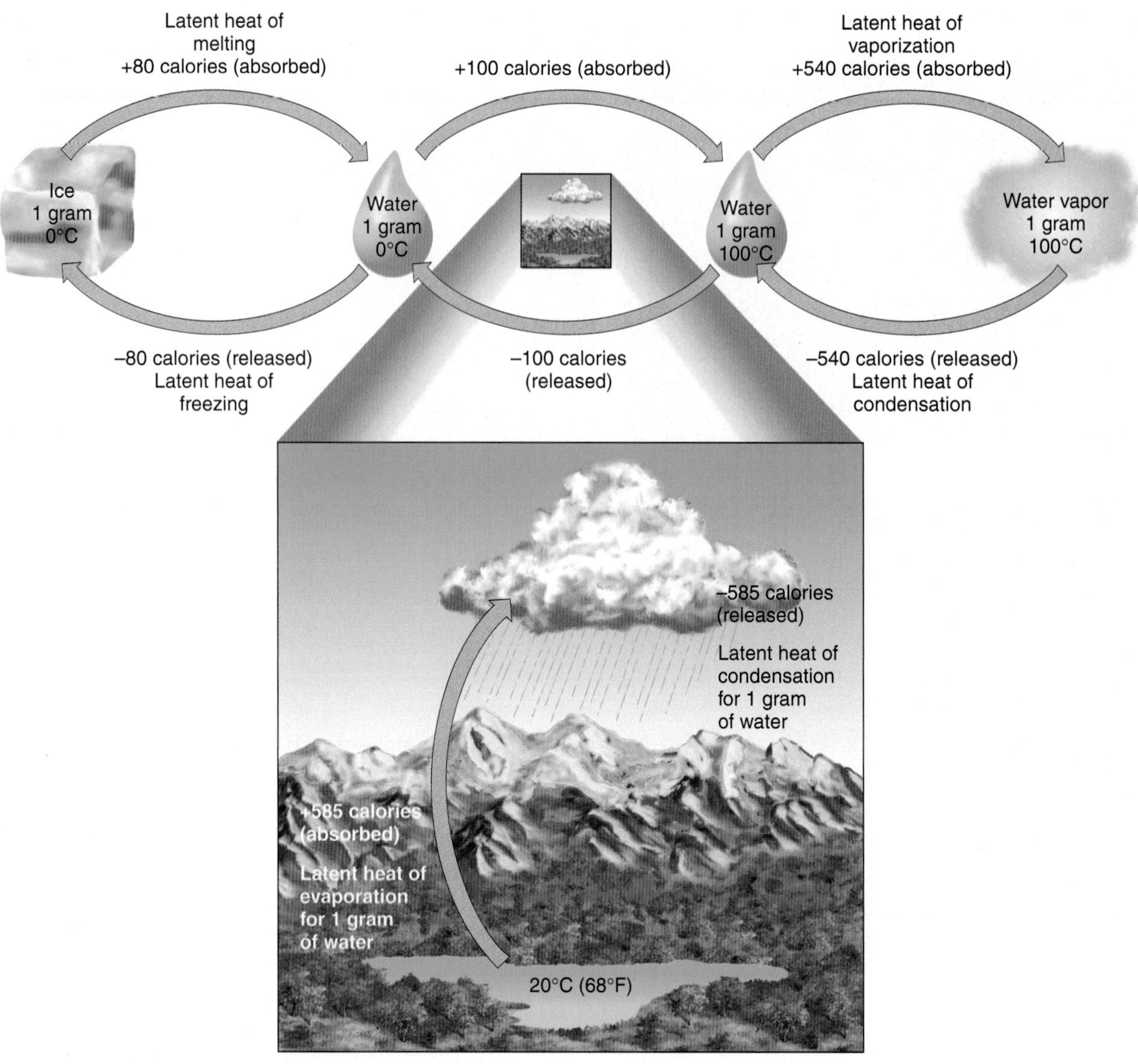

FIGURE 5.6 Water's heat energy characteristics.
A lot of latent heat energy is involved in the phase changes from ice to water and from water to water vapor. To transform 1 g of ice at 0°C to 1 g of water vapor at 100°C requires 720 calories. The inset landscape illustrates average conditions in the environment; phase change between water (lake at 20°C) and water vapor.

breaks away from the surface through evaporation must absorb from the environment approximately 585 calories as the **latent heat of evaporation** (see the natural scene in Figure 5.6). This is slightly more energy than would be required if the water were boiling (540 cal). You can feel this absorption of latent heat as evaporative cooling on your skin when it is wet. This latent heat exchange is the dominant cooling process in Earth's energy budget.

The process reverses when air that contains water vapor is cooled. The vapor eventually condenses back into the liquid state, forming moisture droplets and thus liberating 585 calories as the *latent heat of condensation* for every gram of water. Because of these unique heat properties, water is a major contributor of energy to the atmosphere. Let's now take a look at water vapor in the atmosphere.

Humidity

Humidity refers to water vapor content in the air. The capacity of air to hold water vapor is primarily a function of the temperature of both the air and the water vapor, which usually is the same. Warmer air has a greater capacity for water vapor, whereas cooler air has a lesser capacity.

We are all aware of humidity in the air, for its relationship to air temperature determines our sense of comfort. North Americans spend billions of dollars a year to adjust humidity, either with air conditioning (extracting water vapor and cooling) or with air humidifying (adding water vapor). To determine the energy available to empower weather, it is essential to know the water vapor content of air.

News Report 5.1

Breaking Roads and Pipes and Sinking Ships

Road crews are out in the summer in many parts of the country repairing streets and freeways. Winter damage to highways often is the result of the phase change from water to ice. Rainwater seeps into roadway cracks and then expands as it freezes, thus breaking up the pavement. Perhaps you have noticed that bridges are subjected to the greatest roadbed damage. The reason is that cold air can circulate beneath a bridge and produces more freeze-thaw cycles on the bridge than in the roadbed on rock and soil.

The expansion of freezing water exerts a tremendous force—enough to crack plumbing or an automobile engine block. People living in very cold climates use antifreeze and engine heaters to avoid such damage. Wrapping water pipes with insulation is a common winter task in many places.

A major shipping hazard in higher latitudes is posed by floating ice. Since ice has 0.91 the density of water, an iceberg sits with approximately 10/11 of its mass below water level. The irregular edges of submarine ice can buckle and breach the side of a passing ship, as happened to the RMS *Titanic* in 1912 on its maiden voyage, causing its sinking and a shattering of society's faith in technology.

Historically, this physical property of water was put to useful work in quarrying rock for building materials. Holes were drilled and filled with water before winter, so that when cold weather arrived the water would freeze and expand, cracking the rock into manageable shapes.

Relative Humidity

Next to air temperature and barometric pressure, the most common piece of information in local weather broadcasts is **relative humidity**. Relative humidity is a ratio (expressed as a percentage) that compares the amount of water vapor that is actually in the air (*content*) with the maximum water vapor the air *could hold* at a given temperature (*capacity*). If the air is relatively dry in comparison to its capacity, the percentage is lower; if the air is relatively moist, the percentage is higher; and if the air is holding all the moisture it can for its temperature, the percentage is 100% (Figure 5.7). The formula to calculate relative humidity is

$$\text{Relative humidity} = \frac{\text{actual water vapor } \textit{content} \text{ of the air}}{\text{maximum water vapor } \textit{capacity} \text{ of the air}} \times 100$$

Relative humidity varies because of evaporation, condensation, or temperature changes, all of which affect both the moisture content (the numerator) and the capacity (the denominator) of the air to hold water vapor. Relative humidity is an expression of an ongoing process between air and moist surfaces, for condensation and evaporation operate continuously—water molecules move back and forth between air and water.

Air is said to be **saturated**, or full, if it is holding all the water vapor that it can hold at a given temperature (100% relative humidity). In saturated air, the net transfer of water molecules between a moist surface and air is in equilibrium. Saturation indicates that any further addition of water vapor (increase in content) or any decrease in temperature (reduction in capacity) will result in active condensation (clouds, fog, or precipitation).

The temperature at which a given mass of air becomes saturated is termed the **dew-point temperature**. In other words, *air is saturated when the dew-point temperature and the air temperature are the same.* A cold drink in a glass provides a common example of these conditions. The water droplets that form on the outside of the glass condense from the air because the air layer next to the

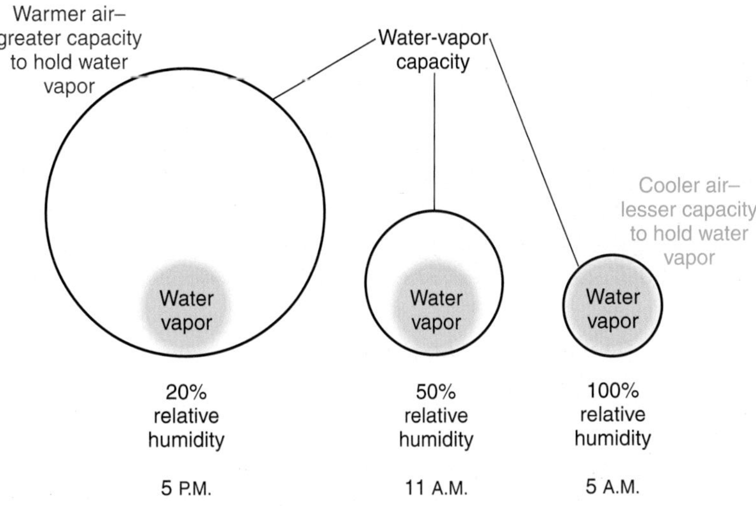

FIGURE 5.7 Water vapor content, capacity, and relative humidity.
The water vapor capacity of warm air is greater than the capacity of cold air, so relative humidity changes with temperature during the day.

glass is chilled to below its dew-point temperature and thus is saturated (Figure 5.8).

Satellites now routinely sense water vapor content of the lower atmosphere. Water vapor absorbs infrared wavelengths, making it possible to distinguish areas of relatively high water-vapor content from areas of low water-vapor content, using infrared sensors. Figure 5.9 portrays the Western Hemisphere showing water-vapor content recorded by satellite. This knowledge is important to forecasting because it shows the energy available (latent heat) and precipitation potential of those systems.

During a typical day, air temperature and relative humidity relate inversely—as temperature rises, relative humidity falls (Figure 5.10). Relative humidity is highest at dawn, when air temperature is lower and the capacity of the air to hold water vapor is less. Relative humidity is lowest in the late afternoon, when higher air temperatures increase the capacity of the air to hold water vapor. The

FIGURE 5.9 Image of water vapor in the atmosphere. Water-vapor content over the full Western Hemisphere seen in a *GOES-8* infrared image. Note subpolar low-pressure circulation. [*GOES* image courtesy of NESDIS Satellite Services Division, NOAA.]

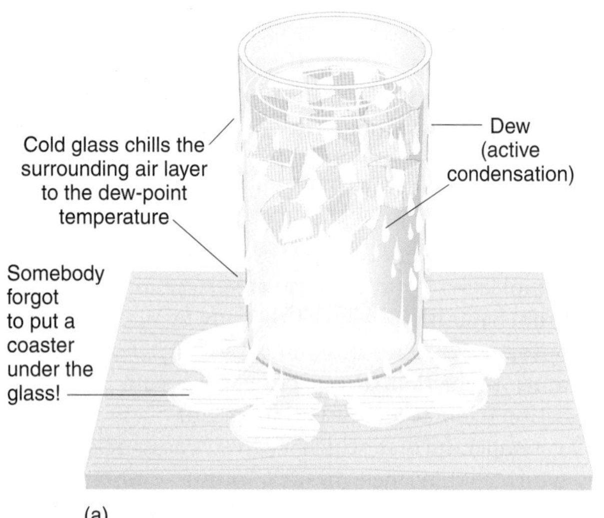

FIGURE 5.8 Dew-point temperature examples. (a) The low temperature of the glass chills the surrounding air layer to the dew point and saturation. Thus, water vapor condenses out of the air and onto the glass as dew. (b) Cold air above the rain-soaked rocks is at the dew point and is saturated. Condensation shrouds the rock in a changing veil of clouds. [Photo by author.]

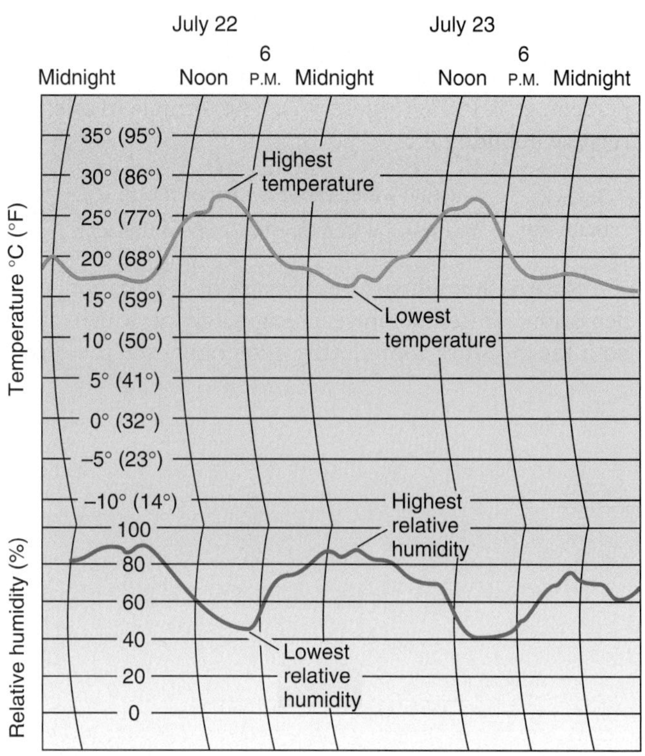

FIGURE 5.10 Typical daily temperature and relative humidity.
The water vapor capacity of warm air is greater than the capacity of cold air, so relative humidity changes with temperature—higher relative humidity in the A.M., lower relative humidity in the P.M.

actual water-vapor content in the air may have remained the same throughout the day, but because the temperature varied, relative humidity percentages dropped from morning to afternoon.

Expressions of Relative Humidity

Humidity and relative humidity may be expressed in several ways, each with its own utility and application; two of these expressions are vapor pressure and specific humidity.

Vapor Pressure. One way of describing humidity is related to air pressure. As free water molecules evaporate from a surface into the atmosphere, they become water vapor, one of the gases in air. That portion of total air pressure that is made up of water vapor molecules is termed **vapor pressure**. Like air pressure, it is expressed in millibars (mb).

Water vapor molecules continue to evaporate from a moist surface, slowly diffusing into the air, until the increasing vapor pressure in the air causes some molecules to return to the surface. Saturation is reached when the movement of water molecules between surface and air is in equilibrium. The maximum capacity of the air at a given temperature is termed the *saturation vapor pressure* and indicates the maximum pressure that water vapor molecules can exert. Any increase or decrease in temperature causes the saturation vapor pressure to change. Figure 5.11 graphs the saturation vapor pressure at varying air temperatures. The graph illustrates that for every temperature increase of 10 C°(18 F°), the vapor pressure capacity of air nearly doubles.

The inset in Figure 5.11 compares saturation vapor pressure over water and over ice surfaces at subfreezing temperatures. You can see that saturation vapor pressure is greater above a water surface than over an ice surface—that is, it takes more water vapor molecules to saturate air above water than it does above ice.

Specific Humidity. A useful humidity measure is one that remains constant as temperature and pressure change. **Specific humidity** refers to the mass of water vapor (in grams) per mass of air (in kilograms) at any specified temperature. The maximum mass of water vapor that a kilogram of air can hold at any specified temperature is termed the *maximum specific humidity* and is plotted in Figure 5.12.

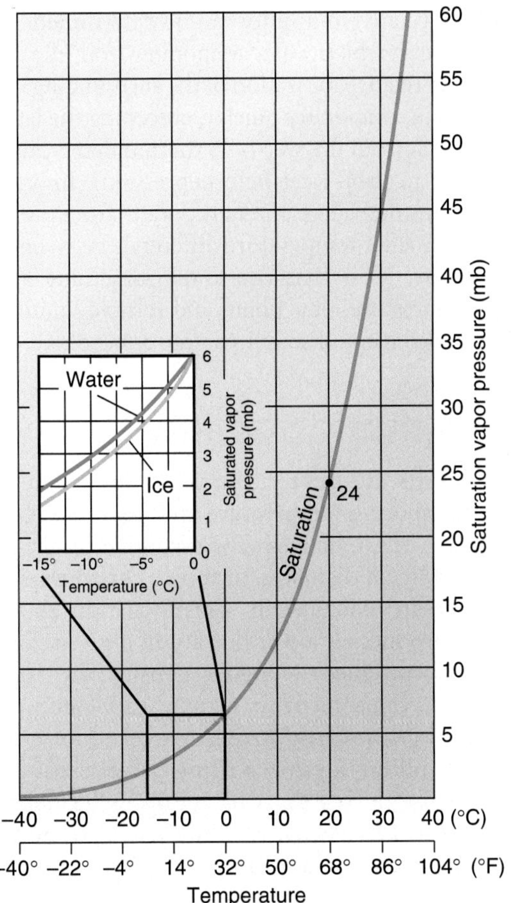

FIGURE 5.11 Saturation vapor pressure.
Saturation vapor pressure of air at various temperatures—the maximum capacity of air to hold water vapor, expressed in vapor pressure. As the graph shows, air at 20°C (68°F) has a saturation vapor pressure of 24 mb; that is, the air is saturated if the water-vapor portion of the air pressure is at 24 mb. Thus, if the water-vapor content actually present is exerting a vapor pressure of only 12 mb in 20°C air, the relative humidity is 50% (12 mb ÷ 24 mb = 0.50 × 100 = 50%).

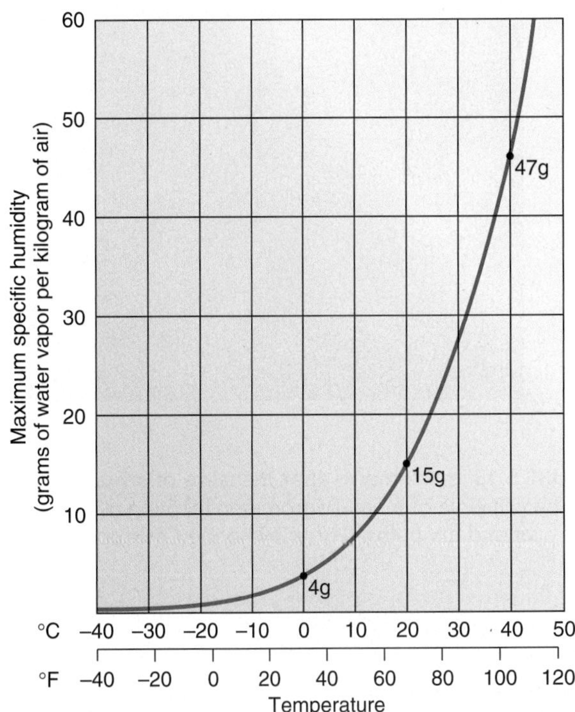

FIGURE 5.12 Maximum specific humidity.
Maximum specific humidity for a mass of air at various temperatures—the maximum capacity of the air to hold water vapor, expressed by weight. The graph shows that a kilogram of air could hold a maximum specific humidity of 47 g of water vapor at 40°C (104°F), 15 g at 20°C (68°F), and about 4 g at 0°C (32°F). Therefore, if a kilogram of air at 40°C has a specific humidity of 12 g, its relative humidity is 25.5% (12 g ÷ 47 g = 0.255 × 100 = 25.5%).

Because it is measured in mass, specific humidity is not affected by changes in temperature or pressure, such as occur when an air parcel rises to higher elevations, and is therefore valuable in forecasting weather and describing the moisture content of large air masses. Specific humidity stays constant despite volume changes.

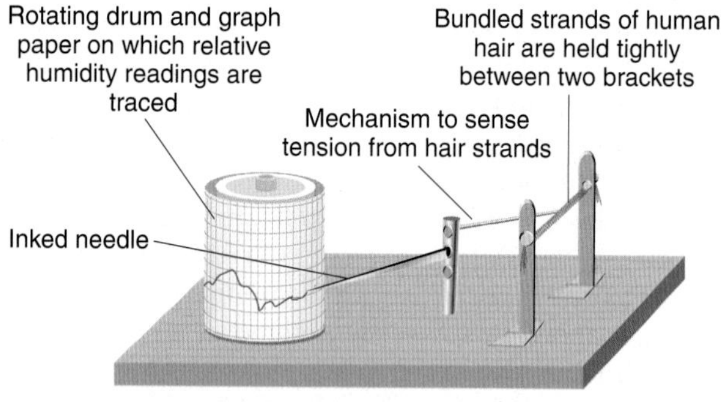

Rotating drum and graph paper on which relative humidity readings are traced

Bundled strands of human hair are held tightly between two brackets

Mechanism to sense tension from hair strands

Inked needle

(a)

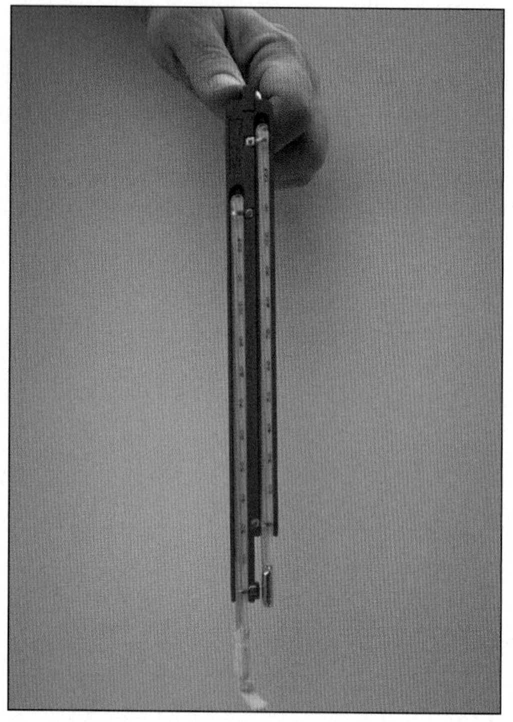

(b)

FIGURE 5.13 Instruments that measure relative humidity. (a) The principle of a hair hygrometer. (b) Sling psychrometer with wet and dry bulbs. [Photo by Bobbé Christopherson.]

Buoyancy force

Air parcel

Gravitational force

Higher density

Lower density

Lower air pressure

Higher air pressure

Earth's surface

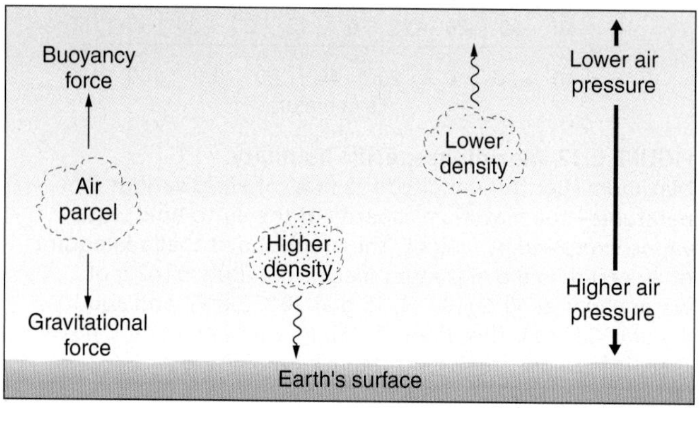

FIGURE 5.14 The forces acting on an air parcel. Buoyancy and gravitational forces work on an air parcel. Different densities produce rising or falling parcels in response to imbalance in these forces.

Instruments for Measurement. Relative humidity is measured with various instruments. The *hair hygrometer* uses the principle in which human hair changes as much as 4% in length between 0 and 100% relative humidity. The instrument connects a standardized bundle of human hair through a mechanism to a gauge. As the hair absorbs or loses water in the air, its change in length indicates relative humidity (Figure 5.13a).

Another instrument used to measure relative humidity is a *sling psychrometer*. This device has two thermometers mounted side-by-side on a metal holder. One is called the *dry-bulb thermometer*; it simply records the ambient (surrounding) air temperature. The other thermometer is called the *wet-bulb thermometer*. This bulb is covered by a cloth wick, which is moistened. The psychrometer is then spun by its handle (Figure 5.13b). (Other versions use a motor to spin the thermometers, or they are set up with a fan.) After a minute or two, the temperature of each bulb is compared on a relative humidity (psychrometric) chart, from which relative humidity can be determined.

The rate at which water evaporates from the wick depends on the relative saturation of the surrounding air. If the air is dry, water evaporates quickly, absorbing the latent heat of evaporation from the wet-bulb thermometer, causing the temperature to drop (wet-bulb depression). In an area of high humidity, much less water evaporates from the wick resulting in a smaller temperature difference between the two thermometers. Now that you know something about atmospheric moisture, dew point, and relative humidity, let's examine the concept of stability in the atmosphere.

Atmospheric Stability

Meteorologists use the term *parcel* to describe a body of air that has specific temperature and humidity characteristics. Think of an air parcel as a volume of air, perhaps 300 m (1000 ft) in diameter, or more. Differences in temperature create changes in density within the parcel. Warm air produces a lower density in a given volume of air; cold air produces a higher density. Two opposing forces work on a parcel of air: an upward *buoyancy force* and a downward *gravitational force*. A parcel of lower density will rise (is more buoyant); a rising parcel expands as external pressure decreases. A parcel of higher density will descend (is less buoyant); a falling parcel compresses as external pressure increases. Figure 5.14 shows air parcels and illustrates these relationships.

Stability refers to the tendency of an air parcel, with its water-vapor cargo, either to remain in place or to change vertical position by ascending (rising) or descending (falling).

Determining the degree of stability involves measuring simple temperature relationships between the air parcel and the surrounding air. Such temperature measurements are made daily with balloon soundings (instrument packages called *radiosondes*) at thousands of weather stations.

Adiabatic Processes

The **normal lapse rate**, as introduced in Chapter 2, is the average decrease in temperature with increasing altitude, a value of 6.4 C° per 1000 m (3.5 F° per 1000 ft). This average for temperature change is for still, calm air; but it can differ greatly under varying weather conditions, and so the actual lapse rate at a particular place and time is labeled the **environmental lapse rate**.

An ascending parcel of air tends to cool by expansion, responding to the reduced pressure at higher altitudes. In contrast, *descending air tends to heat by compression* (Figure 5.15). Both ascending and descending temperature changes are assumed to occur *without any heat exchange between the surrounding environment and the vertically moving parcel of air.*

The warming and cooling rates for a parcel of expanding or compressing air are termed **adiabatic**. (*Diabatic* means occurring with an exchange of heat; *adiabatic* means occurring *without* a loss or gain of heat.) Adiabatic temperatures are measured with one of two specific rates, depending on moisture conditions in the parcel: *dry* adiabatic rate (DAR) and *moist* adiabatic rate (MAR).

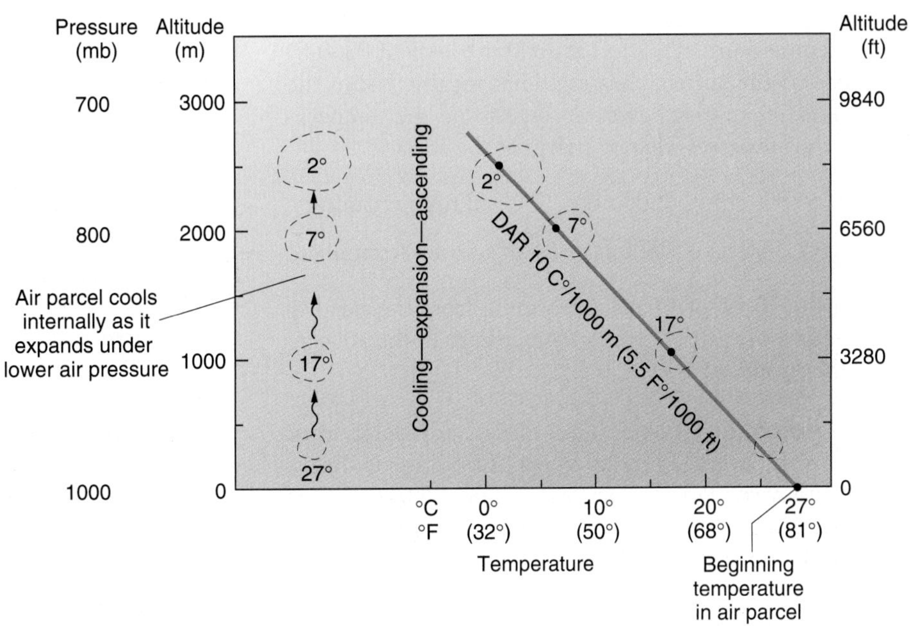

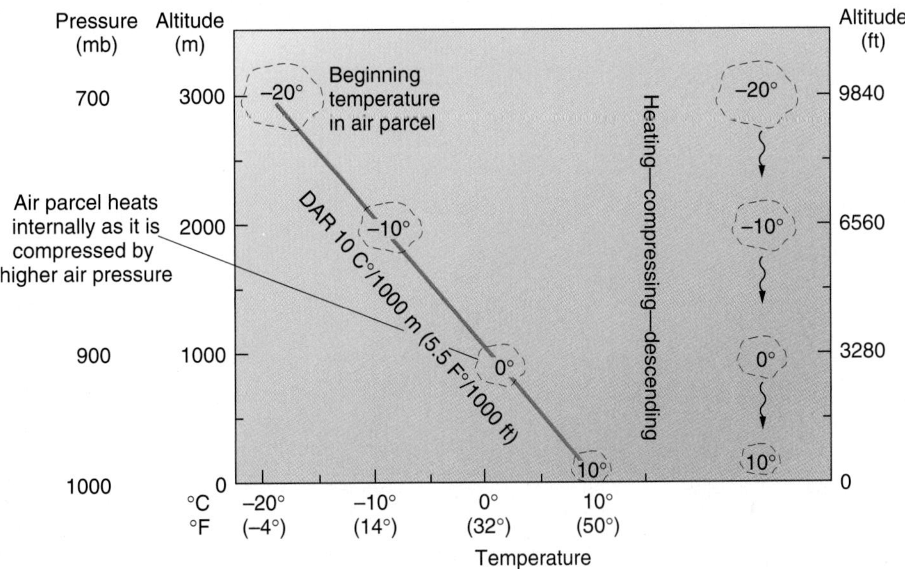

FIGURE 5.15 Adiabatic cooling and heating.
Vertically moving air parcels expand when they rise (because air pressure is less with increasing altitude) and are compressed when they descend. (a) A rising air parcel that is less than saturated cools adiabatically at the dry adiabatic rate (DAR). (b) A descending air parcel that is less than saturated heats adiabatically by compression at the DAR.

Dry Adiabatic Rate. The **dry adiabatic rate (DAR)** is the rate at which "dry" air cools by expansion (if ascending) or heats by compression (if descending). "Dry" air is less than saturated, with a relative humidity less than 100%. The DAR is 10 C° per 1000 m (5.5 F° per 1000 ft), as illustrated in Figure 5.15. For example, in Figure 5.15a, consider an unsaturated parcel of air at the surface, whose temperature measures 27°C (81°F). It rises, expands, and cools at the DAR as it lifts from the ground to 2500 m (approximately 8000 ft). The temperature of the parcel at 2500-m altitude is 2°C (36°F), having cooled by expansion as it rose:

$$(10\ \text{C°}/1000\ \text{m}) \times 2500\ \text{m} = 25\ \text{C° of total cooling}$$

$$(5.5\ \text{F°}/1000\ \text{ft}) \times 8000\ \text{ft} = 44\ \text{F° of total cooling}$$

Subtracting the 25 C° of cooling from the starting temperature of 27°C gives the temperature at 2500 m of 2°C. (Note that the parcel cooled adiabatically, without a loss of heat to the environment.)

In Figure 5.15b, assume that an unsaturated air parcel with a temperature of −20°C at 3000 m (−4°F at 9800 ft) descends to the surface, heating adiabatically. Using the dry adiabatic lapse rate, we can determine the temperature of the air parcel when it arrives at the surface:

$$(10\ \text{C°}/1000\ \text{m}) \times 3000\ \text{m} = 30\ \text{C° of total warming}$$

$$(5.5\ \text{F°}/1000\ \text{ft}) \times 9800\ \text{ft} = 54\ \text{F° of total warming}$$

Adding the 30 C° of adiabatic warming from the starting temperature of −20°C gives a temperature in the air parcel at the surface of 10°C (50°F).

Moist Adiabatic Rate. The **moist adiabatic rate (MAR)** is the average rate at which ascending air that is moist (saturated) cools by expansion. The *average* MAR is 6 C° per 1000 m (3.3 F° per 1000 ft), or roughly 4 C° (2 F°) less than the DAR. However, the MAR varies with moisture content and temperature, and can range from 4 C° to 10 C° per 1000 m (2 F° to 5.5 F° per 1000 ft). The reason is that in a saturated air parcel, latent heat of condensation is liberated as sensible heat, which reduces the adiabatic rate of cooling. The release of latent heat may vary, which affects the MAR. The MAR is much lower than the DAR in warm air, whereas the two rates are more similar in cold air.

Stable and Unstable Atmospheric Conditions. The relationship among the DAR, MAR, and the environmental (actual) lapse rate is complex and determines the stability of the atmosphere over an area. You can see the range of possible relationships in Figure 5.16.

Let's apply this concept to a specific situation. Figure 5.17 illustrates two of these temperature relationships in the atmosphere that lead to different conditions: unstable and stable. For the sake of illustration, both examples begin with a parcel of air at the surface at 25°C (77°F). In each example, examine the temperature relationships between the parcel of air and the surrounding environment. Assume that a lifting mechanism is present (we examine lifting mechanisms shortly).

Under the unstable conditions in Figure 5.17a, the air parcel continues to rise through the atmosphere because it is warmer (and therefore less dense and more buoyant) than the surrounding environment. The air surrounding the lifting parcel is cooler by an environmental lapse rate of 12 C° per 1000 m (6.6 F° per 1000 ft). By 1000 m (3280 ft), the parcel has adiabatically cooled by expansion to 15°C (59°F), but the surrounding air is at 13°C (55.4°F). The temperature in the parcel is 2 C° (3.6 F°) warmer than the surrounding air, and therefore less dense, so it continues to lift, cool, and approach saturation (100% relative humidity and cloud formation).

The stable conditions in Figure 5.17b result from an environmental lapse rate of 5 C° per 1000 m (3 F° per

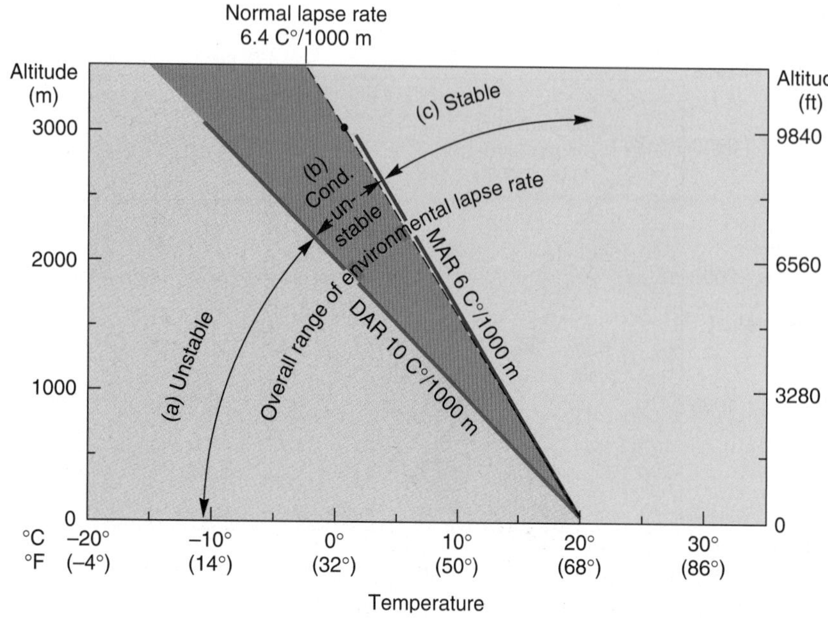

FIGURE 5.16 Temperature relationships and atmospheric stability.
The relationship between dry and moist adiabatic rates and environmental lapse rates produces three atmospheric conditions: (a) unstable (environmental lapse rate exceeds the DAR), (b) conditionally unstable (environmental lapse rate is between the DAR and MAR), and (c) stable (environmental lapse rate is less than DAR and MAR).

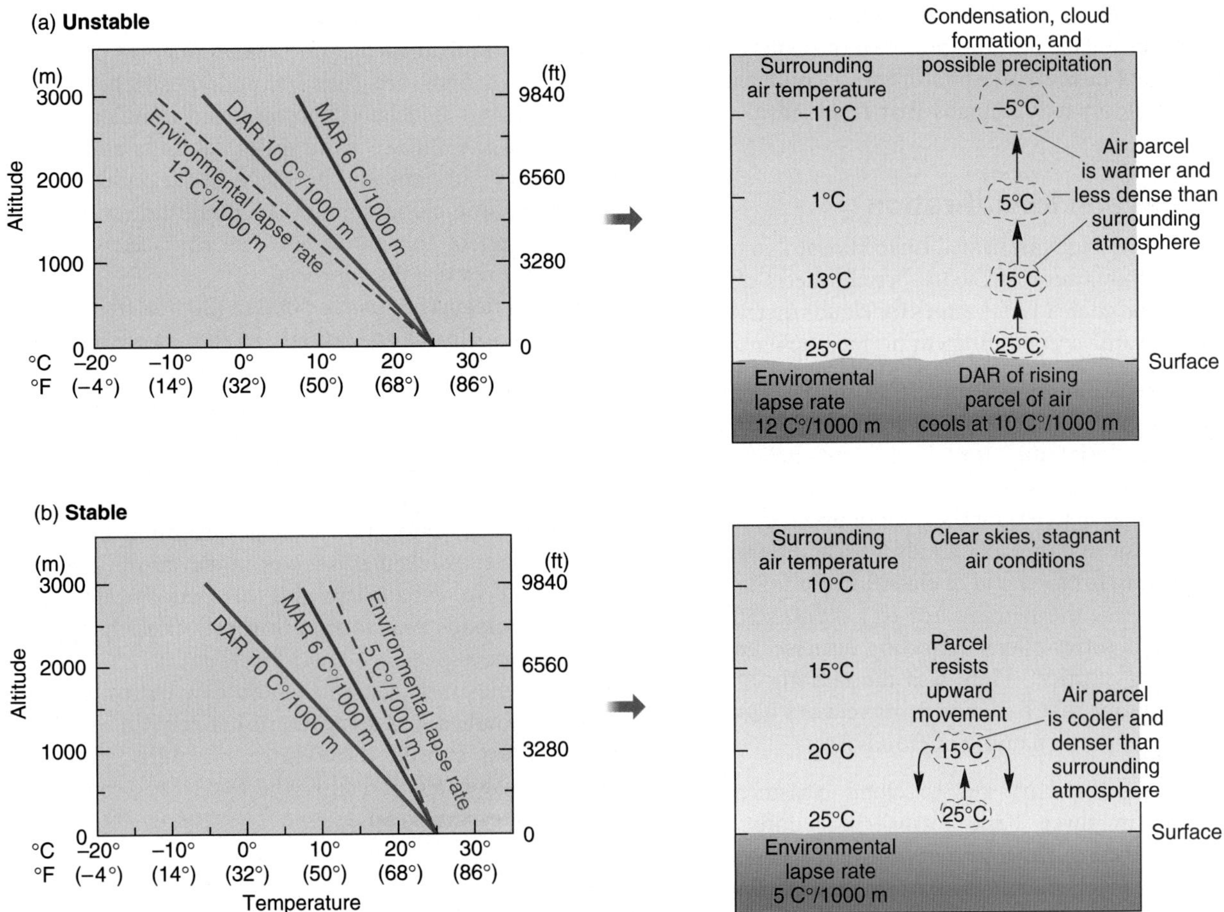

FIGURE 5.17 Unstable and stable conditions.
Specific examples of (a) unstable and (b) stable conditions in the lower atmosphere. Note the environmental lapse rate in each graph: higher than the DAR in (a), lower than the MAR in (b).

1000 ft), which is less than both the DAR and the MAR. Thus, the parcel of air is forced to settle back to its original position because it is cooler (higher density, less buoyant) than the surrounding environment. The parcel lacks a physical motive for liftoff.

With these stability relationships in mind, let's look at the most visible expressions of stability in the atmosphere: clouds and fog.

Clouds and Fog

Clouds are beautiful indicators of atmospheric stability, moisture content, and weather conditions. A **cloud** is an aggregation, or grouping, of moisture droplets and ice crystals that are suspended in air and are great enough in volume and density to be visible to the human eye. Although cloud types are too numerous to fully describe in this text, their classification scheme is simple and summarized here.

Clouds are not initially composed of raindrops. Instead, they are made up of a multitude of **moisture droplets**, each individually invisible to the human eye without magnification. An average raindrop, at 2000μm diameter (0.2 cm or 0.078 in.), is made up of a million or more of these moisture droplets. After more than a century of debate, scientists have determined how these droplets form during condensation and what processes cause them to coalesce into raindrops.

Under unstable conditions, a parcel of air may rise to an altitude where it becomes saturated—that is, the air cools to the dew-point temperature and 100% relative humidity. Further cooling of the air parcel due to lifting produces active condensation of water vapor to water. This condensation requires microscopic particles called **condensation nuclei**, which always are present in the atmosphere.

In continental air masses, which average 10 billion nuclei per cubic meter, condensation nuclei are typically derived from ordinary dust, volcanic and forest-fire soot and ash, and particles from fuel combustion. Given the air composition over cities, great concentrations of such nuclei are available. In maritime air masses, which average one billion nuclei per cubic meter, the nuclei are formed from a high concentration of sea salts derived from ocean sprays. Salts are particularly attracted to moisture and thus are called *hygroscopic nuclei.*

Given the preconditions of saturated air, availability of condensation nuclei, and the presence of cooling (lifting) mechanisms in the atmosphere, active condensation occurs. Let us look at the clouds that result from these processes.

Cloud Types and Identification

In 1803 an English biologist named Luke Howard, in his article "On the Modification of Clouds," established a classification system and coined Latin names for clouds that we still use. About Howard's accomplishment his biographer stated,

> Clouds were no longer exempt from human comprehension, and Howard, in contributing both a system of analysis and a full Latin nomenclature covering their families and genera, had contributed more than anyone to easing the path of understanding.... But the naming of clouds was a different kind of gesture for the hand of classification to have made. Here was the naming not of a solid, stable thing but of a series of self-canceling evanescences [disappearing entities]. Here was the naming of a fugitive presence that hastened to its onward dissolution. Here was the naming of clouds.*

Altitude and *shape* are key to cloud classification. Clouds come in three basic forms—flat, puffy, and wispy—which occur in four primary altitude classes and 10 basic types. Clouds that develop horizontally—flat and layered—are *stratiform* clouds. Those that develop vertically—puffy and globular—are *cumuliform*. Wispy clouds usually quite high in altitude, composed of ice crystals, are *cirroform*. These three basic forms occur in four altitudinal classes: low, middle, high, and those that vertically develop through these altitude classes. Figure 5.18 illustrates the general appearance of the basic classes and types of clouds and includes representative photographs.

Low clouds, ranging from the surface up to 2000 m (6500 ft) in the middle latitudes, are simply called *stratus* or *cumulus* (Latin for "layer" and "heap," respectively). **Stratus** clouds appear dull, gray, and featureless. When they yield precipitation, they are called **nimbostratus** (*nimbo-* denotes precipitation or rain bearing), and their showers typically fall as drizzling rain (Figure 5.18e).

Cumulus clouds appear bright and puffy, like cotton balls (Figure 5.18h). When they do not cover the sky, they float by in infinitely varied shapes. Vertically developed cumulus clouds extend beyond low altitudes into middle and high altitudes and are illustrated to the far right in Figure 5.18.

Stratus and cumulus middle-level clouds are denoted by the prefix *alto-*, as in *altostratus*. They are made of water droplets and, when cold enough, can be mixed with ice crystals. **Altocumulus** clouds, in particular, represent a broad category that occurs in many different styles (Figure 5.18a).

*R. Hamblyn, *The Invention of Clouds, How An Amateur Meteorologist Forged the Language of the Skies* (New York: Farrar, Straus, and Giroux, 2001), pp. 165, 171.

A cumulus cloud can develop into a towering giant called **cumulonimbus** (again, *-nimbus* denotes precipitation). Such clouds are called thunderheads because of their shape and associated lightning and thunder (Figure 5.19). Note the surface wind gusts, updrafts and downdrafts, heavy rain, and the presence of ice crystals at the top of the rising cloud column. High-altitude winds may shear the top of the cloud into the characteristic anvil shape of the mature thunderhead.

Clouds occurring above 6000 m (20,000 ft) are composed principally of ice crystals in thin concentrations. These wispy filaments, usually white except when colored by sunrise or sunset, are **cirrus** clouds (Latin for "curl of hair"), sometimes dubbed mare's tails. Cirrus clouds look as if an artist took a brush and placed delicate feathery strokes high in the sky (Figure 5.18b). Often, cirrus clouds are associated with an oncoming storm, especially if they thicken and lower in elevation. The prefix *cirro-* as in *cirrostratus* and *cirrocumulus*, indicates other high clouds that form a thin veil or puffy appearance, respectively.

Sometimes near the end of the day, lumpy, grayish, low-level clouds called **stratocumulus** may fill the sky in patches. Near sunset, these spreading puffy stratiform remnants may catch and filter the Sun's rays, sometimes indicating clearing weather.

Fog

By international agreement, **fog** is a cloud layer on the ground, with visibility restricted to less than 1 km (3300 ft). The presence of fog tells us that the air temperature and the dew-point temperature at ground level are nearly identical, producing saturated conditions. Generally, an inversion layer caps fog, with as much as 30 C° (50 F°) difference in air temperature between the ground under the fog and the clear, sunny skies above. Two forms of fog related to cooling are advection fog and radiation fog. News Report 5.2 describes fog as a water resource for some.

Advection Fog. As the name implies, **advection fog** forms when air in one place migrates to another place where saturated conditions exist. When warm, moist air overlays cooler ocean currents, lake surfaces, or snow masses, the layer of air directly above the surface is chilled to the dew point. Off all subtropical west coasts in the world, summer fog forms in this manner (Figure 5.20).

Another type of fog forms when cold air flows over the warm water of a lake, ocean surface, or even a swimming pool. An **evaporation fog**, or steam fog, may form as the water molecules evaporate from the water surface into the cold overlying air, effectively humidifying the air. When visible at sea, the term *sea smoke* is applied to this shipping hazard. Figure 5.21 shows an evaporation fog over a lake at sunrise.

A type of advection fog, involving the movement of air, forms when moist air is forced to higher elevations along a hill or mountain. This upslope lifting leads to cooling by expansion as the air rises. The resulting

News Report 5.2

Harvesting Fog

Desert organisms have adapted remarkably to the presence of coastal fog along western coastlines in subtropical latitudes. For example, sand beetles in the Namib Desert in extreme southwestern Africa harvest water from the fog. They hold up their wings so condensation collects and runs down to their mouths. For centuries, coastal villages in the deserts of Oman collected water drips from trees that came from coastal fogs.

In the Atacama Desert of Chile and Peru, residents stretch large nets to intercept the fog; moisture condenses on the netting and drips into trays, flowing through pipes to a 100,000-liter (26,000-gallon) reservoir. Large sheets of plastic mesh along a ridge of the El Tofo mountains harvest water from advection fog (Figure 1). Chungungo, Chile, receives 10,000 liters (2,600 gallons) of water from 80 fog-harvesting collectors in a project developed by Canadian (International Development Research Center) and Chilean interests and made operational in 1993. At least 30 countries across the globe experience conditions suitable for this water resource technology. (See http://www.idrc.ca/nayudamma/fogcatc_72e.html.)

FIGURE 1 Fog harvesting.
In the mountains inland from Chungungo, Chile, polypropylene mesh stretched between two posts captures advection fog for local drinking water supplies. [Photo by Robert S. Schemenauer.]

upslope fog forms a stratus cloud at the level of saturation. Along the Appalachians and the eastern slopes of the Rockies, such fog is common in winter and spring. Another fog associated with topography is **valley fog** (Figure 5.22). Because cool air is denser, it settles in low-lying areas, producing a fog in the chilled, saturated layer near the ground.

Radiation Fog. **Radiation fog** forms when radiative cooling of a surface chills the air layer directly above that surface to the dew-point temperature, creating saturated conditions and fog (Figure 5.23). This fog occurs especially on clear nights over moist ground; it does not occur over water, because water does not cool appreciably overnight. Slight movements of air deliver even more moisture to the cooled area for more fog formation of greater depth.

Specific conditions of humidity, stability, and cloud coverage occur in regional, homogenous masses of air. These air masses interact to produce weather patterns—our next topic.

Air Masses

Each area of Earth's surface imparts its varying characteristics to the air it touches. Such a distinctive body of air is an **air mass**, and it initially reflects the characteristics of its *source region.* For example, weather reporters speak of "a cold Canadian air mass" or "moist tropical air mass."

The longer an air mass remains stationary over a region, the more definite its physical attributes become. Within each air mass there is a homogeneity of temperature and humidity that sometimes extends through the lower half of the troposphere.

Air masses generally are classified according to the moisture and temperature characteristics of their source regions:

1. *Moisture*—designated **m** for maritime (wetter) and **c** for continental (drier).
2. *Temperature* (latitude)—designated **A** (arctic), **P** (polar), **T** (tropical), **E** (equatorial), and **AA** (antarctic).

The principal air masses that affect North America in winter and summer are mapped and described in Figure 5.24.

Continental polar (**cP**) air masses form only in the Northern Hemisphere and are most developed in winter when they dominate cold weather conditions. An area covered by **cP** air in winter experiences cold, stable air, clear skies, high pressure, and anticyclonic wind flow, all visible on the weather map in Figure 5.25 (p. 154). The Southern Hemisphere lacks the necessary continental masses (continentality) at high latitudes to produce continental polar characteristics.

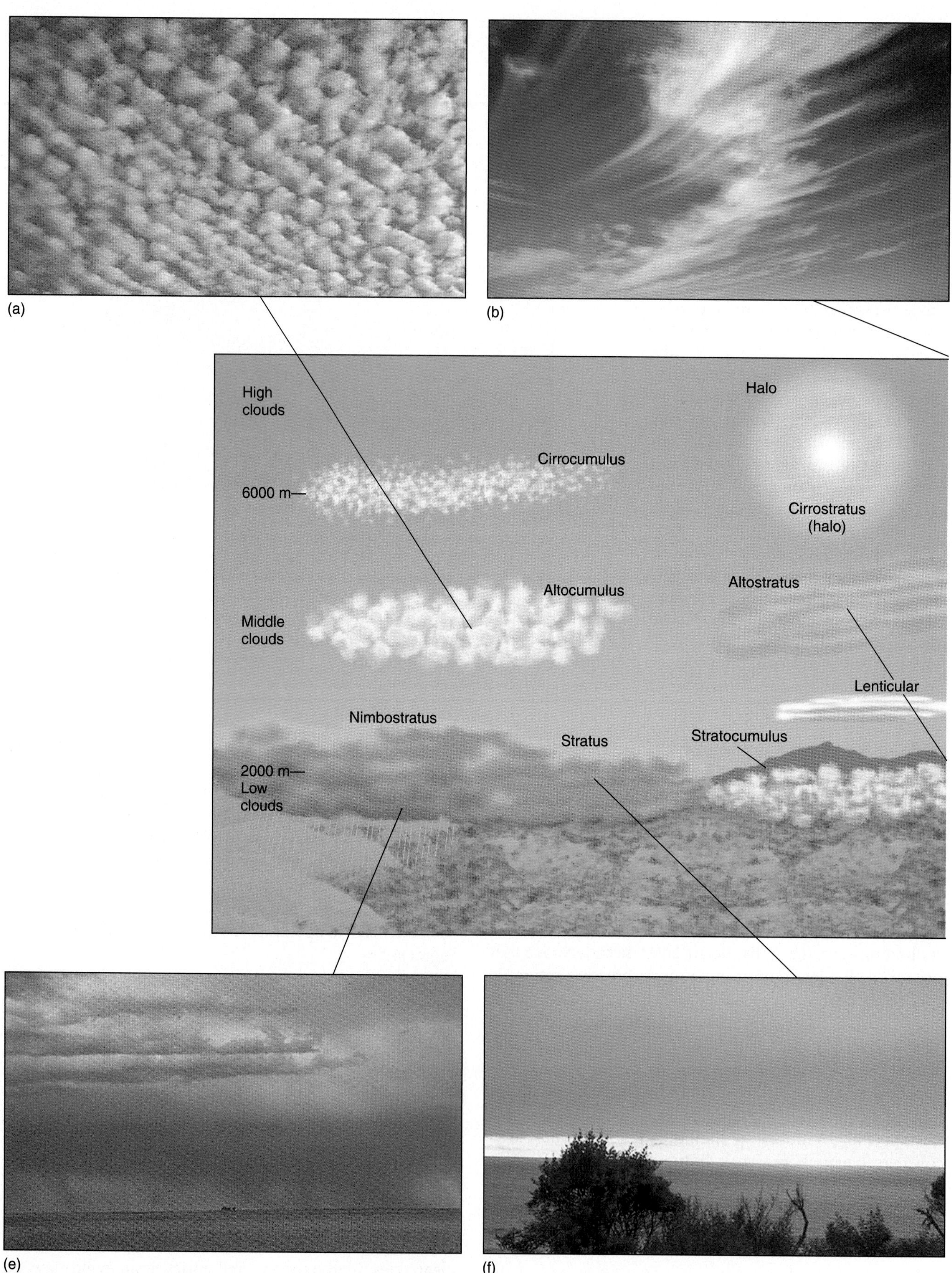

FIGURE 5.18 Principal clouds.
Principal cloud types, classified by form (cirroform, stratiform, and cumuliform) and altitude (low, middle, high, and vertically developed across altitude). (a) altocumulus, (b) cirrus, (c) cirrostratus, (d) cumulonimbus, (e) nimbostratus, (f) stratus, (g) altostratus, and (h) cumulus. [Photos by author, except (b) and (d) photo by Bobbé Christopherson.]

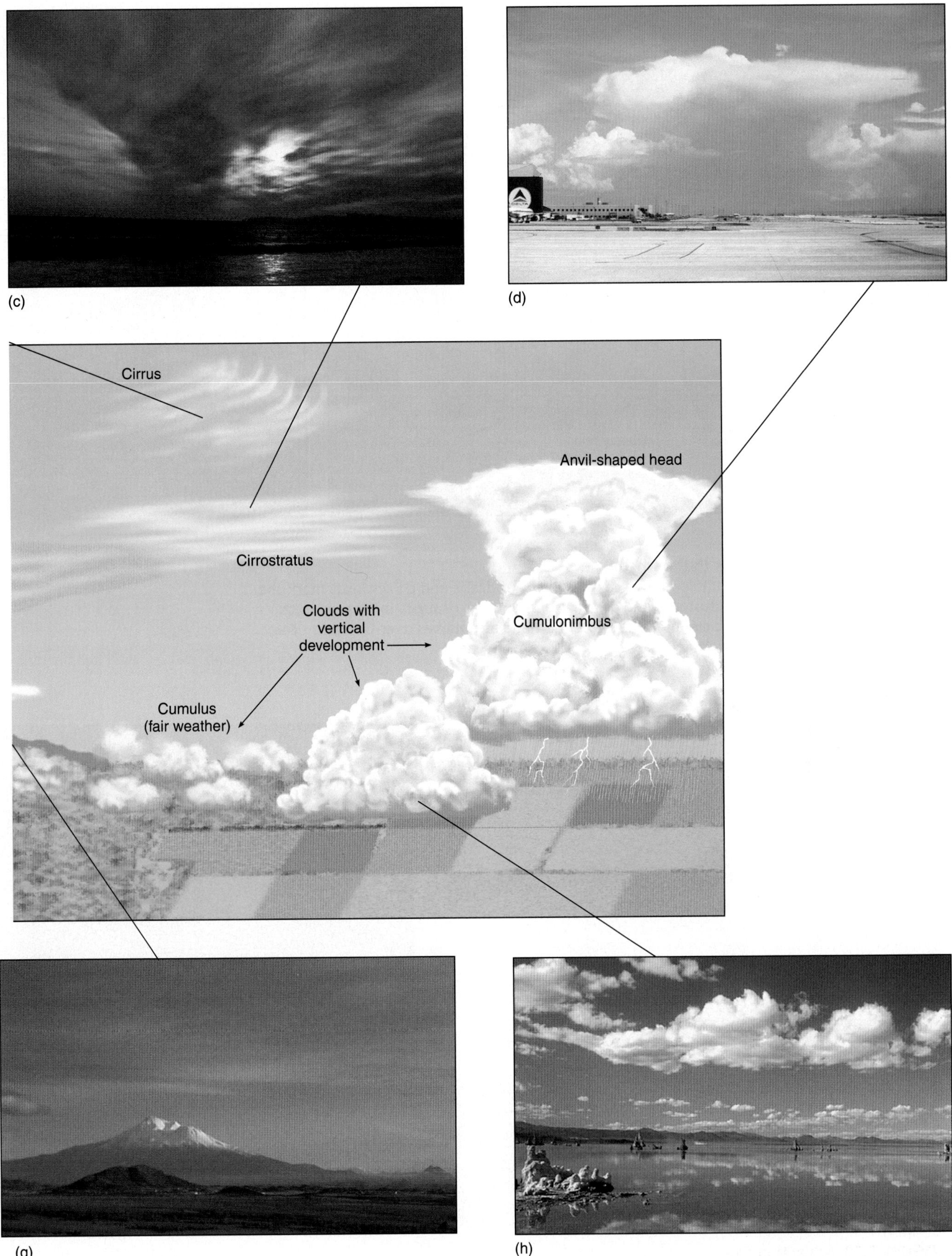
Cirrus
Anvil-shaped head
Cirrostratus
Clouds with vertical development
Cumulonimbus
Cumulus (fair weather)

(c)
(d)
(g)
(h)

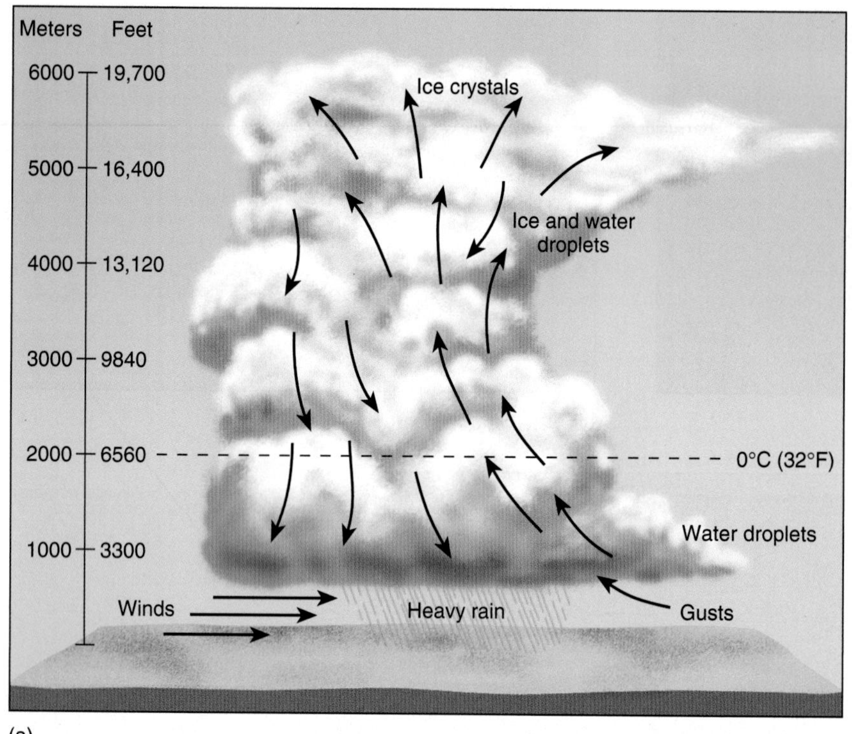

(a)

(b)

FIGURE 5.19 Cumulonimbus thunderhead.
(a) Structure and form of a cumulonimbus cloud. Violent updrafts and downdrafts mark the circulation within the cloud. Blustery wind gusts occur along the ground. (b) Space Shuttle astronauts capture a dramatic cumulonimbus thunderhead as it moves over Galveston Bay, Texas. [Space Shuttle photo from NASA.]

FIGURE 5.20 Advection fog.
San Francisco's Golden Gate Bridge shrouded by an advection fog characteristic of summer conditions along a western coast. [Photo by author.]

FIGURE 5.21 Evaporation fog at sunrise.
Evaporation fog (sea smoke) at dawn on a cold morning at Donner Lake, California. Later that morning as temperatures rose, what do you think happened to the evaporation fog? [Photo by Bobbé Christopherson.]

FIGURE 5.22 Valley fog.
Cold air settles in the valleys of the Appalachian Mountains, chilling the air to the dew point and forming a valley fog. [Photo by author.]

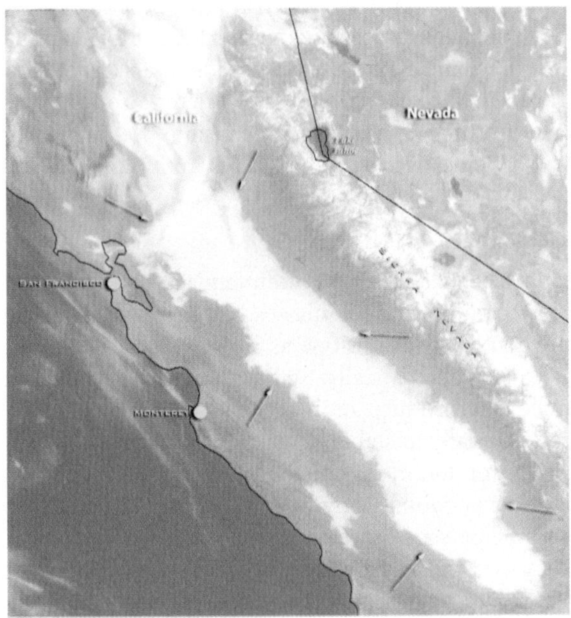

FIGURE 5.23 Radiation fog.
Satellite image of a radiation fog in Central California, June 11, 2001. This fog is locally known as a tule fog (pronounced "toolee") because of its association with the tule (bulrush) plants that line the low-elevation islands and marshes of the Sacramento River and San Joaquin River delta regions. [*GOES-10* image courtesy of NOAA.]

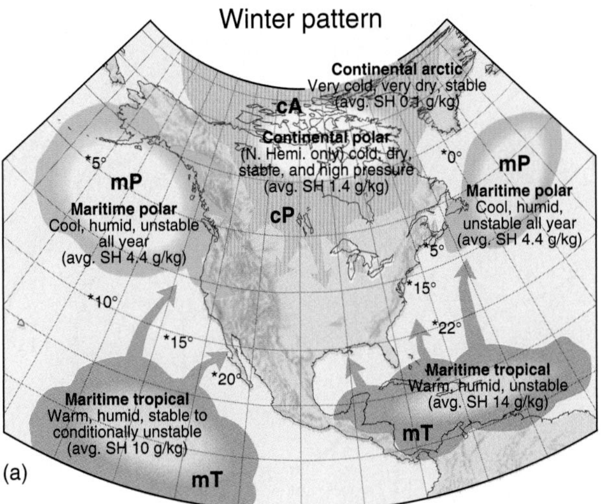

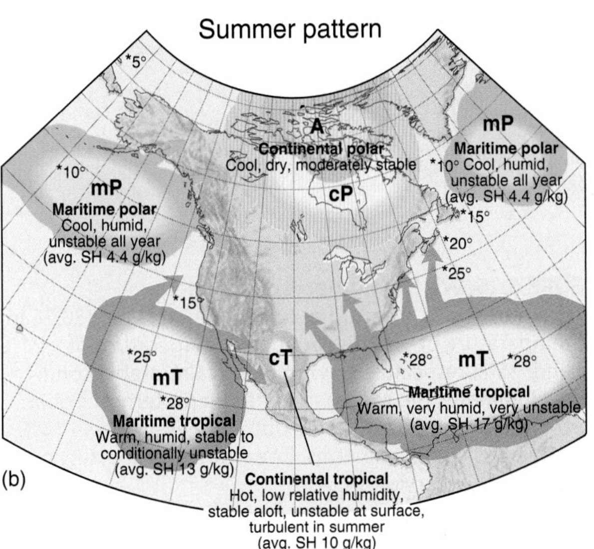

FIGURE 5.24 Principal air masses.
Air masses that influence North America and their characteristics in (a) winter and (b) summer. (*Temperatures shown are sea-surface temperatures [SSTs] in °C.)

Maritime polar (**mP**) air masses in the Northern Hemisphere are northwest and northeast of the North American continent, and within them cool, moist, unstable conditions prevail throughout the year. The Aleutian and Icelandic subpolar low-pressure cells reside within these **mP** air masses, especially in their well-developed winter state.

North America also is influenced by two maritime tropical (**mT**) air masses—the **mT** *Gulf/Atlantic* and the **mT** *Pacific*. The humidity experienced in the East and Midwest is created by the **mT** *Gulf/Atlantic* air mass, which is particularly unstable and active from late spring to early fall. In contrast, the **mT** *Pacific* is stable to conditionally unstable and generally much lower in moisture content and available energy.

Please review Figure 4.14 and the discussion of subtropical high-pressure cells off the coast of North America—the moist, unstable conditions on the western edge of the Atlantic (east coast) and the drier, stable conditions on the eastern edge of the Atlantic (west coast of Europe). These conditions, coupled respectively with warmer and cooler ocean currents, produce the characteristics of each air mass source region.

As air masses migrate from their source regions, their temperature and moisture characteristics are modified. For example, an **mT** *Gulf/Atlantic* air mass may carry humidity to Chicago and on to Winnipeg but gradually will lose its initial characteristics with each day's passage northward.

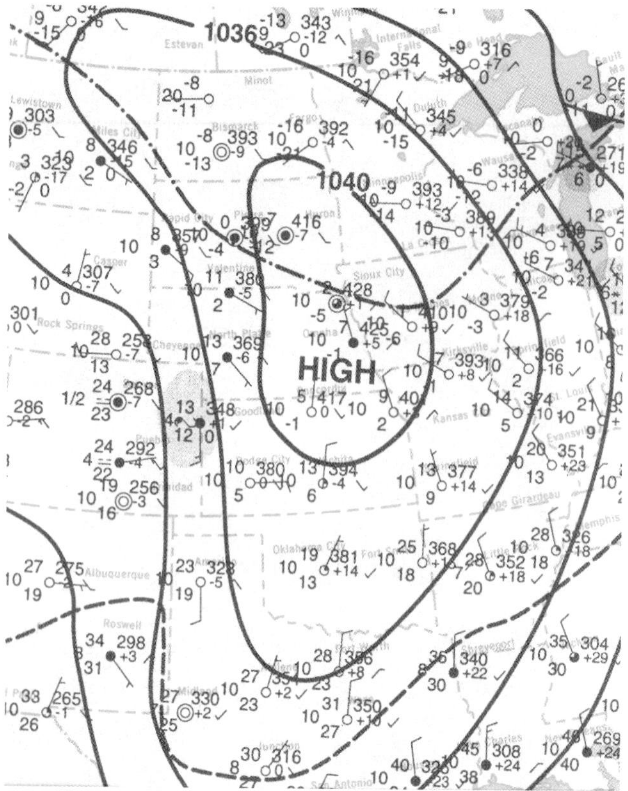

FIGURE 5.25 Winter high-pressure system.
A **cP** air mass with a central pressure of 1042.8 mb (30.76 in.), air temperature of −17°C, dew-point temperature of −21°C (2° and −5°F, respectively), with clear, calm, stable conditions, dominates the Midwest. Note the pattern of isobars portraying the **cP** air mass. The dotted lines are the −18°C (0°F) and 0°C (32°F) isotherms. [Weather map courtesy of National Weather Service, NOAA.]

Atmospheric Lifting Mechanisms

ANIMATION

Cold and warm fronts

For air masses to cool adiabatically (by expansion) and to reach the dew-point temperature and saturate, condense, form clouds, and perhaps precipitate, they must lift and rise in altitude. Four principal lifting mechanisms operate in the atmosphere (Figure 5.26): **convergent lifting** (air flows toward an area of low pressure, such as in the tropics, where convergent lifting of warm, moist air produces heavy precipitation); *convectional lifting* (stimulated by local surface heating); *orographic lifting* (air is forced over a barrier such as a mountain range); and *frontal lifting* (along the leading edges of contrasting air masses). Let us discuss these lifting mechanisms.

Convectional Lifting

When an air mass passes from a maritime source region to a warmer continental region, heating from the warmer land causes lifting in the air mass. Other sources of surface heating might include an urbanized area (heat island) or an area of dark soil in a plowed field—the warmer surfaces produce **convectional lifting**. If conditions are *unstable*, initial lifting sustains and clouds develop. Figure 5.27 illustrates convectional lifting stimulated by local heating, with unstable conditions present in the atmosphere. The rising parcel of air continues its ascent because it is warmer (less dense) than the surrounding environment.

In the figure, the MAR is used above the *lifting condensation level*, which is visible along the cloud base, where the rising air mass becomes saturated. Buoyancy is added at this level through the release of latent heat by condensation. Continued lifting above this altitude produces cooling of the air temperature and the dew-point temperature at the MAR.

Florida's precipitation generally illustrates two of these lifting mechanisms—convergence and convection. Heating of the land by insolation produces uplift and converging onshore winds from the Atlantic and the Gulf of Mexico that collide over Florida. Florida appears highlighted and painted with clouds in a satellite image (Figure 5.28). Convectional showers tend to form in the afternoon and early evening, causing the highest frequency of days with thunderstorms in the United States.

Orographic Lifting

The physical presence of a mountain acts as a topographic barrier to migrating air masses. **Orographic lifting** occurs when air rises upslope as it pushes against a mountain. It cools adiabatically. Stable air forced upward in this manner may produce stratiform clouds, whereas unstable or conditionally unstable air usually forms a line of cumulus and cumulonimbus clouds. An orographic barrier enhances convectional activity and causes additional lifting during the passage of weather fronts and cyclonic systems, thereby extracting more moisture from passing air masses. Figure 5.29 illustrates the operation of orographic lifting.

The wetter intercepting slope is termed the *windward slope*, as opposed to the drier far-side slope, known as the *leeward slope*. Moisture condenses from the lifting air mass on the windward side of the mountain; on the leeward side the descending air mass is heated by compression, causing evaporation of any remaining water in the air. Thus, air can begin its ascent up a mountain warm and moist but finish its descent on the leeward slope hot and dry. In North America, *chinook winds* (called *föhn* or *foehn* winds in Europe) are the warm, downslope air flows characteristic of the leeward side of mountains. Such winds can bring a 20 C° (36 F°) jump in temperature and greatly reduced relative humidity.

The term **rain shadow** is applied to dry regions leeward of mountains. East of the Cascade Range, Sierra Nevada, and Rocky Mountains, such rain-shadow patterns predominate. In fact, the precipitation pattern of windward and leeward slopes persists worldwide, as confirmed by the precipitation maps for the world shown in Chapter 6 and North America in Chapter 7.

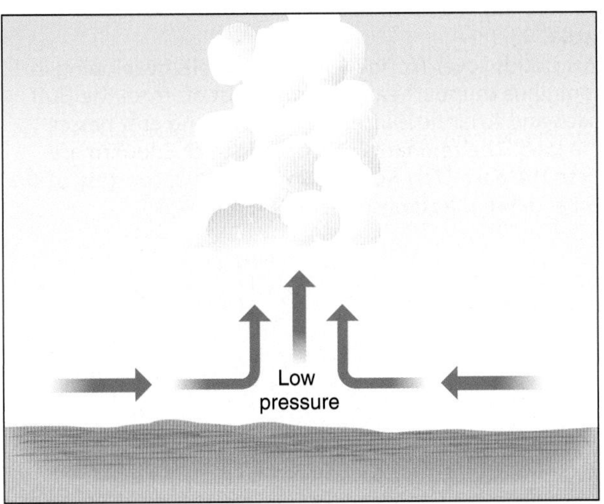

(a) Convergent

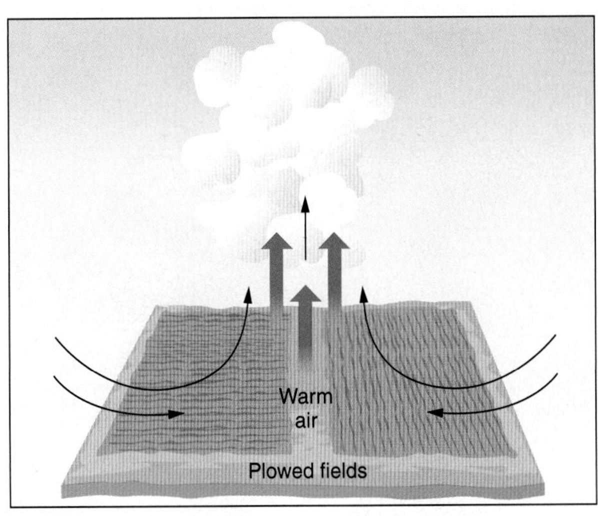

(b) Convectional (local heating)

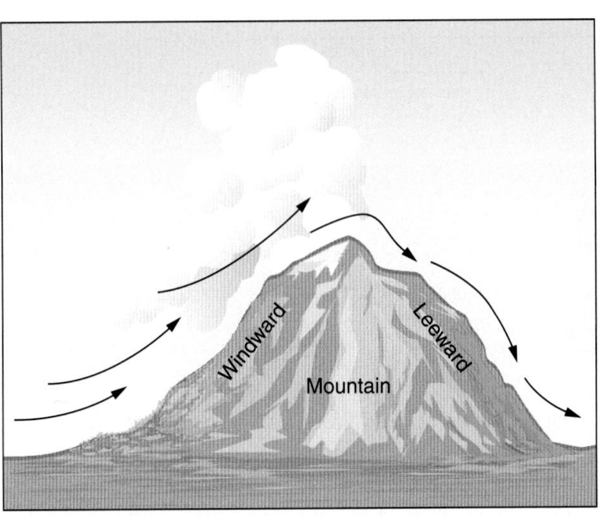

(c) Orographic (barrier)

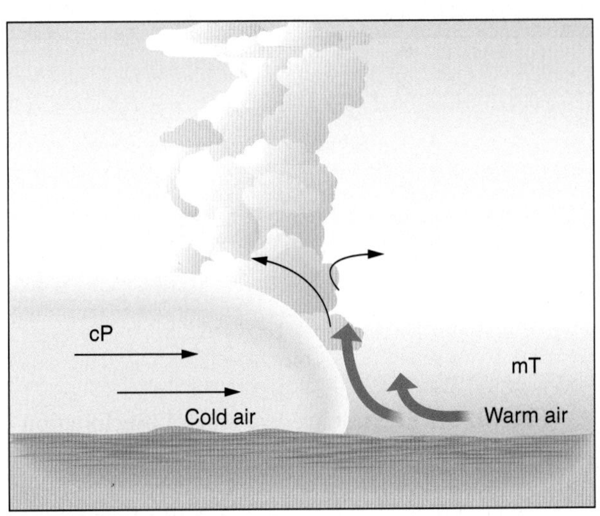

(d) Frontal (e.g. cold front)

FIGURE 5.26 Four atmospheric lifting mechanisms.
(a) Convergent lifting. (b) Convectional lifting. (c) Orographic lifting. (d) Frontal lifting.

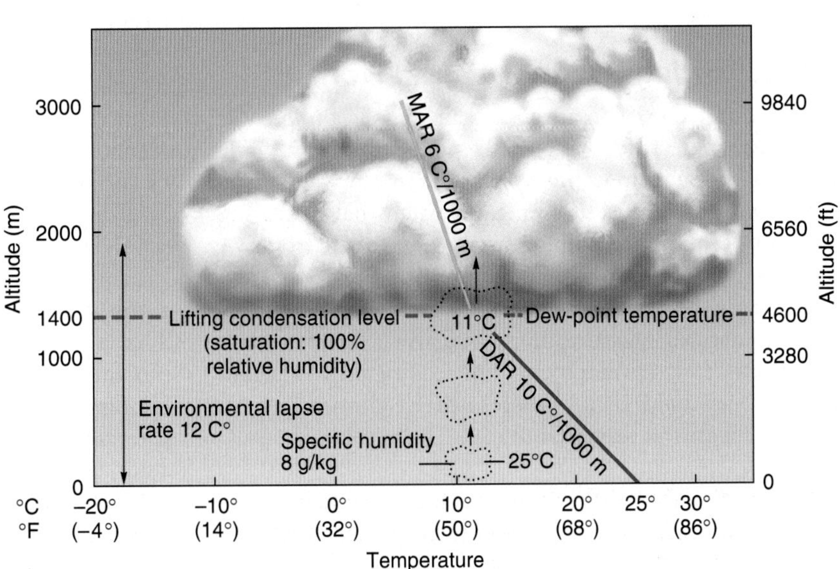

FIGURE 5.27 Local heating and convection.
Local heating and convection under unstable atmospheric conditions. Given specific humidity and temperature conditions within the lifting parcel of air and the unstable conditions in the environment, the dew point is reached at 1400 m (4600 ft). Note that the DAR is used when the air parcel is less than saturated, changing to the MAR above the condensation level.

FIGURE 5.28 Convectional activity over the Florida peninsula.
Cumulus clouds cover Florida with several cells developing into cumulonimbus thunderheads. Warm, moist air from the Gulf of Mexico and Atlantic is lifted by local heating as it passes over the land. The remnants of Tropical Storm Edouard are slightly to the east. [*Terra* image from 9/3/2002, courtesy of the MODIS Land Rapid Response Team, NASA/GSFC.]

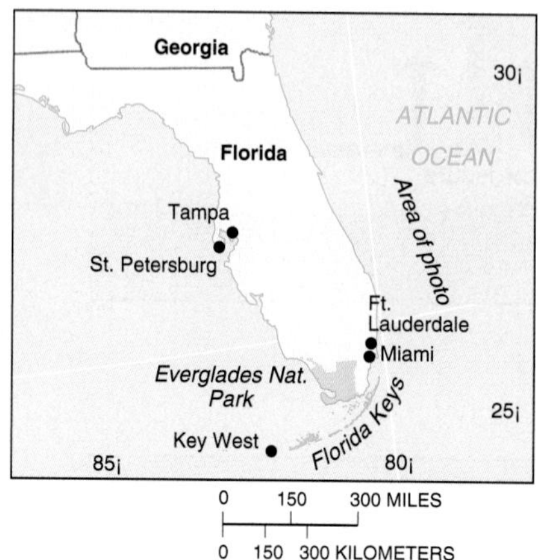

The world's greatest average annual precipitation occurs in the United States on the windward slopes of Mount Wai'ale'ale, on the island of Kaua'i, Hawai'i, which rises 1569 m (5147 ft) above sea level. Rainfall averages 1234 cm (486 in. or 40.5 ft) a year (records for 1941–1992). In contrast, the rain-shadow side of Kaua'i receives only 50 cm (20 in.) of rain annually. If no islands existed at this location, this portion of the Pacific Ocean would receive only an average 63.5 cm (25 in.) of precipitation a year.

Frontal Lifting (Warm and Cold Fronts)

The leading edge of an advancing air mass is a *front.* Vilhelm Bjerknes (1862–1951) first applied the term while working with a team of meteorologists in Norway during World War I, because it seemed to them that migrating air-mass "armies" were doing battle along their fronts. A front is a place of atmospheric discontinuity—a narrow zone that is the contact between two air masses that differ in temperature, pressure, humidity, wind direction and speed, and cloud development. The leading edge of a cold air mass is a **cold front**, whereas the leading edge of a warm air mass is a **warm front**.

Cold Front. On weather maps, such as the one illustrated in Figure 5.32 (p. 160), a cold front is identified by a line marked with triangular spikes pointing in the direction of frontal movement along an advancing **cP** air mass. The steep face of the cold air mass suggests its ground-hugging nature, caused by its density and uniform physical character (Figure 5.30a).

Warmer, moist air in advance of the cold front is lifted upward abruptly and is subjected to the same adiabatic rates and factors of stability/instability that pertain to all lifting air parcels. A day or two ahead of the cold front's passage, high cirrus clouds appear, telling observers that a lifting mechanism is on the way. The cold front's advance is marked by a wind shift, temperature drop, and lowering barometric pressure. Air pressure reaches a local low as the line of most intense lifting passes, usually just ahead of the front itself. Clouds may build up along the cold front into characteristic cumulonimbus types and may appear as an advancing wall of clouds. Precipitation usually is hard, containing large droplets, and can be accompanied by hail, lightning, and thunder. The aftermath usually brings northerly winds in the Northern Hemisphere (southerly winds in the Southern Hemisphere), lower temperatures, increasing air pressure from the invading cooler, denser air, and broken cloud cover.

A fast-advancing cold front can cause violent lifting and create a zone slightly ahead of the front called a

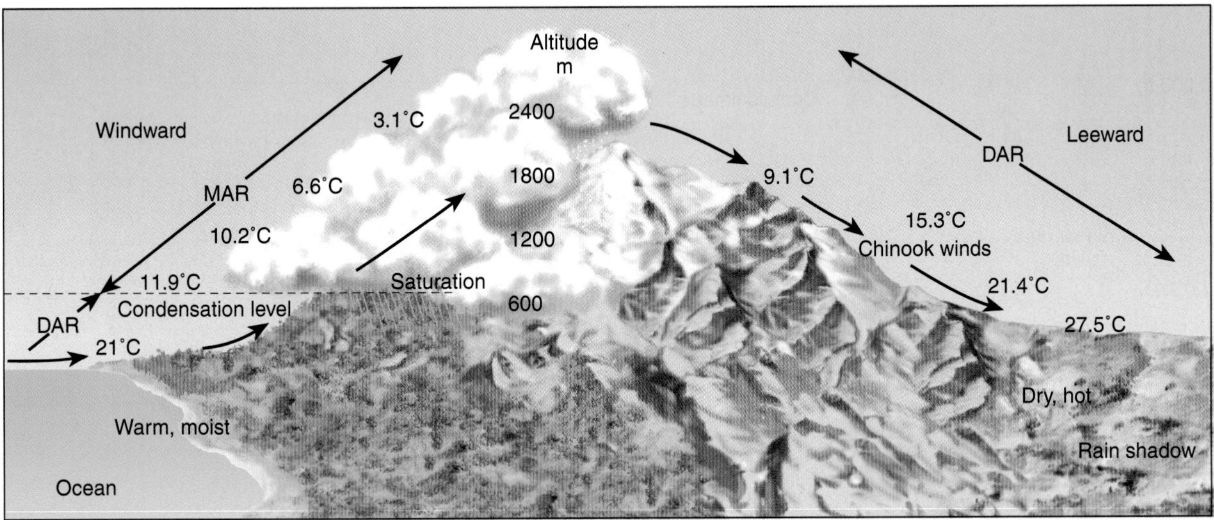

(a)

(b)

FIGURE 5.29 Orographic precipitation.
(a) Orographic barrier and precipitation patterns—unstable conditions assumed. Warm, moist air is forced upward against a mountain range, producing adiabatic cooling and eventual saturation. On the leeward slopes as the air travels downward, compressional heating takes place, producing relatively hot, dry air in the rain shadow of the mountain. (b) A leeward rain shadow is in stark contrast to the clouds of the windward side. Dust is stirred up by downslope winds. [Photo by author.]

squall line. Along a squall line such as the one in the Gulf of Mexico shown in Figure 5.30b, wind patterns are turbulent and wildly changing, and precipitation is intense. The well-defined line of clouds in the photo rises abruptly, with new thunderstorms forming along the distinct front. Tornadoes also may develop along such a squall line.

Warm Front. A warm front appears on weather maps as a line marked with semicircles facing the direction of frontal movement (see Figure 5.32). The leading edge of an advancing warm air mass is unable to displace cooler, passive air, which is denser. Instead, the warm air tends to push the cooler, underlying air into a characteristic wedge shape, with the warmer air sliding up over the

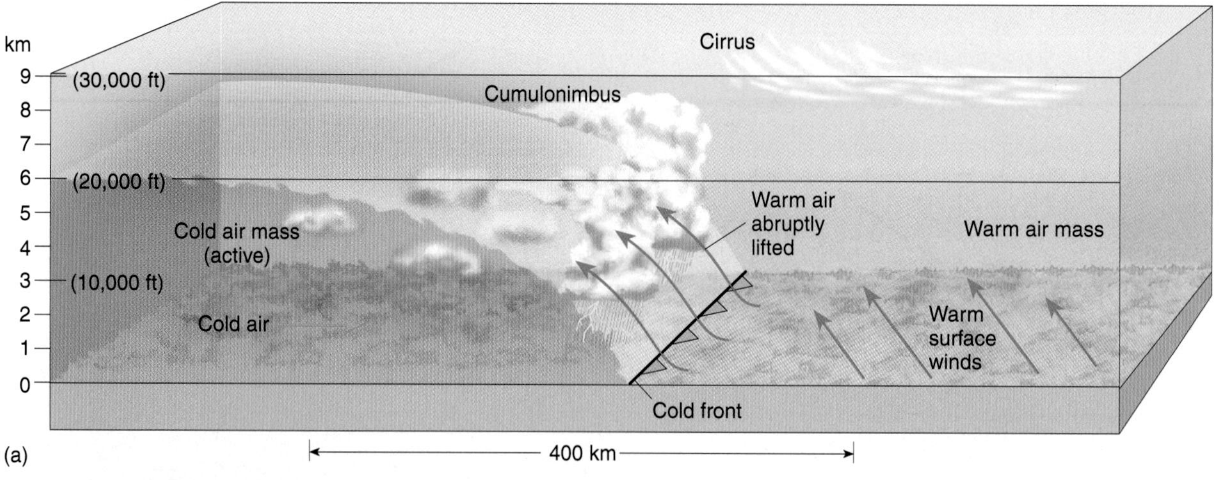

(b)

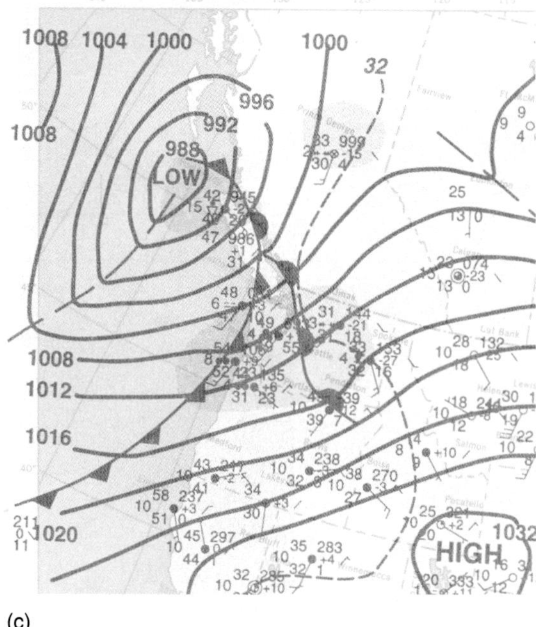

(c)

FIGURE 5.30 A typical cold front.
(a) Warm, moist air is forced to lift abruptly by denser advancing cold air. As the air lifts, it cools by expansion at the DAR, cooling to the dew-point temperature as it rises to a level of active condensation and cloud formation. Cumulonimbus clouds may produce large raindrops, heavy showers, lightning and thunder, and hail. (b) Squall line along a cold front in the Gulf of Mexico is marked by a sharp line of cumulonimbus clouds rising to 16,800 m (55,120 ft). (c) Cold front and strong cyclonic system approach the Pacific coast, February 1, 2000. [Space Shuttle photo from NASA; (c) weather map courtesy of *Daily Weather Maps*, NWS, NOAA.]

cooler air. Thus, in the cooler-air region, a temperature inversion is present, sometimes causing poor air drainage and stagnation.

Figure 5.31 illustrates a typical warm front in which **mT** air is gently lifted, leading to stratiform cloud development and characteristic nimbostratus clouds and drizzly precipitation. High cirrus and cirrostratus clouds announce the advancing warm-front system. Closer to the front, the clouds lower and thicken to altostratus, and then to stratus within several hundred kilometers of the front.

Midlatitude Cyclonic Systems

Midlatitude cyclones

Weather is the short-term condition of the atmosphere, as compared with *climate*, which reflects long-term atmospheric conditions and extremes. Weather is, at the same time, both a “snapshot” of atmospheric conditions and a technical status report of the Earth-atmosphere, heat-

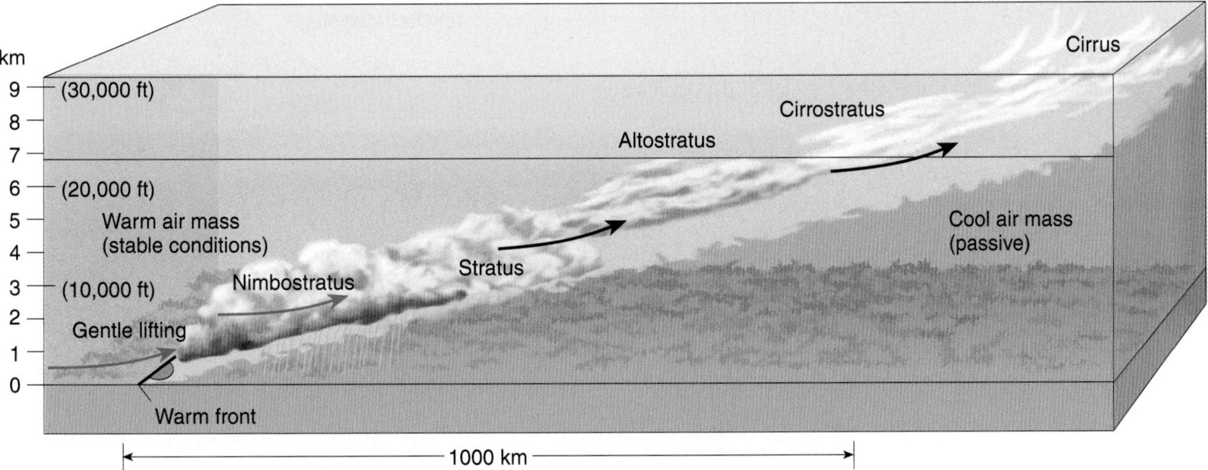

FIGURE 5.31 A typical warm front.
Note the sequence of cloud development as the warm front approaches. Warm air slides upward over a wedge of cooler, passive air near the ground. Gentle lifting of the warm, moist air produces stratus and nimbostratus clouds and drizzly rain showers in contrast to the more dramatic precipitation associated with the passage of a cold front.

energy budget. **Meteorology** is the scientific study of the atmosphere (*meteor* means "heavenly" or "of the atmosphere"). Embodied within this science is a study of the atmosphere's physical characteristics and motions, related chemical, physical, and geological processes, the complex linkages of atmospheric systems, and weather forecasting.

We turn to local media for the day's weather report from the National Weather Service in the United States (http:www.nws.noaa.gov) or Canadian Meteorological Centre (http://www.msc-smc.ec.gc.ca/cmc/index_e.html) to see current satellite images and to hear weather analysis and tomorrow's forecast. Internationally, the World Meteorological Organization coordinates weather information (see http://www.wmo.ch/). Many weather sources and related topics are found in this chapter on the *Elemental Geosystems Home Page*.

The conflict between contrasting air masses along fronts leads to conditions appropriate for the development of a **midlatitude cyclone**, or **wave cyclone**, a migrating center of low pressure, with converging, ascending air, spiraling inward counterclockwise in the Northern Hemisphere (or converging clockwise in the Southern Hemisphere). The pressure gradient force, the Coriolis force, and surface friction generate this cyclonic motion (see Chapter 4).

Wave cyclones form a dominant type of weather pattern in the middle and higher latitudes of both the Northern and Southern Hemispheres and act as a catalyst for air-mass conflict. Such a *midlatitude cyclone*, or extratropical cyclone, can be born along the polar front, particularly in the region of the Icelandic and Aleutian subpolar low-pressure cells in the Northern Hemisphere (see Figures 4.11 and 4.13).

Development and strengthening of a midlatitude wave cyclone is known as *cyclogenesis*. In addition to the polar front, certain other areas are associated with wave cyclone development and intensification: the eastern slope of the Rockies and other north-south mountain barriers, the Gulf coast, and the east coasts of North America and Asia.

Life Cycle of a Midlatitude Cyclone

Figure 5.32 shows the birth, maturity, and death of a typical midlatitude cyclone in four stages along with an idealized weather map. Weather conditions for selected stations are noted using standard weather station symbols. On the average, a midlatitude cyclone takes 3–10 days to progress through these stages and cross the North American continent.

Along the polar front, cold and warm air masses converge and conflict. The polar front forms a discontinuity of temperature, moisture, and winds that establishes potentially unstable conditions. However, for a wave cyclone to form along the polar front, a point of surface air *convergence* must be matched by a compensating area of air *divergence* aloft. Even a slight disturbance along the polar front, perhaps a small change in the path of the jet stream, can initiate the converging, ascending motion of a surface low-pressure system (Figure 5.32a).

Storm Tracks. Cyclonic storms—1600 km (1000 mi) wide—and their attendant air masses move across the continent along *storm tracks*, which shift latitudinally with the Sun, crossing North America farther to the north in summer and farther to the south in winter (Figure 5.33). As the storm tracks shift northward in the spring, **cP** and **mT** air masses are brought into their clearest conflict. This is the time of strongest frontal activity, with associated thunderstorms and tornadoes. Storm tracks follow the path of upper-air winds and jet streams, which direct storm systems across the continent. Figure 5.33b portrays storm tracks for an actual month—March 1991.

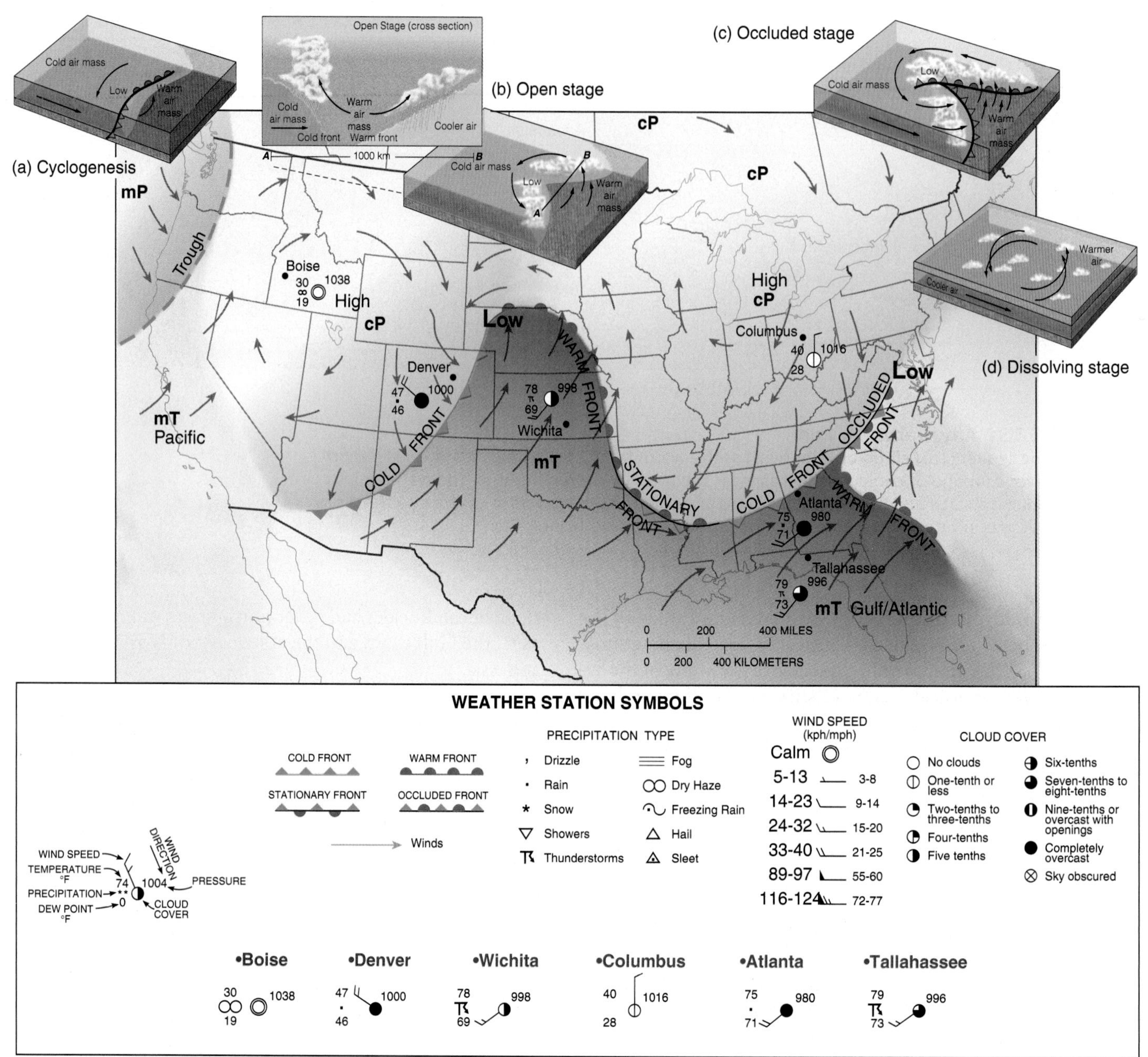

FIGURE 5.32 Idealized stages of a midlatitude wave cyclone. (a) Cyclogenesis is noted where surface convergence and lifting begin. (b) The open stage. (c) The occluded stage. (d) The dissolving stage is reached at the end of the storm track as the cyclone spins down, no longer energized by the latent heat from condensing moisture. Standard weather symbols are in the inset box. After studying the text and the map, can you describe conditions in Boise, Denver, Wichita, Columbus, Atlanta, and Tallahassee depicted on this weather map?

Open Stage. As the midlatitude cyclone matures, the counterclockwise flow (in the Northern Hemisphere) draws the cold air mass from the north and west and the warm air mass from the south. A characteristic open stage forms as shown in Figure 5.32b and in the enlarged cross section (line A–B).

Such an open stage of a midlatitude cyclone occurred April 20, 2000, with low pressure centered over the Iowa-Missouri border. Figure 5.34 shows you a portion of the daily weather map for that day and an image from *SeaWiFS* sensors aboard satellite *OrbView-2*. Note how the isobars portray the cyclone and the low of 997.6 mb (29.45 in.) on the map. Note the cloud pattern stretching along the cold front and the swirling clouds around the low. Compare the cold front and warm front and overall patterns with Figure 5.32.

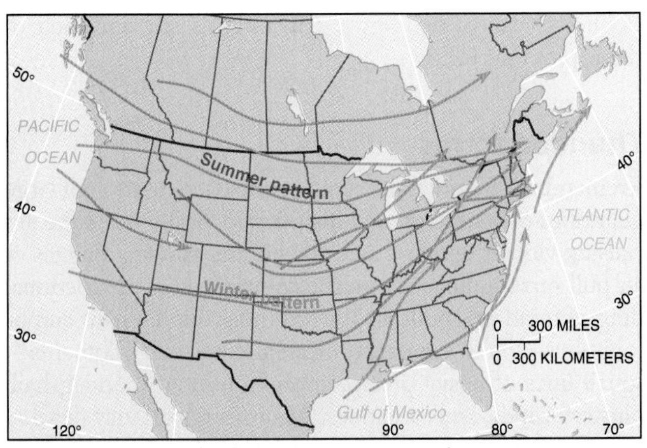

(a) Average storm tracks

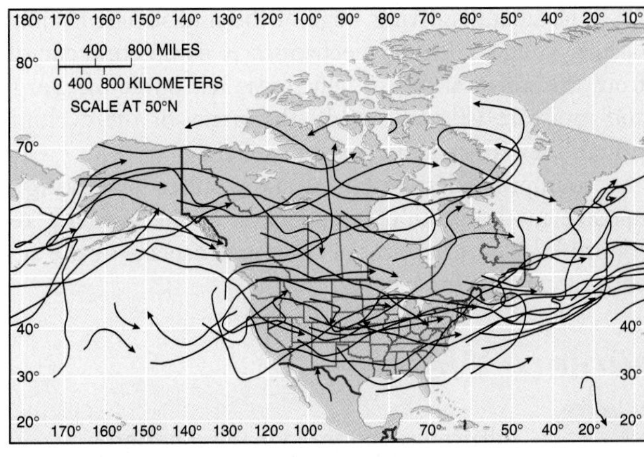

(b) Actual storm tracks in March 1991

FIGURE 5.33 Average and actual storm tracks for North America.
(a) Seasonal storm track patterns vary summer and winter. (b) Actual storm tracks for March 1991. [(b) From *Storm Data* 33, no. 3 (March 1991); Asheville, North Carolina: NESDIS/NCDC/NOAA.]

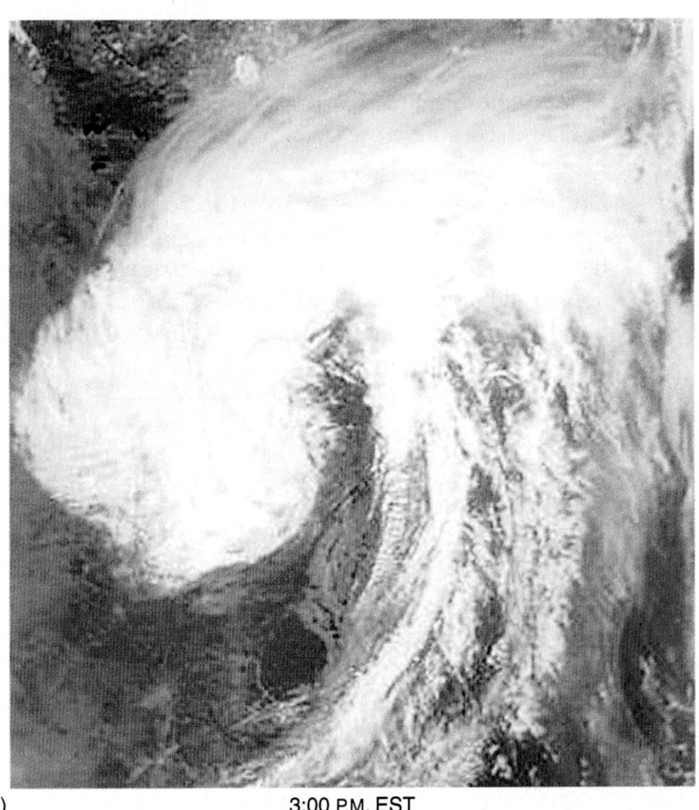

(a) 3:00 P.M. EST

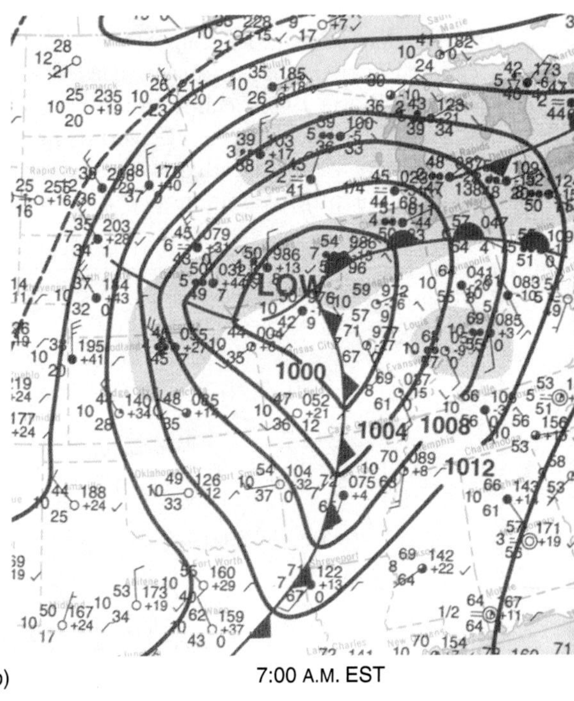

(b) 7:00 A.M. EST

FIGURE 5.34 Open stage of a midlatitude cyclone.
(a) *SeaWiFS* image of a cyclonic system over the Midwest. The cloud patterns are areas of precipitation; clear skies are behind the cold front as **cP** air mass covers the landscape. (b) A segment from the April 20, 2000, daily weather map, about 8 hours earlier than the image, showing the low-pressure system centered on 997.6 mb (29.45 in.). Counterclockwise winds circulate around the low. [(a) *SeaWiFS* image used with permission of ORBIMAGE. All rights reserved. (b) Segment of *Daily Weather Map* courtesy of NWS, NOAA.]

Occluded Stage. Because a **cP** air mass is more homogeneous in temperature and pressure than an **mT** air mass, the cold front leads a denser, more unified mass, and therefore moves faster than the warm front. Cold fronts can average 40 kmph (25 mph), whereas warm fronts average 16–24 kmph (10–15 mph). Thus, a cold front may overrun the cyclonic warm front, producing an **occluded front** within the overall cyclonic circulation (Figure 5.32c). Precipitation may be moderate to heavy initially and then decrease as the warmer air wedge is lifted higher by the advancing cold air mass.

The final dissolving stage of the midlatitude cyclone occurs when its lifting mechanism is completely cut off from the warm air mass, which was its source of energy and moisture (Figure 5.32d). Remnants of the cyclonic system then dissipate in the atmosphere.

To improve data gathering and weather forecasting, a massive modernization is underway in the agencies responsible. News Report 5.3 overviews aspects of the science of weather forecasting.

Violent Weather

ANIMATION **Hurricanes and tornado wind patterns**

SATELLITE LOOP **Hurricane Georges**

Weather provides a continuous reminder that the flow of energy across the latitudes can at times set into motion violent and potentially destructive conditions. Thunderstorms, tornadoes, and tropical cyclones are such forms of violent weather.

Weather is often in the news, as weather-related destruction has risen more than 500% over the past two decades. Weather-related losses exceeded $10 billion per year through the 1990s, eclipsing the previous average of less than $2 billion a year. Government research and monitoring of violent weather is centered at NOAA's National Severe Storms Laboratory in several cities (see http://www.nssl.noaa.gov; consult this site for each of the topics that follow).

Thunderstorms

Tremendous energy is liberated by the condensation of large quantities of water vapor. This process locally heats the air, causing violent updrafts and downdrafts as rising parcels of air pull surrounding air into the column and as the frictional drag of raindrops pulls air toward the ground. Giant cumulonimbus clouds can create dramatic weather moments—squall lines of heavy precipitation, lightning, thunder, hail, blustery winds, and tornadoes. Thunderstorms may develop within an air mass, in a line along a front (particularly a cold front), or where mountain slopes cause orographic lifting.

Thousands of thunderstorms occur on Earth at any given moment. Equatorial regions and the ITCZ experience many of them, as exemplified by the city of Kampala in Uganda, East Africa (north of Lake Victoria). This city sits virtually on the equator and annually averages 242 days of thunderstorms—a record. Figure 5.35 shows the distribution of annual days with thunderstorms across the United States and Canada. You can see that in North America most thunderstorms occur in areas dominated by **mT** air masses.

Lightning and Thunder. An estimated 8 million lightning strikes occur each day on Earth. **Lightning** refers to flashes of light caused by enormous electrical discharges—tens to hundreds of millions of volts—that briefly superheat the air to temperatures of 15,000° to 30,000°C

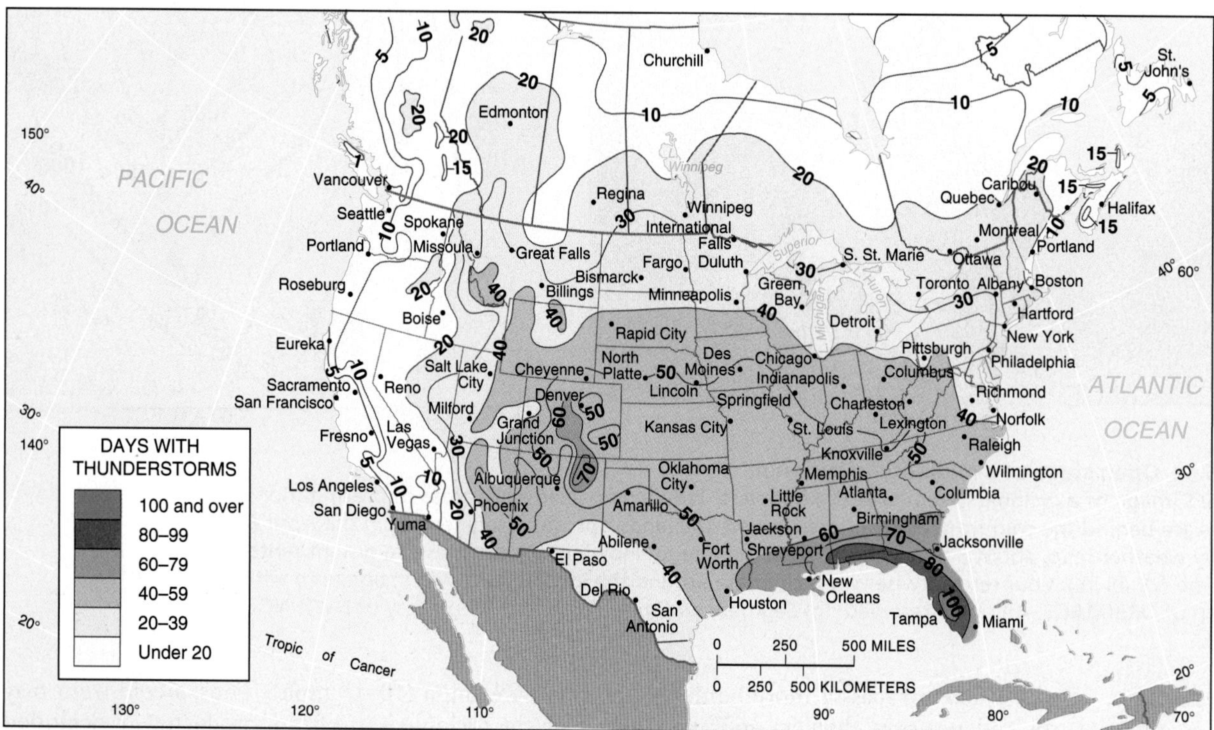

FIGURE 5.35 Thunderstorm occurrence.
Average annual number of days with thunderstorms. [Courtesy of National Weather Service and Map Series 3, *Climatic Atlas of Canada*, Atmospheric Environment Service.]

(27,000°–54,000°F). The violent expansion of this abruptly heated air sends shock waves through the atmosphere—the sonic bangs known as **thunder**. The greater the distance a lightning stroke travels, the longer the thunder echoes. Lightning at great distance from the observer may not be accompanied by thunder and is called *heat lightning*.

Lightning poses a hazard to aircraft and to people, animals, trees, and structures. Individuals should acknowledge that certain precautions are mandatory when the threat of a lightning discharge is near. Nearly 200 deaths and thousands of injuries occur each year in the United States and Canada. Therefore heed severe storm warnings that caution people to remain indoors.

NASA's Lightning Imaging Sensor (LIS) aboard the *Tropical Rainfall Measuring Mission* (*TRMM*) satellite monitors lightning. The LIS can image lightning strikes day or night, within clouds or between cloud and ground. The sensor's data show that about 90% of all strikes occur over land in response to increased convection over relatively warmer continental surfaces, with expected seasonal shifts keyed to the high Sun as shown in Figure 5.36a and b. (See http://thunder.msfc.nasa.gov/lis/.)

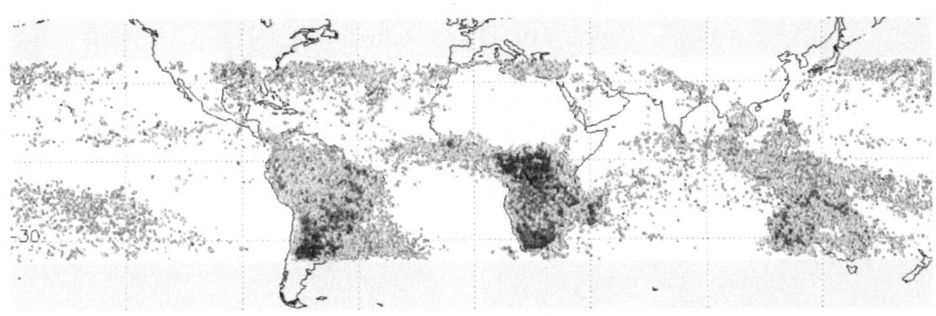

(a) Winter (Dec. 1999, Jan. and Feb. 2000)

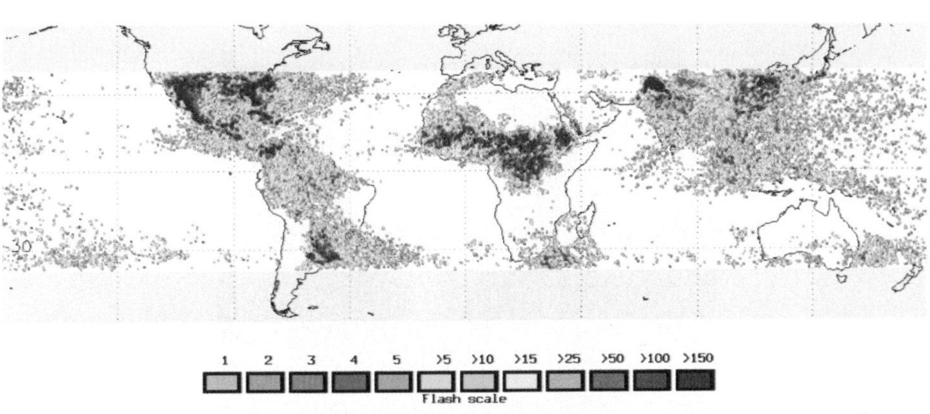

(b) Summer (June, July, August 2000)

(c)

FIGURE 5.36 Seasonal images show global lightning. A composite of 3 months' data derived from NASA's Lightning Imaging Sensor (LIS) records all lightning strikes between 35° N and 35° S latitudes during (a) winter (December 1999–February 2000) and (b) summer (June–August 2000). (c) Multiple cloud-to-ground lightning strikes captured in a time-lapse photo. [Images (a) and (b) courtesy of NASA's Global Hydrology and Climate Center, Marshall Spaceflight Center, Huntsville, Alabama. (c) Photo from C. Clark, NOAA Photo Library, NSSL.]

News Report 5.3

Weather Forecasting

As our knowledge of the interactions that produce weather and instruments and software improve, so too will the accuracy of weather forecasts. *Synoptic analysis* is the evaluation of weather data collected at a selected time. Building a database of wind, pressure, temperature, and moisture conditions is key to *numerical* (computer-based) *weather prediction* and the development of weather-forecasting models. Development of numerical models is a great challenge because the atmosphere operates as a nonlinear (irrational) system, tending toward chaotic behavior. Slight variations in input data or slight changes in the basic assumptions of the model's behavior can produce widely varying forecasts.

Weather data necessary for the preparation of a synoptic map and forecast include:

- Barometric pressure (sea level and altimeter setting)
- Pressure tendency (steady, rising, falling)
- Surface air temperature
- Dew-point temperature
- Wind speed, direction, and character (gusts, squalls)
- Type and movement of clouds
- Current weather
- State of the sky (current sky conditions)
- Visibility, vision obstruction (fog, haze)
- Precipitation since last observation

New developments in supercomputing, an Earth-bound instrument network in the Automated Surface Observing System (ASOS) arrays, orbiting observation systems, and Doppler radar installations are rapidly advancing the science of the atmosphere. By the end of 2001, 155 WSR-88D (*W*eather *S*urveillance *R*adar) Doppler radar systems as part of the NEXRAD (Next Generation Weather Radar) program were operational through the National Weather Service (122), in conjunction with the Federal Aviation Administration (12) and the Department of Defense (21) (Figure 1). In Canada, the National Radar Project will have 30 CWSR-98 radars in service.

ASOS sensor instrument arrays are essential to our surface weather-observing network (Figure 2a). The 883 ASOS installations include: rain gauge (tipping bucket), temperature/dew point sensor, barometer, present weather identifier, wind speed indicator, direction sensor, cloud height indicator, freezing rain sensor, thunderstorm sensor, and visibility sensor, among other items.

NOAA operates the Forecast Systems Laboratory (FSL), Boulder, Colorado, that is developing new forecasting tools (see http://www.fsl.noaa.gov). A few innovations include: wind profilers using radar to profile winds from the surface to high altitudes; a High Performance Computing System (277 CPUs networked to act as one, Figure 2e) to model software for 3-D weather models and other computations; improved international cooperation and weather data dissemination; and a new standard Advanced Weather Interactive Processing System (AWIPS, Figure 2c, d). See Career Link 6.1 in the next chapter and my interview with an FSL researcher.

FIGURE 1 Weather installation.
Doppler radar installation at the Indianapolis International Airport operated by the National Weather Service. The radar antenna is sheltered within the dome structure. [Photo by Bobbé Christopherson.]

(continued)

News Report 5.3 *(continued)*

(a)

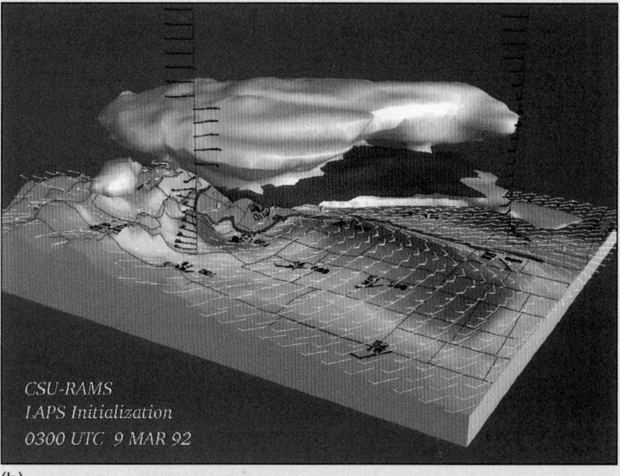

(b)

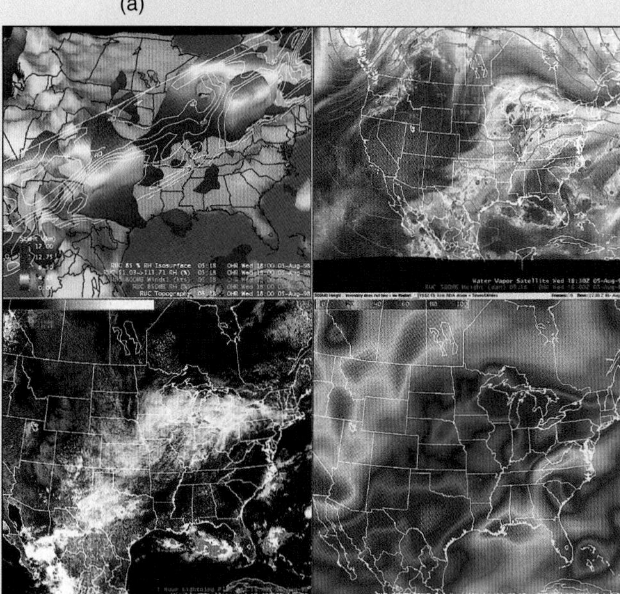

(c)

(d)

(e)

FIGURE 2 ASOS weather instruments and AWIPS workstation display.
(a) Automated Surface Observing System (ASOS) installation, one of 883 ASOS stations in use for data gathering as part of the U.S. primary surface weather observing network. (b) Forecast Systems Laboratory (FSL) scientists develop three-dimensional modeling programs to portray weather and simulate future weather for more accurate forecasting. On the image note the gray area that depicts significant cloud water content and the red layer within the cloud where aircraft icing is forecast. The black wind barbs off the two vertical axes denote wind profiles through the atmosphere. The view is from the southeast. (c) Powerful computers are needed to process the data and produce these 3-D virtual images on the Advanced Weather Interactive Processing System (AWIPS), employing screens with multiple frames. Of the many items that can be displayed, here we see a simulated 3-D pressure analysis, water vapor and visible satellite image—including lightning strikes—and humidity. (d) An AWIPS workstation. (e) A portion of the High Performance Computing System at the FSL. [Photo (a) and (e) by Bobbé Christopherson; (b–d) images courtesy of the FSL, NOAA, Boulder, Colorado.]

Hail. Hail is common in the United States and Canada, although somewhat infrequent at any given location, hitting a specific location perhaps every 1 or 2 years in the highest frequency areas. Annual hail damage in the United States tops $750 million. **Hail** generally forms within a cumulonimbus cloud when a raindrop circulates repeatedly above and below the freezing level in the cloud, adding layers of ice until its weight can no longer be supported by the circulation in the cloud.

Pea-sized and marble-sized hail are common, although hail the size of golf balls and baseballs also is reported at least once or twice a year somewhere in North America. For larger hail to form, the frozen pellets must stay aloft for longer periods. The pattern of hail occurrence across the United States and Canada is similar to that of thunderstorms shown in Figure 5.35, dropping to zero approximately by the 60th parallel.

Atmospheric Turbulence. Most airplane flights experience at least some turbulence—the encountering of air of different densities, or air layers moving at different speeds and directions. This is a natural state of the atmosphere and passengers are asked to keep seat belts fastened when in their seats to avoid injury, even if the seat-belt light is turned off.

Thunderstorms can produce severe turbulence in the form of downbursts, which are exceptionally strong downdrafts. Downbursts are classified by size: a *macroburst* is at least 4.0 km (2.5 mi) wide and in excess of 210 kmph (130 mph), whereas a *microburst* is smaller in size and speed. A microburst causes rapid changes in wind speed and direction, and it produces the dreaded *wind shear* that can bring down aircraft. Such turbulence events are short-lived and hard to detect, although the Forecast Systems Laboratory, among others, is making progress in developing forecasting methods.

Tornadoes

The updrafts associated with a cold-front squall line and cumulonimbus cloud development appear on satellite images as pulsing bubbles of clouds. According to one hypothesis, a spinning, cyclonic, rising column of mid-troposphere-level air forms a **mesocyclone**. A mesocyclone can range up to 10 km (6 mi) in diameter and rotate vertically within a *supercell cloud* (the parent cloud) to a height of thousands of meters (Figure 5.37a). A well-developed mesocyclone most certainly will produce heavy rain, large hail, blustery winds, and lightning; some mature mesocyclones will generate tornado activity.

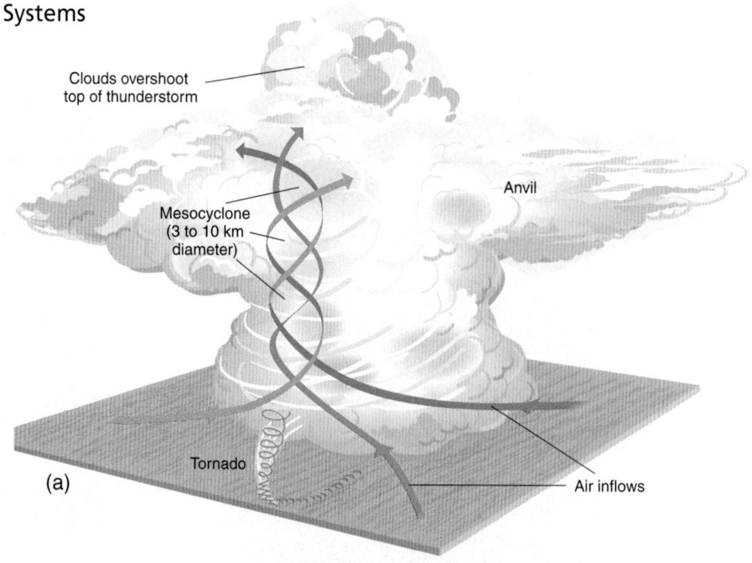

(b)

(c)

FIGURE 5.37 Twister!—tornado terror strikes.
(a) Mesocyclone forms as a rotating updraft within the thunderstorm. If a tornado forms, it will descend from the cloud base and the lower portion of the mesocyclone. (b) A tornado emerges from the base of a wall cloud near Oakfield, Wisconsin. (c) An F5 tornado struck in May 1999. This scene is a few blocks east of I-35. The church visible in the background was used as an emergency center. [Photos (b) by Don Lloyd/AP/Wide World Photos; photo (c) by Todd Lindley.]

As more moisture-laden air is drawn up into the circulation of a mesocyclone, more energy is liberated, and the rotation of air becomes more rapid. The narrower the mesocyclone, the faster the spin of converging parcels of air being sucked into the rotation. The swirl of the mesocyclone itself is visible, as are smaller, dark gray **funnel clouds** that pulse from the bottom side of the parent cloud. The potential of this stage of development is the lowering of a funnel cloud to Earth—a **tornado**. See the chapter-opening photograph for an example of a mesocyclone, wall cloud, and tornado.

A tornado can range from a few meters to a few hundred meters in diameter and can last anywhere from a few moments to tens of minutes. The May 3, 1999, outbreak in Oklahoma featured an F5 tornado that was one-mile (1610-m) wide at its base, traveled 65 km (40 mi), lasted an hour and a half, caused a billion dollars in damage, and killed 36 people!

When tornado circulation occurs over water, a **waterspout** forms, and surface water is drawn up into the funnel some 3–5 m (10–16 ft). The rest of the waterspout funnel is made visible by the rapid condensation of water vapor.

Tornado Measurement and Science. Pressures inside a tornado usually are about 10% less than those in the surrounding air. The inrushing convergence created by such a horizontal pressure gradient causes high wind speeds. The late Theodore Fujita, a noted meteorologist from the University of Chicago, designed a scale for classifying tornadoes according to wind speed and related property damage. The Fujita Scale ranks tornadoes as F0 and F1 (weak, with winds less than 180 kmph, or 112 mph—about 85% of all tornadoes); F2 and F3 (strong, with winds 181–332 kmph, or 113–206 mph—about 13%); F4 (violent, with winds over 333 kmph, or 207 mph—about 2%); and F5 (over 420 kmph, 261 mph—less than 1%). (See http://www.tornadoproject.com/fscale/fscale.htm.)

Tornadoes have struck all 50 states and all the Canadian provinces and territories (Figure 5.38). May and June

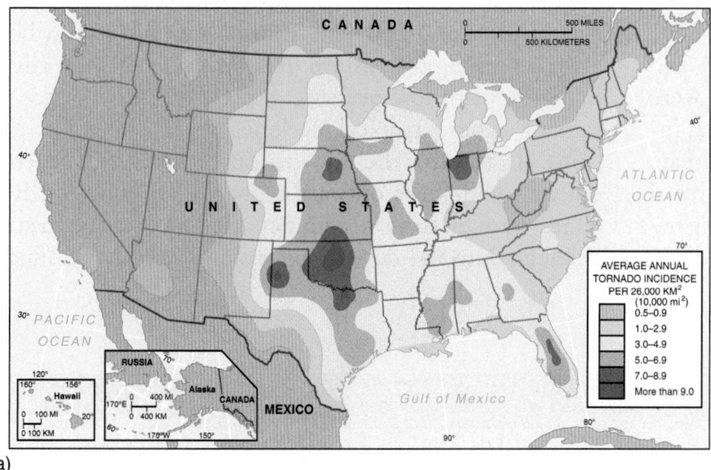

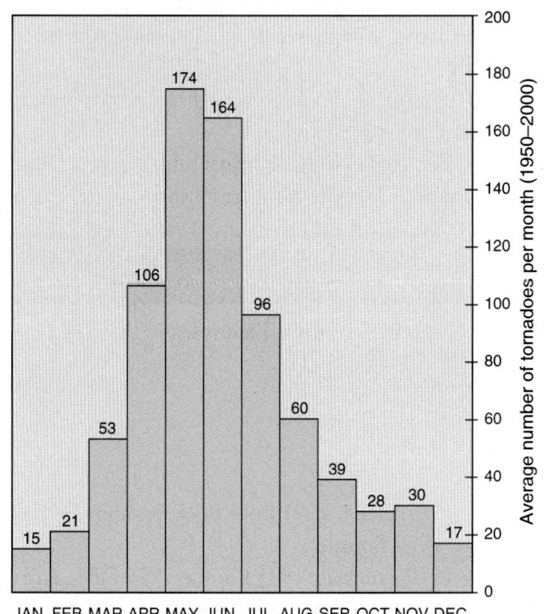

FIGURE 5.38 Tornado occurrence in the United States, 1950–2000.
(a) Average number of tornadoes per 26,000 km² (10,000 mi²) for 1950 to 2000. (b) Average number of tornadoes per month. Remarkably, 155 tornadoes during January 1999 and 70 in one day in November 2002 rendered such monthly averages meaningless. A fully developed tornado locks onto the ground. [Data courtesy of the National Severe Storm Forecast Center, NWS, Kansas City, Missouri. Photo (c) by C. Lavoie/First Light.]

are the peak months. In November 2002, unusually warm temperatures and a fast moving cold front produced 70 tornadoes in one day, or 40 more than the average for the whole month. These exploded through five states in the South and East, killing 40 people and injuring hundreds. A small number of tornadoes are reported on other continents each year, but North America receives the greatest share because its latitudinal position and shape permit contrasting air masses access to one another.

In the United States, 43,257 tornadoes struck between 1950 and 2002, causing 4598 deaths or about 86 deaths per year. Tornadoes took 130 lives in the United States in 1998 alone. Of note, this half-century of tornadoes resulted in more than 78,000 injuries and property damage of more than $25 billion. The yearly average of $475 million in damage is rising at about $50 million a year. Importantly, the 1998–2002 seasons produced 4 F5 and 29 F4 tornadoes.

The annual average number of tornadoes before 1990 was 787. Interestingly, since 1990 the average per year has risen to 1170, with 1992, 1995, 1996, and 2001 surpassing 1200 tornadoes; 1999 had more than 1300; and 1998 with more than 1400. During 2002 numbers were nearer the long-term average. Reasons for the annual increase in tornado frequency range from global climate change to better reporting by a larger population armed with video cameras. (See http://www.srh.noaa.gov/oun/storms/.)

The National Severe Storms Forecast Center in Kansas City, Missouri, is a key forecasting center. As a result of Doppler radar, only about 15% of tornadoes strike without some public warning. Warning times of about 12 minutes are generally possible with current technology (see the Storm Prediction Center at http://www.spc.noaa.gov/climo/).

Tropical Cyclones

A powerful manifestation of Earth's energy and moisture system is the **tropical cyclone**, which originates entirely within tropical air masses. The tropics extend from the Tropic of Cancer at 23.5° N to the Tropic of Capricorn at 23.5° S, including the equatorial zone between 10° N and 10° S. Approximately 80 tropical cyclones occur annually worldwide. Table 5.1 presents the criteria for classification of tropical cyclones based on wind speed.

Meteorologists now think that the cyclonic motion begins with slow-moving easterly waves of low pressure in the trade wind belt of the tropics, such as the Caribbean area. It is along the eastern (leeward) side of these migrating troughs of low pressure, a place of convergence and rainfall, where tropical cyclones form. Surface airflow spins in toward the low-pressure area (convergence), ascends, and flows outward aloft. This important divergence aloft acts as a chimney, pulling more moisture-laden air into the developing system, fueled by abundant

Table 5.1 Tropical Cyclone Classification

Designation	Winds	Features
Tropical disturbance	Variable, low	Definite area of surface low pressure; patches of clouds
Tropical depression	Up to 34 knots (63 kmph, 39 mph)	Gale force, organizing circulation; light to moderate rain
Tropical storm	35–63 knots (63–118 kmph, 39–73 mph)	Closed isobars; definite circular organization; heavy rain; assigned a name
Hurricane (Atlantic and E. Pacific) Typhoon (W. Pacific) Cyclone (Indian Ocean, Australia)	Greater than 65 knots (119 kmph, 74 mph)	Circular, closed isobars; heavy rain, storm surges; tornadoes in right-front quadrant

Saffir-Simpson Hurricane Damage Potential Scale

Category	Wind Speed	Notable Atlantic Examples
1	65–82 knots (74–95 mph)	—
2	83–95 knots (96–110 mph)	—
3	96–113 knots (111–130 mph)	1985 Elena; 1991 Bob; 1995 Roxanne, Marilyn; 1998 Bonnie
4	114–135 knots (131–155 mph)	1979 Frederic; 1985 Gloria; 1995 Felix, Luis, Opal; 1998 Georges, 1999 Floyd
5	> 135 knots (> 155 mph)	1935 No. 2; 1938 No. 4; 1960 Donna; 1961 Carla; 1969 Camille; 1979 David; 1988 Gilbert; 1989 Hugo; 1992 Andrew; 1998 Mitch

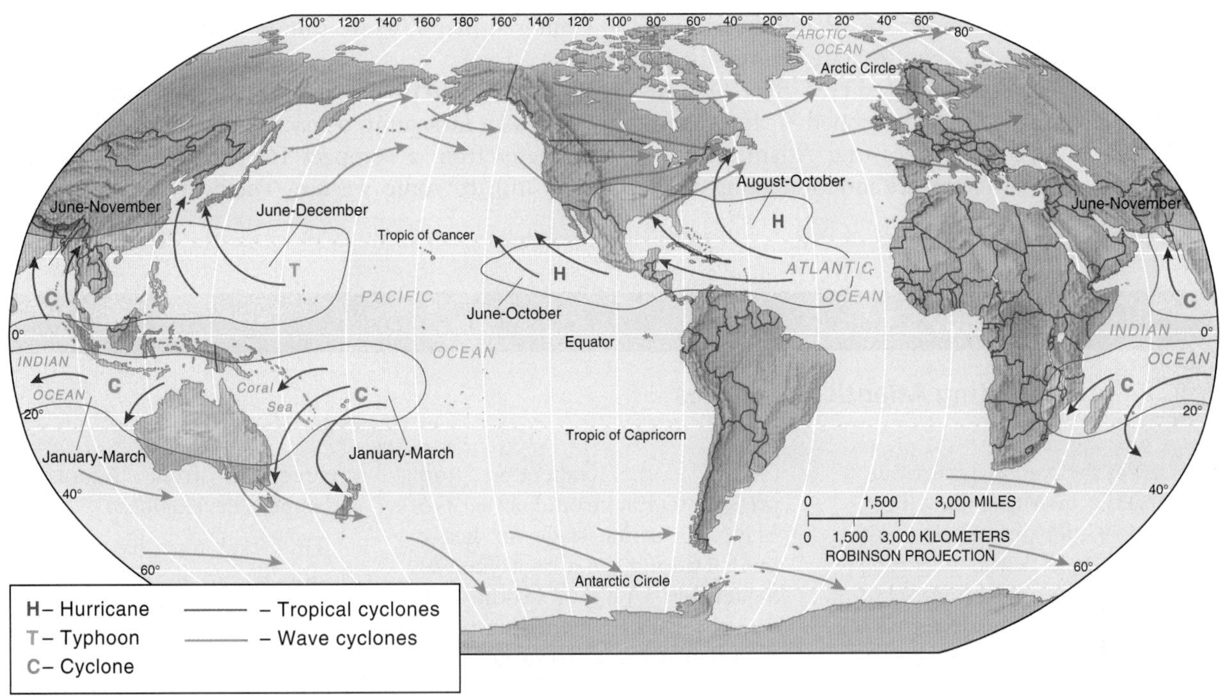

FIGURE 5.39 Worldwide pattern of the most intense tropical cyclones. Typical tropical storm tracks with principal months of occurrence and regional names. For the North Atlantic region from 1871 to 2000, 1084 tropical cyclones (storms and hurricanes) developed. The peak day in this Atlantic region (assuming a 9-day moving average calculation) is September 10, with 64 cyclones for the 129-year period. Also shown for comparison are characteristic storm tracks of midlatitude cyclones.

supply of water vapor and thus the necessary latent heat to fuel these storms.

The map in Figure 5.39 shows areas in which tropical cyclones form and some of the characteristic storm tracks traveled and the range of months during which tropical cyclones are most likely to appear. They tend to occur when the equatorial low-pressure trough (ITCZ) is farthest from the equator, during the months following the summer solstice in each hemisphere. For example, storms that strike the southeastern United States do so mostly between August and October, following the June 21 solstice. About 10% of all tropical disturbances have the right ingredients to intensify into a full-fledged **hurricane** or **typhoon** (see Table 5.1).

Hurricanes and Typhoons. Tropical cyclones are potentially the most destructive storms experienced by humans, costing an average of several hundred million dollars in property damage and thousands of lives each year. The tropical cyclone that struck Bangladesh in 1970 killed an estimated 300,000 people, and the one in 1991 claimed over 200,000. In the United States, death tolls are much lower, but still significant. The Galveston, Texas, hurricane of 1900 killed 6000; Hurricane Audry (1957), 400; Hurricane Gilbert (1988), 318; Hurricane Camille (1969), 256; Hurricane Agnes (1972), 117; and the near-record 1995 Atlantic season of 19 tropical storms, 11 of which became hurricanes, caused 121 deaths and $7.7 billion in damage. In 1998 Hurricanes Bonnie and Georges damaged property and took lives. Hurricane Mitch (October 26–November 4, 1998) was the deadliest Atlantic Hurricane in 2 centuries, killing more than 12,000 people in Central America.

Statistically, the damage caused by tropical cyclones is increasing substantially. Ironically, the relatively mild hurricane seasons from the mid-1960s to mid-1990s encouraged weak zoning and rapid development of vulnerable coastal lowlands. Loss of life is decreasing in most parts of the world owing to better forecasting and understanding of these storms. *Science* offered this "Perspective":

> The risk of human losses is likely to remain low, however, because of a well-established warning and rescue system and ongoing improvements in hurricane prediction. A main concern is the risks of high damage costs (up to $100 billion in a single event) because of ongoing population increases in coastal areas and increasing investment in buildings and extensive infrastructure in general.*

*L. Bengtsson, "Hurricane Threats," *Science* 293 (July 20, 2001), p. 441.

The years 1995–1999 were the most active 5 consecutive years since record keeping began, with 1995 as the second-greatest year for number of Atlantic hurricanes. Atmospheric scientists at Colorado State University, led by William Gray, successfully predicted the 1995 Atlantic hurricane season and developed a forecasting model. Focus Study 5.1, "Forecasting Atlantic Hurricanes," discusses recent hurricanes and forecasting methods. (See the National Hurricane Center at http://www.nhc.noaa.gov/ or the Joint Typhoon Warning Center at http://www.npmoc.navy.mil/jtwc.html.)

Physical Structure. Fully organized tropical cyclones possess an intriguing physical appearance. They range in diameter from a compact 160 km (100 mi), to 960 km (600 mi), to some western Pacific typhoons that reach

Focus Study 5.1

Forecasting Atlantic Hurricanes

Forecasters at the National Hurricane Center (NHC) in Miami have been busy since 1995, for this was the most active period in Atlantic hurricane history, with 103 named tropical storms, including 61 hurricanes (25 intense). This record level of activity was despite the El-Niño-retarded 1997 season. An important innovation assisting the NHC and state and local governments with storm analysis is a new forecasting method developed by a team led by William M. Gray at Colorado State University. (See http://typhoon.atmos.colostate.edu/forecasts/.)

Their analysis of weather records for the period 1900–2001 disclosed a significant causal relationship between Atlantic tropical storms and intense hurricanes that make landfall along the U.S. Gulf and East Coasts and several meteorological variables. Of 16 predictive elements used by the Colorado scientists, five key forecast variables emerge:

1. The presence or absence of warm water off the coast of Peru in the eastern Pacific. Sea-surface temperatures are part of the El Niño and La Niña phenomena (discussed in Chapter 6). A strong El Niño featuring warm surface water in the eastern Pacific Ocean tends to dampen the development of Atlantic hurricanes. The opposite effect occurs during La Niña episodes when cool water occurs in the eastern Pacific Ocean.
2. Rainfall, temperature, and pressure patterns in West Africa, specifically the Sahel along the southern margins of the Sahara (10°–20° N). Temperature and pressure regimes from February to May can set the stage for heavy rainfall during the normal June to September rainy season. A drought in this same region tends to retard Atlantic storm development.
3. Directional flows of equatorial stratospheric winds (20,000 to 23,000 m altitude, 68,000 to 75,000 ft), which tend to shift and reverse roughly every 12 to 16 months. These are called Quasi-Biennial Oscillation (QBO) winds. Winds from the west (westerlies) double Atlantic tropical storm probability and intensity, whereas, when the QBO easterly flow intensifies, storm tops are sheared and storm development is discouraged.
4. Sea-surface temperatures are important in the Atlantic Ocean, between Africa and the Caribbean, and along the North American East Coast. Higher than average sea-surface temperature and salinity levels are related to a more intense tropical storm/hurricane season—the trend since 1994. Lower temperatures and salinity tend to dampen storm activity—characteristic of 1900–1925 and 1970–1993 periods.
5. The status of the Azores high-pressure anticyclone is important (see Figure 4.14 and text discussion). If the circulation system is weak and below average in strength, then the east Atlantic tradewinds are weakened and tropical storm development is favored. A strong Azores high dampens development.

Dr. Gray's team has issued forecasts for 20 years and done better than climatology forecasts (long-term averages), especially when it comes to predicting named storms. The Colorado State team issues their forecast each December (for the next year), and issues updates in April, June, and August.

The 1995 Atlantic Hurricane Season—an Example

As Atlantic hurricane seasons (June 1 to November 1) go, 1995 was the second-most active in tropical storm count since record keeping began in 1870. Of 19 tropical storms, 11 became hurricanes, and 5 of these achieved category 3 status or higher. Only 1933 (21 tropical storms) exceeded this total. The season totaled $8 billion in overall damage, and more than 60 people were killed.

Figure 1 maps all 19 tropical storms, including 11 hurricanes (colored tracks). The season began with Allison and ended with Tanya. Figure 2 is a satellite image from August 30 that shows 4 tropical storms caught in one frame.

Each storm went through stages of tropical depression, tropical storm, hurricane, and extratropical depression. These four stages are based on wind speed (depression, storm, hurricane) and location of the low-pressure center (tropical or extratropical). Top wind speed and lowest central pressure are noted for each hurricane on the map.

(coninued)

Focus Study 5.1 *(continued)*

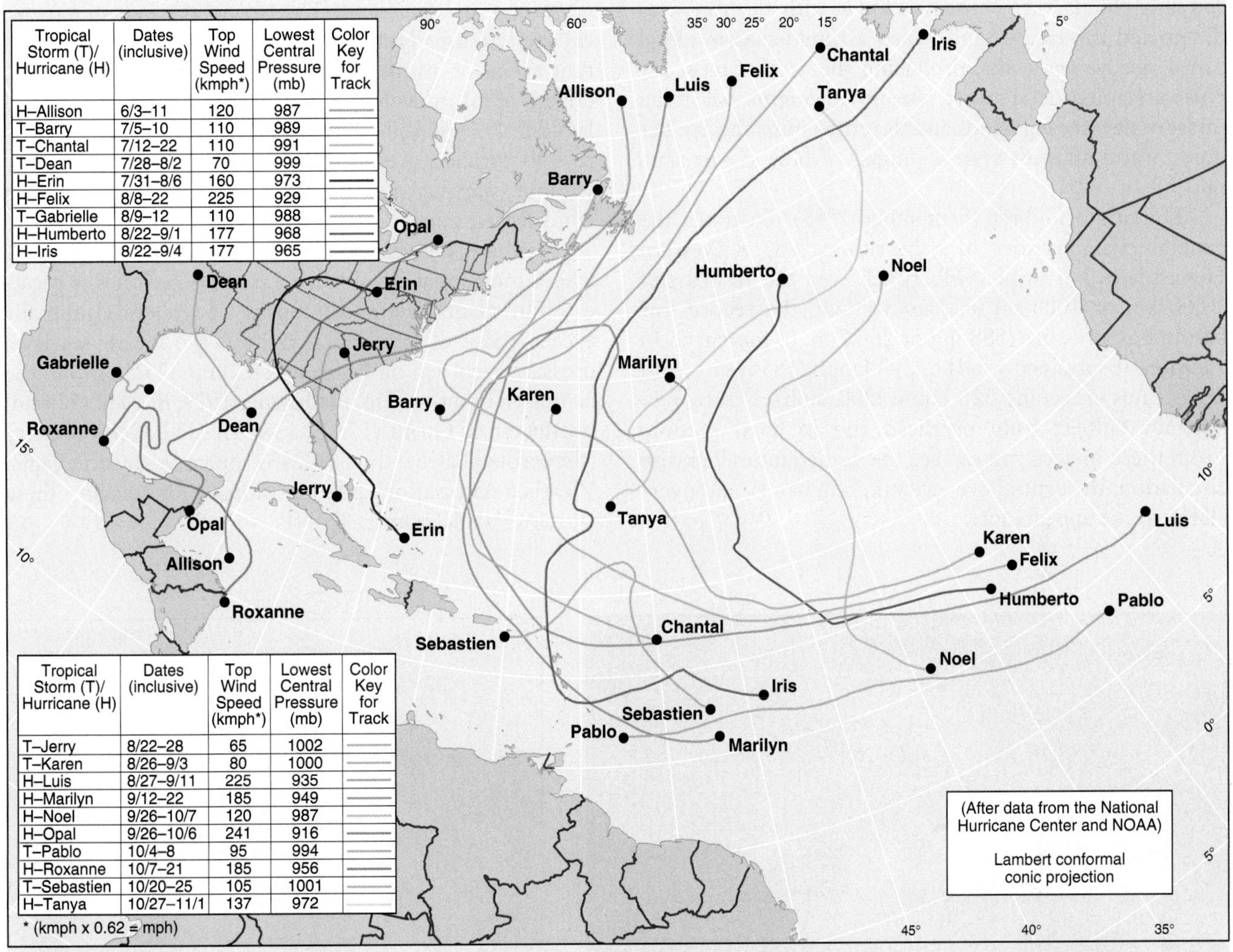

Tropical Storm (T)/ Hurricane (H)	Dates (inclusive)	Top Wind Speed (kmph*)	Lowest Central Pressure (mb)	Color Key for Track
H–Allison	6/3–11	120	987	
T–Barry	7/5–10	110	989	
T–Chantal	7/12–22	110	991	
T–Dean	7/28–8/2	70	999	
H–Erin	7/31–8/6	160	973	
H–Felix	8/8–22	225	929	
T–Gabrielle	8/9–12	110	988	
H–Humberto	8/22–9/1	177	968	
H–Iris	8/22–9/4	177	965	
T–Jerry	8/22–28	65	1002	
T–Karen	8/26–9/3	80	1000	
H–Luis	8/27–9/11	225	935	
H–Marilyn	9/12–22	185	949	
H–Noel	9/26–10/7	120	987	
H–Opal	9/26–10/6	241	916	
T–Pablo	10/4–8	95	994	
H–Roxanne	10/7–21	185	956	
T–Sebastien	10/20–25	105	1001	
H–Tanya	10/27–11/1	137	972	

* (kmph x 0.62 = mph)

(After data from the National Hurricane Center and NOAA)

Lambert conformal conic projection

FIGURE 1 1995 Atlantic storm season.
Atlantic, Caribbean, and Gulf of Mexico hurricane season, 1995. Noted on the map are inclusive dates from each storm's birth as a tropical depression to dissipation, highest wind speed in kmph (kmph × 0.62 = mph), and lowest pressure (mb). [Data courtesy of NOAA and the National Hurricane Center, Miami.]

No matter how accurate storm forecasts become, coastal and lowland property damage will continue to increase until better hazard zoning and development restrictions are in place. The property insurance industry appears to be taking action to promote these improvements. They are requiring tougher building standards to obtain coverage—or, in some cases, they are refusing to insure property along vulnerable coastal lowlands.

FIGURE 2 Satellite observes four storms at once.
Tropical storms Karen and Luis and Hurricanes Humberto and Iris march westward across the Atlantic Ocean during late August 1995. [Image from National Weather Service and NOAA (NESDIS).]

1300–1600 km (800–1000 mi). Vertically, these storms dominate the full height of the troposphere. The inward-spiraling clouds form dense *rain bands*, with a central area designated the *eye*, around which a thunderstorm cloud called the *eyewall* swirls, producing the area of most intense precipitation. The eye remains an enigma, for in the midst of devastating winds and torrential rains the eye has quiet, warm air with even a glimpse of blue sky or stars possible.

Hurricane Gilbert (September 1988) illustrates the general characteristics of a hurricane. For a Western Hemisphere hurricane, Gilbert achieved the record size (1600 km or 1000 mi in diameter) and the record low barometric pressure (888 mb or 26.22 in. of mercury). In addition, it sustained winds of 298 kmph (185 mph), with peak winds exceeding 320 kmph (200 mph). Figure 5.40 portrays Gilbert from overhead and in cross section. From these images you can see the vast counterclockwise circulation, the central eye, eyewall, rain bands, and overall hurricane appearance.

Such a storm travels at 16–40 kmph (10–25 mph). When it moves ashore, or makes *landfall*, it pushes *storm surges* of seawater inland. The strongest winds of a tropical cyclone usually are recorded in its right-front quadrant (relative to the storm's directional path), where dozens of fully developed tornadoes may be at the time of landfall. For example, Hurricane Camille in 1969 had up to 100 tornadoes embedded in that quadrant, striking without warning amidst howling winds and pouring wind-driven rain.

Typhoons form in the vast region of the Pacific, where the annual frequency of tropical cyclones is greatest. The magnitude and intensity of typhoons generally exceed those of hurricanes. In fact, the lowest sea-level pressure recorded on Earth was 870 mb (25.69 in.) in the center of Typhoon Tip, in October 1979, 837 km (520 mi) northwest of Guam (17° N 138° E). The Geostationary Meteorological Satellite (GMS), operated by the Japan Weather Association, provides images for tracking these western Pacific storms.

(a)

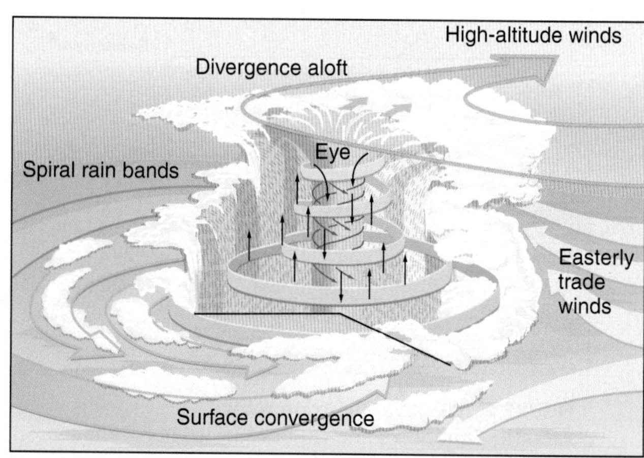

(b)

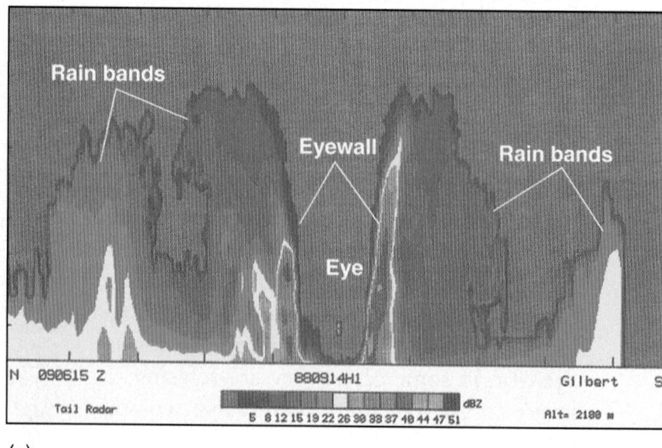

(c)

FIGURE 5.40 Profile of a hurricane.
Category 5 Hurricane Gilbert, September 13, 1988: (a) *GOES-7* satellite image; (b) a stylized portrait of a mature hurricane, drawn from an oblique view (cutaway shows the eye, rain bands, and wind flow patterns; (c) side-view radar image (Side-Looking Airborne Radar—SLAR) taken from an aircraft flying through the central portion of the storm. Rain bands of greater cloud density are false-colored in yellows and reds. [(a) and (c) National Hurricane Center, Miami, and NOAA.]

(a)

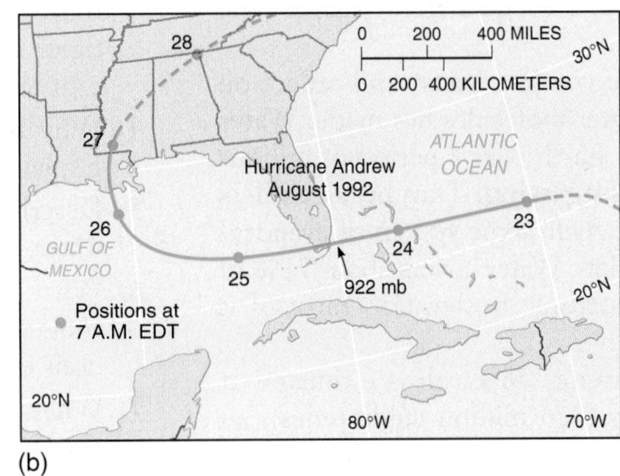

(b)

FIGURE 5.41 Hurricane Andrew.
(a) *NOAA-10* AVHRR image of Hurricane Andrew, August 24, 1992 (4:41 P.M. EDT) right after passage across southern Florida. (b) The track of Hurricane Andrew from August 23 through 28, 1992. [Image from NOAA as supplied by the EROS Data Center.]

Hurricanes of the 1990s. Nature's power and capacity to cause personal, societal, and economic tragedy were displayed in the 1990s. Hurricane Andrew tore through Florida and Louisiana in 1992, Bonnie and Georges crossed the Gulf Coast in 1998, and Mitch devastated Central America in 1999. Hurricane Andrew produced sustained surface winds of 266 kmph (165 mph). In 1999, Hurricanes Dennis, Floyd, and Irene ravaged North Carolina—the fourth year in a row that North Carolina received heavy tropical downpours from hurricanes. Rainfall was measured in feet instead of inches (Figure 5.41).

The tragedy from Andrew in Florida, Mitch in Central America, and the North Carolina hurricanes is the loss of homes, businesses, livestock, and even entire towns. These regions will be decades in recovery. An important lesson is that storms damage human structures more than they damage natural systems such as the Florida Everglades. Remember this perspective: *Urbanization, agriculture, pollution, and water diversion pose a greater ongoing threat to the Everglades and coastal barrier islands than do hurricanes.*

Unfortunately, thoughtful hazard planning is rarely practiced by public or private decision makers, whether along coastal lowlands, river floodplains, or earthquake fault zones. The consequence is that our entire society bears the financial cost of planning failure, whether in Florida, North Carolina, California, or the Midwest, not to mention those victims who directly shoulder the physical, emotional, and economic hardship of the event. This recurrent, yet avoidable cycle—construction, devastation, reconstruction, devastation—was reinforced ironically in *Fortune* magazine more than 30 years ago after the destruction by Hurricane Camille:

> Before long the beachfront is expected to bristle with new motels, apartments, houses, condominiums, and office buildings. Gulf Coast businessmen, incurably optimistic, doubt there will ever be another hurricane like Camille, and even if there is, they vow, the Gulf Coast will rebuild bigger and better after that.*

**Fortune*, October 1969, p. 62.

Summary and Review—Atmospheric Water and Weather

● *Describe* the origin of Earth's waters, the present distribution, and water's unique heat properties.

The next time it rains where you live, pause and reflect on the journey each of those water molecules has made. Water molecules came from within Earth over a period of billions of years, in a process called **outgassing**. Thus began endless cycling of water through the hydrologic system of evaporation-condensation-precipitation. Water covers about 71% of Earth; less than 3% is non-oceanic freshwater—most of it frozen.

The present volume of water on Earth is estimated at 1.36 billion cubic kilometers (326 million cubic miles), an amount achieved roughly 2 billion years ago. This overall steady-state equilibrium might seem in conflict with the many changes in sea level that have occurred over Earth's history, but it is not. Worldwide changes in sea level are called **eustasy**, and are related to the change in volume of water in the oceans. Some of these changes are explained by the amount of water stored in glaciers and ice sheets. At present, sea level is rising because of increases in the temperature of the oceans and the record melting of glacial ice.

Part of Earth's uniqueness is that its water exists naturally in all three states—solid, liquid, and gas—owing to Earth's temperate position relative to the Sun. A change from one state to another is called a **phase change**. The change from solid to vapor is called **sublimation**; from liquid to solid, freezing; from solid to liquid, melting; from vapor to liquid, condensation; and from liquid to vapor, vaporization, or evaporation.

The heat energy required for water to change phase is termed **latent heat**, because, once absorbed, it is hidden within the structure of the water, ice, or water vapor. For 1 g of water to become 1 g of water vapor at boiling requires addition of 540 calories, the **latent heat of vaporization**; at less than boiling the **latent heat of evaporation** describes the energy absorbed during phase change. When this 1 g of water vapor condenses, the same amount of heat energy is liberated, as 540 calories of the **latent heat of condensation**. The *latent heat of sublimation* is the energy exchanged in the phase change from ice to vapor and vapor to ice. Weather is powered by the tremendous amount of latent heat energy involved in the phase changes among the three states of water.

outgassing (p. 136)
eustasy (p. 136)
phase change (p. 138)
sublimation (p. 138)
latent heat (p. 139)
latent heat of vaporization (p. 139)
latent heat of condensation (p. 139)
latent heat of evaporation (p. 140)

1. Approximately when did Earth's water reach its present volume?
2. If the quantity of water on Earth has been quite constant in volume for at least 2 billion years, how can sea level have fluctuated? Explain.
3. Describe the locations of Earth's water, both oceanic and fresh. What is the largest repository of freshwater at this time? In what ways is this distribution of water significant to modern society?
4. Why might you describe Earth as the water planet? Explain.
5. Describe the three states of matter as they apply to ice, water, and water vapor.
6. What happens to the physical structure of water as it cools below 4°C (39°F)? What are some visible indications of these physical changes?
7. What is latent heat? How is it involved in the phase changes of water?
8. Take 1 g of water at 0°C and follow it through to 1 g of water vapor at 100°C, describing what happens along the way. What amounts of energy are involved in the changes that take place?

● *Define* humidity, relative humidity, and dew-point temperature and *illustrate* three atmospheric conditions—unstable, conditionally unstable, and stable.

The amount of water vapor in the atmosphere is termed **humidity**. The ability of air to hold water vapor is principally a function of the temperature of the air and of the water vapor (usually the same). **Relative humidity** is a percentage expression of the humidity content of the air compared with the capacity of the air to hold water vapor at a given temperature—content compared to capacity. Air is said to be **saturated**, or filled to capacity, if it contains all the water vapor it can hold at a given temperature (100% relative humidity). The temperature at which air achieves saturation is called the **dew-point temperature**.

Among the various ways to express humidity and relative humidity are vapor pressure and specific humidity. **Vapor pressure** is that portion of the atmospheric pressure that is produced by the presence of water vapor. A comparison of vapor pressure with the saturation vapor pressure at any moment produces a relative humidity percentage. **Specific humidity** is the mass of water vapor (in grams) per mass of air (in kilograms) at any specified temperature. A comparison of specific humidity with the maximum specific humidity produces a relative humidity percentage.

Meteorologists use the term *parcel* to describe a body of air that has specific temperature and humidity characteristics. Warm air has a lower density in a given volume of air; cold air has a higher density. **Stability** refers to the tendency of an air parcel, with its water-vapor cargo, either to remain in place or to change vertical position by ascending (rising) or descending (falling). An air parcel is *stable* if it resists displacement upward or, when disturbed, it tends to return to its starting place. An air parcel is *unstable* if it continues to rise until it reaches an altitude where the surrounding air has a density (air temperature) similar to its own.

An ascending (rising) parcel of air cools by expansion, responding to the reduced air pressure at higher altitudes. A descending (falling) parcel heats by compression.

The **normal lapse rate**, as introduced in Chapter 2, is the average decrease in temperature with increasing altitude, a value of 6.4 C°/1000 m (3.5 F°/1000 ft). This rate of temperature change is for still, calm air. It can vary greatly under different weather conditions. Consequently, the actual lapse rate at a particular place and time is labeled the **environmental lapse rate**.

The warming and cooling rates for a parcel of expanding or compressing air are termed **adiabatic**. The **dry adiabatic rate (DAR)** is the rate at which "dry" air cools by expansion (if ascending) or heats by compression (if descending). The term *dry* is used when air is less than saturated (relative humidity less than 100%). The DAR is 10 C°/1000 m (5.5 F°/1000 ft). The **moist adiabatic rate (MAR)** is the average rate at which ascending air that is moist (saturated) cools by expansion, or descending air warms by compression. The average MAR is 6 C°/1000 m (3.3 F°/1000 ft). This is roughly 4 C° (2 F°) less than the dry rate. The MAR, however, varies with moisture content and temperature and can range from 4 C° to 10 C° per 1000 m (2 F° to 5.5 F° per 1000 ft).

A simple comparison of the DAR and MAR in a vertically moving parcel of air with that of the environmental lapse rate in the surrounding air determines the atmosphere's stability, that is, whether it is unstable (lifting of air parcels continues) or stable (air parcels resist vertical displacement).

humidity (p. 140)
relative humidity (p. 141)
saturated (p. 141)
dew-point temperature (p. 141)
vapor pressure (p. 143)
specific humidity (p. 143)
stability (p. 144)
normal lapse rate (p. 145)
environmental lapse rate (p. 145)
adiabatic (p. 145)
dry adiabatic rate (DAR) (p. 146)
moist adiabatic rate (MAR) (p. 146)

9. What is humidity? How is it related to the energy present in the atmosphere? To our personal comfort and how we perceive apparent temperatures?
10. Define relative humidity. What does the concept represent? What is meant by the terms *saturation* and *dew-point temperature*?
11. Using different measures of humidity in the air as given in the chapter, derive relative humidity values (vapor pressure/saturation vapor pressure; specific humidity/maximum specific humidity).
12. How do the two instruments described in this chapter measure relative humidity?
13. Differentiate between stability and instability relative to a parcel of air rising vertically in the atmosphere.
14. What are the forces acting on a vertically moving parcel of air? How are they affected by the density of the air parcel?
15. How do the adiabatic rates of heating or cooling in a vertically displaced air parcel differ from the normal lapse rate and environmental lapse rate?
16. What would atmospheric temperature and moisture conditions be on a day when the weather is unstable? When it is stable? Relate in your answer what you would experience if you were outside watching.

● *Explain* cloud formation and *identify* major cloud classes and types, including fog.

A **cloud** is an aggregation of tiny moisture droplets and ice crystals suspended in the air. Clouds are reminders of the powerful heat-exchange system in the environment. **Moisture droplets** in a cloud form when saturated air and the presence of **condensation nuclei** lead to *condensation*.

Low clouds, ranging from the surface up to 2000 m (6500 ft) in the middle latitudes, are called **stratus** (flat clouds, in layers) or **cumulus** (puffy clouds, in heaps). When stratus clouds yield precipitation, they are called **nimbostratus**. Middle-level clouds are denoted by the prefix *alto-*. **Altocumulus** clouds, in particular, represent a broad category that occurs in many different styles. A cumulus cloud can develop into a towering giant called **cumulonimbus** (*-nimbus* denotes precipitation). Such clouds are called *thunderheads* because of their shape and their associated lightning, thunder, surface wind gusts, updrafts and downdrafts, heavy rain, and hail. Clouds at high altitude, principally composed of ice crystals, are called **cirrus**. Sometimes near the end of the day, lumpy, grayish, low-level clouds called **stratocumulus** may fill the sky in patches.

Fog is a cloud that occurs at ground level. **Advection fog** forms when air in one place migrates to another place where conditions exist that can cause saturation—for example, when warm, moist air moves over cooler ocean currents. Another type of advection fog forms when cold air flows over the warm water of a lake, ocean surface, or swimming pool. This **evaporation fog**, or *steam fog*, may form as the water molecules evaporate from the water surface into the cold overlying air. **Upslope fog** is produced when moist air is forced to higher elevations along a hill or mountain. Another fog caused by topography is **valley fog**, formed because cool, denser air settles in low-lying areas, producing fog in the chilled, saturated layer near the ground. **Radiation fog** is caused by radiative cooling of a surface that chills the air layer directly above the surface to the dew-point temperature, creating saturated conditions and fog.

cloud (p. 147)
moisture droplet (p. 147)
condensation nuclei (p. 147)
stratus (p. 148)
cumulus (p. 148)
nimbostratus (p. 148)
altocumulus (p. 148)
cumulonimbus (p. 148)
cirrus (p. 148)
stratocumulus (p. 148)
fog (p. 148)
advection fog (p. 148)
evaporation fog (p. 148)
upslope fog (p. 149)
valley fog (p. 149)
radiation fog (p. 149)

17. Specifically, what is a cloud? Describe the droplets that form a cloud.
18. Explain the condensation process: What are the two essential requirements?
19. What are the basic forms of clouds? Using Figure 5.18, describe how the basic cloud forms vary with altitude.
20. Explain how clouds might be used as indicators of the conditions of the atmosphere. Of expected weather.
21. What type of cloud is fog? List and define the principal types of fog.

● *Describe* air masses that affect North America and *identify* four types of atmospheric lifting mechanisms.

Specific conditions of humidity, stability, and cloud coverage occur in a regional, homogenous **air mass**. The longer an air mass remains stationary over a region, the more definite its physical attributes become. Air masses are categorized by their *moisture content*—**m** for maritime (wetter) and **c** for continental (drier)—and their *temperature* (a function of latitude)—designated **A** (arctic), **P** (polar), **T** (tropical), **E** (equatorial), and **AA** (antarctic).

Air masses can be lifted by **convergent lifting** (air flows conflict, forcing some of the air to lift), **convectional lifting** (air passing over warm surfaces gains buoyancy), **orographic lifting** (passage over a topographic barrier), and *frontal lifting*. In North America, *chinook winds* (called *föhn* or *foehn* winds in Europe) are the warm, downslope air flows characteristic of the leeward side of mountains. Orographic lifting creates wetter windward slopes and drier leeward slopes situated in the **rain shadow** of the mountain. Conflicting air masses at a front produce a **cold front** (and sometimes a zone of strong wind and rain) or a **warm front**.

air mass (p. 149)
convergent lifting (p. 154)
convectional lifting (p. 154)
orographic lifting (p. 154)
rain shadow (p. 154)
cold front (p. 156)
warm front (p. 156)

22. How does a source region influence the type of air mass that forms over it? Give specific examples of each basic classification.
23. Of all the air masses, which are of greatest significance to the United States and Canada? What happens to them as they migrate to locations different from their source regions? Give an example of air-mass modification.
24. Explain why it is necessary for an air mass to rise if there is to be precipitation.
25. What are the four principal lifting mechanisms that cause air masses to ascend, cool, condense, form clouds, and perhaps produce precipitation? Briefly describe each.
26. Differentiate between the structure of a cold front and a warm front.

● *List* the measurable elements of weather and *describe* the life cycle of a midlatitude cyclone.

Weather is the short-term condition of the atmosphere; **meteorology** is the scientific study of the atmosphere. Analyzing and understanding patterns of wind, air pressure, temperature, and moisture conditions portrayed on daily weather maps is key to numerical (computer-based) weather forecasting.

A **midlatitude cyclone**, or **wave cyclone**, is a vast low-pressure system that migrates across the continent, pulling air masses into conflict along fronts. *Cyclogenesis*, the birth of the low-pressure circulation, can occur off the west coast of North America, along the polar front, along the lee slopes of the Rockies, in the Gulf of Mexico, and along the east coast. An **occluded front** is produced when a cold front overtakes a warm front in the maturing cyclone. These systems are guided by the jet streams of the upper troposphere along seasonally shifting *storm tracks*.

weather (p. 158)
meteorology (p. 159)
midlatitude cyclone (p. 159)
wave cyclone (p. 159)
occluded front (p. 162)

27. Differentiate between a cold front and a warm front as types of frontal lifting and what you would experience with each one.
28. How does a midlatitude cyclone act as a catalyst for conflict between air masses?
29. Diagram a midlatitude cyclonic storm during its open stage. Label each of the components in your illustration, and add arrows to indicate wind patterns in the system.
30. What is your principal source of weather data, information, and forecasts? Where does your source obtain its data? Have you used the Internet and World Wide Web to obtain weather information? In what ways will you personally apply this knowledge in the future? What benefits do you see?

● *Analyze* various types of violent weather—thunderstorms, tornadoes, and tropical cyclones.

The violent power of some weather phenomena poses a hazard to society. Thunderstorms produce **lightning** (electrical discharges in the atmosphere), **thunder** (sonic bangs produced by the rapid expansion of air after intense heating by lightning), and **hail** (ice pellets formed within cumulonimbus clouds). A spinning, cyclonic column rising to midtroposphere level—a **mesocyclone**—is sometimes visible as the swirling mass of a cumulonimbus cloud. Dark gray **funnel clouds** may pulse from the bottom side of the parent cloud. A **tornado** is formed when the funnel connects with Earth's surface. A **waterspout** forms when a tornado circulation occurs over water.

Within tropical air masses, large low-pressure centers can form along easterly wave troughs. Under the right conditions, a tropical cyclone is produced. Depending on wind speeds and central pressure, a **tropical cyclone** can become a **hurricane**, **typhoon**, or *cyclone*, when winds exceed 65 knots (119 kmph, 74 mph). As forecasting and the public's perception of weather-related hazards have improved, loss of life has decreased, although property damage continues to increase.

lightning (p. 162)
thunder (p. 163)
hail (p. 166)
mesocyclone (p. 166)
funnel clouds (p. 167)
tornado (p. 167)
waterspout (p. 167)
tropical cyclone (p. 168)
hurricane (p. 169)
typhoon (p. 169)

31. What constitutes a thunderstorm? What type of cloud is involved? What type of air mass would you expect in an area of thunderstorms in North America?
32. Lightning and thunder are powerful phenomena in nature. Briefly describe how they develop.
33. Describe the formation process of a mesocyclone. How is this development associated with that of a tornado?
34. Evaluate the pattern of tornado activity in the United States. What generalizations can you make about the distribution and timing of tornadoes?
35. What are the different classifications for tropical cyclones? List the various names used worldwide for hurricanes.
36. What forecast factors did scientists use to accurately predict the 1995 and other Atlantic hurricane seasons?
37. Relative to improving weather forecasting, what are some of the technological innovations discussed in this chapter?

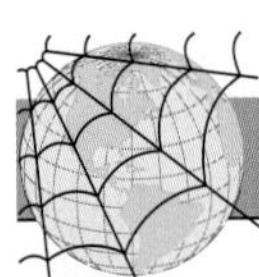

NetWork

The *Elemental Geosystems Home Page* provides on-line resources for this chapter on the World Wide Web. Once on the Home Page, click on this textbook, scroll the Table of Contents menu, select this chapter, and click "Begin." You will find self-tests that are graded, review exercises, specific updates for items in the chapter, and in "Destinations" many links to interesting related pathways on the Internet. *Elemental Geosystems* is found at **http://www.prenhall.com/christopherson**.

Critical Thinking

A. Examine the photograph of an iceberg in Figure 5.5b. The photo was taken in Disko Bay, Greenland. Determine what caused the "shoreline" watermark above the ocean surface around the iceberg. Why do you think the iceberg appears to be riding higher in the water than several weeks earlier? In rough terms, how much ice (in percent) do you think is beneath the surface in comparison to the amount you see above sea level? Explain why this physical trait can be a hazard to shipping.

B. Using Figure 5.18, begin to observe clouds on a regular basis. See if you can relate the cloud type to particular weather conditions at the time of observation. You may want to keep a log in your notebook throughout the course. Seeing clouds in this way will help you to understand weather.

C. Select "Destinations" from the left-hand frame of Chapter 5 on the *Elemental Geosystems Home Page*. Be prepared—this is a fantastic listing of 30 links on the Net. Any item in color and underlined will open and take you directly to that home page. Sample at least five of these weather links and describe what you find. You may want to mark some of these links with your own browser for ready reference when you need weather information.

D. Relative to coastal devastation from tropical weather, the following line appears in the text: "This recurrent, yet avoidable cycle—construction, devastation, reconstruction, devastation. . . . " This describes the ever-increasing dollar losses to property from tropical storms and hurricanes at a time when improved forecasts have resulted in a significant reduction in loss of life. (In Chapter 13, these issues are discussed further in Focus Study 13.1, along with the 1999 North Carolina flooding.) In your opinion, what is the solution to halting these increasing losses, this cycle of destruction? How would you implement your plan?

The Taklimakan Desert is in the Tarim Basin of China. It extends between the Kunlun Mountains and Tibetan Plateau to the south and the Tian Shan Mountains to the north. Shifting sand seas cover more than 80% of the region. Massive dust storms, originating in this region, have traveled as far as North America. [October 27, 2001 Terra MODIS sensor courtesy of MODIS Land Rapid Response Team, NASA/GSFC.]

6 Global Climate Systems

Key Learning Concepts

By knowing and understanding the key learning concepts in this chapter, you should be able to:

- *Define* climate and climatology and *explain* the difference between climate and weather.
- *Review* the role of temperature, precipitation, air pressure, air-mass patterns and sea-surface temperatures used to establish climatic regions.
- *Review* the development of climate classification systems and *compare* genetic and empirical systems as ways of classifying climate.
- *Describe* the principal climate classification categories other than deserts and *locate* these regions on a world map.
- *Explain* the precipitation and moisture efficiency criteria used to determine the arid and semiarid climates and *locate* them on a world map.
- *Outline* future climate patterns from forecasts presented and *explain* the causes and potential consequences.

Earth experiences an almost infinite variety of *weather*—the condition of the atmosphere at any given place and time. This variability, when considered along with the average conditions at a place over time, constitutes **climate**. Climate includes not just the averages but the extremes experienced in a place.

Today climatologists know that intriguing linkages exist in the Earth-atmosphere-ocean system. For example, an El Niño in the Pacific is tied to drought-breaking rains in the West, floods in Louisiana and Northern Europe, and a drought in Australia. And strong monsoonal rains in West Africa correlate with the development of intense Atlantic hurricanes.

Climatologists are concerned about observed changes occurring in the global climate. Climate patterns are changing at a pace not evidenced in the records of the past millennia. Climate and natural vegetation changes during the next 50 years could exceed the total of all changes since the peak of the last ice-age episode, some 18,000 years ago.

In this chapter: Climates are so diverse that no two places on Earth's surface experience exactly the same climatic conditions; in fact, Earth is a vast collection of microclimates. However, broad similarities among local climates permit their grouping into climatic regions.

Early climatologists faced the challenge of what to use as a basis for climate classification. Many of the physical elements of the environment, studied in the first five chapters of this text, link together to explain climates. In this edition I am using a simplified classification system to discuss climates based on physical factors that help uncover the "why" question—why climates are in certain locations. Though imperfect, this method is easily understood and is based on a widely used classification system devised by climatologist Wladimir Köppen (pronounced KUR-pen). For reference, Appendix C details the Köppen climate classification system.

Climatologists use powerful computer models to simulate changing complex interactions in the atmosphere, hydrosphere, lithosphere, and biosphere. This chapter concludes with a discussion of climate change and its vital implications for our global society. The climatic regions we study in this chapter will migrate during this century.

Earth's Climate System and Its Classification

Climatology, the study of climate and its variability, analyzes long-term weather patterns over time or space and the controls that produce Earth's diverse climatic conditions. One type of climatic analysis locates areas of similar weather statistics and groups them into **climatic regions** that contain characteristic weather patterns. Weather measurements and observations, gathered at different locations within a region, are plotted on maps and compared to identify climate types. These observed patterns are at the core of climate classifications. Expected weather and climatic patterns can be disrupted, as is the case with the recurring El Niño phenomenon, discussed in Focus Study 6.1. Note that climate cannot actually be observed and really does not exist at any particular moment. Climate is therefore a conceptual statistical construction from measured weather elements.

Climates may be humid with distinct seasons, dry with consistent warmth, or moist and cool—almost any combination is possible. There are places where rainfall exceeds 20 cm (8 in.) each month and where average monthly temperatures remain above 27°C (80°F) throughout the year. Other places may go without rain for 10 years at a time.

Students reading this text in Singapore experience precipitation every month ranging from 13.1 to 30.6 cm (5.1 to 12.0 in.), or 228.1 cm (89.8 in.) during an average year, whereas students at the university in Karachi, Pakistan, measure only 20.4 cm (8 in.) of rain over an entire year.

Climates greatly influence *ecosystems*, the natural, self-regulating communities formed by plants and animals in their nonliving environment. On land, the basic climatic regions determine to a large extent the location of the world's major ecosystems. These regions, called *biomes*, include forest, grassland, savanna, tundra, and desert. Plant, soil, and animal communities are associated with these biomes. Because climate cycles through periodic change and is never really stable, ecosystems should be thought of as being in a constant state of adaptation and response.

The present global climatic warming trend is producing changes in plant and animal distributions. Figure 6.1 presents a schematic view of Earth's climate system, showing both internal and external processes and linkages that influence climate and thus regulate such changes.

Climate Components: Insolation, Temperature, Air Pressure, Air Masses, and Precipitation

Before we look at climatic regions, let's briefly review key concepts from the first five chapters. Uneven insolation over Earth's surface, varying with latitude, is the energy input for the climate system. Daylength and temperature patterns vary daily and seasonally (Chapter 2). The principal controls of temperature are latitude, altitude, land-water heating differences, and the amount and duration of cloud cover. The pattern of world temperatures and annual temperature ranges is in Chapter 3 and portrayed in Figures 3.24, 3.26, and 3.28.

The hydrologic cycle provides the basic means of transferring energy and mass through Earth's climate system. Figure 6.2 shows the worldwide distribution of precipitation (rain, sleet, snow, and hail). You can identify several patterns, such as the way precipitation decreases from western Europe inland toward the heart of Asia (continentality). However, high precipitation values dominate southern Asia, where the landscape receives moisture-laden monsoonal winds from the Indian Ocean in the summer. Southern South America reflects a pattern of wet western (windward) slopes and dry eastern (leeward) slopes (orographic effects). (See Figure 7.4, which portrays the distribution of average annual precipitation in North America.)

Most of Earth's deserts and regions of permanent drought are located in lands dominated by subtropical high-pressure cells, with bordering lands grading to grasslands and to forests as precipitation increases. The most consistently wet places on Earth straddle the equator in the Amazon region of South America, the Congo (Zaire) region of Africa, and the Indonesian and Southeast Asian area, all of which are influenced by equatorial low pressure and the intertropical convergence zone (Figures 4.11, 4.13).

Simply comparing the two principal climatic components—temperature and precipitation—reveals important

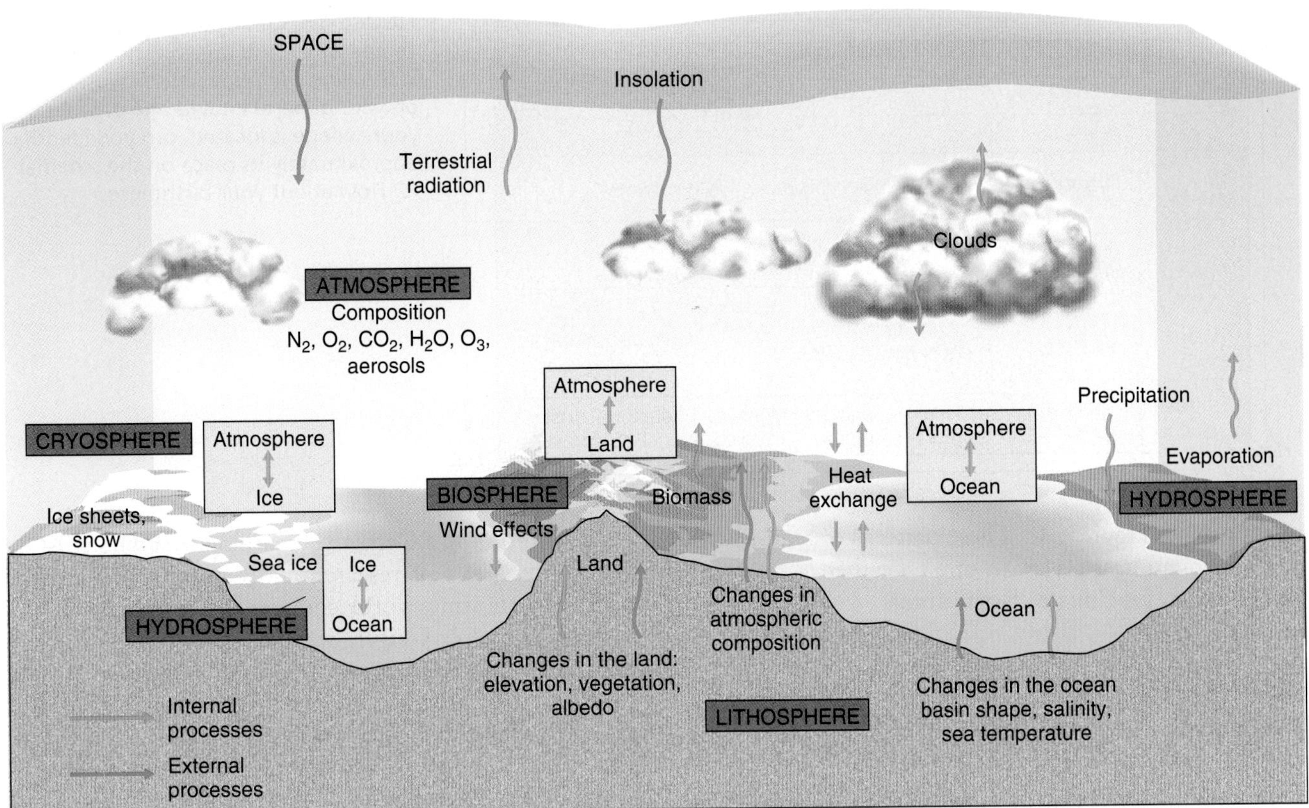

FIGURE 6.1 A schematic of Earth's climate system.
Internal processes that influence climate involve the atmosphere, hydrosphere (streams and ocean), cryosphere (polar ice masses and glaciers), biosphere, and lithosphere (land)—all energized by insolation. All systems interact to produce climatic patterns. *External processes* affect this climatic balance and force climate change. Imagine you are hired to write a computer program to model all these linkages and interactions! [After J. Houghton, *The Global Climate* (Cambridge: Cambridge University Press, 1984), and the Global Atmospheric Research Program.]

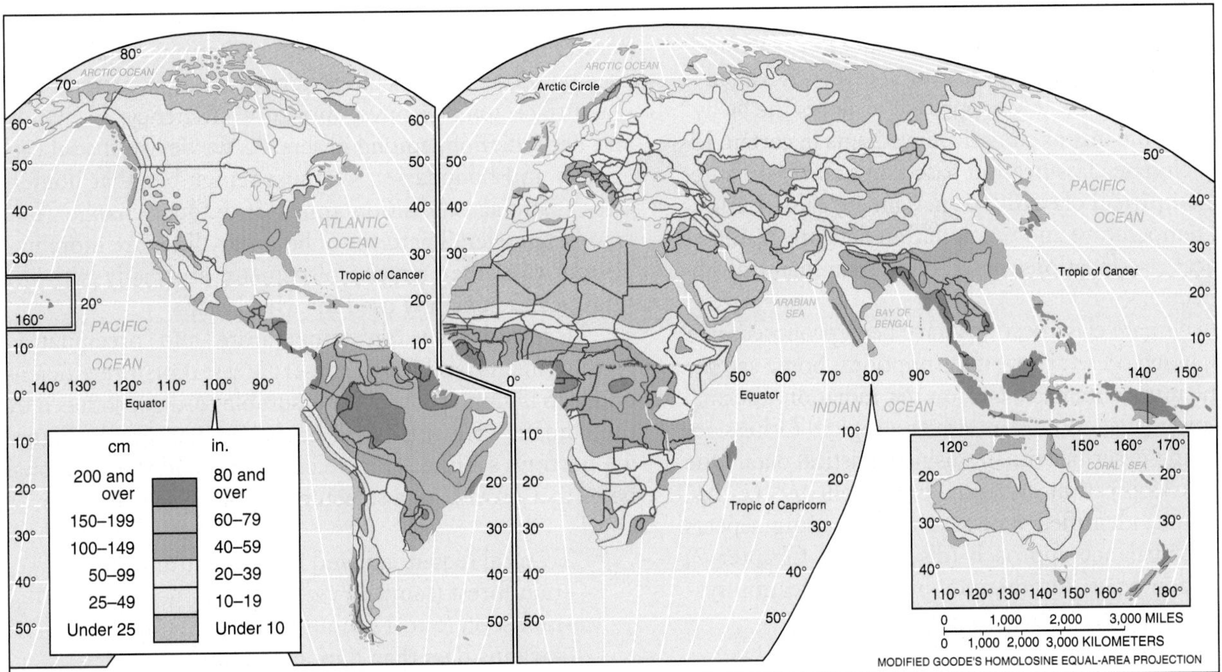

FIGURE 6.2 Worldwide average annual precipitation.
The causes that produce these patterns should be recognizable to you: temperature and pressure patterns; air mass types; convergent, convectional, orographic, and frontal lifting mechanisms; and the general energy availability that decreases toward the poles.

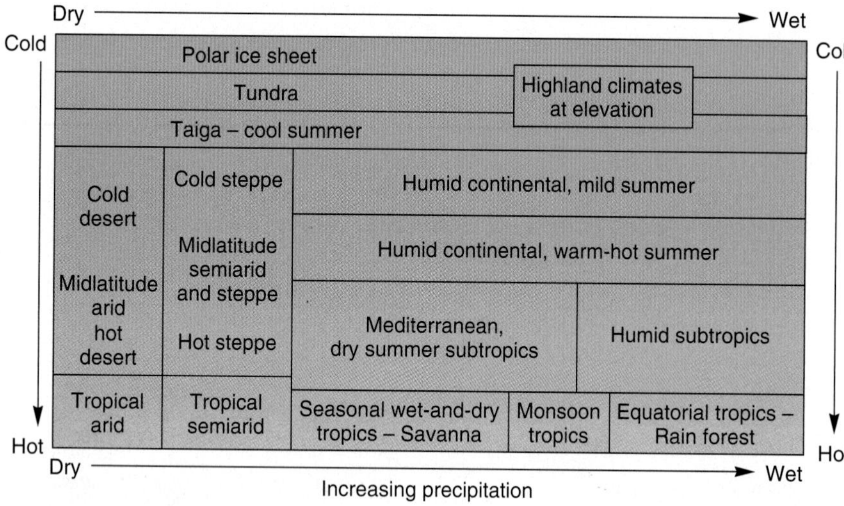

FIGURE 6.3 Climatic relationships. Temperature and precipitation schematic showing climatic relationships. Based on your general knowledge of where your college is located, can you identify approximately its place on the schematic? How about your birthplace?

relationships (Figure 6.3). These patterns of temperature and precipitation, along with other weather factors, provide the basis for climate classification.

Classification of Climatic Regions

The ancient Greeks simplified their view of world climates into three zones: the "torrid zone" referred to warmer areas south of the Mediterranean, the "frigid zone" was to the north, and the area where they lived was labeled the "temperate zone," which they considered the optimum climate. Through exploration and satellite observations, we know that Earth's myriad climatic variations are more complex than the ancients imagined.

Classification is the process of ordering or grouping data or phenomena into related classes. Such generalizations are important organizational tools in science and are especially useful for the spatial analysis of climatic regions. A climate classification based on *causative* factors—for example, the interaction of air masses—is called a **genetic classification**. This approach explores the "why" question as to the mix of climatic ingredients in certain locations. A climate classification based on *statistics* or other data is an **empirical classification**. Climate classifications based on temperature and precipitation data are examples of empirical classifications, based on measurement of observed effects.

Many climate classifications have been proposed over the years, based on a variety of assumptions. Some are genetic, explaining climates based on net radiation, thermal regimes, or air-mass dominance over a region. Others are empirical, and describe climates using statistical data. One empirical classification system, published by C. W. Thornthwaite, identified moisture regions using aspects of the water-balance approach (discussed in Chapter 7) and vegetation types. Another empirical classification system is the Köppen classification system.

A Climate Classification System

The Köppen classification system (pronounced KUR-pen), widely recognized, was designed by Wladimir Köppen (1846–1940), a German climatologist and botanist. In 1928, after years of research and development, his first wall map showing world climates, coauthored with his student Rudolph Geiger, was widely adopted. Köppen continued to refine this system until his death. In Appendix C you find a description of his system and the detailed criteria he used to distinguish climatic regions.

In this chapter I use a system that is somewhat middle ground between genetic and empirical. This allows description of the climatic regions, yet gives you some ideas as to why such climates are found where they occur. The Köppen system provides us an outline and general base map.

Classification Categories. The basis of any classification system is the choice of criteria or causative factors used to draw lines between categories. Some climate elements that could be used include: average monthly temperatures, average monthly precipitation, total annual precipitation, air mass characteristics, ocean currents and sea-surface temperatures, moisture efficiency, insolation and net radiation, among others. As we devise spatial categories and boundaries, we must remember that boundaries really are transition zones of gradual change. The trends and overall patterns of boundary lines are more important than their precise placement, especially with the small scales generally used for world maps.

Here we focus on temperature and precipitation measurements, and for the desert areas, moisture efficiency. Keep in mind these are measurable results produced by the climatic elements listed in the last paragraph. Figure 6.4 portrays six basic climate categories and their regional types that provide us a structure for our discussion.

- Tropical (equatorial and tropical latitudes)
  –rain forest (rainy all year)
  –monsoon (6 to 12 months rainy)
  –savanna (less than 6 months rainy)
- Mesothermal (mild winter)
  –humid subtropical (hot summers)
  –marine west coast (warm to cool summers)
  –Mediterranean (dry summers)

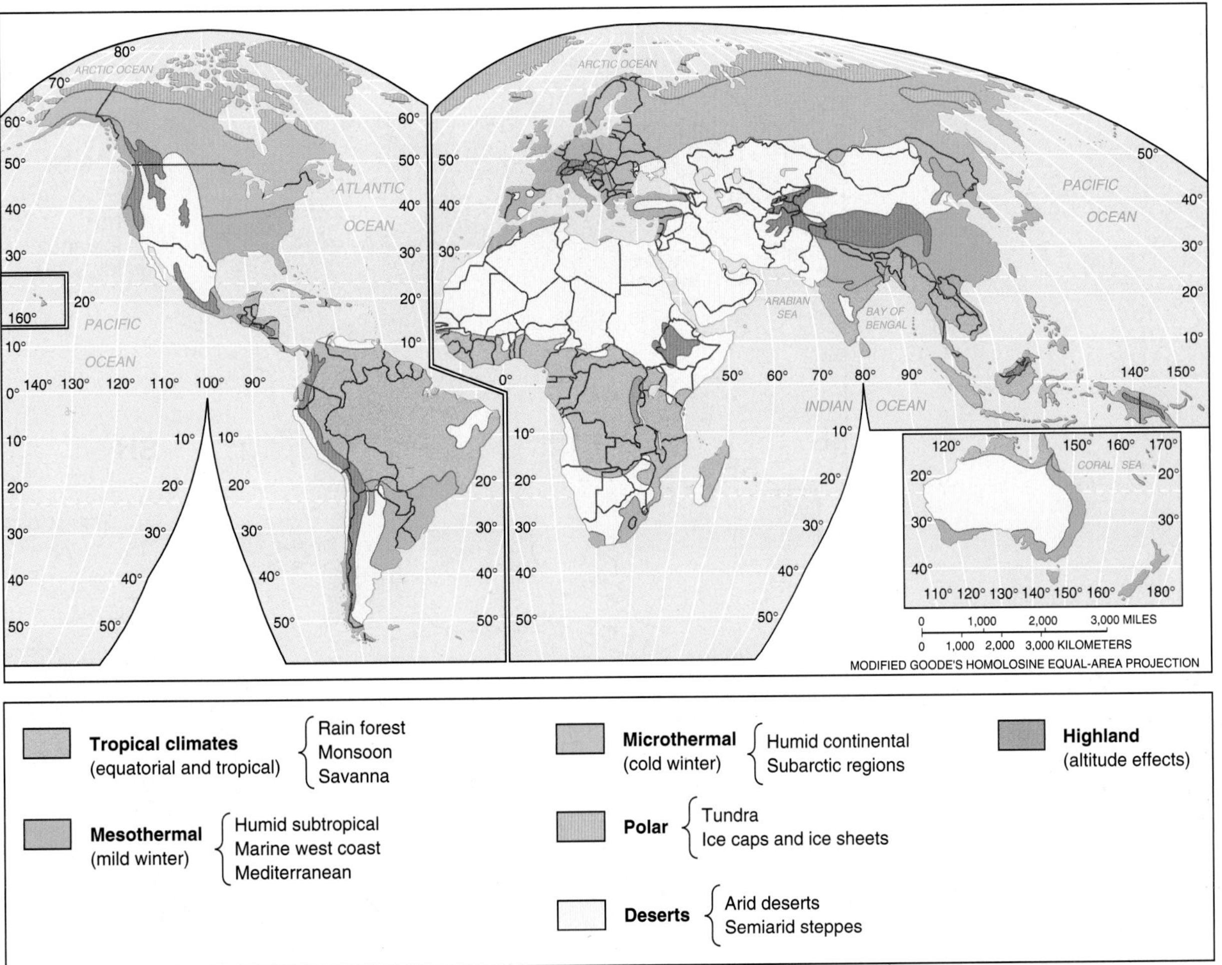

FIGURE 6.4 Climate regions generalized.
Six general climate categories. Relative to the question asked about your campus and birthplace in the caption to Figure 6.3, locate these two places on this map.

- Microthermal (mid- and high-latitude, cold winter)
  –humid continental (hot to warm summers)
  –subarctic regions (cool summers to very cold winters)
- Polar (polar and high latitudes)
  –tundra (high latitude or at altitude)
  –ice caps and ice sheets (perpetually frozen)
- Highland (compared to lowlands at the same latitude, highlands have lower temperatures—recall the normal lapse rate)

Only one climate category is based on moisture efficiency as well as temperature:

- Deserts (permanent moisture deficits)
  –arid deserts (tropical and midlatitude)
  –semiarid steppes (tropical and midlatitude)

Global Climate Patterns. Adding detail to Figure 6.4, we develop the world climate map presented in Figure 6.5. The following sections describe specific climates, organized around each of the main climate categories listed previously. An opening box at the beginning of each climate section gives a simple description of the climate category and causal elements that are in operation. A world map showing distribution and the featured representative cities also is in the introductory box for each climate. In this discussion the names of the climates appear in italics.

Climographs exemplify particular climates for selected cities. A **climograph** is a graph that shows monthly temperature and precipitation, location coordinates, average annual temperature, total annual precipitation, elevation, the local population, annual temperature range, annual hours of sunshine (if available, as an indication of cloudiness), and a location map. Along the top of each climograph, find the dominant weather features that are influential in that climate.

Discussions of soils, vegetation, and major terrestrial biomes that fully integrate these global climate patterns are in Chapters 15 and 16. Table 16.2 synthesizes all this information and will enhance your understanding of this chapter, so please place a tab on that page and refer to it as you read.

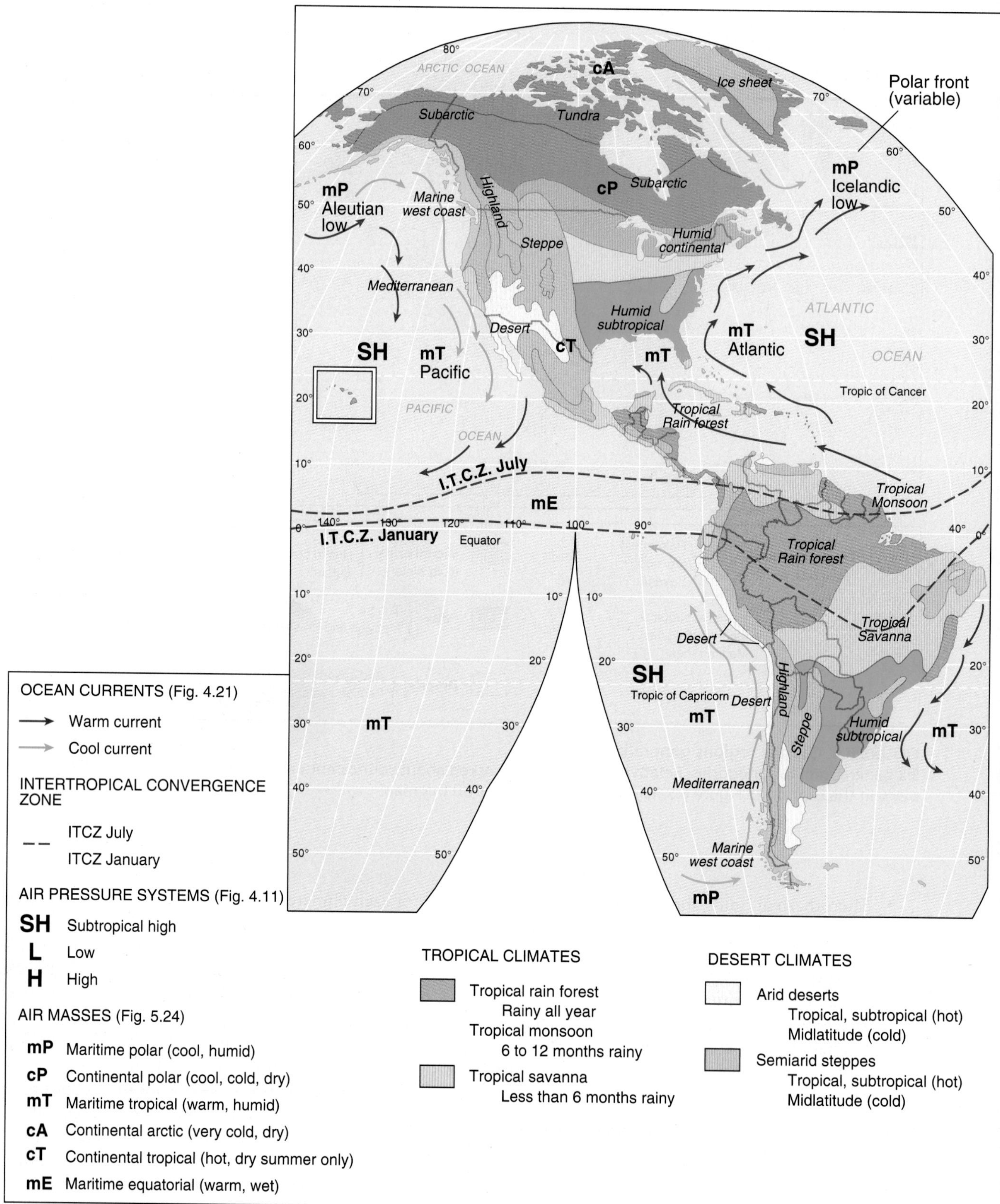

FIGURE 6.5 World climate classifications.
Annotated on this map are selected air masses, near shore ocean currents, pressure systems, and the January and July locations of the ITCZ. Use the colors in the legend to locate various climate types; some labels of the climate names appear in italics on the map to guide you.

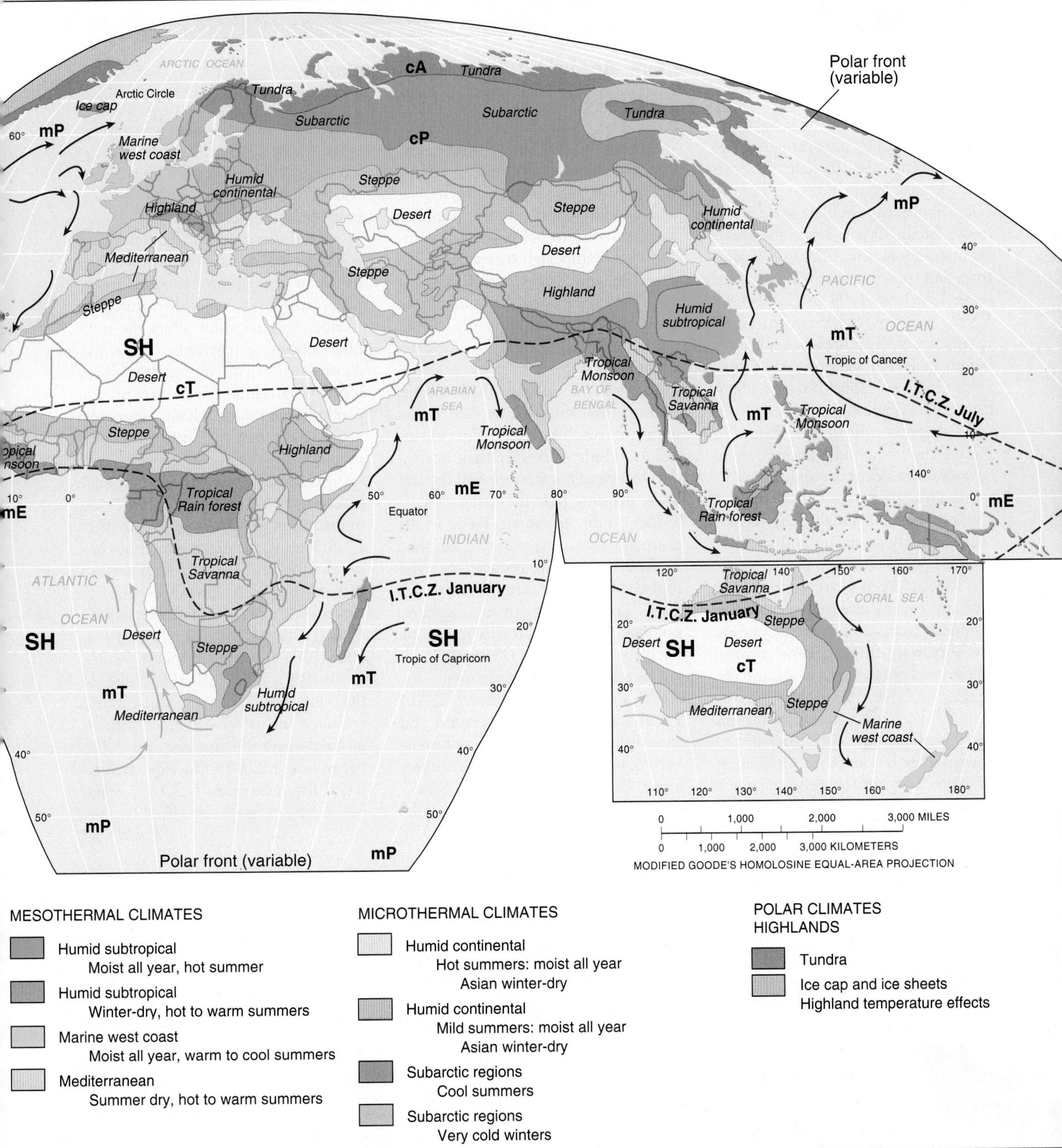
ARCTIC OCEAN
cA
Tundra
Polar front
(variable)
Arctic Circle
Ice cap
Tundra
Subarctic
Subarctic
Tundra
60°
mP
cP
Marine
west coast
Humid
continental
Steppe
Highland
Desert
Steppe
Humid
continental
mP
Desert
Mediterranean
Steppe
Highland
40°
PACIFIC
Steppe
Humid
subtropical
30°
OCEAN
SH
Desert
mT
Tropic of Cancer
Desert
cT
ARABIAN
SEA
Tropical
Monsoon
BAY OF
BENGAL
Tropical
Savanna
20°
I.T.C.Z. July
mT
mT
Tropical
Monsoon
Steppe
Highland
Tropical
Monsoon
10°
140°
10°
0°
50°
60°
mE
70°
80°
90°
mE
mE
Tropical
Rain forest
Equator
Tropical
Rain forest
0°
INDIAN
OCEAN
Tropical
Savanna
10°
ATLANTIC
I.T.C.Z. January
OCEAN
20°
SH
Desert
Steppe
SH
Tropic of Capricorn
mT
mT
30°
Humid
subtropical
Mediterranean
40°
40°
50°
mP
50°
Polar front (variable)
mP
120°
Tropical
Savanna
140°
150°
160°
170°
CORAL SEA
I.T.C.Z. January
Steppe
20°
20°
Desert
SH
Desert
cT
30°
30°
Mediterranean
Steppe
Marine
west coast
40°
40°
110°
120°
130°
140°
150°
160°
180°
0
1,000
2,000
3,000 MILES
0
1,000
2,000
3,000 KILOMETERS
MODIFIED GOODE'S HOMOLOSINE EQUAL-AREA PROJECTION
MESOTHERMAL CLIMATES
Humid subtropical
Moist all year, hot summer
Humid subtropical
Winter-dry, hot to warm summers
Marine west coast
Moist all year, warm to cool summers
Mediterranean
Summer dry, hot to warm summers
MICROTHERMAL CLIMATES
Humid continental
Hot summers: moist all year
Asian winter-dry
Humid continental
Mild summers: moist all year
Asian winter-dry
Subarctic regions
Cool summers
Subarctic regions
Very cold winters
POLAR CLIMATES
HIGHLANDS
Tundra
Ice cap and ice sheets
Highland temperature effects

Focus Study 6.1

The El Niño Phenomenon—Record Intensity, Global Linkages

Climate is the consistent behavior of weather over time, but average weather conditions also include extremes that depart from normal. The El Niño-Southern Oscillation (ENSO) in the Pacific Ocean forces the greatest interannual variability of temperature and precipitation on a global scale. The 1997–1998 and 1982–1983 ENSO events were the two strongest in 120 years. The spring wildflower bloom in Death Valley in 1998 provides visible evidence of the resultant heavy rains (Figure 1).

Peruvians coined the name El Niño ("the boy child") because these episodes may occur around the traditional December celebration time of Christ's birth. Actually El Niños occur as early as spring and summer and may persist through the year.

Revisit ocean currents in Figure 4.21 and see that the region off South America's west coast is dominated by the northward-flowing Peru current. These cold waters move toward the equator and join the westward movement of the south equatorial current.

The Peru current is part of the normal counterclockwise circulation of winds and surface ocean currents around the subtropical high-pressure cell dominating the eastern Pacific in the Southern Hemisphere. As a result, an eastern location such as Guayaquil, Ecuador, normally receives 91.4 cm (36 in.) of precipitation each year under dominant high pressure, whereas western islands in the Indonesian archipelago receive over 254 cm (100 in.) under dominant low pressure. This normal alignment of pressure is shown in Figure 2a.

What is ENSO?

Occasionally, for unexplained reasons, pressure patterns and surface ocean temperatures shift from their usual locations. Higher pressure than normal develops over the western Pacific, and lower pressure over the eastern Pacific (off South America). Trade winds normally moving from east to west weaken and can be reduced or even replaced by an eastward (west-to-east) flow. The shifting of atmospheric pressure and wind patterns across the Pacific is the *Southern Oscillation*.

Sea-surface temperatures increase, sometimes more than 8 C° (14 F°) above normal in the central and eastern Pacific, replacing the normally cold, upwelling, nutrient-rich water along Peru's coastline. Such ocean-surface warming, the "warm pool," may extend to the International Date Line. This surface pool of warm water is known as El Niño. Thus, the designation ENSO is derived (El Niño-Southern Oscillation). This condition is shown in Figure 2b.

The thermocline (boundary of colder, deep ocean water) lowers in depth in the eastern Pacific Ocean. The change in wind direction and warmer surface water slows the normal upwelling currents; upwelling controls nutrient availability. This loss of nutrients affects the phytoplankton and food chain, depriving many fish, marine mammals, and predator birds of nourishment.

Scientists at the National Oceanographic and Atmospheric Administration (NOAA) speculate that ENSO events occurred nine times between 1396 and 1941–1942. They are certain that such ENSO events occurred in 1953, 1957–1958, 1965, 1969–1970, 1972–1973, 1976–1977, 1982–1983 (second strongest event), 1986–1987, and 1991–1993 (one of the longest). The most intense episode, in 1997–1998, disrupted global weather; another El Niño began late in 2002. The expected interval for recurrence is 3 to 5 years, but it may range from 2 to 12 years.

(continued)

(a)

(b)

FIGURE 1 El Niño's impact on the desert.
Death Valley, southeastern California, in (a) full spring bloom following record rains triggered by the 1997–1998 El Niño and (b) the same scene in spring 2002 in its stark desert grandeur. A dramatic effect caused by changes in the distant tropics of the Pacific Ocean. [Photos by Bobbé Christopherson.]

Focus Study 6.1 *(continued)*

(continued)

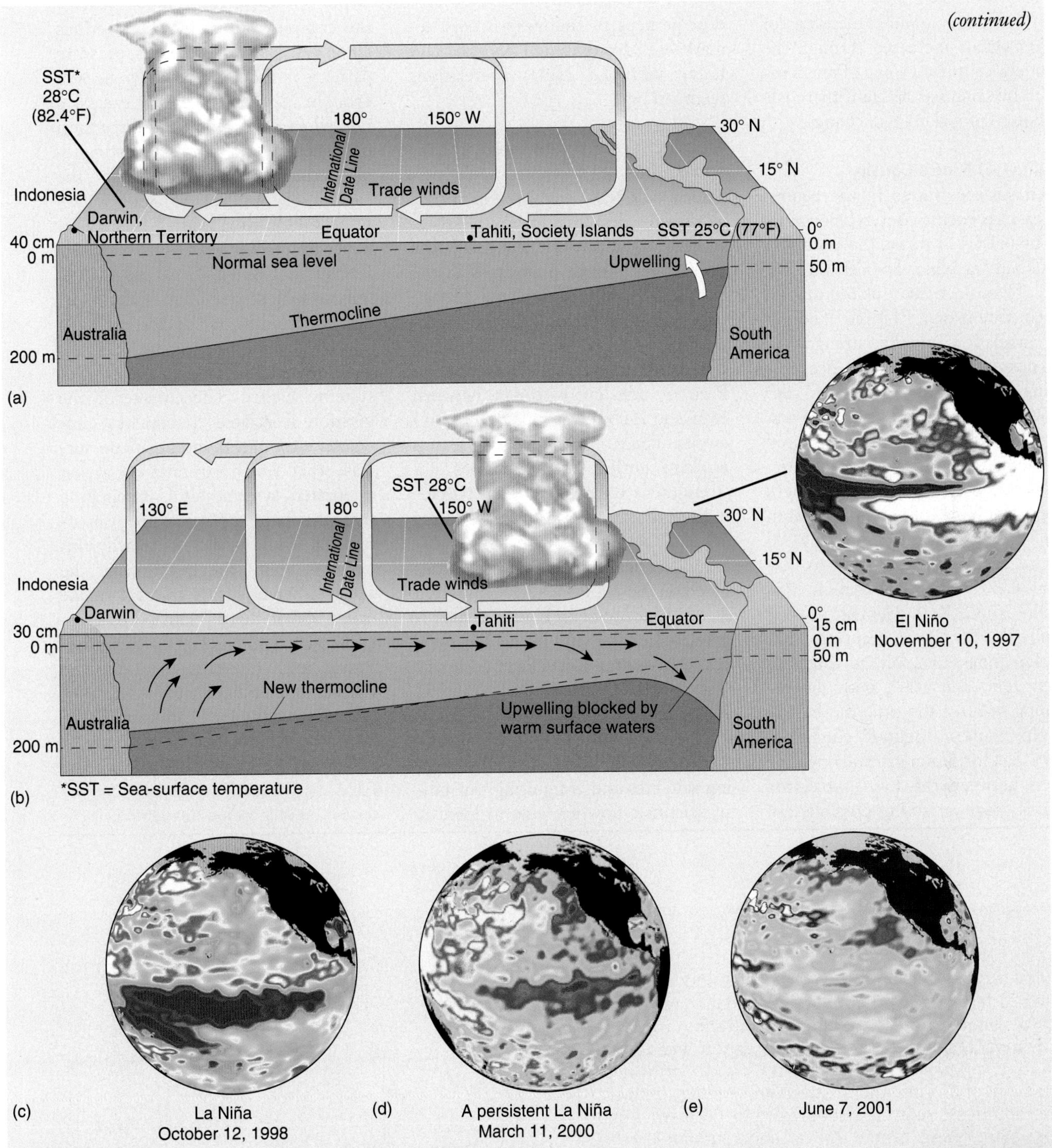

SST*
28°C
(82.4°F)
180°
150° W
30° N
15° N
International Date Line
Indonesia
Trade winds
Darwin,
Northern Territory
Equator
Tahiti, Society Islands
SST 25°C (77°F)
0°
40 cm
0 m
0 m
Normal sea level
Upwelling
50 m
Thermocline
Australia
South America
200 m
(a)

SST 28°C
130° E
180°
150° W
30° N
15° N
International Date Line
Indonesia
Trade winds
Darwin
Tahiti
Equator
0°
30 cm
15 cm
0 m
0 m
50 m
New thermocline
Upwelling blocked by warm surface waters
Australia
South America
200 m
*SST = Sea-surface temperature
(b)

El Niño
November 10, 1997

(c) La Niña
October 12, 1998

(d) A persistent La Niña
March 11, 2000

(e) June 7, 2001

FIGURE 2 Normal, El Niño, and La Niña changes in the Pacific.
(a) Normal patterns in the Pacific; (b) El Niño wind and weather patterns across the Pacific Ocean and *TOPEX/Poseidon* satellite image for November 1997 (white and red colors indicate warmer surface water—a warm pool). (c) *TOPEX/Poseidon* image of La Niña conditions in transition in the Pacific on October 1998 (purple and blue colors for cooler surface water—a cool pool). (d) A persistent La Niña in March 2000 satellite image. (e) Image from June 2001, showing no El Niño as equatorial waters slowly warm with sea-surface temperature near normal. (f) El Niño returns, August 2002, as the warm pool of waters moves eastward toward South America at more than 200 km (125 mi) per day. [(a) and (b) Adapted and author corrected from C. S. Ramage, "El Niño." © 1986 by *Scientific American*, Inc.; (b)–(f) *TOPEX/Poseidon* images courtesy of Jet Propulsion Laboratory, NASA.]

(f) August 13, 2002

Focus Study 6.1 *(continued)*

The frequency and intensity of ENSO events increased through the twentieth century, a topic of much research by scientists to see if there is a relation to global climate change.

La Niña—El Niño's Cousin

When surface waters in the central and eastern Pacific cool to below normal by 0.4 C° or more, the condition is dubbed La Niña, Spanish for "the girl." This is a weaker condition and less consistent than El Niño. There is no correlation in the strength or weakness of each. For instance, following the record 1997–98 ENSO event, the subsequent La Niña was not as strong as predicted and shared the Pacific with lingering warm water.

Between 1900 and 1998 there were 13 La Niñas of note, the latest in 1988, 1995, and late 1998 to 2000 (Figure 2c and d). According to National Center for Atmospheric Research (NCAR) scientist Kevin Trenberth, El Niños occurred 31% and La Niñas 23% of the time between 1950 and 1997, with the remaining 46% of the time the Pacific was in a more neutral condition. Don't look for symmetry and opposite effects between the two events, for there is great variability possible, except perhaps in Indonesia where remarkable drought (El Niño) and heavy rain (La Niña) correlations seem strong.

Global Effects Related to the ENSO and La Niña

Effects of ENSO and La Niña occur worldwide: droughts in South Africa, southern India, Australia, and the Philippines; strong hurricanes in the Pacific, including Tahiti and French Polynesia; and flooding in the southwestern United States and mountain states, Bolivia, Cuba, Ecuador, and Peru. In India, every drought between 1525 and 1900 seems linked to ENSO events. The Atlantic hurricane season weakens during El Niño years and strengthens during La Niñas. (Refer back to the discussion in Focus Study 5.1 and review William Gray's Atlantic hurricane forecasting methods based on these linkages.) Precipitation in the southwestern United States is greater in El Niño than La Niña years. In contrast, the Pacific Northwest is wetter with La Niña than El Niño.

Since the 1982–1983 event, and with the development of remote-sensing satellites and computing capability, scientists now are able to identify the complex global interconnections among surface temperatures, pressure patterns in the Pacific, occurrences of drought in some places and excessive rainfall in others, and the disruption of fisheries and wildlife. Estimates place the overall damage from the 1982–1983 ENSO at more than US$8 billion worldwide. Present estimates of weather-related costs for the ENSO in 1997–1998 exceed US$80 billion, with some 300 million people displaced and 30,000 deaths.

Discovery of these truly Earth-wide relations is at the heart of Earth systems science. The climate of one location is related to climates elsewhere, although it should be no surprise that Earth operates as a vast integrated system: "It is fascinating that what happens in one area can affect the whole world. As to why this happens, that's the question of the century. Scientists are trying to make order out of chaos," said NOAA scientist Alan Strong. (For ENSO monitoring and forecasts, see Climate Prediction Center at http://www.ncep.noaa.gov/ or the Jet Propulsion Laboratory at http://www.jpl.nasa.gov/elnino/ or NOAA's El Niño Theme Page at http://www.pmel.noaa.gov/toga-tao/el-nino/nino-home.html.)

Tropical Climates (equatorial and tropical latitudes)

Tropical climates, the most extensive, occupy about 36% of Earth's surface, including both ocean and land areas. The tropical climates straddle the equator from about 20° N to 20° S, roughly between the Tropics of Cancer and Capricorn, thus the name. Tropical climates stretch northward to the tip of Florida and south-central Mexico, central India, Southeast Asia, and southward to northern Australia, Madagascar, central Africa, and southern Brazil. These climates truly are winterless. Important causal elements include:

- Consistent daylength and insolation input, which produce consistently warm temeratures;
- Intertropical convergence zone (ITCZ), which brings rains as it shifts seasonally with the high Sun;
- Warm ocean temperatures and unstable maritime air masses.

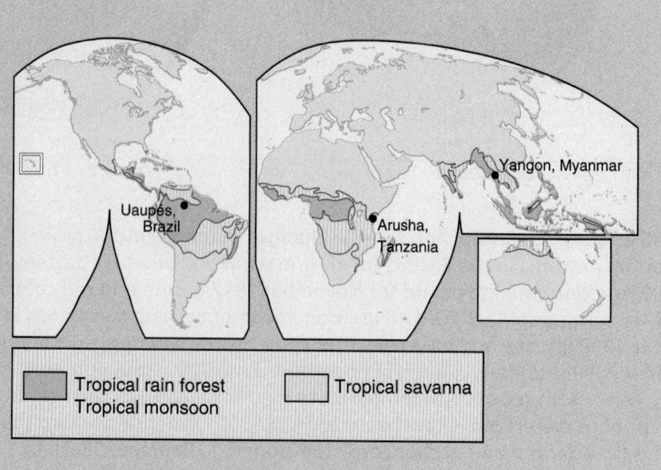

Tropical climates have three distinct regimes: *tropical rain forest* (ITCZ present all year), *tropical monsoon* (ITCZ 6 to 12 months), and *tropical savanna* (ITCZ less than 6 months).

Tropical Rain Forest Climates

The *tropical rain forest* climate is constantly moist and warm. Convectional thunderstorms, triggered by local heating and trade-wind convergence, peak each day from midafternoon to late evening inland. These thunderstorms may hit earlier in the day where marine influence is strong. This precipitation follows the migrating intertropical convergence zone (ITCZ, Chapter 4). Its extreme positions for July and January are plotted in Figure 6.5. The ITCZ shifts northward and southward with the summer Sun throughout the year, but it influences *tropical rain forest* regions during all 12 months. Not surprisingly, water surpluses are enormous, creating the world's greatest stream discharges in the Amazon and Congo (Zaire) Rivers.

High rainfall sustains lush evergreen broadleaf tree growth, producing Earth's equatorial and tropical rain forests. Their leaf canopy is so dense that little light diffuses to the forest floor, leaving the ground surface dim and sparse in plant cover. Dense surface vegetation occurs along riverbanks, where light is abundant (see Figure 6.6b). Widespread deforestation of Earth's rain forest is detailed in Chapter 16.

High temperature promotes energetic bacterial action in the soil so that organic material is quickly consumed. Heavy precipitation washes away certain minerals and nutrients. The resulting soils are somewhat sterile and can support intensive agriculture only if supplemented by fertilizer.

Uaupés, Brazil (Figure 6.6), is characteristic of *tropical rain forest*. On the climograph you see that the lowest-precipitation month receives nearly 15 cm (6 in.), and the annual temperature range is barely 2 C° (3.6 F°). The ITCZ is present all year. In all such climates, the diurnal (day to night) temperature range exceeds the annual average minimum–maximum (coolest to warmest) range: Day–night temperatures can range more than 11 C° (20 F°), more than five times the annual range of monthly averages.

The only interruption of *tropical rain forest* climates is in the highlands of the South American Andes and in East Africa (see Figure 6.5). There, higher elevations produce lower temperatures; Mount Kilimanjaro is less than 4° S of the equator, but at 5895 m (19,340 ft) elevation it has permanent glacial ice on its summit. Such mountainous sites fall within the H (highland) climate designation.

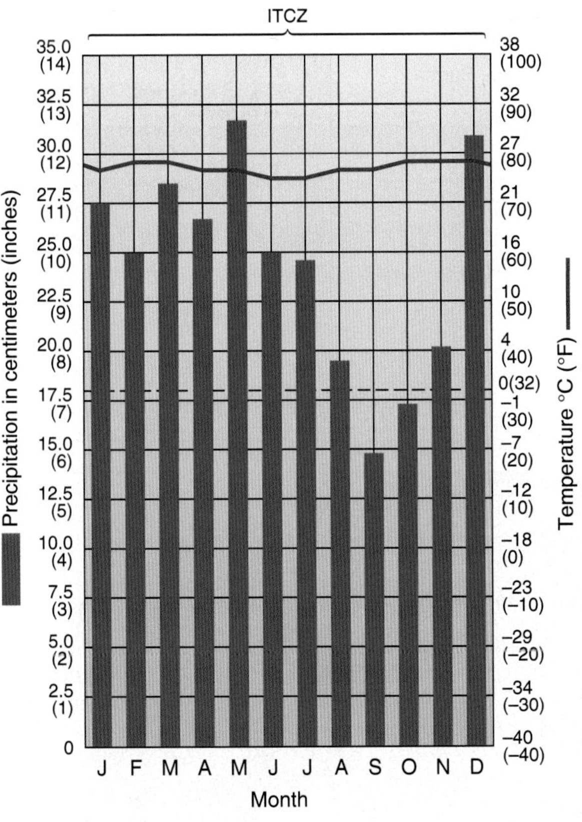

Station: Uaupés, Brazil
Lat/long: 0°08' S 67°05' W
Avg. Ann. Temp.: 25°C (77°F)
Total Ann. Precip.: 291.7 cm (114.8 in.)
Elevation: 86 m (282.2 ft)
Population: 10,000
Ann. Temp. Range: 2 C° (3.6 F°)
Ann. Hr of Sunshine: 2018

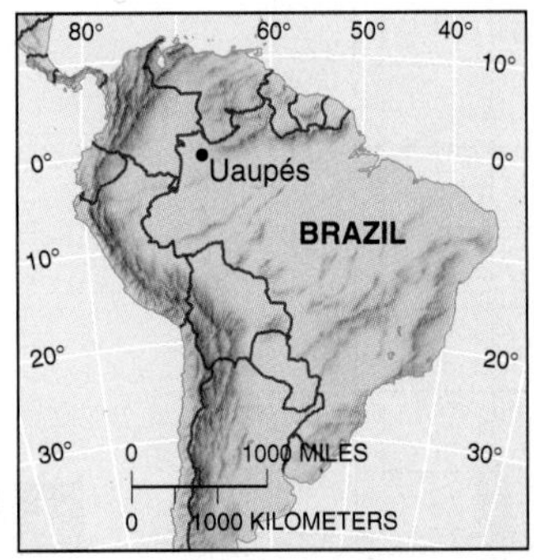

FIGURE 6.6 Tropical rain forest climate.
(a) Climograph for Uaupés, Brazil (*tropical rain forest*). (b) The rain forest near Uaupés along a tributary of the Rio Negro. [Photo by Will and Deni McIntyre/Photo Researchers, Inc.]

Tropical Monsoon Climates

The *tropical monsoon* climates feature a dry season that lasts one or more months. Rainfall brought by the ITCZ affects these areas from 6 to 12 months of the year, with the dry season occurring when the convergence zone is not overhead. Yangon, Myanmar (formerly Rangoon, Burma), is an example of a monsoon climate in the tropics, as illustrated by the climograph and photograph in Figure 6.7.

Tropical monsoon climates lie principally along coastal areas within the tropical rain forest climatic realm and experience seasonal variation of winds and precipitation. Evergreen trees grade into thorn forests on the drier margins near the adjoining tropical savannas.

Tropical Savanna Climates

Tropical savanna climates occur poleward of the *tropical rain forest* climates. The ITCZ dominates these climates for 6 months or less of the year as it migrates with the summer Sun. Summers are wetter than winters because convectional rains accompany the shifting ITCZ when it is overhead. This produces a notable dry condition, when the ITCZ is farthest away and high pressure dominates. Thus, the natural water demand exceeds the natural water supply in winter, causing soil-moisture shortages.

Temperatures vary more in tropical savanna climates than in tropical rain forest regions. The tropical savanna regime can have two temperature maximums because the Sun's direct rays are overhead twice during the year (before and after the summer solstice in each hemisphere as the Sun moves between the equator and the tropic). Dominant grasslands with scattered, drought-resistant trees, able to cope with the highly variable precipitation, characterize the *tropical savanna* regions (Figure 6.8b).

Arusha, Tanzania, is in a characteristic *tropical savanna* city (Figure 6.8). This city of more than 100,000 people is

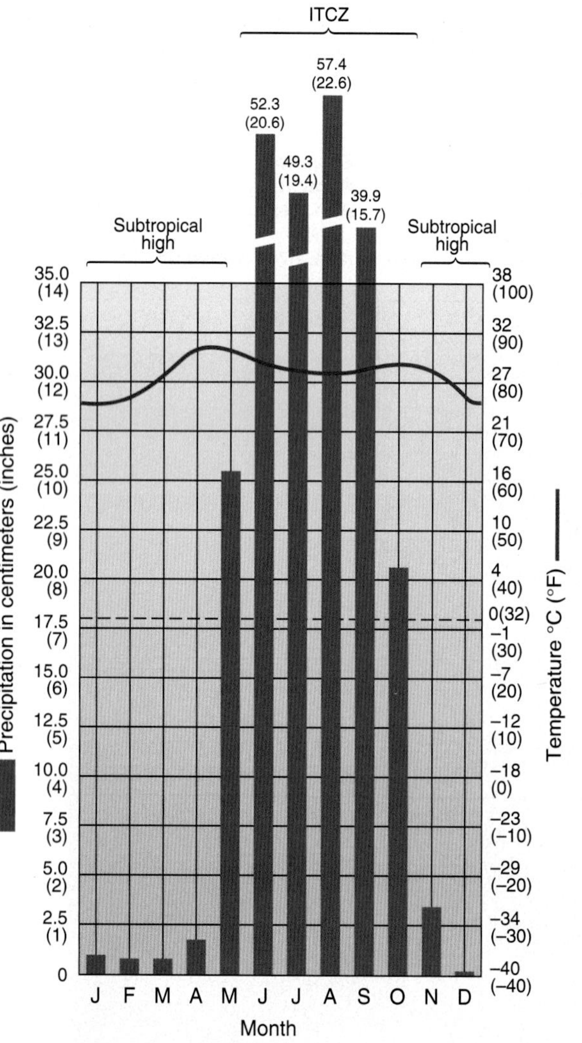

(a)

Station: Yangon, Myanmar*
Lat/long: 16°47' N 96°10' E
Avg. Ann. Temp.: 27.3°C (81.1°F)
Total Ann. Precip.: 252.7 cm (99.5 in.)
*(Formerly Rangoon, Burma)
Elevation: 23 m (76 ft)
Population: 2,500,000
Ann. Temp. Range: 5.5 C° (9.9 F°)

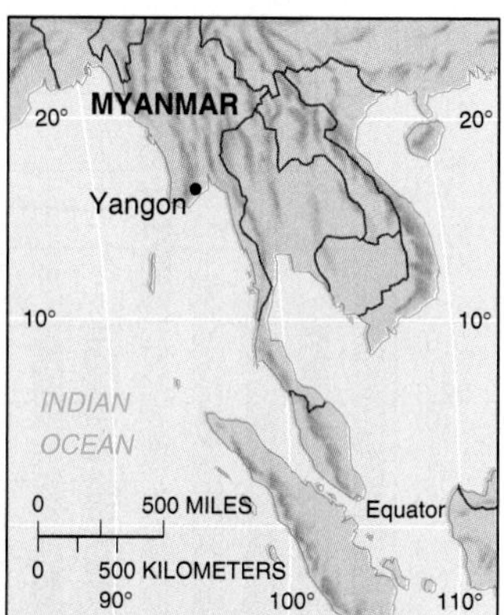

(b)

FIGURE 6.7 Tropical monsoon climate.
(a) Climograph for Yangon, Myanmar (formerly Rangoon, Burma) (*tropical monsoon*). (b) The monsoonal forest near Malang, Java, at the Purwodadi Botanical Gardens. [Photo by Tom McHugh/Photo Researchers, Inc.]

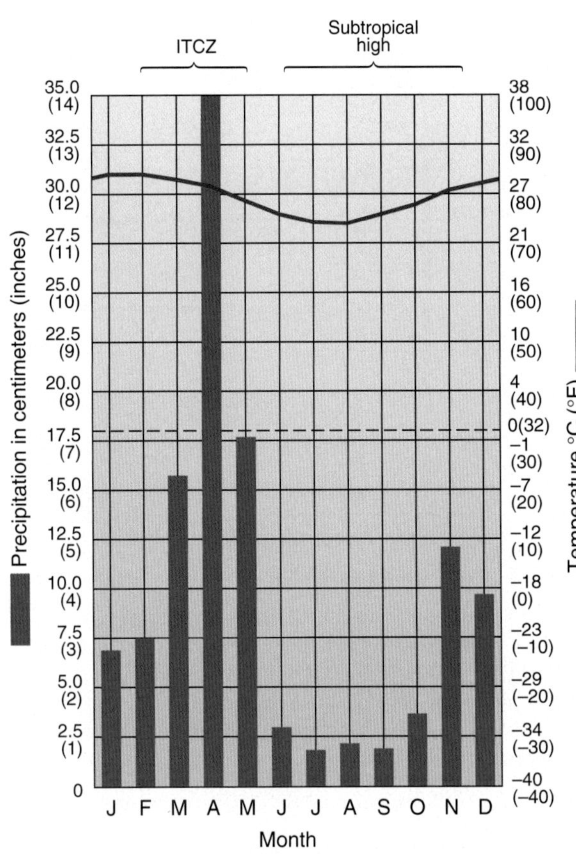

(a)

Station: Arusha, Tanzania
Lat/long: 3°24' S 36°42' E
Avg. Ann. Temp.: 26.5°C (79.7°F)
Total Ann. Precip.: 119 cm (46.9 in.)
Elevation: 1387 m (4550 ft)
Population: 100,000
Ann. Temp. Range: 4.1 C° (7.4 F°)
Ann. Hr of Sunshine: 2600

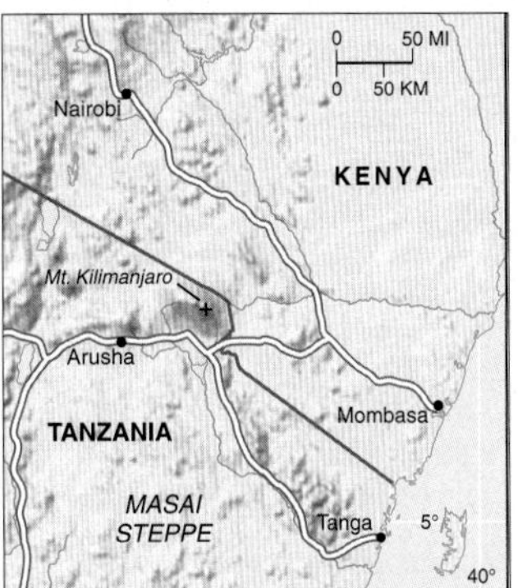

(b)

FIGURE 6.8 Tropical savanna climate.
(a) Climograph for Arusha, Tanzania (*tropical savanna*); note the intense dry period. (b) Characteristic landscape in Kenya, with plants and animals adapted to seasonally dry water budgets. [Photo by Stephen J. Krasemann/DRK Photo.]

east of the famous Serengeti plains savanna grassland and Olduvai Gorge, site of human origins, and north of Tarangire National Park. Temperatures are consistent with tropical climates, despite the elevation of the station. Note the marked dryness from June to October, which defines changing dominant pressure systems rather than annual changes in temperature. This region is near the transition to the dryer hot-desert steppe climates to the northeast.

Mesothermal Climates (midlatitude, mild winters)

Mesothermal, meaning "middle temperature," describes these warm and temperate climates, where true seasonality begins and seasonal contrasts in vegetation, soil, and human lifestyle adaptations are evident. More than half the world's population—approximately 55%—resides in these mesothermal climates. These climates occupy the second-largest percentage of Earth's land and sea surface behind the tropical climates, about 27%. Together, the tropical and mesothermal climates dominate over half of Earth's oceans and about one-third of its land area.

The mesothermal climates, and nearby portions of the microthermal (cold winter) climates, are regions of great weather variability, for these are the latitudes of greatest air mass interaction. Causal elements include:

- Shifting air masses of martime and continental origin are guided by upper air westerly winds and undulating Rossby waves and jet streams.
- Migrating cyclonic (low pressure) and anticyclonic (high pressure) systems bring changeable weather conditions and air mass conflicts.
- Sea-surface temperatures influence air mass strength: cooler along west coasts (weaken) and warmer along east coasts (strengthen).
- Summers transition from hot to warm to cool as you move away from the tropics. Climates are humid, except where subtropical high pressure produces dry-summer conditions.

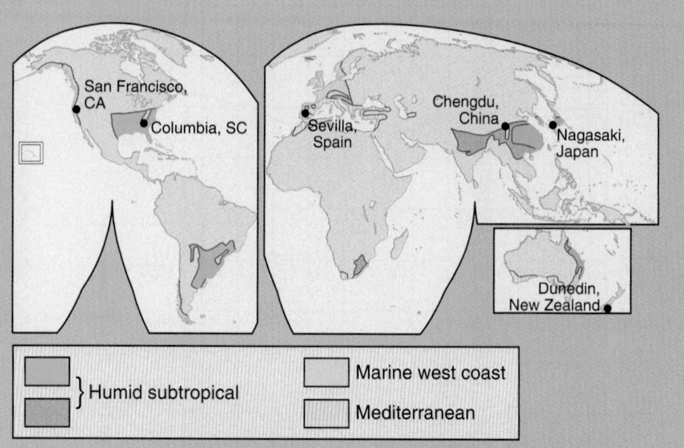

Mesothermal climates have four distinct regimes based on precipitation variability: *humid subtropical* (hot summers, moist all year), *humid subtropical* (hot to warm summers, dry winters in Asia), *marine west coast* (warm to cool summers, moist all year), and *Mediterranean* (warm to hot summers, dry summers).

Humid Subtropical Hot-Summer Climates

The *humid subtropical hot-summer* climates are either moist all year or have a pronounced winter-dry period as occurs in eastern and southern Asia. Maritime tropical air masses generated over warm waters off eastern coasts influence *humid subtropical hot-summer* climates during summer. This warm, moist, unstable air produces convectional showers over land. In fall, winter, and spring, maritime tropical and continental polar air masses interact, generating frontal activity and frequent midlatitude cyclonic storms. Overall, precipitation averages 100–200 cm (40–80 in.) a year. In North America, Columbia, South Carolina, is characteristic (Figure 6.9) of this climate regime, with precipitation totals of 126.5 cm (49.8 in.), hot, humid summers, and mild winters.

In eastern and southern Asia, winter precipitation is a bit less because of the effects of the Asian monsoon, unlike the precipitation of humid subtropical cities in the United States (Atlanta, Memphis, Norfolk, New Orleans). These winter-dry climates relate to the seasonal pulse of the monsoons. They extend poleward from tropical savanna climates and have a summer month that receives 10 times more precipitation than their driest winter month. A representative station is Chengdu, China. Figure 6.10 demonstrates the strong correlation between precipitation and the high-summer Sun (review the monsoonal maps in Figure 4.20.)

The concentration of people in north-central India, the bulk of China's 1.2 billion people, and the many who live in climatically similar portions of the United States, prove the habitability of the humid subtropical climates. The intense summer rains of the Asian monsoon can cause problems, as do occasional dramatic thunderstorms and tornadoes in the southeastern United States (News Report 6.1).

Marine West Coast Climates

Marine west coast climates feature mild winters and cool summers and dominate Europe and other middle-to-high-latitude west coasts (see Figure 6.5). In the United States and Canada, these climates with their cooler summers are in contrast to the hot-summer humid climate of the southeastern United States.

Maritime polar air masses—cool, moist, unstable—control *marine west coast* climates. Weather systems forming along the polar front move into these regions throughout the year, making weather quite unpredictable. Coastal fog, annually totaling 30 to 60 days, is a part of the moderating marine influence. Frosts are possible and tend to shorten the growing season.

Marine west coast climates are unusually mild for their latitude owing to marine influences. They extend along the coastal margins of the Aleutian Islands in the North Pacific, cover the southern third of Iceland in the North Atlantic and coastal Scandinavia, and dominate the British Isles. To have average monthly temperatures that are above freezing throughout the year at such high-latitude locations is hard to imagine.

Unlike marine west coast climates in Europe, these climates in Canada, Alaska, Chile, and Australia are backed by mountains and remain restricted to coastal

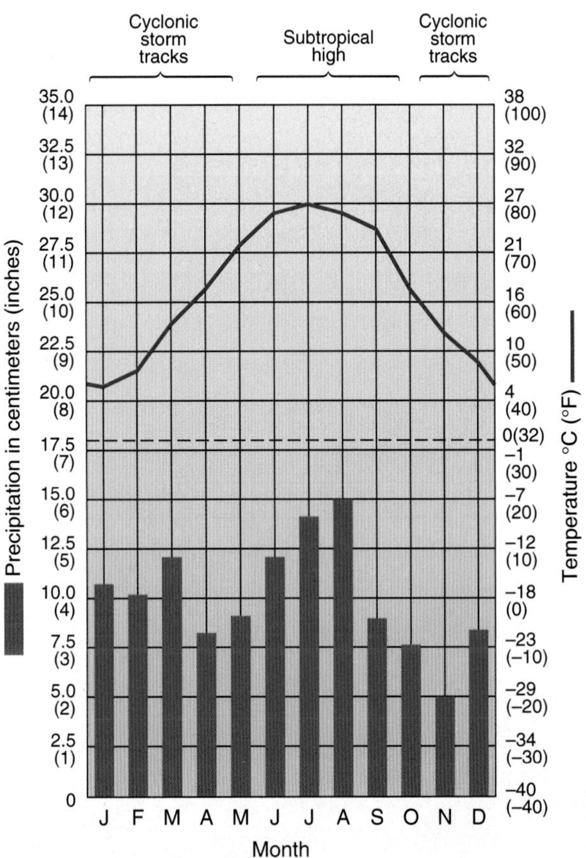

(a)

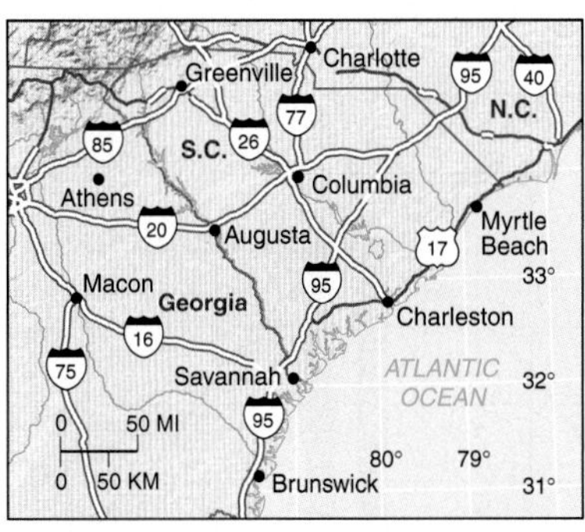

(b)

FIGURE 6.9 Humid subtropical climate, rainy all year.
(a) Climograph for Columbia, South Carolina (*humid subtropical*). Note the more consistent precipitation pattern, as Columbia receives seasonal cyclonic storm activity and summer convection showers within maritime tropical air. (b) The mixed deciduous and evergreen forest south of Atlanta, Georgia, typical of the humid subtropical southeastern United States. [Photo by Bobbé Christopherson.]

environs. The climograph for Vancouver demonstrates the moderate temperature patterns and the annual temperature range for a *marine west coast* city (Figure 6.11). The climograph for Dunedin, New Zealand, demonstrates the moderate temperature patterns and the annual temperature range for a *marine west coast* station in the Southern Hemisphere (Figure 6.12, p. 196).

An interesting anomaly occurs in the eastern United States. In portions of the Appalachian highlands, increased elevation moderates summer temperatures in the surrounding *humid subtropical hot-summer* climate, producing a *marine west coast* cooler summer. The climograph for Bluefield, West Virginia (Figure 6.13, p. 196), reveals marine west coast temperature and precipitation patterns, despite its location in the east. Vegetation similarities between the Appalachians and the Pacific Northwest are quite noticeable and have enticed many emigrants from the East to settle in these climatically familiar environments in the Northwest.

Mediterranean Dry-Summer Climates

Across the planet during summer months, shifting subtropical high-pressure cells block moisture-bearing winds from adjacent regions. For example, the continental tropical air mass over the Sahara in Africa shifts northward in summer over the Mediterranean region and blocks maritime air masses and cyclonic systems. The shifting of stable, warm-to-hot, dry air over an area in summer and away from these regions in the winter creates a unique dry-summer and wet-winter pattern. Mediterranean climates experience at least 70% of their annual precipitation during the winter months. This is in contrast to the majority of the world that experience summer-maximum precipitation.

Added to the high-presure influence, cool offshore currents (the California current, Canary current, Peru current, Benguela current, and West Australian current)

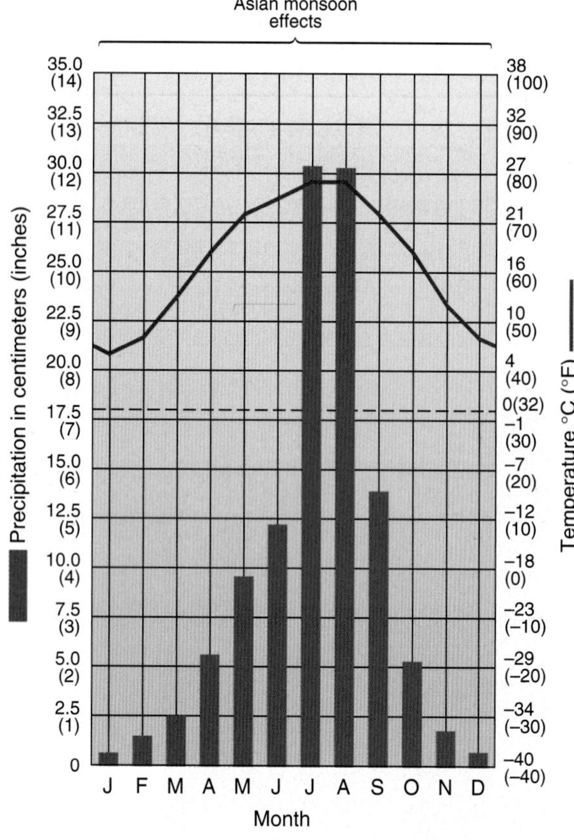

Station: Chengdu, China
Lat/long: 30°40' N 104°04' E
Avg. Ann. Temp.: 17°C (62.6°F)
Total Ann. Precip.: 114.6 cm (45.1 in.)
Elevation: 498 m (1633.9 ft)
Population: 2,260,000
Ann. Temp. Range: 20 C° (36 F°)
Ann. Hr of Sunshine: 1058

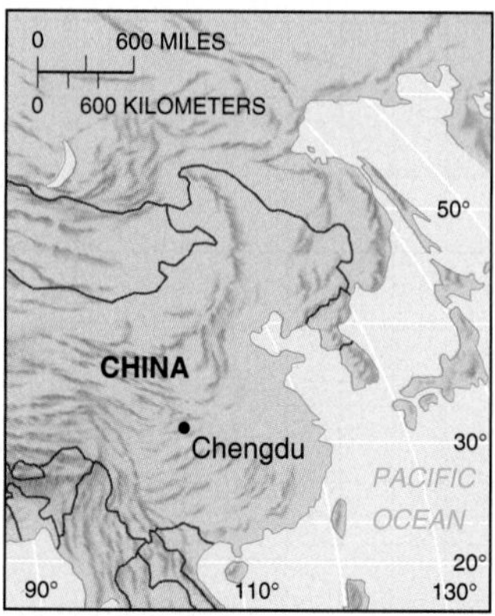

FIGURE 6.10 Humid subtropical winter-dry climate.
(a) Climograph for Chengdu, China. (b) Landscape of southern interior China characteristic of this winter-dry climate. This valley is near Mount Daliang in Sichuan Province. [Photo by Jin Zuqi, Sovfoto/Eastfoto.]

produce stability in overlying air masses along west coasts, poleward of subtropical high pressure. The world climate map (Figure 6.5) shows these currents and Mediterranean dry-summer regions along the western margins of North America, central Chile, and the southwestern tip of Africa, as well as across southern Australia and the Mediterranean Basin—the climate's namesake region.

Figure 6.14 compares the climographs of Mediterranean dry-summer cities of San Francisco, California, and Sevilla (Seville), Spain. Coastal maritime effects moderate San Francisco's climate, producing a cool summer. The transition to hotter summers occurs no more than 24–32 km (15–20 mi) inland from San Francisco.

Along these west coasts, the warm, moist air that contacts cool ocean water produces frequent summer fog, discussed in Chapter 5. However, this type of fog is not associated with the enclosed Mediterranean Sea region itself because offshore water temperatures there are higher than water temperatures near other similar climatic regions during the summer months. Instead, the Mediterranean regime features relatively high humidity values.

The *Mediterranean dry-summer* climate brings natural summer water-resource shortages. Winter precipitation recharges soil moisture, but water usage usually exhausts it by late spring. Large-scale agriculture requires irrigation; some subtropical fruits, nuts, and vegetables are uniquely suited to these conditions. Natural vegetation features a hard-leafed, drought-resistant variety known locally as *chaparral* in the western United States. (Chapter 16 discusses other local names for this type of vegetation in other parts of the world.)

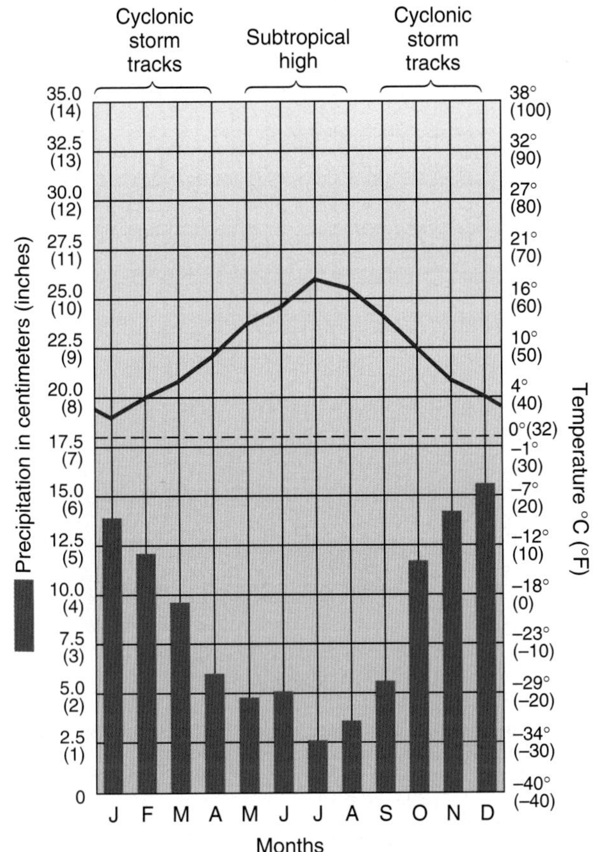

Station: Vancouver, British Columbia
Lat/long: 49°11' N, 123°10' W
Avg. Ann. Temp.: 10° C (50° F) **Cfb**
Total Annual Precipitation: 104.8 cm (41.3 in.)
Elevation: sea level
Population: 1,580,000
Ann. Temp. Range: 16 C° (28.8 F°)
Ann. Hrs. of Sunshine: 1723

(a)

FIGURE 6.11 A marine west coast climate.
(a) Climograph and locator map for Vancouver, British Columbia, a *marine west coast* climate.
(b) The Vancouver waterfront and skyline. [Photo by author.]

News Report 6.1

Record Rains from the Asian Monsoons

The subtropical monsoonal climates hold several precipitation records. Cherrapunji, India, in the Assam Hills south of the Himalayas, is the all-time precipitation holder for a single year and for every other time interval from 15 days to 2 years. Because of the summer monsoons that pour in from the Indian Ocean and the Bay of Bengal, Cherrapunji has received 930 cm (30.5 ft) of rainfall in one month and 2647 cm (86.8 ft) in one year—both world records. Because of extensive deforestation, these rains produce tremendous soil erosion. Sediments generated by this erosion are deposited in the Bay of Bengal as fragile, temporary islands that attract settlements. Floodwaters and heavy monsoonal downpours easily overwhelm these fertile islands. Two massive tropical cyclones struck the nearby coast of India in 1999, further worsening the fragile coastal conditions.

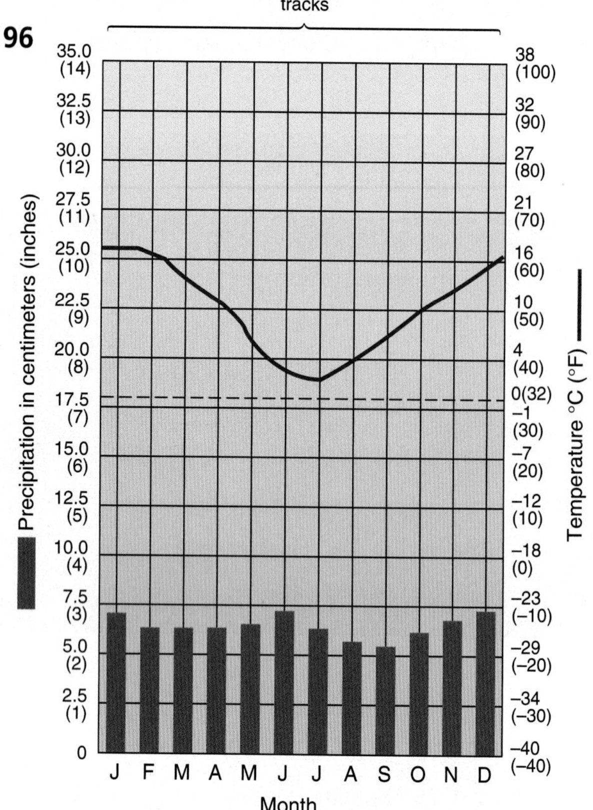

Station: Dunedin, New Zealand
Lat/long: 45°54' S 170°31' E
Avg. Ann. Temp.: 10.2°C (50.3°F)
Total Ann. Precip.: 78.7 cm (31.0 in.)
Elevation: 1.5 m (5 ft)
Population: 120,000
Ann. Temp. Range: 14.2 C° (25.5 F°)

(a)

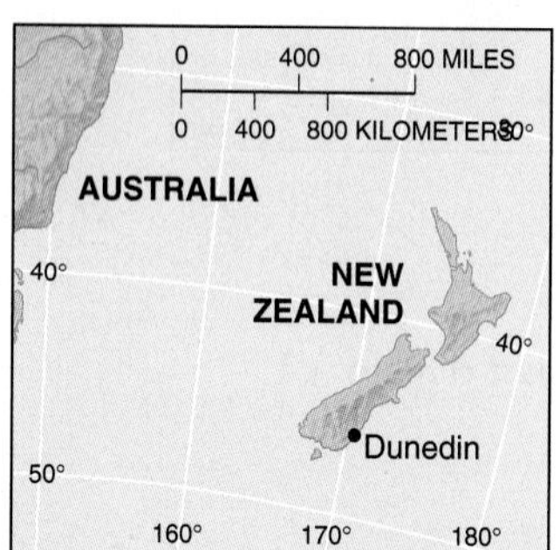

(b)

FIGURE 6.12 A Southern Hemisphere marine west coast climate.
(a) Climograph for Dunedin, New Zealand demonstrates a *marine west coast* climate with a mild summer. (b) Meadow, forest, and mountains on South Island, New Zealand. [Photo by Brian Enting/Photo Researchers, Inc.]

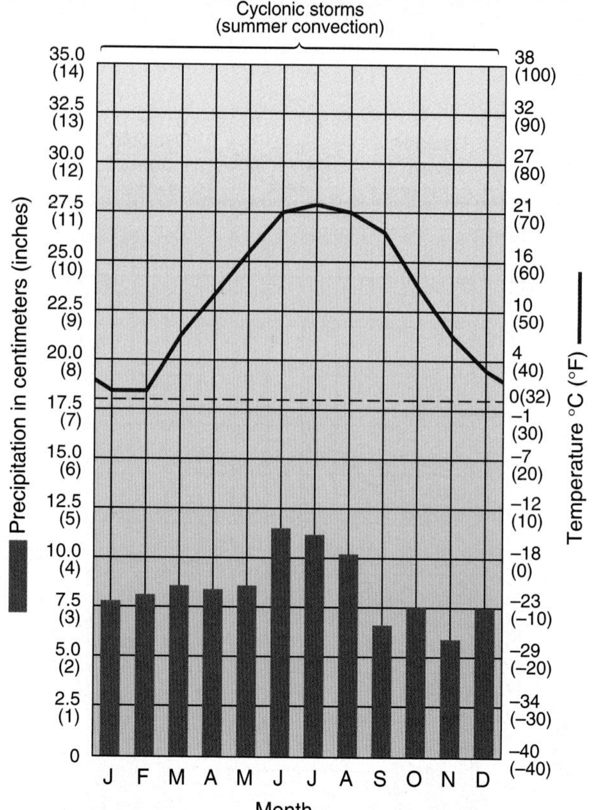

Station: Bluefield, West Virginia
Lat/long: 37°16' N 81°13' W
Avg. Ann. Temp.: 12°C (53.6°F)
Total Ann. Precip.: 101.9 cm (40.1 in.)
Elevation: 780 m (2559 ft)
Population: 16,000
Ann. Temp. Range: 21 C° (37.8 F°)

(a)

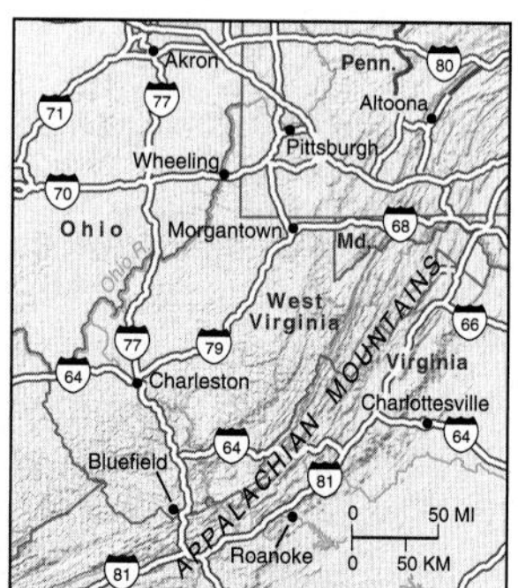

(b)

FIGURE 6.13 Climate in the Appalachians.
(a) Climograph for Bluefield, West Virginia, exhibits *marine west coast* characteristics. (b) Characteristic mixed forest of Dolly Sods Wilderness in the Appalachian highlands. [Photo by David Muench Photography, Inc.]

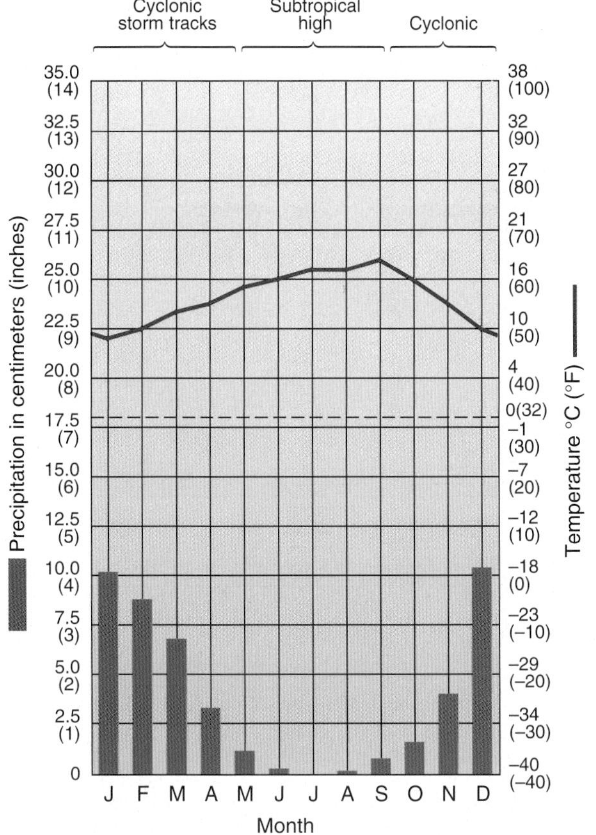

Station: San Francisco, California
Lat/long: 37°37' N 122°23' W
Avg. Ann. Temp.: 14°C (57.2°F)
Total Ann. Precip.: 47.5 cm (18.7 in.)
Elevation: 5 m (16.4 ft)
Population: 777,000
Ann. Temp. Range: 9 C° (16.2 F°)
Ann. Hr of Sunshine: 2975

(a)

Cyclonic storm tracks — Subtropical high — Cyclonic storm tracks

Station: Sevilla, Spain
Lat/long: 37°22' N 6°00' W
Avg. Ann. Temp.: 18°C (64.4°F)
Total Ann. Precip.: 55.9 cm (22 in.)
Elevation: 13 m (42.6 ft)
Population: 683,000
Ann. Temp. Range: 16 C° (28.8 F°)
Ann. Hr of Sunshine: 2862

(b)

(c)

(d)

FIGURE 6.14 Mediterranean climates.
(a) Climographs for San Francisco (*Mediterranean cool-summer*) and (b) Sevilla, Spain (*Mediterranean hot-summer*). (c) Central California landscape of oak savanna. (d) The countryside around Olvera, Andalusia, Spain. [Photos by (c) Bobbé Christopherson; (d) Kaz Chiba/Gamma-Liaison Agency, Inc.]

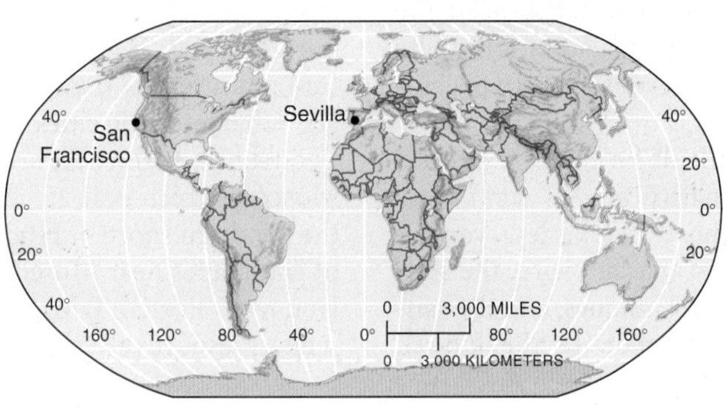

Microthermal Climates (mid- and high-latitude, cold winter)

Humid microthermal climates have a winter season with some summer warmth. Here the term *microthermal* means cool temperate to cold. Approximately 21% of Earth's land surface is influenced by these climates, equaling about 7% of Earth's total surface. These climates occur poleward of the mesothermal climates and experience great temperature ranges related to continentality and air mass conflicts. Temperatures decrease with increasing latitude and toward the interior of continental landmasses and intensely cold winters. In contrast to moist all-year regions (northern tier across the United States and Canada, eastern Europe through the the Ural Mountains) is the winter-dry association with the Asian dry monsoon.

In Figure 6.5, note the absence of microthermal climates in the Southern Hemisphere. Because the Southern Hemisphere lacks substantial landmasses, microthermal climates develop there only in highlands. Important causal elements include:

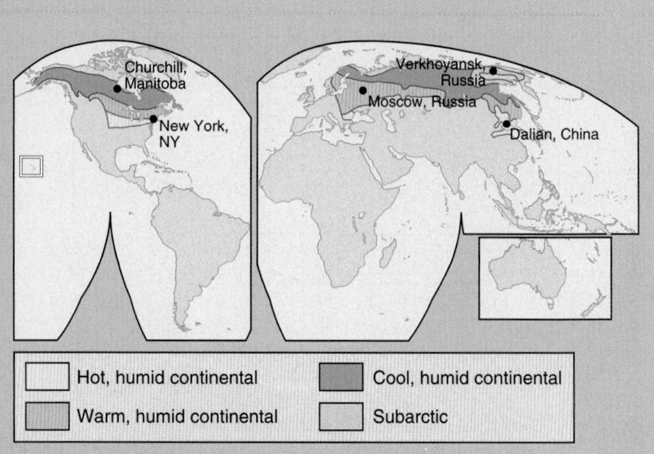

- Increasing seasonality (daylength and Sun altitude), and greater temperature ranges (daily and annually).
- Upper-air westerly winds and undulating Rossby waves, which bring warmer air northward and colder air southward for cyclonic activity; convectional thunderstorms from mT air masses in summer.
- Asian winter-dry pattern for the microthermal climates, increasing east of the Ural Mountains to the Pacific Ocean and eastern Asia.
- Hot summers cooling northward from the mesothermal climates; short spring and fall seasons surrounding winters that are cold to very cold.
- Continental interiors serving as source regions for intense continental polar (cP) air masses that dominate winter, blocking cyclonic storms.

Microthermal climates have four distinct regimes based on increasing cold with latitude and precipitation variability: *humid continental hot-summer* (Chicago, New York); *humid continental warm-summer* (Duluth, Toronto, Moscow); *subarctic cool summers* (Churchill, Manitoba); and the formidable extremes of a frigid *subarctic very cold winter* in Verkhoyansk and northern Siberia.

Humid Continental Hot-Summer Climates

Humid continental hot-summer climates are differentiated by their annual precipitation distribution. In the summer maritime tropical air masses influence both humid continental moist-all-year and winter-dry climates. In North America frequent stormy weather is possible from conflicting air masses—maritime tropical and continental polar—especially in winter. The climographs for New York City, New York, and Dalian, China, illustrate these two humid, hot-summer regimes (Figure 6.15).

Originally, forests covered the *humid continental hot-summer* region of the United States west to the Indiana-Illinois border. Beyond that approximate line, tall-grass prairies extended westward to about the 98th meridian (98° W) and the approximate location of the 51 cm (20 in.) isohyet (line of equal precipitation). The short-grass prairies extended to the west, where precipitation is less.

Deep sod made farming difficult for the first settlers, as did the climate. However, native grasses soon were replaced with domesticated wheat and barley. Various inventions brought from the East (barbed wire, the self-scouring steel plow, well-drilling techniques, and the railroads) helped open the region further. In the United States today, the *humid continental hot-summer* region is the location of corn, soybean, hog, and cattle production (Figure 6.15c).

Humid Continental Mild-Summer Climates

Soils are thinner and less fertile in the cooler microthermal climates, yet agricultural activity is important and includes dairy cattle, poultry, flax, sunflowers, sugar beets, wheat, and potatoes. Frost-free periods range from fewer than 90 days in the north to as many as 225 days in the south. Overall, precipitation is less than in the hot-summer regions to the south. However, notably heavier snowfall is important to soil moisture recharge when it melts. Various snow-capturing strategies are used, including fences and tall stubble (plant stalks left standing after harvest) in fields to create snow drifts and thus more moisture retention on the soil.

Characteristic cities are Duluth, Minnesota, and Saint Petersburg, Russia. Figure 6.16 presents a climograph for Moscow, which is at 55° N, or about the same latitude as the southern shore of Hudson Bay in Canada. The photos of landscapes near Moscow, Russia, and Sebago Lake (inland from Portland, Maine) show summer and late winter scenes, respectively.

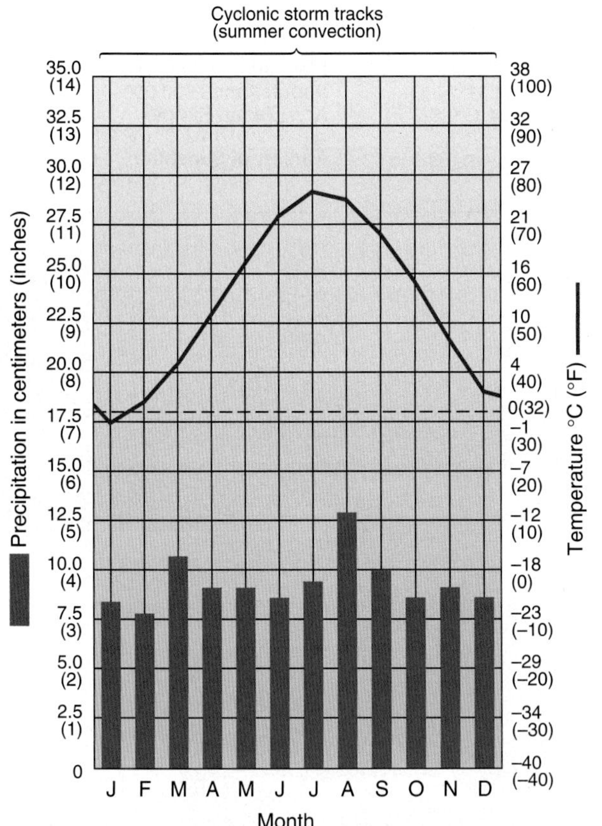

Station: New York, New York
Lat/long: 40°46' N 74°01' W
Avg. Ann. Temp.: 13°C (55.4°F)
Total Ann. Precip.: 112.3 cm (44.2 in.)
Elevation: 16 m (52.5 ft)
Population: 8,008,000
Ann. Temp. Range: 24 C° (43.2 F°)
Ann. Hr of Sunshine: 2564

(a)

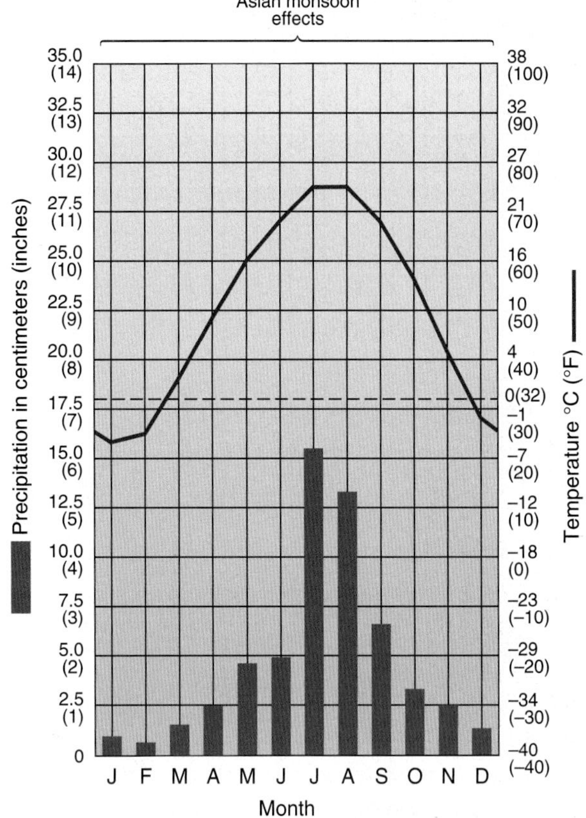

Station: Dalian, China
Lat/long: 38°54' N 121°54' E
Avg. Ann. Temp.: 10°C (50°F)
Total Ann. Precip.: 57.8 cm (22.8 in.)
Elevation: 96 m (314.9 ft)
Population: 5,550,000
Ann. Temp. Range: 29 C° (52.2 F°)
Ann. Hr of Sunshine: 2762

(b)

(c)

(d)

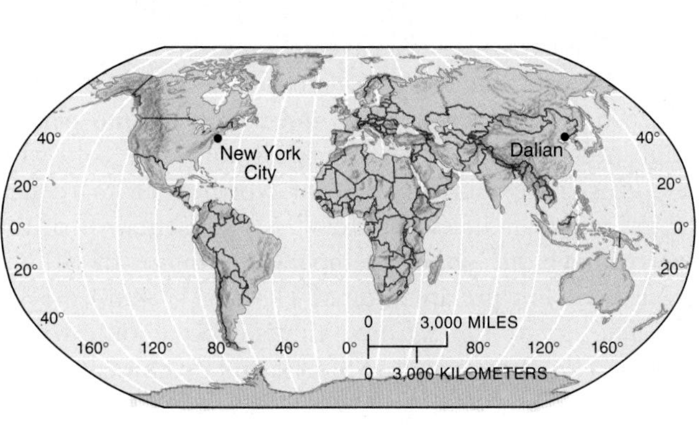

FIGURE 6.15 Humid continental hot-summer climate. Climographs for (a) New York City (*humid continental, moist all year*) and (b) Dalian, China (*humid continental winter-dry*). (c) Typical of this climate region, a deciduous forest and ready-to-harvest soybean field in central Indiana near Zelma. (d) New York City's Central Park emerging from winter just before spring and the return of leaves and warmth. [Photos (c) and (d) by Bobbé Christopherson.]

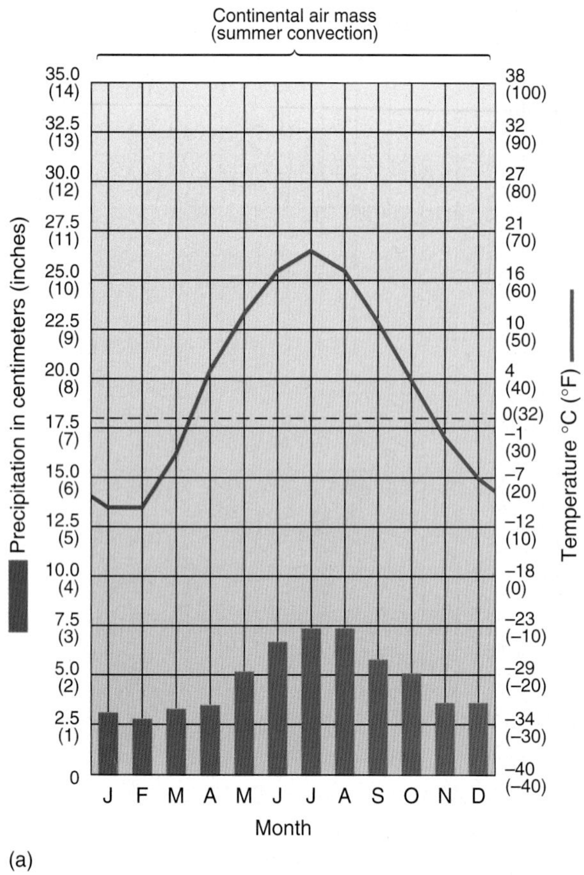

(a)

Station: Moscow, Russia
Lat/long: 55°45' N 37°34' E
Avg. Ann. Temp.: 4°C (39.2°F)
Total Ann. Precip.: 57.5 cm (22.6 in.)
Elevation: 156 m (511.8 ft)
Population: 9,900,000
Ann. Temp. Range: 29 C° (52.2 F°)
Ann. Hr of Sunshine: 1597

(b)

(c)

FIGURE 6.16 Humid continental cool-summer climate.
(a) Climograph for Moscow, Russia. (b) Springtime fields near Saratov, Russia, during the short summer season. (c) Winter scene of Sebago Lake and forests inland from Portland, Maine. [Photos by (b) Wolfgang Kaehler/Gamma, Inc.; (c) by Bobbé Christopherson]

The dry winter associated with the vast Asian landmass, specifically Siberia, is exclusively associated with the extremely dry and frigid winter high-pressure anticyclone that dominates. This system produces the dry monsoons of southern and eastern Asia in the winter months, as winds blow out of Siberia toward the Pacific and Indian Oceans (see Figure 4.20). The intruding dry cold of continental air is a significant winter feature. A representative *humid continental mild-summer* climate is Vladivostok, Russia, on the Sea of Japan, usually one of only two ice-free ports in that country.

Subarctic Climates

Farther poleward, seasonality becomes greater. The short growing season is more intense during long summer days. The cold subarctic climates include vast stretches of Alaska, Canada, northern Scandinavia, and Russia. Discoveries of minerals and petroleum reserves have led to new interest in portions of these regions.

Areas that receive 25 cm (10 in.) or more of precipitation a year on the northern continental margins are covered by the so-called snow forest of fir, spruce, larch, and birch—the *boreal* forests of Canada and the *taiga* of Russia. These forests are in transition to the more open northern woodlands and to the tundra region of the far north. Forests thin out to the north when the warmest summer month drops below an average temperature of 10°C (50°F).

Soils are thin in these lands once scoured by glaciers. Precipitation received and the demand for soil moisture both are low. Soils are generally moist and either partially or totally frozen beneath the surface, a phenomenon known as *permafrost*. The Churchill, Manitoba, climograph (Figure 6.17) shows average monthly temperatures below freezing for at least 7 months of the year, during which time light snow cover and frozen ground persist. High pressure dominates Churchill during its cold winter—this is the source region for North America's continental polar air mass. Churchill is representative of the *subarctic* region: annual temperature range of 40 C° (72 F°) and low precipitation of 44.3 cm (17.4 in.).

The dry-winter *subarctic* climates occur only within Russia. The intense cold of Siberia and north-central and eastern Asia is difficult to comprehend. A severe example of a *subarctic* location is Verkhoyansk, Siberia (Figure 6.18). This area experiences an average temperature lower than freezing for 7 months, with 4 months averaging below −34°C (−30°F). And, as described in Chapter 3, minimum temperatures of below −68°C (−90°F) have been recorded there. Yet summer maximum temperatures in these same areas normally exceed +37°C (+98°F), and average 16°C (60°F)!

Verkhoyansk has probably the world's greatest annual temperature range from winter to summer: a remarkable 63 C° (113.4 F°). Winters feature brittle metals and plastics, triple-thick window panes, and temperatures that render straight antifreeze a solid.

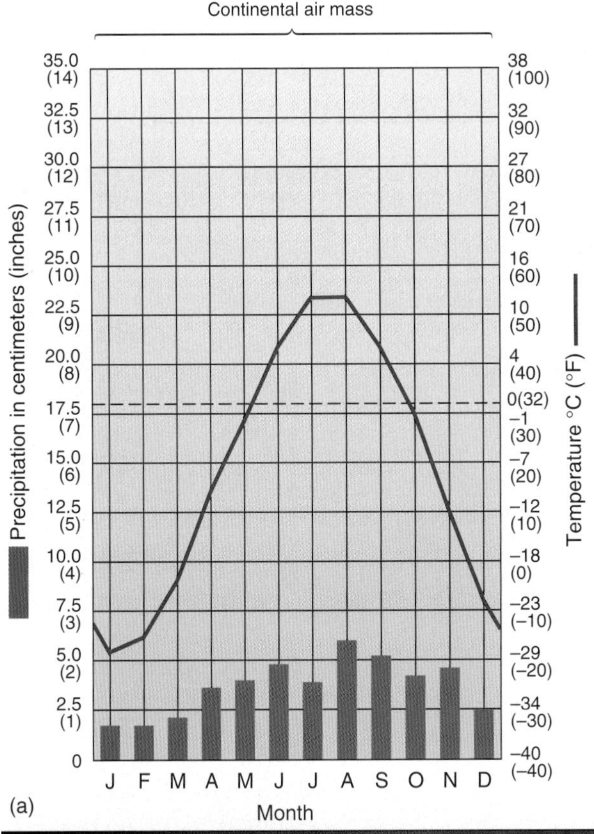

(a)

Station: Churchill, Manitoba
Lat/long: 58°45′ N 94°04′ W
Avg. Ann. Temp.: −7°C (19.4°F)
Total Ann. Precip.: 44.3 cm (17.4 in.)
Elevation: 35 m (114.8 ft)
Population: 1400
Ann. Temp. Range: 40 C° (72 F°)
Ann. Hr of Sunshine: 1732

0 1,000 MILES
0 1,000 KILOMETERS
Churchill
CANADA
50°
40°
30°
30°
120°
90°
70°

(b)

(c)

FIGURE 6.17 Subarctic climate.
(a) Climograph for Churchill, Manitoba (*subarctic* climate). (b) Winter scene with foraging polar bear in the Churchill area near Hudson Bay. (c) Summer scene in Churchill area. [Photos by (b) Art Wolfe/Tony Stone Images; (c) Norbert Rosing/Animals Animals/Earth Scenes.]

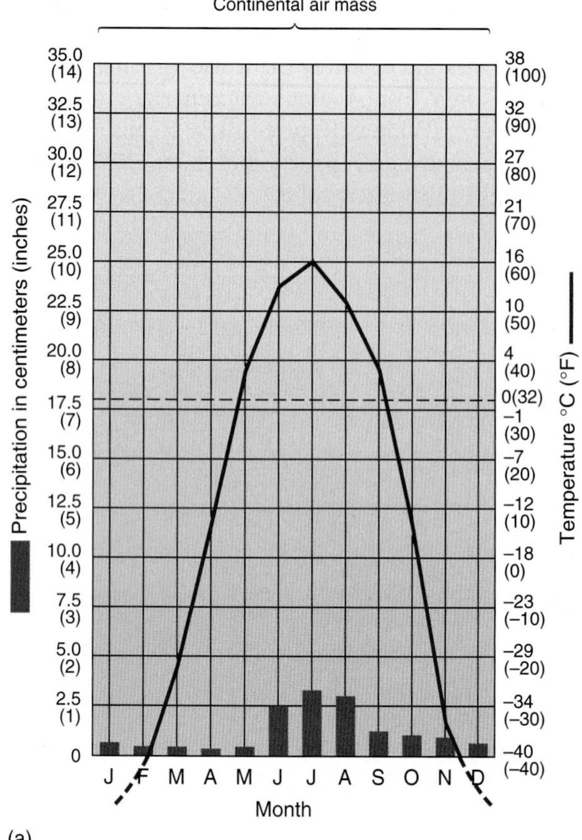

(a)

Station: Verkhoyansk, Russia
Lat/long: 67°35′ N 133°27′ E
Avg. Ann. Temp.: −15°C (5°F)
Total Ann. Precip.: 15.5 cm (6.1 in.)
Elevation: 137 m (449.5 ft)
Population: 1400
Ann. Temp. Range: 63 C° (113.4 F°)

(b)

FIGURE 6.18 Extreme subarctic winter-dry climate.
(a) Climograph for Verkhoyansk, Russia. (b) Scene in the town of Verkhoyansk during their brief summer. [Photo by Dean Conger/National Geographic Society.]

Polar Climates (polar regions)

The polar climates cover about 19% of Earth's total surface and about 17% of its land area. These climates have no true summer like that in lower latitudes. Poleward of the Arctic and Antarctic Circles, daylength increases in summer until daylight becomes continuous, yet average monthly temperatures never rise above 10°C (50°F). These temperature conditions are intolerant to tree growth. Principal climatic factors in these frozen and barren regions are the following:

- Extremes of daylength between winter and summer determine the amount of insolation received.
- Low Sun altitude even during the long summer days is the principal climatic factor.
- Extremely low humidity produces low precipitation amounts—Earth's frozen deserts.

Polar climates have three regimes: *tundra* (high latitude, or altitude); *polar marine* (oceanic association, slight moderation of extreme cold); and *ice caps* and *ice sheet* (perpetually frozen).

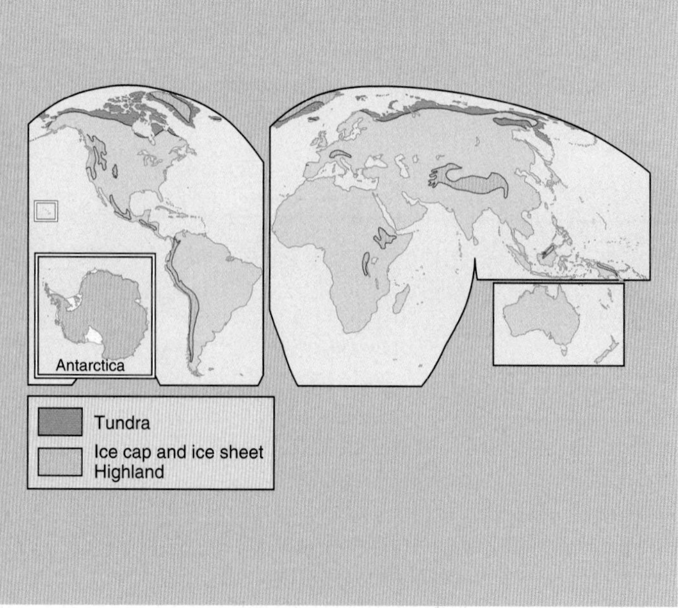

Tundra Climate

In a *tundra* climate, land is under continuous snow cover for 8–10 months, but when the snow melts and spring arrives, numerous plants appear—stunted sedges, mosses, flowering plants, and lichens. Much of the area experiences permafrost (frozen ground) conditions. The tundra also is the summer home of mosquitoes of legend and black gnats. Global climate change is bringing dramatic alterations to the tundra, its plants, animals, and permafrost. In 1998, parts of Canada registered temperatures 5 C° above average, 1.0 C° above normal for all of Canada—the warmest in the 51-year record. A 2002 study of 67 glaciers in Alaska discovered average thickness decreasing at more than a half meter a year, with 131 km^2 (430 mi^2) of land now uncovered and ice free. These conditions are having profound effect in the extent of the tundra regions.

Tundra climates are strictly a Northern Hemisphere occurrence, except for elevated mountain locations in the Southern Hemisphere and a portion of the Antarctic Peninsula, where plants are now growing for the first time. Because of elevation, the summit of Mount Washington in New Hampshire (1914 m, or 6280 ft) statistically qualifies as a highland *tundra* climate of small scale.

Ice Cap and Ice Sheet Climate

Most of Antarctica and Greenland fall within the *ice cap* and *ice sheet* climate, as does the North Pole, with all months averaging below freezing. These regions are dominated by dry, frigid air masses, with vast expanses that never warm above freezing. The area of the North Pole is actually a sea covered by ice, whereas Antarctica is a substantial continental landmass covered by Earth's greatest ice sheet. For comparison, winter minimums at the South Pole (July) frequently drop below the temperature of solid carbon dioxide or "dry ice" (that is, below −78°C, or −109°F).

Antarctica is constantly snow-covered but receives less than 8 cm (3 in.) of precipitation each year. Antarctic ice has accumulated to several kilometers thickness and is the largest repository of freshwater on Earth (Figure 6.19).

This ice is a vast historical record of Earth's atmosphere. Within it, thousands of past volcanic eruptions from all over the world have deposited ash layers, and ancient combinations of atmospheric gases lie trapped in frozen bubbles. An analysis of ice cores taken from Greenland and Antarctica is in Chapter 14.

Polar Marine Climate. *Polar marine* areas are more moderate than other polar climates in winter, with no month averaging below −7°C (20°F), yet they are not as warm as *tundra* climates. Because of marine influences, annual temperature ranges are low. This climate exists along the Bering Sea, the tip of Greenland, northern Iceland, Norway, and in the Southern Hemisphere, generally over oceans between 50° S and 60° S. Precipitation, which frequently falls as sleet, is greater in these regions than in continental polar climates.

FIGURE 6.19 The frozen Antarctic landscape.
The Antarctic landscape; Earth's frozen freshwater reservoir. [Photo by Galen Rowell/Mountain Light Photography, Inc.]

Dry Arid and Semiarid Climates (permanent moisture deficits)

The *dry arid* and *semiarid* regions are where we consider moisture efficiency along with temperature for understanding the climate. Dry climates are the world's arid deserts and semiarid regions, with their unique plants, animals, and physical features. They occupy more than 35% of Earth's land area and clearly are the most extensive climate over land. The mountains, rock strata, long vistas, and the resilient struggle for life are magnified by the dryness. Sparse vegetation leaves the landscape bare; water demand exceeds the precipitation water supply throughout *dry arid* and *semiarid* climates, creating permanent water deficits (water balance is discussed in Chapter 7). The extent of these deficits distinguishes deserts from steppe climatic regions. (See specific annual and daily desert temperature regimes, including the highest record temperatures and surface energy budgets in Chapter 3; desert landscapes in Chapter 12; and desert environments in Chapter 16.) Important causal elements in these dry lands include:

- Dry, subsiding air in subtropical high pressure systems dominates.
- Midlatitude deserts and steppes form in the rain shadow of mountains, those regions to the lee of precipitation-intercepting mountains.

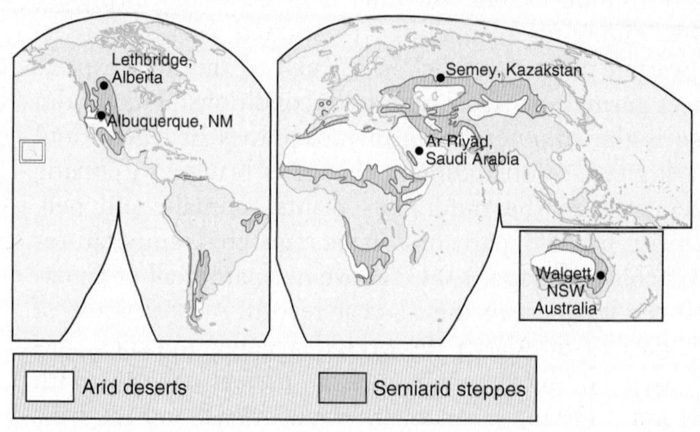

- Continental interiors, particularly central Asia, are far from moisture bearing air masses.
- Shifting subtropical high pressure systems produce semiarid steppe lands around the periphery of arid deserts.

Dry climates are distributed by latitude and the amount of moisture deficits in four distinct regimes: *Arid deserts* (low-latitude hot, midlatitude cold) and *semiarid steppes* (low-latitude hot, midlatitude cold).

Desert Characteristics

Desert vegetation is typically *xerophytic:* drought-resistant, waxy, hard-leafed, or otherwise adapted to aridity and low transpiration loss. Along stream channels, plants called *phreatophytes*, or "water-well plants," have roots that penetrate to great depths for the water they need (Figure 6.20).

The world climate map in Figure 6.5 reveals the pattern of Earth's dry climates, which cover broad regions between 15° and 30° N and S latitudes. In these areas, subtropical high-pressure cells predominate, with subsiding, stable air and low relative humidity. Under generally cloudless skies, these subtropical deserts extend to western continental margins, where cool, stabilizing ocean currents operate offshore and summer advection fog forms. The Atacama Desert of Chile, the Namib Desert of Namibia, the Western Sahara of Morocco, and the Australian Desert each lies adjacent to a coastline.

Extension of dry regions into higher latitudes of North and South America is associated with rain shadows, caused by orographic lifting over western mountains. The isolated interior of Asia, far distant from any moisture-bearing air masses, falls within these *dry arid* and *semiarid* climates.

Major subdivisions include: *deserts* (precipitation supply roughly less than one-half of the natural moisture demand) and *semiarid steppes* (precipitation supply roughly more than one-half of natural moisture demand). Important is whether precipitation falls principally in the winter with a dry summer, in the summer with a dry winter, or is evenly distributed. Winter rains are most effective because they fall at a time of lower moisture demand. Relative to temperature, the lower latitude deserts and steppes tend to be hotter with less seasonal change, than the midlatitude deserts and steppes where mean annual temperatures are below 18°C (64.4°F) and freezing winter temperatues are possible.

Low-Latitude Hot Desert Climates

Low-latitude hot desert climates are Earth's true tropical and subtropical deserts and feature annual average temperatures above 18°C (64.4°F). They generally are concentrated on the western sides of continents, although Egypt, Somalia, and Saudi Arabia also fall within this climate regime. Rainfall is from local summer convectional showers. Some regions receive nearly nothing, whereas others may receive up to 35 cm (14 in.) precipitation a year. A representative *low-latitude hot desert* station is Ar Riyāḑ (Riyadh), Saudi Arabia (Figure 6.21a and b).

Along the Sahara's southern margin is a drought-tortured region. Human populations suffered great hardship in the last several decades as desert conditions gradually expanded over their homelands. The sparse environment sets the stage for a rugged lifestyle and subsistence economies, pictured here near Timbuktu, Mali (Figure 6.21c). Chapter 12 presents the process of desertification (expanding desert conditions).

Midlatitude Cold Desert Climates

Midlatitude cold desert climates cover only a small area: the countries bordering southern Russia, the Gobi Desert, and Mongolia in Asia; the central third of Nevada and areas of the American Southwest, particularly at high elevations; and Patagonia in Argentina. Because of lower temperatures and lower natural moisture demand, rainfall also

(b)

(a)

(c)

FIGURE 6.20 Desert landscapes.
(a) Desert plants are particularly well adapted to the harsh environment of Joshua Tree National Park, part of the Mojave desert in southeastern California. (b) Undependable water flows between the towering sandstone walls surrounding Chinle Wash in Canyon de Chelly, Arizona. (c) The silt-laden Colorado River flows just north of Moab, Utah, cutting through beautiful red sandstone. [Photos (a) and (c) by Bobbé Christopherson; (b) by author.]

must be low for an area to be a *midlatitude cold desert*. Here we find total annual average rainfall at about 15 cm (6 in.).

A representative city is Albuquerque, New Mexico, with 20.7 cm (8.1 in.) of precipitation and an annual average temperature of 14°C (57.2°F) (Figure 6.22). Across central Nevada stretches a characteristic expanse of midlatitude cold desert, greatly modified by more than a century of grazing and other extensive uses (Figure 6.22b). Comparing *midlatitude cold desert* of Albuquerque and *low-latitude hot desert* of Ar Riyāḍ, Saudi Arabia, reveals an interesting similarity in the annual temperature range and a distinct difference in precipitation patterns.

Low-Latitude Hot Steppe Climates

Low-latitude hot steppe climates generally exist around the periphery of hot deserts, where shifting subtropical high-pressure cells create a distinct summer-dry and winter-wet pattern. This climate type is seen around the Sahara's periphery and in Iran, Afghanistan, and the Turkistan, Kazakstan, region. Average annual precipitation in *low-latitude hot steppe* areas usually is below 60 cm (23.6 in.). Walgett, in interior New South Wales, Australia, provides a Southern Hemisphere example of this climate (Figure 6.23, p. 208).

Midlatitude Cold Steppe Climates

The *midlatitude cold steppe* climates occur poleward of about 30° latitude and the midlatitude desert climates. Such midlatitude steppes are not generally found in the Southern Hemisphere. As with other dry climate regions, rainfall in the steppe climate area is variable and undependable, ranging from 20 to 40 cm (7.9 to 15.7 in.). Not all rainfall is convectional, for cyclonic systems penetrate the continents; however, most storms produce little precipitation. Figure 6.24 (p. 209) presents a comparison between Asian and North American *midlatitude cold steppe*; consider Semey (Semipalatinsk) in Kazakstan (greater temperature range, precipitation evenly distributed) and Lethbridge in Alberta (lesser temperature range, summer maximum precipitation).

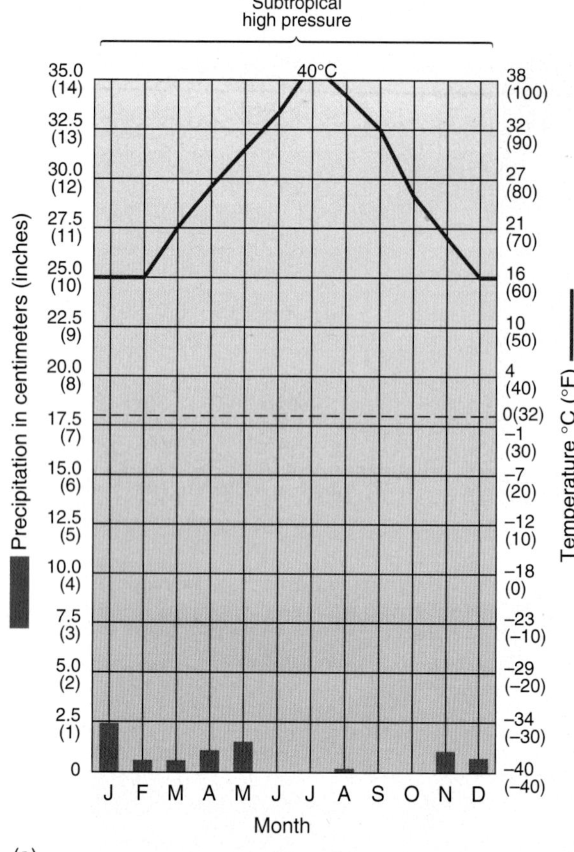

Station: Ar Riyāḍ (Riyadh), Saudi Arabia
Lat/long: 24°42'N 46°43' E
Avg. Ann. Temp.: 26°C (78.8°F)
Total Ann. Precip.: 8.2 cm (3.2 in.)

Elevation: 609 m (1998 ft)
Population: 1,800,000
Ann. Temp. Range: 24 C° (43.2 F°)

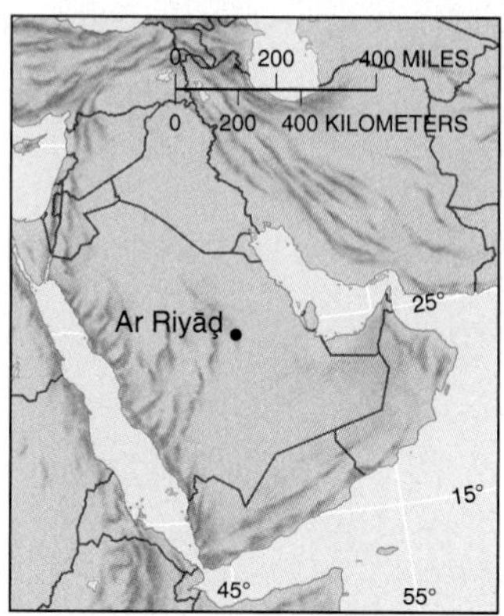

(a)

(b)

(c)

FIGURE 6.21 Low-latitude hot desert climate.
(a) Climograph for Ar Riyāḍ (Riyadh), Saudi Arabia (*low-latitude hot desert*). (b) The Arabian desert sand dunes near Ar Riyāḍ. (c) Herders bring a few cattle to market near Timbuktu, Mali. Precipitation has been below normal in the region since 1966. [Photos by (b) Ray Ellis/Photo Researchers, Inc.; (c) by Betty Press/Woodfin Camp & Associates.]

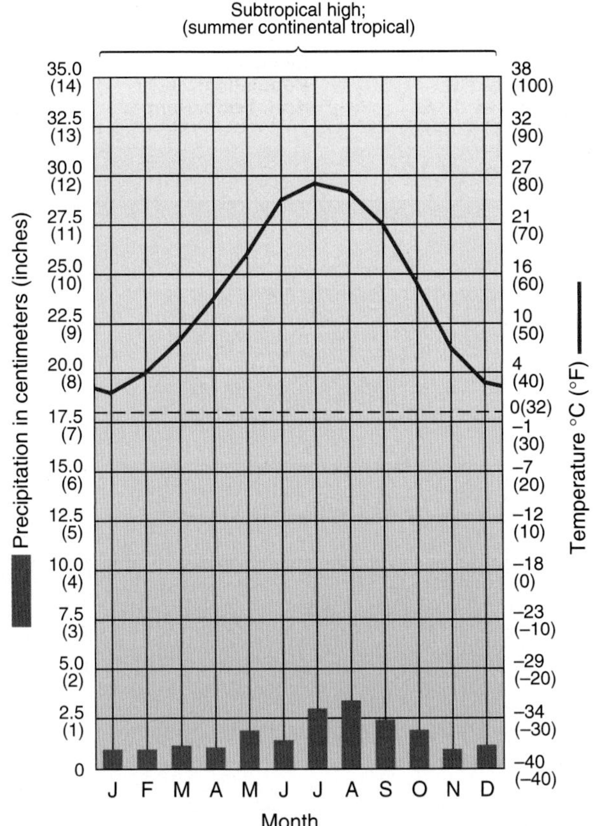

(a)

Station: Albuquerque, New Mexico
Lat/long: 35°03′ N 106°37′ W
Avg. Ann. Temp.: 14°C (57.2°F)
Total Ann. Precip.: 20.7 cm (8.1 in.)

Elevation: 1620 m (5315 ft)
Population: 449,000
Ann. Temp. Range: 24 C° (43.2 F°)
Ann. Hr of Sunshine: 3420

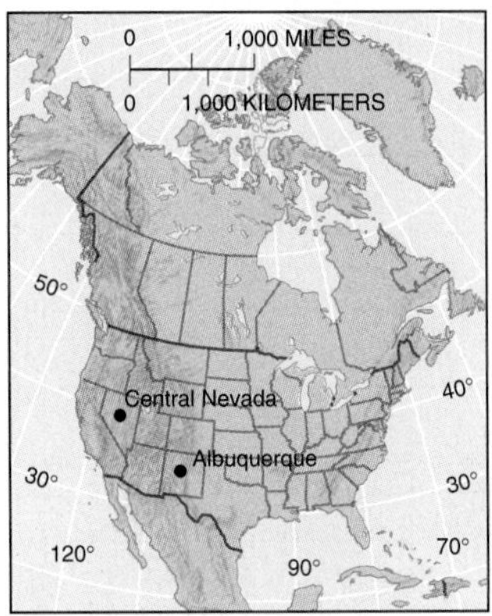

FIGURE 6.22 Midlatitude cold desert climate.
(a) Climograph for Albuquerque, New Mexico (*midlatitude cold desert*). (b) Cold desert landscape of the Basin and Range Province in Nevada, east of Ely along highway U.S. 50 (b). [Photo by Bobbé Christopherson.]

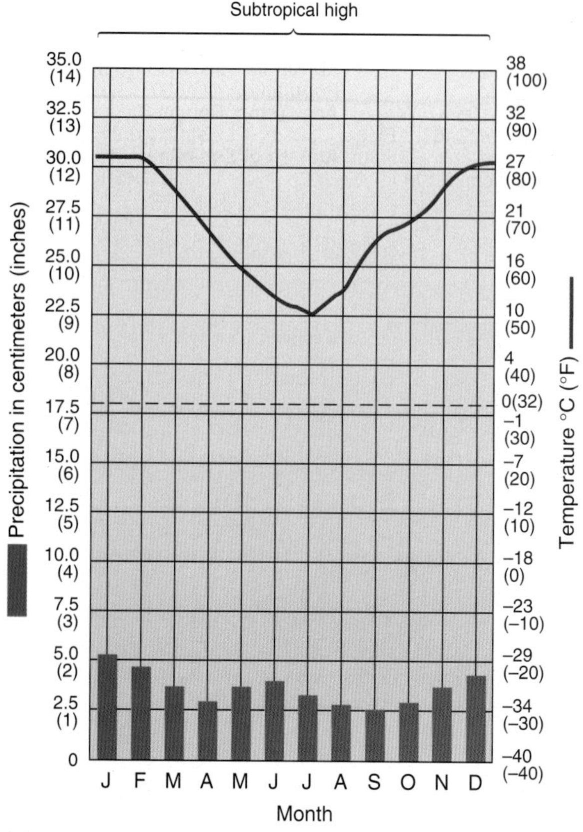

Station: Walgett, New South Wales, Australia
Lat/long: 30°0′ S 148°07′ E
Avg. Ann. Temp.: 20°C (68°F)
Total Ann. Precip.: 45.0 cm (17.7 in.)
Elevation: 133 m (436 ft)
Population: 2160
Ann. Temp. Range: 17 C° (31 F°)

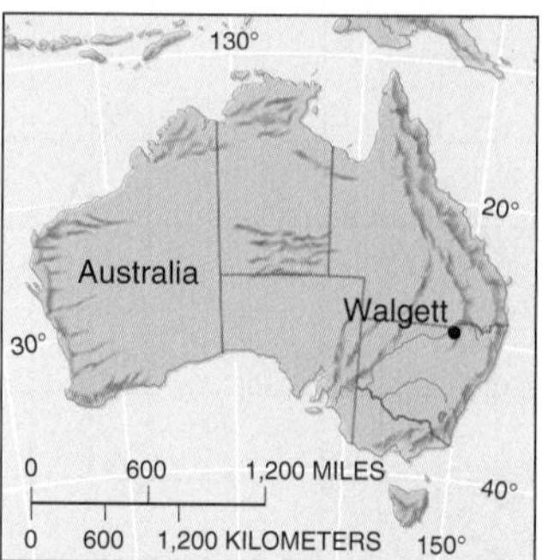

FIGURE 6.23 Low-latitude hot steppe climate.
(a) Climograph for Walgett, New South Wales, Australia (*low-latitude hot steppe*). (b) Vast plains characteristic of north-central New South Wales. [Photo by Otto Rogge/Stock Market.]

Global Climate Change

ANIMATION **Global Warming, climate change**

Significant climatic change has occurred on Earth in the past and most certainly will occur in the future. There is nothing society can do about long-term influences that cycle Earth through swings from ice ages to warmer periods. However, our global society must address short-term changes that are influencing global temperatures within the life span of present generations.

Record-high global temperatures dominated the past two decades, records for both land and ocean and for both day and night. 1998 was the record year for warmth. Understanding this warming and all related impacts is an important applied topic of Earth systems science and a challenge for the spatial analysis ability of physical geography.

Global Warming

More than a decade ago, climatologists Richard Houghton and George Woodwell described the present climatic condition:

> The world is warming. Climatic zones are shifting. Glaciers are melting. Sea level is rising. These are not hypothetical events from a science fiction movie; these changes and others are already taking place, and we expect them to accelerate over the next years as the amounts of carbon dioxide, methane, and other trace gases accumulating in the atmosphere through human activities increase.*

The 2001 Intergovernmental Panel on Climate Change (IPCC) *Third Assessment Report* stated:

> ... there is new and stronger evidence that most of the warming observed over the past 50 years is attributable to human activities.... Both temperature and sea level are projected to continue to rise throughout the 21st century for all scenarios studied.†

*R. Houghton and G. Woodwell, "Global climate change," *Scientific American*, April 1989, p. 36.

†IPCC, Working Group I, *Climate Change 2001, The Scientific Basis* (London: Cambridge Press, 2001), p. ix.

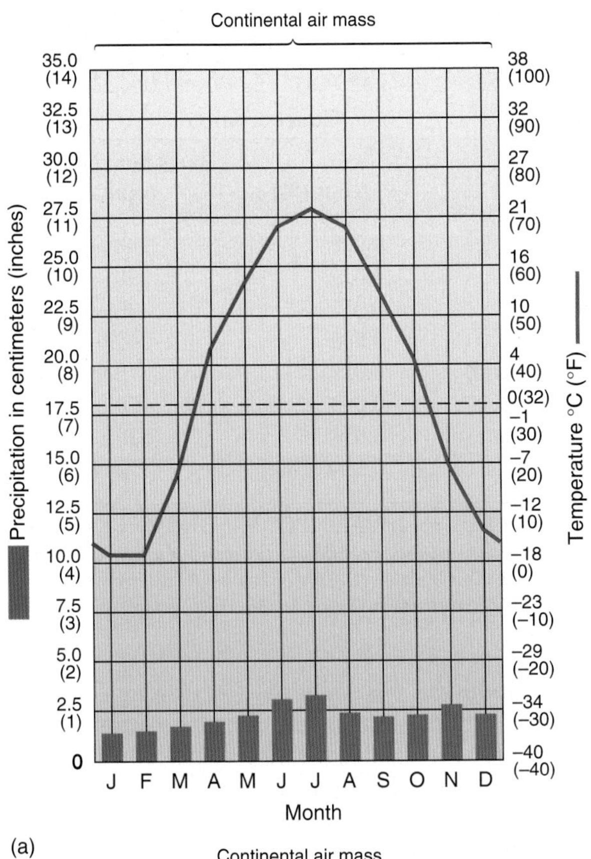

(a)

Station: Semey (Semipalatinsk), Kazakstan
Lat/long: 50°21' N 80°15' E
Avg. Ann. Temp.: 3°C (37.4°F)
Total Ann. Precip.: 26.4 cm (10.4 in.)
Elevation: 206 m (675.9 ft)
Population: 339,000
Ann. Temp. Range: 39 C° (70.2 F°)

(b)

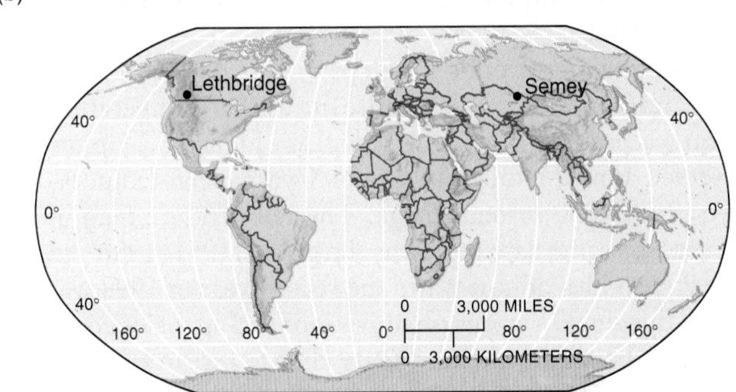

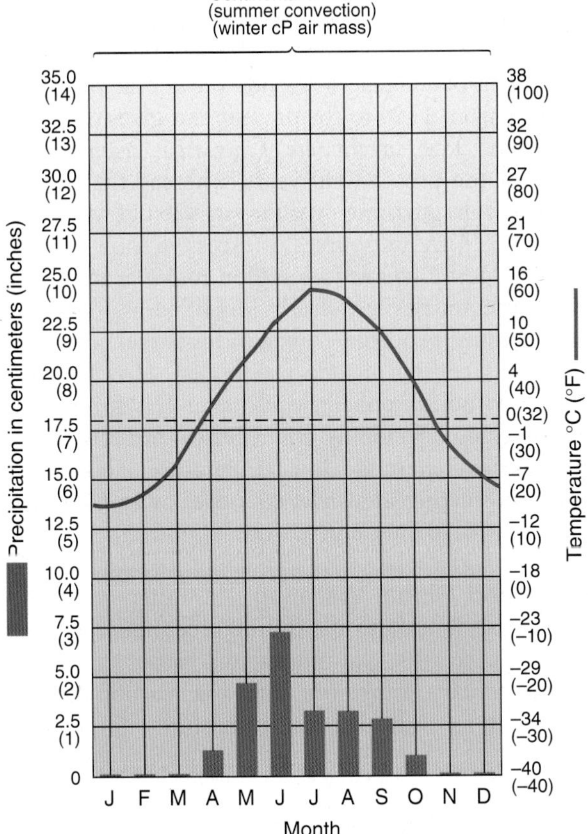

(c)

Station: Lethbridge, Alberta
Lat/long: 49°42' N 110°50' W
Avg. Ann. Temp.: 2.9°C (37.3°F)
Total Ann. Precip.: 25.8 cm (10.2 in.)
Elevation: 910 m (2985 ft)
Population: 58,840
Ann. Temp. Range: 24.3 C° (43.7 F°)

(d)

FIGURE 6.24 Midlatitude cold steppe climate.
(a) Climograph for Semey (Semipalatinsk) in Kazakstan and (b) herders in the region near Semey. (c) Climograph for Lethbridge in Alberta. (d) Canadian prairies and grain elevators of southern Alberta. [Photos by (b) Sovfoto/Eastfoto; (d) author.]

The IPCC, formed in 1988, is an organization operating under the United Nations Environment Programme (UNEP) and the World Meteorological Organization (WMO) and is the scientific organization coordinating global climate change research, climate forecasts, and policy formulation.

Human activities are enhancing Earth's natural greenhouse effect. There is an international scientific consensus that air temperatures are the highest since recordings were begun in earnest more than 140 years ago. This warming trend is *very likely** due to a buildup of greenhouse gases.

In terms of **paleoclimatology**, the science that studies past climates (discussed in Chapter 14), proxy indicators (ice core data, sediments, coral reefs, ancient pollen, tree-ring density, among others) point to the recent past as the warmest in the last 600 years, and further, that the increase in temperature during the twentieth century was very likely the largest in any century over the past 1000 years. Earth is within 1 C° (1.8 F°) of equaling the highest average temperature of the past 125,000 years (based on ice-core data).

The rate of warming in the past 30 years exceeds any comparable period in the entire measured temperature record, according to NASA scientists. Figure 6.25 plots observed annual temperatures and 5-year mean temperatures from 1880 through 2002. The temperature map in (b) uses the same base period as the graph (1951–1980), to give you an idea of how warm the record year of 1998 was across the globe. Various organizations and agencies coordinate and conduct global climate change research. News Report 6.2 offers an overview and contact information.

Scientists are working to determine the difference between *forced fluctuations* (human-caused) and *unforced fluctuations* (natural) as a key to predicting future climate trends. Because the gases that generate temperature changes are human in origin, various management strategies are possible to reduce human-forced changes. *Radiatively active* gases are atmospheric gases, such as carbon dioxide (CO_2), methane (CH_4), nitrous oxide (N_2O), chlorofluorocarbons (CFCs), and water vapor, that absorb and radiate infrared wavelengths. Let's begin by examining the problem at its roots.

Carbon Dioxide and Global Warming. Carbon dioxide and water vapor are the principal radiatively active gases causing Earth's natural greenhouse effect. They are transparent to light but opaque to the infrared wavelengths radiated by Earth, and thus they transmit light from the Sun to Earth but delay heat-energy loss to space. While detained, this heat energy is absorbed and reradiated over and over, warming the lower atmosphere. As concentrations of these infrared-absorbing gases increase, more heat energy remains in the atmosphere and temperatures increase.

*As the standard scientific reference on climate change, the IPCC uses the following words to indicate levels of confidence: *virtually certain* (greater than a 99% chance the result is true), *very likely* (90–99% chance), *likely* (66–90% chance), *medium likelihood* (33–66% chance), *unlikely* (10–33% chance), and *very unlikely* (1–10% chance).

Table 6.1 Carbon Dioxide Concentration in the Lower Atmosphere

Year	Concentration*	
	Percent (%)	Parts per million (ppm)
1750	0.028	280
1888	0.029	290
1970	0.032	320
1985	0.035	345.9
2001 (year end)	0.037	370.9
2020 (estimate)	0.047	470
2050 (estimate)	0.053	530
2100 range (est.)	0.55–0.97	550–970

*See Chapter 3 and pp. 219–24 in IPCC Working Group I, *Climate Change 2001, The Scientific Basis* (Washington: Cambridge University Press, 2001). For update see http://cdiac.esd.ornl.gov/ftp/maunaloa-co2/maunaloa.co2.

The Industrial Revolution, beginning in the mid–1700s, initiated tremendous burning of fossil fuels. Fossil-fuel burning coupled with the destruction and inadequate replacement of harvested forests, continues to increase atmospheric carbon dioxide levels. Carbon dioxide alone is responsible for 64% of the global warming trend. Carbon dioxide levels today are more than 30% higher than they were in the preindustrial era. Table 6.1 shows the increasing percentage of carbon dioxide in the lower atmosphere from 1750 to the present and gives estimates for the future. The current *rate of increase* is faster than at any time in the past 20,000 years, whereas the present *concentration* tops anything over the last 420,000 years.

Figure 6.26 shows sources of excessive (non-natural) carbon dioxide by country or region in 1995 and 2025. Developing countries are clearly identified as the sector with the greatest probable growth in fossil fuel consumption and new carbon dioxide production. However, national and corporate policies could alter this forecast by actively steering developing countries toward alternative energy sources (renewable, low temperature, labor intensive) and redirecting industrial countries away from past practices toward greater efficiencies.

Methane and Global Warming. Another radiatively active gas contributing to the overall greenhouse effect is methane (CH_4), which, at more than 1% per year, is increasing in concentration even faster than carbon dioxide. Air bubbles in ice show that concentrations of methane in

FIGURE 6.25 Global temperature trends.
(a) Global temperature trends (annual and five-year means) from 1880 to 2002. The 0 baseline represents the 1951–1980 global average—which is 14°C (57.2°F). (b) Temperature map shows temperature anomalies for the record year 1998; the coloration represents °C departures from the base period 1951–1980. [Courtesy of GISS/NASA, and NCDC/NOAA.]

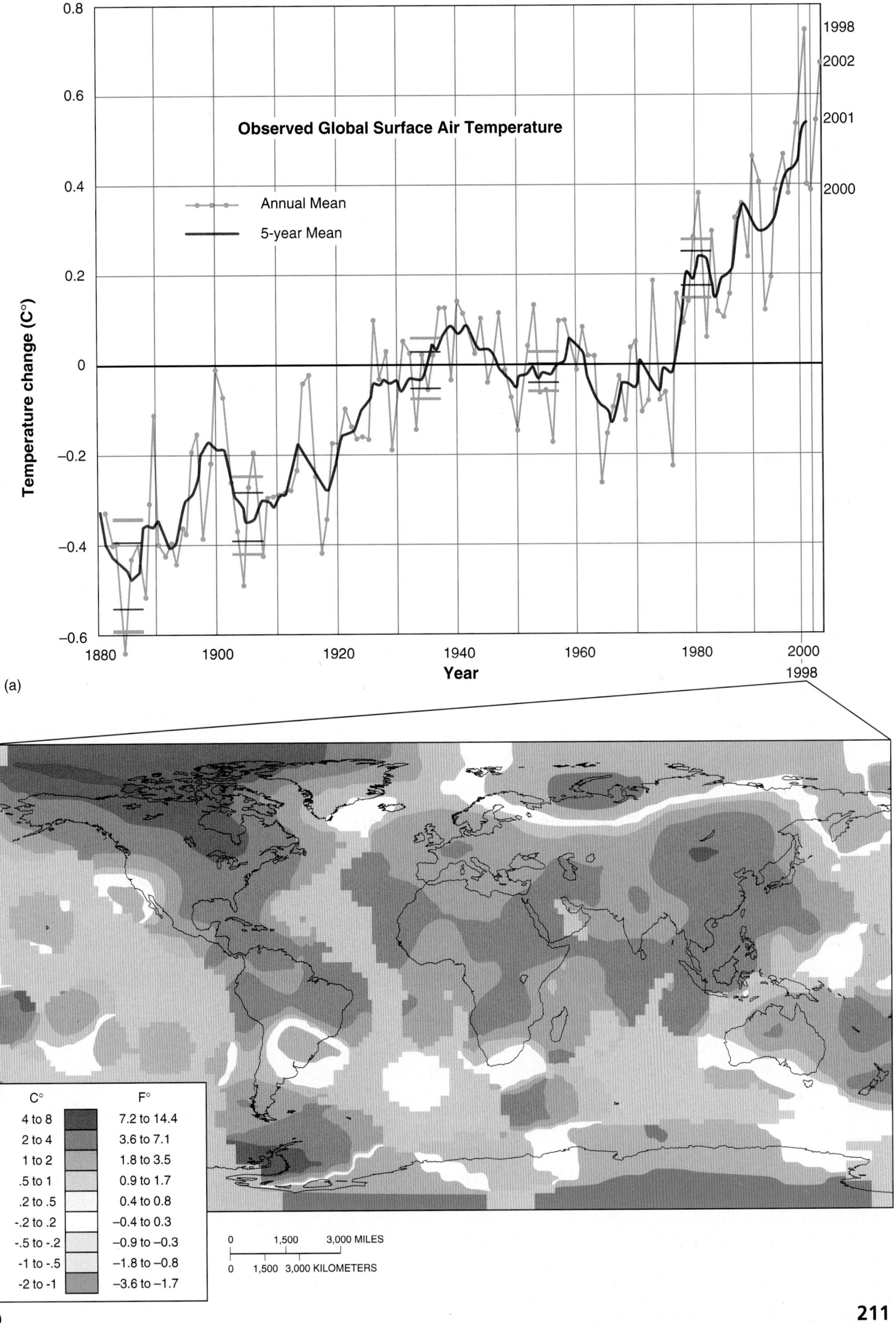

Observed Global Surface Air Temperature

Annual Mean

5-year Mean

Temperature change (C°)

0.8
0.6
0.4
0.2
0
−0.2
−0.4
−0.6

1880 1900 1920 1940 1960 1980 2000

Year

1998
2002
2001
2000

(a)

1998

C°	F°
4 to 8	7.2 to 14.4
2 to 4	3.6 to 7.1
1 to 2	1.8 to 3.5
.5 to 1	0.9 to 1.7
.2 to .5	0.4 to 0.8
-.2 to .2	−0.4 to 0.3
-.5 to -.2	−0.9 to −0.3
-1 to -.5	−1.8 to −0.8
-2 to -1	−3.6 to −1.7

0 1,500 3,000 MILES

0 1,500 3,000 KILOMETERS

(b)

the past, between 500 and 27,000 years ago, were approximately 0.7 ppm (parts per million). Whereas current atmospheric concentrations are 1.745 ppm, or more than double the preindustrial level. We are at an atmospheric concentration of methane that is higher than at any time in the past 420,000 years.

Methane is generated by such organic processes as digestion and rotting in the absence of oxygen (anaerobic processes). About 50% of the excess methane comes from bacterial action in the intestinal tracts of livestock and from organic activity in flooded rice fields. Burning of vegetation causes another 20% of the excess, and bacterial action inside the digestive systems of termite populations also is a significant source. Methane is now believed responsible for at least 19% of the total atmospheric warming, complementing the warming caused by the buildup of carbon dioxide.

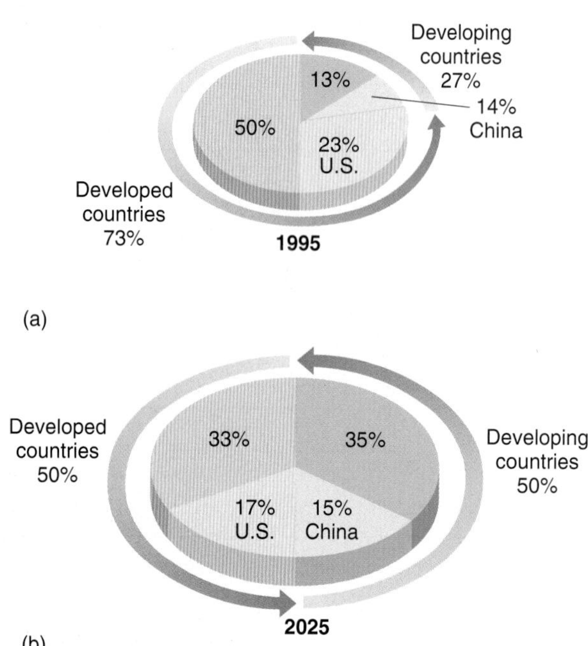

FIGURE 6.26 Origin of excessive carbon dioxide. Countries and regions of origin for excessive carbon dioxide in (a) 1995 and (b) forecast for 2025. [Data from United Nations Environment Programme.]

Other Greenhouse Gases. Nitrous oxide (N_2O) is the third most important greenhouse gas that is being forced by human activity—up 17% in atmospheric concentration since 1750, higher than at any time in the past 1000 years. Fertilizer use increases the processes in soil that emit nitrous oxide, although more research is needed to fully understand the relationships. Chlorofluorocarbons (CFCs) and other halocarbons also contribute to global warming, by absorbing infrared in wavelengths missed by carbon dioxide and water vapor. As radiatively active gases, CFCs enhance the greenhouse effect in the troposphere and are a cause of ozone depletion in the stratosphere.

Climate Models and Future Temperatures

The scientific challenge in understanding climate change is to sense climatic trends in what is essentially a nonlinear (unpredictable), chaotic natural climate

News Report 6.2

Coordinating Global Climate Change Research

A cooperative global network of all United Nation members participate in the United Nations Environment Programme (UNEP, http://www.unep.org) and the World Meteorological Organization (WMO, http://www.wmo.ch/). The World Climate Research Programme (WCRP, http://www.wmo.ch/web/wcrp/wcrp-home.html) and its network under the supervision of the Global Climate Observing System (GCOS, http://www.wmo.ch/web/gcos/gcoshome.html) coordinate data gathering and research. The ongoing climate assessment process within the UNEP is conducted by the Intergovernmental Panel on Climate Change (IPCC, http://www.ipcc.ch/), with completed reports issued by three Working Groups in 1990, a 1992 supplementary report, 1995, and the latest *Third Assessment Report* in 2001.

In the United States coordination is found at the U.S. Global Change Research Program (http://www.usgcrp.gov/). An overall source for information is http://globalchange.gov, which publishes an on-line monthly summary of all related developments. Also important are programs and services at NASA agencies such as Goddard Institute for Space Studies (GISS, http://www.giss.nasa.gov/), Global Hydrology and Climate Center (GHCC, http://www.ghcc.msfc.nasa.gov), and at NOAA agencies at the National Climate Data Center (NCDC, http://www.ncdc.noaa.gov/) and the National Environmental Satellite, Data, and Information Service (NESDIS, http://www.nesdis.noaa.gov/), among others. The Pew Center on Global Climate Change offers credible analysis and overview and has issued several policy reports at http://www.pewclimate.org/.

The multiagency National Ice Center is at http://www.natice.noaa.gov/. Important research is done at the National Center for Atmospheric Research (http://www.ncar.ucar.edu/). For Canada, information and research is coordinated by Environment Canada (http://www.ec.gc.ca/climate/index.html). The effect of global warming on permafrost, which involves half of Canadian land area, is found at http://www.socc.uwaterloo.ca/permafrost/permafrost_current.cfm.

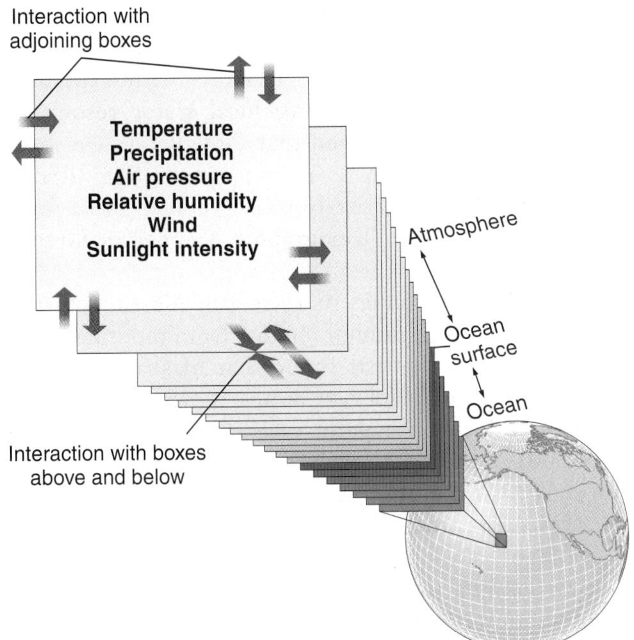

FIGURE 6.27 A general circulation model scheme. Temperature, precipitation, air pressure, relative humidity, wind, and sunlight intensity are sampled in myriad grid boxes. In the ocean, sampling is limited, but temperature, salinity, and ocean current data are considered. The general circulation model program simulates the interactions within a grid layer, and between layers on all six sides.

system. Imagine the tremendous task of building a computer model of all the climatic components we have discussed, and of programming these linkages (suggested in Figure 6.1) over different time frames and at various scales!

The most complex climate model at present, known as a **general circulation model (GCM)**, was developed from mathematical models originally established for forecasting weather. There are at least a dozen established GCMs now operating around the world. Submodel programs for the atmosphere, ocean, land surface, cryosphere, and biosphere operate within the GCM. The most sophisticated models couple atmosphere and ocean submodels and are known as *Atmosphere-Ocean General Circulation Models (AOGCMs)*. The most powerful computer to date became operational in Japan in 2002. The *Earth Simulator* computer is modeling Earth's climate and the planet's interior in ways not previously possible.

The first step in describing a climate is to define a manageable portion of Earth's climatic system for study. Climatologists create dimensional "grid boxes" that extend from beneath the ocean to the tropopause, in multiple layers (Figure 6.27). Resolution of these boxes in the atmosphere is about 250 km in the horizontal and 1 km (155 and 0.6 mi) in the vertical; in the ocean the boxes use the same horizontal resolution and a vertical resolution of about 200 to 400 m (650–1300 ft). Analysts must deal not only with the climatic components within each grid layer but also with the interaction among the layers on all sides.

A comparative benchmark among the operational GCMs is *climatic sensitivity* to doubling of carbon dioxide levels in the atmosphere. GCMs do not predict specific temperatures, but they do offer various scenarios of global warming. The Goddard Institute for Space Studies projects a distinct warming for 2029, using a moderate-case scenario (Figure 6.28). Consistent with other computer studies, unambiguous warming appears over some continental regions, especially China, with the greatest warming assumed to occur over Arctic and Antarctic regions of sea ice. Remarkably, as forecast, Canada had its warmest summer and year, and second warmest winter, in history in 1998. Interestingly, GCM-generated temperature maps correlate well with the observed global warming patterns experienced in the 1990s. Note regions of forecasted cooling on the map.

The 2001 IPCC *Third Assessment Report* predicts a range of average warming between 1990 to 2100 from a "low forecast" of 1.4 C° (2.5 F°), to a "high forecast" of 5.8 C° (10.4 F°). The middle case of 3.6 C° (6.5 F°) represents a significant increase in global land and ocean temperatures and will produce significant consequences.

Consequences of Global Warming

The consequences of uncontrolled atmospheric warming are complex. Regional climate responses are expected as temperature, precipitation, soil moisture, and air mass characteristics change. Although the ability to accurately forecast such regional changes still is evolving, some

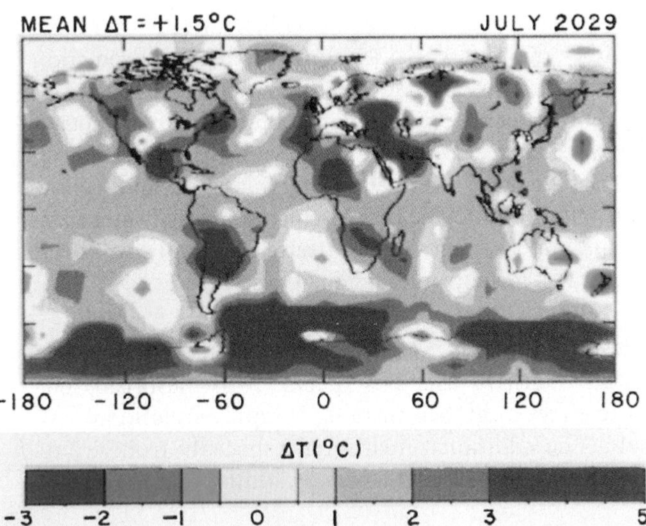

FIGURE 6.28 July 2029 temperature forecast. GISS general circulation model (GCM) forecasts for the year 2029 for a middle-case scenario. Actual temperature measurements for the last dozen years are similar to this middle-case forecast. [From J. Hansen et al., "Global Climate Changes as Forecast by Goddard Institute for Space Studies Three-Dimensional Model," *Journal of Geophysical Research* 43, no. 8 (August 1988): plate 6. Published by the American Geophysical Union.]

consequences of warming are forecast and in several regions already are underway.

Effects on World Food Supply and the Biosphere. Modern single-crop (monoculture) agriculture is delicate and more susceptible to changes in air temperature, water availability, irrigation, and soil chemistry than traditional multicrop agriculture. Available soil moisture is projected to be at least 10% less throughout the midlatitudes 30 years from now as hot, dry weather increases. To maintain managed ecosystems more energy, water, and resources will have to be expended.

Crop patterns, as well as natural habitats of plants and animals, will shift to maintain preferred temperatures. In the midlatitudes, climatic regions could shift poleward by 150 to 550 km (90 to 350 mi) during this twenty-first century, according to climate models. Canadian agricultural harvests would improve according to projections. The possibility exists that billions of dollars in agricultural losses in one region could be countered by corresponding gains in another. The greatest food security risks are with those people who are poor, isolated, landless, or living near sea level along vulnerable coastlines and on islands.

In this century, biosphere models predict that a global average of 30% (varying regionally from 15% to 65%) of the present forest cover will undergo major species redistribution—greatest at high latitudes and least in the tropics. Scientists predict that more than half of Earth's boreal (northern) forests could disappear due to climate change-induced stress on forest ecosystems.

Many plant species are already "on the move" to more favorable locations. Land dwellers also must adapt to changing forage. Several insect species have been found at unexpected higher latitudes, while disappearing from their expected lower latitude habitats. Studies of birds and mammals suggest that the future mix of species will change greatly—some lowland communities will experience a 40% turnover in species mix. Dr. A.T. Peterson, University of Kansas researcher in climate change and biodiversity, stated that no one had suspected that the turnover would be so great.

Studies completed by the U.S. Institute of Public Health, the Netherlands Environmental Protection Agency, the Development Research Centre in India, and the World Health Organization point to health impacts of climate change. Populations previously unaffected by malaria, dengue fever, lymphatic filariasis, and yellow fever (all mosquito vectored), schistosomiasis (water snail vector), and sleeping sickness (tsetse fly vector), will be at greater risk in subtropical and midlatitude areas.

Melting Glaciers, Melting Ice Sheets, and Sea Level. Perhaps the most pervasive (widespread) climatic effect of global warming is rapid escalation of ice melt. The additional meltwater, especially from continental ice masses and glaciers, is adding to a rise in sea level worldwide. Mount Kilimanjaro in Africa, portions of the South American Andes, and the Himalayas will very likely lose most of their glacial ice within the next two decades or less, affecting local water resources. NASA scientists determined that Greenland's ice sheet is thinning by about 1 m per year. See http://visibleearth.nasa.gov/cgi-bin/viewrecord?2172 for a map by the Airborne Topographic Mapper of Greenland's ice loss.

Glacial ice continues its retreat in Alaska. Chapter 14 presents a 2002 study of ice loss from mountain glaciers in Alaska. Scientists found that Alaska's ice losses contributed to sea-level rise, at nearly twice the volume of the entire Greenland Ice Sheet during the 1995–2001 period—Alaskan glacier melt contributed almost 10% of the observed rate of global sea-level rise (see News Report 14.1 for a mass balance graph and Alaska map).

Surrounding the margins of Antarctica, and constituting about 11% of its surface area, are numerous ice shelves, especially where sheltering inlets or bays exist. Covering many thousands of square kilometers, these ice shelves extend over the sea while still attached to continental ice. The loss of these ice shelves does not significantly raise sea level, for they already displace seawater. The concern is for the possible surge of grounded continental ice that the ice shelves hold back from the sea.

Although ice shelves constantly break up to produce icebergs, some large sections have recently broken free. In 1998 an iceberg the size of Delaware (150 km by 35 km, or 90 by 20 mi) broke off the Ronne Ice Shelf, southeast of the Antarctic Peninsula. In March 2000 an iceberg tagged B-15 broke off the Ross Ice Shelf (some 90° longitude west of the Antarctic Peninsula), measuring twice the area of Delaware, 300 km by 40 km, or 190 mi by 25 mi (Figure 6.29a).

Since 1993, six ice shelves have disintegrated in Antarctica. About 8000 km^2 (3090 mi^2) of ice shelf are gone, changing maps, freeing up islands to circumnavigation, and creating thousands of icebergs (Figure 6.29b). The Larsen Ice Shelf, along the east coast of the Antarctic Peninsula, has been retreating slowly for years. Larsen-A suddenly disintegrated in 1995. In only 35 days in early 2002, Larsen-B collapsed into icebergs (Figure 6.29c). (See the multi-agency National Ice Center at http://www.natice.noaa.gov/home.htm for an update.)

This loss of Antarctic ice is a direct result of the 2.5 C° (4.5 F°) temperature increase in the region in the last 50 years. As researchers summarized,

> This breakup [of Larsen-A] followed a period of steady retreat that coincided with a regional trend of atmospheric warming. The observations imply that after an ice shelf retreats beyond a critical limit, it may collapse rapidly as a result of disturbed mass balance.*

*H. Rott, P. Skvarca, and T. Nagler, "Rapid collapse of northern Larsen Ice Shelf, Antarctica," *Science* 271 (February 9, 1996): 788.

(a)

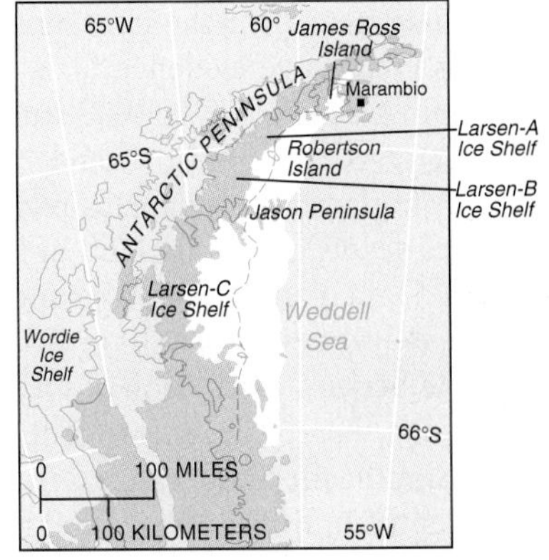

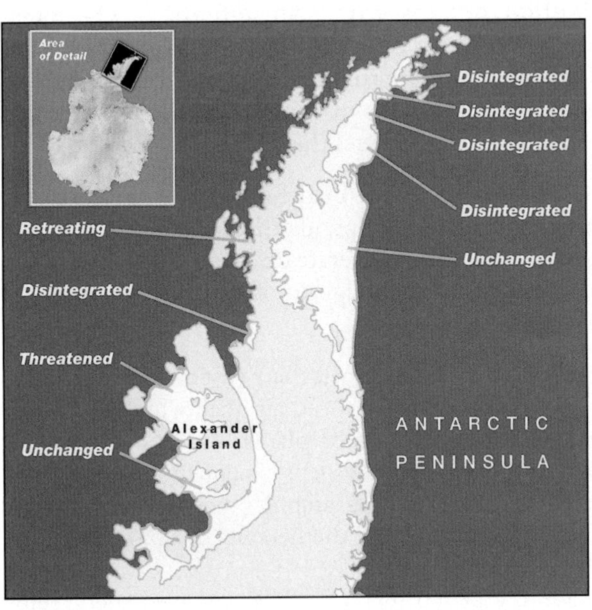

(b)

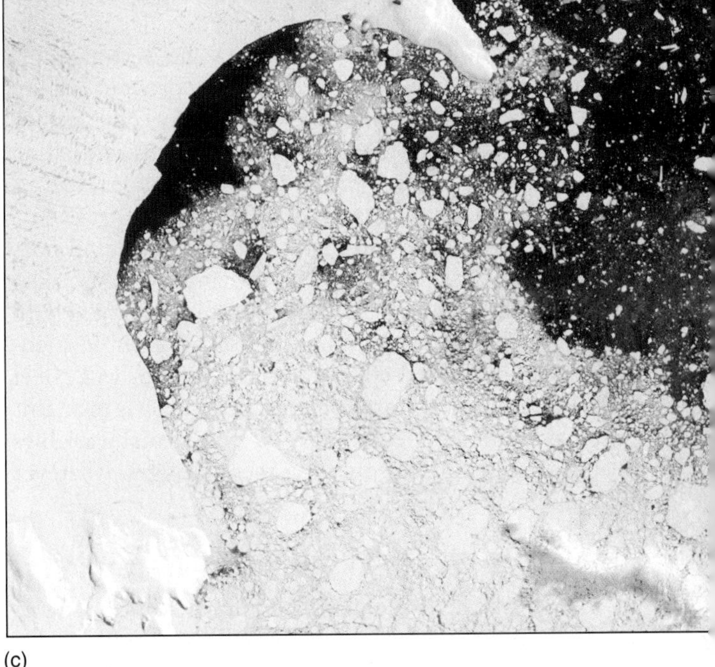

(c)

FIGURE 6.29 Disintegrating ice shelves along the Antarctic peninsula.
(a) March 2000 image of iceberg B-15 as it breaks off the Ross Ice Shelf, near Roosevelt Island, Antarctica. (b) Map of the Antarctic Peninsula showing the status of ice shelves between 1995–1998. Note the location of the Larsen ice shelves. (c) Continued disintegration and retreat of Larsen-B is evident in February, 2000. [(a) *Terra* MODIS sensor courtesy of MODIS Land Science Team, NASA. (b) British Antarctic Survey map. All rights reserved. (c) Satellite image courtesy of *Landsat-7* team, GSFC/NASA.]

In response to the increasing warmth, the Antarctic Peninsula is sporting new vegetation growth, previously not seen there.

In the north polar region, pack ice (floating sea ice and glacial ice) covers the Arctic Ocean. Since 1970 almost half of this ice pack has disappeared due to regional waming, averaging about 1.2 C° (2.2 F°) per decade during this time. According to research published in 2002, ice melt since 1978 is at a 9% per decade rate. This means that the Arctic Ocean could be ice free later this century. Scientists are tracking these cold meltwaters for they may disrupt warmer sea-surface temperatures in the Atlantic, such as deflecting the usual path of the Gulf Stream, thus dramtically changing expected weather and climate patterns.

A loss of polar ice mass, augmented by melting of alpine and mountain glaciers, will affect sea-level rise. The IPCC 2001 median forecast value of 0.48 m (18.9 in.) is two to four times the rate of increase over the last century. The IPCC assessment states that "between one-third to one-half of the existing mountain glacier mass could disappear over the next hundred years." Also, "... there is conclusive evidence for a worldwide recession of mountain glaciers. ... " A 2002 study by researchers at the University of Colorado, Boulder, analyzed new North American glacial-melt data, as related to the overall forecast. Sea-levels could rise 0.89 m (36 in.), or greater than the IPCC median case. For perspective, for every 0.30 m of rise, shorelines retreat 30 m (98 ft) inland.

The Scripps Institution of Oceanography, in La Jolla, California, has kept ocean temperature records since 1916. Significant temperature increases are being recorded to depths of over 300 m (1000 ft) as ocean temperature records are set. Even the warming of the ocean itself will contribute about 25% of sea-level rise, simply because of *thermal expansion* of the water. In addition, any change in ocean temperature has a profound effect on air mass characteristics and weather. A 2002 study by the Australian Institute of Marine Science found present ocean temperatures off Australia the warmest since the record began in 1870. This correlates with a massive bleaching and loss of 60% of the corals in the Great Barrier Reef.

A quick survey of world coastlines shows that even a moderate rise could bring change of unparalleled proportions. At stake are the river deltas, lowland coastal farming valleys, and low-lying mainland areas, all contending with high water, high tides, and higher storm surges. Particularly tragic social and economic consequences will affect small island states. Clearly, physical geography is in an important position to synthesize all the spatial variables needed in planning to cope with these changes, whatever their mildness or severity.

Political Action to Slow Global Warming

A product of the 1992 Earth Summit in Rio de Janeiro was the United Nations Framework Convention on Climate Change (FCCC). The leading body of the Convention is the *Conference of the Parties* (*COP*) operated by the countries that ratified the FCCC, 186 countries by 2000. Meetings were held in Berlin (*COP-1*, 1995) and Geneva (*COP-2*, 1996). These meetings set the stage for *COP-3* in Kyoto, Japan, in December 1997, where 10,000 participants adopted the *Kyoto Protocol* by consensus. The latest gatherings to refine agreements were *COP-6* held in the Hague in late 2000, *COP-7* in Marrakech, Morocco, in 2001, and *COP-8* in New Delhi in 2002. Seventeen national academies of science endorsed the Kyoto Protocol.

The Kyoto Protocol binds more-developed countries to a collective 5.2% reduction in greenhouse gas emissions as measured at 1990 levels for the period 2008 to 2012. Relative to the United States and its goal of a 7% cut below 1990 levels, the United States withdrew in 2001 from the climate treaty process and abandoned emission control goals, alone in its dissent. A 2001 National Research Council report affirmed conclusions about human-induced climate change, saying that the IPCC Third Assessment Report "... accurately reflects the current thinking of the scientific community on this issue" (NRC, *Climate Change Science, An Analysis of Some Key Questions*, Washington: National Academy Press, May 2001). (For updates on the status of the Kyoto Protocol, see http://www.unfccc.int/resource/kpstats.pdf.)

An interesting development finds individual U.S. states adopting their own strategies to combat global warming, independent of the federal government's actions. The list includes, among others, California, Texas, Minnesota, Nebraska, New Jersey, North Carolina, Oregon, Massachusetts, Georgia, and Wisconsin. Initiatives focus on reducing emissions of greenhouse gases by encouraging efficiency in the transportation sector and the generation of more electricity from solar and wind energy sources.

Mitigation Actions with "No Regrets." The Intergovernmental Panel on Climate Change (IPCC) declared that "no regrets" opportunities to reduce carbon dioxide emissions are available in most countries. The IPCC Working Group III defines this as follows:

> No regrets options are by definition greenhouse gas emissions reduction options that have negative net costs, because they generate direct and indirect benefit that are large enough to offset the costs of implementing the options.

Benefits that equal or exceed their cost to society include reduced energy cost, improved air quality and health, reduced need for new power plants, reduction in tanker spills and oil imports, and deployment of renewable and sustainable energy sources, among others. This holds true without even considering the benefits of slowing the rate of climate change.

One key to "no regrets" is the untapped energy-efficiency potential. In Europe, scientists determined that carbon emissions could be reduced to less than half the 1990 level by 2030, at a negative cost (reported by the International Project for Sustainable Energy Paths, http://www.ipsep.org/). In the United States, five Department of Energy national laboratories (Oak Ridge, Lawrence Berkeley, Pacific Northwest, National Renewable Energy, and Argonne) reported that the United States can meet the Kyoto carbon emission reduction targets with negative overall costs (cash benefit savings) ranging from –$7 to –34 billion. (For more, see Working Group III, *Climate Change 2001, Mitigation*, London: Cambridge University Press, 2001, pp. 21, 474–76, and 506–07.)

Summary and Review—Global Climate Systems

Define climate and climatology and *explain* the difference between climate and weather.

Climate is dynamic, not fixed. **Climate** is a synthesis of weather phenomena at many scales, from planetary to local, in contrast to weather, which is the condition of the atmosphere at any given time and place. Earth experiences a wide variety of climatic conditions that can be grouped by general similarities into climatic regions. **Climatology** is the study of climate and attempts to discern similar weather statistics and identify **climatic regions**.

climate (p. 179)
climatology (p. 180)
climatic regions (p. 180)

1. Define climate and compare it with weather. What is climatology?
2. Explain how a climatic region synthesizes climate statistics.
3. How does the El Niño phenomenon produce the largest interannual variability in climate? What are some of the changes and effects that occur worldwide?

Review the role of temperature, precipitation, air pressure, air-mass patterns, and sea-surface temperatures used to establish climatic regions.

Climatic inputs include insolation (pattern of solar energy in the Earth-atmosphere environment); temperature (sensible heat energy content of the air); precipitation (rain, sleet, snow, and hail; the supply of moisture); air pressure (varying patterns of atmospheric density); and air masses (regional-sized homogeneous units of air). Climate is the basic element in ecosystems—self-regulating communities of plants and animals that thrive in specific environments.

4. How do radiation receipts, temperature, air pressure inputs, and precipitation patterns interact to produce climate types? Give an example from a humid environment and one from an arid environment.
5. Evaluate the relationships among a climatic region, ecosystem, and biome.

Review the development of climate classification systems and *compare* genetic and empirical systems as ways of classifying climate.

Classification is the process of ordering or grouping data in related categories. A **genetic classification** is based on causative factors, such as the interaction of air masses. An **empirical classification** is one based on statistical data, such as temperature or precipitation. This text analyzes climate using aspects of both approaches, with a map based on climatological elements.

classification (p. 182)
genetic classification (p. 182)
empirical classification (p. 182)

6. What are the differences between a genetic and an empirical classification system?
7. What are some of the climatological elements used in classifying climates?

Describe the principal climate classification categories other than deserts and *locate* these regions on a world map.

Here we focus on temperature and precipitation measures. Keep in mind these are measurable results produced by interacting elements of weather and climate. These data are plotted on a **climograph** to display the characteristics of the climate.

There are six basic climate categories. Temperature and precipitation considerations form the basis of five climate categories and their regional types:

- Tropical (equatorial and tropical latitudes)
  - rain forest (rainy all year)
  - monsoon (6 to 12 months rainy)
  - savanna (less than 6 months rainy)
- Mesothermal (mild winter)
  - humid subtropical (hot summers)
  - marine west coast (warm to cool summers)
  - Mediterranean (dry summers)
- Microthermal (mid- and high-latitude, cold winter)
  - humid continental (hot to warm summers)
  - subarctic regions (cool summers to very cold winters)
- Polar (polar and high latitudes)
  - tundra (high latitude or at altitude)
  - ice caps and ice sheets (perpetually frozen)
- Highland (compared to lowlands at the same latitude, highlands have lower temperatures—recall the normal lapse rate)

Only one climate category is based on moisture efficiency as well as temperature:

- Deserts (permanent moisture deficits)
  - arid deserts
  - semiarid steppes

climograph (p. 183)

8. List and discuss each of the principal climate categories. In which one of these general types do you live? Which category is the only type associated with the annual distribution and amount of precipitation?
9. What is a climograph, and how is it used to display climatic information?
10. Which of the major climate types occupies the most land and ocean area on Earth?
11. Characterize the tropical climates in terms of temperature, moisture, and location.
12. Using Africa's tropical climates as an example, characterize the climates produced by the seasonal shifting of the ITCZ with the high Sun.
13. Mesothermal (subtropical and midlatitude, mild winter) climates occupy the second largest portion of Earth's entire surface. Describe their temperature, moisture, and precipitation characteristics.
14. Explain the distribution of the *humid subtropical hot-summer* and *Mediterranean dry-summer* climates at similar latitudes and the difference in precipitation patterns

between the two types. Describe the difference in vegetation associated with these two climate types.

15. Which climates are characteristic of the Asian monsoon region?
16. Explain how a *marine west coast* climate type can occur in the Appalachian region of the eastern United States.
17. What role do offshore ocean currents play in the distribution of the *marine west coast* climates? What type of fog is formed in these regions?
18. Discuss the climatic conditions for the coldest places on Earth outside the poles.

• *Explain* the precipitation and moisture efficiency criteria used to determine the arid and semiarid desert climates and *locate* them on a world map.

The dry and semiarid climates are described by precipitation rather than temperature. Dry climates are the world's arid deserts and semiarid regions, with their unique plants, animals, and physical features. The *dry arid* and *semiarid* climates occupy more than 35% of Earth's land area, clearly the most extensive climate over land.

Major subdivisions are *arid deserts* in tropical and midlatitude areas (precipitation—natural water supply—less than one-half of natural water demand) and *semiarid steppes* in tropical and midlatitude areas (precipitation more than one-half of natural water demand).

19. In general terms, what are the differences among the four desert classifications? How are moisture and temperature distributions used to differentiate these subtypes?
20. Relative to the distribution of dry climates, describe at least three locations where they occur across the globe and the reasons for their presence in these locations.

• *Outline* future climate patterns from forecasts presented and *explain* the causes and potential consequences.

Various activities of present-day society are producing climatic changes, particularly a global warming trend. The highest average annual temperatures since instrumental measurements began dominated the 1980s and 1990s. These conditions were confirmed by the 2001 *Third Assessment Report* from the Intergovernmental Panel on Climate Change. IPCC predicted future surface-temperature response to a doubling of carbon dioxide with a range of increase from 1.4 C° (2.5 F°) to 5.8 C° (10.4 F°) between the present and 2100. Natural climatic variability over the span of Earth's history is the subject of **paleoclimatology**. A **general circulation model (GCM)** is used to forecast climate patterns and is evolving to greater capability and accuracy than in the past. People and their political institutions can use GCM forecasts to form policies aimed at reducing unwanted climate change.

paleoclimatology (p. 210)
general circulation model (GCM) (p. 213)

21. Explain climate forecasts. How do general circulation models (GCMs) produce such forecasts?
22. Describe the potential climatic effects of global warming on polar and high-latitude regions. What are the implications of these climatic changes for persons living at lower latitudes?
23. How is climatic change affecting agricultural production? Natural environments? Forests? The possible spread of disease?
24. What are the present actions being taken to delay the effects of global climate change? What is the Kyoto Protocol?

Career Link 6.1

Tracy Smith, Research Meteorologist

After two decades at the Forecast Systems Laboratory (FSL), Tracy Smith is still enthusiastic about her work. Tracy asserted, "Weather is exciting! There is such satisfaction in developing a model and forecasting something and then having it happen—predict it and then find it. We are challenged to improve weather forecasting. If we can add a few minutes to severe weather warning times and narrow the focus of the warning area, or improve aviation safety, lives can be saved."

FSL developed the Advanced Weather Interactive Processing System (AWIPS) to assist forecasters across the country (pictured in News Report 5.3, Figure 2c and d). Tracy works on computer models to improve the "data ingest" of all the information acquired from a growing list of new technologies. As part of her Rapid Update Cycle (RUC) effort, she is working to increase the update frequency, accuracy, and resolution of forecasting models. Tracy said, "This is one of the reasons atmospheric science is such a wide-open field as we attack these challenges."

When I asked about her interest in weather, she responded with a spirited laugh, "I didn't really decide to be a meteorologist until I was 15 or 16 years old!" I reminded her that this was an interesting thing for a teenager to say. She said, "When I was four or five my brother and sister brought a schoolbook home called *Hurricanes and Twisters*. I read it over and over." She was fascinated by the sky and, living in northeastern Ohio, she remembers those "big booming nocturnal thunderstorms in the summer."

When she was seven years old, the great Palm Sunday tornado outbreak of 1965 hit Ohio at Pittsfield (with an F5) and nearby Grafton (with an F4), where she visited her grandparents. She said, "I remember that night, waking up and hearing this great storm outside." Tracy's grandparents' house was slightly damaged but the homes around them were leveled—she walked along the street and saw the devastation. Her childhood

(continued)

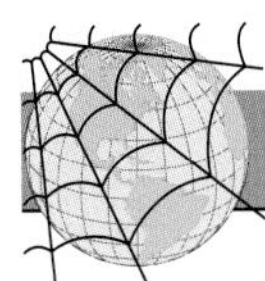

NetWork

The *Elemental Geosystems Home Page* provides on-line resources for this chapter on the World Wide Web. Once on the Home Page, click on this textbook, scroll the Table of Contents menu, select this chapter, and click "Begin." You will find self-tests that are graded, review exercises, specific updates for items in the chapter, and in "Destinations" many links to interesting related pathways on the Internet. *Elemental Geosystems* is found at **http://www.prenhall.com/christopherson**.

Critical Thinking

A. The text asked that you find the climate conditions for your campus and your birthplace and locate these two places on Figures 6.3 and 6.4. Briefly describe your pursuit of the information and the sources you used: library, Internet, teacher, phone calls to state and provincial climatologists. Now, refer to Appendix C to refine your assessment of climate for these two locations. Briefly show how you worked through the Köppen climatic categories to find the climate classification for your two cities.

B. On the *Elemental Geosystems Home Page* and in Focus Study 6.1 there are URL references for El Niño and La Niña. Sample three or four sources to determine the up-to-date status of the El Niño phenomena at present. How are current conditions different from the record El Niño event in 1997–1998? What is the current strength or weakness of any La Niña event? What Internet links were most helpful to you in completing this status report?

C. The chapter discusses the disintegration of six ice shelves on the Antarctic Peninsula and several huge icebergs that are now adrift. Go to the British Antarctic Survey (**http://www.nbs.ac.uk/**), National Ice Center (**http://www.natice.noaa.gov/**), or other source and determine the present status of the ice shelves around Antarctica. Have any changes occurred since the disintegration events described in this chapter took place? What related information and images did you find in your search?

Career Link 6.1 *(continued)*

fascination and experiences blossomed into a great career working with the atmospheric sciences she loves. Tracy admits to sometimes using vacation time to storm chase!

Tracy graduated from Bowling Green University, Ohio, with a degree in geography and a minor in math. She then earned her Master's degree in atmospheric sciences at Colorado State University, Fort Collins. For her thesis, she studied the radiation budget at the top of the atmosphere over the Indian monsoon region, using satellite data. Her committee was made up of both atmospheric science and geography professors. "Geography is important to atmospheric science because we are dealing with spatial elements of the atmosphere, hydrologic cycle, and topography. And, because of the impact on people, geography brings in the human aspect," she states.

Tracy participates in the daily weather briefings at FSL and once a month conducts the briefing. She says it is important to practice communication skills by using these weather briefings in order to be able to explain the forecast to others.

As to the future: "We want to refine the system, gain higher resolution, integrate all new data sources as they develop, and improve accuracy. We are just scratching the surface to get all this operational. The speed and capability of these computers are increasing so fast that this will be the next revolution in forecasting for us. Improving scale and accuracy are our ongoing challenges." And with her infectious laugh she asserts, "I like my job! I am paid to learn and study things and follow my instincts to solve big problems. What better way to spend your day than trying to understand the world around you while helping society!" Tracy's enthusiasm is obvious as she demonstrates the AWIPS workstation to me (Figure 1).

FIGURE 1 Tracy Smith, Research Meteorologist. Tracy works at the Forecast Systems Laboratory, National Weather Service, NOAA. Here she shows this author the workings of an AWIPS workstation, discussed in Chapter 5. [Photo by Bobbé Christopherson.]

El Niño rains and heavy snow in the Rockies produced record-high discharge rates on the Colorado River in 1983. Here, spillways at Hoover Dam operate for the first time in 40 years. The most-regulated river in the U.S. flooded housing subdivisions and resort facilities downstream from these releases. Davis Dam regulates Hoover's discharge, however on this day Davis came within 30 cm (1 ft) of being topped. [Photo by author.]

7 Water Resources

Key Learning Concepts

By knowing and understanding the key learning concepts in this chapter, you should be able to:

- *Illustrate* the hydrologic cycle with a simple sketch and *label* it with definitions for each water pathway.
- *Relate* the importance of the water-budget concept to an understanding of the hydrologic cycle, water resources, and soil moisture for a specific location.
- *Construct* the water-balance equation as a way of accounting for the expenditures of water supply and *define* each of the components in the equation and its specific operation.
- *Describe* the nature of groundwater and *define* the elements of the groundwater environment.
- *Identify* critical aspects of freshwater supplies for the future and *cite* specific issues related to sectors of use, regions and countries, and potential remedies for any shortfalls.

The physical reality of life is defined by water. It is the essence of our existence and is therefore the most critical resource supplied by Earth systems. Our bodies are composed of about 70% water by weight, as are plants and animals. We use water to cook, bathe, wash clothing, and dilute our wastes. We water small gardens and vast agricultural tracts. Most industrial processes would be impossible without water. Fortunately, water is a renewable resource, constantly cycling through the environment, endlessly renewed.

One billion people lacked access to safe water in 2002; some 2.4 billion lacked adequate sanitary facilities. During the first half of this century, water availability per person will drop by 74%, as population increases and adequate water quality decreases.

Meanwhile water use is forecasted to increase by 40% over this same time. Author Peter Gleick of the Pacific Institute summarizes:

> The overall economic and social benefits of meeting basic water requirements far outweigh any reasonable assessment of the costs of providing those needs.... Water-related diseases cost society on the order of $125 billion per year.... Yet the cost of providing new infrastructure needs for all major urban water sectors has been estimated at around $25 to $50 billion per year. While these costs are far below the costs of failing to meet these needs, they are two to three times the average rate of spending for water during the 1980s and 1990s.*

In the previous two chapters we saw how the exchanges of energy between water and the atmosphere drive Earth's weather and climate systems. The flow of water links the atmosphere, ocean, land, and living things through exchanges of energy and matter. In particular, the energy and moisture exchange between plants and the atmosphere is important to the status of the climate system.

In this chapter: The hydrologic cycle and global water balance give us a model for understanding the global plumbing system. Water spends time in the ocean, in the air, on the surface, and underground as groundwater. Water availability to plants from precipitation and from the soil is critical to water-resource issues.

We look at water resources using a water-budget approach—similar in many ways to a money budget—in which we examine water "receipts" and "expenses" at specific locations. Precipitation provides the principal receipt of moisture, whereas evaporation and plant transpiration are the principal expenditures. This budget approach can be applied at any scale, from a small garden, to a farm, to a regional landscape.

Water is not always naturally available where and when we want it. Consequently, we rearrange surface-water resources to suit our needs. We drill wells, build cisterns and reservoirs, and dam and divert streams to redirect water either spatially (geographically, from one area to another) or temporally (over time, from one part of the calendar to another). All of this activity constitutes water-resource management.

This chapter concludes by considering the quantity and quality of the water we withdraw and consume for irrigation, industrial, and municipal uses—our specific water supply. Adequate water supplies in terms of quantity and quality loom as *the resource issue* for many parts of the world in this century. Ismail Serageldin, chair of the World Water Commission, is quoted bluntly in a recent book on water issues: "... the wars of the twenty-first century will be fought over water.... Water is the most critical issue facing human development."†

*Peter Gleick, *The World's Water 2000–2001* (Washington, D.C.: Island Press, 2000), p. 13.

†M. De Villiers, *Water, The Fate of Our Most Precious Resource* (New York: Houghton Mifflin Co., 2000), p. 13–4.

The Hydrologic Cycle

ANIMATION **Global Warming, climate change**

Vast currents of water, water vapor, ice, and energy are flowing about us continuously in an elaborate, open, global plumbing system. Together they form the **hydrologic cycle**, which has operated for billions of years, from the lower atmosphere down to several kilometers beneath Earth's surface. The cycle involves the circulation and transformation of water throughout Earth's atmosphere, hydrosphere, cryosphere, lithosphere, and biosphere. (See NASA's Global Hydrology and Climate Center, http://www.ghcc.msfc.nasa.gov/.)

A Hydrologic Cycle Model

Figure 7.1 is a simplified model of this complex system. Let's use the ocean as a starting point for our discussion. More than 97% of Earth's water is in the ocean, and here most evaporation and precipitation occurs. We can trace 86% of all evaporation to the ocean. The other 14% evaporates from the land, including water moving from the soil into plant roots and passing through their leaves (a process called *transpiration*, described later in this chapter).

In the figure, you can see that of the ocean's evaporated 86%, 66% combines with 12% advected from the land to produce 78% of all precipitation that falls back into the ocean. The remaining 20% of moisture evaporated from the ocean, plus 2% of land-derived moisture, produces the 22% of all precipitation that falls over land. Clearly, the bulk of continental precipitation comes from the oceanic portion of the cycle.

Surface Water

Precipitation that reaches Earth's surface follows two basic pathways—overland flow and movement into the ground. Water soaks in through **infiltration**, or penetration of the soil surface. It then permeates soil or rock through a vertical movement called **percolation**.

The atmospheric advection of water vapor from sea to land and land to sea at the top of Figure 7.1 appears to be unbalanced: 20% moving inland but only 12% moving out to sea. However, this exchange is balanced by 8% as runoff that flows from land to sea. Most of this runoff—about 95%—comes from surface waters washing across land as *overland flow* and *streamflow*. Only 5% of this runoff is attributable to slow-moving subsurface groundwater. These percentages indicate that the small amount of water in rivers and streams is very dynamic, whereas the large quantity of subsurface groundwater is sluggish in comparison and comprises only a small portion of runoff.

The residence time of water in any part of the hydrologic cycle determines its relative importance in affecting Earth's climates. The short time spent by water in transit through the atmosphere (10-day average) is reflected in temporary fluctuations in regional weather patterns.

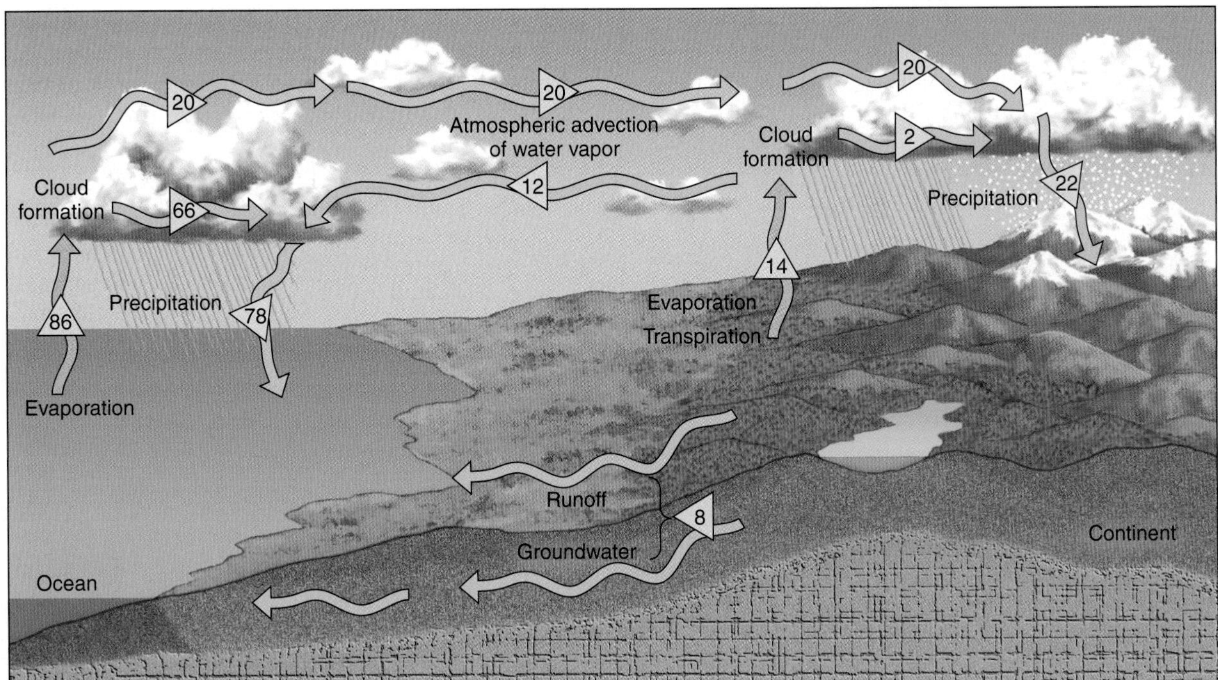

FIGURE 7.1 The hydrologic cycle model.
The hydrologic cycle model with percentages and directional arrows denoting flow paths. Global average values are shown as percentages; note that all evaporation (86% + 14% = 100%) equals all precipitation (78% + 22% = 100%). Locally, various parts of the cycle will vary, creating imbalances.

Long residence times, such as the 3000–10,000 years in deep-ocean circulations, groundwater aquifers, and glacial ice, act to moderate temperatures and climates. These slower parts of the cycle work as a "system memory"; heat is stored and released to buffer change.

To observe the hydrologic cycle and its related energy budgets, scientists established the Global Energy and Water Cycle Experiment (GEWEX). As part of this effort, the GEWEX Continental-Scale International Project is underway (GCIP, see http://www.ogp.noaa.gov/mpe/gapp/gcip/index.htm), a multiscale hydrometeorological investigation of the 13.2 million km^2 (5.1 million mi^2) Mississippi River Basin. The goal is to correlate seasonal, annual, and interannual water resource availability with the climate prediction system. Now that you are acquainted with the hydrologic cycle and surface water, let's examine the concept of the water budget as a method of assessing water resources.

Soil-Water-Budget Concept

A **soil-water budget** can be established for any area of Earth's surface—a continent, nation, region, or field—by calculating the total precipitation input and the total water output. The water budget is a portrait of the hydrologic cycle at a specific site or area for any period of time, including estimation of streamflow. C. W. Thornthwaite (1899–1963), a geographer pioneering in applied water-resource analysis, worked with others to develop a water-balance methodology. They related water-balance concepts to geographical problems, especially to accurately determining irrigation quantity and timing.

The Soil-Water-Balance Equation

The water balance allows us to examine specific aspects of the hydrologic cycle at Earth's surface. Think of a soil-water budget as a money budget: precipitation income must be balanced against expenditures of evaporation, transpiration, and runoff. Soil-moisture storage acts as a savings account, accepting deposits and giving up withdrawals of water. To understand the method we must first examine the terms and concepts in the water-balance equation in Figure 7.2. Refer to the equation in the figure as you read through the next series of headings.

Precipitation (PRECIP) Input. The supply of moisture to Earth's surface is **precipitation** (PRECIP)—rain, sleet, snow, and hail (recall from the hydrologic cycle that 22% of Earth's PRECIP falls on land). In some climates, additional deposits of water at Earth's surface occur as the result of dew and fog. Precipitation is measured with the **rain gauge**, which receives precipitation through an open top so that water can be measured by depth, weight, or volume. Wind can cause an undercatch because the drops or snowflakes are not falling vertically. For example, a wind of 37 kmph (23 mph) produces an undercatch as great as 40%, meaning that an actual 1 in. rainfall might gauge at only 0.6 in. The windshields you see above the gauge's opening reduce this error by catching raindrops that arrive at an angle (Figure 7.3).

According to the World Meteorological Organization, more than 40,000 weather-monitoring stations are operating worldwide, with over 100,000 places measuring precipitation. Precipitation patterns over the United States and Canada are shown in Figure 7.4. A map displaying the pattern of world precipitation is included in the previous chapter as Figure 6.2.

Actual Evapotranspiration (ACTET). **Evaporation** is the net movement of water molecules away from a wet surface into air that is less than saturated. **Transpiration** in plants is the outward movement of water through small

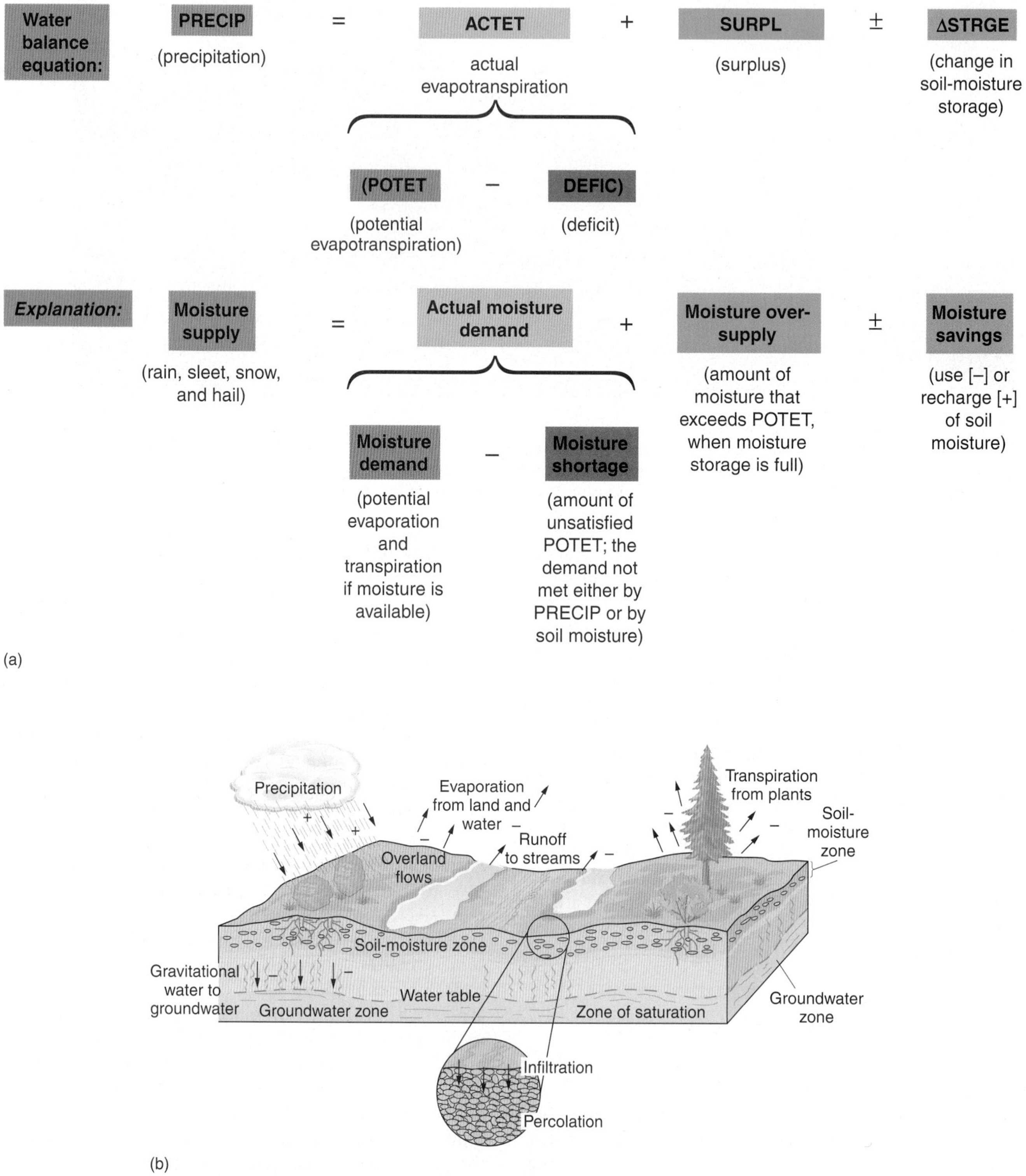

Water balance equation: PRECIP (precipitation) = ACTET (actual evapotranspiration) + SURPL (surplus) ± ΔSTRGE (change in soil-moisture storage)

ACTET = (POTET (potential evapotranspiration) − DEFIC) (deficit)

Explanation: Moisture supply (rain, sleet, snow, and hail) = Actual moisture demand + Moisture over-supply (amount of moisture that exceeds POTET, when moisture storage is full) ± Moisture savings (use [–] or recharge [+] of soil moisture)

Actual moisture demand = Moisture demand (potential evaporation and transpiration if moisture is available) − Moisture shortage (amount of unsatisfied POTET; the demand not met either by PRECIP or by soil moisture)

(a)

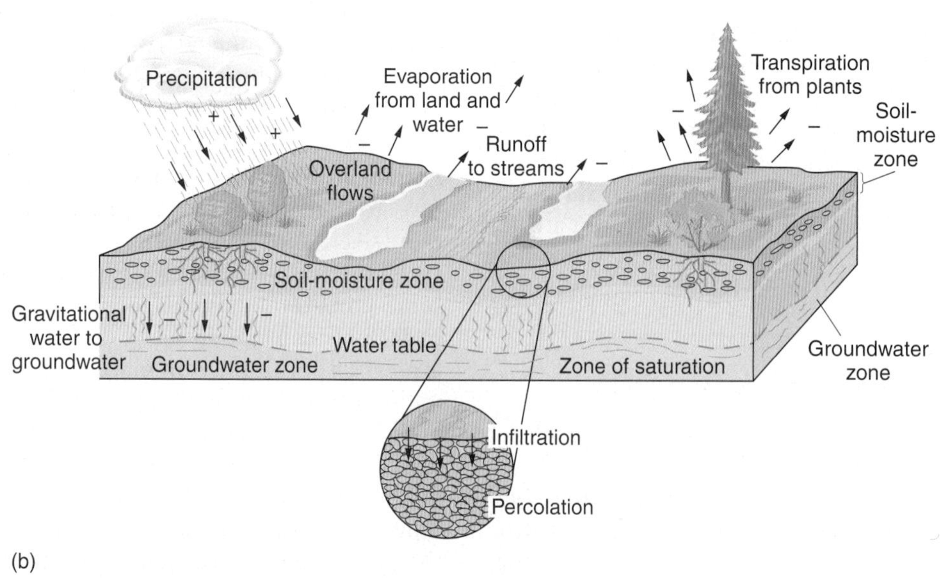

(b)

FIGURE 7.2 The water-balance equation and surface-moisture environment.
(a) The water-balance equation explained. The outputs (components to the right of the equal sign) are an accounting of the precipitation input (left of equal sign). (b) The principal pathways for precipitation through the soil-moisture environment include interception by plants, throughfall to the ground, collection on the surface, evapotranspiration, and water movement through the soil to groundwater.

FIGURE 7.3 A rain gauge.
A standard rain gauge is cylindrical. A funnel guides water into a bucket that rests on an electronic weighing device. The gauge is designed to minimize evaporation, which would cause low readings. The wind shield around the top of the gauge minimizes the undercatch produced by wind. [Photo by Bobbé Christopherson.]

openings (stomata) in the underside of leaves. Both evaporation and transpiration are water-balance outputs, and their rates directly respond to temperature and humidity. In addition, transpiration rates are partially controlled by the plants themselves as they conserve or release water with control cells around their stomata. Transpired quantities can be significant: On a hot day, a single tree may transpire hundreds of liters of water; a forest, millions of liters. Evaporation and transpiration are combined into one term—**evapotranspiration** (14% of evaporation and transpiration occurs from land and plants).

Potential evapotranspiration (POTET) is the water that would evaporate and transpire under optimum moisture conditions (adequate precipitation and full soil-moisture capacity). Filling a bowl with water and letting it evaporate illustrates this concept: When the bowl becomes dry, is there still an evaporation demand? A demand is present, of course, even when the bowl is dry.

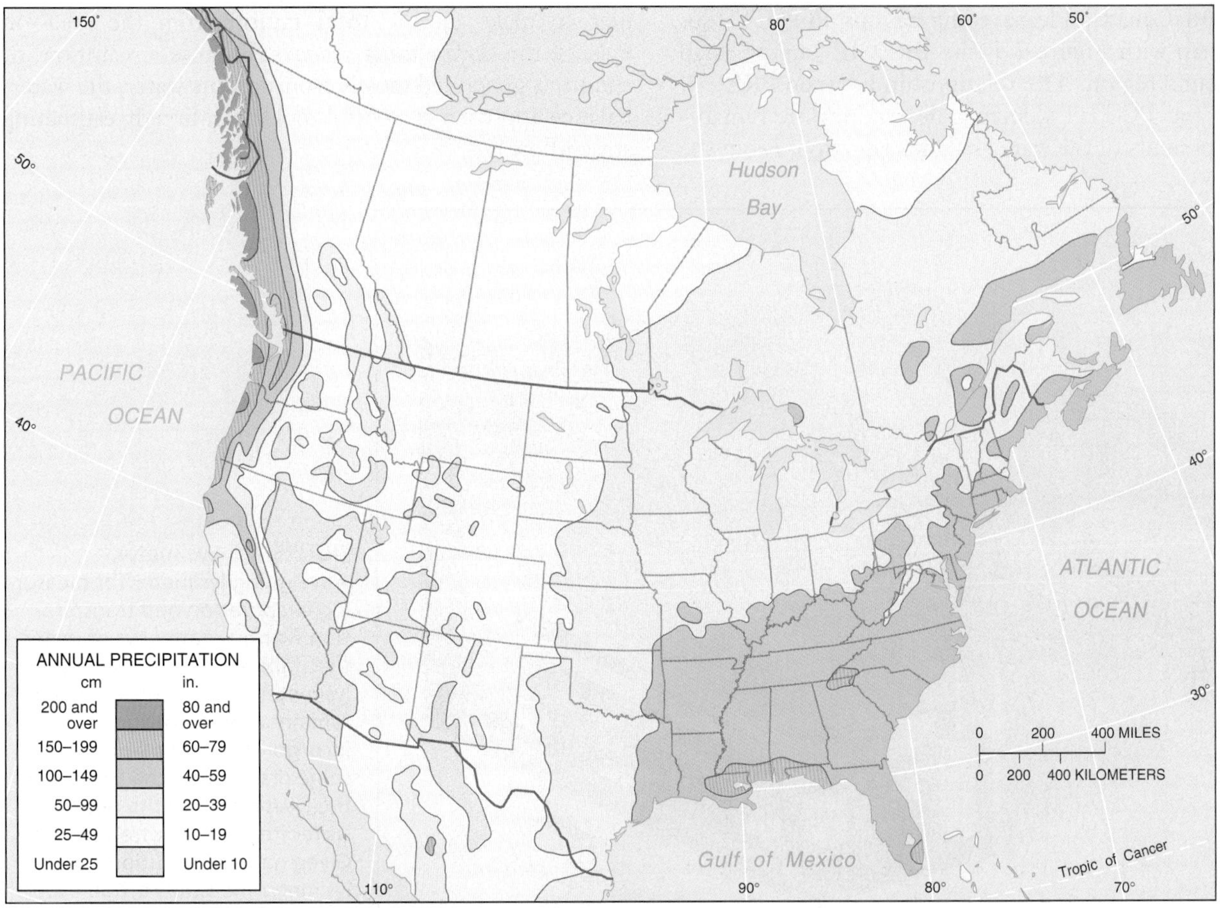

FIGURE 7.4 Precipitation (water supply, PRECIP) in North America.
Annual precipitation (water supply) in the United States and Canada. Compare to Figure 7.6, p. 227. [Adapted from U.S. National Weather Service, U.S. Department of Agriculture, and Environment Canada.]

Therefore, the amount of water that would evaporate and/or transpire if water always was available is the POTET, or the amount that would evaporate from the bowl if it was constantly supplied with water.

Determining POTET. The easiest method for measuring POTET employs an **evaporation pan**, or *evaporimeter*. As evaporation occurs, water in measured amounts is automatically replaced in the pan. Mesh screens over the pan protect against measurement errors created by wind, which accelerates evaporation.

A **lysimeter** is a relatively elaborate device for measuring POTET (Figure 7.5). It is a buried tank, open at the surface and approximately a cubic meter in size. Using an actual portion of a field, a lysimeter isolates soil, subsoil, and plants so that the moisture moving through the sampled area can be measured. Lysimeters and evaporation pans are somewhat limited in availability across North America but provide a database with which to develop ways of calculating POTET. This means that methods of estimating POTET based on meteorological data are widely used and easily implemented for regional applications because of the ready availability of precipitation and temperature data.

Figure 7.6 presents POTET values derived by a method that Thornthwaite developed for the United States and Canada. Please compare this POTET (demand) map with Figure 7.4, the PRECIP (supply) map for the same region. The relationship between PRECIP supplies and POTET demands determines the remaining components of the water-balance equation. From the two maps, can you identify regions where PRECIP is higher than POTET? (Eastern United States.) Where POTET is higher than PRECIP? (Southwestern United States.) Where you live, is the water demand usually met or exceeded by the precipitation supply? Or does your area experience a natural water shortage? How might you find out?

Deficit. POTET can be satisfied by PRECIP or by moisture derived from the soil, or it can be met artificially through irrigation. If these sources are inadequate, the location experiences a moisture shortage. The unsatisfied moisture demand is recorded as a **deficit** (DEFIC). By subtracting this deficit from the POTET demand, we determine the **actual evapotranspiration**, or ACTET, that takes place. Under ideal conditions the potential and actual evapotranspiration rates are about the same, so that plants do not experience a water shortage or prolonged deficits that would lead to drought.

Surplus. If POTET is satisfied and soil moisture is full, then additional water input becomes **surplus** (SURPL), or water oversupply. This excess water may sit on the surface or flow through the soil to groundwater storage. Surplus water that flows across the surface toward stream channels is termed *overland flow*. Such flow combines with other precipitation and subsurface flows into river channels to make up the **total runoff** from the area—in essence the *surface water resource*. Because streamflow, or runoff, is generated mostly from surplus water, the water-balance approach is a useful tool for indirectly estimating streamflow.

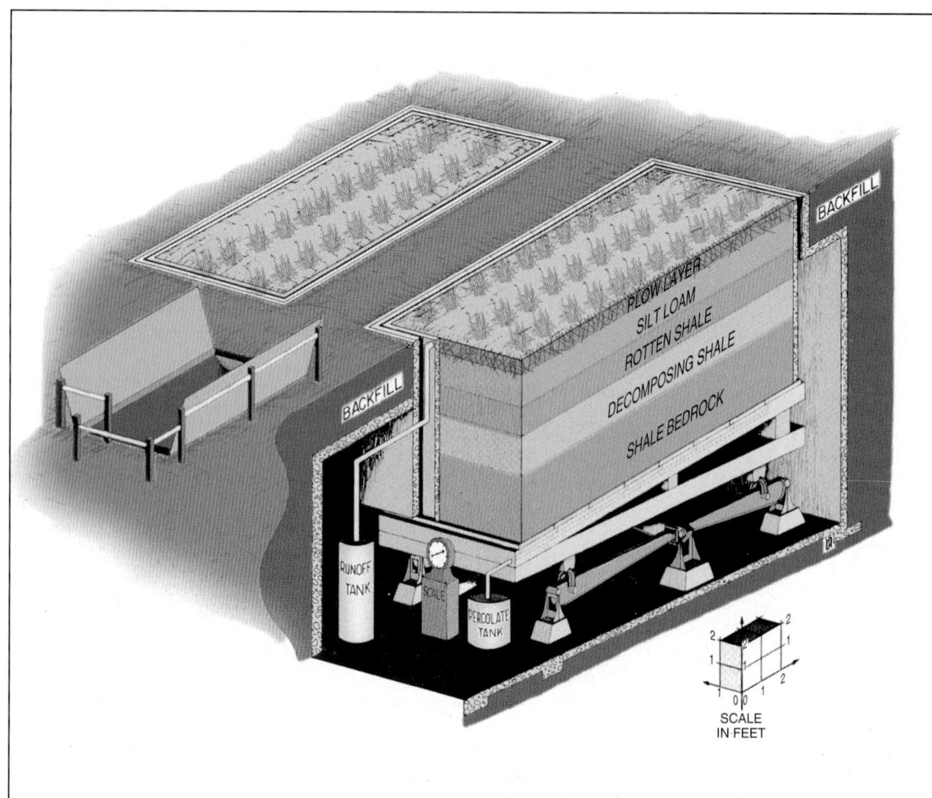

FIGURE 7.5 Lysimeter.
A weighing lysimeter for measuring evaporation and transpiration. Various pathways of water that falls on the soil above the lysimeter are traced: Some water remains as soil moisture, some is incorporated into plant tissues, some drains from the bottom of the lysimeter, and the remainder is credited to evapotranspiration. Given natural conditions, the lysimeter measures actual evapotranspiration. [Adapted from illustration courtesy of Lloyd Owens, Agricultural Research Service, USDA, Coshocton, Ohio.]

Soil-Moisture Storage. **Soil-moisture storage** is a "savings account" of water that can receive deposits and allow withdrawals as conditions change in the water balance. Soil-moisture storage (ΔSTRGE) refers to the amount of water that is stored in the soil and is accessible to plant roots. (Soil moisture is discussed in more detail in Chapter 15.)

Moisture retained in the soil comprises two classes of water, only one of which is accessible to plants. The inaccessible class is *hygroscopic water*, the thin molecular layer that is tightly bound to each soil particle by the natural hydrogen bonding of water molecules (Figure 7.7, left). Such water exists even in the desert but is not available to meet moisture demands. Soil is said to be at the **wilting**

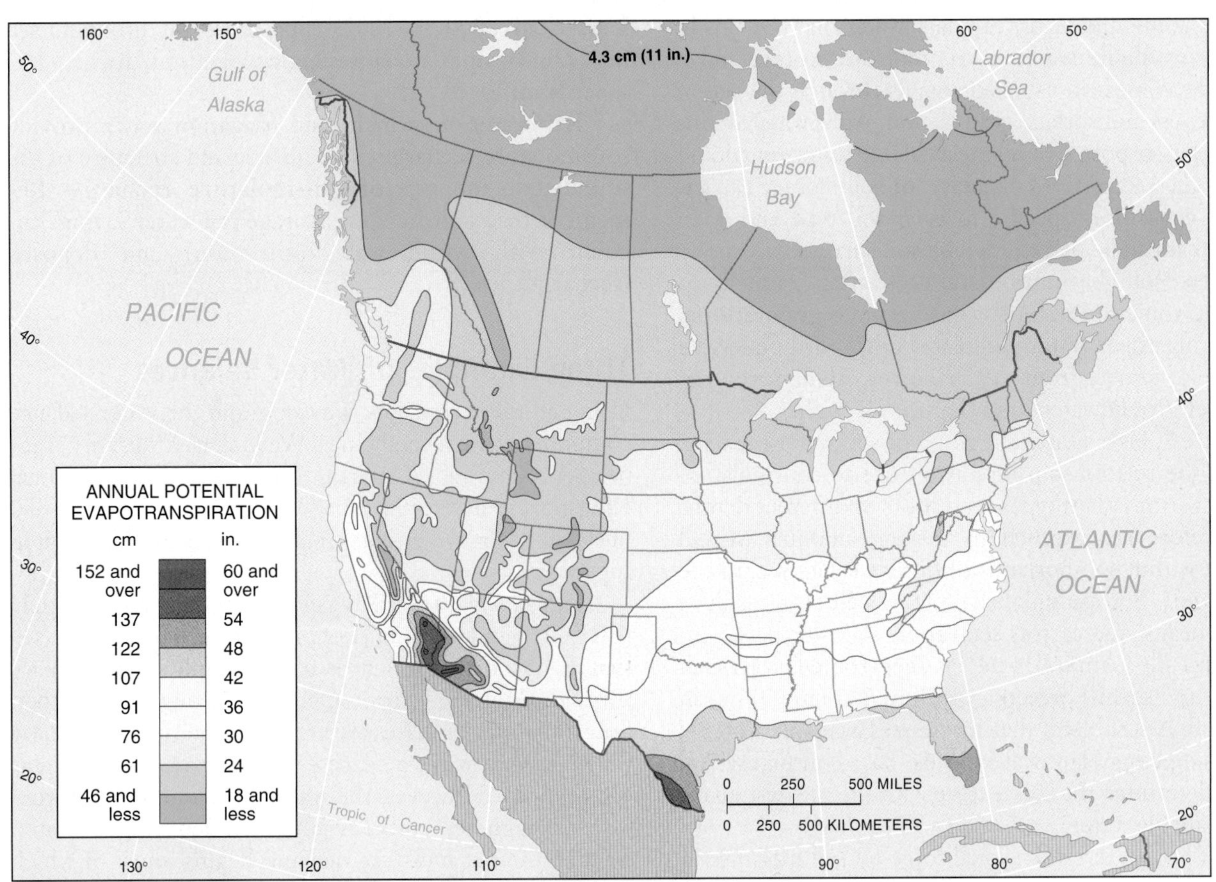

FIGURE 7.6 Potential evapotranspiration (water demand, POTET) for the United States and Canada.
[From Charles W. Thornthwaite, 1948, "An Approach Toward a Rational Classification of Climate," *Geographical Review* 38, p. 64. Adapted by permission from the American Geographical Society. Canadian data adapted from Marie Sanderson, "The Climates of Canada According to the New Thornthwaite Classification," *Scientific Agriculture* 28 (1948): 501–517.]

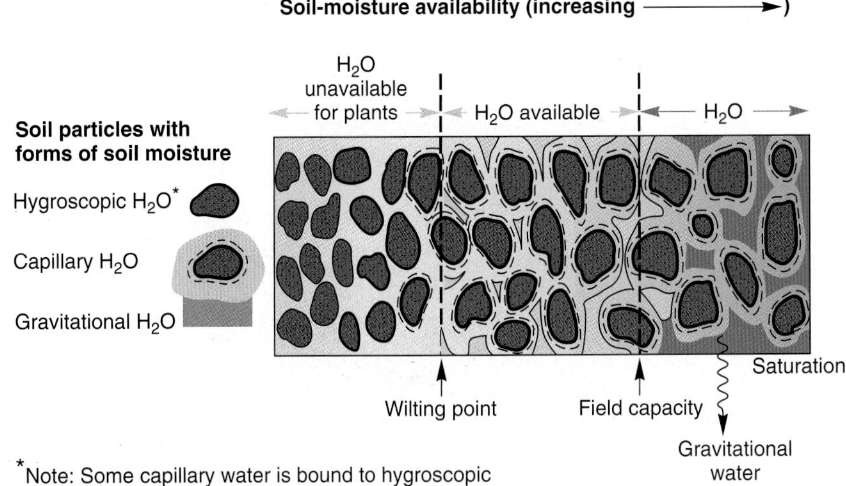

FIGURE 7.7 Types of soil moisture.
Hygroscopic and gravitational water are unavailable to plants; only capillary water is available. [After Donald Steila, *The Geography of Soils*, © 1976, p. 45. Reprinted by permission of Prentice Hall, Inc., Englewood Cliffs, NJ.]

point when all that is left in the soil is unextractable water; the plants wilt and eventually die after a prolonged period of such moisture stress. Unfortunately, most of us have watched a houseplant, or dorm-room plant, go through this process.

The soil moisture that is generally accessible to plant roots is **capillary water**, held in the soil by surface tension and hydrogen bonding between the water and the soil (Figure 7.7, middle). Most capillary water that remains in the soil is **available water** in soil moisture storage and is removable to meet moisture demands through the action of plant roots and surface evaporation. After water drains from the larger pore spaces, the available water remaining for plants is termed **field capacity**, or soil-storage capacity. Field capacity is specific to each soil type and is an amount that can be determined by soil surveys.

When soil becomes saturated after rainfall or snowmelt, some soil water surplus becomes **gravitational water**. This excess water percolates downward under the influence of gravity from the shallower capillary zone to the deeper groundwater zone (Figure 7.7, right).

Figure 7.8 is a simplified graph of soil water capacity, showing the relationship of soil texture to soil moisture content. Various plant types send roots to different depths and therefore are exposed to varying amounts of soil moisture within soil horizons (different layers within the soil column). For example, shallow-rooted crops such as spinach, beans, and carrots send roots down about 65 cm (25 in.) in a silt-loam soil, whereas deep-rooted crops such as alfalfa and shrubs exceed a depth of 125 cm (50 in.) in such a soil. A soil blend that maximizes available water is best for supplying plant water needs. Based on Figure 7.8, can you determine the soil texture with the greatest quantity of available water?

Plant roots exert a tensional force on soil moisture to absorb it. As water transpires through the leaves, a pressure gradient is established throughout the plant, thus "pulling" water into the plant roots. When soil moisture is at field capacity, plants obtain water with less effort. As the soil water is reduced by **soil-moisture utilization**, the plants must exert greater effort to extract the same amount of moisture. As a result, even though a small amount of water may remain in the soil, plants may be unable to exert enough pressure to utilize it. The *unsatisfied demand* is a *deficit*.

Avoiding a deficit and reducing plant growth inefficiencies are the goals of irrigation, for the harder plants must work to get water, the less they will yield and grow. Imagine the impact on soil moisture and agricultural water demand resulting from the widespread drought conditions that affected approximately 30% of the U.S. from 2000 to 2002. (For more on drought, see **http://www.ngdc.noaa.gov/paleo/drought/drght_what.html**.)

Rainwater, snowmelt, and irrigation water provide soil-moisture recharge. The texture and structure of the soil dictate the rate of **soil-moisture recharge**. Remember that soil-moisture storage is a water savings account with withdrawals (utilization) and deposits (recharge).

Three Examples of Water Balances

Using all these concepts, we can graph the water-balance components for several cities. We look at Kingsport, Tennessee; Ottawa, Ontario; and Phoenix, Arizona. Kingsport is at 36.6° N, 82.5° W, with an elevation of 390 m (1280 ft). Its water-balance data appear on the graph in Figure 7.9.

Figure 7.9 shows Kingsport's average water supply (PRECIP) and demand (POTET) for a year. We can see that Kingsport experiences a net supply from January through May and from October through December. Soil-moisture estimates are at field capacity from January through May, with the excess supply providing a surplus each month. However, the warm days and months from June through September result in net demands for water. In June the net moisture demand begins, most of which is withdrawn from soil-moisture storage for shallow-rooted crops.

July, August, and September feature soil-moisture utilization, so that by the end of September soil-moisture for shallow-rooted plants is near the wilting point. During October, however, net supply is generated and is credited as soil-moisture recharge. Recharge occurs

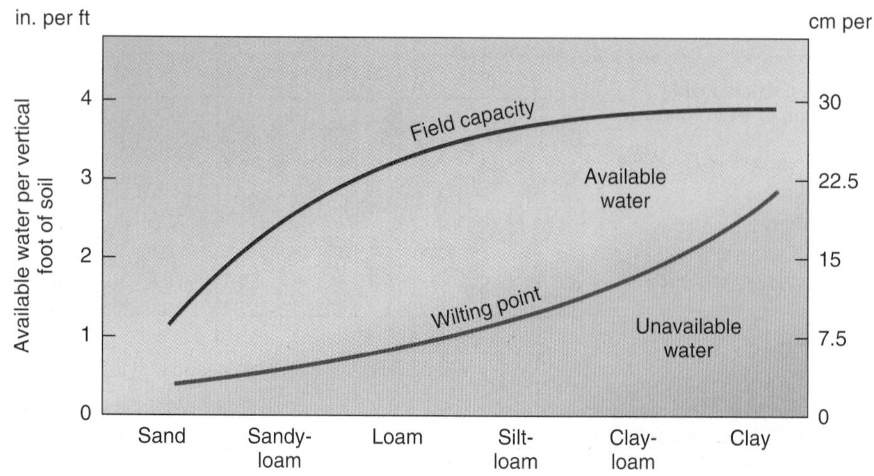

FIGURE 7.8 Soil-moisture availability.
The relation between soil moisture availability and soil texture determines the area between the two curves that show field capacity and wilting point. A loam soil (one-third each of sand, silt, and clay) has roughly the most available water per vertical foot of soil exposed to plant roots. [After U.S. Department of Agriculture, 1955 *Yearbook of Agriculture-Water*, p. 120.]

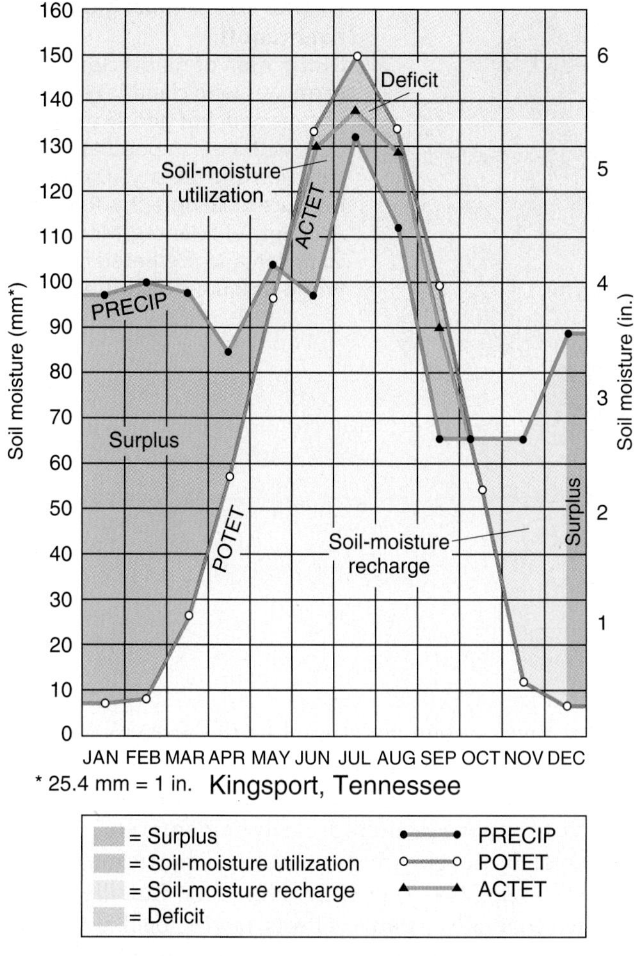

FIGURE 7.9 Sample water budget.
Annual average water-balance components graphed for Kingsport, Tennessee. The comparison of plots for precipitation inputs and potential evapotranspiration outputs determines the condition of the soil-moisture environment. A typical pattern of spring surplus, summer soil-moisture utilization, a small summer deficit, autumn soil-moisture recharge, and ending surplus highlights the year.

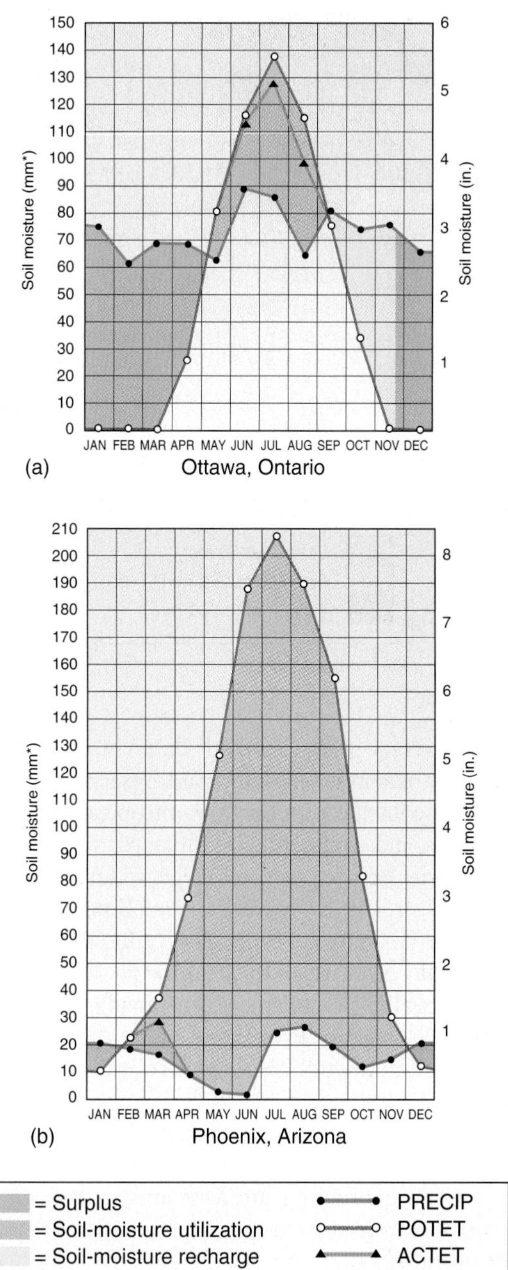

FIGURE 7.10 Sample water budgets for two cities.
Sample water-balance regimes for (a) Ottawa, Ontario, Canada, and (b) Phoenix, Arizona.

again in November and December, bringing the soil moisture back up to field capacity by the end of the year.

Obviously, not all cities experience the surplus moisture patterns of this humid-continental city and surrounding region; different climatic regimes experience different interactions among their water-balance components. Also, medium- and deep-rooted plants produce different results. Figure 7.10 presents water-balance graphs for the moist and cooler climate of Ottawa, Ontario, Canada, and the drier and warmer climate of Phoenix, Arizona.

Water Budget and Water Resources

Water distribution is uneven over space and time. Because we require a steady supply, we build large-scale management projects intended to redistribute the water resource either geographically, by moving water from one place to another, or over time, by storing water from time of receipt until it is needed. Substantial human effort has been expended to override natural water-balance regimes and recurring drought. The analysis of stream contribution to water resources is facilitated using the soil-water-budget approach.

Streams may be perennial (constantly flowing), *perennial* means "through the year" or intermittent. In either case, the total runoff that moves through them comes from surplus surface-water runoff, subsurface throughflow, and groundwater. Figure 7.11 maps annual global river runoff for the world. Highest runoff amounts are along the equator within the tropics, reflecting the continual rainfall along the intertropical convergence zone (ITCZ). Southeast Asia also experiences high runoff, as do northwest coastal mountains in the Northern Hemisphere. In

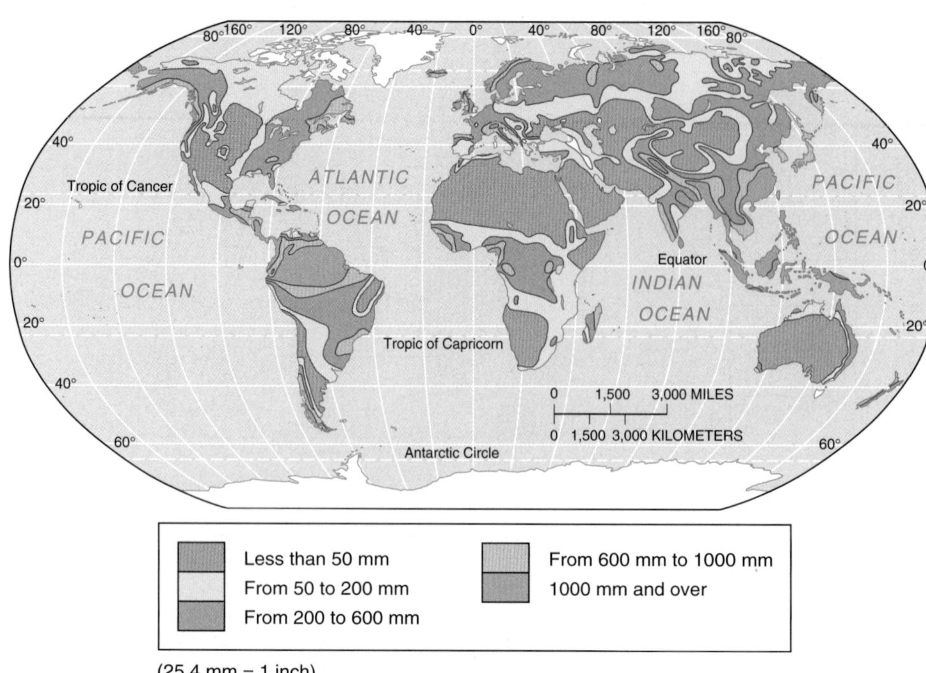

FIGURE 7.11 Annual global river runoff. Distribution of runoff closely correlates with climatic region, as expected, but poorly correlates with human population distribution and density. [Data from Institute of Geography, Russian Academy of Sciences, Moscow, as compiled and presented by World Resources Institute.]

countries having great seasonal fluctuations in runoff, groundwater becomes an important reserve. Regions of lower runoff coincide with Earth's subtropical deserts, rainshadow areas, and continental interiors, particularly in Asia.

In North America, several large water-management projects are already in operation: the Bonneville Power Authority in Washington State, the Tennessee Valley Authority in the Southeast, the California Water Project, and the Central Arizona Project. In Canada, the Churchill Falls and Nelson River Projects of Manitoba and the proposed $12 billion Gull Island and related hydroelectric projects in Québec along the lower Churchill, Saint-Jean, and Romaine Rivers are significant. The best sites for multipurpose hydroelectric projects already are taken, so battles among conflicting interests and analysis of negative environmental impacts invariably accompany a new project proposal.

A particularly ambitious regional water project is the Snowy Mountains Hydroelectric Authority Scheme of Australia, where water-balance variations created the need to relocate water. In the Snowy Mountains, part of the Great Dividing Range in extreme southeastern Australia, precipitation ranges from 100 to 200 cm (40–80 in.) a year, whereas interior Australia receives under 50 cm (20 in.) and drops to less than 25 cm (10 in.) farther inland (see Figure 6.2). Using some of the longest tunnels ever built, vast pumping systems, numerous reservoirs, and power plants, this project—to pump surpluses from one region to eliminate deficits in another—is operational. Today, over 2 million acre-feet of irrigation water is pumping to interior farmlands from the Snowy Mountains (see **http://www.snowyhydro.com.au/**). In 2002, an effort to restore the Snowy River began with the turning of a valve, shutting off one of the water diversions. Benefits from water resource management often come with negative effects elsewhere; in this case, damage to the Snowy River system was caused by the project's 50 years of diversions from its flows.

An interesting application of the water-balance approach to water resources is analyzing an event, such as the Dust Bowl, an urban flood, or a hurricane. Focus Study 7.1 (pp. 234–5) presents Hurricane Camille (1969) and its generally positive effects on regional water resources, in contrast to its coastal damage and flooding.

Groundwater Resources

Groundwater is a part of the hydrologic cycle that lies beneath the surface but is tied to surface supplies. Groundwater is the largest potential source of freshwater in the hydrologic cycle—larger than all surface lakes and streams combined. Between Earth's land surface and a depth of 4 km (13,100 ft) worldwide, some 22% of Earth's freshwater resides. Despite this volume and its obvious importance, groundwater is widely abused. In many areas it is polluted or consumed in quantities beyond natural recharge rates. Remember: *groundwater is not an independent source of water for it is tied to surface supplies for recharge*.

About 50% of the U.S. population derive a portion of its freshwater from groundwater sources. Between 1950 and 1995, annual groundwater withdrawal increased 150%. Figure 7.12 shows potential groundwater resources in the United States and Canada.

Groundwater Profile and Movement

Figure 7.13 illustrates the following discussion. Groundwater is fed by surplus water, which percolates downward from the zone of capillary water as gravitational water. This excess surface water moves through the **zone of aeration** where soil and rock are less than saturated (an un-

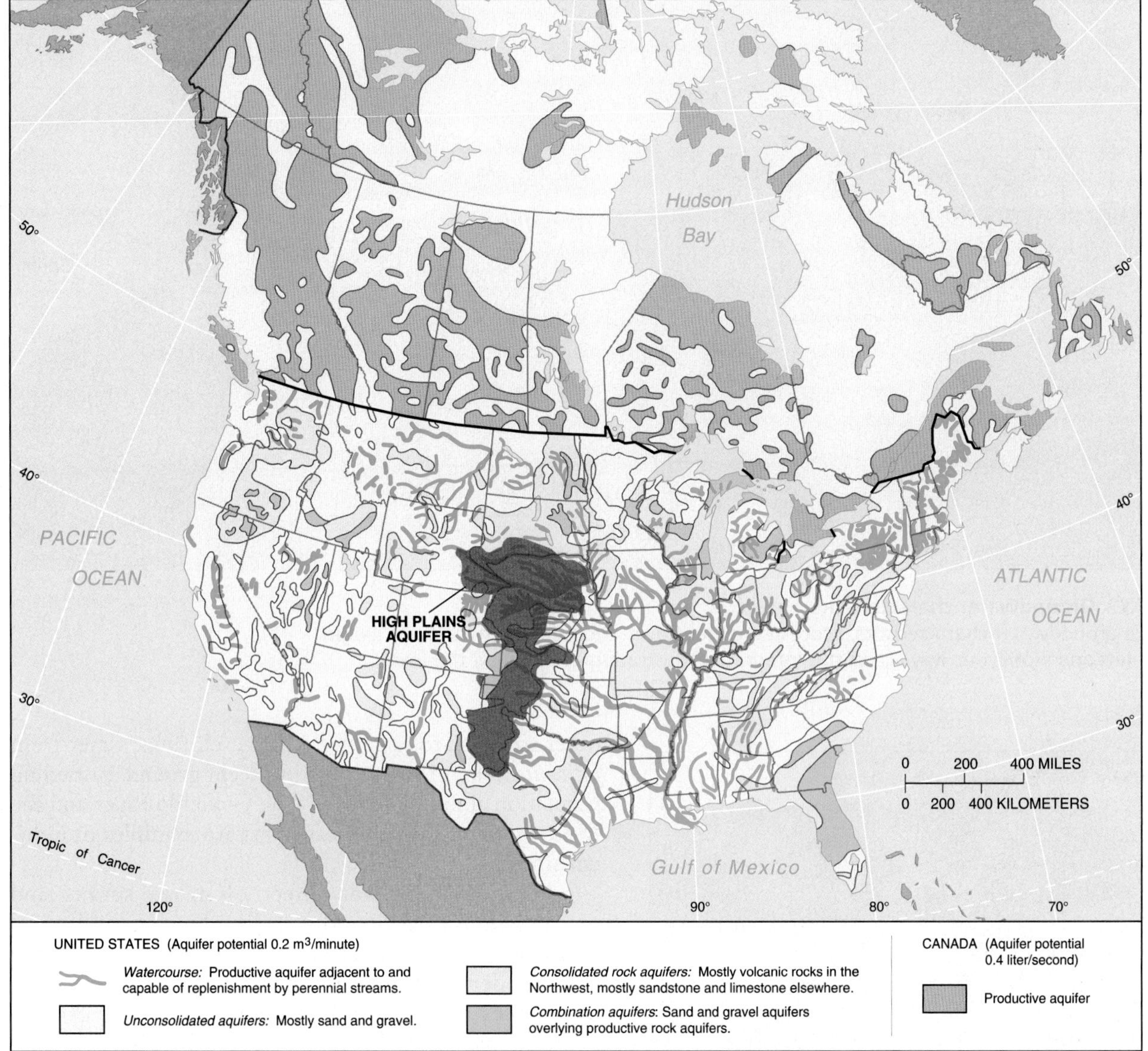

FIGURE 7.12 Groundwater resource potential for the United States and Canada. Highlighted areas of the United States and Canada are underlain by productive aquifers capable of yielding freshwater to wells. [Courtesy of Water Resources Council for the United States and the *Inquiry on Federal Water Policy* for Canada.]

saturated zone), and some pore spaces contain air. Eventually water reaches an area of collected subsurface water known as the **zone of saturation**, where the pores contain only water. Like a sponge made of sand, gravel, and rock, this zone stores water within its structure, filling all available pores and voids.

The **porosity** of this layer is dependent on the arrangement, size, and shape of individual component particles; the cement between them; and their degree of compaction. Subsurface structures are referred to as **permeable** or **impermeable**, depending on whether they permit or obstruct water flows.

An **aquifer** is a rock layer that is permeable to groundwater flow in usable amounts. An **aquiclude** is a body of rock that does not conduct water in usable amounts. The zone of saturation is an **unconfined aquifer**, a water-bearing layer that is not confined by an overlying impermeable layer. The upper limit of the water that collects in the zone of saturation is called the **water table**; it is the contact surface between the zones of saturation and aeration. (See these components on the left section in Figure 7.13.)

The slope of the water table, which generally follows the contours of the land surface, controls groundwater movement. Water wells must penetrate the water table to optimize their potential flow. Where the water table intersects the surface, it creates springs or feeds lakes or riverbeds (Figure 7.14). Ultimately, groundwater may enter stream channels and flow as surface water (near the center in Figure 7.13). During dry periods the water table can act to sustain river flows.

Figure 7.15 illustrates the relation between the water table and surface streams in two different climatic settings. In humid climates, where the water table generally

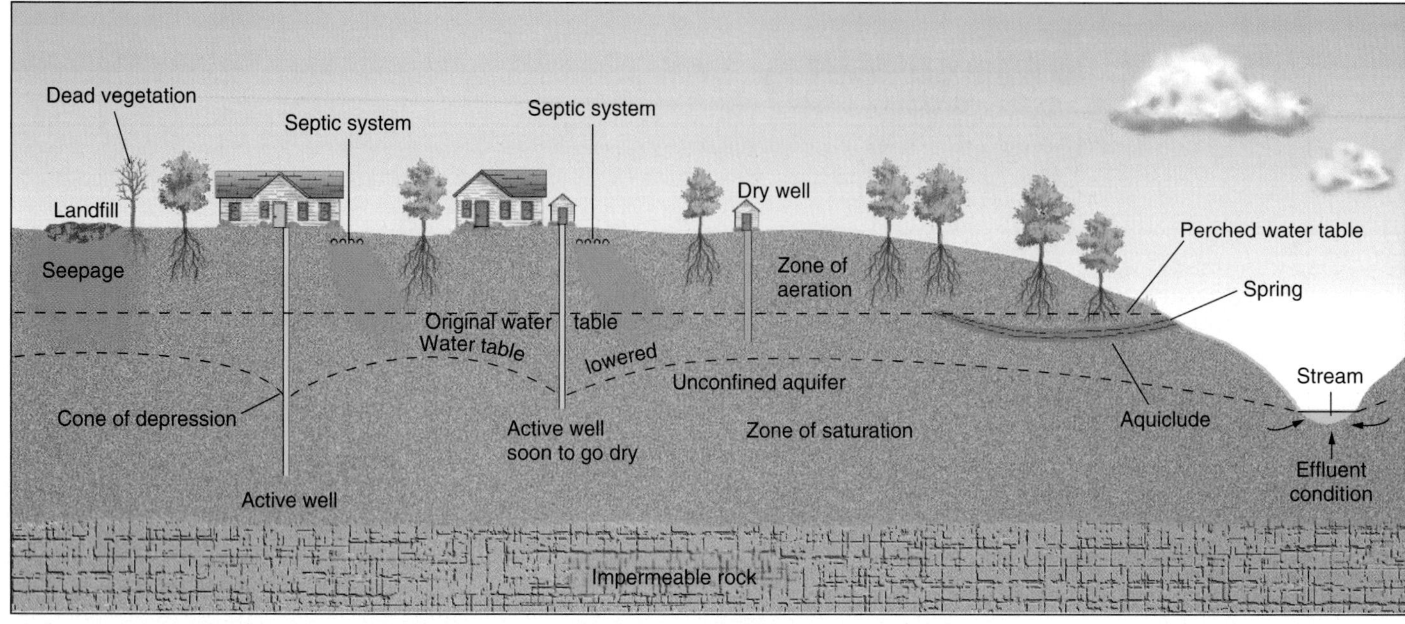

FIGURE 7.13 Groundwater characteristics.
Subsurface groundwater characteristics, processes, water flows, and human interactions. Begin at the far left and work your way across this integrative illustration as you read the text.

FIGURE 7.14 An active spring.
Springs provide evidence of groundwater emerging at the surface, here flowing into Crystal Lake, Salt River Range, Wyoming. [Photo by Bobbé Christopherson.]

supplies a continuous base flow to a stream and is higher than the stream channel, the stream is called *effluent* because it receives the water flowing out (effluent) from the surrounding ground. The Mississippi River and countless other streams are examples. In drier climates, water from *influent* streams flows into the adjacent ground, sustaining vegetation along the stream. The Colorado River and the Rio Grande of the American West are examples of influent streams.

An interesting side effect of urban sprawl and paving land for parking lots, malls, office and industrial parks, and subdivisions is that the process seals the land, rendering it impermeable to water. This covering of surfaces deprives the groundwater system of adequate recharge water, thereby further reducing water tables.

A confined aquifer differs from an unconfined aquifer in several important ways. A **confined aquifer** is bounded above and below by impermeable layers of rock or sediment (Figure 7.13, right), while an *unconfined aquifer* is not. In addition, the two aquifers differ in **aquifer recharge area**, which is the surface area where water enters an aquifer to recharge it. For an unconfined aquifer, that area generally extends above the entire aquifer to the ground surface. But in the case of a confined aquifer, the recharge area is far more restricted (area to the right of the industrial pollution site in Figure 7.13). Once these recharge areas are identified and mapped, the appropriate government body should zone them to prohibit pollution discharges, septic and sewage system installations, or hazardous-material dumping. Of course, an informed population would insist on such action.

Aquifers also differ in water pressure. A well drilled into an unconfined aquifer (Figure 7.13, left) must be pumped to make the water level rise above the water table. In contrast, the water in a confined aquifer is under the pressure of its own weight, creating a pressure level called the *potentiometric surface*, which actually can be

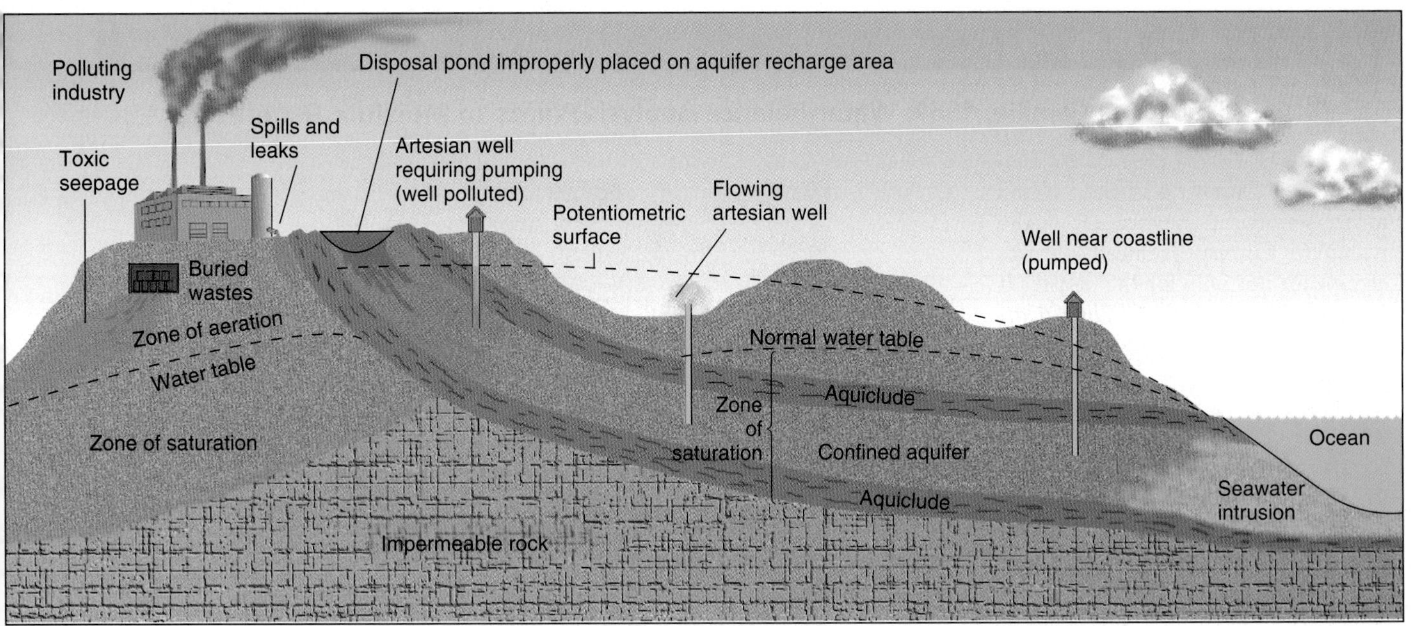

above ground level (to the right in the figure). Under this condition, **artesian water**, or groundwater confined under pressure, may rise up in wells and even flow out at the surface if the head of the well is below the potentiometric surface. In other wells, however, pressure may be inadequate, and the artesian water must be pumped the remaining distance to the surface.

Groundwater Utilization

As water is pumped from a well, the adjacent water table within an unconfined aquifer will experience a **drawdown**, or become lower, if the rate of pumping exceeds the horizontal flow of water in the aquifer. The resultant shape of the water table around the well is called a **cone of depression** (Figure 7.13, left).

Aquifers frequently are pumped beyond their flow and recharge capacities, resulting in **groundwater mining**. Large tracts experience chronic groundwater overdrafts in the Midwest, West, lower Mississippi Valley, Florida, and the Palouse region of eastern Washington State. In many places the water table or artesian water level has declined more than 12 m (40 ft). In the United States, groundwater mining is of special concern in the great High Plains Aquifer, which is the topic of Focus Study 7.2 (pp. 236–7).

About half of India's irrigation water and half of industrial and urban water needs are met by the groundwater reserve. And, in approximately 20% of India's agricultural districts, groundwater mining through more than 17 million wells is beyond recharge rates.

Another possible effect of water removal from an aquifer is that the rock layer loses its internal support—the water in the pore spaces between the rock grains—and overlying rock may *crush the aquifer*, resulting in land subsidence and lower surface elevations. For example, within an 80-km (50-mi) radius of Houston, Texas, land has subsided more than 3 m (10 ft). In the Fresno area of the San

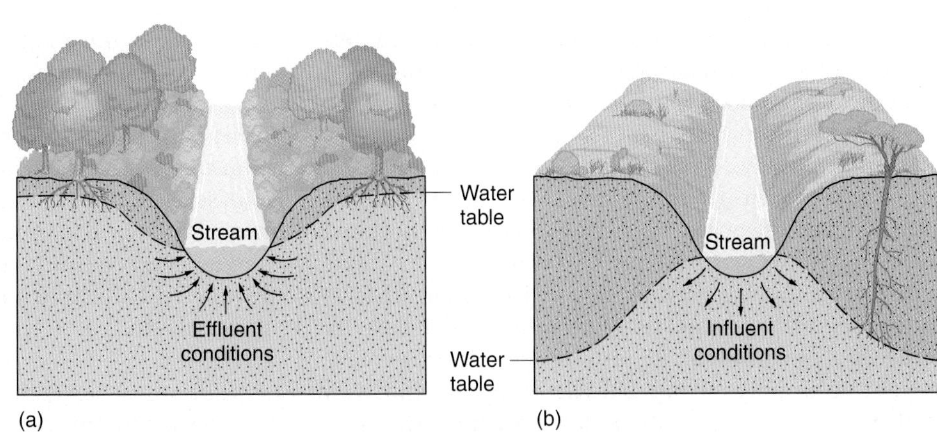

FIGURE 7.15 Groundwater interaction with streamflow. (a) Effluent stream base flow is partially supplied by a high water table, characteristic of humid regions. (b) Influent stream supplies a lower water table, characteristic of drier regions. The water table may drop below the stream channel, effectively drying it out.

Focus Study 7.1

Hurricane Camille, 1969: Water-Balance Analysis Points to Moisture Benefits

Hurricane Camille was one of the most devastating hurricanes of the twentieth century. Ironically, it was significant not only for the disaster it brought (256 dead, $1.5 billion damage) but also for the drought it abated (ended).

Figure 1 shows Camille inland from the Gulf Coast north of Biloxi, Mississippi. Hurricane-force winds sharply diminished after the hurricane made landfall, leaving a vast rainstorm that traveled from the Gulf Coast through Mississippi, western Tennessee, Kentucky, and into central Virginia (Figure 2). Severe flooding drowned the Gulf Coast near landfall and the James River basin of Virginia, where torrential rains produced record floods. But Camille actually had beneficial aspects; it ended a year-long drought along major portions of its storm track.

FIGURE 1 Satellite image of Hurricane Camille. Hurricane Camille made landfall on the night of August 17, 1969, and continued inland on August 18. The storm is progressing northward through Mississippi in the image. [*ESSA-9* satellite image from NOAA.]

Figure 2a maps the precipitation from Camille (moisture supply) and portrays the storm's track from landfall in Mississippi to the coast of Virginia and Delaware. By comparing actual water budgets along Camille's track with water budgets for the same three days with Camille's rainfall totals removed artificially, I analyzed the moisture impact of the storm on the region's water budget. Figure 2b maps the moisture shortages that were avoided because of Camille's rains—termed "deficit abatement." Think of this as drought that did not continue because Camille's rains occurred. Over vast portions of the affected area, Camille reduced dry-soil conditions, restored pastures, and filled low reservoirs.

Thus, Camille's monetary benefits inland outweighed its damage by an estimated 2 to 1 ratio. (Of course, the tragic loss of life does not fit into a financial equation.) Hurricanes should be viewed as normal and natural meteorological events that have terrible destructive potential, mainly to coastal lowlands. However, they contribute to the normal precipitation regimes of the southern and eastern United States. About one-third of all hurricanes making landfall in the United States provide beneficial precipitation to local water budgets, determined by a survey of weather records.

(continued)

Joaquin Valley in central California, after years of intensive pumping of groundwater for irrigation, land levels have dropped almost 10 m (33 ft) because of water removal and soil compaction.

Unfortunately, collapsed aquifers may not be rechargeable even if surplus gravitational water becomes available, because pore spaces may be permanently closed. Surface mining ("strip mining") for coal in eastern Texas, northeastern Louisiana, Wyoming, Arizona, and other locales also destroys aquifers. The water-well fields that serve the Tampa Bay–St. Petersburg, Florida, area are another case in point. Pond and lake levels, swamps, and wetlands in the area are declining and land surfaces are subsiding as the groundwater drawdown for export to coastal cities increases.

An additional problem arises when aquifers are overpumped near the ocean or coastal seas. In such a circumstance the groundwater contact between freshwater and seawater often migrates inland. As a result, wells may become contaminated with saltwater, and the aquifer may become useless as a freshwater source. Figure 7.13 illustrates this *seawater intrusion* on the far right margin.

Like any renewable resource, groundwater can be tapped indefinitely as long as the rate of extrac-

Focus Study 7.1 *(continued)*

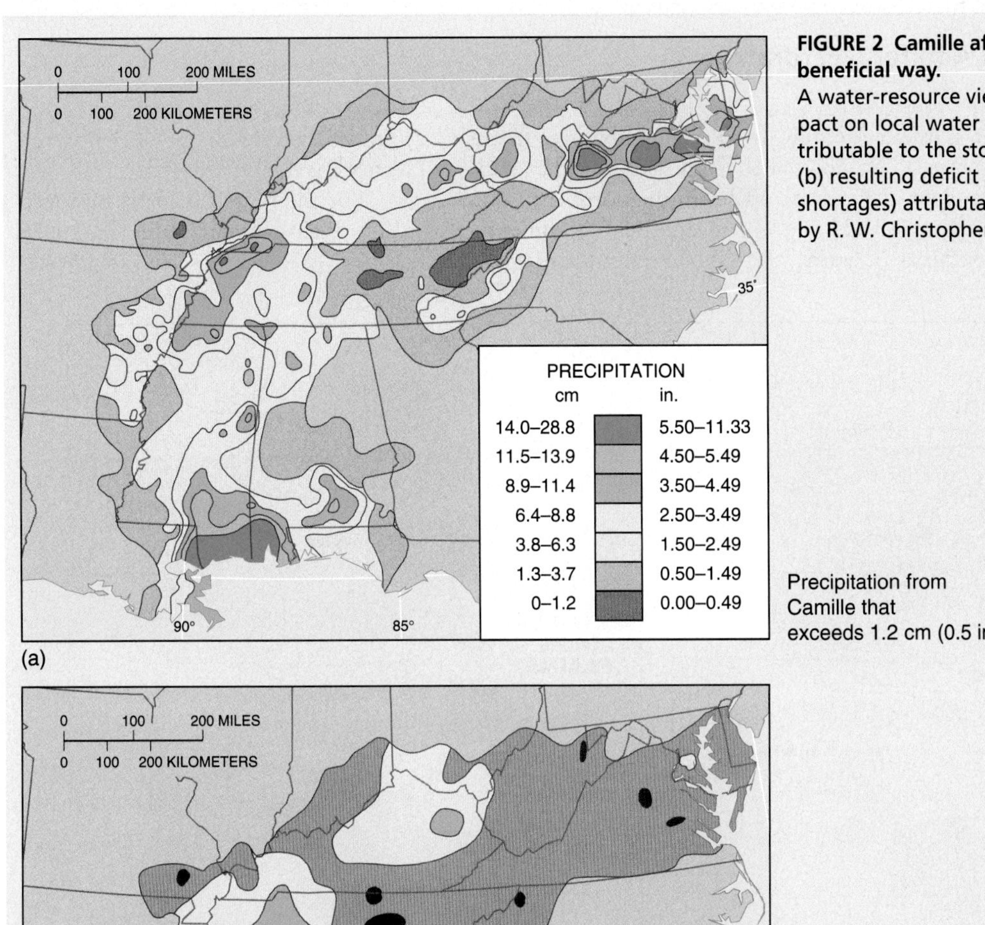

FIGURE 2 Camille affects water budgets in a beneficial way.
A water-resource view of Hurricane Camille's impact on local water budgets: (a) precipitation attributable to the storm (moisture supply); (b) resulting deficit abatement (avoided moisture shortages) attributable to Camille. [Data and maps by R. W. Christopherson. All rights reserved.]

tion does not exceed the rate of replenishment. But just like a bank account, a groundwater reserve will dwindle if withdrawals exceed deposits. Few governments have established and enforced rules and regulations to insure that groundwater sources are exploited at a sustainable rate.... No government has yet adequately tackled the issue of groundwater depletion, but it is at least getting more attention.*

*L. Brown, et al., and the Worldwatch Institute, *Vital Signs, The Environmental Trends That Are Shaping Our Future* (New York: W.W. Norton & Co., 2000), p. 122, 123.

Groundwater depletion and water problems are severe in the Middle East, as detailed in News Report 7.1 (p. 238). The groundwater resource beneath Saudi Arabia accumulated over thousands of years, but the increasing withdrawals are not being naturally recharged to any appreciable degree at present due to the desert climate. Some researchers have suggested that groundwater in the region will be depleted by the year 2007, although worsening water-quality problems will no doubt arise before this date. Desalination of seawater to augment diminishing groundwater supplies is becoming increasingly important as a freshwater source (Figure 7.16, p. 238). A new desalination plant went online in the Tampa-St. Petersburg, Florida, area in 2003.

Focus Study 7.2

High Plains Aquifer Overdraft

Earth's largest known aquifer is the High Plains aquifer. It lies beneath the American High Plains, an eight-state, 450,600 km^2 (174,000 mi^2) area from southern South Dakota to Texas (Figure 1a). Precipitation over the region varies from 30 cm in the southwest to 60 cm in the northeast (12 to 24 in.). For several hundred thousand years, the aquifer's sand and gravel were charged with meltwaters from retreating glaciers.

For the past 100 years, however, High Plains groundwater has been

(continued)

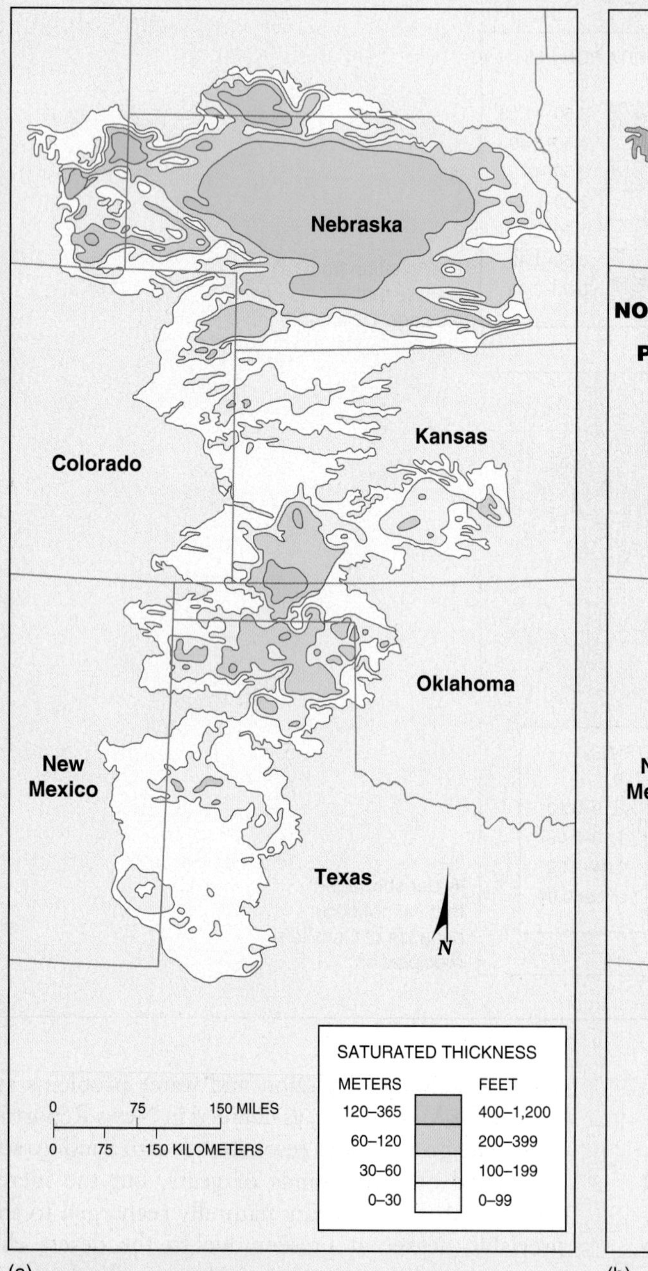

(a)

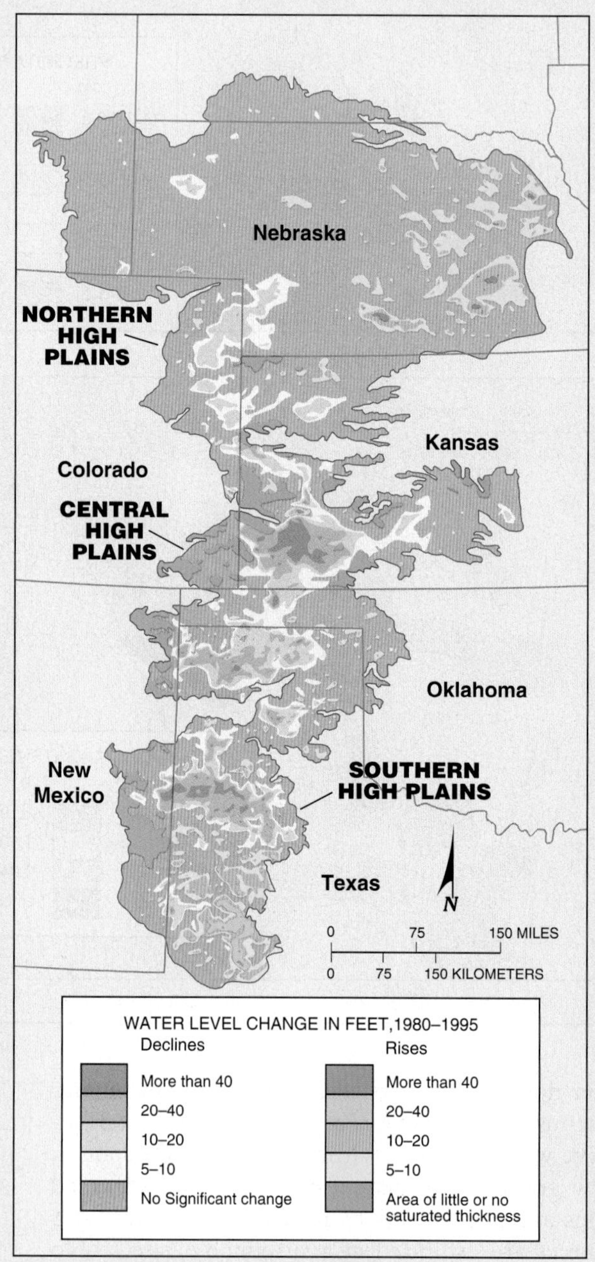

(b)

FIGURE 1 High Plains aquifer.
(a) The average saturated thickness of the High Plains aquifer, the largest known aquifer in the world. (b) Water level changes in the aquifer from 1980 to 1995, given in feet. [(a) After D. E. Kromm and S. E. White, "Interstate groundwater management preference differences: The High Plains region," *Journal of Geography* 86, no. 1 (January–February 1987): 5. (b) USGS "Water-Level Changes in the High Plains Aquifer, 1980 to 1995," Fact Sheet FS-068-97, Lincoln, Nebraska: 1998.]

Focus Study 7.2 *(continued)*

heavily mined; mining intensified after World War II with the introduction of central-pivot irrigation devices (Figure 2). These large circular devices provide vital water to wheat, sorghums, cotton, corn, and about 40% of the grain fed to cattle in the United States. The USGS began monitoring this water mining in 1988. (See http://ne.water.usgs.gov/highplains/hpactivities.html for links or http://webserver.cr.usgs.gov/nawqa/hpgw/HPGW_home.html for a study.)

The High Plains aquifer irrigates about one-fifth of all U.S. cropland: 120,000 wells provide water for 5.7 million hectares (14 million acres). This withdrawal is down from the peak of 170,000 wells in 1978. In 1980, water was being pumped from the aquifer at the rate of 26 billion cubic meters (21 million acre-feet) a year, an increase of more than 300% since 1950. By 1995 withdrawals had decreased 10% due to declining well yields, increasing pumping costs, and implementation of some water efficiency and conservation techniques.

In the past five decades, the water table in the aquifer has dropped more than 30 m (100 ft), and throughout the 1980s it averaged a 2 m (6 ft) drop each year. The USGS estimates that recovery of the High Plains aquifer (those portions that have not been crushed or subsided) would take at least 1000 years if groundwater mining stopped today.

Figure 1b maps changes in water levels from 1980 to 1995. Declining water levels are most severe in northern Texas, where the saturated thickness of the aquifer is least, through the Oklahoma panhandle and into Kansas. Rising water levels are noted in portions of south central Nebraska and a portion of Texas owing to recharge from surface irrigation, a period of above-normal precipitation years, and downward percolation from canals and reservoirs.

FIGURE 2 Central-pivot irrigation.
Myriad central-pivot irrigation systems water crops in north-central Nebraska. A growing season for corn requires from 10 to 20 revolutions of the sprinkler arm, depending on the weather, applying about 3 cm of water per revolution. The High Plains aquifer is at depths of more than 76 m (250 ft) in this part of Nebraska. [Photo by Comstock.]

Obviously, billions of dollars of agricultural activity cannot be abruptly halted, but neither can profligate water mining continue. This issue raises tough questions: How best to manage cropland? Can extensive irrigation continue? Can the region continue to meet the demand to produce commodities for export—principally grain and packaged beef? Should we continue high-volume farming of certain crops that are in chronic oversupply? Should we rethink federal policy on crop subsidies and price supports? What would be the impact on farmers and rural communities of any changes to the system?

Present irrigation practices, if continued, will destroy about half of the High Plains aquifer resource (and two-thirds of the Texas portion) by the year 2020. Add to this the approximate 10% loss of soil moisture due to climatic warming, as forecast by computer models for this region by 2050, and we have a portrait of a major regional water problem in this century and a spatial challenge for society.

FIGURE 7.16 Water desalination.
Freshwater is supplied to Saudi Arabia from the Jubail water desalination plant along the Red Sea. Saudi Arabia obtains a significant amount of its water from desalination of seawater. [Photo by Gamma/Liaison Network.]

Pollution of the Groundwater Resource

When surface water is polluted, groundwater also becomes contaminated because it is recharged from surface-water supplies. Groundwater migrates slowly compared to surface water. Surface water flows rapidly and flushes pollution downstream, but sluggish groundwater, once contaminated, remains polluted virtually forever.

Pollution can enter groundwater from industrial waste injection wells, septic tank outflows, seepage from hazardous-waste disposal sites, industrial toxic-waste dumps, residues of agricultural pesticides, herbicides, fertilizers, and residential and urban wastes in landfills. As an example, there are about 10,000 underground gasoline storage tanks suspected of leaking at active and abandoned gas (filling) stations. Thus, pollution can come either from a point source (about 35% of pollution) or from a large general area (a nonpoint source, 65%), and it can spread over a great distance, such as from an agricultural field or urban runoff, as illustrated in Figure 7.13.

In the face of government inaction, serious contamination of groundwater continues nationwide, as described some 20 years ago.

> One characteristic is the practical irreversibility of groundwater pollution, causing the cost of clean-up to be prohibitively high.... It is a questionable ethical practice to impose the potential risks associated with groundwater contamination on future generations when steps can be taken today to prevent further contamination.*

*J. Tripp, "Groundwater Protection Strategies," in *Groundwater Pollution, Environmental and Legal Problems* (Washington, D.C.: American Association for the Advancement of Science, 1984), p. 137. Reprinted by permission.

News Report 7.1

Middle East Water Crisis: Running on Empty

The Persian Gulf states soon may run out of freshwater. Their vast groundwater resource is over pumped to such an extent that salty seawater is encroaching into aquifers tens of kilometers inland. By the year 2007, groundwater on the Arabian Peninsula may become undrinkable. Imagine the hottest issue in the Middle East to be water, not oil!

Remedies for groundwater overuse are neither easy nor cheap. In the Persian Gulf area, additional freshwater is obtained by desalination of seawater, using desalination plants along the coasts. These processing plants remove salt from seawater by distillation and evaporation processes. In fact, approximately 60% of the world's 4000 desalination plants are presently operating in Saudi Arabia and other Persian Gulf states (see Figure 7.16). (See the links listed under Middle East Water Information Network at http://water1.geol.upenn.edu.)

A planned water pipeline will carry water overland from Turkey in two branches. The Gulf water branch runs southeast through Jordan and Saudi Arabia, with extensions into Kuwait, Abu Dhabi, and Oman; the other line runs south through Syria, Jordan, and Saudi Arabia to the cities of Makkah (Mecca) and Jeddah. This pipeline would import 6 million cubic meters (1.68 billion gallons) of water a day some 1500 km (930 mi), the same distance as from New York City to St. Louis!

Other remedies are possible. Traditional agricultural practices could be modernized to use less water. Urban water use could be made more efficient to reduce groundwater demand. Aquifers and rivers could be shared, although at present no negotiated accords exist for this purpose in the Middle East. Conflicts at various scales pose a grave threat to water pipelines and desalination facilities throughout the region.

> There are severe water shortages in the Middle East.... The resources of the Nile, the Tigris-Euphrates, and the Jordan are overextended owing both to natural causes and to those deriving from human behavior.... In addition to a severe shortage in the quantity of water, there is a growing concern over water quality.*

* N. Kliot, *Water Resources and Conflict in the Middle East* (London: Routledge, 1994), p. 1.

Our Water Supply

Human thirst for adequate water supplies (quantity and quality) is a major issue in the this century. Internationally, increase in per capita water use is double the rate of population growth. Since we are so dependent on water, it seems that humans should cluster where good water is plentiful. But accessible water supplies are not well correlated with population distribution or the regions where population growth is greatest. Just to produce our food requires enormous quantities of water. Yet the quality and quantity of our supply often is taken for granted. See News Report 7.2 for more on personal water use.

Table 7.1 shows estimated world water supplies, present world population, estimated population figures for 2025, population change forecast between 2002 and 2050 at present growth rates, and a comparison of population and global runoff percentages for each continent. The table gives an idea of the unevenness of Earth's water supply (tied to climatic variability) and water demand (tied to level of development, affluence, and per capita consumption).

For example, North America's mean annual water discharge is 5960 km^3 and Asia's is 13,200 km^3. However, North America has only 6.8% of the world's population, whereas Asia has 60.6%, with a population doubling time less than half North America's.

In northern China, 550 million people living in approximately 500 cities lack adequate water supplies. For comparison, note that the 1990 floods cost China $10 billion, whereas water shortages are running at more than $35 billion a year in costs to their economy. The Yellow River, a water resource for many Chinese, runs dry every year and in 1997 it failed to reach the sea for almost 230 days! An analyst for the World Bank stated that China's water shortages pose a more serious threat than floods during this century.

Water resources are different from other resources in that there is no alternative substance to water. Water resource stress related to quantity and/or quality shortfalls will dominate future political agendas. Water shortages increase the probability for international conflict, endanger public health, reduce agricultural productivity, and damage life-supporting ecological systems. New dams and river-management schemes might increase runoff accessibility by 10% over the next 50 years, but population is projected to grow by 46% in the same time period.

Water Supply in the United States

The U.S. water supply is derived from surface and groundwater sources that are fed by an average daily precipitation of 4200 BGD (billion gallons a day). That sum is based on an average annual precipitation value of 76.2 cm (30 in.) divided evenly among the 48 contiguous states (excluding Alaska and Hawai'i). Important to such an assessment are the measuring standards used for water resources.

The 4200 billion gallons of average daily precipitation mentioned is unevenly distributed across the country and unevenly distributed throughout the year. For example, New England's water supply is so abundant that only about 1% of available water is consumed each year. (The same is true in Canada, where the resource greatly exceeds that in the United States.) But in the dry Colorado River Basin mentioned earlier, the discharge is completely consumed. In fact, by treaty and compact agreements, the Colorado River actually is budgeted beyond its average discharge. This paradox results from political misunderstanding of the water resource itself. (See Focus Study 12.1 in Chapter 12 for more on this issue.)

Table 7.1 Estimate of Available Global Water Supply Compared by Region and Population

Region (2002 population in millions)	Land Area in Thousands of km^2	(mi^2)	Share of Mean Annual Discharge in km^3/year	(BGD)	Global Stream Runoff (%)	Global Population, 2025 (%)	Population, 2025 (millions)	Projected Population Change (+) 2002–2050 (%)
Africa (840)	30,600	(11,800)	4,220	(3,060)	11	16.2	1,281	120
Asia (3,766)	44,600	(17,200)	13,200	(9,540)	36	60.2	4,741	41
Australia–Oceania (32)	8,420	(3,250)	1,960	(1,420)	5	0.5	40	47
Europe (728)	9,770	(3,770)	3,150	(2,280)	8.8	9.2	718	−11
North America (421) (Canada, Mexico, U.S.)	22,100	(8,510)	5,960	(4,310)	15	6.6	514	45
Central and South America (429)	17,800	(6,880)	10,400	(7,510)	26	7.3	565	53
Global (6,215) (excluding Antarctica)	134,000	(51,600)	38,900	(28,100)	—	100.00	7,859	46

Note, BGD, billion gallons per day. Population data from "2002 World Population Data Sheet," Washington, D.C.: Population Reference Bureau, 2002.

News Report 7.2

Personal Water Use and Water Measurement

On an individual level, statistics indicate that American and Canadian urban dwellers directly use an average 680 liters (180 gallons) of water per person per day, whereas rural populations average only 265 liters (70 gallons) or less per person per day.

However, each of us indirectly accounts for a much greater water demand, because we all consume food and other products produced with water. For the United States, dividing the total water withdrawal by a population of 287.4 million people (2002) yields a per capita direct and indirect use of approximately 5600 liters (1480 gallons) of water per day. As population and per capita affluence increase, greater demands are placed on the water resource base, which is essentially fixed. To compare water intensity, an average American diet requires 5000 liters (1320 gal) of water a day to produce, whereas an average sub-Saharan diet only requires 1800 liters (475 gal) a day.

Simply providing the variety of food we enjoy requires voluminous amounts of water. For example, 77 g (2.7 oz) of broccoli requires 42 l (11 gal) of water to grow and process; producing 250 ml (8 oz) of milk requires 182 l (48 gal) of water; producing 28 g (1 oz) of cheese requires 212 l (56 gal); producing 1 egg requires 238 l (63 gal).

And then there are the toilets, the majority of which still flush approximately four gallons of water! Imagine the complexity of servicing the desert city of Las Vegas, with 125,000 hotel rooms times the number of toilets times flushes per day—in addition to the usage of a resident population of 1.3 million people! And, to further aggravate water realities in the desert, most all the hotels display extravagant water features.

Clearly, the average North American lifestyle depends on enormous quantities of water and thus is vulnerable to shortfalls or quality problems. Water conservation and efficiency remain as potentially our best water resource—the demand side of the equation.

Water Measures

In most of the United States, hydrologists measure streamflow in cubic feet per second (ft^3/s); Canadians use cubic meters per second (m^3/s). In the eastern United States, or for large-scale assessments, water managers use millions of gallons a day (MGD), billions of gallons a day (BGD), or billions of liters a day (BLD).

In the western United States, where irrigated agriculture is so important, the measure frequently used is acre-feet per year. One acre-foot is an acre of water, 1 foot deep, equivalent to 325,872 gallons (43,560 ft^3, or 1234 m^3, or 1,233,429 liters). An acre is an area that is about 208 feet on a side and is 0.4047 hectares.

For global measurements, 1 km^3 = 1 billion cubic meters= 810 million acre-feet; 1000 m^3 = 264,200 gallons = 0.81 acre-feet. For smaller measures, 1 m^3 = 1000 liters = 264.2 gallons.

Figure 7.17 shows the distribution of the U.S. supply of 4200 BGD. The national water budget has two general outputs: 71% actual evapotranspiration (ACTET), involving 2970 BGD, and 29% surplus (SURPL). This 29% surplus is what we directly use.

Instream, Nonconsumptive, and Consumptive Uses

The surplus 1230 BGD is runoff, available for withdrawal, consumption, and various instream uses.

- *Instream uses* are those that use streamwater in place: navigation, wildlife and ecosystem preservation, waste dilution and removal, hydroelectric power production, fishing resources, and recreation.
- *Nonconsumptive uses*, or *withdrawal*, remove water from the supply, use it, and then return it to the same supply. Nonconsumptive water is used by industry, agriculture, municipalities, and in steam-electric power generation. A portion of water withdrawn is consumed.
- *Consumptive uses* remove water from a stream but do not return it, so it is not available for a second or subsequent use. Some consumptive examples include water that evaporates or is vaporized in steam-electric plants.

When water is returned to the system, water quality usually is altered—water is contaminated chemically with pollutants or waste or thermally with heat energy. In Figure 7.17, this wastewater portion of the budget is returned to runoff and eventually the ocean. The water resource can be extended through recovery, treatment, and reuse of wastewater.

Contaminated or not, returned water becomes a part of all water systems downstream. A good example is New Orleans, the last city to withdraw municipal water from the Mississippi River. New Orleans receives diluted and mixed contaminants added throughout the entire Missouri-Ohio-Mississippi River drainage. Abnormally high cancer rates among citizens living along the Mississippi River between Baton Rouge and New Orleans have led to the ominous label "Cancer Alley" for the region. The nutrient discharge from the wastes and fertilizers in the river creates the "Dead Zone" in the Gulf of Mexico, discussed in Chapter 16.

The estimated U.S. withdrawal of water for 1995 was 402 BGD, an increase of 290% since 1940 but a decrease of 10% since the peak year of 1980 and 2% less than 1990. Total per capita use dropped from 1340 gal/day in

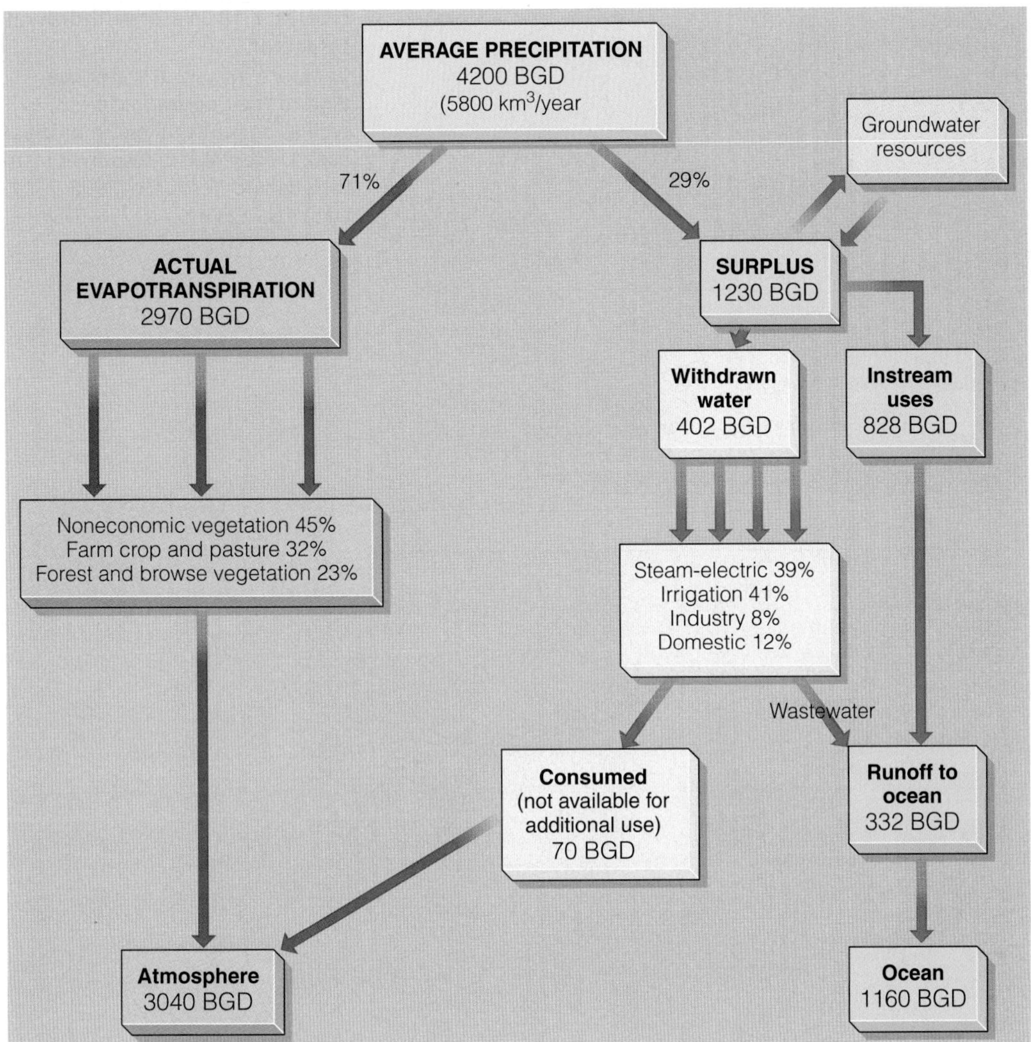

FIGURE 7.17 U.S. water budget.
Daily water budget for the contiguous 48 states in billions of gallons a day (BGD). Approximately 30% of available surplus was withdrawn for agricultural (irrigation-livestock), municipal (domestic-commercial), and industrial (industry-mining-thermoelectric) use. 1995 is the latest year for which data are available. [Data from *Estimated Use of Water in the United States in 1995,* Denver: USGS Circular 1200, 1998.]

1990 to 1280 gal/day in 1998. A shift in emphasis from supply-side (producer) to demand-side (consumer) planning and management has produced these efficiencies.

The four main uses of withdrawn water in the United States in 1995 were steam-electric power (39%) and industry and mining (8%), irrigation-livestock (41%), and domestic-commercial (12%). In contrast, Canada uses only 12% of its withdrawn water for irrigation and 70% for industry. Figure 7.18 compares regions by their use of withdrawn water during 2000. It graphically illustrates the differences between more-developed and less-developed parts of the world. (For studies of water use in the United States see http://water.usgs.gov/public/watuse/.)

Future Water Considerations

Budgeting precipitation input makes apparent the limits of water resources. How can we satisfy the growing demand for water? Water supplies available per person decline as population increases; thus, world population growth since 1970 has reduced per capita water supplies by a third. Also, pollution limits the water-resource base, so that even before *quantity* constraints are felt, *quality* problems may limit the health and growth of a region.

A composite index comparing available water resources to current use patterns, supplies, and national income reveals regional scarcities in Africa, the Middle East, Asia, Peru, and Mexico (Figure 7.19).

> Unlike other important commodities such as oil, copper, or wheat, freshwater has no substitutes for most of its uses. It is also impractical to transport the large quantities of water needed in agriculture and industry more than several hundred kilometers. Freshwater is now scarce in many regions of the world, resulting in severe ecological degradation,

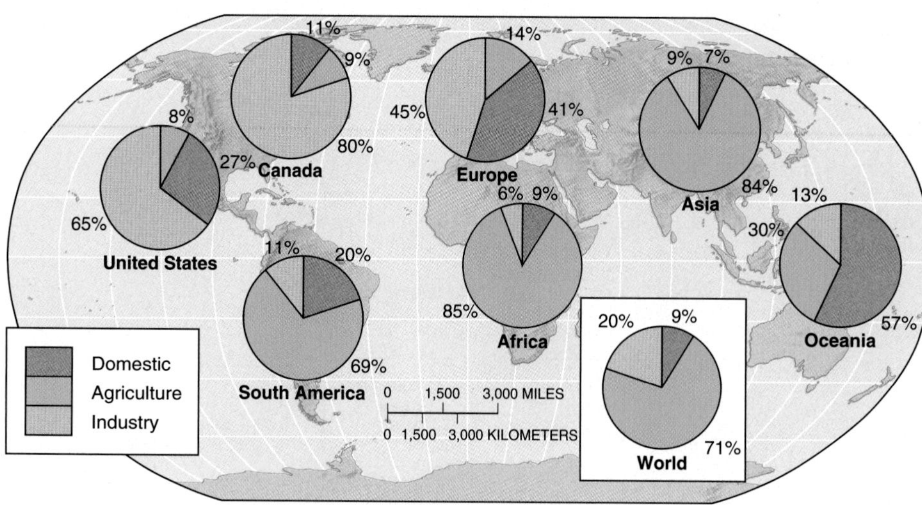

FIGURE 7.18 Water withdrawal by sector.
Compare industrial water use among the geographic areas, as well as agricultural and municipal uses. [After World Resources Institute, *World Resources 2000–2001*, Data Table FW.1, pp. 276–77.]

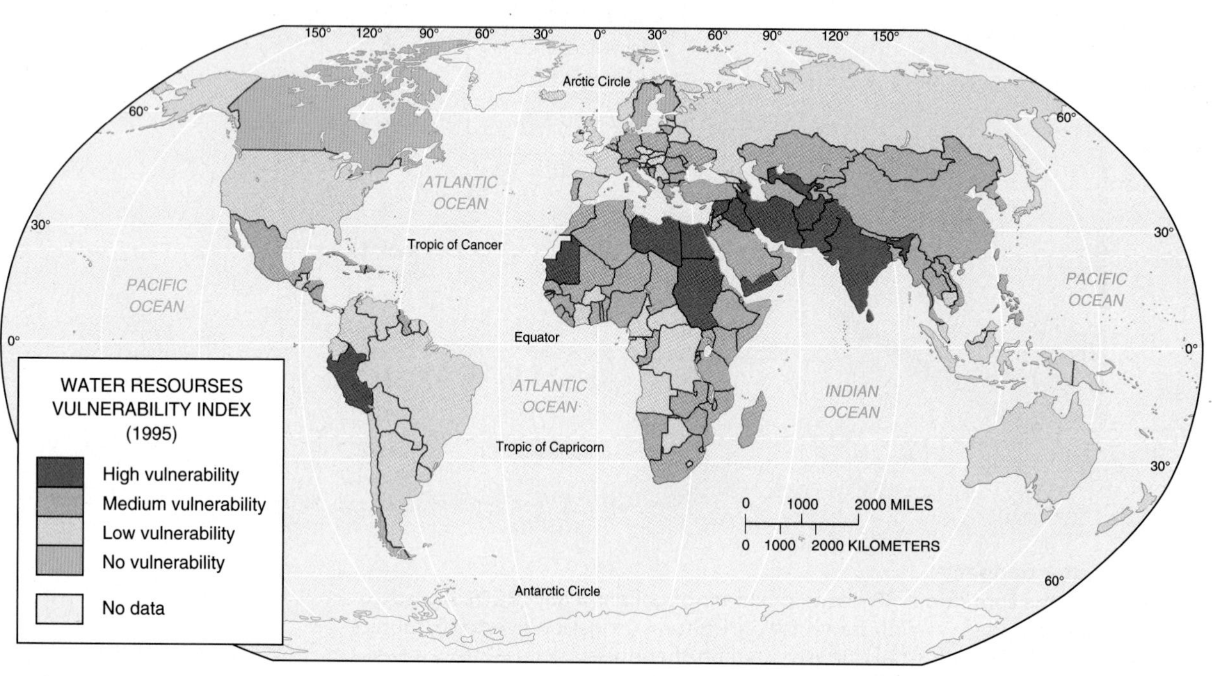

FIGURE 7.19 Global Water Scarcity.
Water resource vulnerability, a composite index. [Compiled by The World Resources Institute from data gathered by the Stockholm Environment Institute, in *Comprehensive Assessment of the Freshwater Resources of the World*, appearing in *World Resources 1998–1999*, New York: The World Resources Institute, 1998, p. 223.]

limits on agriculture and industrial production, threats to human health, and increased potential for international conflict.*

Clearly, cooperation is needed, yet we continue toward a water resource crisis without the guiding concept of a *world water economy* as a frame of reference. When will more international coordination begin, and which country or group of countries will lead the way to sustain future water resources? The author of *The World's Water 2000–2001* summarized the present condition and portrayed two alternatives in the journal *Nature*:

> Not a moment too soon, the world is awakening to the need to rethink fundamentally the way freshwater resources are distributed, managed, and used. In an era of technological breakthroughs and the wonders of the information revolution, millions die each year from preventable water-related diseases, and hundreds of millions more suffer from debilitating illnesses.... Two paths lie before us: a "hard path" that will rely almost exclusively on centralized infrastructure to capture, treat, and deliver water supplies

*S. L. Postal, and others, "Human appropriation of renewable fresh water," *Science* 271, no. 5250 (February 9, 1996), p. 785.

[supply focus, centralized, capital intensive]; and a "soft path" that will complement the former by investing in decentralized facilities, efficient technologies and policies, and human capital [demand focus, decentralized, labor and consumer intensive]. This soft path will seek to improve overall productivity rather than to find new sources of supply.*

*Peter H. Gleick, "Soft Water Paths," *Nature* 418 (July 25, 2002), p. 373.

Summary and Review—Water Resources

● ***Illustrate*** **the hydrologic cycle with a simple sketch and *label* it with definitions for each water pathway.**

The flow of water links the atmosphere, ocean, and land through energy and matter exchanges. The **hydrologic cycle** is a model of Earth's water system, which has operated for billions of years from the lower atmosphere to several kilometers beneath Earth's surface. Water soaks into the subsurface through **infiltration**, or penetration of the soil surface. It further permeates soil or rock through vertical movement called **percolation**.

hydrologic cycle (p. 222)
infiltration (p. 222)
percolation (p. 222)

1. Sketch and explain a simplified model of the complex flows of water on Earth—the hydrologic cycle.
2. What are the possible routes that a raindrop may take on its way to and into the soil surface?
3. Compare precipitation and evaporation volumes from the ocean with those over the land. Describe advection flows of moisture and countering surface and subsurface runoff.

● ***Relate*** **the importance of the water-budget concept to an understanding of the hydrologic cycle, water resources, and soil moisture for a specific location.**

A **soil-water budget** can be established for any area, from region to field, of Earth's surface by measuring the precipitation input and the output of various water demands in the area considered. Water surpluses supply streams and soil moisture storage needs and satisfy potential evapotranspiration demands. Similar to financial accounting, only through budgeting can water resources be truly understood and managed.

soil-water budget (p. 223)

4. How might an understanding of the hydrologic cycle in a particular locale, or a soil-moisture budget of a site, assist you in assessing water resources? Give some specific examples.

● ***Construct*** **the water-balance equation as a way of accounting for the expenditures of water supply and *define* each of the components in the equation and its specific operation.**

The moisture supply to Earth's surface is **precipitation** (PRECIP), arriving as rain, sleet, snow, and hail. Precipitation is measured with the **rain gauge**. **Evaporation** is the net movement of free water molecules away from a wet surface into air. **Transpiration** is the movement of water through plants and back into the atmosphere; it is a cooling mechanism for plants. Evaporation and transpiration are combined into one term—**evapotranspiration**. The ultimate demand for moisture is **potential evapotranspiration** (POTET), the amount of water that *would* evaporate and transpire under optimum moisture conditions (adequate precipitation and adequate soil moisture). Evapotranspiration is measured with an **evaporation pan** (*evaporimeter*) or the more elaborate **lysimeter**.

Unsatisfied POTET is **deficit** (DEFIC). By subtracting DEFIC from POTET, we determine **actual evapotranspiration**, or ACTET. Ideally, POTET and ACTET are about the same, so that plants have sufficient water. If POTET is satisfied and the soil is full of moisture, then additional water input becomes **surplus** (SURPL), which may puddle on the surface, flow across the surface toward stream channels, or percolate through the soil underground. The *overland flow* includes precipitation and groundwater flows into river channels to make up the **total runoff** from the area.

A "savings account" of water that receives deposits and provides withdrawals as water-balance conditions change is the **soil-moisture storage** (Δ STRGE). This is the volume of water stored in the soil that is accessible to plant roots. In soil, *hygroscopic water* is inaccessible because it is a molecule-thin layer that is tightly bound to each soil particle by hydrogen bonding. As available water is utilized, soil reaches the **wilting point** (all that remains is unextractable water). **Capillary water** is generally accessible to plant roots because it is held in the soil by surface tension and hydrogen bonding between water and soil. Almost all capillary water that remains in the soil is **available water** in soil-moisture storage. After water drains from the larger pore spaces, the available water remaining for plants is termed **field capacity**, or storage capacity. When soil is saturated after a precipitation event, surplus water in the soil becomes **gravitational water** and percolates to groundwater. As **soil-moisture utilization** removes soil water, the plants work harder to extract the same amount of moisture. **Soil-moisture recharge** is the rate at which needed moisture enters the soil.

precipitation (p. 223)
rain gauge (p. 223)
evaporation (p. 224)
transpiration (p. 224)
evapotranspiration (p. 225)
potential evapotranspiration (p. 225)

evaporation pan (p. 226)
lysimeter (p. 226)
deficit (p. 226)
actual evapotranspiration (p. 226)
surplus (p. 226)
total runoff (p. 226)
soil-moisture storage (p. 227)
wilting point (p. 227)
capillary water (p. 228)
available water (p. 228)
field capacity (p. 228)
gravitational water (p. 228)
soil-moisture utilization (p. 228)
soil-moisture recharge (p. 228)

5. Discuss the meaning of this statement: "The soil-water budget is an assessment of the hydrologic cycle at a specific site."
6. What are the components of the water-balance equation? Construct the equation and place each term's definition below its abbreviation in the equation.
7. Compare the annual water-balance data for Kingsport, Tennessee; Ottawa, Ontario; and Phoenix, Arizona. What differences do you see in the components? What similarities? What does this tell you about local water resources in each area?
8. Explain how to derive actual evapotranspiration (ACTET) in the water-balance equation.
9. What is potential evapotranspiration (POTET)? How do we go about estimating this potential rate? What factors did Thornthwaite use to determine this value?
10. Explain the operation of soil-moisture storage, soil-moisture utilization, and soil-moisture recharge. Include discussion of field capacity, capillary water, and wilting point concepts.
11. In the case of silt-loam soil from Figure 7.8, roughly, what is the available water capacity? How is this value derived?
12. For water resources, what does it mean to move water geographically or across the calendar?
13. In terms of water balance and water management, explain the logic behind the Snowy Mountains Scheme in southeastern Australia.

● ***Describe*** **the nature of groundwater and *define* the elements of the groundwater environment.**

Groundwater is a part of the hydrologic cycle, but it lies beneath the surface beyond the soil-moisture root zone. Groundwater replenishment is tied to surface surpluses because groundwater does not exist independently. Excess surface water moves through the **zone of aeration**, where soil and rock are less than saturated. Eventually, the water reaches the **zone of saturation**, where the pores are completely filled with water.

The texture and the structure of the soil dictate available pore spaces, or **porosity**. **Permeable** is the degree to which water can flow through soil; **impermeable** describes the condition where flow is restricted. Permeability depends on particle sizes and the shape and packing of soil grains. An **aquifer** is a rock layer that is permeable to groundwater flow in usable amounts. An **aquiclude** (aquitard) is a body of rock that does not conduct water in usable amounts.

The upper limit of the water that collects in the zone of saturation is called the **water table**; it is the contact surface between the zones of saturation and aeration. An **unconfined aquifer** has a permeable layer on top and an impermeable one beneath. A **confined aquifer** is bounded above and below by impermeable layers of rock or sediment. The **aquifer recharge area** extends over an entire unconfined aquifer. Water in a confined aquifer is under the pressure of its own weight, creating a pressure level to which the water can rise on its own, called the *potentiometric surface*, which can be above ground level. Groundwater confined under pressure is **artesian water**; it may rise up in wells and even flow out at the surface without pumping, if the head of the well is below the potentiometric surface.

As water is pumped from a well, the surrounding water table within an unconfined aquifer will experience a **drawdown**, or become lower, if the rate of pumping exceeds the horizontal flow of water in the aquifer around the well. This excessive pumping causes a **cone of depression**. Aquifers frequently are pumped beyond their flow and recharge capacities, a condition known as **groundwater mining**.

groundwater (p. 230)
zone of aeration (p. 230)
zone of saturation (p. 231)
porosity (p. 231)
permeable (p. 231)
impermeable (p. 231)
aquifer (p. 231)
aquiclude (p. 231)
water table (p. 231)
unconfined aquifer (p. 231)
confined aquifer (p. 232)
aquifer recharge area (p. 232)
artesian water (p. 233)
drawdown (p. 233)
cone of depression (p. 233)
groundwater mining (p. 233)

14. Are groundwater resources independent of surface supplies, or are the two interrelated? Explain your answer.
15. Make a simple sketch of the subsurface environment, labeling zones of aeration and saturation and the water table in an unconfined aquifer. Then add a confined aquifer to the sketch.
16. At what point does groundwater utilization become groundwater mining? Use the High Plains aquifer example to explain your answer.
17. What is the nature of groundwater pollution? Can contaminated groundwater be cleaned up easily? Explain.

● ***Identify*** **critical aspects of freshwater supplies for the future and *cite* specific issues related to sectors of use, regions and countries, and potential remedies for any shortfalls.**

Nonconsumptive uses, or water *withdrawal*, remove water from the supply, use it, and then return it to the stream. *Consumptive* uses remove water from a stream but do not return it, so the water is not available for a second or subsequent use. Americans in the 48 contiguous states withdraw approximately one-third of the available surplus runoff for irrigation, industry, and municipal uses. Water-resource planning regionally and globally, using water-budget principles, is essential if Earth's societies are to have enough water of adequate quality.

18. Describe the principal pathways involved in the water budget of the contiguous 48 states. What is the difference between withdrawal and consumptive use of water resources? Compare these with instream uses.

19. Characterize each of the sectors withdrawing water: irrigation, industry, and municipalities. What are the present usage trends in more-developed and less-developed nations?

20. Briefly assess the status of world water resources. What challenges are there in meeting future needs of an expanding population and growing economies? How do you think world-water-economy thinking and planning might begin?

21. How does acid deposition (acid rain) tie together the issues of air pollution, energy production and consumption from Chapter 2 and Focus Study 2.2 and water resource quality in this chapter?

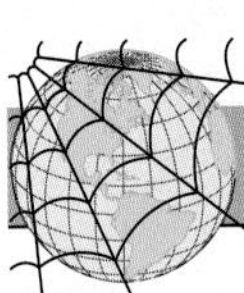

Network

The *Elemental Geosystems Home Page* provides on-line resources for this chapter on the World Wide Web. Once on the home page, click on this textbook, scroll the Table of Contents menu, select this chapter, and click "Begin." You will find self-tests that are graded, short essay and review exercises, specific updates for items in the chapter, and many links to interesting related pathways on the Internet under "Destinations." *Elemental Geosystems* is at http://www.prenhall.com/christopherson.

Critical Thinking

A. Select your campus, yard, or perhaps a houseplant, and apply soil-water-budget concepts. What is the supply of water? Estimate the ultimate water supply (PRECIP) and ultimate demand (POTET) for the area selected. Estimate water needs and how they vary with season as components of the water budget change.

If you are maintaining a house- or dorm-room plant this can be an interesting process. You meter out water to satisfy the water demands created by room temperature and the placement of the plant. You might find that work with the soil in the container improves capillary water function and soil moisture storage. You are doing the water budget!

B. You have, no doubt, consumed several glasses of water since you woke this morning. Where did this water originate? Obtain the following information: the name of the water supplier or agency, whether the supplier is using surface or groundwater to meet demands, and how the water is metered and billed. If your state or province requires water quality reporting, obtain a copy of the analysis of your tap water.

What about the water on your campus? If the campus has its own wells, how is the quality tested? Who on campus is in charge of supervising these wells?

Lastly, what is your subjective assessment of your water today: taste, smell, hardness, clarity? Compare these perceptions with others in your class.

PART THREE

Earth's Changing Landscape Systems

Dramatic Hawaiian landscape where active lava flows plunge into the ocean, forming the Kamoamoa bench and representing some of the newest land surface on Earth. Look carefully [illegible] of red-glowing molten [illegible] eruptions, beginning in 198[illegible], [illegible] the longest continuous period of active eruptions in history. Photo by Bobbé Christopherson.]

Earth is a dynamic planet whose surface is actively shaped by physical agents of change. Part 3 is organized around two broad systems of these agents—endogenic and exogenic. The **endogenic system** (Chapters 8 and 9) encompasses internal processes that produce flows of heat and material from deep below the crust, powered by radioactive decay—*the solid realm* of Earth. Earth's surface responds by moving, warping, and breaking, sometimes in dramatic episodes of earthquakes and volcanic eruptions.

At the same time, the **exogenic system** (Chapters 10–14) involves external processes that set into motion air, water, and ice, all powered by solar energy—*the fluid realm* of Earth's environment. These media are sculpting agents that carve, shape, and reduce the landscape. One such process—weathering—breaks up and dissolves the crust, making materials available for erosion, transport, and deposition by rivers, winds, wave action, and flowing glaciers. Thus, Earth's surface is the interface between two systems, one that builds the landscape and one that reduces it.

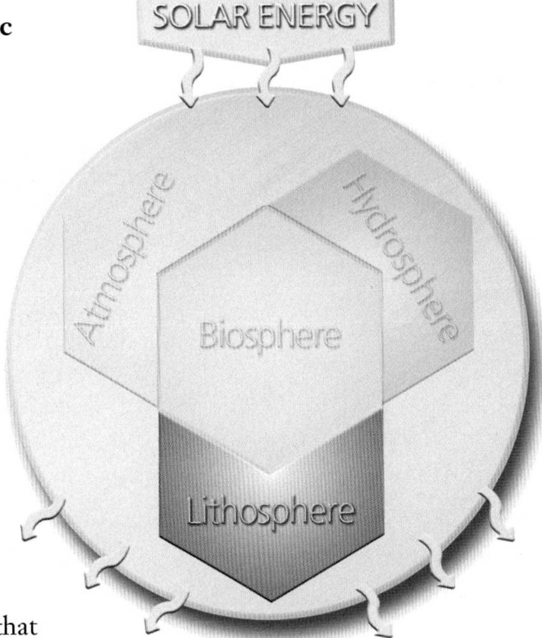

Mount Etna erupts in view of International Space Station (ISS). Visible is the ash and steam from a vigorous eruption. Gas emissions are along the north slope (lower left) through a series of vents. Smoke on the lower slope is from lava-flow set wildfires. View is to the southeast across the island of Sicily. Ashfall from this episode reached North Africa. [October 30, 2002, ISS photo courtesy of Earth Science and Image Analysis Laboratory, JSC, NASA..]

8 The Dynamic Planet

Key Learning Concepts

By knowing and understanding the key learning concepts in this chapter, you should be able to:

- *Distinguish* between the endogenic and exogenic systems, *determine* the driving force for each, and *explain* the pace at which these systems operate.
- *Diagram* Earth's interior in cross section and *describe* each distinct layer.
- *Illustrate* the geologic cycle and *relate* the rock cycle and rock types to endogenic and exogenic processes.
- *Describe* Pangaea and its breakup and *relate* several physical proofs that crustal drifting is continuing today.
- *Portray* the pattern of Earth's major plates and *relate* this pattern to the occurrence of earthquakes, volcanic activity, and hot spots.

The twentieth century was a time of great discovery about Earth's internal structure and dynamic crust, yet much remains undiscovered. We are in a time of revolution in our understanding of how the present arrangement of continents and oceans evolved. One task of physical geography is to explain the *spatial implications* of all this new knowledge and its effect on Earth's surface and society.

A new era of Earth-systems science is emerging, combining various sciences within the study of physical geography. Remember that geography is an *approach*, a way of looking at things over physical space, the *spatial science* that examines place, location, human-Earth relationships, the uniqueness of regions, and movement (see Figure 1.1). The geographic essence of geology, geophysics, paleontology (fossil study), seismology (earthquake study), and geomorphology (landform study) are all integrated to produce an overall picture of Earth's surface environment.

In this chapter: Earth's interior is organized, with a core surrounded by roughly concentric shells of material from the mantle to the crust. The interior is unevenly heated by the radioactive decay of unstable elements. Physical convection flows rise in some regions and sink in others. A rock cycle produces three classes of rocks through igneous, sedimentary, and metamorphic processes. Internal processes coupled with the rock cycle, hydrologic cycle, and tectonic cycle result in a varied crustal surface.

Earth features extensive mountain ranges both on land and the ocean floor, drifting continental and oceanic crust, and frequent earthquakes and volcanic events. All of this movement of material results from endogenic forces within Earth—the subject of this chapter.

The Pace of Change

ANIMATION **Applying relative dating principles**

The **geologic time scale** is a summary timeline of all Earth history. Figure 8.1 reflects currently accepted names of time intervals that encompass Earth's history, from vast *eons* through briefer *eras*, *periods*, and *epochs*. The sequence in this scale is based on the relative positions of rock strata above or below each other—*relative time*. An important general principle is that of *superposition*, which states that rock and sediment always are arranged with the youngest beds "superposed" near the top of a rock formation and the oldest at the base, if they have not been disturbed.

The absolute ages on the scale, determined by scientific methods such as dating by radioactive isotopes, help refine the time-scale sequence—*absolute time*. Thus, the geologic time scale is a product of both relative and absolute time measures. See News Report 8.1 for more on this absolute dating method. Also on the geologic time scale, see the labels denoting the six major extinctions of life forms in Earth history. These range from 440 million years ago (m.y.a.) to the present one caused by modern civilization. For more on the geologic time scale, see **http://www.ucmp.berkeley.edu/exhibit/geology.html**.

The guiding principle of Earth science is uniformitarianism, first proposed by James Hutton in his *Theory of the Earth* (1795) and later amplified by Charles Lyell in his *Principles of Geology* (1830). **Uniformitarianism** assumes that *the same physical processes active in the environment today have been operating throughout geologic time*. "The present is the key to the past" describes this principle. Evidence unfolding from modern scientific exploration and from the landscape record of volcanic eruptions, earthquakes, and Earth's processes support uniformitarianism.

However, geologic time is punctuated by dramatic events, such as massive landslides, earthquakes, volcanic episodes, and extraterrestrial asteroid impacts. Within the principle of uniformitarianism, these localized catastrophic events occur as small interruptions in the generally uniform processes that shape the slowly evolving landscape. Here, the *punctuated equilibrium* concept (interruptions in the flow of events, system jumps to new operation levels) studied in the life sciences and paleontology might apply to aspects of Earth's long developmental history.

We start our journey deep within the planet. A knowledge of Earth's internal structure and energy is key to understanding the surface.

News Report 8.1

Radioactivity: Earth's Time Clock

The age of Earth and the age of the earliest known crustal rock are astounding, for we think in terms of Earth's annual trips around the Sun and the pace of our own lives. We need something greater than human time to measure the vastness of geologic time. Nature has provided a way: *radiometric dating*. It is based on the steady decay of certain atoms.

An atom is composed of protons and neutrons in its nucleus. Certain forms of atoms (isotopes) have unstable nuclei; that is, the protons and neutrons do not remain together indefinitely. As particles break away and the nucleus disintegrates, radiation is emitted and the atom decays into a different element—this process is *radioactivity*.

Radioactivity provides the steady time clock needed to measure the age of ancient rocks. It works because the decay rates for different isotopes are determined precisely, and they do not vary. These rates are expressed as *half-life*, the time required for one-half of the unstable atoms in a sample to decay. Some examples of unstable elements that become stable elements and their half-lives include uranium-238 to lead-206 (4.5 billion years); thorium-232 to lead-208 (14.1 billion years); potassium-40 to argon-40 (1.3 billion years); and, in organic (living) materials, carbon-14 to nitrogen-14 (5730 years).

The presence of these decaying elements and stable end products in sediment or rock allows scientists to read the radiometric "clock." They compare the amount of original isotope in the sample with the amount of decayed end product in the sample. If the two are in a ratio of 1:1 (equal parts), one half-life has passed. Errors can occur if the sample has been disturbed or subjected to natural weathering processes that might alter its radioactivity. To increase accuracy, investigators may check a sample using more than one radiometric measurement. Calibration of the past 10,000 years is accomplished using tree-ring analysis; the past 45,000 years using fossils found in lake sediments; and the past 250,000 years using ice-core records from Greenland and Antarctica, among several tools available. Using several radiometric methods, Earth's oldest known rocks are absolutely dated.

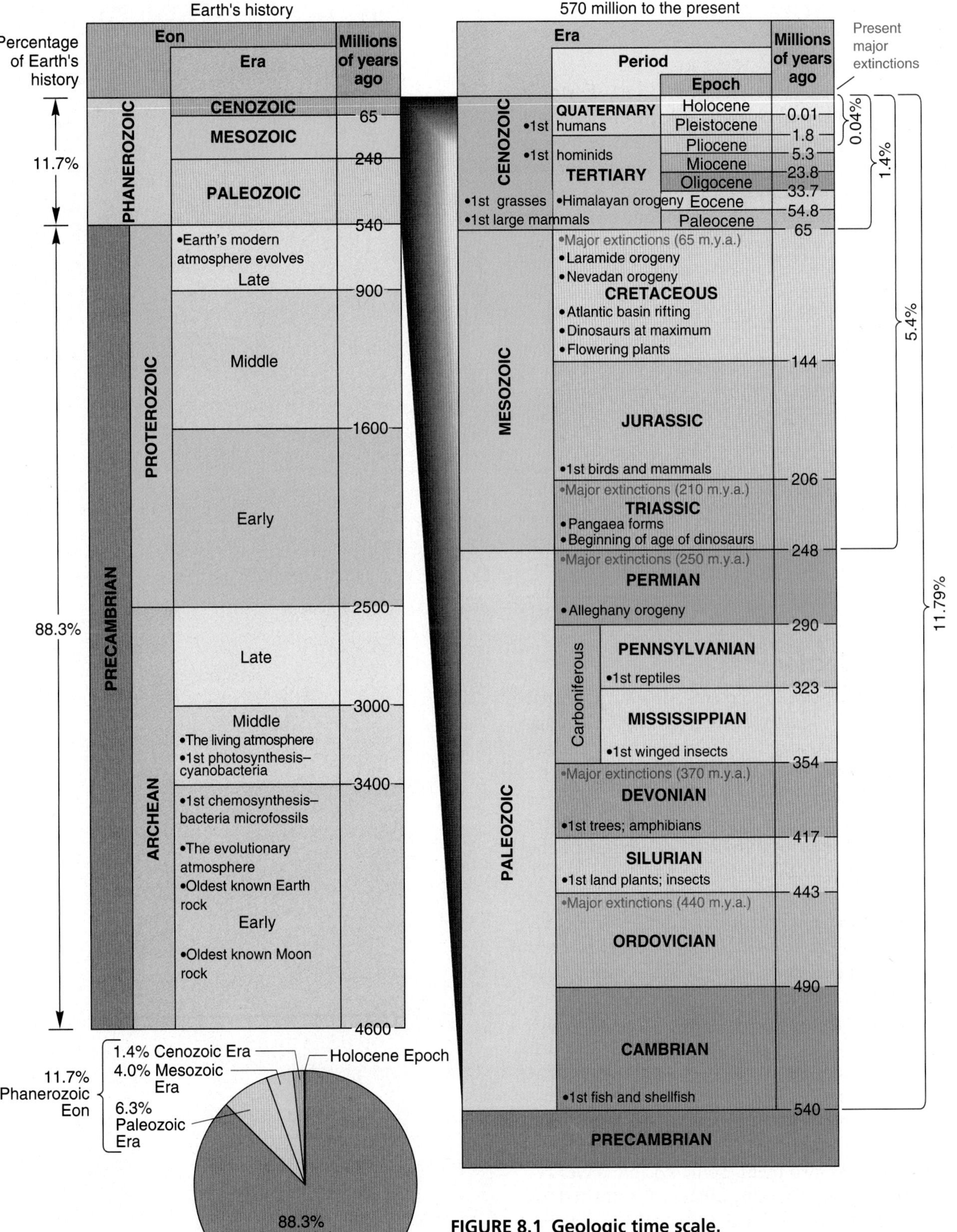

FIGURE 8.1 Geologic time scale.
Both relative and absolute dating methods calibrate the geologic time scale. Relative dating determines the sequence of events and time intervals between them. Technological means, especially radiometric dating, determine absolute dates. In the column at the left, note that more than 88% of geologic time occurred during the Precambrian Era. Highlights of Earth's history also are shown in the figure as bulleted items, including the six major extinctions of life forms (the sixth extinction episode is underway at the present time). Dates appear in m.y.a., million years ago. [Data for updating time scale from Geological Society of America.]

Earth's Structure and Internal Energy

Along with the other planets and the Sun, Earth is thought to have condensed and congealed from a nebula of dust, gas, and icy comets about 4.6 billion years ago (Chapter 2). Until recently, the oldest known surface rocks on Earth were in northwestern Canada. The Acasta Gneiss is a 3.96 billion-year-old formation; nearby the Slave Province rocks date to 4.03 billion. Rocks from Greenland date back to 3.8 billion years old.

Recent research found detrital zircons (particles of preexisting zirconium silica oxides forming rock) in Western Australia dating between 4.2 and 4.4 billion years old; these are possibly the oldest materials in Earth's crust. These discoveries tell us something significant: Earth was forming continental crust at least 4 billion years ago, during the Archean Eon.

Throughout Earth's formation process, heat was accumulating. As the protoplanet compacted into a smaller volume, energy was transformed into heat through compression. In addition, significant heat was trapped inside Earth from the decay of isotopes (unstable forms) of uranium, thorium, potassium, and other radioactive elements.

Earth in Cross Section

As Earth solidified, heavier elements slowly gravitated toward the center, and lighter elements slowly welled upward to the surface, concentrating in the crust. Consequently, Earth's interior is arranged roughly in concentric layers, each one distinct either in chemical composition or temperature, with heat radiating outward from the center by conduction and then by physical convection in the more plastic levels nearer the surface.

Our knowledge of Earth's *internal differentiation* into these layers is acquired entirely through indirect evidence, because we are unable to drill more than a few kilometers into Earth's crust. There are several physical properties of Earth materials that enable us to approximate the nature of the interior. For example, when an earthquake sends shock waves through the planet, the cooler areas, which generally are more rigid, transmit these **seismic waves** at a higher velocity than do the hotter areas.

Density also has an effect on seismic velocities, and fluid or plastic zones simply do not transmit some types of seismic waves; they absorb them. Some seismic waves reflect off interior density changes; others are refracted, or bent, as they travel through Earth. Thus, the distinctive ways in which seismic waves pass through Earth and the time they take to travel between two surface points help seismologists deduce the structure of Earth's interior (Figure 8.2).

Figure 8.3 illustrates the dimensions of Earth's interior in comparison to surface distances in North America to give you a sense of scale. An airplane flying from Halifax, Nova Scotia, to San Francisco would travel the same distance as that from Earth's center to its surface. Note, on

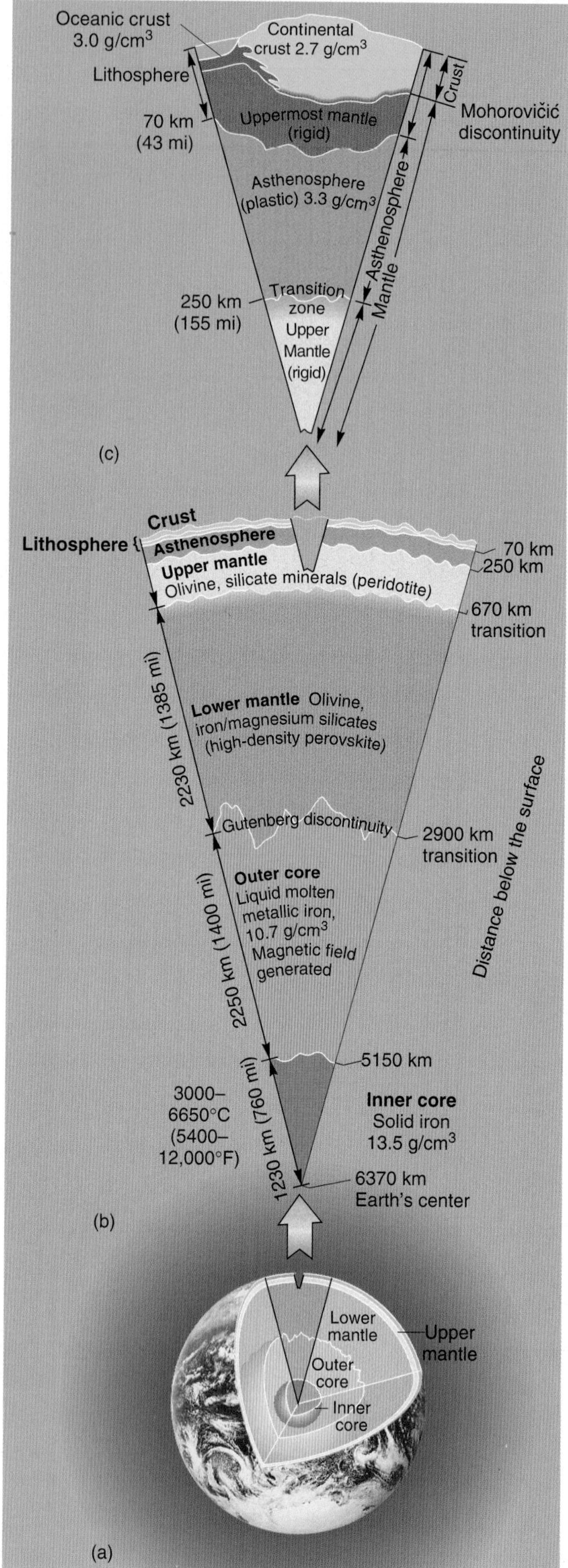

FIGURE 8.2 Earth in cross section.
(a) Cutaway showing Earth's interior. (b) Earth's interior in cross section, from inner core to the crust. (c) Detail of the structure of the lithosphere and its relation to the asthenosphere. [Photo from NASA.]

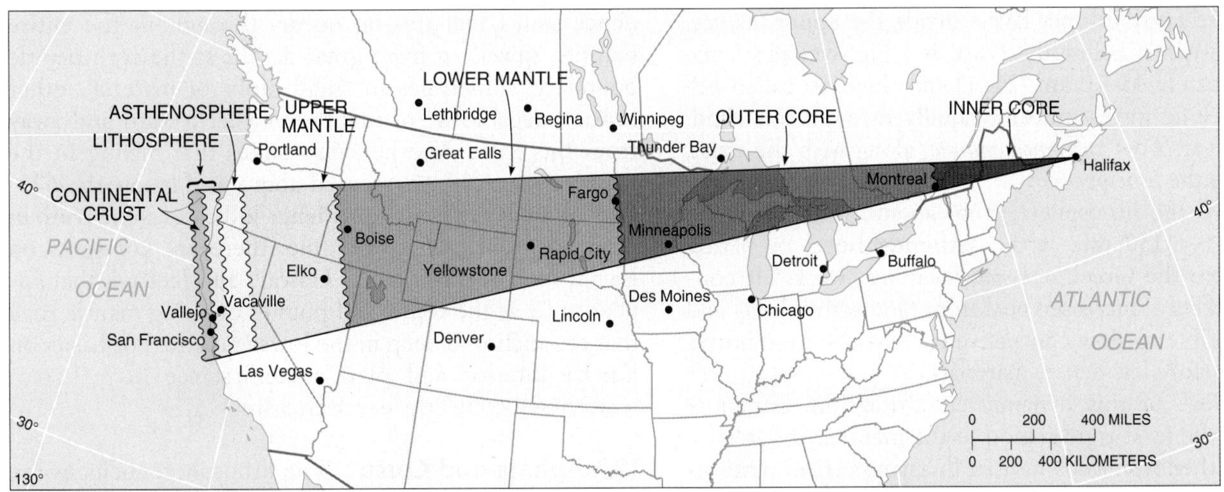

FIGURE 8.3 Distances from core to crust.
Surface map, using the distance from Halifax, Nova Scotia, to San Francisco to compare the distance from Earth's center to the surface. The continental crust thickness is the same as the distance between Vallejo (in the eastern portion of San Francisco Bay) and the city of San Francisco.

the West Coast, the thinness of the crust, extending only from Vallejo to San Francisco.

Earth's Core. A third of Earth's entire mass, but only a sixth of its volume, lies in its dense core. The **core** is differentiated into two regions—*inner core* and *outer core*—divided by a transition zone of several hundred kilometers (Figure 8.2). Estimated core temperatures range from 3000°C (5400°F) to as high as 6650°C (12,000°F). The inner core is thought to be solid iron and well above the melting temperature of iron at the surface. It remains solid because of tremendous pressures. The iron is not pure but probably is combined with silicon, and possibly oxygen and sulfur. The outer core is molten, metallic iron with lighter densities than the inner core.

Earth's Magnetism. The fluid outer core generates at least 90% of Earth's magnetic field and the magnetosphere that surrounds and protects the surface from the solar wind. One hypothesis for Earth's magnetic field describes circulation patterns in the outer core that are influenced by Earth's rotation. A new discovery suggests that Earth's solid inner core rotates slightly faster than the rest of the planet. This circulation generates electric currents, which in turn induce the magnetic field.

An intriguing feature of Earth's magnetic field is that its polarity sometimes fades to zero and then returns to full strength with north and south magnetic poles reversed. The field does not blink on and off but instead oscillates slowly to low intensity and then rapidly regains its full strength. This **geomagnetic reversal** has taken place nine times during the past 4 million years and hundreds of times over Earth's history. The average period of a magnetic reversal is 500,000 years; occurrences possibly vary from as short as several thousand years to as long as tens of millions of years.

There appears to be a trend in recent geologic time toward more frequent reversals with shorter intervals. When Earth is without polarity in its magnetic field, a random pattern of magnetism results. The effects of these low-intensity episodes on life are still speculative, but without a magnetosphere, the surface environment is unprotected from cosmic radiation and solar wind.

The reasons for these magnetic reversals are unknown; however, the spatial patterns they create at Earth's surface give us important evidence to understand the evolution of landmasses and the movement of the continents. When new iron-bearing rocks solidify from molten material at Earth's surface, small magnetic particles in the rocks align according to the orientation of the magnetic poles at that time. This pattern is then locked in place as the rocks cool and harden. All across Earth, rocks of comparable types and ages bear this identical record of magnetic reversals. In addition, these magnetic patterns record the migration of the continents over time in relation to Earth's magnetic poles.

Earth's Mantle. A transition zone of several hundred kilometers divides the outer core from the mantle. Scientists at the California Institute of Technology analyzed more than 25,000 earthquakes and determined that this transition area is bumpy and uneven, with ragged peak-and-valley-like formations. Some of the motions in the mantle may be created by this rough texture at what is called the *Gutenberg discontinuity* (Figure 8.2b). A *discontinuity* is a place where physical differences occur between adjoining regions in Earth's interior or surface rock, such as between the outer core and lower mantle.

The **mantle** is about 80% of Earth's total volume. It is rich in oxides and silicates of iron and magnesium (FeO, MgO, and SiO_2) and is denser and tightly packed at depth, grading to lesser densities toward the surface. Of the mantle's volume, 50% is in the lower mantle, which is composed of rock that probably has never been at Earth's surface. A broad transition zone of several hundred kilometers separates the upper mantle from the lower mantle. The entire mantle experiences a gradual temperature increase with depth.

Three fairly distinct layers divide the upper mantle. Outermost, just below the crust, is a high-velocity zone approximately 45–70 km (25–43 mi) thick, so called because seismic waves transmit rapidly through this rigid, cooler layer. This *uppermost mantle*, along with the crust, makes up the *lithosphere*.

Below the lithosphere, from about 70 km down to 250 km (43–155 mi), is the **asthenosphere**, or plastic layer, from the Greek *asthenos*, meaning "weak." It contains pockets of increased heat from radioactive decay and is susceptible to slow convective currents in these hotter (and therefore less dense) materials.

Because of this dynamic condition, the asthenosphere is the least rigid region of the mantle. About 10% of the asthenosphere is molten in asymmetrical patterns and hot spots. The resulting slow movement in this zone disturbs the overlying crust and creates tectonic activity—folding, faulting, and general deformation of surface rocks. Below the asthenosphere resides the third layer, the rest of the upper mantle. Here the rocks are solid again, a zone of greater density, high seismic velocity, and increasing pressures.

The depth affected by convection currents is the subject of much scientific research. One body of evidence states that mixing occurs throughout the entire mantle, upwelling from great depths at the core-mantle boundary, sometimes in small blobs of material, other times "megablobs" of mantle convect toward and away from the crust. Another view states that mixing in the mantle is layered, segregated above and below the 670-km boundary. Presently, evidence indicates some truth in both positions. As an example, there are hot spots on Earth, such as those under Hawai'i and Iceland, that appear to be at the top of tall plumes of rising mantle rock that are anchored deep in the lower mantle. For basics on Earth's interior and plate tectonics, see http://www.solarviews.com/eng/earthint.htm.

Lithosphere and Crust. The lithosphere includes the entire **crust**, which represents only 0.01% of Earth's mass, and the uppermost mantle down to a depth of about 70 km (43 mi). The boundary between the crust and the high-velocity portion of the lithospheric upper mantle is another discontinuity called the **Mohorovičić discontinuity**, or **Moho** for short, named for the Yugoslavian seismologist who determined that seismic waves change at this depth, owing to sharp contrasts of materials and densities.

News Report 8.2

Drilling the Crust to Record Depths

Scientists wanting to sample mantle material directly have unsuccessfully tried for decades to penetrate Earth's crust to the Moho discontinuity. The longest-lasting deep-drilling attempt is on the northern Kola Peninsula near Zapolyarny, Russia, 250 km north of the Arctic Circle—the *Kola Borehole* (*KSDB*). Twenty years of high-technology drilling (1970–1989) produced a hole 12,262 km deep (7.6 mi, or 40,230 ft), purely for exploration and science. Crystalline rock 1.4 billion years old at 180°C (356°F) was reached. The site has other active boreholes, and the fifth is underway. (See an analysis log at http://www.icdp-online.de/.) A record holder for depth of a gas well is in Oklahoma; it was stopped at 9750 m (32,000 ft) when the drill bit ran into molten sulfur.

FIGURE 1 Ocean drilling ship. A modern ocean-floor drilling ship, the *JOIDES Resolution*. The International Ocean Drilling Program operates the research ship. Since it began operations in 1984, *JOIDES* has spent more than 5000 days at sea, traveled more than half a million kilometers, drilled 1555 holes into the seafloor, and recovered almost 200,000 m (650,000 ft) of core for analysis. [Photo courtesy of ODP/Texas A & M University.]

Oceanic crust is thinner than continental crust and is the object of several drilling attempts. The International Ocean Drilling Program (ODP), a cooperative effort directed by Texas A & M University, drilled a 2.5 km (1.6 mi) deep hole in an oceanic rift near the Galápagos Islands from 1975 to 1993, with further drilling planned at the site (Figure 1). An ocean floor of distinct layers of lava and deeper magma chambers was found. But the Moho and Earth's mantle remain untapped. See http://www-odp.tamu.edu/ or http://www.oceandrilling.org/ for the ODP home page and information. Construction of a new largest deep-ocean drilling ship is ongoing in Okayama, Japan, with completion scheduled for 2006. The new ODP ship will be able to drill to 7 km, more than three times the *JOIDES Resolution*.

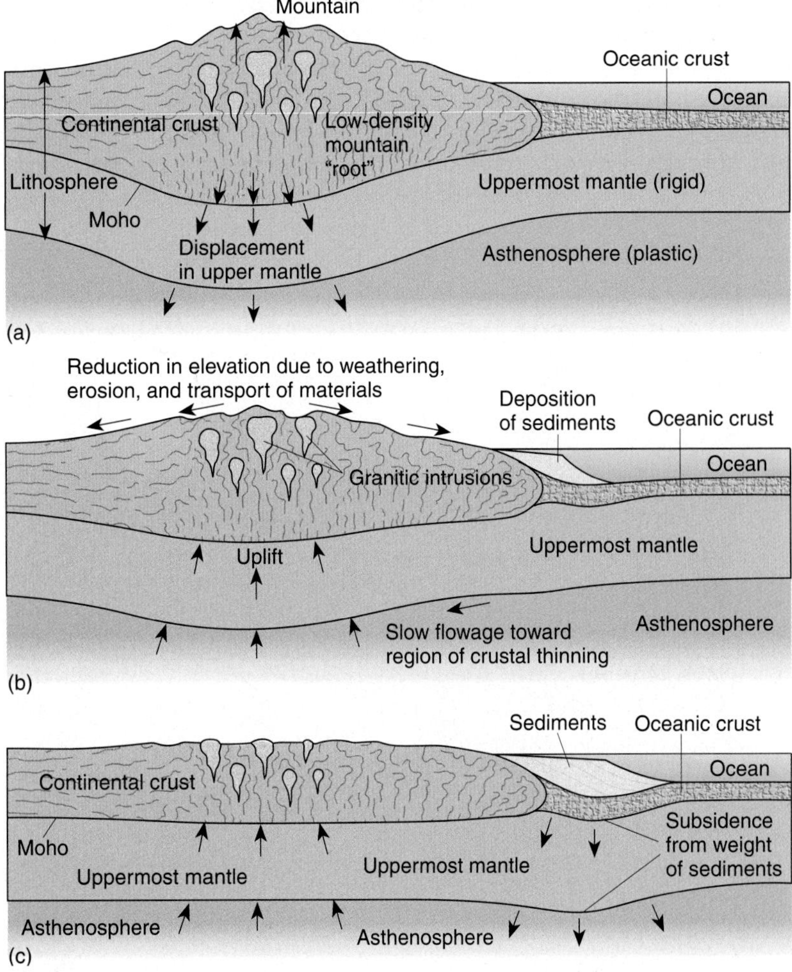

FIGURE 8.4 Isostatic adjustment of the crust. Earth's entire crust is in a constant state of compensating adjustment as suggested by these three sequential stages: (a) The mountain mass slowly sinks, displacing mantle material. (b) Because of the loss of mass through erosion and transportation, the crust isostatically adjusts upward, and sediments accumulate in the ocean. (c) As the continent thins, the heavy sediment load offshore begins to deform the lithosphere beneath the ocean.

Figure 8.2c illustrates the relationship of the crust to the rest of the lithosphere and the asthenosphere below. Crustal areas below mountain masses extend lower into the asthenosphere, perhaps to 50–60 km (31–37 mi), whereas the crust beneath continental interiors averages about 30 km (19 mi) in thickness, and oceanic crust averages only 5 km (3 mi). Drilling into the mantle through the crust remains an elusive scientific goal (News Report 8.2).

The composition, density, and texture of continental and oceanic crust are quite different. Continental crust is basically **granite**. It is crystalline and high in silica, aluminum, potassium, calcium, and sodium. (Sometimes continental crust is called *sial*, shorthand for *si*lica and *al*uminum.) Oceanic crust is **basalt**. It is granular and high in silica, magnesium, and iron. (Sometimes oceanic crust is called *sima*, shorthand for *si*lica and *ma*gnesium.)

Buoyancy is the principle that something less dense, such as wood, floats in something denser, such as water. The principles of buoyancy and balance were combined in the 1800s into the important principle of **isostasy**. Isostasy explains certain vertical movements of Earth's crust.

Think of Earth's outer crust as floating on the denser layers beneath, much as a boat floats on water. With a greater load (for example, glaciers, sediment, mountains), the crust tends to ride lower in the asthenosphere. Without that load (for example, when a glacier melts), the crust rides higher—an uplift known as an *isostatic rebound*. For example, with the rapid loss of glacial ice across southern Alaska since the mid-1990s, scientists are measuring relatively rapid isostatic uplift of the region—some 36 mm (1.4 in.) a year. Thus, the entire crust is in a constant state of compensating adjustment, or isostasy, slowly rising and sinking in response to its own weight; pushed and dragged about by currents in the asthenosphere (Figure 8.4).

Earth's crust is the outermost irregular shell that resides restlessly on a dynamic and diverse interior. Let us now examine the processes at work on this crust and the variety of rock types that compose the landscape.

The Geologic Cycle

The rock cycle

Earth's crust is in an ongoing state of change, being formed, deformed, moved, and broken by physical, chemical, and biological processes. The endogenic (internal) system is at work building landforms while the exogenic (external) system is wearing them down. This vast give-and-take at the Earth-atmosphere interface is the **geologic cycle**. The overall cycle is fueled by Earth's internal heat and by solar energy, influenced by the leveling force of

Earth's gravity. Figure 8.5 illustrates the geologic cycle, combining many of the elements discussed in this text.

As you can see in the figure, the geologic cycle is composed of three principal cycles—hydrologic, rock, and tectonic. The *hydrologic cycle*, along the top of Figure 8.5, through erosion, transportation, and deposition, processes Earth materials with the chemical and physical action of water, ice, and wind. The *rock cycle* produces the three basic rock types found in the crust—igneous, metamorphic, and sedimentary. The *tectonic cycle* brings heat energy and new material to the surface and recycles material, creating movement and deformation of the crust.

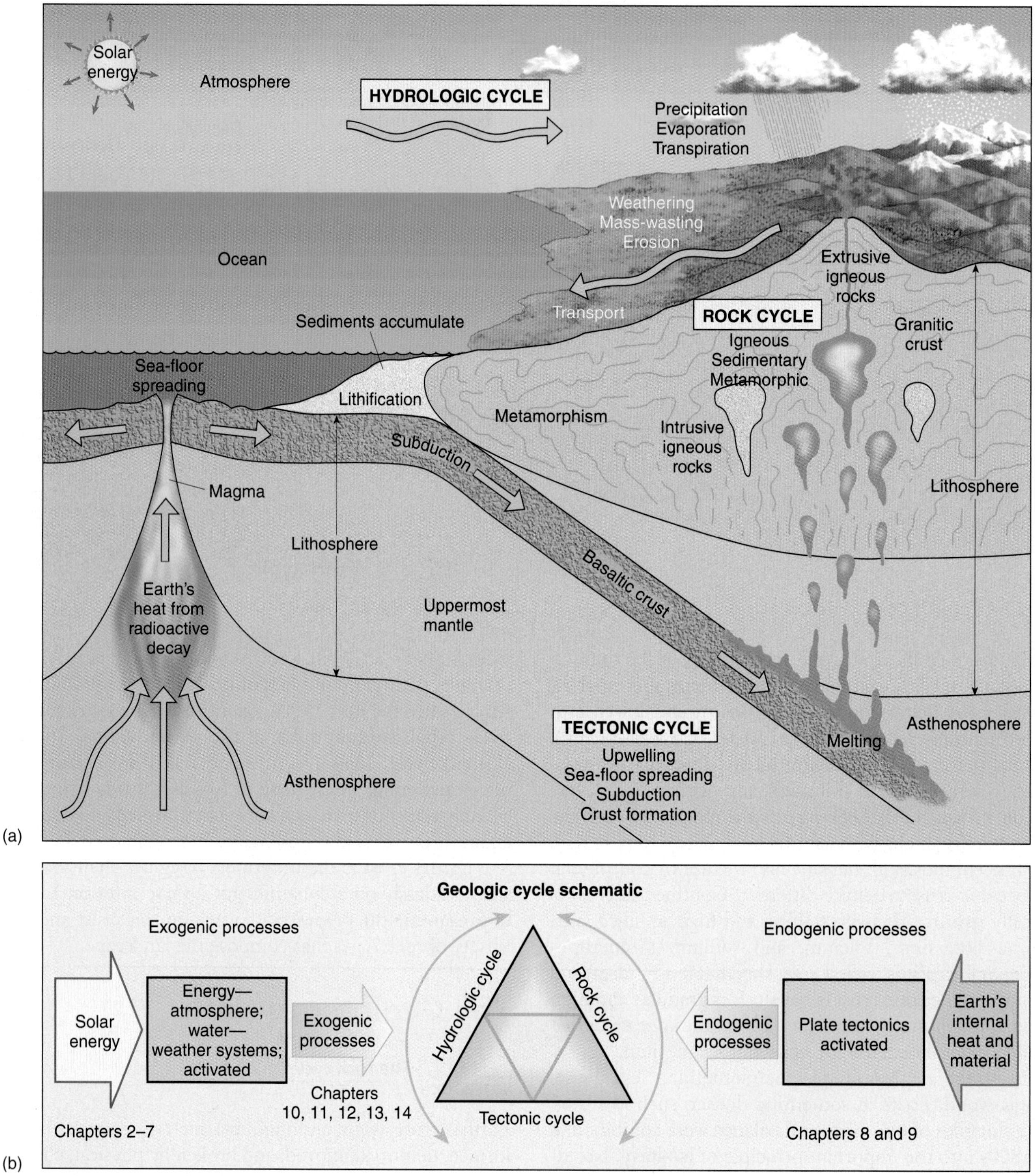

(a)

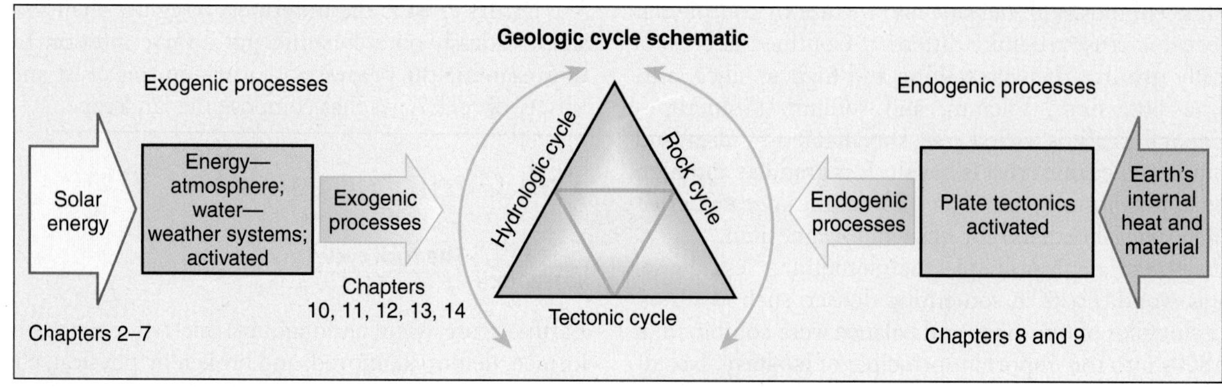

(b)

FIGURE 8.5 The geologic cycle.
(a) The geologic cycle, showing the interactive relationship of the rock cycle, tectonic cycle, and hydrologic cycle, as endogenic (internal) and exogenic (external) processes operate at or near Earth's surface. (b) Note these process-systems on the schematic diagram and the related coverage of chapters in this book.

Rock Cycle

ANIMATION Formation of intrusive ingneous features
Foliation (metamorphic rock)

Only eight natural elements compose 99% of Earth's crust, and only two of these—oxygen and silicon—account for 74.3% (Table 8.1). Oxygen is the most reactive gas in the lower atmosphere, readily combining with other elements. For this reason, the percentage of oxygen is higher in the crust than the 21% concentration in the atmosphere. The internal differentiation process in which less dense elements migrated toward the surface explains the relatively large percentages of lightweight elements such as silicon and aluminum in the crust.

A **mineral** is an element or combination of elements that forms an inorganic natural compound. A mineral can be described with a specific symbol or chemical formula and possesses specific qualities, including a crystalline structure. One of the most widespread mineral families on Earth is the *silicates*, because silicon (Si) readily combines with oxygen and other elements. The common mineral *quartz* is silicon dioxide, and has a distinctive six-sided crystal (see photo in Table 8.1). This mineral family also includes feldspar, clay minerals, and numerous gemstones.

Another important mineral family is the *carbonate* group, which features carbon in combination with oxygen and other elements such as calcium, magnesium, and potassium. An example is the mineral calcite ($CaCO_3$), a form of calcium carbonate. Of the nearly 4200 minerals, only 30 are common, with just 10 of those making up 90% of the minerals in the crust. *Mineralogy* is the study of the composition, properties, and classification of minerals (see **http://webmineral.com/** or **http://www.minsocam.org/MSA/Research_Links.html**).

A **rock** is an assemblage of minerals bound together (such as granite, a rock containing three minerals), or it may be a mass of a single mineral (such as rock salt). Literally thousands of rocks have been identified, all the result of three rock-forming processes: *igneous* (melted), *sedimentary* (from settling out), and *metamorphic* (altered). Figure 8.6 illustrates the interrelationships between these three processes in the *rock cycle*. The next three sections examine each process.

Igneous Processes. Rocks that solidify and crystallize from a molten state are **igneous rocks**. They form from **magma**, which is molten rock beneath the surface (hence the name *igneous*, which means "fire-formed" in Latin). Magma is fluid, highly gaseous, and under tremendous pressure. It is either *intruded* into crustal rocks, known as country rock, or *extruded* onto the surface as **lava**.

The cooling history of the rock—how fast it cooled and how steadily the temperature dropped—determines its texture and degree of crystallization. Textures range from coarse-grained (slower cooling, with more time for larger crystals to form) to fine-grained or glassy (faster cooling). Most rocks in the crust are igneous, although sedimentary rocks, soil, or the ocean frequently cover them. Familiar igneous examples are granite, basalt, and rhyolite. Figure 8.7 illustrates the variety of occurrences of igneous rocks, both on and beneath Earth's surface.

Intrusive igneous rock that cools slowly in the crust forms a **pluton**, a general term for any intrusive igneous rock body, regardless of size or shape. It is named after the Roman god of the underworld, Pluto. The largest pluton form is a **batholith** (Figure 8.7), defined as an irregular-shaped mass with a surface exposure greater than 100 km^2 (40 mi^2) that has invaded layers of crustal rocks. Batholiths form the mass of many large mountain ranges, for example, the Sierra Nevada batholith in California, the Idaho batholith, and the Coast Range batholith of Canada and extreme northern Washington State (Figure 8.8a).

Basalt is the most common fine-grained extrusive igneous rock, here on Earth and in the Solar System. It makes up the bulk of the ocean floor, constituting 71% of

Table 8.1 Common Elements in Earth's Crust

Element	Percentage of Earth's Crust by Weight
Oxygen (O)	46.6
Silicon (Si)	27.7
Aluminum (Al)	8.1
Iron (Fe)	5.0
Calcium (Ca)	3.6
Sodium (Na)	2.8
Potassium (K)	2.6
Magnesium (Mg)	2.1
All others	1.5
Total	100.00

Note. A quartz crystal (SiO_2) consists of Earth's two most abundant elements, silicon (Si) and oxygen (O). Inset photo from *Laboratory Manual in Physical Geology*, 3rd ed., R. M. Busch, ed. © 1993 by Macmillan Publishing Co.

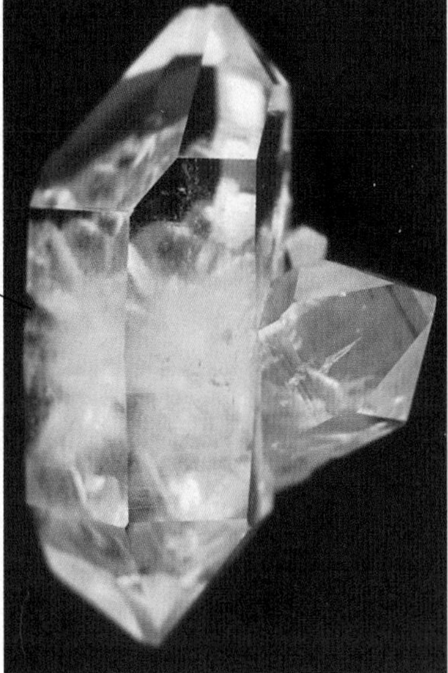

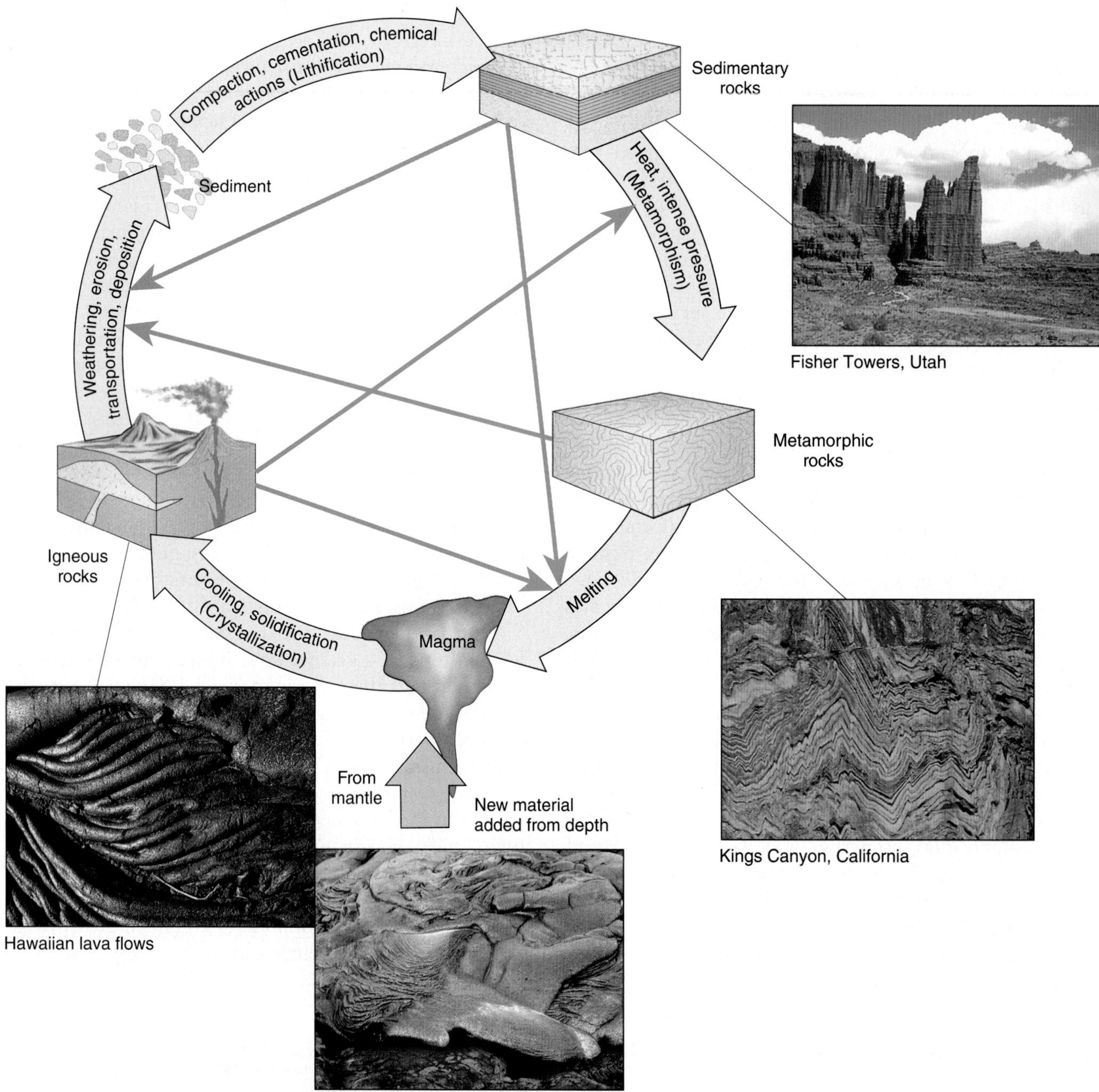

FIGURE 8.6 The rock cycle.
A rock-cycle schematic demonstrates the relation among igneous, sedimentary, and metamorphic processes. The arrows indicate that each rock type can enter the cycle at various points and be transformed into other rock types. [Adapted by permission from R. M. Busch, ed., *Laboratory Manual in Physical Geology,* 3rd ed., © 1993 by Macmillan Publishing Co. Photos by Bobbé Christopherson.]

Earth's surface. Lava flows such as those on the big island of Hawai'i cool to form basalt (Figure 8.8b).

Sedimentary Processes. Most sedimentary rocks are derived from existing rocks or from organic materials such as bone and shell. The exogenic processes of weathering and erosion provide the raw materials needed to form sedimentary rocks. Bits and pieces of former rocks—principally quartz, feldspar, and clay minerals—are eroded and then mechanically transported (by water, ice, wind, and gravity) to other sites where they are deposited. In addition, some minerals go into solution and form sedimentary deposits by precipitating from those solutions; this is an important process in the oceanic environment.

Various sedimentation forms are created in different environments—deserts, glaciers, beaches, the tropics, and so on. Characteristically, sedimentary rocks are laid down by wind, water, and ice in horizontally layered beds.

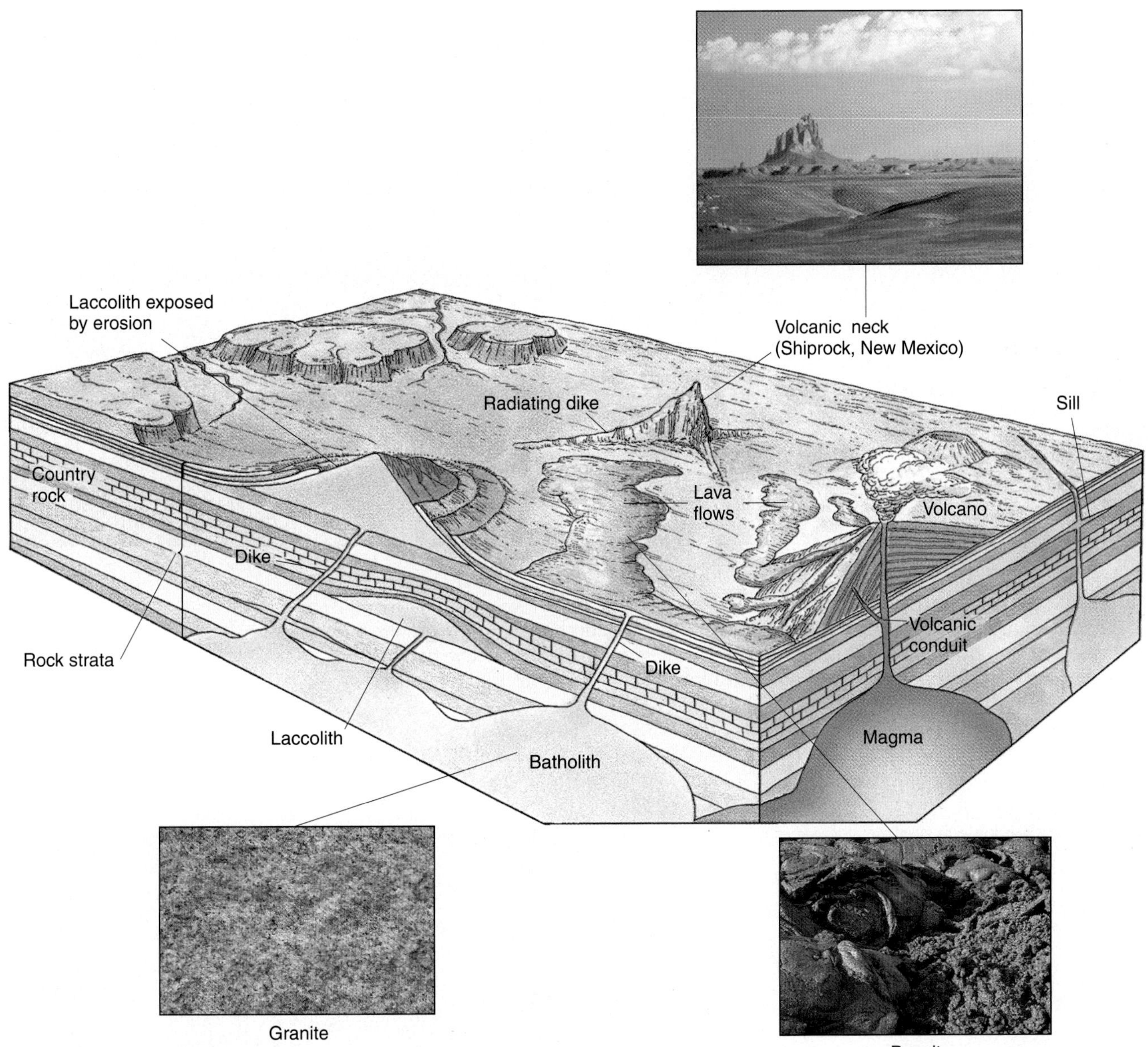

FIGURE 8.7 Igneous rock types.
The variety of occurrences of igneous rocks, both intrusive (below the surface) and extrusive (on the surface). Photographs show samples of White Mountain Conway granite (intrusive), Hawaiian basalt (extrusive), and the Shiprock volcanic neck. [Granite and basalt photos by Bobbé Christopherson. Photo of Shiprock volcanic neck by author.]

These layered strata are important records of past ages, and **stratigraphy** is the study of their sequence (superposition), thickness, and spatial distribution—clues to the age and origin of the rocks. Figure 8.9a is a sandstone sedimentary rock in a desert landscape. Note the various layers in the formation and how differently they resist erosional processes. The various layers visible in the rock formation tell the story of past floods and wetter times, and past droughts and drier times. Ancient sand dunes are hardened near the top of the rock.

The two primary sources of sedimentary rocks are the mechanically transported bits and pieces of former rock, called *clastic sediments*, and the minerals in solution, called *chemical sediments*. Clastic sedimentary rocks, such as sandstone, are derived from weathered and fragmented rocks that are further worn in transport—everything from boulders to microscopic clay particles. The process of cementation, compaction, and hardening of sediments into **sedimentary rocks** is called **lithification**. Various cements fuse rock particles together; lime ($CaCO_3$, or calcium carbonate) is the most common, followed by iron oxides (Fe_2O_3) and silica (SiO_2). Drying (dehydration), heating, or chemical reactions also unite particles.

Chemical sedimentary rocks are not formed from physical pieces of broken rock but instead are dissolved, transported in solution, and chemically precipitated out of solution (they are essentially nonclastic). The most common chemical sedimentary rock is **limestone**, which is lithified *calcium carbonate*, $CaCO_3$. A similar form is *dolomite*, which is lithified calcium-magnesium carbonate,

(a)

(b)

FIGURE 8.8 Intrusive and extrusive rocks.
(a) Exposed granites of the Sierra Nevada batholith. The boulders sitting on the granite are erratics left behind by glacial ice that melted thousands of years ago. (b) Active basaltic lava flows on Hawai'i in 2002. [Photos by Bobbé Christopherson.]

(a)

(b)

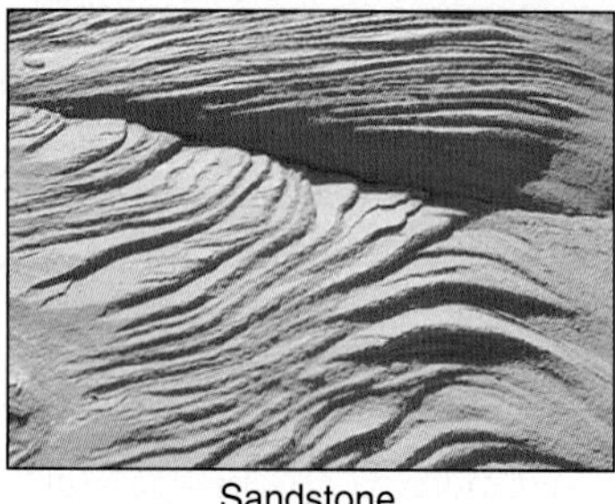
Sandstone

Limestone

FIGURE 8.9 Sedimentary rock types.
Two types of sedimentary rock. (a) Sandstone formation with sedimentary strata subjected to differential weathering; note weaker underlying siltstone. (b) A limestone landscape, formed by chemical sedimentary processes, in south-central Indiana. [(a) Photos by author; (b) photo by Bobbé Christopherson; inset photo by Richard M. Busch.]

(a)

(b)

FIGURE 8.10 Death Valley, wet and dry.
A Death Valley landscape (a) one day after a record rainfall event when the valley was covered by several square kilometers of water only a few centimeters deep. (b) One month later the water evaporated and the same playa (lake bed) is coated with evaporites (borated salts). [Photos by author.]

$CaMg(CO_3)_2$. About 90% of all limestone comes from organic sources; it is biochemical—derived from shell and bone produced by biological activity. Once formed, these rocks are vulnerable to chemical weathering that produces unique landforms, as discussed in the weathering section of Chapter 10 and as shown in Figure 8.9b.

Chemical sediments also form from inorganic sources when water evaporates and leaves behind a residue of salts. These **evaporites** may exist as common salts, such as gypsum or sodium chloride (table salt), to name only two, and often appear as flat, layered deposits across a dry landscape. This process is dramatically demonstrated in the pair of photographs in Figure 8.10, taken in Death Valley National Park one day and one month after a record 2.57-cm (1.01-in.) rainfall.

Chemical deposition also occurs in the water of natural hot springs from chemical reactions between minerals and oxygen; this is a sedimentary process related to *hydrothermal activity.* A deposit of travertine, a form of calcium carbonate, at Mammoth Hot Springs in Yellowstone National Park is an example (Figure 8.11).

Metamorphic Processes. Any rock, either igneous or sedimentary, may be transformed into a **metamorphic rock** by going through profound physical and/or chemical changes under pressure and increased temperature. (The name *metamorphic* comes from the Greek, meaning to "change form.") Metamorphic rocks generally are more compact than the original rock and therefore are harder and more resistant to weathering and erosion (Figure 8.12a).

The ancient roots of mountains are composed predominantly of metamorphic rocks. Exposed at the bottom of the inner gorge of the Grand Canyon in Arizona, a Precambrian metamorphic rock called the Vishnu Schist is a remnant of such an ancient mountain root (Figure 8.12b).

Several conditions can cause metamorphism, particularly when subsurface rock is subjected to high temperatures and high compressional stresses occurring over millions of years. Igneous rocks are compressed during collisions of portions of Earth's crust (described under the section Plate Tectonics later in this chapter), or they may be sheared and stressed along earthquake fault zones. Sometimes they simply are crushed under great weight when a crustal area is thrust beneath other crust.

Another metamorphic condition occurs when sediments collect in broad depressions in Earth's crust and, because of their own weight, create enough pressure in the bottommost layers to transform the sediments into metamorphic rock, termed *regional metamorphism.* Also, molten magma rising within the crust may *cook* adjacent rock, a process called *contact metamorphism.*

FIGURE 8.11 Hydrothermal deposits.
Mammoth Hot Springs, Yellowstone National Park, is an example of a hydrothermal deposit principally composed of travertine ($CaCO_3$), deposited as a chemical precipitate from heated spring water as it evaporated. [Photo by author.]

(a)

(b)

FIGURE 8.12 Metamorphic rocks.
(a) A metamorphic rock outcrop in Greenland, the Amitsoq Gneiss, at 3.8 billion years old, is the second-oldest rock found on Earth. (b) Precambrian Vishnu Schist composes these rock cliffs in the inner gorge of the Grand Canyon in northern Arizona. Despite the fact that the schist is harder than steel, the Colorado River has cut down into the uplifted Colorado Plateau, exposing and eroding these ancient rocks. [(a) Photo by Kevin Schafer/Peter Arnold, Inc.; (b) photo by author.]

Table 8.2 Metamorphic Rocks

Parent Rock	Metamorphic Equivalent	Texture
Shale (clay minerals)	Slate	Foliated
Granite, slate, shale	Gneiss	Foliated
Basalt, shale, peridotite	Schist	Foliated
Limestone, dolomite	Marble	Nonfoliated
Sandstone	Quartzite	Nonfoliated

Metamorphic rocks may be changed both physically and chemically from the original rocks. If the mineral structure demonstrates a particular alignment after metamorphism, the rock is *foliated*, and some minerals may appear as wavy striations (streaks or lines) in the new rock. In contrast, parent rock with a more homogeneous (evenly mixed) makeup may produce a *nonfoliated* rock. Table 8.2 lists some metamorphic rock types and their parent rocks. The two inset photographs demonstrate foliated and nonfoliated textures.

You have seen how the rock-forming processes yield the igneous, sedimentary, and metamorphic materials of Earth's crust. Next we look at how vast *tectonic processes* press, push, and drag portions of the crust in large-scale movements, causing the continents to *drift*.

Plate Tectonics

Have you ever looked at a world map and noticed that a few of the continental landmasses appear to have matching shapes like pieces of a jigsaw puzzle—particularly South America and Africa? The incredible reality is that the continental pieces once did fit together! Continental landmasses not only migrated to their present locations but continue to move at speeds up to 6 cm (2.4 in.) per year.

We say that the continents are *adrift* because convection currents in the asthenosphere and upper mantle are dragging them around. The key point is that the arrangement of continents and oceans we see today is not permanent but is in a continuing state of change.

A continent such as North America is actually a collage of crustal pieces and fragments that have migrated to form the landscape we know. Through a historical chronology, let us trace the discoveries that produced this major revolution in science, called *plate tectonics*. The facts that the continents drift and the young seafloor spreads from rifts are revolutionary discoveries in the tectonic cycle that came to light only in the twentieth century.

A Brief History

As early mapping gained accuracy, some observers noticed a symmetry to the coastlines. Abraham Ortelius (1527–1598) described the apparent fit of some continental coastlines in his *Thesaurus Geographicus* (1596). In 1620, the English philosopher Sir Francis Bacon noted gross similarities between the shapes of Africa and South America (although he did not suggest that they had drifted apart). Others wrote—unscientifically—about such apparent relationships, but it was not until much later that a valid explanation was proposed. In 1912, German geophysicist and meteorologist Alfred Wegener publicly presented in a lecture his idea that Earth's landmasses migrate. His book *Origin of the Continents and Oceans* appeared in 1915. Wegener today is regarded as the father of this concept, which he called *continental drift.* But scientists at the time, knowing little of Earth's interior structure and bound by an inertial mindset as to how continents and mountains were formed, were unreceptive to Wegener's revolutionary proposal and, in a nonscientific spirit, rejected it outright.

Wegener thought that all landmasses were united in one supercontinent approximately 225 million years ago, during the Triassic period—see Figure 8.15b (p. 266). (As you read this, it may be helpful to refer to the geologic time scale in Figure 8.1.) This one landmass he called **Pangaea**, for "all Earth." Although his initial model kept the landmasses together far too long, and his idea about the driving mechanism was incorrect, Wegener's overall configuration of Pangaea was right.

To come up with his Pangaea fit, he began studying the research of others, specifically the geologic record, the fossil record, and the climatic record for the continents. He concluded that South America and Africa were related in many complex ways. He also concluded that the large midlatitude coal deposits, which stretch from North America to Europe to China and date to the Permian and Carboniferous periods (245–360 million years ago), existed because these regions once were more equatorial in location and therefore covered by lush vegetation that became coal.

As modern scientific capabilities built the case for continental drift, the 1950s and 1960s saw a revival of interest in Wegener's concepts and, finally, confirmation. Aided by an avalanche of discoveries, the theory today is universally accepted as an accurate model of the way Earth's surface evolves, and virtually all Earth scientists accept the fact that continental masses move about in dramatic ways. *Tectonic*, from the Greek *tektonikùs*, meaning "building" or "construction," refers to changes in the configuration of Earth's crust as a result of internal forces. **Plate tectonic** processes include upwelling of magma, crustal plate movements, earthquakes, volcanic activity, subduction, warping, folding, and faulting of the crust.

Sea-floor Spreading and Production of New Crust

ANIMATION Sea-floor spreading, subduction

The key to establishing the fact of continental drift was a better understanding of the seafloor. The seafloor has a remarkable feature: an interconnected mountain chain (ridge) some 64,000 km (40,000 mi) in extent and averaging more than 1000 km (600 mi) in width. A striking map of this great undersea mountain chain opens Chapter 9. How did this mountain chain get there?

In the early 1960s, geophysicists Harry H. Hess and Robert S. Dietz proposed **sea-floor spreading** as the mechanism that builds this mountain chain and drives continental movement. Hess said that these submarine mountain ranges, called the **mid-ocean ridges**, were the direct result of upwelling flows of magma from hot areas in the upper mantle and asthenosphere and perhaps deeper sources. When mantle convection brings magma up to the crust, the crust is fractured and the magma spills out and cools to form new seafloor, building the ridges and spreading laterally. Figure 8.13 suggests that the ocean floor is rifted and scarred along mid-ocean ridges, with arrows indicating the direction in which the seafloor spreads.

As new crust is generated and the seafloor spreads, the alignment of the magnetic field in force at the time dictates the alignment of magnetic particles in the cooling rock, creating a kind of magnetic tape recording in the seafloor. Each magnetic reversal and reorientation of Earth's polarity is recorded in the oceanic crust. Figure 8.14 illustrates just such a recording, from the Mid-Atlantic ridge near Iceland. Note the mirror images that develop on either side of the sea-floor rift as a result of these magnetic reversals recorded in the rock.

If you measure the same distance from the rift to either side, you find the same orientation of magnetic content in the rock. Thus, the periodic reversals of Earth's magnetic field have proven a valuable clue to understanding sea-floor spreading. In essence, scientists were provided with a global pattern of magnetic stripes to use in fitting together pieces of Earth's crust.

As the age of the seafloor was determined from these sea-floor recordings of Earth's magnetic-field reversals and other measurements, the complex harmony of the two concepts of continental drift and sea-floor spreading became clearer. The youngest crust anywhere on Earth is at the spreading centers of the mid-ocean ridges. With increasing distance from these centers, Earth's surface gets older. The Pigafetta Basin, the oldest seafloor, is in the western Pacific near Japan, coincidentally farthest from a spreading center.

Overall, the seafloor is relatively young—nowhere does it exceed 200 million years in age, remarkable when you remember that Earth's age is 4.6 billion years. The reason is that oceanic crust is not permanent—it is slowly plunging beneath continental crust along Earth's deep oceanic trenches. The discovery that the seafloor is so young demolished earlier geologic thinking that the oldest rocks would be found there.

Subduction of Crust

In contrast to the upwelling zones along the mid-ocean ridges are the areas of descending crust elsewhere. We must remember that the basaltic ocean crust has a greater average density (3.0 g/cm^3) than continental crust (2.7 g/cm^3). As a result, when continental crust and oceanic crust slowly collide, the denser ocean floor dives beneath the lighter continent. This action forms a descending **subduction zone** (see Figure 8.13). The world's **oceanic trenches** are subduction zones and are the deepest features on Earth's surface. The deepest is the

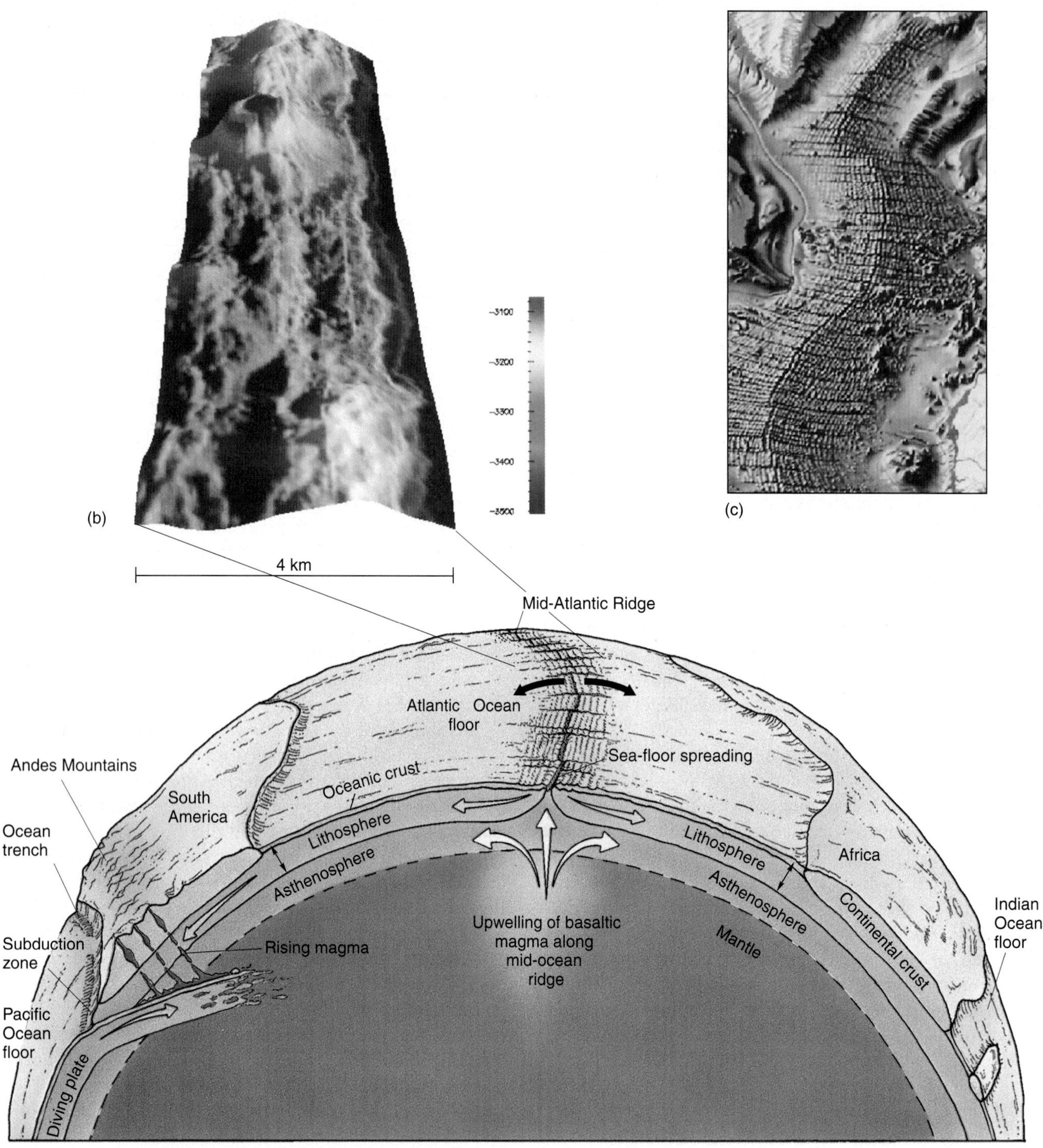

FIGURE 8.13 Crustal movements.
(a) Sea-floor spreading, upwelling currents, subduction, and plate movements, shown in cross section. Arrows indicate the direction of the spreading. (b) A 4-km-wide image of the Mid-Atlantic Ridge, showing linear faults, a volcanic crater, a rift valley, and ridges. The image was taken by the TOBI (towed ocean-bottom instrument) at approximately 29° N latitude. (c) Detail from Tharp's ocean floor map. [(a) After P. J. Wyllie, *The Way the Earth Works,* © 1976, by John Wiley & Sons. Adapted by permission; (b) image courtesy of D. K. Smith, Woods Hole Oceanographic Institute, Woods Hole, Massachusetts. All rights reserved; (c) inset derived from Marie Tharp's *Floor of the Oceans.*]

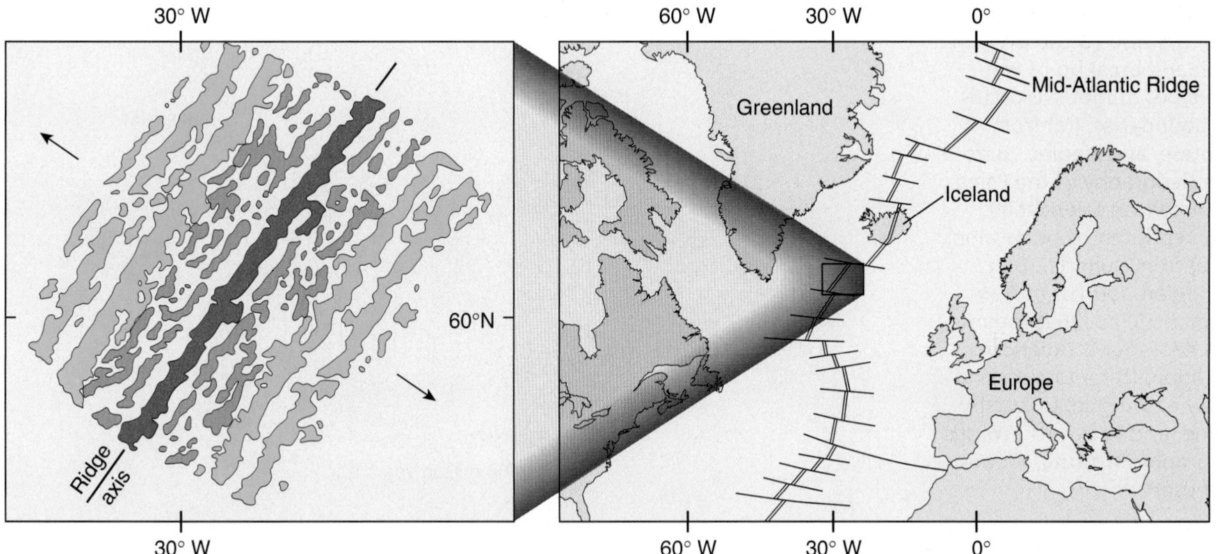

FIGURE 8.14 Magnetic reversals recorded.
Magnetic reversals recorded in the seafloor south of Iceland along the Mid-Atlantic Ridge. [Magnetic reversals reprinted from *Deep-Sea Research* 13, J. R. Heirtzler, S. Le Pichon, and J. G. Baron. Copyright © 1966, Pergamon Press, p. 247.]

Mariana Trench near Guam, which descends to −11,033 m (−36,198 ft) below sea level.

The subducted portion of crust is dragged down into the mantle, where it remelts, is recycled, and eventually migrates back toward the surface through deep fissures and cracks in crustal rock (Figure 8.13a, left). Volcanic mountains like the Andes in South America and the Cascade Range from northern California to the Canadian border form inland of these subduction zones. The fact that spreading ridges and subduction zones are areas of earthquake and volcanic activity provide important clues for the study of the drifting continents.

Sea-floor spreading, subduction, and mantle convection all are phenomena of plate tectonics, which by 1968 had become the all-encompassing term for the concepts that began with Wegener's original continental drift proposal. (For an overview, see http://geology.usgs.gov/index.shtml.) Using current scientific findings, let us go back and reconstruct the past and Pangaea.

The Formation and Breakup of Pangaea

ANIMATION **Pangaea breakup, plate movements**

The supercontinent of Pangaea and its subsequent breakup into today's continents represent only the last 225 million years of Earth's 4.6 billion years, or only the most recent 1/23 of Earth's existence. During the other 22/23 of geologic time, other things were happening, including the formation and breakup of previous supercontinents. The landmasses as we know them were unrecognizable through most of Earth's history.

Figure 8.15a begins with the pre-Pangaea arrangement of 465 million years ago (during the middle Ordovician Period). Figure 8.15b illustrates an updated version of Wegener's Pangaea, 225–200 million years ago (Triassic–Jurassic Periods). The movement of plates that occurred by 135 million years ago (the beginning of the Cretaceous Period) is in Figure 8.15c. Figure 8.15d presents the arrangement from 65 million years ago (beginning of the Tertiary Period). Finally, the present arrangement in modern geologic time (the late Cenozoic Era) is in Figure 8.15e.

Earth's present crust is divided into at least 14 plates, of which about half are major and half are minor, in terms of area (Figure 8.16). Each of these broad plates are composed of literally hundreds of smaller pieces and perhaps dozens of microplates that have migrated together. The arrows in the figure indicate the direction in which each plate is presently moving, and the length of the arrows suggests the rate of movement during the past 20 million years. A plate motion calculator by the University of Tokyo is at http://triton.ori.u-tokyo.ac.jp/~intridge/pmc/nuvel1.html.

Compare Figure 8.16 with the image of the ocean floor in Figure 8.17 (both on p. 268). In the image, satellite radar-altimeter measurements have recorded the sea-surface height to a remarkable accuracy of 0.03 m, or 1 in. The sea-surface elevation is far from uniform; it is higher or lower in direct response to the mountains, plains, and trenches of the ocean floor beneath it. The small bumps and dips in the sea surface are due to slight differences in Earth's gravity caused by sea-floor topography.

For example, a massive mountain on the ocean floor exerts a high gravitational field and attracts water to it, producing a sea-surface bump; over a trench there is less gravity, resulting in a sea-surface dip. An ocean-floor mountain that is 2000 m (6500 ft) high and 20 km (12 mi) across its base creates a 2 m rise in the sea surface. Until this use of satellite technology was developed at Scripps Institution of Oceanography and NOAA, these

FIGURE 8.15 Continents adrift, from 465 million years ago to the present. The formation and breakup of Pangaea and the types of motions occurring at plate boundaries. [(a) from Bambach, Scotese, and Ziegler, "Before Pangaea: The Geography of the Paleozoic World," *American Scientist* 68 (1980): 26–38, reprinted by permission; (b) through (e) from Robert S. Dietz and John C. Holden, *Journal of Geophysical Research* 75, no. 26 (September 10, 1970): 4939–56, © American Geophysical Union; (f) remote sensing image courtesy of S. Tighe, University of Rhode Island, and R. Detrick, Woods Hole Oceanographic Institute, Woods Hole, Massachusetts.]

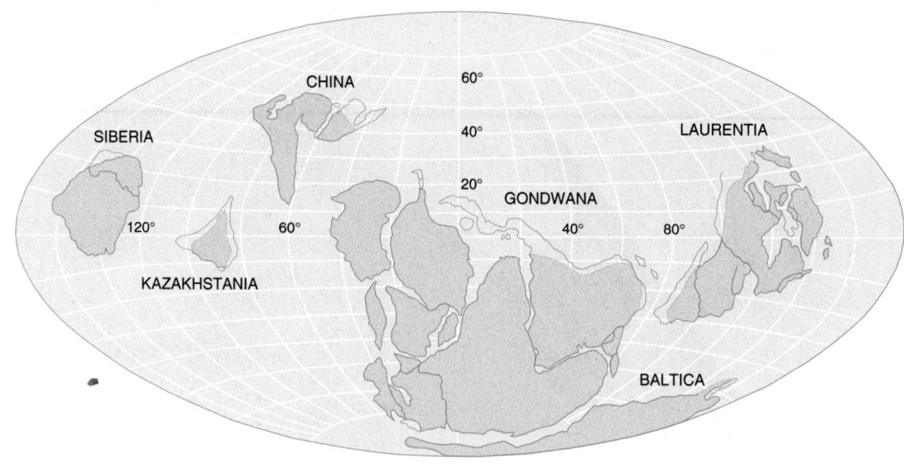

(a) 465 million years ago

Pangaea—all Earth. Panthalassa ("all seas"), became the Pacific Ocean, and the Tethys Sea (partly enclosed by the African and the Eurasian plates), became the Mediterranean Sea, and trapped portions formed the present-day Caspian Sea. The Atlantic Ocean did not exist.

Africa shared a common connection with both North and South America. Today the Appalachian Mountains in the eastern United States and the Atlas Mountains of northwestern Africa reflect this common ancestry; they are, in fact, portions of the same mountain range, torn thousands of kilometers apart!

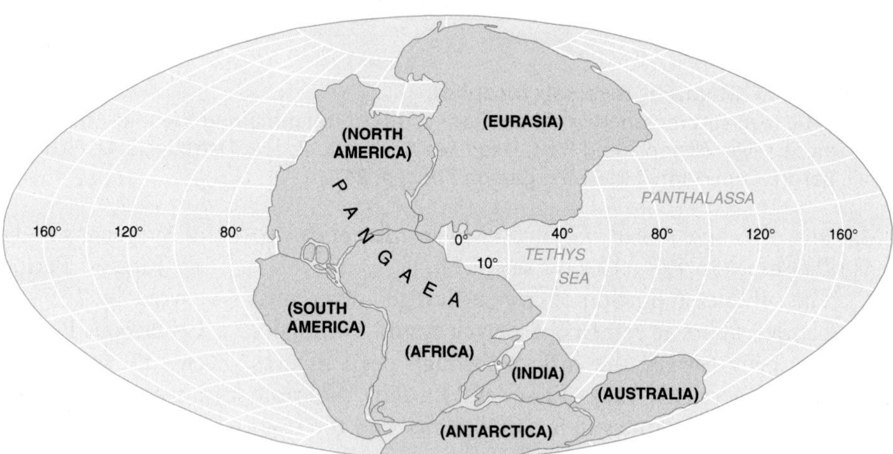

(b) 225 million years ago

New seafloor that formed since the last map is highlighted (shaded areas). An active spreading center rifted North America away from landmasses to the east, shaping the coast of Labrador. India was farther along in its journey, with a spreading center to the south and a subduction zone to the north; the leading edge of the India plate was diving beneath Eurasia.

The outlines of South America, Africa, India, Australia, and southern Europe appear within the continent called *Gondwana* in the Southern Hemisphere. North America, Europe, and Asia comprise *Laurasia*, the northern portion.

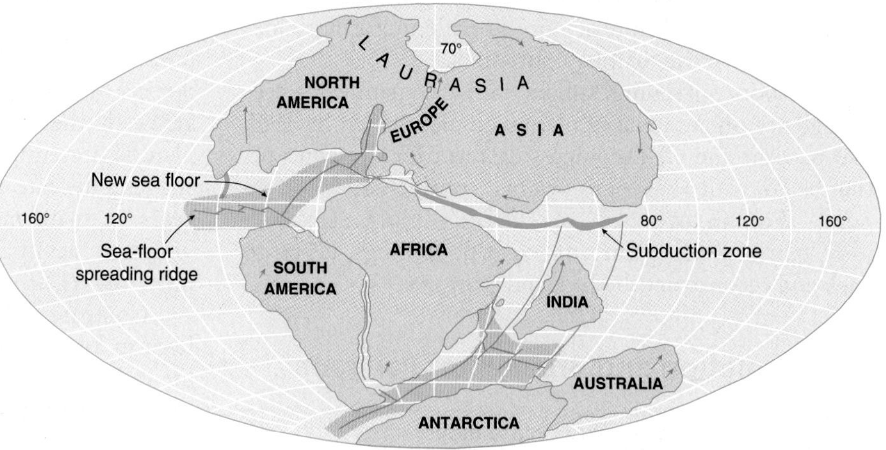

(c) 135 million years ago

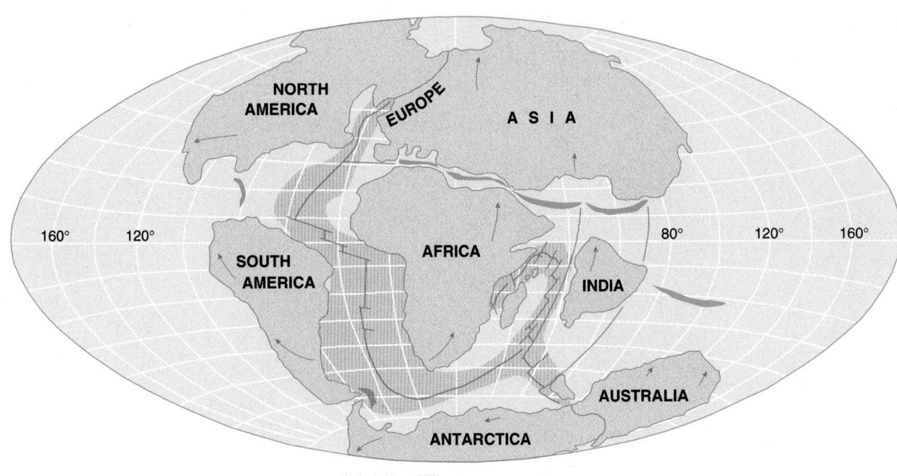

(d) 65 million years ago

Sea-floor spreading along the Mid-Atlantic Ridge had grown some 3000 km (almost 1900 mi) in 70 million years. Africa moved northward about 10° in latitude, leaving Madagascar split from the mainland and opening up the Gulf of Aden. The rifting along what would be the Red Sea began. The India plate moved three-fourths of the way to Asia, as Asia continued to rotate clockwise. Of all the major plates, India traveled the farthest—almost 10,000 km.

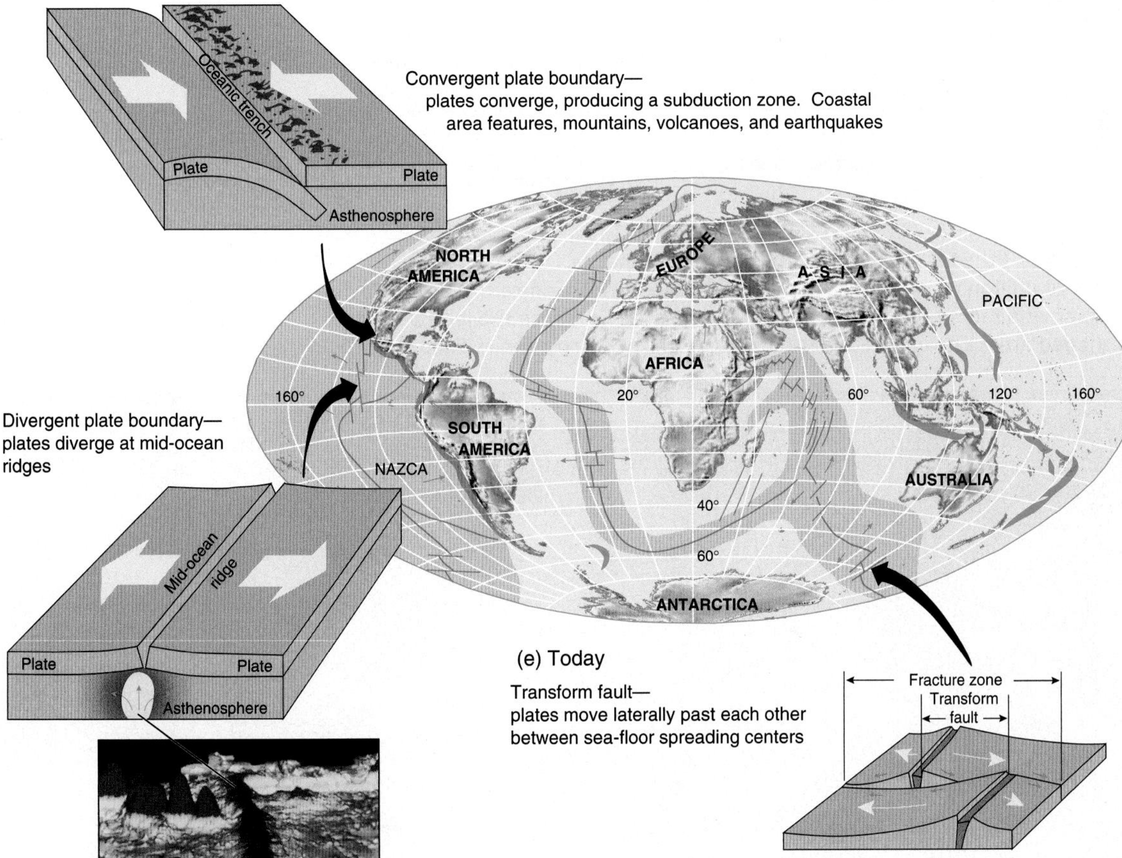

Oceanic trench
Plate
Plate
Asthenosphere

Convergent plate boundary—
plates converge, producing a subduction zone. Coastal area features, mountains, volcanoes, and earthquakes

NORTH AMERICA
EUROPE
ASIA
PACIFIC
AFRICA
160°
20°
60°
120°
160°
SOUTH AMERICA
NAZCA
AUSTRALIA
40°
60°
ANTARCTICA

Divergent plate boundary—
plates diverge at mid-ocean ridges

Mid-ocean ridge
Plate
Plate
Asthenosphere

(e) Today

Transform fault—
plates move laterally past each other between sea-floor spreading centers

Fracture zone
Transform fault

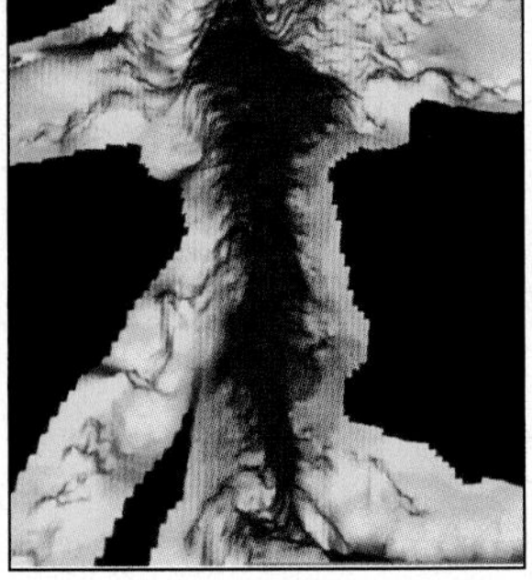

(f) East-Pacific rise, sea-floor spreading ridge (center of image at 9° N)

From 65 million years ago until the present, more than half of the ocean floor was renewed. The northern reaches of the India plate underthrust the southern mass of Asia through subduction, forming the Himalayas in the upheaval created by the collision. Plate motions continue to this day.

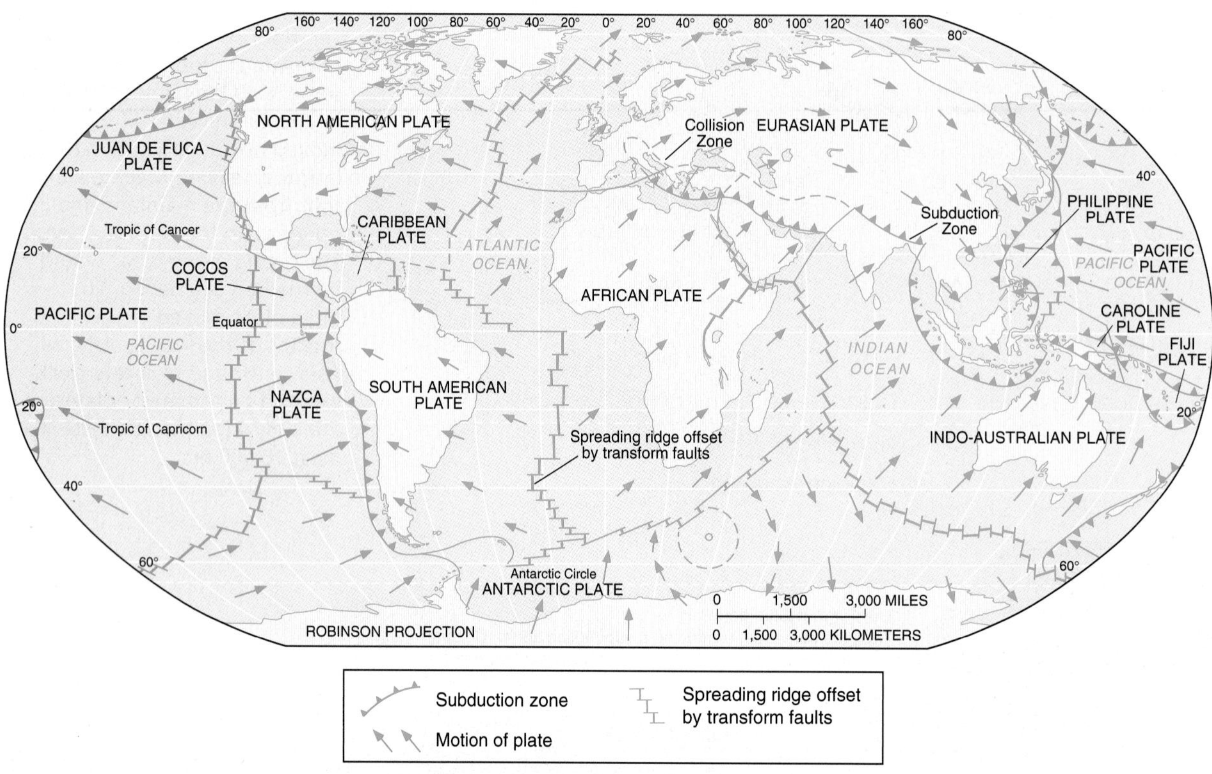

FIGURE 8.16 Earth's major lithospheric plates and their movements.
Each arrow represents 20 million years of movement. The longer arrows indicate that the Pacific and Nazca plates are moving more rapidly than the Atlantic plates. Compare the length of these arrows with those shaded areas on Figure 8.15. [Adapted from U.S. Geodynamics Committee.]

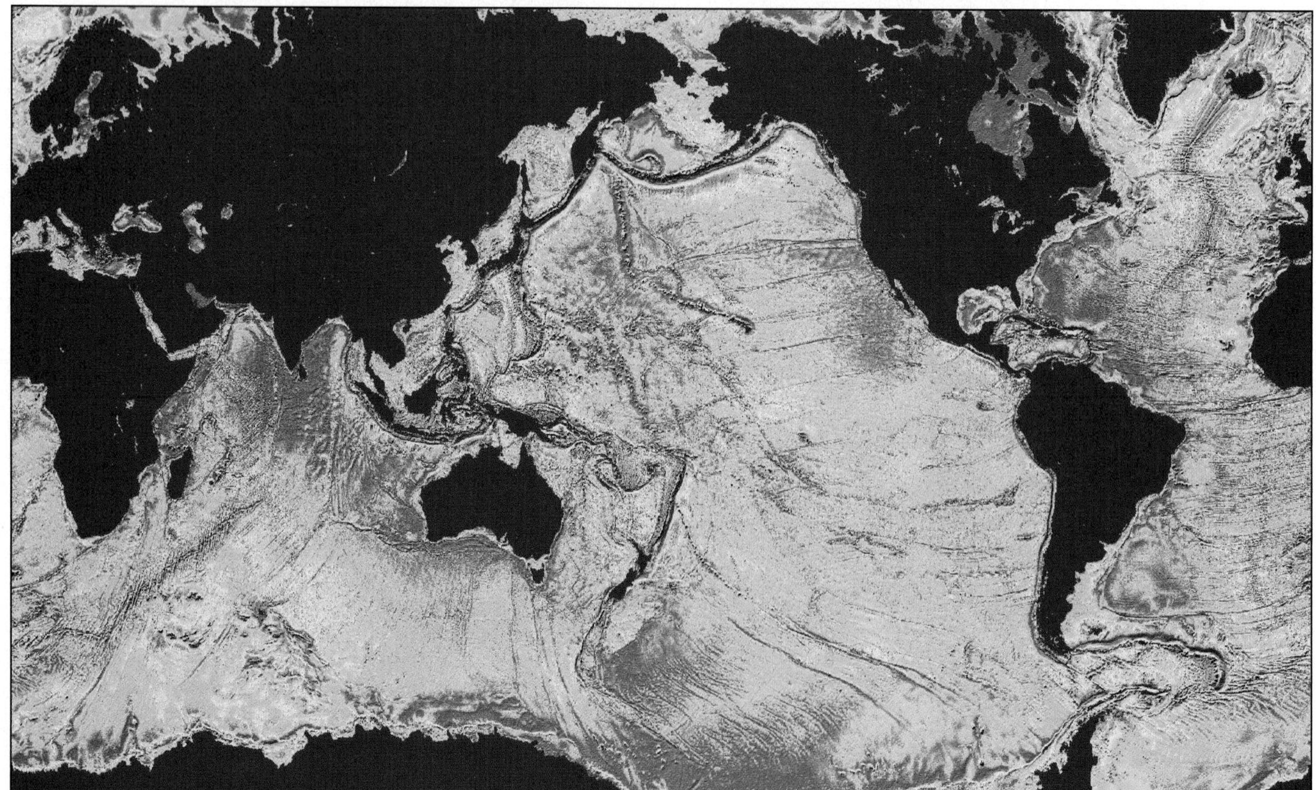

FIGURE 8.17 The ocean floor revealed.
A global gravity anomaly map derived from *Geosat* and *ERS-1* altimeter data. The radar altimeters measure sea-surface heights. Variation in sea-surface elevation is a direct indication of the topography of the ocean floor. [Global gravity anomaly map image courtesy of D. T. Sandwell, Scripps Institution of Oceanography. All rights reserved, 1995.]

small changes in height were not visible. Now we have a comprehensive map of the ocean *floor*, derived from remote sensing of the ocean *surface*.

The map of the seafloor that begins Chapter 9 also may be helpful in identifying the plate boundaries in the following discussion. Try correlating the features illustrated in Figures 8.16, 8.17, and the opening map for Chapter 9—interesting relationships emerge.

Plate Boundaries

NOTEBOOK **Plate boundaries**

ANIMATION **India collision with Asia**
Transform faults, plate margins

The boundaries where plates meet clearly are dynamic places. The block diagram inserts in Figure 8.15e show the three general types of motion and interaction that occur along the boundary areas:

- *Divergent boundaries* (lower left) are characteristic of sea-floor spreading centers, where upwelling material from the mantle forms new seafloor, and crustal plates are spread apart ("constructional"). These are zones of tension. An example noted in the figure is the divergent boundary along the East Pacific rise, which gives birth to the Nazca plate and the Pacific plate. Figure 8.14 illustrates patterns of magnetic reversals on such a divergent plate boundary south of Iceland along the Mid-Atlantic Ridge. Although most divergent boundaries occur at mid-ocean ridges, there are a few within continents themselves. An example is the Great Rift Valley of East Africa, where continental crust is being pulled apart.
- *Convergent boundaries* (upper left) are characteristic of collision zones, where areas of continental and/or oceanic crust collide. These are zones of compression and crustal loss ("destructional"). Examples include the subduction zone off the west coast of South and Central America (noted in Figure 8.15e) and the area along the Japan and Aleutian trenches. Along the western edge of South America, the Nazca plate collides with and is subducted beneath the South American plate, creating the Andes Mountains chain and related volcanoes. The collision of India and Asia is another example of a convergent boundary.
- *Transform boundaries* (lower right) occur where plates slide laterally past one another at right angles to a sea-floor spreading center, neither diverging nor converging, and usually with no volcanic eruptions. These are the right-angle fractures stretching across the mid-ocean ridge system worldwide (visible in Figure 8.16 and the Chapter 9 opening illustration).

Related to plate boundaries, another piece of the tectonic puzzle fell into place in 1965 when University of Toronto geophysicist Tuzo Wilson first described the nature of transform boundaries and their relation to earthquake activity. All the spreading centers on Earth's crust feature these perpendicular scars, which stretch out on both sides. Some transform faults are a few hundred kilometers long; others, such as those along the East Pacific rise, stretch out 1000 km or more (over 600 mi). Across the entire ocean floor, spreading-center boundaries are the location of *transform faults*. The faults generally are parallel to the direction in which the plate is moving; and the fracture zone, which follows these breaks in Earth's crust, is active only along the fault section between ridges of spreading centers.

Along transform faults the motion is one of horizontal displacement—no new crust is formed or old crust subducted. The famous San Andreas fault system in California, where continental crust has overridden a transform system, is related to this type of motion. The fault that triggered the 1995 Kobe earthquake in Japan has this type of horizontal motion, as does the North Anatolian fault system and its 1999 quake that struck Turkey (see Figure 9.13).

Earthquakes and Volcanoes

Plate boundaries are the primary location of earthquake and volcanic activity, and the correlation of these phenomena is an important aspect of plate tectonics. The next chapter discusses earthquakes and volcanic activity in more detail, but their general relationship to the tectonic plates is important to point out here. Figure 8.18 identifies earthquake zones and volcanic sites. The "ring of fire" surrounding the Pacific basin, named for the frequent incidence of volcanoes, is most evident. The subducting edge of the Pacific plate thrusts deep into the crust and upper mantle, producing molten material that makes its way back toward the surface, causing active volcanoes along the Pacific Rim. Such processes occur similarly at plate boundaries throughout the world.

Hot Spots

A dramatic aspect of Earth's internal dynamics is the estimated 50–100 hot spots across Earth's surface. **Hot spots** are individual sites of upwelling material, noted on Figure 8.18, that arrives at the surface in tall plumes from the mantle, sometimes producing thermal effects in groundwater and the crust. Some of these sites are developed for geothermal power (Focus Study 8.1). Hot spots occur beneath both oceanic and continental crust and are anchored deep in the stiff lower mantle, tending to remain fixed relative to migrating plates, but not always. Thus, the area of a plate that is above a hot spot is locally heated for the brief geologic time it is there (a few hundred thousand or million years).

An example of an isolated hot spot is the one that has formed the Hawaiian-Emperor Islands chain (Figure 8.19). The Pacific plate has moved across this hot, upward-erupting plume for almost 80 million years, with the resulting string of volcanic islands moving northwestward away from the hot spot. Thus, the age of each island or

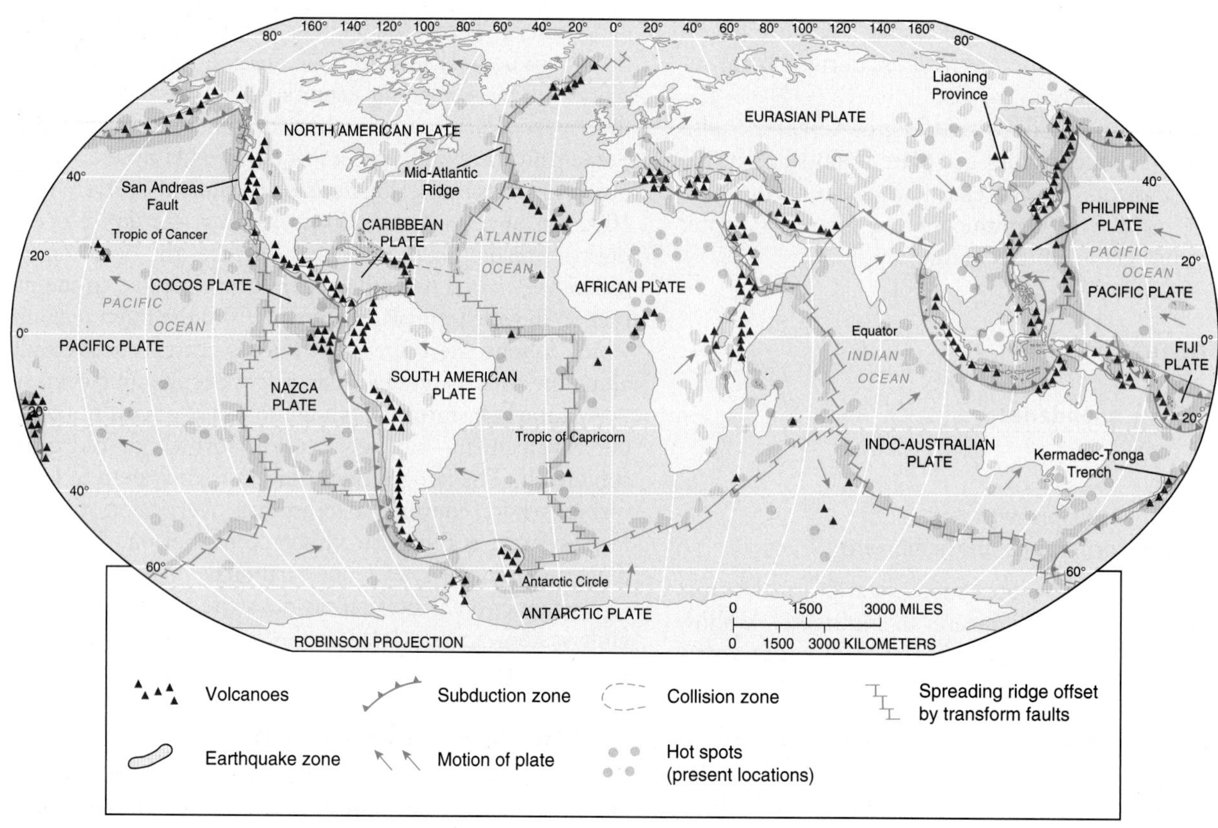

FIGURE 8.18 Earthquake and volcanic activity locations.
Earthquake and volcanic activity in relation to major tectonic plate boundaries, and principal hot spots. [Earthquake and volcano data from *Earthquakes,* B. A. Bolt, © 1988 W. H. Freeman and Company; reprinted with permission. Hot spots adapted from USGS data.]

seamount in the chain increases northwestward from the island of Hawai'i, as you can see from the ages marked in the figure.

The big island of Hawai'i actually took less than 1 million years to build to its present stature. The oldest island in the Hawaiian part of the chain is Kaua'i, at approximately 5 million years old; it is weathered, eroded, and deeply etched with canyon and valley terrain. The youngest island in the chain is still a *seamount*, a submarine mountain that does not reach the surface. It rises 3350 m (11,000 ft) from the ocean floor but is still 975 m (3200 ft) beneath the ocean surface. Even though this new island will not experience the Sun for about 10,000 years, it is already named Lō'ihi.

To the northwest of this active hot spot in Hawai'i, the island of Midway rises as a part of the same system. From there the Emperor Seamounts stretch northwestward until they reach about 40 million years of age. At that point, this linear island chain shifts direction northward, evidently reflecting a change in the movement of the Pacific plate at that earlier time. At the northernmost extreme, the seamounts that formed about 80 million years ago are now approaching the Aleutian Trench, where they eventually will be subducted beneath the Eurasian plate. Recent research suggests that the change in direction of the chain you see in Figure 8.19a (approximately 43 million years ago) could be related to movement of the hot spot itself. Evidently the mantle flows disrupt the upwelling plume that forms the hot spot.

Iceland is an example of an active hot spot sitting astride a mid-ocean ridge (Figure 8.14). It also is an example of a segment of mid-ocean ridge rising above sea level. This hot spot has generated enough material to form Iceland and continues to cause volcanic eruptions. As a result, Iceland still is growing in area and volume. The youngest rocks are near the center of Iceland, with rock age increasing toward the eastern and western coasts.

Take a moment to locate these features (mid-ocean ridges, subduction zones, Hawaiian and Icelandic hot spots) on the Chapter 9 opening map illustration.

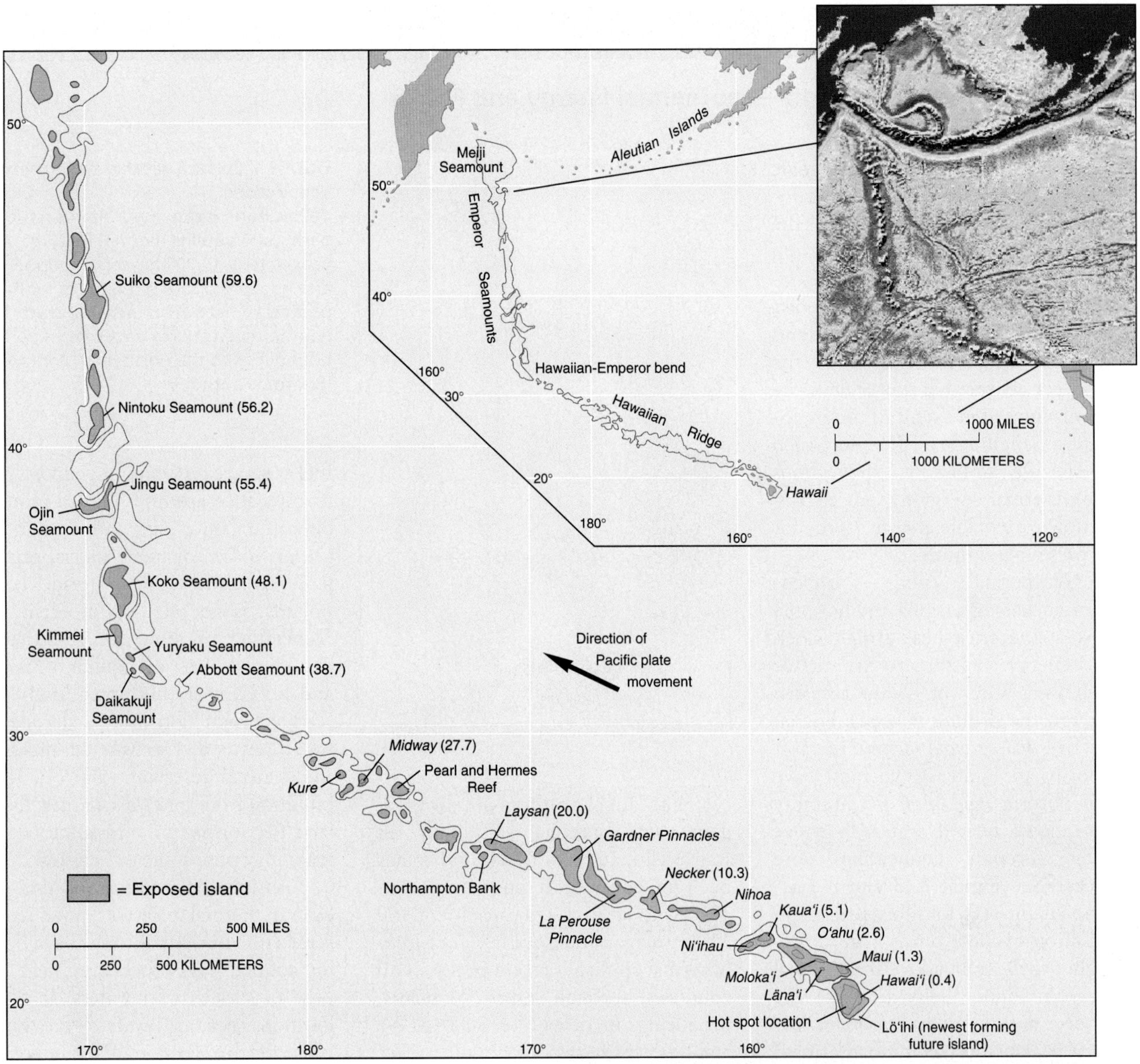

(a)

FIGURE 8.19 Hot spot tracks across the North Pacific.
Hawai'i and the linear volcanic chain of islands known as the Emperor Seamounts. (a) The islands and seamounts in the chain are progressively younger toward the southeast. Ages, in millions of years, are shown in parentheses. Note that Midway Island is 27.7 million years old, meaning that the site was over the plume 27.7 million years ago. (b) Lō'ihi is forming 975 m (3200 ft) beneath the Pacific Ocean; presently an undersea volcano (seamount), it will continue to grow into the next Hawaiian island. This view is from South Point, Hawai'i, Lō'ihi is east of this headland. [(a) After D. A. Clague, "Petrology and K–Ar (Potassium–Argon) Ages of Dredged Volcanic Rocks from the Western Hawaiian Ridge and the Southern Emperor Seamount Chain," *Geological Society of America Bulletin* 86 (1975): 991; inset from global gravity anomaly map image, Scripps Institution of Oceanography. All rights reserved. (b) photo by Bobbé Christopherson.]

(b)

Focus Study 8.1

Heat from Earth—Geothermal Energy and Power

A tremendous amount of endogenic energy flows from Earth's interior toward the surface. Temperatures at the base of Earth's crust range from 200°–1000°C (392°–1800°F). Convection and conduction transports this geothermal energy in enormous quantities from the mantle to the crust, yet geothermal power development is limited in extent to certain locations (Figure 1). Despite such site-specific limitations, applications of geothermal energy for power production are a viable energy source for the present and future.

Geothermal energy happens when pockets of magma and hot portions of the crust heat groundwater. This energy transmits to the surface by heated water or steam accessed through the drilling of wells. For an effective *underground thermal reservoir* to form, an aquifer strata must have high porosity and high permeability that allows heated water to move freely through connecting pore spaces. (See Figure 7.13 and the accompanying text for illustration and definition of these terms.) An impermeable rock strata above the thermal reservoir aquifer helps preserve the resource and describes a *hydrothermal reservoir*, the most common form of thermal repository.

Geothermal energy literally refers to heat from Earth's interior, whereas *geothermal power* relates to specific applied strategies of geothermal electric or geothermal direct applications. *Geothermal electric* uses steam, or hot water that flashes to steam, to drive a turbine-generator. In *geothermal direct*, hot water is used to heat buildings, to cool buildings using a heat exchange system, to heat greenhouses, to heat soil, for manufacturing processes, for aquaculture, and to heat swimming pools, among many uses.

For either electrical production or direct use, the end-use facilities must be fairly near the well head because of heat losses in piping the hot water or steam over distance. Of course, once electricity is generated, it can feed into the regional power grid for transmission to distant markets.

FIGURE 1 Surface geothermal activity in Yellowstone.
Yellowstone is the most famous national park that contains thermal features. Approximately 10,000 geysers, mudpots, and hot springs are evidence of molten pockets of heat in an unstable crust. This geothermal resource area crosses park boundaries to the north into Montana. [Photo by author.]

The first geothermal electrical generating station was built in Larderello, Italy, in 1904, and it has been in continuous operation since 1913 with an installed capacity of 360 MWe (megawatts electric). Today geothermal applications are in use in 30 countries, through some 200 plants, producing an output of 8000 MWe and direct heat of 12,000 MWt (megawatts thermal).

In Reykjavik, Iceland, the majority of space heating is geothermal. In the Paris Basin, France, some 25,000 residences are heated using geothermal direct 70°C (158°F) water. In the Philippines almost 30% of total electrical production is generated with geothermal energy. The top six countries for installed geothermal electrical generation are the United States (2850 MWe), Philippines (1848 MWe), Italy (769 MWe), Mexico (743 MWe), Indonesia (590 MWe), and Japan (530 MWe). Worldwide potential is estimated to be in excess of 80,000 MWe and hundreds of thousands of MWt.

Other than the limiting factors of site-specific production and having the power plant or end-use near the well, there are several other considerations with which to deal: (1) Although fuel costs are low, equipment costs and maintenance are high due to mineral deposition on piping and equipment. (2) Corrosion of metal alloys, principally by sulfide and chloride compounds, is an ongoing problem. (3) Depending on how the resource is maintained, loss of aquifer pressure and lowering groundwater levels can reduce production after a decade or so, requiring new wells to be sunk into the thermal reservoir. And (4), local pollution from water discharges, localized fog formation, and some winter icing may pose some difficulties.

An important mitigation that can extend thermal reservoir potential is reinjection of water into the heat-bearing aquifer. This process can help dispose of surface wastewaters. In some locations reclaimed water (gray water) is used for reinjection. The Geysers in northern California, presently the world's largest-capacity geothermal electrical generation installation, uses this reinjection of wastewater.

The Geysers Geothermal Field (the name despite the lack of any geysers in the area) began production in 1960, increased to a peak in 1989 at 1967 MWe, and lowered to a present capacity of 1070 MWe (Figure 2). By the mid-1990s some 600 wells had been drilled. The average well depth is 2500 m (8200 ft). Declining production relates to the expected problems of decreased yields. The Geysers are operating below the capacity of the field potential at the time of this writing.

Some 2800 MWe of geothermal power capacity is operating in California, Hawai'i, Nevada, and Utah. Figure 3 maps low-, medium-, and

(continued)

Focus Study 8.1 *(continued)*

high-temperature known geothermal resource areas (KGRA) in the United States. About 300 cities are within 8 km (5 mi) of a KGRA. The Department of Energy has identified about 9000 potential development sites. In Canada, a proposed project at Mount Meager, British Columbia, will produce electricity; geothermal direct heats buildings at Carleton University, Ottawa; and some direct operations are in process in Nova Scotia.

As with other non-fossil fuel, potentially renewable energy resources, energy industry politics plays a role in the slow implementation of geothermal power. In times past, the price of geothermal steam was coupled with fossil fuel prices, so events completely unrelated to geothermal power could dictate its competitive stance. Geothermal plant owners in California have had to sue utilities for timely payments during the energy distribution crisis of 2000–2001. Apparently the potential of heat from Earth will have to wait a little longer to become a more significant player. (For more information see http://www.eren.doe.gov/geothermal/, or http://geothermal.marin.org/, or http://www.geothermal.org/, or http://www.smu.edu/geothermal/.)

FIGURE 2 The Geysers Geothermal Field, California. The Geysers is the largest geothermal electric installation in the world. One of the 21 generating stations is Eagle Rock, Unit 11, with a 73 MWe capacity, in commercial operation since 1975. [Photo courtesy of Calpine, San Jose, California, http://www.geysers.com.]

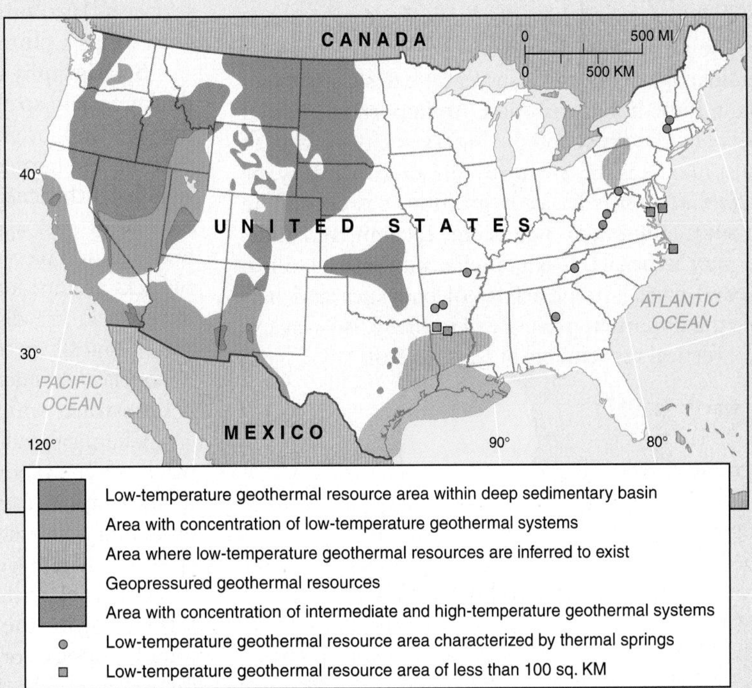

FIGURE 3 Geothermal resources. Geothermal resources in the conterminous United States. [Map courtesy of USGS, from Duffield, Sass, and Sorey, *Tapping the Earth's Natural Heat*, USGS Circular 1125, 1994, p. 35.]

Summary and Review—The Dynamic Planet

● ***Distinguish*** **between the endogenic and exogenic systems,** ***determine*** **the driving force for each, and** ***explain*** **the pace at which these systems operate.**

The Earth-atmosphere interface is where the **endogenic system** (internal), powered by heat energy from within the planet, interacts with the **exogenic system** (external), powered by insolation and influenced by gravity. These systems work together to produce Earth's diverse landscape. The **geologic time scale** is an effective device for organizing the vast span of geologic time. It depicts the sequence of Earth's events (relative time) and the approximate actual dates (absolute time).

The most fundamental principle of Earth science is **uniformitarianism**. Uniformitarianism assumes that the *same physical processes active in the environment today have been operating throughout geologic time.*

endogenic system (p. 247)
exogenic system (p. 247)
geologic time scale (p. 251)
uniformitarianism (p. 251)

1. To what extent is Earth's crust active at this time in its history?
2. Define the endogenic and the exogenic systems. Describe the driving forces that energize these systems.
3. How is the geologic time scale organized? What is the basis for the time scale in relative and absolute terms? What era, period, and epoch are we living in today?

4. Describe uniformitarianism in Earth's development. How can this flow of events and time be interrupted?

Diagram Earth's interior in cross section and *describe* each distinct layer.

We have learned about Earth's interior from indirect evidence—the way its various layers transmit **seismic waves.** The **core** is differentiated into an *inner core* and an *outer core*—divided by a transition zone. Earth's magnetic field is generated almost entirely within the outer core. Polarity reversals in Earth's magnetism are recorded in cooling magma that contains iron minerals. The pattern of **geomagnetic reversals** frozen in rock helps scientists piece together the story of Earth's mobile crust.

Outside Earth's core lies the **mantle**, differentiated into lower mantle and upper mantle. It experiences a gradual temperature increase with depth and stiffening due to increased pressures. The upper mantle is divided into three fairly distinct layers. The uppermost mantle, along with the crust, makes up the *lithosphere*. Below the lithosphere is the **asthenosphere**, or *plastic layer*. It contains pockets of increased heat from radioactive decay and is susceptible to slow convective currents in these hotter materials. An important internal boundary between the **crust** and the high-velocity portion of the uppermost mantle is the **Mohorovičić discontinuity**, or **Moho**. *Continental crust* is basically **granite**; it is crystalline and high in silica, aluminum, potassium, calcium, and sodium. *Oceanic crust* is **basalt**; it is granular and high in silica, magnesium, and iron. The principles of buoyancy and balance produce the important principle of **isostasy**. Isostasy explains certain vertical movements of Earth's crust.

seismic waves (p. 252)
core (p. 253)
geomagnetic reversal (p. 253)
mantle (p. 253)
asthenosphere (p. 254)
crust (p. 254)
Mohorovičić discontinuity (Moho) (p. 254)
granite (p. 255)
basalt (p. 255)
isostasy (p. 255)

5. Make a simple sketch of Earth's interior, label each layer, and list the physical characteristics, temperature, composition, and range of size of each on your drawing.
6. What is the present thinking on how Earth generates its magnetic field? Is this field constant, or does it change? Explain the implications of your answer.
7. Describe the asthenosphere. Why is it also known as the plastic layer? What are the consequences of its convection currents?
8. What is a discontinuity? Describe the principal discontinuities within Earth.
9. Define isostasy and isostatic rebound, and explain the crustal equilibrium (balance between buoyancy and gravity) concept.
10. Diagram the uppermost mantle and crust. Label the density of the layers in grams per cubic centimeter. What two types of crust were described in the text in terms of rock composition?

Illustrate the geologic cycle and *relate* the rock cycle and rock types to endogenic and exogenic processes.

The **geologic cycle** is a model of the internal and external interactions that shape the crust. A **mineral** is an inorganic natural compound having a specific chemical formula and possessing a crystalline structure. A **rock** is an assemblage of minerals bound together (such as granite, a rock containing three minerals), or it may be a mass of a single mineral (such as rock salt). Thousands of different rocks have been identified.

The geologic cycle comprises three cycles: the *hydrologic cycle*, the *tectonic cycle*, and the *rock cycle*. The rock cycle describes the three principal rock-forming processes and the rocks they produce. **Igneous rocks** form from **magma**, which is molten rock beneath the surface. Magma is fluid, highly gaseous, and under tremendous pressure. It either *intrudes* into crustal rocks, cools, and hardens, or it *extrudes* onto the surface as **lava**. Intrusive igneous rock that cools slowly in the crust forms a **pluton**. The largest pluton form is a **batholith**.

Stratigraphy is the study of the sequence (superposition), thickness, and spatial distribution of strata that yield clues to the age and origin of the rocks. The cementation, compaction, and hardening of sediments into **sedimentary rocks** is called **lithification**. These layered strata form important records of past ages. Clastic sedimentary rocks are derived from bits and pieces of weathered rocks. Chemical sedimentary rocks are not formed from physical pieces of broken rock, but instead are dissolved minerals, transported in solution, and chemically precipitated out of solution (they are essentially nonclastic). The most common chemical sedimentary rock is **limestone**, which is lithified calcium carbonate, $CaCO_3$.

Chemical sediments also form from inorganic sources when water evaporates and leaves behind a residue of salts. These **evaporites** may exist as common salts, such as gypsum or sodium chloride (table salt), to name only two, and often appear as flat, layered deposits across a dry landscape.

Any rock, either igneous or sedimentary, may be transformed into a **metamorphic rock** by going through profound physical or chemical changes under pressure and increased temperature.

geologic cycle (p. 255)
mineral (p. 257)
rock (p. 257)
igneous rock (p. 257)
magma (p. 257)
lava (p. 257)
pluton (p. 257)
batholith (p. 257)
stratigraphy (p. 259)
sedimentary rock (p. 259)
lithification (p. 259)
limestone (p. 259)
evaporite (p. 261)
metamorphic rock (p. 261)

11. Illustrate the geologic cycle and define each component: rock cycle, tectonic cycle, and hydrologic cycle.
12. What is a mineral? A mineral family? Name the most common minerals on Earth. What is a rock?

13. Describe igneous processes. What is the difference between intrusive and extrusive types of igneous rocks?
14. Briefly explain how the rock formation in Figure 8.9a demonstrates through its layers a record of past climates.
15. Briefly describe sedimentary processes and lithification. Describe the sources and particle sizes of sedimentary rocks.
16. What is metamorphism, and how are metamorphic rocks produced? Name some original parent rocks and metamorphic equivalents.

- ***Describe*** **Pangaea and its breakup and *relate* several physical proofs that crustal drifting is continuing today.**

The present configuration of the ocean basins and continents are the result of *tectonic processes* involving Earth's interior dynamics and crust. Alfred Wegener coined the phrase *continental drift* to describe his idea that the crust moves in response to vast forces within the planet. **Pangaea** was the name he gave to a single assemblage of continental crust some 225 million years ago that subsequently broke apart. Earth's crust is fractured into huge slabs or plates, each moving in response to flowing currents in the mantle. The all-encompassing theory of **plate tectonics** includes **sea-floor spreading** along **mid-ocean ridges** and denser oceanic crust diving beneath lighter continental crust along **subduction zones**—places where deep **oceanic trenches** form. Supporting evidence includes: the pattern of magnetic reversals in rocks worldwide, matching animal and plant fossil record in adjoining continents, patterns of relative and absolute dating of crust rock, and deposits of fossil fuels, among many items.

Pangaea (p. 263)
plate tectonics (p. 263)
sea-floor spreading (p. 263)
mid-ocean ridge (p. 263)
subduction zone (p. 263)
oceanic trenches (p. 263)

17. Briefly review the history of the theory of continental drift, sea-floor spreading, and the all-inclusive plate tectonics theory. What was Alfred Wegener's role?
18. Define upwelling, and describe related features on the ocean floor. Define subduction and explain the process.
19. What was Pangaea? What happened to it during the past 225 million years?
20. Characterize the three types of plate boundaries and the actions associated with each type.

- ***Portray*** **the pattern of Earth's major plates and *relate* this pattern to the occurrence of earthquakes, volcanic activity, and hot spots.**

Occurrences of often damaging earthquakes and volcanic eruptions are correlated with plate boundaries. Three types of plate boundaries form: divergent, convergent, and transform. Along the offset portions of mid-ocean ridges, horizontal motions produce *transform faults*.

As many as 50 to 100 **hot spots** exist across Earth's surface, where tall plumes of magma, anchored in the lower mantle, remain fixed as eruptions penetrate drifting plates. **Geothermal energy** literally refers to heat from Earth's interior, whereas geothermal power relates to specific applied strategies of geothermal electric or geothermal direct applications.

hot spots (p. 269)
geothermal energy (p. 272)

21. What is the relation between plate boundaries and volcanic and earthquake activity?
22. What is the nature of motion along a transform fault? Name a famous example of such a fault.
23. How is the Hawaiian-Emperor chain of islands and seamounts an example of plate motion and hot spot activity? Correlate your answer with the ages of the Hawaiian Islands.

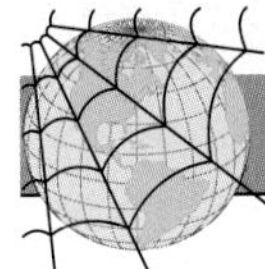

NetWork

The *Elemental Geosystems Home Page* provides on-line resources for this chapter on the World Wide Web. Once on the Home Page, click on this textbook, scroll the Table of Contents menu, select this chapter, and click "Begin." You will find self-tests that are graded, review exercises, specific updates for items in the chapter, and in "Destinations" many links to interesting related pathways on the Internet. *Elemental Geosystems* is found at http://www.prenhall.com/christopherson.

Critical Thinking

A. Using the maps in this chapter, determine your present location relative to Earth's crustal plates. Now, using Figure 8.15b, approximately locate where your present location was 225 million years ago; express it in a rough estimate using the equator and the longitudes noted on the map.

B. Relative to the motion of the Pacific plate shown in Figure 8.19 (note the scale in lower-left corner), the island of Midway formed 27.7 million years ago over the hot spot that is active under the southeast coast of the big island of Hawai'i today. Given the scale of the map, roughly determine the average annual speed of the Pacific plate in cm per year for Midway to have traveled this distance.

Floor of the Ocean, 1975 by Bruce C. Heezen and Maria Tharp.
The scarred ocean floor is clearly visible: sea-floor spreading centers marked by oceanic ridges that stretch over 64,000 km (40,000 mi), subduction zones indicated by deep oceanic trenches, and transform faults slicing across oceanic ridges.

Follow the East Pacific rise (an ocean ridge and spreading center) northward as it trends beneath the west coast of the North American plate, disappearing under earthquake-prone California. The continents, offshore continental shelves, and the expanse of the sediment-covered abyssal plain all are identifiable on this illustration.

On the floor of the Indian Ocean, you can see the wide track along which the India plate traveled northward to its collision with the Eurasian plate. Vast deposits of sediment cover the Indian Ocean floor, south of the Ganges River to the east of India. Sediments derived from the Himalayan Range blanket the floor of the Bay of Bengal (south of Bangladesh) to a depth of 20 km (12.4 mi). These sediments result from centuries of soil erosion in the land of the monsoons.

Isostasy is in action in central and west-central Greenland, where the weight of the ice sheet depresses portions of the land far below sea level—now visible on the map with the ice artificially removed by the cartographer (see the extreme upper-left corner of the map). In contrast, the region around Canada's Hudson Bay became ice-free about 8000 years ago and has isostatically rebounded 300 m (1100 ft).

In the area of the Hawaiian Islands, the hot-spot track is marked by a chain of islands and seamounts that you can follow along the Pacific plate from Hawai'i to the Aleutians. Subduction zones south and east of Alaska and Japan, as well as along the western coast of South and Central America, are visible as dark trenches. Along the left margin of the map you can find Iceland's position on the Mid-Atlantic Ridge.

9 Earthquakes and Volcanoes

Key Learning Concepts

By knowing and understanding the key learning concepts in this chapter, you should be able to:

- *Describe* first, second, and third orders of relief and *relate* examples of each from Earth's major topographic regions.
- *Describe* the several origins of continental crust and *define* displaced terranes.
- *Explain* compressional processes and folding; *describe* four principal types of faults and their characteristic landforms.
- *Relate* the three types of plate collisions associated with orogenesis and *identify* specific examples of each.
- *Explain* the nature of earthquakes, their measurement, and the nature of faulting.
- *Distinguish* between an effusive and an explosive volcanic eruption and *describe* related landforms, using specific examples.

Earth's physical systems move to the front pages when an earthquake strikes or a volcanic eruption threatens a city. Some 50 volcanic eruptions occur every year, although most are in remote locations or beneath the ocean surface. The drama of Mount Etna's recent activity is shown in the Chapter 8 opening photo.

Of the 13 magnitude 7.0 quakes in the world during 2002, a magnitude 7.9 near Denali National Park, Alaska, was notable as the largest that year (Figure 9.1). The remote location prevented loss of life but dramatic effects took place: cracked roads and bridges, huge boulders and rocks that fell onto glacial surfaces, offsets in glacial crevasses, many avalanches, and 150 km (92 mi) of the Alaska pipeline supports broken. The fault rupture extended eastward from the epicenter for over 210 km (130 mi), with horizontal offsets of nearly 9 m (26 ft).

Such tectonic activity has repeatedly deformed, recycled, and reshaped Earth's crust during its 4.6-billion-year existence, sometimes in dramatic episodes. The

(a)

(b)

FIGURE 9.1 Earthquakes strike Alaska and Taiwan. (a) A magnitude 7.9 earthquake strikes central Alaska, 66 km (41 mi) southeast of Denali National Park. A portion of the Trans-Alaska oil pipeline was knocked off its supports. (b) The damage in Chi-Chi, Nantou County, Taiwan, that killed over 2000 people and left 100,000 homeless, was caused by an earthquake in 1999. Because of extensive instrumentation across in the region, a wealth of data on this quake was made available for analysis. Damage and building failures were severe as shown here in Yulin. [Photos by (a) Earthquake Engineering Research Institute; (b) Reuters/Simon Kwang/Archive Photos.]

principal tectonic and volcanic zones lie along plate boundaries. The arrangement of continents and oceans, the origin of mountain ranges, topography, and the locations of earthquake and volcanic activity are all the result of dynamic Earth processes.

In this chapter: We examine processes that construct Earth's surface and create world structural regions. Tectonic processes deform, recycle, and reshape Earth's crust. These processes occur sometimes in dramatic episodes but most often in slow, deliberate motions that build the landscape. Continental crust has been forming throughout most of Earth's 4.6-billion-year existence. The arrangement of continents and oceans, the origin of mountain ranges, the topography of the land and the seafloor, and the locations of earthquake and volcanic activity are all evidence of our dynamic Earth. Principal seismic and volcanic zones occur along plate boundaries and hot spot locations, thus linking plate tectonics to the devastation of major earthquakes and local threats of major volcanic eruptions and their attendant potential global climatic impact.

We begin our look at tectonics and volcanism on the ocean floor, hidden from direct view. The illustration that opens this chapter is a striking representation of Earth with its blanket of water removed, as revealed to us through decades of direct and indirect observation. Careful examination of this portrait gives us a helpful review of the concepts learned in the previous chapter, laying the foundation for this and subsequent chapters. Be sure to take the quick tour presented with this chapter-opening map. Try to correlate this sea-floor illustration with the maps of crustal plates and plate boundaries shown in Figures 8.16 and the gravity anomaly image in Figure 8.17.

Earth's Surface Relief Features

Relief refers to vertical elevation differences in the landscape. Examples include the low relief of Iowa and Nebraska, medium relief in foothills along mountain ranges, and high relief in the Rockies and Himalayas. The undulating form of Earth's surface is called **topography**—portrayed effectively on topographic maps—the lay of the land.

The relief and topography of Earth's crustal landforms have played a vital role in human history: High mountain passes have both protected and isolated societies; ridges and valleys have dictated transportation routes; and vast plains have encouraged better methods of communication and travel. Earth's topography has stimulated human invention and adaptation.

The U.S. Geological Survey has prepared a digitized shaded-relief map of the United States. The entire map is made up of 12 million spot elevations, with less than a kilometer between any two. A Web site prepared by the USGS, "A Tapestry of Time and Terrain," uses this data, colored geologic regions, and detailed topography in a useful virtual map: http://tapestry.usgs.gov.

Crustal Orders of Relief

For convenience, geographers group the landscape's topography into three *orders of relief*. Orders of relief classify landscapes by scale, from enormous ocean basins and continents down to local hills and valleys.

First Order of Relief. The broadest category of landforms includes huge continental platforms and ocean basins. **Continental platforms** are the masses of crust that reside above or near sea level, including the undersea continental shelves along the coastlines. The **ocean basins** are entirely below sea level and are portrayed in the chapter-opening illustration. Approximately 71% of Earth is covered by water, with only 29% of its surface appearing as continents and islands.

Second Order of Relief. Continental masses, mountain masses, plains, and lowlands are continental features that are in the second order of relief category. A few examples are the Alps, Rocky Mountains (both Canadian and American), west Siberian lowland, and Tibetan Plateau. The great rock cores ("shields") that form the heart of each continental mass are of this second order. In the ocean basins, second order of relief includes continental rises, slopes, abyssal plains, mid-ocean ridges, submarine canyons, and subduction trenches—all visible in the sea-floor illustration that opens this chapter.

Third Order of Relief. The third and most-detailed order of relief includes individual mountains, cliffs, valleys, hills, and other landforms of smaller scale. These features are identifiable as local landscapes.

Hypsometry. Figure 9.2 is a hypsographic curve, which shows the distribution of Earth's surface by area and elevation (the Greek *hypsos* means "height"). Relative to Earth's diameter of 12,756 km (7926 mi), the surface generally is of low relief; for comparison, Mount Everest is only 8.8 km (5.5 mi) above sea level, and Mauna Kea in Hawai'i is 10.0-km (6.2-mi) tall measured from the seafloor. Note the new GPS measurement for the summit of Mount Everest announced in 1999—8850 m (29,035 ft). See News Report 9.1 for more on this geographic milestone.

The average elevation of Earth's solid surface is actually under water: −2070 m (−6790 ft) below mean sea level. The average elevation just for exposed land is +875 m (+2870 ft). For the ocean depths, the average elevation is −3800 m (−12,470 ft). From these statistics you can see that, on the average, the oceans are much deeper than continental regions are high.

Earth's Topographic Regions

The three orders of relief can be further generalized into six topographic regions: plains, high tablelands, hills and low tablelands, mountains, widely spaced mountains, and depressions (Figure 9.3). An arbitrary elevation or descriptive limit in common use defines each type of topography (see the legend in the figure). Four of the continents possess extensive *plains*, which are identified as areas with local relief of less than 100 m (325 ft) and slope angles of 5° or less. Some plains have high elevations of over 600 m (2000 ft); in the United States the high plains achieve elevations above 1220 m (4000 ft). The Colorado Plateau and Antarctica are notable *high tablelands*, with elevations exceeding 1520 m (5000 ft). *Hills and low tablelands* dominate Africa.

Mountain ranges are characterized by local relief exceeding 600 m (2000 ft) and appear on each continent. Earth's relief and topography are undergoing constant change as a result of processes that form crust.

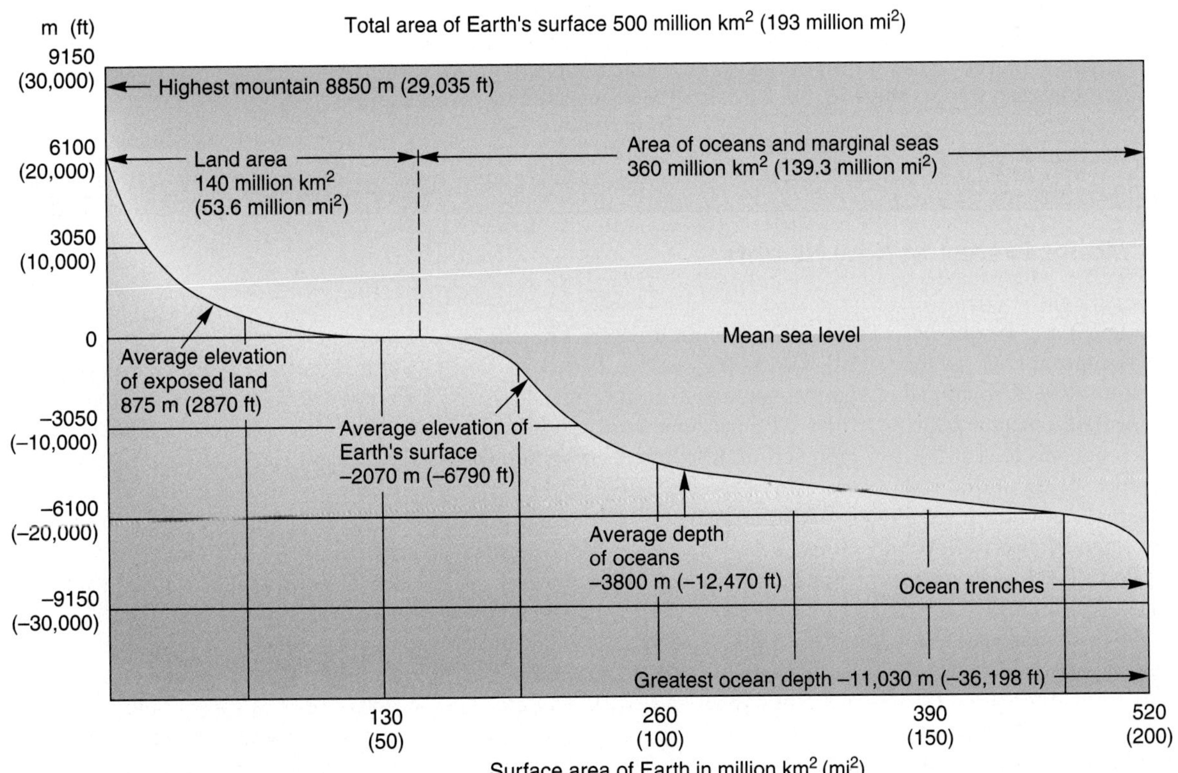

FIGURE 9.2 Earth's hypsometry.
Hypsographic curve of Earth's surface, charting elevation as related to mean sea level. From the highest point above sea level (Mount Everest) to the deepest oceanic trench (Mariana Trench) Earth's overall relief is almost 20 km (12.5 mi). The new height given for Mount Everest was announced in 1999.

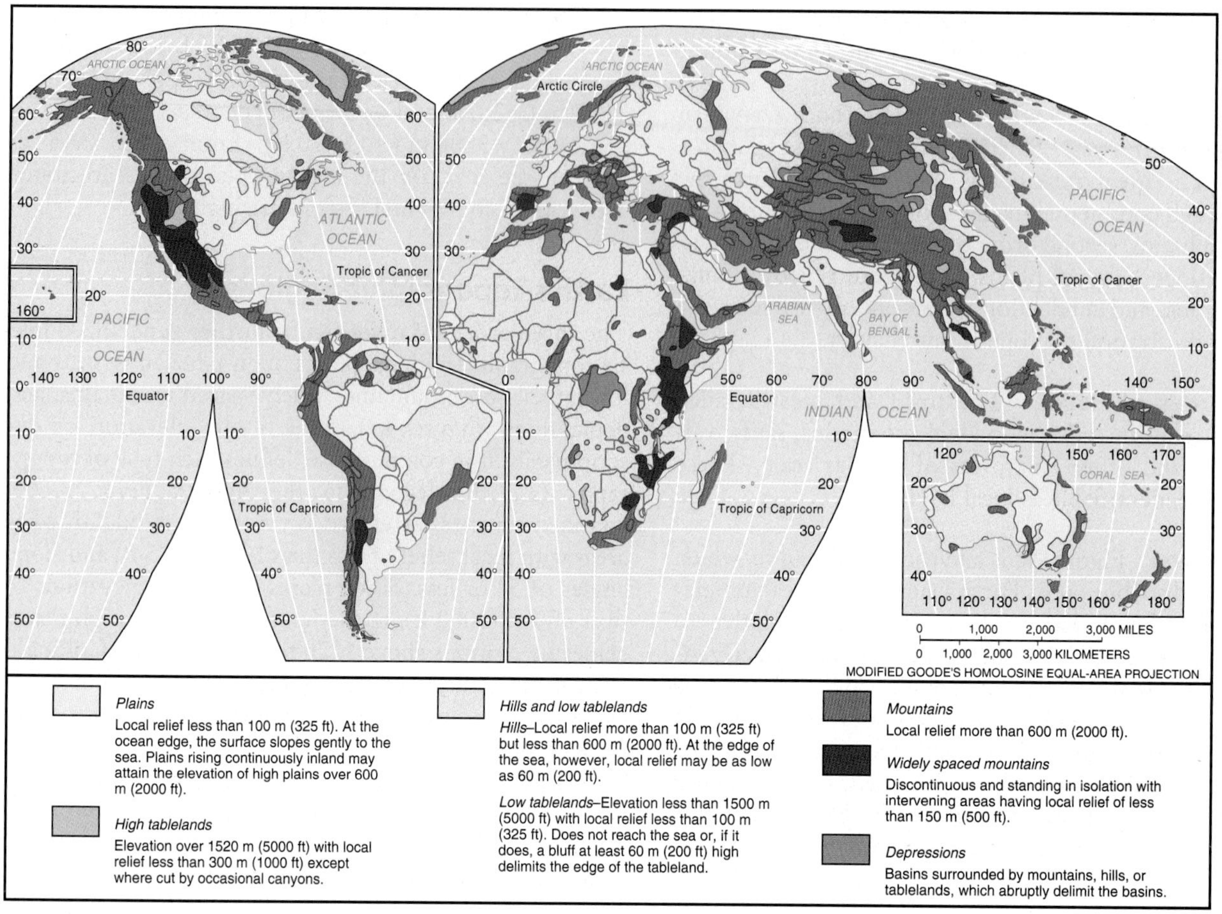

Plains
Local relief less than 100 m (325 ft). At the ocean edge, the surface slopes gently to the sea. Plains rising continuously inland may attain the elevation of high plains over 600 m (2000 ft).

High tablelands
Elevation over 1520 m (5000 ft) with local relief less than 300 m (1000 ft) except where cut by occasional canyons.

Hills and low tablelands
Hills–Local relief more than 100 m (325 ft) but less than 600 m (2000 ft). At the edge of the sea, however, local relief may be as low as 60 m (200 ft).
Low tablelands–Elevation less than 1500 m (5000 ft) with local relief less than 100 m (325 ft). Does not reach the sea or, if it does, a bluff at least 60 m (200 ft) high delimits the edge of the tableland.

Mountains
Local relief more than 600 m (2000 ft).

Widely spaced mountains
Discontinuous and standing in isolation with intervening areas having local relief of less than 150 m (500 ft).

Depressions
Basins surrounded by mountains, hills, or tablelands, which abruptly delimit the basins.

MODIFIED GOODE'S HOMOLOSINE EQUAL-AREA PROJECTION

FIGURE 9.3 Earth's topographic regions.
Earth's topography is characterized as plains, high and low tablelands, hills, mountains, and depressions. [After Richard E. Murphy, "Landforms of the World," Map Supplement No. 9, *Annals of the Association of American Geographers* 58, no. 1 (March 1968). Adapted by permission.]

News Report 9.1

Mount Everest at New Heights

The announcement came November 11, 1999, in Washington, D.C., at the opening reception of the 87th annual meeting of the American Alpine Club. The press conference was held at National Geographic Society headquarters. Mount Everest has a new revised elevation determined by direct Global Positioning System (GPS) placement on the mountain's icy summit!

In May 1999, mountaineers Pete Athans and Bill Crouse reached the summit with five Sherpas. The climbers operated Global Positioning System satellite equipment on the very top of Mount Everest and determined the precise height of the world's tallest mountain. A photo in Chapter 1, News Report 1.1, shows a GPS unit placed by climber Wally Berg 18 m below the summit in 1998. This site is still Earth's highest benchmark with a permanent metal marker installed in rock. GPS instruments around the region permit differential calibration for exact measurement (Figure 1). This is a continuation of a measurement effort begun in 1995 by Bradford Washburn, renowned mountain photographer/explorer and honorary director of Boston's Museum of Science. Washburn announced the new measurement of 8850 m (29,035 ft).

Astronaut Thomas D. Jones made the photo of Mount Everest in Figure 2 during Space Shuttle flight STS-80 in 1996 aboard Space Shuttle *Columbia*. This startling view caught the mountain's triangular East Face bathed in morning light (east is to the lower left, north is to the lower right, south at the upper left; see locator map direction arrow). The North Face is in shadow. Several glaciers are visible flowing outward from the mountains, darkened by rock and debris.

The National Geographic Society press release stated: "This latest measurement stemmed from Washburn's desire to use lightweight Trimble GPS receivers, along with lithium batteries that work in severe temperature conditions, to establish the highest bedrock survey station. Then, running two receivers simultaneously, one at the South Col and the other at the summit, an extremely accurate altitude was established for the top of the mountain." Wally Berg's preliminary

(continued)

News Report 9.1 *(continued)*

placement of a GPS unit was an important step in meeting this goal.

Washburn stated, "The reading of 29,035 ft (8850 m) showed no measurable change in the height of Everest calculated since GPS observations began 4 years ago. But from these GPS readings ... it appears that the horizontal position of Everest seems to be moving steadily and slightly northeastward; between 3 and 6 mm a year (up to 0.25 in. a year)." The mountain range is being driven further into Asia by plate tectonics—the continuing collision of Indian and Asian landmasses.

FIGURE 1 GPS installation measures Mount Everest.
A global positioning system (GPS) installation near Namche Bazar in the Khumbu (Everest) region of Nepal. The station is at latitude 27.8° N, longitude 86.7° E, at 3523 m (11,558 ft). The summit of Mount Everest is 30 km (18.6 mi) distant—visible above the right side of the weather dome. A network of such GPS installations correlated with the unit the climbers took to the summit. [Photo by Charles Corfield, science manager to the 1998 and 1999 expeditions.]

FIGURE 2 Mount Everest in morning light from orbit aboard Space Shuttle *Columbia*.
[Photo courtesy of NASA, made by Dr. Thomas Jones, astronaut, 1996.]

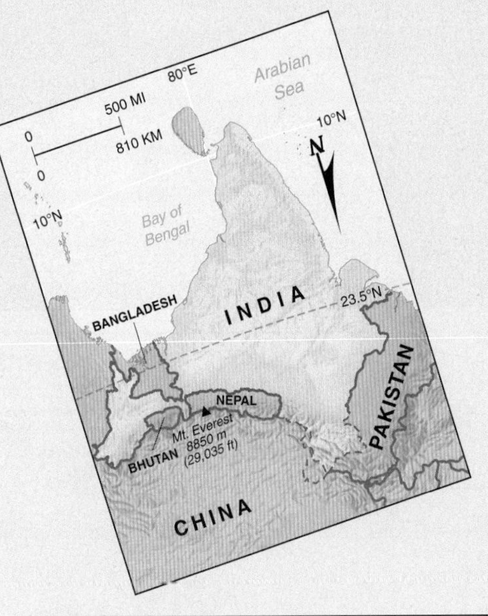

Crustal Formation Processes

How did Earth's continental crust form? What gives rise to the three orders of relief just discussed? Ultimately, the answer is tectonic activity, driven by our planet's internal energy, and the exogenic processes of weathering and erosion, powered by the Sun through the actions of air, water, ice, and waves.

Tectonic activity produces continental crust that is quite varied. Nonetheless, continental crust generally is thought of in three categories: (1) residual mountains and continental cores ("shields") that are inactive remnants of ancient tectonic activity; (2) tectonic mountains and landforms produced by active folding and faulting movements that deform the crust; and (3) volcanic features formed by the surface accumulation of molten rock from eruptions of subsurface materials. Thus, several distinct processes operate together to produce the continental crust.

Continental Shields

All continents have a nucleus of crystalline rock on which the continent "grows" with additions of other crust and sediments. The nucleus is the *craton*, or heartland region, of the continental crust. A region where a craton is exposed at the surface is called a **continental shield**. Geophysical observations of thousands of seismic wave patterns portray these shields as cold and stiff portions of the lithosphere, extending in depth to some 250 km (average lithosphere is 70 km, Figure 8.2). These deep craton roots are the *craton keels* of the continents, as in a keel on a boat. These are regions where the old, deep lithosphere has remained locked to the overlying crust, resistant to the push and pull of tectonic forces. The low iron content and lesser density than deeper rocks have kept these shields from sinking into the mantle.

Cratons generally have been eroded to a low elevation and relief and are old (most exceed 2 billion years in age, and all are Precambrian, or older than 570 million years). Portions of cratons are covered with layers of sedimentary rock that are quite stable over time. An example of such a stable *platform* is the region that stretches from the Rockies to the Appalachians and northward into central Canada. Figure 9.4 shows the principal areas of exposed shields and a photo of the Canadian shield.

Building Continental Crust

Continental crust results from a complex process that involves sea-floor spreading and formation of oceanic crust, its subduction, remelting, and subsequent rise of magma, as summarized in Figure 8.13 and Figure 9.5.

To understand this process, begin with the magma that originates in the asthenosphere and mantle and wells up along the mid-ocean ridges. Magma is less than 50%

(b)

(a)

FIGURE 9.4 Continental shields.
(a) Portions of major continental shields that have been exposed by erosion. Adjacent portions of these shields remain covered. (b) Canadian shield landscape in central Labrador interior, stable for hundreds of millions of years, stripped by past glaciations, and marked by intrusive igneous dikes (magmatic intrusions). [(a) After Richard E. Murphy, "Landforms of the World," Map Supplement No. 9, *Annals of the Association of American Geographers* 58, no. 1 (March 1968). Adapted by permission; (b) Photo by R. S. Hildebrand, Geological Survey of Canada.]

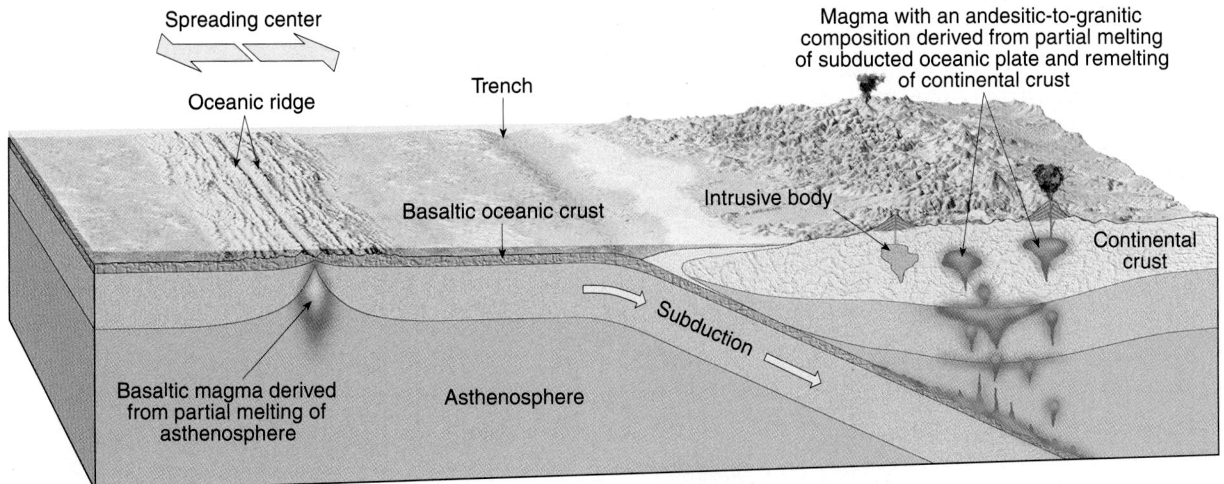

FIGURE 9.5 Crustal formation.
Material from the asthenosphere upwells along sea-floor spreading centers. Basaltic ocean floor is subducted beneath lighter continental crust, where it melts, along with its cargo of sediments, water, and minerals. This melting generates magma, which makes its way up through the crust to form igneous intrusions and extrusive eruptions. [From E. J. Tarbuck and F. K. Lutgens, *Earth, An Introduction to Physical Geology*, 5th ed., art by Dennis Tasa, © Prentice Hall, 1996, Fig. 20.20, p. 502.]

silica, is rich in iron and magnesium, and has a low-viscosity (thin) texture. This material rises at the spreading centers, cools to form ocean floor, spreads outward, and collides with continental crust along its far edges. This denser oceanic crust plunges beneath the lighter continental crust, into the mantle, where it remelts. The new magma then rises and cools, forming more continental crust, in the form of intrusive granitic igneous rock.

As the subducting oceanic plate works its way under a continental plate, it takes with it sediment and trapped water, melting and incorporating various elements from the crust into the mixture. As a result, the magma (generally called a "melt") that migrates upward from a subducted plate contains 50%–75% silica, is high in aluminum, and has a high-viscosity (thick) texture. Note that this melt is different in composition from the magma that arose directly from the asthenosphere at spreading centers to form new seafloor.

Bodies of the silica-rich magma may reach the surface in explosive volcanic eruptions, or they may stop short and become subsurface intrusive bodies in the crust, cooling slowly to form crystalline plutons such as batholiths (see Figures 8.7 and 8.8).

Terranes. A surprising discovery is that each of Earth's major plates actually is a collage of many crustal pieces acquired from a variety of sources. Accretion, or accumulation, has occurred as crustal fragments of ocean floor, curved chains (or arcs) of volcanic islands, and other pieces of continental crust have been forced against the edges of continental shields and platforms. These migrating crustal pieces, which became attached to the plates, are called **terranes** (not to be confused with "terrain," which refers to the topography of a tract of land).

Displaced terranes have histories different from those of the continents that capture them, are usually framed by fractures, and are distinguished in rock composition and structure from their new continental homes. As an example, accreted terranes are particularly prevalent in the region surrounding the Pacific. At least 25% of the growth of western North America can be attributed to the accretion of terranes since the early Jurassic period (190 million years ago). A good example is the Wrangell Mountains, which lie just east of Prince William Sound and the city of Valdez, Alaska. The *Wrangellia terranes*—a former volcanic island arc and associated marine sediments from near the equator—migrated approximately 10,000 km (6200 mi) to form these mountains (Figure 9.6).

The Appalachian Mountains, extending from Alabama to the Maritime Provinces of Canada, possess bits of land once attached to ancient portions of Europe, Africa, South America, Antarctica, and various oceanic islands. The discovery of terranes, made only in the 1980s, demonstrates one of the ways continents are assembled.

Crustal Deformation Processes

Rocks, whether igneous, sedimentary, or metamorphic, are subject to powerful stress by tectonic forces, gravity, and the weight of overlying rocks. The three types of stress: *tension* (stretching), *compression* (shortening), and *shear* (tearing and twisting), are shown in Figure 9.7. *Strain* is how rocks respond to these stresses as is expressed in *folding* (bending) or *faulting* (breaking). Whether a rock bends or breaks depends on several factors, including composition and how much pressure is on the rock. An important quality is whether the rock is

FIGURE 9.6 The Wrangell Mountains.
Snow-covered Wrangell Mountains north of the Chugach Mountains in this November 7, 2001, image, in east-central Alaska and across the Canadian border. Note Mount McKinley (Denali) at 6194 m, 20,322 ft, in upper-left part of the image, north of Cook Inlet, highest elevation in North America. [*Terra* image from MODIS sensor, courtesy of MODIS Land Rapid Response Team, NASA/GSFC.]

Wrangell Mountains, Alaska

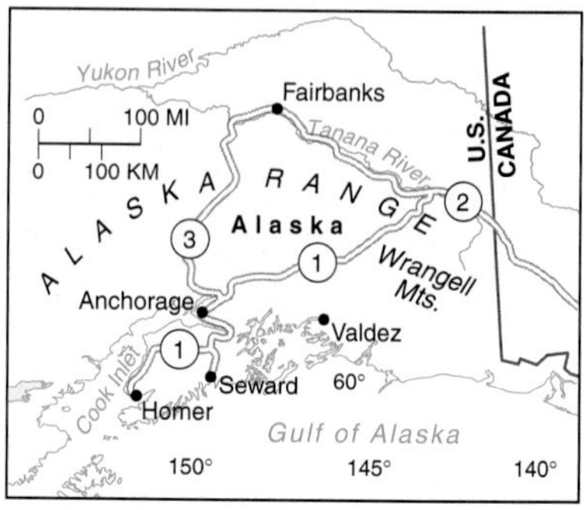

Yukon River
Fairbanks
0 100 MI
0 100 KM
Tanana River
U.S.
CANADA
ALASKA RANGE
Alaska
Wrangell Mts.
Anchorage
Valdez
Cook Inlet
Homer
Seward
60°
Gulf of Alaska
150° 145° 140°

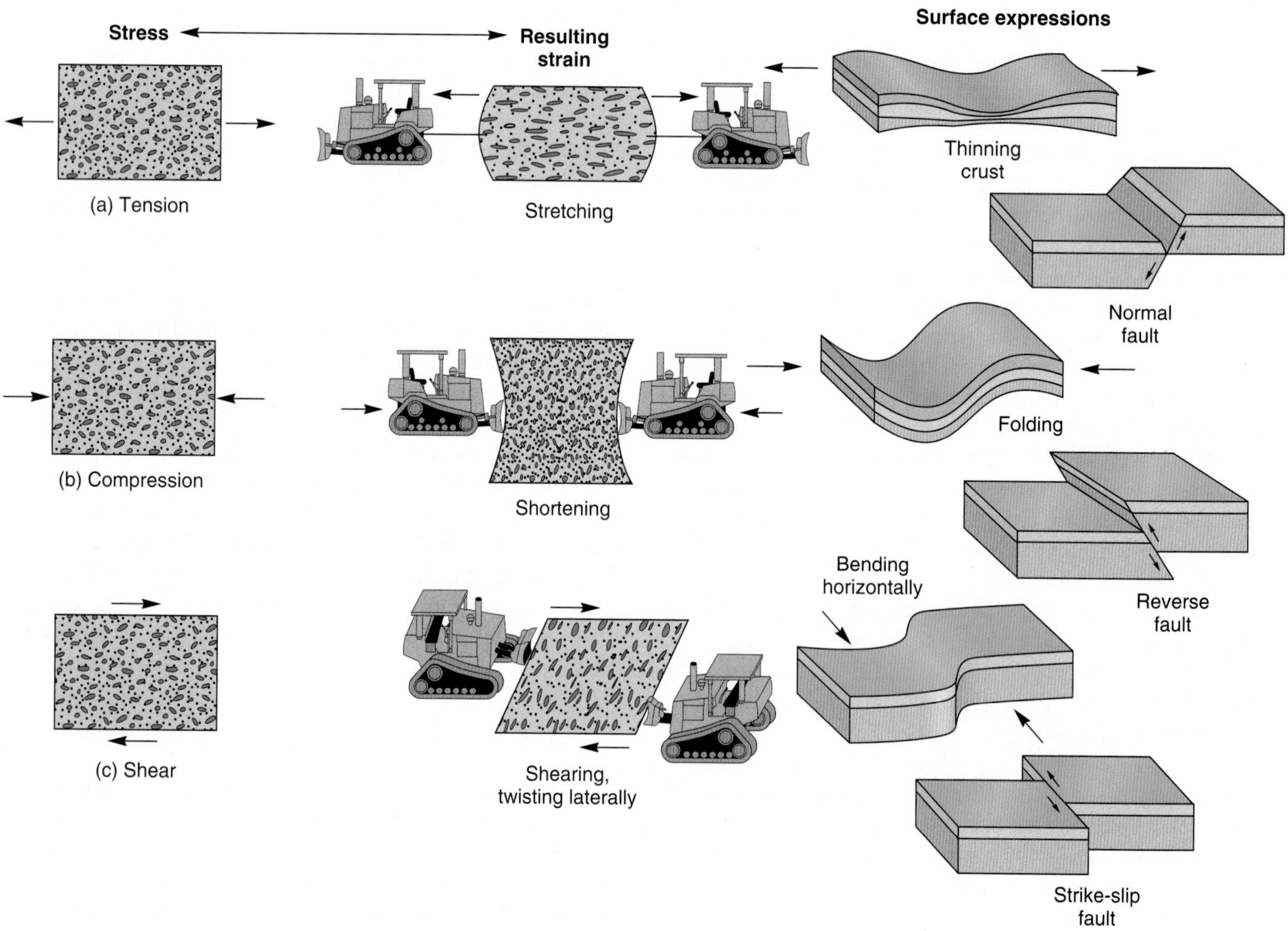

Stress ← → Resulting strain

Surface expressions

(a) Tension

Stretching

Thinning crust

Normal fault

(b) Compression

Shortening

Folding

Reverse fault

(c) Shear

Shearing, twisting laterally

Bending horizontally

Strike-slip fault

FIGURE 9.7 Three kinds of stress, strain, and resulting surface expressions.
The bulldozers represent the stress, or force, on the rock strata. (a) Tension stress produces a stretching and thinning of the crust, and a *normal fault.* (b) Compression produces shortening and folding, and *reverse faulting* of the crust. In a back-and-forth horizontal motion, (c) shear stress produces a bending of the crust, and on breaking, a *strike-slip fault.*

brittle or *ductile*. The patterns created by these processes are evident in the landforms we see today.

Folding and Broad Warping

Folds, anticlines, and synclines

When layered rock strata are subjected to compressional forces, they become deformed (Figure 9.8). Convergent plate boundaries intensely compress rocks, deforming them in a process known as **folding**. If we take pieces of thick fabric, stack them flat on a table, and then slowly push on opposite ends of the stack, the cloth layers will bend and form similar folds (Figure 9.8a).

If we then draw a line down the center axis of a resulting ridge and trough, we are able to see how the names of the folds are assigned. Along the *ridge* of a fold, layers *slope downward away from the axis*, resulting in an **anticline**. In the trough of a fold, however, layers *slope downward toward the axis*, producing a **syncline** (Figure 9.8a, left).

If the axis of either type of fold is not "level" (horizontal, or parallel to Earth's surface), the layers then *plunge* (dip down) at an angle. Relative to resource recovery it is important to know the angle of folds to Earth's surface and their location. Petroleum, for example, collects in the upper portions of folds in permeable rock layers such as sandstone.

A residual ridge may form within a syncline, due to the weathering response of rock strata of different resistances (a "synclinal ridge," Figure 9.8a, right). An interstate

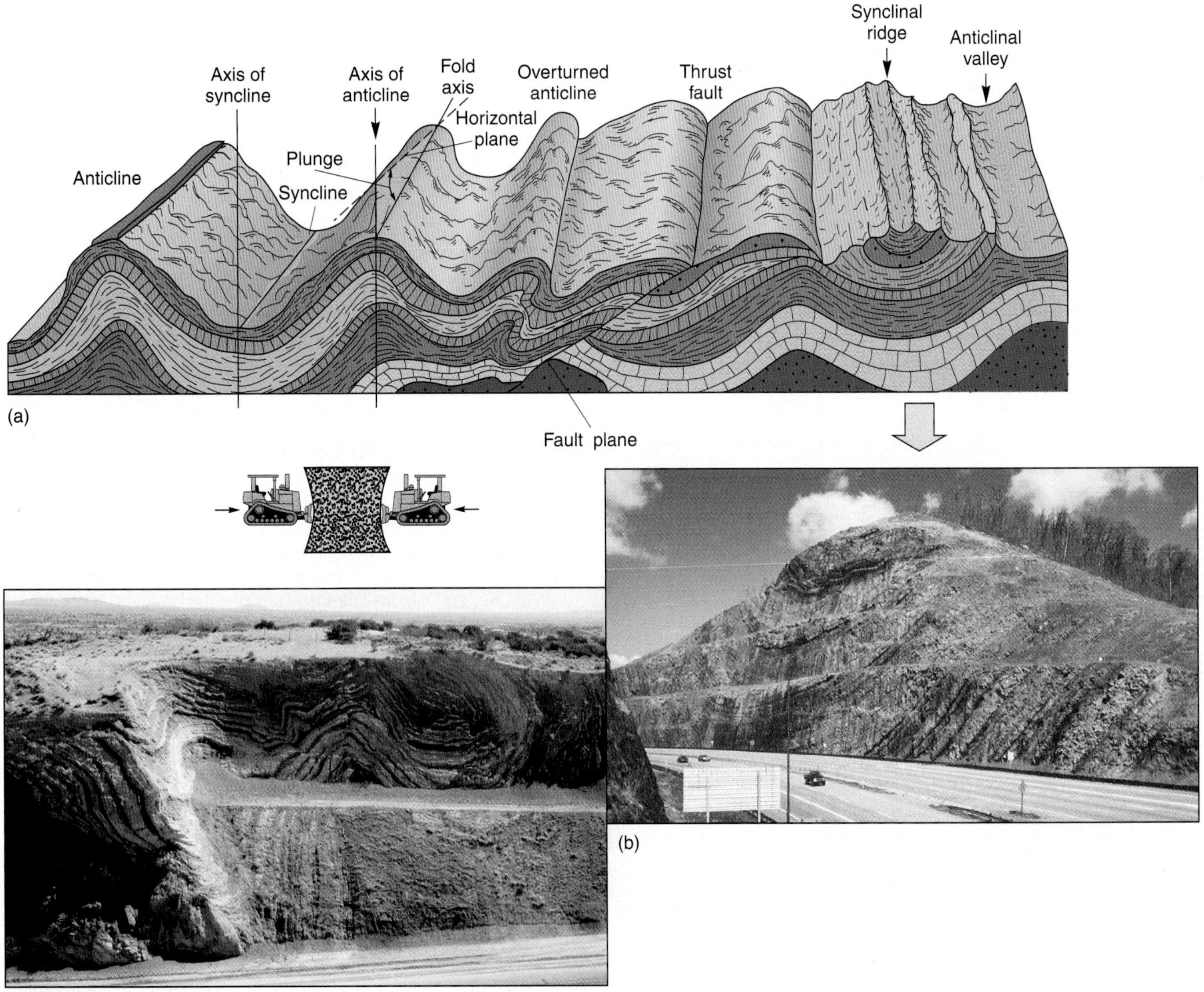

FIGURE 9.8 Folded landscapes.
(a) Folded landscape and the basic types of fold structures. (b) A roadcut exposes a synclinal ridge in western Maryland. This syncline is a natural outdoor classroom, and the state of Maryland built an interpretive center with a walkway above the highway to permit study of the syncline. (c) Compressional folding and faulting in a roadcut along the San Andreas fault zone in Southern California. [Photos by (b) Mike Boroff/Photri-Microstock; (c) Bobbé Christopherson.]

highway roadcut dramatically exposes such a synclinal ridge in western Maryland (Figure 9.8b). Compressional forces often push folds far enough that they actually overturn on their own strata ("overturned anticline," near center of figure). Further stress eventually fractures the rock strata along distinct lines, and some overturned folds are thrust up, causing a considerable shortening of the original strata ("thrust fault"). Areas of intense stress, compressional folding, and faulting are visible in a roadcut where the San Andreas fault in Southern California passes beneath the Antelope Valley Freeway (Figure 9.8c).

The Canadian Rocky Mountains and the Appalachian Mountains illustrate well the complexity of the resulting folded landscape. Satellites let us view many of these structures from an orbital perspective, as in Figure 9.9. Northwest of the Strait of Hormuz, just north of the Persian Gulf, are the Zagros Mountains of Iran. In the satellite image, anticlines form the parallel ridges and active weathering and erosion processes expose underlying strata.

In addition to the rumpling of rock strata just discussed, Earth's continental crust also is subjected to broad warping actions. These actions produce similar up-and-down bending of strata, but the bends are far greater in extent than those produced by folding. Warping forces include mantle convection, isostatic adjustment such as from the loss of weight from previous ice loads across northern Canada, or swelling from an underlying hot spot. Warping features include small individual foldlike structures called *basins* and *domes* (Figure 9.10a, b). Or they can range up to regional features the size of the Ozark Mountain complex in Arkansas and Missouri, the Colorado Plateau in the West, the Richat dome in Mauritania, or the Black Hills of South Dakota (Figure 9.10c, d).

Faulting

ANIMATION **Fault types**

A freshly poured concrete sidewalk is smooth and strong. But stress the sidewalk by driving heavy equipment over it, and the resulting strain might cause a fracture. Pieces on either side of the fracture may move up, down, or horizontally, depending on the direction of stress. When rock strata are strained beyond their ability to remain a solid unit, they express the strain as a fracture. Displacement of rock formations on either side of the fracture is the process known as **faulting**.

Thus, *fault zones* are areas of crustal movement. At the moment of fracture the fault line shifts and a sharp release of energy occurs, called an **earthquake** or *quake* (Figure 9.11). The fracture surface along which the two sides of a fault move is the *fault plane*. The tilt and orientation of the fault plane provide the basis for naming the three basic types of faults.

Normal Fault. A **normal fault**, or tension fault, occurs when rocks pull apart. They move vertically along an inclined fault plane so that one "block" of rock ends up lower than the other (Figure 9.11a). The downward-shifting side is the *hanging wall*; it drops relative to the *footwall block*.

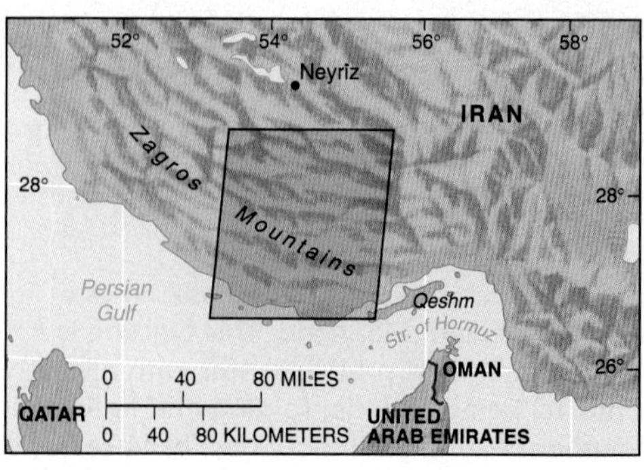

FIGURE 9.9 Folding in the Zagros crush zone, Iran.
Southern Zagros Mountains in the Zagros crush zone between the Arabian and Eurasian plates. This area was a dispersed terrane (migrating crustal piece) that separated from the Eurasian plate. However, the collision produced by the northward push of the Arabian block is now shoving this terrane back into Eurasia and forming the folded mountains shown. [NASA image.]

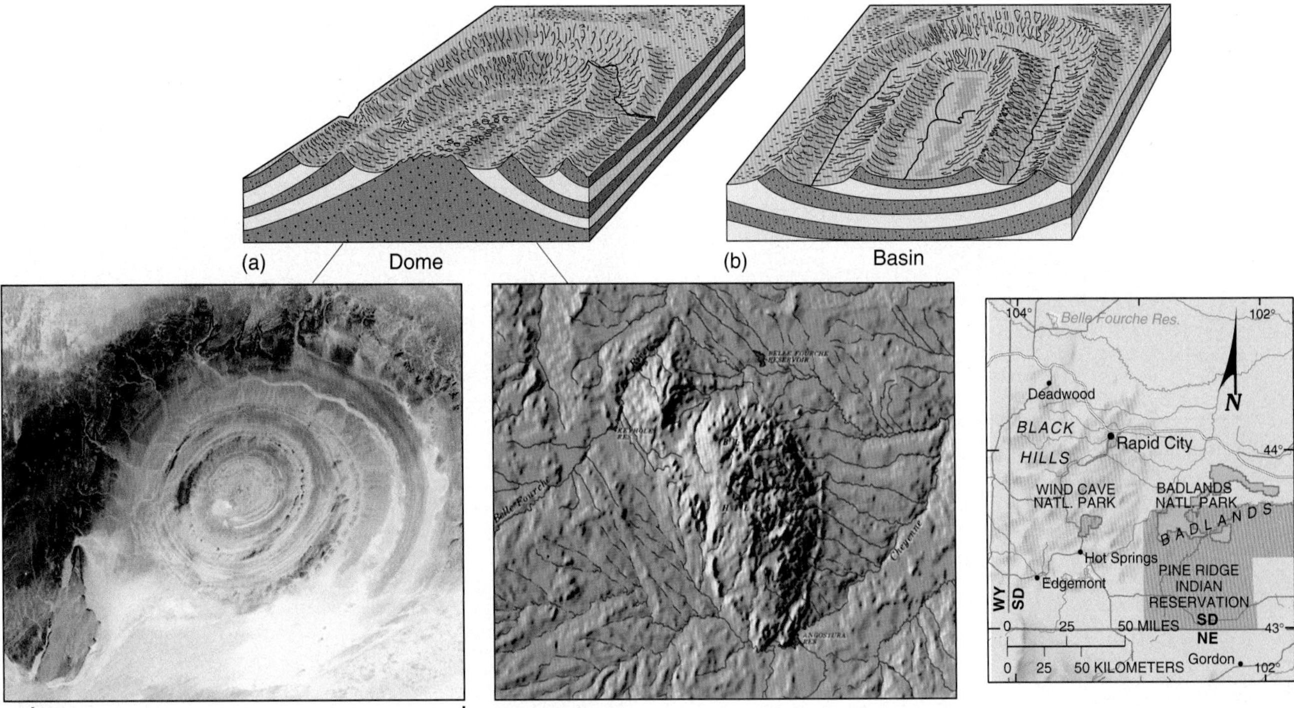

FIGURE 9.10 Domes and basins.
(a) An upwarped dome. (b) A structural basin. (c) The Richat dome structure of Mauritania. (d) The Black Hills of South Dakota is a dome structure, shown here on a digitized relief map. [(c) Image from *Terra* ASTER sensor, October 7, 2000, courtesy of NASA/GSFC/MITI and U.S./Japan ASTER Science Team; (d) from USGS digital terrain map I-2206, 1992.]

The exposed fault plane sometimes is visible along the base of faulted mountains, where individual ridges truncate owing to the movements of the fault and appear as triangular facets at the ends of the ridges. A cliff formed by faulting is a *fault scarp*, or *escarpment.*

Reverse (Thrust) Fault. Compressional forces associated with converging plates force rocks to move *upward* along the fault plane, producing a **reverse fault**, or compression fault (Figure 9.11b). On the surface it appears similar in form to a normal fault. If the fault plane forms a low angle relative to the horizontal, the fault is termed a **thrust fault**, or *overthrust fault*, indicating that the overlying block has shifted far over the underlying block (Figure 9.8a, "thrust fault"). Place one hand palm-down on the back of the other and move your hands past each other—this is the motion of a low-angle thrust fault with one side pushing over the other.

In the Alps, several such overthrusts result from compressional forces of the ongoing collision between the African and Eurasian plates. Beneath the Los Angeles Basin, such overthrust faults produce a high risk of earthquakes and caused many quakes in the twentieth century, including the $30 billion 1994 Northridge earthquake. These blind (unknown until they rupture) thrust faults beneath the Los Angeles region are a major earthquake threat in the future.

Strike-Slip Fault. If movement along a fault plane is horizontal, as produced by a *transform fault*, it is called a **strike-slip fault** (Figure 9.11c). The movement is *right-lateral* or *left-lateral*, depending on the motion perceived when you observe movement on one side of the fault relative to the other side.

Although strike-slip faults do not produce cliffs (scarps), they can create linear *rift valleys*, and sag ponds where disruptions of groundwater flows occur, as is the case with the San Andreas fault system of California. The rift is clearly visible in the photograph where the edges of the North American and Pacific plates are grinding past each other as a result of transform faults associated with a former sea-floor spreading center. The evolution of this fault system is shown in Figure 9.12.

Note in the figure how in (1) the East Pacific rise developed as a spreading center with associated transform faults, while the North American plate was progressing westward after the breakup of Pangaea. (2) Forces then shifted the transform faults toward a northwest-southeast alignment along a weaving axis. (3) Finally, the western margin of North America overrode those shifting transform faults. The map notes the average rates of fault motions at several locations.

In *relative* terms, the motion along this series of faults is right-lateral, whereas in *absolute* terms the North American plate still is moving westward. Consequently, the San Andreas system is a series of faults that are *transform* (associated with a former spreading center), *strike-slip* (horizontal in motion), and *right-lateral* (one side is moving to the right relative to the other side).

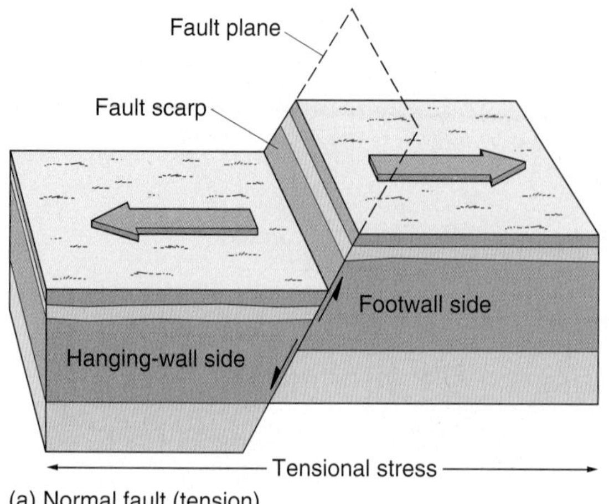

(a) Normal fault (tension)

(a)

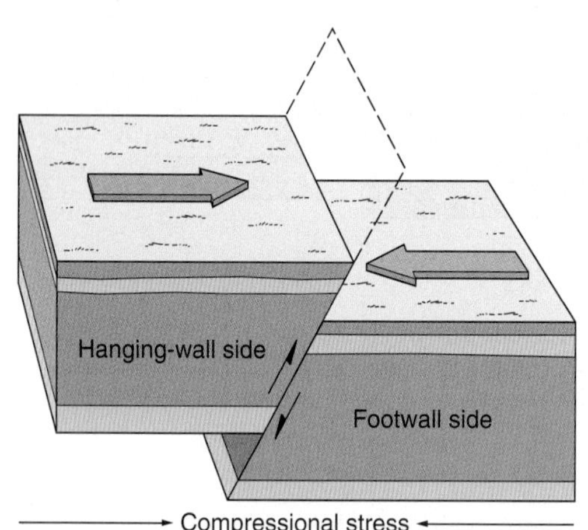

(b) Thrust or reverse fault (compression)

(b)

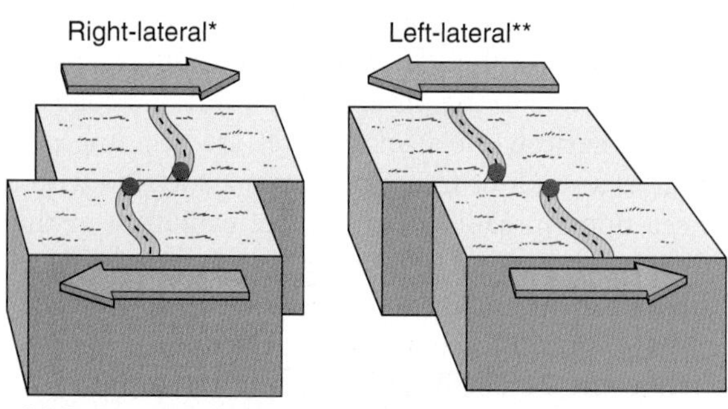

(c) Strike-slip fault (lateral shearing)

* Viewed from either dot on each road, movement to opposite side is *to the right.*
** Viewed from either dot on each road, movement to opposite side is *to the left.*

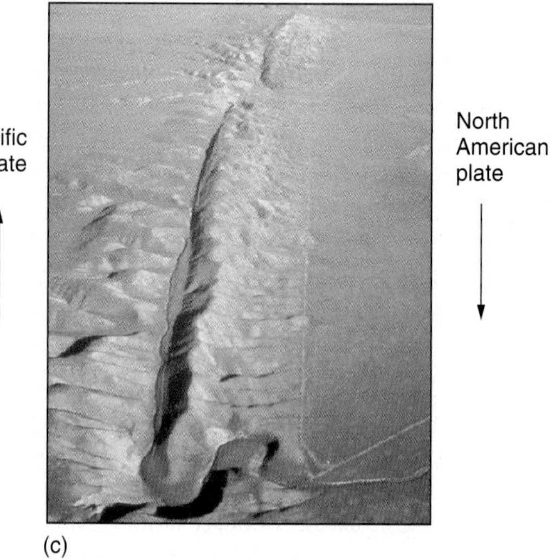

(c)

FIGURE 9.11 Types of faults.
(a) A normal fault produced by tension in the crust, visible along the edge in Tashkent, Uzbekistan. (b) A thrust, or reverse fault, produced by compression in the crust, visible in these offset strata in coal seams and volcanic ash in British Columbia. (c) A strike-slip fault produced by lateral shearing, clearly seen looking north along the San Andreas fault rift zone on the eastern edge of the Coast Ranges in California–Pacific plate to the left, North American plate to the right. [(a) Photo by Fred McConnaughey/Photo Researchers, Inc.; (b) photo by Fletcher and Baylis/Photo Researchers, Inc.; (c) Kevin Schafer/Peter Arnold, Inc.]

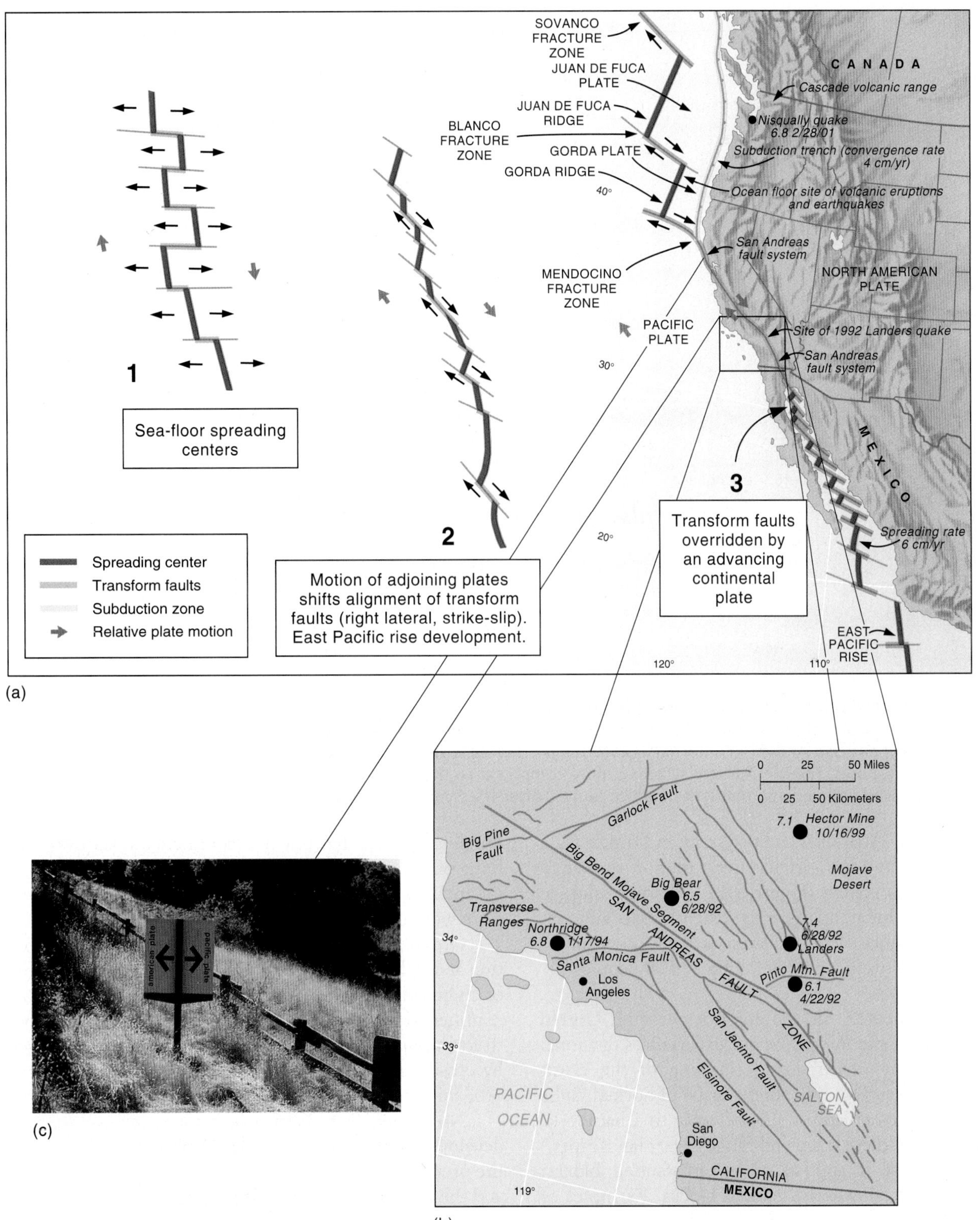

FIGURE 9.12 San Andreas fault formation.
(a) Formation of the San Andreas fault system as a series of transform faults in three successive stages. (b) Enlargement shows the portion of southern California where the 1992 Landers, Big Bear, the 1994 Northridge (Reseda), and 1999 Hector Mine earthquakes occurred. Magnitude ratings are shown for six quakes. Note the epicenter location of the January 2001 magnitude 6.8 Nisqually quake, Washington state, related to the subduction zone offshore. (c) Trail sign near the epicenter of the 1906 San Francisco earthquake roughly marks the Pacific–North American plate boundary, Marin County, Point Reyes National Seashore, California. [(c) Photo by author.]

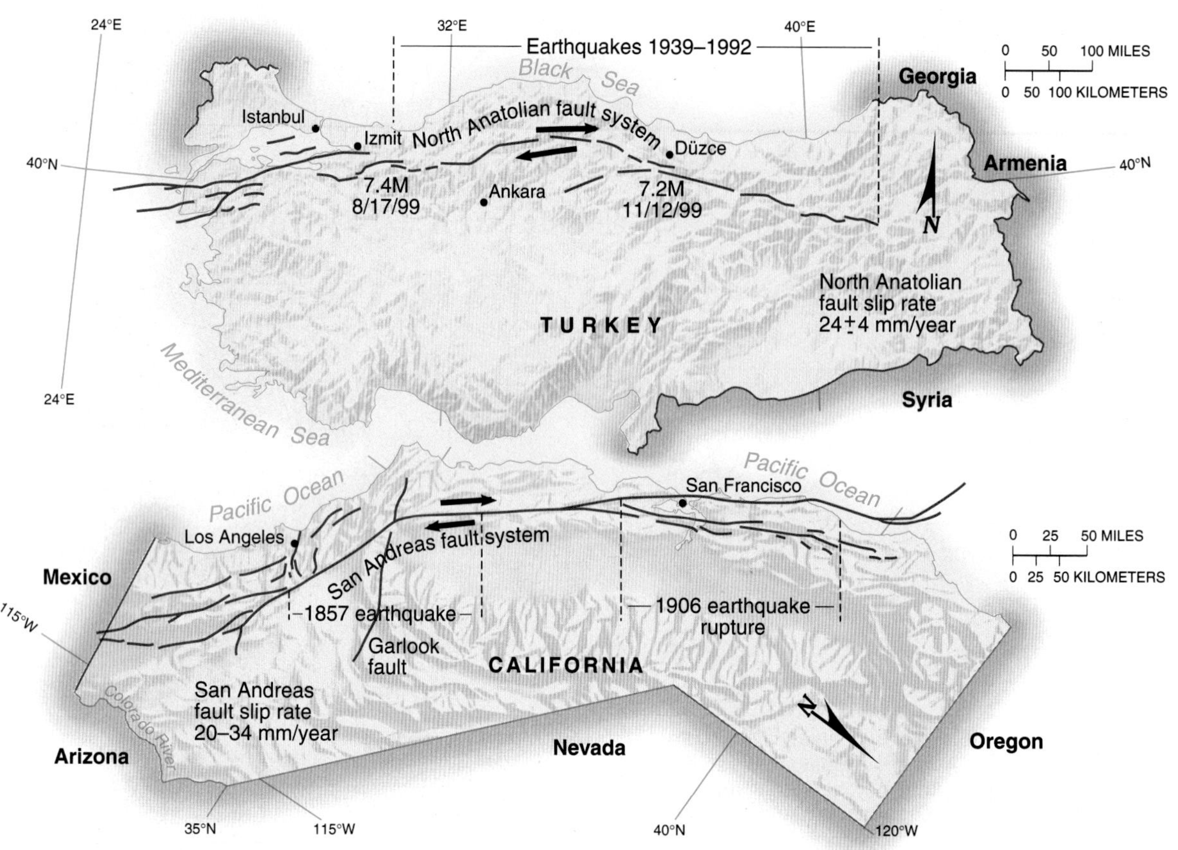

FIGURE 9.13 Strike-slip faults in Turkey and California.
Note these two strike-slip, right-lateral faults and the dates of activity along each—the North Anatolian (Turkey) and San Andreas (California). California is oriented with north to the right for comparison purposes. The San Andreas system is more complex than the North Anatolian because it is involved with a dense network of active faults. [After USGS map by Ross S. Stein.]

The North Anatolian fault system in Turkey is a strike-slip fault and has a right-lateral motion similar to the San Andreas system (Figure 9.13). A progression of earthquakes has hit along the entire extent of the fault system in Turkey since 1939; the latest devastation occurred in August and November 1999, killing 20,000 people.

Faults in Concert. In the interior western United States, the *Basin and Range Province* experiences tensional forces caused by uplifting and thinning of the crust, which crack the surface into aligned pairs of normal faults and a distinctive landscape. (Please refer to Chapter 12 and a section on these basin and range desert landscapes.) The term **horst** is applied to upward-faulted blocks; **graben** refers to downward-faulted blocks. Examples of horst-and-graben landscapes also include the Great Rift Valley of East Africa (associated with crustal spreading), which extends northward to the Red Sea, and the Rhine graben through which the Rhine River flows in Europe (Figure 9.14).

Orogenesis (Mountain Building)

Orogenesis literally is the birth of mountains (*oros* comes from the Greek for "mountain"). An *orogeny* is a mountain-building episode that thickens continental crust over millions of years. Earth's major chains of folded and faulted mountains, called *orogens*, correlate remarkably with the plate tectonics model.

Orogenesis can occur through large-scale deformation and uplift of the crust. It also may include the capture of migrating terranes and cementation of them to the continental margins, or the intrusion of granitic magmas to form plutons. These granite masses often are exposed by erosion following uplift. Uplift is the final act of the orogenic cycle.

No orogeny is a simple event; many involve previous developmental stages dating back millions of years, and the processes are ongoing today. Major mountain ranges, and the orogens that caused them, include:

- Rocky Mountains of North America (*Laramide orogeny*, 40–80 million years ago)
- Sierra Nevada of California (*Sierra Nevadan orogeny*, 35 million years ago, with older batholithic intrusions dating back 130–160 million years)
- Appalachian Mountains and the Ridge and Valley Province (nearly parallel ridges and valleys) of the eastern United States (*Alleghany orogeny*, 250–300 million years ago, preceded by at least two earlier

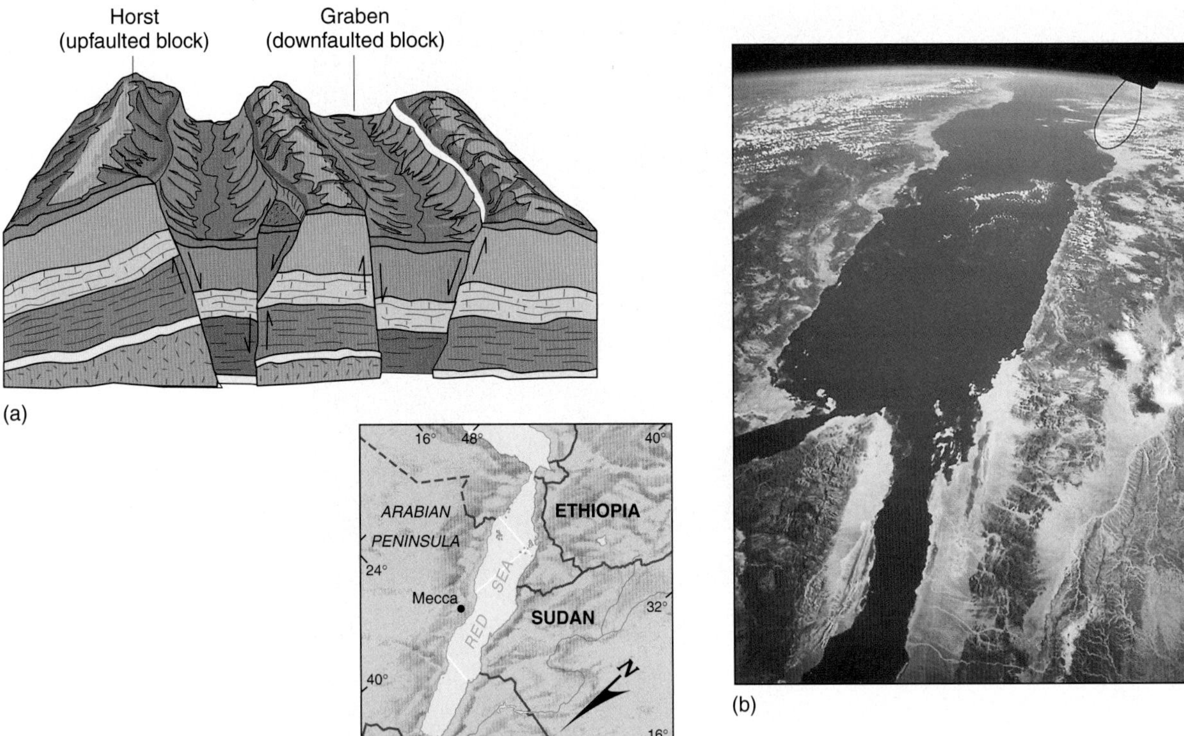

FIGURE 9.14 Faulted landscapes.
(a) Pairs of faults produce a horst and graben landscape characteristic of the Basin and Range Province in the western United States. (b) The Red Sea occupies a down-dropped block that is part of the rift system that runs through East Africa. [(b) *Gemini* photo from NASA.]

orogenies; in Europe this is contemporary with the *Hercynian orogeny*)

- Alps of Europe (*Alpine orogeny*, 20–120 million years ago and continuing to the present, with many earlier episodes) shown in Figure 9.15
- Himalayas of Asia (*Himalayan orogeny*, 45–54 million years ago, beginning with the collision of the India plate and Eurasia plate and continuing to the present)

Types of Orogenies

Figure 9.16 illustrates convergent plate collisions associated with orogenesis:

(a) *Oceanic plate–continental plate collision orogenesis.* This type of convergence is now occurring along the Pacific coast of the Americas and has formed the Andes, the Sierra of Central America, the Rockies, and other western mountains. We see folded sedimentary formations, with intrusions of magma forming granitic plutons at the heart of these mountains. Their buildup has been augmented by capturing of displaced terranes, cemented during their collision with the continental mass. Also, note the associated volcanic activity inland from the subduction zone.

(b) *Oceanic plate–oceanic plate collision orogenesis.* Such collisions can produce either simple volcanic island arcs or more complex arcs, such as Indonesia and Japan, which include deformation and metamorphism of rocks, and granitic intrusions. These processes formed the chains of island arcs and volcanoes that continue from the southwestern Pacific to the western Pacific, the Philippines, the Kurils, and on through portions of the Aleutians.

Both collision types, (a) oceanic–continental and (b) oceanic–oceanic, are active around the Pacific Rim. Both are thermal in nature, because the diving plate melts and migrates back toward the surface as molten rock. The region of active volcanoes and earthquakes around the Pacific is known as the *circum-Pacific belt* or, more popularly, the *ring of fire*. The third type of plate convergence is different.

(c) *Continental plate–continental plate collision orogenesis.* Here the orogenesis is mechanical; large masses of continental crust are subjected to intense folding, overthrusting, faulting, and uplifting. The converging plates crush and deform both marine sediments and basaltic oceanic crust. The formation of the European Alps is a result of such compression forces and includes considerable crustal shortening,

FIGURE 9.15 European Alps.
Western (France), Central (Italy), and Eastern (Austria) segments comprise the crescent shape of the Alps. Complex overturned faults and crustal shortening due to compressional forces occur along convergent plates. The Alps are some 1200 km (750 mi) in length, occupying 207,000 km^2 (80,000 mi^2). Note the aerosols and other pollutants concentrated south of the Alps in northern Italy. [*Terra* MODIS image courtesy of MODIS Land Rapid Response Team, NASA/GSFC.]

forming great overturned folds, called *nappes* (see Figure 9.15).

As mentioned earlier, the collision of India with the Eurasian landmass produced the Himalayan Mountains. That collision is estimated to have shortened the overall continental crust by as much as 1000 km (about 600 mi) and to have produced telescoping sequences of thrust faults at depths of 40 km (25 mi). The Himalayas feature the tallest above-sea-level mountains on Earth, including Mount Everest at 8850 m elevation (29,035 ft), and all 10 of Earth's highest peaks, as measured from sea level.

The disruption created by this plate collision has reached far under China, and frequent earthquakes there signal the continuation of this rapid-paced collision. As evidence of this ongoing strain, the January 2001 quake in Gujarat, India, was along a shallow, east-west tending thrust fault that gave way under the pressure of the northward-pushing India plate. More than one million buildings were destroyed or damaged, and 20,023 people died.

Grand Tetons and the Sierra Nevada. The Sierra Nevada of California and the Grand Tetons of Wyoming are in later stages of mountain-building orogenesis. Each is a *tilted-fault block* mountain range, in which a normal fault on one side of the range produces a tilted landscape of dramatic relief (Figure 9.17). Slowly cooling magma intruded into those blocks and formed granitic cores of coarsely crystalline rock. After tremendous tectonic uplift and the removal of overlying material through weathering, erosion, and transport, those granitic masses are now exposed in each mountain range. In some areas, the overlying material originally covered these batholiths by more than 7500 m (25,000 ft).

Recent research in the Sierra Nevada disclosed that some of the uplift was isostatic, caused by the erosion of overburden and loss of melting ice mass (that is, loss in overlying weight) following the last ice age some 18,000 years ago. The accumulation of sediments in the adjoining valley depressed the crust, thus enhancing relief in the landscape.

The Appalachian Mountains. The old, eroded, fold-and-thrust belt of the eastern United States contrasts with the younger mountains of the western portions of North America. As noted, the *Alleghany orogeny* followed at least two earlier orogenic cycles of uplift and the accretion of several captured terranes.

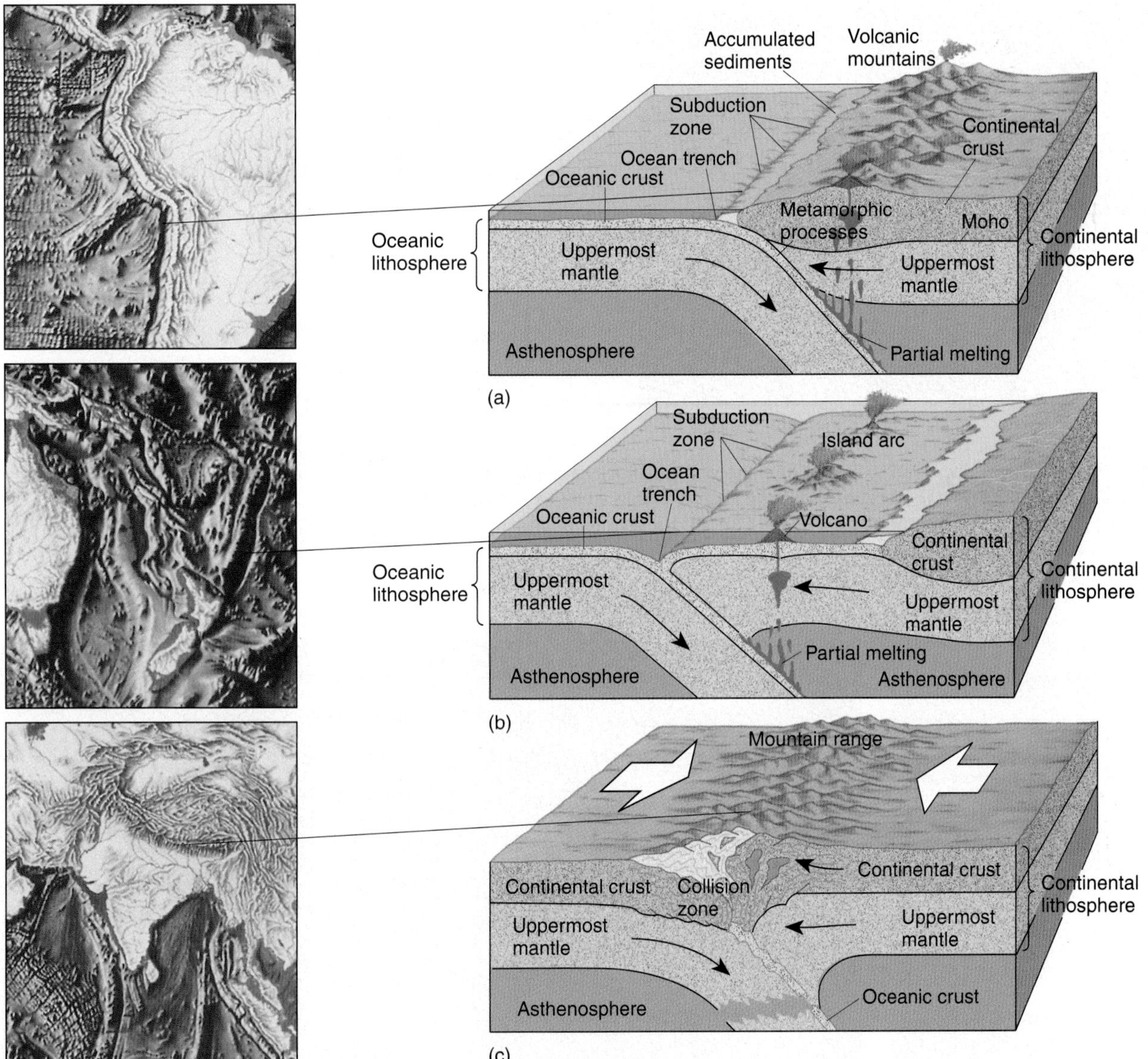

FIGURE 9.16 Three types of plate convergence.
Real-world examples illustrate three types of crustal collisions. (a) Oceanic–continental (example: Nazca plate–South American plate collision and subduction). (b) Oceanic–oceanic (example: New Hebrides Trench near Vanuatu, 16° S 168° E). (c) Continental–continental (example: India plate and Eurasian landmass collision and resulting Himalayan Mountains). Can you identify more of these plate convergence areas on the chapter-opening map? [Inset illustrations derived from *Floor of the Oceans*, 1975, by Bruce C. Heezen and Marie Tharp. © 1980 by Marie Tharp.]

The original material for the Appalachian Mountains resulted from the collisions that produced Pangaea. In fact, the Atlas Mountains of northwestern Africa were connected to the Appalachians at some time in the past, but the Atlas Mountains, embedded in the African plate, rafted apart from the Appalachians.

The Appalachian Mountain region comprises several landscape subregions: the Ridge and Valley Province (elongated sequences of folded sedimentary rock); the Blue Ridge Province (principally of crystalline rock, highest where North Carolina, Virginia, and Tennessee converge); the Piedmont (hilly to gentle terrain along most of the eastern and southern margins of the mountains); and the east coastal plain (from gentle hills to flat plains that extend to the coast).

Figure 9.18 displays the linear folds of the Appalachian system. Note how these dissected ridges are cut through by rivers, forming *water gaps*. These important breaks in the rugged ridges greatly influenced migration, settlement patterns, and the diffusion of cultural traits in the 1700s. The initial flow of people, goods, and ideas was guided by this topography.

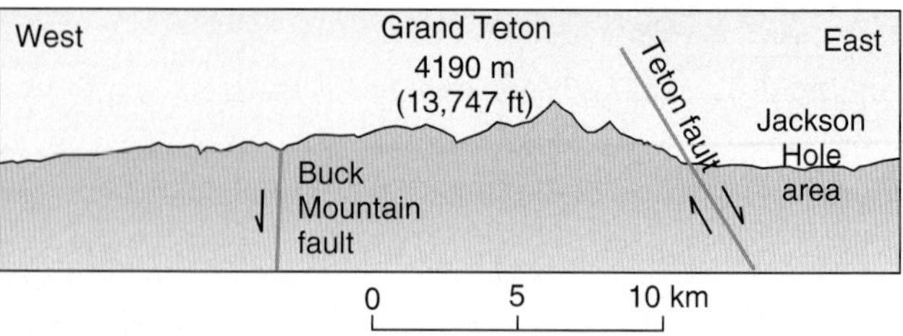

FIGURE 9.17 Tilted-fault block.
The Teton Range in Wyoming is an example of a tilted-fault block, a range of scenic beauty featuring 2130 m (7,000 ft) rugged relief between Jackson Hole and the summits. The Grand Teton is the highest peak in the range at 4190 m (13,747 ft). [Photo by author.]

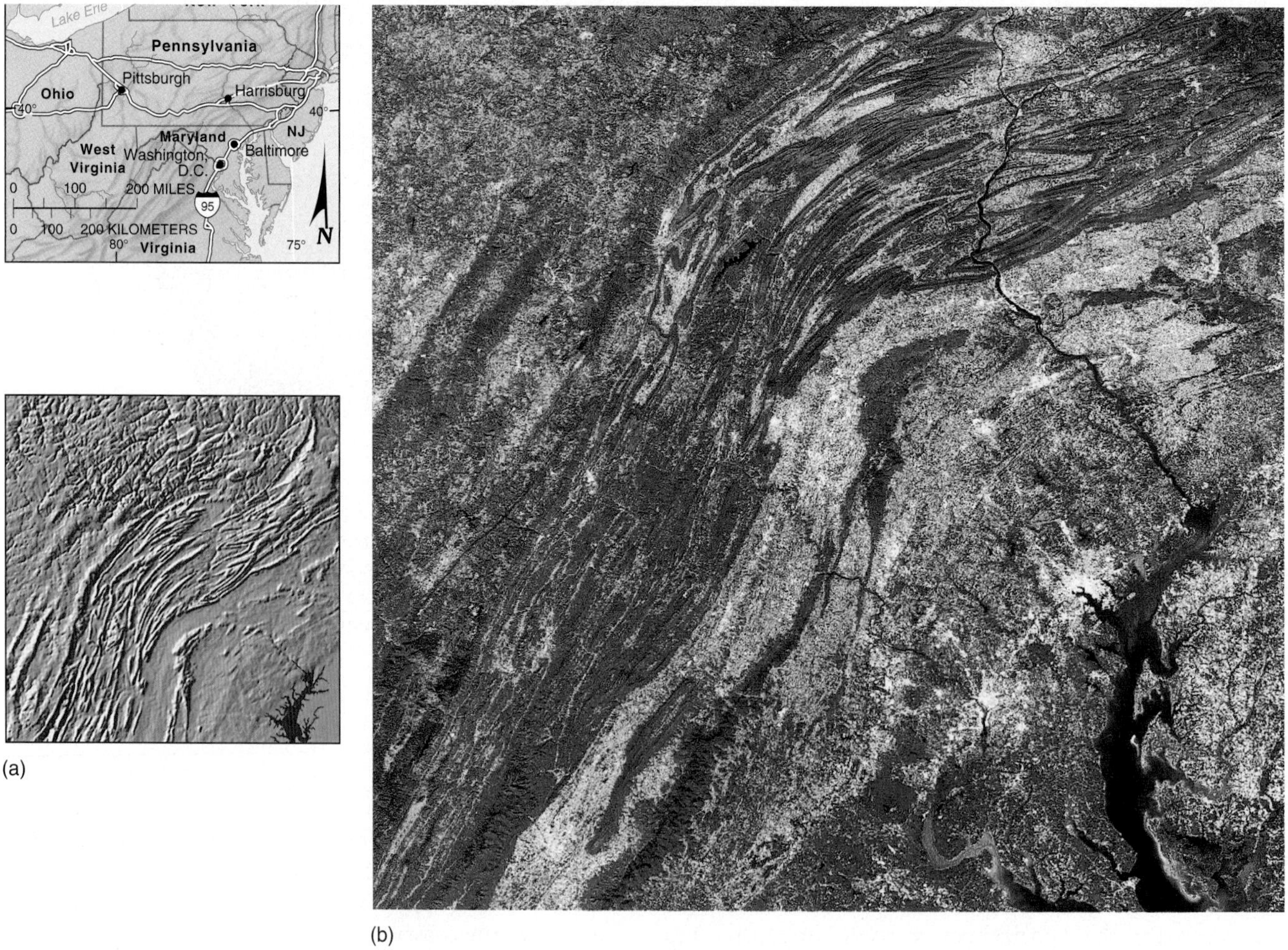

(a)

(b)

FIGURE 9.18 The Appalachian Mountains.
(a) USGS digital terrain map segment. (b) Appalachian Mountain region of Pennsylvania south through Maryland, Virginia, and West Virginia in a *Terra* MODIS sensor image from October 11, 2000, showing true fall colors. The highly folded nature of this entire region is clearly visible in the satellite image. [(a) USGS digital terrain map I-2206, 1992; (b) *Terra* image courtesy of MODIS Land Science Team, NASA/GSFC.]

World Structural Regions

Figure 9.19 defines seven essential structural regions, highlights Earth's two major continental mountain systems, and allows interesting comparison with the chapter-opening illustration of Earth's continental crust. Looking at the distribution of world structural regions helps summarize the information presented about the three rock-forming processes (igneous, sedimentary, metamorphic), plate tectonics, landform construction, and orogenesis.

The map reveals two large mountain chains—Earth's *alpine system*. The relatively young mountains along the western margins of the North and South American plates stretch from the tip of Tierra del Fuego to the massive peaks of Alaska, forming the *Cordilleran system*. The mountains of southern Asia, China, and northern India continue in a belt through the upper Middle East to Europe and the European Alps, constituting the *Eurasian-Himalayan system*.

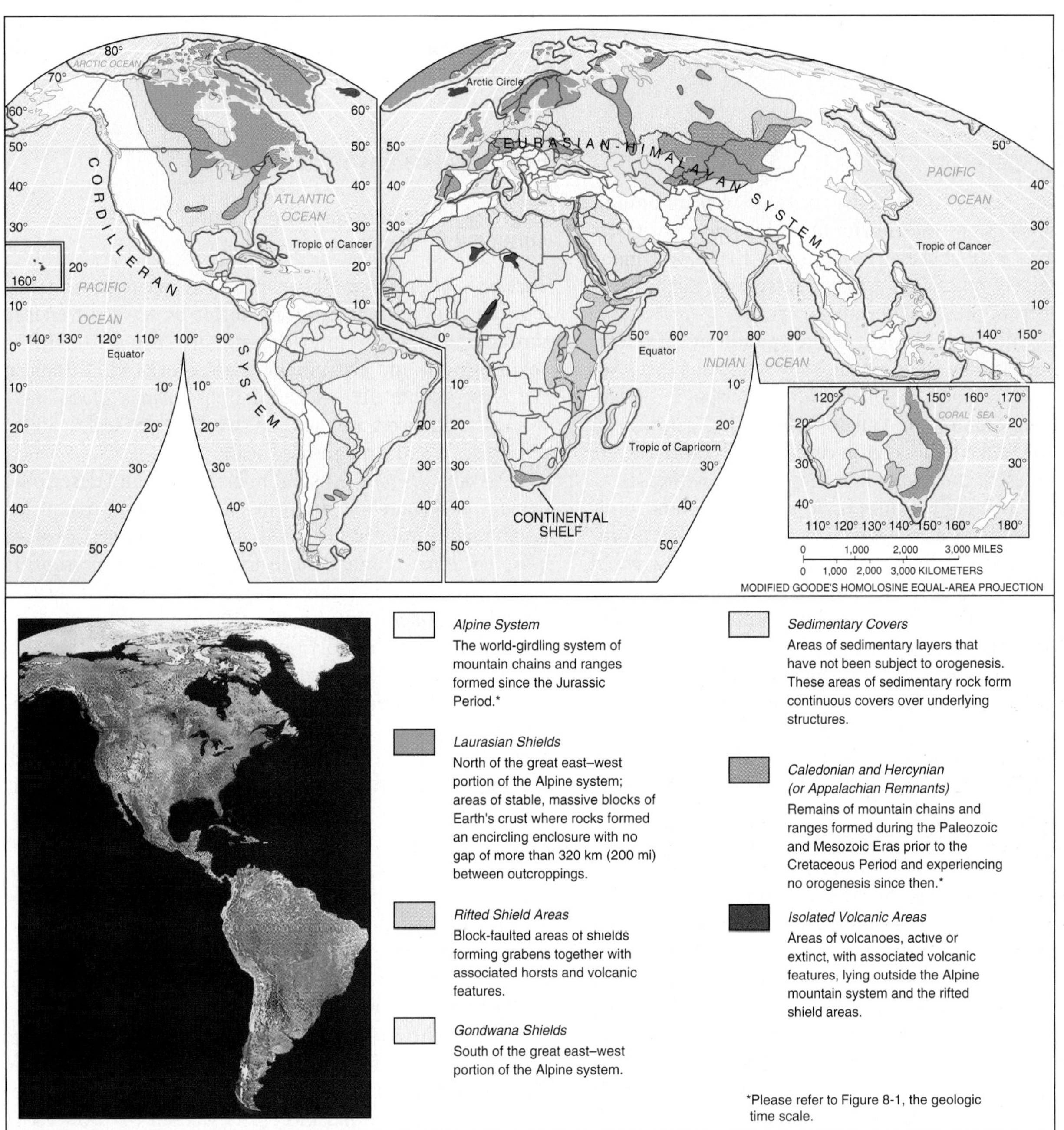

FIGURE 9.19 World structural regions and major mountain systems.
Because each structural region in this figure includes related landforms adjacent to the central feature of the region, some of the regions appear larger than the structures themselves. Structural regions in the Western Hemisphere are visible on this composite *Landsat* image inset. [After Richard E. Murphy, "Landforms of the World," Map Supplement No. 9, *Annals of the Association of American Geographers* 58, no. 1 (March 1968). Adapted by permission. Image courtesy of EROS Data Center and the National Geographic Society.]

As you examine the map, identify the continental shields at the heart of each landmass. These areas are surrounded by continental platforms composed of sedimentary deposits. Various mountain chains, rifted regions, and isolated volcanic areas are portrayed. On the continent of Australia, you see older mountain sequences to the east, sedimentary layers covering basement rocks west of these ranges, and portions of the original Gondwana, an ancient landscape in the central and western region. (Remember that Gondwana was a landmass that included Antarctica, Australia, South America, Africa, and the southern portion of India; it broke away from Pangaea some 200 million years ago.)

Earthquakes

Crustal plates do not move smoothly past one another. Instead, stress builds strain in the rocks along plate boundaries until the sides release and lurch into new positions. Earthquakes may strike at any time on Earth, but many are in sparsely populated areas. Devastating earthquakes can result from these rapid shifts of crust.

This reality was brought home to millions of television viewers as they watched the 1989 baseball World Series from Candlestick Park near San Francisco. Less than half an hour before the call to "play ball," a powerful earthquake rocked the region and turned sportscasters into newscasters and sports fans into disaster witnesses.

In the Liaoning Province of northeastern China, ominous indications of tectonic activity began in 1970. Foreboding symptoms included land uplift and tilting, increased numbers of minor tremors, and changes in the region's magnetic field—all of this after almost 120 years of quiet.

These precursors of tectonic events continued for almost 5 years before Chinese scientists took the bold step of forecasting an earthquake. Finally, on February 4, 1975, at 2:00 P.M., some 3 million people evacuated in what turned out to be a timely manner; the quake struck at 7:36 P.M., within the predicted time frame. Ninety percent of the buildings in the city of Haicheng were destroyed, but thousands of lives were saved, and success was proclaimed—an earthquake had been forecast and preparatory action taken, for the first time in history.

In contrast, only 17 months later, at Tangshan in the northeastern province of Hebei (Hopei), an earthquake occurred on July 28, 1976, without warning, killing a quarter of a million people. (This is the official death toll; other estimates range as high as 650,000 dead.) The earthquake also destroyed 95 percent of the buildings and was strong enough to throw people against the ceilings of their homes. How is it possible that these two tectonic events produced such different human consequences—one forecasted and one a surprise?

Students at California State University–Northridge, in the San Fernando Valley of southern California, need no reminder of the power of earthquakes. Their campus was near the epicenter of the most devastating earthquake in U.S. history in terms of property damage—$30 billion in destruction across the region. The January 17, 1994, Northridge (Reseda) earthquake, a 6.8 magnitude, and over 10,000 aftershocks (through 2000), caused approximately $350 million damage to campus buildings 2 weeks before the beginning of spring classes. Amazingly, the semester began just 3 weeks late in 450 temporary trailers and graduation still was held in May!

In 1999 more than 30,000 people perished in quakes in Turkey, Greece, Taiwan, and Mexico. Quake-related deaths worldwide in 2000 dropped to 231. Fifteen earthquakes greater than magnitude 7.0 struck in 2001, killing 21,400 people; 20,023 deaths alone were in Gujarat state, eastern India. Thousands died in the Hindu Kush region of war-torn Afghanistan in a 2002 earthquake.

Earthquake Essentials

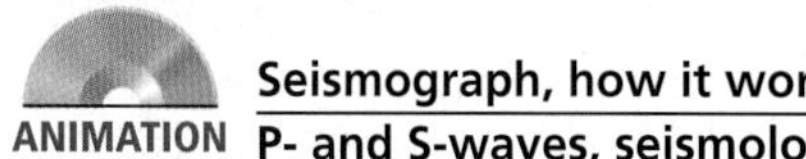

Tectonic earthquakes are those quakes associated with faulting. Their vibrations are transmitted as waves of energy throughout Earth's interior and are detected with a **seismograph**, an instrument that records vibrations in the crust—some 4000 seismographs form a global network. Earthquakes are rated on two different scales, an intensity scale and a magnitude scale.

An *intensity scale* is useful in classifying and describing damage to terrain and structures following an earthquake. Earthquake intensity is rated on the arbitrary *Mercalli scale*, a Roman numeral scale from I to XII representing "barely felt" to "catastrophic total destruction." It was designed in 1902 and modified in 1931 to be more applicable to conditions in North America (Table 9.1). Table 9.2 presents a sampling of significant earthquakes.

A system designed by Charles Richter in 1935 estimates earthquake magnitude. In this method, a seismograph located at least 100 km (62 mi) from the origin of the quake (epicenter) records the amplitude of seismic waves. The *Richter scale* is then estimated from the seismogram chart. The scale is open-ended and logarithmic; that is, each whole number on the scale represents a 10-fold increase in the measured *wave amplitude*. Translated into energy, each whole number demonstrates a 31.5-fold increase in the amount of energy released. Thus, a 3.0 on the Richter scale represents 31.5 times more energy than a 2.0 and 992 times more energy than a 1.0.

Today, the Richter scale is improved and made more quantitative. The revision was necessary because the scale did not properly measure or differentiate between quakes of high magnitude. Seismologists wanted to know more about what they call the *seismic moment* to understand a broader range of possible motions during an earthquake.

The **moment-magnitude scale** in use since 1993 is more accurate than Richter's *amplitude-magnitude* scale for large earthquakes. Moment magnitude considers the amount of fault slippage produced by the earthquake, the size of the surface (or subsurface) area that ruptured, and

Table 9.1 Earthquake Characteristics and Frequency Expected Each Year

Effects in Populated Areas	Approximate Intensity (Modified Mercalli Scale)	Approximate Magnitude (Moment Magnitude Scale)	Number per Year
Nearly total damage	XII	>8.0	1 every few years
Great damage	X–XI	7–7.9	18
Considerable-to-serious damage to buildings: railroad tracks bent	VIII–IX	6–6.9	120
Felt-by-all, with slight damage to buildings	V–VII	5–5.9	800
Felt-by-some to felt-by-many	III–IV	4–4.9	6,200
Not felt, but recorded	I–II	2–3.9	500,000
Recorded, minor	—	less than 2.0	3,000,000

Source: USGS, Earthquake Information Center.

Table 9.2 A Sampling of Significant Earthquakes*

Year	Date	Location	Number of Deaths	Mercalli Intensity	Moment Magnitude (Richter)
1556	Jan. 23	Shaanxi Province, China	830,000	*	?
1737	Oct. 11	Calcutta, India	300,000	*	?
1812	Feb. 7	New Madrid, Missouri	Several	XI–XII	?
1857	Jan. 9	Fort Tejon, California	?	X–XI	?
1870	Oct. 21	Montreal to Québec, Canada	?	IX	?
1886	Aug. 31	Charleston, South Carolina	?	IX	6.7
1906	Apr. 18	San Francisco, California	3,000	XI	7.7 (8.25)
1923	Sept. 1	Kwanto, Japan	143,000	XII	7.9 (8.2)
1939	Dec. 27	Erzincan, Turkey	40,000	XII	7.6 (8.0)
1960	May 22	Southern Chile	5,700	XII	9.5 (8.6)
1964	Mar. 28	Southern Alaska	131	X–XII	9.2 (8.6)
1970	May 31	Northern Peru	66,000	X-XII	7.9 (7.8)
1971	Feb. 9	San Fernando, California	65	VII–IX	6.7 (6.5)
1972	Dec. 23	Managua, Nicaragua	5,000	X–XII	6.2 (6.2)
1976	Jul. 28	Tangshan, China	250,000	XI–XII	7.4 (7.6)
1978	Sept. 16	Iran	25,000	X–XII	7.8 (7.7)
1985	Sept. 19	Mexico City, Mexico	7,000	IX–XII	8.1 (8.1)
1988	Dec. 7	Armenia-Turkey border	30,000	XII	6.8 (6.9)
1989	Oct. 17	Loma Prieta (near Santa Cruz, California)	67	VII–IX	7.0 (7.1)
1991	Oct. 20	Uttar Pradesh, India	1,700	IX–XI	6.2 (6.1)
1994	Jan. 17	Northridge (Reseda), California	66	VII–IX	6.8
1995	Jan. 17	Kobe, Japan	5,500	XII	6.9
1996	Feb. 17	Indonesia	110	X	8.1
1997	Feb. 28	Armenia–Azerbaijan	1,100	XII	6.1
1997	May 10	Northern Iran	1,600	XII	7.3
1998	May 30	Afghanistan–Tajikistan	4,000	XII	6.9
1998	Jul. 17	Papua, New Guinea	2,200	X	7.1
1999	Jan. 26	Armenia, Colombia	1,000	VIII–IX	6.0
1999	Aug. 17	Izmit, Turkey	17,100	VIII–XI	7.4
1999	Sept. 7	Athens, Greece	150	VI–VIII	5.9
1999	Sept. 20	Chi-Chi, Taiwan	2,500	VI–X	7.6
1999	Nov. 12	Düzce, Turkey	700	VI–X	7.2
2001	Jan. 26	Gujarat state, India	20,023	X–XII	7.7
2002	Nov. 3	Near Denali National Park, Alaska	1	X	7.9

*Note: earthquakes are not on an increase in recent years; merely, the list reflects a greater sample for recent years.

the nature of the materials that faulted, including how resistant they were to failure.

A reassessment of past quakes increased the rating of some and decreased others. The 1964 earthquake at Prince William Sound in Alaska had an amplitude magnitude of 8.6, but on the moment-magnitude scale its rating increases to a 9.2. The 1906 San Francisco quake goes from an 8.25 down to a 7.7; the Tangshan quake from a 7.6 to a 7.4.

The subsurface area along a fault plane where the motion of seismic waves is initiated is called the *focus*, or *hypocenter* (see Figure 9.20). The area at the surface directly above the focus is the **epicenter**. Shock waves produced by an earthquake radiate outward from both the focus and epicenter areas. An *aftershock* may occur after the main shock, sharing the same general area of the epicenter. A *foreshock* also is possible, preceding the main shock. Usually, the greater the distance from an epicenter, the less severe the shock. At a distance of 40 km (25 mi), only 1/10 of the full effect of an earthquake normally is felt. However, if the ground is unstable, distant effects can be magnified, as they were in Mexico in 1985 and in San Francisco in 1989.

The National Earthquake Information Center in Golden, Colorado, reports earthquake epicenters. For references, see http://wwwneic.cr.usgs.gov/ and the National Geophysical Data Center at http://www.ngdc.noaa.gov/seg/. As part of the "Learning from Earthquakes Project," reconnaissance teams examine the sites of major quakes; see http://www.eqnet.org. For a real time listing of earthquakes worldwide, go to http://wwwneic.cr.usgs.gov/neis/bulletin/bulletin.html. In southern California the USGS is coordinating a Southern California Seismographic Network (SCSN at http://pasadena.wr.usgs.gov/ and specifically http://www.trinet.org/scsn/scsn.html) to correlate 350 instruments for immediate earthquake location analysis and disaster coordination.

The Nature of Faulting

We earlier described specific types of faults and faulting motions. How a fault actually breaks still is under investigation, but the basic process is described by the **elastic-rebound theory**. Generally, two sides along a fault appear to be locked by friction, resisting any movement despite the powerful forces acting on adjoining pieces of crust. This stress continues to build strain along the fault surfaces, storing elastic energy like a wound-up spring.

As the strain builds and exceeds the frictional lock, movement occurs, releasing a burst of mechanical energy, returning both sides of the fault to a condition of less strain. This type of sudden movement rocked the region around Kobe, Japan, in 1995, with a magnitude 6.9 earthquake. See News Report 9.2 for an account of this event and similarities to conditions in the San Francisco Bay area.

Think of the fault plane as a surface with irregularities that act as sticking points, preventing movement, similar to two pieces of wood held together by drops of glue rather than an even coating of glue. Research scientists at the USGS and the University of California have identified these small areas of high strain as *asperities*. They are the points that break and release the sides of the fault.

If the fracture along the fault line is isolated to a small asperity break, the quake will be small in magnitude. Clearly, as some asperities break (perhaps recorded as small foreshocks), the strain increases on surrounding asperities that remain intact. Thus, small earthquakes in an area may be precursors to a major quake. However, if the break involves the release of strain along several asperities, the quake will be greater in extent and will involve the shifting of massive amounts of crust. The latest seismic evidence points to a wavelike pattern, as rupturing spreads along the fault plane, rather than the entire fault surface giving way at once.

The San Francisco Earthquakes. In 1906 an earthquake devastated San Francisco, then a city of 400,000 people (several million people live in the metropolitan region today). A magnitude 7.7 (8.25 Richter) tremor rocked the city, felling buildings and initiating a firestorm fed by broken gas pipes. Movement along the San Andreas fault was evident over 435 km (270 mi). This quake prompted intensive research to discover the nature of faulting. The elastic-rebound theory developed as a result. (Realize that this occurred 6 years before Wegener's continental drift hypothesis was proposed.)

The earthquake that disrupted the 1989 World Series, mentioned earlier, involved a portion of the San Andreas fault approximately 16 km (9.9 mi) east of Santa Cruz and 95 km (59 mi) south of San Francisco (Figure 9.20). The quake registered 7.0 on the moment magnitude scale. A fault ruptured at a focus unusually deep for the San Andreas system—more than 18 km (11.5 mi) below the surface.

Unlike previous earthquakes—such as the one in 1906, when the plates shifted a maximum of 6.4 m (21 ft) relative to each other—there was no evidence of a fault plane or rifting at the surface in the Loma Prieta area. Instead, the fault plane suggested in Figure 9.20 shows the two plates moving horizontally approximately 2 m (6 ft) past each other deep below the surface, with the Pacific plate thrusting 1.3 m (4.3 ft) upward. This vertical motion is unusual for the San Andreas fault and perhaps is a clue that this portion of the San Andreas system is more complex than previously thought.

The Southern California Earthquakes. Since the mid-1980s, in an area east of Los Angeles seven earthquakes greater than 6.0 magnitude have struck. After a magnitude 6.1 quake in 1992, a magnitude 7.4 tremor rocked the lightly populated area near Landers, California (see the enlargement map of southern California in Figure 9.12b). Although it was the single largest quake in California in 30 years, the remote location kept injuries and damage slight. Involved in this series were four different faults and some unknown segments. The main faulting caused a displacement of 6.1 m (20 ft).

For yet unexplained reasons, related earthquakes over the next few weeks struck in Mammoth Lakes, about 645

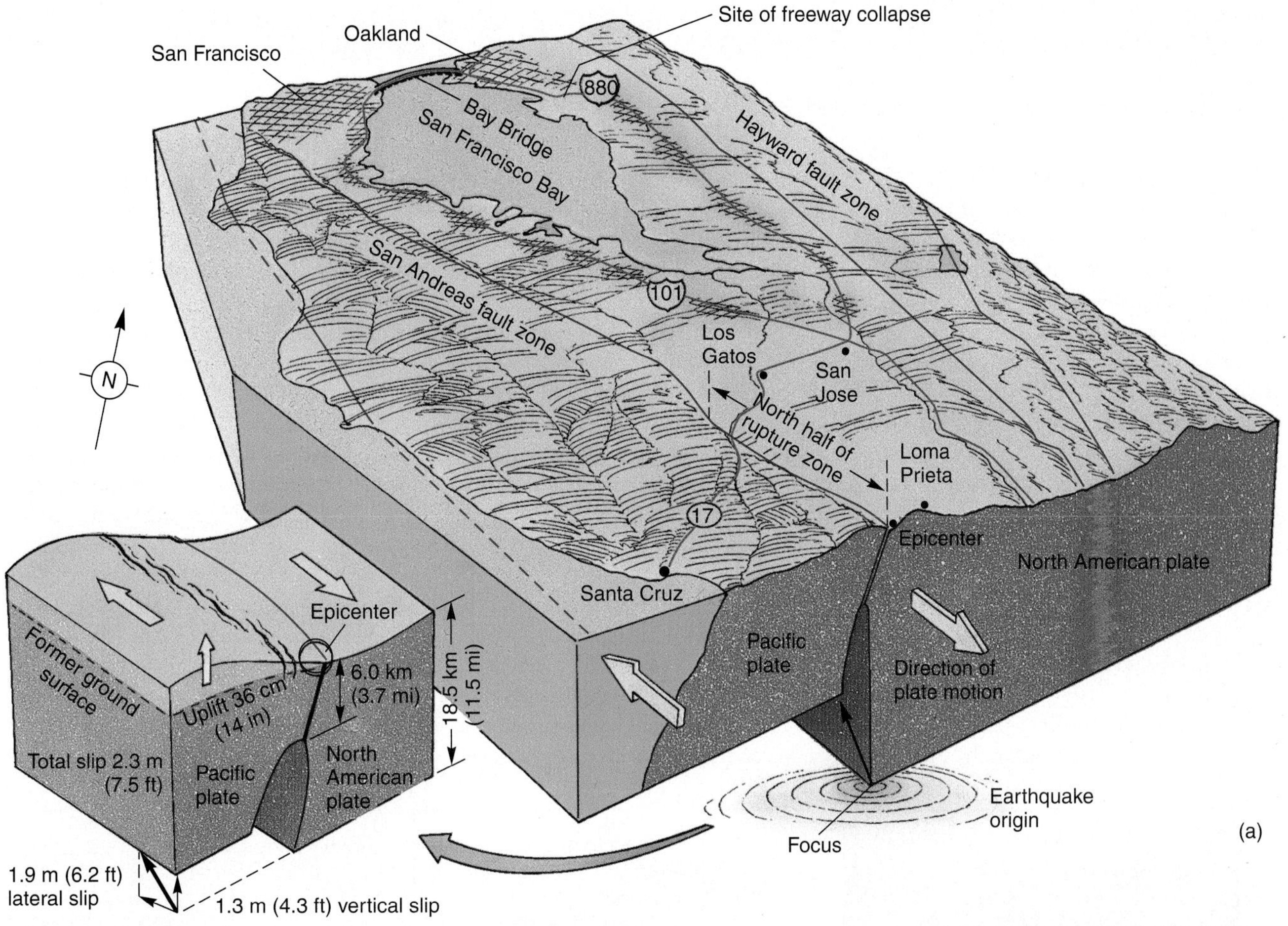

(b)

(c)

FIGURE 9.20 Anatomy of an earthquake.
(a) The fault-plane solution for the 1989 Loma Prieta, California, earthquake shows the lateral and vertical (thrust) movements occurring at depth. There was no surface expression of this fault plane. Damage totaled $8 billion, 14,000 people were displaced from their homes, 4000 were injured, and 67 killed. (b) In only 15 seconds, more than 2 km (1.2 mi) of the Route 880 Cypress Freeway collapsed. (c) Section failure in the San Francisco–Oakland Bay Bridge actually is the way the bridge handles strain without completely failing. [(a) After P. J. Ward and R. A. Page, *The Loma Prieta Earthquake of October 17, 1989* (Washington, D.C.: U.S. Geological Survey, November 1989), p. 1; (b) and (c) courtesy of California Department of Transportation.]

km (400 mi) to the north; at Mount Shasta in northern California; in southern Nevada and Utah; and 1810 km (1125 mi) distant in Yellowstone National Park, Wyoming.

The 1971 San Fernando, 1987 Whittier, 1988 Pasadena, 1991 Sierra Madré, and 1994 Northridge (Reseda) earthquakes are a few of the many quakes associated with deeply buried thrust faults. Earthquakes roughly originate at foci 18 km (11 mi) deep in the crust. Although not directly aligned with the San Andreas, scientists think that southern California will be affected by more of these thrust-fault actions as strain continues to build along the nearby San Andreas system of faults.

News Report 9.2

A Tragedy in Kobe, Japan—the Hyogo-ken Nanbu Earthquake

Early on the morning of January 17, 1995, a year to the day after the Northridge quake, a section of the Nojima fault zone, approximately 16 km (10 mi) below the surface, became the focus of a devastating earthquake near Kobe, Japan. On the moment-magnitude scale the quake registered 6.9; it killed 5500 people, injured 26,000, and caused over $100 billion in damage. About 200,000 homes, or 10% of all the housing in the metropolitan area, were damaged; some 80,000 buildings collapsed completely; major transportation arteries were severed; and the busy port was brought to a standstill (Figure 1).

For Japan, this was the worst quake loss since the 1923 temblor that hit Tokyo. The Kobe region had large earthquakes in A.D. 868, 1596, and 1916, but after almost 80 years of relative calm, the people did not expect this surprise—despite the fact the fault and the earthquake mechanism are well understood. The fault moved in a right-lateral, strike-slip motion, similar to the San Andreas fault in California. This event moved the land along three segments, where displacement of 1.0 to 1.5 m occurred in a rupture zone that stretched over 30 km (18 mi) in length.

The lessons are many, especially the occurrence of liquefaction and the failure of filled land. Areas of Osaka Bay reclaimed by landfill taken from the sea liquefied quickly in the shaking. Soil surfaces failed, collapsing the structures they supported (Figure 1b). Importantly, these human-made landfills and islands are characteristic of the San Francisco Bay region, where, since the early 1900s, about half the original surface area of the bay is reclaimed with fill and occupied with buildings.

A positive note was the good performance of landfill sites in Kobe and Osaka where drains had been installed and where pilings were in place to support structures. The cost of such retrofits in the United States could easily exceed several hundred billion dollars. In the meantime, the resemblance between the geology of the Kobe region and the San Francisco Bay region—specifically, the menacing Hayward fault system—is troubling. The Kobe lessons raise great concern and a call to action.

(a)

(b)

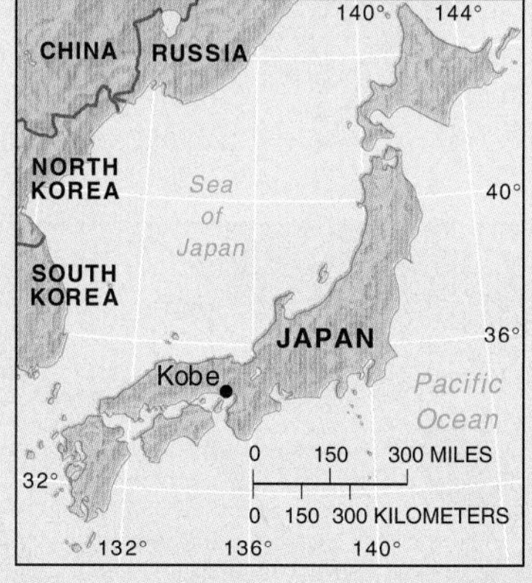

FIGURE 1 1995 earthquake in Kobe, Japan. (a) Catastrophic failure to an elevated freeway; note the failure of the supporting pillars. (b) Ground failure, mainly liquefaction, tilts this building to near collapse; note the space between the building and the ground on the right side. [Photos by Haruyoshi Yamaguchi/Sygma.]

Earthquake Forecasting and Planning

A major challenge for seismologists is how to predict the specific time and place for a quake. One approach is to examine the history of each plate boundary and determine the frequency of earthquakes in the past. Seismologists then construct maps that provide an estimate of expected earthquake activity. Areas that are quiet and overdue for an earthquake are termed *seismic gaps*; such an area forms a gap in the earthquake occurrence record and is therefore a place that possesses accumulated strain.

The areas around San Francisco and northeast of Los Angeles represent such gaps where the fault system appears to be locked by friction and is accumulating strain. The 1989 Loma Prieta earthquake was predicted in 1988 by the U.S. Geological Survey (USGS) as having a 30% chance of occurring with a 6.5 magnitude, within 30 years. The actual quake filled a portion of the seismic gap in that region. The 1999 earthquakes in Turkey filled a similar gap in the earthquake record for that region.

The USGS issued an earthquake probability estimate for the San Francisco Bay Region in 1999. According to scientists, there is a 70% (±10%) probability of at least a magnitude 6.7, or greater, quake, capable of damage, before 2030. (See USGS Fact Sheet 151-99; **http://quake.usgs.gov/**.)

Actual implementation of an action plan to reduce death, injury, and property damage from earthquakes is very difficult. The political environment adds complexity; sadly, an accurate earthquake prediction would be viewed as a threat to a region's economy (Figure 9.21). If we examine the potential socioeconomic impact of earthquake prediction on

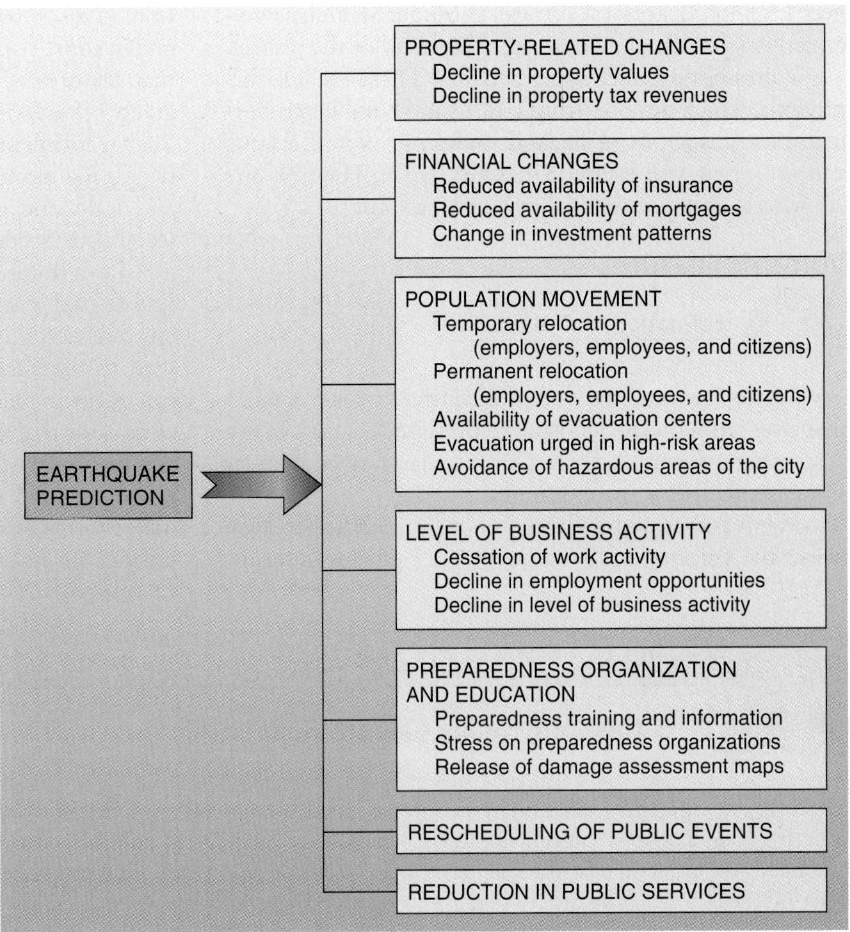

FIGURE 9.21 Socioeconomic impacts of an earthquake prediction. Business, political, and monetary interests are at stake as scientists get closer to being able to predict earthquakes. [After J. E. Haas and D. S. Mileti, *Socioeconomic Impact of Earthquake Prediction on Government, Business, and Community* (Boulder: University of Colorado, Institute of Behavioral Sciences). Used by permission.]

an urban community, we find surprising negative economic impacts in the period before the quake hits. Imagine a chamber of commerce, bank, real estate agent, tax assessor, or politician who would privately welcome an earthquake prediction and such negative publicity for their city.

A valid and applicable generalization seems to be that *humans and their institutions are unable or unwilling to perceive hazards in a familiar environment*. Such an axiom of human behavior certainly helps explain why large populations continue to live and work in earthquake-prone settings in developed countries. Similar questions also can be raised about populations in areas vulnerable to floods, droughts, hurricanes and coastal storm surge, and settlements on barrier islands. Hazard perception studies and risk management represents a growing discipline within geography. The Career Link at the end of this chapter presents more on this subject. (See the Natural Hazards Observer at **http://www.colorado.edu/hazards/o/o.html** as part of the Natural Hazards Center at the University of Colorado at **http://www.colorado.edu/hazards/index.html**.)

Volcanism

Volcanic eruptions across the globe remind us of Earth's internal energy. Ongoing eruption activity matches plate tectonic activity as shown on the map in Figure 8.18. A few recent eruptions include Kīlauea in Hawai'i, Soufriere Hills on Montserrat (Lesser Antilles), Mount Etna (Sicily, Italy), several on the Kamchatka Peninsula (Russia), Rabaul (Papua, New Guinea), Grímsvötn (Iceland), Axial Summit and the Jackson Segment (sea-floor spreading center off the coast of Oregon), Shishaldin (Unimak Island, Alaska; see Figure 9.31), Lascar (Chile), White Island (New Zealand; see Figure 9.27), Nyiragongo (Congo, Africa), Mayon (Philippines), among many.

Over 1300 identifiable volcanic cones and mountains exist on Earth, although fewer than 600 are active (have had at least one eruption in recorded history). In an average year, about 50 volcanoes erupt worldwide, varying from modest activity to major explosions. (An index to the world's volcanoes is at **http://vulcan.wr.usgs.gov/Volcanoes/framework.html**, and the National Museum of Natural History Global Volcanism Program lists information for more than 8500 eruptions at **http://www.nmnh.si.edu/gvp/**. For more volcanic eruption listings, see **http://volcano.und.nodak.edu/vwdocs/current_volcs/current.html**.)

Eruptions in remote locations and at depths on the seafloor go largely unnoticed, but the occasional eruption of great magnitude near a population center makes headlines. North America has about 70 volcanoes (mostly inactive) along the western margin of the continent. Mount St. Helens in Washington State is a famous active example, and over 1 million visitors a year travel to Mount St. Helens Volcanic National Monument to see volcanism for themselves.

Volcanoes produce some benefits. These include fertile soils, which develop from lava, as in Hawai'i; geothermal energy, such as in Iceland, Italy, and New Zealand; and even new real estate in Iceland, Japan, Hawai'i, and elsewhere, as lava extends shorelines seaward.

Volcanic Features

ANIMATION **Formation of Crater Lake**

A **volcano** forms at the end of a central vent or conduit that rises from the asthenosphere through the crust into a volcanic mountain. A **crater**, or circular surface depression, usually forms at the summit.

Magma rises and collects in a magma chamber deep below the volcano until conditions are right for an eruption. This subsurface magma carries tremendous heat, and in some areas it boils groundwater, as seen in the thermal features of Yellowstone National Park. The steam given off is a potential energy source if it is accessible. Such **geothermal energy** has provided heating and electricity for more than 90 years in Iceland, New Zealand, and Italy, and power production in northern California (see Focus Study 8.1).

Lava (molten rock), gases, and **pyroclastics**, or *tephra* (pulverized rock and clastic materials of various sizes ejected violently during an eruption) pass through the vent to the surface and build a volcanic landform. Lava can occur in many different textures and forms, which accounts for the varied behavior of volcanoes and the different landforms they build. Here we look at five volcanic landforms and their origins: cinder cones, calderas, the major forms of shield volcanoes, plateau basalts, and composite volcanoes.

News Report 9.3

Is the Long Valley Caldera Next?

Trees are dying in the forests on Mammoth Mountain, in a portion of the Long Valley Caldera, near the California-Nevada border. The oval caldera is about 15 by 30 km (9 by 19 mi) long in a north-south direction, and sits at 2000 m (6500 ft) elevation, rising to 2600 m (8500 ft) along its western side.

Tons of carbon dioxide are coming up through the soil in the old caldera each day. Carbon dioxide levels reach 30% to 96% in some soil samples, thus killing the forest in an area of heaviest concentration. The source: active and moving magma at some 3 km (1.86 mi) depth. This and other gas emissions are used elsewhere in the world to forecast a rebirth of volcanic activity. At Mammoth Mountain, these gases signify magmatic activity and may be a warning of things to come (Figure 1).

The Long Valley Caldera was formed by a powerful volcanic eruption 730,000 years ago. The eruption exceeded the volume output of the 1980 Mount St. Helens event by more than 500 times. Volcanic activity of a lesser extent has occurred in the area over the past several hundred thousand years, most recently between A.D. 1720 and 1850 in the Mono Lake area north of Long Valley.

Since the late 1970s episodes of earthquakes shook Long Valley—thousands of quakes overall, some three dozen greater than magnitude 3.0. The quakes are focused at about 11 km beneath the region. As of this writing, the surface has shown some lifting or deformation by this activity. The combination of volcanic gas production and earthquakes is drawing scientific and public attention to this area because of the potential for a massive eruption sometime in this century. The region is a popular tourist mecca and recreational center, with second-home residential development and a growing year-round population.

By the end of 2002 earthquake activity remained at 5 to 10 earthquakes, less than magnitude 2.0 per day, or slightly below the average through the 1990s. The swelling of the uplifting central dome continued uplift through the year at a rate of about 2 cm per year (0.8 in. per year), comparable to the previous decade. For the latest information see http://quake.wr.usgs.gov/VOLCANOES/LongValley/.

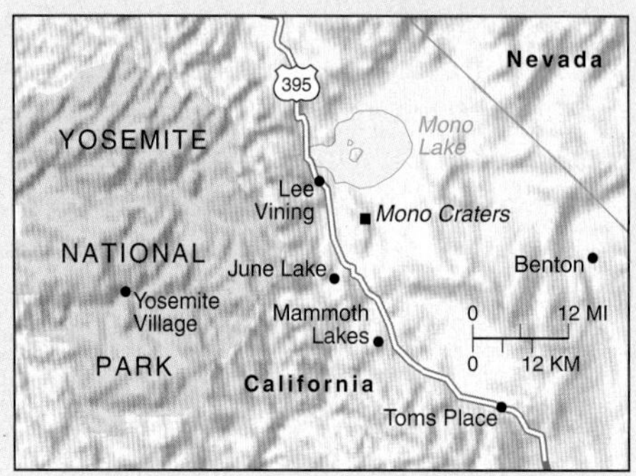

FIGURE 1 Trees are dying. Forest killed by carbon dioxide and a warning sign about the danger. The hazard has killed one person to date. The CO_2 is gassing up through soils from subsurface magma and related volcanic activity. Rates of CO_2 emissions in this forest continue to average levels of 50 to 150 tons per day. [Photo by Bobbé Christopherson.]

A **cinder cone** is a small cone-shaped hill usually less than 450 m (1500 ft) high, with a truncated top formed from cinders that accumulate during moderately explosive eruptions. Cinder cones are made of tephra and scoria (cindery rock full of air bubbles, or vesicular), as exemplified by Paricutín in southwestern Mexico.

Another distinctive landform is a large basin-shaped depression called a **caldera** (Spanish for "kettle"). A caldera forms when summit material on a volcanic mountain collapses inward after an eruption or other loss of magma. A caldera may fill with rainwater, as it did in beautiful Crater Lake in southern Oregon. News Report 9.3 discusses the Long Valley Caldera in California.

In the Cape Verde Islands (15° N 24.5° W), Fogo Island has such a caldera in the midst of a volcanic cone, opening to the sea on the eastern side. The best farmland is in the caldera, so the people choose to live and work there. Unfortunately, eruptions began in 1995, destroying farmland and forcing the evacuation of 5000 people. Previously, eruptions occurred between A.D. 1500 and 1750, followed by six more in the next century, each destroying houses and crops.

Location and Types of Volcanic Activity

The location of volcanic mountains on Earth is a function of plate tectonics and hot-spot activity. Volcanic activity occurs in three areas:

1. Along subduction boundaries at continental plate–oceanic plate convergence (such as Mount St. Helens and Kliuchevskoi, Siberia) or oceanic plate–oceanic plate convergence (Philippines and Japan);
2. Along sea-floor spreading centers on the ocean floor (Iceland on the Mid-Atlantic ridge, or off the coast of Oregon and Washington) and areas of rifting on continental plates (the rift zone in east Africa);
3. At hot spots, where individual plumes of magma rise to the crust (such as Hawai'i and Yellowstone National Park).

Figure 9.22 illustrates these three types of volcanic activity, which you can compare to the active volcano sites and plate boundaries shown in Figure 8.18. Figures 9.23 to 9.28 are included with Figure 9.22 to illustrate various aspects of volcanic activity.

The variety of forms among volcanoes makes them hard to classify; most fall in transition between one type and another. Even during a single eruption, a volcano may behave in several different ways. The primary factors in determining an eruption type are (1) the chemistry of its magma, which is related to the magma's source, and (2) its viscosity. Viscosity is magma's thickness (resistance to flow, or degree of fluidity); it ranges from low viscosity (very fluid) to high viscosity (thick and flowing slowly). We consider two types of eruptions—effusive and explosive—and the characteristic landforms they build.

Effusive Eruptions. Effusive eruptions are the relatively gentle eruptions that produce enormous volumes of lava annually on the seafloor and in places such as Hawai'i and Iceland. Direct eruptions from the asthenosphere produce a low-viscosity magma that is very fluid and cools to form a dark, basaltic rock (less than 50% silica and rich in iron and magnesium). Gases readily escape from this magma because of its texture, causing an **effusive eruption** that pours out on the surface, with relatively small explosions and little tephra. However, dramatic fountains of basaltic lava sometimes shoot upward, powered by jets of rapidly expanding gases (Figure 9.25a).

An effusive eruption may come from a single vent or from a flank eruption through side vents in surrounding slopes. If such vents form a linear opening, they are *fissures*; eruptions from fissures sometimes create a dramatic *curtain of fire* (sheets of molten rock spraying into the air). In Iceland, active fissures are spread throughout the plateau landscape. In Hawai'i, rift zones capable of erupting tend to converge on the central crater, or vent. The interior of such a crater is often a sunken caldera, which may fill with low-viscosity magma during an eruption, forming a molten lake that may then overflow lava downslope.

On the island of Hawai'i, the continuing Kīlauea eruption is the longest in recorded history, active since January 3, 1983. During 1989–1990, lava flows from Kīlauea actually consumed several visitor buildings in the Hawai'i Volcanoes National Park and two housing subdivisions in Kalapana, Kapa'ahu, and Kaimu. Kīlauea has produced more lava than any other vent on Earth in recorded history (see Part 3's opening photo, Figures 9.25 and 9.26). Pu'u O'o is the active crater at Kīlauea (Figure 9.29).

Beginning on Mothers Day 2002, an aggressive eruption sequence intensified (9.25c), flowing over Chain of Craters Road, and continuing (2003) to send streams of lava to the ocean. A Career Link interview with a geographer/park ranger working at this site is at the end of this chapter.

A typical mountain landform built from effusive eruptions is gently sloped, gradually rising from the surrounding landscape to a summit crater. The shape is similar in outline to a shield of armor lying faceup on the ground, and therefore is called a **shield volcano**. The shield shape and size of Mauna Loa in Hawai'i are distinctive when compared with Mount Rainier in Washington, which is a different type of volcano and the largest in the Cascade Range (Figure 9.30). The height of the shield is the result of successive eruptions, flowing one on top of another. Mauna Loa is one of five shield volcanoes that make up the island of Hawai'i, and it took at least 1 million years to accumulate its mass. Mauna Kea is slightly taller, but Mauna Loa is the most massive single mountain on Earth.

In rifting areas, effusive eruptions send material out through elongated fissures, forming extensive sheets on the surface. The Columbia Plateau of the northwestern United States represents the eruption of **plateau basalts**, or *flood basalts* (Figure 9.24).

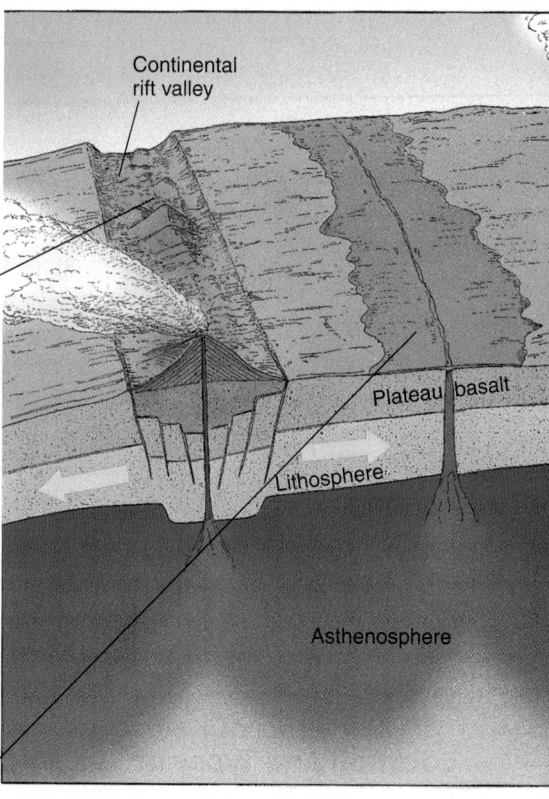

FIGURE 9.22 Tectonic settings of volcanic activity.
Magma rises and lava erupts from rifts, through crust above subduction zones, and where thermal plumes at hot spots break through the crust. [Adapted from U.S. Geological Survey, *The Dynamic Planet* (Washington, D.C.: Government Printing Office, 1989).]

FIGURE 9.23 Rift Valley.
This rift is in Iceland, where the mid-Atlantic ridge system is seen above sea level. [Photo by Nada Pecnik/Visuals Unlimited.]

FIGURE 9.24 Two expressions of volcanic activity.
In the foreground are plateau (flood) basalts characteristic of the Columbia Plateau in Oregon. Mount Hood, in the background to the west of the plateau, is a composite volcano, part of the Cascade Range of active and dormant volcanoes that result from subduction of the Pacific plate beneath the North American plate. [Photo by author.]

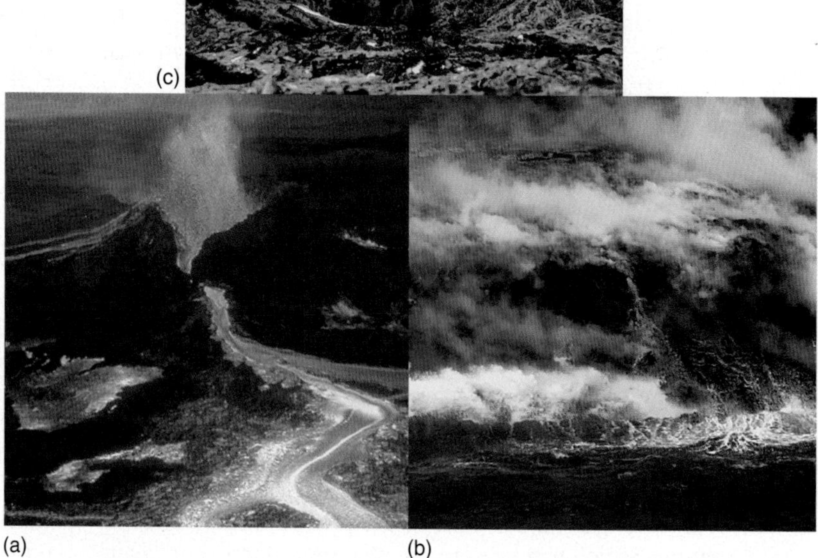

FIGURE 9.25 Volcanic fountaining and flows in Hawai'i.
(a) Kīlauea erupts a fountain of lava from its East Rift spatter cone on the big island of Hawai'i. (b) Lava flows reach the sea, quenched in the boiling surf. (c) Continuing flows from Earth's most active volcano form spectacular lava falls in 2002. [(a) Photo by Kepa Maly/U.S. Geological Survey; (b) and (c) photos by Bobbé Christopherson.]

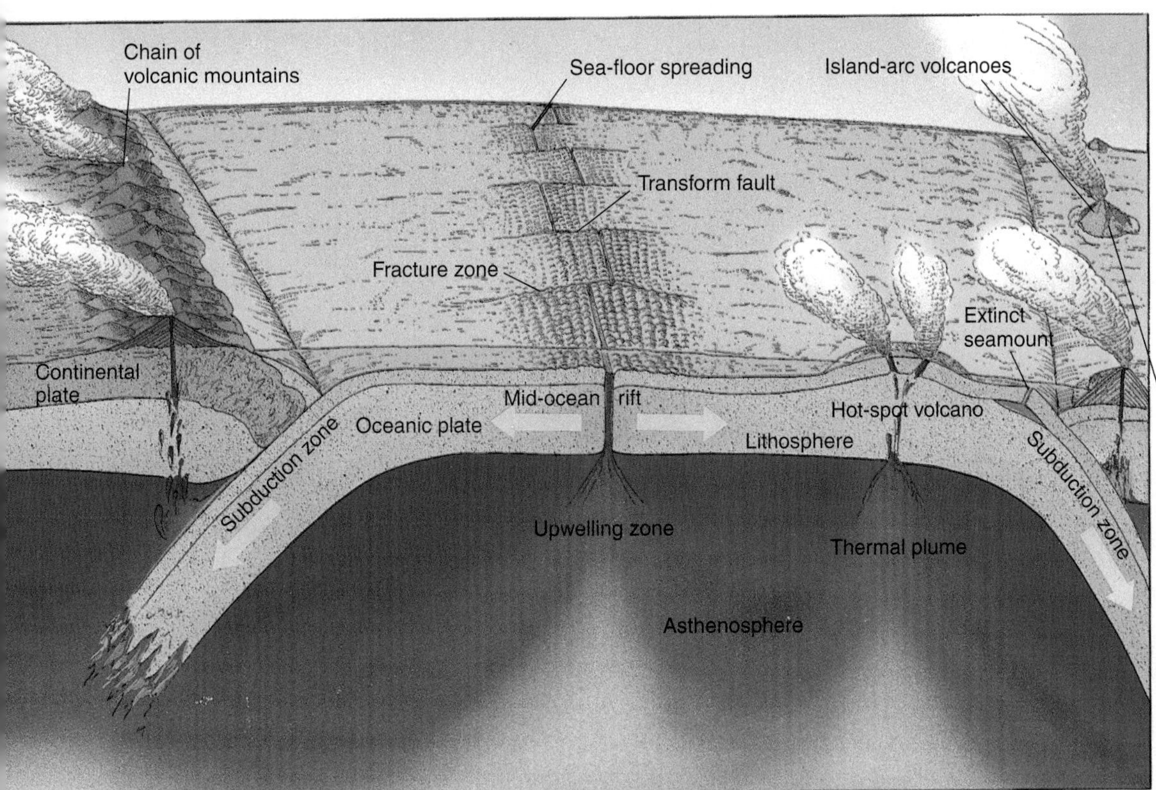

FIGURE 9.28 Extrusive igneous rocks compared.
A dacite volcanic bomb (lighter rock on left) from Mount St. Helens and a basalt lava rock (darker rock on right) from Hawai'i. What does this coloration tell you about the rocks' history and composition? [Photos by Bobbé Christopherson.]

FIGURE 9.26 Kīlauea landscape.
Aerial photo of the newest land on the planet produced by the massive flows of basaltic lavas from the Kīlauea volcano, Hawai'i Volcano National Park. In the distance, ocean water reluctantly swallows the intense 1200°C (2200°F) lava, producing steam and hydrochloric acid mist. [Photo by Bobbé Christopherson.]

FIGURE 9.27 Volcanic island off the coast of New Zealand.
White's Isle, New Zealand (37.5° S 177° E) actively erupted through 2001, sometimes gas and steam, other times bursts of pyroclastics and ash without warning. [Photo by Harvey Lloyd/Stock Market.]

FIGURE 9.29 Scientists monitor the active Pu'u O'o, at Kīlauea, Hawai'i.
The cracked and faulted crater of Pu'u O'o Vent, part of the Kīlauea volcano, Hawai'i, is a risky environment for volcanic research; note the helicopter and two people at the crater rim! Direct measurement and monitoring give scientists data to forecast volcanic eruptions. Active lava from the crater continues to make its way to the ocean. [Photo by Bobbé Christopherson.]

Extensive sheet eruptions also episodically flow from the mid-ocean ridges, forming new seafloor in the process. Worldwide, the volume of material produced in effusive fissure eruptions along spreading centers is far greater than that of the continental eruptions. Iceland continues to form in this effusive manner, with upwelling basaltic flows creating new Icelandic real estate.

Explosive Eruptions. Volcanic activity along subduction zones produces the well-known explosive volcanoes. Magma produced by the melting of subducted oceanic plate and other materials is thicker (more viscous) than magma from effusive volcanoes; it is 50%–75% silica and high in aluminum. Consequently, it tends to block the magma conduit inside the volcano, trapping and compressing gases, causing pressure to build and setting the stage for an **explosive eruption**. This type of magma forms a lighter rock at the surface as illustrated in the comparison in Figure 9.28.

The term **composite volcano** is used to describe explosively formed mountains. (They are sometimes called *stratovolcanoes*, but because shield volcanoes also can exhibit a stratified structure, *composite* is the preferred term.) *Composite* volcanoes tend to have steep sides; they are more conical than shield volcanoes. Their alternating layers of lava and ash and cone shape give them their name: *composite cones* (Figure 9.31). If a single summit vent remains stationary, a remarkable symmetry may develop as the mountain grows in size, as demonstrated by Mount Orizaba in Mexico, Mount Shishaldin in Alaska, Mount Fuji in Japan, the pre-1980-eruption shape of Mount St. Helens in Washington, and Mount Mayon in the Philippines. The beautifully symmetrical summit of Mount Mayon erupted in 2001 and 2002.

As the magma in a composite volcano forms plugs near the surface, blocked-off passages build tremendous pressure, keeping the trapped gases compressed and liquefied. When the blockage can no longer hold back this pressurized inferno, explosions equivalent to megatons of TNT blast the tops and sides off these mountains. Much less lava is produced than in shield eruptions, but composite volcanoes eject larger amounts of *pyroclastics*, which include volcanic *ash* (< 2 mm in diameter), dust, cinders, *lapilli* (up to 32 mm in diameter), *scoria* (volcanic slag), pumice, and *aerial bombs* (explosively ejected blobs of incandescent lava). Table 9.3 presents a few notable

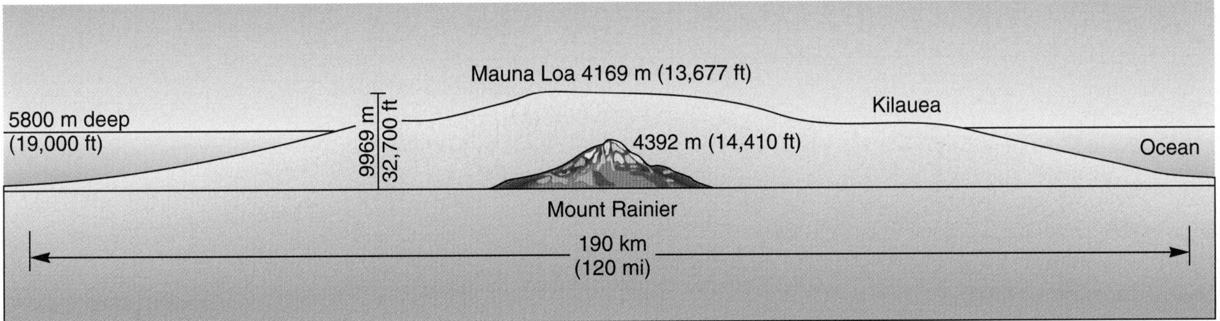

FIGURE 9.30 Shield and composite volcanoes compared. Comparison of Mauna Loa in Hawai'i (a shield volcano) and Mount Rainier in Washington State (a composite volcano). Their strikingly different profiles signify different tectonic origins. [After USGS, *Eruption of Hawaiian Volcanoes* (Washington, D.C.: Government Printing Office, 1986). Inset photo of Mauna Loa by Bobbé Christopherson.]

composite volcano eruptions. Focus Study 9.1 details the much-publicized 1980 eruption of Mount St. Helens.

Unlike the volcanoes in Hawai'i Volcanoes National Park, where tourists gather at observation platforms to watch the relatively calm effusive eruptions, composite volcanoes do not invite close inspection and can explode with little warning.

Mount Pinatubo Eruption. In June 1991, after 600 years of dormancy, Mount Pinatubo in the Philippines erupted. The summit of the 1460-m (4795-ft) volcano exploded, devastating many surrounding villages and permanently closing the U.S. Clark Air Force Base. Fortunately, scientists from the U.S. Geological Survey, working with local scientists, accurately predicted the eruption. A resulting timely evacuation of the surrounding countryside saved thousands of lives, however 800 people were killed.

Although volcanoes are regional events, their spatial implications can be worldwide. The single volcanic eruption of Mount Pinatubo was significant to the global environment and energy budget, as discussed in Chapters 1, 2, 3, 4, and 6 and summarized in Figure 1.5:

- 15–20 million tons of ash and sulfuric acid mist were blasted into the atmosphere, concentrating at 16–25 km (10–15.5 mi) altitude.

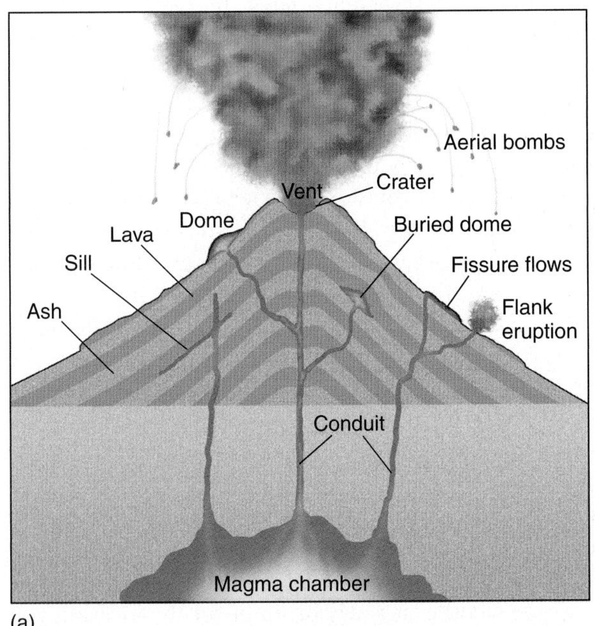

(a)

(b)

FIGURE 9.31 A composite volcano.
(a) A typical composite volcano with its cone-shaped form. (b) An eruption of Mount Shishaldin, Alaska, a composite volcano, in December 1995. It was active through 2001. [(b) Photo courtesy of AeroMap U.S., Inc., Anchorage, Alaska.]

Table 9.3 Notable Composite Volcano Eruptions

Date	Name of Locations	Deaths	Amount Extruded (Mostly Pyroclastics) in km^3 (mi^3)
Prehistoric	Yellowstone, Wyoming	Unknown	2400 (576)
4600 B.C.	Mount Mazama (Crater Lake, Oregon)	Unknown	50–70 (12–17)
1900 B.C.	Mount St. Helens	Unknown	4 (0.95)
A.D. 79	Mount Vesuvius, Italy	20,000	3 (0.7)
1500	Mount St. Helens	Unknown	1 (0.24)
1815	Tambora, Indonesia	66,000	80–100 (19–24)
1883	Krakatau, Indonesia	36,000	18 (4.3)
1902	Mont Pelée, Martinique	29,000	Unknown
1912	Mount Katmai, Alaska	Unknown	12 (2.9)
1943–1952	Paricutín, Mexico	0	1.3 (0.30)
1980	Mount St. Helens	54	4 (0.95)
1985	Nevado del Ruiz, Colombia	23,000	1 (0.24)
1991	Mount Unzen, Japan	10	2 (0.5)
1991	Mount Pinatubo, Philippines	800	12 (3.0)

- 12 km^3 (3 mi^3) of material was ejected and extruded by the eruption (12 times the volume from Mount St. Helens).
- 60 days after the eruption, about 42% of the globe was affected (from 20° S to 30° N) by the thin, spreading aerosol cloud in the atmosphere.
- Colorful twilight and dawn skies were observed worldwide.
- An increase in atmospheric albedo of 1.5% (4.3 W/m^2) occurred.
- An increase in the atmospheric absorption of insolation followed (2.5 W/m^2).
- A decrease in net radiation at the surface and a lowering of Northern Hemisphere average temperatures of 0.5 C° (0.9 F°) were measured.
- Atmospheric scientists and volcanologists were able to study the eruption aftermath using satellite-borne orbiting sensors and general circulation model computer simulations.

Volcano Forecasting and Planning

The USGS and the Office of Foreign Disaster Assistance (U. S. AID) operate the Volcano Disaster Assistance Program (VDAP; see **http://vulcan.wr.usgs.gov/Vdap/framework.html**). The need for such a program is evident in that, over the past 15 years, volcanic activity has killed 29,000 people, forced more than 800,000 to evacuate their homes, and caused more than $3 billion in damage. The program was established after 23,000 people died in the eruption of Nevado del Ruiz, Colombia, in 1985.

As an example of the need for accurate forecasting, a volcano watch continues at Soufriere Hills, an erupting composite volcano on the island of Montserrat in the Lesser Antilles (17° N 62° W). The Montserrat Volcano Observatory reports a dozen evacuations since volcanic activity began in 1995 and strengthened through 1998. Are these precursors to a main event? Or just routine rumblings that will grow quiet again? Although the U.S. VDAP is in place to help local scientists with eruption forecasts, residents remain skeptical after many false alerts, which only heightens the risk. VDAP is at work at more than two dozen volcanoes in the world. An effort such as this led to the life-saving evacuation of 60,000 people hours before Mount Pinatubo exploded.

The key to the effort is to set up mobile volcano-monitoring systems (seismometers, gas-emission sampling) at the most-threatened sites. In addition, satellite remote sensing is helping VDAP to monitor eruption cloud dynamics, atmospheric emissions and climatic effects, and lava and thermal measurements, among other items. GPS technology adds accuracy to topographic measurements never before possible.

For example, at the South Sister volcano near Bend, Oregon, dormant for 1500 years, measurements disclosed a 3-cm per year growing bulge, now almost 20 cm (8 in.) over an area of 100 km^2 (62 mi^2). This size area tells scientists that the magma causing the bulging is large. GPS and seismic monitoring are tracking the changes in the mountain. Integrated seismographic networks and monitoring are making possible early warning systems.

In this era of the Internet you can access "volcano cams" positioned around the world, giving you 24-hour surveillance of many volcanoes. For an exciting visual adventure, go to the following URL and add it to your bookmarks (note whether it is day or night for the location you are checking): **http://vulcan.wr.usgs.gov/Photo/volcano_cams.html**.

Focus Study 9.1

The 1980 Eruption of Mount St. Helens

Probably the most studied and photographed composite (explosive) volcano on Earth is Mount St. Helens, located 70 km (45 mi) northeast of Portland, Oregon, and 130 km (80 mi) south of the Tacoma-Seattle area of Washington. Mount St. Helens is the youngest and most active of the Cascade Range of volcanoes, which form a line from Mount Lassen in California to Mount Baker in Washington (Figure 1). The Cascade Range is the product of the Juan de Fuca sea-floor spreading center off the coast of northern California, Oregon, Washington, and British Columbia, and the plate subduction that occurs offshore, as identified in the upper right of Figure 9.12.

The mountain had been quiet since 1857. New activity began in March 1980, with a sharp earthquake registering a magnitude 4.1. The first eruptive outburst occurred one week later, beginning with a 4.5 quake and continuing with a thick black plume of ash and the development of a small summit crater. Ten days later, the first volcanic earthquake, a *harmonic tremor*, registered on the many instruments that had been hurriedly placed around the volcano. Harmonic tremors are slow, steady vibrations, unlike the sharp releases of energy associated with earthquakes and faulting. Harmonic tremors told scientists that magma was on the move within the mountain.

Also developing was a massive bulge on the north side of the mountain. This bulge indicated the direction of the magma flow within the volcano. A bulge represents the greatest risk from a composite volcano, for it could signal a potential lateral burst through the bulge and across the landscape.

Early on Sunday, May 18, the area north of the mountain was rocked by a magnitude 5.0 quake, the strongest to date. The mountain, with its distended 245 m (800 ft) bulge, was shaken, but nothing else happened. Then a second quake (5.1) hit at 8:32 A.M., loosening the bulge and launching the eruption. David Johnston, a volcanologist with the U.S. Geological Survey, was only 8 km (5 mi) from the mountain, servicing instruments, when he saw the eruption begin. He radioed headquarters in Vancouver, Washington, saying, "Vancouver, Vancouver, this is it!" He perished in the eruption. For a continuously updated live picture of Mount St. Helens from Johnston's approximate observation point, go to http://www.fs.fed.us./gpnf/mshnvm/volcanocam/. The camera is mounted below the roof line at the Johnston Ridge Observatory 8 km (5 mi) from the mountain. The observatory is open to the public and is named for this brave scientist.

As the contents of the mountain exploded, a surge of hot gas (about 300°C, or 570°F), a *nuée ardente* (rapidly moving, very hot, steam-filled, explosive ash, pyroclastics, and incandescent gases) moved northward, hugging the ground and traveling at speeds up to 400 kmph (250 mph) for a distance of 28 km (17 mi).

The slumping north face of the mountain produced the greatest landslide in recorded history; about 2.75 km^3 (0.67 mi^3) of rock, ice, and trapped air, all fluidized with steam, surged at speeds approaching 250 kmph (155 mph). Landslide materials traveled for 21 km (13 mi) into the valley, blanketing the forest, covering a lake, and filling rivers. A series of photographs, taken at 10-second intervals from the east looking west, records this sequence (Figure 2). The eruption continued with intensity for 9 hours, first clearing out old rock from the throat of the volcano and then blasting new material.

Before the eruption, Mount St. Helens was 2950 m (9677 ft) tall; the eruption blew away 418 m (1370 ft).

(continued)

(a)

(b)

(c)

FIGURE 1 Mount St. Helens before and after the eruption.
(a) Mount St. Helens prior to the 1980 eruption. (b) The devastated and scorched land shortly after the eruption in 1980. The scorched earth and tree blowdown area covered some 38,950 hectares (95,000 acres). (c) A landscape in recovery as life moves back in and takes hold to establish new ecosystems in 1999 along the Toutle River and debris flow. [(a) Photo by Pat and Tom Lesson/Photo Researchers; (b) photo by Krafft-Explorer/Photo Researchers, Inc.; (c) photo by Bobbé Christopherson.]

But today, Mount St. Helens is building a lava dome within its crater. As destructive as such eruptions are, they also are constructive, for this is the way in which a volcano eventually builds its height. The thick lava rapidly and repeatedly plugs and breaks in a series of lesser dome eruptions that may continue for several decades. The dome already is over 300 m (1000 ft) high; a new mountain is being born from the eruption of the old.

An ever-resilient ecology is recovering as plants and animals reclaim the devastated landscape. In Chapter 16, matching photographs from 1983 and 1999 document the slow successional recovery near the volcano (see Fig. 16.19). More than two decades have passed, yet strong interest in the area continues; scientists and millions of tourists visit the Mount St. Helens Volcanic National Monument every year. For more information see the Cascades Volcano Observatory at **http://vulcan.wr.usgs.gov/** and specifically Mount St. Helens at **http://vulcan.wr.usgs.gov/Volcanoes/MSH/framework.html**.

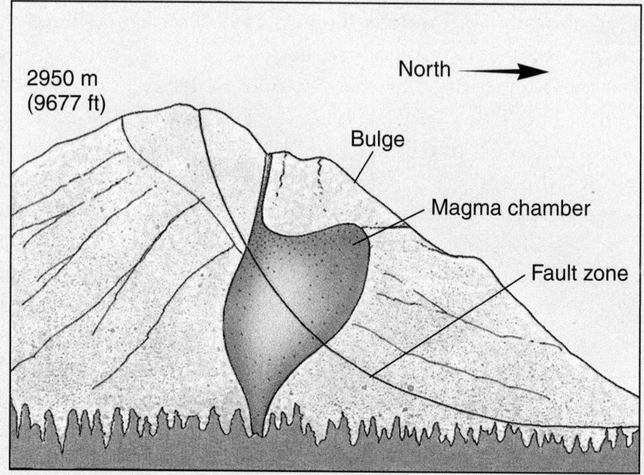

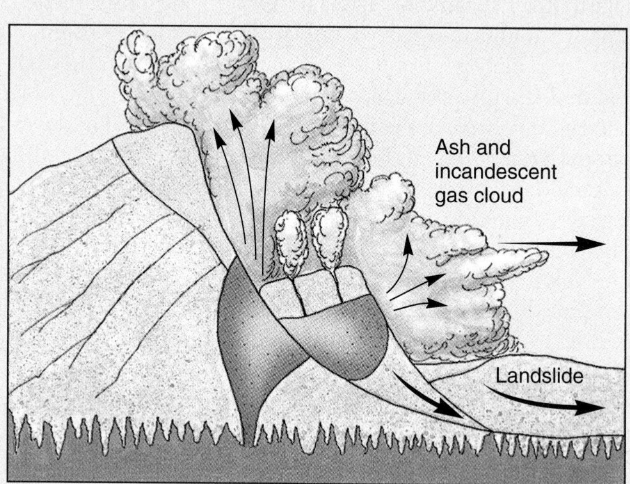

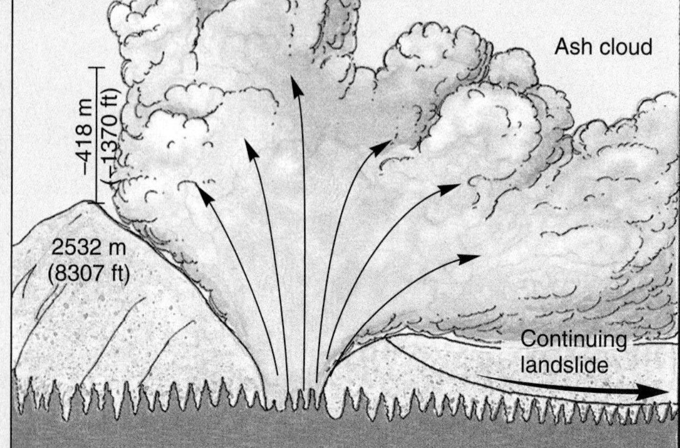

FIGURE 2 The Mount St. Helens eruption sequence and corresponding schematics.
[Photo sequence by Keith Ronnholm. All rights reserved.]

Summary and Review—Earthquakes and Volcanoes

● *Describe* first, second, and third orders of relief and *relate* examples of each from Earth's major topographic regions.

Earth's surface is dramatically shaped by tectonic forces generated within the planet. **Relief** is the vertical elevation difference in a local landscape. The undulating physical surface of Earth, including relief, is called **topography**. Convenient descriptive categories are termed *orders of relief*. The coarsest level of landforms includes the **continental platforms** and **ocean basins**; the finest comprises local hills and valleys.

relief (p. 278)
topography (p. 278)
continental platforms (p. 278)
ocean basins (p. 278)

1. How does the map of the ocean floor (chapter-opening illustration) exhibit the principles of plate tectonics? Briefly analyze.
2. What is meant by an "order of relief"? Give an example from each order.
3. Explain the difference between relief and topography.

● *Describe* the several origins of continental crust and *define* displaced terranes.

The continents are formed as a result of several processes, including upwelling material from below the crust and migrating portions of crust. The source of material generated by plate tectonic processes determines the behavior and the composition of the crust that results.

A continent has a nucleus of ancient crystalline rock called a *craton*. A region where a craton is exposed is termed a **continental shield**. As continental crust forms, it is enlarged through accretion of dispersed **terranes**. An example is the *Wrangellia terrane* of the Pacific Northwest and Alaska.

continental shield (p. 282)
terranes (p. 283)

4. What is a craton? Relate this structure to continental shields and platforms, and describe these regions in North America.
5. What is a migrating terrane, and how does it add to the formation of continental masses?
6. Briefly describe the journey and destination of the Wrangellia Terrane.

● *Explain* compressional processes and folding; *describe* four principal types of faults and their characteristic landforms.

The crust is deformed by folding, broad warping, and faulting, which produce characteristic landforms. Compression causes rocks to deform in a process known as **folding**, during which rock strata bend and may overturn. Along the *ridge* of a fold, layers *slope downward away from the axis*, which is an **anticline**. In the *trough* of a fold, however, layers *slope downward toward the axis*, called a **syncline**.

When rock strata are stressed beyond their ability to remain a solid unit, they express the strain as a fracture. Rocks on either side of the fracture are displaced relative to the other side in a process known as **faulting**. Thus, *fault zones* are areas where fractures in the rock demonstrate crustal movement. At the moment of fracture, a sharp release of energy, called an **earthquake**, or *quake*, occurs.

When forces pull rocks apart, the tension causes a **normal fault**, sometimes visible on the landscape as a scarp, or escarpment. Compressional forces associated with converging plates force rocks to move upward, producing a **reverse fault**. A low-angle fault plane is referred to as a **thrust fault**. Horizontal movement along a fault plane that produces a linear rift valley is called a **strike-slip fault**. In the U.S. interior West, the Basin and Range Province is an example of aligned pairs of normal faults and a distinctive horst-and-graben landscape. The term **horst** is applied to upward-faulted blocks; **graben** refers to downward-faulted blocks.

folding (p. 285)
anticline (p. 285)
syncline (p. 285)
faulting (p. 286)
earthquake (p. 286)
normal fault (p. 286)
reverse fault (p. 287)
thrust fault (p. 287)
strike-slip fault (p. 287)
horst (p. 290)
graben (p. 290)

7. Diagram a simple folded landscape in cross section, and identify the features created by the folded strata.
8. Define the four basic types of faults. How are faults related to earthquakes and seismic activity?
9. How did the Basin and Range Province evolve in the western United States? What other examples exist of this type of landscape?

● *Relate* the three types of plate collisions associated with orogenesis and *identify* specific examples of each.

Orogenesis is the birth of mountains. An orogeny is a mountain-building episode, occurring over millions of years, that thickens continental crust. It can occur through large-scale deformation and uplift of the crust. It also may include the capture of migrating terranes and cementation of them to the continental margins, and the intrusion of granitic magmas to form plutons.

Oceanic plate–continental plate collision orogenesis is now occurring along the Pacific coast of the Americas and has formed the Andes, the Sierra of Central America, the Rockies, and other western mountains. *Oceanic plate–oceanic plate collision* orogenesis produces either simple volcanic island arcs or more complex arcs such as Japan, the Philippines, the Kurils, and portions of the Aleutians. The region around the Pacific contains expressions of each type of collision in the *circum-Pacific belt*, or the *ring of fire.*

Continental plate–continental plate collision orogenesis is quite mechanical; large masses of continental crust, such as the Himalayan Range, are subjected to intense folding, overthrusting, faulting, and uplifting.

orogenesis (p. 290)

10. Define orogenesis. What is meant by the birth of mountain chains?
11. Name some significant orogenies.
12. Identify on a map Earth's two large mountain chains. What processes contributed to their development?
13. How are plate boundaries related to episodes of mountain building? Explain how different types of plate boundaries produce differing orogenic episodes and different landscapes.
14. Relate tectonic processes to the formation of the Appalachians and the Alleghany orogeny.

● *Explain* the nature of earthquakes, their measurement, and the nature of faulting.

Earthquakes generally occur along plate boundaries; major ones can be disastrous. Earthquakes result from faults, which are under continuing study to learn the nature of faulting, stress and the buildup of strain, irregularities along fault plane surfaces, the way faults rupture, and the relationship among active faults. Earthquake prediction and improved planning are active concerns of *seismology*, the study of earthquake waves and Earth's interior. Seismic motions are measured with a **seismograph**.

Charles Richter developed the *Richter scale*, a measure of earthquake magnitude. A more precise and quantitative scale that assesses the *seismic moment* is now used, especially for larger quakes; this is the **moment-magnitude scale**. The area at the surface directly above the focus (where motion initiates seismic waves) is the **epicenter**. The specific mechanics of how a fault breaks are under study, but the **elastic-rebound theory** describes the basic process. In general, two sides along a fault appear to be locked by friction, resisting any movement. This stress continues to build strain along the fault surfaces, storing elastic energy like a wound-up spring. When energy is released abruptly as the rock breaks, both sides of the fault return to a condition of less strain.

seismograph (p. 296)
moment magnitude scale (p. 296)
epicenter (p. 298)
elastic-rebound theory (p. 298)

15. Describe the differences in human response between the two earthquakes that occurred in China in 1975 and 1976.
16. Compare the North Anatolian, Turkey, and San Andreas, California, fault systems. Discuss their similarities and differences.
17. Differentiate between the Mercalli and moment magnitude and amplitude scales. How are these used to describe an earthquake? Why has the Richter scale been updated and modified?
18. What is the relationship between an epicenter and the focus of an earthquake? Give examples from the Loma Prieta, California, and Kobe, Japan, earthquakes.
19. What local soil and surface conditions in San Francisco severely magnify the energy felt in earthquakes?
20. How do the elastic-rebound theory and asperities help explain the nature of faulting?
21. Describe the San Andreas fault and its relationship to ancient sea-floor spreading movements along transform faults.
22. How is the seismic gap concept related to expected earthquake occurrences? Are any gaps correlated with earthquake events in the recent past? Explain.
23. What do you see as the biggest barrier to effective earthquake prediction?

● *Distinguish* between an effusive and an explosive volcanic eruption and *describe* related landforms, using specific examples.

Volcanoes offer direct evidence of the makeup of the asthenosphere and uppermost mantle. A **volcano** forms at the end of a central vent or pipe that rises from the asthenosphere through the crust into a volcanic mountain. A **crater**, or circular surface depression, usually forms at the summit. Areas where magma is near the surface may heat groundwater, producing **geothermal energy**.

Eruptions produce **lava** (molten rock), gases, and **pyroclastics** (pulverized rock and clastic materials ejected violently during an eruption) that pass through the vent to openings and fissures at the surface and build volcanic landforms, such as a **cinder cone**, a small hill, or a large basin-shaped depression called a **caldera**.

Volcanoes are of two general types, based on the chemistry and the viscosity of the magma involved. **Effusive eruption** produces a **shield volcano** (such as Kīlauea in Hawai'i) and extensive deposits of **plateau basalts**, or flood basalts. Magma of higher viscosity leads to an **explosive eruption** (such as Mount Pinatubo in the Philippines), producing a **composite volcano**. Volcanic activity has produced some destructive moments in history but constantly creates new seafloor, land, and soils.

volcano (p. 302)
crater (p. 302)
geothermal energy (p. 302)
lava (p. 302)
pyroclastics (p. 302)
cinder cone (p. 303)
caldera (p. 303)
effusive eruption (p. 303)
shield volcano (p. 303)
plateau basalts (p. 303)
explosive eruption (p. 306)
composite volcano (p. 306)

24. What is a volcano? In general terms, describe some related features.
25. Where do you expect to find volcanic activity in the world? Why?
26. Compare effusive and explosive eruptions. Why are they different? What distinct landforms are produced by each type? Give examples of each.
27. Describe several recent volcanic eruptions.

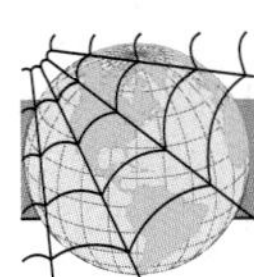

NetWork

The *Elemental Geosystems Home Page* provides on-line resources for this chapter on the World Wide Web. Once on the Home Page, click on this textbook, scroll the Table of Contents menu, select this chapter, and click "Begin." You will find self-tests that are graded, review exercises, specific updates for items in the chapter, and in "Destinations" many links to interesting related pathways on the Internet. *Elemental Geosystems* is found at http://www.prenhall.com/christopherson.

Critical Thinking

A. Using the topographic regions map in Figure 9.3, assess the region within 100 km (62 mi) and beyond to 1000 km (620 mi) of your campus. Describe the topographic character and the variety of relief within these two regional scales. You may want to consult local maps and atlases in your analysis. Do you perceive that this type of topographic region influences lifestyles? Economic activities? Transportation? History of the region?

B. Determine from library, Internet, and local agencies the seismic potential, or volcanic potential, if any, for the region in which your campus is located. If a hazard exists, does it influence regional planning? Availability of property insurance? Explain any other significant issues you found in working on this critical thinking challenge.

C. Go to one of the several URLs in this chapter covering active volcanoes in the world. Select one volcano that is currently erupting and prepare a profile of it, including its history, present activity, and overall risk assessment.

Career Link 9.1

Travis Heggie, Geographer/Park Ranger/Ph.D. Candidate

My wife and I arrived at the end of Chain of Craters Road in Hawai'i Volcanoes National Park (HAVO) on the Big Island to witness heightened lava flow activity, which by July 2002 had overrun the end of the road. This is part of the ongoing 1983 eruption of the Pu'u O'o Vent of the Kīlauea volcano. Our excitement built as we scrambled across the uneven wreckage of basaltic lava from earlier flows—glassy, crunchy sounds beneath our feet. We could feel the intense heat of the lava a kilometer distant, as it flowed over the pali (steep slope) and made its way along and beneath the surface toward the ocean. There was Travis, helping, guiding, instructing anyone in need or with questions. His enthusiasm glowed like the lava in the distance. Travis is a park ranger on the volcano. And this man is a geographer with an excitement and joy that is contagious (Figure 1).

Travis was born in Alberta, Canada, about 60 mi (96 km) from Waterton-Glacier International Peace Park. Travis and his family soon moved to Cedar City, Utah, where his father was attending Southern Utah University. His dad began his teaching career there. "My first six years were in Utah. I have fond memories of the canyons and red rock landscapes of southern Utah." The family moved back to Alberta when Travis was seven. "My dad was a teacher. I come from a family of teachers. I want to teach."

After high school, Travis went to South Korea on a two-year volunteer mission for his church, which required him to learn the Korean language. Travis said, "My interest in geography came during my mission. I was a small-town kid, thrown out into the world. I wanted to know where places were, why they were there, and how to communicate with local people." Following his mission, he met and married his wife Tracy and they returned to South Korea to work at a training institute. His facility with

(continued)

Career Link 9.1 *(continued)*

Korean language was important in helping diplomats prepare for overseas assignments.

The couple returned to the states in 1994 and enrolled at Rick's College, Idaho. A world regional geography class changed the course of his life. "When I took that class, I knew geography was my calling; that class got me! This was what I wanted to learn, what I wanted to be. I realized I could combine my interests in world events, climate studies, biology, ecology, and geology into one discipline called 'geography,' for here I could examine the 'big picture,' because geography seemed to have it all, through the lens of spatial analysis. One course was an Earth science course for teachers. The teacher had spent time at the Hawai'i Volcanoes National Park (HAVO). He showed a lot of slides and got me interested in Earth processes—this set up my next move after Rick's College. We went to Hawai'i. I think the geography department at the University of Hawai'i at Hilo is one of the best places in the world to do geography! Which I did for the next four years."

Travis said, "The professors at Hilo were so dynamic, such inspiring teachers and leaders. The work was rigorous but also rewarding." Travis started working with the National Park Service (NPS) while at the University of Hawai'i. He answered an ad for a student-training program and was hired. He wants to always keep this connection in his career.

He looks to history for other role models. "Captain Cook in the 18th century was a born geographer with a desire to explore and a quest to understand this globe. The geographic quest is what I refer to as the 'call of the wild.' I surely caught this quest in Hilo." Travis feels strongly that geography is an exploration and an adventure.

Travis completed a senior thesis titled, "Environmental Impacts of Polynesian Colonization in Pacific-Oceania." He said, "I read Captain Cook's journals at night when I was staying in American Samoa doing my master's research. I was fascinated with his observations. Cook talked about the use of fire in clearing landscapes on these islands and the way people transformed landscapes."

Next Travis went on to graduate school at Texas A & M, majoring in physical geography, emphasizing biogeography. His Master's thesis was "Rainforest composition and succession on a South Pacific Island (Island of Tutuila, American Samoa)," completed in 2001. Travis explained, "I have permanent research plots on the island and continue to gather data on several forest communities from sea level up to 245 m (800 feet) elevation." Travis analyzed abandoned agricultural land and old plantations, looking at what species were regenerating. He analyzed 18 years of land-use data and put together a broad picture of succession and regeneration.

Now Travis is working on his Ph.D. at Texas A & M, although he still finds time to work on the lava flows for the NPS in Hawai'i. His wife Tracy completed a Master's in geography at Texas A & M as well, researching water quality in Colorado. She is enrolled in the doctorate program.

(continued)

FIGURE 1 Travis Heggie, Park Ranger, Geographer. Park ranger Travis standing at a warning barricade, placed to keep the public at a safe distance from active lava flows in Hawai'i Volcanoes National Park, June 2002. [Photo by Bobbé Christopherson.]

Career Link 9.1 *(continued)*

Travis' dissertation topic is "Natural Hazards, Risk, and Emergency Management: A Case Study of Hawai'i Volcanoes National Park." Travis has a risk/hazard analysis advisor on his Ph.D. committee, a geographer who specializes in environmental impacts, and a professor from the Urban Planning Department. And he is now in the Recreation, Parks, and Tourism/Environmental Hazard Management Program. Travis explained this diversity, "I am a geographer housed in another department. I feel it is important to keep the big picture in mind. It is a real benefit to be a geographer, because we are trained in human and physical geographies, and the geosciences."

For his dissertation Travis obtained accident and injury data back to 1992 and he interviewed more than 800 tourists in the HAVO. He found that people not familiar with this environment are most susceptible to injury; not adjusting hazard perception is common in human behavior. Injuries range from abrasions and blisters, to sprains, sunburns, cuts, and fractures, with 5% suffering thermal burns where the lava enters the ocean. The holiday mind set is powerful, "... they are in Hawai'i and have to see the 'red glowing stuff,' no matter what! And local business and politics encourage this, because it represents tourism dollars."

Travis said, "I work on risk communication. People have to get the message about hazards." He said, "Hazard perception does not seem to be affected by signs. I found that only 24% of people actually stop and read at least one sign. If there is a sign that will work psychologically, I recommend a big STOP sign, with a warning message and posted accident reports."

Travis observed, "People tend to feel safe in a national park, whether its lava in Hawai'i or grizzly bears in Yellowstone, people don't readily perceive hazards in a national park—a prevailing attitude according to my perception surveys. The reality is that national parks are inherently dangerous places, because they tend to be in areas of dramatic expressions of nature and Earth systems. For instance, on July 29, 2002, it was like a war zone on the Kīlauea volcano. Fumes and hellish lava moving at a 1000-feet an hour, with methane explosions all about, and the burning asphalt of the highway giving off toxic fumes (Figure 2). People needed to heighten their hazard perception, yet we had difficulty getting everyone to respond to warnings."

As to the future and his goals, Travis volunteered, "My whole aim in life is to be a teaching professor, do research, and travel. I have a passion for all three and see them related. I want to maintain contacts with the Hawai'i Volcano National Park, and take students there and to the National Park of American Samoa for field work."

Travis' own words say it best as to what he wants readers of this Career Link to remember. Without hesitation, he implored, "*Do Geography!* I see people with geography degrees who sold themselves short because they narrowly specialized, missing the true scope of the discipline. So when I say *Do geography!* I literally mean 'do it all.' You need to have an understanding of both human and physical geography, as well as, climatology, economic geography, and regional courses, and an ability with analytical tools, such as statistics, field methods, and GIS."

He gave an example of the geographic advantage. "When I applied to Hawai'i Volcanoes National Park, my boss told me I was hired over students from other disciplines because it was obvious to him that a geographer could talk the same geology as the geologists, and also discuss the human and cultural aspects, biogeography, climatology, and geomorphology. Geography has it all."

This man is an enthusiastic geographer and skilled lifelong student of the world. *Do geography!*

FIGURE 2 Lava flows run over highway, July 2002. The Chain of Craters Road is engulfed in 1200°C (2200°F) lava, incinerating the asphalt, releasing toxic smoke. The former ranger station hut was moved a kilometer from the end of the road to avoid the inferno. [Photo by Travis Heggie.]

The rugged Alabama Hills are rounded granite outcrops that demonstrate spheroidal weathering—the scene of many movies and television commercials. You can imagine a movie chase scene raising clouds of dust! Mount Whitney, the highest peak in the contiguous states, tops the Sierra Nevada Range in the background. [Photo by Bobbé Christopherson.]

10

Weathering, Karst Landscapes, and Mass Movement

Key Learning Concepts

By knowing and understanding the key learning concepts in this chapter, you should be able to:

- *Define* the science of geomorphology.
- *Illustrate* the forces at work on materials residing on a slope.
- *Define* weathering and *explain* the importance of the parent rock and joints and fractures in rock.
- *Describe* frost action, crystallization, hydration, pressure-release jointing, and the role of freezing water as physical weathering processes.
- *Describe* the susceptibility of different minerals to the chemical weathering processes called hydrolysis, oxidation, carbonation, and solution.
- *Review* the processes and features associated with karst topography.
- *Portray* the various types of mass movements and *identify* examples of each in relation to moisture content and speed of movement.

One of the benefits you receive from a physical geography course is a new appreciation of the scenery. Whether you go by car, train, plane, or on foot, travel is an opportunity to experience Earth's varied landscapes and to witness the active processes that produce them. Warning: Seeing Earth's landscapes with understanding may make each trip take longer!

In the last two chapters, we discussed the endogenic (internal) processes of our planet and how they produce landforms. However, as the landscape is formed, several exogenic (external) processes simultaneously wear and waste it. Perhaps you noticed rough and broken highways in areas that experience freezing temperatures. The pavement breaks into chunks each winter and develops potholes. Or maybe you have seen older marble structures such as tombstones that rainwater has etched and dissolved. Perhaps after a rainstorm you noticed mud flowing from a hillslope or maybe large segments of a hillside ready to let loose in a slide or flow of debris. A news broadcast may bring you word of an avalanche in Colombia, mudflows in Los Angeles, a landslide in Turkey, or hot mudflows surging down the slopes of a volcano in Japan.

Here we begin a five-chapter examination of exogenic processes at work on the landscape. This chapter examines weathering and mass movement of the lithosphere. The next four chapters look at specific exogenic agents and their handiwork—river systems, wind-influenced landscapes, deserts in water-deficit regions, coastal processes and landforms, and regions worked by ice and glaciers. Whether you enjoy time along a river, love the desert or the waves along a coastline, or live in a place where glaciers once carved the land, you will find something of interest in these chapters.

In this chapter: We look at physical (mechanical) and chemical weathering processes that break up, dissolve, and generally reduce the landscape. Such weathering releases essential minerals from bedrock for soil formation or enrichment. Perhaps you live in a region that has caves and caverns. Water has dissolved enormous underground worlds of mystery and darkness, yet many caves remain undiscovered. In addition, we examine mass movement processes that continually operate in and upon the landscape. Mass movement of surface materials rearranges landforms, providing dramatic reminders of the power of nature.

Landmass Denudation

Geomorphology is the science of landforms—their origin, evolution, form, and spatial distribution. This discipline is an important aspect of physical geography. **Denudation** is any process that wears away or rearranges landforms. The principal denudation processes affecting surface materials include *weathering*, *mass movement*, *erosion*, *transportation*, and *deposition*, as produced by the agents of moving water, air, waves, ice, and the pull of gravity.

Interactions between the structural elements of the land and denudation processes are complex. They represent a continuing struggle between Earth's internal and external processes, between the resistance of materials and weathering and erosional processes. The 15-story-tall Delicate Arch in Utah is dramatic evidence of this struggle (Figure 10.1). Differing resistances of the rocks, coupled with variations in the processes at work on the rock, are carving this delicate sculpture—an example of **differential weathering**, where a more resistant cap rock protects supporting strata below.

Dynamic Equilibrium Approach to Landforms

A landscape is an open system, with highly variable inputs of energy and materials. Uplift creates the *potential energy of position* above sea level and therefore a disequilibrium between relief and energy. The Sun provides radiant energy that is converted into *heat energy*. The hydrologic cycle imparts *kinetic energy* through mechanical motion. *Chemical energy* is made available from the atmosphere and various reactions within the crust.

As physical factors fluctuate in an area, the surface constantly responds in search of equilibrium. Every change produces compensating actions and reactions. The balancing act between tectonic uplift and reduction by weathering and erosion, between the resistance of rocks and the ceaseless attack of weathering and erosion, is the **dynamic equilibrium model**.

FIGURE 10.1 Delicate weathering in the desert. Delicate Arch—a dramatic example of differential weathering in Arches National Park, Utah. Resistant rock strata at the top of the structure have helped preserve the arch beneath them as surrounding rock was eroded away. The arch stands 15 stories tall! In the distance are the snow-covered Manti LaSal Mountains, extinct volcanics. [Photo by author.]

According to current thinking, landscapes in a dynamic equilibrium show ongoing adaptations to the ever changing conditions of rock structure, climate, local relief, and elevation. Endogenic events, such as faulting or a lava flow, or exogenic events, such as a heavy rainfall or a forest fire, may provide new sets of relationships for the landscape. Following such destabilizing events, a landform system arrives at a **geomorphic threshold**—the point at which there is enough energy to overcome the resistance against movement. At this threshold the system breaks through to a new set of equilibrium relationships as the landform and its slopes enter a period of adjustment and realignment. The pattern over time follows a sequence: (1) equilibrium stability (fluctuating around some average), (2) a destabilizing event, (3) a period of adjustment, and (4) development of a new and different condition of equilibrium stability.

The disturbed hillslope in Figure 10.2 is in the midst of compensating adjustment. The failure of saturated slopes caused a landslide into the river. As a consequence, the new dam of material threw the stream into a disequilibrium condition between its flow and sediment load.

Slow, continuous-change events, such as soil development and erosion, tend to maintain an approximate equilibrium condition. Less frequent, dramatic events such as a major landslide or dam collapse require longer recovery times before an equilibrium is reestablished.

FIGURE 10.2 A slope in disequilibrium.
Unstable, saturated soils gave way, leaving a debris dam partially blocking the river. [Photo by author.]

Slopes. Material loosened by weathering is susceptible to erosion and transportation. However, if it is to move downslope, the forces of erosion must overcome other forces: friction, inertia (the resistance to movement), and the cohesion of particles to one another (Figure 10.3a). If the angle is steep enough for gravity to overcome frictional forces, or if the impact of raindrops or moving animals or even wind dislodges material, then erosion of particles and transport downslope can occur.

Slopes, or hillslopes, are curved, inclined surfaces that form the boundaries of landforms. Figure 10.3b illustrates essential slope forms that may vary with conditions of rock structure and climate. Slopes generally feature an *upper waxing* slope near the top ("waxing" here means increasing). This convex surface curves downward and grades into the *free face* below. The presence of a free face indicates an outcrop of resistant rock that forms a steep scarp or cliff.

Downslope from the free face is a *debris slope* that receives rock fragments and materials from above. In humid climates, constantly moving water carries away material as it arrives; in arid climates, graded debris slopes persist and accumulate. A debris slope grades into a *waning slope*, a concave surface along the base of the slope that forms a *pediment*, or broad, gently sloping erosional surface.

Slopes are open systems and seek an *angle of equilibrium* among the forces described here. Conflicting forces work together on slopes to establish an optimum compromise incline that balances these forces. A geomorphic threshold (change point) is reached when any of the conditions in the balance is altered. All the forces on the slope then compensate by adjusting to a new dynamic equilibrium.

Slopes, then, are shaped by the relationship between rates of weathering and breakup of slope materials and the rates of mass movement and erosion of those materials. A slope is thought of as stable if its strength is greater than these denudation processes, and unstable if materials are weaker than these processes.

Now, with the concepts of landmass denudation and slope development in mind, let's examine specific processes that operate to reduce landforms.

Weathering Processes

Rocks at or just below Earth's surface are exposed to both physical and chemical weathering processes. **Weathering** encompasses a group of processes by which surface and subsurface rock disintegrates into mineral particles or dissolves in water. Weathering does not transport the weathered materials; it simply generates them for transport by the agents of water, wind, waves, and ice—all influenced by gravity. In most areas, the upper surface of bedrock undergoes continual weathering to broken-up rock called **regolith**. Loose surface material comes from further weathering of regolith and from transported and deposited regolith (Figure 10.4a). In some areas regolith may be missing or undeveloped, thus exposing an outcrop of unweathered bedrock.

Bedrock is the *parent rock* from which weathered regolith and soils develop. The bedrock is traceable through similarities in composition. For example, the sand and soils in Figure 10.4c derive their color and character from the parent rock in the background. Fragmented material, known as **sediment**, combines with weathered rock to form the **parent material** from which soil evolves.

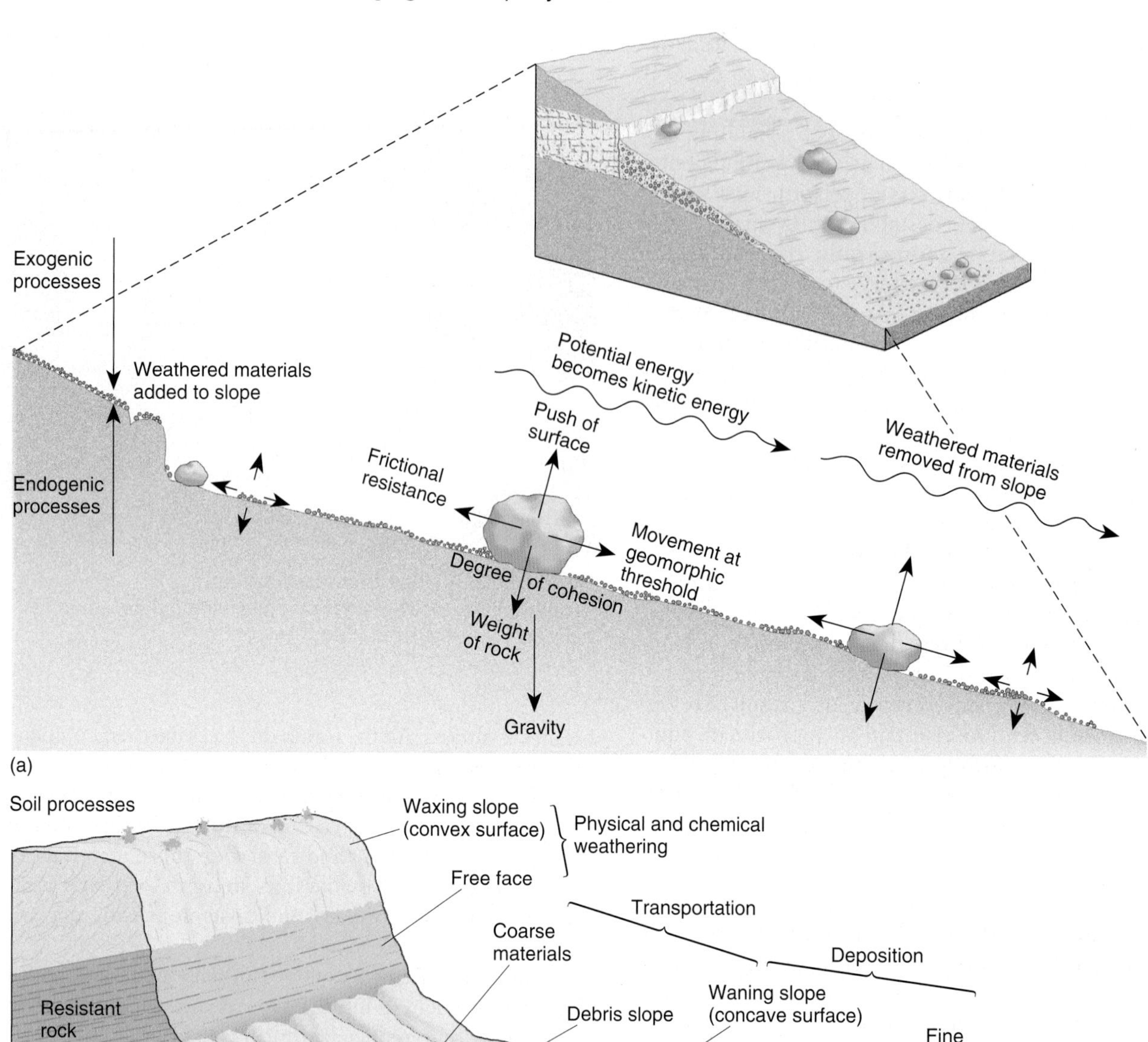

FIGURE 10.3 Slope mechanics and form.
(a) Directional forces (noted by arrows) act on materials along an inclined slope. (b) The principal elements of a slope.

Factors Influencing Weathering Processes

Weathering is greatly influenced by the character of the bedrock: hard or soft, soluble or insoluble, broken or unbroken. *Jointing* in rock is important for weathering processes. **Joints** are fractures or separations in rock that occur without displacement of the sides (as would be the case in faulting). The presence of these usually plane (flat) surfaces increases the surface area of rock exposed to both physical and chemical weathering.

Important controls on weathering rates are climatic elements—precipitation, temperature, and freeze-thaw cycles. Also significant is the position of the water table and water movement (hydraulics).

Another control over weathering rates is the *geographic orientation* of a slope—whether it faces north, south, east, or west. Orientation controls the slope's exposure to Sun, wind, and precipitation. Slopes facing away from the Sun's rays tend to be cooler, moister, and more vegetated than

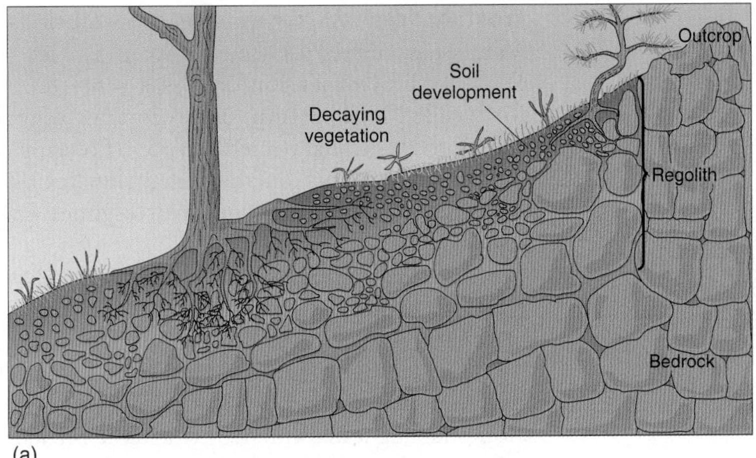

(a)

(c)

(b)

FIGURE 10.4 Regolith, soil, and parent materials. (a) A cross section of a typical hillside. (b) A cliff exhibits hillside components. (c) These reddish-colored dunes in the Navajo Tribal Park, near the Utah-Arizona border, derive their color from the red-sandstone parent materials in the background. [Photos by author.]

slopes in direct sunlight. This effect of orientation is especially noticeable in the middle and higher latitudes.

Vegetation is also a factor in weathering. Although vegetative cover can protect rock by shielding it from raindrop impact and providing roots to stabilize soil, it also produces organic acids from the partial decay of organic matter; these acids contribute to chemical weathering. Plant roots can enter crevices and break up a rock, exerting enough pressure to drive rock segments apart, thereby exposing greater surface area to other weathering processes (Figure 10.5). You may have observed how tree roots can heave the sections of a sidewalk or driveway sufficiently to raise and crack the concrete.

The role of the climatic factor warrants further discussion. There is a relation among climate (annual precipitation and temperature), physical weathering, and chemical weathering processes. In general, physical weathering dominates in drier, cooler climates, whereas chemical weathering dominates in wetter, warmer climates. Extreme dryness reduces weathering rates, as is experienced in desert climates (*low-latitude arid desert*). In the hot, wet tropical and equatorial rain forest climates (*tropical rain forest*), most rocks weather rapidly, and the weathering extends deep below the surface.

The scale at which we analyze weathering processes is important. Research at *microscale* levels reveals greater complexity in the relation of climate and weathering. At the small scale of actual reaction sites on the rock surface, both physical and chemical weathering processes can occur across varied climate types. Hygroscopic water (a

(a)

(b)

FIGURE 10.5 Physical weathering examples.
(a) Frost action shattered this granite rock; ice expansion forced the rock segments apart. (b) Roots exert a force on the sides of this joint in the rock. [Photos by (a) Bobbé Christopherson; (b) author.]

molecule-thin water layer on soil particles) and capillary water (soil water) activate chemical weathering processes, even in the driest landscape. (Review sections in Chapter 7 for these water types.)

Imagine all the factors that influence weathering rates as operating in concert: climatic influence, soil water and groundwater, rock composition and structure (jointing), slope orientation, vegetation, and microscopic boundary-layer conditions at reaction sites. We separate these processes here for convenience of study. Of course in all this, *time* is the crucial factor, for these processes require long periods of time to operate.

Physical Weathering Processes

When rock is broken and disintegrated without any chemical alteration, the process is **physical weathering**, or *mechanical weathering*. By breaking up rock, physical weathering greatly increases the surface area on which chemical weathering may operate. In the complexity of nature, physical and chemical weathering usually operate together in a combination of processes.

Frost Action. Water expands by as much as 9% of its volume as it freezes (see Chapter 5, "Ice, the Solid Phase"). This expansion creates a powerful mechanical force called **frost action**, or freeze-thaw action, that can exceed the tensional strength of rock. Freezing actions are important in humid microthermal climates (*humid continental*) and in subarctic and polar regimes (*subarctic* and *polar* climates).

Repeated cycles of water freezing and thawing break rock segments apart. The work of ice begins in small openings, gradually expanding until rocks are cleaved (split) as demonstrated in Figure 10.5. Figure 10.6 shows this action on blocks of rock, causing *joint-block separation* along existing joints and fractures in the rock. This weathering action, called *frost-wedging*, pushes portions of the rock apart. Cracking and breaking can be in any shape, depending on the rock structure. The softer supporting rock underneath the slabs in the photo already has weathered physically—an example of differential weathering.

Several cultures have used this principle of frost action to quarry rock. Pioneers in the early American West drilled holes in rock, poured water in the holes, and then plugged them. During the cold winter months, expanding ice broke off large blocks along lines determined by the hole patterns. In spring the blocks were hauled to cities for use as construction material. Frost action also produces unwanted fractures that damage pavement and split water pipes.

Spring can be a risky time to venture into mountainous terrain. As rising temperatures melt the winter's ice, newly fractured rock pieces fall without warning and may even start rock slides. The falling rock pieces may physically shatter on impact—another form of physical weathering (Figure 10.7). One such rockfall in Yosemite National Park (1996) involved a crashing drop of 670 m, at 260 kmph (2200 ft, 160 mph), of a 162,000-ton granite slab. The impact shattered the rock, covering 50 acres in powdered rock and felling 500 trees. Yet another rock fall dropped off Glacier Point in 1999, killing one person.

FIGURE 10.6 Physical weathering.
Physical weathering along joints in rock produces discrete blocks in the backcountry of Canyonlands National Park in Utah. [Photo by author.]

FIGURE 10.7 Rockfall.
Shattered rock debris, the product of a large rockfall in Yosemite National Park. A rockfall involving 162,000 tons of granite shocked Yosemite in July 1996; another occurred in 1999. [Photo by author.]

Crystallization. Especially in arid climates, dry weather draws moisture to the surface of rocks. As the water evaporates, crystals form from dissolved minerals. As this process continues over time and the crystals grow and enlarge, they exert a force great enough to spread apart individual mineral grains and begin breaking up the rock. Such *crystallization*, or *salt-crystal growth*, is a form of physical weathering.

In the Colorado Plateau of the Southwest, deep indentations develop in sandstone cliffs, especially above impervious layers, where salty water slowly flows out of the rock strata. Crystallization then loosens the sand grains, and subsequent erosion and transportation by water and wind complete the sculpting process. Over 1000 years ago, Native Americans built entire villages in these weathered niches at several locations, including Mesa Verde in Colorado and Arizona's Canyon de Chelly (pronounced "canyon duh shay," Figure 10.8).

Hydration. Another process involving water, but little chemical change, is **hydration**. Hydration is a physical weathering process in which water is absorbed by a mineral (such a hydrate is gypsum, which is hydrous calcium sulfate: $CaSO_4 \cdot 2H_2O$). This addition of water initiates swelling and stress within the rock, mechanically forcing grains apart. Hydration can lead to granular disintegration and further susceptibility of the rock to chemical weathering, especially by oxidation and carbonation. Hydration also is at work on the sandstone niches shown in the cliff-dwelling photo.

Pressure-Release Jointing. To review from Chapter 8, rising magma that is deeply buried and subjected to high pressure forms intrusive igneous rocks called plutons, which cool slowly and produce coarse-grained, crystalline granitic rocks (see illustration in Figure 8.8a). As the landscape is subjected to uplift, the weathering regolith that covers such a pluton can be eroded and transported away, eventually exposing the pluton as a mountainous batholith.

As the tremendous weight of the overburden is removed from the granite, the pressure of deep burial is relieved. The batholith responds with a slow but enormous heave; and in a process known as *pressure-release jointing*, layer after layer of rock peels off in curved slabs or plates, thinner at the top of the rock structure and thicker at the sides. As these slabs weather, they slip off in a process called **sheeting** (Figure 10.9a). This exfoliation process creates arch-shaped and dome-shaped features on the exposed landscape. An **exfoliation dome** is the largest single weathering feature on Earth (Figure 10.9b, c).

Chemical Weathering Processes

Chemical weathering refers to actual decomposition and decay of the constituent minerals in rock due to chemical alteration of those minerals, always in the presence of water. The chemical breakdown becomes more intense as both temperature and precipitation increase. Although individual minerals vary in susceptibility, no rock-forming minerals are completely unresponsive to chemical weathering. A familiar example of chemical weathering is the eating away of cathedral façades and the etching of tombstones caused by increasing acid precipitation.

Spheroidal weathering is a chemical process that occurs when water penetrates joints and fractures in rock and dissolves the rock's cementing materials. The resulting rounded edges and corners resemble exfoliation but do not result from pressure-release jointing (Figure 10.10a). This type of weathering and rock decay in turn opens spaces for frost action.

A boulder can be attacked from all sides, shedding spherical shells of decayed rock. As the rock breaks down, more and more surface area is exposed for further weathering. This type of weathering is visible in the photograph

(a)

Sandstone

Niche

Shale

(b)

FIGURE 10.8 Physical weathering in sandstone. (a) Cliff dwelling site in Canyon de Chelly, Arizona, used by the Anasazi people until about 900 years ago. The niche in the rock was partially formed by crystallization, which forced apart mineral grains and broke up the sandstone. The dark streaks on the rock are desert varnish, composed of iron oxides with traces of manganese and silica. (b) Water and an impervious sandstone layer helped concentrate weathering processes in the niche. [Photo by author.]

of a weathered rock from the Alabama Hills in Figure 10.10b; note the spheroidal rock form.

Hydrolysis. When minerals chemically combine with water, the process is called **hydrolysis**. Compared with *hydration*, a physical process in which water is simply absorbed, the hydrolysis process involves active participation of water in chemical reactions to produce different minerals.

When weaker minerals in rock are changed by hydrolysis, the interlocking crystal network in the rock fails, and *granular disintegration* takes place. Such disintegrating granite may appear etched, corroded, and softened. For example, the weathering of feldspar minerals in granite can be caused by a reaction to the normal mild acids dissolved in precipitation:

$$\text{feldspar (K, Al, Si, O)} + \text{carbonic acid and water} \rightarrow \text{residual clays} + \text{dissolved minerals} + \text{silica}$$

So the by-products of chemical weathering of feldspar in granite include clay (such as kaolinite) and silica. As clay is formed from some minerals in the granite, quartz (SiO_2) particles are left behind. The resistant quartz may wash downstream, eventually becoming sand on some distant beach. Clay minerals become a major component in soil and in shale, a common sedimentary rock.

Oxidation. One type of chemical weathering occurs when the oxygen dissolved in water oxidizes (combines with) certain metallic elements to form oxides. This is a chemical weathering process known as **oxidation**. Perhaps most familiar is the "rusting" of iron in rocks or soil that produces a reddish brown stain of iron oxide (Fe_2O_3). Such iron-bearing rocks then weather easily because these oxides are physically weaker than the original iron-bearing mineral.

Carbonation and Solution. The simplest form of chemical weathering occurs when a mineral dissolves into *solution*—for example, when sodium chloride (common table salt) dissolves in water. Water is called the *universal solvent* because it is capable of dissolving at least 57 of the natural elements and many of their compounds. It readily dissolves carbon dioxide, thereby yielding precipitation containing carbonic acid (H_2CO_3). This acid is strong enough to react with many minerals, especially limestone, in a process called **carbonation**. Carbonation simply means reactions whereby *carbon* combines and transforms

(a)

(b)

(c)

FIGURE 10.9 Exfoliation in granite.
Exfoliation processes loosen slabs of granite, freeing them for further weathering and downslope movement: (a) Great arches form in the White Mountains of New Hampshire. (b) Exfoliated layers of rock are visible in characteristic dome formations in granites. The loosened slabs of rock are susceptible to further weathering and downslope movement. This view is from the east side of Half Dome in Yosemite National Park, California. (c) Half Dome perspective from the west; relief is approximately 1500 m (5000 ft) from the top of the dome to the glaciated valley below. [Photos by (a) Bobbé Christopherson, (b) and (c) by author.]

minerals containing calcium, magnesium, potassium, and sodium into *carbonates*.

When rainwater attacks formations of *limestone* (which is calcium carbonate, $CaCO_3$), the constituent minerals are dissolved and wash away with the rainwater. Walk through an old cemetery and you can observe the carbonation of marble, a metamorphic form of limestone. Limestone and marble, whether in tombstones or in rock formations, appear pitted and weathered wherever adequate water is available for carbonation. In this era of human-induced increases of acid precipitation, carbonation processes are greatly enhanced (see Focus Study 2.2, "Acid Deposition: A Continuing Blight on the Landscape").

The chemical weathering process of carbonation dominates entire landscapes composed of limestone. These are the regions of karst topography, which we look at next.

Karst Topography and Landscapes

NOTEBOOK **Karst Farm Park, Indiana**

Limestone is so abundant on Earth that many landscapes are composed of it. These areas are quite susceptible to chemical weathering. Such weathering creates a specific landscape of pitted and bumpy surface topography, poor surface drainage, and well-developed channels underground. Remarkable labyrinths of underworld caverns also may develop.

These are the hallmarks of **karst topography**, originally named for the Krš Plateau in Yugoslavia, where these processes were first studied. Approximately 15% of Earth's land area has some developed karst, with outstanding examples found in southern China, Japan, Puerto Rico, Cuba, the state of Yucatán in Mexico, Kentucky, Indiana, New Mexico, and Florida.

Formation of Karst

For a limestone landscape to develop into karst topography, there are several necessary conditions:

- The limestone formation must contain 80% or more calcium carbonate for solution processes to proceed effectively.

(a)

(b)

FIGURE 10.10 Chemical weathering and spheroidal weathering.
(a) Chemical weathering processes act on the joints in granite to dissolve weaker minerals, leading to a rounding of the edges of the cracks. (b) Rounded granite outcrop demonstrates spheroidal weathering and the disintegration of rock.
[Photos by Bobbé Christopherson.]

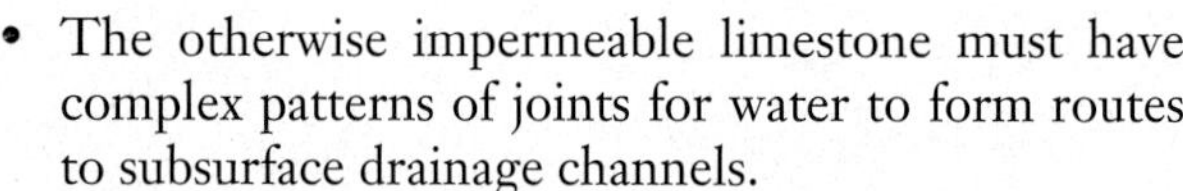

- The otherwise impermeable limestone must have complex patterns of joints for water to form routes to subsurface drainage channels.
- There must be an aerated (containing air) zone between the ground surface and the water table.
- Vegetation cover must supply varying amounts of organic acids that enhance the solution process.

The role of climate in providing optimum conditions for karst processes remains in debate, although the amount and distribution of rainfall appears important. Karst occurs in arid regions, but it is primarily due to former climatic conditions of greater humidity. Karst is rare in the Arctic and Antarctic because the water, although present, is generally frozen.

As with all weathering processes, time is a factor. Early in the twentieth century, karst landscapes were thought to progress through evolutionary stages of development, as if they were aging. Today, these landscapes are thought to be locally unique, a result of specific conditions, and there is little evidence that different regions evolve in some predictable sequence. Nonetheless, mature karst landscapes do display certain characteristic forms.

Lands Covered with Sinkholes

The weathering of limestone landscapes creates many **sinkholes**, which form in circular depressions. (Traditional studies may call a sinkhole a *doline*.) If a solution sinkhole collapses through the roof of an underground cavern, a *collapse sinkhole* is formed. A gently rolling limestone plain might be pockmarked by sinkholes with depths of 2 to 100 m (7 to 330 ft) and diameters of 10 to 1000 m (33 to 3300 ft), as shown in Figure 10.11a. Through continuing solution and collapse, sinkholes may coalesce to form a *karst valley*, an elongated depression up to several kilometers long.

The area southwest of Orleans, Indiana, has over 1000 sinkholes in just 2.6 km^2 (1 mi^2). In this area, the Lost River, a "disappearing stream," flows more than 13 km (8 mi) underground before it resurfaces at its Lost River rise (near the Orangeville rise shown in Figure 10.11e); its flow is diverted from the surface through sinkholes and solution channels. Its dry bed can be seen on the lower left of the topographic map in Figure 10.11b.

In Florida, several sinkholes have made news because lowered water tables (lowered by pumping) caused their collapse into underground solution caves, taking with them homes, businesses, and even new cars from an auto dealership. One such sinkhole collapsed in a suburban area in 1981, and others in 1993 and 1998 (Figure 10.12a, b).

A complex landscape in which sinkholes intersect is called a *cockpit karst*. The sinkholes can be symmetrically shaped in certain circumstances; one at Arecibo, Puerto Rico, is perfectly shaped for a radio telescope installation (Figure 10.12c).

Caves and Caverns

Most caves are formed in limestone rock, because it is so easily dissolved by carbonation. The largest limestone caverns in the United States are Mammoth Cave in Kentucky, Carlsbad Caverns in New Mexico, and Lehman Cave in Nevada. Limestone formations in which the Carlsbad Caverns formed were themselves deposited 200 million years ago when the region was covered by shallow seas. Regional uplifts associated with the Rockies (the Laramide orogeny 40–80 million years ago) then brought the region above sea level, leading to active cave formation (Figure 10.13).

FIGURE 10.11 Features of karst topography in Indiana. (a) Idealized features of karst topography in southern Indiana. (b) Karst topography southwest of the town of Orleans, Indiana. On average, 1022 sinkholes occur per 1 mi^2 in this area. On the map, note the contour lines: Depressions are indicated with small hachures (tick marks) on the downslope side of contour lines. (c) Gently rolling karst landscape and cornfields near Orleans, Indiana. (d) This pond is in a sinkhole depression near Palmyra, Indiana. (e) The Orangeville rise, near Orangeville, Indiana, just north of the Lost River rise. During periods of high rainfall, this rise is almost filled with water. [(a) Adapted from W. D. Thornbury, *Principles of Geomorphology,* illustration by W. J. Wayne, p. 326; © 1954 by John Wiley & Sons. (b) Mitchell, Indiana quadrangle, USGS. Photos c, d, and e by Bobbé Christopherson.]

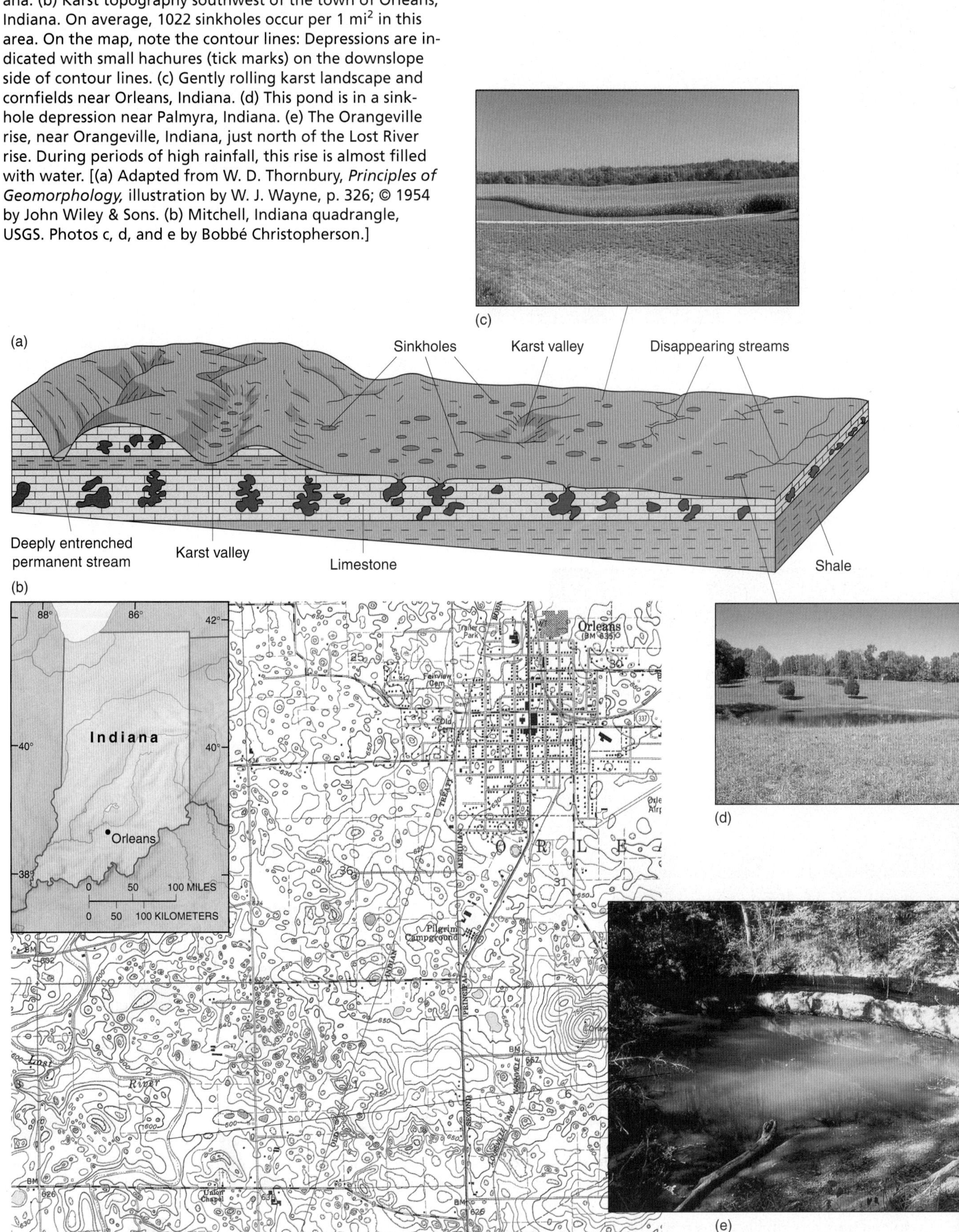

(a)

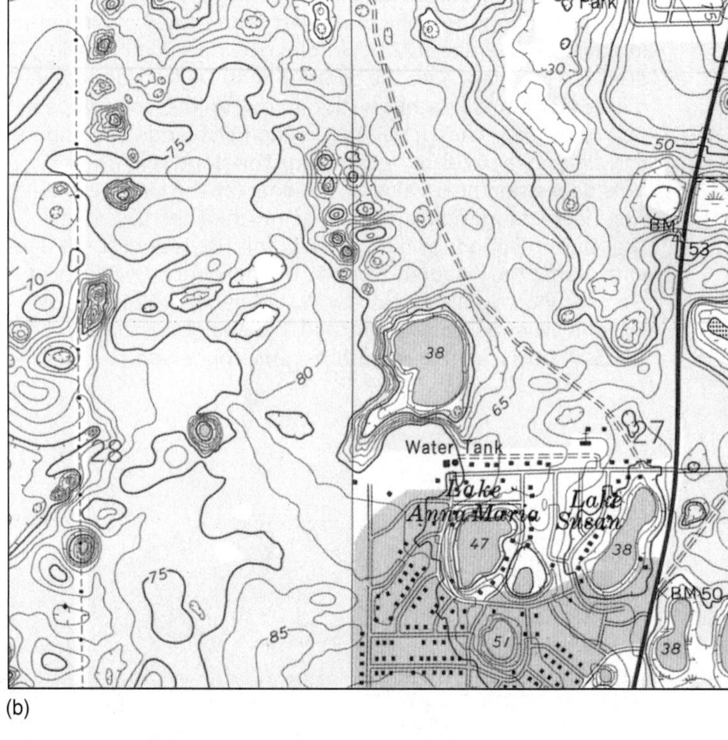

(b)

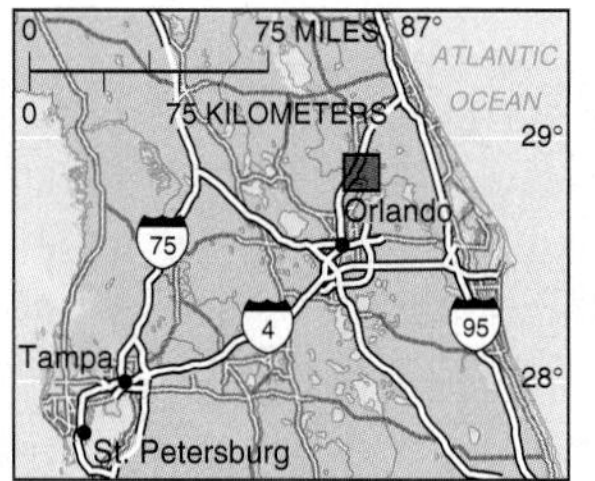

(c)

FIGURE 10.12 Sinkhole landscapes.
(a) Florida sinkhole, formed in 1981 in Winter Park, a suburb of Orlando. (b) Karst area 25 km (15.5 mi) north of Winter Park depicted on a topographic map. Note the depressions marked by small hachures (tick marks). (c) Cockpit karst topography near Arecibo, Puerto Rico, provides a natural depression for the dish antenna of a giant radio telescope. The Arecibo Observatory is part of the National Astronomy and Ionosphere Center, which is operated by Cornell University under contract with the National Science Foundation. [(b) Orange City quadrangle, USGS; photos by (a) Jim Tuten/Black Star; (c) courtesy of Cornell University.]

Caves generally form just beneath the water table, where a later lowering of the water level exposes them to further development. The *dripstones* that form under such conditions are produced as water containing dissolved minerals slowly drips from the cave ceiling. Calcium carbonate precipitates out of the evaporating solution, literally a molecular layer at a time, and accumulates at a point below on the cave floor. Depositional features called *stalactites* grow down from the ceiling and *stalagmites* build up from the floor, sometimes growing until they connect and form a continuous *column*. A dramatic subterranean world is thus created. This is an

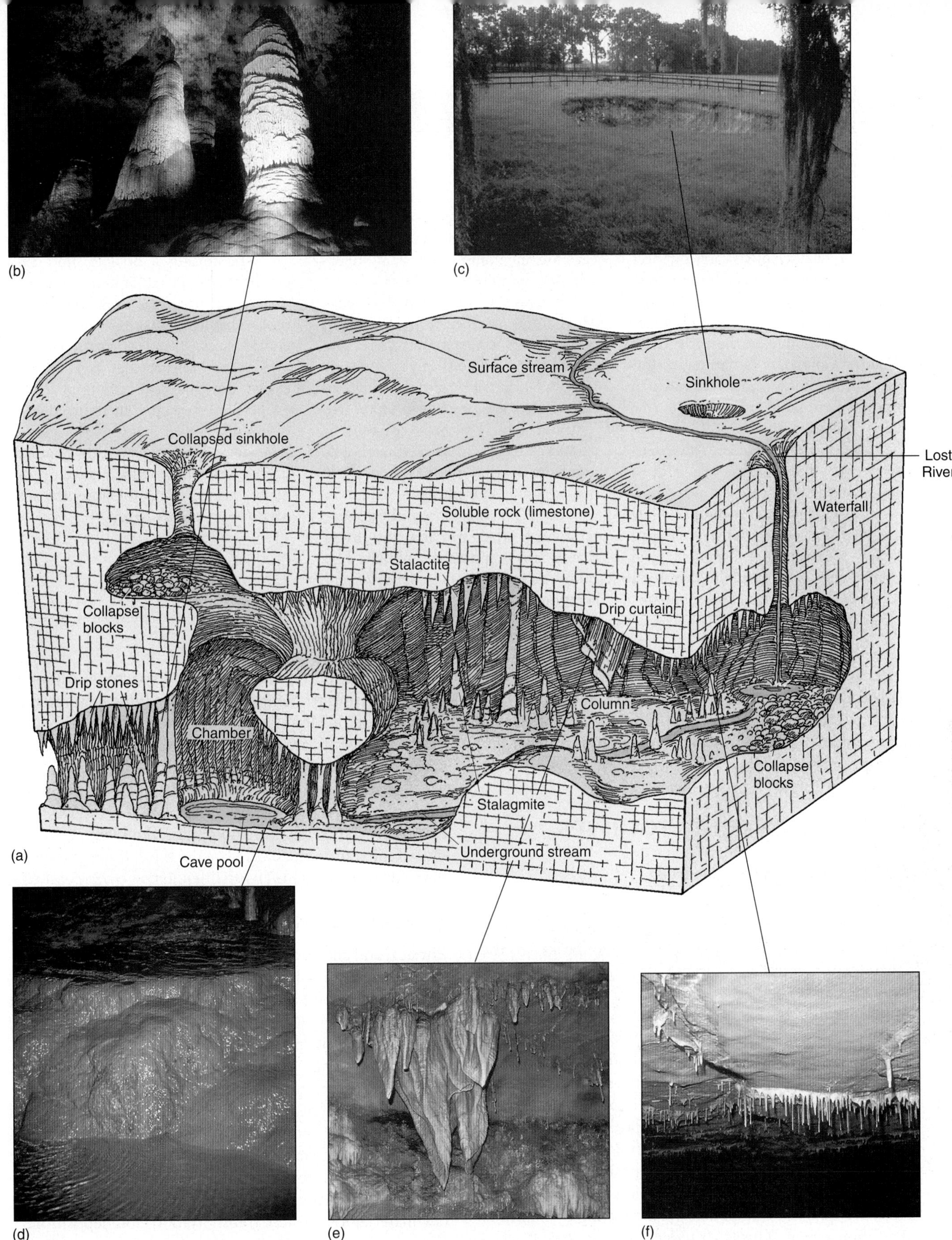

FIGURE 10.13 Cavern features.
(a) An underground cavern and related forms in limestone. (b) A column in Carlsbad Caverns, New Mexico, where a series of underground caverns includes rooms more than 1200 m (4000 ft) long and 190 m (625 ft) wide. Although a national park since 1930, unexplored portions remain. (c) A sinkhole in a Florida pasture. From Marengo Caves, Marengo, Indiana: (d) A flowstone and pool of water. (e) Dripstone drapery formations. (f) Soda straws hanging from ceiling cracks, forming one molecular layer at a time. [(b) Photo by author; (c) photo by Thomas M. Scott, Florida Geological Survey; photos (d, e, f) by Bobbé Christopherson.]

News Report 10.1

Amateurs Make Cave Discoveries

The exploration and scientific study of caves is called *speleology*. Physical and biological scientists carry out professional investigations of caves; however, amateur cavers, or "spelunkers," have made many important discoveries. As an example, in the early 1940s George Colglazier, a farmer southwest of Bedford, Indiana, awoke to find his farm pond at the bottom of a deep collapsed sinkhole. This sinkhole is now the entrance to an extensive cave system that includes a subterranean navigable stream.

Cave habitats are unique. They are nearly closed systems—self-contained ecosystems with simple food chains and great stability. In total darkness, bacteria synthesize inorganic elements and produce organic compounds that sustain many types of cave life, including algae, small invertebrates, amphibians, and fish.

In a cave discovered in 1986, near Movile in southeastern Romania, cave-adapted invertebrates were discovered after millions of years of sunless isolation. Thirty-one of these organisms were previously unknown. Without sunlight, the ecosystem in Movile is based on sulfur-metabolizing bacteria that synthesize organic matter using energy from oxidation processes. These chemosynthetic bacteria feed other bacteria and fungi that in turn support cave animals. The sulfur bacteria produce sulfuric acid compounds that may prove to be important in the chemical weathering of some caves.

The mystery, intrigue, and excitement of cave exploration lie in the variety of dark passageways, enormous chambers that narrow to tiny crawl spaces, strange formations, and underwater worlds that can be accessed only by cave diving. Private property owners and amateur adventurers discovered many of the major caves, a fact that keeps this popular science/sport very much alive. (For many related links of interest see http://www.caves.org/.)

aspect of geomorphology in which amateur cavers have made important discoveries about these unique habitats (see News Report 10.1). For more on caves and related formations, see http://www.goodearthgraphics.com/virtcave/virtcave.html.

Mass Movement Processes

NOTEBOOK

Madison River landslide

Nevado del Ruiz, northernmost of two dozen dormant (not extinct, sometimes active) volcanic peaks in the Cordilleran Central of Colombia, had erupted six times during the past 3000 years, killing 1000 people during its last eruption in 1845. On November 13, 1985, at 11 P.M. after a year of earthquakes and harmonic tremors, a growing bulge on its northeast flank, and months of small summit eruptions, Nevado del Ruiz violently erupted in a lateral explosion. The mountain was back in action.

On this night, the familiar pyroclastics, lava, and blast were not the worst problem. The hot eruption quickly melted ice on the mountain's snowy peak, liquefying mud and volcanic ash, sending a hot mudflow downslope. Such a flow is a *lahar*, an Indonesian word referring to mudflows of volcanic origin. This lahar moved rapidly down the Lagunilla River toward the villages below. The wall of mud was at least 40 m (130 ft) high as it approached Armero, a regional center with a population of 25,000. The city slept as the lahar buried its homes: 23,000 people were killed there and in other afflicted river valleys; thousands were injured; 60,000 were left homeless. The volcanic debris flow is now a permanent grave for its victims. Although less devastating to human life, the mudflow generated by the eruption of Mount St. Helens also was a lahar. Not all mass movements are this destructive, but such processes play a major role in the denudation of the landscape.

For more on mass movement hazards, including landslides, see the Web site of the Natural Hazards Center at the University of Colorado, Boulder, at http://www.Colorado.edu/hazards/ or the USGS Geologic Hazards page at http://landslides.usgs.gov/index.html.

Mass Movement Mechanics

Physical and chemical weathering processes create an overall weakening of surface rock, making the rock more susceptible to the pull of gravity. The term **mass movement** applies to all downward movements of materials propelled and controlled by gravity, such as the lahar just described. These movements can range from dry to wet, slow to fast, small to large, and from free-falling to gradual or intermittent in motion. The term *mass movement* sometimes is used interchangeably with *mass wasting*, which is the general process involved in mass movements and erosion of the landscape. To combine the concepts, we can say that mass movement of material works to waste slopes and provide raw material for erosion, transportation, and deposition.

The Role of Slopes. All mass movements occur on slopes. If we try to pile dry sand on the beach, the grains will flow downward until an equilibrium is achieved. The steepness of the slope that results when loose sand comes to rest depends on the size and texture of the grains; this is called the *angle of repose*. This angle represents a balance

FIGURE 10.16 Rockfall and talus cone.
Rockfall and talus cone at the base of a steep slope. [Photo by author.]

FIGURE 10.17 A debris avalanche in Alaska.
A debris avalanche covers portions of the Cascade Glacier, west of the Saint Elias Range in Alaska. The one pictured here was the largest of several dozen triggered by a 1979 earthquake. [Photo by George Plafker.]

10.17). The photo reveals characteristic grooves, lobes, and large rocks associated with these fluid avalanches.

Landslide. A sudden rapid movement of a cohesive mass of regolith or bedrock that is not saturated with moisture is a **landslide**—a large amount of material failing simultaneously. Surprise is the danger, for the downward pull of gravity dominates the struggle for equilibrium in an instant. One such surprise event struck near Longarone, Italy, in 1963, and is described in Focus Study 10.1.

To eliminate the surprise element, scientists are using the global positioning system (GPS) to monitor landslide

Focus Study 10.1

Vaiont Reservoir Landslide Disaster

Place: Northeastern Italy, Vaiont Canyon in the Italian Alps, rugged scenery, 680 m (2200 ft) above sea level.

Location: 46.3° N 12.3° E, near the border of the Veneto and Friuli-Venezia Giulia regions.

Situation: Centrally located in the region, an ideal situation for hydroelectric power production.

Site: Steep-sided, narrow glaciated canyon, opening to populated lowlands to the west; ideal site for a narrow-crested, high dam, and deep reservoir.

The Plan: Build the second highest dam in the world, using a new thin-arch design, 262 m (860 ft) high, 190 m (623 ft) crest length, impounding a reservoir capacity of 150 million cubic meters (5.3 billion cubic feet) of water (third largest reservoir in the world), and generate hydroelectric power.

Geologic Analysis:

1. Steep canyon walls composed of interbedded limestone and shale; badly cracked and deformed structures; open fractures in the shale are inclined toward the future reservoir body. The steepness of the canyon walls enhances the strong driving forces (gravitational) at work on rock structures.
2. High potential for bank storage (water absorption by canyon walls) into the groundwater system, which will increase water pressure on all rocks in contact with the reservoir.
3. The nature of the shale beds is such that cohesion will be reduced as its clay minerals become saturated.

(continued)

Focus Study 10.1 *(continued)*

4. Evidence of ancient rockfalls and landslides on the north side of the canyon.
5. Evidence of creep activity along the south side of the canyon.

Political/Engineering Decision: Begin design and construction of a thin-arch dam at this site.

Events During Construction and Filling: Large volumes of concrete were injected into the bedrock as "dental work" in an attempt to strengthen the fractured rock. During reservoir filling in 1960, 700,000 cubic meters (2.5 million cubic feet) of rock and soil slid into the reservoir from the south side. A slow creep of slope materials along the entire south side began shortly thereafter, increasing to about 1 cm per week by January 1963. Creep rate increased as the level of the reservoir rose. By mid-September 1963, the creep rate exceeded 40 cm (16 in.) a day.

This sequence of events set the stage for one of the worst dam disasters in history. Heavy rains began on September 28, 1963. Alarmed at the runoff from the rain into the reservoir and the increase in creep rate along the south wall, engineers opened the outlet tunnels on October 8 in an attempt to bring down the reservoir level—but it was too late.

The next evening, in only 30 seconds, a landslide of 240 million cubic meters (8.5 billion cubic feet) crashed into the reservoir, shaking seismographs across Europe. A 150 m thick (500 ft) slab of mountainside gave way (2 km by 1.6 km in area, 1.2 mi by 1 mi), sending shock waves of wind and water up the canyon 2 km (1.2 mi) and splashing a 100 m (330 ft) wave over the dam. The former reservoir was effectively filled with bedrock, regolith, and soil that almost entirely displaced its water content (Figure 1). Amazingly, the experimental dam design held.

Downstream, in the unsuspecting town of Longarone, near the mouth of Vaiont Canyon on the Piave River, people heard the distant rumble and were quickly drowned by the 69 m (226 ft) wave of water that came out of the canyon. That night 3000 people perished. As for a lesson or recommendation, we need only to reread the preconstruction geologic analysis that began this focus study. The Italian courts eventually prosecuted the persons responsible.

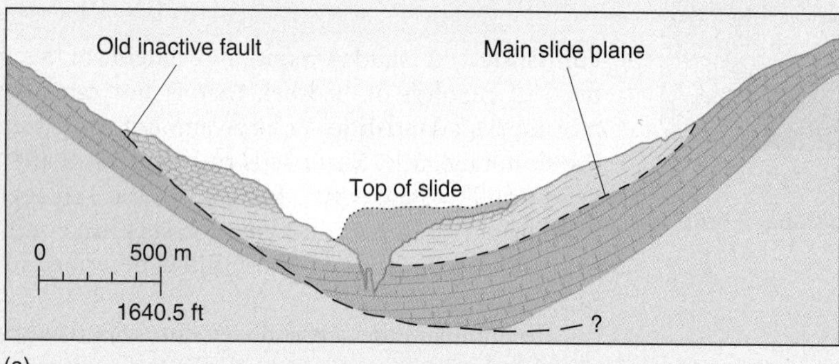

(a)

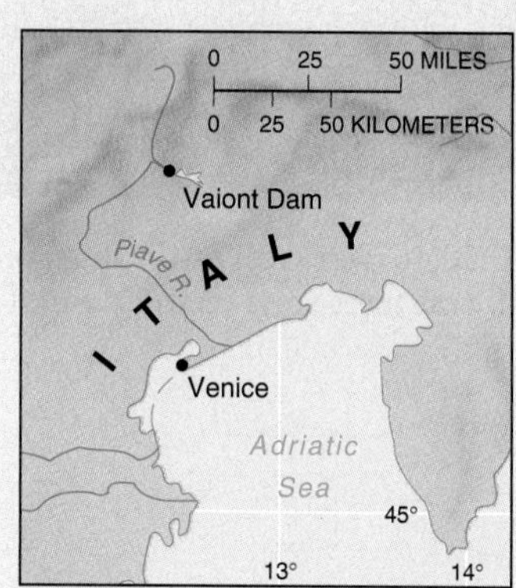

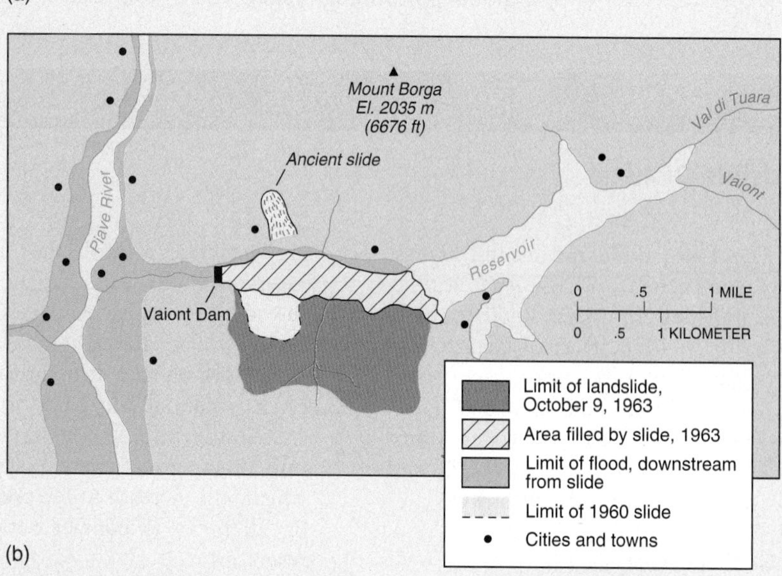

(b)

FIGURE 1 Vaiont Reservoir disaster.
(a) Cross section of the Vaiont River Valley. (b) Map of the disaster site noted with evidence of previous activity. [After G. A. Kiersch, "The Vaiont Reservoir Disaster," California Division of Mines and Geology, Mineral Information Service, vol. 18, no. 7, pp. 129–138.]

movement. With GPS, scientists measure slight land shifts in suspect areas for clues to possible mass wasting. GPS was applied in two cases in Japan and effectively identified prelandslide movements of 2–5 cm per year, providing information to expand the area of hazard concern.

Slides occur in one of two basic forms—translational or rotational (Figure 10.15). *Translational slides* involve movement along a planar (flat) surface roughly parallel to the angle of the slope. The Madison Canyon landslide described earlier was a translational slide. Flow and creep patterns also are considered translational in nature.

Rotational slides occur when surface material moves along a concave surface. Frequently, underlying clay presents an impervious surface to percolating water. As a result, water flows along the clay surface, undermining the overlying block. The surface may rotate as a single unit, or it may present a stepped appearance.

A landslide presentation and solution diagrams may be viewed at **http://www.kingston.ac.uk/~ce_s011/slides.htm**. For a comprehensive look at an ongoing landslide situation in southern California, see the Anaheim Hills Landslide Update at **http://anaheim-landslide.com/**. This site includes general background on mass movements, including many diagrams, and before and after photos.

Flow. When the moisture content of moving material is high, the suffix *-flow* is used, as in *earthflow* and *mudflow*, illustrated in Figure 10.15. An estimated 20,000 people died in Venezuela as the result of massive mudflows in 1999. Heavy rains frequently saturate barren mountain slopes and set them moving, as was the case east of Jackson Hole, Wyoming, in the spring of 1925. A slope above the Gros Ventre River (pronounced "grow vaunt") broke loose and slid downward as a unit. The slide occurred because sandstone formations rested on weak shale and siltstone, which became moistened and soft and offered little resistance to the overlying strata.

Because of melted snow and rain, the water content of the Gros Ventre landslide was high enough to classify it as an earthflow. About 37 million m^3 (1.3 billion ft^3) of wet soil and rock moved down one side of the canyon and surged 30 m (100 ft) up the other side (Figure 10.18). The earthflow dammed the river and formed a lake. Two years later, the lake water broke through the temporary dam, transporting a tremendous quantity of debris over the region downstream.

Creep. A persistent mass movement of surface soil is called **soil creep**. Individual soil particles are lifted and disturbed by the expansion of soil moisture as it freezes, by cycles of moistness and dryness, by daily temperature variations, or even by grazing livestock or digging animals.

In the freeze-thaw cycle, particles are lifted at right angles to the slope by freezing soil moisture, as shown in Figure 10.19. However, when the ice melts, the particles are pulled straight downward by the force of gravity. As the process is repeated, the soil cover gradually creeps its way downslope.

The overall wasting of a slope may cover a wide area and may cause fence posts, utility poles, and even trees to lean downslope. Various strategies are used to arrest the mass movement of slope material—grading the terrain, building terraces and retaining walls, planting ground cover—but the persistence of creep always prevails.

Human-Induced Mass Movements

Every highway roadcut or surface-mining activity that exposes a slope can hasten mass wasting because the oversteepened surfaces are thrown into a new search for equilibrium. Imagine the disequilibrium in slope relationships created by a highway roadcut.

Large open-pit strip mines—such as the Bingham Copper Mine west of Salt Lake City, the Berkeley Pit in Butte, Montana, and the extensive strip mining for coal in

FIGURE 10.18 The Gros Ventre slide near Jackson, Wyoming. Evidence of the 1925 landslide is still visible after more than 75 years. [Photo by Steven K. Huhtala.]

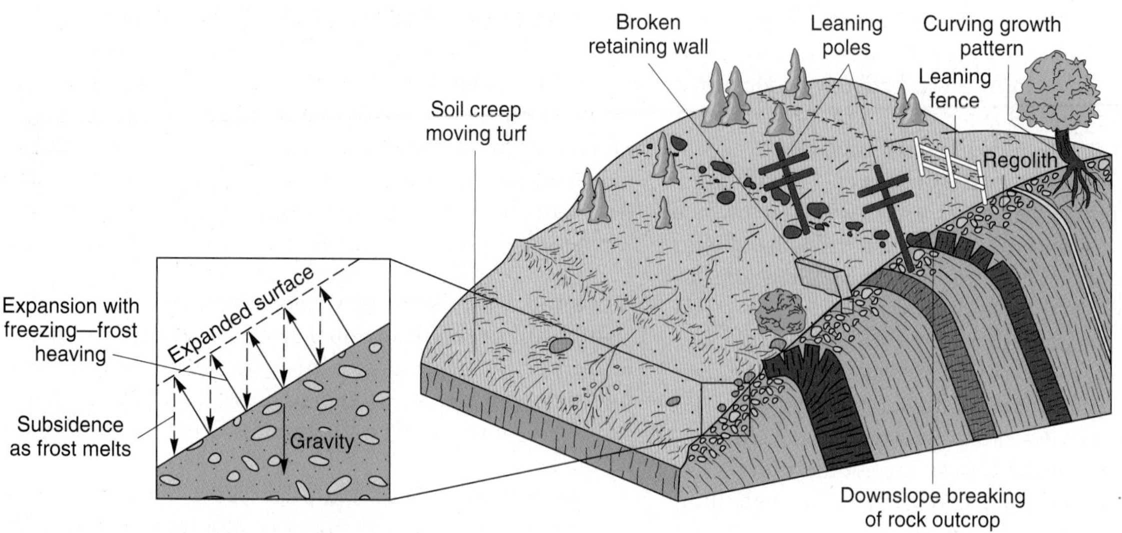

FIGURE 10.19 Soil creep and its effects.

the East and West—are examples of human-induced mass movements, generally called **scarification** (Figure 10.20). At the Bingham Copper Mine, a mountain literally was removed. The disposal of *tailings* (mined ore of little value) and waste material is a significant problem with such large excavations because the tailing piles prove unstable and susceptible to further weathering, mass wasting, or wind dispersal. Wind dispersal is a particular problem with uranium tailings in the West due to their radioactivity.

Land subsidence and collapse in mined areas produce further mass movements of land. This is a major problem in portions of the Appalachians where buildings, utility poles, and streets, as well as drainage patterns, are affected.

Scientists can informally quantify the scale of human-induced scarification for comparison with natural denudation processes. R. L. Hooke, Department of Geology and Geophysics, University of Minnesota, used estimates of U.S. excavations for new housing, mineral production (including the three largest—stone, sand and gravel, and coal), and highway construction. He then prorated these quantities of moved earth for all countries; each was rated using its gross domestic product (GDP), energy consumption, and agriculture's effect on river sediment loads. From these, he calculated a global estimate for human earth moving.

Hooke estimated that humans, as a geomorphic agent, annually move 40 to 45 billion tons (40–45 Gt/yr) of the planet's surface. Therefore, humans exceed natural river sediment transfer (14 Gt/yr), movement through stream meandering (39 Gt/yr), haulage by glaciers (4.3 Gt/yr), movement due to wave action and erosion (1.25 Gt/yr), wind transport (1 Gt/yr), sediment movement by continental and oceanic mountain building (34 Gt/yr), or deep-ocean sedimentation (7 Gt/yr). As Hooke stated,

> *Homo sapiens* has become an impressive geomorphic agent. Coupling our earth-moving prowess with our inadvertent adding of sediment load to rivers and the visual impact of our activities on the landscape, one is compelled to acknowledge that, for better or for worse, this biogeomorphic agent may be the premier geomorphic agent of our time.*

*R. L. Hooke, "On the efficacy of humans as geomorphic agents," *GSA Today*, The Geological Society of America, 4, no. 9 (September 1994): 217–226.

Summary and Review—Weathering, Karst Landscapes, and Mass Movement

● *Define* the science of geomorphology.

Geomorphology is the science that analyzes and describes the origin, evolution, form, and spatial distribution of landforms. The exogenic system, powered by solar energy and gravity, tears down the landscape through processes of landmass **denudation** involving weathering, mass movement, erosion, transportation, and deposition. Different rocks offer differing resistance to these weathering processes and produce a pattern of **differential weathering** on the landscape. This balancing act between tectonic uplift and reduction by weathering and erosion, between the resistance of rocks and the ceaseless attack of weathering and erosion, is summarized in the **dynamic equilibrium model**.

geomorphology (p. 318)
denudation (p. 318)
differential weathering (p. 318)

FIGURE 10.20 Scarification.
(a) Black Mesa, Arizona, strip-mining for coal. (b) Bingham Canyon, Utah, west of Salt Lake City, strip-mining for copper and other minerals. (c) Abandoned Berkeley Pit copper mine, Butte, Montana. (d) Spoil banks in a West Virginia coal-mining area. [Photos by author.]

dynamic equilibrium model (p. 318)

1. Define geomorphology, and describe its relationship to physical geography.
2. Define landmass denudation. What processes are included in the concept?
3. What is the interplay between the resistance of rock structures and weathering variabilities?
4. Explain how Figure 10.1 is an example of differential weathering.
5. What are the principal considerations in the dynamic equilibrium model?

- ***Illustrate* the forces at work on materials residing on a slope.**

Slopes are shaped by the relation between rate of weathering and breakup of slope materials and the rate of mass movement and erosion of those materials. A slope is considered stable if it is stronger than these denudation processes; it is unstable if it is weaker. In this struggle against gravity, a slope may reach a **geomorphic threshold**—the point at which there is enough energy to overcome resistance against movement. **Slopes** that form the boundaries of landforms have several general components: *waxing slope*, *free face*, *debris slope*, and *waning slope*. Slopes seek an angle of equilibrium among the operating forces.

geomorphic threshold (p. 319)
slopes (p. 319)

6. Describe conditions on a hillslope that is right at the geomorphic threshold. What factors might push the slope beyond this point?
7. Given all the interacting variables, do you think a landscape ever reaches a stable, old-age condition? Explain.
8. What are the general components of an ideal slope?
9. Relative to slopes, what is meant by an "angle of equilibrium"? Can you apply this concept to the photograph in Figure 10.2?

● *Define* weathering and *explain* the importance of parent rock and joints and fractures in rock.

Weathering processes disintegrate both surface and subsurface rock into mineral particles or dissolve them in water. The upper layers of surface material undergo continual weathering and create broken-up rock called **regolith**. Weathered **bedrock** is the *parent rock* from which regolith forms. The unconsolidated, fragmented material that develops after weathering is called **sediment**, which, along with weathered rock, forms the **parent material** from which soil evolves.

Important in weathering processes are **joints**, the fractures and separations in the rock. Jointing opens up rock surfaces on which weathering processes operate. Factors that influence weathering include: character of the bedrock (hard or soft, soluble or insoluble, broken or unbroken), climatic elements (temperature, precipitation, freeze-thaw cycles), position of the water table, slope orientation, surface vegetation and its subsurface roots, and time. Agents of change include moving air, water, waves, and ice.

weathering (p. 319)
regolith (p. 319)
bedrock (p. 319)
sediment (p. 319)
parent material (p. 319)
joints (p. 320)

10. Describe weathering processes operating on an open expanse of bedrock. How does regolith develop? How is sediment derived?
11. Describe the relationship between climatic conditions and rates of weathering activities. Is the scale at which we consider weathering processes important?
12. What is the relation among parent rock, parent material, regolith, and soil?
13. What role do joints play in the weathering process? Give an example from one of the illustrations in this chapter.

● *Describe* frost action, crystallization, hydration, pressure-release jointing, and the role of freezing water as physical weathering processes.

Physical weathering refers to the breakup of rock into smaller pieces with no alteration of mineral identity. The physical action of water when it freezes (expands) and thaws (contracts) is a powerful agent in shaping the landscape. This **frost action** may break apart any rock. Working in joints, expanded ice can produce *joint-block separation* through the process of *frost-wedging*.

Another process of physical weathering is *crystallization*; as crystals in rock grow and enlarge over time they force apart mineral grains and break up rock. **Hydration** occurs when a mineral absorbs water and expands, thus creating a strong mechanical force that stresses rocks.

As overburden is removed from a granitic batholith, the pressure of deep burial is relieved. The granite slowly responds with *pressure-release jointing*,with layer after layer of rock peeling off in curved slabs or plates. As these slabs weather, they slip off in a process called **sheeting**. This *exfoliation process* creates a dome-shaped feature on the exposed landscape, called an **exfoliation dome**.

physical weathering (p. 322)
frost action (p. 322)
hydration (p. 323)
sheeting (p. 323)
exfoliation dome (p. 323)

14. What is physical weathering? Give an example.
15. Why is freezing water such an effective physical weathering agent?
16. What weathering processes produce a granite dome? Describe the sequence of events.

● *Describe* the susceptibility of different minerals to the chemical weathering processes called hydrolysis, oxidation, carbonation, and solution.

Chemical weathering is the chemical decomposition of minerals in rock. **Spheroidal weathering** is chemical weathering that occurs in cracks in the rock. As cementing and binding materials are removed, the rock begins to disintegrate, and sharp edges and corners become rounded.

Hydrolysis breaks down silicate minerals in rock, as in the chemical weathering of feldspar into clays and silica. Water is not just absorbed, as in hydration, but actively participates in chemical reactions. **Oxidation** is the reaction of oxygen with certain metallic elements, the most familiar example being the rusting of iron, producing iron oxide. *Solution* is chemical weathering. For instance, a mild acid such as carbonic acid in rainwater will cause **carbonation**, wherein carbon combines with certain minerals, such as calcium, magnesium, potassium, and sodium.

chemical weathering (p. 323)
spheroidal weathering (p. 323)
hydrolysis (p. 324)
oxidation (p. 324)
carbonation (p. 324)

17. What is chemical weathering? Contrast this set of processes to physical weathering.
18. What is meant by the term spheroidal weathering? Describe the spheroidal weathering process.
19. What is hydrolysis? How does it affect rocks?
20. Iron minerals in rock are susceptible to which form of chemical weathering? What characteristic color is associated with this type of weathering?
21. With what kind of minerals do carbon compounds react, and under what circumstances does this reaction occur? What is this weathering process called?

● *Review* the processes and features associated with karst topography.

Karst topography refers to distinctively pitted and weathered limestone landscapes. Surface circular **sinkholes** form and may combine to form a *karst valley*. A sinkhole may collapse through the roof of an underground cavern, forming a *collapse sinkhole*. The formation of caverns is part of karst processes and groundwater erosion.

Limestone caves feature many unique erosional and depositional features, producing a dramatic subterranean world.

karst topography (p. 325)
sinkholes (p. 326)

22. Describe the development of limestone topography. What is the name applied to such landscapes? From what area was this name derived?
23. Differentiate among sinkholes, karst valleys, and cockpit karst. Within which form is the radio telescope at Arecibo, Puerto Rico?
24. In general, how would you characterize the region southwest of Orleans, Indiana?
25. What are some of the unique erosional and depositional features you find in a limestone cavern?

Portray the various types of mass movements and *identify* examples of each in relation to moisture content and speed of movement.

Any movement of a body of material, propelled and controlled by gravity, is called **mass movement**, also called *mass wasting*. The *angle of repose* of loose sediment grains represents a balance of driving and resisting forces on a slope. Mass movement of Earth's surface produces some dramatic incidents, including **rockfall** (a volume of rock that falls), **debris avalanche** (mass of tumbling, falling rock, debris, and soil at high speed), **landslide** (a large amount of material failing simultaneously), *mudflow* (material in motion with a high moisture content), and **soil creep** (a persistent movement of individual soil particles that are lifted by the expansion of soil moisture as it freezes, cycles of wetness and dryness, and temperature variations, or the impact of grazing animals). In addition, human mining and construction activities have created massive **scarification** of landscapes.

mass movement (p. 330)
rockfall (p. 332)
debris avalanche (p. 332)
landslide (p. 333)
soil creep (p. 335)
scarification (p. 336)

26. Define the role of slopes in mass movements, using the terms *angle of repose*, *driving force*, *resisting force*, and *geomorphic threshold*.
27. What events occurred in the Madison River Canyon in 1959?
28. What are the classes of mass movement? Describe each briefly and differentiate among these classes.
29. Name and describe the type of mudflow associated with a volcanic eruption.
30. Describe the difference between a landslide and what happened on the slopes of Nevado Huascarán.
31. What is scarification, and why is it considered a type of mass movement? Give several examples of scarification. Why are humans a significant geomorphic agent?

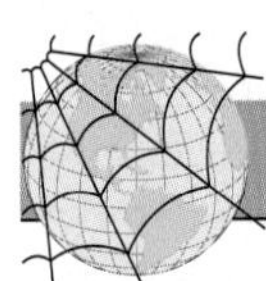

NetWork

The *Elemental Geosystems Home Page* provides on-line resources for this chapter on the World Wide Web. Once on the Home Page, click on this textbook, scroll the Table of Contents menu, select this chapter, and click "Begin." You will find self-tests that are graded, review exercises, specific updates for items in the chapter, and in "Destinations" many links to interesting related pathways on the Internet. *Elemental Geosystems* is found at http://www.prenhall.com/christopherson.

Critical Thinking

A. Locate a slope, possibly near campus, near your home, or at a local roadcut. Using Figure 10.3a and b, can you identify the forces and forms of a hillslope at your site? How would you go about assessing the stability of the slope? Is there any evidence of the mass wasting of materials, soil creep, or other processes discussed in this chapter?

B. The USGS has completed a landslide hazards potential map for the conterminous United States. You can check out this resource at http://landslides.usgs.gov/landslide.html. How does the map portray landslide incidences? In what way does the map rate landslide susceptibility to an area? Are you able to find a location that you have visited and determine its vulnerability?

C. Go to the *Elemental Geosystems Home Page* and to Critical Thinking in Chapter 10. Study the two remote sensing images. The Himalayan Mountains are the primary source of the sediment transported by the rivers. Imagine yourself as a piece of rock near the top of Mount Everest. Describe a scenario by which you join a collection of sediments in the Bay of Bengal. Discuss the changes in weathering processes acting upon you between the high, cold reaches of the Himalayas and the hot, humid mud flats of the Ganges Delta.

Career Link 10.1

Gregory A. Pope, Ph.D., Associate Geography Professor

Gregory Pope lists physical geography, weather, geomorphology, weathering processes, GIS, and global environmental change as areas of special interest. He assumed his teaching post at Montclair State in 1996 and is the advisor for students in the geography program. Greg's master's and Ph.D. degrees were completed at Arizona State University. His dissertation developed a boundary-layer model for interpreting spatial variation in weathering. He has published several articles derived from his dissertation.

Earlier in this chapter you can see Greg's influence in my mention of the importance of scale in analyzing weathering processes. He found both chemical and physical weathering at small-scale reaction sites on rock, even in the driest landscapes.

FIGURE 1 Gregory A. Pope, Associate Geography Professor and Geography Department Chair at Montclair State University, Upper Montclair, New Jersey. Here, leading a field trip in New York City's Central Park. [Photo by Bobbé Christopherson.]

Greg did his undergraduate work at the University of Colorado in his hometown of Colorado Springs. As a child, he camped with his family in Colorado and other parts of the West. Especially vivid to him are camping trips to the Rockies: "I didn't know what geomorphology or geology were, but the rocks and mountain formations fascinated me." Beginning when he was only 4 or 5, during automobile trips to Illinois, he was the family navigator, using his own highway map collection.

Greg remembers, "We had a geographic encyclopedia set that I read and used for homework throughout my lower grades. I also read all the geography and geology books in my junior high before I finished eighth grade; they called it social studies, not geography, but it was really geography. I convinced my teacher to let me do an urban planning project about the unplanned growth in Colorado Springs. I asked the teacher to let me design a new city plan, as if Colorado Springs wasn't there. I did this one whole semester instead of what the rest of the class did."

"Following graduation from UCCS, I received a National Geographic Society (NGS) internship and worked in Washington, D.C. for half a year. I received a lot of valuable research experience working on *World*, the NGS children's magazine." He also worked as a contributing writer for the NGS National Geography Bee.

As a teaching assistant in graduate school at Arizona State, Greg, along with his advisor Ronald Dorn, began studying the White Mountains of eastern California, which fostered his interest in desert geomorphology. He knew he wanted to teach. "When I taught the labs, something all teaching assistants do, I felt comfortable. I knew teaching had to be part of my future."

Greg is studying the weathering of ancient to modern architectural structures in Portugal and collaborating with geoarcheologists who study weathering of petroglyphs. Also, he is looking at weathering in some stone work, specifically isolating pre-Clean Air Act (1970) contamination and weathering, examining tombstones near pollution sources. With his students he examines the rates of building-stone weathering around New York City—his take on urban geology—and then maps the spatial results. In addition, Greg is doing research on the tectonic geomorphology of the

(continued)

Career Link 10.1 *(continued)*

Rocky Mountains, a lifelong subject of interest. And under an NSF grant he is studying the effects of the recent wildfires, and the intense temperatures they generate, on weathering processes and soils in Colorado.

In 2001, Greg, along with geographer Dr. Patricia Beyer of Bloomsburg University, led us on a field trip through Central Park in New York City—called "Cultural Stones of Central Park: Outcrops, Arches, and Statues." They produced a wonderful field guide for our expedition. The popular park went through many stages of design and construction over the last 150 years including a variety of arches and bridges to accommodate different forms of traffic. Thirty-six bridges remain of the original thirty-nine, built from a great variety of rock.

Greg said, "The stones used by the designers and builders in the art and architecture of the Park are truly 'cultural stones.' Sources for all this building stone cover a wide geography, from many locales across the Northeast and into Canada." Greg and Patricia showed us how these materials presented a diverse array of surfaces on which physical and chemical weathering processes could work (Figure 2).

What lies ahead for geography? "Global change studies will be important; the integration of environmental studies is another possible avenue. Some discipline must emerge to pull all the disparate parts together—geography should be it, Greg feels. "In the direction I am going—weathering studies, geoarcheology, and work in building my department's program—I am having fun! I am inventing stuff and combining subjects that interest me. And being a geographer, I get to synthesize all this into coherent patterns! I get to work with people from so many different fields. My advice is to have fun with it, have joy in what you choose to do."

FIGURE 2 Weathered sandstone in Central Park, New York City.
Here on the sides of one of the many arches, constructed in 1859, we see the ravages of both physical and chemical weathering. The patterns etched into this Alberta sandstone, found in New Brunswick, are completely gone in several squares. Encrusted salts, leading to crystallization processes are visible. [Photo by Bobbé Christopherson.]

The Elbe River floods Zwinger Castle and neighborhoods in Dresden, Germany, August 16, 2002. The Elbe reached 9 m (29 ft) above flood stage, slighly higher than the all-time record. Such extreme events appear related to more anomolous weather patterns. [Photo © AP/Wide World Photos.]

11 River Systems and Landforms

Key Learning Concepts

By knowing and understanding the key learning concepts in this chapter, you should be able to:

- *Define* fluvial and *outline* the fluvial processes: erosion, transportation, and deposition.
- *Construct* a basic drainage basin model and *identify* different types of drainage patterns, with examples.
- *Describe* the relation among velocity, depth, width, and discharge, and *explain* the various ways that a stream erodes and transports its load.
- *Develop* a model of a meandering stream, including point bar, undercut bank, and cutoff, and *explain* the role of stream gradient in these flow characteristics.
- *Define* a floodplain and *analyze* the behavior of a stream channel during a flood.
- *Differentiate* the several types of river deltas and *detail* each.
- *Explain* flood probability estimates and *review* strategies for mitigating flood hazards.

Earth's rivers and waterways form vast arterial networks that both shape and drain the continents, transporting the by-products of weathering, mass movement, and erosion. To call them "Earth's lifeblood" is no exaggeration, inasmuch as rivers redistribute mineral nutrients important for soil formation and plant growth and serve society in many ways.

Rivers not only provide us with essential water supplies, they also receive, dilute, and transport wastes, provide critical cooling water for industry, and form one of the world's most important transportation networks. Rivers have been significant throughout human history. The devastating record floods in Europe in 2002 demonstrate the interrelationship between rivers and society (see the chapter-opening photo).

At any one moment approximately 1250 km^3 (300 mi^3) of water flows through Earth's waterways. Even though this volume represents only 0.003% of all freshwater, the work performed by this energetic flow makes it the dominant agent of landmass denudation. Of the world's rivers, those with the greatest *discharge* (stream flow rate) are the Amazon (Figure 11.1), the Congo (Zaire) of Africa, the Chàng Jiang (Yangtze) of Asia, and the Orinoco of South America (Table 11.1). In North America, the greatest discharges are from the Mississippi-Missouri-Ohio, Saint Lawrence, and Mackenzie river systems.

Hydrology is the science of water, its global circulation, distribution, and properties, specifically water at and below Earth's surface. For hydrology links on the Web, see http://terrassa.pnl.gov:2080/hydroweb.html or the Global Hydrology and Climate Center at http://www.ghcc.msfc.nasa.gov/; also see the Amazon River at http://boto.ocean.washington.edu/eos/index.html.

In this chapter: We begin with a look at the largest rivers on Earth. Essential fluvial concepts of base level, drainage basin and watershed, and drainage patterns follow. Factors that affect streamflow characteristics and the work performed by flowing water, including erosion, transport, and deposition, are discussed. Human response to floods and floodplain development are important aspects of river management and conclude the chapter.

Fluvial Processes and Landscapes

Stream-related processes are termed **fluvial** (from the Latin *fluvius*, meaning "river"). Geographers seek to describe recognizable stream patterns and the fluvial processes that created them. Fluvial systems, like all natural systems, have characteristic processes and produce predictable landforms. Yet a stream system can behave with randomness, unpredictability, and disorder.

Insolation and gravity are the driving forces of fluvial systems because they power the hydrologic cycle. Individual streams vary greatly, depending on their region's climate, the variety of surface composition and topography

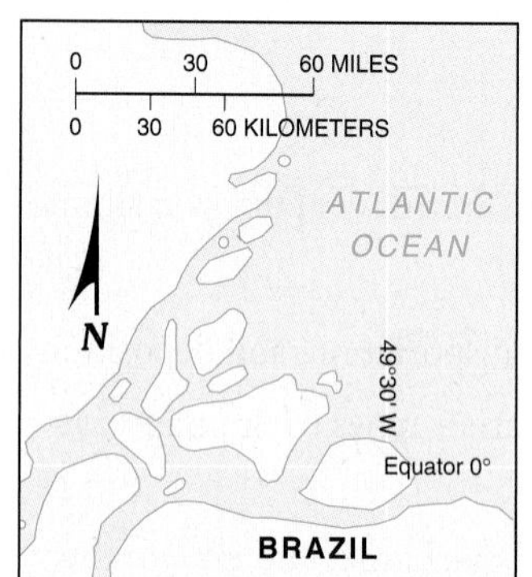

FIGURE 11.1 Mouth of the Amazon River.
The mouth of the Amazon River discharges a fifth of all the freshwater that enters the world's oceans. The mouth of the Amazon is 160 km (100 mi) wide. Amazon's drainage basin, which is as large as the Australian continent, produces millions of tons of sediments. Large islands of sediment are left where the river's discharge leaves the mouth and flows into the Atlantic Ocean. [*Terra* MISR sensor image courtesy of NASA/GSFC/JPL and the MISR Team.]

Table 11.1 Largest Rivers on Earth Ranked by Discharge Volume

Rank by Volume	Average Discharge at Mouth in Thousands of m³/s (cfs)	River (with Tributaries)	Outflow/Location	Length km (mi)	Rank by Length
1	180 (6350)	Amazon (Ucayali, Tambo, Ene, Apurimac)	Atlantic Ocean/ Amapá-Pará, Brazil	6570 (4080)	2
2	41 (1460)	Congo, also known as the Zaire (Lualaba)	Atlantic Ocean/ Angola, Congo	4630 (2880)	10
3	34 (1201)	Yangtze (Chàng Jiang)	East China Sea/Kiangsu, China	6300 (3915)	3
4	30 (1060)	Orinoco	Atlantic Ocean/Venezuela	2737 (1700)	27
5	21.8 (779)	La Plata estuary (Paraná)	Atlantic Ocean/Argentina	3945 (2450)	16
6	19.6 (699)	Ganges (Brahmaputra)	Bay of Bengal/India	2510 (1560)	23
7	19.4 (692)	Yenisey (Angara, Selenga or Selenge, Ider)	Gulf of Kara Sea/Siberia	5870 (3650)	5
8	18.2 (650)	Mississippi (Missouri, Ohio, Tennessee, Jefferson, Beaverhead, Red Rock)	Gulf of Mexico/Louisiana	6020 (3740)	4
9	16.0 (568)	Lena	Laptev Sea/Siberia	4400 (2730)	11
17	9.7 (348)	St. Lawrence	Gulf of St. Lawrence/Canada and U.S.	3060 (1900)	21
36	2.83 (100)	Nile (Kagera, Ruvuvu, Luvironza)	Mediterranean Sea/Egypt	6690 (4160)	1

over which they flow, the nature of vegetation and plant cover, and the length of time they have existed in a specific setting.

Wind, water, and ice dislodge, dissolve, or remove surface material in the process of **erosion**. Thus, streams produce fluvial erosion, which supplies weathered sediment for **transport** to new locations, where it is laid down in a **deposition** process. A stream is a mixture of water and solids—carried in solution, in suspension, and by mechanical transport. **Alluvium** is the general term for the clay, silt, and sand transported by running water.

Base Level of Streams

American geologist and ethnologist John Wesley Powell (1834–1902) was an early director of the U.S. Geological Survey, an explorer of the Colorado River, and a pioneer in understanding the landscape (Figure 11.2). In 1875 he put forward the idea of **base level**, or a level below which a stream cannot erode its valley further. The hypothetical *ultimate base level* is sea level. You can imagine base level as a surface extending inland from sea level, inclined gently upward under the continents (Figure 11.3). Ideally, this is the lowest practical level for all denudation processes.

Powell also recognized that not every landscape degraded down to sea level, for other base levels seemed to be in operation. A *local base level*, or a temporary one, may control the lower limit of local streams. That local base level might be a river, a lake, a hard and resistant rock structure, or a human-made dam. In arid landscapes, with

FIGURE 11.2 Powell's journey to base level. John Wesley Powell developed his base level concept during extensive exploration of the West. He first ventured down the Colorado River in 1869 in heavy oak boats, drifting by awe-inspiring red cliffs, incomprehensible in scale at the time. [Photo by Bobbé Christopherson.]

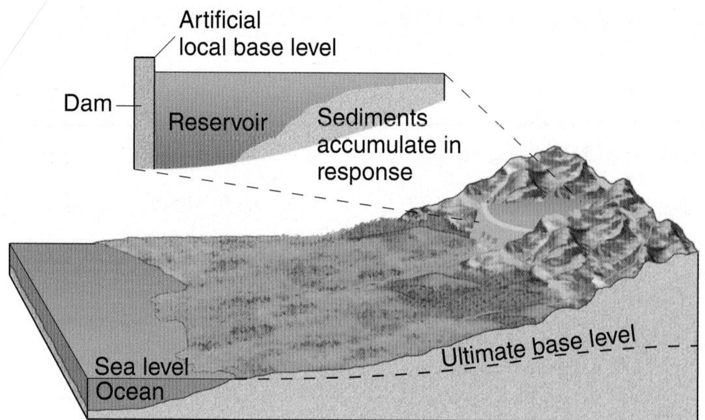

FIGURE 11.3 Ultimate and local base level concepts. The concepts of ultimate base level (sea level) and local base level (natural, such as a lake, or artificial, such as a dam). Note how base level curves gently upward from the sea as it is traced inland; this is the theoretical limit for stream erosion.

their intermittent precipitation, valleys, plains, or other low points determine local base level.

Ideally, endogenic processes build and create *initial* landscapes, whereas exogenic processes work toward low relief, an ultimate condition of little change, and the stability of *sequential* landscapes. Various theories have been proposed to model these denudation processes and account for the appearance of the landscape at different developmental stages, not only for humid temperate climates but also for coastal, arid desert, and equatorial landscapes.

A *dynamic equilibrium*—a balance among force, form, and process—is the preferred model used by many contemporary geomorphologists. Slope and landform stability are considered a function of the resistance of rock materials to the attack of denudation processes.

Over time, the work of streams modifies the landscape dramatically. Landforms are produced by two basic processes: (1) erosive action of flowing water, and (2) deposition of stream-transported materials. Let us begin our study by examining a basic fluvial unit—the drainage basin.

The Drainage Basin System

Every stream has a drainage basin. A **drainage basin** is the spatial geomorphic unit occupied by a river system (Figure 11.4). Ridges that form *drainage divides* are the dividing lines that control into which basin precipitation drains. Drainage divides define a **watershed**, the catchment area of the drainage basin. In any drainage basin, water initially moves downslope in a thin film called **sheetflow**, or *overland flow.* High ground that separates one valley from another, and directs sheetflow, is an *interfluve.* Surface runoff concentrates in *rills*, or small-scale downhill grooves, which may develop into deeper *gullies* and then a stream course in a valley.

Drainage Divides and Basins. Several high drainage divides, called **continental divides**, are situated in the

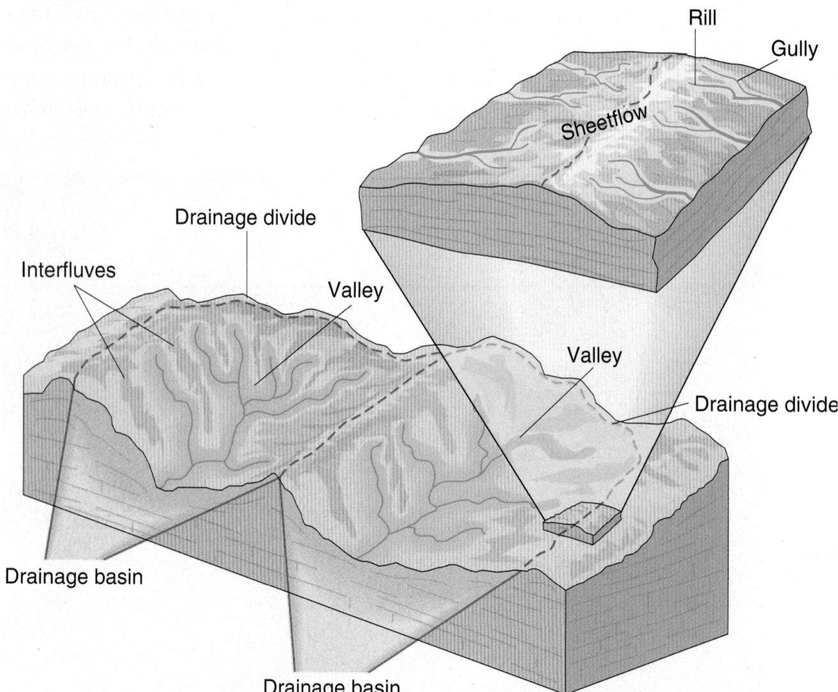

FIGURE 11.4 A drainage basin. A drainage divide separates a drainage basin and its watershed from other basins. Note the high ground that separates one valley from another—an interfluve.

United States and Canada. These are extensive mountain and highland regions that separate drainage basins, sending flows to the Pacific, the Gulf of Mexico, the Atlantic, Hudson Bay, or the Arctic Ocean. The principal drainage basins and continental divides in the United States and Canada are on the map in Figure 11.5. These drainage basins provide a spatial framework for water studies and water planning.

A major drainage basin system, such as the one created by the Mississippi-Missouri-Ohio river system, is made up of many smaller drainage basins, which in turn comprise even smaller basins, each divided into specific watersheds. Each drainage basin gathers and delivers its precipitation and sediment to a larger basin, concentrating the volume in the main stream. For example, consider the travels of rainfall in north-central Pennsylvania. This water feeds hundreds of small streams that flow into the Allegheny River (noted on Figure 11.5). The Allegheny River then joins with the Monongahela River at Pittsburgh to form the Ohio River. The Ohio flows southwest and connects with the Mississippi River at Cairo, Illinois, and eventually flows on past New Orleans to the Gulf of

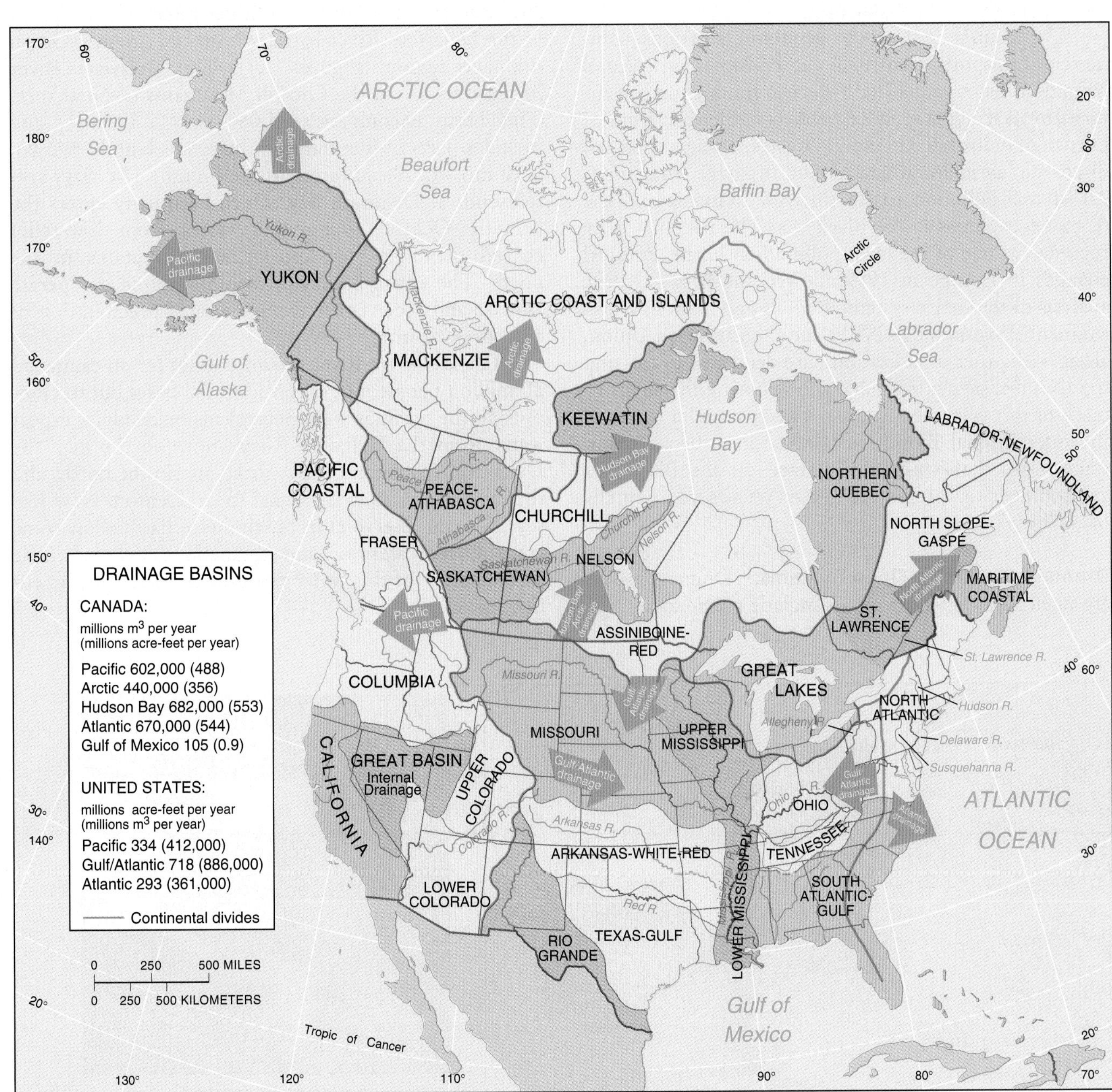

FIGURE 11.5 Drainage basins and continental divides.
Continental divides (blue lines) separate the major drainage basins that empty into the Pacific, Atlantic, and Gulf of Mexico, and to the north through Canada into Hudson Bay and the Arctic Ocean. Subdividing these large-scale basins are major river basins. [After U.S. Geological Survey; *The National Atlas of Canada,* 1985, Energy, Mines, and Resources Canada; and Environment Canada, *Currents of Change—Inquiry on Federal Water Policy—Final Report 1986.*]

exico. Each contributing tributary adds its discharge and sediment load to the larger river.

In our example, sediment weathered and eroded in north-central Pennsylvania is transported thousands of kilometers and accumulates as the Mississippi River delta on the floor of the Gulf of Mexico.

Imagine the complexity of an international drainage basin, such as the Danube River in Europe, that flows 2850 km (1770 mi) from western Germany's Black Forest to the Black Sea. The river crosses or forms the borders between nine countries (Figure 11.6). A total of area of 817,000 km^2 (315,000 mi^2) fall within the drainage basin, including some 300 tributaries.

The Danube serves many economic functions: commercial transport, municipal water source, agricultural irrigation, fishing, and hydroelectric production. An international struggle is underway to save the river from its burden of industrial and mining wastes, sewage, chemical discharge, agricultural runoff, and drainage from ships. All of this pollution passes through Romania and the deltaic ecosystems in the Black Sea. The river is widely regarded as one of the most polluted on Earth. Political changes in Europe in 1989 allowed the first scientific analysis of the entire system. The United Nations Environment Programme (UNEP) and the European Union, along with other organizations, are dedicated to clearing the Danube, saving the delta, and restoring the environment of this valuable resource. The river delta is one of the international Biosphere Reserves; see http://www.unesco.org/mab/wnbr.htm. (More on the Black Sea environment is posted at http://www.grid.unep.ch/bsein/index.html.)

Drainage Basins as Open Systems. Drainage basins are open systems whose inputs include precipitation, the minerals and rocks of the regional geology, and changes of energy with both the uplift and subsidence provided by tectonic activities. System outputs of water and sediment leave through the mouth of the river. Change that occurs in any portion of a drainage basin can affect the entire system, because the stream adjusts to carry the appropriate load of sediment relative to discharge (its volume of flow) and velocity. A stream drainage system exhibits a constant struggle toward an equilibrium among interacting variables of discharge, transported sediment load, channel shape, and channel slope.

An Example: the Delaware River Basin. Let us look at the Delaware River basin, within the Atlantic Ocean drainage region (Figure 11.7). The Delaware River headwaters are in the Catskill Mountains of New York. This basin encompasses 33,060 km^2 (12,890 mi^2) and includes parts of five states in the river's length, 595 km (370 mi) from headwaters to the mouth. The river system ends at Delaware Bay, which eventually enters the Atlantic Ocean. Topography varies from low-relief coastal plains to the Appalachian Mountains in the north. The entire basin lies within a humid, temperate climate and receives an average annual precipitation of 120 cm (47.2 in.).

The Delaware River provides water for an estimated 20 million people, not only within the basin but to cities outside the basin as well. Several major conduits export water from the Delaware River. Note on the map the Delaware Aqueduct to New York City (in the north) and the Delaware & Raritan Canal (near Trenton, New Jersey). Several reservoirs in the drainage basin allow some control over water flow and storage for dry periods. The sustainability of this water resource is critical to the entire region.

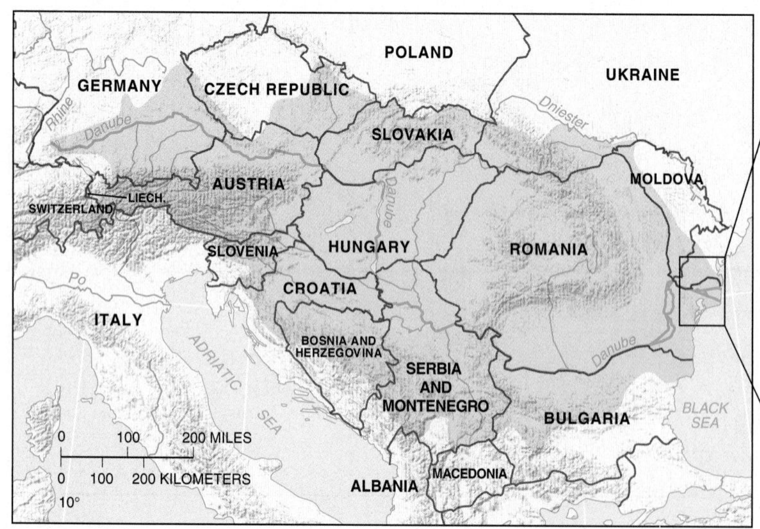

Danube River delta

FIGURE 11.6 An international drainage basin—the Danube River.
The Danube crosses or forms the border of nine countries as it flows across Europe to the Black Sea. The river spews polluted discharge into the Black Sea through its arcuate form delta. [*Terra* MODIS image, June 15, 2002, courtesy of MODIS Land Rapid Response Team, NASA/GSFC.]

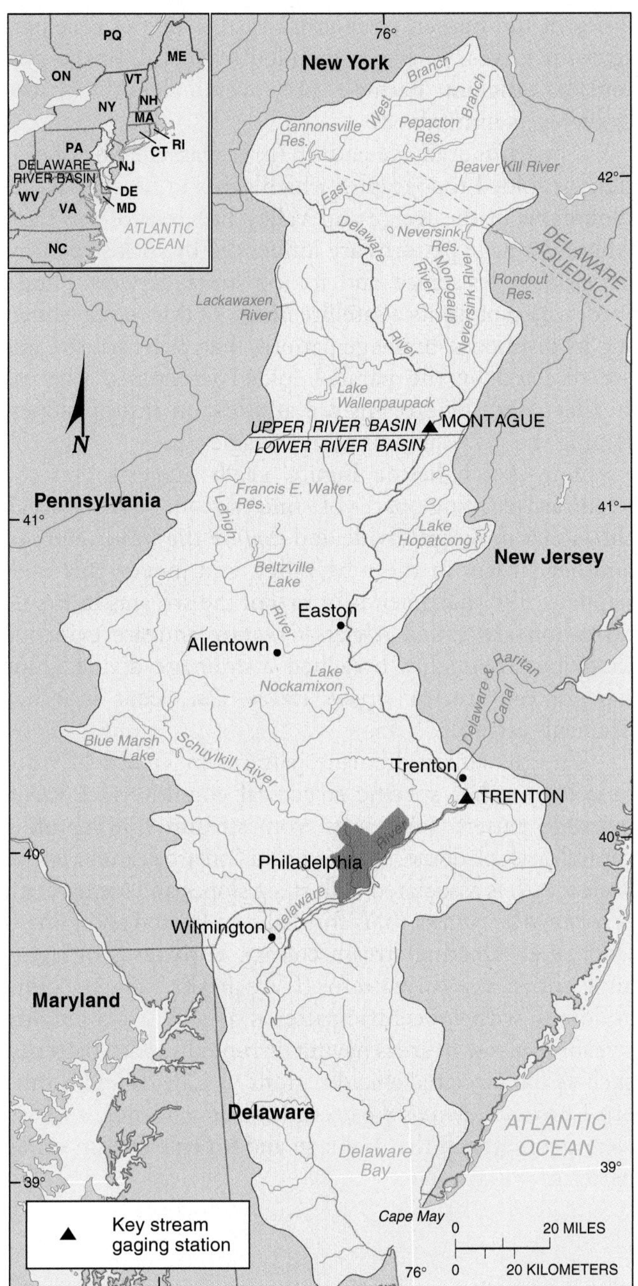

FIGURE 11.7 The Delaware River Basin.
The Delaware River drains portions of five states. The basin was studied by the USGS to assess the potential impact of global warming on a river system. [After U.S. Geological Survey, 1986, "Hydrologic events and surface water resources," *National Water Summary 1985,* Water Supply Paper 2300 (Washington, D.C.: Government Printing Office), p. 30.]

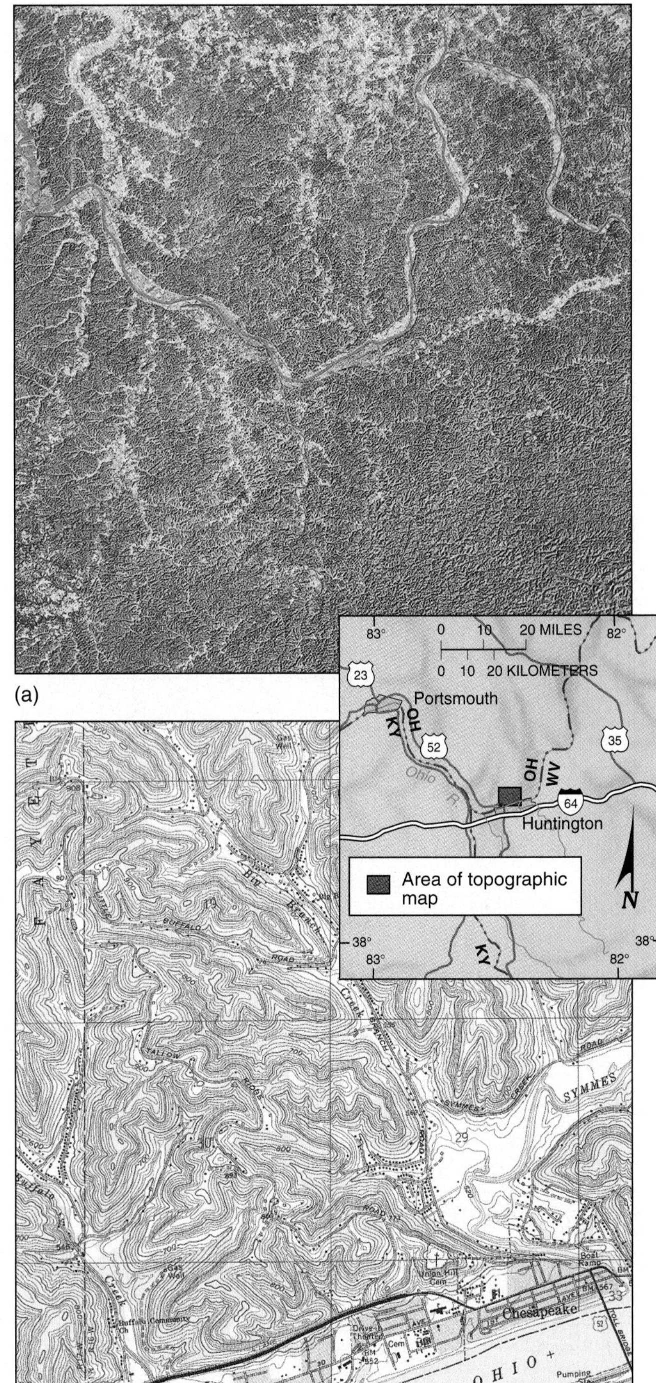

FIGURE 11.8 A stream-dissected landscape.
(a) A highly dissected topography (featuring a dendritic drainage pattern shown in Figure 11.9) around the junction of West Virginia, Ohio, and Kentucky. All of these borders are formed by rivers. (b) A portion of the USGS topographic map covering the area north of Huntington, West Virginia, reveals the intricate complexity of dissected landscapes. (The image and map are not at the same scale.) [(a) *Landsat* image from NASA; (b) Huntington Quadrangle, USGS.]

Drainage Patterns

The **drainage pattern** is an arrangement of channels that is determined by slope, differing rock resistance, climatic and hydrologic variability, relief of the land, and structural controls imposed by the landscape. Consequently, the drainage pattern of any land area on Earth is a remarkable visual summary of every characteristic—geologic and climatic—of that region.

The *Landsat* image in Figure 11.8 is of the Ohio River area near the junction of the West Virginia, Ohio, and Kentucky borders. The high-density drainage pattern and intricate dissection of the land occur because of the region's generally level sandstone, siltstone, and shale strata, which are easily eroded, given the humid mesothermal climate. Fluvial action, along with other denudation processes, is responsible for this dissected topography.

The seven most common drainage patterns encountered in nature are represented in Figure 11.9. The most familiar pattern is *dendritic* (a); this treelike pattern is similar to that of many natural systems, such as capillaries in the human circulatory system or the vein patterns in leaves. Energy expended by this drainage system is efficient because the overall length of the branches is minimized.

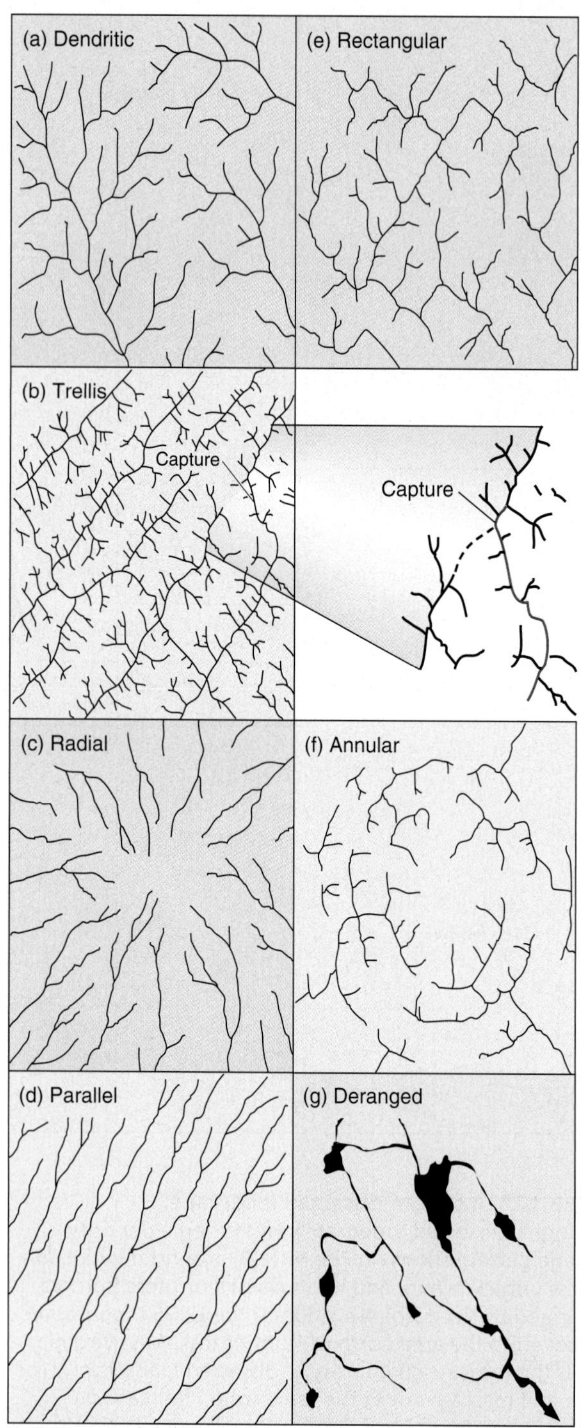

FIGURE 11.9 The seven most common drainage patterns. Each pattern is a visual summary of all the geologic and climatic conditions of its region. [After A. D. Howard, "Drainage Analysis in Geological Interpretation: A Summation," *Bulletin of American Association of Petroleum Geologists* 51, 1967, p. 2248. Adapted by permission.]

The *trellis* drainage pattern (b) is characteristic of dipping or folded topography, which exists in nearly parallel mountains of the Ridge and Valley Province of the East, where drainage patterns are influenced by rock structures of variable resistance and folded strata. Review Figure 9.18, which presents a satellite image of this region showing its distinctive drainage pattern. The principal streams are directed by the parallel folded structures, whereas smaller dendritic streams are at work on nearby slopes, joining the main streams at right angles.

The sketch beside Figure 11.9b suggests that the headward-eroding part of one stream could break through a drainage divide and *capture* the headwaters of another stream in the next valley, and indeed this does happen. The sharp bends in two of the streams in the illustration are called *elbows of capture* and are evidence that one stream has breached a drainage divide. This type of capture, or *stream piracy*, can occur in other drainage patterns.

The remaining drainage patterns in Figure 11.9 are caused by other specific structural conditions. A *radial* drainage pattern (c) results from streams flowing off a central peak or dome, such as occurs on a volcano. *Parallel drainage* (d) is associated with steep slopes and some relief. A *rectangular pattern* (e) is formed by a faulted and jointed landscape, directing stream courses in patterns of right-angle turns. *Annular* patterns (f) are produced by structural domes, with concentric patterns of rock strata guiding stream courses. In areas having disrupted surface patterns, such as the glaciated shield regions of Canada and northern Europe, a *deranged* pattern (g) is in evidence, with no clear geometry in the drainage and no true stream valley pattern.

Streamflow Characteristics

A mass of water positioned above base level in a stream has *potential energy*. As the water flows downstream under the influence of gravity, this energy becomes *kinetic energy*. The rate of this potential energy conversion depends on the steepness of the stream channel.

Stream channels vary in *width* and *depth*. The streams that flow in them vary in *velocity* and in the *sediment load* they carry. All these factors may increase with increasing discharge. **Discharge** is calculated by multiplying the velocity of the stream by its width and depth for a specific cross section of the channel as stated in the simple expression:

$$Q = wdv$$

where Q = discharge; w = channel width; d = channel depth; and v = stream velocity. As Q increases, some combination of channel width, depth, and stream velocity increases. Discharge is expressed either in cubic meters per second (m^3/s) or cubic feet per second (cfs).

Given the interplay of channel width, depth, and stream velocity, the cross section of a stream varies over time, especially during heavy floods. Figure 11.10 shows changes in the San Juan River channel in Utah that occurred during a flood. The increase in discharge increases the velocity and therefore the carrying capacity of the river as the flood progresses. As a result, the river's ability to scour materials from its bed is enhanced. Such scouring represents a powerful clearing action, especially in the excavation of alluvium.

You can see in the figure that the San Juan River's channel was deepest on October 14, when floodwaters were highest (blue outline). Then, as the discharge returned to normal, the kinetic energy of the river was reduced, and the bed again filled as new sediments were deposited. The process depicted in Figure 11.10 moved a depth of about 3 m (10 ft) of sediment from this cross section of the stream channel. A scouring experiment intended to provide new recreational areas and restore wildlife habitats in the Grand Canyon is in News Report 11.1.

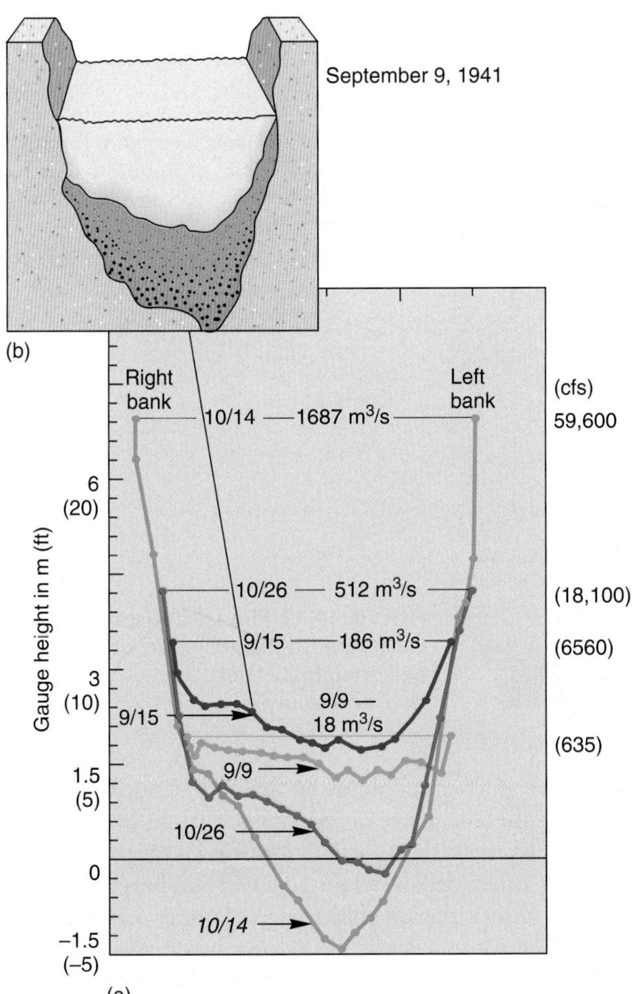

FIGURE 11.10 A flood affects a stream channel. (a) Stream channel cross sections showing the progress of a 1941 flood on the San Juan River near Bluff, Utah. (b) Detail of stream-channel profile on September 9, 1941. [After L. Leopold and T. Maddock, Jr., *The Hydraulic Geometry of Stream Channels and Some Physiographic Implications*, U.S. Geological Survey Professional Paper 252 (Washington, D.C.: Government Printing Office, 1953), p. 32.]

Stream Erosion

The erosional work of a stream carves and shapes the landscape through which it flows. Several types of erosional processes are operative. **Hydraulic action** is the work of water moving materials through the exertion of pressure and shearing force. Running water causes hydraulic squeeze-and-release action that loosens and lifts rocks. As this debris moves along, it mechanically erodes the streambed further through **abrasion**, the process whereby rock particles grind and carve the streambed.

The upstream tributaries in a drainage basin usually have small and irregular discharges, with most of the stream energy expended in turbulent eddies. As a result, hydraulic action in these upstream sections is at maximum, even though the coarse-textured load of such a stream is small. The downstream portions of a river, however, move much larger volumes of water past a given point and carry larger suspended loads of sediment. Thus, both volume and velocity are important determinants of the amount of energy expended in the erosion and transportation of sediment (Figure 11.11).

Stream Transport

You may have watched a river or creek after a rainfall, the water colored brown by the heavy sediment load being transported. The amount of material available to a stream is dependent on topographic relief, the nature of rock and materials through which the stream flows, climate, vegetation, and the types of processes at work in a drainage basin. *Competence*, which is a stream's ability to move particles of specific size, is a function of stream velocity. The total possible load that a stream can transport is its *capacity*. Four transportation processes erode materials: solution, suspension, saltation, and traction (Figure 11.12).

Solution refers to the **dissolved load** of a stream, especially the chemical solution derived from minerals such as limestone or from soluble salts. The main contributor of material in solution is chemical weathering. Sometimes the undesirable salt content that hinders human use of rivers comes from dissolved rock formations or from springs in the stream channel; as an example, the San Juan and Little Colorado Rivers that flow into the Colorado River near the Utah-Arizona border add dissolved salts to the system.

The **suspended load** consists of fine-grained, clastic particles (bits and pieces of rock) physically held aloft in the stream, with the finest particles held in suspension until the stream velocity slows to near zero. Turbulence in the water, with random upward motions, is an important mechanical factor in holding a load of sediment in suspension.

The **bed load** refers to those coarser materials that are dragged along the bed of the stream by **traction** or are rolled and bounced along by **saltation** (from the Latin *saltim*, which means "by leaps or jumps"). With increased kinetic energy, parts of the bed load are rafted up

(a)

(b)

FIGURE 11.11 Stream velocity and discharge increase together.
(a) A low-volume, low-velocity (but turbulent) mountain stream in the White Mountains, New Hampshire; and, (b) the high-discharge, high-velocity (but laminar, or smooth water flow) portion of the Ocmulgee River near Jacksonville, Georgia. [Photos by Bobbé Christopherson.]

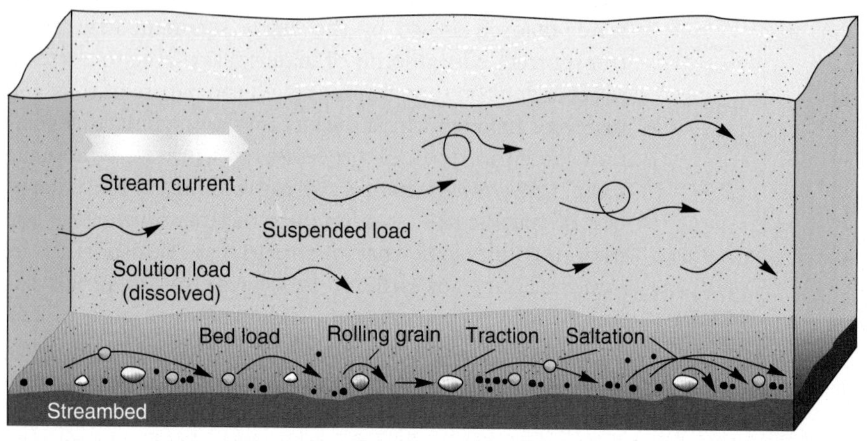

FIGURE 11.12 Fluvial transport.
Fluvial transportation of eroded materials through saltation, traction, suspension, and solution.

and become suspended load. This fact was demonstrated in the flood-induced channel deepening of the San Juan River shown in Figure 11.10 and the Grand Canyon scouring in News Report 11.1. Saltation is also a process in the transportation of materials by wind (see Chapter 12).

The first Spanish explorers to visit the Grand Canyon reported in their journals that they were kept awake at night by the thundering sound of boulders tumbling along the Colorado River bed. Such sounds today are substantially lessened because of the reduced velocity of the Colorado resulting from the many dams and control facilities that now trap sediments and reduce bed load capacity.

If the load in a stream exceeds its capacity, sediments accumulate as **aggradation** (the opposite of degradation) as the stream channel builds up through deposition. With excess sediment, a stream becomes a maze of interconnected channels that form a **braided stream** pattern. Braiding often occurs when reduced discharge affects a stream's transportation ability such as under seasonal conditions, or when a landslide occurs upstream, or where weak banks of sand or gravel exist, producing excess supplies of sediment. Locally, braiding also may result from a new sediment load, which frequently is associated with meltwaters, such as in the glacial materials, such as rock, sediment, and glacial flour (ground up rock), that exceed stream capacity in the 15-km (9.3-mi) stretch of the Brahmaputra River about 35 km (22 mi) south of Lhasa, Tibet, in south-central Asia (Figure 11.13).

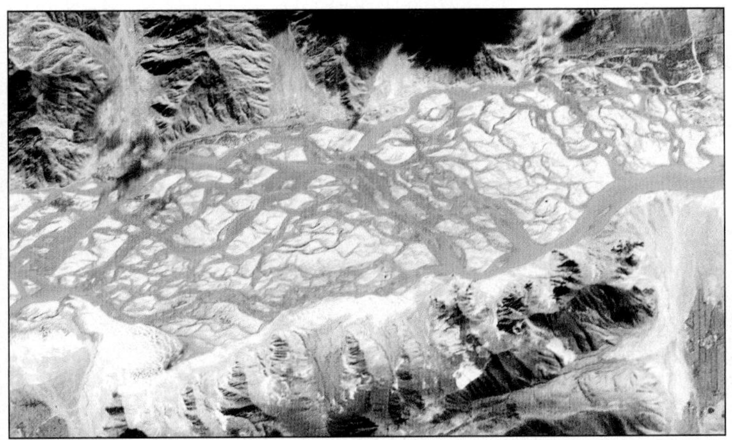

FIGURE 11.13 A braided stream.
A braided Brahmaputra River channel some 35 km (22 mi) south of Lhasa, Tibet, in a narrow valley south of the Tibetan Plateau. These streams reflect excessive sediment load associated with glacial meltwaters filled with fine sediments, or "glacial flour." [Photo by International Space Station astronaut courtesy of Earth Science and Image Analysis Lab, JSC/NASA.]

Flow and Channel Characteristics

ANIMATION **Stream processes, floodplains**
Oxbow lake formation

Flow characteristics of a stream are best seen in a cross-section view. The greatest velocities in a stream are near the surface at the center, corresponding with the deepest part of the stream channel (Figure 11.14). Velocities decrease closer to the sides and bottom of the channel because of the frictional drag on the water flow. In a curving stream, the maximum velocity line migrates from side to side along the channel, deflected by the curves.

Where slopes are gradual, stream channels assume a sinuous (snakelike) form weaving across the landscape. This action produces a **meandering stream**, from the Greek *maiandros*, after the ancient Maiandros River in Asia Minor (the present-day Menderes River in Turkey) which had a meandering channel pattern. The outer portion of each meandering curve is subject to the greatest erosive action and can be the site of a steep bank called an **undercut bank.** In contrast, the inner portion of a meander receives sediment fill, forming a deposit called a **point bar.**

As meanders develop, the scour-and-fill features gradually work their way downstream. As a result, the landscape near a meandering river bears meander scars of residual deposits from the previous river channels.

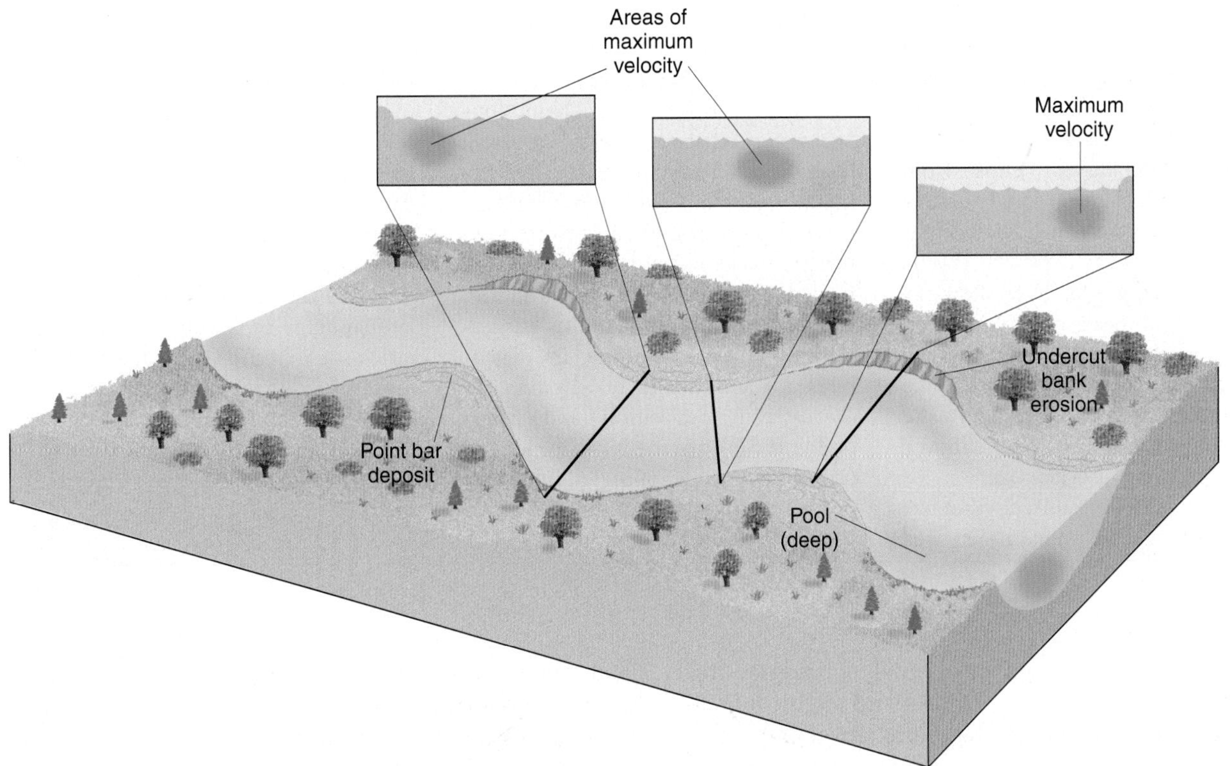

FIGURE 11.14 Meandering stream profile.
Aerial view and cross sections of a meandering stream, showing the location of maximum flow velocity, point-bar deposits, and areas of undercut bank erosion.

News Report 11.1

Scouring the Grand Canyon for New Beaches and Habitats

Glen Canyon Dam sealed the Colorado River gorge north of the Grand Canyon, near Lees Ferry and the Utah-Arizona border (see the map in Focus Study 12.1, Figure 1). Impoundment of water began in 1963, as did the inundation of many upstream canyons from the dam by the new Lake Powell. The lake began collecting the tremendous sediment load of the Colorado River, reducing the volume of water in the river below the dam and eliminating the natural seasonal fluctuations in river discharge. The production of hydroelectricity at Glen Canyon further affected the Grand Canyon with highly variable water releases keyed to electrical turbine operations. Water in the canyon rose and fell as much as 4.3 m (14 ft) as the distant lights and air conditioning of Las Vegas and Phoenix went on and off.

FIGURE 1 Glen Canyon Dam floods the Grand Canyon. Bypass tubes dramatically discharge water from Lake Powell in a 1996 experiment to restore the damaging effects of the dam on the Grand Canyon. Note the workers on the walkway for scale. [Photo by Tom Smart, *Deseret News*.]

All of these changes affected the canyon. Over the years, the beaches were starved for sand, channels filled with sediment, fisheries disrupted, and backwater channels depleted of nutrients. In 1996, an unprecedented experiment took place. The Grand Canyon was artificially flooded with a 1260 m^3/s (45,000 cfs) release from Glen Canyon (Figure 1). The flow lasted for seven days and then was reduced to 224 m^3/s (8000 cfs). Flows were decreased gradually to allow the newly formed beaches to drain, in contrast to the rapid opening of the pipes that initiated the flood. Lake Powell dropped 1 m (3.3 ft), and Lake Mead, downstream behind Hoover Dam, rose 0.8 m (2.6 ft).

The results were initially thought to be positive: 35% more beach area was created by the scouring of sand and sediments from the channel. Approximately 80% of the aggradation (building up of beaches) took place in the first 40 hours and was completed by the 100-hour mark. Numerous backwater channels were created, flush with fresh nutrients for the humpback chub and other endangered fish species. In contrast to these benefits were a few negatives in the form of existing ecosystem disruption. Present thinking is that this success was limited because of the lack of sediment to supply the floodwaters for downstream deposition and beach building. Some of the initial beach deposits quickly dissappeared. For more on this experiment, see http://water.usgs.gov/pubs/FS/FS-060-99/.

As of 2003, hydrologists were preparing for a second test, this time with a difference. Scientists are going to time the artifical flood with natural peak runoff that supplies fresh sediment to the system. Also, instead of seven days, the artificial flood releases from Glen Canyon will last only 2.5 days. It is thought that tributaries will deliver enough sediment to accomplish the task of rebuilding beaches downstream. This requires a test in late summer or autumn when tributaries such as the Paria River deliver more than three million tonnes of sediment to the Colorado River. (See Grand Canyon Monitoring and Research Center for progress reports at http://www.gcmrc.gov/.)

Former point-bar deposits leave low-lying ridges, creating a bar-and-swale relief (ridges and slight depressions). The photograph of the Itkillik River in Alaska, Figure 11.15a, shows both meanders and meander scars.

Figure 11.15b shows (in four stages) how a stream meander can form a cutoff. The stream erodes its outside bank as the curve migrates downstream (1); the neck of land created by the looping meander (2) eventually erodes through and forms a *cutoff* (3). When the former meander becomes isolated from the rest of the river, the resulting **oxbow lake** (4) may gradually fill with silt or may again become part of the river when it floods. The Mississippi River is many miles shorter today than it was in the 1830s because of artificial cutoffs that were dredged across these necks of land to improve navigation and safety.

(a)

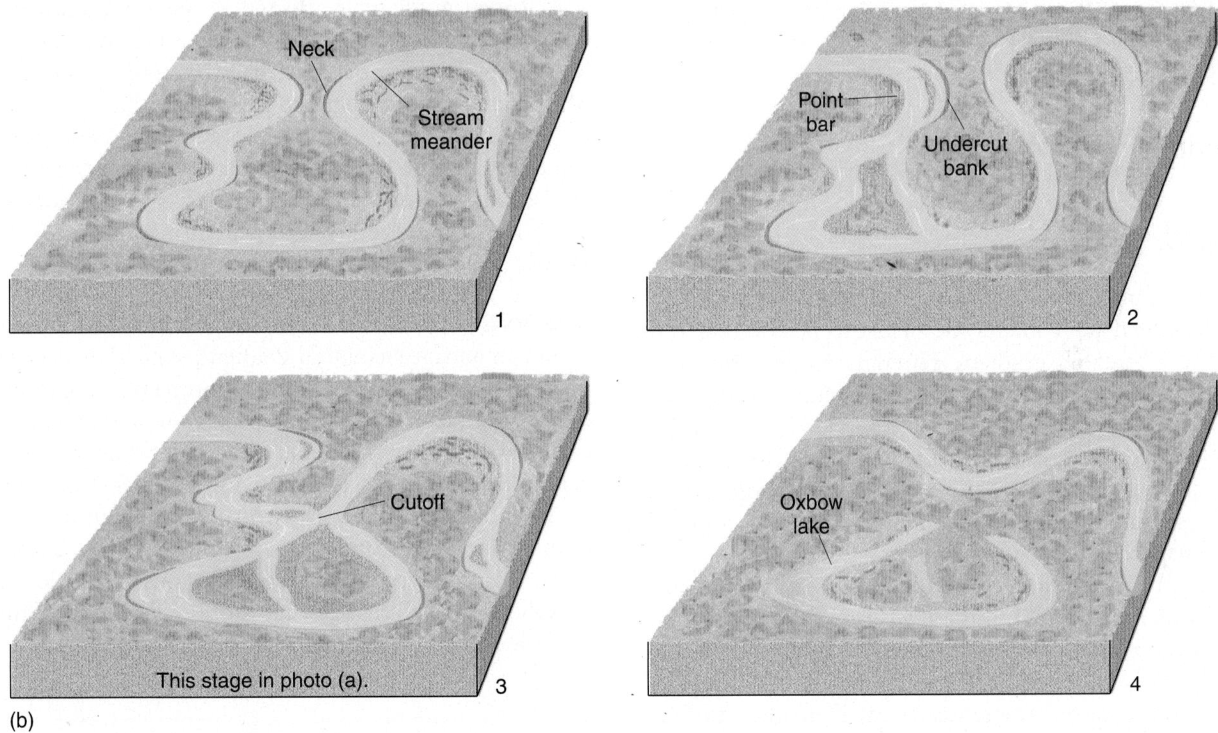

(b)

FIGURE 11.15 Meandering stream development.
(a) Itkillik River in Alaska. (b) Development of a river meander and oxbow lake simplified in four stages. [(a) U.S. Geological Survey photo.]

Streams often are used as natural political boundaries, as we saw with the Danube River earlier, but it is easy to see how disagreements might arise when boundaries are based on river channels that change course. For example, the Ohio, Missouri, and Mississippi Rivers can shift their positions quite rapidly during times of flood and, therefore, make confusing the boundaries based on them. Carter Lake, Iowa, provides us with a case in point (Figure 11.16). The Nebraska-Iowa border was originally placed midchannel in the Missouri River. In 1877, the river cut off the meander loop around the town of Carter Lake, leaving the town "captured" by Nebraska. The new oxbow lake was called Carter Lake and still is the state boundary.

Boundaries should always be fixed by surveys independent of river locations, because border disputes result when river channels change course. Such surveys have been completed along the Rio Grande near El Paso, Texas, and along the Colorado River between Arizona and California, permanently establishing political boundaries separate from changing river locations.

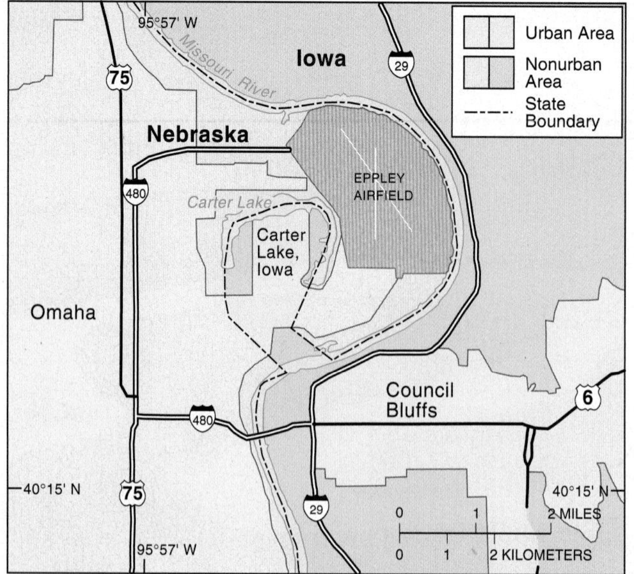

FIGURE 11.16 Rivers make poor political boundaries. Carter Lake, Iowa, sits within the curve of a former meander that was cut off by the Missouri River. The city and oxbow lake remain part of Iowa even though they are stranded within Nebraska.

Stream Gradient

NOTEBOOK Niagara Falls

Every stream has a degree of inclination or **gradient**, which is the decline in elevation from its headwaters to its mouth. A stream's gradient generally forms a concave-shaped slope (Figure 11.17). Characteristically, the *longitudinal profile* of a stream (a side view) has a steeper slope upstream and a more gradual slope downstream. This curve assumes its shape for complex reasons related to the stream's ability to do enough work to accomplish the transport of the load it receives.

A **graded stream** condition occurs when the load carried by the stream and the landscape through which it flows become mutually adjusted (balanced); it is in a state of *dynamic equilibrium* among erosion, transported load, deposition, and the stream's capacity. Dynamic equilibrium infers that the stream and landscape work together to maintain this balance.

Both high-gradient and low-gradient streams can achieve a graded condition. The difference in gradient is the result of variation in each stream's discharge and the nature of the transported load. A stream's profile tells geographers about characteristics of slope, discharge, and load.

> A graded stream is one in which, over a period of years, *slope* is delicately adjusted to provide, with available *discharge* and with prevailing *channel characteristics*, just the velocity required for transportation of the load supplied from the drainage basin.*

Attainment of a graded condition does not mean that the stream is at its lowest gradient, but rather that it represents a balance among erosion, transportation, and deposition over time along a specific portion of the stream.

One problem with applying the graded stream concept is that an individual stream can have both graded and ungraded portions. It may have graded sections without having an overall graded slope. In fact, variations and interruptions are the rule rather than the exception.

Stream gradient is affected by tectonic uplift of the landscape, or a change in base level. Such changes stimulate renewed erosional activity. If tectonic forces slowly lift the landscape, the stream gradient will increase. Imagine this happening to the Colorado River landscape in Figure 11.2. The region and stream flowing through the landscape become *rejuvenated.* With rejuvenation, river meanders actively return to downcutting and become *entrenched meanders* in the landscape (Figure 11.18).

Nickpoints. When the longitudinal profile of a stream shows an abrupt change in gradient, such as at a waterfall or an area of rapids, the point of interruption is termed a **nickpoint** (also spelled *knickpoint*). At a nickpoint, the conversion of potential energy to kinetic energy is concentrated and works to eliminate the nickpoint. Figure 11.19 shows a stream with two such interruptions. Nickpoints can result from a stream flowing across a zone of hard, resistant rock, or from various tectonic uplift episodes, such as might occur along a fault line. Temporary blockage in a channel, caused by a landslide or a log-

*J. H. Mackin, "Concept of the Graded River," *Geological Society of America Bulletin*, 59 (1948): 463.

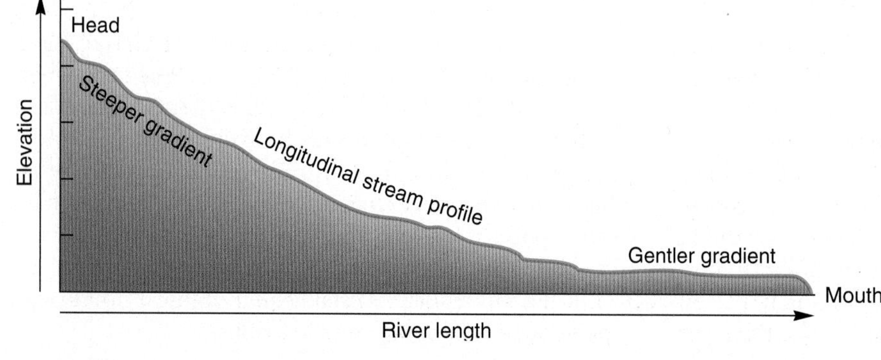

FIGURE 11.17 An ideal longitudinal profile. Idealized cross section of the longitudinal profile of a stream showing its gradient. Upstream segments have a steeper gradient; downstream the gradient is more gentle. The middle and lower portions in the illustration appear graded, or in dynamic equilibrium.

FIGURE 11.18 Entrenched meanders.
The San Juan River near Mexican Hat, Utah, cuts down into the uplifted Colorado Plateau landscape producing entrenched meanders. [(a) Photo by Betty Crowell; (b) photo by Randall M. Christopherson.]

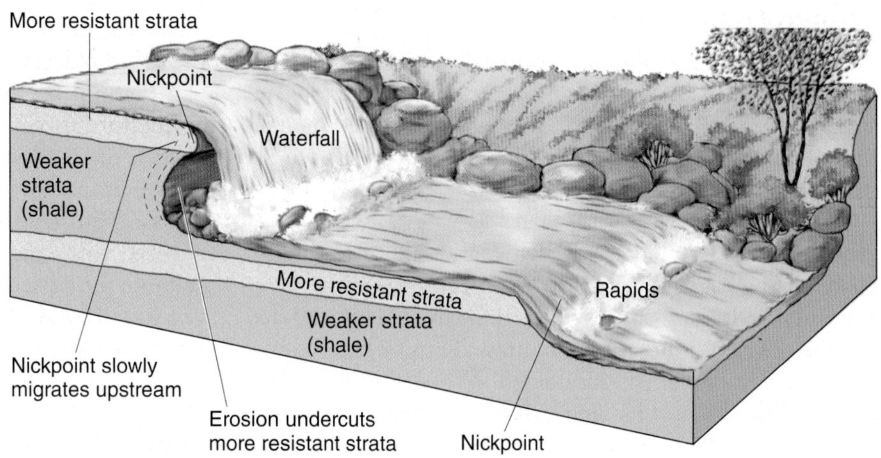

FIGURE 11.19 Nickpoints interrupt a stream profile.
Longitudinal stream profile showing nickpoints produced by resistant rock strata. Potential energy is converted into kinetic energy and concentrated at the nickpoint, accelerating erosion, which eventually eliminates the feature.

jam, also could be considered a nickpoint; when the logjam breaks, the stream quickly readjusts its channel to its former grade.

One of the more interesting and beautiful gradient breaks is a waterfall. At its edge, a stream becomes free falling, moving at high velocity and causing increased abrasion on the channel below. This action generally undercuts the waterfall, and eventually the rock ledge at the lip of the fall collapses, causing the waterfall to shift a bit farther upstream. Thus, nickpoints migrate upstream. The height of the waterfall is reduced gradually as debris accumulates at its base.

At Niagara Falls on the Ontario-New York border, glaciers advanced and receded over the region, exposing resistant rock strata underlain by less-resistant shales. As the less-resistant material continues to weather away, the overlying rock strata collapse, allowing the falls to erode farther upstream toward Lake Erie (Figure 11.20).

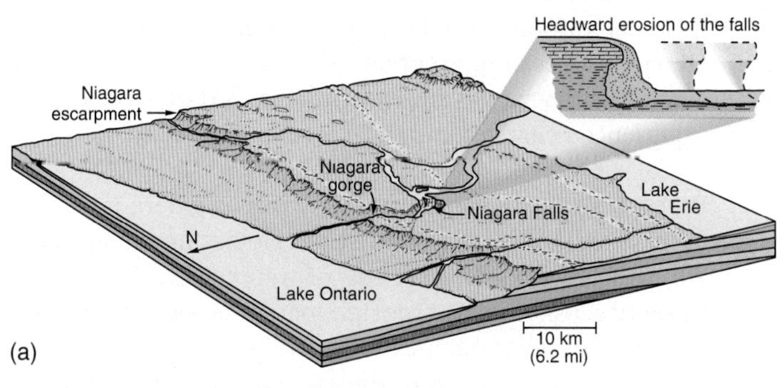

FIGURE 11.20 Retreat of Niagara Falls.
(a) Headward retreat of Niagara Falls from the Niagara escarpment. The falls retreat at about 1.3 m (4.3 ft) per year. (b) Niagara Falls, with the American Falls portion almost completely shut off by upstream controls for inspection. Horseshoe Falls in the background still is flowing. Such inspections allow engineers to assess the progress of natural processes that are working to eliminate the Niagara Falls nickpoint. [(a) After W. K. Hamblin, *Earth's Dynamic Systems*, 6th ed., (Upper Saddle River, N.J.: Macmillan Publishing, an imprint of Prentice Hall, Inc. © 1992), Fig. 12.15, p. 246. Used by permission. (b) Photo courtesy of the New York Power Authority.]

Niagara Falls is a place where natural processes labor to eliminate a nickpoint, reducing this portion of the river to a series of mere rapids. In fact, the falls have retreated more than 11 km (6.8 mi) from the steep face of the Niagara escarpment (cliff) during the past 12,000 years. In the past engineers have used control facilities upstream to reduce flows over the American Falls at Niagara for inspection of the cliff (Figure 11.20b). A nickpoint is a mobile feature on the landscape. As this example demonstrates, a nickpoint should be thought of as a relatively temporary feature.

Stream Deposition

Deposition is the next logical event after weathering, mass movement, erosion, and transportation. In *deposition*, a stream deposits alluvium, or unconsolidated sediments, thereby creating specific depositional landforms, such as floodplains, terraces, and deltas.

Floodplains. The flat, low-lying area along a stream channel that is subjected to recurrent flooding is a **floodplain**. It is formed when the river overflows its channel during times of high flow. Thus, when floods occur, the floodplain is inundated. When the water recedes, alluvial deposits generally mask the underlying rock. Figure 11.21 illustrates a characteristic floodplain, with the present river channel embedded in the plain's alluvial deposits.

On either bank of most streams, **natural levees** develop as by-products of flooding. When flood waters arrive, the river overflows its banks, loses velocity as it spreads out, and drops a portion of its sediment load to form the levees. Larger sand-sized particles drop out first, forming the principal component of the levees, with finer silts and clays deposited farther from the river. Successive floods increase the height of the levees (*levée* is French for "raising"). The levees may grow in height until the river channel becomes elevated, or *perched* above the surrounding floodplain.

Notice on Figure 11.21 an area labeled *backswamp* and a stream called a *yazoo tributary*. The natural levees and elevated channel of the river prevent this tributary from joining the main channel, so it flows parallel to the river and through the backswamp area. (The name comes from the Yazoo River in the southern part of the Mississippi floodplain.) On the topographic map (Figure 11.21b), you can see the natural levees represented by several contour lines that run immediately adjacent to the Tallahatchie River. These contour lines (5-ft interval) denote a height of 10–15 ft (3–4.5 m) above the river and the adjoining floodplain. Next time you have an opportunity to see a river and its floodplain, look for levees (they may be low and subtle).

People build cities on floodplains because they are nearly level and next to the water, despite the threat of flooding. Government assurances of artificial protection from floods or of disaster assistance often encourage people to settle there. Government assistance may include building artificial levees on top of natural levees. Artificial levees do increase the capacity in the channel, but they also lead to even greater floods when they are topped or when they fail. The catastrophic floods along the Mississippi River and its tributaries in 1993 illustrated the risk of floodplain settlement. Total damage from these floods exceeded $30 billion (News Report 11.2). In 2001 some two-thirds of disaster losses were attributable to floods. Tropical storm Allison left $6 billion in damage in its wandering visit to Texas in June 2001; most damage occurred from flooding along occupied floodplains. The 2002 floods in Europe (see chapter-opening photo) produced estimated losses exceeding US $50 billion, although the degree of damage is still unfolding in 2003.

Perhaps the best use of some floodplains is for crop agriculture, because inundation generally delivers nutrients to the land with each new alluvial deposit. A significant example is the Nile River in Egypt, where annual flooding enriches the soil. However, floodplains that are covered with coarse sediment—sand and gravel—are less suitable for agriculture. Are there river floodplains where you live? If so, how would you assess present land-use patterns, people's hazard perception, and local planning and zoning?

Stream Terraces. As explained earlier, several factors may rejuvenate stream energy and stream-landscape relationships so that a stream again scours downward with increased erosion. The resulting entrenchment of the river into its own floodplain produces **alluvial terraces** on both sides of the valley, which look like topographic steps above the river. Alluvial terraces generally appear paired at similar elevations on the sides of the valley (Figure 11.22). If more than one set of paired terraces is present, the valley probably has undergone more than one episode of rejuvenation.

If the terraces on the sides of the valley do not match in elevation, then entrenchment actions must have been continuous as the river meandered from side to side, with each meander cutting a terrace slightly lower in elevation. Thus, alluvial terraces represent an original depositional feature, a floodplain, which is subsequently eroded by a stream that experienced changes in stream load and capacity.

River Deltas. The mouth of a river is where it reaches its base level. The river's forward velocity rapidly decelerates as it enters a larger body of water, with the reduced velocity causing its transported load to exceed its capacity. Coarse sediments such as sand and gravel drop out first, with finer clays carried to the extreme end of the deposit. The depositional plain formed at the mouth of a river is called a **delta**, named after the Greek letter *delta*: Δ, the triangular shape which was perceived by Herodotus in ancient times to be similar to the shape of the Nile River delta.

Each flood stage deposits a new layer of alluvium over the surface of the delta so that it grows outward. At the same

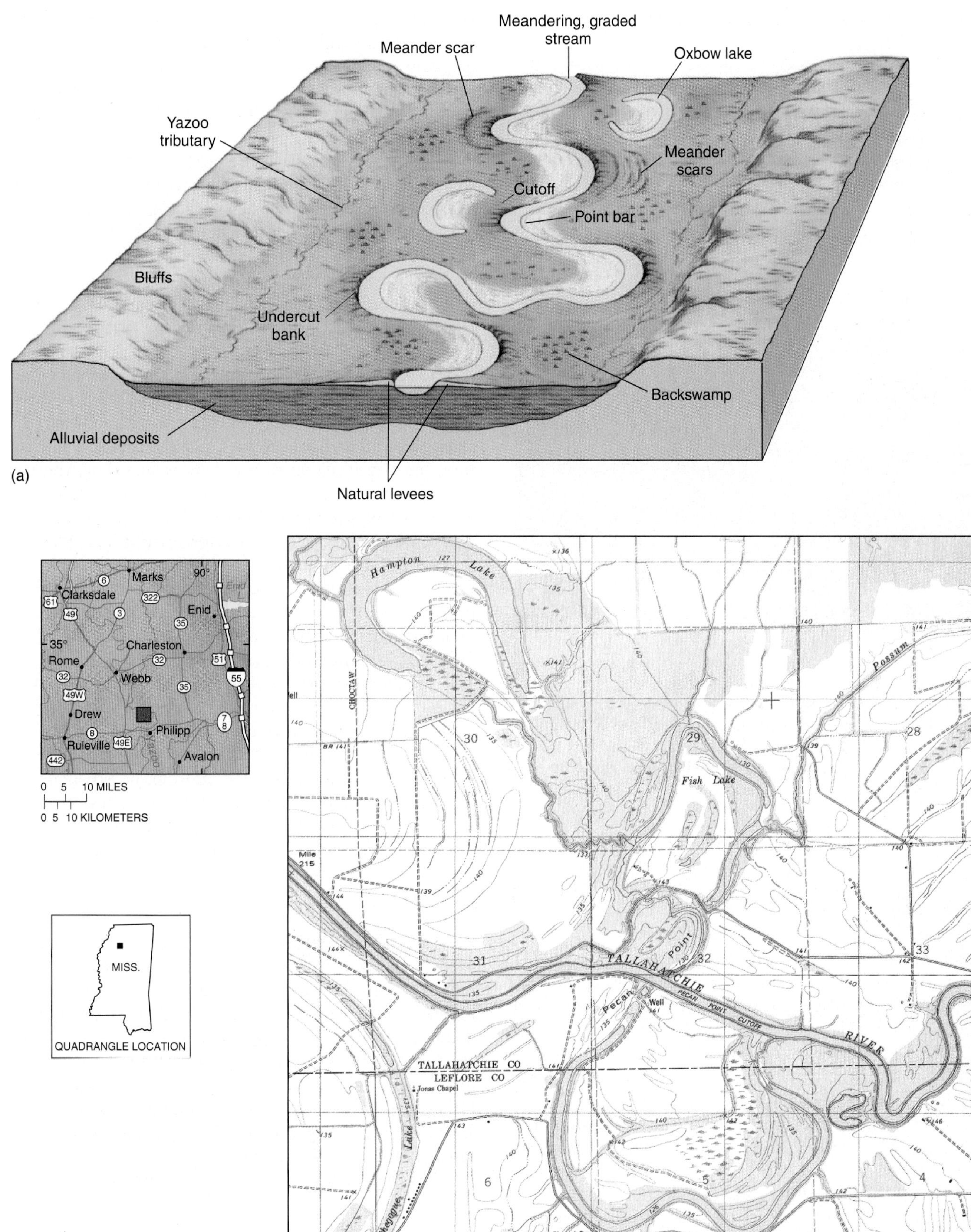

FIGURE 11.21 A floodplain.
(a) Typical floodplain landscape and related landscape features. (b) A portion of the Philipp, Mississippi, topographic map quadrangle. [(b) Topographic map from USGS.]

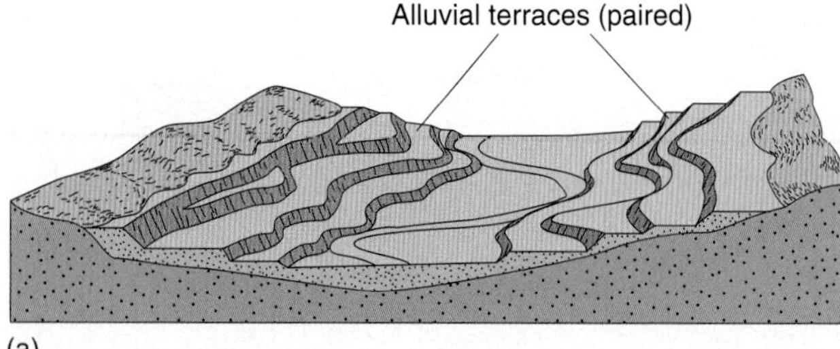

(a)

(b)

FIGURE 11.22 Alluvial stream terraces.
(a) Alluvial terraces are formed as a stream cuts into a valley. (b) Alluvial terraces along the Rakaia River, in New Zealand. [(a) After W. M. Davis, *Geographical Essays* (New York: Dover, 1964 [1909]), p. 515; (b) photo by Bill Bachman/Photo Researchers, Inc.]

time river channels divide into smaller channels known as *distributaries*, which appear as a reverse of the dendritic drainage pattern discussed earlier. The Ganges River delta features an intricate braided pattern of distributaries. Alluvium carried from deforested slopes upstream is the excess sediment that forms the many deltaic islands (Figure 11.23).

The Nile River forms an *arcuate* (arc-shaped) *delta* (Figure 11.24). Also arcuate in form is the Danube River delta in Romania as it enters the Black Sea (see Figure 11.6), and the Ganges and Indus River deltas. In another distinct form, the Seine River in France and the Tiber River in Italy have *estuarian deltas*, or one that is in the process of filling an **estuary**, which is the seaward mouth of a river where the river's freshwater encounters seawater.

Mississippi River Delta. The Mississippi River has a *bird-foot delta*, a long channel with many distributaries and sediments carried beyond the tip of the delta into the Gulf of Mexico. Over the past 120 million years the Mississippi deposited sediments downstream all the way from southern Minnesota and Illinois. During the past 5000 years, the river formed seven distinct deltaic complexes along the Louisiana coast. The seventh and current subdelta has been building for at least 500 years. Each lobe reflects distinct course changes in the Mississippi River, probably where the river broke through its natural levees. In 1966, Kolb and Lopik, two engineering geologists, prepared a map of this recent deltaic history, which shows the relatively smaller size and difference in configuration of the present delta (Figure 11.25a).

The Mississippi River delta clearly is dynamic over time as sediments accumulate on the floor of the Gulf of Mexico and the distributaries shift (Figure 11.25c). The main channel persists because of much effort and expense directed at maintaining the artificial levee system. The 3.25 million km^2 (1.25 million mi^2) Mississippi drainage basin produces enough sediment to extend the Louisiana coast 90 m (295 ft) a year—550 million metric tonnes a year.

To further complicate this situation, the tremendous weight of the sediments in the Mississippi River is creating isostatic adjustments in Earth's crust. In addition, the removal of oil and gas reserves deflated the land. Therefore, the entire region of the delta is subsiding, thereby placing ever increasing stress on natural and artificial levees and other structures along the lower Mississippi. Severe problems are a certainty for existing and planned settlements unless further intervention or relocation efforts take place. Past protection and reclamation efforts by the U.S. Army Corps of Engineers apparently have only worsened the flood peril, as demonstrated by the 1993 floods in the Midwest.

An additional problem for the lower Mississippi Valley is the possibility, in a worst-case flood, that the river could break from its existing channel and seek a new route to the Gulf of Mexico. An obvious alternative is the Atchafalaya River, now blocked off from the Mississippi at 320 km (200 mi) from its mouth. It would be less than half the distance to the Gulf (see arrow in Figure 11.25b, inset). Occurrence of a major flood is only a matter of time, and residents should prepare for the river to change channel.

Rivers Without Deltas. The Amazon River, which exceeds 180,000 m^3/s (6.35 million cfs) discharge and carries sediments far into the Atlantic, lacks a true delta. Its mouth, 160 km (100 mi) wide, formed an underwater deltaic plain deposited on a sloping continental shelf. As a result, the Amazon's mouth is braided into a broad maze of islands and channels (see satellite image in Figure 11.1). Other rivers lack deltaic formations if they lack significant sediment, or if they discharge into strong erosive currents. The Columbia River of the U.S. Northwest is one that lacks a delta because of offshore currents.

Floods and River Management

Throughout history, civilizations have settled floodplains and deltas, especially since the agricultural revolution of 10,000 years ago when the fertility of floodplain soils was discovered. Early villages generally were built away from the area of flooding, or on stream terraces, because the

FIGURE 11.23 The Ganges River enters the Bay of Bengal.
The complex distributary pattern in the "many mouths" of the Ganges River delta in Bangladesh and extreme eastern India from the *Terra* satellite. [*Terra* MODIS sensor image courtesy of MODIS Land Team, NASA.]

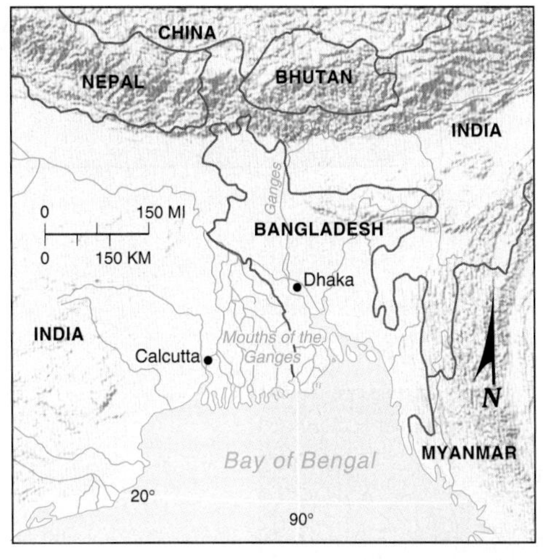

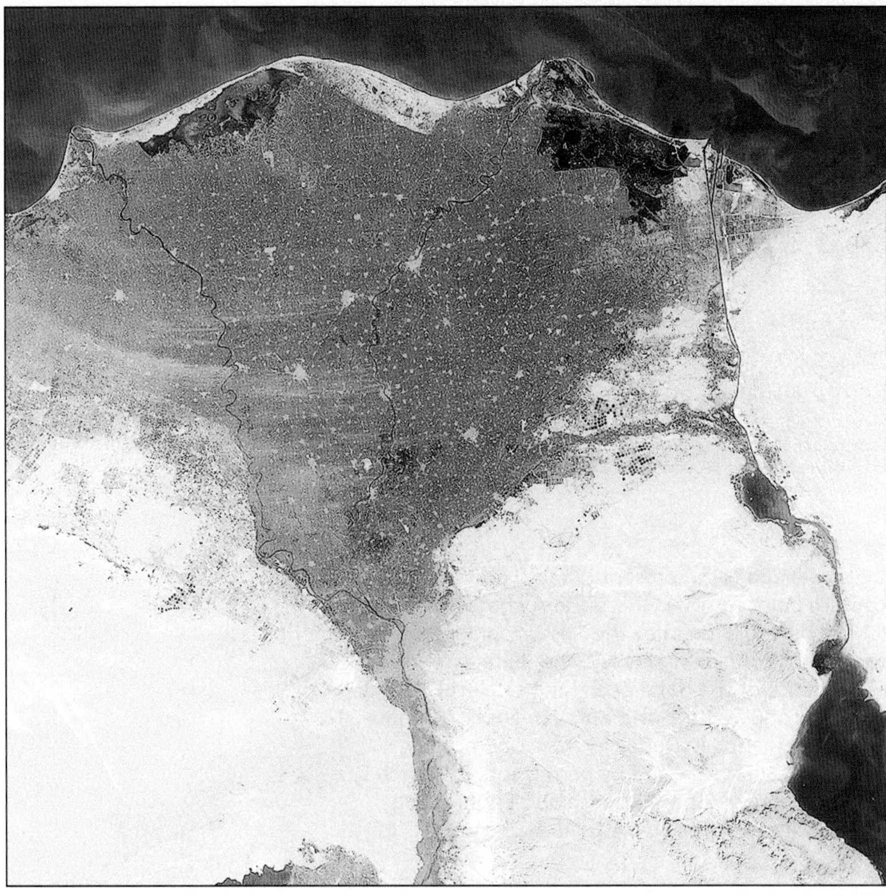

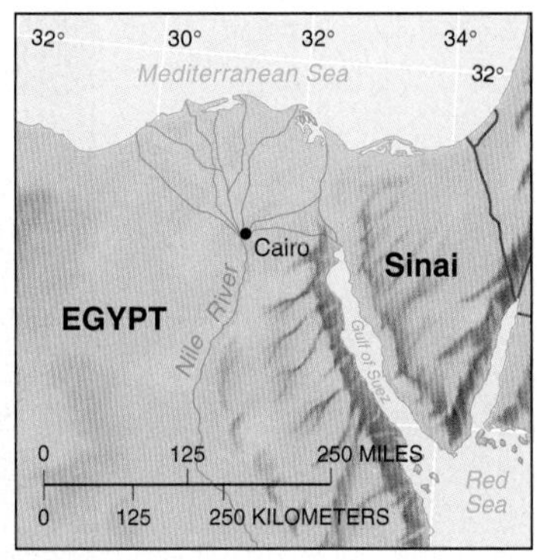

FIGURE 11.24 The Nile River delta—an arcuate delta.
The arcuate Nile River delta. Intensive agricultural activity and small settlements are visible on the delta and along the Nile River floodplain in this true color image. Cairo is at the apex of the delta. You can see the two main distributaries: Damietta to the east and Rosetta to the west. Because of the disruption of sediment flow into the delta, the Nile Delta is receding from the coast at an alarming 50 to 100 m (165 to 330 ft) per year. [January 30, 2001, *Terra* MISR sensor image courtesy of MISR Team, NASA/GSFC/JPL.]

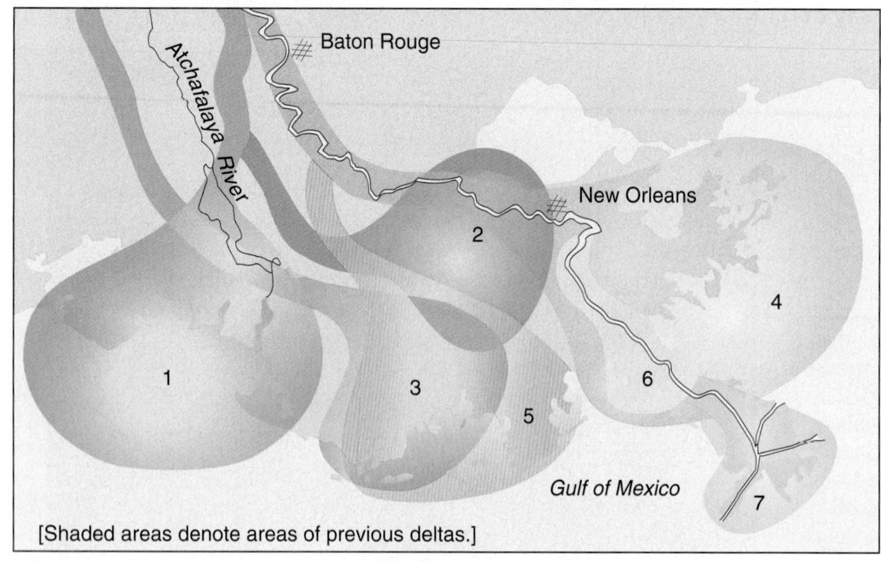

(a)

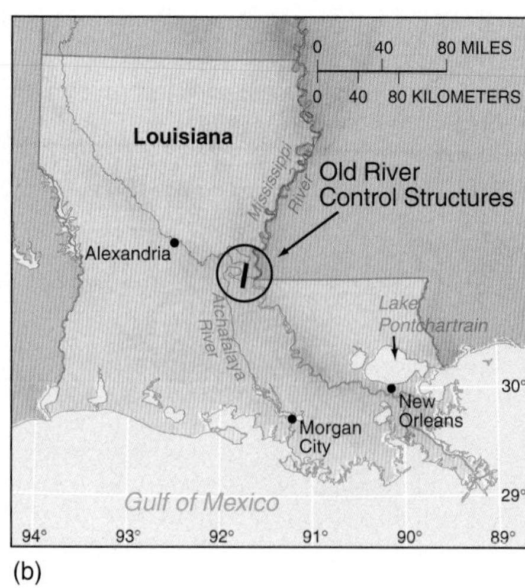

(b)

(c)

FIGURE 11.25 The Mississippi River delta.
(a) Evolution of the present delta, from 5000 years ago (1) to present (7). (b) Location map of the Old Control Structures and potential capture point (arrow) where the Atchafalaya River may one day divert the present channel. (c) The bird-foot delta of the Mississippi River receives a continuous supply of sediments, focused by controlling levees. The delta extends ever farther into the Gulf of Mexico, although subsidence of the delta and rising sea level have diminished the overall surface area. [(a) Adapted from C. R. Kolb and J. R. Van Lopik, "Depositional environments of the Mississippi River deltaic plain," in *Deltas in Their Geologic Framework* (Houston: Houston Geological Society, 1966). Adapted by permission. (c) March 5, 2001, *Terra* MODIS sensor image courtesy of Liam Gumley, Space Science and Engineering Center, University of Wisconsin, and the MODIS Science Team, NASA.]

News Report 11.2

The 1993 Midwest Floods

The Mississippi and Missouri Rivers are no strangers to flooding. The damage a flood might cause increases as more people settle on the vulnerable floodplains. The widespread floods of 1993 in the upper Mississippi and lower Missouri River basins exceeded peak discharge records at nearly 100 gaging (stream-measuring) stations, making it one of the greatest floods in U.S. history. Despite the passage of more than 10 years, these floods are still the topic of much discussion.

In spring 1993, a series of low-pressure weather systems (with their counterclockwise winds) stalled in the West, and high pressure (with its clockwise winds) dominated the Eastern Seaboard. These systems combined to produce a region of sustained convergence, instability, and thundershowers over the Midwest. Precipitation was 150% to 200% above normal for most cities in the flooded area (Figure 1a). By late June, many reservoirs were filled, and soils were saturated by record rainfall. Then came July, when some stations received 75 cm (30 in.) of precipitation in the first 3 weeks! Three remarkable aspects of "The Great Flood of 1993" were that the flood stage along some rivers was sustained for over a month, flood crests set new historical records, and many locations experienced multiple crests.

As 10,000 km (6200 mi) of levees were overtopped, more than 1000 levee breaches (broken through) happened. Water reached almost 10 m (32 ft) above flood stage in some areas, flooding many cities and towns, and 6 million hectares (14 million acres) of forest and farmland in the Missouri and Mississippi river drainage basins (Figure 1b). The result was a Presidential Disaster Declaration for 487 counties in Illinois, Iowa, Kansas, Minnesota, Missouri, the Dakotas, Nebraska, and Wisconsin. Overall, damage estimates top $30 billion.

Some cities had prepared for the disaster through improved levee construction and hazard zoning of susceptible floodplains. Many others had postponed raising local taxes for such action and had done little to prevent flood damage. The unevenness between areas created political friction. This flood event was a painful reminder of the power of nature in our lives and the need for improved hazard perception, preparation, and avoidance strategies. For imagery of these floods, see http://rsd.gsfc.nasa.gov/rsd/images/Flood_cp.html, and for a brief paper, see http://www.nwrfc.noaa.gov/floods/papers/oh_2/great.htm.

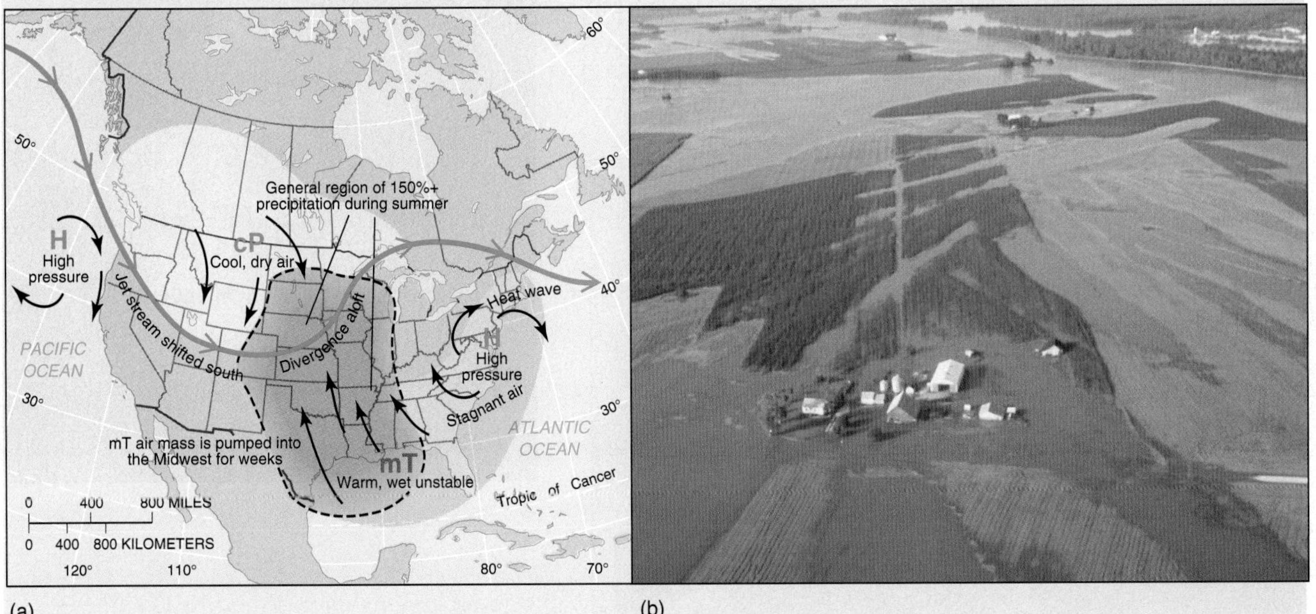

FIGURE 1 The great flood of 1993.
(a) Weather pattern summarized for the summer of 1993. Atmospheric conditions produced a steady flow of moisture-rich, unstable, maritime tropical air from the Gulf of Mexico for a prolonged period of time. (b) A farm 32 km (20 mi) south of Des Moines, Iowa, is inundated during the Midwest floods of 1993. [Photo by Les Stone/Sygma.]

floodplain was dedicated exclusively to farming. However, as commerce grew, sites near rivers became important for transportation. Port and dock facilities were built, as were river bridges. Also, because water is a basic industrial raw material used for cooling and for diluting and removing wastes, waterside industrial sites became desirable. However, all these human activities on vulnerable flood-prone lands require planning, zoning, and restrictions to avoid disaster. Unfortunately, this hazard perception is not a general rule.

Floods and Floodplains

A **flood** is a high-water level that overflows the natural (or artificial) levees along any portion of a stream. Understanding flood patterns of a drainage basin is as complex as understanding the weather, for floods and weather are equally variable, and both include a level of unpredictability. The abuse and misuse of river floodplains brought catastrophe to North Carolina in 1999. In short succession during September and October, Hurricanes Dennis, Floyd, and Irene delivered several feet of precipitation to the state, each storm falling on already saturated ground. Some 50,000 people were left homeless and at least 50 died; over 4000 homes were lost and an equal amount badly damaged (Figure 11.26a). The dollar estimate for the ongoing disaster through 1999 approached $10 billion. However, the real tragedy will be unfolding for years to come.

Hogs, in factory farms, outnumber humans in North Carolina. More than 10 million hogs, each producing two tons of waste per year, were located in about 3000 agricultural factories. These generally unregulated operations

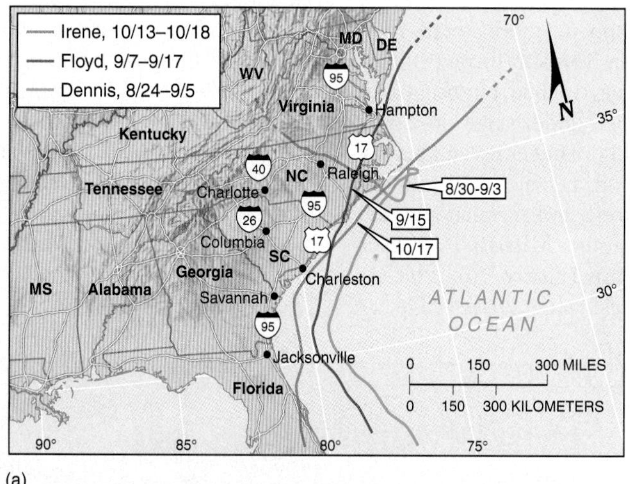

(a)

(b)

(b)

FIGURE 11.26 North Carolina 1999 floodplain disaster.
(a) Three hurricanes deluged North Carolina with several feet of rain during September and October 1999, Hurricane Floyd being the worst. (b) Hundreds of thousands of livestock were killed in the floods, which also washed out hundreds of animal sewage lagoons into wetlands, streams, and the ocean. Many of these factory farms and lagoons were sited on floodplains. (c) Hogs fight to stay alive as the Neuse River floodwaters rise around a hog farm a few miles from New Bern, North Carolina. Those animals too tired to swim drown in the polluted water. [Photos by (b) Mel Nathanson and (c) Chris Seward, both from the *Raleigh News & Observer*.]

collect millions of tons of manure into open lagoons, many set on river floodplains. The hurricane downpour flushed out these waste lagoons, spewing hundreds of millions of gallons of untreated sewage into wetlands, streams, and eventually Pamlico Sound and the ocean—a spreading "dead zone." Added to this waste were hundreds of thousands of hog, poultry, and other livestock carcasses, industrial toxins, floodplain junkyard oil, and municipal waste, creating an environmental catastrophe (Figure 11.26b). Final assessment of overall damage will take a decade, if not longer.

Catastrophic floods continue to be a threat in poor nations. Tropical storms off the Bay of Bengal devastated India in 1999—river floodplains and coastal lowlands were most vulnerable to destruction. In Bangladesh, intense monsoonal rains and tropical cyclones in 1988 and 1991 created devastating floods over the country's vast alluvial plain (130,000 km^2 or 50,000 mi^2). One of the most densely populated countries on Earth, Bangladesh was more than three-fourths covered by floodwaters.

Excessive forest harvesting in the upstream portions of the Ganges-Brahmaputra River watersheds increased runoff and added to the severity of the flooding. Over time the increased load carried by the river was deposited in the Bay of Bengal, creating new islands (see Figure 11.23). These islands, barely above sea level, became sites of new settlements and farming villages. When the floodwaters finally did recede, the lack of freshwater—coupled with crop failures, disease, and pestilence—led to famine and the death of tens of thousands of people. About 30 million people were left homeless, and many of the alluvial islands disappeared. (For information on worldwide floods, see **http://www.dartmouth.edu/artsci/geog/floods/** or a daily flood summary at **http://www.nws.noaa.gov/oh/hod/handbook_products.htm**.)

Floods and floodplains are rated statistically for the expected intervals between floods. A *10-year flood* is expected to occur once every 10 years (that is, it has a 10% probability of occurring in any one year). Such a frequency labels a floodplain as one of moderate threat. A 50-year or 100-year flood is of greater and perhaps catastrophic consequence, but it is also less likely to occur in a given year. These statistical estimates are probabilities that events will occur randomly during any single year of the specified period. Of course, two decades might pass without a 10-year flood, or 10-year flood volumes could occur 3 years in a row. The record-breaking Mississippi River Valley floods in 1993, or the North Carolina floods of 1999 easily exceeded a 1000-year flood probability of occurrence. Focus Study 11.1 presents a closer look at floodplain strategies.

Focus Study 11.1

Floodplain Strategies

Detailed measurements of streamflows and flood records have been kept rigorously in the United States only for about 100 years, in particular since the 1940s. At any selected location along any given stream, the *probable maximum flood* (PMF) is a hypothetical flood of such a magnitude that there is virtually no possibility it will be exceeded. Because the collection and concentration of rainfall produces floods, hydrologists speak of a corollary, the *probable maximum precipitation* (PMP) for a given drainage basin, which is an amount of rainfall so great that it will never be exceeded.

These parameters are used by hydrologic engineers to establish a *design flood* against which to take protective measures. For urban areas near creeks, planning maps often include survey lines for a 50-year or a 100-year floodplain; such maps have been completed for most U.S. urban areas. The design flood usually is used to enforce planning restrictions and special insurance requirements. Restrictive zoning using these floodplain designations is an effective way of avoiding potential damage. (The floodplain managers' organization at **http://www.floods.org/** is a respected source of information.)

Unfortunately, such political action is not generally implemented, and the scenario all too often goes like this: (1) Minimal zoning precautions are not carefully supervised; (2) a flooding disaster occurs; (3) the public is outraged at being caught off guard; (4) businesses and homeowners are surprisingly resistant to stricter laws and enforcement; (5) eventually another flood refreshes the memory and promotes more knee-jerk planning. As strange as it seems, *there is little indication that our risk perception improves as the risk increases.*

A Few Planning Strategies

A planning strategy used in some large river systems is to develop artificial floodplains, by constructing *bypass channels* to accept seasonal or occasional floods. When not flooded, the bypass channel can serve as farmland, often benefiting from occasional soil-replenishing inundations. When the river reaches flood stage, large gates called weirs are opened, allowing the water to enter the bypass channel. This alternate route relieves the main channel of the burden of carrying the entire discharge.

Dams (an artificial structure placed in a river channel) and *reservoirs* (an impoundment of water behind a dam; a human-made lake) are common streamflow control methods within a watershed. For conservation purposes, a dam holds back seasonal peak flows for distribution during low-water periods. In this way, streamflows are regulated to assure year-round water supplies. Dams also are constructed for flood control, to hold back excess flows for later release at more moderate discharge levels. Adding hydroelectric power production to these

(continued)

Focus Study 11.1 *(continued)*

functions of conservation and flood control can define a modern multipurpose reclamation project.

The function of reservoir impoundment is to provide flexible storage capacity within a watershed to regulate river flows, especially in a region with variable precipitation. Figure 1 shows one reservoir during drought conditions and during a time of wetter weather 4 years later.

(a)

(b)

FIGURE 1 Reservoir extremes.
Four years separate these comparative photographs of the New Hogan reservoir, central California, during (a) dry and (b) wet weather conditions. During intense storms in 1996, this reservoir came very close to spilling water over the top of the dam. [Photos by author.]

Reservoir Considerations

Unfortunately, the multipurpose benefits of reservoir construction are countered by some negative consequences. The area upstream from a dam becomes permanently drowned. In mountainous regions, this may mean loss of white water rapids and recreational sections of a river. In agricultural areas, the ironic end result may be that a hectare of farmland is inundated upstream to preserve a hectare of farmland downstream.

Furthermore, dams built in warm and arid climates lose substantial water to evaporation, compared with the free-flowing streams they replace. Reservoirs in the southwestern United States can lose 3–4 meters (10–13 ft) of water a year. Also, sedimentation can reduce the effective capacity of a reservoir and can shorten a dam's life span, as mentioned earlier regarding Glen Canyon Dam.

Vast environmental disruption for the sake of economic gain no longer appears popular with the public. The James Bay Project in central and northern Québec is a case in point. Launched in 1970 by Hydro-Québec and only one-third complete at this time, the project might eventually include 215 dams, 25 power stations, and 20 river diversions. Many unexpected environmental problems have arisen because no environmental impact studies were completed at the outset. The early stages remain a huge experiment with fragile ecosystems. Well into the planning phase, the public learned that corporate interests sought inexpensive, publicly generated electric power, and that much of it was for export to the United States—all at public expense.

The second major phase of the James Bay Project, the Great Whale project (named for a local river), may never be completed because of the success of conservation programs begun by utilities in the northeastern United States and court challenges to assess impacts before further construction. In addition, the state of New York in 1992 withdrew its offer to buy 1000 megawatts of power from the Hydro-Québec project. For more on this project from a local perspective see the First Nations 1999 resolution at **http://www.afn.ca/resolutions/1990/sca/res14.htm**.

A Final Thought about Floods

The benefit of any levee, bypass, or other project intended to prevent flood destruction is measured in avoided damage and is used to justify the cost of the protection facility. Thus, ever increasing damage leads to the justification of ever increasing flood control structures. All such strategies are subjected to cost-benefit analysis, but bias is a serious drawback because such an analysis usually is prepared by an agency or bureau with a vested interest in building more flood control projects.

As suggested in a half-century old article titled "Settlement Control Beats Flood Control,"* there are

(continued)

Focus Study 11.1 ***(continued)***

other ways to protect populations than with enormous, expensive, sometimes environmentally disruptive projects. Strictly zoning the floodplain is one approach. However, the flat, easily developed floodplains near pleasant rivers are desirable for housing, and thus weaken political resolve. A reasoned zoning strategy would set aside the floodplain for farming or passive recreation, such as a riverine park, golf course, or plant and wildlife sanctuary, or for other uses that are not hurt by natural floods. This study concluded that "urban and industrial losses would be largely obviated [avoided] by set-back levees and zoning and thus cancel the biggest share of the assessed benefits which justify big dams."

*Walter Kollmorgen, *Economic Geography* 29, no. 3 (July 1953): 215.

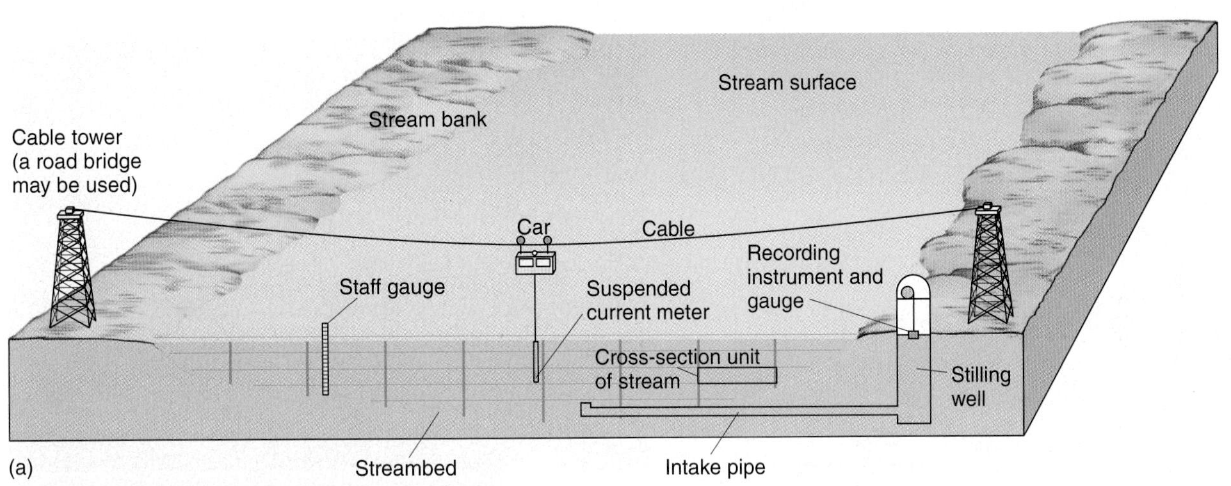

FIGURE 11.27 Streamflow measurement.
(a) A typical streamflow measurement installation: staff gauge, stilling well with recording instrument, and suspended current meter. (b) An automated hydrographic station sends telemetry to a satellite for collection by the USGS. [Photo courtesy of the California Department of Water Resources.]

Streamflow Measurement

To develop the best possible flood management, the behavior of each large watershed and stream is measured and analyzed. Unfortunately, such data often are not available for small basins or for the changing landscapes of urban areas where urban creek flooding is an increasing problem.

The key is to measure *streamflow*—the height and discharge of a stream (Figure 11.27). A *staff gauge*, a pole placed in a stream bank and marked with water heights, is used to measure stream level. With a fully measured cross section, stream level can be used to determine discharge (discharge is equal to width times depth times velocity). A *stilling well* is sited on the stream bank and a gauge is mounted in it to measure stream level. A movable current meter can be used to sample velocity at various locations.

Approximately 11,000 stream gaging stations (hydrologists use this spelling) are used in the United States (an average of over 200 per state). Of these, 7000 are operated by the U.S. Geological Survey and have continuous recorders for stage (level) and discharge (see http://water.usgs.gov/pubs/circ/circ1123/). Many of these stations automatically send telemetry data to satellites, from which information is retransmitted to regional centers (Figure 11.27b). Environment Canada's Water Survey of Canada maintains more than 3000 gaging stations. A National Weather Service hydrologist and forecaster profile appears in Career Link 11.1 at the end of this chapter.

If we study streamflow measurements, we can understand channel characteristics as conditions vary. A graph of stream discharge over a time period for a specific place is called a **hydrograph**. The hydrograph in Figure 11.28a shows the relation between precipitation input and stream discharge. During dry periods at low-water stages the flow is described as *base flow* and is largely maintained by input from local groundwater. When rainfall occurs in some portion of the watershed, the runoff collects and is concentrated in streams and tributaries. The amount, location, and duration of the rainfall episode determine the *peak flow*. Also important is the nature of the surface in a watershed; for example, a hydrograph for a specific portion of a stream changes after a forest fire or urbanization of the watershed.

Human activities have enormous impact on water flow in a basin. The effects of urbanization are quite dramatic, both increasing and hastening peak flow and causing damage. In fact, urban areas produce runoff patterns quite similar to those of deserts. The sealed surfaces of the city drastically reduce infiltration and soil moisture recharge, behaving much like the hard, nearly barren surfaces of the desert, producing flash flood events.

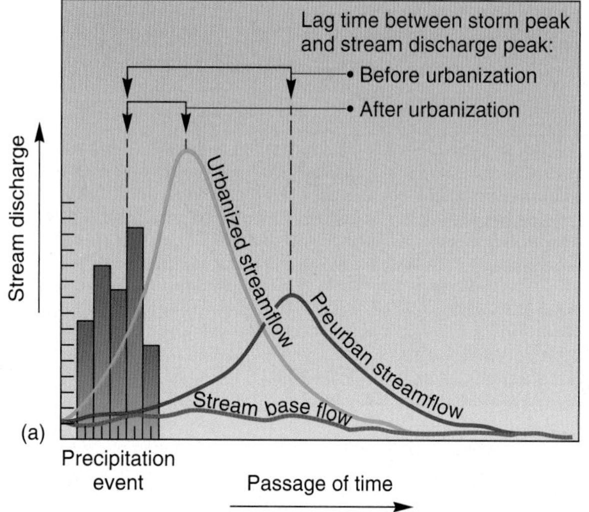

FIGURE 11.28 Urban flood profile.
(a) Effect of urbanization on a typical stream hydrograph. Normal base flow is indicated with a dark blue line. The purple line indicates discharge after a storm, prior to urbanization. Following urbanization, stream discharge dramatically increases and peaks earlier, as shown by the light blue line. (b) Severe flooding of an urban area in Linda, California, after a levee break on the Sacramento River—second time (1986, 1996) for this subdivision in 10 years! [Photo from California Department of Water Resources.]

Summary and Review—River Systems and Landforms

• *Define* fluvial and *outline* the fluvial processes: erosion, transportation, and deposition.

River systems, fluvial processes and landscapes, floodplains, and river control strategies are important to human populations as demands for limited water resources increase. **Hydrology** is the science of water, its global circulation, distribution, and properties, specifically water at and below Earth's surface. Stream-related processes are called **fluvial**. Water dislodges, dissolves, or removes surface material in the process called **erosion**. Streams produce *fluvial erosion*, in which weathered sediment is picked up for **transport**, and movement to new locations. Sediments are laid down by another process, **deposition**. **Alluvium** is the general term for the clay, silt, and sand deposited by running water. **Base level** is a level below which a stream cannot erode its valley further.

hydrology (p. 344)
fluvial (p. 344)
erosion (p. 345)
transport (p. 345)
deposition (p. 345)
alluvium (p. 345)
base level (p. 345)

1. What role is played by rivers in the hydrologic cycle?
2. What are the five largest rivers on Earth in terms of discharge? Relate these to the weather patterns in each area and to regional potential evapotranspiration (moisture demand) and precipitation (moisture supply).
3. Define fluvial. What is a fluvial process?
4. What is the difference between a local and an ultimate base level?
5. According to Figure 11.5, in which drainage basin are you located? Where are you in relation to the various continental divides?

• *Construct* a basic drainage basin model and *identify* different types of drainage patterns, with examples.

The basic fluvial system is a **drainage basin**, which is an open system. *Drainage divides* define the **watershed** catchment (water receiving) area of the drainage basin. In any drainage basin, water initially moves downslope in a thin film called **sheetflow**, or *overland flow*. This surface runoff concentrates in *rills*, or small-scale downhill grooves, which may develop into deeper *gullies* and a stream course in a valley. High ground that separates one valley from another and directs sheetflow is termed an *interfluve*. Extensive mountain and highland regions act as **continental divides** that separate major drainage basins. **Drainage pattern** refers to the arrangement of channels in an area as determined by the steepness, variable rock resistance, variable climate, hydrology, relief of the land, and structural controls imposed by the landscape. There are seven basic drainage patterns generally found in nature: dendritic, trellis, radial, parallel, rectangular, annular, and deranged.

drainage basin (p. 346)
watershed (p. 346)
sheetflow (p. 346)
continental divides (p. 346)
drainage pattern (p. 349)

6. What is the spatial geomorphic unit of an individual river system? How is it determined on the landscape? Define the several relevant key terms used.
7. On Figure 11.5, follow the Allegheny-Ohio-Mississippi river systems to the Gulf of Mexico, analyze the pattern of tributaries, and describe the channel. What role do continental divides play in this drainage?
8. Describe drainage patterns. Define the various patterns that commonly appear in nature. What drainage patterns exist in your hometown? Where you attend school?

- ***Describe* the relation among velocity, depth, width, and discharge, and *explain* the various ways that a stream erodes and transports its load.**

Stream channels vary in *width* and *depth*. The streams that flow in them vary in velocity and in the *sediment load* they carry. All of these factors may increase with increasing discharge. **Discharge** is calculated by multiplying the velocity of the stream by its width and depth for a specific cross section of the channel.

Hydraulic action is the work of *turbulence* in the water. Running water causes hydraulic squeeze-and-release action to loosen and lift rocks and sediment. As this debris moves along, it mechanically erodes the streambed further, through a process of **abrasion**.

Solution refers to the **dissolved load** of a stream, especially the chemical solution derived from minerals such as limestone or dolomite or from soluble salts. The **suspended load** consists of fine-grained, clastic particles held aloft in the stream, with the finest particles held in suspension until the stream velocity slows nearly to zero. **Bed load** refers to coarser materials that are dragged along the stream bed by **traction** or are rolled and bounced along by **saltation**. If the load in a stream exceeds its capacity, **aggradation** occurs, or the accumulation of excess sediment, as deposition fills the stream channel. With excess sediment, a stream becomes a maze of interconnected channels that form a **braided stream** pattern.

discharge (p. 350)
hydraulic action (p. 351)
abrasion (p. 351)
dissolved load (p. 351)
suspended load (p. 351)
bed load (p. 351)
traction (p. 351)
saltation (p. 351)
aggradation (p. 352)
braided stream (p. 352)

9. What was the impact of flood discharge on the channel of the San Juan River near Bluff, Utah? Why did these changes take place?
10. How does stream discharge complete its erosive work? What are the processes at work in the channel?
11. Differentiate between stream competence and stream capacity.
12. How does a stream transport its sediment load? What processes are at work?

- ***Develop* a model of a meandering stream, including point bar, undercut bank, and cutoff, and *explain* the role of stream gradient in these flow characteristics.**

Where the slope is gradual, stream channels develop a sinuous form called a **meandering stream**. The outer portion of each meandering curve is subject to the fastest water velocity and can be the site of a steep **undercut bank**. On the other hand, the inner portion of a meander experiences the slowest water velocity and forms a **point bar** deposit. When a meander neck is cut off as two undercut banks merge, the meander becomes isolated and forms an **oxbow lake**.

Every stream develops its own **gradient** and establishes a longitudinal profile. A portion of the stream is designated a **graded stream** when the stream is adjusted among available discharge, channel characteristics, its velocity, and the load supplied from the drainage basin. An interruption in a stream's longitudinal profile is called a **nickpoint**. A nickpoint can develop as the stream flows across hard resistant rock or after tectonic uplift episodes.

meandering stream (p. 353)
undercut bank (p. 353)
point bar (p. 353)
oxbow lake (p. 354)
gradient (p. 356)
graded stream (p. 356)
nickpoint (p. 356)

13. Describe the flow characteristics of a meandering stream. What is the pattern of flow in the channel? What are the erosional and depositional features and the typical landforms created?
14. Explain these statements: (a) All streams have a gradient, but not all streams are graded. (b) Graded streams may have ungraded segments.
15. Why is Niagara Falls an example of a nickpoint? Without human intervention, what do you think would eventually take place at Niagara Falls?
16. Apply these concepts (gradient, graded stream, meandering stream, nickpoint), where appropriate to a stream in your area. Explain and discuss.

- ***Define* a floodplain and *analyze* the behavior of a stream channel during a flood.**

Floodplains have been an important site of human activity throughout history. Rich soils, bathed in fresh nutrients by floodwaters, attract agricultural activity and urbanization. Despite our knowledge of historical devastation by floods, floodplains are settled, raising issues of human hazard perception. The flat, low-lying area along a stream channel that is subjected to recurrent flooding is a **floodplain**. It is formed when the river overflows its channel during times of high flow. On either bank of most streams, **natural levees** develop as byproducts of flooding. On the floodplain, backswamps and yazoo tributaries may develop. **Alluvial terraces** are formed by the entrenchment of a river into its own floodplain.

floodplain (p. 358)
natural levees (p. 358)
alluvial terraces (p. 358)

17. Describe the formation of a floodplain. How are natural levees, oxbow lakes, backswamps, and yazoo tributaries produced?
18. How is it possible to travel fewer kilometers on the Mississippi River between St. Louis and New Orleans today than 100 years ago? Explain.
19. Describe any floodplains near where you live or where you go to college. Have you seen any of the floodplain features discussed in this chapter? If so, which ones?

● *Differentiate* the several types of river deltas and *detail* each.

A depositional plain formed at the mouth of a river is called a **delta**. When the mouth of a river enters the sea and is inundated by the sea in a mix with freshwater it is called an **estuary**.

delta (p. 358)
estuary (p. 360)

20. What is a river delta? What are the various deltaic forms? Give some examples.
21. How might life in New Orleans change in the this century? Explain.
22. Describe the Ganges River delta. What factors upstream explain its form and pattern? Assess the consequences of settlement on this delta.
23. What is meant by the statement "the Nile Delta is receding" (caption Figure 11.24)?

● *Explain* flood probability estimates and *review* strategies for mitigating flood hazards.

A **flood** occurs when high water overflows the natural or artificial levees of a stream. Both floods and the floodplains they might occupy are rated statistically for the expected time interval between floods. A 10-year flood is the greatest level of flooding that is likely once every 10 years. A graph of stream discharge over time for a specific place is called a **hydrograph**.

Collective efforts by government agencies undertake to reduce flood probability. Such management attempts include the construction of artificial levees, bypasses, straightened channels, diversions, dams, and reservoirs. Society still is learning how to live in a sustainable way with Earth's dynamic river systems.

flood (p. 364)
hydrograph (p. 368)

24. Specifically, what is a flood? How are such flows measured and tracked?
25. Differentiate between a hydrograph from a natural terrain and one from an urbanized area.
26. What do you see as the major consideration regarding floodplain management? How would you describe the general attitude of society toward natural hazards and disasters?
27. What do you think the author of the article "Settlement Control Beats Flood Control" meant by the title? Explain your answer, using information presented in the chapter.

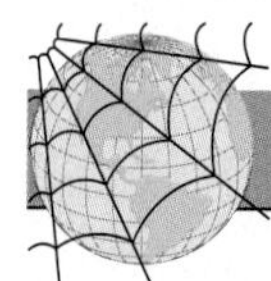

NetWork

The *Elemental Geosystems Home Page* provides on-line resources for this chapter on the World Wide Web. Once on the Home Page, click on this textbook, scroll the Table of Contents menu, select this chapter, and click "Begin." You will find self-tests that are graded, review exercises, specific updates for items in the chapter, and in "Destinations" many links to interesting related pathways on the Internet. *Elemental Geosystems* is found at http://www.prenhall.com/christopherson.

Critical Thinking

A. Determine the name of the river drainage basin within which your campus is located. Where are its headwaters? Where is the river's mouth? If you are in the United States or Canada, use Figure 11.5 to locate the larger drainage basins and divides for your region. Is there any regulatory organization that oversees planning and coordination for this drainage basin?

B. Relative to the drainage basin you determined in A, see if there is a topographic map on file in the library, geography department, or at a local outdoor recreation store that covers the portion of the basin near campus. After examining the map, can you discern a prominent drainage pattern for the area (Figure 11.9)?

C. Under "Destinations" in Chapter 11 of the *Elemental Geosystems Home Page*, there are links to many sources of flood information. Relative to the discussion of the 1993 Midwest flood in the chapter, what related information do you find?

Career Link 11.1

Julie Dian-Reed, Service Hydrologist and Weather Forecaster

Julie reminiscences, "Growing up, I lived in northwestern Indiana. When I was in high school I would go sailing on Lake Michigan with my older sister and her husband. I was fascinated by the winds, clouds, and thunderstorms over the lake and how they affected sailing. Even a trip to the beach for the day allowed me to watch the thunderstorms over the lake."

Julie did her undergraduate work at the University of Indiana in geography, with a major in meteorology, although she entered college intent on following biology and music. She moved to the University of Illinois for her master's work in climatology, also in the geography department. Her advisor was in agricultural meteorology. "I studied climatological aspects of integrated pest management, the theory that by using a variety of methods the reliance on chemical pesticides can be reduced. I found that knowledge of weather forecasting and weather patterns aided farmers in knowing when to use pesticides. Wind and weather conditions dictate migration patterns of pests, and I tried to track down these relationships."

After graduate school, Julie accepted a position at the Illinois State Water Survey, and later worked at the Midwest Regional Climate Center, where she compiled climate data and studied climate trends. In 1992 she joined the National Weather Service (NWS) in Cincinnati as a meteorological intern. She gathered surface and radar observations, issued severe storm warnings, and honed her forecasting skills. Also, Julie was heavily involved in public education and NWS outreach programs. The great Midwest floods hit in 1993 (see News Report 11.2), so her first full year with NWS was eventful.

Julie's next job in the NWS was at the Ohio River Forecast Center (OHRFC) as a hydrology intern. There she completed additional hydrology coursework, pulling together various specialties into her expertise as a weather forecasting hydrologist. She became the Service Hydrologist at the Weather Forecast Office, Wilmington, Ohio, in May 1995. "I try to get everyone thinking like a hydrologist. Once the rain hits the ground it shouldn't pass from one area of responsibility to another, especially during flash flood and river flood situations. The geographer in me sees the spatial relations among rainfall events, hydrology, and watershed characteristics. Our staff is good at this and everyone is tuned into this approach."

FIGURE 1 Julie Dian-Reed, Service Hydrologist and Weather Forecaster
National Weather Service, Wilmington, Ohio. [Photo by National Weather Service.]

Her office is co-located with the River Forecast Center—"just across the hall, so it is nice to have that interaction. I can see what goes on regarding river forecasting and integrate this with the weather forecast. Then, from this mix, we put out a comprehensive warning or advisories to the public."

Julie works with Advanced Weather Interactive Processing System (AWIPS), installed in their office in 1998, to put out forecasts (Figure 1). "It is a UNIX-based system, where all the data are accessed in multiframe screens on several monitors. This kind of integration makes the warning and forecasting process very efficient. We have also utilized instant messaging software to coordinate with neighboring NWS forecast offices. This helps us create a more seemless forecast," she said. She also operates the Doppler radar at Wilmington on some work shifts. "Once the input data arrives, we work at the terminal to develop the derived data product."

Careers in this field are bright and involve the GIS revolution. "My background in geography helps me bring together diverse subjects to complement our weather and river forecasting. We are beginning to incorporate GIS information as part of our suite of tools for both weather and river forecasting and warning.

"I enjoy the variety of weather we receive in this region, and following the weather is a hobby of mine. Even while not working, I track the weather and try to take photos of a flooding event." Julie is enthusiastic about her integrated vocation and avocation!

Lower Sonoran Desert near Tucson, Arizona. North America's hottest desert in the southwestern U.S., northern Mexico, and most of Baja, California. Many plants and animals inhabit this region in a rich biodiversity despite the low average annual precipitation of 14.0 cm (5.5 in.). The Saguaro cactus (*Carnegiea gigantea*) can reach 18 m (60 ft) in height and be several hundred years old, taking almost 60 years to reach 2.5 m (8 ft) and begin flowering. [Photo by author.]

12 Wind Processes and Desert Landscapes

Key Learning Concepts

By knowing and understanding the key learning concepts in this chapter, you should be able to:

- *Characterize* the unique work accomplished by wind and eolian processes.
- *Describe* eolian erosion, including deflation, abrasion, and the resultant landforms.
- *Describe* eolian transportation and *explain* saltation and surface creep.
- *Identify* the major classes of sand dunes and *present* examples within each class.
- *Define* loess deposits, their origins, locations, and landforms.
- *Portray* desert landscapes and *locate* these regions on a world map.

Wind is an agent of geomorphic change. Like moving water, moving air (wind) causes erosion, transportation, and deposition of materials. Like moving water, moving air is a fluid, and it behaves similarly, although it has a lower viscosity (it is "thinner") than water. Although lacking the lifting ability of water, wind processes can modify and move quantities of materials in deserts and along coastlines.

Wind contributes to soil formation in distant places, bringing fine particles from regions where glaciers once were active. Fallow fields (those not planted) give up their soil resource to destructive wind erosion. Scientists are only now getting an accurate picture of the amount of windblown dust that fills the atmosphere and crosses the oceans between continents. Chemical fingerprints and satellites trace windblown dust from African soils to South America and Asian landscapes, to Europe, and to North America.

Earth's dry lands stand out in stark contrast on the water planet. In desert environments, an overall lack of moisture and stabilizing vegetation allow wind to create extensive sand seas and dunes of infinite variety. The polar regions are deserts as well, so there should be no surprise that weather monitoring stations in Yuma, Arizona, and in Antarctica receive the same amount of annual precipitation. These polar and high-latitude deserts possess unique features related to their cold, dry environment—aspects covered in Chapter 14.

Arid landscapes display distinctive landforms and life forms: "instead of finding chaos and disorder the observer never fails to be amazed at a simplicity of form, an exactitude of repetition and a geometric order. . . ."*

In this chapter: We examine the work of wind, associated erosion, transport, and depositional processes and resulting landforms. Most deserts are rocky and covered with desert pavement, whereas some dry landscapes are covered in sand dunes. Windblown fine particles form vast loess deposits, the basis for rich agricultural soils. For convenience of organization, we include the discussion of arid lands in this chapter. In desert environments, an overall lack of moisture sets the landscapes in sharp, Sun-baked relief. Water is the major erosional agent in the desert, yet water is the limiting resource for human development. The Colorado River is the subject of an important focus study.

The Work of Wind

The work of wind—erosion, transportation, and deposition—is called **eolian** (also spelled *aeolian*; for Aeolus, ruler of the winds in Greek mythology). Ralph Bagnold, a British major stationed in Egypt in 1925, was a pioneer in desert research. He was an engineering officer who spent much of his time in the deserts west of the Nile. Bagnold measured, sketched, and developed hypotheses about the wind and desert forms. His often-cited work, *The Physics of Blown Sand and Desert Dunes*, was published in 1941 following the completion in London of wind-tunnel simulations of windy desert conditions.

The actual ability of wind to move materials is small compared with that of other transporting agents such as water and ice, because air is so much less dense than these other media. Yet over time, wind accomplishes enormous work. Bagnold studied the ability of wind to transport sand over the surface of a dune. Wind of 50 kmph (30 mph) can move approximately a half ton of sand per day over a 1-m-wide section of dune! Consistent local wind can prune and shape vegetation (Figure 12.1).

*R. A. Bagnold, *The Physics of Blown Sand and Desert Dunes* (London: Methuen, 1941).

FIGURE 12.1 The work of wind.
Wind-sculpted tree near South Point, Hawai'i. Nearly constant tradewinds keep this tree naturally pruned. [Photo by Bobbé Christopherson.]

Eolian Erosion

Two principal wind-erosion processes are **deflation**, the removal and lifting of individual loose particles, and **abrasion**, the grinding of rock surfaces with a "sandblasting" action by particles captured in the air. Deflation and abrasion produce a variety of distinctive landforms and landscapes.

Deflation. Deflation literally blows away loose or noncohesive sediment. **Desert pavement** is formed from pebbles and gravel left behind after wind deflation and water wash away fine materials and concentrate and cement remaining rock pieces. Resembling a cobblestone street, desert pavement protects underlying sediment from further deflation (Figure 12.2). Desert pavements are so common that many provincial names have been used for them—for example, *gibber plain* in Australia, *gobi* in China, and, in Africa, *lag gravels* or *serir* (or *reg* desert, if some fine particles remain).

Heavy recreational activity damages fragile desert landscapes. Over 14 million off-road vehicles (ORVs) are in use in the United States. Such vehicles crush plants and animals, disrupt desert pavement, promote deflation losses to wind, and create ruts that easily concentrate sheetwash to form gullies. Measures to restrict ORV use to specific areas, preserving the remaining desert, are controversial.

Military activities can also threaten delicate desert landscapes. A serious environmental impact of the 1991 Persian Gulf War was the disruption of desert pavement. Thousands of square kilometers of stable desert pavement were shattered by more than 68,000 tons of explosives and disrupted by the movement of heavy vehicles. The resulting loosened sand and silt became available for deflation, plaguing cities and farms with increased dust and sand accumulations. People living in desert regions, where frequent sand storms occur, contend with the sandblasting of painted surfaces and etched window glass.

Wherever wind encounters loose sediment, deflation also may form basins. Called **blowout depressions**, these range from small indentations of less than a meter up to areas hundreds of meters wide and many meters deep. Chemical weathering, although it operates slowly in the desert, is important in the formation of a blowout, for it removes the cementing materials that give particles their cohesiveness. Large depressions in the Sahara are at least partially formed by deflation. The enormous Munkhafad

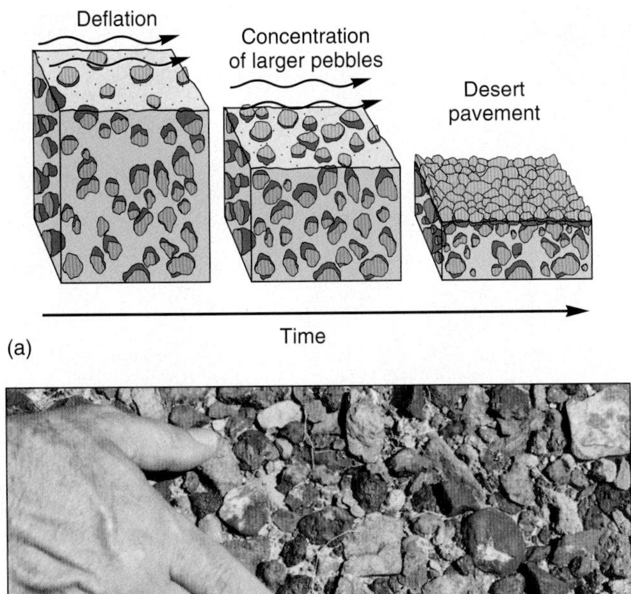

(b)

FIGURE 12.2 Desert pavement.
(a) Desert pavement is formed from larger rocks and fragments left after deflation and sheetwash remove loose material; (b) a typical desert pavement. [Photo by Bobbé Christopherson.]

el Qaṭṭâra (Qaṭṭâra Depression) just inland from the Mediterranean Sea in the Western Desert of Egypt, which covers 18,000 km^2 (6950 mi^2), is now about 130 m (427 ft) below sea level at its lowest point.

Abrasion. Sandblasting is commonly used to clean stone surfaces on buildings or to remove unwanted markings from streets. Abrasion by windblown particles is nature's version of sandblasting and is especially effective at polishing exposed rocks when the abrading particles are hard and angular. Variables that affect the rate of abrasion include the hardness of vulnerable surface rocks and the wind velocity and constancy. Abrasive action is restricted to the area immediately above the ground, usually no more than a meter or two in height, because sand grains are lifted only a short distance.

Rocks exposed to eolian abrasion appear pitted, grooved, or polished, and usually are aerodynamically shaped in a specific direction, according to the flow of airborne particles. Rocks that bear such evidence of eolian erosion are called **ventifacts**. On a larger scale, deflation and abrasion are capable of streamlining rock structures that are aligned parallel to the most effective wind direction, leaving behind distinctive, elongated ridges called **yardangs**. Yardangs can range from meters to kilometers in length and up to many meters in height. Abrasion is concentrated on the windward end of each yardang, with deflation operating on the leeward portion. A wind-sculpted form in Snow Canyon, Utah, is shown in Figure 12.3.

On Earth, some yardangs are large enough to be detected on satellite imagery. The Ica Valley of southern Peru contains yardangs reaching 100 m (330 ft) in height and several kilometers in length, and yardangs in the Lūt Desert of Iran attain 150 m (490 ft) in height. The Sphinx in Egypt perhaps partially formed as a yardang, suggesting a head and body to the ancients. Some scientists think

FIGURE 12.3 A yardang.
A small wind-sculpted rock formation in Snow Canyon outside St. George, Utah. [Photo by Bobbé Christopherson.]

this shape led them to complete the bulk of the sculpture artificially with masonry.

Eolian Transportation

As mentioned in Chapter 4, atmospheric circulation is capable of transporting fine material such as volcanic debris worldwide within days. The distance that wind is capable of transporting particles varies greatly with particle size. Wind exerts a drag or frictional pull on surface materials. Only the finest dust particles travel significant distances (Figure 12.4). The finer material suspended in a *dust storm* is lifted much higher than the coarser particles of a *sand storm*, which may be lifted only about 2 m (6.5 ft). People living in areas of frequent dust storms face the infiltration of very fine particles into their homes and businesses through even the smallest cracks. (Figure 2.23 illustrates such dust storms in Nevada and over the eastern Mediterranean Sea and blowing alkali dust from the exposed shorelines of Mono Lake, California.)

Deflation and wind transport produced a catastrophe called the American Dust Bowl of the 1930s. (See News

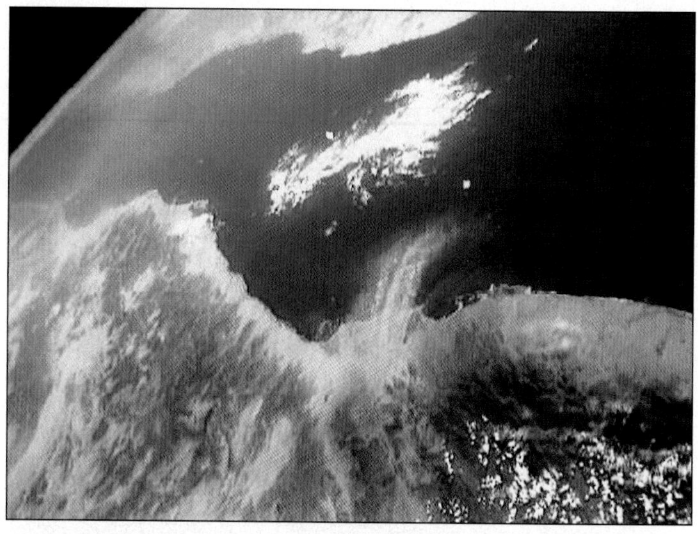

FIGURE 12.4 Dust over the Red Sea.
A dust storm is blowing off the Sudan coast at 21° N over the Red Sea. (In this photo, north is to the left, south to the right.) [Astronaut photo from Shuttle STS-43 courtesy of JSC Digital Image Collection, NASA.]

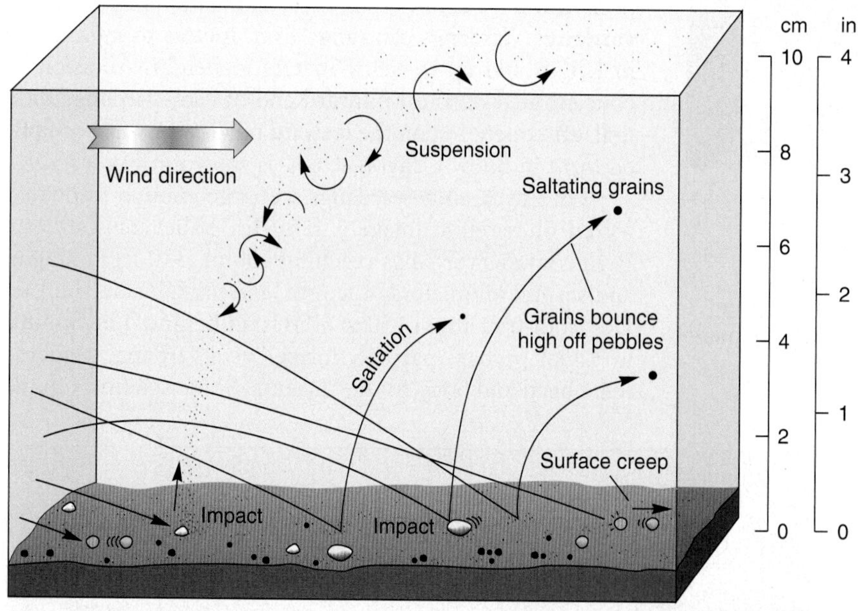

FIGURE 12.5 Wind moves sand.
(a) Eolian saltation and surface creep are forms of sediment transportation. (b) Sand grains saltating along the surface in the Stovepipe Wells dune field, Death Valley. [Photo by author.]

Report 12.1.) Overgrazing and poor agricultural practices left soil vulnerable. The transported dust darkened the skies of midwestern cities and the lives of millions.

Through processes of weathering, erosion, and transportation, mineral grains are removed from parent rock and redistributed. In Figure 10.4c, you can see the relationship between the composition and color of the sandstone in the background and the derived sandy surface in the foreground. Wind action is not significant in the weathering process that frees individual grains of sand from the parent rock, but it is active in relocating the weathered grains.

The term *saltation* was used in Chapter 11 to describe the movement of particles by water. The term also describes the wind transport of grains along the ground—grains usually larger than 0.2 mm (0.008 in.). Skipping and bouncing actions account for about 80% of wind transport of particles (Figure 12.5). Compared with fluvial transport, in which saltation is accomplished by hydraulic lift, eolian saltation is executed by aerodynamic lift, elastic bounce, and impact (compare Figures 12.5 and 11.12).

Saltating particles crash into other particles, knocking them both loose and forward. This action causes **surface creep**, which is the sliding and rolling of particles too large for saltation. Surface creep affects about 20% of the material being transported. Once in motion, particles can continue moving at lower wind velocities. In a desert or along a beach, you can hear the myriad saltating grains of sand produce a slight hissing sound, almost like steam escaping, as they bounce along and collide with surface particles.

Sand erosion and transport from a beach may be slowed by conservation measures such as the introduction of stabilizing native plants, the use of fences, and the restriction of pedestrian traffic to walkways (Figure 12.6).

FIGURE 12.6 Preventing sand transport.
Further erosion and transport of coastal dunes can be controlled by stabilizing strategies, planting native plants, and confining pedestrian traffic to walkways, as at Popham Beach State Park, Maine. [Photo by Bobbé Christopherson.]

Eolian Depositional Landforms

The smallest features shaped by individual saltating grains are *ripples* (Figure 12.7). Ripples form in crests and troughs positioned transversely (at a right angle) to the direction of the wind. The length of time particles are airborne influences their formation patterns. Eolian ripples are different from fluvial ripples because the impact of saltating grains is very slight in water.

A common assumption is that sand covers most deserts. Instead, desert pavements predominate across most subtropical arid landscapes; sand covers only about 10% of desert areas. Transient ridges or hills of sand grains are known as dunes. A **dune** is a wind-sculpted accumulation of sand. An extensive area of dunes, such as that found in North Africa, is characteristic of an **erg desert**, which means **sand sea**.

The Grand Erg Oriental in the central Sahara exceeds 1200 m (4000 ft) in thickness and covers 192,000 km^2 (75,000 mi^2). This sand sea has been active for more than 1.3 million years and has average dune heights of 120 m (400 ft). Sahara Marzūq sand sea shown in Figure 12.8a is wider than 300 km (185 mi). Similar sand seas, such as the Grand Ar Rub'al Khālī Erg, are active in Saudi Arabia, and in the Taklimakan Desert featured in the Chapter 6 opening satellite image. Eolian processes are at work on the Martian surface forming similar parallel ridges along the floor of Melas Chasma in Figure 12.8b.

Dune Movement and Form. Dune fields, whether in arid regions or along coastlines, tend to migrate in the direction of effective, sand-transporting winds. In this regard, stronger seasonal winds or winds from a passing storm may prove more effective than average prevailing

FIGURE 12.7 Sand ripples.
Myriad sand ripple patterns may become lithified into fixed patterns in rock. The area in the photo is approximately 1 m (3.3 ft) wide. [Photo by author.]

(a)

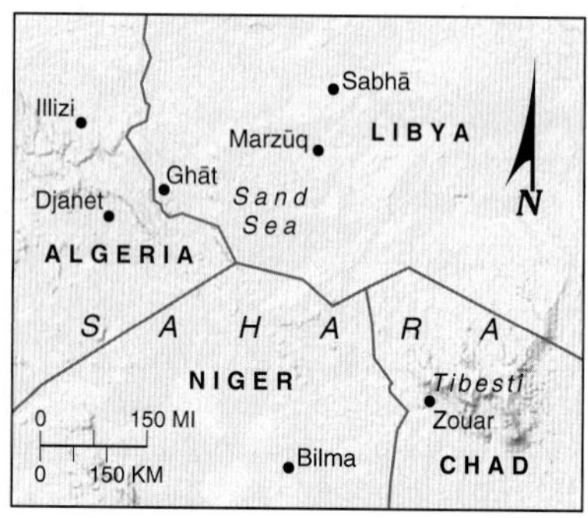

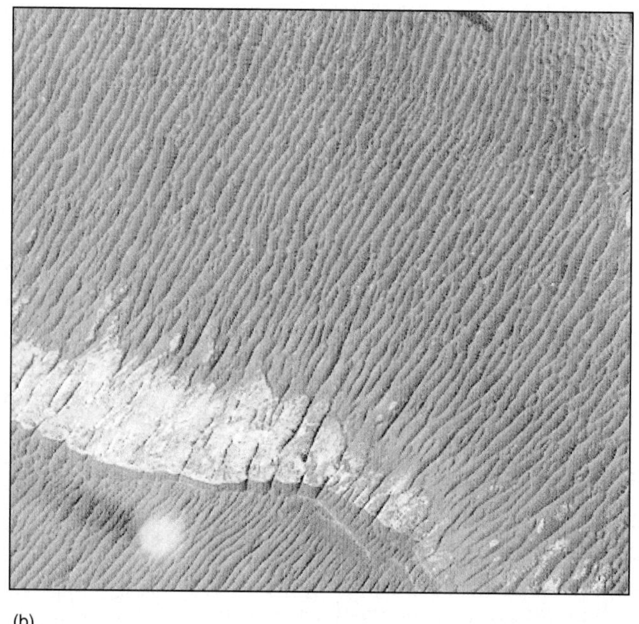

(b)

FIGURE 12.8 A sand sea.
(a) The Sahara Marzūq, an erg desert that dominates southwestern Libya. Effective northwesterly winds shape the pattern and direction of the transverse and barchanoid (series of connected barchans) dunes. This sand sea exceeds 300 km (185 mi) across. (b) A similar pattern of dunes appears on the Martian surface in the southern area of Melas Chasma in Valles Marineris. Note the dust devil in the lower left. Area covered is about 2 km (1.2 mi) wide. [(a) *Terra* MODIS sensor image courtesy of MODIS Land Rapid Response Team, NASA/GSFC, November 9, 2001. (b) Mars Global Surveyor, Mars Orbiter Camera image courtesy of NASA/JPL/Malin Space Science Systems, July 11, 1999.]

News Report 12.1

The Dust Bowl

Deflation and wind transport of loess soils produced a catastrophe in the American Great Plains in the 1930s—the Dust Bowl. More than a century of overgrazing and intensive agriculture left soil susceptible to drought and eolian processes. The deflation of many centimeters of soil occurred in southern Nebraska, Kansas, Oklahoma, Texas, and eastern Colorado.

Fine sediments were lifted by winds forming severe dust storms. The transported dust darkened the skies of Midwestern cities and drifted over farmland. Streetlights were left on throughout the day in Kansas City, St. Louis, and other midwestern cities and towns. Such episodes can devastate economies, cause tremendous loss of topsoil, and even bury farmsteads. For more on the Dust Bowl, see **http://www.pbs.org/wgbh/amex/dustbowl/**. Southeastern Australia experienced severe dust storms in 1993 that included consequences similar to those of the American Dust Bowl. A continuing drought through 2002 and 2003 threatens more severe consequences for Australia.

winds. When saltating sand grains encounter small patches of sand, their kinetic energy (motion) is dissipated and they accumulate. As height increases above 30 cm (12 in.), a *slipface* and characteristic dune features form.

Study the dune in Figure 12.9 and you can see that winds characteristically create a gently sloping *windward side* (stoss side), and a more steeply sloped slipface on the *leeward side*. A dune usually is asymmetrical in one or more directions. The angle of a slipface is the angle at which loose material is stable—its *angle of repose*. Thus, the constant flow of new material makes a slipface a type of *avalanche slope*. As sand moves over the crest of the dune to the brink, it builds up and avalanches as the slipface continually adjusts, seeking its angle of repose (usually 30°–34°). In this way, a dune migrates downwind with the effective wind, as suggested by the successive dune profiles in Figure 12.9.

Dunes that move actively are called *freedunes* and reflect most dynamically the interaction between fluid atmospheric winds and moving sand. However, because the sand is moving close to the ground, it may encounter an obstruction such as stabilizing vegetation or a rock outcrop, resulting in a *tied dune*, or one that is fixed in place. The ever changing form of these eolian deposits is part of their beauty, eloquently described by one author: "I see hills and hollows of sand like rising and falling waves. Now at midmorning, they appear paper white. At dawn they were fog gray. This evening they will be eggshell brown."* Their many wind-shaped styles make classification difficult. We can simplify dune forms into three classes—*crescentic*, *linear*, and *star* dunes (Figure 12.10).

*J. E. Bowers, *Seasons of the Wind* (Flagstaff, Ariz.: Northland Press, 1985), p. 1.

Crescentic dunes are crescent-shaped ridges of sand that form in response to a fairly unidirectional wind pattern. The crescentic group is most common, with related forms including *barchan dunes* (limited sand), *transverse dunes* (abundant sand), *parabolic dunes* (vegetation controlled), and *barchanoid ridges* (rows of coalesced barchans).

Linear dunes generally form long parallel ridges separated by sheets of sand or bare ground. Linear dunes characteristically are much longer than they are wide; some exceed 100 km (60 mi) in length. Winds producing these dunes are principally bidirectional, so the slipface alternates from side to side. Related forms include *longitudinal dunes* and the *seif* (Arabic for "sword"), which is a sharper, narrower linear dune with a more sinuous crest.

Star dunes are the mountainous giants of the sandy desert. They form in response to complicated, changing wind patterns and have multiple slipfaces. They are pinwheel-shaped, with several radiating arms rising and joining to form a common central peak. The best examples of star dunes are in the Sahara and the Namib Desert, where they approach 200 m (650 ft) in height (Figure 12.11).

These same dune-forming principles and terms (for example, dune, barchan, slipface) apply to snow-covered landscapes. *Snow dunes* form as wind deposits snow in drifts. In semiarid farming areas, fences and tall stubble are left in the fields to capture drifting snow, which contributes significantly to soil moisture on melting.

Figure 12.12a correlates active sand regions with deserts (tropical, continental interior, and coastal). Active sand dunes cover only about 10% of all continental land between 30° N and 30° S. Also noted on the map are dune fields in humid climates: along coastal Oregon, the south shore of Lake Michigan (Figure 12.12b), along the Gulf and Atlantic coastlines, in Europe, and elsewhere. The remarkable coastal desert sands of Namibia are pictured in Figure 12.12c.

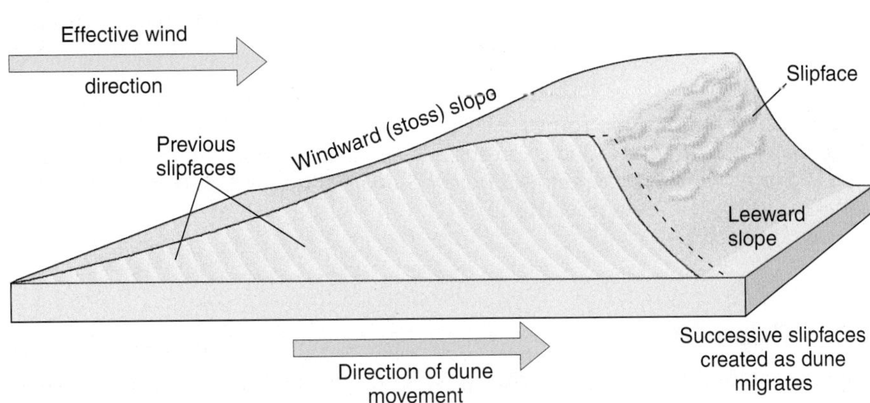

FIGURE 12.9 Dune cross section. Cross section of a dune, showing the pattern of successive slipfaces as the dune migrates in the direction of the effective wind.

Class	Type	Description
Crescentic	Barchan	Crescent-shaped dune with horns pointed downwind. Winds are constant with little directional variability. Limited sand available. Only one slipface. Can be scattered over bare rock or desert pavement or commonly in dune fields.
	Transverse	Asymmetrical ridge, transverse to wind direction (right angle). Only one slipface. Results from relatively ineffective wind and abundant sand supply.
	Parabolic	Role of anchoring vegetation important. Open end faces upwind with U-shaped "blow-out" and arms anchored by vegetation. Multiple slipfaces, partially stabilized.
	Barchanoid ridge	A wavy, asymmetrical dune ridge aligned transverse to effective winds. Formed from coalesced barchans; look like connected crescents in rows with open areas between them.

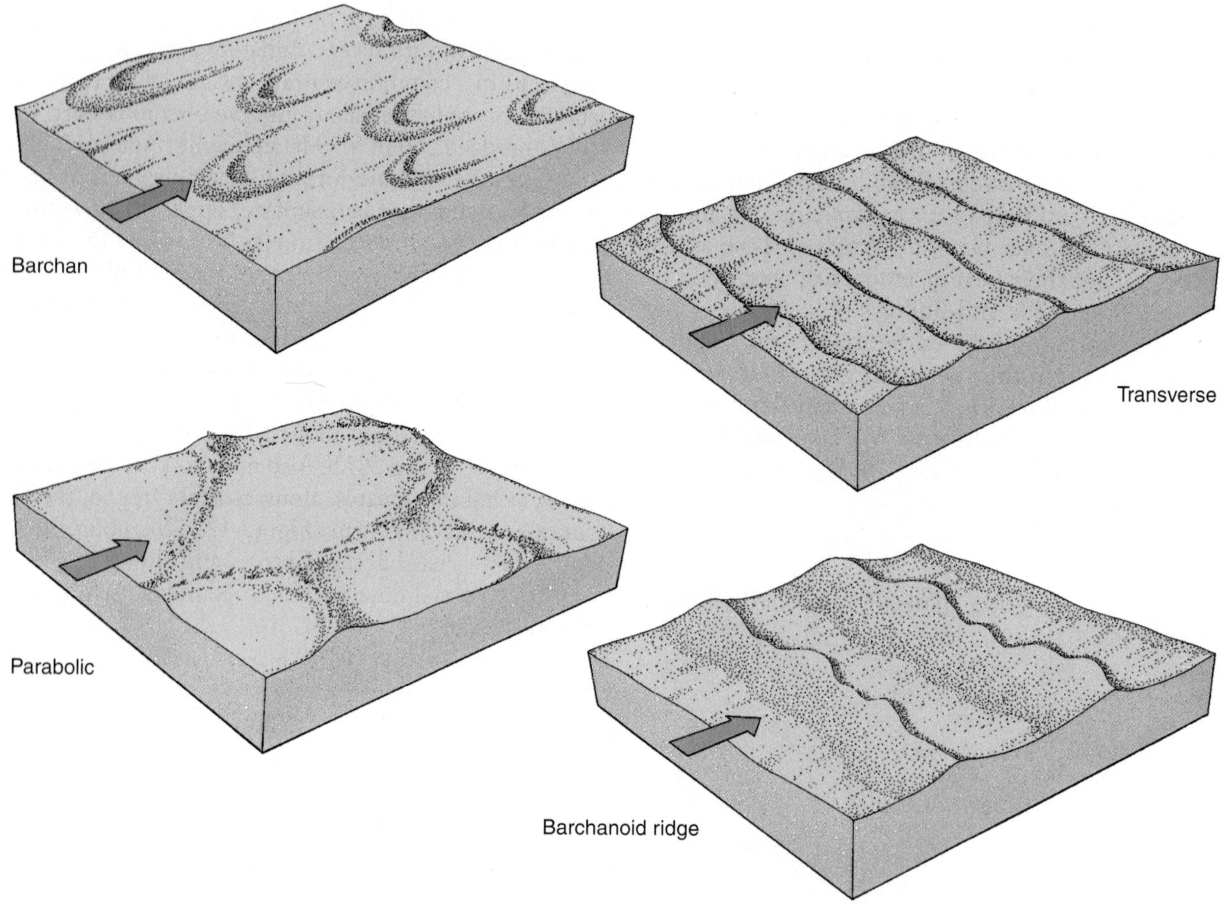

FIGURE 12.10 Major dune forms.
Three major dune classes include crescentic, linear, and star dune. Arrows show wind direction. [Adapted from E. D. McKee, *A Study of Global Sand Seas*, U.S. Geological Survey Professional Paper 1052. Washington, DC: 1979.]

Loess Deposits

Approximately 15,000 years ago, in several episodes, Pleistocene glaciers retreated in many parts of the world, leaving behind large glacial outwash deposits of fine-grained clays and silts (<0.06 mm or 0.0023 in.). These materials were blown great distances by the wind and redeposited in unstratified, homogeneous deposits, named **loess** by peasants working along the Rhine River Valley in Germany.

No specific landforms were created; instead, loess covered existing landforms with a thick blanket of material that assumed the general topography of the existing

Class	Type	Description
Linear	Longitudinal	Long, slightly sinuous, ridge-shaped dune, aligned parallel with the wind direction; two slipfaces. Can be 100 m high and 100 km long, and the "draas" at the extreme can reach to 400 m high. Results from strong effective winds varying in one direction.
	Serif	After Arabic word for "sword"; a more sinuous crest and shorter than longer longitudinal dunes. Rounded toward upwind direction and pointed downwind. (Not illustrated.)
Star dune		The giant of dunes. Pyramidal or star-shaped with three or more sinuous radiating arms extending outward from a central peak. Slipfaces in multiple directions. Results from effective winds shifting in all directions. Tend to form isolated mounds in high effective winds and connected sinuous arms in low effective winds.
Other	Dome	Circular or elliptical mound with no slipface. Can be modified into barchanoid forms.
	Reversing	Asymmetrical ridge form intermediate between star dune and transverse dune. Wind variability can alter shape between forms.

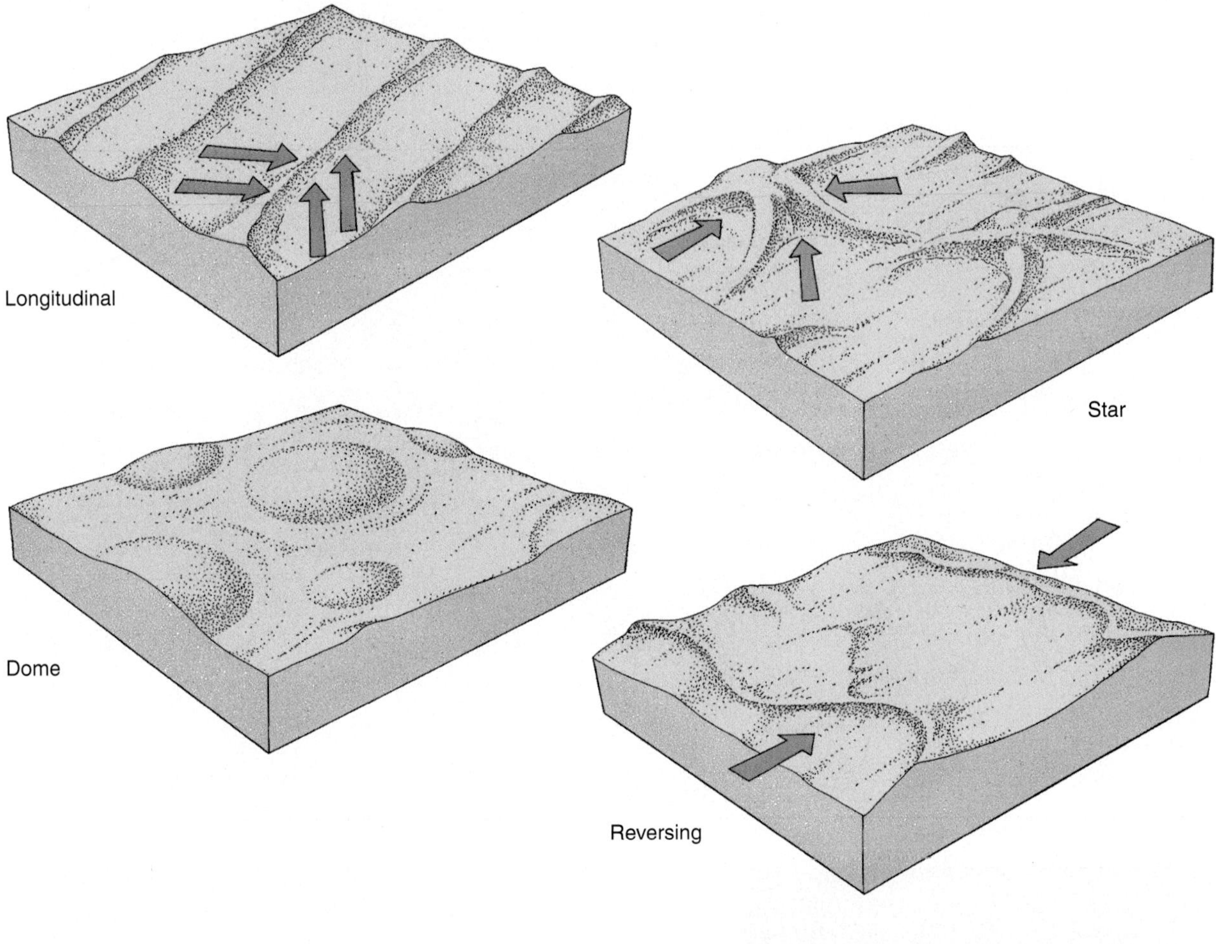

Longitudinal

Star

Dome

Reversing

FIGURE 12.11 Mountains of the desert.
Star dune in the Namib Desert of Namibia in southwestern Africa. [Photo by Comstock, Inc.]

landscape. Because of its own binding strength, loess weathers and erodes into steep bluffs, or vertical faces. At Xi'an, Shaanxi, in China, a loess wall was excavated for dwelling space (Figure 12.13a). When a bank is cut into a loess deposit, it generally will stand vertically, although it can fail if saturated (Figure 12.13b).

Figure 12.14 shows the worldwide distribution of loess deposits. Significant accumulations of loess throughout the Mississippi and Missouri valleys form continuous deposits 15–30 m (50–100 ft) thick. Loess deposits also occur in eastern Washington State and Idaho. The soils in these regions are fertile, for loess deposits are well drained, deep, and have excellent moisture retention. Loess deposits also cover much of Ukraine, central Europe, China, the Pampas-Patagonia regions of Argentina, and lowland New Zealand. These soils derived from loess are some of Earth's "bread basket" farming regions. Transport of loess soils occurred in the catastrophic Dust Bowl in the 1930s in the United States; see News Report 12.1.

In Europe and North America, loess originates mainly from glacial sources. The vast deposits of loess in China, covering more than 300,000 km^2 (115,800 mi^2), are thought to come from windblown desert sediment rather than glacial sources. Accumulations in the Loess Plateau of China exceed 300 m (1000 ft) thickness, forming some complex weathered badlands and some good agricultural land. These windblown deposits are interwoven with much of Chinese history. New research found that plumes of windblown dust from African deserts have moved across the Atlantic Ocean to enrich soils of the Amazon rain forest of South America, the southeastern United States, and the Caribbean islands.

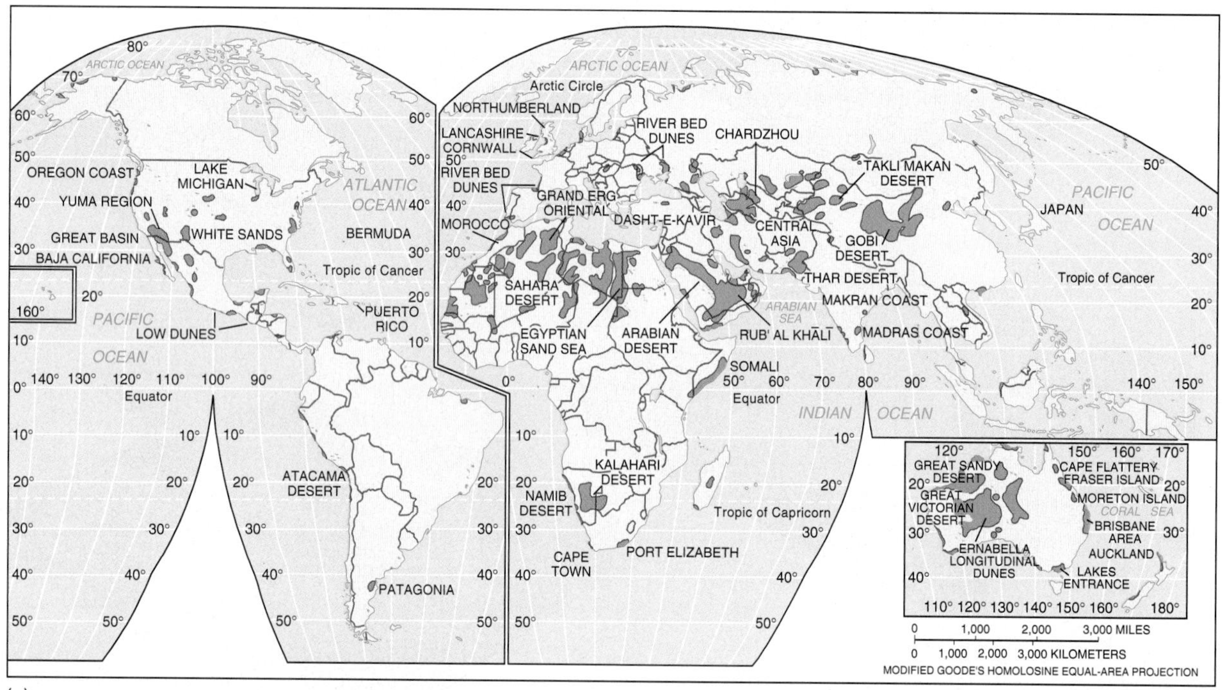

(a)

(b)

(c)

FIGURE 12.12 Sandy regions of the world.
(a) Worldwide distribution of active and stable sand regions. (b) Sand dunes along the shore of Lake Michigan in Indiana Dunes State Park in Indiana. (c) Sandy area in the Namib Desert, Namibia, Africa. [(a) After R. E. Snead, *Atlas of World Physical Features*, p. 134, © 1972 by John Wiley & Sons. Adapted by permission of John Wiley & Sons, Inc. Photos by (b) Bobbé Christopherson, (c) by Nigel J. Dennis/Photo Researchers, Inc.]

(a)

(b)

FIGURE 12.13 Example of loess deposits.
(a) Loess formation in Xi'an, Shaanxi, China, has strong enough structure to be excavated for dwelling rooms. (b) Loess bluffs along the Arikaree River in extreme northwestern Cheyenne County, Kansas. [Photos by (a) Betty Crowell; (b) Steve Mulligan Photography.]

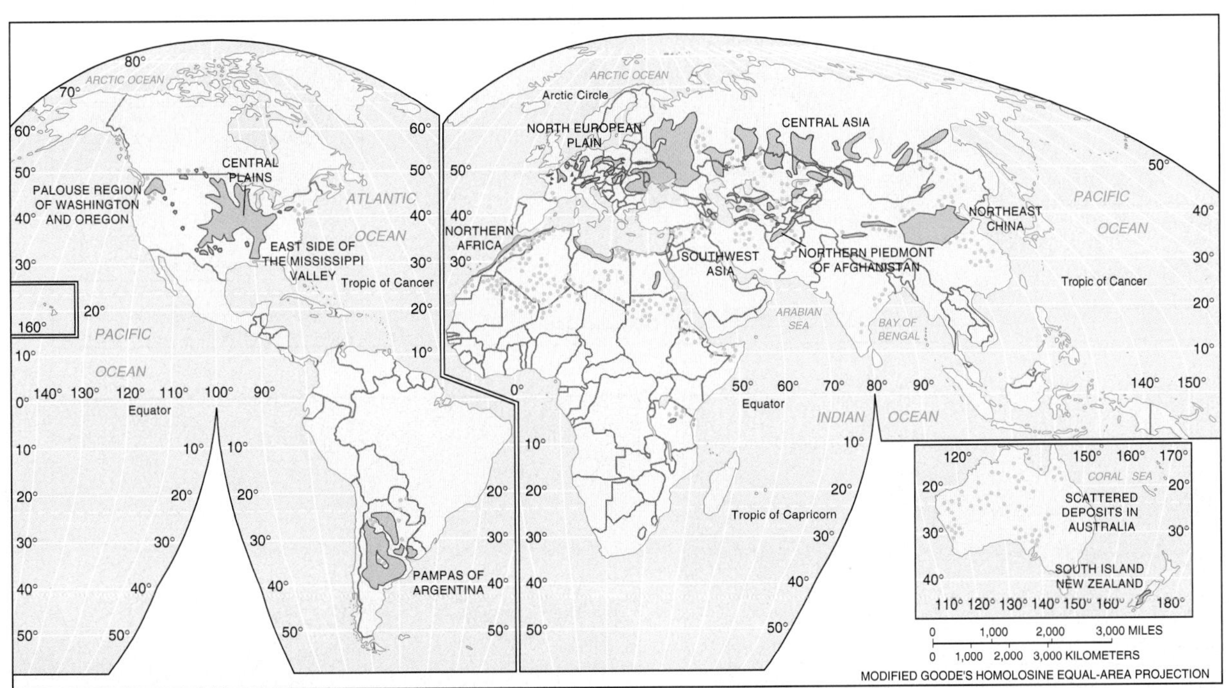

FIGURE 12.14 Worldwide loess deposits.
Dots represent small scattered loess formations. [After R. E. Snead, *Atlas of World Physical Features*, p. 138, copyright © 1972 by John Wiley & Sons. Adapted by permission of John Wiley & Sons, Inc.]

Overview of Desert Landscapes

Arid and semiarid dry climates cover as much as 35% of all land, constituting the largest single climatic region on Earth. See Figures 6.4 and 6.5 for the location of these *arid deserts* and *semiarid steppe* climate regions, and the world biome map in Chapter 16 for the distribution of these desert environments. And Earth's deserts are expanding, as we discuss shortly. Now let us look at the link between climate and Earth's deserts.

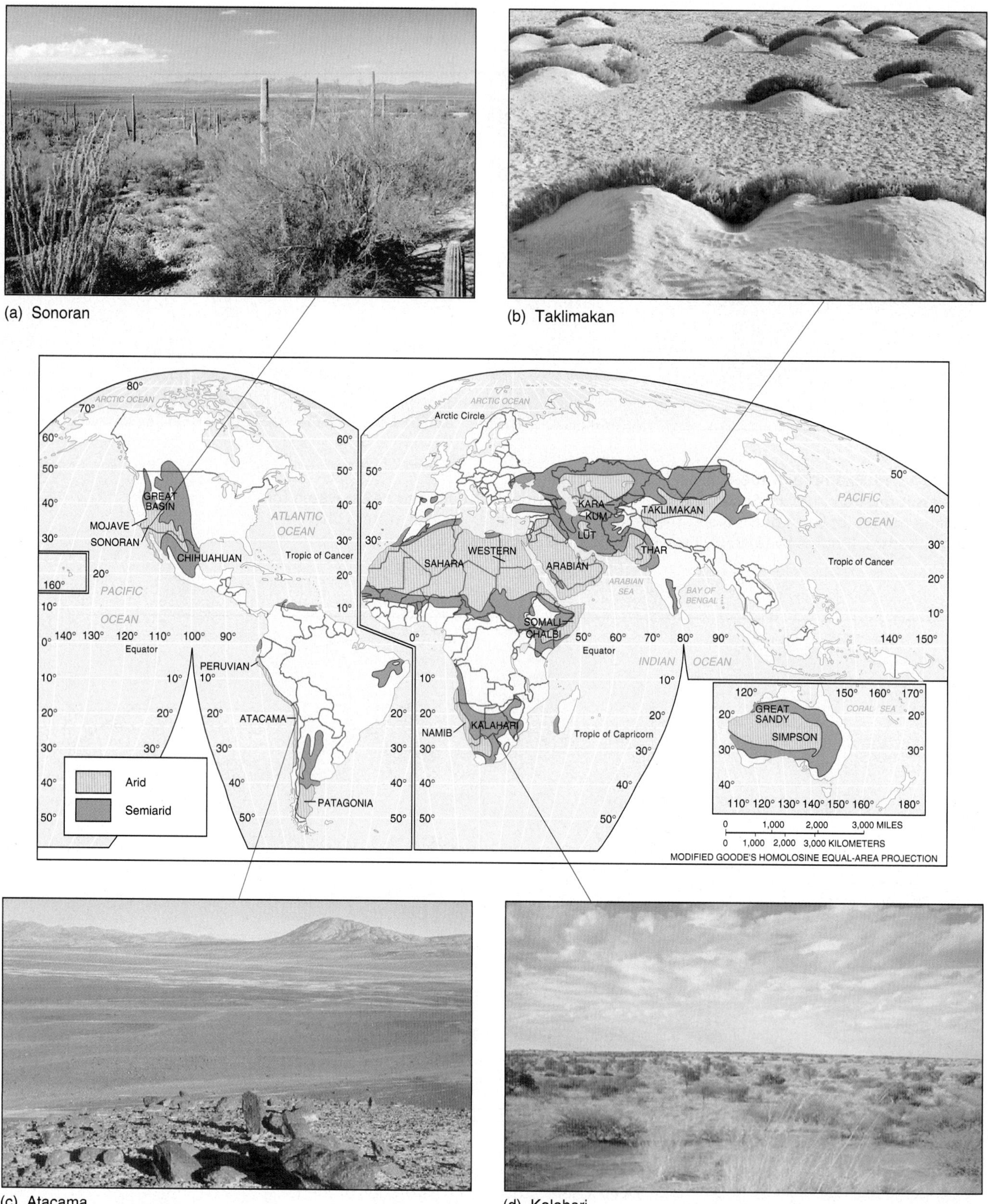

FIGURE 12.15 The world's dry regions.
Worldwide distribution of arid and semiarid lands—Earth's deserts and steppes. (a) Sonoran Desert in the American southwest. (b) Taklimakan Desert in central Asia. (c) Atacama Desert in subtropical Chile near Baquedano. (d) Kalahari Desert in south-central Africa. [Photos by (a) author; (b) Wu Chunzhan/New China Pictures/East Photo; (c) Jacques Jangoux/Photo Researchers, Inc.; and (d) Nigel J. Dennis/Photo Researchers, Inc.]

Desert Climates

Physical conditions explain the spatial distribution of these dry lands: subtropical high-pressure cells between 15° and 35° N and S (see Figures 4.11 and 4.13), rain shadows on the lee side of mountain ranges (see Figure 5.29), and areas at great distance from moisture-bearing air masses, such as central Asia. Figure 12.15 portrays this distribution according to the climate classification used in this text and presents photographs of four major desert regions. These areas possess unique landscapes created by the interaction of intermittent precipitation events, weathering processes, and wind. Rugged, hard-edged desert landscapes of cliffs and scarps contrast sharply with the vegetation-covered, rounded and smoothed slopes characteristic of humid regions.

Desert environments experience high sensible heat conditions and intense ground heating. Such areas receive a high input of insolation through generally clear skies and experience high radiative heat losses at night. A typical desert water balance shows high potential evapotranspiration (moisture demand), low precipitation (moisture supply), and prolonged seasonal moisture deficits (see, for example, Figure 7.10 for Phoenix, Arizona). Fluvial processes in the desert feature intermittent running water, with hard, poorly vegetated desert pavement yielding high runoff during rainstorms—the infamous flash flood.

Desert Fluvial Processes

Precipitation events in a desert may be rare—indeed, a year or two apart—but when they do occur, a dry streambed fills with a torrent called a **flash flood**. Such channels may fill in a few minutes and surge briefly during and after a storm. Depending on the region, such a dry streambed is known as a **wash**, an *arroyo* (Spanish), or a *wadi* (Arabic). A desert highway that crosses a wash usually is posted to warn drivers not to proceed if rain is in the vicinity, for a flash flood can suddenly sweep away anything in its path.

When washes fill with surging flash floodwaters, a unique set of ecological relationships quickly develops. Crashing rocks and boulders break open seeds. The seeds respond to the timely moisture and germinate. Other plants and animals also spring into brief life cycles as the water irrigates their limited habitats. Please see the desert comparison photos in Focus Study 6.1 for a look at the possible remarkable contrasts fostered by moisture.

At times of intense rainfall, remarkable scenes fill the desert. Figure 12.16 shows two photographs made just 1 month apart in a sand dune field in Death Valley, California. The rainfall event that occurred produced 2.57 cm (1.01 in.) of precipitation in 1 day, in a place that receives only 4.6 cm (1.83 in.) in an average year. The runoff flowed for hours and then collected in low spots on the hard, underlying clay surfaces. The water was quickly consumed by the high evaporation demand, so that in just a month these short-lived watercourses were dry and covered with accumulations of alluvial materials.

As runoff water evaporates, salt crusts may be left behind on the desert floor. An intermittently wet and dry low area in a region of closed drainage is called a **playa**, the site of an *ephemeral lake* when water is present. Accompanying our earlier discussion of evaporites, Figure 8.10 shows such a playa in Death Valley, covered with salt precipitate just 1 month after the record rainfall event mentioned here. Figure 3.27 is a photograph of instruments recording weather conditions on the playa in Death Valley.

Permanent lakes and continuously flowing rivers are uncommon features in the desert, although the Nile River and the Colorado River are notable exceptions. Both these rivers are *exotic streams* with their headwaters in a wetter region and the majority of their channels in a

(a)

(b)

FIGURE 12.16 An improbable river in Death Valley.
The Stovepipe Wells dune field of Death Valley, California, shown (a) the day after a 2.57 cm (1.01 in.) rainfall and (b) one month later (identical location). [Photos by author.]

desert region. Focus Study 12.1 describes the Colorado River and its problem of overuse.

Alluvial Fans. In arid climates, a prominent landform is the **alluvial fan**, which occurs at the mouth of a canyon where it exits into a valley. Flowing water produces the fan when it loses velocity as it leaves the constricted channel of the canyon. It drops layer upon layer of sediment along the base of the mountain block. Water then flows over the surface of the fan and produces a braided drainage pattern, shifting from channel to channel with each precipitation event (Figure 12.17). A continuous apron, or **bajada** (Spanish for "slope") may form if individual alluvial fans coalesce into one sloping surface (see Figure 12.21). Fan formation of any sort is reduced in humid climates because perennial streams constantly carry away much of the alluvium.

An interesting aspect of an alluvial fan is the natural sorting of materials by size. Near the mouth of the canyon at the apex of the fan, coarser materials are deposited, grading slowly to pebbles and finer gravels with distance out from the mouth. Next, deposits of sands and silts drop, with the finest clays and salts carried in suspension and solution all the way to the valley floor. Minerals dissolved in solution accumulate there as evaporite deposits.

Well-developed alluvial fans also can be a major source of groundwater. Some cities—San Bernardino, California, for example—are built on alluvial fans and extract their municipal water supplies from them. However, because water resources in an alluvial fan are recharged from surface supplies, and these fans are in arid regions, groundwater mining and overdraft beyond recharge rates are common. In other parts of the world, such water-bearing alluvial fans are known as *qanat* (Iran), *karex* (Pakistan), and *foggara* (Western Sahara).

Desert Landscapes

Contrary to popular belief, deserts are not wastelands, for they abound in specially adapted plants and animals. Moreover, the limited vegetation, intermittent rainfall, intense insolation, and distant vistas produce starkly beautiful landscapes. And all deserts are not the same: For example, North American deserts have more vegetation cover than do the generally barren Saharan expanses.

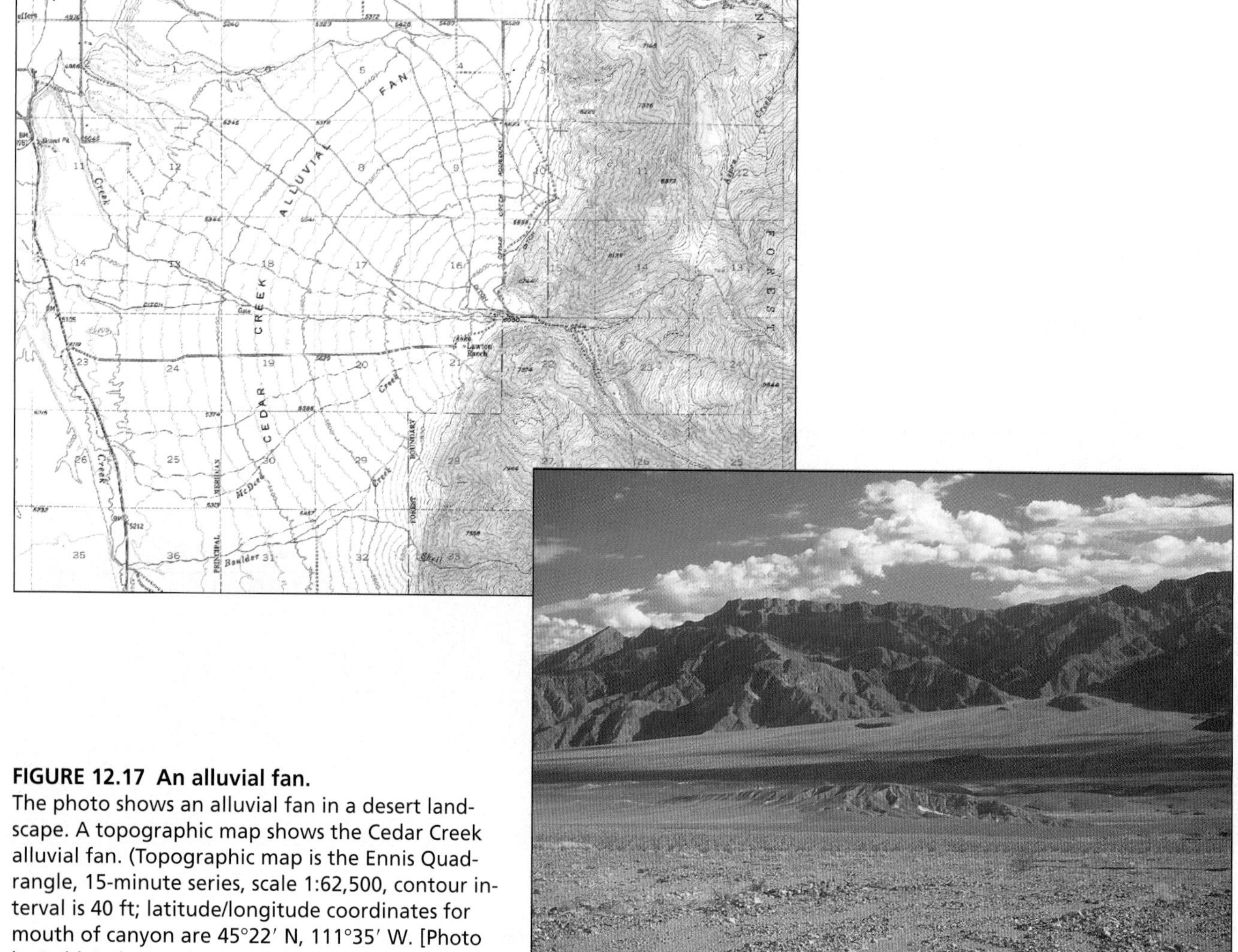

FIGURE 12.17 An alluvial fan.
The photo shows an alluvial fan in a desert landscape. A topographic map shows the Cedar Creek alluvial fan. (Topographic map is the Ennis Quadrangle, 15-minute series, scale 1:62,500, contour interval is 40 ft; latitude/longitude coordinates for mouth of canyon are 45°22′ N, 111°35′ W. [Photo by Bobbé Christopherson; USGS map.]

Deserts occur worldwide as topographic plains, such as the Great Sandy and Simpson Deserts of Australia, the Arabian and Kalahari Deserts of Africa, and portions of the extensive Taklimakan Desert, which cover some 270,000 km^2 (105,000 mi^2) in the central Tarim Basin of China (see the opening image for Chapter 6). Deserts also are found in mountainous regions: interior Asia, from Iran to Pakistan, and in China and Mongolia. In South America, lying between the ocean and the Andes, is the rugged Atacama Desert.

The shimmering heat waves and related mirage effects in the desert are products of light refraction through layers of air that have developed a temperature gradient near the hot ground. The book *Desert Solitaire* captures the desert's enchantment:

> Around noon the heat waves begin flowing upward from the expanses of sand and bare rock. They shimmer like transparent, filmy veils between my sanctuary in the shade and all the sun-dazzled world beyond. Objects and forms viewed through this tremulous flow appear somewhat displaced or distorted.... the great Balanced Rock floats a few inches above its pedestal, supported by a layer of superheated air. The buttes, pinnacles, and fins in the windows area bend and undulate beyond the middle ground like a painted backdrop stirred by a draft of air.*

Resistant horizontal rock strata in buttes, pinnacles, and mesas of arid landscapes exhibit differential weathering. Removal of the less-resistant sandstone strata produces unusual desert sculptures—arches, pedestals, and delicately balanced rocks (Figure 12.18). Specifically, the upper layers of sandstone along the top of an arch or butte are more resistant to weathering and protect the sandstone rock beneath (another example is in Figure 10.1).

Desert landscapes are places where stark erosional remnants stand above the surrounding terrain as knobs, hills, or "island mountains." Such a bare, exposed rock, called an *inselberg*, or island mountain, is exemplified by Uluru (Ayers) Rock in Australia (Figure 12.19).

In a desert area, weak surface material may weather to a complex, rugged low topography, called a **badland**, probably so-named because it offered little economic value and was difficult to traverse in nineteenth-century wagons. The Badlands region of the Dakotas is of this form, as are portions of Death Valley, California.

Sand dunes that existed in some ancient deserts lithified, forming sandstone structures that bear the imprint of cross-stratification. When such a dune was accumulating, sand cascaded down its slipface, and distinct bedding planes (layers) were established that remained after the dune lithified (Figure 12.20, p. 391). Ripple marks, animal tracks, and fossils also are found preserved in these sandstones, which originally were eolian-deposited sand dunes.

*E. Abbey, *Desert Solitaire*, © 1968 by Edward Abbey (New York: McGraw-Hill, 1968), p. 154.

FIGURE 12.18 A balanced rock—differential weathering. Balanced Rock in Arches National Park, Utah, where writer/naturalist Edward Abbey (quoted in the text) worked as a ranger years before it became a park. The overall feature is 39 m (128 ft) tall and composed of Entrada sandstone. The balance-rock portion is 17 m (55 ft) tall and weighs 3255 metric (3577 short) tons. [Photo by author.]

FIGURE 12.19 Australian landmark. Uluru (Ayers) Rock in Northern Territory, Australia, is an isolated mass of weathered rock. The formation, 348 m (1145 ft) high and 2.5 km long and 1.6 km wide (1.5 by 1.0 mi), is sacred to Aboriginal peoples. It is protected in Uluru National Park, established in 1950. [Photo by Porterfield/Chickering.]

Focus Study 12.1

The Colorado River: a System Out of Balance

An exotic stream has headwaters in a humid region of water surpluses, then flows mostly through arid lands for the rest of its journey to the sea. Exotic streams have few incoming tributaries. Consequently, an exotic stream has a discharge pattern that is opposite that of a typical stream: Instead of discharge increasing downstream, it decreases. The Nile and Colorado Rivers are prominent examples.

In the case of the Nile, the East African mountains and plateaus provide a humid source area. The Nile first rises in remote headwaters as the Kagera River in the eastern portion of the Lake Plateau country of East Africa. On its way to Lake Victoria, it forms the partial boundary of Tanzania, Rwanda, and Uganda. The Nile itself then flows from the lake and continues on its 6650 km (4132 mi) course to its mouth on the Mediterranean Sea near Cairo.

The Colorado River Basin

The Colorado rises on the high slopes of Mount Richthofen (3962 m, or 13,000 ft) in Rocky Mountain National Park, Colorado (Figure 1a) and flows almost 2317 km (1440 mi). The river ends as a trickle of water that disappears in the sand, kilometers short of its former mouth in the Gulf of California.

Orographic precipitation totaling 102 cm (40 in.) per year (mostly snow) falls in the Rockies, feeding the Colorado headwaters. But at Yuma, Arizona, near the river's end, annual precipitation is a scant 8.9 cm (3.5 in.), an extremely small amount when compared with the annual potential evapotranspiration demand in the Yuma region of 140 cm (55 in.).

From its source region, the Colorado River quickly leaves the humid Rockies and spills out into the arid desert of western Colorado and eastern Utah. At Grand Junction, Colorado, near the Utah border, annual precipitation is only 20 cm (8 in.; Figure 1b). After carving its way through the intricate labyrinth of canyonlands in Utah, the river enters Lake Powell, 945 m (3100 ft) lower in elevation than the river's source area upstream in the Rockies (Figure 1c). The Colorado then flows through the Grand Canyon chasm, formed by its own erosive power.

West of the Grand Canyon, the river turns southward, tracing its final 644 km (400 mi) as the Arizona-California border. Along this stretch of the river we find: Hoover Dam, just east of Las Vegas (Figure 1d); Davis Dam (Figure 1e), built to control the releases from Hoover; Parker Dam for the water needs of Los Angeles; three more dams for irrigation water (Figure 1f), Palo Verde, Imperial, and Laguna; and, finally, Morelos Dam at the Mexican border (Figure 1g). Mexico owns the end of the river and whatever water is left, although the river no longer reaches its mouth into the Gulf of California (Figure 1i).

Figure 1j shows the annual water discharge and suspended sediment load for the Colorado River at Yuma, Arizona, from 1905 to 1964. The completion of Hoover Dam in the 1930s dramatically reduced suspended sediment. Addition of Glen Canyon Dam upstream from Hoover Dam in 1963 further reduced stream flows. In fact, Lake Powell, which formed upstream behind the artificial base level of Glen Canyon Dam, is forecast to fill with sediment by the end of this century.

Overall, the drainage basin encompasses 641,025 km^2 (247,500 mi^2) of mountain, basin-and-range, plateau, canyon, and desert landscapes, in parts of seven states and two countries. A discussion of the Colorado River is included in this chapter because of its crucial part in the history of the Southwest and its role in the future of this dry region.

Dividing Up the Colorado's Dammed Water

John Wesley Powell (1834–1902), the first person of record to successfully navigate the Colorado River through the Grand Canyon, was the first director of the U.S. Bureau of Ethnology and later director of the U.S. Geological Survey (1881–1892). Powell perceived that the challenge of the West was too great for individual efforts and believed that solutions to problems such as water availability could be met only through private cooperative efforts. His 1878 study (reprinted 1962), *Report of the Lands of the Arid Region of the United States*, is a conservation landmark.

Today Powell probably would be skeptical of the intervention by

(continued)

FIGURE 1 The Colorado River Basin.
The Colorado River basin, showing division of the upper and lower basins near Lees Ferry in northern Arizona. (a) Headwaters of the Colorado River near Mount Richthofen in the Colorado Rockies. (b) The river near Moab, Utah. (c) Glen Canyon Dam, a regulatory, administrative facility near Lees Ferry, Arizona. (d) Hoover Dam spillways in rare operation during 1983 floods. (e) Davis Dam in full release during flood. (f) Irrigation canal and cropland. (g) Morelos Dam at the Mexican border is the final stop as the river dwindles to a mere canal. (h) Central Arizona Project aqueduct west of Phoenix. (i) The Colorado River stops short of its former delta in the Gulf of California. Note the ecological contrast between California and Mexico sides of the border. (j) Dam construction affects river discharge and sediment yields. [(a) Photo by Bobbé Christopherson; (b through g) photos by author; (h) photo by Tom Bean/DRK Photo; (i) August 25, 2002, *Terra* MODIS image courtesy of MODIS Rapid Response Team, NASA/GSFC; and, (j) data from USGS, 1985, *National Water Summary 1984*, Water Supply Paper 2275 (Washington, D.C.: Government Printing Office), p. 55).]

Focus Study 12.1 *(continued)*

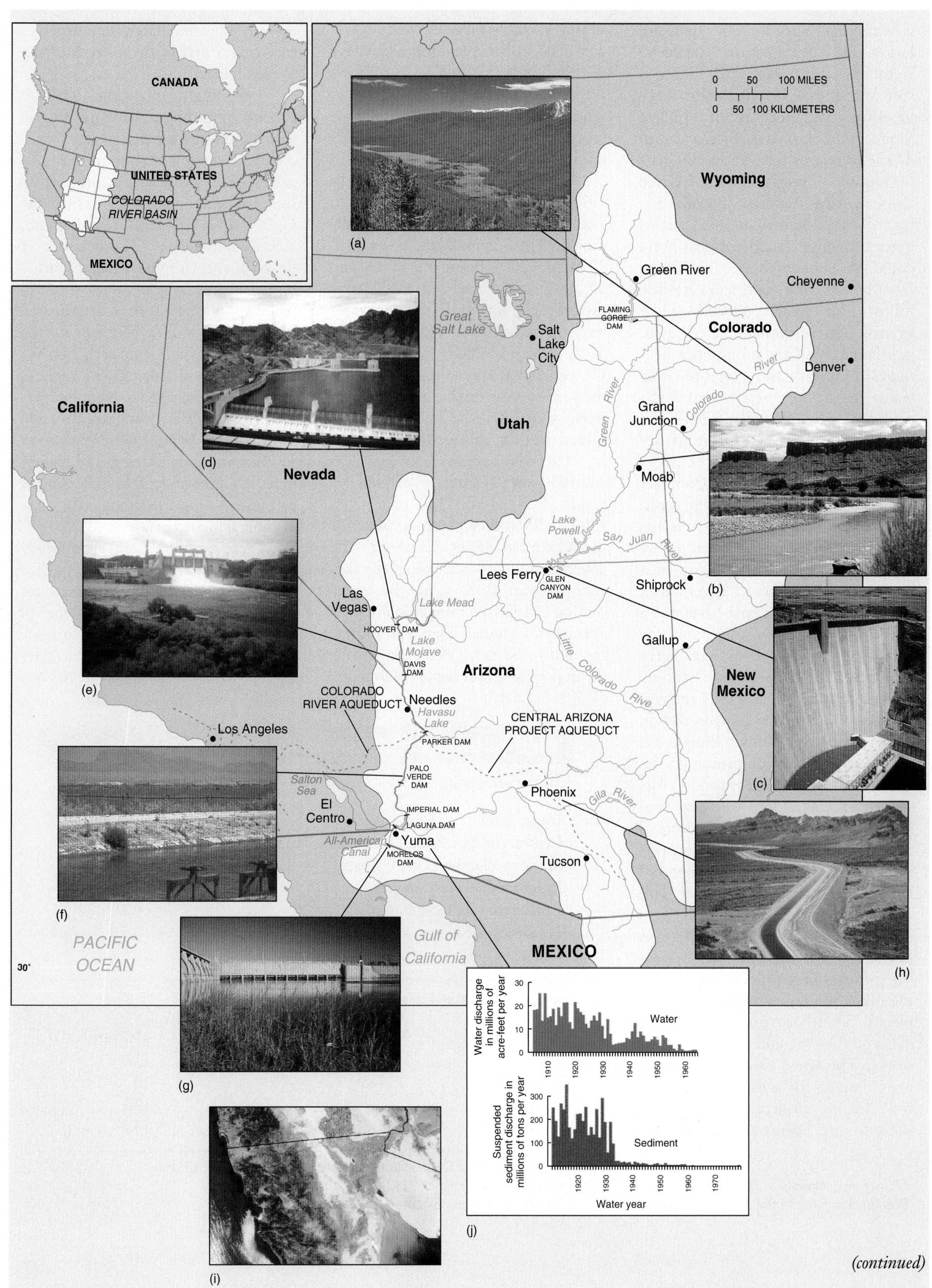

(continued)

Focus Study 12.1 *(continued)*

government agencies in building large-scale reclamation projects. Lake Powell is named after him, despite his probable opposition were he alive. An anecdote in Wallace Stegner's book *Beyond the Hundredth Meridian* relates that, at an 1893 international irrigation conference held in Los Angeles, Powell observed development-minded delegates bragging that the entire West could be conquered and reclaimed from nature, that "rain will certainly follow the plow." Powell spoke against that sentiment: "I tell you, gentlemen, you are piling up a heritage of conflict and litigation over water rights, for there is not sufficient water to supply the land."* He was booed from the hall. But history has shown Powell to be correct.

The Colorado River Compact was signed by six of the seven basin states in 1923. (The seventh, Arizona, signed in 1944, the same year as the Mexican Water Treaty.) With this compact, the Colorado River basin was divided into an upper basin and a lower basin, arbitrarily separated for administrative purposes at Lees Ferry near the Utah-Arizona border (noted on the map in Figure 1). Congress adopted the Boulder Canyon Act in 1928, authorizing Hoover Dam as the first major reclamation project on the river. Also authorized was the All-American Canal into the Imperial Valley, which required an additional dam. Los Angeles then began its project to bring Colorado River water 390 km (240 mi) from still another dam and reservoir to their city.

Shortly after Hoover Dam was finished and downstream enterprises were thus offered flood protection, the other projects were quickly completed. There are now eight major dams on the river and many irrigation works. The latest effort to redistribute Colorado River water is the Central Arizona Project, which carries water to the Phoenix area (Figure 1h).

* W. Stegner, *Beyond the Hundredth Meridian* (Boston: Houghton Mifflin Co., 1954), p. 343.

Highly Variable River Flows

The flaw in this planning and water distribution is that exotic streamflows are highly variable, and the Colorado is no exception. In 1917, the discharge measured at Lees Ferry totaled 24 million acre-feet (maf), whereas in 1934 it dropped to only 5.03 maf. In 1977, the discharge dropped again to 5.02 maf, but in 1984 it rose to an all-time high of 24.5 maf. Yet, in contrast, 2000 and 2001 flows fell to 11 and 10.7 maf, respectively, and 2002 was exceptionally low in discharge. Water levels in Lake Powell were the lowest in its history in 2003. In addition to this variability, approximately 70% of the year-to-year discharge is between April and July; the other 30% is spread over the balance of the year.

The average flows between 1906 and 1930 were 17.7 maf a year, but averages dropped to 13.25 maf from 1990 to 2002. As a planning basis for the Colorado River Compact, the government used average river discharges from 1914 up to the treaty signing in 1923, an exceptionally high average of 18.8 maf. That amount was thought to be more than enough for the upper and lower basins each to receive 7.5 maf and, later, for Mexico to receive 1.5 maf in the 1944 Mexican Water Treaty.

We might question whether proper long-range planning should rely on the providence of high variability. Tree-ring analyses of past climates disclosed that the only other time Colorado discharges were at the high 1914–1923 level was between A.D. 1606 and 1625. The dependable flows of the river have been consistently overestimated. This shortfall problem is shown in an estimated budget of 20-million acre feet for the river's water (Table 1); clearly, the situation is out of balance, for there is not enough discharge to meet budgeted demands. A few wet years here and there (1997, 1995, 1984–1986) seem to spark more confidence and trigger more disputes on how to divvy up the surpluses—only delaying the inevitable deficit crisis.

Presently, the seven states ideally want rights that total as high as 25.0 maf. When added to the guarantee for Mexico this comes up to 26.5 maf of water wants. And six states share one opinion: California's right to the water

(continued)

Table 1 Estimated Colorado River Budget through 2002

Water Demand	Quantity (maf)[a]
Upper Basin (7.5) Lower Basin (7.5)[b]	15.0
Central Arizona Project (rising to 2.8 maf)	1.0
Mexican allotment (1944 Treaty)	1.5
Evaporation from reservoirs	1.5
Bank storage at Lake Powell	0.5
Phreatophytic losses (water-demanding plants)	0.5
Budgeted total demand	20.0 maf
Average flows at Lees Ferry	
1990–2002	13.25 maf
2000	11.06
2001	10.75
1906–1930	17.7
1931–2002	14.8

Source: Bureau of Reclamation, and states of Arizona and California.
[a] 1 million acre-feet = 325,872 gallons; 1.24 million liters.
[b] In-basin consumptive uses 75% agricultural.

Focus Study 12.1 ***(continued)***

must be limited to a court-ordered 4.4 maf, which it exceeds every year—1998 saw California take 4.9 maf; 2001, 5.2 maf. California's access to surplus Colorado River water ceased in 2002.

Water Loss at Glen Canyon Dam and Lake Powell

Glen Canyon Dam was completed and began water impoundment (Lake Powell) in 1963, 27.4 km (17 mi) north of the basin division point at Lees Ferry. The advancing water flooded the deep, fluted, inner gorges of many canyons, including the Glen, Navajo, Labyrinth, and Cathedral.

Glen Canyon Dam's primary purpose, according to the Bureau of Reclamation, is to regulate flows between the upper and lower basins. An additional benefit is the production of hydroelectric power, which is sold at a low wholesale rate to utilities across the Southwest. Also, there is a growing recreation and tourism industry around Lake Powell, as previously inaccessible desert scenery now is reachable by boat.

Many thought that Lake Mead, behind Hoover Dam, could serve the primary administrative function of flow regulation, especially considering the serious water-loss problems with the Glen Canyon Dam. Porous Navajo sandstone underlies most of the Lake Powell reservoir at Glen Canyon Dam. This sandstone absorbs an estimated 0.5 maf of the river's overall annual discharge as bank storage. The higher the lake level, the greater the loss into the sandstone. Secondly, Lake Powell is an open body of water in an arid desert, where hot, dry winds accelerate evaporative losses. Another 0.5 maf of the Colorado's overall discharge is lost annually from the Lake Powell reservoir in this manner. Thirdly, now-permanent sandbars and banks have stabilized along the regulated river, allowing water-demanding plants called phreatophytes to establish and extract an additional 0.5 maf of the river flow.

Flood Control in 1983

Intense precipitation and heavy snowpack in the Rockies, attributable to the 1982–1983 El Niño (see Focus Study 6.1), led to record-high discharge rates on the Colorado, testing the controllability of one of the most dammed rivers in the world. Federal reservoir managers were not prepared for the high discharge, since they had set aside Lake Mead's primary purpose (flood control) in favor of competing water and power interests. What followed was a human-caused flood on the most regulated river in the world!

The only time the spillways at Hoover Dam had ever operated was more than 40 years earlier, when the reservoir capacity was artificially raised for a test; now they were opened to release the floodwaters (Figure 1d; Chapter 7's opening photo illustrates this event). Davis Dam, which regulates releases from Hoover Dam, was within 30 cm (1 ft) of overflow, a real problem for a structure made partially of earth fill (Figure 1e). In addition, Glen Canyon Dam was over capacity and at risk, and was damaged by the volume of discharge tearing through its spillways. The decision to increase releases doomed towns and homeowners along the river to flooding, especially in subdivisions near Needles, California.

We might wonder what John Wesley Powell would think if he were alive today to witness such errant attempts to control the mighty and variable Colorado. He foretold such a "heritage of conflict and litigation."

Basin and Range Province. A *province* is a large region that shares several geologic or physiographic traits. The **Basin and Range Province** of the western United States consists of alternating basins and mountain ranges that lie in the rain shadow of mountains to the west. Thus, the province has a dry climate, few permanent streams, and *interior drainage patterns*—drainage basins that lack any outlet to the ocean (see Figure 11.5). The Basin and Range Province—almost 800,000 km^2 (300,000 mi^2)—was a major barrier to early settlers in their migration westward (Figure 12.21a, b). The desert climate and north-south-trending mountain ranges were harsh challenges.

FIGURE 12.20 Cross bedding in sedimentary rocks. The bedding pattern, called cross-stratification, in these sandstone rocks tells us about conditions that were established in the dunes before lithification (hardening into rock). [Photo by Bobbé Christopherson.]

As the North American plate moved westward, it overrode former oceanic crust and hot spots at such a rapid pace that slabs of subducted material literally were run over. This motion stretched the crust, creating a landscape fractured by many faults. The present landscape consists of nearly parallel sequences of *horsts* (upward-faulted blocks, which are the "ranges") and *grabens* (downward-faulted blocks, which are the "basins" or valleys). Figure 12.21c illustrates this pattern. Note the **bolson**, a slope-and-basin area between the crests of two adjacent ridges in a dry region of interior drainage. The figure also identifies a *playa* (central salt pan), and a *bajada* (coalesced alluvial fans). A *pediment* is an area of bedrock that is covered with a thin veneer, or coating, of alluvium. It is an erosional surface, as opposed to the depositional surface of the bajada, and is created as a mountain front retreats through weathering and erosion.

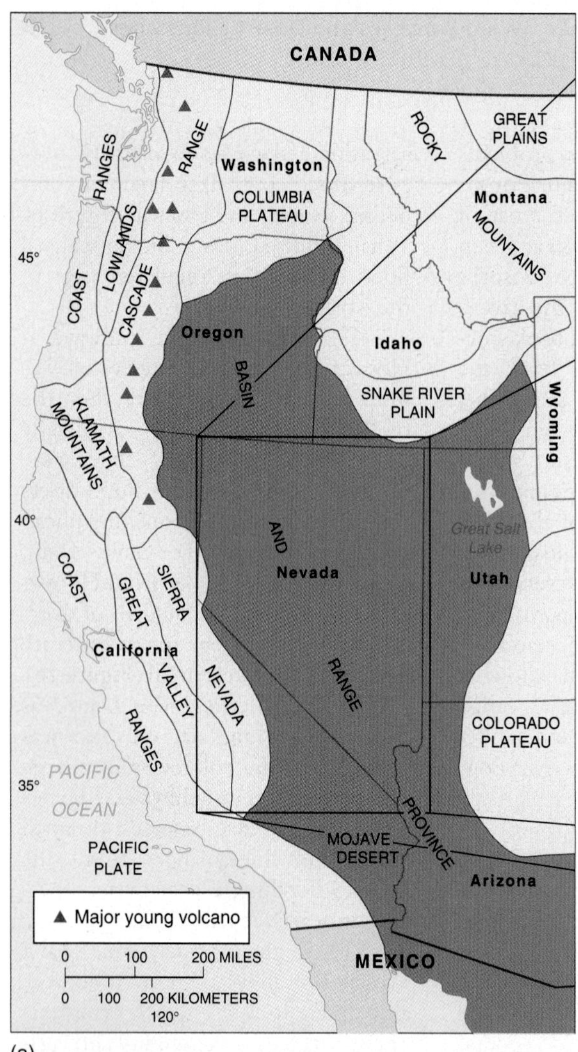

(a)

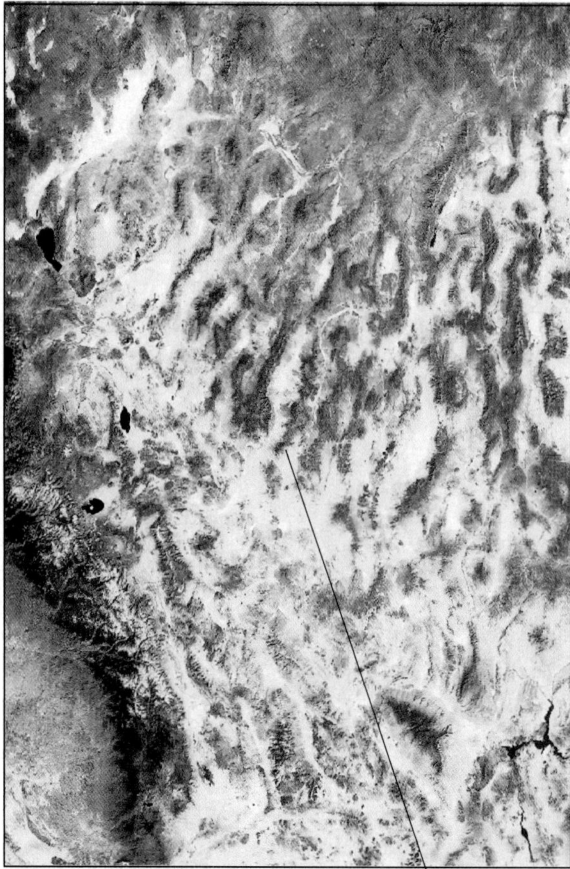

(b)

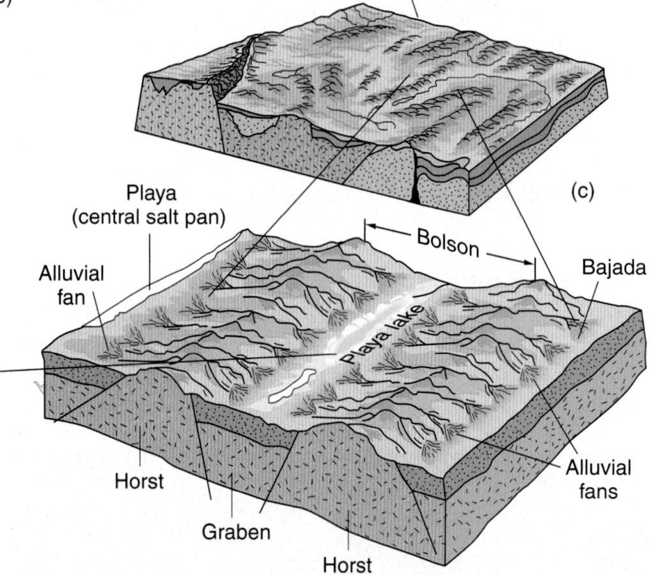

(c)

(d)

(e)

FIGURE 12.21 Basin and Range Province in the western United States.
Recent scientific discoveries demonstrate that this province extends south through northern and central Mexico. (a) Map and (b) *Landsat* image of the area. (c) A bolson in the mountainous desert landscape of the Basin and Range Province. Parallel normal faults produce a series of horsts (ranges) and grabens (basins). (d) Death Valley features a central playa, parallel mountain ranges, alluvial fans, and bajada along the base of the ranges. (e) "The Loneliest Road in America," Highway U.S. 50, slashes westward through the province. [(b) Image from NASA; photos by (d) author; (e) Bobbé Christopherson.]

John McPhee captured the feel of this desert province in his book *Basin and Range*:

> Supreme over all is silence. Discounting the cry of the occasional bird, the wailing of a pack of coyotes, silence—a great spatial silence—is pure in the Basin and Range. It is a soundless immensity with mountains in it. You stand . . . and look up at a high mountain front, and turn your head and look fifty miles down the valley, and there is utter silence.*

Basin-and-range relief is abrupt, and rock structures are angular and rugged. As the ranges erode, transported materials accumulate to great depths in the basins, gradually producing extensive desert plains. The basin's elevation averages between 1220 and 1525 m (4000–5000 ft) above sea level, with mountain crests rising higher by some 915 to 1525 m (3000–5000 ft). Death Valley is an extreme example (Figure 12.21d); its lowest basin has an elevation of −86 m (−282 ft). However, to the west of the valley, the Panamint Range rises to 3368 m (11,050 ft) at Telescope Peak—over 3 vertical kilometers (2 mi) of desert relief.

When you traverse U.S. Highway 50 across Nevada, you cross five passes (horsts) of over 1950 m (6400 ft) and numerous basins (grabens). Throughout the drive you are reminded where you are by prideful signs that plainly state: "The Loneliest Road in America." It is difficult to imagine crossing this topography with wagons and oxen in the 19th century (Figure 12.21e).

*J. McPhee, *Basin and Range* (New York: Farrar, Straus, Giroux, 1981), p. 46.

Desertification

We are witnessing an unwanted expansion of Earth's desert lands in a process known as **desertification**. This now is a worldwide phenomenon along the margins of semiarid and arid lands. Desertification is due principally to poor agricultural practices (overgrazing and agricultural activities that abuse soil structure and fertility), improper soil-moisture management, erosion and salinization, deforestation, and the ongoing global climatic change that is shifting temperature and precipitation patterns.

The southward expansion of Saharan conditions through portions of the *Sahel region* has left many African peoples on land that no longer experiences the rainfall of just two decades ago. Other regions at risk of desertification stretch from Asia and central Australia to portions of North and South America. The United Nations estimates that degraded lands have covered some 800 million hectares (2 billion acres) since 1930; many millions of additional hectares are added each year (Figure 12.22). Urgently needed is an improved database for a more accurate accounting of the problem and a better understanding of what is occurring.

Figure 12.22 is drawn from a map prepared for a U.N. conference on desertification. Desertification areas are ranked: A moderate hazard area has an average 10%–25% drop in agricultural productivity; a high hazard area has a 25%–50% drop; and a very high hazard area has more than a 50% decrease. Because human activities and economies, especially unwise grazing practices, appear to be the major cause of desertification, actions to slow the process are available. The severity of this problem is magnified by the poverty in many of the affected regions.

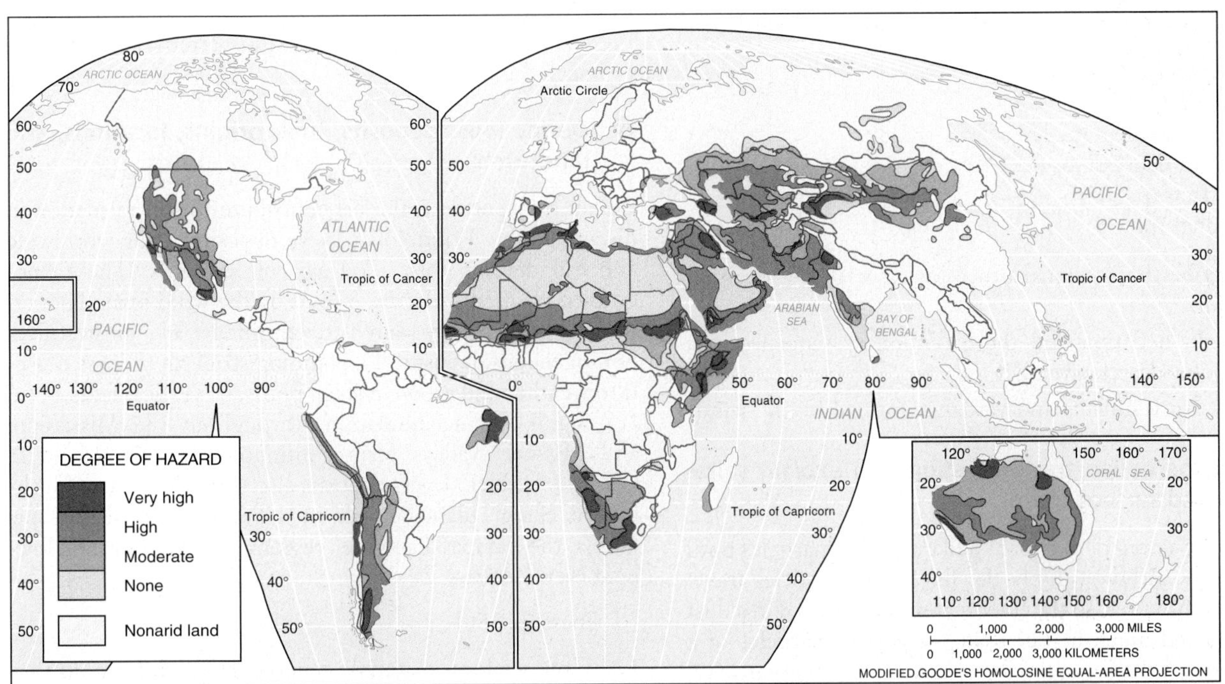

FIGURE 12.22 The desertification hazard.
Worldwide desertification estimates by the United Nations. [Data from U.N. Food and Agricultural Organization (FAO), World Meteorological Organization (WMO), United Nations Educational, Scientific, and Cultural Organization (UNESCO), Nairobi, Kenya; as printed in J. M. Rubenstein, *An Introduction to Human Geography*, Figure 14–16, p. 509. © 1999 Prentice Hall, Inc. Used by permission.]

Summary and Review—Wind Processes and Desert Landscapes

● *Characterize* the unique work accomplished by wind and eolian processes.

The movement of the atmosphere in response to pressure differences produces wind. Wind is a geomorphic agent of erosion, transportation, and deposition. **Eolian** processes modify and move sand accumulations along coastal beaches and deserts. Wind's ability to move materials is small compared with that of water and ice.

eolian (p. 374)

1. Who was Ralph Bagnold? What was his contribution to eolian studies?
2. Explain the term *eolian* and its application in this chapter. How would you characterize the ability of wind to move material?

● *Describe* eolian erosion, including deflation, abrasion, and the resultant landforms.

Two principal wind-erosion processes are **deflation**, the removal and lifting of individual loose particles, and **abrasion**, the "sandblasting" of rock surfaces with particles captured in the air. Fine materials are eroded by wind deflation and moving water, leaving behind concentrations of pebbles and gravel called **desert pavement**. Wherever wind encounters loose sediment, deflation may remove enough material to form basins. Called **blowout depressions**, they range from small indentations less than a meter wide up to areas hundreds of meters wide and many meters deep. Rocks that bear evidence of eolian erosion are called **ventifacts**. On a larger scale, deflation and abrasion are capable of streamlining rock structures, leaving behind distinctive rock formations or elongated ridges called **yardangs**.

deflation (p. 374)
abrasion (p. 374)
desert pavement (p. 374)
blowout depressions (p. 374)
ventifacts (p. 375)
yardangs (p. 375)

3. Describe the erosional processes associated with moving air.
4. Explain deflation and the evolutionary sequence that produces desert pavement.
5. How are ventifacts and yardangs formed by the wind?

● *Describe* eolian transportation and *explain* saltation and surface creep.

Wind exerts a drag or frictional pull on surface particles until they become airborne. Only the finest dust particles travel significant distances, so the finer material suspended in a dust storm is lifted much higher than the coarser particles of a sand storm. Saltating particles crash into other particles, knocking them both loose and forward. The motion called **surface creep** slides and rolls particles too large for saltation.

surface creep (p. 377)

6. Differentiate between a dust storm and a sand storm.
7. What is the difference between eolian saltation and fluvial saltation?
8. Explain the concept of surface creep.

● *Identify* the major classes of sand dunes and *present* examples within each class.

In arid and semiarid climates and along some coastlines where sand is available, **dunes** accumulate. A dune is a wind-sculpted accumulation of sand. An extensive area of dunes, such as that found in North Africa, is characteristic of an **erg desert**, or **sand sea**. When saltating sand grains encounter small patches of sand, their kinetic energy (motion) is dissipated and they start to accumulate into a dune. As height increases above 30 cm (12 in.), a steeply sloping *slipface* on the lee side and characteristic dune features are formed. Dune forms are broadly classified as *crescentic*, *linear*, and *star*.

dunes (p. 377)
erg desert (p. 377)
sand sea (p. 377)

9. What is the difference between an erg desert and desert pavement? Which type is a sand sea? Are all deserts covered by sand? Explain.
10. What are the three classes of dune forms? Describe the basic types of dunes within each class. What do you think is the major shaping force for sand dunes?
11. Which form of dune is the mountain giant of the desert? What are the characteristic wind patterns that produce such dunes?

● *Define* loess deposits, their origins, locations, and landforms.

Eolian transported materials contribute to soil formation in distant places. Windblown **loess** deposits occur worldwide and can develop into good agricultural soils. These fine-grained clays and silts are moved by the wind many kilometers, where they are redeposited in unstratified, homogeneous deposits. The binding strength of loess causes it to weather and erode in steep bluffs, or vertical faces.

Significant accumulations throughout the Mississippi and Missouri valleys form continuous deposits 15–30 m (50–100 ft) thick. Loess deposits also occur in eastern Washington State, Idaho, much of Ukraine, central Europe, China, the Pampas-Patagonia regions of Argentina, and lowland New Zealand.

loess (p. 380)

12. How are loess materials generated? What form do they assume when deposited?
13. Name a few examples of significant loess deposits on Earth.

- ***Portray* desert landscapes and *locate* these regions on a world map.**

Dry and semiarid climates occupy about 35% of Earth's land surface. The spatial distribution of these dry lands is related to subtropical high-pressure cells between 15° and 35° N and S, to rain shadows on the lee side of mountain ranges, or to areas at great distance from moisture-bearing air masses, such as central Asia.

Water events are rare, yet running water is still the major erosional agent in deserts. Precipitation events may be rare, but when they do occur, a dry streambed may fill with a torrent called a **flash flood**. Depending on the region, such a dry streambed is known as a **wash**, an *arroyo* (Spanish), or a *wadi* (Arabic). As runoff water evaporates, salt crusts may be left behind on the desert floor. An intermittently wet and dry low area in a region of closed drainage is called a **playa**, site of an *ephemeral lake* when water is present.

In arid climates, a prominent landform is the **alluvial fan** at the mouth of a canyon where it exits into a valley. The fan is produced by flowing water that abruptly loses velocity as it leaves the constricted channel of the canyon and deposits a layer of sediment along the mountain block. A continuous apron, or **bajada**, may form if individual alluvial fans coalesce. In a dry region, weak surface materials may weather to a complex, rugged low topography, called a **badland.** A *province* is a large region that is characterized by several geologic or physiographic traits. The **Basin and Range Province** of the western United States consists of alternating basins and mountain ranges. A slope-and-basin area between the crests of two adjacent ridges in a dry region of interior drainage is termed a **bolson. Desertification** is the process that leads to unwanted expansion of the Earth's desert lands.

flash flood (p. 385)
wash (p. 385)
playa (p. 385)
alluvial fan (p. 386)
bajada (p. 386)
badland (p. 391)
Basin and Range Province (p. 391)
bolson (p. 391)
desertification (p. 393)

14. Characterize desert energy and water balance regimes. What are the significant patterns of occurrence for arid landscapes in the world?
15. How would you describe the water budget of the Colorado River? What was the basis for agreements regarding distribution of the river? Why has thinking about the river's discharge been so optimistic?
16. Describe a desert bolson from crest to crest. Draw a simple sketch with the components of the landscape labeled.
17. Where is the Basin and Range Province? Briefly describe its appearance and character.
18. What is meant by desertification? Using the map in Figure 12.22, and the text description, locate several of the affected regions of desert expansion.

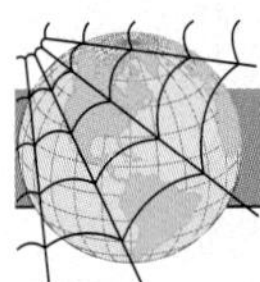

NetWork

The *Elemental Geosystems Home Page* provides on-line resources for this chapter on the World Wide Web. Once on the Home Page, click on this textbook, scroll the Table of Contents menu, select this chapter, and click "Begin." You will find self-tests that are graded, review exercises, specific updates for items in the chapter, and in "Destinations" many links to interesting related pathways on the Internet. *Elemental Geosystems* is found at **http://www.prenhall.com/christopherson.**

Critical Thinking

A. "Water is the major erosion and transport medium in the desert." Respond to this quotation. How is this possible? What factors have you learned from this chapter that prove this statement true?

B. Where are the nearest eolian features (coastal, lakeshore, desert dunes, or loess deposits) to your present location? Which causative factors discussed in this chapter explain the features you identified?

C. Given the estimated Colorado River budget in Table 1, Focus Study 12.1, analyze what this means for the future of the region. How is it possible for such a budget to be sustainable? In your opinion, should such realities be part of the planning and development process in the region?

Sand and shell beach along the New Jersey shore. An ongoing sand nourishment program is necessary to maintain the sand inventory, which is under assault from storms and rising sea levels. [Photo by Bobbé Christopherson.]

13 The Oceans, Coastal Processes, and Landforms

Key Learning Concepts

By knowing and understanding the key learning concepts in this chapter, you should be able to:

- *Describe* the chemical composition of seawater and the physical structure of the ocean.
- *Identify* the components of the coastal environment and *list* the physical inputs to the coastal system, including tides and mean sea level.
- *Describe* wave motion at sea and near shore and *explain* coastal straightening as a product of wave refraction.
- *Identify* characteristic coastal erosional and depositional landforms.
- *Describe* barrier islands and their hazards as they relate to human settlement.
- *Assess* living coastal environments: corals, wetlands, salt marshes, and mangroves.
- *Construct* an environmentally sensitive model for settlement and land use along the coast.

Walk along a shoreline and you witness the dramatic interaction of Earth's vast oceanic, atmospheric, and lithospheric systems. At times, the ocean attacks the coast in a stormy rage of erosive power; at other times, the moist sea breeze, salty mist, and repetitive motion of the water are gentle and calming. Few have captured the confrontation between land and sea as well as Rachel Carson:

> The edge of the sea is a strange and beautiful place. All through the long history of Earth it has been an area of unrest where waves have broken heavily against the land, where the tides have pressed forward over the continents, receded, and then returned. For no two successive days is the shoreline precisely the same. Not only do the tides advance and retreat in their eternal rhythms, but the level of the sea itself is never at rest. It rises or falls as the glaciers melt

> or grow, as the floors of the deep ocean basins shift under its increasing load of sediments, or as the Earth's crust along the continental margins warps up or down in adjustment to strain and tension. Today a little more land may belong to the sea, tomorrow a little less. Always the edge of the sea remains an elusive and indefinable boundary.*

Commerce and access to sea routes, fishing, and tourism prompt many people to settle near the ocean. A 1995 scientific assessment estimates that about 40% of Earth's population live within 100 km (62 mi) of coastlines. In the United States, about 50% of the people live in areas designated as *coastal* (this includes the Great Lakes). Therefore, an understanding of coastal processes and landforms is important to humanity. And because these processes along coastlines often produce dramatic change, they are essential to consider in planning and development. A World Resources Institute study found as much as half of the world's coastlines at some risk of loss or disruption (see "Coastlines at Risk" at http://www.wri.org/wri/indictrs/coastrsk.htm).

The ocean is a vast ecosystem, intricately linked to life on the planet and to life-sustaining systems in the atmosphere, the hydrosphere, and the lithosphere. In the recent *Atlas of the Oceans—The Deep Frontier*, Jean-Michel Cousteau commented on "The Future of the Ocean":

> Today, we are coming to better appreciate the extent to which our actions affect an ecosystem—and the people who depend on it—thousands of miles away. The reef fisherman in Fiji is not undone by the local poacher, but by global warming intensified by the driving of a car in downtown Toronto, Canada. Yet these connections are not all bad news. The web of interdependence is built with strands of responsibility and hope. Our ever-expanding ability to communicate across borders and oceans is helping to drive a truly global dialogue about the planet's most pressing environmental challenges.†

In this chapter: We begin the chapter with a brief look at our global oceans and seas—1998 was celebrated as The International Year of the Ocean by all United Nations countries (see http://www.yoto98.noaa.gov/). The physical and chemical properties of the ocean distinguish it from the waters of the continent. The coastlines are areas of dynamic change and beauty, where oceans and seas confront the land. Coverage includes discussions of tides, waves, coastal erosional and depositional landforms, beaches, barrier islands, and organic processes, including corals, wetlands, salt marshes, and mangroves. A systems framework of specific inputs (components and driving forces), actions (movements and processes), and outputs (results and consequences) organize our discussion of coastal processes. We conclude with a look at the considerable human impact on coastal environments.

*Rachel Carson, "The Marginal World," in *The Edge of the Sea*. © 1955 by Rachel Carson, © renewed 1983 by R. Christie (Boston: Houghton Mifflin), p. 11.

†Sylvia Earle, *Atlas of the Oceans—The Deep Frontier*. © 2001 by National Geographic Society, text © 2001 by Sylvia Earle (Washington: National Geographic Society), p. 171.

Global Oceans and Seas

The ocean is one of Earth's last great scientific frontiers and is of great interest to geographers. Remote sensing from orbiting spacecraft, aircraft, surface vessels, a network of bouys, and submersibles is providing a wealth of data and a new understanding of the oceanic system. The pattern of sea-surface temperatures is presented in Figure 3.22 and ocean currents in Figure 4.21. The world's oceans, their area, volume, and depth, are listed in a table in Figure 13.1 along with the locations of oceans and major seas.

Chemical Composition of Seawater

Water is called the "universal solvent," dissolving at least 57 of the 92 elements found in nature. In fact, most natural elements and the compounds they form are found in the seas as dissolved solids, or *solutes*. Thus, seawater is a solution, and the concentration of dissolved solids is called **salinity**.

The oceans are a remarkably homogeneous mixture. The ratio of individual salts does not change, despite minor fluctuations in overall salinity. In 1874 the British *HMS Challenger* sailed around the world, taking surface and depth measurements and collecting samples of seawater. Analyses of those samples first demonstrated the uniform composition of seawater.

Ocean Chemistry. Ocean chemistry is a result of complex exchanges among seawater, the atmosphere, minerals, bottom sediments, and living organisms. In addition, significant flows of mineral-rich water enter the ocean through hydrothermal (hot water) vents in the ocean floor. These vents are called "black smokers" for the dense, black, mineral-laden water that spews from them. The uniformity of seawater results from complementary chemical reactions and continuous mixing—after all, the ocean basins interconnect and water circulates among them.

Seven elements account for more than 99% of the dissolved solids in seawater. They are (with their ionic form): chlorine (as chloride, Cl^-), sodium (as Na^+), magnesium (as Mg^{2+}), sulfur (as sulfate, SO_4^{2-}), calcium (as Ca^{2+}), potassium (as K^+), and bromine (as bromide, Br^-). Seawater also contains dissolved gases (such as carbon dioxide, nitrogen, and oxygen), suspended and dissolved organic matter, and a multitude of trace elements.

Commercially, only sodium chloride (common table salt), magnesium, and bromine are extracted in any significant amount from the ocean. Future mining of minerals from the seafloor is technically feasible, although it remains uneconomical.

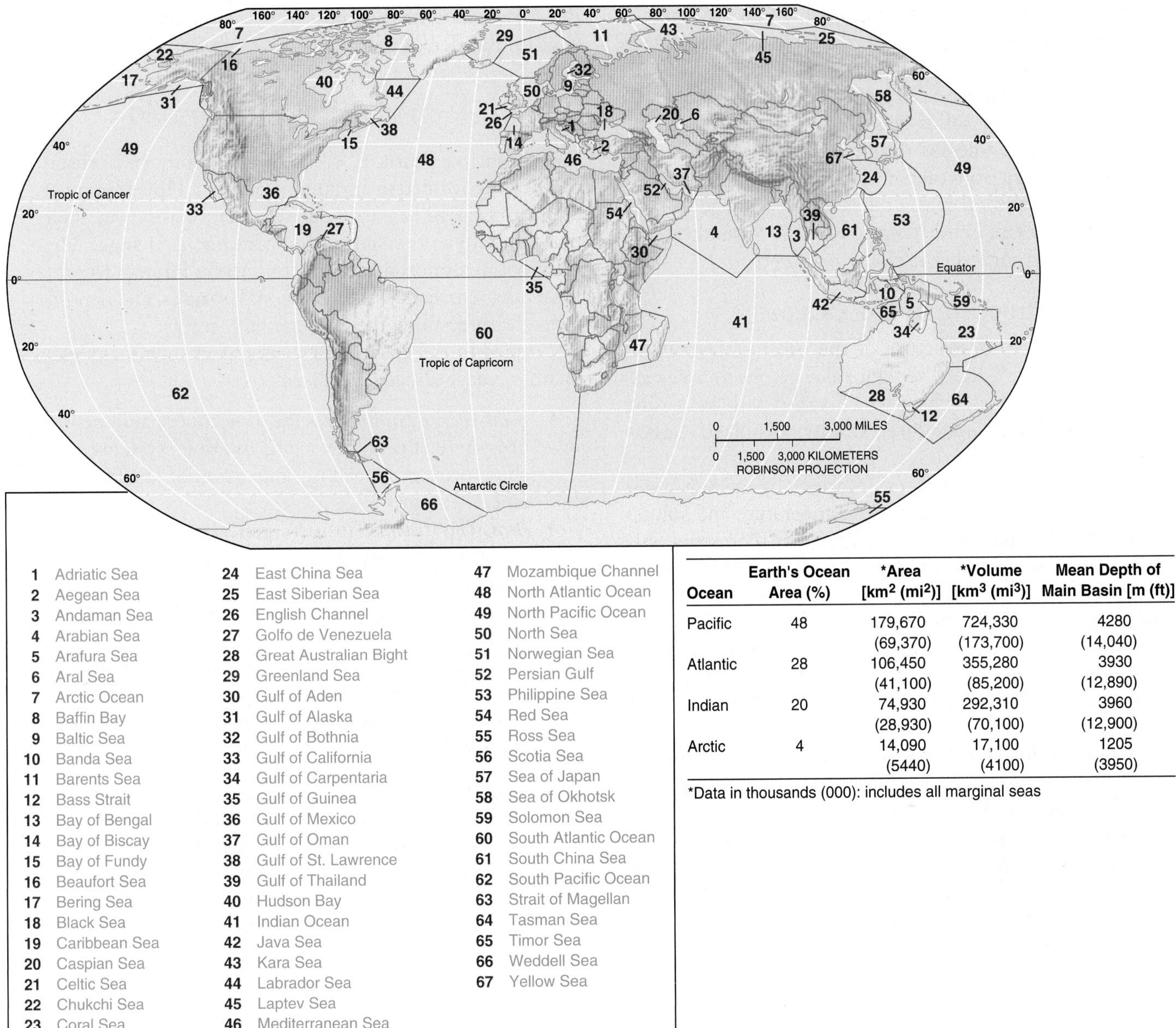

Ocean	Earth's Ocean Area (%)	*Area [km^2 (mi^2)]	*Volume [km^3 (mi^3)]	Mean Depth of Main Basin [m (ft)]
Pacific	48	179,670 (69,370)	724,330 (173,700)	4280 (14,040)
Atlantic	28	106,450 (41,100)	355,280 (85,200)	3930 (12,890)
Indian	20	74,930 (28,930)	292,310 (70,100)	3960 (12,900)
Arctic	4	14,090 (5440)	17,100 (4100)	1205 (3950)

*Data in thousands (000): includes all marginal seas

FIGURE 13.1 Principal oceans and seas of the world.
A sea is generally smaller than an ocean and is near a landmass; sometimes the term refers to a large, inland, salty body of water. A profile of Earth's four oceans is in the inset table.

Average Salinity: 35‰. There are several ways to express salinity (dissolved solids by volume) in seawater, using the worldwide average value:

- 35‰ (‰ = parts per thousand), the most common notation
- 3.5% (% = parts per hundred)
- 35,000 ppm (parts per million)
- 35,000 mg per liter
- 35 g/kg

Salinity worldwide normally varies between 34‰ and 37‰; variations are attributable to atmospheric conditions above the water and to the volume of freshwater inflows. In equatorial water, precipitation is great throughout the year, diluting salinity values to slightly lower than average (34.5‰). In subtropical oceans—where evaporation rates are greatest because of the influence of hot, dry subtropical high-pressure cells—salinity is more concentrated, increasing to 36‰.

The term **brine** is applied to water that exceeds the average of 35‰ salinity. **Brackish** applies to water that is less than 35‰ salts. In general, oceans are lower in salinity near landmasses because of freshwater runoff and river discharges. Extreme examples include the Baltic Sea (north of Poland and Germany) and the Gulf of Bothnia (between Sweden and Finland), which average 10‰ or less salinity because of heavy freshwater runoff and low evaporation rates.

On the other hand, the Sargasso Sea, within the North Atlantic subtropical gyre (circulation pattern), averages

38‰. The Persian Gulf has a salinity of 40‰ as a result of high evaporation rates in a nearly enclosed basin. Deep pockets, or "brine lakes," along the floor of the Red Sea and the Mediterranean Sea register up to a salty 225‰.

The dissolved solids remain in the ocean, but the water recycles endlessly through the hydrologic cycle, driven by energy from the Sun. The water you drink today may have water molecules in it that not long ago were in the Pacific Ocean, in the Yangtze River, in groundwater in Sweden, or airborne in the clouds over Peru.

Physical Structure of the Ocean

The basic physical structure of the ocean is layered, as shown in Figure 13.2. The figure also graphs four key aspects of the ocean, each of which varies with increasing depth: average temperature, salinity, dissolved carbon dioxide, and dissolved oxygen level.

The ocean's surface layer is warmed by the Sun and is wind-driven. Variations in water temperature and solutes are blended rapidly in a *mixing zone* that represents only 2% of the oceanic mass. Below the mixing zone is the *thermocline transition zone*, a more than kilometer-deep region of decreasing temperature gradient that lacks the motion of the surface. Friction at these depths dampens the effect of surface currents. In addition, colder water temperatures at the lower margin tend to inhibit any convective movements.

From a depth of 1–1.5 km (0.6–0.9 mi) to the ocean floor, temperature and salinity values are quite uniform. Temperatures in this *deep cold zone* are near 0°C (32°F). Water in the deep cold zone does not freeze, however, because of its salinity and intense pressures at those depths; seawater freezes at about −2°C (28.4°F) at the surface. The coldest water is at the bottom, except near the poles, where the coldest water may be near or at the surface. Now let us shift to the edge of the sea and examine Earth's coastlines.

Coastal System Components

Earth's surface features, such as mountains and crustal plates, were formed over millions of years. However, most of Earth's coastlines are relatively new, existing in their present state as the setting for continuous change. The land, ocean, atmosphere, Sun, and Moon interact to produce tides, currents, waves, erosional features, and depositional features along the continental margins.

Inputs to the coastal environment include many elements we have already discussed:

- *Solar energy* input drives the atmosphere and the hydrosphere. Prevailing winds, weather systems, and climate are produced by conversion of insolation to kinetic energy.
- *Atmospheric winds*, in turn, generate ocean currents and waves, key inputs to the coastal environment.
- *Climatic regimes*, which result from insolation and moisture, strongly influence coastal geomorphic processes.
- *The nature of coastal rock* is important in determining rates of erosion and sediment production.
- *Human activities* are an increasingly significant input to coastal change.

All these inputs occur within the ever present influence of gravity's pull, not only from Earth but also from the Moon and the Sun. Gravity provides the potential energy of position for materials in motion and produces the tides. A dynamic equilibrium among all these components produces coastline features of infinite variety and beauty.

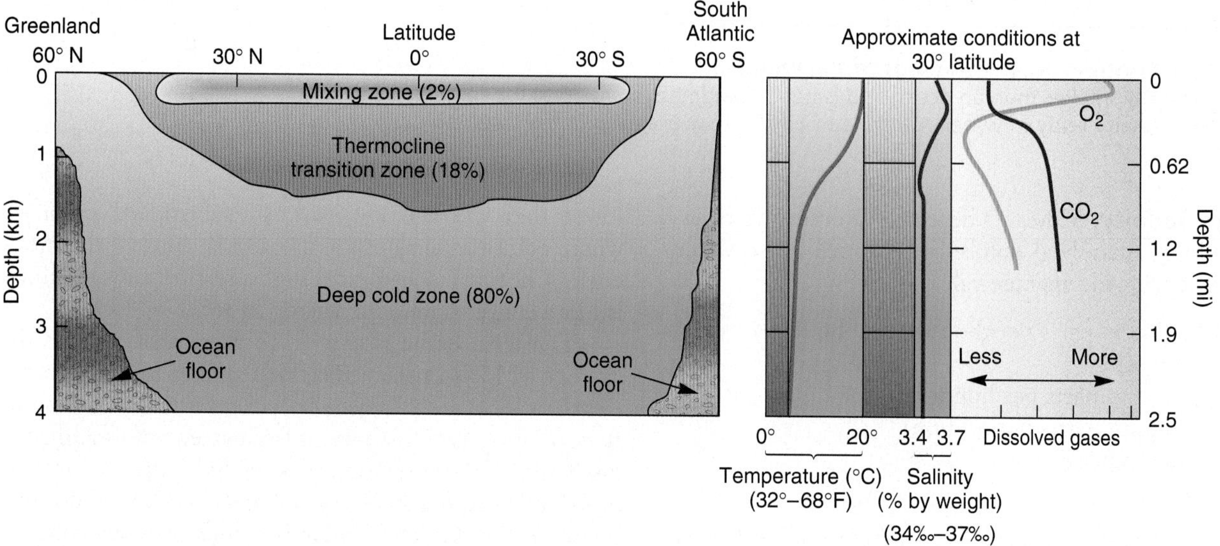

FIGURE 13.2 The ocean's physical structure.
Schematic of average physical structure observed throughout the ocean's vertical profile as sampled along a line from Greenland to the South Atlantic. Temperature, salinity, and dissolved gases are plotted by depth.

The Coastal Environment and Sea Level

The coastal environment is called the **littoral zone.** (*Littoral* comes from the Latin word for "shore.") The littoral zone spans some land as well as water. Landward, it extends to the highest water line that occurs on shore during a storm. Seaward, it extends to the point at which storm waves can no longer move sediments on the seafloor (usually at depths of approximately 60 m or 200 ft). The specific contact line between the sea and the land is the *shoreline*, and adjacent land is considered the *coast*. Figure 13.3 illustrates the littoral zone and includes specific components discussed later in the chapter.

Because the level of the ocean varies, the littoral zone naturally shifts position from time to time. A rise in sea level causes submergence of land, whereas a drop in sea level produces coastline emergence. Uplift and subsidence of the land also cause changes in the littoral zone.

Sea level is a relative term and reflects changes in both water quantity and land elevations. Because sea level is not a straightforward measurement, there exists no international system to determine exact sea level. The Global Sea Level Observing System (GLOSS) is an international group actively working on sea level issues. (For copies of their newsletters and other scientific discussion, see http://www.pol.ac.uk/psmsl/gb.html. GLOSS is part of the larger Permanent Service for Mean Sea Level, which you can find at http://www.pol.ac.uk/psmsl/programmes/. The National Ocean Service is found at http://www.nos.noaa.gov/.)

Mean sea level (MSL) is a value based on average tidal levels recorded hourly at a given site over a period of at least 19 years, which is one full lunar tidal cycle. Mean sea level varies because of ocean currents and waves, tides, air temperature and pressure differences, ocean temperature variations, slight variations in Earth's gravity, and changes in oceanic volume.

At present, the overall mean sea level for the United States is calculated at approximately 40 locations along the coastal margins of the continent. These sites are being upgraded with new equipment in the Next Generation Water Level Measurement System, specifically along the U.S. and Canadian Atlantic coasts, Bermuda, and the Hawaiian Islands. These measurements are augmented by remote-sensing technology, including the *TOPEX/Poseidon* satellite launched in August 1992 (see http://topex-www.jpl.nasa.gov/).

This satellite has two radar altimeters that measure changes in mean sea level at any one location every 10 days between 66° N and 66° S latitudes. These measurements are made to an astonishing precision of 4.2 cm, 1.7 in. (Figure 13.4)! Spectacular *TOPEX/Poseidon* portraits of El Niño and La Niña in the Pacific are shown in Focus Study 6.1 in Chapter 6.

Of great assistance are the *NAVSTAR* satellites that make up the Global Positioning System (GPS), which allow correlation from a network of ground- and ocean-based measurements. See News Report 13.1 for more information about sea level.

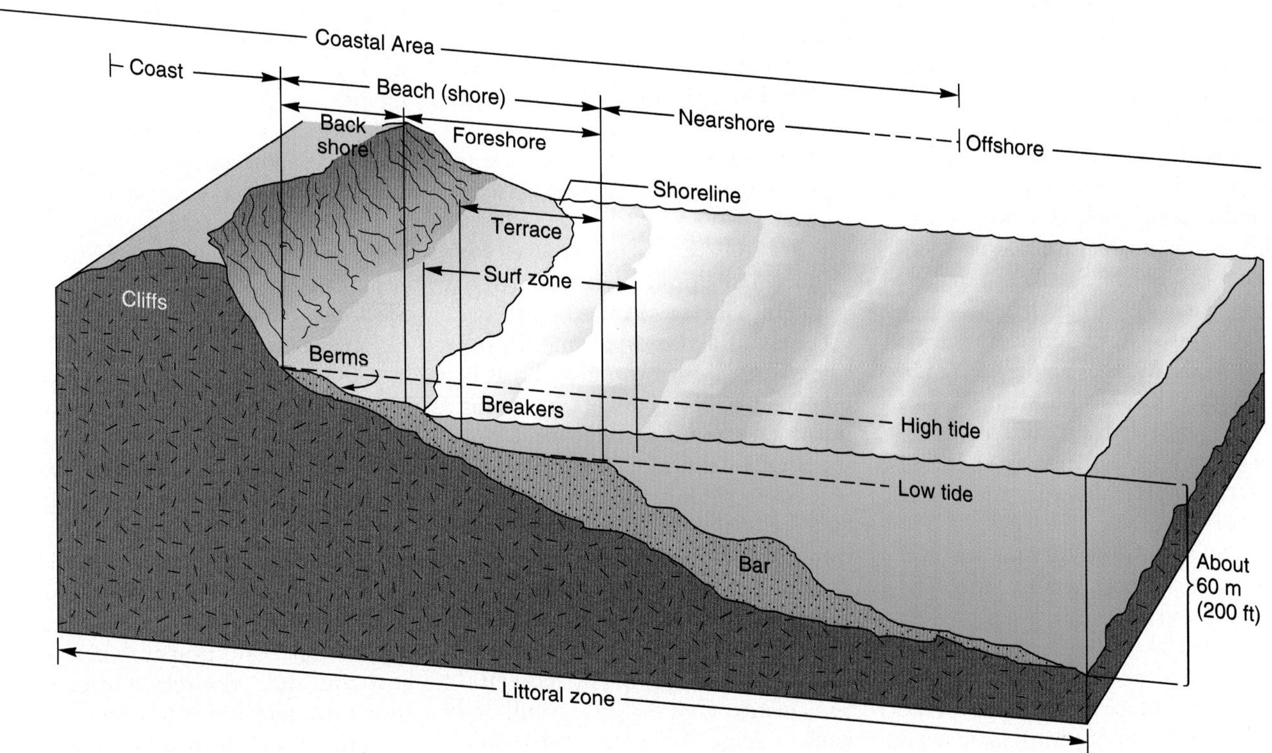

FIGURE 13.3 The littoral zone.
The littoral zone includes the coast, beach, and nearshore environments.

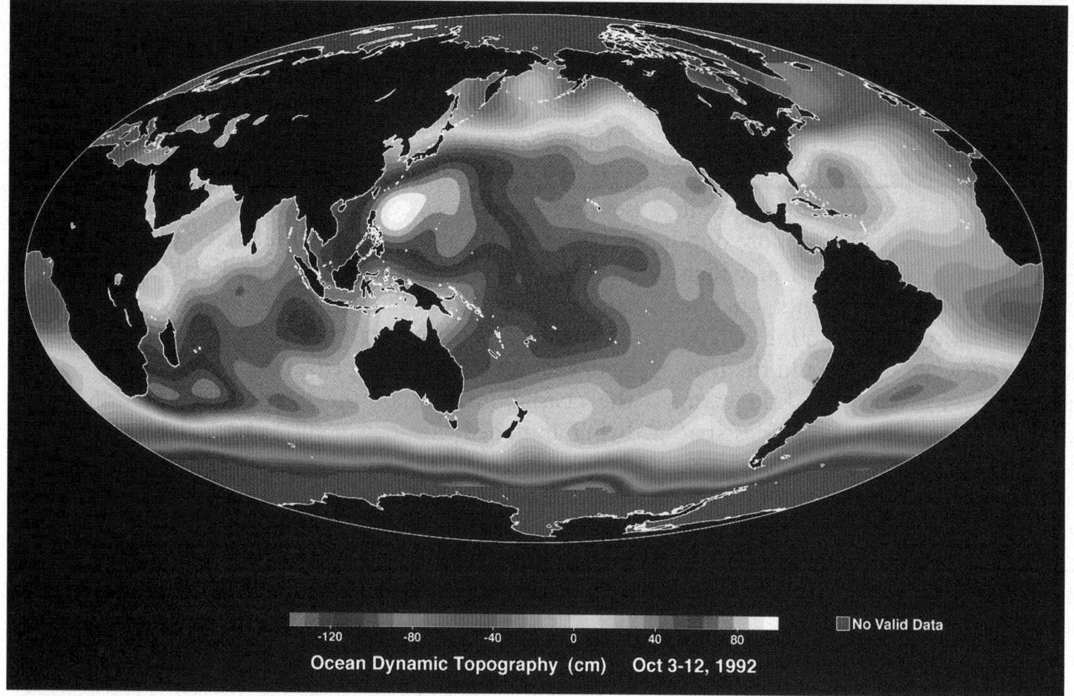

FIGURE 13.4 Ocean topography as revealed by satellite.
Sea-level data recorded by the radar altimeter aboard the *TOPEX/Poseidon* satellite, October 3–12, 1992. The color scale is given in centimeters above or below Earth's geoid. The overall relief portrayed in the image is about 2 m (6.6 ft). The maximum sea level is in the western Pacific Ocean (white). The minimum is around Antarctica (blue and purple). [Image from JPL and NASA/GSFC. *TOPEX/Poseidon* is a joint U.S.-French Mission.]

News Report 13.1

Sea Level Variations and the Present Rise

Sea level varies along the extent of North American shorelines. The mean sea level (MSL) of the U.S. Gulf coast is about 25 cm (10 in.) higher than that of Florida's east coast, which is the lowest in North America. MSL rises northward along the eastern coast, to 38 cm (15 in.) higher in Maine than in Florida. Along the U.S. western coast, MSL is higher than Florida by about 58 cm (23 in.) in San Diego and by about 86 cm (34 in.) in Oregon.

Overall, North America's Pacific coast MSL averages about 66 cm (26 in.) higher than on the Atlantic coast. MSL is affected by differences in ocean currents, air pressure and wind patterns, water density, and water temperature.

Over the long term, sea-level fluctuations expose a range of coastal landforms to tidal and wave processes. As average global temperatures cycle through cold or warm climatic spells, the quantity of ice locked up in the ice sheets of Antarctica and Greenland and in hundreds of mountain glaciers can increase or decrease and result in sea level changes accordingly. At the peak of the Pleistocene glaciation about 18,000 B.P. (years before the present), sea level was about 130 m (430 ft) lower than it is today. On the other hand, if Antarctica and Greenland ever became ice-free (ice sheets fully melted), sea level would rise at least 65 m (215 ft) worldwide.

Just 100 years ago sea level was 38 cm (15 in.) lower along the coast of southern Florida. Venice, Italy, has experienced a rise of 25 cm (10 in.) since 1890. Venice is under threat of innundation. Sea levels are rising today at a rate not in evidence during the last 3000 years.

Given these trends and the predicted climatic change, sea level will continue to rise and be potentially devastating for many coastal locations. A rise of only 0.3 m (1 ft) would cause shorelines worldwide to move inland an average of 30 m (100 ft). This elevated sea level would inundate some valuable real estate along coastlines worldwide. Some 20,000 km^2 (7800 mi^2) of land along North American shores alone would be drowned, at a staggering loss of US $650 billion. A 95 cm (3.1 ft) sea-level rise could inundate 15% of Egypt's arable land, 17% of Bangladesh, and many island countries and communities. However, uncertainty exists in these forecasts.

In 2001 the Intergovernmental Panel on Climate Change (IPCC, discussed in Chapter 6) forecast global MSL rise for this century, given regional variations, as a range from 0.11 to 0.88 m (4.3 to 34.6 in.). The median value of 0.48 m (18.9 in.) is two to four times the rate of increase over the last century. These increases would continue beyond 2100 even if greenhouse gas concentrations are stabilized. Despite any uncertainty, planning should start now along coastlines worldwide because preventive strategies are cheaper than recovery costs from possible destruction. Insurance underwriters have begun the process by refusing coverage for shoreline properties vulnerable to rising sea level.

Coastal System Actions

Beachdrift, coastal erosion

The coastal system is the scene of complex tidal fluctuation, winds, waves, ocean currents, and the occasional impact of storms. These forces shape landforms ranging from gentle beaches to steep cliffs, and they sustain delicate ecosystems.

Tides

Tides are complex daily oscillations in sea level, ranging worldwide from barely noticeable to several meters. They are experienced to varying degrees along every ocean shore around the world. As tides flood (rise) and ebb (fall), the daily migration of the shoreline landward and seaward causes significant changes that affect sediment erosion and transportation. Figure 13.5 illustrates the relationship among the Moon, the Sun, and Earth and the generation of variable tidal bulges on opposite sides of the planet.

Tides also are important in human activities, including navigation, fishing, and recreation. Tides are especially important to ships because the entrance to many ports is limited by shallow water, and thus high tide is required for passage. Tall-mast ships may need a low tide to clear overhead bridges. Tides also exist in large lakes; but because the tidal range is small, tides are difficult to distinguish from changes caused by wind. Lake Superior, for instance, has a tidal variation of only about 5 cm (2 in.).

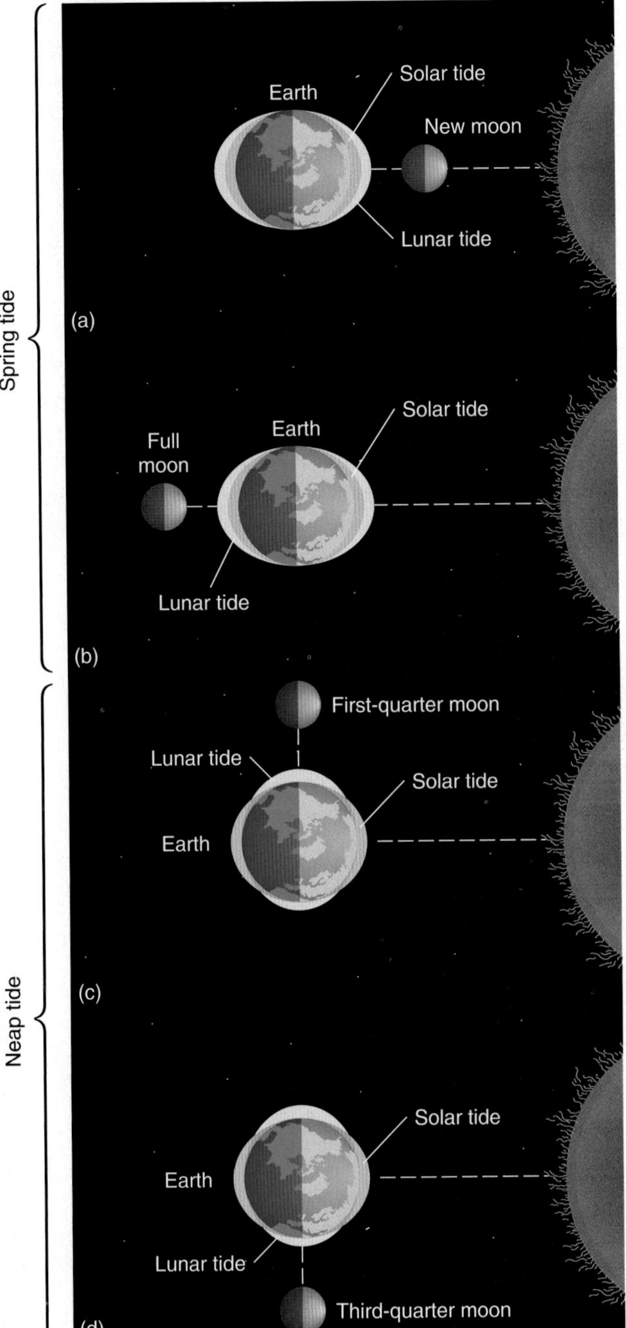

FIGURE 13.5 The cause of tides.
Gravitational relations of Sun, Moon, and Earth combine to produce spring tides (a, b) and neap tides (c, d). (Tides are greatly exaggerated for illustration.)

Causes of Tides. Earth's relation to the Sun and the Moon and the reasons for the seasons are covered in Chapter 2. These astronomical relationships also produce the pattern of tides, the complex daily oscillations in sea level that are experienced to varying degrees around the world. Tides also are influenced by the size, depth, and topography of ocean basins, by latitude, and by shoreline shape.

Tides are produced by the gravitational pull exerted on Earth and its oceans by both the Sun and the Moon. The Sun's influence is only about half that of the Moon's because of the Sun's greater distance from Earth, although it still is a significant force.

The gravitational pull of both bodies on Earth actually pulls Earth's atmosphere, oceans, and lithosphere on the side that is facing the Sun or Moon at the moment. Earth's solid and fluid surfaces all experience some stretching as a result of this gravitational pull. The stretching raises large *tidal bulges* in the atmosphere (which we can't see), smaller tidal bulges in the ocean, and very slight bulges in Earth's rigid crust. Our concern here is the tidal bulges in the ocean. For complex reasons, gravitational force and inertia also create a bulge on the opposite side of Earth. These opposing bulges are always there in the alignments of Earth, Sun, and Moon, as shown in Figure 13.5.

Every 24 hours and 50 minutes, any given point on Earth rotates through these two bulges. Thus, every day, most coastal locations experience two high (rising) tides known as *flood tides*, and two low (falling) tides known as *ebb tides*. The difference between consecutive high and low tides is the *tidal range*.

Spring and Neap Tides. When Earth, Sun, and the Moon are aligned as shown in Figure 13.5 (a) and (b), their combined gravitational pull creates a greater *spring tide* (meaning to "spring forth," not the season). When positioned as in (c) and (d), their weaker pulls create a lesser *neap tide* (*neap* means "without the power of advancing").

Tides also are influenced by other factors, causing a great variety of tidal ranges. For example, some locations may experience almost no difference between high and low tides. The highest tides occur where open water is forced into partially enclosed gulfs or bays. The Bay of Fundy in Nova Scotia records the greatest tidal range on Earth, a difference of 16 m (52.5 ft) (Figure 13.6a and b). (For more on tides and tide prediction, contact the Scripps Institution of Oceanography library at **http://scilib.ucsd.edu/sio/tide/**.)

Tidal Power. The fact that sea level changes daily with the tides suggests an opportunity: Could these predictable flows be harnessed to produce electricity? The answer is yes, given the right conditions. The bay or estuary under consideration must have a narrow entrance suitable for the construction of a dam with gates and locks, and it must experience a tidal range of flood and ebb tides large enough to turn turbines, at least a 5 m (16 ft) range.

About 30 locations in the world are suited for tidal power generation, although at present only three of them are actually producing electricity. Two are outside North America—a 4-megawatt (MW) station in Russia at Kislaya-Guba Bay, on the White Sea, since 1968, and a facility in the Rance River estuary on the Brittany coast of France, since 1967. The tides in the Rance estuary fluctuate up to 13 m (43 ft), and power production has been almost continuous there, providing an electrical generating capacity of a moderate 240 megawatts (about 20% of the capacity of Hoover Dam).

The third site is on the Bay of Fundy. According to the Canadian government, the present cost of tidal power at ideal sites is economically competitive with that of fossil fuels, although certain environmental concerns must be addressed. At one of several favorable sites on the Bay of Fundy, the Annapolis Tidal Generating Station was built in 1984. Nova Scotia Power Incorporated operates this 20-MW plant (Figure 13.6c).

Waves

Wind friction on the surface of the ocean generates undulations of water called **waves**. They travel in *wave trains*, or groups of waves. On a small scale, a moving boat creates a wake, which consists of waves; at a larger

(a)

(b)

(c)

FIGURE 13.6 Tidal range and tidal power.
Tidal range is great in some bays and estuaries, such as in the Bay of Fundy at flood tide (a) and ebb tide (b). The flow of water between high and low tide is ideal for turning turbines and generating electricity, as is done at the Annapolis Tidal Generating Station (c), near the Bay of Fundy in Nova Scotia, in operation since 1984. [Photos by Jeff Newbery.]

scale, storms around the world generate large groups of wave trains. A stormy area at sea is a *generating region* for these waves, which radiate outward in all directions. Intricate patterns of waves traveling in all directions crisscross the ocean. The waves seen along a coast may be the product of a storm center thousands of kilometers away.

Regular patterns of smooth, rounded waves are **swells**—these are the mature undulations of the open ocean. As waves leave the generating region, wave energy continues to run in these swells, which can range from small ripples to very large flat-crested waves. A deep-water wave leaving a generating region tends to extend its wavelength many meters (see Figure 13.7 noting crest, trough, wavelength, and wave height).

Water within a wave in the open ocean is not really migrating but is transferring energy from molecule to molecule through the water in simple cyclic undulations, or *waves of transition* (Figure 13.7). Individual water particles move forward only slightly, forming a vertically circular pattern. The diameter of the paths formed by the orbiting water particles decreases with depth. As a deep-ocean wave approaches the shoreline and enters shallower water (10–20 m or 30–65 ft), the water particle orbits are vertically reduced. The restriction causes more elliptical (flattened) orbits to form near the bottom. This change from circular to elliptical orbits of the water particles slows the entire wave, although more waves continue arriving. As the peak of each wave rises, a point is reached when the wave height exceeds its vertical stability and the

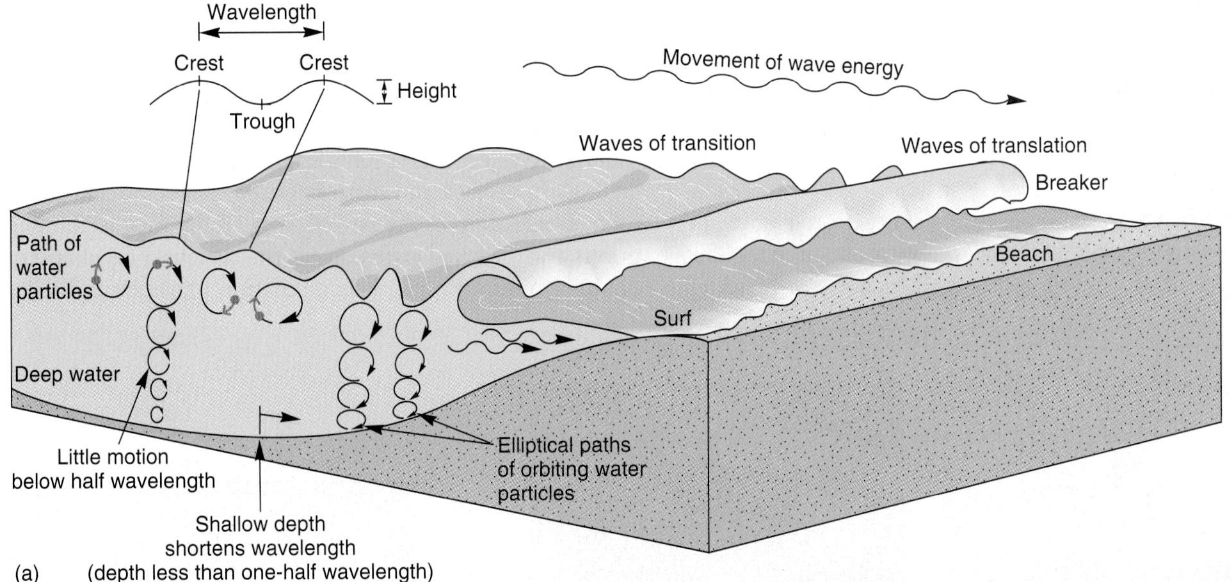

FIGURE 13.7 Wave formation and breakers.
(a) Wave structure and the orbiting tracks of water particles change from circular motions in deep waves of transition to more elliptical orbits in shallow waves of translation. (b) Cascades of waves attack the shore near Pacific Grove, California. Storm-driven wave heights range between 3 to 6 m (10 to 20 ft). [Photo by Bobbé Christopherson.]

wave falls into a characteristic **breaker**, crashing onto the beach (Figure 13.7b).

In a breaker the orbital motion of transition gives way to form *waves of translation*, in which energy and water both move forward toward shore as water cascades down from the wave crest. Both the energy of arriving waves and the slope of the shore determine wave style: Plunging breakers indicate a steep bottom profile, whereas spilling breakers indicate a gentle, shallow bottom profile.

As various wave trains move along in the open sea, they interact by *interference*. These interfering waves sometimes align so that the wave crests and troughs from one wave train are in phase with those of another. When this in-phase condition occurs, the height of the waves is increased, sometimes dramatically. The resulting "killer waves" or "sleeper waves" can sweep in unannounced and overtake unsuspecting victims. Signs along portions of the California, Oregon, Washington, and British Columbia coastline warn beachcombers to watch for "killer waves." On the other hand, out-of-phase wave trains will dampen wave energy at the shore. The changing beat of the surf along a beach actually is produced by the patterns of *wave interference* that occurred in far-distant areas of the ocean.

Wave Refraction. Generally, wave action results in coastal straightening. As waves approach an irregular coast, they bend around *headlands*, which are protruding landforms generally composed of resistant rocks (Figure 13.8). The submarine topography refracts (bends) approaching waves into patterns that focus energy around headlands, and dissipate energy in coves and the submerged coastal valleys between them. Thus, headlands represent a specific focus of wave attack along a coastline. This **wave refraction** (wave bending) along a coastline redistributes wave energy so that different sections of the coastline are subject to variations in erosion potential.

When waves approach a coast, they usually arrive at some angle other than parallel to it (Figure 13.9). As the waves enter shallow water, they bend and generate a current parallel to the coast, zigzagging in the prevalent direction of the incoming waves. (Think of the end of the wave nearer the shore slowing, while the part of the wave farther from shore in deeper water is moving faster.) This **longshore current**, or *littoral current*, depends on wind direction and wave direction. A longshore current generates only in the surf zone and works in combination with wave action to transport large amounts of sand, gravel, sediment, and debris along the shore as *longshore drift*.

Particles on the beach also are moved along as *beach drift*, shifting back and forth between water and land with each swash and backwash of surf. Individual sediment grains trace arched paths along the beach. The chapter-opening photo shows an accumulation of sand and shells,

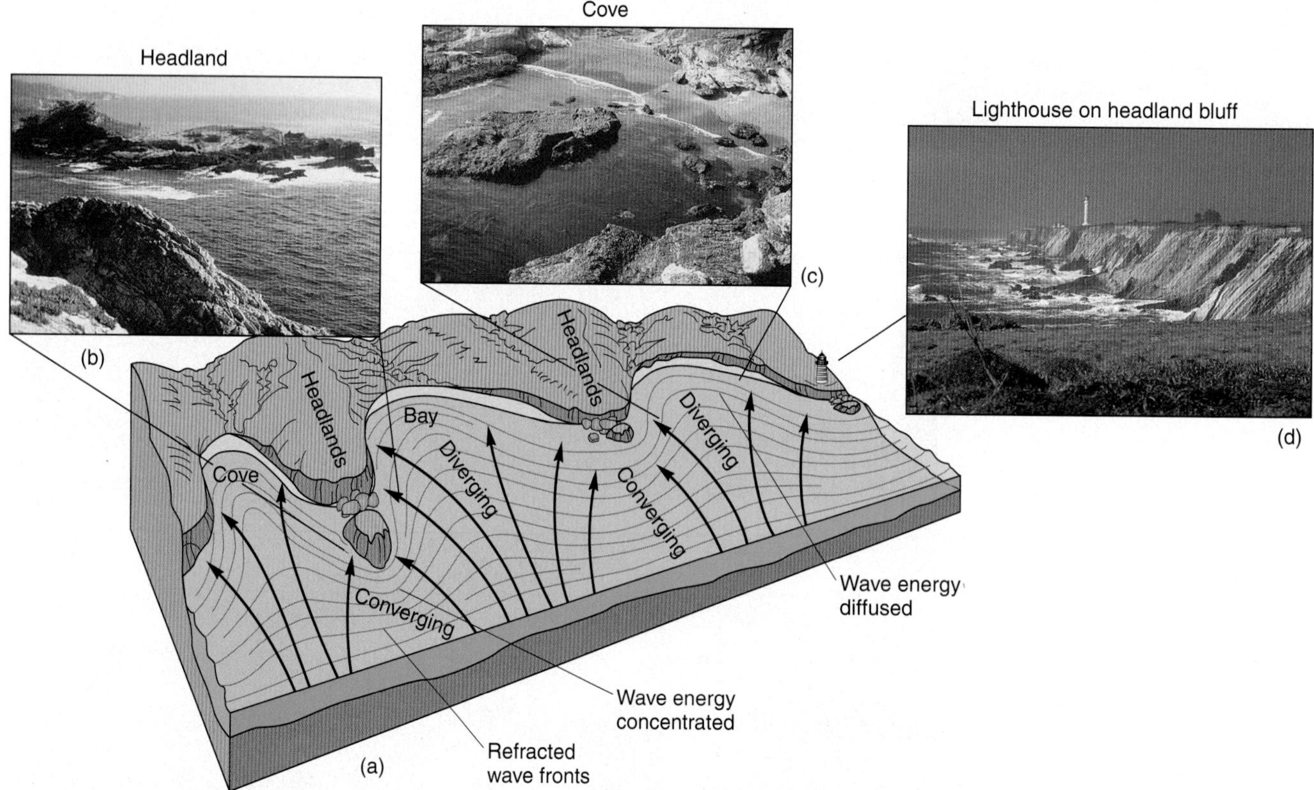

FIGURE 13.8 Coastal straightening.
(a) The process of coastal straightening is brought about by wave refraction. (b) Wave energy is concentrated as it converges on headlands and (c) is diffused as it diverges in coves and bays. (d) Headlands are frequent sites for lighthouses such as the light at Point Arena, California. [Photos by (b) and (c) author; (d) Bobbé Christopherson.]

FIGURE 13.9 Longshore current and beach drift. (a) Longshore currents are produced as waves approach the surf zone and shallower water. Longshore and beach drift results as substantial volumes of material are moved along the shore. (b) Processes at work along Point Reyes Beach, Point Reyes National Seashore, California. [Photo by author.]

in motion along a beach. These dislodged materials are available for transport and eventual deposition in coves and inlets and can represent a significant volume.

Tsunami, or Seismic Sea Wave. A type of wave that can greatly influence coastlines is the **tsunami**. Tsunami is Japanese for "harbor wave," named for its devastating effect in harbors. Tsunami are incorrectly referred to as "tidal waves," but they have no relation to the tides. They are formed by sudden, sharp motions in the seafloor, caused by earthquakes, submarine landslides, or eruptions of undersea volcanoes. Thus, they properly are called *seismic sea waves*.

A large undersea disturbance usually generates a solitary wave of great wavelength. Tsunami generally exceed 100 km (60 mi) in wavelength (crest to crest) but only a meter (3 ft) or so in height. They travel at great speeds in deep-ocean water—velocities of 600–800 kmph (375–500 mph) are not uncommon—but often pass unnoticed on the open sea because their great wavelength makes the slow rise and fall of water hard to observe.

As a tsunami approaches a coast, the shallow water forces the wavelength to shorten. As a result, the wave height may increase up to 15 m (50 ft) or more. Such a wave has the potential for coastal devastation, resulting in property damage and death. For example, in September 1992 the citizens of Casares, Nicaragua, were surprised by a 12 m (39 ft) tsunami that took 270 lives, and another hit Indonesia and killed 1000. Two tsunami in 1994 killed 250 in Java, Indonesia, and 62 people in the Philippines. Earthquakes triggered all these waves.

Hawai'i is vulnerable to tsunami because of its position in the open central Pacific, surrounded by the ring of fire that outlines the Pacific Basin. As an example, a tsunami produced near the Philippines would take 10 hours to reach Hawai'i. The U.S. Army Corps of Engineers has reported 41 damaging tsunami in Hawai'i during the past 142 years—statistically, 1 every 3.5 years.

Because tsunami travel so quickly and are undetectable in the open ocean, accurate forecasts are difficult. A warning system now is in operation for nations surrounding the Pacific, where the majority of tsunami occur. Warnings always should be heeded, despite the many false alarms, for the causes lie beneath the ocean and are difficult to monitor in any consistent manner. (For the tsunami research program, see http://www.pmcl.noaa.gov/tsunami/. The tsunami home page is at http://www.geophys.washington.edu/tsunami/welcome.html.)

Coastal System Outputs

Coastlines are active portions of continents, with energy and sediment continuously delivered to a narrow environment. The action of tides, currents, wind, waves, and changing sea level produces a variety of erosional and depositional landforms. We look first at erosional coastlines such as the U.S. West Coast, then at depositional coastlines as found along the East and Gulf Coasts. In this era of rising sea level, coastlines become more dynamic.

Erosional Coastal Processes and Landforms

The active margins of the Pacific Ocean along the North and South American continents are characteristic coastlines affected by erosional landform processes. *Erosional coastlines* tend to be rugged, of high relief, and tectonically active, as expected from their association with the leading edge of drifting lithospheric plates (see plate tectonics discussion in Chapter 8). Figure 13.10 presents features commonly observed along an erosional coast.

Sea cliffs form by the undercutting action of the sea. As indentations are produced at water level, such a cliff becomes notched, leading to subsequent collapse and retreat of the cliff. Other erosional forms evolve along cliff-dominated coastlines, including *sea caves*, *sea arches*, and *sea stacks*. As erosion continues, arches may collapse, leaving isolated stacks in the water (Figure 13.10b, c). The coasts of southern England and Oregon are prime examples of such erosional landscapes.

Wave action can cut a horizontal bench in the tidal zone, extending from a sea cliff out into the sea. Such a structure is called a **wave-cut platform**, or *wave-cut terrace*. If the relationship between the land and sea-level changes over time, multiple platforms or terraces may rise like stair steps back from the coast. These marine terraces are remarkable indicators of an emerging coastline, with some terraces more than 365 m (1200 ft) above sea level. A tectonically active region, such as the California coast, has many examples of multiple wave-cut platforms (Figure 13.10e) and the many hazards associated with building there (Figure 13.10d).

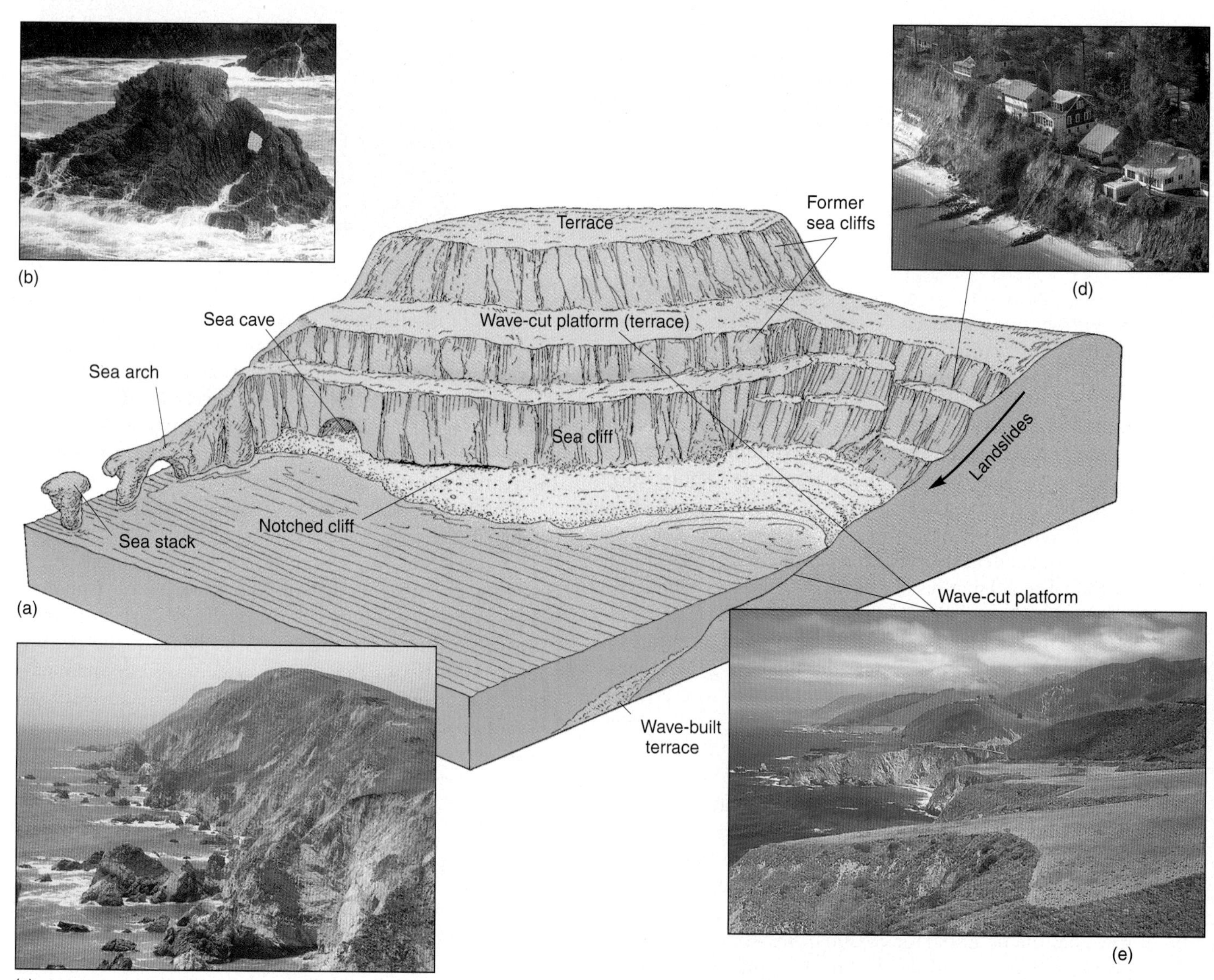

FIGURE 13.10 Erosional coastal features.
(a) Characteristic coastal erosional landforms: caves, (b) arches, (c) stacks, (d) collapsing cliffs, and (e) wave-cut platforms near Bixby Bridge along the Cabrillo Highway. [Photos by (b), (c) author; (d) Lowell Georgia/Photo Researchers, Inc.; (e) Bobbé Christopherson.]

Depositional Coastal Processes and Landforms

Depositional coasts generally are located along land of gentle relief, where sediments are available from many sources. Such is the case with the Atlantic and Gulf coastal plains of the United States, which lie along the relatively passive, trailing edge of the North American lithospheric plate. Erosional processes and inundation, particularly during storm activity, influence these depositional coasts.

Characteristic wave- and current-deposited landforms are illustrated in Figure 13.11. A **barrier spit** consists of material deposited in a long ridge extending out from a coast attached at one end; it partially crosses and blocks the mouth of a bay. Classic examples include Sandy Hook, New Jersey (south of New York City), and Cape Cod, Massachusetts. The Little Sur River in California (13.11a) and Prion Bay in Southwestern National Park in Tasmania (13.11b) show barrier spits forming partway across the mouths of their respective rivers.

A spit becomes a **bay barrier**, sometimes referred to as a *baymouth bar*, if it completely cuts off the bay from the ocean and forms an inland **lagoon**. Spits and barriers are made up of materials that have been eroded and transported by *littoral drift* (beach and longshore drift combined). For much sediment to accumulate, offshore currents must be weak. Tidal flats and salt marshes are characteristic low-relief features wherever tidal influence is greater than wave action.

A **tombolo** occurs when sediment deposits connect the shoreline with an offshore island or sea stack (Figure 13.11c). A tombolo forms when sediments accumulate on an underwater wave-built terrace.

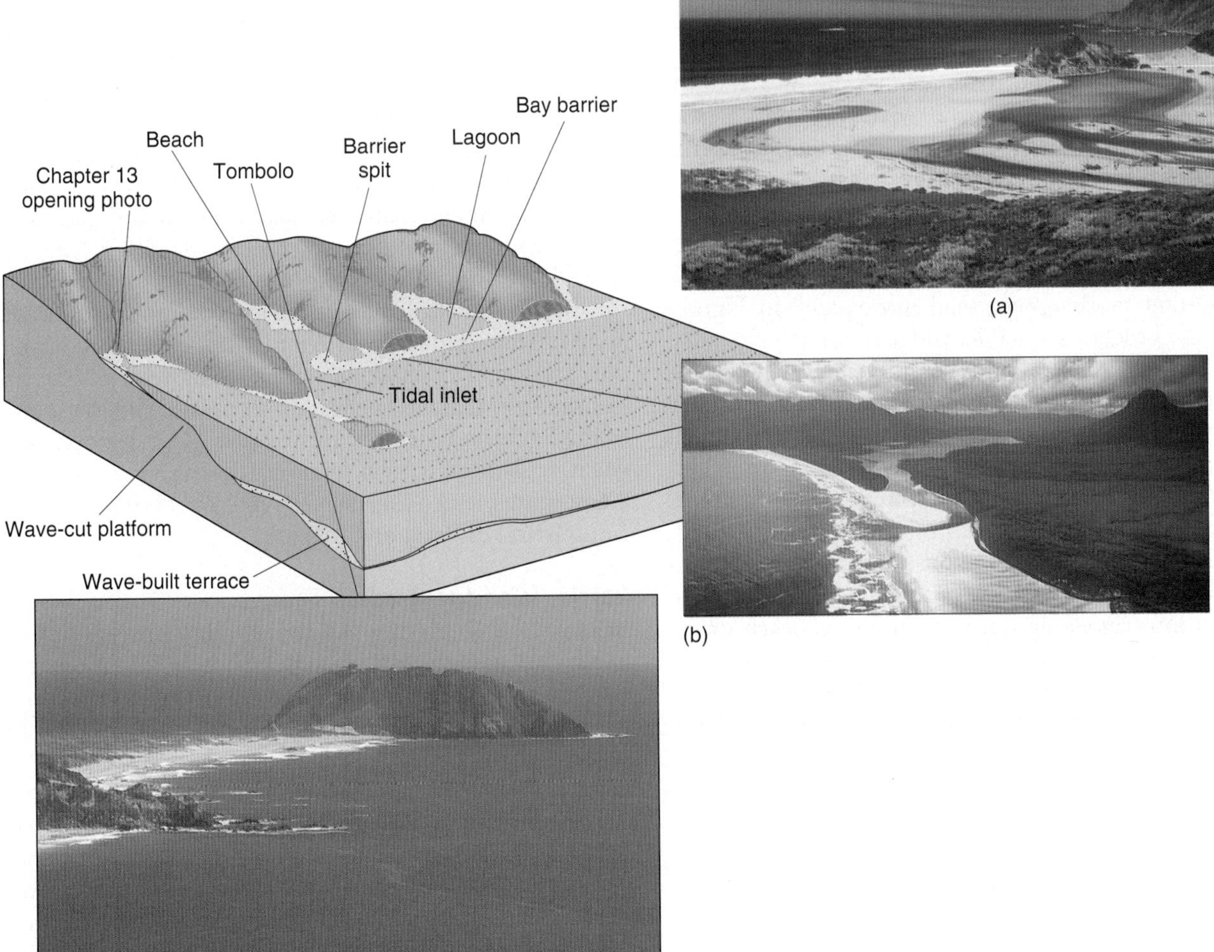

FIGURE 13.11 Characteristic depositional coastal features.
Characteristic coastal depositional landforms: beaches, spits, barriers, and lagoons. (a) Little Sur River enters the Pacific Ocean, a barrier spit nearly blocks its way, and a characteristic summer fog bank lies offshore. (b) A barrier spit developing across the mouth of the New River, forming the New River Lagoon and Prion Bay at Southwestern National Park in Tasmania. (c) A tombolo at Point Sur along the central California coast where sediment deposits connect the shore with an island. [Photos by (b) Reg Morrison/Auscape International Pty. Ltd.; (a) and (c) Bobbé Christopherson.]

Beaches. Of all the features associated with a depositional coastline, beaches probably are the most familiar. Beaches vary in type and permanence, especially along coastlines dominated by wave action. Technically, a **beach** is that place along a coast where sediment is in motion, deposited by waves and currents. Material from the land temporarily resides there while it is in active transit along the shore. You may have experienced a beach at some time, along a seacoast, a lakeshore, or even a stream. Perhaps you have even built your own "landforms" in the sand, only to see them washed away by the waves.

The beach zone ranges, on average, from 5 m (16 ft) above high tide to 10 m (33 ft) below low tide (see Figure 13.3). The specific definition varies greatly along individual shorelines. Worldwide, sands of quartz (SiO_2) dominate beaches because it resists weathering, and therefore remains after other minerals are removed. In volcanic areas, beaches are derived from wave-processed lava. Hawai'i and Iceland feature examples of these black-sand beaches.

Many beaches, such as those in southern France and western Italy, are of pebbles and cobbles—a type of "shingle beach." Some shores have no beaches at all; scrambling across boulders and rocks may be the only way to move along the coast. The coast of Maine and portions of the Atlantic provinces of Canada are classic examples, composed of resistant granite rock that is scenically rugged but with few beaches.

A beach acts to stabilize a shoreline by absorbing wave energy, as is evident by the amount of material that is in almost constant motion (see "sand movement" in Figure 13.9). Some beaches are stable, whereas others cycle seasonally. Protected areas along a coastline tend to accumulate sand. They accumulate during the summer, are moved offshore by winter storm waves, forming a submerged bar, and are redeposited onshore the following summer.

Changes in coastal sediment transport can disrupt human activities—beaches are lost, harbors closed, and coastal highways and beach houses can be inundated with sediment. Consequently, various strategies are employed to interrupt longshore currents and beach drift. The goal of intervention is usually to halt sand accumulation or to force accumulation in a desired way through construction of engineered structures—"hard" shoreline protection.

Figure 13.12 illustrates common approaches: a *jetty* to block material from harbor entrances; a *groin* to slow drift action along the coast, and a *breakwater* to create a zone of still water near the coastline. However, interrupting the coastal drift, which naturally replenishes beaches, may lead to unwanted changes in sediment distribution downcurrent. Careful planning and impact assessment should be part of any strategy for preserving or altering a beach.

Beach nourishment refers to the artificial placement of sand along a beach. Through such efforts, a beach that normally experiences a net loss of sediment will instead show a net gain. In contrast to hard structures, this hauling of sand to replenish a beach is considered "soft" shoreline protection. (See News Report 13.2, p. 416.)

Barrier Beach Formations. Barrier chains are long, narrow, depositional features, generally of sand, that form offshore roughly parallel to the coast. Common forms are **barrier beaches**, or the broader, more extensive landform, **barrier islands**. Tidal variation in the area usually is moderate to low, with adequate sediment supplies coming from nearby coastal plains. Figure 13.13 illustrates the many features of barrier chains, using North Carolina's famed Outer Banks as an example, including Cape Hatteras, across Pamlico Sound from the mainland. The area presently is designated as one of three national seashore reserves supervised by the National Park Service.

On the landward side of a barrier formation are tidal flats, marshes, swamps, lagoons, coastal dunes, and beaches. Barrier beaches appear to adjust to sea level and may naturally shift position from time to time in response to wave action and longshore currents. Breaks in the barrier, visible in the image, are inlets, connecting the bay with the ocean. Their name "barrier" is appropriate, for they take the brunt of storm energy and actually act as protection for the mainland. Because of the continuing loss to a barrier island, the famous Cape Hatteras lighthouse was moved inland in 1999 to safer ground (Figure 13.13c). Its new position is about 488 m (1600 ft) from the ocean, or approximately the distance it was in 1870 when it was built—that much sand was lost back to the sea. (See **http://www.ncsu.edu/coast/chl/** for details of the extraordinary effort to save this landmark.)

Barrier beaches and islands are common worldwide, lying offshore of nearly 10% of Earth's coastlines. Examples are found offshore of Africa, India (east coast), Sri Lanka, Australia, the north slope of Alaska, and the shores of the Baltic and Mediterranean Seas. The most extensive chain of barrier islands is along the Atlantic and Gulf Coast states, extending some 5000 km (3100 mi) from Long Island to Texas and Mexico.

Barrier Island Hazards. Because many barrier islands seem to be migrating landward, they are an unwise choice for homesites or commercial building. Nonetheless, they are a common choice, even though they take the brunt of storm energy. The hazard represented by the settlement of barrier islands was made graphically clear when Hurricane Hugo assaulted South Carolina in 1989. The storm attacked the Grand Strand barrier islands off the northern half of South Carolina's coastline, most affecting Charleston and the South Strand portion of the islands. The storm made landfall as the worst hurricane to strike there in 35 years—beachfront houses, barrier-island developments, and millions of tons of sand were swept away.

The barrier islands off the Louisiana shore are disappearing at rates approaching 20 m (65 ft) per year. Hurricanes have taken their toll on these barrier islands. Also, they are affected by subsidence through compaction of Mississippi delta sediments and a changing sea level that is rising at 1 cm (0.4 in.) per year in the region. Hurricane

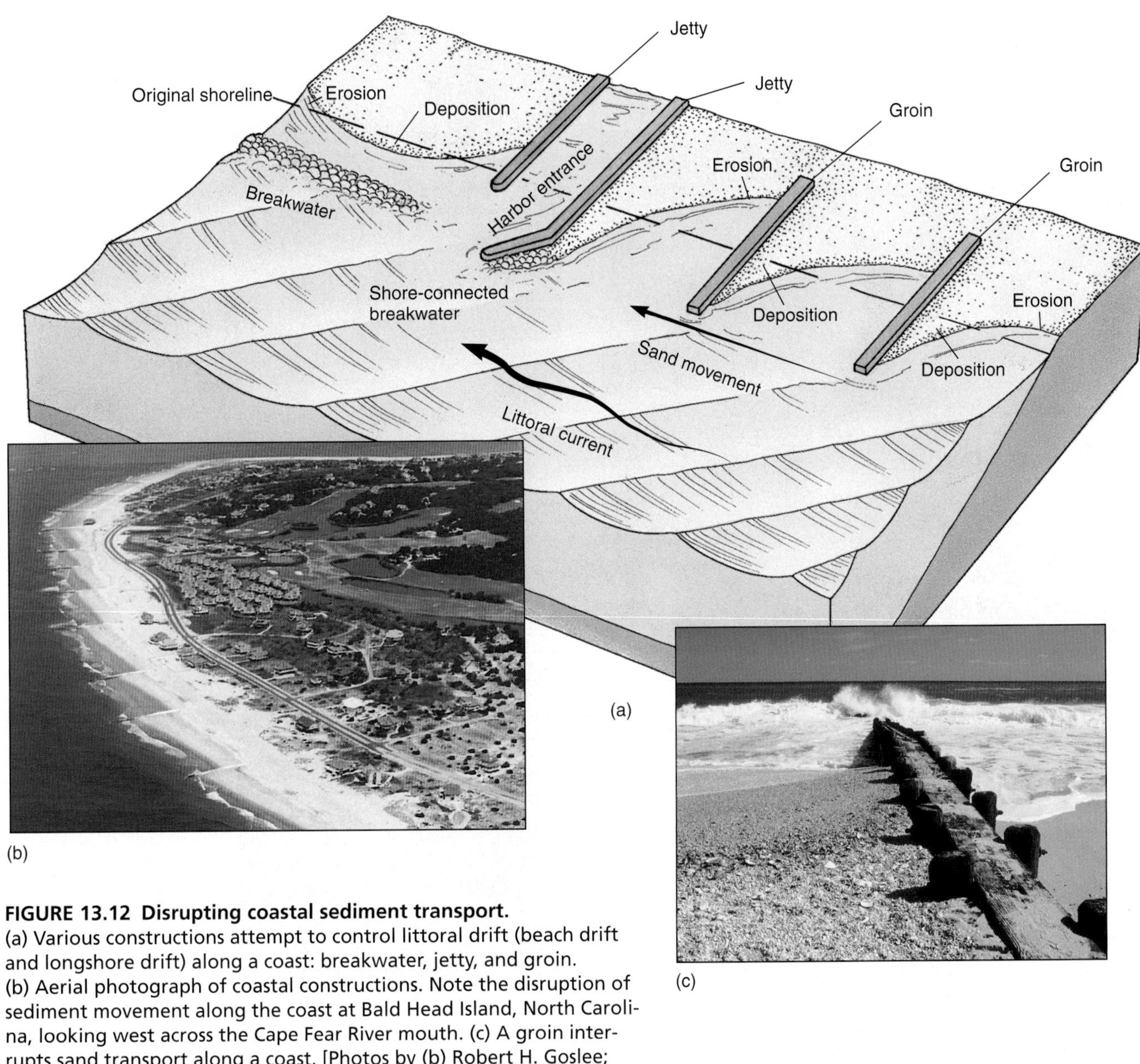

FIGURE 13.12 Disrupting coastal sediment transport. (a) Various constructions attempt to control littoral drift (beach drift and longshore drift) along a coast: breakwater, jetty, and groin. (b) Aerial photograph of coastal constructions. Note the disruption of sediment movement along the coast at Bald Head Island, North Carolina, looking west across the Cape Fear River mouth. (c) A groin interrupts sand transport along a coast. [Photos by (b) Robert H. Goslee; and (c) Bobbé Christopherson.]

Andrew eroded sand from 70% of barrier islands in 1992. In 1998, Hurricane Georges destroyed large tracts of the Chandeleur Islands (30 to 40 km, 19 to 25 mi, from the Louisiana and Mississippi Gulf Coast), leaving the mainland more exposed (Figure 13.14). The regional office of the USGS predicts that in a few decades the barrier islands may be gone. Louisiana's increasingly exposed wetlands are disappearing at rates of 40 km^2 (15 mi^2) per year.

Biological Processes: Coral Formations

Not all coastlines form due to purely physical processes. Some form as the result of organic processes, such as coral growth. A **coral** is a simple marine animal with a small, cylindrical, saclike body (polyp); it is related to other marine invertebrates, such as anemones and jellyfish. Corals secrete calcium carbonate ($CaCO_3$) from the lower part of their bodies, forming a hard external skeleton. Corals function in a *symbiotic* (mutually helpful) relationship with algae—that is, they live in close association with the algae, and each depends on the other for survival. Algae photosynthesize some of the food for the coral, and in turn the corals provide some nutrients and shelter for the algae.

Coral reefs are the most diverse among marine ecosystems. Preliminary estimates of coral species range upward to nine million worldwide, yet, as in most ecosystems in water or on land, biodiversity is declining in these communities.

Figure 13.15 shows the distribution of currently living coral formations. Corals thrive in warm tropical oceans, so the difference in ocean temperature between the western coasts and eastern coasts of continents is critical to their distribution. Western coastal waters tend to be cooler, thereby discouraging coral activity, whereas eastern coastal currents are warmer and thus enhance coral growth. Living

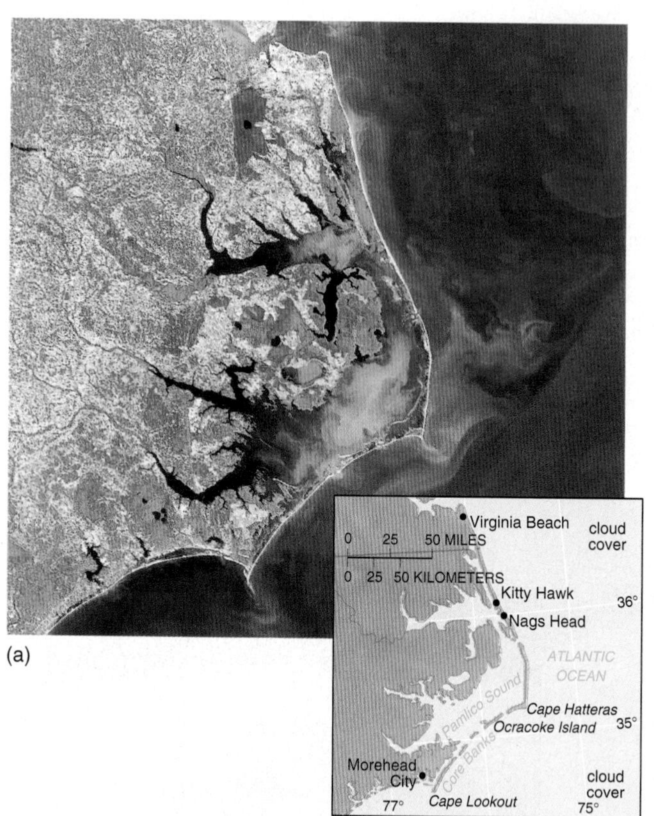

(a)

(b)

(c)

FIGURE 13.13 Barrier island chain.
(a) *Terra* image of barrier island chain along the North Carolina coast. Hurricane Emily swept past Cape Hatteras in August 1993, causing damage and beach erosion. You can see key depositional forms: spit, island, beach, lagoon, and inlet. A sound is a large inlet of the ocean; Pamlico Sound forms an ideal example. (b) View from Cape Hatteras lighthouse illustrates the narrow strand of sand that stands between the ocean and the mainland. (c) Cape Hatteras lighthouse is moved inland in 1999 to safer ground, away from the receding shoreline. [(a) *Terra* MODIS image courtesy of NASA/GSFC; (b) photo by Eric Horan/Liaison Agency, Inc.; (c) © AP/Wide World Photos.]

colonial corals range in distribution from about 30° N to 30° S and occupy a very specific ecological zone: 10–55 m (30–180 ft) depth, 27‰–40‰ (parts per thousand) salinity, and 18° to 29°C (64°–85°F) water temperature.

Corals require clear, sediment-free water and consequently do not locate near the mouths of sediment-charged freshwater tributaries (note the lack of these formations along the U.S. Gulf Coast). Increasing ocean water temperature is evidently causing a worldwide bleaching of corals. See News Report 13.3 (p. 417) for more on this important problem.

Coral Reefs. Some corals are colonial, and their skeletons accumulate in enormous structures. Through many generations, live corals near the ocean's surface build on the foundation of older coral skeletons, which in turn may rest upon a volcanic seamount or some other submarine feature built up from the ocean floor. *Coral reefs* form by this process. Thus, a coral reef is a biologically derived sedimentary rock. It can assume one of several distinctive shapes.

The principal shapes of reefs are fringing reefs, barrier reefs, and atolls. In 1842 Charles Darwin presented a hypothesis for the evolution of reef formation: As reefs develop around a volcanic island and the island itself gradually subsides, equilibrium is maintained between the subsidence of the island and the upward growth of coral. This generally accepted idea is shown in Figure 13.16 with specific examples of each reef stage: *fringing reefs* (platforms of surrounding coral rock), *barrier reefs* (forming enclosed lagoons), and *atolls* (circular, ring-shaped).

Wetlands, Salt Marshes, and Mangrove Swamps

Some coastal areas have great *biological productivity* (plant growth, spawning ground for fish, shellfish, and other organisms) stemming from trapped organic matter and sediments. Such a rich coastal marsh environment can greatly outproduce a wheat field in raw vegetation per acre.

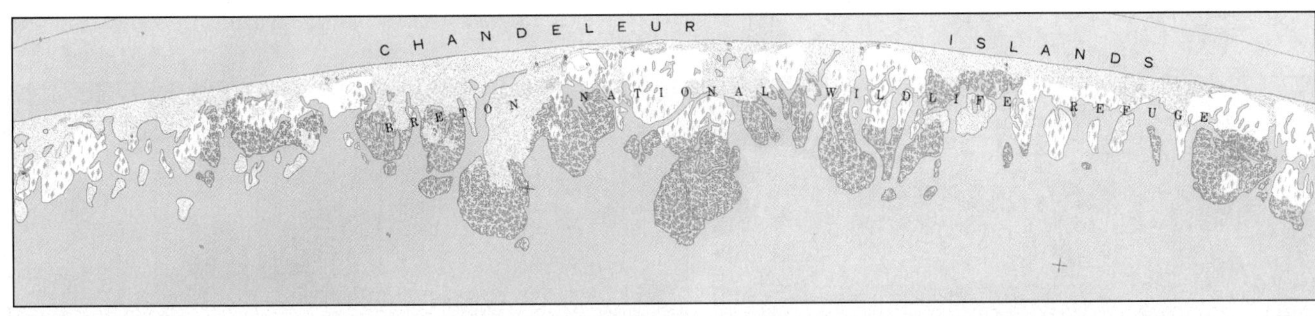

(a)

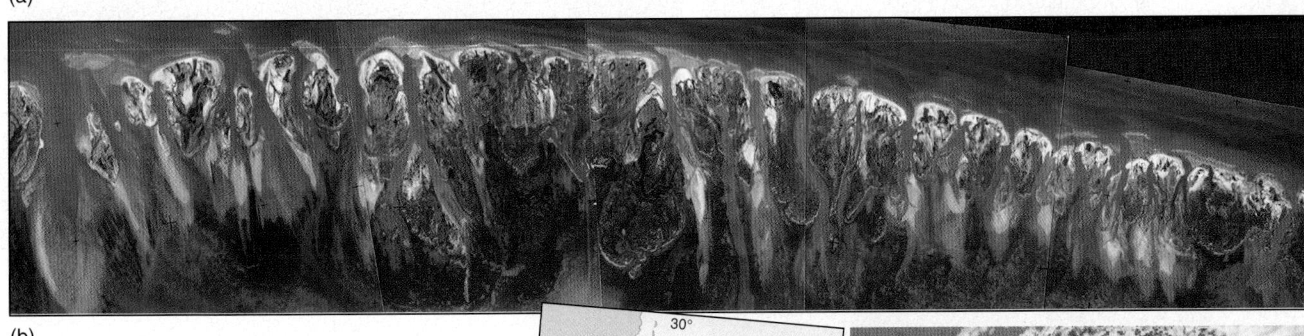

(b)

SATELLITE LOOP

Hurricane Georges

FIGURE 13.14 Hurricane Georges takes its toll (1998).
Hurricane Georges eroded huge amounts of sand from the Chandeleur Islands, just off the Louisiana and Mississippi Gulf Coast. Compare the *before* topographic map (a) with the *after* aerial photo composite (b). (c) *GOES-8* image of Hurricane Georges; note the Chandeleur Islands just south of the hurricane's eye. [(b) Photo by Aerial Data Service, Earth Imaging, a USGS service company; topographic map provided by USGS; (c) Image courtesy of NOAA.]

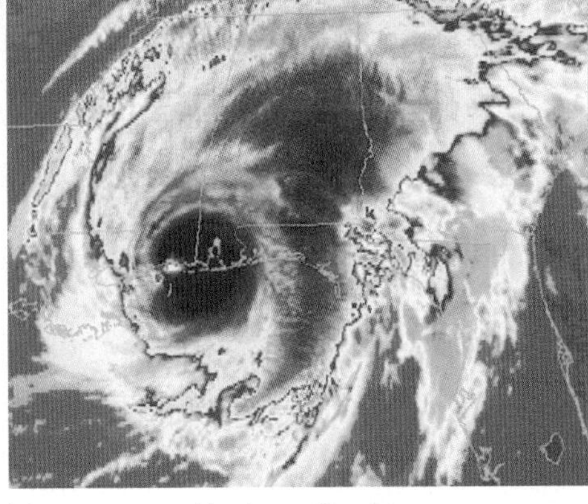

(c) Hurricane Georges

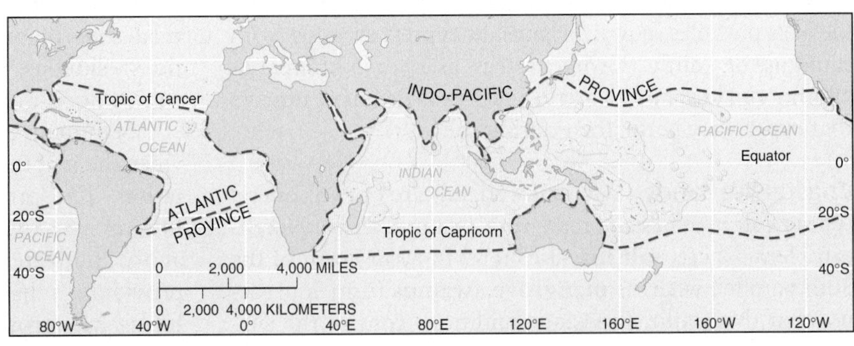

FIGURE 13.15 Worldwide distribution of living coral formations.
Yellow areas include prolific reef growth and atoll formation. The red dotted line marks the geographical limits of coral activity. Colonial corals range in distribution from about 30° N to 30° S. [After J. L. Davies, *Geographical Variation in Coastal Development* (Essex, England: Longman House, 1973). Adapted by permission.]

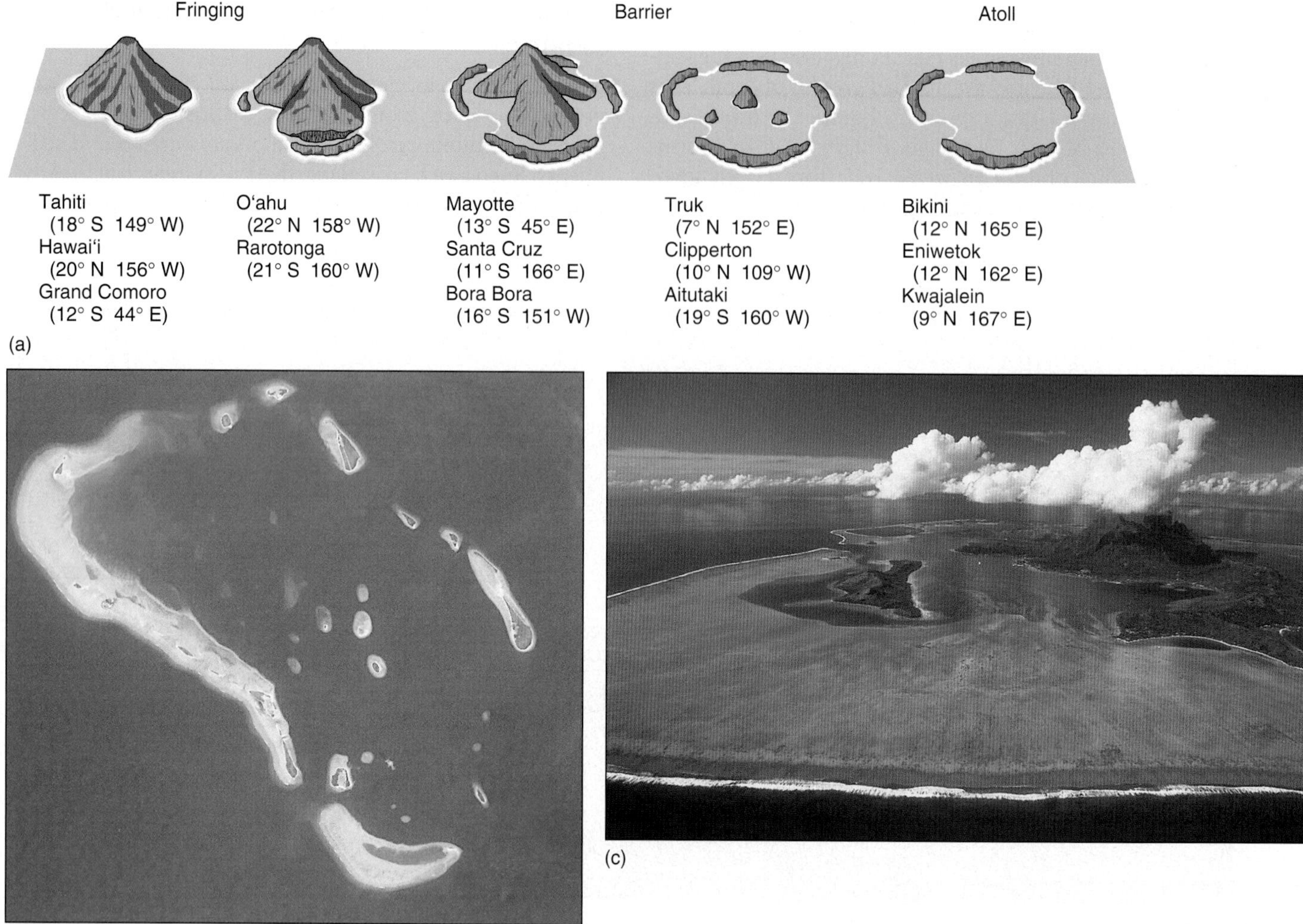

FIGURE 13.16 Coral formations.
(a) Common coral formations in a sequence of reef growth formed around a subsiding volcanic island: fringing reefs, barrier reefs, and an atoll. (b) Satellite image of a portion of the Maldive Islands, in the Indian Ocean (5° N 75° W). (c) Aerial photograph of the atolls in Bora Bora, Society Islands (16° S 152° W). [(a) After D. R. Stoddart, *The Geographical Magazine* 63 (1971): 610; (b) *Landsat-7* image courtesy of NASA; (c) photo by Harvey Lloyd/Stock market.]

These wetland ecosystems are quite fragile and are threatened by human development.

Wetlands are saturated with water enough of the time to support *hydrophytic vegetation* (plants that grow in water or very wet soil). Wetlands usually occur on poorly drained soils. Geographically, they occur not only along coastlines but also as bogs (peatlands with high water tables), as potholes in prairie lands, as cypress swamps (with standing or gently flowing water), as river bottomlands and floodplains, and as arctic and subarctic environments that experience permafrost during the year.

Coastal Wetlands. Coastal wetlands are of two general types—salt marshes and mangrove swamps. In the Northern Hemisphere, **salt marshes** tend to form north of the 30th parallel, whereas **mangrove swamps** form equatorward of that point. Freezing conditions control the survival of mangrove seedlings and thus dictate their distribution. Roughly the same latitudinal limits apply in the Southern Hemisphere.

Salt marshes usually form in estuaries and behind barrier beaches and spits. An accumulation of mud produces a site for the growth of *halophytic* (salt-tolerant) plants. This vegetation then traps additional alluvial sediments and adds to the salt marsh area. Because salt marshes are in the intertidal zone (between the farthest reaches of high and low tides), sinuous, branching channels are produced as tidal waters flood into and ebb from the marsh (Figure 13.17).

Sediment accumulation on tropical coastlines provides the site for mangrove trees, shrubs, and other vegetation. The prop roots of the mangrove are constantly finding new anchorages (Figure 13.18b). They are visible above the water line but reach below the water surface, providing a habitat for a multitude of specialized life forms. Mangrove swamps often secure enough material to form islands.

FIGURE 13.17 Coastal salt marsh.
Salt marshes are productive ecosystems commonly occurring poleward of 30° latitude in both hemispheres. This is Gearheart Marsh, part of the Arcata Marsh system on the Pacific Coast in northern California. [Photo by Bobbé Christopherson.]

(a)

(b)

(c)

FIGURE 13.18 Mangroves.
Mangroves tend to grow equatorward of 30° latitude. (a) Mangroves along the East Alligator River (12° S latitude) in Kakadu National Park, Northern Territory, Australia. (b) Mangroves retain sediments and can form anchors for island formations such as Aldabra Island (9° S), Seychelles. (c) Mangroves in the Florida Keys. [Photos by (a) Belinda Wright/ DRK Photo; (b) Wolfgang Kaehler Photography; (c) author.]

News Report 13.2

Engineers Nourish a Beach

The city of Miami, Florida, and surrounding Dade County have spent almost $70 million since the 1970s in a continuing effort to rebuild their beaches. Sand is transported to the replenishment area from a source area. To maintain a 200 m wide beach (660 ft), planners determine net sand loss per year and set a schedule for replenishment. In Miami Beach, an 8-year replenishment cycle is maintained. During Hurricane Andrew, in 1992, the replenished Miami Beach is thought to have prevented millions of dollars in shoreline structural damage.

Unforeseen environmental impact may accompany the addition of sand to a beach, especially if the sand is from an unmatched source, in terms of its ecological traits. If the new sands do not match (physically and chemically) the existing varieties, disruption of coastal marine life is possible. The U.S. Army Corps of Engineers, which operates the Miami replenishment program, is running out of "borrowing areas" for sand that matches the natural sand of the beach. A proposal to haul a different type of sand from the Bahamas is being studied as to possible environmental consequences.

The constant assault of the sea requires fortification of eroding sand. Because houses are built so close to the beach on what was the primary dune, they are vulnerable to innundation. In Figure 1 hauled-in sand is piled to form a temporary protective berm. The sign warns beachcombers to stay off the sand because traffic will scatter the effort.

In contrast to the term *hard structures*, this hauling of sand to replenish a beach is considered "soft" shoreline protection. Enormous energy and material must be committed to counteract the relentless energy that nature expends along the coast. For more on beach nourishment through Duke University, see **http://www.env.duke.edu/psds/nourishment.htm**.

FIGURE 1 Beach sand replenishment. Efforts to create protection for the beach and former primary dune must be continuous. Barrier islands are a dynamic system of rapid change; increasing storm activity and rising sea level place these coastal environments at risk. Note the sign to help keep the sand in place. One bad storm can wipe out these efforts. [Photo by Bobbé Christopherson.]

The World Resources Institute and the U.N. Environment Programme estimate mangrove losses at between 40% (example, Cameroon and Indonesia) and nearly 80% (example, Bangladesh and Philippines) since pre-agricultural times. Deliberate removal was a common practice by many governments because of false fears of disease or pestilence in these swamplands. Today, clearing for commercial and residential development is common.

Human Impact on Coastal Environments

The development of modern societies is closely linked to estuaries, wetlands, barrier beaches, and coastlines. Estuaries are important sites of human settlement because they provide natural harbors, a food source, and convenient sewage and waste disposal. Society depends on daily tidal flushing of estuaries to dilute pollution. Thus, the estuarine and coastal environment is vulnerable to abuse and destruction if development is not carefully planned. We are approaching a time when most barrier islands will be developed and occupied (Figure 13.19).

Barrier islands and coastal beaches do migrate over time. Houses that were more than a mile from the sea in the Hamptons along the southeastern shore of Long Island, New York, are now within only 30 m (100 ft) of it (Figure 13.20). The time remaining for these homes can be quickly shortened by a single hurricane or by a sequence of storms with the intensity of the series that occurred in 1962 or the severe storm that struck the U.S. and Canadian Atlantic coast in December 1992.

Despite our understanding of beach and barrier-island migration, the effect of storms, and warnings from scientists and government agencies, coastal development proceeds. Society behaves as though beaches and barrier

FIGURE 13.19 Barrier island subdivisions.
The town of Barnegat Light abruptly stops where Long Beach Island is protected in Barnegat Lighthouse State Park, New Jersey. The protrected area is almost entirely composed of replenished sand. [Photo by Bobbé Christopherson.]

FIGURE 13.20 Beach erosion dooms houses and public works on Long Island, New York.
[Photo by Mark Wexler/Woodfin Camp & Associates.]

islands are stable, fixed features, or as though they can be engineered to be permanent. Experience shows us that severe erosion generally cannot be prevented. We witness the degradation of the cliffs of southern California, the shores of the Great Lakes, the New Jersey shore, the Gulf coast, the devastation of coastal South Carolina by Hurricane Hugo, and the destruction brought by five recent hurricanes to North Carolina—three in 1999 alone. Shoreline planning is the topic of Focus Study 13.1.

The key to protective environmental planning and zoning is to *allocate responsibility and cost in the event of a disaster*. An ideal system places a hazard tax on land, based on assessed risk, and restricts the government's responsibility to fund reconstruction or an individual's right to reconstruct on frequently damaged sites. Comprehensive mapping of erosion-hazard areas would help avoid the ever increasing costs from recurring disasters.

News Report 13.3

Worldwide Coral Bleaching Worsens

Life in the ocean is under stress from a combination of rising ocean temperatures, human activity, and pollution. Increased deaths and many new diseases, never before reported, are appearing in everything from corals to sardines to seals. A troubling phenomenon is occurring among corals around the world as normally colorful corals turn stark white by expelling their nutrient-supplying, colorful algae (red-brown to green), a phenomenon known as *bleaching*. Exactly why the corals eject their symbiotic partner is unknown, for without algae the corals die. Scientists are tracking their unprecedented bleaching and dying worldwide. Locations in the Caribbean Sea and Indian Ocean, as well as off the shores of Australia, Indonesia, Japan, Kenya, Florida, Texas, and Hawai'i are experiencing this phenomenon.

Possible causes include local pollution, disease, sedimentation, and changes in salinity. Another possible cause is the 1 to 2 C° (1.8 to 3.6 F°) warming of sea-surface temperatures, as stimulated by greenhouse warming of the atmosphere. In a report, the *Status of Coral Reefs of the World: 2000* from the Global Coral Reef Monitoring Network, a finding was made that warmer water is a greater threat to corals than local pollution or other environmental assaults (see http://coral.aoml.noaa.gov/gcrmn/).

Coral bleaching worldwide is continuing as average ocean temperatures climb higher, thus linking the issue of climate change to the health of all living coral formations. By the end of 2000, approximately 30% of reefs were lost, especially following the record El Niño event of 1998. (For more information and Internet links, see http://www.usgs.gov/coralreef.html.) The *World Atlas of Coral Reefs* in 2001 summarized:

Humans are thus bringing new pressures to bear on the world's coral reefs and driving more profound changes, more rapidly, than any natural impact has ever done. Overfishing has become so widespread that there are few, if any, reefs in the world which are not threatened.... From onshore a much greater suite of damaging activities is taking place. Often remote from reefs, deforestation, urban development, and intensive agriculture are now producing vast quantities of sediments and pollutants which are pouring into the sea and rapidly degrading coral reefs.... A further specter overshadowing the world of coral reefs is that of global climate change.*

*M. D. Spalding, C. Ravilious, E. P. Green, and UNEP/WCMC, *World Atlas of Coral Reefs* (Berkeley: University of California Press, 2001), p. 11.

Focus Study 13.1

An Environmental Approach to Shoreline Planning

Coastlines are places of wonderful opportunity. They also are zones of specific constraints. Poor understanding of this resource and a lack of environmental analysis often go hand in hand. Ecologist and landscape architect Ian McHarg, in his book *Design with Nature*, discusses the New Jersey shore. He shows how proper understanding of a coastal environment could have avoided problems from major storms and coastal development. Much of what is presented here applies to coastal areas elsewhere. Figure 1 illustrates the New Jersey shore from ocean to back bay along with photographs illustrating the natural setting in the small preserved portions of the shore and those same landforms in their developed state. Let us walk across this landscape and discover how it should be treated under ideal conditions.

Beaches and Dunes—Where to Build?

We begin our walk at the water's edge and proceed across the beach. Sand beaches are the primary natural defense against the ocean; they act as bulwarks against the pounding of a stormy sea. The shoreline tolerates recreation, but not construction, because of its shifting, changing nature during storms, daily tidal fluctuations, and the potential effects of rising sea level. Beaches are susceptible to pollution and require environmental protection to control nearshore dumping of dangerous materials. In recent years, high water levels in the Great Lakes and along the southern California coast attest to the vulnerability of shorelines to erosion and inundation. Wave erosion attacks cliff formations and undermines, bit by bit, the foundations of houses and structures built too close to the edge (Figure 2).

Walking inland, we encounter the primary dune. The primary dune is even more sensitive than the beach. It is fragile, easily disturbed, and vulnerable to erosion; it cannot tolerate the passage of people trekking to the beach. Delicate plants struggle to hold the sand in place. Primary dunes are like human-made dikes in the Netherlands: They are the next defense against the sea, so disturbance by development or heavy traffic should be limited. Carefully controlled access points to the beach should be designated, and restricted access should be enforced, for even foot traffic can cause destruction. (See the beach stabilization example in Figure 12.6.)

The trough behind the primary dune is relatively tolerant of limited recreation and building. The plants that affix themselves to the surface send roots down to fresh groundwater reserves and anchor the sand in the process. Thus, if construction should inhibit the surface recharge of that water supply, the natural protective ground cover could die and destabilize the environment. Or subsequent saltwater intrusion might contaminate well water for human use. Clearly, groundwater resources and the location of recharge aquifers must be considered in planning.

Behind the trough is the secondary dune, another line of defense against the sea. It, too, is tolerant of some use yet is vulnerable to destruction. Next is the backdune, more suitable for development than any zone between it and the sea. Further inland are the bayshore and the bay, where no dredging or filling and only limited dumping of treated wastes and toxics should be permitted. This zone is tolerant to intensive recreation. In reality, the opposite of such careful assessment and planning prevails.

A Scientific View and a Political Reality

Before an intensive government study of the New Jersey shore was completed in 1962, no analysis of coastal hazards had been done outside of academic circles. What is common knowledge to geographers, botanists, biologists, and ecologists in the classroom and laboratory still has not filtered through to the general planning and political processes. As a result, on the New Jersey shore and along much of the Atlantic and Gulf coasts, improper development of the fragile coastal zone (primary dunes, trough, and secondary dunes) led to extensive destruction during storms in 1962, 1992, and numerous other times.

Scientists estimate that the coastlines will continue their retreat, in some cases tens of meters, in a few decades. Such estimates persist along the entire East Coast around to the Gulf coast. Along Long Beach Island, New Jersey, street flooding is common following rain storms, as storm drains back up during high tide. Society must reconcile ecology and economics if these coastal environments are to be sustained.

South Carolina enacted the Beach Management Act of 1988 (modified 1990) to apply some of McHarg's principles. In its first 2 years, more than 70 lawsuits protested the act as an invalid seizure of private property without compensation. Similar measures in other states have faced the same difficult path.

Implementation of any planning process is problematic; political pressure is intense, and results are mixed. Ian McHarg voices an optimistic hope: "May it be that these simple ecological lessons will become known and incorporated into ordinance [law] so that people can continue to enjoy the special delights of life by the sea."*

*Ian McHarg, *Design with Nature*. © 1969 by Ian L. McHarg. (Bantam Doubleday Dell Publishing Group, Inc.), p. 17.

(continued)

Focus Study 13.1 *(continued)*

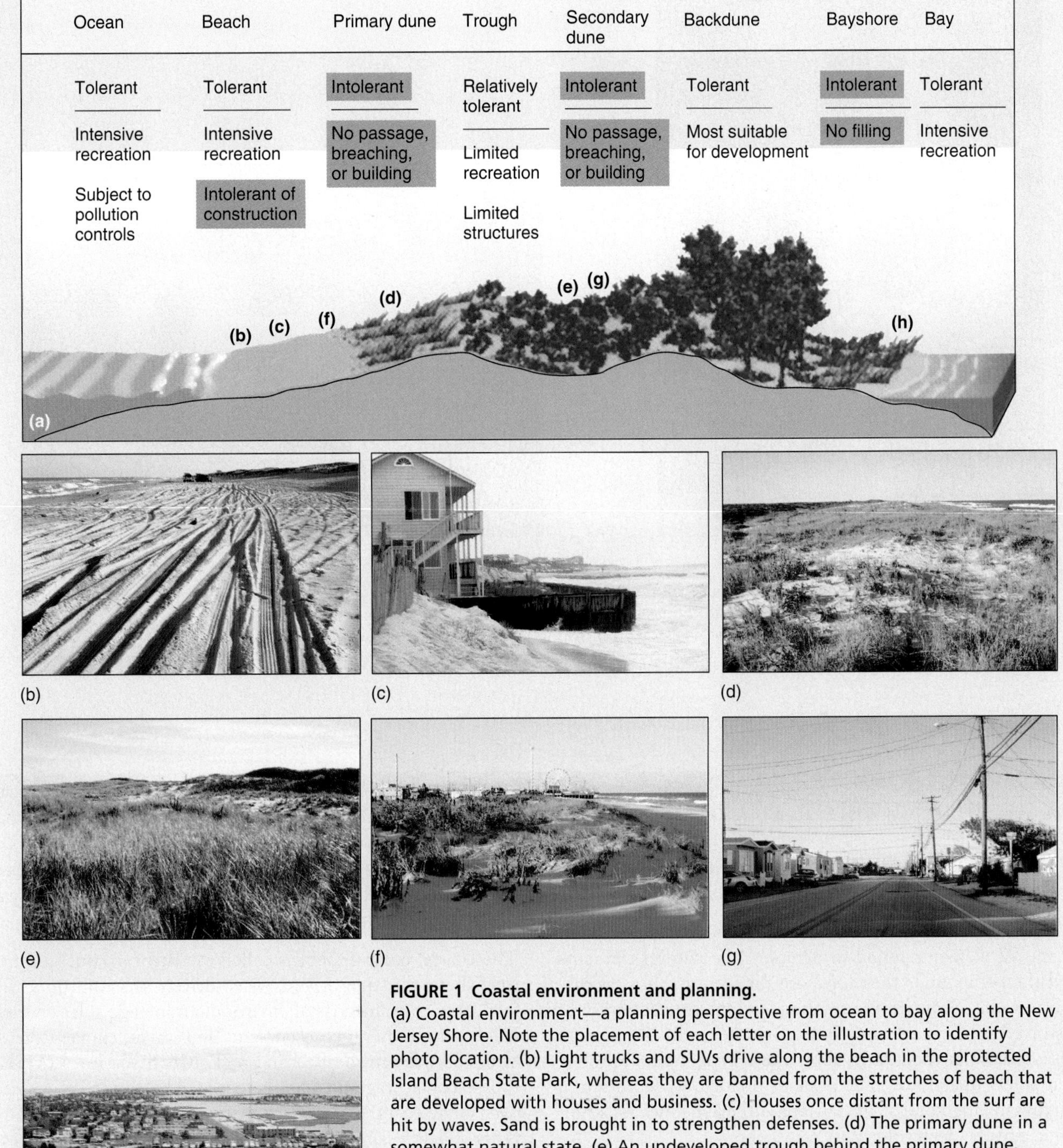

FIGURE 1 Coastal environment and planning.
(a) Coastal environment—a planning perspective from ocean to bay along the New Jersey Shore. Note the placement of each letter on the illustration to identify photo location. (b) Light trucks and SUVs drive along the beach in the protected Island Beach State Park, whereas they are banned from the stretches of beach that are developed with houses and business. (c) Houses once distant from the surf are hit by waves. Sand is brought in to strengthen defenses. (d) The primary dune in a somewhat natural state. (e) An undeveloped trough behind the primary dune. (f) Development along the primary dunes challenges the natural setting—an amusement park. (g) The former trough, paved and developed, barely above high tide levels. (h) Development and commerce in the bayshore and bay. [(a) After *Design with Nature* by Ian McHarg. Copyright © 1969 by Ian L. McHarg. Adapted by permission of Ian McHarg. Photos (b) through (h) by Bobbé Christopherson.]

(continued)

Focus Study 13.1 *(continued)*

(a)

(b)

FIGURE 2 Coastal erosion and a failed house foundation.
(a) Coastal erosion and a failed house foundation. (b) Seemingly futile efforts to halt collapse of coastal cliffs. [Photos by (a) author; (b) Bobbé Christopherson.]

Summary and Review—The Oceans, Coastal Processes, and Landforms

● *Describe* the chemical composition of seawater and the physical structure of the ocean.

Water is called the "universal solvent," dissolving at least 57 of the 92 elements found in nature. Most natural elements and the compounds they form are found in the seas as dissolved solids. Seawater is a solution, and the concentration of dissolved solids is called **salinity**. **Brine** exceeds the average 35‰ (parts per thousand) salinity; **brackish** applies to water that is less than 35‰. The ocean is divided by depth into a narrow mixing zone at the surface, a thermocline transition zone, and the deep cold zone.

salinity (p. 398)
brine (p. 399)
brackish (p. 399)

1. Describe the salinity of seawater: its composition, amount, and distribution.
2. Review the table of ocean information in Figure 13.1. Locate each of the four oceans and the Southern Ocean on a map.
3. What are the three general zones relative to physical structure within the ocean? Characterize each by temperature, salinity, dissolved oxygen, and dissolved carbon dioxide.

● *Identify* the components of the coastal environment and *list* the physical inputs to the coastal system, including tides and mean sea level.

The coastal environment is called the **littoral zone** and exists where the tide-driven, wave-driven sea confronts the land. Inputs to the coastal environment include solar energy, wind and weather, ocean currents and waves, climatic variation, and the nature of coastal rock. **Mean sea level (MSL)** is based on average tidal levels recorded hourly at a given site over many years. MSL varies spatially because of ocean currents and waves, tidal variations, air temperature and pressure differences, ocean temperature variations, slight variations in Earth's gravity, and changes in oceanic volume.

Tides are complex daily oscillations in sea level, ranging worldwide from barely noticeable to many meters. Tides are produced by the gravitational pull of both the Moon and the Sun. Most coastal locations experience two high (rising) *flood tides*, and two low (falling) *ebb tides* every day. The difference between consecutive high and low tides is the tidal range. *Spring tides* exhibit the greatest tidal range, when the Moon and Sun are either in conjunction or opposition. *Neap tides* produce a lesser tidal range.

littoral zone (p. 401)

mean sea level (MSL) (p. 401)
tide (p. 403)

4. What are the key terms used to describe the coastal environment?
5. Define mean sea level. How is this value determined? Is it constant or variable around the world? Explain.
6. What interacting forces generate the pattern of tides?
7. What characteristic tides are expected during a new Moon or a full Moon? During the first-quarter and third-quarter phases of the Moon? What is meant by a flood tide? An ebb tide?
8. Is tidal power being used anywhere to generate electricity? Explain briefly how such a plant would utilize the tides to produce electricity. Are there any sites in North America? Where are they?

● *Describe* wave motion at sea and near shore and *explain* coastal straightening as a product of wave refraction.

Friction between moving air (wind) and the ocean surface generates undulations of water that we call **waves**. Wave energy in the open sea travels through water, but the water itself stays in place. Regular patterns of smooth, rounded waves, called **swells**, are the mature undulations of the open ocean. Near shore, the restricted depth of water slows the wave, forming *waves of translation*, in which both energy and water actually move toward shore. As the crest of each wave rises, the wave falls into a characteristic **breaker**.

Wave refraction redistributes wave energy so that different sections of the coastline vary in erosion potential. Headlands are eroded, whereas coves and bays receive materials, with the long-term effect of straightening the coast. As waves approach a shore at an angle, refraction produces a **longshore current** of water moving parallel to the shore, producing the *longshore drift*. This longshore drift of sand, sediment and gravel, and assorted materials along the beach transport as *beach drift*—together these materials make up the overall littoral drift.

A **tsunami** is a seismic sea wave triggered by an undersea landslide or earthquake. It travels at great speeds in the open sea and gains height as it comes ashore, posing a coastal hazard.

wave (p. 404)
swell (p. 405)
breaker (p. 406)
wave refraction (p. 406)
longshore current (p. 406)
tsunami (p. 407)

9. What is a wave? How are waves generated, and how do they travel across the ocean? Does the water travel with the wave? Discuss the process of wave formation and transmission.
10. Describe the refraction process that occurs when waves reach an irregular coastline. How is the coastline straightened?
11. Define the components of beach drift and the longshore current and longshore drift.
12. Explain how a seismic sea wave attains such tremendous velocities. Why is it given a Japanese name?

● *Identify* characteristic coastal erosional and depositional landforms.

An *erosional coast* features wave action that cuts a horizontal bench in the tidal zone, extending from a sea cliff out into the sea. Erosional features include *stacks*, *arches*, and *cliffs*. Such a structure is called a **wave-cut platform**, or *wave-cut terrace*. In contrast, *depositional coasts* generally are located along land of gentle relief, where depositional sediments are available from many sources. Characteristic landforms deposited by waves and currents are: a **barrier spit** (material deposited in a long ridge extending out from a coast); a **bay barrier**, or *baymouth bar* (a spit that cuts off the bay from the ocean and forms an inland **lagoon**); a **tombolo** (where sediment deposits connect the shoreline with an offshore island or sea stack); and a **beach** (temporary land along the shore where sediment is in motion, deposited by waves and currents). A beach helps to stabilize the shoreline, although it may be unstable seasonally.

wave-cut platform (p. 408)
barrier spit (p. 409)
bay barrier (p. 409)
lagoon (p. 409)
tombolo (p. 409)
beach (p. 410)

13. What is meant by an erosional coast? What are the expected features of such a coast?
14. What is meant by a depositional coast? What are the expected features of such a coast?
15. How do people attempt to modify littoral drift? What strategies are used? What are the positive and negative impacts of these actions?
16. Describe a beach—its form, composition, function, and evolution.
17. What success has Miami had with beach replenishment? Is it a practical strategy?

● *Describe* barrier islands and their hazards as they relate to human settlement.

Barrier chains are long, narrow, depositional features, generally of sand, that form offshore roughly parallel to the coast. Common forms are **barrier beaches**, and the broader, more extensive **barrier islands**. Barrier formations are transient coastal features, constantly on the move, and they are a poor, but common, choice for development.

barrier beach (p. 410)
barrier island (p. 410)

18. On the basis of the information in the text and any other sources at your disposal, do you think barrier islands and beaches should be used for development? If so, under what conditions? If not, why not?
19. After Hurricane Hazel destroyed the Grand Strand off South Carolina in 1954, settlements were rebuilt, only to be hit by Hurricane Hugo 35 years later, in 1989. Why do these recurring events happen to human populations?

- ***Assess* living coastal environments: corals, wetlands, salt marshes, and mangroves.**

A **coral** is a simple marine invertebrate that forms a hard, calcified external skeleton. Over generations, corals accumulate in large reef structures. Corals live in a *symbiotic* (mutually helpful) relationship with algae; each is dependent on the other for survival.

Wetlands are lands saturated with water that support specific plants adapted to wet conditions. They occur along coastlands and inland in bogs, swamps, and river bottomlands. Coastal wetlands form as **salt marshes** poleward of the 30th parallel in each hemisphere and **mangrove swamps** equatorward of these parallels.

coral (p. 411)
wetlands (p. 414)
salt marsh (p. 414)
mangrove swamp (p. 414)

20. How are corals able to construct reefs and islands?
21. Describe a trend in corals that is troubling scientists, and discuss some possible causes.
22. Why are the coastal wetlands poleward of 30° N and S latitude different from those that are equatorward? Describe the differences.

- ***Construct* an environmentally sensitive model for settlement and land use along the coast.**

Coastlines are zones of specific constraints. Poor understanding of this resource and a lack of environmental analysis often go hand in hand in producing frequent disasterous disruption of coastal ecosystems property losses. Society must reconcile ecology and economics if these coastal environments are to be sustained.

23. Describe the condition of the Chandeleur Islands in the Gulf of Mexico before and after Hurricane Georges in 1998 (Figure 13.14). What does this tell us about the fate of other barrier islands?
24. What type of environmental analysis is needed for rational development and growth in a region like the New Jersey shore? Examine, analyze, and explain the photographs in Focus Study 13.1, Figure 1. Evaluate South Carolina's approach to coastal hazards and protection.

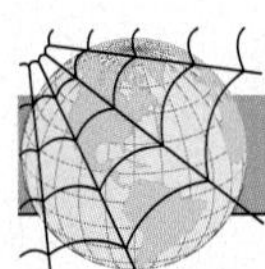

NetWork

The *Elemental Geosystems Home Page* provides on-line resources for this chapter on the World Wide Web. Once on the Home Page, click on this textbook, scroll the Table of Contents menu, select this chapter, and click "Begin." You will find self-tests that are graded, review exercises, specific updates for items in the chapter, and in "Destinations" many links to interesting related pathways on the Internet. *Elemental Geosystems* is found at **http://www.prenhall.com/christopherson**.

Critical Thinking

A. This chapter includes the following statement, "The key to protective environmental planning and zoning is to *allocate responsibility and cost in the event of a disaster*. An ideal system places a hazard tax on land, based on assessed risk, and restricts the government's responsibility to fund reconstruction or an individual's right to reconstruct on frequently damaged sites." What do you think about this as a policy statement? How would you approach implementing such a strategy? In what way could you use geographic information systems (GIS), as described in Chapter 1, to survey, assess, list owners, and follow taxation status for a vulnerable stretch of coastline?

B. Under "Destinations" in Chapter 13 of the *Elemental Geosystems Home Page* there is a link called "Coral Reefs." Sample some of the buttons on this page. Do you find any information about the damage to and bleaching of coral reefs reported in 1998 or 1999? Which places in the world were affected? Are there some suspect causes presented in "Bleaching Hot Spots" or any of the "Coral Reef Alliance" references?

Moreno Glacier in Santa Cruz Province of extreme southern Argentina (49.5° south latitude). The glacier blocks off the Rico Arm of Lake Argentina as a 60-m (197-ft) high ice dam. Every few years the water breaks through in a spectacular display. Beech trees frame this view of crevasses and tumbled glacial ice. Can you see the dark streak of debris that forms a moraine on the glacier? [Photo by Galen Rowell/Mountain Light Photography, Inc.]

14 Glacial and Periglacial Landscapes

Key Learning Concepts

By knowing and understanding the key learning concepts in this chapter, you should be able to:

- *Differentiate* between alpine and continental glaciers and *describe* their principal features.
- *Describe* the process of glacial ice formation and *portray* the mechanics of glacial movement.
- *Describe* characteristic erosional and depositional landforms created by alpine glaciation and continental glaciation.
- *Analyze* the spatial distribution of periglacial processes and *describe* several unique landforms and topographic features related to permafrost and frozen ground phenomena.
- *Explain* the Pleistocene ice age epoch and related glacials and interglacials and *describe* some of the methods used to study paleoclimatology.

A large measure of the freshwater on Earth is frozen, with the bulk of that ice sitting restlessly in just two places—Greenland and Antarctica. The remaining ice covers various mountains and fills some alpine valleys. More than 29 million cubic kilometers (7 million cubic miles) of water is tied up as ice, or about 77% of all freshwater. These deposits of ice provide an extensive frozen record of Earth's climatic history over the past several million years and perhaps some clues to its climatic future. This is Earth's **cryosphere**, the portion of the hydrosphere and ground that is perennially frozen, generally at high latitudes and elevations.

Earth's cryosphere is in a state of dramatic change, as worldwide glacial ice retreats. Present melt rates exceed anything in the ice record. In the European Alps alone some 75% of the glaciers have receded in the past 50 years, losing more than 50% of their ice mass since 1850. At this rate, the European Alps will have only 20% of their pre-industrial glacial ice left by 2050. Mount Kilimanjaro in Africa, portions of the South American Andes, and the Himalayas could lose their glacial ice within the next several decades, affecting local water resources. Later in this chapter the loss of Alaska's glacial ice is documented along with its contribution to rising sea levels.

In this chapter: We focus on Earth's extensive ice deposits—their formation, movement, and the ways in which they produce various erosional and depositional landforms. Glaciers, transient landforms themselves, leave in their wake a variety of landscape features. The fate of glaciers is intricately tied to change in global temperature, which ultimately concerns us all. We discuss the methods used to decipher past climates—the science of *paleoclimatology*—and the clues for understanding the future.

We examine the cold, near-glacial world of permafrost and periglacial processes. Approximately 20% of Earth's land area is subject to freezing conditions and frost action characteristic of periglacial regions. Higher latitudes are experiencing greater increases in average temperatures than are lower latitudes. These areas near former and existing glaciers are reminders of the last ice age. The chapter ends at Earth's poles, the Arctic Ocean and Antarctica.

Rivers of Ice

A **glacier** is a large mass of ice, resting on land or floating shelflike in the sea adjacent to land. Glaciers are not frozen lakes or groundwater ice but form by the continual accumulation of snow that recrystallizes into an ice mass. Glaciers move under the pressure of their own great weight and the pull of gravity. In fact, they move slowly in streamlike patterns, merging as tributaries into rivers of ice, as you can see in Figure 14.1.

Today, about 11% of Earth's land area is dominated by these slowly flowing streams of ice. During colder episodes in the past, as much as 30% of continental land was covered by glacial ice. Through these "ice ages," below-freezing temperatures prevailed at lower latitudes, allowing snow to accumulate year after year.

Glaciers form in areas of permanent snow, both at high latitudes and high elevations. A *snowline* is the lowest elevation where snow remains year-round—specifically, the lowest line where winter snow accumulation persists throughout the summer. Glaciers exist on some high mountains along the equator, such as in the Andes Mountains of South America and on Mount Kilimanjaro in Tanzania. In equatorial mountains, the snowline is around 5000 m (16,400 ft); on midlatitude mountains, such as the European Alps, snowlines average 2700 m (8850 ft); and in southern Greenland snowlines are down to 600 m (1970 ft).

(For Internet links to glaciers, see the Glacier Page at http://southpole.rice.edu/, Global Land Ice Measurements from Space at http://wwwflag.wr.usgs.gov/GLIMS/glimshome.html, or the National Snow and Ice Data Center at http://www-nsidc.colorado.edu/. *Landsat*-7 images are listed at http://www.emporia.edu/earthsci/gage/glacier7.htm.)

Glaciers are as varied as the landscape itself. They fall within two general groups, based on their form, size, and flow characteristics: alpine glaciers and continental glaciers.

Alpine Glaciers

With few exceptions, a glacier in a mountain range is called an **alpine glacier**, or *mountain glacier*. The *snowfield* that continually feeds the glacier with new snow is at a higher elevation. The name comes from the Alps of central Europe, where such glaciers abound. Alpine glaciers form in several subtypes. One prominent type is a *valley glacier*, an ice mass confined within a valley that originally formed by stream action. Such glaciers range in length from only 100 m (325 ft) to over 100 km (60 mi). In the satellite image in Figure 14.1, how many valley glaciers join the main glacier?

In Figure 14.2, at least a dozen valley glaciers are identifiable in the high-altitude photograph of the Eldridge and Ruth Glaciers. They fill valleys as they flow from source areas near Mount McKinley.

FIGURE 14.1 Rivers of ice.
Alpine glaciers merge from adjoining glacial valleys in the northeast region of Ellesmere Island (80° N 75° W) in the Canadian Arctic. [*Terra* ASTER image courtesy of University of Alberta, NASA/GSFC/ERSDAC/JAROS and the U.S./Japan ASTER Science Team, July 31, 2000.]

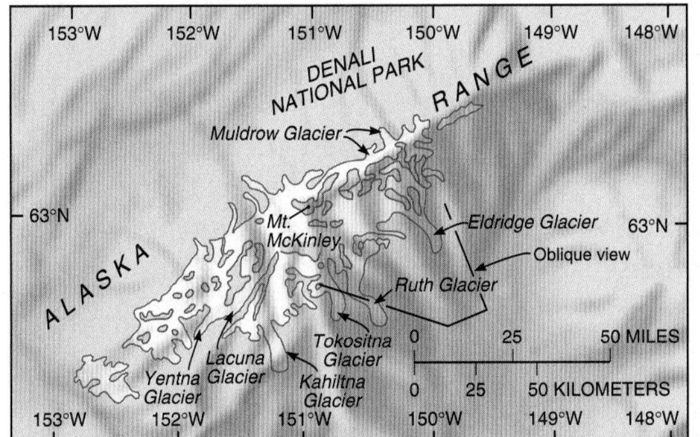

FIGURE 14.2 Glaciers in south-central Alaska.
Oblique infrared (false-color) photo of Eldridge and Ruth Glaciers, with Mount McKinley at upper left, in the Alaska Range of Denali National Park. Photo made at 18,300 m (60,000 ft). [Alaska High Altitude Aerial Photography from EROS Data Center, USGS.]

As a valley glacier flows slowly downhill, the mountains, canyons, and river valleys beneath its mass are profoundly altered by its erosive passage. Some of the debris created by the glacier's excavation is transported in and on the ice, visible as dark streaks of material being carried for deposition elsewhere (see Figures 14.1 and 14.5).

Most alpine glaciers originate in a mountain snowfield that is confined in a bowl-shaped recess. This scooped-out erosional landform at the head of a valley is called a **cirque**. A glacier that forms in a cirque is called a *cirque glacier*. Several cirque glaciers may jointly feed a valley glacier. Numerous cirques and cirque glaciers appear in the high-altitude photo in Figure 14.2.

Wherever several valley glaciers pour out of their confining valleys and coalesce at the base of a mountain range, a *piedmont glacier* is formed and spreads freely over the nearby lowlands. A *tidal glacier*, such as the Columbia Glacier on Prince William Sound in Alaska, ends in the sea, *calving* (breaking off) to form floating ice, known as **icebergs** (Figure 14.3). (See Glaciers of Prince William Sound at http://www.alaska.net/~sea/glacier.html.)

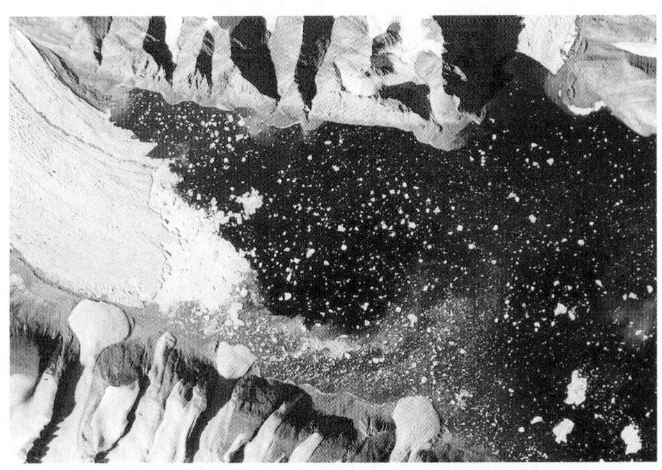

FIGURE 14.3 Icebergs forming.
Eugenie Glacier floats out on the water and breaks up into icebergs in Dobbin Bay, Ellesmere Island, Canada. [*Terra* ASTER sensor image courtesy of University of Alberta, NASA/GSFC/MITI/ERSDAC/JAROS, and U.S./Japan ASTER Science Team.]

Continental Glaciers

On a larger scale than individual alpine glaciers, a continuous mass of ice is known as a **continental glacier** and in its most extensive form is an **ice sheet**. Most glacial ice exists in the snow-covered ice sheets that blanket 80% of Greenland and 90% of Antarctica. Antarctica alone has 91% of all the glacial ice on the planet.

These two ice sheets represent such an enormous mass that large portions of each landmass have been isostatically depressed (under the weight of the ice) more than 2000 m (6500 ft) below sea level. Each ice sheet is more than 3000 m (9800 ft) deep, burying all but the highest peaks.

Ice caps and *ice fields* are two additional types of continuous ice cover associated with mountain locations. An **ice cap** is roughly circular and covers an area of less than 50,000 km^2 (19,300 mi^2). An ice cap completely buries the underlying landscape. The volcanic island of Iceland features several ice caps (Figure 14.4a). Iceland's Grímsvötn Volcano lies beneath the ice cap and erupted in 1996, producing large quantities of melted glacial water and floods, a flow called *jökulhlaup* by Icelanders.

An **ice field** extends in a characteristic elongated pattern in a mountainous region with ridges and peaks visible above the buried terrain. Figure 14.4b shows the southern Patagonian ice field in the Andes, one of Earth's largest,

(a)

(b)

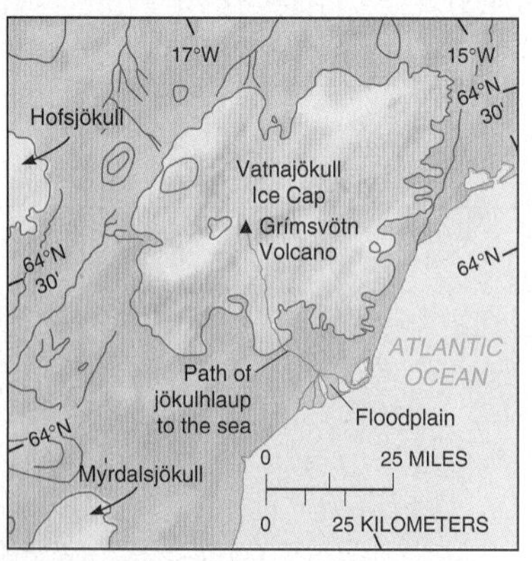

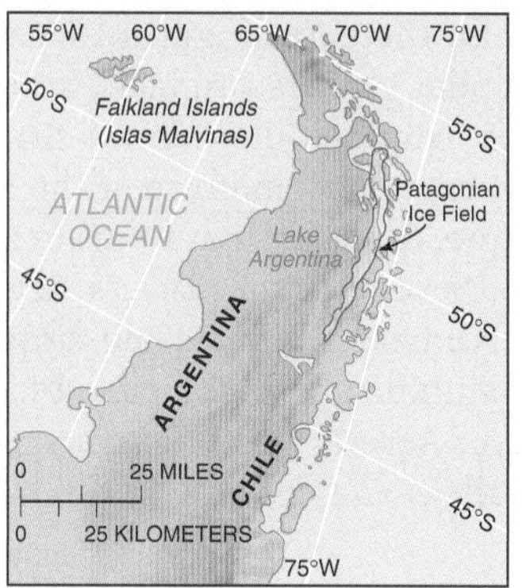

FIGURE 14.4 Ice cap and ice field.
(a) The Vatnajökull ice cap in southeastern Iceland (jökull means ice cap in Danish). A volcanic eruption beneath this ice cap in 1996 triggered massive flooding through an unpopulated portion of Iceland. Note the location of the Grímsvötn Volcano on the map, and the path of the floods to the sea. (b) The southern Patagonian ice field of Argentina. [(a) *Landsat* image from NASA; (b) photo by Cosmonauts G. M. Greshko and Yu V. Romanenko, *Salyut 6*.]

stretching 360 km (224 mi) in length between 46° S and 51° S latitude and up to 90 km (56 mi) in width. An ice field is not extensive enough to form the characteristic dome of an ice cap.

Glacial Processes

ANIMATION

Budget of a glacier, mass balance

A glacier is a dynamic body, moving relentlessly downslope at rates that vary within its mass, excavating the landscape through which it flows. The mass is dense ice that forms from snow and water through a complex process over time. A glacier's *mass budget* consists of net gains or losses of this glacial ice, which determine whether the glacier expands or retreats. Let us now look at glacial ice formation, mass balance, movement, and erosion before we discuss the fascinating landforms produced by these processes.

Formation of Glacial Ice

Dense ice comprises a glacier, formed from snow and water through a process of compaction, recrystallization, and growth. The essential input to a glacier is precipitation that accumulates in a *snowfield*, a glacier's accumulation zone (Figure 14.5a and b). Snowfields usually are at the highest elevation of an ice sheet, ice cap, or head of a valley glacier—usually in a cirque. Highland snow accumulation sometimes is the product of orographic processes, as discussed in Chapter 5. Avalanches from surrounding mountain slopes can add to the snowfield.

As the snow accumulation deepens in sedimentary-like layers, the increasing thickness results in increased weight and pressure on underlying portions. Rain and summer snowmelt then contribute water, which stimulates further melting, and that meltwater seeps down into the snowfield and refreezes.

Snow that survives the summer and into the following winter begins a slow transformation into glacial ice. Air spaces among ice crystals are pressed out as snow packs to a greater density. The ice crystals recrystallize and consolidate under pressure. In a transition step to glacial ice, snow becomes **firn**, which has a compact, granular texture.

As this process continues, many years pass before dense glacial ice is produced. Formation of **glacial ice** is analogous to formation of metamorphic rock: Sediments (snow and firn) are pressured and recrystallized into a dense metamorphic rock (glacial ice). In Antarctica, glacial ice formation may take 1000 years because of the dryness of the climate (minimal snow input), whereas in wet climates this time is reduced to just a few years because of the volume of new snow constantly being added to the system.

Glacial Mass Balance

A glacier is an open system, with *inputs* of snow and *outputs* of ice, meltwater, and water vapor. At its upper end, a glacier is fed by snowfall and other moisture in the *accumulation zone* (Figure 14.5a). This area ends at the **firn line**, indicating where the winter snow and ice accumulation survived the summer melting season.

Toward a glacier's lower end, it is wasted (reduced) through several processes: melting on the surface, internally, and at its base; ice removal by deflation (wind); the calving of ice blocks; and sublimation (recall from Chapter 5 that this is the direct evaporation of ice). Collectively, these losses are **ablation**.

The zone where accumulation gain balances ablation loss is the **equilibrium line** (Figure 14.5b). This area of a glacier generally coincides with the firn line. A glacier achieves *positive net balance* of mass—grows larger—during cold periods with adequate precipitation. In warmer times, the equilibrium line migrates up-glacier, and the glacier retreats—grows smaller—because of its *negative net balance*. Internally, gravity continues to move a glacier forward even though its lower terminus might be in retreat owing to ablation. The mass-balance losses from the South Cascade glacier and 67 Alaskan glaciers provide us two examples in News Report 14.1, p. 432.

Highways of ice that flow from accumulation areas high in the mountains are marked by trails of transported debris called *moraines* (Figure 14.5a and c). A *lateral moraine* accumulates along the sides. A *medial moraine* forms down the middle when two glaciers merge and their lateral moraines combine.

Glacial Movement

We generally think of ice as those small, brittle cubes from the freezer, but glacial ice is quite different. In fact, glacial ice behaves in a plastic (pliable) manner, for it distorts and flows in its lower portions in response to weight and pressure from above and the degree of slope below. In contrast, the glacier's upper portions are quite brittle. Rates of flow range from almost nothing to a kilometer or two per year on a steep slope. The rate of accumulation of snow in the formation area is critical to the speed.

The movement of a glacier is not like a block of rock sliding downhill. The greatest movement within a valley glacier occurs *internally*, below the rigid surface layer, which fractures as the underlying zone moves forward (Figure 14.6a). At the same time, the base creeps and slides along, varying its speed with temperature and the presence of any lubricating water beneath the ice. This *basal slip* usually is much less rapid than the internal plastic flow of the glacier, so the upper portion of the glacier flows ahead of the lower portion.

Unevenness in the landscape beneath the ice may cause the pressure to vary, melting some of the basal ice by compression at one moment, only to have it refreeze later; this process is an important factor in glacial downslope movement. This melting/refreezing action incorporates rock debris into the glacier. Consequently, the basal ice layer, which can extend tens of meters above the base of the glacier, has a much greater debris content than the ice above.

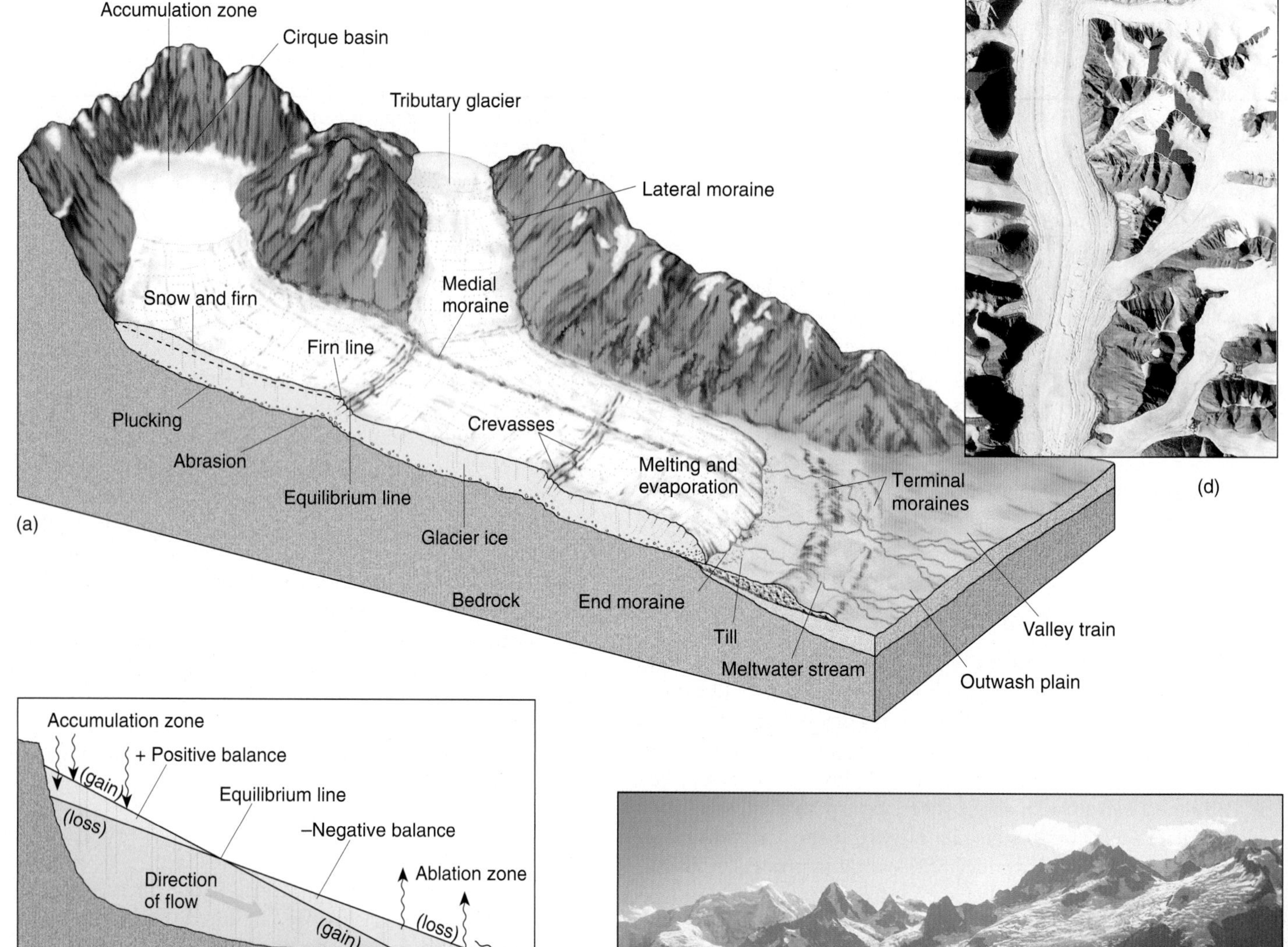

FIGURE 14.5 Mass balance and retreating alpine glacier. (a) Cross section of a typical retreating alpine glacier. (b) Annual mass balance of a glacial system, showing how the relation between accumulation and ablation controls the location of the equilibrium line. (c) Johns Hopkins Glacier, Glacier Bay National Park, Alaska, and, (d) a valley glacier off the Agassiz ice cap, Ellesmere Island, Canada, demonstrates many of the features in the illustration. Note the formation of the medial moraine where two glaciers merge. [(c) Photo by Frank S. Balthis. (d) *Terra* ASTER sensor image courtesy of University of Alberta, NASA/GSFC/MITI/ERSDAC/JAROS, and U.S./Japan ASTER Science Team.]

A flowing glacier can develop vertical cracks known as **crevasses** (Figure 14.6). These crevasses result from friction with valley walls, or tension from stretching as the glacier passes over convex slopes, or compression as the glacier passes over concave slopes. Traversing a glacier, whether an alpine glacier or an ice sheet, is dangerous because a thin veneer of snow sometimes masks the presence of a crevasse.

Tributary valley glaciers can merge to form a *compound valley* glacier. The flowing movement of a compound valley glacier is different from the flow of a river with tributaries. Tributary glaciers become compound glaciers when they merge alongside one another, rather than blending as do rivers. This concept is visible on the John Hopkins Glacier in Figure 14.5c, as each tributary maintains its own pattern of transported debris (dark streaks).

Glacier Surges. Although glaciers flow plastically and predictably most of the time, some will lurch forward with little or no warning in a **glacial surge**. This is not quite as abrupt as it sounds; in glacial terms, a surge can be tens of

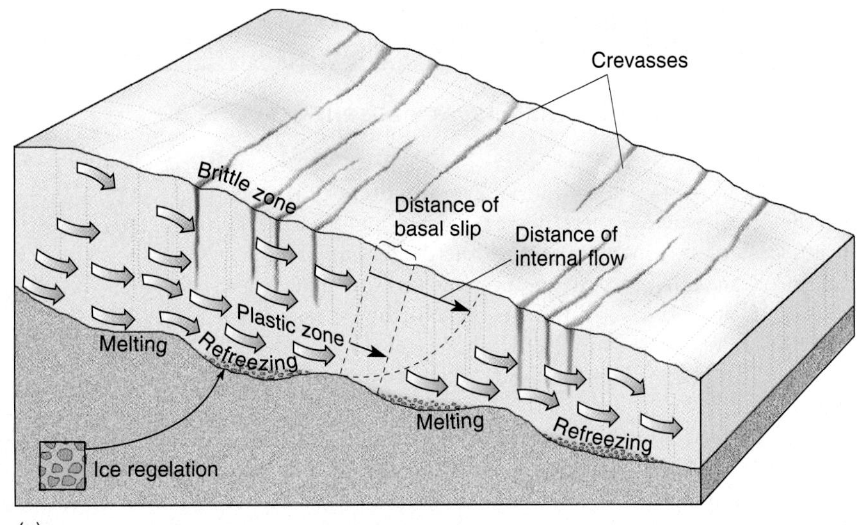

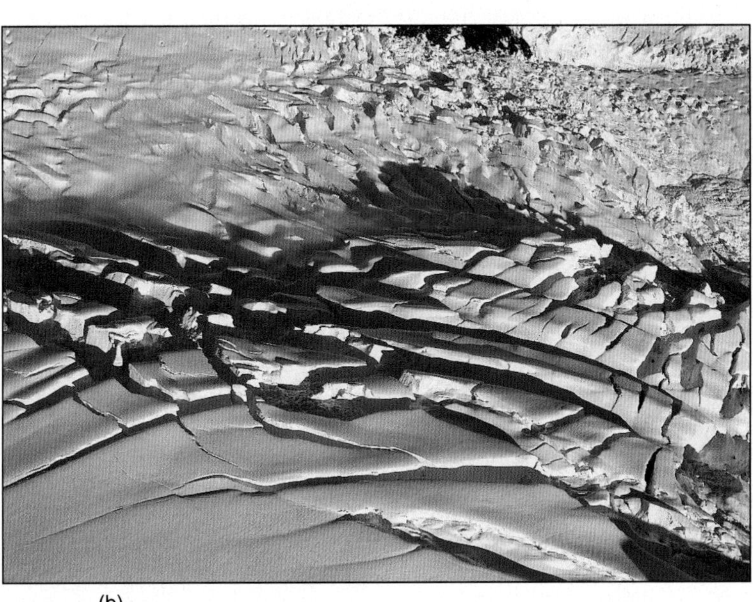

FIGURE 14.6 Glacial movement.
(a) Cross section of a glacier, showing its forward motion and brittle cracking at the surface and flow along its basal layer. (b) Surface crevasses and cracks are evidence of a glacier's forward motion (near the Don Sheldon Amphitheater, Denali National Park, Alaska). (c) Numerous crevasses in the Pine Island Glacier, off the West Antarctic Ice Sheet near 75° S 100° W. [(b) Photo by Michael Collier; (c) *Terra* ASTER sensor image courtesy of NASA/GSFC/MITI/ERSDAC/JAROS, and the U.S./Japan ASTER Science Team.]

meters per day. The Jakobshavn Glacier in Greenland, for example, is known to move between 7 and 12 km (4.3 and 7.5 mi) a year.

In the spring of 1986 Hubbard Glacier and its tributary Valerie Glacier surged across the mouth of Russell Fjord in Alaska, cutting it off from contact with Yakutat Bay. This area in southeastern Alaska is fed by annual snowfall in the Saint Elias mountain range that averages more than 8.5 m (28 ft) a year. But the rapidity of the surge was surprising. The fjord was dubbed Russell Lake for the time it remained dammed from the ocean. The glacier's movement exceeded 34 m (112 ft) per day during the peak surge, an enormous increase over its normal rate of 15 cm (6 in.) per day. Hubbard was still in surge mode through 2002.

As a surge begins, icequakes are detectable, and ice faults develop. The exact cause of such a glacial surge is being studied. Some surge events result from a buildup of water pressure under the glacier, sometimes enough to actually float the glacier slightly. Another cause of glacial surges could be the presence of a water-saturated layer of sediment, a so-called *soft bed*, beneath the glacier. A soft bed is a deformable layer that cannot resist the tremendous sheer stress produced by the moving ice of the glacier. Scientists examining cores taken from several ice streams now accelerating through the West Antarctic Ice Sheet think they identified this cause.

News Report 14.1

South Cascade and Alaskan Glaciers Lose Mass

As part of a global trend, glacial mass budgets are running in the negative on average with significant ice losses in Alaska, the Andes, the European Alps, and the Himalayas. The Columbia Glacier in southern Alaska retreated 12 km (7.5 mi) and reduced its thickness by more than 400 m (1312 ft) between 1982 and 2000. Its flow speed increased from 5 to 30 m (16.4 to 98.4 ft) per day in this time frame. This appears part of a regional trend (see R.M. Krimmel, *Photogrammetric Data Set, 1957–2000, and Bathymetric Measurements for Columbia Glacier, Alaska*, USGS Water-Resources Report 01-4089, Tacoma, Washington, 2001).

The net mass balance of the South Cascade Glacier in Washington State demonstrated significant losses between 1955 and 2001. In just one year (September 1991 to October 1992), the terminus of the glacier retreated 38 m (125 ft), resulting in major changes to the surface and sides of the glacier. More than a 2% loss of the glacier's mass occurred in just that one year. The *net accumulation* and *net wastage* are illustrated in Figure 1.

There are multiple causes for these losses. Increasing average air temperature and decreasing precipitation are possible causes. The heavy snowfall in the Northwest associated with the La Niña episode is reflected in the positive mass balance for 1999 and 2000. A comparison of the trend of this glacier's mass balance with that of others in the world shows that temperature changes apparently are causing widespread reductions in middle- and lower-elevation glacial ice. The present wastage (ice loss) from alpine glaciers worldwide is thought by some to contribute over 25% to the rise in sea level.

Research covering almost 50 years of 67 glaciers in Alaska was published in 2002. The study estimated volume changes in these glaciers. The glaciers were grouped into seven regions shown in Figure 2a. The rate of glacier-wide thickness loss for 67 glaciers between the mid-1950s and 1995 and for 28 glaciers between 1995 and 2001 are in Figure 2b.

A total volume change was estimated to be -52 ± 15 km^3 per year (-12.5 ± 3.6 mi^3), equaling a water equivalent of -96 ± 35 km^3 per year (-23 ± 8.4 mi^3). These losses alone account for a rise in global sea level of 0.14 ± 0.04 and 0.27 ± 0.10 mm per year (0.55 and 1.06 in. per year). This is twice the volume estimated to be coming from the entire Greenland ice sheet during the 1995 to 2001 time period. Therefore, about 9% of global sea level rise is coming from the Alaskan meltdown.

The reasons for these losses are complex. Thinning and wastage are not solely a result of climatic warming. The individual dynamics of each glacier must be considered. Although these changes appear initiated by negative mass balances. Relative to the contribution of the Alaskan meltdown to sea level rise, the scientists summarized,

> Compared with estimated inputs from the Greenland ice sheet and other sources, Alaskan glaciers have, over the past 50 years, made the largest single glaciological contribution to rising sea

(continued)

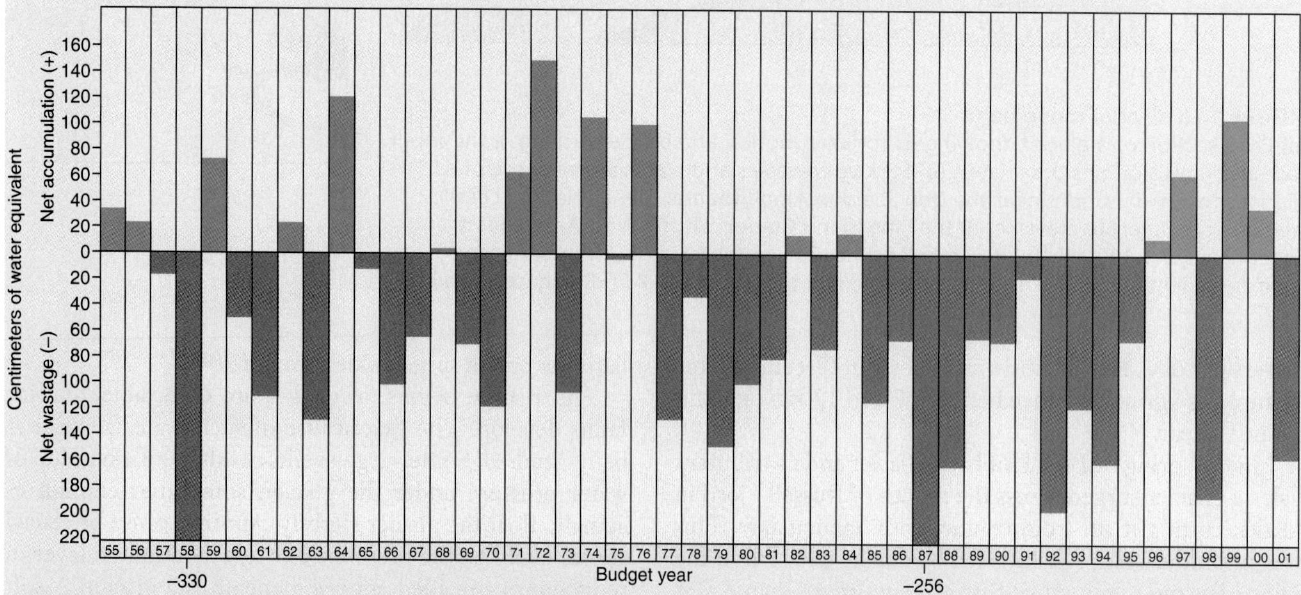

FIGURE 1 South Cascade Glacier net mass balance, 1955–2001.
A negative net mass balance has dominated this shrinking glacier since 1955. Data in centimeters (2.54 cm per in.). [Data from R. M. Krimmel, *Water, Ice, Meteorological, and Speed Measurements at South Cascade Glacier, Washington, 2001 Balance Year*. USGS Water Resources Report, Tacoma, Washington, 2001; and personal communication.]

News Report 14.1 *(continued)*

level yet measured.... the different rates of thinning observed in the various Alaskan regions may be important in characterizing patterns of climate change.*

*A.A. Arendt, *et. al.*, "Rapid Wastage of Alaska Glaciers and Their Contribution to Rising Sea Level," *Science* 297, (July 19, 2002): 382–86.

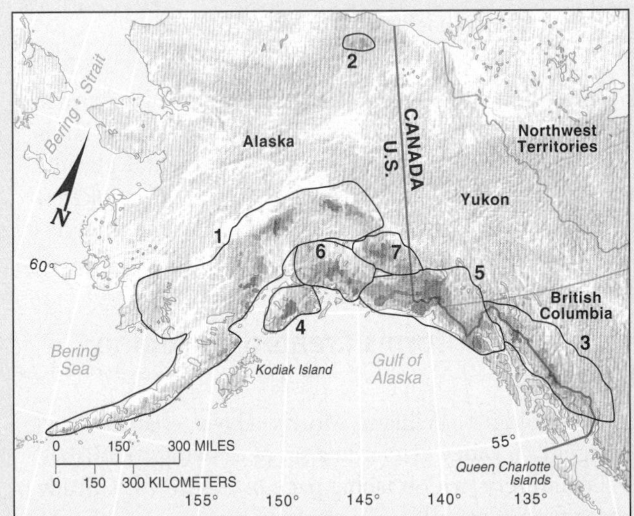

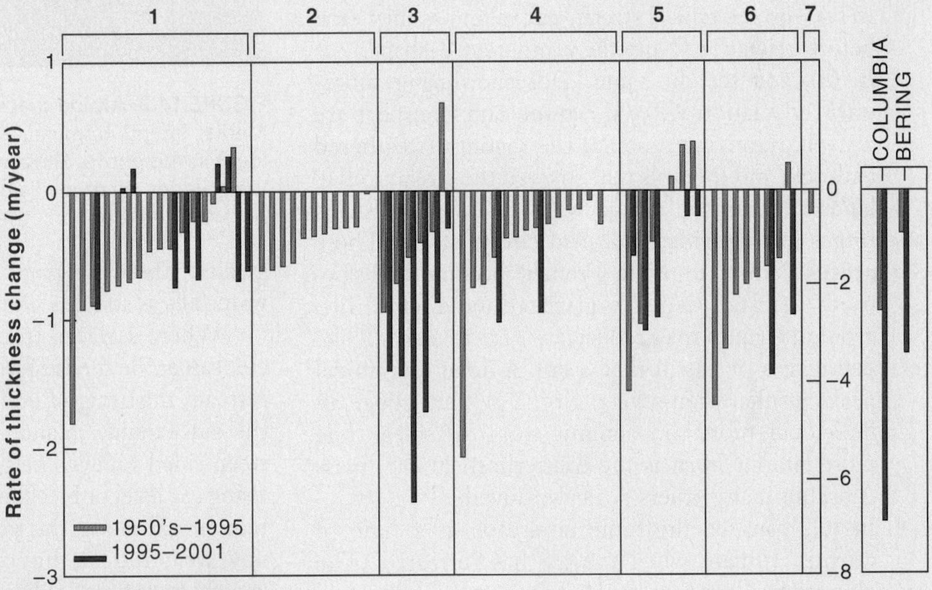

FIGURE 2 Negative mass balances for Alaska's glaciers. (a) Sixty-seven Alaskan glaciers divided into seven regions: 55 in Alaska, 11 cross the Alaskan-Canadian border, and one is in the Yukon. (b) Rate of change (meters per year) in glacier-wide thickness for 67 glaciers in the earlier period (1950s to 1995, solid brown bars) and 28 glaciers in the recent period (1995 to 2001, red bars). Note the grouping of glaciers into seven regions noted on the map of Alaska in (a). [Adapted by permission of the American Association for the Advancement of Science from Anthony A. Arendt, *et. al.*, Figures 1 and 3, "Rapid Wastage of Alaska Glaciers and Their Contribution to Rising Sea Level," *Science* 297, (July 19, 2002): 382–86.]

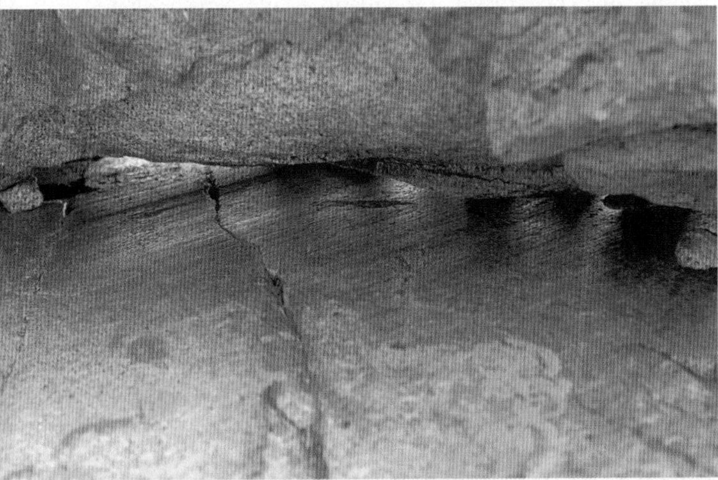

FIGURE 14.7 Glacial sandpapering polishes rock. Glacial polish and striations are evidence of glacial abrasion and erosion. The polished, marked surface is seen sheltered beneath a glacial erratic—rock left behind by a retreating glacier. [Photo by Bobbé Christopherson.]

Glacial Erosion. Glacial erosion is similar to a large excavation project, with the glacier hauling debris from one site and depositing it in another. The passing glacier mechanically plucks rock material and carries it away. Rock pieces actually freeze to the basal layers of the glacier in this *glacial plucking* process. Once embedded in the ice, this debris scours and sandpapers the landscape as the glacier moves—a process called **abrasion**. Abrasion and gouging produce a smooth surface on exposed rock, which shines with *glacial polish* when the glacier retreats. Larger rocks in the glacier act much like chisels, working the underlying surface to produce glacial striations (scratches) in the flow direction (Figure 14.7).

Glacial Landforms

Teton Glacier Notebook

Glacial erosion and deposition produce unique landforms. You might expect all glaciers to create the same landforms, but this is not so. Alpine and continental glaciers generate their own characteristic landscapes. We look first at erosional landforms created by alpine glaciers, then depositional landforms. Finally, we examine the results of continental glaciation.

Erosional Landforms Created by Alpine Glaciation

Geomorphologist William Morris Davis characterized the stages of a valley glacier in a set of drawings published in 1906 and redrawn on facing page in Figure 14.8. Study of these figures reveals ice as the sculptor:

- In (a), you see typical stream-cut valleys as they exist before glaciation. Note the prominent **V**-shape.
- In (b), you see the same landscape during subsequent glaciation. Glacial erosion and transport are actively removing much of the regolith (weathered bedrock) and the soils that covered the stream valley landscape. As the cirque walls erode away, sharp ridges form, dividing adjacent cirque basins. These **arêtes** ("knife-edge" in French) become the sawtooth, serrated ridges in glaciated mountains. Two eroding cirques may reduce an arête to a saddlelike depression or pass, called a **col**. A **horn** (pyramidal peak) results when several cirque glaciers gouge an individual mountain summit from all sides. The most famous horn is the Matterhorn in the Swiss Alps, but many others occur worldwide.
- In (c), you see the same landscape at a time of warmer climate when the ice has retreated. The glaciated valleys now are **U**-shaped, greatly changed from their previous stream-cut **V** form. You can see the steep sides and the straightened course of the valleys. Physical weathering from the freeze-thaw cycle has loosened rock along the steep cliffs, where it has fallen to form *talus slopes* along the valley sides.

A small mountain lake, especially one that collects in a cirque basin behind risers of rock material, is called a **tarn** (Figure 14.8). Small, circular, stair-stepped lakes in a series are called **paternoster lakes** because they look like a string of rosary (religious) beads forming in individual rock basins aligned down the course of a glaciated valley. Such lakes may have formed by the differing resistance of rock to glacial processes or by the damming effect of glacial deposits.

The valleys carved by tributary glaciers are left stranded high above the glaciated valley floor following the removal of previous slopes and stream courses by the glacier. These *hanging valleys* are the sites of spectacular waterfalls as streams plunge down the steep cliffs.

FIGURE 14.9 Alpine glacier depositional features. Medial, lateral, terminal, and ground moraine deposits are evident in this photo. These were produced by the Hole-in-the-Wall Glacier, Wrangell-Saint Elias National Park, Alaska. Also, note the kettle pond in the foreground. [Photo by Tom Bean.]

Where a glacial trough intersects the ocean, the glacier can erode the landscape below sea level. As the glacier retreats, the trough floods and forms a deep **fjord** in which the sea extends inland, filling the lower reaches of the steep-sided valley. The fjord may be flooded further by a rising sea level or by changes in the elevation of the coastal region. All along the glaciated coast of Alaska, glaciers now are in retreat, thus opening many new fjords that previously were blocked by ice. Coastlines with notable fjords include those of Norway, Greenland, Chile, South Island of New Zealand, Alaska, and British Columbia.

Depositional Landforms Created by Alpine Glaciation

You see how glaciers excavate tremendous amounts of material and create fascinating landforms in the process.

FIGURE 14.10 A valley train deposit. Peyto Glacier in Alberta, Canada. Note valley train, braided stream, and milky-colored glacial meltwater. [Photo by author.]

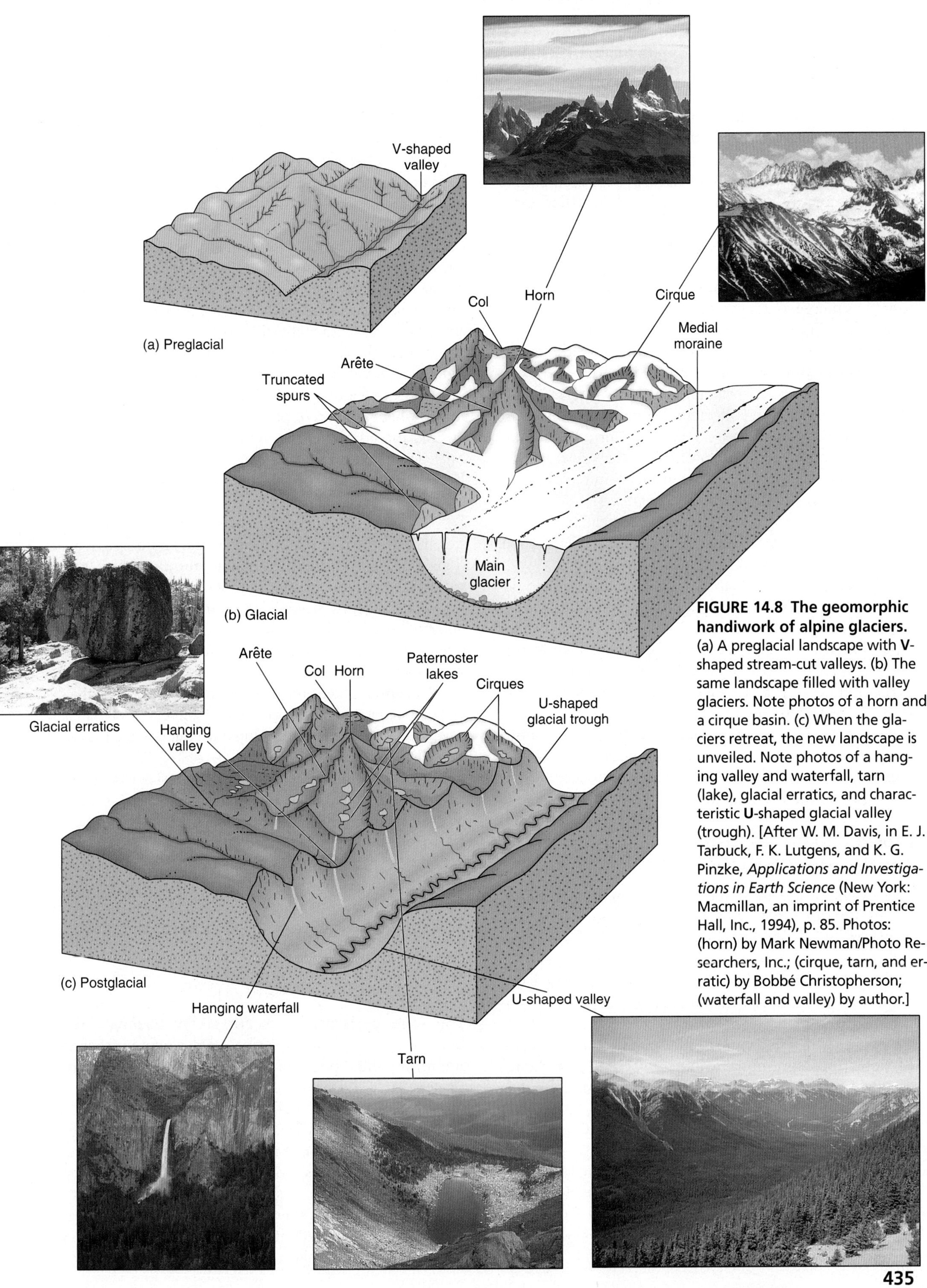

FIGURE 14.8 The geomorphic handiwork of alpine glaciers. (a) A preglacial landscape with **V**-shaped stream-cut valleys. (b) The same landscape filled with valley glaciers. Note photos of a horn and a cirque basin. (c) When the glaciers retreat, the new landscape is unveiled. Note photos of a hanging valley and waterfall, tarn (lake), glacial erratics, and characteristic **U**-shaped glacial valley (trough). [After W. M. Davis, in E. J. Tarbuck, F. K. Lutgens, and K. G. Pinzke, *Applications and Investigations in Earth Science* (New York: Macmillan, an imprint of Prentice Hall, Inc., 1994), p. 85. Photos: (horn) by Mark Newman/Photo Researchers, Inc.; (cirque, tarn, and erratic) by Bobbé Christopherson; (waterfall and valley) by author.]

Glaciers also produce another set of landforms when they melt and deposit their eroded and transported debris cargo. At the toe of the glacier where the glacier retreats, accumulated debris marks its former margins. This is the end, or *terminus*, of the glacier (Figure 14.5a). **Glacial drift** is the general term for *all* glacial deposits, both unsorted and sorted. Direct ice deposits are unstratified and unsorted and are called **till**. Sediments deposited by glacial meltwater are sorted and are termed **stratified drift**.

Moraine is the name for specific landforms produced by the deposition of these sediments. A **lateral moraine** forms along each side of a glacier. If two glaciers with lateral moraines join, a **medial moraine** may form as lateral moraines merge. (Can you identify these elements on the active glacier in Figure 14.5c?)

Eroded debris that is dropped at the glacier's farthest extent is called a *terminal moraine*. However, there also may be *end moraines*, formed wherever a glacier pauses after reaching a new equilibrium. Both of these forms are clearly visible in Figure 14.9. Lakes may form behind terminal and end moraines following a glacier's retreat, with the moraine acting as a dam. If a glacier is in retreat, individual deposits are called *recessional moraines*. A deposition of till generally spread across a surface is called a *ground moraine*, or *till plain*, and may hide the former landscape. Such plains are found in portions of the Midwest.

All these types of till are unsorted and unstratified debris. In contrast, streams of glacial meltwater can deposit sorted and stratified drift beyond a terminal moraine. This type of meltwater-deposited material down valley from a glacier is called a *valley train deposit*. Peyto Glacier in Alberta produces such a valley train that continues into Peyto Lake (Figure 14.10). Distributary stream channels appear braided across its surface. The picture also shows the milky meltwater associated with glaciers laden with finely ground "rock flour." Meltwater is produced by glaciers at all times, not just when they are retreating.

Erosional and Depositional Features of Continental Glaciation

The extent of the most recent continental glaciation in North America and Europe, 18,000 years ago, is portrayed in Figure 14.21. Because continental glaciers form across broad, open landscapes under different circumstances than alpine glaciers, the intricately carved alpine features, lateral moraines, and medial moraines all are lacking in continental glaciation.

Figure 14.11 illustrates some of the most common erosional and depositional features associated with the retreat of a continental glacier. A **till plain** forms behind an end moraine, featuring *unstratified* coarse till, low, rolling relief, and deranged drainage patterns (see Chapter 11, Figure 11.9g). Beyond the morainal deposits, **outwash plains** of *stratified drift* feature stream channels that are meltwater-fed, braided, and overloaded with debris deposited across the landscape.

Figure 14.11a shows a sinuously curving, narrow ridge of coarse sand and gravel called an **esker**. An esker forms along the channel of a meltwater stream that flows beneath a glacier, in an ice tunnel, or between ice walls beneath the glacier. As a glacier retreats, the steep-sided esker is left behind in a pattern roughly parallel to the path of the glacier. The ridge may not be continuous, and in places may even appear branched, following the course set by the subglacial watercourse. Commercially valuable deposits of sand and gravel are quarried from some eskers.

Sometimes an isolated block of ice, perhaps more than a kilometer (0.6 mi) across, persists in a ground moraine, an outwash plain, or a valley floor after a glacier retreats. Perhaps 20 to 30 years are required for it to melt. In the interim, sediment continues to accumulate around the melting ice block. When the block finally melts, it leaves behind a steep-sided hole that frequently fills with water. This feature is called a **kettle** (Figure 14.11b). Thoreau's Walden Pond (quotation in Chapter 5), is such a kettle.

Another feature of outwash plains is a **kame**—a small hill, knob, or mound of poorly sorted sand and gravel deposited directly by water, by ice in crevasses, or in ice-caused indentations in the ground surface (Figure 14.11c). Kames also can be found in deltaic forms and in terraces along valley walls.

Glacial action also forms streamlined hills—one erosional, called a roche moutonnée, and the other depositional, called a drumlin. A **roche moutonnée** ("sheep rock" in French) is an asymmetrical hill of exposed bedrock. Its gently sloping upstream side (stoss side) was polished smooth by glacial action, whereas its downstream side (lee side) is abrupt and steep where rock pieces were plucked by the glacier (Figure 14.12).

A **drumlin** is deposited till that was streamlined in the direction of continental ice movement, blunt end upstream and tapered end downstream. Multiple drumlins (called *swarms*) occur across the landscape in portions of New York and Wisconsin, among other areas. Sometimes their shape is that of an elongated teaspoon bowl, lying face down. Drumlins may attain lengths of 100–5000 m (330 ft to 3.1 mi) and heights up to 200 m (650 ft). Figure 14.13 shows a portion of a topographic map for the area south of Williamson, New York, which experienced continental glaciation. In studying the map, can you identify the numerous drumlins? In what direction do you think the continental glaciers moved across this region?

Periglacial Landscapes

The term **periglacial** was coined by scientist W. V. Lozinski in 1909 to describe cold-climate processes, landforms, and topographic features along the margins of glaciers, past and present. These periglacial regions occupy over 20% of Earth's land surface (Figure 14.14). Periglacial landscapes are either near permanent ice or at high elevation, and are seasonally snow free. Under these conditions, a unique set of periglacial processes operate, including permafrost, frost action, and ground ice.

Climatologically, these regions are in *subarctic* and *polar* climates (especially the *tundra* climate region). Such climates occur either at high latitude (tundra and boreal forest environments) or high elevation in lower-latitude mountains (alpine environments). These periglacial regions are dominated by processes that relate to physical weathering and mass movement (Chapter 10), climate (Chapter 6), and soil (Chapter 15).

Geography of Permafrost

When soil or rock temperatures remain below 0°C (32°F) for at least 2 years, a condition of **permafrost** develops. An area that has permafrost but is not covered by glaciers is considered periglacial. Note that this criterion is based solely on

FIGURE 14.11 Continental glacier depositional features.
Common depositional landforms produced by glaciers: (a) an esker through farmland near Campbellsport, Wisconsin; (b) Walden Pond is a kettle surrounded by mixed forest in northeastern Massachusetts; (c) a kame covered by a woodlot near Campbellsport, Wisconsin. [Photos by (a) and (c) Tom Bean/DRK Photos; (b) Bobbé Christopherson.]

(a)

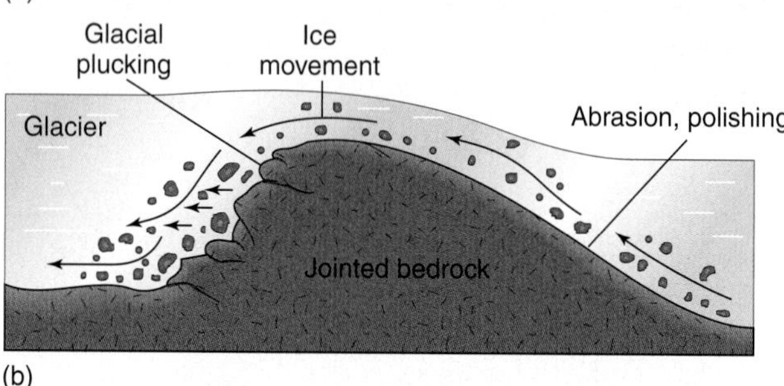

(b)

FIGURE 14.12 Glacially eroded streamlined rock. Roche moutonnée, as exemplified by Lembert Dome in the Tuolumne Meadows area of Yosemite National Park, California. [Photo by author.]

temperature and not on how much water is present. Other than high latitude and low temperatures, two other factors contribute to permafrost: The presence of fossil permafrost from previous ice-age conditions and the insulating effect of snow cover or vegetation that inhibits heat loss.

Continuous and Discontinuous Zones. Permafrost regions are divided into two general categories, continuous and discontinuous, that merge along a general transition zone. *Continuous permafrost* is a perennial region of the most severe cold (purple areas in Figure 14.14). Continuous permafrost affects all surfaces except those beneath deep lakes or rivers and may exceed 1000 m (over 3000 ft) in depth, averaging approximately 400 m (1300 ft).

Discontinuous permafrost occurs in unconnected patches that gradually merge poleward toward the continuous permafrost zone. Equatorward, along lower-latitude margins, permafrost becomes scattered or sporadic until it gradually disappears (dark blue area on the map). Permafrost affects as much as 50% of Canada and 80% of Alaska. In central Eurasia, the effects of continentality and elevation produce discontinuous permafrost that extends equatorward to the 50th parallel. In addition to these two types, zones of high-altitude alpine permafrost extend to lower latitudes, as shown on the map.

Behavior of Permafrost. We have looked at the spatial distribution of permafrost; let us examine how permafrost behaves. Figure 14.15 (p. 440) is a stylized cross section from approximately 75° N to 55° N, using the three sites located on the map in Figure 14.14. The **active layer** is the zone of seasonally frozen ground that exists between the subsurface permafrost layer and the ground surface.

The active layer is subjected to consistent daily and seasonal freeze-thaw cycles. This cyclic melting of the active layer affects as little as 10 cm (4 in.) depth in the north (Ellesmere Island, 78° N), up to 2 m (6.6 ft) in the southern margins (55° N) of the periglacial region, and 15 m (50 ft) in the alpine permafrost of the Colorado Rockies (40° N).

The depth and thickness of the active layer and permafrost zone change slowly in response to climatic change. Higher temperatures degrade (reduce) permafrost and increase the thickness of the active layer; lower temperatures gradually aggrade (increase) permafrost depth and reduce active layer thickness. Although somewhat sluggish in response, the active layer is a dynamic open system driven by energy gains and losses in the subsurface environment. As you might expect, most permafrost exists in disequilibrium with environmental conditions and therefore actively adjusts to inconstant climatic conditions. With the incredibly warm temperatures recorded in the Canadian Arctic since 1990, more disruption of surfaces is occurring—leading to highway, railway, and building damage.

A *talik* is unfrozen ground that may occur above, below, or within a body of discontinuous permafrost or beneath a water body in the continuous region. Taliks occur beneath deep lakes and may extend to bedrock and noncryotic (unfrozen) soil beneath large deep lakes (Figure 14.15). Taliks form connections between the active layer and groundwater, whereas in continuous permafrost groundwater is essentially cut off from surface water. In this way, permafrost disrupts aquifers and taliks, leading to water supply problems.

Ground Ice and Frozen-Ground Phenomena

In regions of permafrost, subsurface water that is frozen is termed **ground ice**. The moisture content of areas with ground ice may vary from nearly none in regions of drier permafrost to almost 100% in saturated soils.

Frost-Action Processes. Frozen water in the soil initiates *frost action*. The 9% expansion of water as it freezes produces strong mechanical forces that fracture rock and disrupt soil at and below the surface. If sufficient water freezes, the saturated soil and rocks are subject to *frost-heaving* (vertical movement) and *frost-thrusting* (horizontal motions). Boulders and slabs of rock are thrust to the surface. Layers of soil (soil horizons) may appear disrupted as if stirred or churned. Frost action processes also produce a contraction in soil and rock volume, opening up cracks for ice wedges to form.

An *ice wedge* develops when water enters a crack in the permafrost and freezes. Thermal contraction in ice-rich soil forms a tapered crack—wider at the top, narrowing toward the bottom. Repeated seasonal freezing and melting pro-

(a)

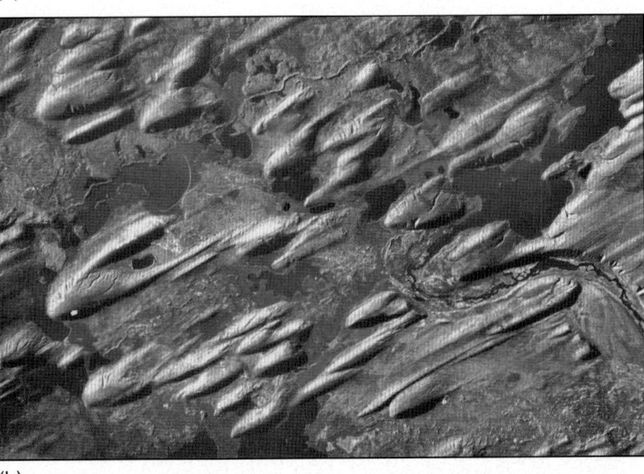

(b)

FIGURE 14.13 Glacially deposited streamlined features. (a) Topographic map south of Williamson, New York, featuring numerous drumlins (7.5-minute series quadrangle map, originally produced at a 1:24,000 scale, 10-ft contour interval). (b) A drumlin field in Snare Lake, Canada. [(a) USGS map Courtesy of U.S. Geological Survey; (b) photo by National Air Photo Library, Ottawa, Canada.]

gressively expands the wedge, which may widen from a few millimeters to 5–6 m and up to 30 m (100 ft) in depth (Figure 14.16). Widening may be small each year, but after many years the wedge can become significant.

The expansion and contraction of frost action results in the transport of stones and boulders. As the water-ice volume changes and the ice wedge deepens, coarser particles are moved toward the surface. An area with a system of ground ice and frost action develops sorted and unsorted accumulations of rock at the surface that take the shape of polygons called patterned ground (Figure 14.17, p. 441).

Hillslope Processes. Soil drainage is poor in areas of permafrost and ground ice. When the surface thaws seasonally, the upper layer of soil and regolith saturates with soil moisture, and the whole layer flows if the landscape is sloped. Such soil flows are in general called *solifluction*; in the presence of ground ice, the more specific term *gelifluction* applies.

The cumulative effect of this landflow can be an overall flattening of a rolling landscape, with sagging surfaces and scalloped and lobed patterns downslope. Periglacial mass movement processes are related to slope dynamics and processes discussed in Chapter 10.

Humans and Periglacial Landscapes

Human populations in areas that experience permafrost and frozen-ground phenomena face various difficulties. Because thawed ground above the permafrost zone fre-

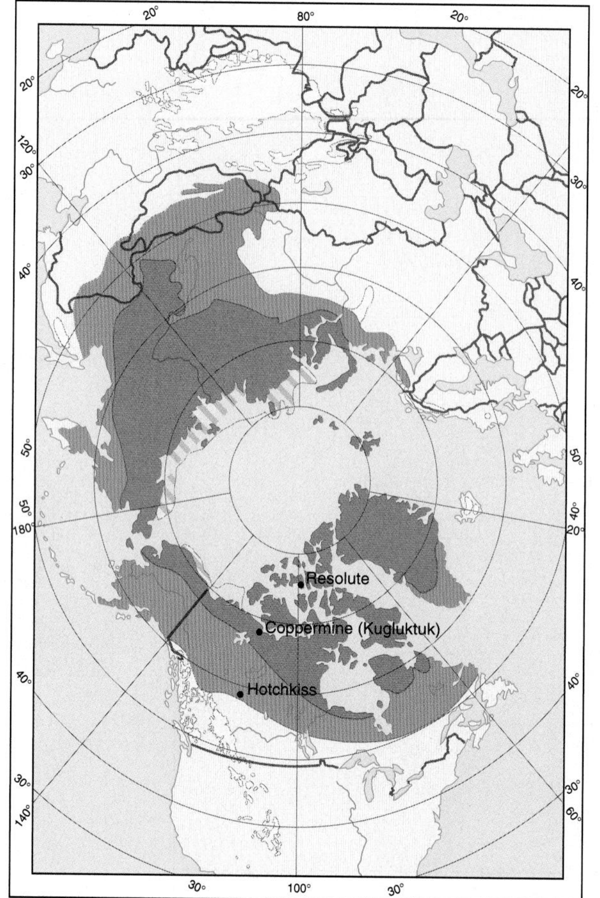

FIGURE 14.14 Permafrost distribution.
Distribution of permafrost in the Northern Hemisphere. Alpine permafrost is noted except for small alpine occurrences in Hawai'i, Mexico, Europe, and Japan that are too small to show on the map. Subsea permafrost occurs in the ground beneath the Arctic Ocean along the margins of the continents noted. Note the towns of Resolute and Coppermine (now Kugluktuk in Nunavut, formerly part of the NWT) and Hotchkiss in Alberta. A cross section of the permafrost beneath these towns is shown in Figure 14.15. [Adapted from Troy L. Péwé, "Alpine Permafrost in the Contiguous United States: A Review," *Arctic and Alpine Research* 15, no. 2 (May 1983): 146. © Regents of the University of Colorado. Used by permission.]

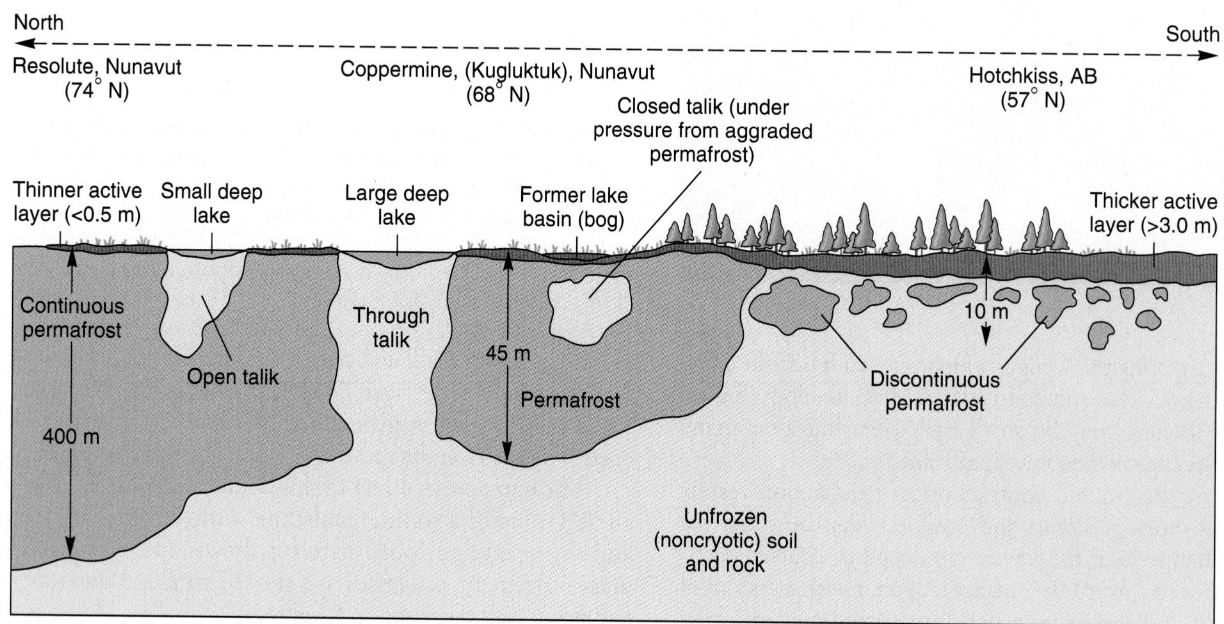

FIGURE 14.15 Periglacial environments.
Cross section of a periglacial region in northern Canada, showing typical forms of permafrost, active layer, talik, and ground ice. The three towns noted are shown on the map in Figure 14.14.

FIGURE 14.16 Ice wedge.
An ice wedge and ground ice in northern Canada. [Photo by Hugh M. French.]

(a)

(b)

FIGURE 14.17 Patterned ground phenomena.
(a) Aerial view of polygonal nets formed in Alaska, a fairly widespread frozen-ground phenomenon. (b) Patterned ground in the Dry Valleys of Antarctica. [(a) Photo from the Marbut Collection, Soil Science Society of America, Inc.; (b) photo by Galen Rowell/Mountain Light Photography, Inc.]

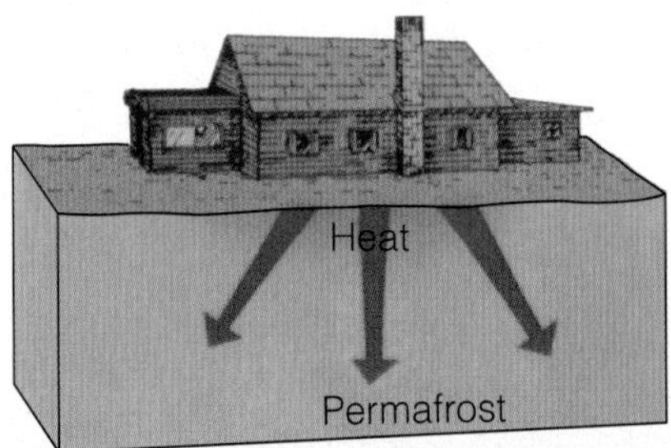

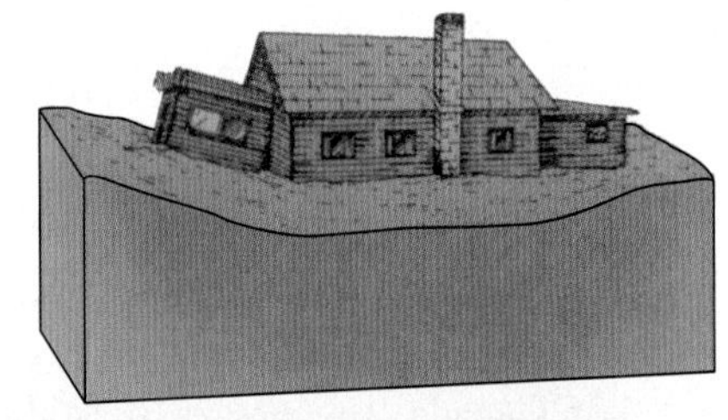

FIGURE 14.18 Permafrost melting and structure collapse.
Building failure due to the melting of permafrost south of Fairbanks, Alaska. [Adapted from U.S. Geological Survey. Photo by Steve McCutcheon; illustration based on U.S. Geological Survey pamphlet "Permafrost" by Louis L. Ray.]

(a)

(b)

FIGURE 14.19 Special structures for permafrost.
(a) Proper construction in periglacial environments requires placement of buildings above ground, and water and sewage lines in elevated "utilidors," Inuvik, Nunavut. (b) Supporting the Trans-Alaska oil pipeline on racks protects the permafrost from heat in the oil (compare with Figure 9.1a). [Photos by (a) Joyce Lundberg; (b) Galen Rowell/Mountain Light Photography, Inc.]

quently shifts, highways and rail lines become warped and twisted and fail, and utility poles are tilted. In addition, any building placed directly on frozen ground literally will melt itself into the defrosting soil (Figure 14.18).

Construction in periglacial regions dictates placing structures above the ground to allow air circulation beneath. This airflow allows the ground to cycle through its normal annual temperature pattern. Utilities such as water and sewer lines must be built above ground in "utilidors" to protect them from freezing and thawing ground (Figure 14.19). Likewise, the Alaskan oil pipeline was constructed above ground on racks to avoid melting the frozen ground, causing shifting that could rupture the line. Earthquakes remain a threat as happened in November 2002. See Figure 9.1a for a photo of this pipeline with broken support brackets.

The Pleistocene Ice Age Epoch

Imagine almost a third of Earth's land surface buried beneath ice sheets and glaciers thousands of meters thick—most of Canada, the northern Midwest, England, and northern Europe, including many mountain ranges. This is how it was at the height of the Pleistocene Epoch of the late Cenozoic Era.

The most recent episode of cold climatic conditions began about 1.65 million years ago, launching the *Pleistocene Epoch*. At the height of the Pleistocene, ice sheets and glaciers covered 30% of Earth's land area, amounting to more than 45 million km^2 (17.4 million mi^2). Periglacial regions during the last ice age covered about twice their present areal extent. The Pleistocene is thought to have been one of the more prolonged cold periods in Earth's history. It featured not just one glacial advance and retreat but at least 18 expansions of ice over Europe and North America, each obliterating and confusing the evidence from the one before.

The term **ice age** is applied to any such extended period of cold (not a single cold spell). An ice age is a period of generally cold climate that includes one or more *glacials*, interrupted by warm spells known as *interglacials*. Each glacial and interglacial is given a name that is usually based on the location where evidence of the episode is prominent—for example, "Wisconsinan glacial."

Modern research techniques include the examination of ancient ratios of oxygen isotopes, depths of coral growth in the tropics, analysis of sediments worldwide, and analysis of the latest ice cores from Greenland and Antarctica. These techniques opened the way for a new chronology and understanding of past climates. Glaciologists currently recognize the Illinoian glacial and Wisconsinan glacial periods, with the Sangamon interglacial between them. These events span the 300,000-year period prior to our present Holocene Epoch (Figure 14.20).

The continental ice sheets over Canada, the United States, Europe, and Asia, some 18,000 years ago, are illustrated on the polar map projection in Figure 14.21a . In North America, the Ohio and Missouri River systems mark the southern terminus of continuous ice. The edge of an ice sheet is not even; instead it expands and retreats in ice lobes. Such lobes were positioned where the Great Lakes are today and contributed to the formation of these lakes when the ice retreated from their gouged, isostatically (weighted down by ice) depressed basins. (For Web links on glaciers and Pleistocene geology, see **http://research.umbc.edu/~miller/geog111/glacierlinks.htm**.)

As these ice sheets and alpine glaciers retreated (see map in Figure 14.21c), they exposed a drastically altered landscape: the rocky soils of New England, the polished and scarred surfaces of Canada's Atlantic Provinces, the sharp crests of the Sawtooth Range and Tetons of Idaho and Wyoming, the scenery of the Canadian Rockies and the Sierra Nevada, the Great Lakes of the United States and Canada, and much more. Study of these glaciated landscapes is important, for we can better understand paleoclimatology (past climates) and discover the mechanisms that produce ice ages and climatic change.

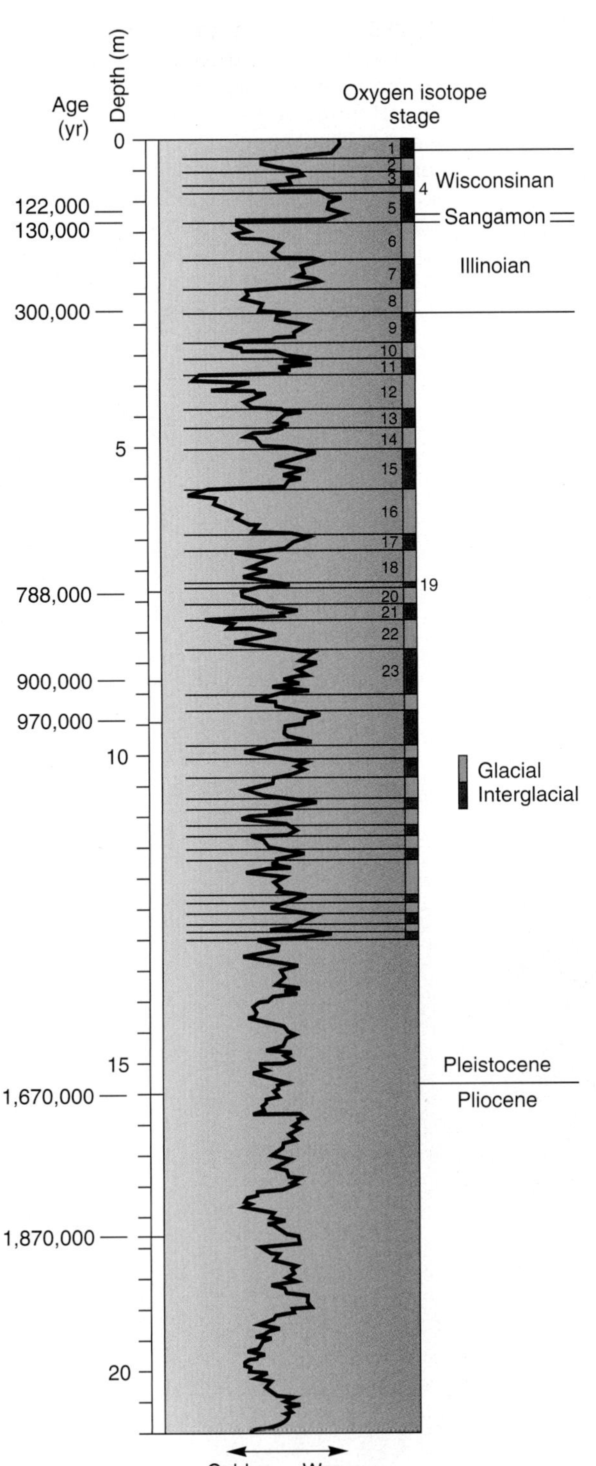

FIGURE 14.20 Temperature record of the past 2 million years.
Pleistocene temperatures, determined by oxygen-isotope fluctuations in fossils found in deep-sea cores. Twenty-three stages cover 900,000 years, with names assigned for the past 300,000 years. [After N. J. Shackelton and N. D. Opdyke, "Oxygen-Isotope and Paleomagnetic Stratigraphy of Pacific Core V28-239, Late Pliocene to Latest Pleistocene," *Geological Society of America Memoir* 145. Copyright © 1976 by the GSA. Adapted by permission.]

Pluvial Periods and Paleolakes

Figure 14.22 portrays the West dotted with large lakes 12,000–30,000 years ago. Except for the Great Salt Lake in Utah (a remnant of the former Lake Bonneville noted on the map) and a few smaller lakes, only dry basins, ancient shorelines, and lake sediments remain today. These ancient lakes are **paleolakes**. The three photos in Figure 14.22 show how these paleolake sites look today: the Bonneville Salt Flats in Utah, Mono Lake and its tufa towers in California, and Severe Dry Lake, Utah.

The term *pluvial* (Latin "rain") describes this period of wetter conditions, principally during the Pleistocene epoch. During pluvial periods, lake levels increased in arid regions. The drier periods between pluvials are called *interpluvials*. Interpluvials are marked with **lacustrine deposits**—lake sediments that form terraces along former shorelines.

Early researchers attempted to correlate pluvial and glacial ages, given their coincidence during the Pleistocene. However, few sites actually demonstrate such a simple relationship. For example, in the western United States, the estimated volume of melted ice is only a small portion of the water volume of pluvial lakes of the region. Also, these lakes tend to predate glacial times and correlate instead with periods of wetter climate, or periods thought to have had lower evaporation rates. In North America, the two largest late Pleistocene paleolakes were in the Basin and Range Province of the West. Figure 14.22 portrays these—Lake Bonneville and Lake Lahontan—and other paleolakes at their highest levels. Pluvial lakes also occurred in Mexico, South America, Africa, Asia, and Australia.

The Great Salt Lake, near Salt Lake City, Utah, and the Bonneville Salt Flats in western Utah are remnants of Lake Bonneville, which at its greatest extent covered more than 50,000 km^2 (19,500 mi^2) and reached depths of 300 m (1000 ft), spilling over into the Snake River drainage to the north. Today the area is a closed basin with no drainage except an artificial outlet to the west where excess water from the Great Salt Lake is pumped to a basin called the Newfoundland Evaporation Basin during floods.

Deciphering Past Climates: Paleoclimatology

We observe glacials and interglacials because Earth's climate has fluctuated in and out of warm and cold ages. Documentation for this fluctuation now is traced in ice cores from Greenland and Antarctica, in layered deposits of silts and clays, in the extensive pollen record from ancient plants, and in the relation of past coral growth to sea level. This evidence is analyzed with radioactive dating methods and other techniques. One especially interesting fact is emerging from these studies: We humans (*Homo erectus* and *H. sapiens* of the last 1.9 million years) have never experienced Earth's normal (more moderate, less extreme) climate, most characteristic of Earth's entire 4.6 billion year span.

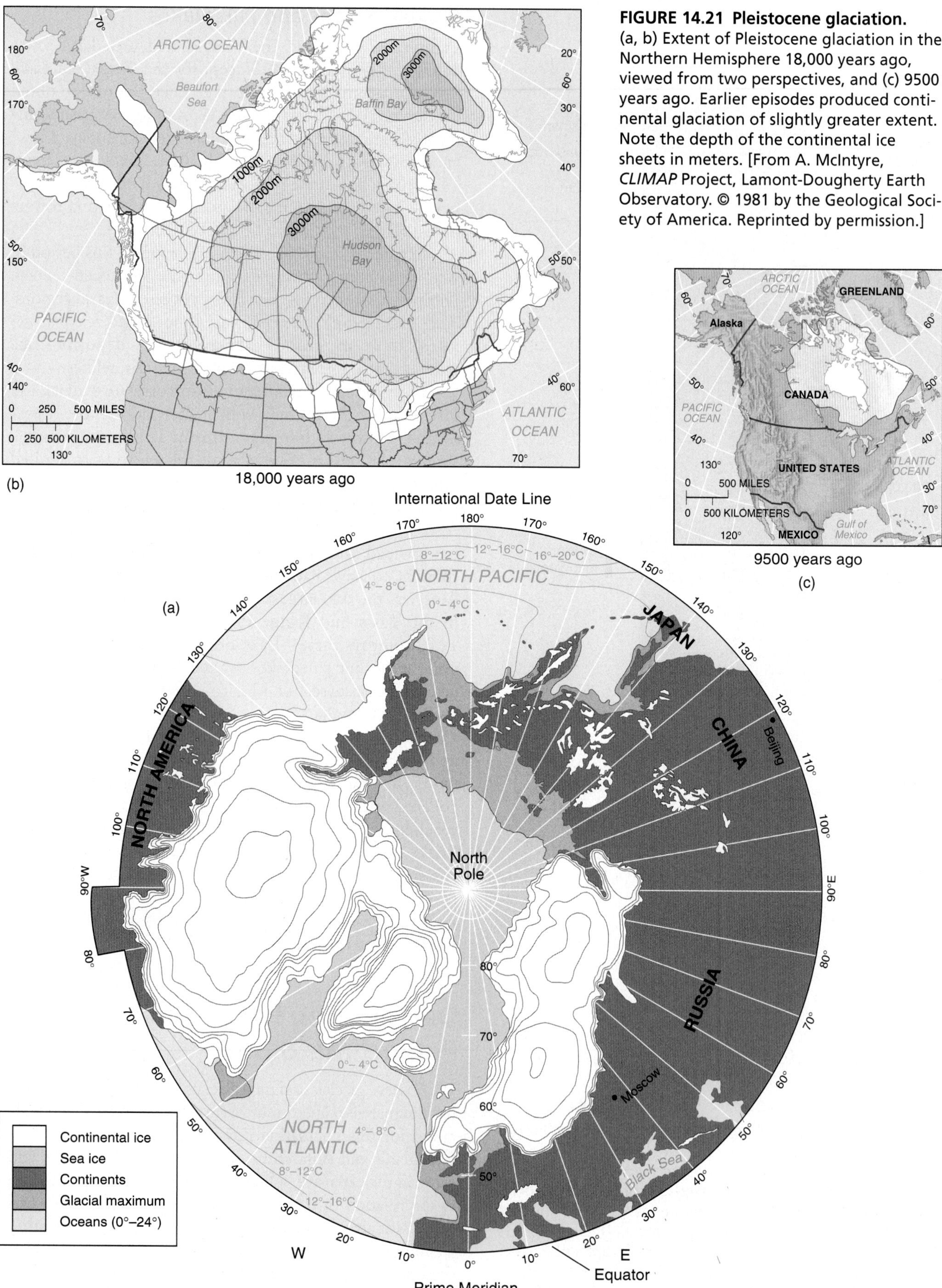

FIGURE 14.21 Pleistocene glaciation. (a, b) Extent of Pleistocene glaciation in the Northern Hemisphere 18,000 years ago, viewed from two perspectives, and (c) 9500 years ago. Earlier episodes produced continental glaciation of slightly greater extent. Note the depth of the continental ice sheets in meters. [From A. McIntyre, *CLIMAP* Project, Lamont-Dougherty Earth Observatory. © 1981 by the Geological Society of America. Reprinted by permission.]

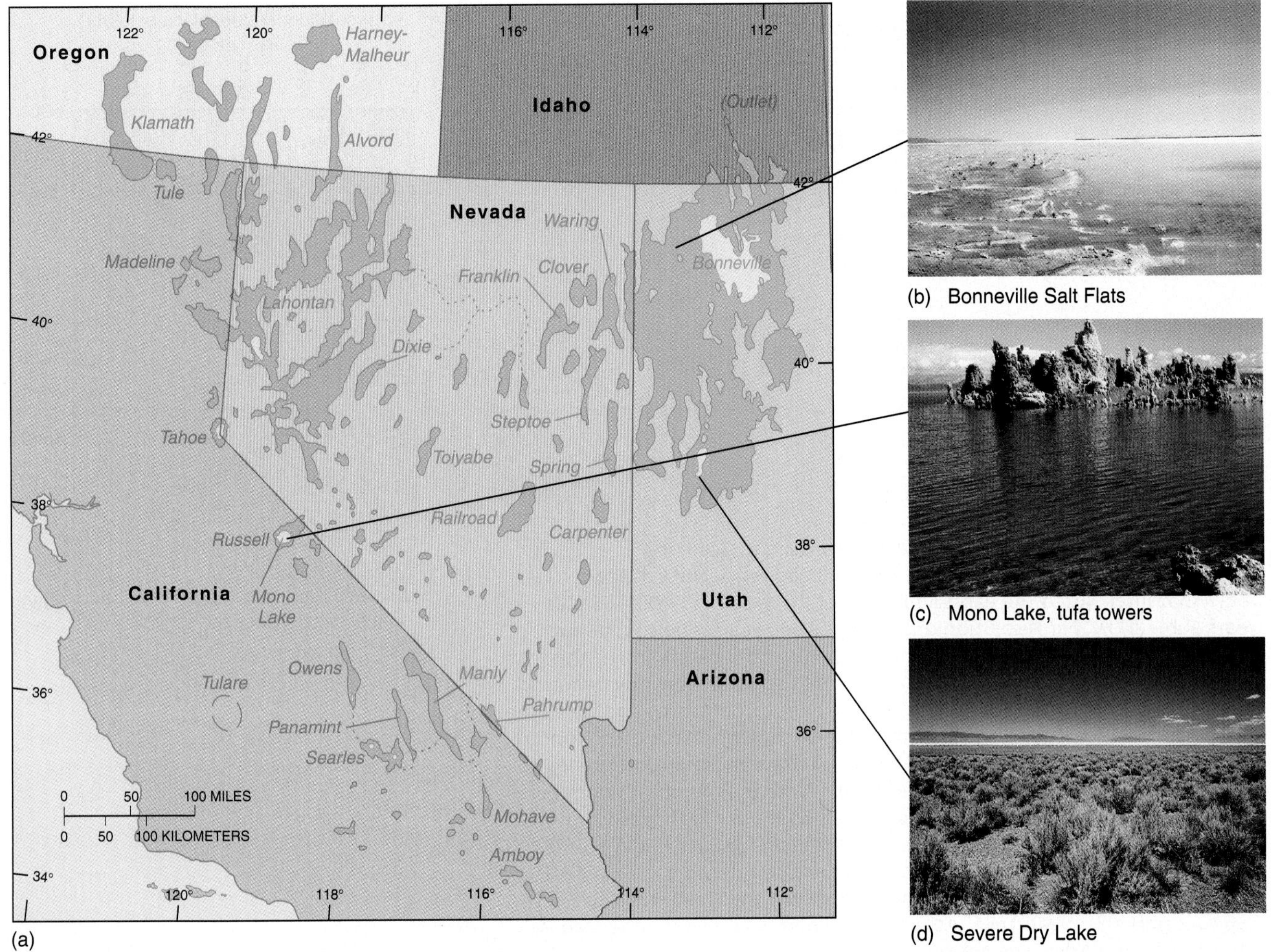

(a)

(b) Bonneville Salt Flats

(c) Mono Lake, tufa towers

(d) Severe Dry Lake

FIGURE 14.22 Paleolakes in the western United States.
Paleolakes of the western United States at their greatest extent 12,000 to 30,000 years ago, a recent pluvial period. Lake Lahontan and Lake Bonneville were the largest. The Great Salt Lake and Severe Dry Lake in Utah are remnants of Lake Bonneville; Mono Lake in California is what remains of pluvial Lake Russell. Modern-day photos of some paleolake locations are in (b) through (d). [After R. F. Flint, *Glacial and Pleistocene Geology*. © 1957 by John Wiley & Sons, Inc. Adapted by permission. Photos (b) and (c) by author; (d) by Bobbé Christopherson.]

Apparently, Earth's climates slowly fluctuated until the past 1.2 billion years, when temperature patterns with cycles of 200–300 million years became more pronounced. The most recent cold episode was the Pleistocene Epoch, which began in earnest 1.65 million years ago and through which we may still be progressing. The Holocene Epoch began approximately 10,000 years ago, when average temperatures abruptly increased 6 C° (11 F°). The period we live in may represent an end to the Pleistocene, or it may be merely a mild interglacial. Figure 14.23 details the climatic record of the past 160,000 years.

Medieval Warm Period and Little Ice Age

In A.D. 1001, Leif Eriksson inadvertently ventured onto the North American continent, perhaps the first European to do so. He and his fellow Vikings were favored by a medieval warming episode as they sailed the less-frozen North Atlantic to settle Iceland and Greenland.

The mild climatic episode that lasted from about A.D. 800 to 1200 is known as the *Medieval Warm Period*. During the warmth, grape vineyards were planted far into England some 500 km (310 mi) north of present-day commercial plantings. Oats and barley were planted in Iceland, and wheat was planted as far north as Trondheim, Norway. The shift to warmer, wetter weather influenced migration and settlement northward in North America, Europe, and Asia.

However, from approximately 1200–1350 through 1800–1900, a *Little Ice Age* took place. Parts of the North Atlantic froze, and expanding glaciers blocked many key mountain passes in Europe. Snowlines in Europe lowered about 200 m (650 ft) in the coldest years. The Greenland colonies were deserted. Cropping patterns changed, and northern forests declined, along with human population in those regions. In the winter of 1779–80, New York's Hudson and East Rivers and the entire Upper Bay froze over. People walked and hauled heavy loads across the ice between Staten and Manhattan Islands!

However, the Little Ice Age was not consistently cold throughout its 700-year reign. Reading the record of the

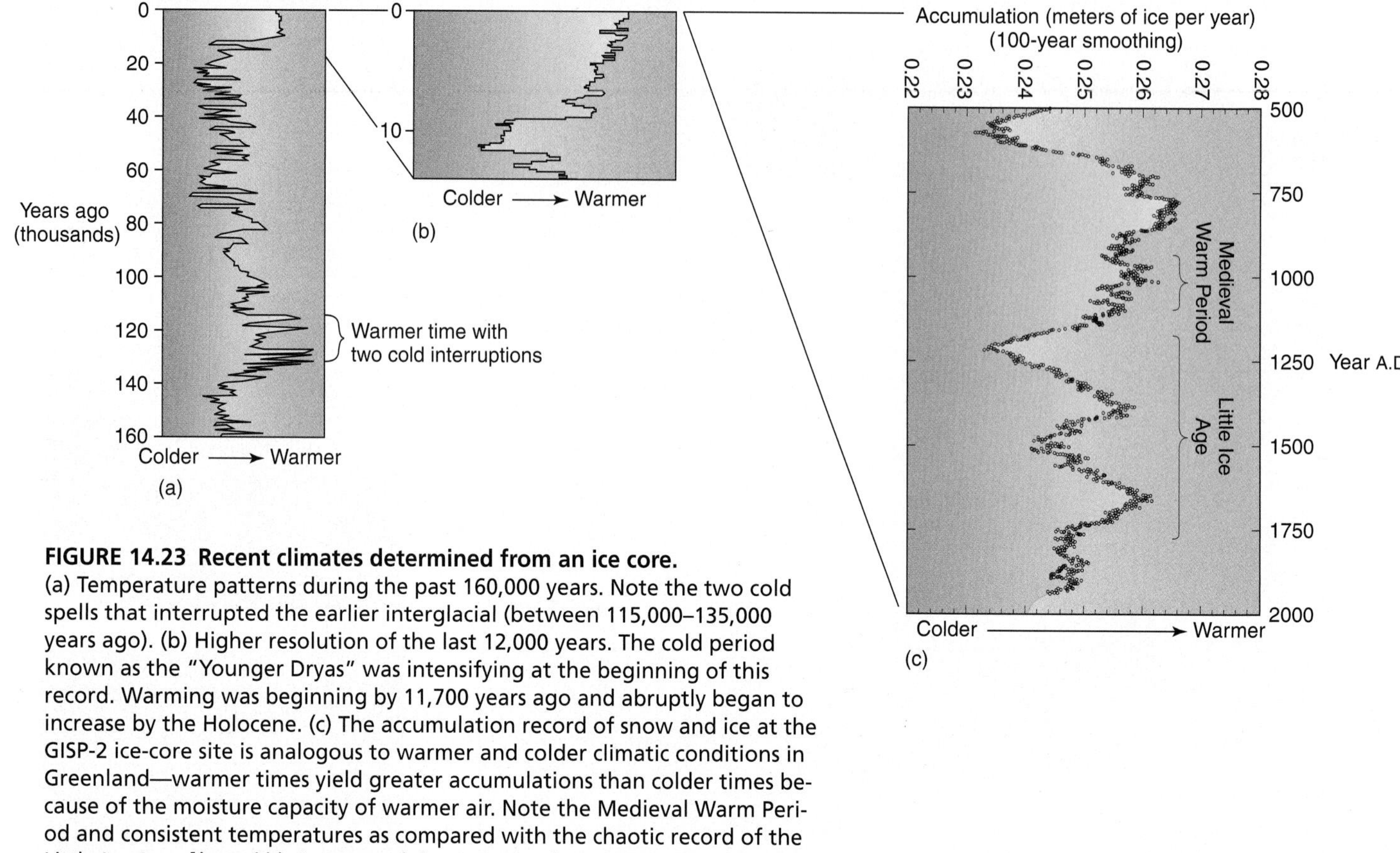

FIGURE 14.23 Recent climates determined from an ice core. (a) Temperature patterns during the past 160,000 years. Note the two cold spells that interrupted the earlier interglacial (between 115,000–135,000 years ago). (b) Higher resolution of the last 12,000 years. The cold period known as the "Younger Dryas" was intensifying at the beginning of this record. Warming was beginning by 11,700 years ago and abruptly began to increase by the Holocene. (c) The accumulation record of snow and ice at the GISP-2 ice-core site is analogous to warmer and colder climatic conditions in Greenland—warmer times yield greater accumulations than colder times because of the moisture capacity of warmer air. Note the Medieval Warm Period and consistent temperatures as compared with the chaotic record of the Little Ice Age. [(a and b) Courtesy of the Greenland Ice Core Project (GRIP); (c) From D. A. Meese and others, "The accumulation record from the GISP-2 core as an indicator of climate change throughout the Holocene," *Science* 266 (December 9, 1994): 1681. Adapted with permission from American Association for the Advancement of Science.]

Greenland ice cores, scientists have found many mild years among the harsh. More accurately, this was a time of rapid, short-term climate fluctuations that lasted only decades.

Ice cores drilled in Greenland have revealed a record of annual snow and ice accumulation that, when correlated with other aspects of the core sample, is indicative of air temperature. Figure 14.23c presents this record since A.D. 500. The warmth of the Medieval Warm Period is evident, whereas the Little Ice Age appears mixed with colder conditions around 1200, 1500, and after about 1800.

Mechanisms of Climate Fluctuation

What mechanisms cause short-term fluctuations? And why is Earth pulsing through long-term climatic changes that span several hundred million years? The ice age concept is being researched and debated with an unprecedented intensity for three principal reasons: (1) Continuous ice cores from Greenland and Antarctica are providing a new, detailed record of weather and climate patterns, volcanic eruptions, and trends in the biosphere (discussed in News Report 14.2). (2) To understand present and future climate change and to refine general circulation models, we must understand the natural variability of the atmosphere and climate. (3) Global warming, and its relation to ice ages, is a major concern.

Because past occurrences of low temperature appear to have followed a pattern, researchers have looked for causes that also are cyclic in nature. They have identified a complicated mix of interacting variables that appear to influence long-term climatic trends. Let us take a look at several of them.

Climate and Celestial Relations. As our Solar System revolves around the distant center of the Milky Way, it crosses the plane of the galaxy approximately every 32 million years. At that time, Earth's plane of the ecliptic aligns parallel to the galaxy's plane, and we pass through regions in space of increased interstellar dust and gas, which may have some climatic effect.

Milutin Milankovitch (1879–1954), a Yugoslavian astronomer who studied Earth-Sun orbital relations, developed other possible astronomical factors. Milankovitch wondered whether the development of an ice age relates to seasonal astronomical factors—Earth's revolution around the Sun, rotation, and tilt (Figure 14.24). In summary:

- Earth's elliptical orbit about the Sun is not constant. The shape of the ellipse varies by more than 17.7 million km (11 million mi) during a 100,000-year cycle, from nearly circular to an extreme ellipse (Figure 14.24a).

- Earth's axis "wobbles" through a 26,000-year cycle, in a movement much like that of a spinning top winding down. Earth's wobble is called precession. As you can see in Figure 14.24b, precession changes the orientation of hemispheres and landmasses to the Sun.
- Earth's present axial tilt of 23.5° varies from 22° to 24° during a 40,000-year period (Figure 14.24c).

Milankovitch calculated, without the aid of today's computers, that the interaction of these Earth-Sun relations creates a 96,000-year climatic cycle. His glaciation model assumes that changes in astronomical relations affect the amounts of insolation received.

Milankovitch died in 1954, his ideas still not accepted by a skeptical scientific community. Now, in the era of computers, remote-sensing satellites, and worldwide efforts to decipher past climates, Milankovitch's valuable work has stimulated much research attempting to explain climatic cycles and experienced some degree of confirmation. A roughly 100,000-year climatic cycle is confirmed in such diverse places as ice cores in Greenland and the accumulation of sediment in Lake Baykal, Siberia.

Climate and Tectonics. Major glaciations also can be associated with plate tectonics because some landmasses have migrated to higher, cooler latitudes. Chapters 8 and 9 explain that the shape and orientation of landmasses and ocean basins have changed greatly during Earth's history. Continental plates drifted from equatorial locations to the polar regions and vice versa, thus exposing the land to a gradual change in climate. Gondwana (the southern half of Pangaea) experienced extensive glaciation that left its mark on the rocks of parts of present-day Africa, South America, India, Antarctica, and Australia. Landforms in the Sahara, for example, bear the markings of even earlier glacial activity. These markings partly are explained by the fact that portions of Africa were centered near the South Pole during the Ordovician Period, 465 million years ago (see Figure 8.15a).

Episodes of mountain building over the past billion years forced mountain summits above the snowline, where snow remains after the summer melt. Mountain chains influence downwind weather patterns and jet stream circulation, which in turn guides weather systems. More dust was present during glacial periods, suggesting drier weather and more extensive deserts beyond the frozen regions.

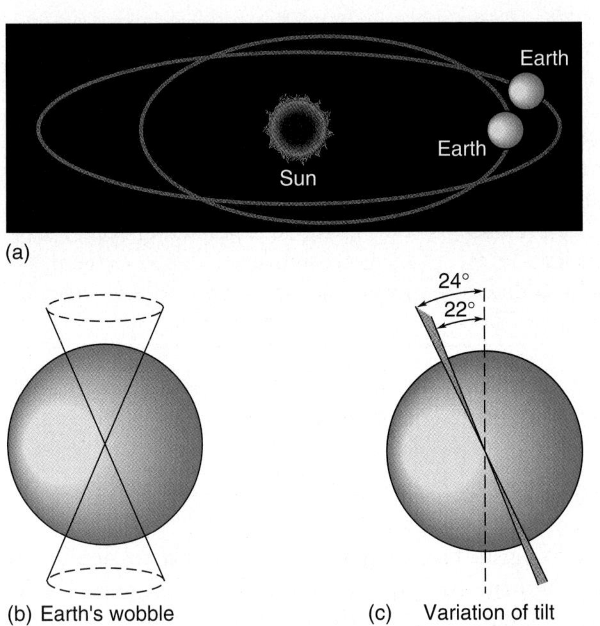

FIGURE 14.24 Astronomical factors that may affect broad climatic cycles.
(a) Earth's elliptical orbit varies widely during a 100,000-year cycle, stretching out to an extreme ellipse. (b) Earth's 26,000-year axial wobble. (c) Variation in Earth's axial tilt every 40,000 years.

Climate and Atmospheric Factors. Some events alter the atmosphere and produce climate change. A volcanic eruption might produce lower temperatures for a year or two. The lower temperatures could initiate a buildup of long-term snow cover at high latitudes. These high-albedo snow surfaces then would reflect more insolation away from Earth, to further enhance cooling in a positive feedback system. The eruption of Mount Pinatubo in the Philippines in 1991 is closely studied, for the eruption caused a temporary cooling and possibly other climatic effects.

The fluctuation of atmospheric greenhouse gases could trigger higher or lower temperatures. An ice core taken at Vostok, the Russian research station near the geographic center of Antarctica, contained trapped air samples from more than 200,000 years in the past. The sample showed carbon dioxide levels varying from more than 290 ppm to a low of nearly 180 ppm (Figure 14.25). Each interglacial, or warmer period, appears to be correlated to higher levels of carbon dioxide. During 2002, atmospheric carbon dioxide exceeded 370 ppm—higher than at any time in the past 200,000 years, principally owing to anthropogenic (human-created) sources.

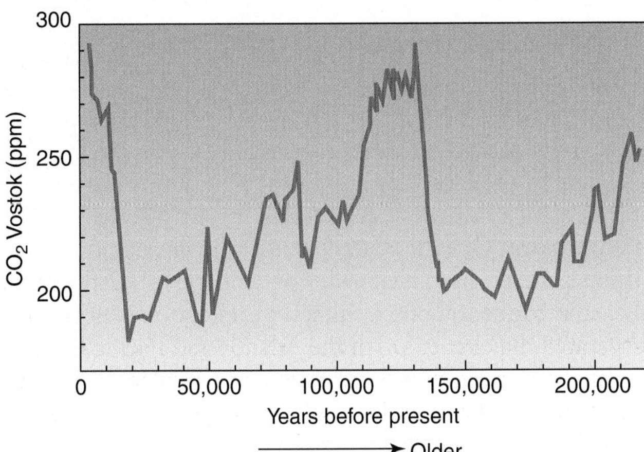

FIGURE 14.25 Vostok ice core and past carbon dioxide concentrations.
Carbon dioxide, trapped in air bubbles in Antarctic ice, measured from the Vostok ice core. [Adapted by permission from J. Jouzel and others, "Extending the Vostok ice-core record of paleoclimate to the penultimate glacial period," *Nature* 364 (July 29, 1993): 411. © 1993 Macmillan Magazines Ltd.]

News Report 14.2

GRIP and GISP-2: Boring Ice for Exciting History

The Greenland Ice Core Project (GRIP) was launched in 1989. A site was selected near the summit of the Greenland ice sheet at 3200 m (10,500 ft) so that the maximum thickness of ice history would be drilled (Figure 1). After 3 years, drilling hit bedrock 3030 m (9940 ft) below the site—or, in terms of time, 250,000 years into the past. The core is 10 cm (4 in.) in diameter.

In 1990, about 32 km (20 mi) west of the summit, the Greenland Ice Sheet Project (GISP-2) began to bore back through time (Figure 1). GISP-2 reached bedrock in 1993. This core is slightly larger in diameter, at 13.2 cm (5.2 in.), and collects about twice the data as GRIP. The existence of a second core helps scientists compensate for any folds or disturbed sections they encountered below 2700 m, or about 115,000 years, in the first core.

FIGURE 1 Greenland ice core locations. GRIP is at 72.5° N 37.5° W. GISP-2 is at 72.6° N 38.5° W. [Map courtesy of GISP-2 Science Management Office.]

What is being discovered from a 3030-m ice core? Locked into the core are air bubbles of past atmospheres, which indicate ancient gas concentrations. Of special interest are the greenhouse gases, carbon dioxide and methane. Chemical and physical properties of the atmosphere and the snow that accumulated each year are frozen in place.

Pollutants are locked into the core record. For example, during cold periods, high concentrations of dust were present, brought by winds from distant dry lands. An invaluable record of past volcanic eruptions is included in the layers, as if on a calendar. Even the exact beginning of the Bronze Age is recorded in the ice core—about 3000 B.C. When the Greeks and later the Romans began smelting copper, they produced ash and smoke that the winds carried all the way to Greenland. The presence of ammonia indicates ancient forest fires at lower latitudes. In addition, the ratio between stable forms of oxygen is an important analog of past temperatures on the ice-sheet surface.

GRIP and GISP-2 greatly refined the paleoclimatology of the late Cenozoic Era, helping us to understand Earth's dynamic climate system. Our chances of predicting future patterns are improved, thanks to these efforts and the core drilling that is underway in Antarctica.

Climate and Oceanic Circulation. Finally, oceanic circulation patterns have changed. For example, the Isthmus of Panama formed about 3 million years ago and effectively separated the circulation of the Atlantic and Pacific Oceans. Changes in ocean basin configuration, surface temperatures, and salinity and in upwelling and downwelling rates affect air mass formation and air temperature.

Our understanding of Earth's climate—past, present, and future—is unfolding. We are learning that climate is a multicyclic system controlled by an interacting set of cooling and warming processes, all founded on celestial relations, tectonic factors, atmospheric variables, and changes in oceanic circulation. In the era of industrialization and consumption of fossil fuels, we add human activity to the climate change mix.

Arctic and Antarctic Regions

Climatologists use environmental criteria to define the Arctic and the Antarctic regions. The *Arctic region* is defined in Figure 14.26 by the green line. This is the 10°C (50°F) isotherm for July (the Northern Hemisphere summer). This line coincides with the visible treeline—the boundary between the northern forests and tundra. News Report 14.3 discusses the possibility of an Arctic ice sheet during the last ice age.

The Arctic Ocean is covered by two kinds of ice: *floating sea ice* (frozen seawater) and *glacier ice* (frozen freshwater). This ice pack thins in the summer months and sometimes breaks up. A 1999 study reported that 43% of the Arctic Ocean ice has melted since 1970, apparently due to global warming. Research published in 2002 reported that Arctic sea ice is being lost at a rate of 9% per decade. During the summer of 2002 the melt rate in Greenland exceeded all previous years, and greatly topped the 1988–2000 average. The fabled Northwest Passage across the Arctic from the Atlantic to the Pacific may soon be a reality as the Arctic ice continues to melt.

The *Antarctic region* is defined by the *antarctic convergence*, a narrow zone that extends around the continent as a boundary between colder antarctic water and warmer water at lower latitudes. This boundary follows roughly the 10°C (50°F) isotherm for February (Southern Hemisphere summer) and is located near 60° S latitude. The Antarctic region that is covered with sea ice represents an area greater than North America, Greenland, and Western Europe combined! (For more information on polar region ice, see the National Ice Center at http://www.natice.noaa.gov/ and the Canadian Ice Service at http://www.cis.ec.gc.ca/.)

The Antarctic Ice Sheet

Antarctica is a continent-sized landmass and therefore is much colder overall than the Arctic, which is an ocean. In simplest terms, Antarctica can be thought of as a continent covered by a single enormous glacier, although it

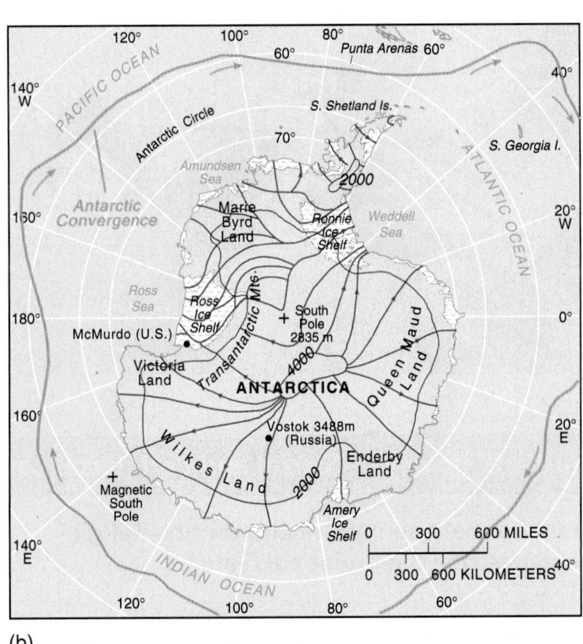

FIGURE 14.26 The Arctic and Antarctic regions.
(a) Note the 10°C (50°F) isotherm in midsummer, which designates the Arctic region. (b) The Antarctic convergence designates the Antarctic region. Arrows on the ice sheet show the general direction of ice movement on Antarctica.

News Report 14.3

An Arctic Ice Sheet?

Questions remain about the nature of the ice in the Arctic Ocean during the Pleistocene Ice Age. Some scientists believe that the ice was similar to the thin sea ice we see today. Others think a deep ice cover blanketed the region. The latter hypothesis received a boost when sonar images of the ocean floor, published in 1994, showed deep gouges and grooves in bottom sediments on the floor of the Fram Strait between Greenland and Norway.

Imagine icebergs twice the size of the largest found in Antarctic water, extending up to 300 m (1000 ft) high and as much as 700 m (2300 ft) deep. These enormous ice vessels scarred the ocean floor with their deep, icy keels. In another area, the sonar disclosed submerged ridges that were scraped smooth. These super-icebergs could have originated from a floating ice sheet or from glaciers surrounding the Arctic Basin. Now the question: To what extent did the North American and European ice sheets join across the Arctic Ocean? Scientists are investigating this question.

contains distinct regions such as the East Antarctic and West Antarctic ice sheets, which respond differently to slight climatic variations.

Ice sheet edges that enter coastal bays form extensive ice shelves, with sharp ice cliffs rising up to 30 m (100 ft) above the sea. Large tabular islands of ice are formed when sections of the shelves break off and move out to sea. These ice islands can be very large; several in the late 1980s and again in 1995, 1998, 2001, and 2002 exceeded the area of Rhode Island, and one the area of Delaware. The disintegration of some of these ice shelves caused by higher temperatures is discussed in Chapter 6 and illustrated in Figure 6.29.

The scientific importance of Antarctica told in the words of the people who work there is in the New South Polar Times at **http://www.spotsylvania.k12.va.us/nspt/home.htm**. A place so remote from civilization is an excellent laboratory for sampling past and present human and natural variables that are transported by atmospheric and oceanic circulation. Far from pollution sources, so cold and dark in winter, and its high altitude make it an ideal location for certain astronomical observations. Glacial processes, landforms, and Earth's frozen climatic record in the ice sheet all provide clues for further scientific exploration (Figure 14.27).

FIGURE 14.27 Scientific base at the South Pole. Aerial view of the Amundsen-Scott South Pole Station made in 2001. The geodesic dome completed in 1975, 50 m wide and 16 m high (165 ft by 52 ft), shelters buildings from the frigid cold and winds. Twenty-eight scientists and support people work through the winter (February to October) and 130 personnel, or more, research and work there in the brief summer. On the left side you see the first of the new modular buildings, which stand 3 m (10 ft) above the surface to accommodate snow accumulation and protect the ice under the building from melting. Built entirely from materials airlifted to the base, they will be completed in 2006. Above this module is the ceremonial South Pole; slightly to the right is the geographic South Pole. This station is at an elevation of 2835 m (9301 ft). [Photo courtesy of Kristan Hutchison, U.S. Antarctic Program.]

Summary and Review—Glacial and Periglacial Landscapes

● *Differentiate* between alpine and continental glaciers and *describe* their principal features.

More than 77% of Earth's freshwater is frozen. Ice covers about 11% of Earth's surface, and periglacial features occupy another 20% of ice-free but cold-dominated landscapes. Earth's **cryosphere** is the portion of the hydrosphere and ground that is perennially frozen, generally at high latitudes and elevations. A **glacier** is a mass of ice sitting on land or floating as an ice shelf in the ocean next to land. Glaciers form in areas of permanent snow. A *snowline* is the lowest elevation where snow occurs year-round; its altitude varies by latitude—higher near the equator, lower poleward.

A glacier in a mountain range is an **alpine glacier**. If confined within a valley, it is termed a *valley glacier*. The area of origin is a snowfield, usually in a bowl-shaped erosional landform called a **cirque**. Where alpine glaciers flow down to the sea, they calve and form **icebergs**. A **continental glacier** is a continuous mass of ice on land. Its most extensive form is an **ice sheet**; a smaller, roughly circular form is an **ice cap**; and the least extensive form, usually in mountains, is an **ice field**.

cryosphere (p. 425)
glacier (p. 426)
alpine glacier (p. 426)
cirque (p. 427)
iceberg (p. 427)
continental glacier (p. 428)
ice sheet (p. 428)
ice cap (p. 428)
ice field (p. 428)

1. Describe the location of most freshwater on Earth today. What is the cryosphere?
2. What is a glacier? What is implied about existing climate patterns in a glacial region?
3. Differentiate between an alpine glacier and a continental glacier.
4. Name the three types of continental glaciers. What is the basis for dividing continental glaciers into types? Which type covers Antarctica?

● *Describe* the process of glacial ice formation and *portray* the mechanics of glacial movement.

Snow becomes glacial ice through accumulation, increasing thickness, pressure on underlying layers, and recrystallization. Snow progresses through transitional steps from **firn** (compact, granular) to a denser **glacial ice** after many years.

A glacier is an open system with inputs and outputs that can be analyzed through observation of the growth and wasting of the glacier itself. A **firn line** is the lower extent of a fresh snow-covered area. A glacier is fed by snowfall and is wasted by **ablation** (losses from its upper and lower surfaces and along its margins). Accumulation and ablation achieve a mass balance in each glacier. The zone where accumulation gain balances ablation loss is the **equilibrium line**.

As a glacier moves downhill, vertical **crevasses** (cracks and openings) may develop. Sometimes a glacier will move rapidly in a **glacier surge**. The presence of water along the basal layer appears to be important in glacial movements. As a glacier moves, it plucks rock pieces and debris, incorporating them into the ice, and this debris scours and sandpapers underlying rock through **abrasion**.

firn (p. 429)
glacial ice (p. 429)
firn line (p. 429)
ablation (p. 429)
equilibrium line (p. 429)
crevasse (p. 430)
glacier surge (p. 430)
abrasion (p.433)

5. Trace the evolution of glacial ice from fresh fallen snow.
6. What is meant by glacial mass balance? What are the basic inputs and outputs underlying that balance?
7. What is meant by a glacier surge? What do scientists think produces surging episodes?

● *Describe* characteristic erosional and depositional landforms created by alpine glaciation and continental glaciation.

Extensive valley glaciers have profoundly reshaped mountains worldwide, carving **V**-shaped stream valleys into **U**-shaped glaciated valleys, producing many distinctive erosional and depositional landforms. As cirque walls erode away, sharp **arêtes** (sawtooth, serrated ridges) form, dividing adjacent cirque basins. Two eroding cirques may reduce an arête to a saddlelike **col**. A **horn** results when several cirque glaciers gouge an individual mountain summit from all sides, forming a pyramidal peak. An ice-carved rock basin left as a glacier retreats may fill with water to form a **tarn**; a string of tarns separated by moraines is called **paternoster lakes**. Where a glacial valley trough joins the ocean, and the glacier retreats, the sea extends inland to form a **fjord**.

All glacial deposits, whether ice-borne or meltwater-borne, constitute **glacial drift**. Direct deposits from ice, called **till**, are unstratified and unsorted. Glacial meltwater deposits are sorted and are called **stratified drift**. A specific landform produced by the deposition of drift is a **moraine**. A **lateral moraine** forms along each side of a glacier; merging glaciers with lateral moraines form a **medial moraine**; and eroded debris dropped at the glacier's terminus is a *terminal moraine*.

Continental glaciation leaves different features than does alpine glaciation. A **till plain** forms behind end moraines, featuring unstratified coarse till, low and rolling relief, and deranged drainage. Beyond the morainal deposits, **outwash plains** of stratified drift feature stream channels that are meltwater-fed, braided, and overloaded with debris that is sorted and deposited across the landscape. An **esker** is a sinuously curving, narrow ridge of coarse sand and gravel that forms along the channel of a meltwater stream beneath a glacier. An isolated block of ice left by a retreating glacier becomes surrounded with debris; when the block finally melts, it leaves a steep-sided **kettle**. A **kame** is a small hill, knob, or mound of poorly sorted sand and gravel that is deposited directly by water or by ice in crevasses.

Glacial action may form two types of streamlined hills: the erosional **roche moutonnée** is an asymmetrical hill of exposed bedrock, gently sloping upstream and abruptly sloping downstream; the depositional **drumlin** is deposited till, streamlined in the direction of continental ice movement (blunt end upstream and tapered end downstream).

arête (p. 433)
col (p. 433)
horn (p. 433)
tarn (p. 434)
paternoster lakes (p. 434)
fjord (p. 434)
glacial drift (p. 434)
till (p. 434)
stratified drift (p. 434)
moraine (p. 434)
lateral moraine (p. 434)
medial moraine (p. 434)
till plain (p. 436)
outwash plain (p. 436)
esker (p. 436)
kettle (p. 436)
kame (p. 436)
roche moutonnée (p. 436)
drumlin (p. 436)

8. How does a glacier accomplish erosion?
9. Describe the evolution of a **V**-shaped stream valley to a **U**-shaped glaciated valley. What features are visible after the glacier retreats?
10. How is an iceberg generated?
11. Differentiate between two forms of glacial drift—till and outwash.
12. What is a morainal deposit? What specific moraines are created by alpine and continental glaciers?
13. What are some common depositional features encountered in a till plain?
14. Contrast a roche moutonnée and a drumlin regarding appearance, orientation, and the way each forms.

● *Analyze* the spatial distribution of periglacial processes and *describe* several unique landforms and topographic features related to permafrost and frozen ground phenomena.

The term **periglacial** describes cold-climate processes, landforms, and topographic features that exist along the margins of glaciers, past and present. Periglacial regions occupy over 20% of Earth's land surface. These areas are either near-permanent ice or are at high elevation, and the ground is seasonally snow-free. When soil or rock temperatures remain below 0°C (32°F) for at least 2 years, **permafrost** ("permanent frost") develops. An area of permafrost that is not covered by glaciers is considered periglacial. Note that this criterion is based solely on temperature and has nothing to do with how much or how little water is present. The **active layer** is the zone of seasonally frozen ground that exists between the subsurface permafrost layer and the ground surface. In regions of permafrost, frozen subsurface water is termed **ground ice**.

periglacial (p. 436)
permafrost (p. 436)
active layer (p. 438)
ground ice (p. 438)

15. In terms of climatic types, describe the areas on Earth where periglacial landscapes occur. Include both higher latitude and higher altitude climate types.
16. Define two types of permafrost, and differentiate their occurrence on Earth. What are the characteristics of each?
17. Describe the active zone in permafrost regions, and relate the degree of development to specific latitudes.
18. What is a talik? Where might you expect to find taliks, and to what depth do they occur?
19. What is the difference between permafrost and ground ice?
20. Describe the role of frost action in the formation of various landforms in the periglacial region.
21. Relate some of the specific problems humans encounter in developing periglacial landscapes.

- ***Explain* the Pleistocene ice age epoch and related glacials and interglacials and *describe* some of the methods used to study paleoclimatology.**

An **ice age** is any extended period of cold. The late Cenozoic Era has featured pronounced ice-age conditions in an epoch called the Pleistocene. During this time, alpine and continental glaciers covered about 30% of Earth's land area in at least 18 glacials, punctuated by interglacials of milder weather. Beyond the ice, **paleolakes** formed because of wetter conditions. Evidence of ice-age conditions is gathered from ice cores drilled in Greenland and Antarctica, from ocean sediments, from coral growth in relation to past sea levels, and from rock.

The apparent pattern followed by these low-temperature episodes indicates cyclic causes. A complicated mix of interacting variables appears to influence long-term climatic trends: celestial relations, solar variability, tectonic factors, atmospheric variables, and oceanic circulation. **Lacustrine deposits** are lake sediments that form terraces along former shorelines. The study of past climates is *paleoclimatology*.

ice age (p. 442)
paleolake (p. 443)
lacustrine deposits (p. 443)

22. What is paleoclimatology? Describe Earth's past climatic patterns. Are we experiencing a normal climate pattern in this era, or have scientists noticed any significant climate trends?
23. Define an ice age. When was the most recent? Explain "glacial" and "interglacial" in your answer.
24. Summarize what science has learned about the causes of ice ages by listing and explaining at least four possible factors in climate change.
25. Describe the role of ice cores in deciphering past climates. What record do they preserve? Where were they drilled?
26. Explain the relationship between the criteria defining the Arctic and Antarctic regions. Is there any coincidence in these criteria and the distribution of Northern Hemisphere forests on the continents?

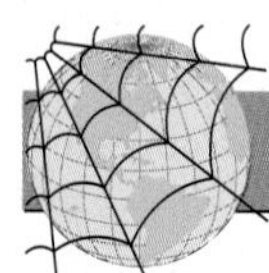

NetWork

The *Elemental Geosystems Home Page* provides on-line resources for this chapter on the World Wide Web. Once on the Home Page, click on this textbook, scroll the Table of Contents menu, select this chapter, and click "Begin." You will find self-tests that are graded, review exercises, specific updates for items in the chapter, and in "Destinations" many links to interesting related pathways on the Internet. *Elemental Geosystems* is found at **http://www.prenhall.com/christopherson**.

Critical Thinking

A. After checking back issues of the *New South Polar Times* at **http://www.spotsylvania.k12.va.us/nspt/home.htm**, imagine duty there for yourself. Of the 28 people who winter over at the station, some serve as scientists, technicians, and support staff. Remember, the last airplane leaves mid-February and the first airplane lands mid-October—such is the isolation. What do you see as the positives and negatives to such service? How would you combat the elements? The isolation? The cold and dark conditions? During the 1999 winter, a doctor who was working at the South Pole station detected her own breast cancer. An unusual aerial winter drop of medications was delivered so she could self-administer chemotherapy. The isolation is unimaginable.

B. Hypothetically, speculate on the relationship between the Alaskan meltdown discussed in this chapter and the occurrence of earthquakes, such as the magnitude 7.9 quake near Denali National Park, Alaska, the largest on Earth during 2002. The rapid melting of glacial ice is causing an isostaic rebound of the landscape (see Figure 8.4). Vertical motion in Alaska is averaging 36 mm (1.42 in.) per year of uplift—the fastest rate anywhere on Earth in 2002. Scientists see a tie between the rebounding crust and glacial ice loss over the past 150 years and especially over the past 30 years. How might isostatic rebound, in response to the unloading of glacial ice, produce strain along faults?

Career Link 14.1

Karl Birkeland, Avalanche Scientist, Forest Service National Avalanche Center

When snow scientist Karl Birkeland was in second grade, he wrote a little essay called "How Glaciers Form." It reads (spelling corrected): "Glaciers form by large snows, where snow does not melt. The reason glaciers are covered with dirt is because the glaciers shake the canyon and the dirt falls on them. Glaciers must be bigger than the Empire State Building. If they are near the water big icebergs might fall in." Karl added, "So I was thinking about this way back then. I have this framed above my desk, misspellings and all."

Karl completed his undergraduate work at the University of Colorado in environment, population, and biology, and even thought about medical school. His father was a professor of geology at Colorado. A skier since childhood, while finishing his degree he was a member of the ski patrol. He began to realize that studying the environment, specifically snow science, was what he wanted to do: "Snow science became a focus. Drawing on my ski patrol work where we had to deal with avalanches, I felt I had a good start."

Karl finished his master's at Montana State University, where his thesis topic was "Spatial Variability of Snow Resistance on Potential Avalanche Slopes." He began work as an avalanche specialist in the Gallatin National Forest, Bozeman, Montana, in 1990, where he established the Avalanche Center. He completed his Ph.D. at Arizona State in 1997. In his dissertation, he looked at slopes throughout the Bridger Mountains, Montana, and developed models for assessing variations in snow stability. This was the first time that anyone had tried to map avalanche conditions over a mountain range in one day.

Karl added, "In the Rockies, we have better preserved weak layers within the snowpack. These weak layers can keep the snow packs unstable for long periods of time, even if there hasn't been fresh snowfall." This regional characteristic makes the location of his National Avalanche Center ideal.

FIGURE 1 Karl Birkeland, Avalanche Scientist. Karl works at the U.S. Forest Service National Avalanche Center. [Photo by Ginger Birkeland.]

As to his present responsibilities, Karl said, "I'm in charge of transferring new and emerging technologies to the avalanche specialists at the regional avalanche centers around the country. I'm also involved in keeping track of any recent scientific developments in our field that should be conveyed to our avalanche workers." Karl assists the director of the Forest Service National Avalanche Center in coordinating all the avalanche centers.

Karl and his team continue to examine synoptic climatology and analyze air masses to understand resulting snow accumulations. Reconstructing past weather patterns is key. These efforts will produce a classification scheme that categorizes past data into coastal, continental, or intermountain characteristic types. "We can tie this in with weather patterns and predict the kind of avalanches we might expect on a given day." He said the big challenge is to "take scientific work and distill out the day-to-day practical significance for avalanche forecasters." In other words, what is included in this research that can help avalanche forecasters right now?

I kept hearing "spatial" in his work, getting a sense of the geographer in him. "That's me!" he answers, "There is such a need to see things spatially, regionally, with maps as the medium of communication." Geography is in the family, as Karl's wife is a Ph.D. geographer studying rivers and riparian vegetation across the Colorado Plateau and southwestern deserts.

Karl participates in the biennial International Snow Science Workshop (see http://www.avalanche.org). The key challenge is merging theory and practice. More than 600 scientists attended the 2000 meeting in Montana. The group is working to increase the global database, accident statistics, and forecasting efforts of many scientists in this growing field of interest. The future involves utilization of GIS modeling techniques to better tie together snowpack variables for spatial analysis.

"We want to bring together the operational avalanche workers and the avalanche researchers. The Internet has done a great job at linking this small community of individuals," Karl said. "This communication is needed because the avalanche risk is growing, owing to the increased development and construction in the mountains that are avalanche prone. All in all, the future is quite exciting in snow science!" And, Karl declared, "I continue to be very pro-geography and I like calling myself a geographer! Much of my current research focuses on spatial variations of snowpack properties, a clearly geographic topic that has critical real-world applications for avalanche forecasters."

PART FOUR

Biogeography Systems

A misty morning near Heather Meadows and the Nooksack River below the Mount Baker volcano, Washington state. Earth's physical systems interact to produce a unique assemblage of soils and ecosystems in the cool, moist, Northwest. [Photo by Bobbé Christopherson.]

Earth is the home of the only known biosphere in the Solar System—a unique, complex, and interactive system of abiotic (nonliving) and biotic (living) components working together to sustain a tremendous diversity of life. Energy enters the system through conversion of solar energy by means of photosynthesis in the leaves of plants. Soil is the essential link among the lithosphere, plants, and the rest of Earth's physical systems. Thus, soil helps sustain life.

Life is organized into a feeding hierarchy from producers to consumers, and on through to decomposers. Taken together, the soils, plants, animals, and nonliving components produce aquatic and terrestrial ecosystems, known as biomes. The resilience of the biosphere, as we know it, is being tested in a real-time, one-time experiment. Today we face crucial issues—preservation of the diversity of life in the biosphere and the survival of the biosphere itself are important applied topics considered in Part 4.

The Palouse of eastern Washington state is a major wheat producing region. Rich Mollisols form the agricultural soils of the region. [Photo by Bobbé Christopherson.]

15 The Geography of Soils

Key Learning Concepts

By knowing and understanding the key learning concepts in this chapter, you should be able to:

- *Define* soil and soil science and *describe* a pedon, polypedon, and typical soil profile.
- *Describe* soil properties of color, texture, structure, consistence, porosity, and soil moisture.
- *Explain* basic soil chemistry, including cation-exchange capacity, and *relate* these concepts to soil fertility.
- *Evaluate* principal soil formation factors, including the human element.
- *Describe* the 12 soil orders of the Soil Taxonomy classification system and *explain* their general occurrence.

Earth's landscape generally is covered with soil. **Soil** is a dynamic natural material composed of fine particles in which plants grow, and that contains both mineral and organic matter. If you have ever planted a garden, tended a houseplant, or been concerned about famine, this chapter will interest you. A knowledge of soil is at the heart of agriculture and food production.

You kneel down and scoop a handful of prairie soil, compressing it and breaking it apart with your fingers. You are holding a historical object—one that bears the legacy of the last 15,000 years, or more. This lump of soil contains information about the last ice age and intervening warm periods, about distinct and distant source materials, and about several physical processes. We are using and abusing this legacy at rates much faster than it formed. Soils do not reproduce, nor can they be recreated.

Soil science is interdisciplinary, involving physics, chemistry, biology, mineralogy, hydrology, taxonomy, climatology, and cartography. Physical geographers are interested in the spatial patterns formed by soil types and the physical factors that interact to produce them. As an integrative science, physical geography is well suited for this task. *Pedology* deals with the origin, classification, distribution, and description of soils (*ped* is from the Greek *pedon*, meaning "soil" or "earth"). *Edaphology* specifically focuses on the study of soil as a medium for sustaining the growth of higher plants (*edaphos* means "soil" or "ground").

Soil science deals with a complex substance whose characteristics vary from kilometer to kilometer, and even centimeter to centimeter. In many locales, an *agricultural extension service* can provide specific information and perform a detailed analysis of local soils. Soil surveys and local soil maps are available for most counties in the United States and for the Canadian provinces. Your local phone book may list the U.S. Department of Agriculture, Natural Resources Conservation Service (http://www.nrcs.usda.gov/), or Agriculture Canada's Soil Information System(http://sis.agr.gc.ca/cansis/intro.html); or you may contact the appropriate department at a local college or university. See the National Soil Survey Center's site at http://soils.usda.gov/, and for links related to world soils, see http://www.metla.fi/info/vlib/soils/old.htm.

In this chapter: The geography of soils deals spatially with a complex substance, the characteristics of which vary from kilometer to kilometer, and even over smaller areas. We begin with soil characteristics and the basic soil sampling and soil mapping units. The soil profile is a dynamic structure, mixing and exchanging materials and moisture across its horizons. Properties of soil include texture, structure, porosity, moisture, and chemistry—all integrating to form soil types. The chapter discusses both natural and human factors that affect soil formation. A global concern exists over the loss of soils to erosion, mistreatment, and conversion to other uses. The chapter concludes with a brief examination of the Soil Taxonomy, the 12 principal soil orders, and their spatial distribution.

Soil Characteristics

Classifying soils is similar to classifying climates. Both involve complex interacting variables, however, simple theoretical models are the basis for soil classifications. The following section presents general assumptions about soil structure, soil properties, and soil formation processes.

Soil Profiles

Just as a book should not be judged by its cover, so soils should not be evaluated at the surface layer only. Instead, a soil profile should extend from the surface to the deepest extent of plant roots, or to where regolith or bedrock is encountered. Such a profile, a **pedon**, is conceptually a hexagonal column from 1 to 10 m^2 in surface area (Figure 15.1). *The pedon is the basic sampling unit used in soil surveys.*

Many pedons together in one area make up a polypedon, which has distinctive characteristics differentiating it from surrounding polypedons. A **polypedon** is an essential soil individual, constituting an identifiable *series* of soils in an area. *The polypedon is the soil unit used in preparing local soil maps.*

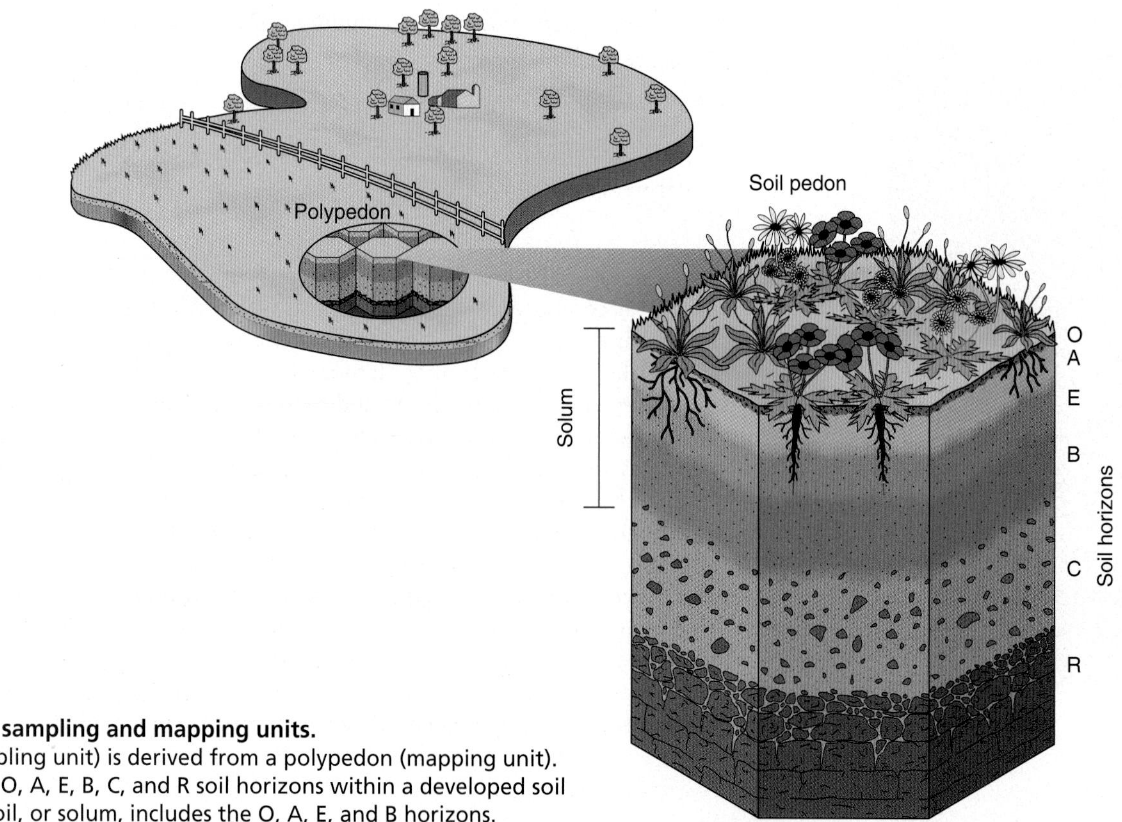

FIGURE 15.1 Soil sampling and mapping units.
A soil pedon (sampling unit) is derived from a polypedon (mapping unit). Shown are typical O, A, E, B, C, and R soil horizons within a developed soil pedon. The true soil, or solum, includes the O, A, E, and B horizons.

Soil Horizons

Each layer exposed in a pedon is a **soil horizon**. A horizon is roughly parallel to the pedon's surface and has characteristics distinctly different from horizons directly above or below. The boundary between horizons usually is visible in the field, using the properties of color, texture, structure, consistence (meaning soil consistency), porosity, and the presence or absence of certain minerals, moisture, and chemical processes (Figure 15.2). Soil horizons are the building blocks of soil classification. However, because they can vary in an almost endless number of ways, a simple model is useful. The enlarged column in Figure 15.1 presents an ideal pedon and *soil profile*, with letters assigned to each soil horizon for identification. A soil pedon may have all these, some better developed than others, or it may have poorly developed horizons.

At the top of the soil profile is the *O* (organic) *horizon*, named for its organic composition, derived from plant and animal litter that was deposited on the surface and transformed into humus. **Humus** is a mixture of decomposed organic materials and is usually dark in color. The O horizon is 20%–30% or more organic matter, important because of its water-absorbing ability and its nutrients.

At the bottom of the soil profile is the *R* (rock) *horizon*, consisting of either unconsolidated (loose) material or consolidated bedrock. The A, E, B, and C horizons mark important mineral strata between O and R. These middle layers are composed of sand, silt, clay, and other weathered by-products.

In the *A horizon*, the presence of humus and clay particles is particularly important, for they provide essential chemical links between soil nutrients and plants. The A horizon usually is darker and richer in organic content than lower horizons. It grades into the *E horizon*, made up of coarse sand, silt, and resistant minerals. From the E horizon, clays and oxides of aluminum and iron are leached (removed by water) and are carried to lower horizons with water as it percolates through the soil. This process in which water rinses upper horizons and removes fine particles and minerals is termed **eluviation**; thus the designation E for this horizon. The greater the precipitation in an area, the higher the rate of eluviation.

In contrast to the A and E horizons, *B horizons* accumulate clays, aluminum, and iron. B horizons are dominated by **illuviation**, a depositional process. (Eluviation is an erosional removal process; illuviation is depositional.) These horizons may exhibit reddish or yellowish hues because of the illuviated presence of mineral and organic oxides. In the humid tropics, these layers often develop to some depth.

Some materials occurring in the B horizon may have formed in place from weathering processes rather than arriving there by *translocation*, or migration. Likewise, clay losses in an A horizon may be caused by destructive processes and not eluviation. Research to better understand erosion and deposition of clays between soil horizons is one of the challenges in soil science.

The combination of the A and E horizons with their *eluviation* removals and the B horizon with its *illuviation* accumulations is designated the **solum**, considered the true definable soil of the pedon. The A, E, and B horizons experience the most active soil processes (labeled on Figure 15.1).

The *C horizon* is weathered bedrock or weathered parent material, excluding the bedrock itself. The C horizon is not much affected by soil operations in the solum and lies outside the biological influences in the shallower horizons. Plant roots and soil microorganisms are rare in the C horizon. In dry climates, calcium carbonates commonly form the cementing material of these hardened layers.

FIGURE 15.2 A typical soil profile.
A Mollisol pedon in southeastern South Dakota. The parent material is glacial till, and the soil is well drained. The dark O and A horizons above the #1 transitions into an E horizon. Distinct carbonate nodules are visible in the lower B and upper C horizons. [Photo from Marbut Collection, Soil Science Society of America, Inc.]

Soil Properties

Observing a real soil profile will help you identify color, texture, structure, and other soil properties. A good place to see a soil profile is at a construction site or excavation, perhaps on your campus, or at a road cut.

Soil Color. Color is important, for it sometimes suggests composition and chemical makeup. If you look at exposed soil, color may be the most obvious trait. Among the many possible hues are the red and yellow soils of the southeastern United States, which are high in ferric (iron) oxides; the black-prairie, richly organic soils of portions of the U.S. grain-growing regions and Ukraine; or the white-to-pale soils attributable to the presence of aluminum oxides and silicates. Color can be deceptive too, for soils of high humus content are often dark, yet clays of warm-temperate and tropical regions with less than 3% organic content are some of the world's blackest soils.

Soil Texture. Soil texture, perhaps a soil's most permanent attribute, refers to the size and organization of particles in the soil. Individual mineral particles are called *soil separates.* Particles range from the finest clays, to silts, to coarser sands, to larger pebbles and gravels.

Figure 15.3 is a soil triangle, showing the relation of sand, silt, and clay concentrations distributed along each side. Each corner of the triangle represents a soil consisting solely of the particle size noted (rarely are true soils composed of a single separate). Every soil on Earth is defined somewhere in this triangle.

Figure 15.3 includes the common designation **loam,** which is a balanced mixture of sand, silt, and clay beneficial to plant growth. A sandy loam with clay content below 30% (lower left) usually is considered ideal by farmers because of its water-holding characteristics and cultivation ease. Soil texture is important in determining water retention and water transmission traits.

Consider a soil type in Indiana called *Miami silt loam.* Samples from it are plotted on the soil texture triangle as points 1, 2, and 3. A sample taken near the surface in the A horizon is recorded at point 1; in the B horizon at point 2; and in the C horizon at point 3. Textural analyses of these samples are summarized in pie diagrams and a table in Figure 15.3. Note that silt dominates the surface, clay the B horizon, and sand the C horizon.

Soil Structure. *Soil structure* refers to the *arrangement* of soil particles, whereas *soil texture* describes the size of soil particles. Structure can partially modify the effects of soil texture. The term *ped* describes an individual unit of soil

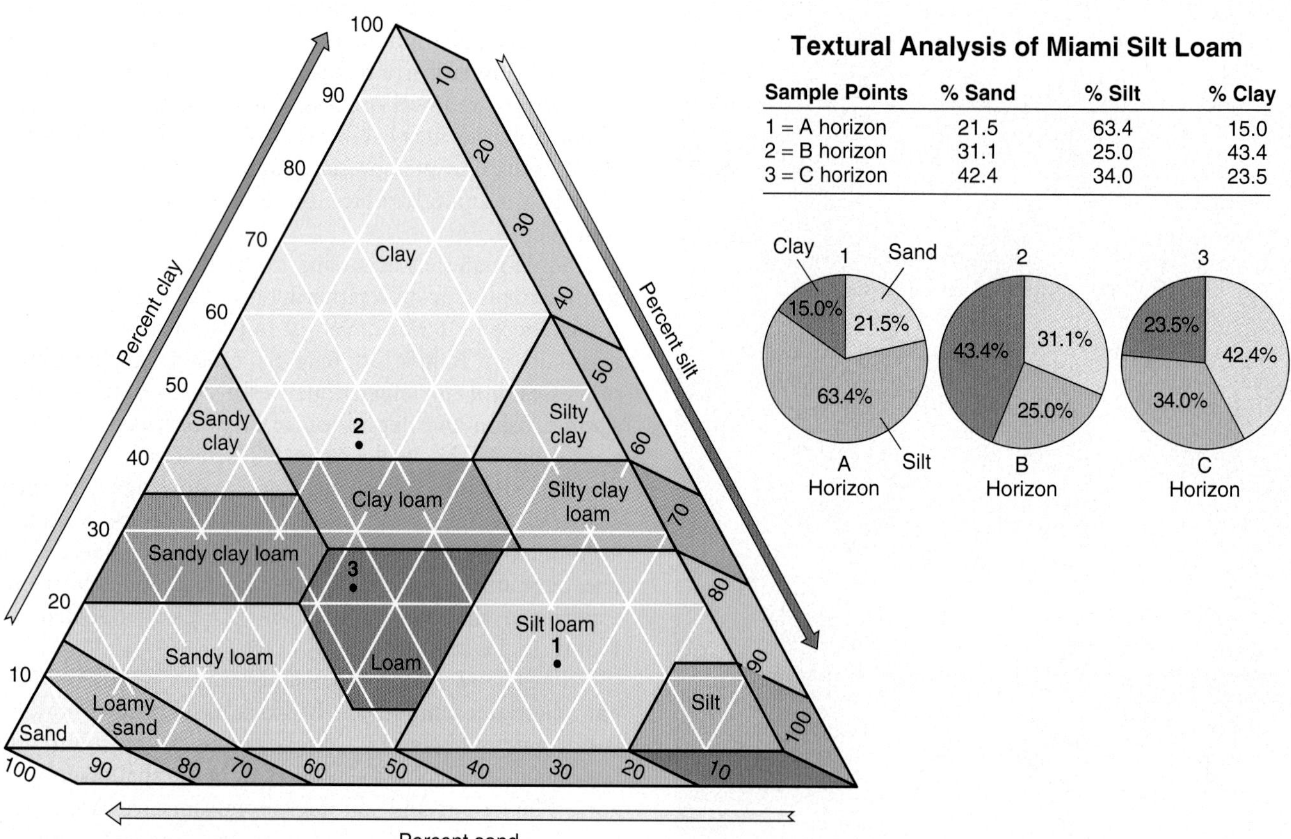

Textural Analysis of Miami Silt Loam

Sample Points	% Sand	% Silt	% Clay
1 = A horizon	21.5	63.4	15.0
2 = B horizon	31.1	25.0	43.4
3 = C horizon	42.4	34.0	23.5

FIGURE 15.3 Soil texture triangle.
Measures of the ratio of clay, silt, and sand determine soil texture. As an example, points 1 (horizon A), 2 (horizon B), and 3 (horizon C) designate samples taken at three different horizons in the Miami silt loam in Indiana. Note that silt dominates the surface, clay the B horizon, and sand the C horizon. Note also the ratio of sand to silt to clay as displayed in the table and in the three pie diagrams. [After U.S. Department of Agriculture, Natural Resources Conservation Service, *Soil Survey Manual*, Agricultural Handbook No. 18, p. 138 (Washington, D.C.: U.S. Government Printing Office, 1993).]

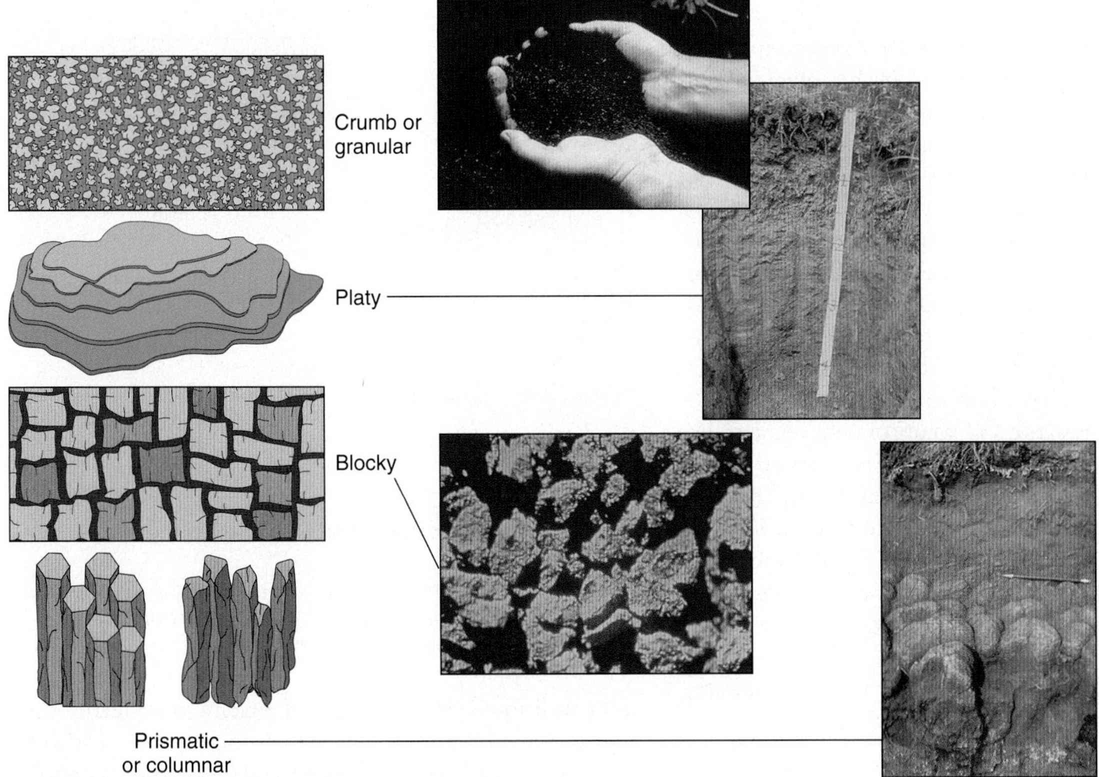

FIGURE 15.4 Types of soil structure.
Structure is important because it controls drainage, rooting of plants, and the delivery of nutrients to plants. The shape of individual peds shown here controls a soil's structure. [Photos by National Soil Survey Center, Natural Resources Conservation Service, Soil Survey Staff.]

particles; it is a tiny natural lump or cluster of particles held together. The shape of these peds determines which of the structural types the soil exhibits (Figure 15.4).

Peds separate from each other along zones of weakness, creating voids (pores) that are important for moisture storage and drainage. More rounded peds have more pore space and greater permeability. They are therefore more productive for plant growth than are coarse, blocky, prismatic, or platy peds, despite comparable fertility.

Soil Consistence. In soil science, the term *consistence* is used to describe the consistency of soil particles. It refers to cohesion in soil, a product of texture and structure. Consistence reflects a soil's resistance to breaking and manipulation under varying moisture conditions.

- A *wet soil* is sticky between the thumb and forefinger, ranging from a little adherence to either finger, to sticking to both fingers, to stretching when the fingers are moved apart. *Plasticity*, the quality of being moldable, is roughly measured by rolling a piece of soil between your fingers and thumb to see whether it rolls into a thin strand.
- A *moist soil* is filled to about half of field capacity (the usable water capacity of soil), and its consistence grades from loose (noncoherent), to *friable* (easily pulverized), to firm (not crushable between thumb and forefinger).
- A *dry soil* is typically brittle and rigid, with consistence ranging from loose, to soft, to hard, to extremely hard.

Soil Porosity. Soil porosity, permeability, and moisture storage are discussed in Chapter 7. Pores in the soil horizon control the flow of water, its intake and drainage, and air ventilation. Important porosity factors are *pore size*, *continuity* (whether the pores are interconnected), *shape* (whether pores are spherical, irregular, or tubular), *orientation* (whether pore spaces are vertical, horizontal, or random), and *pore location* (whether pores are within or between soil peds).

Porosity is improved by the biotic actions of plant roots, animal activity such as the tunneling of gophers or worms, and human intervention through soil manipulation (plowing, adding humus or sand, or planting soil-building crops). Much of a farmer's soil preparation work is done to improve soil porosity.

Soil Moisture. Reviewing Figures 7.7 and 7.8 will help you understand this section. Plants operate most efficiently when the soil is at *field capacity*, that is, maximum water availability for plant use. Field capacity is determined by soil type. The effective rooting depth of a plant determines the amount of soil moisture to which the plant's roots are exposed. If soil moisture is drawn below field capacity, plants use available water inefficiently until the plant's wilting point is reached, beyond which plants cannot extract the water they need.

Soil Chemistry

The soil atmosphere is mostly nitrogen, oxygen, and carbon dioxide. Nitrogen concentrations in the soil are about the same as in the atmosphere. There is less oxygen and more carbon dioxide in soil than in the atmosphere because of ongoing respiration processes.

Water present in soil pores is called *soil solution.* It is the medium in the soil for chemical reactions. Soil solution is critical to plants as their source of nutrients and is the foundation of *soil fertility.* Carbon dioxide and various organic materials combine with water to produce carbonic and organic acids, respectively. These acids are then active in soil processes, as are dissolved alkalis and salts.

To understand how the soil solution behaves, let's go through a quick chemistry review. An *ion* is an atom, or group of atoms, that carries an electrical charge. An ion has either a positive charge or a negative charge. For example, when NaCl (sodium chloride) is dissolved in solution, it separates into two ions: Na^+, a *cation* (positively charged ion), and Cl^-, an *anion* (negatively charged ion). Some ions in soil carry single charges, whereas others carry double or even triple charges (for example, sulfate, SO_4^{2-}; aluminum, Al^{3+}).

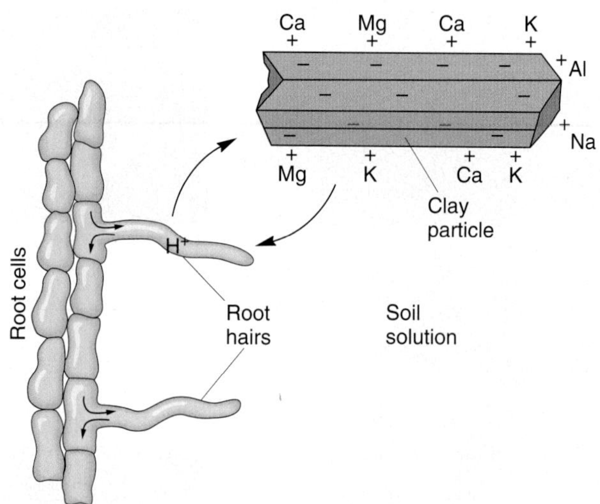

FIGURE 15.5 Soil colloids and cation-exchange capacity (CEC).
Typical soil colloids retain mineral ions by adsorption to their surface (opposite charges attract). This attraction holds the ions until they are absorbed by root hairs.

Soil Colloids and Mineral Ions. Soil colloids retain ions in soil. **Soil colloids** are tiny particles of clay and organic material that carry a negative electrical charge and consequently attract any positively charged ions in the soil (Figure 15.5). The positive ions, many metallic, are critical to plant growth. If it were not for the soil colloids, the positive ions would be leached away by the soil solution and thus would be unavailable to plant roots.

Individual clay colloids are thin and platelike, with parallel surfaces that are negatively charged. Cations attach to the surfaces of the colloids by **adsorption** (not *absorption*). The metallic cations are adsorbed onto the soil colloids. Colloids can exchange cations between their surfaces and the soil solution because of their **cation-exchange capacity (CEC)**. A high CEC means good soil fertility (unless there is a complicating factor, such as high soil acidity).

Soil fertility is the ability of soil to sustain plants. Soil has fertility when it contains organic substances and clay minerals that absorb water and certain elements needed by plants. Billions of dollars are expended to create fertile soil conditions, yet the future of Earth's most fertile soils is threatened because soil erosion is increasing worldwide. The impact on society is potentially disastrous as population and food demands increase.

Soil Acidity and Alkalinity. A soil solution may contain significant hydrogen ions (H^+), cations that stimulate acid formation. The result is an *acid soil.* In contrast, a soil high in base cations (calcium, magnesium, potassium, and sodium) is a *basic* or *alkaline soil.* Acidity and alkalinity are expressed on the pH scale (Figure 15.6).

Pure water is nearly *neutral*, with a pH of 7.0. Readings below 7.0 represent increasing acidity. Readings above 7.0 represent increasing alkalinity. Acidity usually is regarded as strong at 5.0 or lower on the pH scale, whereas 10.0 or above is considered strongly alkaline.

Today the major contributor to soil acidity is acid precipitation (rain, snow, fog). Acid rain actually has been measured below pH 3.0—an incredibly low value for natural precipitation, as acidic as lemon juice. Because most crops are sensitive to specific pH levels, acid soils below pH 6.0 require treatment to raise the pH. The pH is elevated by the addition of bases in the form of minerals that are rich in base cations, usually lime (calcium carbonate).

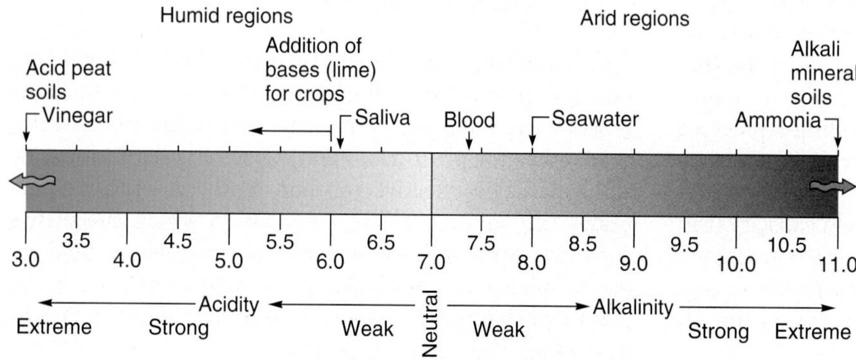

FIGURE 15.6 pH scale.
Soil pH scale, measuring acidity (lower pH) and alkalinity (higher pH). The complete pH scale ranges between 0 and 14.

Increased acidity in the soil solution accelerates the chemical weathering of mineral nutrients and increases their depletion rates, as in soils in the region of the northern forests.

Soil Formation Factors and Management

Three soil-forming factors are important: two natural factors are *dynamic* (climatic and biologic) and *passive* (parent material, topography and relief, and time), and the *human* component is a third. These three factors work together as a system for soil formation. The role of parent material in providing weathered minerals to form soils was discussed in Chapter 10.

Natural Factors. Climate types correlate closely with soil types worldwide. The moisture, evaporation, and temperature regimes of climates determine the chemical reactions, organic activity, and eluviation rates of soils. Not only is the present climate important but many soils exhibit the imprint of past climates, sometimes over thousands of years. Notable is the effect of glaciations, as described and mapped in Chapter 14.

The organic content of soil is determined by vegetation and by animal and bacterial activity. The chemical makeup of vegetation contributes to acidity or alkalinity of the soil solution. For example, broadleaf trees tend to increase alkalinity, whereas needleleaf trees tend to produce higher acidity. Thus, when civilization moves into new areas and alters the natural vegetation by logging or plowing, the affected soils are likewise altered. In many cases they are permanently changed.

Topography also affects soil formation, mainly through slope and orientation. Slopes that are too steep cannot have full soil development because gravity and erosional processes remove materials. Slopes that are nearly level inhibit soil drainage. In the Northern Hemisphere, a south-facing slope is warmer because it receives direct sunlight. North-facing slopes are colder, causing slower snowmelt and lower evaporation rates, which result in more moisture for plants than is available on south-facing slopes. Temperature affects water-balance relationships.

The Human Factor. Human intervention has a major impact on soils. Millennia ago, farmers in every culture on Earth learned to plow slopes "on the contour"—to plow around a slope at the same elevation, not vertically up and down the slope. Contour plowing reduces sheet erosion (which removes topsoil) and simultaneously improves water retention in the soil. However, the demands of today's large farm equipment have forced farmers to sometimes alter their practice of contouring because the equipment cannot make the tight turns that such proper plowing requires. Other effective soil-holding practices also are being eliminated to accommodate the larger equipment. Protective farming methods in declining use include terracing (similar to contour plowing), planting tree rows as windbreaks, and planting shelter belts.

All the identified factors in soil development (climate, biological activity, parent material, landforms and topography, and human activity) require time to operate. A few centimeters' thickness of prime farmland soil may require *500 years* to mature. Yet these same soils are being lost at a few centimeters per year to sheet flow and gullying when the soil-holding vegetation is removed. Soils exposed to precipitation may be completely leached of needed cations, thereby losing their fertility. Figure 15.7 maps regions of soil losses and related concerns, and News Report 15.1 discusses the same issues. Cooperative international action is needed to save Earth's productive soils.

Soil Classification

Classification of soils is complicated by the variety of interactions that create thousands of distinct soils—well over 15,000 soil types in the United States and Canada alone. Not surprisingly, a number of different classification systems are in use worldwide. The United States, Canada, the United Kingdom, Germany, Australia, Russia, and the United Nations Food and Agricultural Organization (FAO) have their own soil classification systems. Each system reflects the environment of its country. For example, the National Soil Survey Committee of Canada developed a system suited to its great expanses of boreal forest, tundra, and cool climatic regimes (detailed in Appendix B).

Soil Taxonomy

The U.S. soil classification system was published in 1975 (*Soil Taxonomy—A Basic System of Soil Classification for Making and Interpreting Soil Surveys*) and revised in a new edition in 1999. Soil scientists refer to it as **Soil Taxonomy**. Over the years various revisions and clarifications in the system were published in *Keys to the Soil Taxonomy*, now in its 8th edition (1998), which includes all the revisions to the 1975 Soil Taxonomy. Major revisions include the addition of two new soil orders: Andisols (volcanic soils) in 1990 and Gelisols (cold and frozen soils) in 1998. Much of the information in this chapter is derived from these two keystone publications. See http://soils.usda.gov/ for a copy of this publication and much more regarding soils.

The U.S. Soil Taxonomy system is based on soil properties actually seen in the field. Thus, it is open to addition, change, and modification as the sampling database grows. An important aspect of the system is that it recognizes the interaction of humans and soils and the changes that humans have introduced, both purposely and inadvertently.

The classification system divides soils into six categories, creating a hierarchical sorting system. Each soil series (the smallest, most detailed category numbering 15,000) ideally includes only one polypedon, but may include portions continuous with adjoining polypedons in the field. In sequence from smallest to largest categories, including the number of occurrences within each, the Soil

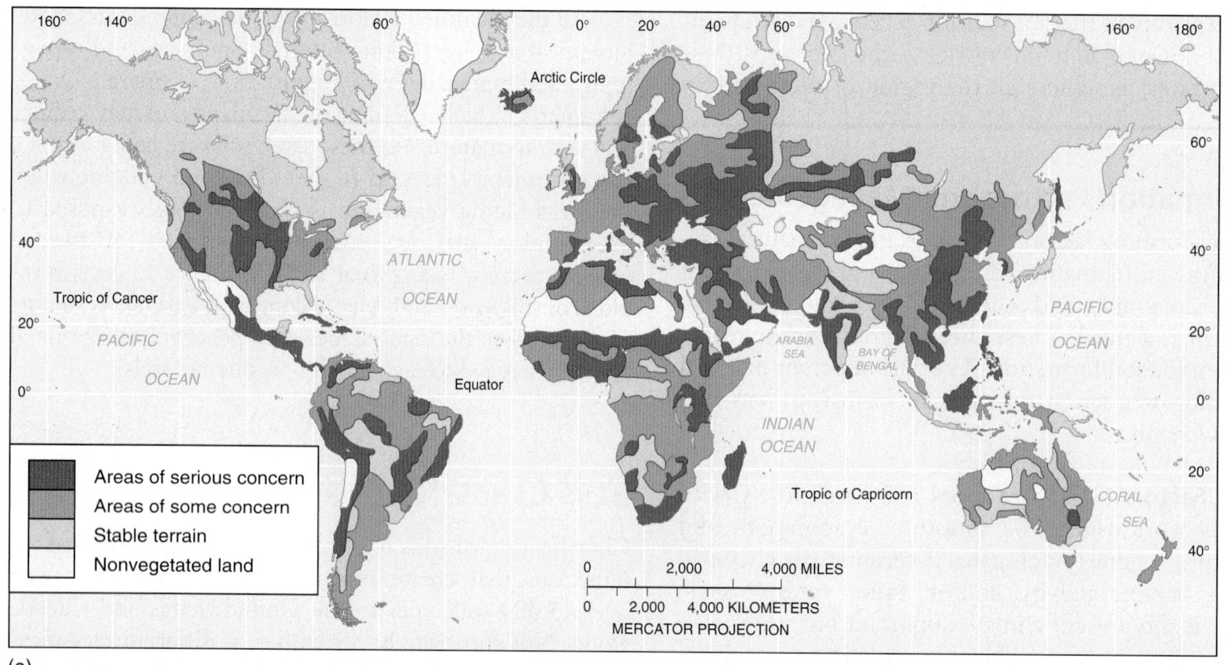

(a)

(b)

FIGURE 15.7 Soil degradation.
(a) Approximately 1.2 billion hectares (3.0 billion acres) of Earth's soils suffer some degree of degradation through erosion caused by human misuse and abuse. (b) Typical soil loss through sheet erosion on a northeastern Wisconsin farm. One millimeter of soil lost from an acre weighs about 5 tons. [(a) *A Global Assessment of Soil Degradation*, adapted from United Nations Environment Programme, International Soil Reference and Information Centre, "Map of Status of Human-Induced Soil Degradation," Sheet 2, Nairobi, Kenya, 1990; (b) photo by D. P. Burnside/Photo Researchers, Inc.]

News Report 15.1

Soil Is Slipping Through Our Fingers

- The U.S. General Accounting Office estimates that from 3 to 5 million acres of prime farmland are lost each year in the United States through mismanagement or conversion to nonagricultural uses. About half of all cropland in the United States and Canada is experiencing excessive rates of soil erosion—these two countries being among the few that monitor loss of topsoil.
- The Canadian Environmental Advisory Council estimates that the organic content of cultivated prairie soils has declined by as much as 40% compared with noncultivated native soils. In Ontario and Québec, losses of organic content increased as much as 50%, and losses are even higher in the Atlantic Provinces, which were naturally low in organic content before cultivation.
- A 1995 study completed at Cornell University concluded that soil erosion is a major environmental threat to the sustainability and productive capacity of agriculture worldwide.
- Since 1950, as much as 38% of the world's farmable land has been lost to soil erosion, and the rate continues at 5 to 6 million hectares (about 12 to 15 million acres) per year—560 million hectares, 1380 million acres, to date (World Resources Institute and UNEP, 1997).
- The causes for degraded soils are, in order of severity: overgrazing, vegetation removal, agricultural activities, overexploitation, and industrial and bioindustrial use (UNEP, 1997).
- The world's human population is growing at the rate of 6.9 million people a month (net increase), increasing the demand for food and agricultural productivity. Especially significant is that proportion of the global population that is adopting a meat-centered diet as American tastes and food outlets spread

(continued)

News Report 15.1 ***(continued)***

worldwide. As we see in the next chapter, meat production is an inefficient use of the grain supply.

Some 35% of farmlands are losing soil faster than it can form—a loss exceeding 23 billion metric tons (25 billion tons) per year. Soil depletion and loss are at record levels from Iowa to China, Peru to Ethiopia, the Middle East to the Americas. The impact on society is potentially disastrous as population and food demands increase.

Soil erosion can be compensated for in the short run by using more fertilizer, increasing irrigation, and by planting higher-yielding strains. But the potential yield from prime agricultural land will drop by as much as 20% over the next 20 years if only moderate erosion continues.

One 1995 study tabulated the market value of lost nutrients and other variables in the most comprehensive soil-erosion study to date. The sum of direct damage (to agricultural land) and indirect damage (to streams, society's infrastructure, and human health) was estimated at more than $25 billion a year in the United States and hundreds of billions of dollars worldwide. (Of course, this is a controversial assessment in the agricultural industry.) The cost to bring erosion under control in the United States is estimated at approximately $8.5 billion, or about 30 cents on every dollar of damage and loss.

Taxonomy recognizes *soil families* (6000), *soil subgroups* (1200), *soil great groups* (230), *soil suborders* (47), and *soil orders* (12).

Pedogenic Regimes. Before the Soil Taxonomy system was devised, *pedogenic regimes* were used. These regimes keyed specific soil-forming processes to climatic regions. Although each pedogenic process may be active in several soil orders and in different climates, we discuss them within the soil order where they commonly occur. Such climate-based regimes are convenient for relating climate and soil processes. However, *the Soil Taxonomy system recognizes the great uncertainty and inconsistency in basing soil classification on such climatic variables.* The principal pedogenic regimes are:

- **laterization**: a leaching process in humid and warm climates, discussed with Oxisols;
- **salinization**: a process that concentrates salts in soils in climates with excessive potential evapotranspiration (moisture demand) rates, discussed with Aridisols;
- **calcification**: a process that produces an illuviated accumulation of calcium carbonates in continental climates, discussed with Mollisols and Aridisols;
- **podzolization**: a process of soil acidification associated with forest soils in cool climates, discussed with Spodosols;
- **gleization**: a process that includes an accumulation of humus and a thick, water-saturated gray layer of clay beneath, usually in cold, wet climates and poor drainage conditions.

Diagnostic Soil Horizons

To identify specific soil types within the Soil Taxonomy, the U.S. Soil Conservation Service describes *diagnostic horizons*. A diagnostic horizon reflects a distinctive physical property (color, texture, structure, consistence, porosity, moisture), or a dominant soil process (discussed with the soil types).

In the solum, two diagnostic horizons may be identified:

- The **epipedon** (literally, "over the soil") is the diagnostic horizon at the surface. It is visibly darkened by organic matter and sometimes is leached of minerals.
- The **diagnostic subsurface horizon** is located below the surface at varying depths.

The presence or absence of either diagnostic horizon usually distinguishes a soil for classification. The Natural Resources Conservation Service publications identify many different types within each diagnostic horizon using a myriad of terminology.

The 12 Soil Orders of the Soil Taxonomy

At the heart of the Soil Taxonomy are 12 general soil orders, which are described in Table 15.1 (p. 468). Their worldwide distribution is shown in Figure 15.8. Please consult this table and map as you read the following descriptions. Because the taxonomy evaluates each soil order on its own characteristics, there is *no priority* to the classification. You will find the 12 orders loosely arranged here by latitude, beginning along the equator as is done in Chapter 6, "Global Climate Systems," and Chapter 16, "Ecosystems and Biomes."

Oxisols. The persistent moisture, intense temperature, and uniform daylength of equatorial latitudes greatly affect soils. These generally old landscapes are deeply developed and have greatly altered minerals (Figure 15.9, p. 469). These soils are called **Oxisols** (tropical soils) because they have a distinctive horizon with a mixture of iron and aluminum oxides. Related vegetation is the luxuriant and diverse tropical and equatorial rain forest.

Heavy precipitation leaches soluble minerals and soil constituents (such as silica) from the A horizon. Typical Oxisols are reddish and yellowish from the iron and aluminum oxides left behind, with a weathered claylike

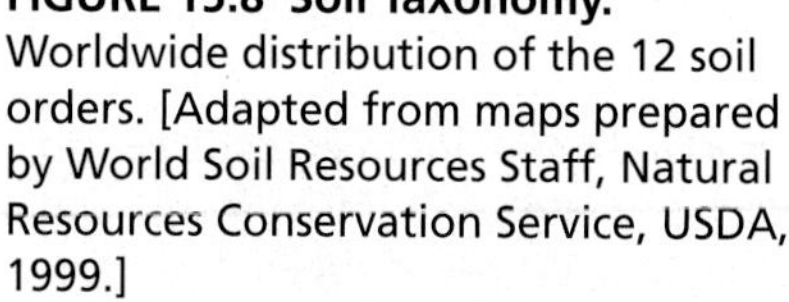

FIGURE 15.8 Soil Taxonomy. Worldwide distribution of the 12 soil orders. [Adapted from maps prepared by World Soil Resources Staff, Natural Resources Conservation Service, USDA, 1999.]

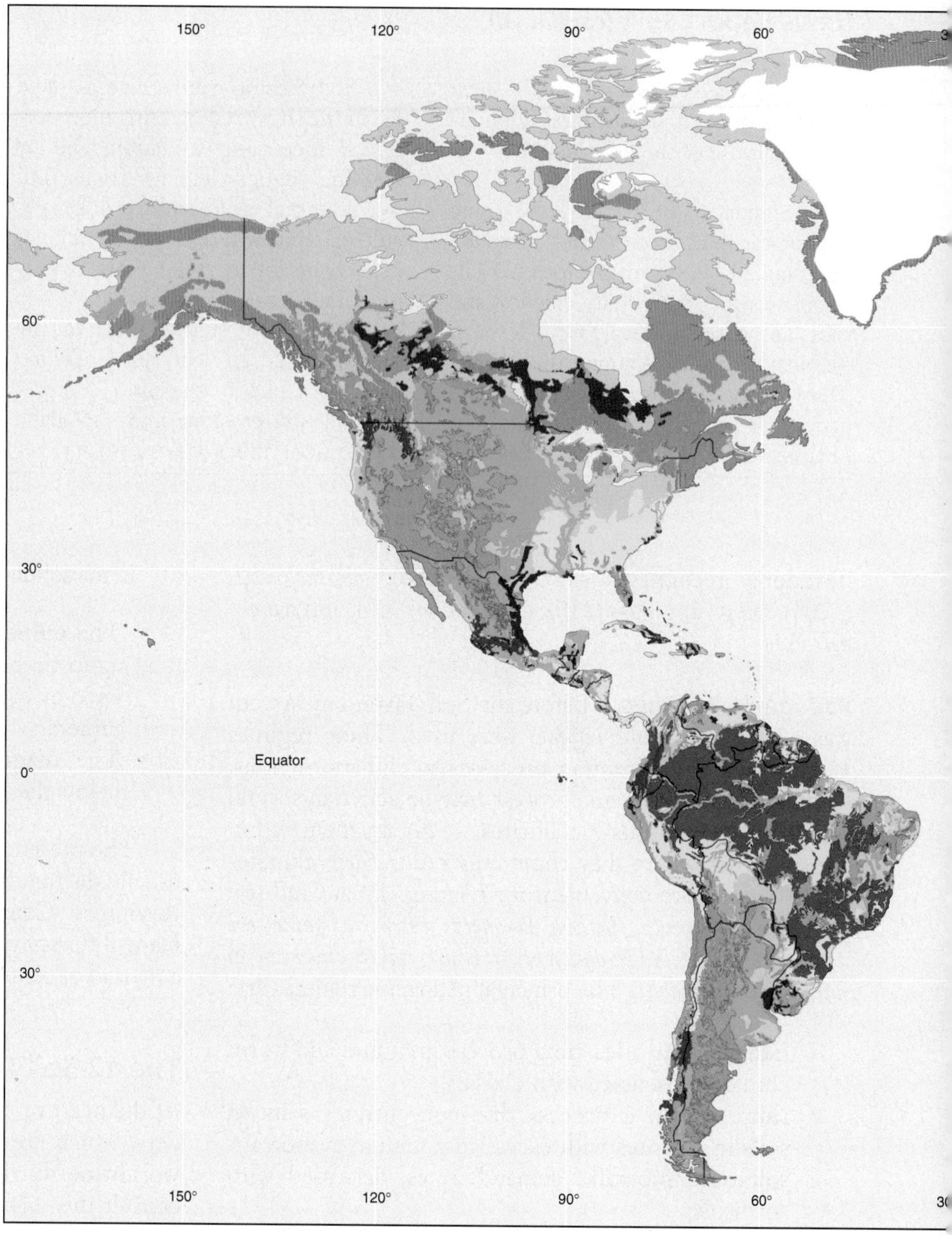

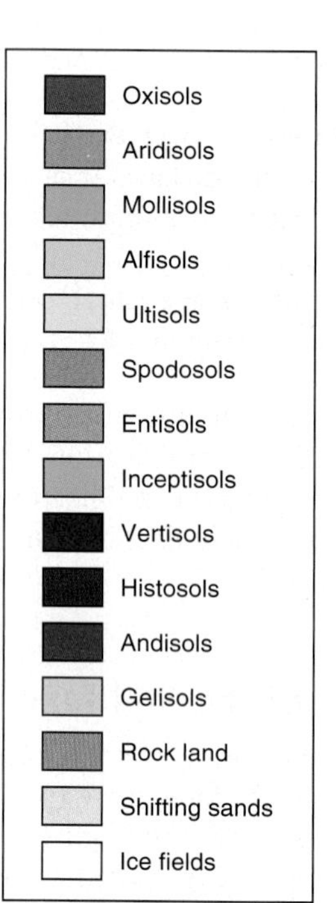

texture, sometimes in a granular soil structure that is easily broken apart. The high degree of eluviation removes basic cations and colloidal material to lower illuviated horizons. Thus, Oxisols are low in CEC (cation-exchange capacity) and fertility, except in regions augmented by alluvial (water-borne sediment) or volcanic materials. Figure 15.10 illustrates *laterization*, the leaching process that operates in well-drained soils in warm, humid tropical and subtropical climates.

The subsurface diagnostic horizon in an Oxisol is highly weathered, containing iron and aluminum oxides, at least 30 cm (12 in.) thick and within 2 m (6.5 ft) of the surface (Figure 15.9). If these horizons are subjected to repeated wetting and drying, a *hardpan* (hardened soil layer) develops, called a **plinthite** (from the Greek *plinthos*, meaning "brick"). This form of soil, also called a laterite, can be quarried in blocks and used as a building material (Figure 15.11).

Simple agricultural activities can be conducted with care in these soils. Early *slash-and-burn* shifting cultivation practices were adapted to these soil conditions and formed a unique style of crop rotation. The scenario went like this: People in the tropics cut down (slashed) and burned the rain forest in small tracts and then cultivated the land with stick and hoe. After several years the soil lost fertility and the people shifted cultivation to another tract and repeated the process. After many years of moving from tract to tract, the group returned to the first patch to begin the cycle again. This practice protected the limited fertility of the soils somewhat, allowing periods of recovery.

However, the invasion of foreign plantation interests, development by local governments, vastly increased

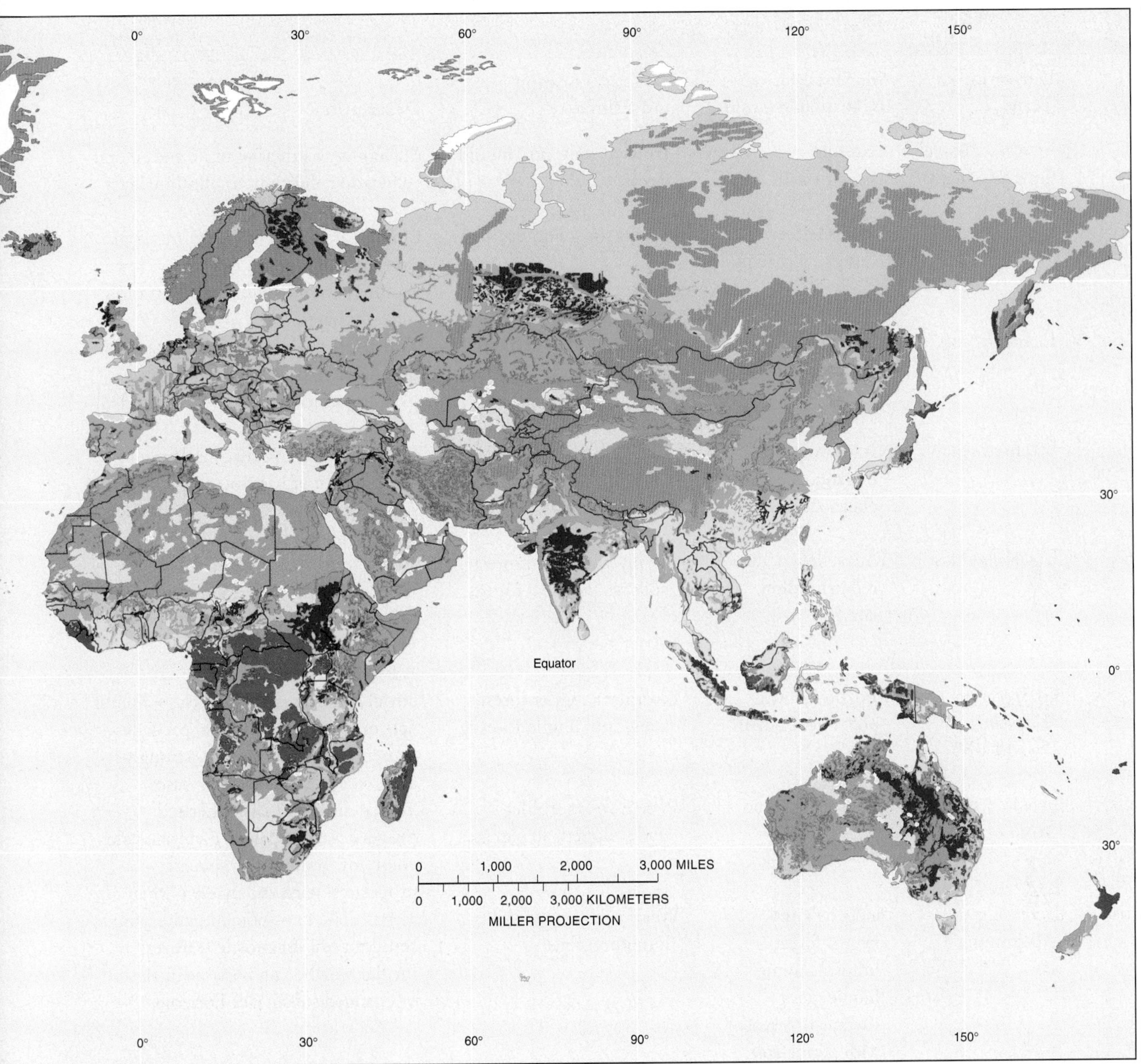

population pressures, and conversion of vast new tracts to pasturage halted this orderly native pattern of land rotation. Permanent tracts of cleared land put tremendous pressure on the remaining forest and brought disastrous consequences. When Oxisols are disturbed, soil loss can exceed a thousand tons per square kilometer per year, not to mention the greatly increased rate of extinction of plant and animal species that accompanies such destruction. The regions dominated by the Oxisols and rain forests are rightfully the focus of much worldwide environmental attention.

Aridisols. The largest single soil order occurs in the world's dry regions. **Aridisols** (desert soils) occupy approximately 19% of Earth's land surface and some 12% of the land in the United States (Figure 15.8). A pale soil color near the surface is diagnostic (Figure 15.12a, p. 470).

Not surprisingly for desert soils, the water balance in Aridisol regions has marked periods of soil moisture deficit and generally inadequate soil moisture for plant growth. High potential evapotranspiration (moisture demand) and low precipitation produce very shallow soil horizons. Usually there is no period greater than 3 months when the soils have adequate moisture. Aridisols also lack organic matter of any consequence. Although the soils are not leached often because of the low precipitation, they are leached easily when exposed to excessive water, for they lack a significant colloidal structure.

Salinization is a soil process that occurs in Aridisols. Salinization results from high potential evapotranspiration

Table 15.1 Soil Taxonomy Soil Orders

Order	Derivation of Term	Marbut 1935 (Canadian System)	General Location and Climate	Description
Oxisols	Fr. *oxide*, "oxide" Gr. *oxide*, "acid or sharp"	Latosols lateritic soils	Tropical soils, hot, humid areas	Maximum weathering of Fe and Al and eluviation, continuous plinthite layer
Aridisols	L. *aridos*, "dry"	Reddish desert, gray desert, sierozems	Desert soils, hot, dry areas	Limited alteration of parent material, low climate activity, light color, low humus content subsurface illuviation of carbonates
Mollisols	L. *mollis*, "soft"	Chestnut, chernozem (Chernozemic)	Grassland soils; subhumid, semiarid lands	Noticeably dark with organic material, humus rich, base saturation high, friable surface with well-structured horizons
Alfisols	Invented syllable	Gray-brown podzolic, degraded chernozem (Luvisol)	Moderately weathered forest soils, humid temperate forests	B horizon high in clays, moderate to high degree of base saturation, illuviated clay accum., no pronounced color change with depth
Ultisols	L. *ultimus*, "fast"	Red-yellow podzolic, reddish yellow lateritic	Highly weathered forest soils, subtropical forests	Similar to Alfisols, B horizon high in clays, generally low amount of base saturation, strong weathering in subsurface horizons, redder than Alfisols
Spodosols	Gr. *spodos* or L. *spodus*, "wood ash"	Podzols, brown podzolic (Podzol)	Northern conifer forest soils, cool humid forests	Illuvial B horizon of Fe/Al clays, humus accum.; without structure, partially cemented; highly leached, strongly acid; coarse texture of low bases
Entisols	Invented syllable from *recent*	Azonal soils, tundra	Recent soils, profile undeveloped, all climates	Limited development; inherited properties from parent material; pale color, low humus, few specific properties; hard and massive when dry
Inceptisols	L. *inceptum*, "beginning"	Ando, subarctic brown forest lithosols, some humic gleys (Brunisol, Cryosol with permafrost, Gleysol wet)	Weakly developed soils, humid regions	Intermediate development; embryonic soils, but few diagnostic features; further weathering possible in altered or changed subsurface horizons
Vertisols	L. *verto*, "to turn"	Grumusols (1949) tropical black clays	Expandable clay soils; subtropics, Tropics; sufficient dry period	Forms large cracks on drying, self-mixing action, contains >30% in swelling clays, light color, low humus content
Histosols	Gr. *histos*, "tissue"	Peat, muck, bog (Organic)	Organic soils, wet places	Peat or bog, >20% organic matter, much with clay >40 cm thick, surface, organic layers, no diagnostic horizons
Andisols	L. *ando*, "volcanic ash"	—	Areas affected by frequent volcanic activity (formerly within Inceptisols and Entisols)	Volcanic parent materials, particularly ash and volcanic glass; weathering and mineral transformation important; high CEC and organic content, generally fertile
Gelisols		Formerly Inceptisols and Entisols (Cryosols, some Brunisols)	High latitudes in Northern Hemisphere, southern limits near tree line	Permafrost within 100 cm of the soil surface. Evidence of cryoturbation (frost churning) and/or an active layer. Patterned-ground.

(a)

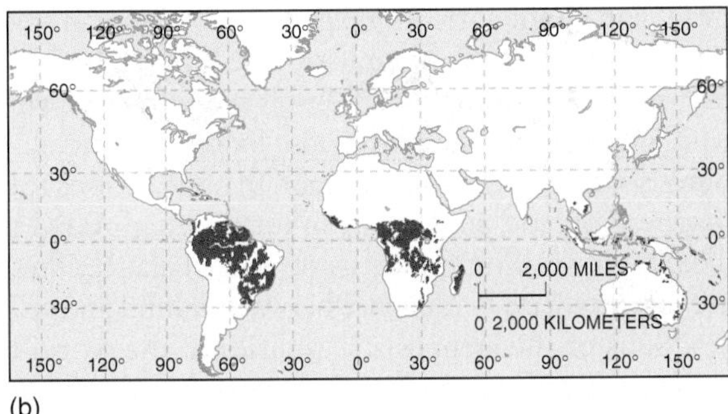

(b)

FIGURE 15.9 Oxisols.
(a) Deeply weathered Oxisol profile in central Puerto Rico. (b) General map of worldwide distribution. [(a) Photo from the Marbut Collection, Soil Science Society of America.]

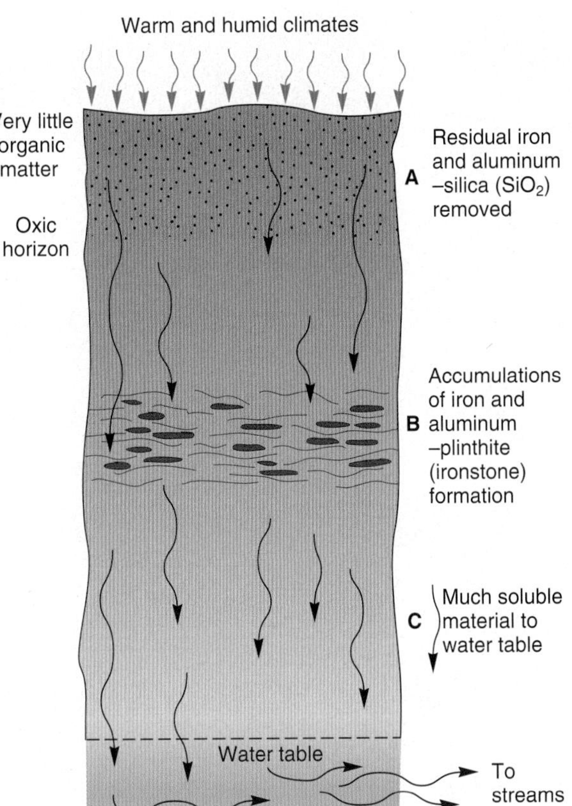

FIGURE 15.10 Laterization.
Laterization process characteristic of Oxisols in tropical and subtropical climatic regimes.

FIGURE 15.11 Oxisols are used for building materials.
Here plinthite blocks are quarried in India. [Photo by Henry D. Foth.]

(a)

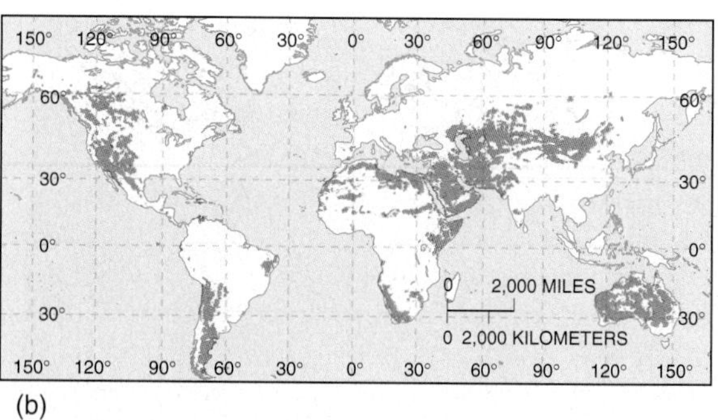

(b)

FIGURE 15.12 Aridisols.
(a) Soil profile from central Arizona. (b) General map showing worldwide distribution of these desert soils. [Photo from the Marbut Collection, Soil Science Society of America.]

rates in the deserts and semiarid regions of the world. Salts dissolved in soil water are brought to surface horizons and deposited as surface water evaporates. These deposits will kill plants when the salts accumulate near the root zone.

Obviously, salinization complicates farming in Aridisols. The introduction of irrigation water may either waterlog poorly drained soils or lead to salinization. Nonetheless, vegetation does grow where soils are well drained and kept low in salt content. If large capital investments are made in water, drainage, and fertilizers, Aridisols possess much agricultural potential (Figure 15.13).

In the Nile and Indus River valleys, for example, Aridisols are intensively farmed with a careful balance of these environmental factors, although thousands of acres of once-productive land, not so carefully treated, now sit idle and salt-encrusted. In California, the Kesterson Wildlife Refuge was reduced to a toxic waste dump in the early 1980s by contaminated agricultural drainage. Focus Study 15.1 (pp. 476–477) elaborates on the Kesterson tragedy and the farming of semiarid lands.

FIGURE 15.13 Agriculture in arid lands.
Agriculture in arid lands of the Coachella Valley of southeastern California. Providing drainage for excessive water application in such areas often is necessary to prevent salinization of the rooting zone. The irrigation water is the subject of much squabbling between agricultural interests and the cities of southern California. [Photo by author.]

Mollisols. **Mollisols** (grassland soils) are some of Earth's most significant agricultural soils. The dominant diagnostic horizon is a dark, organic surface layer some 25 cm (10 in.) thick (Figure 15.14). Mollisols are soft, even when dry, with granular or crumbly peds, loosely arranged. These humus-rich organic soils are high in basic cations (calcium, magnesium, and potassium) and have a high CEC (therefore, high fertility). In terms of water balance, these soils are intermediate in moisture between humid and arid soils.

Soils of the steppes and prairies of the world belong to this soil group: the North American Great Plains, the Palouse of Washington state (Figure 15.15), the Pampas of Argentina, and the region from Manchuria in China through to Europe. Agriculture ranges from large-scale commercial grain farming to grazing along the drier portions of the soil order. With fertilization or soil-building practices, high crop yields are common. The "fertile triangle" of Ukraine, Russia, and western portions of the former Soviet Union is of this soil type.

(a)

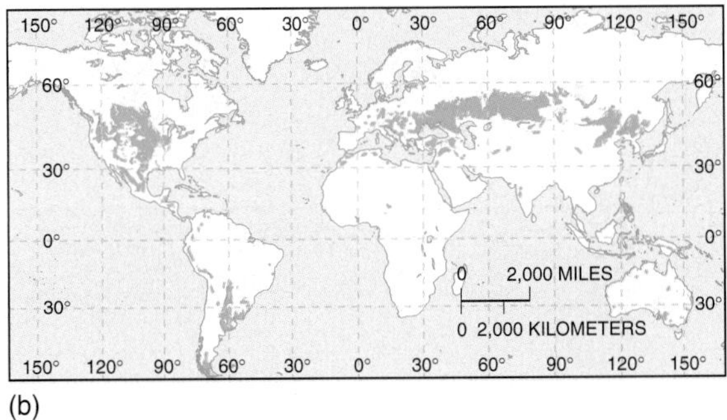

(b)

FIGURE 15.14 Mollisols.
(a) Profile from central Iowa and (b) general map of worldwide distribution. [(a) Photo from Marbut Collection, Soil Science Society of America, Inc.]

In North America, the Great Plains straddle the 98th meridian, which is coincident with the 51-cm (20-in.) isohyet of annual precipitation—wetter to the east and drier to the west. The Mollisols here mark the historic division between the short- and tall-grass prairies (Figure 15.16).

A soil process characteristic of some Mollisols, and adjoining areas of Aridisols, is *calcification*. Calcification is an illuviated accumulation of calcium carbonate or magnesium carbonate in the B and C horizons (Figure 15.17). When cemented or hardened, these deposits are called *caliche*, or *kunkur*. They occur in widespread soil formations in central and western Australia, the Kalahari region of interior southern Africa, and the High Plains of the west-central United States, among other places.

Alfisols. **Alfisols** (moderately weathered forest soils) are the most widespread of the soil orders, extending from near the equator to high latitudes. Representative Alfisol areas include Boromo and Burkina (interior western Africa); Fort Nelson, British Columbia; the states near the Great Lakes, and the valleys of central California. Most Alfisols have a pale, grayish brown-to-reddish epipedon and are considered moist versions of the Mollisol soil group. Moderate eluviation is present, as well as a subsurface horizon of illuviated white clays because of increased precipitation (Figure 15.18).

Alfisols have moderate-to-high reserves of basic cations and are fertile. However, productivity depends on moisture and temperature. Alfisols usually are supplemented by a moderate application of lime and fertilizers in areas of active agriculture. Some of the best farmland in the United States stretches from Illinois, Wisconsin, and Minnesota through Indiana, Michigan, and Ohio to Pennsylvania and New York. This land produces grains, hay, and supports a dairy industry. These naturally productive soils are farmed intensively with subtropical fruits, nuts, and special crops that can grow only in a few locales worldwide. (For example, California grows grapes, citrus, artichokes, almonds, and figs, among others; see Figure 15.18c.)

Ultisols. Farther south in the United States are the **Ultisols**, highly weathered forest soils. An Alfisol might degenerate into an Ultisol, given time and exposure to increased weathering under moist conditions. These soils

FIGURE 15.15 Fertile soils of the Palouse.
Mollisols of eastern Washington state form the basis of a productive agricultural landscape. Wheat, grown here, when added to maize (corn) and rice, provide more than 50% of the world's grain supply. These rich soils will become increasingly valuable in the future. [Photo by Bobbé Christopherson.]

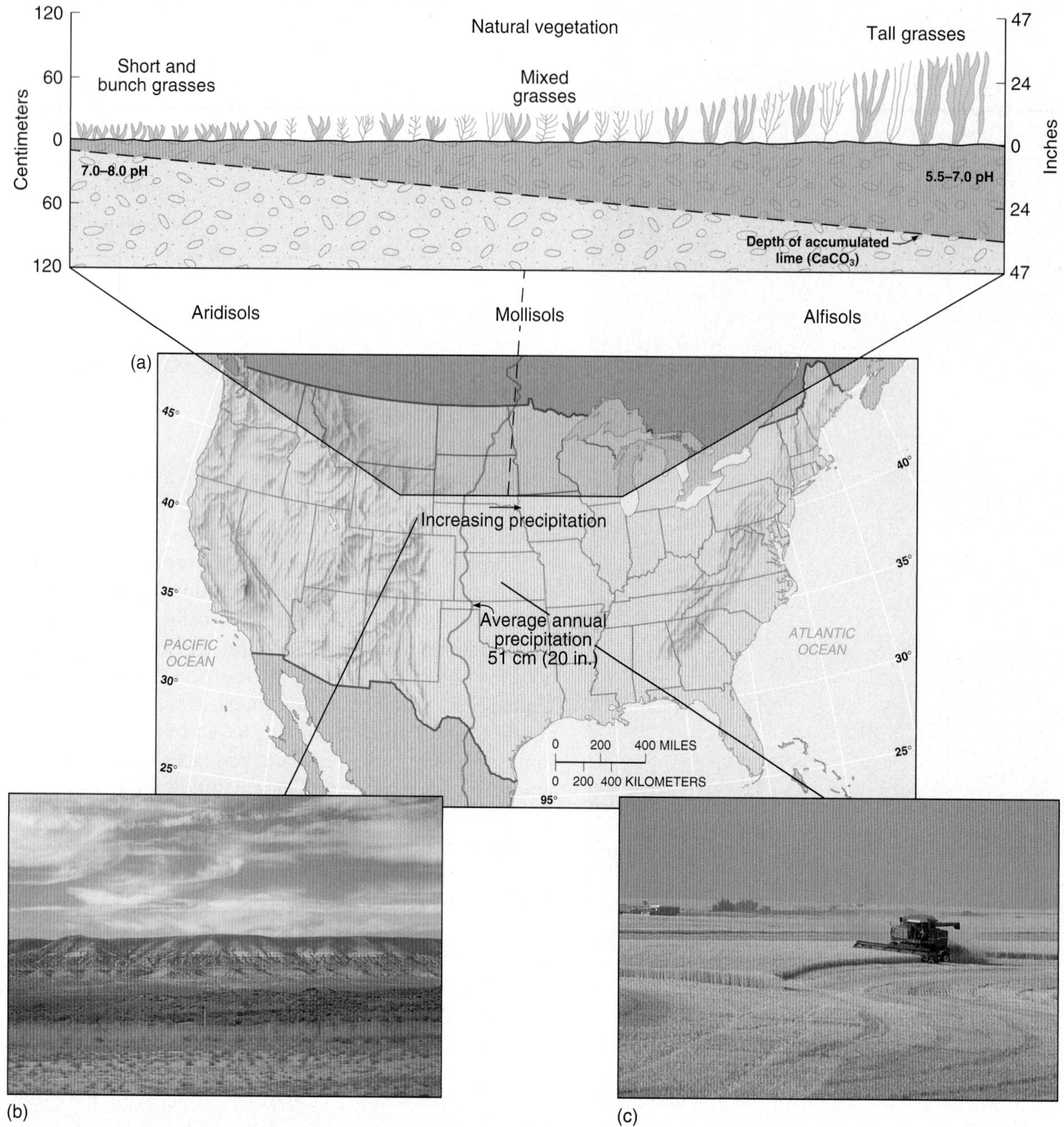

FIGURE 15.16 Soils of the Midwest.
(a) Aridisols (to the west), Mollisols (central), and Alfisols (to the east)—a soil continuum in the north-central United States and southern Canadian prairies. Graduated changes that occur in soil pH and the depth of accumulated lime are shown. (b) Bunch grasses and shallow soils of Wyoming. (c) Wheat harvest from rich Mollisols near Hardtner, Kansas. [(a) Illustration adapted from N. C. Brady, *The Nature and Properties of Soils*, 10th ed., © 1990 by Macmillan Publishing Company, adapted by permission; (b) photo by author; (c) photo by Garry D. McMichael/Photo Researchers, Inc.]

tend to be reddish because of residual iron and aluminum oxides in the A horizon (Figure 15.19).

The increased precipitation in Ultisol regions means greater mineral alteration, more eluvial leaching, and therefore a lower level of basic cations, leading to infertility. Certain agricultural practices and the effect of soil-damaging crops such as cotton and tobacco reduce fertility, depleting nitrogen and exposing soil to erosion. However, these soils respond well if subjected to good management—for example, crop rotation that restores nitrogen and cultivation practices that prevent sheetwash and soil erosion. Figure 15.19c shows peanut plantings that assist in nitrogen restoration. Much needs to be done to achieve sustainable management of these soils.

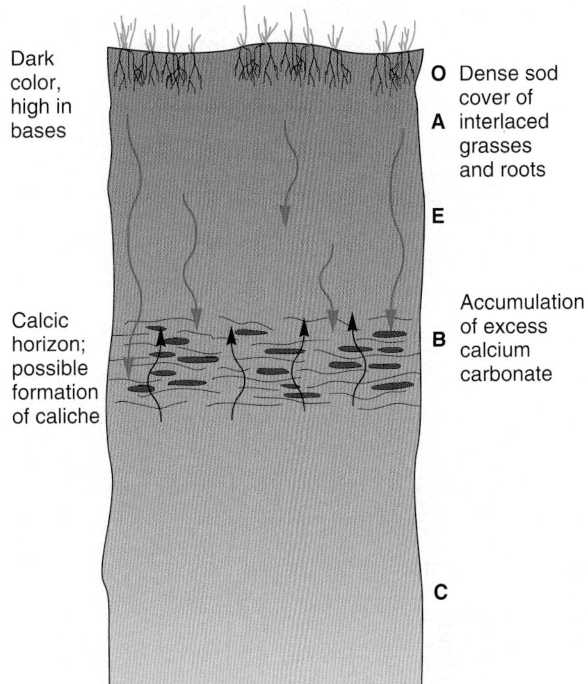

FIGURE 15.17 Calcification in soil.
Calcification process in Aridisol/Mollisol soils in climatic regimes that have a potential evapotranspiration equal to or greater than precipitation.

Spodosols. The **Spodosols** (northern coniferous forest soils) occur generally to the north and east of the Alfisols. They are in cold and forested moist regimes (*humid continental mild summer* climates) in northern North America and Eurasia, Denmark, the Netherlands, and southern England. Because there are no comparable climates in the Southern Hemisphere, this soil type is not identified there. Spodosols form from sandy parent materials, shaded under evergreen forests of spruce, fir, and pine. Spodosols with more moderate properties form under mixed or deciduous forests (Figure 15.20).

Spodosols lack humus and clay in the A horizons. An eluviated white horizon, sandy and leached of clays and irons, lies in the A horizon instead and overlies a horizon of illuviated organic matter and iron and aluminum oxides. The surface horizon receives organic litter from base-poor, acid-rich trees, which contribute to acid accumulations in the soil. The low pH (acidic) soil solution effectively

(a)

(b)

(c)

FIGURE 15.18 Alfisols.
(a) Soil profile from central California. (b) General map showing worldwide distribution of these moderately weathered forest soils. (c) Cultivated Alfisols farmland near Lompoc, California. [(a) Photo from the Marbut Collection, Soil Science Society of America, Inc. (c) Photo by Bobbé Christopherson.]

(a)

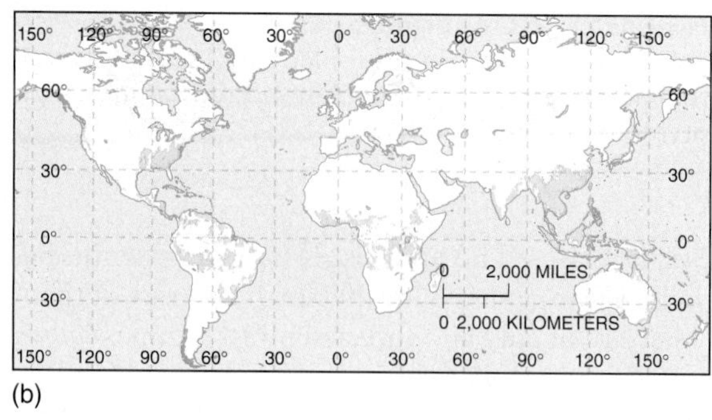

(b)

(c)

FIGURE 15.19 Ultisols.
(a) Profile from the upper coastal plain of central North Carolina and (b) general map of worldwide distribution. (c) A type of Ultisol in central Georgia planted with rows of peanuts, near Plains in west-central Georgia, bears its characteristic reddish color. [(a) Photo from Marbut Collection, Soil Science Society of America, Inc. (c) Photo by Bobbé Christopherson.]

removes clays, iron, and aluminum, passing to the upper diagnostic horizon. An ashen gray color is common in these subarctic forest soils and is characteristic of a formation process called *podzolization* (Figure 15.20c). In the Canadian system, Spodosols fall within the Podzolic Great Group, as seen in the temperate rain forests of Vancouver Island, British Columbia (Figure 15.20d; and Appendix B, Table 3).

When agriculture is attempted, the low base-cation content of Spodosols requires the addition of nitrogen, phosphate, and potash (potassium carbonate), and perhaps crop rotation as well. A soil amendment such as limestone can significantly increase crop production by raising the pH of these acidic soils. For example, the yields of several crops (corn, oats, wheat, and hay) grown in specific Spodosols in New York State were increased up to a third with the application of limestone in a 6-year rotation program.

Entisols. The **Entisols** (recent, undeveloped soils) lack vertical development of horizons. The presence of Entisols is not climate dependent, for they occur in many climates worldwide. Entisols are true soils that have not had sufficient time to generate the usual horizons.

Entisols generally are poor agricultural soils, although those formed from river silt deposits are quite fertile. The conditions that have inhibited complete development also have prevented adequate fertility—too much or too little water, poor structure, and insufficient accumulation of weathered nutrients. Active slopes, alluvium-filled floodplains, poorly drained tundra, tidal mud flats, dune sands and erg (sandy) deserts, and plains of glacial outwash all are characteristic regions that have these soils. Figure 15.21 shows an Entisol in a desert climate where shales formed the parent material.

(a)

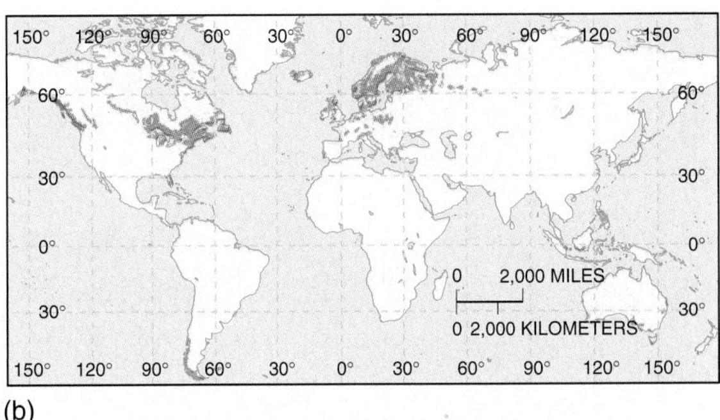

(b)

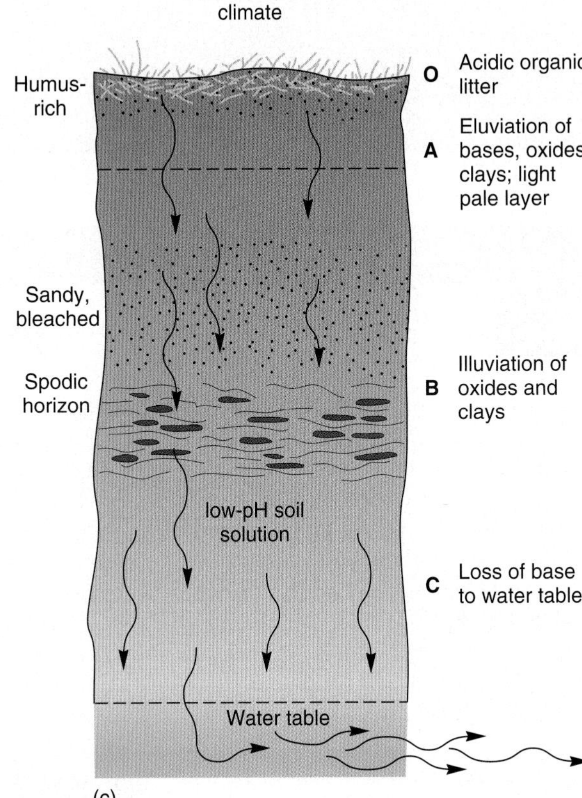

(c)

(d)

FIGURE 15.20 Spodosols. (a) Profile from northern New York and (b) general map of worldwide distribution; (c) podzolization process, typical in cool and moist climatic regimes. (d) Characteristic temperate forest and Spodosols in the cool moist climate of central Vancouver Island. [(a) Photos from (a) Marbut Collection, Soil Science Society of America, Inc. (d) Photo by Bobbé Christopherson.]

FIGURE 15.21 Entisols.
A characteristic Entisol forming from a shale parent material in the desert near Zabriskie Point, Death Valley. A young, poorly developed soil, lacking mature horizons. [Photo by author.]

Focus Study 15.1

Selenium Concentration in Western Soils

Irrigated agriculture has increased greatly since 1800, when only 8 million hectares (about 20 million acres) were irrigated worldwide. Today, approximately 255 million hectares (about 630 million acres) are irrigated, and this figure is on the increase (USDA, FAO, 1996). Representing about 16% of Earth's agricultural land, irrigated land accounts for nearly 36% of the harvest.

Two related problems common in irrigated lands are salinization and waterlogging, especially in arid lands that are poorly drained. In many areas, production has decreased and even ended because of salt buildup in the soils. Examples include areas along the Tigris and Euphrates Rivers, the Indus River valley, sections of South America and Africa, and the western United States.

Irrigation in the West

About 95% of the irrigated acreage in the United States lies west of the 98th meridian. This region is increasingly troubled with salinization and waterlogging problems. In addition, at least nine sites in the West, particularly California's western San Joaquin Valley, are experiencing related contamination of a more serious nature—increasing selenium concentrations. Toxic effects of selenium were reported during the 1980s in some domestic animals grazing on grasses grown in selenium-rich soils in the Great Plains. In California, as parent materials weathered, selenium-rich alluvium washed into the semiarid valley, forming the soils that needed only irrigation water to become productive.

Drainage of agricultural wastewater poses a particular problem in semiarid and arid lands, where river discharge is inadequate to dilute and remove field runoff. One solution to prevent salt accumulations and waterlogging is to place field drains beneath the soil to collect gravitational water from fields that have been purposely overwatered to keep salts away from the effective rooting depth of the crops. But agricultural drain water must go somewhere, and for the San Joaquin Valley of central California this problem triggered a 15-year controversy (Figure 1).

Death of Kesterson

Central California's potential outlets for agricultural runoff are to the ocean, or San Francisco Bay, or the Central Valley. But all these suggested destinations failed to pass environmental impact assessments under Environmental Policy Act requirements. Nonetheless, about 80 miles of a drain were finished by the late 1970s, even though no formal plan or adequate funding was in place—a drain was

FIGURE 1 Soil drainage canal collects contaminated water from field drains. [Photo by author.]

built with no outlet. In the absence of any plan, large-scale irrigation continued, supplying the field tile outlets with salty, selenium-laden runoff that made its way to the Kesterson National Wildlife Refuge in the northern portion of the San Joaquin Valley east of San Francisco. The unfinished drain abruptly stopped at the boundary to the refuge.

In only 3 years the selenium-tainted drainage destroyed the wildlife refuge, which was officially declared a contaminated toxic waste site (Figure 2). Aquatic life forms (e.g., marsh plants, plankton, and insects) had taken in the selenium, which then made its way up the food chain and into the diets of other life forms in the refuge. According to U.S. Fish and

(continued)

Focus Study 15.1 *(continued)*

Wildlife Service scientists, the toxicity moved through the food chain and genetically damaged and killed wildlife, including all varieties of birds that nested at Kesterson; approximately 90 % of the exposed birds perished or were injured. Because this wildlife refuge was a major migration flyway and stopover point for birds from throughout the Western Hemisphere, the destruction of this refuge also violated several multinational wildlife protection treaties.

Such damage to wildlife presents a real warning to human populations—remember where we are in the food chain. The field drains were sealed and removed in 1986, following a court order that forced the federal government to uphold existing laws. Irrigation water then immediately began backing up in the corporate farmlands, producing both waterlogging and selenium contamination.

Since 1985, more than 0.6 million hectares (1.5 million acres) of irrigated Aridisols and Alfisols have gone out of production in California, marking the end of several decades of irrigated farming in climatically marginal lands. In a 2002 settlement, the federal government paid out over $100 million to buy damaged land and take it permanently out of production. The drainage problem in the Westlands Irrigation District is still unresolved. Severe cutbacks in irrigated acreage no doubt will continue, underscoring the need to preserve prime farmlands in wetter regions and to understand essential soil processes.

Frustrated agricultural interests have asked the federal government to finish the drain, either to San Francisco Bay or to the ocean. However, neither option appears capable of passing an environmental impact analysis. Another strategy is to allow irrigated lands to start pumping again into the former wildlife refuge because it is now a declared toxic dump site anyway! There are nine such threatened sites in the West; Kesterson was simply the first to fail.

FIGURE 2 Salt-encrusted soil and plants at the contaminated Kesterson National Wildlife Refuge. [Photo by Gary R. Zahm.]

Inceptisols. **Inceptisols** (weakly developed soils) are inherently infertile. They are weakly developed young soils, although more developed than the Entisols. Inceptisols include a wide variety of different soils, all having in common a lack of maturity with evidence of weathering just beginning. Inceptisols are associated with moist soil regimes and are regarded as eluvial because they demonstrate a loss of soil constituents throughout their profile but retain some weatherable minerals. This soil group has no distinct illuvial horizons.

Inceptisols include glacially derived till and outwash materials from New York down through the Appalachians, and alluvium on the Mekong and Ganges floodplains.

Gelisols. **Gelisols** (cold and frozen soils) and their three suborders are new to the Soil Taxonomy (1998) and represent inclusion of high latitude (Canada, Alaska, Russia, and Antarctica) and high elevation (mountains) soil conditions (Figure 15.22). Temperatures in these regions are at or below 0°C (32°F), making soil development a slow process and disturbances of the soil long lasting. Gelisols can develop organic diagnostic horizons even though cold temperatures slow decomposition of materials. Tundra vegetation, such as lichens, mosses, sedges, and other plants adapted to the harsh cold, are characteristic of Gelisols.

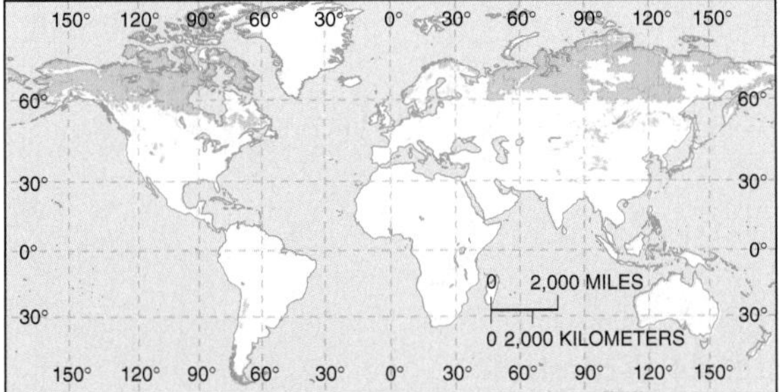

FIGURE 15.22 Gelisols.
General map of worldwide distribution.

Gelisols are subject to *cryoturbation* (frost churning and mixing in soil) in the freeze-thaw cycle in the active layer of periglacial regions (see Chapter 14). This process disrupts soil horizons, pulling organic material to lower layers and lifting rocky C-horizon material to the surface. Patterned-ground phenomena are possible under such conditions.

As we saw in Chapter 14, periglacial processes occupy about 20% of Earth's land surface, including permafrost (frozen ground) that underlies some 13% of these lands. Previously these soils were included in Inceptisols, Entisols, and Histosols soil orders. In the Canadian System of Soil Classification these are included in the *Crysolic* soil order (see Appendix B).

Andisols. **Andisols** (volcanic parent materials) occur in areas of volcanic activity. These soils formerly were classified as Inceptisols and Entisols, but in 1990 they were placed in this new order. Andisols are derived from volcanic ash and volcanic glass. Previous soil horizons frequently are found buried by ejecta from repeated volcanic eruptions. Andisols are unique in their mineral content and in their recharge by eruptions.

Weathering and mineral transformations are important in this soil order. Volcanic glass weathers readily into

FIGURE 15.23 Andisols in agricultural production. Fertile Andisols planted with sugar cane on Kaua'i, Hawai'i. Hawai'i is an important producer of sugar cane in the United States. [Photo by Wolfgang Kaehler Photography.]

a clay colloid and oxides of aluminum and iron. Andisols feature a high CEC and high water-holding ability and develop moderate fertility. The fertile fields of Hawai'i produce sugar cane, pineapple, and other important cash crops in Andisols (Figure 15.23). Andisol distribution is small in areal extent; however, such soils are locally important around the volcanic "ring of fire" around the Pacific Rim.

Vertisols. **Vertisols** (expandable clay soils) are heavy clay soils. They contain more than 30% swelling clays (which swell significantly when they absorb water). They are located in regions experiencing highly variable soil moisture balances through the seasons. These soils occur in areas of subhumid-to-semiarid moisture and moderate-to-high temperature. Vertisols frequently form under savanna and grassland vegetation in tropical and subtropical climates and are sometimes associated with a distinct dry season following a wet season. Although widespread, individual Vertisol units are limited in extent.

(a)

150° 120° 90° 60° 30° 0° 30° 60° 90° 120° 150°
60° 30° 0° 30°
0 2,000 MILES
0 2,000 KILOMETERS

(b)

FIGURE 15.24 Vertisols.
(a) Profile in the Lajas Valley of Puerto Rico and (b) general map of worldwide distribution. [(a) Photo from Marbut Collection, Soil Science Society of America, Inc.]

FIGURE 15.25 Vertisols in production.
Vertisols in the Texas coastal plain, northeast of Palacios near the Tres Palacios River, planted with a commercial sorghum crop. Note the dark soil color indicative of Vertisols, wet and shiny from the rains of Tropical Storm Allison, June 2001. [Photo by Bobbé Christopherson.]

Vertisol clays are black when wet (not because of organics, but because of specific mineral content) and range to brown and dark gray. These deep clays swell when moistened and shrink when dried. In the process, vertical cracks form, which widen and deepen as the soil dries, producing cracks 2–3 cm (0.8–1.2 in.) wide and up to 40 cm (16 in.) deep. Loose material falls into these cracks, only to disappear when the soil again expands and the cracks close. After many such cycles, soil contents tend to invert or mix vertically, bringing lower horizons to the surface (Figure 15.24).

Despite the fact that clay soils are plastic and heavy when wet, with little available soil moisture for plants, Vertisols are high in bases and nutrients and thus are some of the better farming soils wherever they occur. For example, they occur in a narrow zone along the coastal plain of Texas and in a section along the Deccan region of India (Figure 15.25). Vertisols often are planted with grain sorghums, corn, and cotton.

Histosols. **Histosols** (organic soils) are formed from accumulations of thick organic matter. In the midlatitudes, when conditions are right, beds of former lakes may turn into Histosols where water gradually is replaced by organic material, forming a bog and layers of peat. (Lake succession and bog/marsh formation are discussed in Chapter 16.) Histosols also form in small, poorly drained depressions where conditions are ideal for the formation of significant deposits of *sphagnum peat* (Figure 15.26a). This material can be cut, baled, and sold as a soil amendment. Dried peat has served for centuries as a low-grade fuel. The area southwest of Hudson Bay in Canada is typical of Histosol formation (Figure 15.26b).

(a)

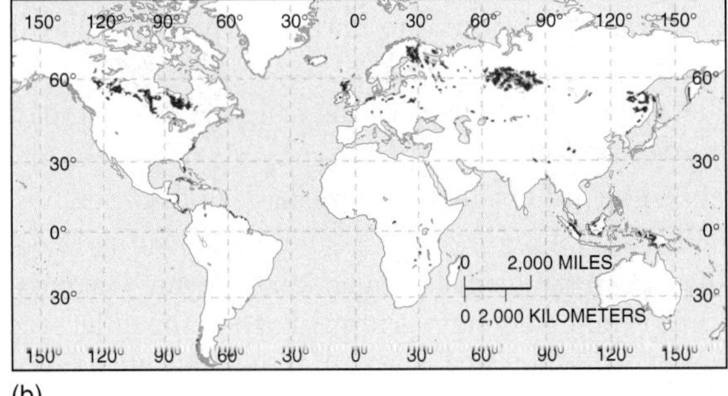

(b)

FIGURE 15.26 Histosols.
(a) A Histosol bog in coastal Maine, near Popham Beach State Park. (b) General map of worldwide distribution. [Photo by Bobbé Christopherson.]

Summary and Review—The Geography of Soils

Define soil and soil science and *describe* a pedon, polypedon, and typical soil profile.

Soil is the portion of the land surface in which plants can grow. It is a dynamic natural body composed of fine materials and contains both mineral and organic matter. **Soil science** is the interdisciplinary study of soils involving physics, chemistry, biology, mineralogy, hydrology, taxonomy, climatology, and cartography. *Pedology* deals with the origin, classification, distribution, and description of soil. *Edaphology* specifically focuses on the study of soil as a medium for sustaining the growth of plants.

The basic sampling unit used in soil surveys is the **pedon**. The **polypedon** is the soil unit used to prepare local soil maps and may contain many pedons. Each discernible layer in an exposed pedon is a **soil horizon**. The horizons are designated: O (contains **humus**, a complex mixture of decomposed and synthesized organic materials); A (rich in humus and clay, darker); E (zone of **eluviation**, the removal of fine particles and minerals by water); B (zone of **illuviation**, the deposition of clays ands minerals translocated from elsewhere); C (regolith, weathered bedrock); and R (bedrock). Soil horizons A, E, and B experience the most active soil processes and together are designated the **solum**.

soil (p. 457)
soil science (p. 457)
pedon (p. 458)
polypedon (p. 458)
soil horizon (p. 459)
humus (p. 459)
eluviation (p. 459)
illuviation (p. 459)
solum (p. 459)

1. Soils provide the foundation for animal and plant life and therefore are critical to Earth's ecosystems. Why is this true?
2. What are the differences among soil science, pedology, and edaphology?
3. Define polypedon and pedon, the basic units of soil.
4. Characterize the principal aspects of each soil horizon. Where does the main accumulation of organic material occur? Where does humus form? Explain the difference between the eluviated layer and the illuviated layer. Which horizons constitute the solum?

Describe soil properties of color, texture, structure, consistence, porosity, and soil moisture.

We use several physical properties to classify soils. *Soil color* suggests composition and chemical makeup. *Soil texture* refers to the size of individual mineral particles and the proportion of different sizes. For example, **loam** is a balanced mixture of sand, silt, and clay. *Soil structure* refers to the arrangement of soil *peds*, which are the smallest natural cluster of particles in a soil. The cohesion of soil particles to each other is called *soil consistence*. *Soil porosity* refers to the size, alignment, shape, and location of spaces in the soil. *Soil moisture* refers to water in the soil pores and its availability to plants.

loam (p. 460)

5. How can soil color be an indication of soil qualities? Give a couple of examples.
6. Define a soil separate. What are the various sizes of particles in soil? What is loam? Why is loam regarded so highly by agriculturists?
7. What is a quick, hands-on method for determining soil consistence?
8. Summarize the role of soil-moisture in mature soils.

Explain basic soil chemistry, including cation-exchange capacity, and *relate* these concepts to soil fertility.

Particles of clay and organic material form negatively charged **soil colloids** that attract and retain positively charged mineral ions in the soil. **Adsorption** is a process whereby cations attach to soil colloids. The capacity to exchange ions between colloids and roots is an ability called the **cation-exchange capacity (CEC)**. CEC is a measure of **soil fertility**, the ability of soil to sustain plants. Fertile soil contains organic substances and clay minerals that absorb water and retain certain elements needed by plants.

soil colloids (p. 462)
adsorption (p. 462)
cation-exchange capacity (CEC) (p. 462)
soil fertility (p. 462)

9. What are soil colloids? How are they related to cations and anions in the soil? Explain cation-exchange capacity.
10. What is meant by the concept of soil fertility?

Evaluate principal soil formation factors, including the human element.

Environmental factors that affect soil formation include parent materials, climate, vegetation, topography, and time. Human influence is having great impact on Earth's prime soils. Mismanagement, destruction, and conversion to other uses threatens essential soil fertility for agriculture. Much soil loss is preventable through the application of known technologies, improved agricultural practices, and government policies.

11. Briefly describe the contribution of the following factors and their effect on soil formation: parent material, climate, vegetation, landforms, time, and humans.
12. Explain some of the details that support the concern over loss of our most fertile soils. What cost estimates have been placed on soil erosion?

Describe the 12 soil orders of the Soil Taxonomy classification system and *explain* their general occurrence.

The **Soil Taxonomy** classification system is used in the United States and is built around an analysis of various diagnostic horizons and 12 soil orders, as actually seen in the field. The system divides soils into six hierarchical categories: series, families, subgroups, great groups, suborders, and orders.

Specific soil-forming processes keyed to climatic regions (not a basis for classification) are: *pedogenic regimes*: **laterization** (leaching in warm and humid climates), **salinization** (collection of salt residues in surface horizons in hot, dry climates), **calcification** (accumulation of carbonates in the B and C horizons in drier continental climates), **podzolization** (soil acidification in forest soils in cool climates), and **gleization** (humus and clay accumulation in cold, wet climates with poor drainage).

The Soil Taxonomy system uses two diagnostic horizons to identify soil: the **epipedon**, or the surface soil, and the **diagnostic subsurface horizon**, or the soil below the surface at various depths. (For an overview and definition of the 12 soil orders, please refer to Table 15.1, "Soil Taxonomy Soil Orders.") The 12 soil orders are: **Oxisols** (tropical soils, including a hardpan **plinthite**, a building material), **Aridisols** (desert soils), **Mollisols** (grassland soils), **Alfisols** (moderately weathered, temperate forest soils), **Ultisols** (highly weathered, subtropical forest soils), **Spodosols** (northern coniferous forest soils), **Entisols** (recent, undeveloped soils), **Inceptisols** (weakly developed, humid region soils), **Gelisols** (cold soils underlain by permafrost), **Andisols** (soils formed from volcanic materials), **Vertisols** (expandable clay soils), and **Histosols** (organic soils).

Soil Taxonomy (p. 463)
laterization (p. 465)
salinization (p. 465)
calcification (p. 465)
podzolization (p. 465)
gleization (p. 465)
epipedon (p. 465)
diagnostic subsurface horizon (p. 465)
Oxisols (p. 465)
plinthite (p. 466)
Aridisols (p. 467)
Mollisols (p. 470)
Alfisols (p. 471)
Ultisols (p. 471)
Spodosols (p. 473)
Entisols (p. 474)
Inceptisols (p. 477)
Gelisols (p. 477)
Andisols (p. 477)
Vertisols (p. 478)
Histosols (p. 479)

13. Summarize what led soil scientists to develop the new Soil Taxonomy classification system?
14. What is the basis of the Soil Taxonomy system? How many orders, suborders, great groups, subgroups, families, and soil series are there?
15. Define an epipedon and a subsurface diagnostic horizon. Give a simple example of each.
16. Locate each soil order on the world map and on the U.S. map as you give a general description of it.
17. How was slash-and-burn shifting cultivation, as practiced in the past, a form of crop and soil rotation and conservation of soil properties?
18. Describe the salinization process in arid and semiarid soils. What associated soil horizons develop?
19. Which of the soil orders are associated with Earth's most productive agricultural areas?
20. What is the significance to plants of the 51 cm (20 in.) isohyet in the Midwest relative to soils, pH, and lime content?
21. Describe the podzolization process associated with northern coniferous forest soils. What characteristics are associated with the surface horizons? What strategies might enhance these soils?
22. What former Inceptisols now form a new soil order named in 1998? Describe these soils as to location, nature, and formation processes. Why do you think they were separated into their own order?
23. Why has a selenium contamination problem arisen in western U.S. soils? Explain the impact of agricultural practices, and tell why you think this is or is not a serious problem.

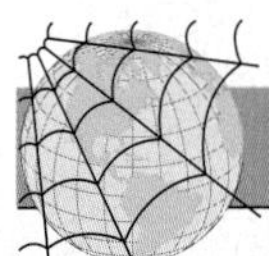

NetWork

The *Elemental Geosystems Home Page* provides on-line resources for this chapter on the World Wide Web. Once on the Home Page, click on this textbook, scroll the Table of Contents menu, select this chapter, and click "Begin." You will find self-tests that are graded, review exercises, specific updates for items in the chapter, and in "Destinations" many links to interesting related pathways on the Internet. *Elemental Geosystems* is found at **http://www.prenhall.com/christopherson**.

Critical Thinking

A. Select a small soil sample from your campus or near where you live. Using the sections in this chapter on soil characteristics, properties, and formation, describe this sample as completely as possible, within these constraints. Using the general soil map and any other sources available, are you able to roughly place this sample in one of the soil orders?

B. Using the local phone directory or Internet, see if you can locate an agency that provides information about local soils and advice for soil management. Is there a place where you can have soil tested?

C. Refer to "Critical Thinking" for Chapter 15 on the *Geosystems Home Page* item 1. Please complete the analysis of the three photographs: deciduous forests of eastern North America, evergreen forests, and arctic tundra environments. What soil order would you expect in these regions? Why?

An ancient temperate rain forest in the Pacific Northwest in the Gifford Pinchot National Forest features huge old-growth Douglas fir, redwoods, cedars, and a mix of deciduous trees, ferns, and mosses. Only a small percentage of these old-growth forests remain. [Photo by Bobbé Christopherson.]

16

Ecosystems and Biomes

Key Learning Concepts

By knowing and understanding the key learning concepts in this chapter, you should be able to:

- *Define* ecology, biogeography, and the ecosystem, community, habitat, and niche concepts.
- *Explain* photosynthesis and respiration and *derive* net photosynthesis and the world pattern of net primary productivity.
- *List* abiotic and biotic ecosystem components and *relate* those components to ecosystem operations and trophic relationships.
- *Define* succession and *outline* the stages of general ecological succession in both terrestrial and aquatic ecosystems.
- *Define* the concepts of terrestrial ecosystem, biome, ecotone, and formation classes.
- *Describe* 10 major terrestrial biomes and *locate* them on a world map.
- *Relate* human impacts, real and potential, to several of the biomes.

Diversity is an impressive feature of the living Earth. The diversity of organisms is a response to the interaction of the atmosphere, hydrosphere, and lithosphere, which produce a variety of conditions within which the biosphere exists. The diversity of life also results from the intricate interplay of living organisms themselves. We, as part of this vast natural complex, seek to find our place and understand the feelings nature stimulates within us.

The physical beauty of nature is certainly among its most powerful appeals to the human animal. The complexity of the aesthetic response is suggested by its wide-ranging expression from the contours of a mountain landscape to the ambient colors of a setting Sun to the fleeting vitality of a breaching whale. Each exerts a powerful aesthetic impact on most people, often accompanied by feelings of awe at the extraordinary physical appeal and beauty of the natural world.*

This sphere of life and organic activity extends from the ocean floor to about 8 km (5 mi) in the atmosphere. The biosphere includes myriad ecosystems from simple to complex, each operating within general spatial boundaries. An **ecosystem** is a self-regulating association of living plants and animals and their nonliving physical environment. In an ecosystem, a change in one component causes changes in others as systems adjust to new conditions.

Earth's surface environment itself is the largest ecosystem within the natural boundary of the atmosphere. Natural ecosystems are open systems for both energy and matter, with almost all ecosystem boundaries functioning as transition zones rather than as sharp demarcations.

Ecology is the study of the relationships between organisms and their environment and among the various ecosystems in the biosphere. The word *ecology*, developed by German naturalist Ernst Haeckel in 1869, is derived from the Greek *oikos* (household, or place to live) and *logos* (study of). **Biogeography**, essentially a spatial ecology, is the study of the distribution of plants and animals, the diverse spatial patterns they create, and the physical and biological processes, past and present, that produce this distribution across Earth.

The degree to which modern society understands ecosystems will help determine our success as a species and the long-term survival of a habitable Earth:

> The time is ripe to step up and expand current efforts to understand the great interlocking systems of air, water, and minerals nourishing the Earth.... Moreover, without vigorous action toward that goal, nations will be seriously handicapped in trying to cope with proven and suspected threats to ecosystems and to human health and welfare resulting from alterations in the cycles of carbon, nitrogen, phosphorus, sulfur, and related materials.... Society depends upon this life-support system of planet Earth.†

Humans are the most influential biotic (living) agents of change. This is not arrogance; it is fact, because we powerfully influence every ecosystem on Earth. Since life arose on the planet, there have been six major extinctions (see Figure 8.1). The fifth one was 65 million years ago, whereas the sixth is going on right now. Of all these extinction episodes, the current one is the only one of biotic origin, for it is being caused by humans.

In this chapter: We explore ecosystems, and the community, habitat, and niche concepts. Plants are the essential living component in the biosphere, translating solar energy into usable forms to energize life. We examine the role of nonliving systems, including biogeochemical cycles. We cover the organization of living ecosystems along complex food chains and webs. Topics include ecosystem stability and resilience, the effects of global change on ecosystems and rates of succession, and consideration of global biodiversity.

We explore Earth's major terrestrial ecosystems, conveniently grouped into 10 biomes. Table 16.1 brings together nearly all the information about Earth's physical systems that is presented through the pages of this text. Through biomes we synthesize the physical geography of place and region.

*S. R. Kellert, "The Biological Basis for Human Values of Nature," in *The Biophilia Hypothesis*, S.R. Kellert and E. O. Wilson, eds. (Washington, D.C.: Island Press, 1993), p. 49.

†G. F. White and M. K. Tolba, *Global Life Support Systems*, United Nations Environment Programme Information, No. 47 (Nairobi, Kenya: United Nations, 1979), p. 1.

Ecosystem Components and Cycles

An ecosystem is a complex of many variables, all functioning independently yet in concert, with complicated flows of energy and matter (Figure 16.1). An ecosystem includes both biotic (living) and abiotic (nonliving) components. Nearly all depend on the input of solar energy; the few limited ecosystems that exist in dark caves or on the ocean floor depend on chemical reactions (chemosynthesis).

Ecosystems are divided into subsystems, with the biotic portion composed of producers, consumers, and detritivores. The abiotic flows in an ecosystem include gaseous and sedimentary nutrient cycles. Figure 16.2 illustrates the essential elements of an ecosystem.

Communities

A convenient biotic subdivision within an ecosystem is a **community**, which is formed by interactions among populations of living animals and plants. Therefore, an ecosystem is the interaction of many communities with the abiotic physical components of its environment.

For example, in a forest ecosystem, a specific community may exist on the forest floor while another community functions in the canopy of leaves high above. Similarly, within a lake ecosystem, the plants and animals that flourish in the bottom sediments form one community, whereas those near the surface form another. A community is identified in several ways—by its physical appearance, the number of species and the abundance of each, the complex patterns of their interdependence, and the trophic (feeding) structure of the community.

Within a community are two important concepts: habitat and niche. **Habitat** is the specific physical location of an organism, the type of environment in which it resides or is biologically adapted to live. In terms of physical and natural factors, most species have specific habitat requirements with definite limits and a specific regimen of sustaining nutrients.

Niche (French *nicher*, "to nest") refers to the function, or occupation, of a life form within a given community. It is the way an organism obtains and sustains the physical, chemical, and biological factors it needs to survive. An individual species must satisfy several aspects in its niche. Among these are a *habitat* niche, a *trophic* (food) niche, and

FIGURE 16.1 The web of life. The intricate web of a spider mirrors the web of life.

> Life devours itself: everything that eats is itself eaten; everything that can be eaten is eaten; every chemical that is made by life can be broken down by life; all the sunlight that can be used is used.... The web of life has so many threads that a few can be broken without making it all unravel, and if this were not so, life could not have survived the normal accidents of weather and time, but still the snapping of each thread makes the whole web shudder, and weakens it.... You can never do just one thing: the effects of what you do in the world will always spread out like ripples in a pond.

[From Friends of the Earth and Amory Lovins, The United Nations Stockholm Conference: *Only One Earth*. London: Earth Island Limited, 1972, p. 20. Photo by author.]

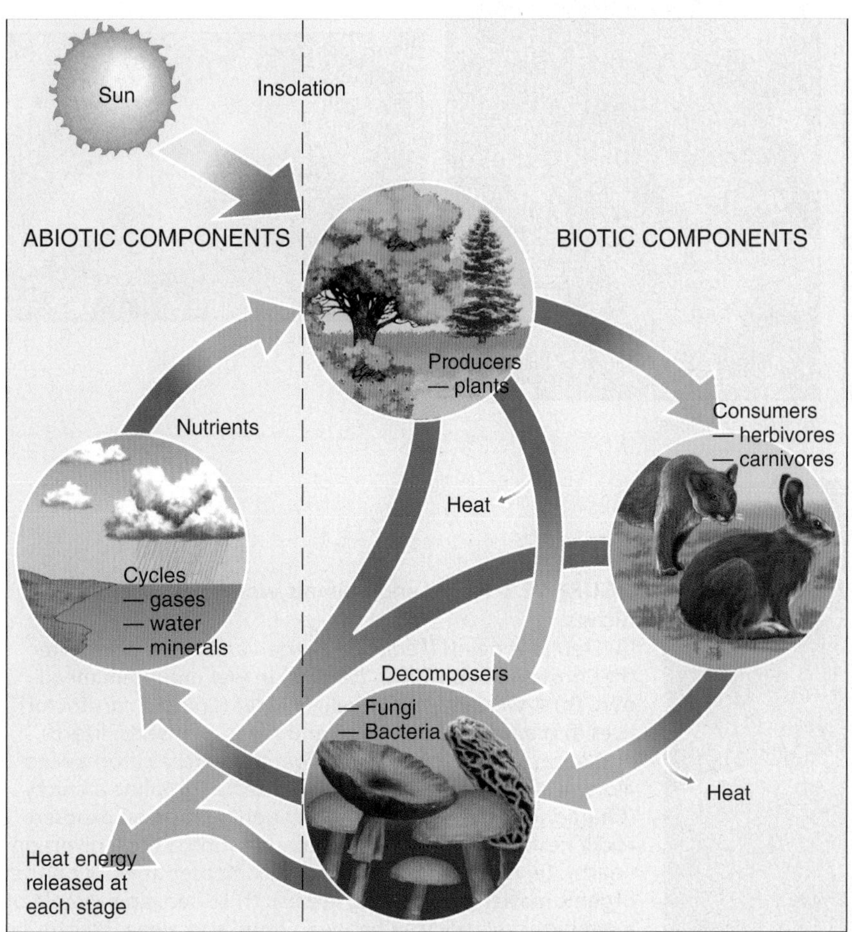

(a)

(b)

FIGURE 16.2 Abiotic and biotic components of ecosystems.
(a) Solar energy is the input that drives the biosphere. Heat energy and biomass are the outputs from the biosphere.
(b) Abiotic and biotic ingredients operate together to sustain this subtropical swamp (low, waterlogged ground) and basking turtles at Juniper Springs Recreation Area, Florida. [(b) Photo by Bobbé Christopherson.]

a *reproductive* niche. For example, the Red-winged Blackbird (*Agelaius phoeniceus*) occurs throughout the United States and most of Canada in habitats of meadow, pastureland, and marsh where it nests. Its trophic niche is weed seeds and cultivated seed crops throughout the year, adding insects to its diet during the nesting season.

Similar habitats produce comparable niches. In a stable community, no niche is left unfilled. The *competitive exclusion principle* states that no two species can occupy the same niche (food or space) successfully in a stable community. Thus, closely related species are separated spatially from one another. In other words, each species operates to reduce competition. This strategy in turn leads to greater diversity as species shift and adapt (Figure 16.3).

Some species are *symbiotic*, an arrangement where two or more species exist together in an overlapping relationship. One type of symbiosis, *mutualism*, occurs when each organism benefits and is sustained over an extended period by the relation. For example, lichen (pronounced "liken") is made up of algae and fungi living together.

(a) Elephant heads

(b) Western yellow-bellied racer

(c) Dragonfly

(d) Killdeer chicks

(e) Mushrooms (wild)

(f) Lichen

FIGURE 16.3 Plants and animals work to fit specific niches.
(a) Elephant heads (*Pedicularis groenlandica*), a wildflower that grows above 1800 m (6000 ft) in wet mountain meadows. (b) A western yellow-bellied racer (*Coluber constrictor*) lives in prairies and meadows and feeds on insects, lizards, and mice. (c) A dragonfly (*Libellula* sp.) perches atop a reed along a stream, feeding on flying insects. (d) Killdeer chicks (*Charadrius vociferus*) begin life roughing it on an exposed rocky nest, with protective parents creating a noisy diversion nearby. (e) Mushrooms, a fungal decomposer at work on the organic matter (composting leaves). (f) Lichen, an example of a symbiotic relationship between fungi and algae. [Photos by (a), (c), (e), (f) Bobbé Christopherson; (b), (d) author.]

FIGURE 16.4 How plants live and grow.
The balance between photosynthesis and respiration determines net photosynthesis and plant growth. [Temperate rain forest photo by Bobbé Christopherson.]

The alga is the producer and food source for the fungus, and the fungus provides structure and physical support. Their mutualism allows the two to occupy a niche in which neither could survive alone. Lichen developed from an earlier parasitic relationship in which the fungi broke into the alga cells. Today, the two organisms have evolved into a supportive harmony and symbiotic relationship (Figure 16.3f). The partnership of corals and algae discussed in Chapter 13 is another example of a symbiotic relationship.

By contrast, another form of symbiosis is a *parasitic* relationship, which may eventually kill the host, thus destroying the parasite's own niche and habitat. An example is mistletoe (*Phoradendron* sp.), which lives on and may kill various kinds of trees. Some scientists are questioning whether our human society and the physical systems of Earth constitute a global-scale symbiotic relationship: mutualism (sustainable) or a parasitic one (nonsustainable).

Plants: The Essential Biotic Component

Plants are the critical biotic link between life and solar energy. *Ultimately, the fate of the biosphere rests on the success of plants and their ability to capture sunlight.*

Land plants and animals became common about 430 million years ago, according to fossilized remains. In the present day, about 270,000 species of plants, mostly vascular (have conductive tissue for fluid and nutrient transport and true roots), are known to exist. Many more species have yet to be identified. Plants are a major source of new medicines and chemical compounds that benefit humanity. Plants also are the core of healthy, functioning ecosystems.

Leaf Activity. Leaves are solar-powered chemical factories. Flows of carbon dioxide, water, light, and oxygen enter and exit the surface of each leaf (see Figure 1.4). They flow in and out through small pores called **stomata** (singular: stoma). Each stoma is surrounded by small guard cells that open and close the pore, depending on the plant's needs at the moment. Veins in each leaf connect to the stems and branches of the plant and thus to its main circulation system. The veins bring in water and nutrients and carry away the sugars produced by photosynthesis.

Water that moves through a plant exits the leaves through the stomata and evaporates from leaf surfaces, thereby assisting heat regulation within the plant. As water evaporates from the leaves, a pressure deficit is created that allows atmospheric pressure to push water up through the plant all the way from the roots in the same manner that a soda straw works. We can only imagine the complex operation of a 100-m (330-ft) tree.

Photosynthesis and Respiration. Under the influence of certain wavelengths of visible light, **photosynthesis** unites carbon dioxide and hydrogen (derived from water in the plant). The process releases oxygen and produces energy-rich organic material. The name of this process is descriptive: *photo-* refers to sunlight, and *-synthesis* describes the combining reaction of materials within plant leaves.

The largest concentration of light-responsive, photosynthetic structures (known as *organelles*) in a leaf rests below the leaf's upper layers. These organelle units within cells are called *chloroplasts*, and within each resides a green, light-sensitive pigment called **chlorophyll**. Within this pigment, light stimulates photochemistry. Consequently, competition for light is a dominant factor in the formation of plant communities. This competition is expressed in the height, orientation, distribution, and structure of plants.

Photosynthesis essentially follows this equation:

$$\underset{\text{(carbon dioxide)}}{6CO_2} + \underset{\text{(water)}}{6H_2O} + \underset{\text{(solar energy)}}{\text{Light}} \rightarrow \underset{\text{(glucose, carbohydrate)}}{C_6H_{12}O_6} + \underset{\text{(oxygen)}}{6O_2}$$

From the equation, you can see that photosynthesis removes carbon (in the form of CO_2) from Earth's atmosphere (Figure 16.4). Carbohydrates, the organic result of

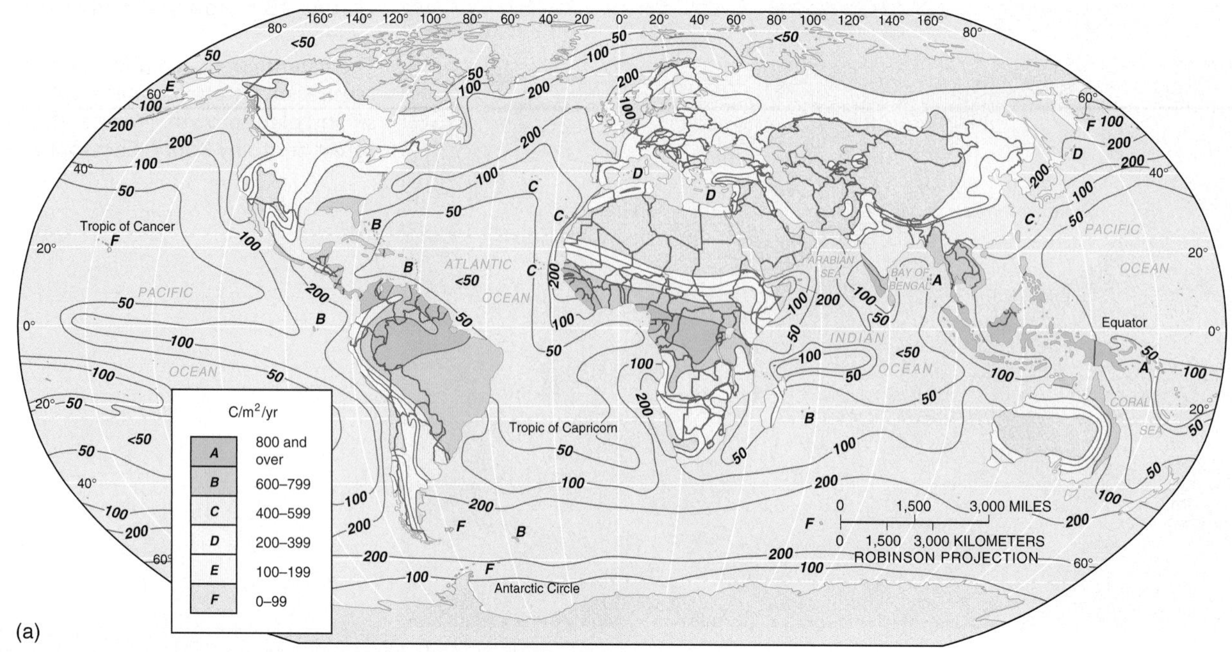

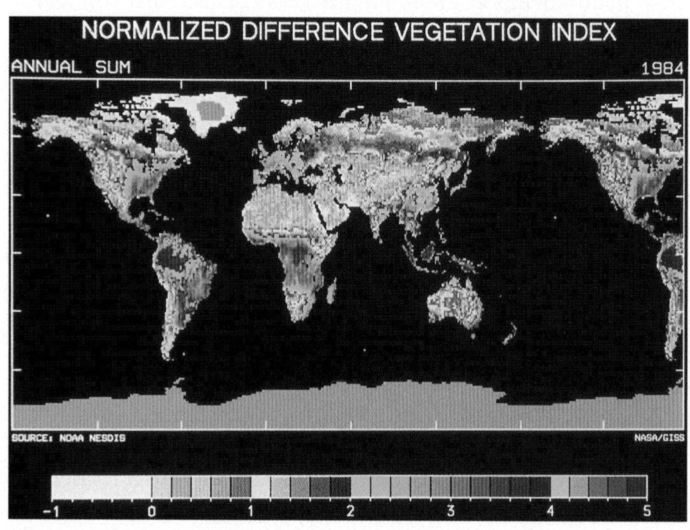

FIGURE 16.5 Net primary productivity.
(a) Worldwide net primary productivity in grams of carbon per square meter per year (approximate values). (b) Normalized difference vegetation index during 1984. False coloration indicates bare ground in browns, dense vegetation in blues. [(a) After D. E. Reichle, *Analysis of Temperate Forest Ecosystems* (Heidelberg, Germany: Springer-Verlag, 1970). Adapted by permission; (b) Goddard Space Flight Center and NOAA.]

the photosynthetic process, are combinations of carbon, hydrogen, and oxygen. They form simple sugars, such as glucose ($C_6H_{12}O_6$). Glucose, in turn, is used by plants to build starches, which are more complex carbohydrates and the principal food storage substance in plants. *Primary productivity* refers to the rate at which energy is stored in such organic substances.

Plants not only store energy; they also must consume some of this energy by converting carbohydrates for their other operations. Thus, **respiration** is essentially a reverse of the photosynthetic process:

$$\underset{\text{(glucose, carbohydrate)}}{C_6H_{12}O_6} + \underset{\text{(oxygen)}}{6O_2} \rightarrow \underset{\text{(carbon dioxide)}}{6CO_2} + \underset{\text{(water)}}{6H_2O} + \underset{\text{(heat energy)}}{\text{energy}}$$

In respiration, plants oxidize carbohydrates, releasing carbon dioxide, water, and energy as heat. The overall growth of a plant depends on a surplus of carbohydrates beyond what is lost through plant respiration. Figure 16.4 presents a simple schematic of this process, which yields plant growth. The difference between photosynthetic production and respiration loss is called *net photosynthesis*.

Net Primary Productivity. The net photosynthesis for an entire plant community is its **net primary productivity**. This is the amount of stored chemical energy (biomass) that the community generates for the ecosystem. **Biomass** is the net dry weight of organic material; it is biomass that feeds the food chain.

Net primary productivity is mapped in terms of *fixed carbon per square meter per year.* ("Fixed" means chemically bound into plant tissues.) You can see in Figure 16.5 that on land, net primary production tends to be highest in the tropics and decreases toward higher latitudes. Even though deserts receive high amounts of solar radiation,

Table 16.1 Net Primary Productivity and Plant Biomass on Earth

Ecosystem	Area (10^6 km²)[a]	Net Primary Productivity per Unit Area (g/m²/yr)[b] Normal Range	Mean	World Net Biomass (10^9/tons/yr)[c]
Tropical rain forest	17.0	1000–3500	2200	37.4
Tropical seasonal forest	7.5	1000–2500	1600	12.0
Temperate evergreen forest	5.0	600–2500	1300	6.5
Temperate deciduous forest	7.0	600–2500	1200	8.4
Boreal forest	12.0	400–2000	800	9.6
Woodland and shrubland	8.5	250–1200	700	6.0
Savanna	15.0	200–2000	900	13.5
Temperate grassland	9.0	200–1500	600	5.4
Tundra and alpine region	8.0	10–400	140	1.1
Desert and semidesert scrub	18.0	10–250	90	1.6
Extreme desert, rock, sand, ice	24.0	0–10	3	0.07
Cultivated land	*14.0*	*100–3500*	*650*	*9.1*
Swamp and marsh	2.0	800–3500	2000	4.0
Lake and stream	2.0	100–1500	250	0.5
Total continental	**149**	—	**773**	**115.17**
Open ocean	332.0	2–400	125	41.5
Upwelling zones	0.4	400–1000	500	0.2
Continental shelf	26.6	200–600	360	9.6
Algal beds and reefs	0.6	500–4000	2500	1.6
Estuaries	1.4	200–3500	1500	2.1
Total marine	**361.0**	—	**152**	**55.0**
Grand total	**510.0**	—	**333**	**170.17**

Source: R. H. Whittaker, *Communities and Ecosystems* (Heidelberg, Germany: Springer-Verlag, 1975), p. 224. Reprinted by permission.
[a]1 km² = 0.39 mi².
[b]1 g per m² = 8.92 lb per acre.
[c]1 metric ton(10^6 g) = 1.1023 tons.

other controlling factors are more important, namely, water availability and soil conditions. But precipitation also affects productivity, as evidenced on the map by the correlations of abundant precipitation with high productivity (tropics) and reduced precipitation with low productivity (subtropical deserts).

In the oceans, productivity is limited by differing nutrient levels. Regions with nutrient-rich upwelling currents generally are the most productive (off western coastlines). The map shows that the tropical ocean and areas of subtropical high pressure are quite low in productivity.

In temperate and high latitudes, the rate at which carbon dioxide is fixed by vegetation varies seasonally. The rate increases in spring and summer as plants flourish with increasing solar input and, in some areas, more available (nonfrozen) water, and decreases in late fall and winter. Rates in the tropics are high throughout the year, and turnover in the photosynthesis-respiration cycle is faster, exceeding by many times the rates experienced in a desert environment or in the far northern limits of the tundra. A lush hectare (2.5 acres) of sugar cane in the tropics might fix 45 metric tons (50 tons) of carbon in a year, whereas desert plants in an equivalent area might achieve only 1% of this amount.

Table 16.1 lists various ecosystems, their net primary productivity per year, and an estimate of net total biomass worldwide. World net primary productivity is estimated at 170 billion metric tons (189 billion tons) of dry organic matter per year. Compare the various ecosystems, especially cultivated land, with the natural communities.

Abiotic Ecosystem Components

Critical in each ecosystem are the flow of energy and the cycling of nutrients and water. Nonliving abiotic components set the stage for ecosystem operations.

Light, Temperature, Water, and Climate. The pattern of solar energy receipt is crucial in both terrestrial and aquatic ecosystems. As explained, solar energy enters an ecosystem by way of photosynthesis, with heat dissipated from the system at many points. The duration of Sun exposure is the *photoperiod.* Along the equator, days are almost always 12 hours in length; however, with increasing distance from the equator, seasonal effects become pronounced, as discussed in Chapter 2. Plants have adapted their flowering and seed germination to seasonal changes in insolation.

Air and soil temperatures determine the rates at which chemical reactions proceed. Significant temperature factors are seasonal variation, duration and pattern of minimum and maximum temperatures, and average temperature (Chapter 3).

The hydrologic cycle and its output of water depend on precipitation/evaporation rates and their seasonal distribution (see Chapters 5 and 7). Water quality is essential—its mineral content, salinity, and levels of pollution determine its quality. Also, daily weather patterns over time create regional climates (see Chapter 6), which in turn affect the pattern of vegetation and ultimately influence soil development. All these factors work together to establish the parameters (limits) for ecosystems that may develop in a given location.

Figure 16.6 illustrates the general relationship among temperature, precipitation, and vegetation. In general terms, can you identify the characteristic vegetation type and related temperature and moisture relationship that fits the area of your home or school?

Beyond these general conditions, each ecosystem further produces its own *microclimate*, specific to individual sites. For example, in forests the insolation reaching the ground is reduced. A pine forest cuts light by 20%–40%, whereas a birch-beech forest reduces it by as much as 50%–75%. Forests also are about 5% more humid than nonforested landscape, have warmer winters and cooler summers, and experience reduced winds.

Life Zones. Alexander von Humboldt (1769–1859), an explorer, geographer, and scientist, deduced that plants and animals recur in related groupings wherever conditions in the abiotic environment are similar. After several years of study in the Andes Mountains of Peru, he described a distinct relationship between altitude and plant communities, his **life zone** concept. As he climbed the mountains, he noticed that the experience was similar to that of traveling away from the equator toward higher latitudes (Figure 16.7, p. 492).

This zonation of plants with altitude is noticeable on any trip from lower valleys to higher elevations. Each life zone possesses its own temperature, precipitation, and insolation relationships and therefore its own biotic communities.

Elemental Cycles

The most abundant natural elements in living matter are hydrogen (H), oxygen (O), and carbon (C). Together, these elements make up more than 99% of Earth's biomass; in fact, all life (organic molecules) contains hydrogen and carbon. In addition, nitrogen (N), calcium (Ca), potassium (K), magnesium (Mg), sulfur (S), and phosphorus (P) are essential *nutrients*, elements necessary for the growth and development of a living organism.

Several key chemical cycles function in nature. Oxygen, carbon, and nitrogen all have *gaseous cycles*, part of which are in the atmosphere. Other elements have *sedimentary cycles*, which principally involve the mineral and solid phases (major ones include phosphorus, calcium, and sulfur). Some elements combine gaseous and sedimentary cycles. These recycling processes are called **biogeochemical cycles**, because they involve chemical reactions in both living and nonliving systems.

Oxygen and Carbon Cycles. We consider oxygen and carbon together because they are so closely intertwined through photosynthesis and respiration (Figure 16.8, p. 493). The atmosphere is the principal reserve of available oxygen. Larger reserves of oxygen exist in Earth's crust, but they are unavailable, chemically bound to other elements.

As for carbon, the greatest pool of it is in the ocean—about 39,000 billion tons, or about 93% of Earth's total carbon. However, all this carbon is bound chemically in carbon dioxide, calcium carbonate, and other compounds. The ocean absorbs carbon dioxide through photosynthesis by phytoplankton. On land, carbon is stored in certain carbonate minerals, such as limestone.

The atmosphere, which is the integrating link in the cycle, contains only about 700 billion tons of carbon (as carbon dioxide) at any moment. This is far less carbon than is stored in fossil fuels and oil shales (13,200 billion metric tons, as hydrocarbon molecules) or in living and dead organic matter (2500 billion metric tons, as carbohydrate molecules). Carbon dioxide in the atmosphere is produced by the respiration of plants and animals, volcanic activity, and fossil fuel combustion by industry and transportation.

The carbon dumped into the atmosphere by human activity constitutes a vast geochemical experiment, using the real-time atmosphere as a laboratory. Since 1970, we have added to the atmospheric pool an amount of carbon equivalent to more than 25% of the total amount added since 1880, a 400% increase over 1950 levels.

The Nitrogen Cycle. The *nitrogen cycle* is the major constituent of the atmosphere, 78.084% of each breath we take. Nitrogen also is important in the makeup of organic molecules, especially proteins, and therefore is essential to living processes. A simplified view of the nitrogen cycle is

FIGURE 16.6 Temperature and precipitation affect ecosystems.
(a) Climate controls (wetness, dryness; warmth, cold) and ecosystem types; (b) subtropical desert in Namibia; (c) Sonoran desert of Arizona; (d) cold desert of north-central Nevada; (e) dry tundra of Nunavut Canada; (f) El Yunque tropical rain forest of Puerto Rico; (g) deciduous trees in fall colors in Ohio; (h) needleleaf forest near Independence Pass, Colorado; and (i) moist tundra of northern Québec. [Photos by (b) NigelDennis/Photo Researchers, Inc.; (c, d) author; (e) John Eastcott/Yva Momatiuk/The Image Works; (f) Tom Bean; (g, h) Bobbé Christopherson; (i) John Eastcott/Yva Momatiuk/Stock Boston.]

Polar and Alpine
Dry tundra
Moist tundra
Cool
Cool desert
Pine-Spruce-Fir forest
Temperate
Temperate desert
Short grass
Tall grass
Deciduous forest
Hot
Hot desert
Grasslands
Tropical forest
Temperature
Precipitation
Wet
Dry

(a) (b) (c) (d) (e) (f) (g) (h) (i)

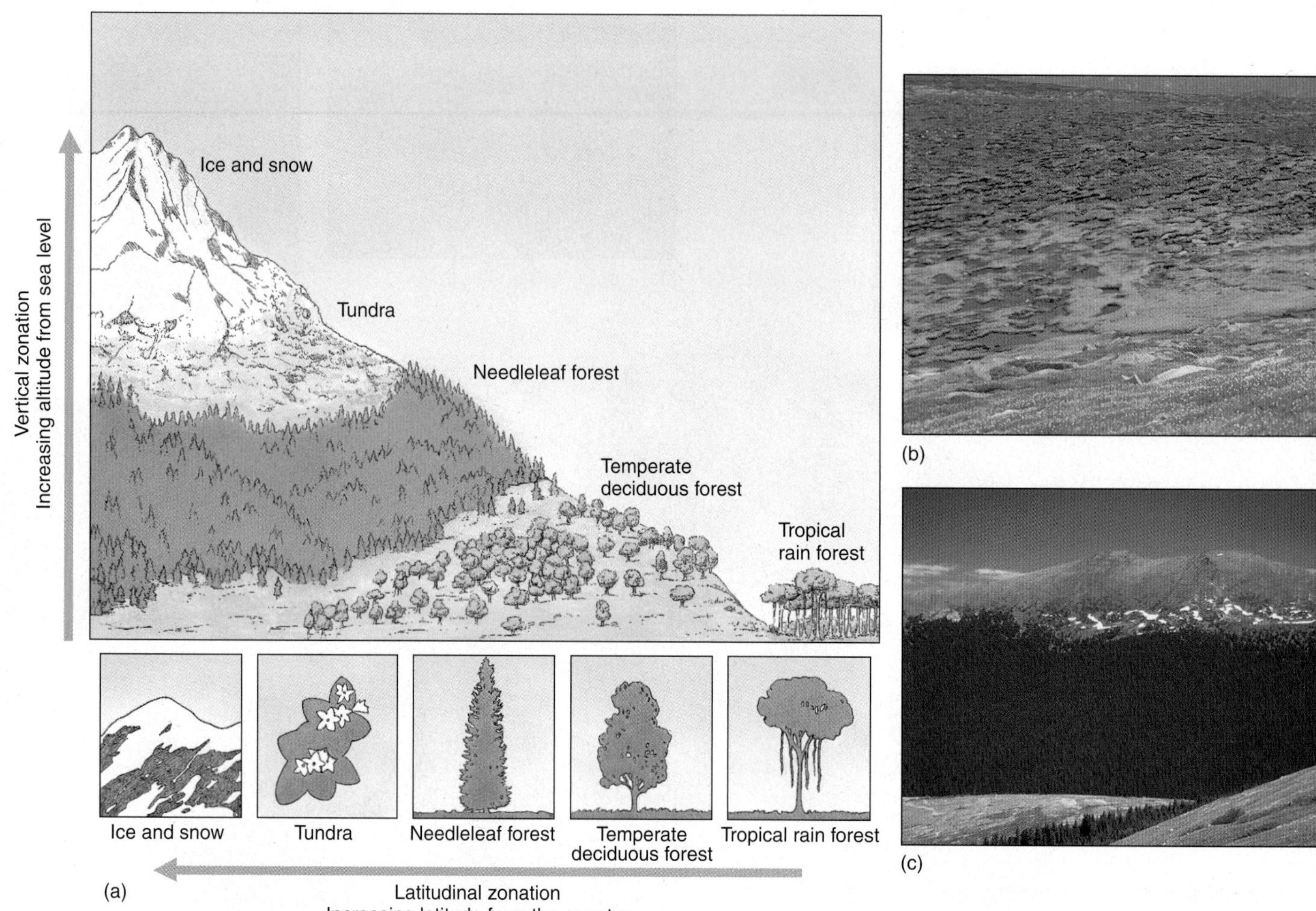

FIGURE 16.7 Vertical and latitudinal zonation of plant communities. (a) Progression of plant community life zones with changing altitude or latitude. (b) Alpine tundra ecosystem in the Colorado Rockies. (c) The timberline for a needleleaf forest in the Rockies. The forest line marks the highest continuous forest, whereas the tree line above it marks the zone in which no trees grow. [Photos by Bobbé Christopherson.]

portrayed in Figure 16.9 (p.494). This vast atmospheric reservoir is inaccessible directly to most organisms.

Nitrogen-fixing bacteria, which live principally in the soil and are associated with the roots of certain plants, are the key link to life. The roots of legumes such as clover, alfalfa, soybeans, peas, beans, and peanuts have such bacteria. Bacteria colonies reside in nodules on the legume roots and chemically combine the nitrogen from the air in the form of nitrates (NO_3) and ammonia (NH_3). Plants use these chemically bound forms of nitrogen to produce their own organic matter. Anyone or anything feeding on the plants thus ingests the nitrogen. Finally, the nitrogen in the organic wastes of the consuming organisms is freed by denitrifying bacteria, which recycle it to the atmosphere.

To improve agricultural yields, many farmers use synthetic inorganic fertilizers, as opposed to soil-building organic fertilizers (manure and compost). Inorganic fertilizers are chemically produced through artificial nitrogen fixation at factories. Humans presently fix more nitrogen as synthetic fertilizer per year than all terrestrial sources combined!

The surplus of usable nitrogen accumulates in Earth's ecosystems. This excess nitrogen load begins a water pollution process that feeds an excessive growth of algae and phytoplankton, increases biochemical oxygen demand, diminishes dissolved oxygen reserves, and eventually disrupts the aquatic ecosystem.

Figure 16.10 (p.495) maps and images the dead zone, a region of oxygen-depleted (hypoxia) water off the coast of Louisiana in the Gulf of Mexico. The Mississippi River carries agricultural fertilizers, farm sewage, and other wastes to the Gulf, causing huge spring blooms of phytoplankton. The Mississippi drainage system handles the runoff for 41% of the continental United States. By summer the biological oxygen demand of bacteria feeding on the decay exceeds the dissolved oxygen; hypoxia develops, killing any fish that ventures into the area.

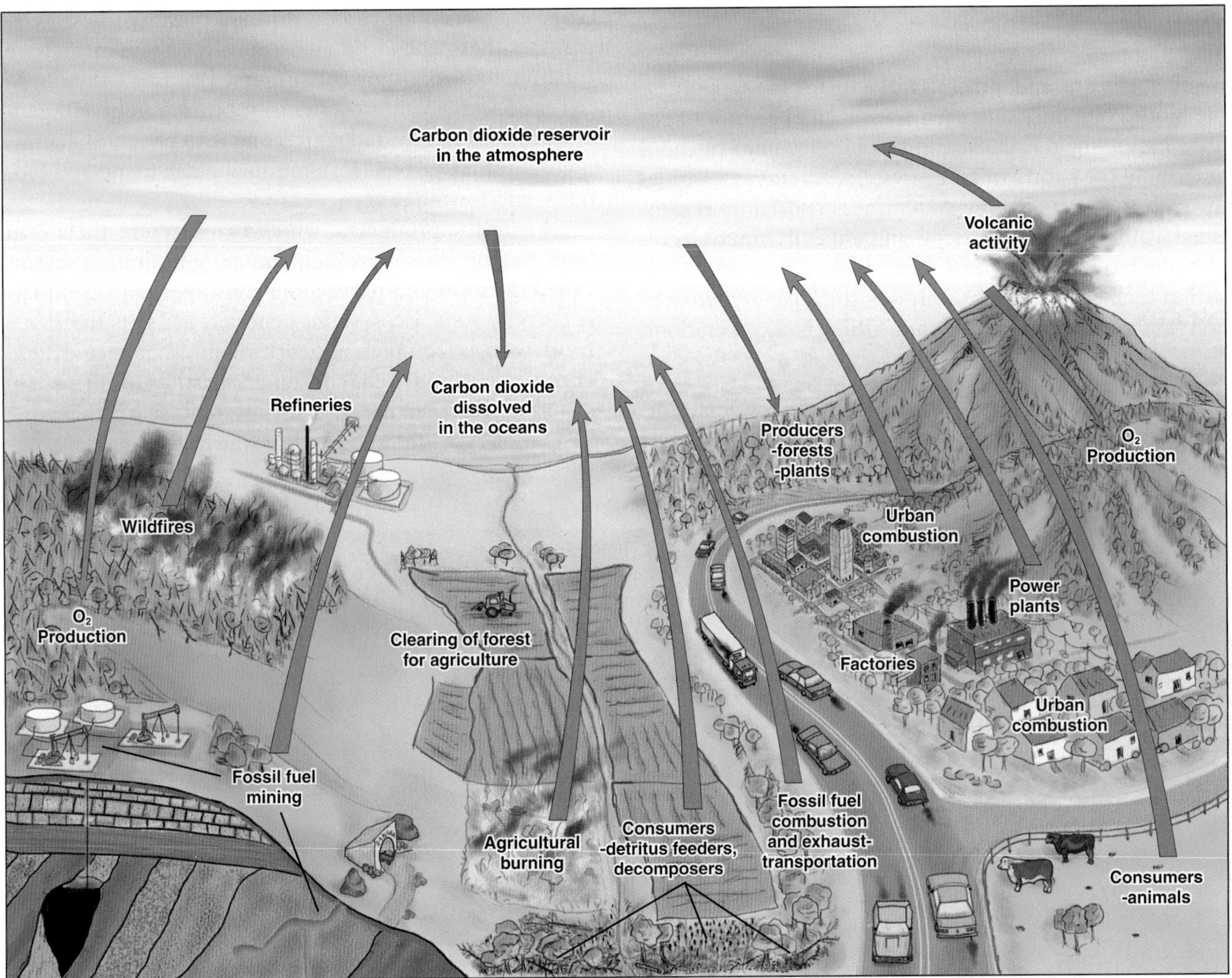

FIGURE 16.8 The carbon and oxygen cycles, simplified.
Carbon is fixed through photosynthesis and begins its passage through the ecosystem. Respiration by living organisms, burning of forests and grasslands, and the combustion of fossil fuels releases carbon to the atmosphere. These cycles are greatly influenced by human activities.

These low-oxygen conditions act as a limiting factor on marine life.

When the 1993 Midwest floods hit, the river's nutrient discharge was such that the size of the dead zone doubled to 17,500 km^2 (6755 mi^2); the worst year was 2002, when it expanded to 22,000 km^2 (8500 mi^2). The agricultural, feedlot, and fertilizer industries dispute the connection between their nutrient input and the dead zone.

Limiting Factors

The term **limiting factor** identifies the one physical or chemical abiotic component that most inhibits biotic operations, through its lack or excess. Here are a few examples:

- Low temperatures limit plant growth at high elevations or high latitudes.
- Lack of water limits growth in a desert.
- Excess water limits growth in a bog.
- Changes in salinity levels limit aquatic ecosystems.
- Lack of iron in ocean surface environments limits photosynthetic production.
- Low phosphorus content of eastern U.S. soils limits plant growth.
- General lack of active chlorophyll above 6100 m (20,000 ft) limits primary productivity.

Each organism possesses a range of tolerance for each limiting factor in its environment. Limiting factors are illustrated vividly in Figure 16.11, which shows the geographic range for two tree species and two bird species. The coast redwood (*Sequoia sempervirens*) is limited to a narrow section of the Coast Ranges in California, covering barely 9500 km^2 (3670 mi^2) and concentrated in areas that receive necessary summer advection fog. The red maple (*Acer rubrum*), on the other hand, thrives over a large area under varying conditions of moisture and temperature, thus demonstrating a broader tolerance to environmental variations.

Biotic Ecosystem Operations

The abiotic components of energy, atmosphere, water, weather, climate, and minerals make up the life support for the biotic components of each ecosystem. The flow of energy, cycling of nutrients, and trophic (feeding) relations determine the nature of an ecosystem. As energy cascades through this process-flow system, it is constantly replenished by the Sun. But nutrients and minerals cannot be replenished from an external source, so they constantly cycle within each ecosystem and through the biosphere in general (Figure 16.12). Let us examine these biotic operations.

Producers, Consumers, and Detritivores. Organisms that are capable of using carbon dioxide as their sole source of carbon are called *autotrophs*, or **producers**. They chemically fix carbon through photosynthesis. Organisms that depend on autotrophs for their carbon are called heterotrophs, or **consumers**. Plants are the essential producers in an ecosystem—capturing light and generating heat energy and converting it to chemical energy, incorporating carbon, forming new plant tissue and biomass, and freeing oxygen.

From the producers, which manufacture their own food, energy flows through the system along a circuit called the **food chain**, reaching consumers and eventually decomposers. Ecosystems generally are structured in a **food web**, a complex network of interconnected food chains. In a food web, consumers participate in several different food chains. Organisms that share the same basic foods are said to be at the same *trophic* (feeding, nutrition) *level*.

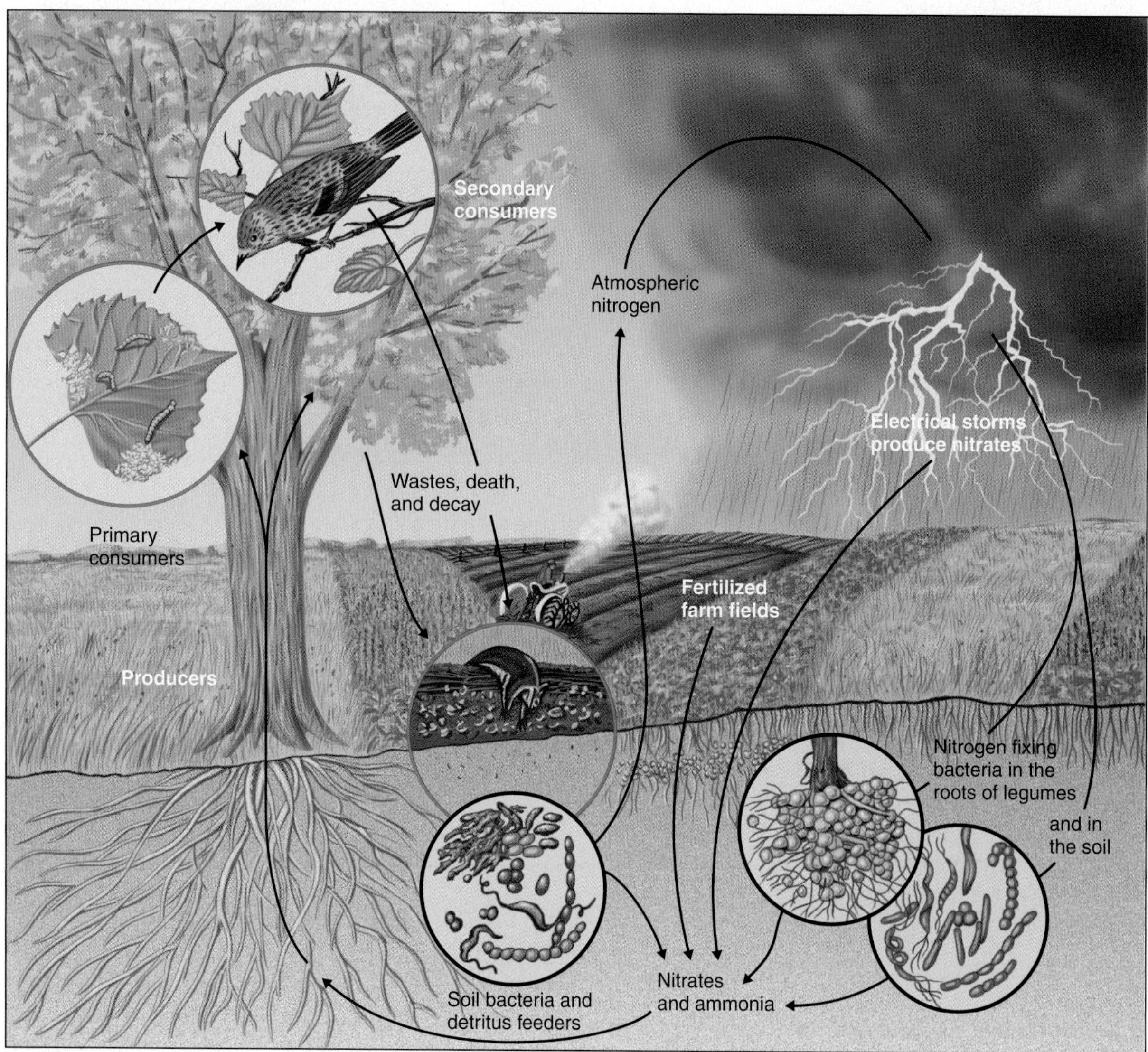

FIGURE 16.9 The nitrogen cycle.
The atmosphere is the essential reservoir of gaseous nitrogen. Atmospheric nitrogen gas is chemically fixed by bacteria to produce ammonia. Lightning and forest fires produce nitrates, and fossil fuel combustion forms nitrogen compounds that are washed from the atmosphere by precipitation. Plants absorb nitrogen compounds and produce organic material. [Adapted from Gerald Audesirk and Teresa Audesirk, *Biology, Life On Earth*, 4th ed., Figure 45-10, p. 903. Copyright © 1996. Adapted by permission of Prentice-Hall, Inc., Upper Saddle River, N.J.]

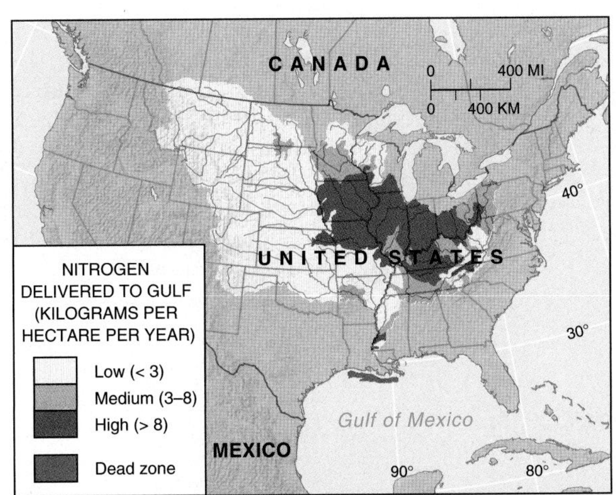

(a)

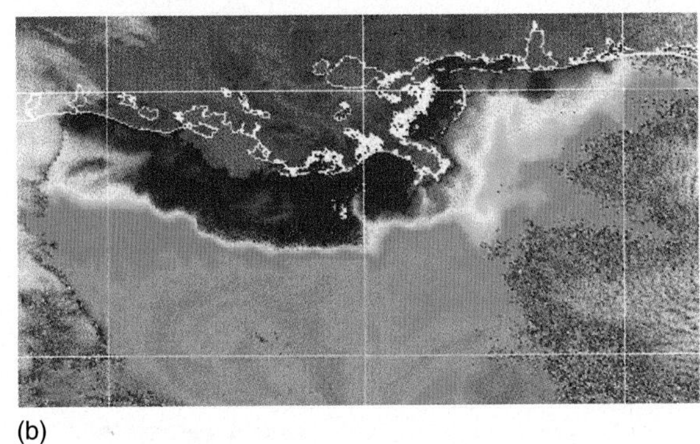

(b)

(c)

FIGURE 16.10 The dead zone.
(a) Nitrogen originates from agricultural activities in the upstream watershed that feeds the Mississippi River system. (b) The flush of nutrients from the Mississippi River drainage basin enriches the offshore waters in the Gulf of Mexico, causing huge phytoplankton blooms in early spring (red areas). (c) By late summer, oxygen-depleted waters (hypoxic) dominate what is known as the *dead zone*, expanding to a record extent in 2002. [(a) Adapted from R. B. Alexander, R. A. Smith, and G. E. Schwarz, "Effect of stream channel size on the delivery of nitrogen to the Gulf of Mexico," *Nature* (February 17, 2000), 761; as presented in *Environment*, Jan/Feb 2001, 16. (b) *SeaWiFS* image from the Goddard Space Flight Center, NASA, courtesy of Omnimage, Dulles, Virginia.]

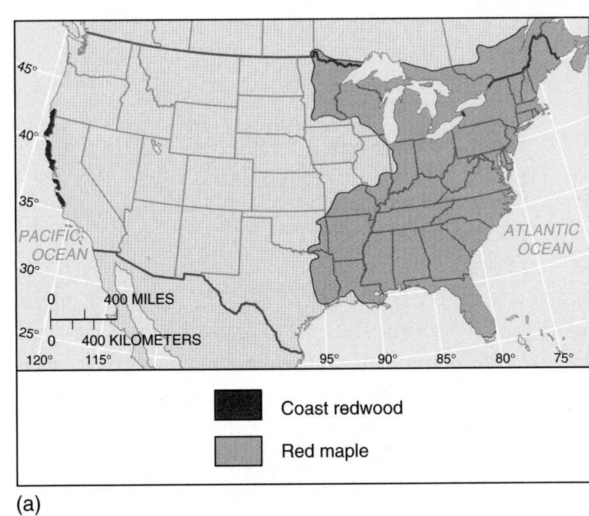

(a)

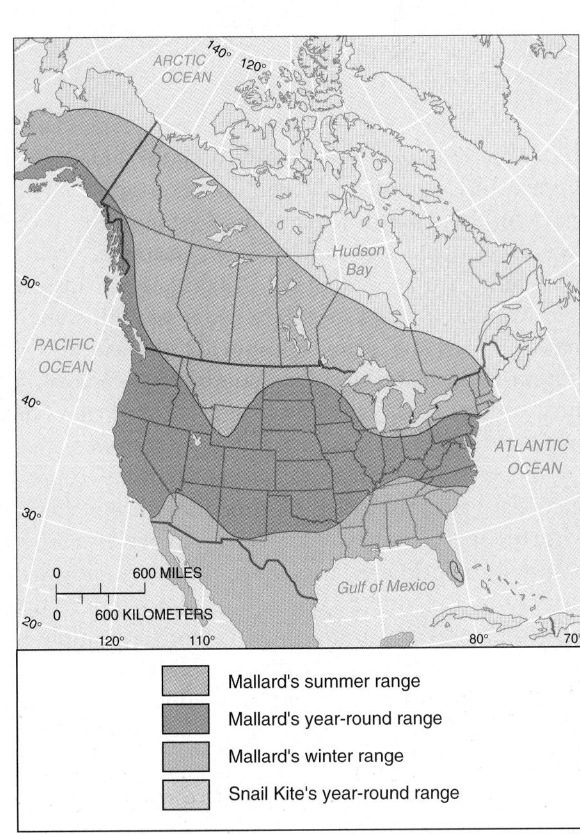

(b)

FIGURE 16.11 Limiting factors affect the distribution of plant and animal species.
(a) Biotic distribution of coast redwood and red maple illustrates limiting factors. (b) The Mallard duck and Snail Kite demonstrate the effect of limiting factors. The Mallard is a generalist and feeds widely. In contrast, the Snail Kite is limited to a single type of snail for food.

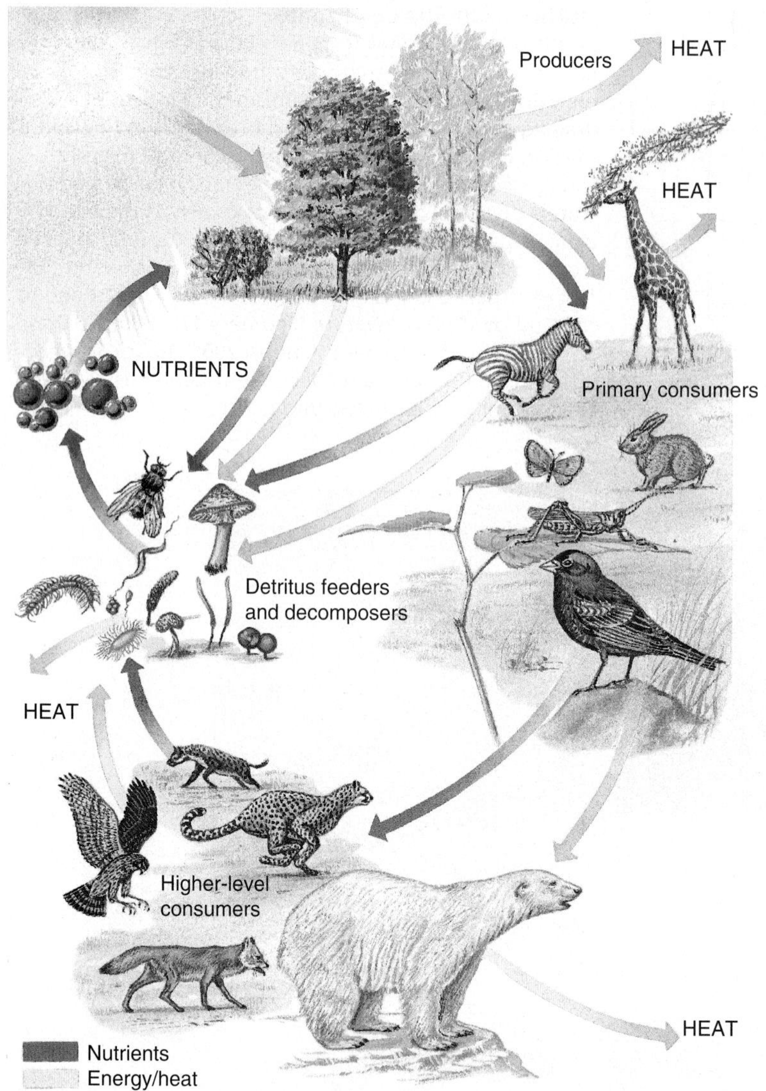

FIGURE 16.12 Energy, nutrient, and food pathways in the environment.
The flow of energy, cycling of nutrients, and trophic (feeding) relationships in a generalized ecosystem. The operation is fueled by radiant energy supplied by sunlight and first captured by the plants. [From T. Audesirk and G. Audesirk, *Biology: Life on Earth*, 4th ed., figure 45-1, p. 892. © 1996 Prentice Hall Inc. Used by permission.]

As an example, look at the food web based on krill (Figure 16.13). Krill is a shrimplike crustacean about 7.5 cm (3 in.) long that is a major food for a diverse group of organisms, including whales, fish, seabirds, seals, and squid in the Antarctic region. All these organisms participate in numerous other food chains as well, some consuming and some being consumed. Because krill are a protein-rich, plentiful food, commercial factory ships, such as those from Japan and Russia, hunt krill. The annual harvest currently approaches a million tons.

Primary consumers feed on producers. Because producers are always plants, the primary consumer is called a **herbivore**, or plant eater. A **carnivore** is a *secondary consumer* and primarily eats meat. A *tertiary consumer* eats primary and secondary consumers and is referred to as the "top carnivore" in the food chain; an example is the sperm whale in the krill food web. A consumer that feeds on both producers (plants) and consumers (meat) is called an **omnivore**—a role occupied by humans, among others.

Any assessment of world food resources depends on the level of consumer being studied. Using humans as an example, many can be fed from the amount of wheat harvested from an acre of land. An acre of land (0.41 hectare) produces about 810 kg (1800 pounds) of grain. However, if herbivores, such as cattle, eat that grain, only 82 kg (180 pounds) of biomass is produced, which can feed far fewer people (Figure 16.14).

In terms of energy, *only about 10% of the calories in plant matter survive from the primary to the secondary trophic level.* When humans, or any carnivore or omnivore, consume meat there is a loss of biomass to the system. More energy is lost to the environment at each progressive step in the food chain. Today approximately half of the cultivated acreage in the United States and Canada is planted for animal consumption—beef and dairy cattle, hogs, chickens, and turkeys. You can see that an omnivorous diet—our dietary pattern—is quite expensive in terms of biomass and energy.

Detritivores (detritus feeders and decomposers) are the final link in the endless chain. Detritivores renew the entire system by releasing simple inorganic compounds and nutrients with the breaking down of organic materials. Detritus refers to all the dead organic debris, remains, fallen leaves, and wastes, that living processes leave. *Detritus feeders*—worms, mites, termites, centipedes,

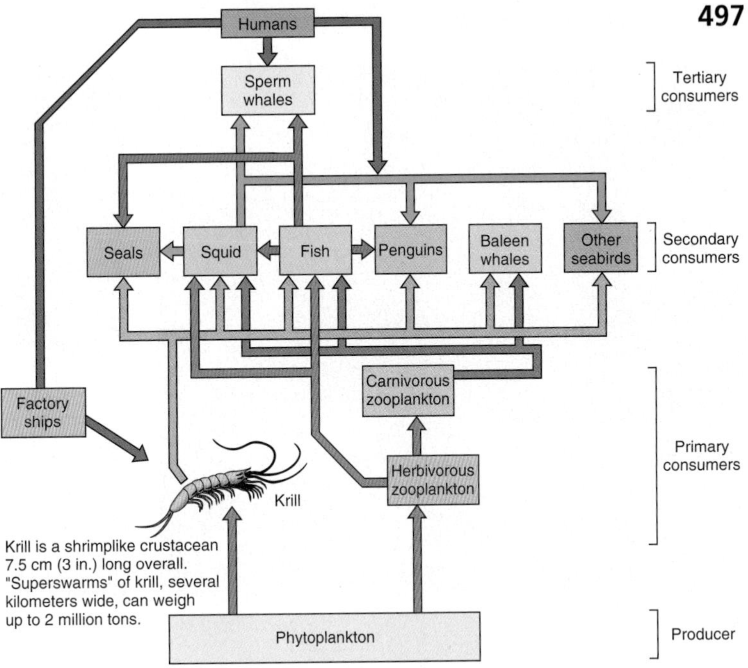

FIGURE 16.13 An Antarctic food web.
The food web of the krill in Antarctic waters, from phytoplankton producers (bottom) through various consumers. Phytoplankton begin this chain by using solar energy in photosynthesis. The krill, along with other organisms, feed on the phytoplankton. Krill in turn are fed upon by the next trophic level.

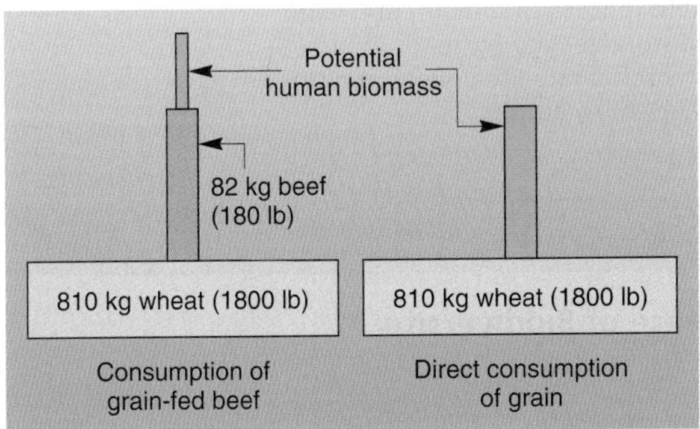

FIGURE 16.14 Pyramids show us efficiency and inefficiency.
Biomass pyramids illustrating the difference in efficiency between direct and indirect consumption of grain.

snails, crabs, even vultures, among others—work like an army to consume detritus and excrete nutrients that fuel an ecosystem. **Decomposers** are primarily bacteria and fungi that digest organic debris outside their bodies, and absorb and release nutrients in the process. This metabolic work of microbial decomposers produces the rotting that breaks down detritus. Detritus feeders and decomposers, although different in operation, have a similar function in an ecosystem.

Stability and Succession

Far from being static, Earth's ecosystems are dynamic (vigorous, energetic) and ever changing. Over time, communities of plants and animals adapted to such variation, evolved, and in turn shaped their environments. Each ecosystem operates in dynamic equilibrium, constantly adjusting to changing conditions to maintain stability. The concept of change is key to the study of ecosystem stability.

For most of the last century, scientists thought that an undisturbed ecosystem—whether forest, grassland, aquatic, or other—would progress to a stage of equilibrium, a stable point, with maximum chemical storage and biomass. Modern research has determined, however, that it is in the intermediate stages of succession when the greatest mineral and biomass inventories are in place. Ecosystems do not progress to some static equilibrium conclusion; such is the great diversity of nature.

Ecosystem Stability and Diversity

Any ecosystem moves toward maximum biomass and stability to survive. However, the tendency for birth and death rates to balance and the composition of species to remain stable, *inertial stability*, does not necessarily foster the ability to recover from change, *resilience*. Examples of stable communities include a redwood forest, a pine forest at a high elevation, and a tropical rain forest near the equator. Yet cleared tracts in a stable community recover slowly (if ever) and therefore have poor resilience. Figure 16.15 shows how clear-cut tracts of former forest have altered drastically the microclimatic conditions, making regrowth of the same species difficult. In contrast, a midlatitude grassland is low in stability; yet when burned, its resilience is high because the community recovers rapidly.

A critical aspect of ecosystem stability is **biodiversity**, or species richness of life (a combination of *bio*logical and *diversity*). The more diverse the species population (both in number of different species and quantity of each species), the species genetic diversity (number of genetic characteristics), and ecosystem and habitat diversity, the better the risk is spread over the entire community. In other words, *greater biodiversity in an ecosystem results in greater stability and greater productivity*. News Report 16.1 presents exciting confirmation of this principle. The importance of biodiversity is discussed further in Focus Study 16.1.

The loss of biodiversity is irreversible, and yet here

(a)

(b)

FIGURE 16.15 Human disruption of stable communities.
(a) Bowron Clearcut, British Columbia—an example of clear-cut timber harvesting that disrupted a stable community and produced drastic changes in microclimatic conditions. Full exposure to sunlight and wind is different from former shade conditions. (b) A logged pine forest and disruptive logging road. About 10% of the Northwest's old-growth forests remain, as identified through orbital satellite images and GIS analysis. [Photos by (a) Galen Rowell/Mountain Light Photography, Inc; and (b) by author.]

News Report 16.1

Experimental Prairies Confirm the Importance of Biodiversity

Field experiments have confirmed an important scientific assumption: that greater biological diversity in an ecosystem leads to greater stability, productivity, and soil nutrient use within that ecosystem. For instance, during a drought, some species of plants will be damaged. In a diverse ecosystem, however, other species with deeper roots and better water-obtaining ability will thrive.

Ecologist David Tilman at the University of Minnesota tracked the operation of 147 grassland plots, each 3 m (9.8 ft) square. Species diversity was carefully controlled on each plot (Figure 1). Plots were sown with different numbers of native North American prairie plant seeds and cared for by a team of 50 people. The results demonstrated that the plots with a more diverse plant community were able to retain and use nutrients more efficiently than the plots with less diversity. This efficiency reduced the loss of soil nitrogen through leaching, thus increasing soil fertility.

Greater plant diversity led to both higher productivity and better resource utilization. Also, total plant cover was found to increase with species richness. Tilman and his colleagues affirm:

> This extends the earlier results to the field, providing direct evidence that the current rapid loss of species on Earth, and management practices that decrease local biodiversity, threaten ecosystem productivity and the sustainability of nutrient cycling. Observational, laboratory, and now field experimental evidence, supports the hypothesis that biodiversity influences ecosystem productivity, sustainability, and stability.*

FIGURE 1 Biodiversity experiment.
Experimental plots are planted with native North American prairie grasses. A total of 147 plots were sown with random seed selections of 1, 2, 4, 6, 8, 12, or 24 species. Experimental results confirm many assumptions regarding the value of rich biodiversity. [Photo by David Tilman/University of Minnesota, Twin Cities Campus.]

*D. Tilman, D. Wedin, and J. Knops, "Productivity and sustainability influenced by biodiversity in grassland ecosystems," *Nature* 379 (February 22, 1996): 720. Also see the summary in *Science* 271 (March 15, 1996): 1497.

we are in the midst of the biosphere's sixth major extinction episode (see Figure 8.1). E. O. Wilson, the famed biologist, stated:

> Biological diversity—the full sweep from ecosystems to species within ecosystems, thence to genes within species—is in trouble. Mass extinctions are commonplace, especially in tropical regions where most of the biodiversity occurs. Among the more recent are more than half the exclusively freshwater fishes of peninsular Asia, half of the fourteen birds of the Philippines island of Cebu, and more than ninety plant species growing on a single mountain ridge in Ecuador. In the United States an estimated 1 percent of all species have been extinguished; another 32 percent are imperiled.*

Climate Change. Obviously, the distribution of plant species is affected by climate change, though many species have survived wide climate swings in the past. Consider the beginning of the Tertiary Period, 75 million years ago. Warm, humid conditions and tropical forests dominated the land to southern Canada, pines grew in the Arctic, and deserts were few. Then, between 15 and 50 million years ago, deserts began developing in the southwestern United States. Mountain-building processes created higher elevations, causing rain-shadow aridity and affecting plant distribution.

Recall, too, that the movement of Earth's tectonic plates created climate changes important in the evolution and distribution of plants and animals (see Figure 8.15). Europe and North America were joined in Pangaea and positioned near the equator, where vast swamps formed (the site of coal deposits today). The southern mass of Gondwana was extensively glaciated as it drifted at high latitudes in the Southern Hemisphere. This glaciation left matching glacial scars and specific distributions of plants and animals across South America, Africa, India, and Australia. The diverse and majestic dinosaurs dispersed with the drifting continents.

The key question for our future is: As temperature patterns change, how fast can plants either adapt to new conditions or migrate through succession (location change) to remain within their shifting specific habitats? Rapid changes in vegetation patterns in northern latitudes occurred in the last two decades, a conclusion based on satellite observations and field surveys. The global warming during spring is reducing snow cover, allowing spring greening to occur significantly earlier and delaying the onset of fall by one to two weeks. The changing distribution of plants, animals, and insects is evidence that this is happening.

Nina Leopold Bradley is studying the Wisconsin setting her father Aldo Leopold made famous in his book *A Sand County Almanac* (Oxford University Press, 1949). He kept detailed records of all things dealing with nature and seasonal change on his tract of land. The new findings show that more than one-third of animal and plant species start spring activities 4 weeks earlier than they did 50 years before.

Adaptation to conditions is key to evolution. Through mutation and natural selection, species have either adapted or failed to adapt to changing environmental conditions over millions of years. Current global change is occurring rapidly, at the rate of decades instead of millions of years. Thus, we see die-out and succession of different species along disadvantageous habitat margins. The displaced species may colonize new regions made more hospitable by climate change. Also shifting are current agricultural climates that produced wheat, corn, soybeans, and other commodities. Society will have to adapt to crops growing in different places. Rapid environmental change can lead to outright extinction of plants and animals that are unable to adapt or disperse.

A study completed by biologist Margaret Davis on North American forests suggests that trees will have to respond quickly if temperatures increase. Changes in the climate inhabited by certain species could shift 100–400 km (60–250 mi) during the next 100 years. Some species, such as the sugar maple, are migrating northward, disappearing from the United States except for Maine, and moving into eastern Ontario and Québec. Davis prepared a map showing the possible impact of increasing temperatures on the distribution of beech and hemlock trees (Figure 16.16).

> Changes in the geographical distributions of plant and animal species in response to future greenhouse warming threaten to reduce biotic diversity.... The risk posed by CO_2-induced warming depends on the distances that regions of suitable climate are displaced northward [in the Northern Hemisphere] and on the rate of displacement.... If the change occurs too rapidly for colonization of newly available regions, population sizes may fall to critical levels, and extinction will occur.†

Ecological Succession

Ecological succession occurs when more complex communities replace older, usually simpler, communities of plants and animals. Each successive community of species modifies the physical environment in a manner suitable for the establishment of a later community of species. Changes apparently move toward a more stable and mature condition, although disturbances are common.

Traditionally, it was assumed that plants and animals eventually formed a *climax community*—a stable, self-sustaining, and symbiotically functioning community with balanced birth, growth, and death—but this notion has been mostly abandoned by scientists. Contemporary conservation biology, biogeography, and ecology assume nature to be in constant adaptation and nonequilibrium. Rather than thinking of an ecosystem as a uniform set of communities, think of an ecosystem as a patchwork mosa-

*E.O. Wilson, *Consilience, The Unity of Knowledge* (New York: Knopf, 1998), p. 292.

†M. B. Davis and C. Zabinski, "Changes in the Geographical Range Resulting from Greenhouse Warming: Effects on Biodiversity in Forests," in R. L. Peters and T. E. Lovejoy, eds., *Global Warming and Biological Diversity* (New Haven, CT: Yale University Press, 1992), p. 297.

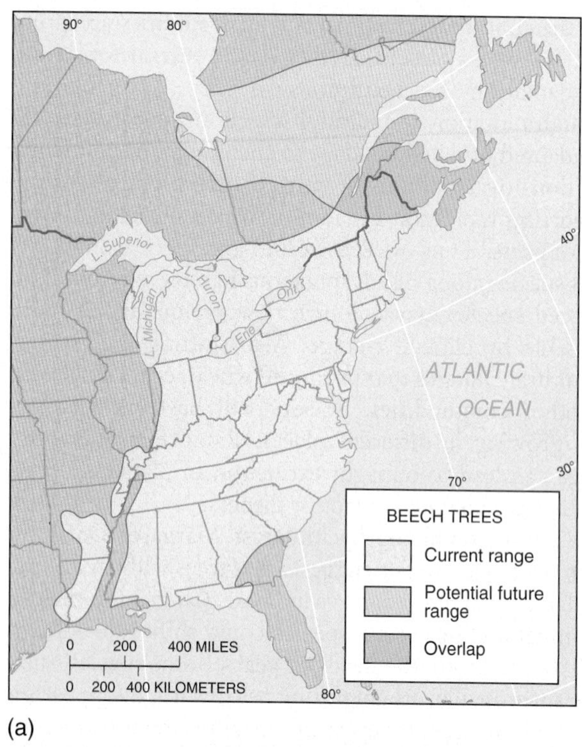

(a)

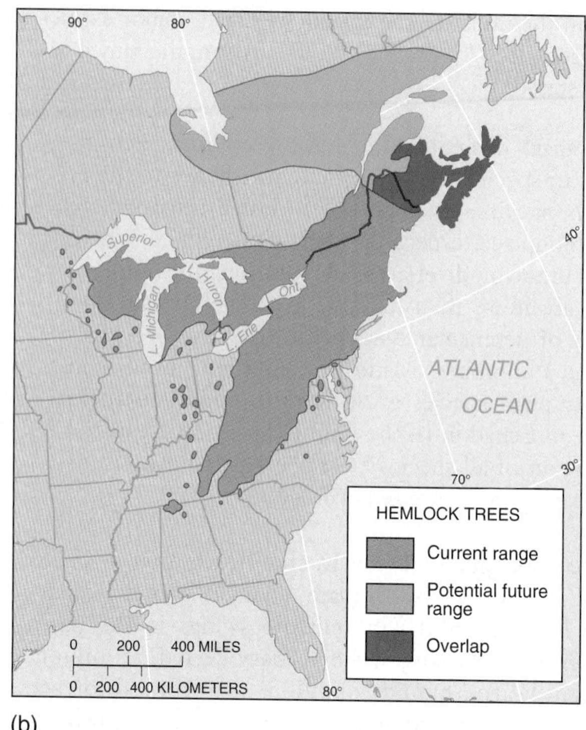

(b)

FIGURE 16.16 Present and predicted distribution of beech trees in North America. Changes in growth patterns due to climate change resulting from a doubling of CO_2 using the Goddard Fluid Dynamics Laboratory (GFDL) general circulation model. [After M. B. Davis and C. Zabinski, "Changes in the Geographical Range Resulting from Greenhouse Warming: Effects on Biodiversity in Forests," in R. L. Peters and T. E. Lovejoy, eds., *Global Warming and Biological Diversity* (New Haven: Yale University Press, 1992), p. 301.]

ic of habitats—each striving to achieve an optimal range and low environmental stress.

Patch dynamics is the study of disturbed portions of habitats. Within an ecosystem, individual patches may arise only to fail later. The complexity of succession is in the interactions of structure and function among patches. Most ecosystems are actually made up of patches of former landscapes.

Given the complexity of natural ecosystems, it is obvious that real succession involves much more than a series of predictable stages ending with a specific monoclimax community. Instead, there may be several stages, or a polyclimax condition, with adjoining ecosystems, or patches within ecosystems, at different stages in the same environment. Mature communities are properly thought of as being in dynamic equilibrium. Or at times, these communities may be out of phase and in nonequilibrium with the immediate physical environment because of the usual lag time in their adjustment.

Succession often requires an initiating disturbance. External examples include windstorm, severe flooding, a volcanic eruption, a devastating wildfire, an agricultural practice such as prolonged overgrazing, climate change, or an insect infestation (Figure 16.17). When existing organisms are disturbed or removed, new communities can emerge. At such times of nonequilibrium transition, the interrelationships among species produce elements of chance, and species having an adaptive edge will succeed in the competitive struggle for light, water, nutrients, space, time, reproduction, and survival.

In Figure 16.17d you see dead and dying Coulter pines (*Pinus coulteri*) in southern California mountains. These drought-stressed trees can die within a week of infestation by opportunistic bark beetles. Thus, the succession of plant and animal communities is an intricate process with many interactive variables, both in space and time and from internal and external processes.

Land and water experience different forms of succession. We first look at terrestrial succession, which is characterized by competition for sunlight, and then at aquatic succession, characterized by progressive changes in nutrient levels. Succession is a dynamic process of unpredictable outcomes—the basis of *dynamic ecology*.

Terrestrial Succession. An area of bare rock or soil with no vestige of a former community can be a site for **primary succession**, the beginning of development of an ecosystem. Examples are any new surface created by mass movement of land, areas exposed by a retreating glacier, cooled lava flows, or lands disturbed by surface mining, clear-cut logging, land development, or volcanic eruption. This primary succession is illustrated in the lava flows of Hawai'i in Figure 16.18 Primary succession often begins with lichens and mosses growing on bare rock (see Figure 16.3f).

More common is **secondary succession**, which begins if the vestiges of a previously functioning community are present. An area where the natural community has been destroyed or disturbed, but where the underlying soil remains

(a)

(b)

(c)

(d)

FIGURE 16.17 Disturbances in ecosystems alter succession patterns. (a) River flooding. (b) Succession progresses in a burned forest, perhaps progressing by chance, as species with an adaptive edge succeed, such as fireweed (*Epilobium* sp.). (c) Unwise practices cause extensive soil erosion on a hog farm in Iowa, 1999; a disrupted prairie landscape. (d) Drought-stressed Coulter pines are attacked by bark beetles, altering relationships in the ecosystem. [Photos by (a, b, d) Bobbé Christopherson; (c) photo by USDA Natural Resources Conservation Service.]

intact, may experience secondary succession. In terrestrial ecosystems, secondary succession begins with early successional species that form a **pioneer community**.

As succession progresses, soil develops and a different set of plants and animals with different niche requirements may adapt. Further niche expansion follows as the community matures. Succession, which is really the developmental process forming new communities, is a set of interactions with sometimes unpredictable outcomes, steered by sometimes random, events that trigger a *threshold* leap to a new set of relations as in a dynamic ecology.

FIGURE 16.18 Pioneer species begin primary succession. Succession and recovery of pioneer species as ferns take hold on new lava flows coming from Kīlauea volcano, Hawai'i. [Photo by Bobbé Christopherson.]

Examples of secondary succession are most of the areas affected by the Mount St. Helens eruption and blast in 1980 (Figure 16.19). Some soils, young trees, and plants were protected under ash and snow, so community development began almost immediately after the event. About 38,450 hectares (95,000 acres) of trees were blown down and burned, however; and as the photos show, the effects of the eruption were devastating and lingering. Of course, the areas completely destroyed near the Mount St. Helens volcano or those buried beneath the massive landslide north of the mountain became candidates for primary succession.

Wildfire and Fire Ecology. Fire and its effects are one of Earth's significant ecosystem and economic processes. In the United States during 2000, about 7.4 million acres, more than double the annual average, burned in wildfires—an incredible 92,000 fires (Figure 16.20a). Over the past 50 years, the role of fire in ecosystems has been the subject of much scientific research and experimentation.

Modern society's demand for fire prevention to protect property goes back to European forestry of the 1800s. Fire prevention became an article of faith for forest managers in North America.

(a) 1979

(b) 1980

Landsat-7 8/22/1999

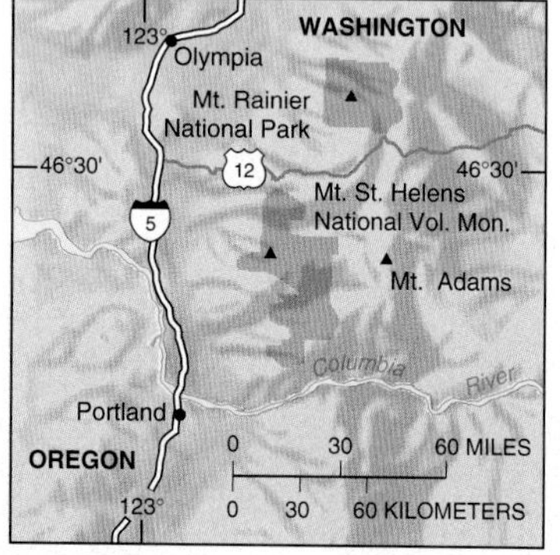

(c) 1983

1999

(d) 1983

1999

(e) 1983

1999

(f) 1983

1999

FIGURE 16.19 The pace of change in the region of Mount St. Helens. (a) The area north of the volcano before the eruption (1979) and (b) after the eruption (1980) experienced profound change in all ecosystems. A series of four photo comparisons made in 1983 and 1999 of matching scenes illustrate the slow recovery of secondary succession: (c) and (d) at Meta Lake; (e) Spirit Lake panorama view (note the mat of floating logs after 19 years); (f) detail of the destroyed forest community. [(a) and all 1983 photos by author; (b) and all 1999 photos by Bobbé Christopherson; *Landsat-7* 8/22/99 satellite image courtesy of NASA/GSFC.]

(a)

(b)

FIGURE 16.20 Wildfires affect environments.
(a) The drama of wildfire unfolds in the Bitterroot National Forest in Montana, August 2000, as elk stand in the Bitterroot River attempting to survive the conflagration. (b) Fire-adapted chaparral vegetation a few months after a wildfire as recovery begins, in the San Jacinto Mountains of southern California. Here we see stump sprouts from chamise, or greasewood, (*Adenostoma fasciculatum*) in abundance after the recent fire. [Photos by (a) John McColgan, BLM Alaska Fire Service; (b) Bobbé Christopherson.]

Today fire is recognized as a natural component of most ecosystems and not the enemy of nature that was once thought (Figure 16.20b). In fact, in many forests, undergrowth and surface litter is purposely burned in controlled "cool fires" to remove fuel that could enable a catastrophic and destructive "hot fire." Forestry experts have learned that when fire-prevention strategies are rigidly followed, they can lead to the abundant undergrowth accumulation that fuels major fires.

In studies of the longleaf pine forest that stretches in a wide band from the Atlantic coastal plain to Texas, fire was discovered to be an integral part of regrowth following lumbering. In fact, seed dispersal of some pine species, such as the knobcone pine, does not occur unless assisted by a forest fire! Heat from the fire opens the cones, releasing seeds so they can fall to the ground for germination. Also, these fire-disturbed areas quickly recover with protein-rich woody growth, young plants, and a stimulated seed production that provides abundant food for animals.

The science of **fire ecology** imitates nature by recognizing fire as a dynamic ingredient in community succession. The U.S. Forest Service first recognized the principle of fire ecology in the early 1940s and formally implemented the practice in 1972. Controlled ground fires, deliberately set to prevent accumulation of forest undergrowth, now are widely regarded as a wise forest management practice and are used across the country.

Nonetheless, after more than a decade of intense forest fires in the western United States, especially those that charred portions of the highly visible Yellowstone National Park (Figure 16.21), an outcry was heard from forestry and recreational interests. The demand was for the Forest Service and ecologists to admit they were wrong and to

(a)

(b)

FIGURE 16.21 Yellowstone burns and is recovering.
(a) A view of the great fire of 1988 in Yellowstone National Park in northwestern Wyoming.
(b) Trees and shrubs return to Yellowstone. [Photos by (a) Joe Peaco/National Park Service; (b) Bobbé Christopherson.]

abandon fire ecology. Critics called fire ecology practice the government's "let it burn" policy.

In its final report on the 1988 Yellowstone fire, a government interagency task force concluded, "an attempt to exclude fire from these lands leads to major unnatural changes in vegetation ... as well as creating fuel accumulation that can lead to uncontrollable, sometimes very damaging wildfire." Thus, participating federal land managers and others reaffirmed their stand that fire ecology is a fundamentally sound concept.

Aquatic Succession. Lakes, ponds, and oxbows exhibit another form of ecological succession. A lake experiences successional stages as it fills with nutrients and sediment and as aquatic plants take root and grow, capturing more sediment and adding organic debris to the system (Figure 16.22a). This gradual enrichment of water bodies is known as **eutrophication**, from the Greek *eutrophos* meaning "well nourished."

For example, in moist climates, a floating mat of vegetation grows outward from the shore to form a bog

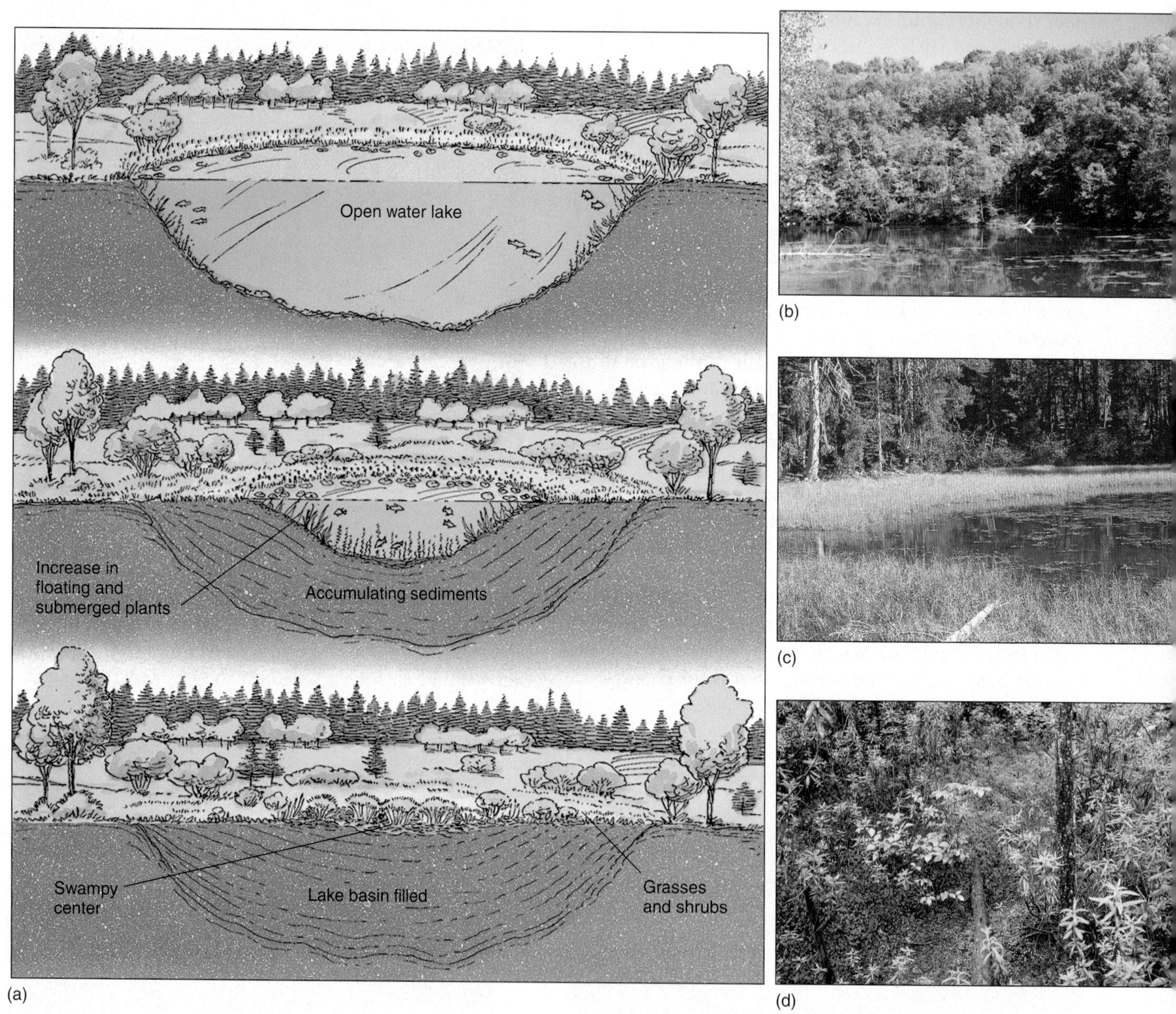

FIGURE 16.22 Lake-bog-meadow succession.
(a) What begins as a lake gradually fills with organic and inorganic sediments, which successively shrink the area of the pond. A bog forms, then a marshy area, and finally a meadow completes the successional stages. (b) Spring Mill Lake, Indiana. (c) Aquatic succession in a mountain lake. (d) The Richmond bog was ocean just 8000 years ago. Frasier River sediment created the mudflats and set the stage for the evolution of this sphagnum bog with acidic soil that "quakes" when you walk on it. Plants in the bog include moss, shore pine, hemlock, blueberry, salal shrubs, and Labrador tea, among others. [Photos by Bobbé Christopherson.]

(Figure 16.22b). Cattails and other marsh plants become established, and partially decomposed organic material accumulates in the basin, with additional vegetation bordering the remaining lake surface. A meadow may form as water is displaced by the peat bog; willow trees follow, and perhaps cottonwood trees. Eventually, the lake may evolve into a forest community.

Now that we have examined ecosystem operations, let us look at terrestrial ecosystem patterns—Earth's terrestrial biomes.

Earth's Major Terrestrial Biomes

A **terrestrial ecosystem** is a self-regulating association of plants and animals and their abiotic environment characterized by specific plant formations. In their growth, form, and distribution, plants reflect Earth's physical systems: its energy patterns; atmospheric composition; temperature and winds; air masses; water quantity, quality, and seasonal timing; soils; regional climates; geomorphic processes; and ecosystem dynamics. We know from history that the effect of climate change on ecosystems will be significant. Two Harvard researchers assessed the impact of these alterations on ecosystems:

> Based on more than a decade of research, it is obvious that the CO_2-rich atmosphere of our future will have direct and dramatic effects on the composition and operation of ecosystems. According to the best scientific evidence, we see no reason to be sanguine [optimistic] about the response of these habitats to our changing environment.*

Terrestrial Ecosystem Concepts

A **biome** is a large, stable terrestrial ecosystem characterized by specific plant and animal communities. Each biome usually is named for its *dominant vegetation*. We can generalize Earth's wide-ranging plant species into these broad biomes: *forest*, *savanna*, *grassland*, *shrubland*, *desert*, and *tundra*. Because plant distributions respond to environmental conditions and reflect variations in climate and soil, the biome-related world climate map (Figure 6.5) is a helpful reference for this chapter.

A boundary transition zone between adjoining ecosystems or biomes is an **ecotone**. Because ecotones are defined by different physical factors, they vary in width. Climatic ecotones usually are more gradual than physical ecotones, because differences in soil or topography sometimes form abrupt boundaries. An ecotone between prairies and northern forests may occupy many kilometers of land. The ecotone is an area of tension as similar species of plants and animals compete for the resource base.

These general biomes are divided into more specific **formation classes**, vegetation units that refer to the dominant plants in a terrestrial ecosystem. Examples are equatorial rain forest, northern needleleaf forest, Mediterranean shrubland, and arctic tundra. Each formation class includes numerous plant communities, and each community includes innumerable plant habitats. Within those habitats, Earth's diversity is expressed in 270,000 plant species. Despite this intricate complexity, we can generalize Earth's numerous formation classes into 10 global terrestrial biome regions.

Few natural communities of plants and animals remain; most biomes have been greatly altered by human intervention. Thus, the "natural vegetation" identified on many biome maps reflects ideal mature vegetation potential, given the physical factors in a region. Even though human practices have greatly altered these ideal forms, it is valuable to study the natural (undisturbed) biomes to better understand the natural environment and to assess the extent of human-caused alteration.

In addition, knowing the ideal in a region helps us approximate natural vegetation in the plants we introduce. In the United States and Canada, we humans are perpetuating a type of transition community, somewhere between a grassland and a forest. We plant trees and lawns and then must invest energy, water, and capital to sustain such artificial modifications of nature. Some irreversible damage might occur when plants, animals, and organisms are knowingly or accidentally brought into an ecosystem or biome in which they are not native. These exotic species pose an increasing concern to scientists and society—a subject of News Report 16.2.

The global distribution of Earth's 10 major terrestrial biomes, based on vegetation formation classes, is portrayed in Figure 16.23 (pp. 508–509). Table 16.2 describes each biome on the map and summarizes other pertinent information from throughout this text—a compilation from many chapters, for Earth's biomes are a synthesis of physical geography.

Equatorial and Tropical Rain Forest

Earth is girdled with a lush biome—the **equatorial and tropical rain forest**. In a climate of consistent daylength (12 hours), high insolation, and average annual temperatures around 25°C (77°F), plant and animal populations have responded with the most diverse body of life on the planet. The Amazon region is the largest tract of equatorial and tropical rain forest, also called the *selva*. Rain forests also cover equatorial regions of Africa, Indonesia, the margins of Madagascar and Southeast Asia, the Pacific coast of Ecuador and Colombia, and the east coast of Central America, with small discontinuous patches elsewhere. The cloud forests of western Venezuela are such tracts of rain forest at high elevation, perpetuated by high humidity and cloud cover. Undisturbed tracts of rain forest are rare.

Rain forests feature ecological niches distributed vertically rather than horizontally, because of the competition for light. The canopy is filled with a rich variety of plants and animals. Lianas (vines) stretch from tree to tree, entwining them with cords that can reach 20 cm (8 in.) in diameter. Epiphytes flourish there, too: such plants

*F. A. Bazzaz and E. D. Fajer, "Plant Life in a CO_2-Rich World," *Scientific American* 256, no. 1 (January 1992): 74.

Table 16.2 Major Terrestrial Biome and Their Characteristics

Biome and Ecosystems (map symbol)	Vegetation Characteristics	Soil Orders (Soil Taxonomy and CSSC)	Climate Type	Annual Precipitation Range	Temperature Patterns	Water Balance
Equatorial and Tropical Rain Forest (ETR) Evergreen broadleaf forest Selva	Leaf canopy thick and continuous; broadleaf evergreen trees, vines (lianas), epiphytes, tree ferns, palms	Oxisols Ultisols (on well-drained uplands)	Tropical	180–400 cm (>6 cm/mo)	Always warm (21–30°C; avg. 25°C)	Surpluses all year
Tropical Seasonal Forest and Scrub (TrSF) Tropical monsoon forest Tropical deciduous forest Scrub woodland and thorn forest	Transitional between rain forest and grasslands; broadleaf, some deciduous trees; open parkland to dense undergrowth; acacias and other thorn trees in open growth	Oxisols Ultisols Verdsols (in India) Some Alfisols	Tropical monsoon savanna	130–200 cm (<40 rainy days during 4 driest months)	Variable, always warm (>18°C)	Seasonal surpluses and deficits
Tropical Savanna (TrS) Tropical grassland Thorn tree scrub Thorn woodland	Transitional between seasonal forests, rain forests, and semiarid tropical steppes and desert; trees with flattened crowns, clumped grasses, and bush thickers; fire association	Alfisols (dry; Ultalfs) Ultisols Oxisols	Tropical savanna	9–150 cm, seasonal	No cold-weather limitations	Tends toward deficits, therefore fire- and drough-susceptible
Midlatitude Broadleaf and Mixed Forest (MBME) Temperate broadleaf Midlatitude deciduous Temperate needleaf	Mixed broadleaf and needleaf trees; deciduous broadleaf, losing leaves in winter; southern and eastern evergreen pines demonstrate fire association	Ultisols Some Alfisols (Podzols, red and yellow)	Humid subtropical warm summer Humid continental warm summer	75–150 cm	Temperate, with cold season	Seasonal pattern with summer maximum PRECIP and POTET (PET); no irrigation needed
Needleleaf Forest and Montane Forest (NF/MF) Taiga Boreal forest Other montane forests and highlands	Needleleaf conifers, mostly evergreen pine, spruce, fir; Russian larch, a deciduous needleleaf	Spodosols Histosols Inceptisols Alfisols (Boralfs: cold) (Glcysols) (Podzols)	Subarctic Humid continental cool summer	30–100 cm	Short summer, cold winter	Low POTET (PET), moderate PRECIP, moist soils, some waterlogged and frozen in winter, no deficits
Temperate Rain Forest (TeR) West coast forest Coast redwoods (U.S.)	Narrow margin of lush evergreen and deciduous trees on windward slopes; redwoods, tallest trees on Earth	Spodosols Inceptisols (mountainous environs) (Podzols)	Marine west coast	150–500 cm	Mild summer and mild winter for latitude	Large surpluses and runoff
Mediterranean Shrubland (MSh) Sclerophyllous shrubs Australian eucalyptus forest	Short shrubs, drought adapted, tending to grassy woodlands chaparral	Alfisols (Xeralfs) Mollisols (Xerolls) (Luvisols)	Mediterranean dry summer	25–65 cm	Hot, dry summers, cool winters	Summer deficits, winter surpluses
Midlatitude Grasslands (MGr) Temperate grassland Sclerophyllous shrub	Tallgrass prairies and shortgrass steppes, highly modified by human activity; major areas of commercial grain farming; plains, pampas, and veld	Mollisols Aridisols (Chernozemic)	Humid subtropical Humid continental hot summer	25–75 cm	Temperate continental regimes	Soil moisture utilization and recharge balanced; irrigation and dry farming in drier areas
Warm Desert and Semidesert (DBW) Subtropical desert and scrubland	Bare ground graduating into xerophytic plants including succulents, cacti, and dry shrubs	Aridisols Entisols (sand dunes)	Desert Arid	<2 cm	Average annual temperature, around 18°C, highest tempera-tures on Earth	Chronic deficits, irregular precipitation events, PRECIP $<\frac{1}{2}$ POTET (PET)
Cold Desert and Semidesert (DBC) Midlatitude desert, scrubland, and steppe	Cold desert vegetation includes short grass and dry shrubs	Aridisols Entisols	Steppe Semiarid	2–25 cm	Average annual temperature around 18°C	PRECIP $>\frac{1}{2}$ POTET (PET)
Arctic and Alpine Tundra (AAT)	Treeless; dwarf shrubs, stunted sedges, mosses, lichens, and short grasses; alpine, grass meadows	Gelisols Histosols Entisols (permafrost) (Organic) (Cryosols)	Tundra Subarctic very cold	15–180 cm	Warmest months <10°C, only 2 or 3 months above freezing	Not applicable most of the year, poor drainage in summer
Ice			Ice sheet, ice cap			

News Report 16.2

Exotic Species Invasions

The above title sounds like that of a "B" science fiction movie; instead, it refers to a real problem in the integrity of many ecosystems, both aquatic and terrestrial. Animal and plant species brought or somehow getting into biomes outside their native home is an important subject in biogeography and invasion biology. Niche takeovers by nonnative species can prove damaging to otherwise healthy ecosystems. These intruders are known as *exotic species*, or *alien* or *nonnative species*.

The "Biological Invasions" Web site from Oregon State University, at http://seagrant.orst.edu/colloquium/index.html, is dedicated to these invasive plants, animals, and other organisms. Some states post warning signs at their borders (Figure 1a), while others offer Web sites such as Minnesota's "Prohibited and Noxious Plants by Scientific Name" (http://www.dnr.state.mn.us/ecological_services/exotics.html).

The African "killer bees" are frequently in the headlines, as are brown tree snakes in Guam, zebra mussels in the Great Lakes, and Eurasian cheatgrass in the Utah desert, to name but a few examples. Brought from Africa to the central coast of Brazil in 1957 to increase honey production, killer bees now have interbred with native bees and range to southern California, Arizona, New Mexico, Texas, and Puerto Rico. More than 1000 people have died from their attacks, although the dangers are in their mass response to disturbances and that some people are allergic to even one bee sting. This is a classic case of geographical diffusion of an exotic species.

Probably 90% of exotic species fail when they try to move into established niches in a community. The 10% that succeed, some 4500 species documented in a 1993 Office of Technology Assessment report, prove damaging to as much as one-fifth of ecosystems they invade. Figures 1b and c show two such alien invaders: purple loosestrife (*Lythrum salicaria*) and kudzu (*Pueraria montana*).

Purple loosestrife was introduced from Europe in the 1800s as a desired ornamental and had some medicinal applications. The plant's seeds also arrived in ships that used soil for ballast. A hardy perennial, it got loose and invaded wetlands across the eastern portions of the United States and Canada, through the upper Midwest and as far west as Vancouver Island, British Columbia, replacing plants on which native wildlife depend (Figure 1b). The plant is known to infect drier landscapes as well and poses a potential threat to agriculture.

The infamous kudzu was brought from Japan in 1876 for cattle feed, erosion control, and as an ornamental. It has spread to east Texas, southern Pennsylvania and across the South into Florida. The plant can grow 0.3 m (1 ft) on a hot, humid summer day and overtake forests, homes, or anything in its path. Serious effort is going into finding some practical use for the prodigious plant, from pulp (treeless paper), to recipes for cooking, to the telling of new kudzu jokes.

In learning about biomes, we find out about native plant, animal, and organism locations and the unique place each species occupies. We must avoid knowingly or perhaps unconsciously adding to this escalating problem by bringing exotic species into nonnative sites.

(a)

(b)

(c)

FIGURE 1 Exotic species.
(a) A sign in Wallowa County, Washington, pleads for awareness and help in this agricultural country, listing noxious (nonnative, often problematic) weeds at the county line. (b) Invasive purple loosestrife near Lake Michigan and the Indiana Dunes National Lakeshore, Indiana (foreground and left center). (c) Kudzu overruns pasture and forest in western Georgia. [Photos by Bobbé Christopherson.]

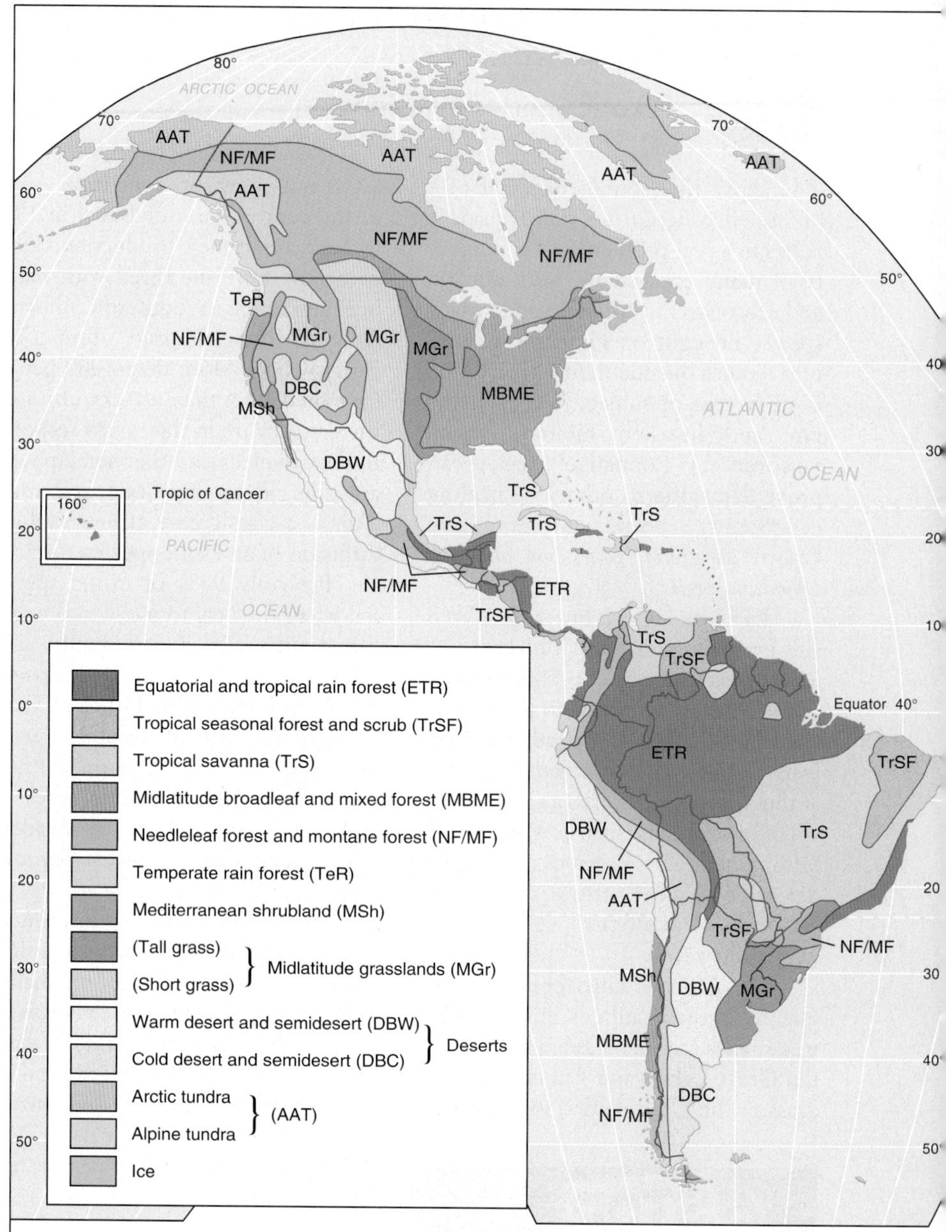

FIGURE 16.23 The 10 major global terrestrial biomes.

as orchids, bromeliads, and ferns that live entirely above ground, supported physically but not nutritionally by the structures of other plants. Windless conditions on the forest floor make pollination difficult—insects, other animals, and self-pollination predominate.

The rain forest canopy forms three levels—see Figure 16.24. The upper level is not continuous but features tall trees whose high crowns rise above the middle canopy. The middle canopy is the most continuous, with its broad leaves blocking much of the light and creating a darkened forest floor. The lower level is composed of seedlings, ferns, bamboo, and the like, leaving the litter-strewn ground level in deep shade and fairly open.

A look at aerial photographs of a rain forest, or views along river banks covered by dense vegetation, aided by the false Hollywood-movie imagery of the jungle, make it difficult to imagine the shadowy environment of the actual rain forest floor (Figure 16.25a). The forest floor receives only about 1% of the sunlight arriving at the canopy. The constant moisture, rotting fruit and moldy odors, strings of thin roots and vines dropping down from above, windless air, and echoing sounds of life in the trees together create a unique environment.

The smooth, slender trunks of rain forest trees are covered with thin bark and buttressed by large wall-like flanks that grow out from the trees to brace the trunks (Figure 16.25b). These buttresses form angular open enclosures, a ready habitat for various animals. There are usually no branches for at least the lower two-thirds of the tree trunks.

Varieties of trees include mahogany, ebony, and rosewood. Logging is difficult because individual species are

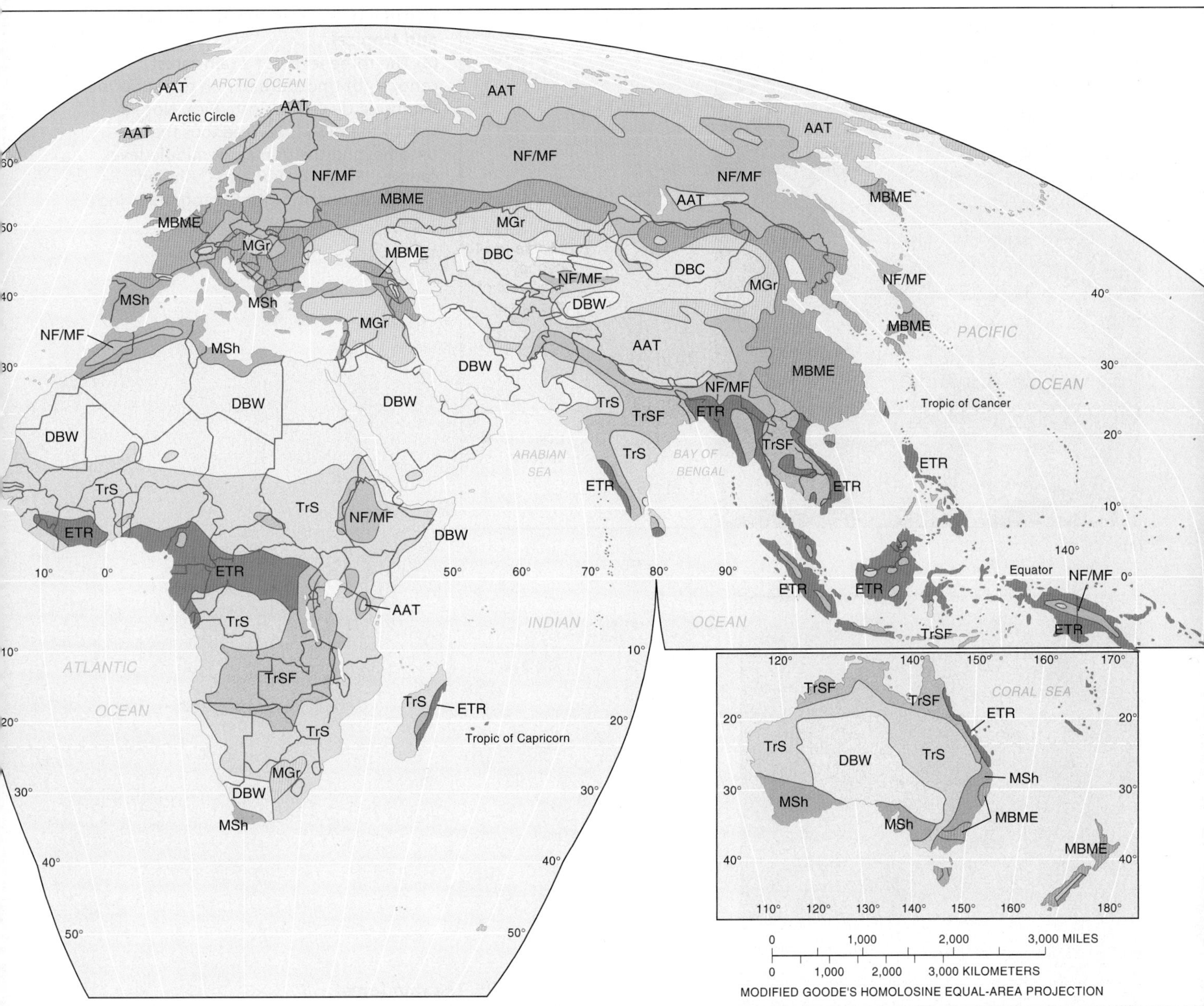

widely scattered; a species may occur only once or twice per square kilometer. Selective cutting is required for species-specific logging, whereas pulpwood production takes everything. Conversion of the forest to pasture usually is accomplished by setting destructive fires.

Rain forests represent approximately half of Earth's remaining forests, occupying about 7% of the total land area worldwide. This biome is stable in its natural state, resulting from the long-term residence of these continental plates near equatorial latitudes and their escape from glaciation. The soils, principally Oxisols, are essentially infertile, yet they support rich vegetation. The trees have adapted to these soil conditions with root systems able to capture nutrients from litter decay at the soil surface. With much investment in fertilizers, pesticides, and machinery, these soils can be productive.

The animal and insect life of the rain forest is diverse, ranging from small decomposers (bacteria) working the surface to many animals living exclusively in the upper stories of the trees. These tree dwellers are referred to as *arboreal*, from the Latin for "tree," and include sloths, monkeys, lemurs, parrots, and snakes. Beautiful birds of many colors, tree frogs, lizards, bats, and a rich insect life that includes over 500 species of butterflies are found in rain forests. Surface animals include pigs (bushpigs and the giant forest hog in Africa, wild boar, bearded pig in Asia, and peccary in South America), species of small antelopes (bovids), and mammal predators (tiger in Asia, jaguar in South America, and leopard in Africa and Asia).

Deforestation of the Tropics. More than half of Earth's original rain forest is gone, cleared for pasture,

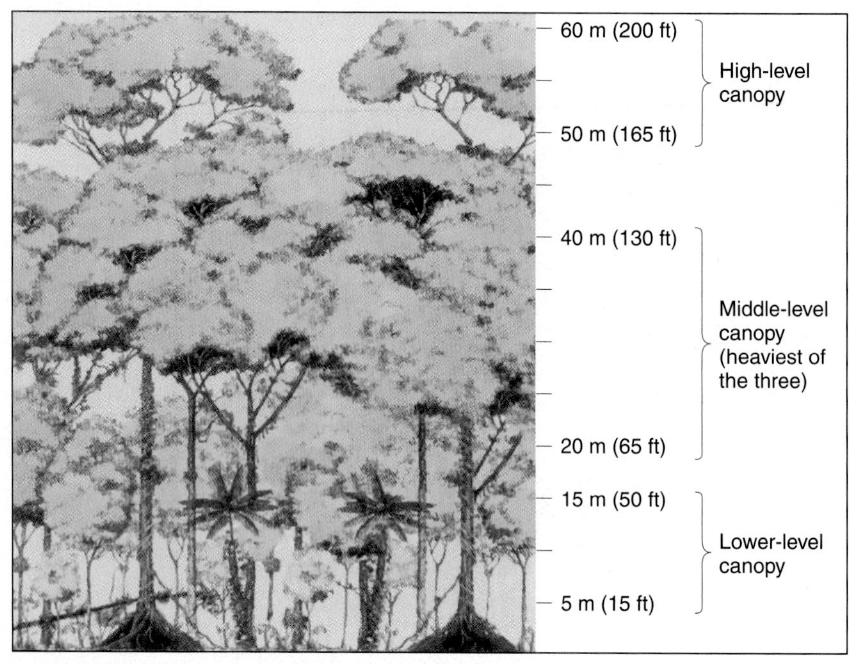

(a)

(b)

FIGURE 16.24 Rain forest—equatorial and tropical.
(a) The three levels of a rain forest canopy. (b) Undisturbed rain forest on the Peninsula de Osa in Costa Rica. Note a few of the high-level tree tops that extend beyond the dominant middle-level canopy. [Photo by Barbara Cushman Rowell/Mountain Light Photography, Inc.]

(a)

(b)

FIGURE 16.25 The rain forest.
(a) Equatorial rain forest is thick along the Amazon River where light breaks through to the surface, producing a rich gallery of vegetation. (b) The rain forest floor in Corcovado, Costa Rica, with typical buttressed trees and lianas. [Photos by (a) Wolfgang Kaehler; (b) Frank S. Balthis.]

timber, fuel wood, and farming. Worldwide an area nearly the size of Wisconsin is lost each year and about a third more is disrupted by selective cutting of canopy trees, where damage occurs along the *edges* of newly deforested areas (Figure 16.26).

When orbiting astronauts look down on the rain forests at night, they see thousands of human-set fires. During the day, the lower atmosphere in these regions is choked with smoke. These fires are used to clear land for agriculture, which is intended to feed the domestic population as well as to produce cash exports of beef, rubber, coffee, and other commodities. Because of the poor soil fertility, the cleared lands are quickly exhausted under intensive farming and are

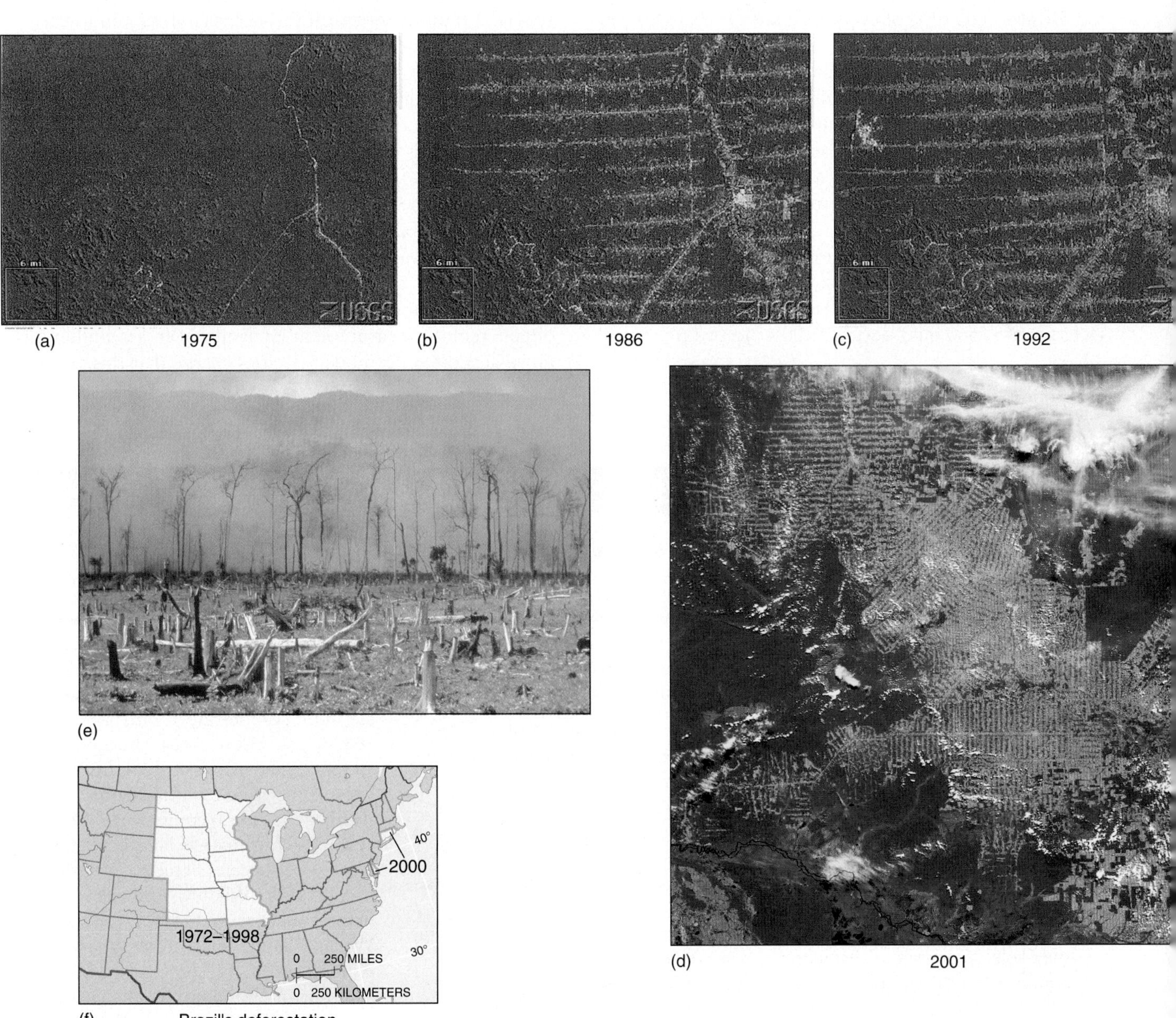

FIGURE 16.26 Rain forest losses.
Satellite images of a portion of western Brazil called Rondônia, recorded in false color in (a) June 1975 (*Landsat 2*), (b) August 1986 (*Landsat 5*), (c) June 1992 (*Landsat 4*), and in true color (d) June 1, 2001 (*Terra* MODIS sensor; the area of the other three images is in the upper left). Highway BR364, the main artery in the region, passes through the towns of Jaru and Ji-Paraná near the center of image (d) at about 62° W 11° S. The branching pattern of feeder roads encroach on the rain forest. (e) Surface view of burning rain forest in Guatemala. (f) Deforested areas in Brazil, cleared between 1972 and 2000, and for 2000 only, are compared to North America. [*Landsat* images courtesy of NASA/USGS EROS Data Center; *Terra* MODIS image courtesy of NASA/MODIS Land Rapid Response Team; (e) photo by George Holton/Photo Researchers, Inc.]

then generally abandoned in favor of newly burned and cleared lands, unless the lands are artificially maintained.

Figures 16.26a, b, c shows satellite false-color images of a portion of western Brazil called Rondônia, recorded in June 1975 (*Landsat 2*), August 1986 (*Landsat 5*), and June 1992 (*Landsat 4*), and in (d) a true-color June 2001 image (*Terra*, MODIS). These images give a sense of the level of rain forest destruction in progress. You can clearly see encroachment along new roads branching from highway BR364. The *edges* of every road and cleared area represent a significant portion of the species habitat disturbance, population dynamic changes, and carbon losses to the atmosphere—significant impact occurs along these edges next to tracts of clear-cut logging. The hotter, drier, and windier edge conditions penetrate the forest up to 100 m. Figure 16.26e shows a tract of former rain forest that has just been burned to begin the clearing and road-building process.

In 1998, despite all the efforts to introduce sustainable forestry practices, deforestation increased 27% over 1997—some 16,800 km^2 (6500 mi^2)—and rose again to 19,836 km^2 (7660 mi^2) lost in 2000, according to the Brazilian Environment Ministry in Brasilia (an area roughly equivalent to Connecticut, Rhode Island, and Delaware combined). Since 1972 this has brought total deforestation to more than 15% of the entire Amazon Basin—more than the area of the combined north-central states of Minnesota, Iowa, Missouri, North and South Dakota, Nebraska, and Kansas (Figure 16.26f), or equivalent to the area of France in just 30 years!

The United Nations Food and Agricultural Organization (FAO) estimates that if this destruction to rain forests continues unabated, these forests will be completely removed by about A.D. 2050! By continent, forest losses are estimated at more than 50% in Africa, over 40% in Asia, and 40% in Central and South America.

To slow this continuing catastrophe of deforestation, in 1985 the FAO, the UN Development Programme, the World Bank, the World Resources Institute, and several nongovernmental organizations initiated the Tropical Forestry Action Plan. This plan, armed with US$8 billion in development aid money, instituted an accurate worldwide survey of the rate of deforestation, using satellites to obtain remote measurement of land cover for building more-accurate GIS models. Critics charge that this effort

Focus Study 16.1

Biodiversity and Biosphere Reserves

When the first European settlers landed on the Hawaiian Islands in the late 1700s, 43 species of birds were counted. Today 15 of those species are extinct, and 19 more are threatened or endangered, with only one-fifth of the original species populations relatively healthy. In most of Hawai'i there are no native species below 1220-m (4000-ft) elevation because of an introduced avian virus. Humans have great impact on animals, plants, and global biodiversity.

> Relatively pristine habitats around the world are being lost at unprecedented rates as an expanding human population converts them to agriculture, forestry, and urban centers. As these habitats are altered, untold numbers of species are disappearing before they have been recognized, much less studied, and the functioning of entire ecosystems is threatened. This loss of biodiversity, at the very time when the value of biotic resources is becoming widely recognized, has made it strikingly clear that current strategies for conservation are failing dismally.*

As more is learned about Earth's ecosystems and their related communities, more is known of their value to civilization and our interdependence with them. Natural ecosystems are a major source of new foods, new chemicals, new medicines, and specialty woods, and of course they are indicators of a healthy, functioning biosphere.

International efforts are underway to study and preserve specific segments of the biosphere, including the following: the UN Environment Programme, World Resources Institute, The World Conservation Union, Rain Forest Action Network, Natural World Heritage Sites, Wetlands of International Importance, and the Nature Conservancy. An important part of the effort is the setting aside of biosphere reserves, and also the focus provided by the World Conservation Monitoring Centre and its IUCN Red List of global endangered species. (See IUCN Red List at **http://www.wcmc.org.uk/data/database/rl_anml_combo.html**, the IUCN at **http://www.iucn.org/**, the endangered species home page of the Fish and Wildlife Service at **http://endangered.fws.gov/**, and the World Wildlife Fund at **http://www.panda.org/**.)

The motivation to set aside natural sanctuaries is directly related to concern over the increase in the rate of species extinctions. We are facing a loss of genetic diversity that may be unparalleled in Earth's history, even compared with the major extinctions over the geologic record.

*T.D. Sisk, A. E. Launer, K. R. Switky, and P. R. Ehrlich, "Identifying extinction threats," *Bioscience* 44 (October 1994): 592.

Species Threatened

Black rhinos (*Diceros bicornis*) and white rhinos (*Ceratotherium simum*) in

(continued)

Focus Study 16.1 *(continued)*

FIGURE 1 The rhinoceros in Africa.
Black rhino and young in Tanzania, escorted by oxpecker birds. These rhinos are quite nearsighted (they can see clearly only up to 10 m); the birds act as the rhino's early-warning system for disturbances in the distance. [Photo by Stephen J. Krasemann/DRK Photo.]

Africa exemplify species in jeopardy. Rhinos once grazed over much of the savanna grasslands and woodlands. Today they survive only in protected districts in heavily guarded sanctuaries and are threatened even there (Figure 1). In 1996, 2408 black rhinos remained, 96% less than the 70,000 in 1960 (source: IUCN/SSC/African Rhino Specialist Group, 1998). Populations have remained statistically stable since 1992. South Africa's government must guard about 50% of the remaining population from poaching.

There were 11 white rhinos (northern subspecies) in existence in 1984, increasing to an estimated 29 by 1998 (Congo 25, Côte d'Ivoire 4). Political unrest in the Congo and the area of the Garamba National Park has led to the deaths of an unknown number of these few. The southern white rhino population reached 8431 by 1998.

Rhinoceros horn sells for US$29,000 per kilogram as an aphrodisiac! These large land mammals are nearing extinction and will survive only as a dwindling zoo population. The limited genetic pool that remains complicates further reproduction.

Table 1 summarizes the numbers of known and estimated species on Earth (see http://www.wri.org/biodiv/gba-unpr.html). Scientists have classified only 1.75 million species of plants and animals of an estimated 13.6 million overall; this figure represents an increase in what scientists once thought to be the diversity of life on Earth. And those yet-to-be-discovered species represent a potential future resource for society. The wide range of estimates places the expected species count between a low of 3.6 million and a high of 111.7 million.

Estimates of annual species loss range between 1000 and 30,000, although this range might be conservative. The possibility exists that over half of Earth's present species could be extinct within the next 100 years. The effects of pollution, loss of wild habitat, excessive grazing, poaching, and collecting are at the root of such devastation. Approximately 60% of the extinctions are attributable to the clearing and loss of rain forests alone.

The Question of Medicine and Food

Wheat, maize (corn), and rice—just three grains—fulfill about 50% of our planetary food demands. About 7000 plant species have been gathered for food throughout human history, but over 30,000 plant species have edible parts. Undiscovered potential food resources are in nature waiting to be

(continued)

Table 1 Known and Estimated Species on Earth

Categories of Living Organisms	Number of Known Species	Estimated Numbers of Species High (000)	Estimated Numbers of Species Low (000)	Working Estimate (000)	Accuracy
Viruses	4000	1000	50	400	Very poor
Bacteria	4000	3000	50	1000	Very poor
Fungi	72,000	27,000	200	1500	Moderate
Protozoa	40,000	200	60	200	Very poor
Algae	40,000	1000	150	400	Very poor
Plants	270,000	500	300	320	Good
Nematodes	25,000	1000	100	400	Poor
Arthropods					
Crustaceans	40,000	200	75	150	Moderate
Arachnids	75,000	1000	300	750	Moderate
Insects	950,000	100,000	2000	8000	Moderate
Mollusks	70,000	200	100	200	Moderate
Chordates	45,000	55	50	50	Good
Others	115,000	800	200	250	Moderate
Total	**1,750,000**	**111,655**	**3635**	**13,620**	**Very poor**

Source: United Nations Environment Programme, *Global Biodiversity Assessment* (Cambridge: Cambridge University Press, 1995), Table 3.1–2, p. 118.

Focus Study 16.1 *(continued)*

found and developed. Biodiversity, if preserved, provides us a potential cushion for all future food needs, but only if species are identified, inventoried, and protected. The same is true for pharmaceuticals.

Nature's biodiversity is like a full medicine cabinet. Between 1959 and 1990, 25% of all prescription drugs were originally derived from plants. In preliminary surveys, 3000 plants have been identified as having anticancer properties. The rosy periwinkle *(Catharanthus roseus)* of Madagascar contains two alkaloids that combat two forms of cancer. (Alkaloids are compounds found in certain plants that help the plant defend against insects and are potentially significant as human medicines; examples are atropine, quinine, and morphine.) Yet less than 3% of flowering plants have been examined for alkaloid content. It defies common sense to throw away the medicine cabinet before we even open the door to see what is inside.

> Genetic material from the developing world is already making an immense contribution to the food tables and medicine cabinets of the industrialized Northern Hemisphere. Agriculture in developed countries reaps an estimated $5 billion a year from Third World genetic material.... One Peruvian variety of tomato has been worth $8 million annually to U.S. tomato processors because it added soluble solid content to tomatoes.... And one in four prescription drugs is derived at least in part from plants.... Their value to the $60-billion-a-year U.S. pharmaceutical industry is considerable.*

*J. W. King, "Breeding uniformity, Will global biotechnology threaten global biodiversity?" *The Amicus Journal* 15, (Spring 1993): 26.

Biosphere Reserves

Formal natural reserves are a possible strategy for slowing the loss of biodiversity and protecting this resource base. Setting up such a biosphere reserve involves principles of *island biogeography*. Island communities are special places for study because of their spatial isolation and the relatively small number of species present. Islands resemble natural experiments, because the impact of individual factors, such as civilization, can be more easily assessed on islands than over larger continental areas.

Studies of islands also can assist in the study of mainland ecosystems, for in many ways a park or biosphere reserve, surrounded by modified areas and artificial boundaries, is like an island. Indeed, a biosphere reserve is conceived as an ecological island in the midst of change. The intent is to establish a core in which genetic material is protected from outside disturbances, surrounded by a buffer zone that is, in turn, surrounded by a transition zone and experimental research areas. An important variable to consider in setting aside a biosphere reserve is any change that might occur in temperature and precipitation patterns as a result of global change. A carefully considered reserve could end up outside its natural range as ecotones (climatic/ecosystem boundaries) shift.

Scientists predict that new, undisturbed reserves will not be possible after A.D. 2010, because pristine areas will be gone. The Man and the Biosphere (MAB) Programme of UNESCO coordinates the biosphere reserve program (http://www.unesco.org/mab/). Nearly 300 such biosphere reserves, covering some 12 million hectares (30 million acres), are now operated voluntarily in 76 countries. Not all protected areas are ideal bioregional entities. Some are simply imposed on existing park space and some remain in the planning stage, although they are officially designated. In Canada there are six biosphere reserve sites; the Canada/MAB Programme is http://www.eman-rese.ca/partners/mab/. The United States has 90 biosphere reserves under various jurisdictions. Also, check another Web site at the University of California at Davis for MAB fauna and MAB flora databases at http://www.ice.ucdavis.edu/mab/.

The ultimate goal is to establish at least one reserve in each of the 194 distinctive biogeographic communities presently identified. UNESCO and the World Conservation Monitoring Centre report the United Nations' list of national parks and protected areas. Presently, 6930 areas covering 657 million hectares (1.62 billion acres) are designated in some protective form, representing about 4.8% of land area on the planet. Even though these are not all set aside to the degree of biosphere reserves, they do demonstrate progress toward preservation of our planet's plant and animal heritage.

> The preservation of species diversity is a problem that must today be confronted by one species, *Homo sapiens*.... The diversity of species is worth preserving because it represents a wealth of knowledge that cannot be replaced. Moreover, today's extinctions are unlike those in previous eras, in which long periods of recovery could follow extinctions. The present situation is an inexorably irreversible one in which human overpopulation will destroy most species unless we plan for protection immediately. Accepting that the goal is worthwhile requires that more energy be devoted to planning and priorities and less to emotionalism and indignation [on all sides].†

†D. E. Koshland Jr., "Preserving biodiversity," *Science* 253, no. 5021 (August 16, 1991): 717.

has yet to show progress. (Among many references, see Tropical Rainforest Coalition at http://www.rainforest.org/, World Resources Institute publications on forests at http://www.wri.org/wri/cat-frst.html, the Rainforest Information Centre at http://www.rainforestinfo.org.au/welcome.htm, or the Rainforest Action Network at http://www.ran.org/.)

Focus Study 16.1 looks more closely at efforts to curb increasing rates of species extinction and loss of Earth's biodiversity, much of which is directly attributable to the loss of rain forests.

Tropical Seasonal Forest and Scrub

A varied biome on the margins of the rain forest is the **tropical seasonal forest and scrub**, which occupies regions of low and erratic rainfall. The shifting intertropical convergence zone (ITCZ) brings precipitation with the seasonally shifting high Sun and dryness with the low Sun, producing a seasonal pattern of moisture deficits, some leaf loss, and dry-season flowering. The term *semideciduous* applies to some of the broadleaf trees that lose their leaves during the dry season.

Areas of this biome have fewer than 40 rainy days during their 4 consecutive driest months, yet heavy monsoon downpours characterize their summers (see Chapter 4, especially Figure 4.20). The climates *tropical monsoon* and *tropical savanna* apply to these transitional communities between rain forests and tropical grasslands.

Portraying such a varied biome is difficult. In many areas, humans disturb the natural biome so that the savanna grassland adjoins the rain forest directly. The biome does include a gradation from wetter to drier areas: monsoonal forests, to open woodlands and scrub woodland, to thorn forests, to drought-resistant scrub species (Figure 16.27). An area of transitional tropical deciduous forest surrounds an area of savanna in southeastern Brazil and portions of Paraguay.

The monsoonal forests average 15 m (50 ft) high with no continuous canopy of leaves, graduating into open orchardlike parkland with grassy openings or into areas choked by dense undergrowth. In more open tracts, a common tree is the acacia, with its flat-topped appearance and usually thorny stems. These trees have branches that look like an upside-down umbrella (spreading and open skyward), as do trees in the tropical savanna.

Local names are given to these communities: the *caatinga* of the Bahia State of northeastern Brazil, the *chaco* area of Paraguay and northern Argentina, the *brigalow* scrub of Australia, and the *dornveld* of southern Africa. The world map in Figure 16.23 shows this biome in Africa, extending from eastern Angola through Zambia to Tanzania; in southeast Asia and portions of India, from interior Myanmar through northeastern Thailand; and in parts of Indonesia.

The trees throughout most of this biome make poor lumber, but some, especially teak, may be valuable for fine cabinetry. In addition, some of the plants with dry-season adaptations produce usable waxes and gums, such as carnauba and palm-hard waxes. Animal life includes the koalas and cockatoos of Australia and the elephants, large cats, rodents, and ground-dwelling birds in other occurrences of this biome.

FIGURE 16.27 Open thorn scrub and forest.
Open thorn forest and savanna in the Samburu Reserve, Kenya. [Photo by Gael Summer-Hebden.]

Tropical Savanna

Large expanses of grassland interrupted by trees and shrubs describe the **tropical savanna**. This is a transitional biome between the tropical forests and semiarid tropical steppes and deserts. The savanna biome also includes treeless tracts of grasslands. The trees of the savanna woodlands are characteristically flat-topped.

Savannas covered more than 40% of Earth's land surface before human intervention but were especially modified by human-caused fire. Fires occur annually throughout the biome. The timing of these fires is important. Early in the dry season they are beneficial and increase tree cover; if late in the season they are very hot and kill trees and seeds. Savanna trees are adapted to resist the "cooler" fires. Elephant grasses averaging 5 m (16 ft) high and forests once penetrated much farther into the dry regions, for they are known to survive there when protected. Savanna grasslands are much richer in humus than the wetter tropics and are better drained, thereby providing a better base for agriculture and grazing. Sorghums, wheat, and groundnuts (peanuts) are common commodities.

Tropical savannas receive their precipitation during less than 6 months of the year, when they are influenced by the ITCZ. The rest of the year they are under the drier influence of shifting subtropical high-pressure cells. Savanna shrubs and trees are frequently *xerophytic*, or drought resistant, with various adaptations to protect them from the dryness: small thick leaves, rough bark, or waxy leaf surfaces.

FIGURE 16.28 Animals and plants of the savanna. Savanna landscape of Samburu, Kenya, with Grevy's zebra and reticulated giraffe. [Photo by Galen Rowell/Mountain Light Photography, Inc.]

Africa has the largest region of this biome, including the famous Serengeti Plains and the Sahel region. Sections of Australia, India, and South America also are part of the savanna biome. Some of the local names for these lands include the *Llanos* in Venezuela, stretching along the coast and inland east of Lake Maricaibo and the Andes; the *Campo Cerrado* of Brazil and Guiana; and the *Pantanal* of southwestern Brazil. Particularly in Africa, savannas are the home of large land mammals that graze on savanna grasses or feed upon the grazers themselves: lion, cheetah, zebra, giraffe, buffalo, gazelle, wildebeest, antelope, rhinoceros, and elephant (Figure 16.28).

However, in our lifetime we may see the reduction of these animal herds to zoo stock only, because of poaching and habitat losses. The loss of both the black and white rhino is discussed in Focus Study 16.1. Establishment of large tracts of savanna as biosphere reserves is critical for the preservation of this biome and its associated fauna.

Midlatitude Broadleaf and Mixed Forest

Moist continental climates support mixed broadleaf and needleleaf forest in areas of warm-to-hot summers and cool-to-cold winters. The **midlatitude broadleaf and mixed forest** biome includes several distinct communities in North America, Europe, and Asia. Relatively lush evergreen broadleaf forests occur along the Gulf of Mexico. Northward are mixed deciduous and evergreen needleleaf stands associated with sandy soils and burning. When areas are given fire protection, broadleaf trees quickly take over.

Pines predominate in the southeastern and Atlantic coastal plains. Into New England and westward in a narrow belt to the Great Lakes, white and red pines and eastern hemlock are the principal evergreens, mixed with deciduous varieties of oak, beech, hickory, maple, elm, chestnut, and many others.

(a)

(b)

FIGURE 16.29 Mixed forest community. (a) Pioneer Mothers' Memorial Forest near Paoli, Indiana, is 36 hectares (88 acres) of old-growth forest, virtually undisturbed since around 1816, featuring black walnut, white oak, yellow poplar, white ash, and beech trees, among others. (b) A young white-tailed deer grazes in the undergrowth of a mixed forest. [Photos by (a) Bobbé Christopherson; (b) John Shaw/Tom Stack & Associates.]

These mixed stands contain valuable timber, but their distribution has been greatly altered by human activity. Native stands of white pine in Michigan and Minnesota were removed before 1910 (Figure 16.29a); only later reforestation sustains their presence today. In northern China these forests have almost disappeared as a result of centuries of occupation. Deforestation in China, as elsewhere, was principally for agricultural purposes, as well as for construction materials and fuel.

The midlatitude broadleaf and mixed forest biome is quite consistent in appearance from continent to continent; at one time it represented the principal vegetation of the *humid subtropical hot-summer* climates, *marine west coast*, and *cool-summer*, *winter drought* climatic regions of North America, Europe, and Asia. To the north of this biome, poorer soils and colder climates favor stands of coniferous trees and a gradual transition to the needleleaf forests.

A wide assortment of mammals, birds, reptiles, and amphibians is distributed throughout this biome. Representative animals (some migratory) include red fox, white-tailed deer, southern flying squirrel, opossum, bear, and a great variety of birds, including tanager and cardinal (Figure 16.29b). A white-tailed deer is characteristic of the mixed forest (Figure 16.29b).

Needleleaf Forest and Montane Forest

Stretching from the east coast of Canada and the Atlantic provinces westward to Alaska and continuing from Siberia across the entire extent of Russia to the European Plain is the **needleleaf forest** biome, also called the **boreal forest** (Figure 16.30). A more open form of boreal forest, transitional to arctic and subarctic regions, is termed the **taiga**. The Southern Hemisphere, lacking *humid microthermal* (at least one month averaging below freezing) climates except in mountainous locales, has no such biome. However, *montane forests* of needleleaf trees exist worldwide at high elevation.

Boreal forests of pine, spruce, and fir occupy most of the subarctic climates on Earth that are dominated by trees. Although these forests are similar in formation, individual species vary between North America and Eurasia. Certain regions of the northern needleleaf biome experience permafrost, discussed in Chapter 14. When coupled with rocky and poorly developed soils, these conditions generally limit the existence of trees to those with shallow root systems.

The Sierra Nevada, Rocky Mountains, Alps, and Himalayas have similar forest communities occurring at lower latitudes. Douglas and white fir grow in the western mountains in the United States and Canada. Economically, these forests are important for lumbering, with saw timber occurring in the southern margins of the biome and pulpwood throughout the middle and northern portions. Present logging practices and whether these yields are sustainable are issues of increasing controversy.

Representative fauna include the wolf, moose (the largest deer), bear, lynx, beaver, wolverine, marten, small rodents, and migratory birds during the brief summer season. Birds include hawks and eagles, several species of grouse, Pine Grosbeak, Clark's Nutcracker, and several owls. About 50 species of insects particularly adapted to the presence of coniferous trees inhabit the biome.

A study released in 1993 by the U.S. Forest Service noted the failing ecology of these forest ecosystems and suggested that timber management plans should include ecosystem preservation as a priority. The government held a "forestry summit" in Portland, Oregon, in 1993, to bring together the opposing sides of the logging-versus-

FIGURE 16.30 Boreal forest of Canada.
[Photo by author.]

Temperate Rain Forest

The **temperate rain forest** biome is recognized by its lush forests at middle and high latitudes, occurring only along narrow margins of the Pacific Northwest in North America (see the chapter-opening photo and Figure 16.31). This biome contrasts with the diversity of the equatorial and tropical rain forest in that only a few species make up the bulk of the trees. The rain forest of the Olympic Peninsula in Washington State is a mixture of broadleaf and needleleaf trees, huge ferns, and thick undergrowth. Precipitation approaching 400 cm (160 in.) per year on the western slopes, moderate air temperatures, summer fog, and an overall maritime influence produce this moist, lush vegetation community. Animals include bear, badger, deer, wild pig, wolf, bobcat, and fox. The trees are home to numerous bird species (Figure 16.31b).

The tallest trees in the world occur in this biome—the coastal redwoods (*Sequoia sempervirens*). Their distribution is shown on the map in Figure 16.11a. These trees can exceed 1500 years of age and typically range in height from 60 to 90 m (200–300 ft), with some exceeding 100 m (330 ft). Virgin stands of other representative trees—Douglas fir, spruce, cedar, and hemlock—have been reduced to a few remaining valleys in Oregon and Washington, less than 10% of the original forest.

(a)

(b)

FIGURE 16.31 Temperate rain forest.
(a) The temperate rain forest of Macmillan Provincial Park, central Vancouver Island, British Columbia—an old-growth Douglas fir forest, with western red cedar and hemlock, sword and bracken ferns, and hanging mosses. (b)The Northern Spotted Owl is characteristic of the temperate rain forest. [Photos by (a) Bobbé Christopherson; (b) Greg Vaughn/Tony Stone Images.]

environment conflict. Despite political actions in 2002 to abandon such understanding, the ultimate solution must be one of economic-ecological synthesis rather than continuing conflict and forest losses.

Mediterranean Shrubland

The **Mediterranean shrubland** biome occupies those regions poleward of the shifting subtropical high-pressure cells. As those cells shift poleward with the high Sun, they cut off available storm systems and moisture. Their stable high-pressure presence produces the characteristic dry summer climate—*Mediterranean dry-summer*—and establishes conditions conducive to fire. Plant ecologists think that this biome is well adapted to frequent fires, for many of its characteristically deep-rooted plants have the ability to resprout from their roots after a fire. Earlier stands of evergreen woodlands no longer dominate (review Figure 16.20).

The dominant shrub formations that occupy these regions are stunted and tough in their ability to withstand hot-summer drought. The vegetation is called *sclerophyllous* (from *sclero* or "hard" and *phyllos* for "leaf"); it averages a meter or two in height and has deep, well-developed roots, leathery leaves, and uneven low branches.

Typically, the vegetation varies between woody shrubs covering more than 50% of the ground and grassy woodlands with 25%–60% coverage. In California, the Spanish word *chaparro* for "scrubby evergreen" gives us the word **chaparral** for this vegetation type (Figure 16.32). A counterpart to chaparral in the Mediterranean region, called *maquis*, includes live and cork oak trees (source of cork), as well as pine and olive trees. In Chile, such a region is called *mattoral*; in southwest Australia, *mallee scrub*. Of course, in Australia the bulk of the eucalyptus species is sclerophyllous in form and structure in whichever climate it occurs.

As described in Chapter 6, Mediterranean climates are important in commercial agriculture for subtropical fruits, vegetables, and nuts, with many food types produced only in this biome (for example, artichokes, olives,

FIGURE 16.32 Chaparral.
Mixed chaparral vegetation associated with the Mediterranean dry-summer climate in southern California. [Photo by Bobbé Christopherson.]

FIGURE 16.33 Farming in the grasslands of North America.
An autumn sunset bathes these Alberta, Canada, prairie farms. Warmer average temperatures are expanding the growing season in Canada and melting permafrost soil in the far north. Conditions of drought are a concern. [Photo by John Eastcott/Yva Momatiuk, DRK Photo.]

almonds). Larger animals, such as different types of deer, are grazers and browsers, with coyote, wolf, and bobcat as predators. Many rodents, other small animals, and a variety of birds also proliferate.

Midlatitude Grasslands

Of all the natural biomes, the **midlatitude grasslands** are most modified by human activity. Here are the world's "breadbaskets"—regions of grain and livestock production (Figure 16.33). In these regions, the only naturally occurring trees were deciduous broadleafs along streams and other limited sites. These regions are called grasslands because of the original predominance of grasslike plants:

> In this study of vegetation, attention has been devoted to grass because grass is the dominant feature of the Plains and is at the same time an index to their history. Grass is the visible feature which distinguishes the Plains from the desert. Grass grows, has its natural habitat, in the transition area between timber and desert.... The history of the Plains is the history of the grasslands.*

In North America, tall-grass prairies once rose to heights of 2 m (6.5 ft) and extended westward to about the 98th meridian, with short-grass prairies farther west. The 98th meridian is roughly the location of the 51-cm (20-in.) isohyet, with wetter conditions to the east and drier to the west (see Figure 15.16).

The deep sod of those grasslands posed problems for the first settlers, as did the climate. The self-scouring steel plow, introduced in 1837 by John Deere, allowed the interlaced grass sod to be broken apart, freeing the soils for agriculture. Other inventions were critical to opening this region and solving its unique spatial problems: barbed wire (the fencing material for a treeless prairie), well-drilling techniques developed by Pennsylvania oil drillers but used for water wells, windmills for pumping, and railroads to conquer the distances.

Few patches of the original prairies (tall grasslands) or steppes (short grasslands) remain within this biome. For prairies alone, the reduction of natural vegetation went from 100 million hectares (250 million acres) down to a few areas of several hundred hectares each. The Tallgrass Prairie National Preserve is a 4410-hectare (10,894-acre) section of prairie located in the Flint Hills of Kansas. This is the only national park unit preserving this vestige of the former grasslands that covered much of the North American Great Plains. A state-designated 10-hectare (25-acre) remnant of the "prairie" is located 8 km north of Ames, Iowa. This is a patch of original grassland that never has felt the plow.

Outside North America, the Pampas of Argentina and Uruguay and the grasslands of Ukraine are characteristic midlatitude grassland biomes. In each region of the world where these grasslands have occurred, human development of them has been critical to territorial expansion.

This biome is the home of large grazing animals, including deer, antelope, pronghorn, and bison (the almost complete annihilation of the latter is part of American history). Grasshoppers and other insects feed on the grasses and crops as well, and gophers, prairie dogs, ground squirrels, Turkey Vulture, grouse, and prairie chickens are on the land. Predators include coyote, the nearly extinct black-footed ferret, badger, and birds of prey, including hawks, eagles, and owls.

Desert Biomes

Earth's **desert biomes** cover more than one-third of its land area. In Chapter 12 we examined desert landscapes and in Chapter 6, desert climates. On a planet with such a rich biosphere, the deserts stand out as unique regions of fascinating adaptations for survival. Much as a group of humans in the desert might behave with short supplies, plant communities also compete for water and site advantage. Some desert plants, called *ephemerals*, wait years for a rainfall event, at which time their seeds quickly germinate, and the plants develop, flower, and produce new seeds, which then rest again until the next rainfall event. The

*From W. P. Webb, *The Great Plains* © 1931 by Walter Prescott Webb Needham Heights, Mass., Ginn and Company, p. 32. Used by permission.

seeds of some xerophytic species open only when fractured by the tumbling, churning action of flash floods cascading down a desert arroyo, and of course such an event produces the moisture that a germinating seed needs.

Perennial desert plants employ other adaptive features to cope with the desert, such as long, deep tap roots (for example, the mesquite); succulence (thick, fleshy, water-holding tissue such as that of cacti); spreading root systems to maximize water capture; waxy coatings and fine hairs on leaves to retard water loss; leafless conditions during dry periods (for example, palo verde and ocotillo); reflective surfaces to reduce leaf temperatures; and tissue that tastes bad to discourage herbivores.

The creosote bush (*Larrea divaricata*, sometimes dominated by the species *tridentata*) sends out a wide pattern of roots and contaminates the surrounding soil with toxins that prevent the germination of other creosote seeds, possible competitors for water. When a creosote bush dies and is removed, surrounding plants or germinating seeds work to occupy the abandoned site but must rely on infrequent rains to remove the toxins.

The faunas of both warm and cold deserts are limited by the extreme conditions and include few resident large animals. Exceptions are the desert bighorn sheep (in nearby mountains) and the camel, which can lose up to 30% of its body weight in water without suffering (for humans, a 10%–12% loss is dangerous). Some representative desert animals are the ringtail cat, kangaroo rat, lizards, scorpions, and snakes. Most of these animals are quite secretive and become active only at night, when temperatures are lower. In addition, various birds have adapted to desert plants and other available food sources—for example, Roadrunners, thrashers, ravens, wrens, hawks, grouse, and nighthawks.

We are witnessing an unwanted expansion of the desert biome. This expansion is due principally to agricultural practices (overgrazing and inappropriate agricultural activities that abuse soil structure and fertility), improper soil-moisture management, erosion and salinization, deforestation, and the ongoing climatic change. A process known as desertification is now a worldwide phenomenon along the margins of semiarid and arid lands (see Figure 12.22). The United Nations estimates that desertified lands have covered some 800 million hectares (2 billion acres) since 1930; many millions of additional hectares are added each year.

FIGURE 16.34 Sonoran desert scene.
Lower Sonoran desert west of Tucson, Arizona (32.3°N latitude; elevation 900 m or 2950 ft). Note adaptive features of the plants pictured: tall, fleshy, saguaro cactus; thorny, sparsely leafed ocotillo (right); barrel and prickly pear cactus, among others. [Photo by author.]

Warm Desert and Semidesert. Earth's **warm desert and semidesert** biomes are caused by the presence of dry air and low precipitation from subtropical high-pressure cells. These areas are very dry, as evidenced by the Atacama Desert of northern Chile, where only a minute amount of rain has ever been recorded—a 30-year annual average of only 0.05 cm (0.02 in.)! As in the Atacama, the true deserts of Earth are under the influence of the descending, drying, and stable air of high-pressure systems from 8 to 12 months of the year.

Vegetation ranges from almost none in the arid deserts to numerous xerophytic shrubs, succulents, and thorn tree forms. The lower Sonoran Desert of southern Arizona is a warm desert (Figure 16.34). This desert landscape features the unique saguaro cactus (*Carnegiea gigantea*) which grows to many meters in height and up to 200 years in age if undisturbed. First blooms do not appear until it is 50 to 75 years old.

A few of the subtropical deserts—such as those in Chile, Western Sahara, and Namibia—are right on the sea coast and are influenced by cool offshore ocean currents. As a result, these true deserts experience summer fog that mists the plant and animal populations with needed moisture.

The equatorward margin of the subtropical high-pressure cell is a region of transition to savanna, thorn tree, scrub woodland, and tropical seasonal forest. Poleward of the warm deserts, the subtropical cells shift to produce the Mediterranean dry-summer regime along west coasts and may grade into cool deserts elsewhere.

Cold Desert and Semidesert. The **cold desert and semidesert** biomes tend to occur at higher latitudes, where seasonal shifting of the subtropical high is of some influence less than 6 months of the year. Specifically, interior locations are dry because of their distance from moisture sources or their location in rain shadow areas on the lee side of mountain ranges such as the Sierra Nevada of the western United States, the Himalayas in Asia, and the Andes in South America. The combination of interior location and rain shadow positioning produces the cold deserts of the Great Basin of western North America.

Winter snows occur in the cold deserts, but are generally light. Summers are hot, with highs from 30° to 40°C (86° to 104°F). Nighttime lows, even in the summer, can cool 10° to 20 C° (18° to 36 F°) from the daytime high. The dryness, generally clear skies, and sparse vegetation lead to high radiative heat loss and cool evenings. Many areas of these cold deserts that are covered by sagebrush and scrub vegetation were actually dry short-grass regions in the past, before extensive grazing altered their appearance. The look of the deserts in the upper Great Basin is the result of more than a century of such cultural practices (see Figure 6.22b).

Arctic and Alpine Tundra

The **arctic tundra** is found in the extreme northern area of North America and Russia, bordering on the Arctic Ocean and generally north of the 10°C (50°F) isotherm for the warmest month. Daylength varies greatly throughout the year, seasonally changing from almost continuous day to continuous night. The region, except for a few portions of Alaska and Siberia, was covered by ice during all of the Pleistocene glaciations.

Winters in this biome, a *tundra* climate classification, are long and cold; cool summers are brief. Intensely cold continental polar and arctic air masses and stable high-pressure anticyclones dominate winter. A growing season of sorts lasts only 60–80 days, and even then frosts can occur at any time. Vegetation is fragile in this flat, treeless world; soils are poorly developed periglacial surfaces that are underlain by permafrost. In the summer months only the surface horizons thaw, thus producing a mucky surface of poor drainage. Roots can penetrate only to the depth of thawed ground, usually about a meter (3 ft). The surface is shaped by freeze-thaw cycles that create the frozen ground phenomena discussed in Chapter 14.

Tundra vegetation is low, ground-level herbaceous plants such as sedges, mosses, arctic meadow grass, and snow lichen, and some woody species such as dwarf willow (Figure 16.35a). Owing to the short growing season, some perennials form flower buds one summer and open them for pollination the next. The tundra supports musk ox, caribou, reindeer, rabbit, ptarmigans, lemmings, and other small rodents (important food for the larger carnivores), wolf, fox, weasel, Snowy Owl, polar bears, and, of course, mosquitoes. The tundra is an important breeding ground for geese, swans, and other waterfowl.

Alpine tundra is similar to arctic tundra but can occur at lower latitudes because it is associated with high elevation. This biome usually is described as above the timberline—that elevation above which trees cannot grow. Timberlines increase in elevation equatorward in both hemispheres. Alpine tundra communities occur in the Andes near the equator, the White Mountains of California, the Rockies, the Alps, and Mount Kilimanjaro of equatorial Africa, as well as mountains from the Middle East to Asia. Alpine meadows feature grasses and stunted shrubs, such as willows and heaths (Figure 16.35b). Characteristic fauna include mountain goats, bighorn sheep, elk, and voles.

(a)

(b)

FIGURE 16.35 Arctic and alpine tundra.
(a) Tundra on the Kamchatka Peninsula, Russia. The Uzon Caldera is in the center-background mountain range. (b) An alpine tundra and mountain goats near Mount Evans, Colorado, at 3660 m (12,000 ft). [Photos by (a) Wolfgang Kaehler; (b) Bobbé Christopherson.]

Summary and Review—Ecosystems and Biomes

Define ecology, biogeography, and the ecosystem, community, habitat, and niche concepts.

Earth's biosphere is unique in the Solar System; its ecosystems are the core of life. **Ecosystems** are self-sustaining associations of living plants and animals and their nonliving physical environment. **Ecology** is the study of the relationships between organisms and their environment and among the various ecosystems in the biosphere. **Biogeography** is the study of the distribution of plants and animals and the diverse spatial patterns they create. A **community** is formed by the interactions among populations of living animals and plants at a particular time. Within a community, a **habitat** is the specific physical location of an organism—its address. A **niche** is the function or operation of a life form within a given community—its profession.

ecosystem (p. 484)
ecology (p. 484)
biogeography (p. 484)
community (p. 484)
habitat (p. 484)
niche (p. 484)

1. What is the relationship between the biosphere and an ecosystem? Define ecosystem and give some examples.
2. What does biogeography include? Describe its relationship to ecology.
3. Briefly summarize what ecosystem operations imply about the complexity of life.
4. Define a community within an ecosystem.
5. What do the concepts of habitat and niche involve? Relate them to some specific plant and animal communities.
6. Describe symbiotic and parasitic relationships in nature. Draw an analogy between these relationships and human societies on our planet. Explain.

Explain photosynthesis and respiration and *derive* net photosynthesis and the world pattern of net primary productivity.

As plants evolved, the *vascular plants* developed conductive tissues. **Stomata** on the underside of leaves are the portals through which the plant participates with the atmosphere and hydrosphere. Plants (primary producers) perform **photosynthesis** as sunlight stimulates a light-sensitive pigment called **chlorophyll**. This process produces the food sugars and oxygen that drive biological processes. **Respiration** is essentially the reverse of photosynthesis and is the way the plant derives energy by oxidizing carbohydrates. **Net primary productivity** is the net photosynthesis (photosynthesis minus respiration) of an entire community. This stored chemical energy is the **biomass** that the community generates, or the net dry weight of organic material.

stomata (p. 487)
photosynthesis (p. 487)
chlorophyll (p. 487)
respiration (p. 488)
net primary productivity (p. 488)
biomass (p. 488)

7. How many plant species are there on Earth?
8. How do plants function to link the Sun's energy to living organisms? What is formed within the light-responsive cells of plants?
9. Compare photosynthesis and respiration and the derivation of net photosynthesis. What is the importance of knowing the net primary productivity of an ecosystem and how much biomass an ecosystem has accumulated?
10. Briefly describe the global pattern of net primary productivity.

List abiotic and biotic ecosystem components and *relate* those components to ecosystem operations and trophic relationships.

Light, temperature, water, and the cycling of gases and nutrients constitute the life-supporting abiotic components of ecosystems. Elevation and latitudinal position on Earth create a variety of physical environments. The zonation of plants with altitude is called **life zones** and is visible as you travel between different elevations.

Life is sustained by **biogeochemical cycles**, through which circulate the gases and sedimentary materials (nutrients) necessary for growth and development of living organisms. The environment may inhibit biotic operations, either through lack or excess. These **limiting factors** may be physical or chemical in nature.

Trophic refers to the feeding and nutrition relations in an ecosystem. It represents the flow of energy and cycling of nutrients. **Producers**, which fix the carbon they need from carbon dioxide, are the plants, including phytoplankton in aquatic ecosystems. **Consumers**, generally animals including zooplankton in aquatic ecosystems, depend on the producers as their carbon source. Energy flows from the producers through the system along a circuit called the **food chain**. Within ecosystems, the feeding interrelationships are complex and arranged in a **food web**.

The primary consumer is an **herbivore**, or plant eater. A **carnivore** (meat eater) is a secondary consumer. A consumer that eats both producers and consumers is called an **omnivore**—a role occupied by humans. **Detritivores** (detritus feeders and decomposers) renew the entire system by releasing simple inorganic compounds and nutrients by breaking down organic materials. *Detritus feeders* include worms, mites, termites, centipedes, snails, crabs, and even vultures, among others. **Decomposers**, primarily bacteria and fungi, digest organic debris outside their bodies, and absorb and release nutrients in the process. This metabolic work of microbial decomposers produces the rotting that breaks down detritus. Biomass and population pyramids characterize the flow of energy and the numbers of producers, consumers, and decomposers operating in an ecosystem.

life zone (p. 490)
biogeochemical cycles (p. 490)

limiting factor (p. 493)
producers (p. 494)
consumers (p. 494)
food chain (p. 494)
food web (p. 494)
herbivore (p. 496)
carnivore (p. 496)
omnivore (p. 496)
detritivores (p. 496)
decomposers (p. 497)

11. What are the principal abiotic components in terrestrial ecosystems?
12. Describe what Alexander von Humboldt found that led him to propose the life-zone concept. What are life zones? Explain the interaction among altitude, latitude, and the types of communities that develop.
13. What are biogeochemical cycles? Describe several of the essential cycles.
14. What is a limiting factor? How does it function to control the spatial distribution of plant and animal species?
15. What roles do producers and consumers play in an ecosystem? What are the detritivores? Describe their "job" in an ecosystem.
16. Describe the relationship among producers, consumers, decomposers, and detritus feeders in an ecosystem. What is the trophic nature of an ecosystem? What is the place of humans in a trophic system?

• *Define* succession and *outline* the stages of general ecological succession in both terrestrial and aquatic ecosystems.

A critical aspect of ecosystem stability is **biodiversity**, or species richness of life (a combination of *bio*logical and *diversity*). The more diverse the species population (both in number of different species and quantity of each species), the species genetic diversity (number of genetic characteristics), and ecosystem and habitat diversity, the better the risk is spread over the entire community. The greater the biodiversity within an ecosystem, the more stable and resilient it is, and the more productive it will be.

Biogeographers trace communities across the ages, considering plate tectonics and ancient dispersal of plants and animals. Past glacial and interglacial climatic episodes have created a long-term succession. Climate change is forcing accelerated succession in ecosystems.

Ecological succession occurs when more complex communities replace older, usually simpler, communities of plants and animals. An area of bare rock or soil with no trace of a former community can be a site for **primary succession. Secondary succession** begins in an area that has a vestige of a previously functioning community in place. The initial community that occupies an area of early succession is a **pioneer community**. Rather than progressing smoothly to a definable stable point, ecosystems tend to operate in a dynamic condition, with succeeding communities overlapping in time and space. Wildfire is one external factor that disrupts a successional community. Aquatic ecosystems also experience community succession, as exemplified by the eutrophication of a lake ecosystem.

The science of **fire ecology** has emerged in an effort to understand the natural role of fire in ecosystem maintenance and succession. A natural process in which lakes receive nutrients and sediment and become enriched is **eutrophication**, a natural aging of water bodies.

biodiversity (p. 497)
ecological succession (p. 499)
primary succession (p. 500)
secondary succession (p. 500)
pioneer community (p. 501)
fire ecology (p. 503)
eutrophication (p. 504)

17. What is meant by ecosystem stability?
18. How does ecological succession proceed? What are the relationships between existing communities and new, invading communities?
19. Discuss the concept of fire ecology in the context of the Yellowstone National Park fires of 1988. What were the findings of the government task force?
20. Summarize the process of succession in a body of water. What is meant by cultural eutrophication?

• *Define* the concepts of terrestrial ecosystem, biome, ecotone, and formation classes.

Earth is the only planet in the Solar System with a biosphere. An impressive feature of the living Earth is its diversity, which biogeographers categorize into discrete spatial biomes for analysis and study. The interplay among supporting physical factors within Earth's ecosystem determines the distribution of plant and animal communities. A **terrestrial ecosystem** is a self-sustaining association of plants and animals and their abiotic environment that is characterized by specific plant formation classes. A **biome** is a large, stable ecosystem characterized by specific plant and animal communities. Biomes carry the name of the dominant vegetation because it is the most easily identified feature: forest, savanna, grassland, shrubland, desert, tundra. A boundary transition zone adjoining ecosystems is an **ecotone**. Biomes are divided into more specific vegetation units called **formation classes**. The structure and appearance of the vegetation units are described as rain forest, needleleaf forest, Mediterranean shrubland, and arctic tundra, as examples. Specific life-form designations include trees, lianas, shrubs, herbs, bryophytes, epiphytes (plants growing above ground on other plants), and thallophytes (lacking leaves, stems, or roots, including bacteria, fungi, algae, and lichens).

terrestrial ecosystem (p. 505)
biome (p. 505)
ecotone (p. 505)
formation class (p. 505)

21. Describe a transition zone between two ecosystems. How wide is an ecotone? Explain.
22. Define biome. What is the basis of the designation?
23. Distinguish between formation classes and life-form designations as a basis for spatial classification.

● *Describe* 10 major terrestrial biomes and *locate* them on a world map.

Biomes are Earth's major terrestrial ecosystems, each named for its dominant plant community. The 10 major biomes are generalized from numerous formation classes that describe vegetation. Ideally, a biome represents a mature community of natural vegetation. In reality, few undisturbed biomes exist in the world, for most have been modified by human activity. Many of Earth's plant and animal communities are experiencing an accelerated rate of change that could produce dramatic alterations within our lifetime.

For an overview of Earth's 10 major terrestrial biomes and their vegetation characteristics, soil orders, climate designation, annual precipitation range, temperature patterns, and water balance characteristics, please review Table 16.1.

equatorial and tropical rain forest (p. 505)
tropical seasonal forest and scrub (p. 515)
tropical savanna (p. 515)
midlatitude broadleaf and mixed forest (p. 516)
needleleaf forest (p. 517)
boreal forest (p. 517)
taiga (p. 517)
temperate rain forest (p. 517)
Mediterranean shrubland (p. 518)
chaparral (p. 518)
midlatitude grasslands (p. 518)
desert biome (p. 519)
warm desert and semidesert (p. 520)
cold desert and semidesert (p. 520)
arctic tundra (p. 521)

24. Using the integrative chart in Table 16.1 and the world map in Figure 16.23, select any two biomes and study the correlation of vegetation characteristics, soil, moisture, and climate in their spatial distribution. Then contrast the two using each characteristic.
25. Describe the equatorial and tropical rain forests. Why is the rain forest floor somewhat clear of plant growth? Why are logging activities for specific species so difficult there?
26. What issues surround the deforestation of the rain forest? What is the impact of these losses on the rest of the biosphere? What new threat to the rain forest has emerged?
27. What do caatinga, chaco, brigalow, and dornveld refer to? Explain.
28. Describe the role of fire or fire ecology in the tropical savanna biome and the midlatitude broadleaf and mixed forest biome.
29. Why does the northern needleleaf forest biome not exist in the Southern Hemisphere? Where is this biome located in the Northern Hemisphere, and what is its relationship to climate type?
30. In which biome do we find Earth's tallest trees? Small, stunted plants, lichens, and mosses dominate which biome?
31. What type of vegetation predominates in the Mediterranean dry summer climates? Describe the adaptations necessary for these plants to survive.
32. What is the significance of the 98th meridian in terms of North American grasslands? What types of inventions were necessary for humans to cope with the grasslands?
33. Describe some of the unique adaptations found in a desert biome.
34. What is desertification (review from Chapter 12 and this chapter)? Explain its impact.
35. What physical weathering processes are specifically related to the tundra biome? What types of plants and animals are found there?

● *Relate* human impacts, real and potential, to several of the biomes.

The equatorial and tropical rain forest biome is undergoing rapid deforestation. Because the rain forest is Earth's most diverse biome and is important to the climate system, such losses are creating great concern among citizens, scientists, and nations. Efforts are under way worldwide to set aside and protect remaining representative sites within most of Earth's principal biomes. The Man and the Biosphere (MAB) Programme of UNESCO coordinates these biosphere reserve programs. Nearly 300 such biosphere reserves, covering some 12 million hectares (30 million acres), are now operated voluntarily in 76 countries.

36. What is the relationship between island biogeography and biosphere reserves? Describe a biosphere reserve. What are the goals?
37. Compare the map in Figure 16.23 with the global climate map in Chapter 6. Select several regions. What correlations can you make between them?

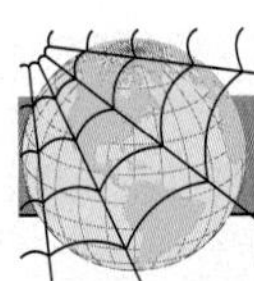

NetWork

The *Elemental Geosystems Home Page* provides on-line resources for this chapter on the World Wide Web. Once on the Home Page, click on this textbook, scroll the Table of Contents menu, select this chapter, and click "Begin." You will find self-tests that are graded, review exercises, specific updates for items in the chapter, and in "Destinations" many links to interesting related pathways on the Internet. *Elemental Geosystems* is found at **http://www.prenhall.com/christopherson.**

Critical Thinking

A. This chapter states, "Some scientists are questioning whether our human society and the physical systems of Earth constitute a global-scale symbiotic relationship (sustainable) or a parasitic one (nonsustainable)." Referring to the definition of these terms, what is your response to the statement? How do you equate our planetary economic system with the need to sustain life-supporting natural systems?

B. Over the next several days as you travel between home, apartment or dorm, the campus, a job, or other journey, observe the landscape. What types of ecosystem disturbances do you see? Imagine that several acres in the same area escaped any disruptions for a century or more, and reflect original formation classes and life-form types. What do you think several of these ecosystems and communities would be like?

C. Given the information presented in this chapter about deforestation in the tropics, assess the present situation for yourself. What are the main issues? What natural assets are at stake? Natural resources? How are global biodiversity and greenhouse warming related to these issues? What is the perspective of the less-developed countries that possess most of the rain forest? What is the perspective of the more-developed countries and their transnational corporations and millions of environmentally active citizens? What kind of action plan would you cast to accommodate all parties? How would you proceed?

D. Using Figure 16.23 (biomes), Figure 6.5 (climates), Figure 7.4 (precipitation), and Figure 5.24 (air masses), and the printed graphic scales on these four maps, consider the following hypothetical. Assume a northward climatic shift in the United States and Canada of 500 km (310 mi). Describe your analysis of conditions through the Midwest from Texas to the prairie provinces of Canada. Describe your analysis of conditions from New York, through New England, and into the Maritime provinces. What economic dislocations and relocations do you envision? How might a GIS (geographic information system) program help in your analysis?

The bristlecone pine (*Pinus longaeva*) provides us a lesson in sustainability and resilience in nature. Here on the subalpine slopes of the White Mountains of California at 3050-m to 3550-m (10,000 to 11,650-ft) elevation, these trees grow to more than 4500 years old. A young bristlecone pine stands in the foreground, mature trees are in the background, and long-dead bristlecones are still standing in the photo. [Photo by Bobbé Christopherson.]

17

Earth and the Human Denominator

Key Learning Concepts

By knowing and understanding the key learning concepts in this chapter, you should be able to:

- *Determine* an answer for Carl Sagan's question, "Who speaks for Earth?"
- *Describe* the growth in human population and *speculate* on possible future trends.
- *Analyze* "An Oily Bird" and *relate* your analysis to energy consumption patterns in the United States and Canada.
- *List* environmental agreements from the Earth Summit and *relate* them to physical geography and Earth systems science (geosystems).
- *List* the 12 paradigms for the twenty-first century.
- *Appraise* your place in the biosphere and *realize* your physical identity as an Earthling.

> During my space flight, I came to appreciate my profound connection to the home planet and the process of life evolving in our special corner of the Universe, and I grasped that I was part of a vast and mysterious dance whose outcome will be determined largely by human values and actions.*

Earth can be observed from profound vantage points, as this astronaut experienced on the 1969 *Apollo IX* mission. Since November 2000, the International Space Station (ISS) has operated at an orbital altitude of about 400 km (250 mi), in the upper reaches of Earth's thermosphere (see Figure 2.17). Just imagine, a 16-nation coalition has built a scientific work station, in orbit, that is 44.5-m long, 73-m wide, and 27.5-m tall (240 ft by 146 ft by 90 ft) and it is not yet finished

*Rusty Schweickart, "Our backs against the bomb, our eyes on the stars," *Discovery*, July 1987, p. 62. Reprinted by permission.

(Figure 17.1)! The 182-tonne (400,000 lb) complex has had 112 visitors through January 2003.

Sixty-five scientific investigations and experiments are completed or underway, to help us better understand Earth and life systems through research in the unique space environment. When completed the ISS will weigh 455 tonnes (1 million lbs) and have the cabin volume of a 747 jumbo jet! Isn't this a fulfillment of our human nature: to question, to explore, to discover, to test the limits of the known, and push beyond to the unknown? (See **http://spaceflight.nasa.gov/station/** for more information and updates.)

Our vantage point in this book is that of physical geography. We examine Earth's many systems: its energy, atmosphere, winds, ocean currents, water, weather, climate, endogenic and exogenic systems, soils, ecosystems, and biomes. Throughout this text you have seen how physical geographers use the latest scientific instruments, remote sensing, GPS, and GIS. This exploration has led us to an examination of the planet's most abundant large animal, *Homo sapiens*.

We stand in the first few years of the twenty-first century. The twenty-first century will be an adventure for the global society, historically unparalleled in experimenting with Earth's life-supporting systems. What preparations and "future thinking" are we doing to understand all that is to occur?

In his 1980 book and public television series *Cosmos*, the late astronomer Carl Sagan asked:

> What account would we give of our stewardship of the planet Earth? We have heard the rationales offered by the nuclear superpowers. We know who speaks for the nations. But who speaks for the human species? Who speaks for Earth?*

*C. Sagan, *Cosmos* (New York: Random House, 1980), p. 329. Reprinted by permission.

FIGURE 17.1 Our orbital perspective—Earth's Space Station.
The International Space Station (ISS) began scientific operations November 2000. More than 300 hours of human space walks went into the construction of the ISS through January 2003. All life-support supplies must be brought from Earth. Water recycling and purification techniques provide the bulk of the water needed. Oxygen is electrolytically derived from water and supplemented by compressed oxygen. Enormous solar-panel arrays generate electricity. Without plants, carbon dioxide must be scrubbed from the air the astronauts breathe. [Space Shuttle *Atlantis* photo made October 16, 2002, courtesy of NASA.]

Indeed, who does speak for Earth? Perhaps we physical geographers, and other scientists who have studied Earth and know the operations of the global ecosystem, should speak for Earth. However, some might say that questions of technology, environmental politics, and future thinking belong outside of science, and that our job is merely to learn how Earth's processes work and to leave the spokesperson's role to others. Biologist-ecologist Marston Bates addressed this line of thought:

> Then we came to humans and their place in this system of life. We could have left humans out, playing the ecological game of "let's pretend humans don't exist." But this seems as unfair as the corresponding game of the economists, "let's pretend nature doesn't exist." The economy of nature and the ecology of humans are inseparable and attempts to separate them are more than misleading, they are dangerous. Human destiny is tied to nature's destiny and the arrogance of the engineering mind does not change this. Humans may be a very peculiar animal, but they are still a part of the system of nature.*

A fact of life is that Earth's more-developed countries (MDCs), through their economic dominance, speak for the billions who live in less-developed countries (LDCs). The reality in 2002 was that the U.S. defense budget alone was greater than the gross domestic product (GDP) of many countries. The gross state product of California's 33 million people was slightly greater than the GDP of China and its 1.28 billion people. The scale of disparity on our home planet is difficult to comprehend.

The fate of traditional modes of life may rest in some distant financial capital. But economics aside, the reality is that the remote lands of Siberia are linked by Earth systems to the Pampas of Argentina to the Great Plains in North America and to those harvesting grain by hand in the remote Pamirs of Tajikistan (Figure 17.2).

To understand these linkages among Earth's myriad systems is our quest in *Elemental Geosystems* (Figure 17.3). We explored the atmosphere, hydrosphere, lithosphere, and biosphere, synthesizing operating systems into the web of life. We now see global linkages among physical and living systems and how actions in one place can affect change elsewhere. In this sense Earth can be compared to a spaceship. In the same way the members of a crew aboard the ISS are inextricably linked to each other's lives and survival, we too are each connected through the operation of planetary systems. Admittedly, the pace and noise of popular culture makes these connections difficult to perceive.

Growing international environmental awareness in the public sector is gradually prodding governments into action. People, when informed, generally favor environmental protection and public health progress over economic interests. A survey by Louis Harris and Associates confirmed these opinion trends. Majorities in 22 nations are willing to "endorse environmental protection at the risk of slowing down economic growth." The public seems to know they are saving money and improving their health if things are done right with the environment.

*M. Bates, *The Forest and the Sea* (New York: Random House, 1960), p. 247. Reprinted by permission.

FIGURE 17.2 Lands distant from the global centers of power.
These lands were depopulated under Stalin when they were part of the former USSR. Only in the last decade or so have people returned to this traditional rural landscape in the Pamirs of Tajikistan, beginning life again, so distant from the more-developed countries. [Photo by Stephen F. Cunha.]

The Human Count and the Future

Because human influence is so pervasive, we consider the totality of our impact the *human denominator*—thus, the chapter title. Just as the denominator in a fraction tells how many parts a whole is divided into, so the growing human population, the increasing demand for resources, and rising planetary impact, must be divided into Earth's relatively fixed resource base.

The human population of Earth passed the 6 billion mark in 1999. Over the span of human history these billion-mark milestones are occurring at closer intervals (Figure 17.4). More people now are alive than at any previous time in the planet's history, unevenly arranged in 192 countries and numerous colonies. Each year more than 83,000,000 more people are added to Earth's population—that is 227,000 a day, or a new United States population every three years. Thirty-seven percent of Earth's 6.215 billion population (2002) live in just two countries (China, 21%, and India, 16%—2.34 billion people combined). Moreover, we are a young planetary population with some 30% of those now alive under the age of 15 (U.S. Bureau of Census *PopClock Projection:* **http://www.census.gov/cgi-bin/popclock**).

The present annual global *natural increase* in population is 1.3%—the annual difference between the *crude birth rate* (per thousand population) and *crude death rate* (per thousand population). This natural increase, if unchanged, will produce a 46 % increase in world population by 2050. Growth is

Energy-Atmosphere System

PART 1: Chapters 2—4

Water, Weather and Climate Systems

PART 2: Chapters 5—7

Geosystems: Our Sphere of Contents

Solar energy

Atmosphere

Hydrosphere

Lithosphere

Soils, Ecosystems, and Biomes

PART 4: Chapters 15—17

Heat energy

FIGURE 17.3 The spheres within *Elemental Geosystems.*
Through this text we covered the atmosphere, hydrosphere, lithosphere, and the synthesizing biosphere—the culmination of life-sustaining interactions. The systems approach shows us the flow of energy and matter and the sequence of events through time, within each sphere and among the spheres through their complex linkages. Ultimately, *Elemental Geosystems* describes the support systems of life. [Atmosphere and hydrosphere photos by author; lithosphere and biosphere photos by Bobbé Christopherson.]

Earth's Changing Landscape Systems

PART 3: Chapters 8—14

not uniformly distributed. *Virtually all new population growth is in the less-developed countries* (LDCs) that now possess 80%, or 4.94 billion, of the total population; the MDCs have the other 20%, or 1.2 billion. Table 17.1 illustrates the natural increase and the differences in growth found in the MDCs and LDCs.

If you consider only the population count, the MDCs do not have a population growth problem. In fact, Japan and some European countries are actually declining in growth or are near replacement levels. However, people in the developed world produce a greater impact on the planet per person and therefore constitute a population impact crisis. An equation expresses this concept:

$$I \text{ (planetary impact)} = P \cdot A \cdot T$$

where P is *population*, A is *affluence*, or consumption per person, and T is *technology*, or the level of environmental impact per unit of production. The United States and Canada, with about 5% of the world's population, produce over 50% of the world's gross domestic product (over $10 trillion a year), and use more than double the energy per capita of Europeans, more than 7 times Latin Americans, 10 times Asians, and 20 times Africans. The United States and Canada produce 22 tons of carbon dioxide emissions per person per year, more than 7 times Latin Americans, 18 times Africans, 2.3 times Europeans, and 10 times Asians.

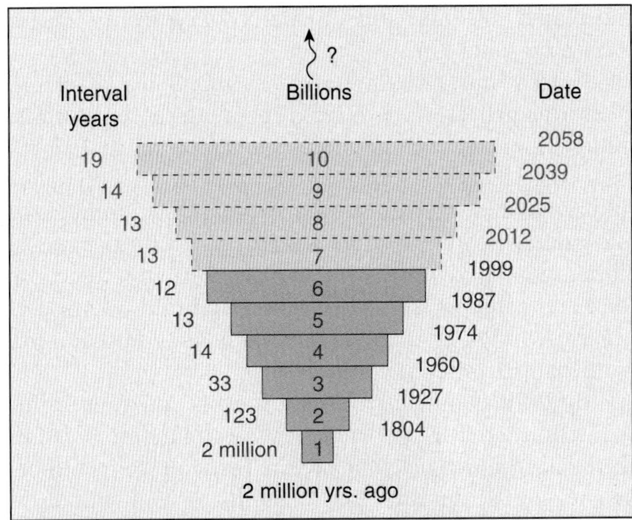

FIGURE 17.4 Human population growth.
The number of years required for the human population to add one billion to the count grows smaller and smaller. The new century should mark a slowing of this history of growth if policy actions are taken. Note the population forecasts for the next half century.

Therefore, consideration by people in the MDCs of the state of Earth systems, natural resources, and sustainability of current practices is critical, responsible leadership is mandatory, and guidance of transnational corporate activity becomes a necessity. Earth systems science is here to provide a high level of planetary monitoring and analysis to assist this process. Not only must the MDCs effect change in their own planetary impact but they also should provide direction for the LDCs to move along the road of progress and improve living conditions, without repeating some of the mistakes being made by the MDCs.

An Oily Bird

At first glance, the chain of events that exposes wildlife to oil contamination seems to stem from a technological problem. An oil tanker splits open at sea and releases its petroleum cargo, which is moved by ocean currents toward shore, where it coats coastal waters, beaches, and animals. In response, concerned citizens mobilize and try to save as much of the spoiled environment as possible (Figure 17.5). But the real problem goes far beyond the physical facts of the spill.

The immediate problem of cleaning oil off a bird symbolizes a national and international concern with far reaching significance. And while we search for answers, oil slicks continue their contamination. In just one year

FIGURE 17.5 An oily bird.
A Western Grebe contaminated with oil from the *Exxon Valdez* tanker accident in Prince William Sound, Alaska. This oily bird is the result of a long chain of events and mistakes. [Photo by Geoffrey Orth/© Sipa Press.]

Table 17.1 Planetary Natural Increase Rates

	Birth Rate	Death Rate	Natural Increase	Population Projection Change Change 2002–2050 (%)
World–2002	**21**	**9**	**1.3**%	46
More developed:	**11**	**10**	**0.1**% (dbl. 548 yrs.)	3
United States	15	9	0.6%	44
Japan	9	8	0.2%	−21
United Kingdom	11	10	0.1%	0.1
Less developed:	**24**	**8**	**1.6**% (dbl. 40 yrs.)	57
LDC exclu. China:	**27**	**9**	**1.9**% (dbl. 38 yrs.)	73
Mexico	26	5	2.1%	48
Nigeria	41	14	2.7%	134
China	13	6	0.7%	9
India	26	9	1.7%	55

Source: World Population Data Sheet 2002, Population Reference Bureau, Washington, D.C. http://www.prb.org.

following the 1989 *Exxon Valdez* oil-spill disaster, nearly 76 million liters (20 million gallons) of oil were spilled in 10,000 accidents worldwide, shown in the startling map in Figure 17.6a. A massive 2002 oil-tanker breakup spilled 67,000 tons (20 million gallons) of oil into the ocean off Galicia, Spain. In addition to oceanic oil spills, some people improperly dispose of crankcase oil from their automobiles in a volume that annually exceeds tanker spills!

In Prince William Sound off the southern coast of Alaska in clear weather and calm seas, the *Exxon Valdez*, a single-hulled supertanker operated by Exxon Corporation, struck a reef. The tanker spilled 42 million liters (11 million gallons) of oil. It took only 12 hours for the *Exxon Valdez* to empty its contents, yet a complete cleanup is impossible and costs and private claims exceed $15 billion.

Eventually, over 2400 km (about 1500 mi) of sensitive coastline was ruined, affecting three national parks and eight other protected areas. For perspective, had this spill occurred farther south, enough oil spilled to blacken every beach and bay along the Pacific Coast from southern Oregon to the Mexican border.

The death toll of animals was massive: At least 5000 sea otters died, or about 30% of resident otters, about 300,000 birds, and uncounted fish, shellfish, plants, and aquatic microorganisms. Sublethal effects, namely mutations, now are appearing in fish. The Pacific herring still is in significant decline, as are the harbor seals; other species such as the Bald Eagle and Common Murre are in recovery. More than a decade later, oil remains in mudflat and marsh soils and still can be scooped from beneath rocks along shorelines.

A great many factors influence our demand for oil. Well over half of our imported oil is burned in vehicles. Improvement of automobile efficiency began in 1975 as ordered by federal regulations. But the 1980s saw a major rollback of auto efficiency standards, a reduction in gasoline prices, slowing of domestic conservation programs, elimination of research for energy alternatives such as solar and wind power, and the continuing slow demise of America's railroad network for passengers and freight. These negative trends continue.

By 1999, comparatively inefficient sports utility vehicles represented half of new car sales. These SUVs are classified as light trucks and are thus exempt from auto-efficiency and some pollution standards. A combination of waste, low prices, the popularity of less-efficient vehicles, and a lack of alternatives has spurred the demand for petroleum. In addition, our land-use policies continue to foster a diffuse sprawl of our population, thereby adding stress to transportation systems. All of these manipulations increased fossil fuel demand to about 8 billion barrels a year for oil alone.

Yet hypocrisy is apparent in our outrage over political events and accidents and waste. We continue to consume gasoline at record levels in inefficient vehicles, thus creating the demand for oil imports. The task of physical geography is to analyze all the spatial aspects of these events in the environment and the ironies that are symbolized by a little oily bird.

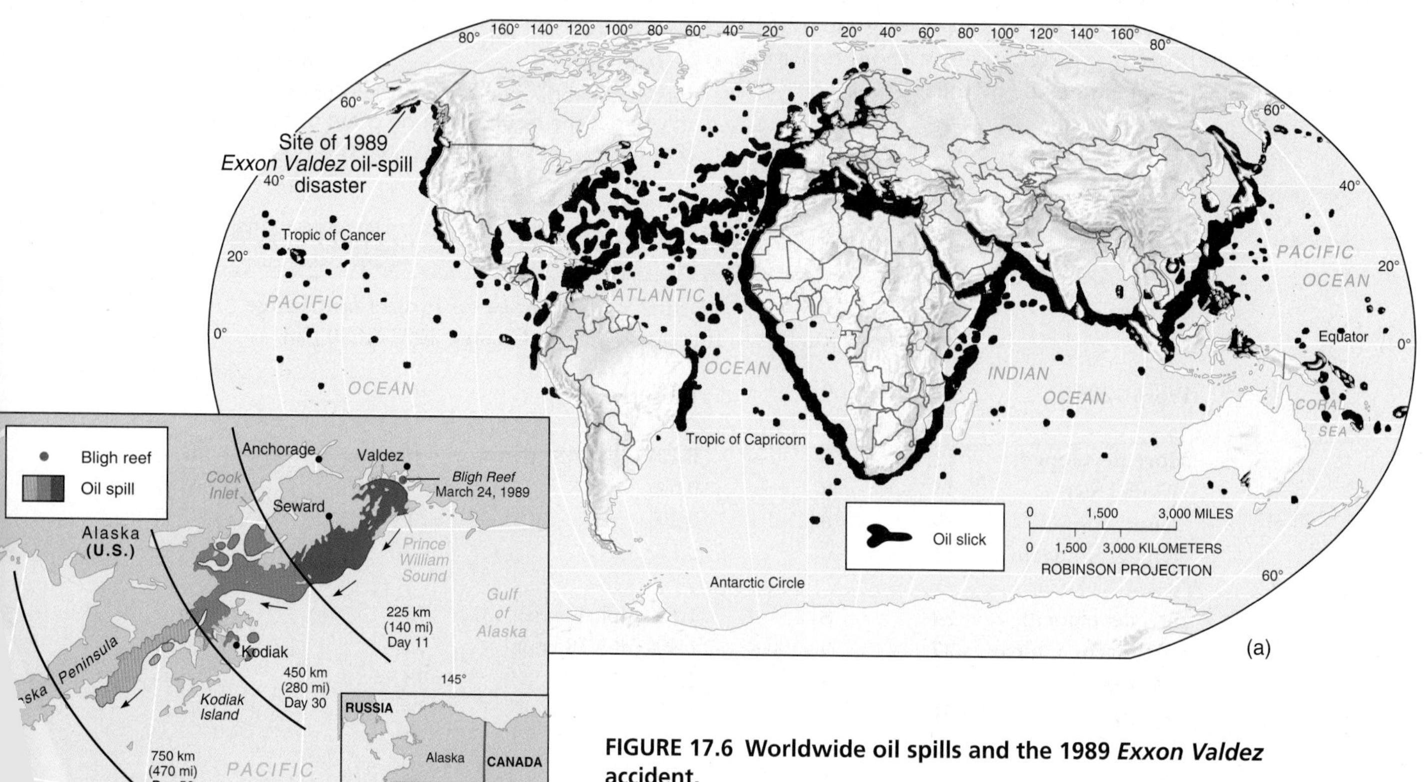

FIGURE 17.6 Worldwide oil spills and the 1989 *Exxon Valdez* accident.
(a) Location of visible oil slicks worldwide. (b) Track of spreading oil for the first 56 days of the Alaskan spill. [(a) Data from Organization for Economic Cooperation and Development.]

The Need for International Cooperation

The idea for a global meeting on the environment was put forward at the 1972 U.N. Conference on the Human Environment held in Stockholm. The U.N. General Assembly in 1987 achieved a landmark in global planning by agreeing to hold the Earth Summit. The first Earth Summit was held in 2002. *Our Common Future* set the tone:

> The Earth is one but the world is not. We all depend on one biosphere for sustaining our lives. Yet each community, each country, strives for survival and prosperity with little regard for its impact on others. Some consume the Earth's resources at a rate that would leave little for future generations. Others, many more in number, consume far too little and live with the prospect of hunger, squalor, disease, and early death.*

The setting in Rio de Janeiro for the first Earth Summit was ironic, for many of the very problems discussed at the conference were evident on the city's streets, with their abundant air pollution, water pollution, toxics, noise, wealth and poverty, and daily struggle for health and education.

Five agreements were written at the Earth Summit: the *Climate Change Framework*; *Biological Diversity Treaty*; the *Management, Conservation, and Sustainable Development of All Types of Forests*; the *Earth Charter* (a nonbinding statement of 27 environmental and economic principles); and *Agenda 21 for Sustainable Development*. A product of the Earth Summit was the United Nations *Framework Convention on Climate Convention* (FCCC), which led to a series of *Convention of the Parties* (COP) meetings. The *Kyoto Protocol*, agreed to in 1997 to reduce global carbon emissions, was finalized in Marrakech, Morocco, in 2001 at *COP-7* (Figure 17.7). As of 2002 the United States was the lone major holdout to the Kyoto Rulebook.

Asking whether the Earth Summit "succeeded" or "failed" is the wrong question. The occurrence of this largest-ever official gathering is a remarkable accomplishment. From the Earth Summit emerged a new organization—the U.N. Commission on Sustainable Development—to oversee the promises made in the five agreements. This momentum lead to a second Earth Summit in 2002 (http://www.earthsummit2002.org/) in Johannesburg, South Africa, with an agenda including climate change, freshwater, gender issues, global public goods, HIV/AIDS, sustainable finance, and the five Rio Conventions.

Members of society must work to move the solutions for environmental and developmental problems off the bench and into play. We need international cooperation to consider our symbiotic relations with each other and with Earth's resilient, yet fragile, life-support systems (News Report 17.1, p. 534).

A critical corollary to these international efforts is the linkage of academic disciplines. A positive step in that direction is the Earth systems science (geosystems) approach, which synthesizes content from across the disciplines to create a holistic perspective. Exciting progress toward an integrated understanding of Earth's physical and biological systems is in progress.

More than a decade ago, *Time* magazine gave Earth an interesting honor and is today a time capsule of activist suggestions (Figure 17.8). We wonder if that Earth awareness is to be a victim of short-lived fame, or how it fits into the global scheme of things today? Is there a follow up in pop culture to making the cover of *Time*? In part, *Time* asked,

> Sooner or later the Earth's inhabitants, so use to adapting the environment to suit their needs, will be forced to adapt themselves to the environment's demands. When that day comes, how will societies respond? How well will the world cope with the long-term changes that are likely to be in store?

[January 2, 1989, p. 70, © 1989 The *Time* Inc. Magazine Company.]

FIGURE 17.7 The Marrakech, Morocco, Climate Summit, November 2001.
The Marrakech Climate Summit (*COP-7*) voted to accept the terms of the Kyoto Protocol. A consensus was reached on the reduction of global carbon dioxide emissions and the Kyoto Rulebook. [Photo courtesy of IISD/ENB-Leila Mead.]

*World Commission on Environment and Development, *Our Common Future* (Oxford, UK: Oxford University Press, 1987), p. 27.

News Report 17.1

Gaia Hypothesis Triggers Debate

Some view Earth as one vast, self-regulating organism. The concept is one of global symbiosis, or mutualism. The *Gaia hypothesis* is the name of this controversial concept (Gaia was the Earth Mother goddess in ancient mythology). First proposed in 1979, British astronomer and inventor James Lovelock and American biologist Lynn Margulis promoted this concept.

Gaia is the ultimate synergistic relationship, in which the whole greatly exceeds the sum of the individual interacting components. The hypothesis contends that life processes control and shape Earth's inorganic physical and chemical processes with the ecosphere so interactively that a very small mass can affect a very large mass. Thus, Lovelock and Margulis think that the material environment and the evolution of species are tightly joined; as species evolve through natural selection, they (including us) in turn affect their environment. The present oxygen-rich composition of the atmosphere is given as proof of this coevolution of living and nonliving systems.

From the perspective of physical geography, the Gaia hypothesis permits a view of all Earth and the spatial interrelations among systems. In fact, such a perspective is necessary for analyzing specific environmental issues. Many variables interact synergistically, producing both wanted and unwanted results.

One disturbing aspect of this unity is that any biotic threat to the operation of an ecosystem tends to move toward extinction itself. This trend preserves the system overall. Earth-systems operation and feedback naturally tend to eliminate offensive members. The degree to which humans represent a planetary threat, then, becomes a topic of great concern, for Earth (Gaia) will prevail, regardless of the outcome of the human experiment.

> The maladies of Gaia do not last long in terms of her life span. Anything that makes the world uncomfortable to live in tends to induce the evolution of those species that can achieve a new and more comfortable environment. It follows that, if the world is made unfit by what we do, there is the probability of a change in regime to one that will be better for life but not necessarily better for us.*

The debate is vigorous regarding the true applicability of this hypothesis to nature, or whether it is true science at all. Regardless, it remains philosophically intriguing in its portrayal of the relationship between humans and Earth.

*J. Lovelock, *The Ages of Gaia—A Biography of Our Living Earth* (New York: W. W. Norton, 1988), p. 178. Used with permission.

FIGURE 17.8 ***Time*** **magazine cover for January 2, 1989—a time capsule.**
More than a decade ago, global concerns about environmental impacts prompted *Time* magazine to deviate from its 60-year tradition of naming a prominent citizen as its person of the year, instead naming Earth the "Planet of the Year." The magazine devoted 33 pages to Earth's physical and human geography. Importantly, *Time* also offered positive policy strategies for consideration—an interesting time capsule for comparison with real events over the years. [January 2, 1989, issue, Copyright © 1989 The *Time* Inc. Magazine Company. Reprinted by permission.]

Twelve Paradigms for the Twenty-first Century

As we conclude, it seems appropriate to consider a brief list of the dominant themes and patterns of concern for the twenty-first century. I hope these *paradigms* provide a useful framework for brainstorming and discussing the critical issues that will affect us.

Twelve Paradigms for the Twenty-first Century

1. Population increases in the less-developed countries
2. Planetary impact per person (on the biosphere and resources; $I = P \cdot A \cdot T$)
3. Feeding the world's population
4. Global and national disparities of wealth and resource allocation
5. Status of women and children (health, welfare, rights)
6. Global climate change (temperatures, sea level, weather and climate patterns, disease, and biomes)
7. Energy supplies and energy demands; renewables and demand management
8. Loss of biodiversity (habitats, genetic wealth, and species richness)
9. Pollution of air, surface water (quality and quantity), groundwater, oceans, and land
10. The persistence of wilderness (biosphere reserves and biodiversity hot spots)
11. Globalization versus cultural diversity
12. Conflict resolution

FIGURE 17.9 Fifth Annual International Geography Olympiad, 2001.
Students from across the globe met in Vancouver, British Columbia, August 2001, for the fifth Geography Olympiad, sponsored by the National Geographic Society, with Alex Trebek of *Jeopardy!* fame as moderator. Here Alex congratulates the Canadian team on their second-place finish—the U.S placed first, the Hungarian team third in the final standings. The Olympiad is an indicator of growing international geographic awareness. [Photo by O. Louis Mazzatenta, 2001 National Geographic Society.]

Who Speaks for Earth?

Geographic awareness and education is an increasingly positive force on Earth. The National Geographic Society conducts an annual National Geography Bee to promote geography education and global awareness to millions of 6th to 8th graders, their parents, and teachers. The International Geography Olympiad is now an annual event (Figure 17.9). There are presently 60 geographic alliances in 47 states, which coordinate geographic education among teachers and students at all levels: K–12, community college, college, and university. People are learning more about Earth-human relations.

Yet ideological and ethical differences remain within society. This dichotomy was addressed by the respected biologist Edward O. Wilson:

> The evidence of swift environmental change calls for an ethic uncoupled from other systems of belief. Those committed by religion to believe that life was put on Earth in one divine stroke will recognize that we are destroying the Creation; and those who perceive biodiversity to be the product of blind evolution will agree. Across the other great philosophical divide, it does not matter whether species have independent rights or, conversely, that moral reasoning is uniquely a human concern. Defenders of both premises seem destined to gravitate toward the same position on conservation.... For what, in the final analysis, is morality but the command of conscience seasoned by a rational examination of consequences? ... An enduring environmental ethic will aim to preserve not only the health and freedom of our species, but access to the world in which the human spirit was born.*

United Nations Secretary-General Kofi Annan, recipient of the 2001 Nobel Peace Prize, spoke to the Association of American Geographers annual meeting (Figure 17.10) on March 1, 2001, and offered us this thought:

FIGURE 17.10 Secretary-General Kofi Annan addresses the AAG.
U.N. Secretary-General and Nobel Prize winner speaks to more than 3000 geographers at the Association of American Geographers Annual Meeting in New York City, March 1, 2001. [Photo courtesy of the Association of American Geographers, by Kevin J. McCormick.]

*E. O. Wilson, *The Diversity of Life* (Cambridge: Harvard University Press, 1992), p. 351.

The idea of interdependence is old hat to geographers, but for most people it is a new garment they are only now trying on for size. Getting it to fit—and getting it imprinted on the mental maps that guide our voices and our choices—is one of the crucial projects of human geography for the 21st century. I look forward to working with you in that all-important journey.

The late Carl Sagan asked, "Who speaks for Earth?" He answered with this perspective:

We have begun to contemplate our origins: starstuff pondering the stars; organized assemblages of ten billion billion billion atoms considering the evolution of atoms; tracing the long journey by which, here at least, consciousness arose. Our loyalties are to the species and the planet. We speak for Earth. Our obligation to survive is owed not just to ourselves but also to that Cosmos, ancient and vast, from which we spring.†

May we all perceive our spatial importance within Earth's ecosystems and do our part to maintain a life-supporting and sustaining Earth for ourselves and countless generations in the future.

Review Questions and Critical Thinking

1. What part do you think technology, politics, and thinking about the future should play in science courses?
2. Assess population growth issues: the count, the impact per person, and future projection. What strategies do you see as important?
3. According to the discussion in the chapter, what worldwide factors led to the *Exxon Valdez* accident? Describe the complexity of that event from a global perspective. In your analysis, examine both supply-side and demand-side issues, as well as environmental and strategic factors.
4. What is meant by the Gaia hypothesis? Describe several concepts from this text that might pertain to this hypothesis.
5. Relate the content of the various chapters in this text to the integrative Earth systems science concept. Which chapters help you to better understand Earth-human relations and human impacts?
6. After examining the list of 12 paradigm issues for the twenty-first century, suggest items that need to be added to the list, omitted from the list, or expanded in coverage. Rearrange the list items as needed to match your concerns.
7. This chapter states that we already know many of the solutions to the problems we face. Why do you think these solutions are not being implemented at a faster pace?
8. Who speaks for Earth?

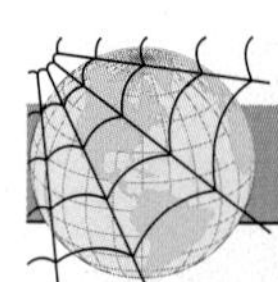

NetWork

The *Elemental Geosystems Home Page* provides on-line resources for this chapter on the World Wide Web. Once on the Home Page, click on this textbook, scroll the Table of Contents menu, select this chapter, and click "Begin." You will find self-tests that are graded, review exercises, specific updates for items in the chapter, and in "Destinations" many links to interesting related pathways on the Internet. *Elemental Geosystems* is found at **http://www.prenhall.com/christopherson**.

†C. Sagan, *Cosmos* (New York: Random House, 1980), p. 345. Reprinted by permission.

Appendix A

Maps in This Text and Topographic Maps

Maps Used in This Text

Elemental Geosystems uses several map projections: Goode's homolosine, Robinson, and Miller cylindrical, among others. Each was chosen to best present specific types of data. **Goode's homolosine projection** is an interrupted world map designed in 1923 by Dr. J. Paul Goode of the University of Chicago. Rand McNally *Goode's Atlas* first used it in 1925. Goode's homolosine equal-area projection (Figure 1) is a combination of two oval projections (*homolo*graphic and *sin*usoidal projections).

Two equal-area projections are cut and pasted together to improve the rendering of landmass shapes. A *sinusoidal projection* is used between 40° N and 40° S latitudes. Its central meridian is a straight line; all other meridians are drawn as sinusoidal curves (based on sine-wave curves) and parallels are evenly spaced. A *Mollweide projection*, also called a *homolographic projection*, is used from 40°N to the North Pole and from 40° S to the South Pole. Its central meridian is a straight line; all other meridians are drawn as elliptical arcs and parallels are unequally spaced—farther apart at the equator, closer together poleward. This technique of combining two projections preserves areal size relationships, making the projection excellent for mapping spatial distributions when interruptions of oceans or continents do not pose a problem.

We use Goode's homolosine projection throughout this book. Examples include the world climate map and smaller climate type maps in Chapter 6, topographic regions and continental shields maps (Figures 9.3 and 9.4), world sand regions and loess deposits (Figures 12.12 and 12.14), and the terrestrial biomes map in Chapter 16.

Another projection we use is the **Robinson projection**, designed by Arthur Robinson in 1963 (Figure 2). This projection is neither equal area nor true shape, but is a compromise between the two. The North and South Poles appear as lines slightly more than half the length of the equator; thus higher latitudes are exaggerated less than on other oval and cylindrical projections. Some of the Robinson maps employed include the latitudinal geographic zones map in Chapter 1 (Figure 1.11), daily net radiation map (Figure 2.10), the world temperature range map in Chapter 3 (Figure 3.28), the maps of lithospheric plates of crust and volcanoes and earthquakes in Chapter 8 (Figures 8.16 and 8.18), and the global oil spills map in Chapter 17 (Figure 17.6).

Another compromise map, the **Miller cylindrical projection**, is used in this text (Figure 3). Examples of this projection include the world time zone map (Figure 1.14), global temperature maps in Chapter 3 (Figures 3.24 and 3.26), two global pressure maps in Figure 4.11, and the global soil maps and soil order maps in Chapter 15 (Figure 15.8). This projection is neither true shape nor true area but is a compromise that avoids the severe scale distortion of the Mercator. The Miller projection frequently appears in world atlases. The American Geographical Society presented Osborn Miller's map projection in 1942.

Mapping, Quadrangles, and Topographic Maps

The westward expansion across the vast North American continent demanded a land survey for the creation of accurate maps. Maps were needed to subdivide the land and to guide travel, ex-

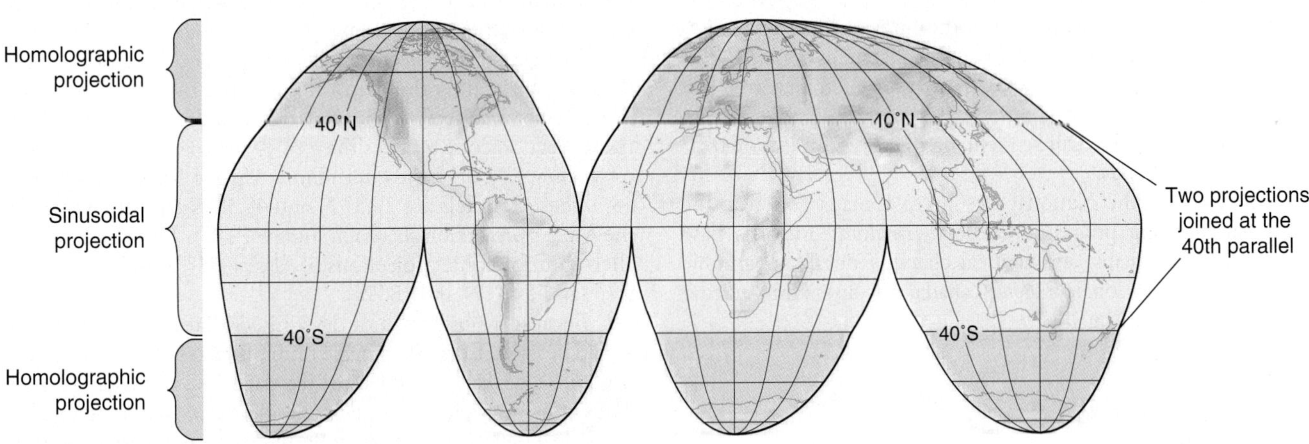

Figure 1 Goode's homolosine projection.
An equal-area map. [Copyright by the University of Chicago. Used by permission of the University of Chicago Press.]

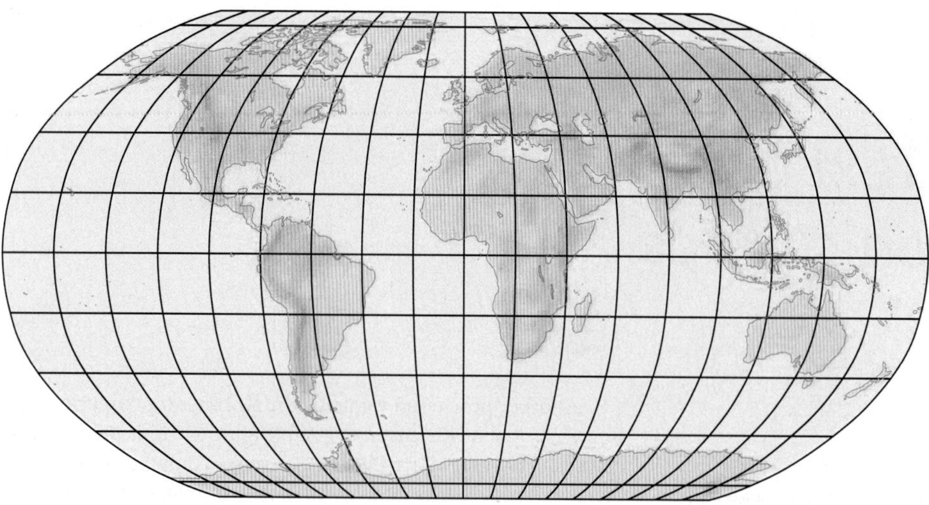

Figure 2 Robinson projection.
A compromise between equal area and true shape. [Developed by Arthur H. Robinson, 1963.]

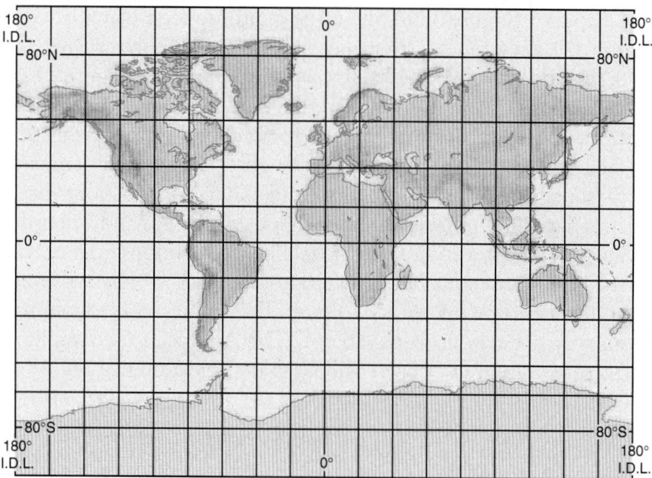

Figure 3 Miller cylindrical projection.
A compromise map projection between equal area and true shape. [Developed by Osborn M. Miller, American Geographical Society, 1942.]

ploration, settlement, and transportation. In 1785 the Public Lands Survey System began surveying and mapping government land in the United States. In 1836 the Clerk of Surveys in the Land Office of the Department of the Interior directed public-land surveys. The Bureau of Land Management replaced this Land Office in 1946. The actual preparation and recording of survey information fell to the U.S. Geological Survey (USGS), also a branch of the Department of the Interior (see **http://mapping.usgs.gov/**).

In Canada, the National Resources Canada conducts the national mapping program. Canadian mapping includes base maps, thematic maps, aeronautical charts, federal topographic maps, and the National Atlas of Canada, now in its fifth edition (see **http://atlas.gc.ca/**).

Quadrangle Maps

The USGS depicts survey information on quadrangle maps, so called because they are rectangular maps with four corner angles. The angles are junctures of parallels of latitude and meridians of longitude rather than political boundaries. These quadrangle maps utilize the Albers equal-area projection, from the conic class of map projections.

The accuracy of conformality (shape) and scale of this base map is improved by the use of not one but two standard parallels. (Remember from Chapter 1 that standard lines are where the projection cone touches the globe's surface, producing greatest accuracy.) For the conterminous United States (the "lower 48"), these parallels are 29.5° N and 45.5° N latitude (noted on the Alber's projection shown in Figure 1.18c). The standard parallels shift for conic projections of Alaska (55° N and 65° N) and for Hawai'i (8° N and 18° N).

Because a single map of the United States at 1:24,000 scale would be more than 200 m wide (more than 600 ft), some system had to be devised for dividing the map into manageable size. Thus, a quadrangle system using latitude and longitude coordinates developed. Note that these maps are not perfect rectangles, because meridians converge toward the poles. The width of quadrangles narrows noticeably as you move north (poleward).

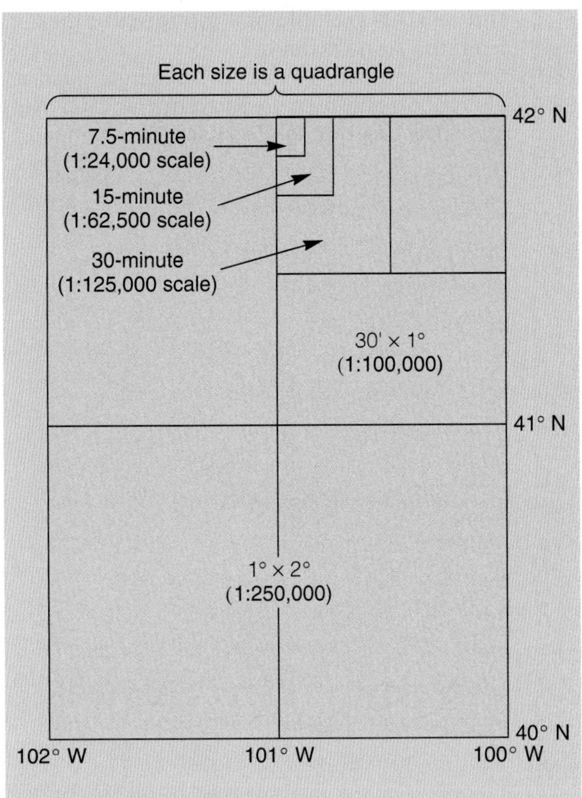

Figure 4 Quadrangle system of maps used by the USGS.

Quadrangle maps are published in different series, covering different amounts of Earth's surface at different scales. You see in Figure 4 that each series is referred to by its angular dimensions, which range from 1° to 2° (1:250,000 scale) to 7.5′ × 7.5′ (1:24,000 scale). A map that is one-half a degree on each side is called a "30-minute quadrangle," and a map one-fourth of a degree on each side is a "15-minute quadrangle" (this was the USGS standard size from 1910 to 1950). A map that is one-eighth of a degree on each side is a 7.5-minute quadrangle, the most widely produced of all USGS topographic maps, and the standard since 1950. The progression toward more-detailed maps and a larger-scale map standard through the years reflects the continuing refinement of geographic data and new mapping technologies.

The USGS National Mapping Program recently completed coverage of the entire country (except Alaska) on 7.5-minute maps (1 in. to 2000 ft, a large scale). It takes 53,838 separate 7.5-minute quadrangles to cover the lower 48 states, Hawai'i, and the U.S. territories. A series of smaller-scale, more-general 15-minute topographic maps offer Alaskan coverage.

In the United States, most quadrangle maps remain in English units of feet and miles. The eventual changeover to the metric system requires revision of the units used on all maps, with the 1:24,000 scale eventually changing to a scale of 1:25,000. However, after completing only a few metric quads, the USGS halted the program in 1991. In Canada, the entire country is mapped at a scale of 1:250,000, using metric units (1.0 cm to 2.5 km). About half the country also is mapped at 1:50,000 (1.0 cm to 0.50 km).

Topographic Maps

The most popular and widely used quadrangle maps are **topographic maps** prepared by the USGS. An example of such a map is a portion of the Cumberland, Maryland, quad shown in Figure 5. You will find topographic maps throughout *Elemental Geosystems* because they portray landscapes so effectively. As examples, see Figures 10.11 and 10.12, karst landscapes and sinkholes near Orleans, Indiana, and Winter Park, Florida; Figure 11.8, river drainage patterns; Figure 11.21, river meander scars; Figure 12.17, an alluvial fan in Montana; and Figure 14.13, drumlins in New York.

A **planimetric map** shows the horizontal position (latitude/longitude) of boundaries, land-use aspects, bodies of water, and economic and cultural features. A highway map is a common example of a planimetric map.

A topographic map adds a vertical component to show topography (configuration of the land surface), including slope and relief (the vertical difference in local landscape elevation). These fine details are shown through the use of elevation contour lines (Figure 6). A *contour line* connects all points at the same elevation. Elevations are shown above or below a vertical datum, or reference level, which usually is mean sea level. The contour interval is the vertical distance in elevation between two adjacent contour lines (20 ft, or 6.1 m in Figure 6b).

The topographic map in Figure 6b shows a hypothetical landscape, demonstrating how contour lines and intervals depict slope and relief, which are the three-dimensional aspects of terrain. The pattern of lines and the spacing between them indicates slope. The steeper a slope or cliff, the closer together the contour lines appear—in the figure, note the narrowly spaced contours that represent the cliffs to the left of the highway. A wider spacing of these contour lines portrays a more gradual slope, as you can see from the widely spaced lines on the beach and to the right of the river valley.

In Figure 7 are the standard symbols commonly used on these topographic maps. These symbols and the colors used are standard on all USGS topographic maps: black for human constructions, blue for water features, brown for relief features and contours, pink for urbanized areas, and green for woodlands, orchards, brush, and the like.

The margins of a topographic map contain a wealth of information about its concept and content. In the margins of topographic maps, you find the quadrangle name, names of adjoining quads, quad series and type, position in the latitude-longitude and other coordinate systems, title, legend, magnetic declination (alignment of magnetic north) and compass information, datum plane, symbols used for roads and trails, the dates and history of the survey of that particular quad, and more.

Topographic maps may be purchased directly from the USGS or Centre for Topographic Information, NRC (**http://maps.nrcan.gc.ca/main.html**). Many state geological survey offices, national and state park headquarters, outfitters, sports shops, and bookstores also sell topographic maps to assist people in planning their outdoor activities.

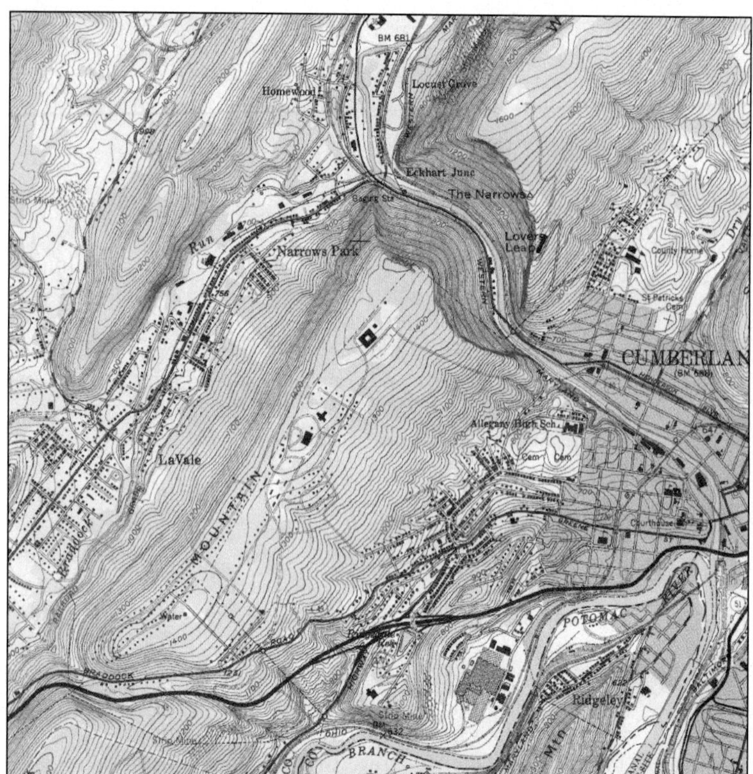

Figure 5 An example of a topographic map from the Appalachians.
Cumberland, Md., Pa., W.Va. 7.5-minute quadrangle topographic map prepared by the USGS. Note the water gap through Haystack Mountain.

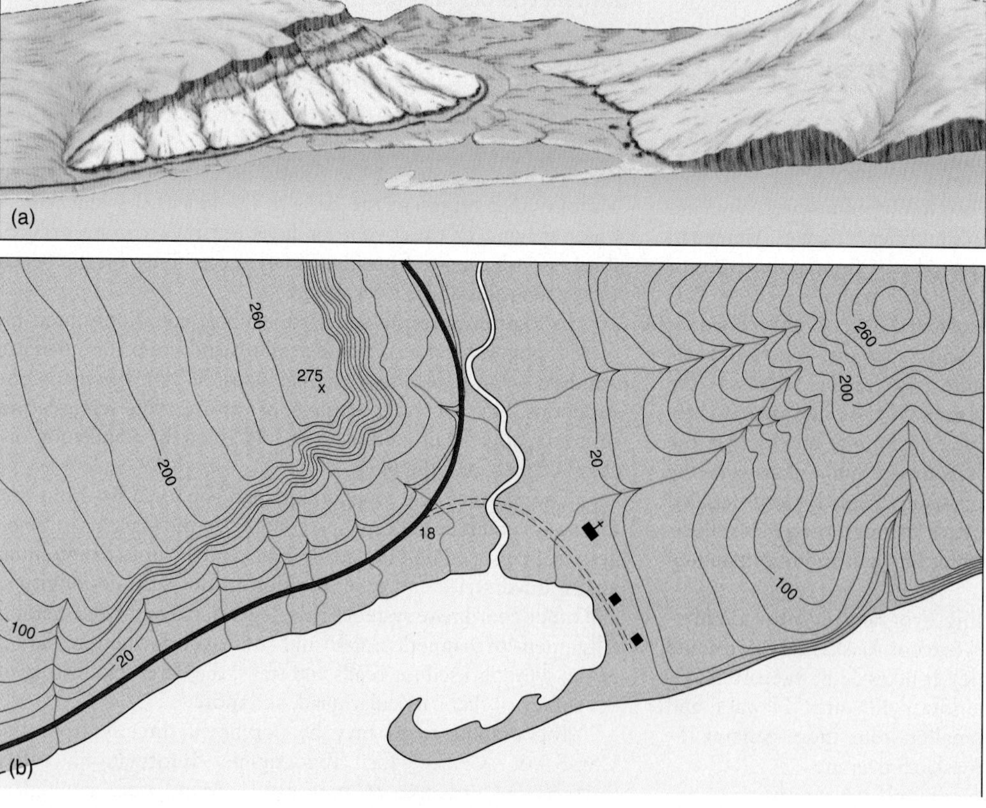

Figure 6 Topographic map of a hypothetical landscape.
(a) Perspective view of a hypothetical landscape. (b) Depiction of that landscape on a topographic map. The contour interval on the map is 20 feet (6.1 m). [After the U.S. Geological Survey.]

Control data and monuments

Vertical control

Third order or better, with tablet	BM×16.3
Third order or better, recoverable mark	×120.0
Bench mark at found section corner	BM 118.3
Spot elevation	×5.3

Contours

Topographic

- Intermediate
- Index
- Supplementary
- Depression
- Cut; fill

Bathymetric

- Intermediate
- Index
- Primary
- Index primary
- Supplementary

Boundaries

- National
- State or territorial
- County or equivalent
- Civil township or equivalent
- Incorporated city or equivalent
- Park, reservation, or monument

Surface features

Levee	Levee
Sand or mud area, dunes, or shifting sand	Sand
Intricate surface area	Strip mine
Gravel beach or glacial moraine	Gravel
Tailings pond	Tailings pond

Mines and caves

Quarry or open pit mine	
Gravel, sand, clay, or borrow pit	
Mine dump	Mine dump
Tailings	Tailings

Vegetation

Woods	
Scrub	
Orchard	
Vineyard	
Mangrove	Mangrove

Glaciers and permanent snowfields

- Contours and limits
- Form lines

Marine shoreline

Topographic maps

- Approximate mean high water
- Indefinite or unsurveyed

Topographic-bathymetric maps

- Mean high water
- Apparent (edge of vegetation)

Coastal features

Foreshore flat	Mud
Rock or coral reef	Reef
Rock bare or awash	
Group of rocks bare or awash	
Exposed wreck	
Depth curve; sounding	3
Breakwater, pier, jetty, or wharf	
Seawall	

Rivers, lakes, and canals

Intermittent stream	
Intermittent river	
Disappearing stream	
Perennial stream	
Perennial river	
Small falls; small rapids	
Large falls; large rapids	
Masonry dam	
Dam with lock	
Dam carrying road	
Perennial lake; Intermittent lake or pond	
Dry lake	Dry lake
Narrow wash	
Wide wash	Wide wash
Canal, flume, or aquaduct with lock	
Well or spring; spring or seep	

Submerged areas and bogs

Marsh or swamp	
Submerged marsh or swamp	
Wooded marsh or swamp	
Submerged wooded marsh or swamp	
Rice field	Rice
Land subject to inundation	Max pool 431

Buildings and related features

- Building
- School; church
- Built-up area
- Racetrack
- Airport
- Landing strip
- Well (other than water); windmill
- Tanks
- Covered reservoir
- Gaging station
- Landmark object (feature as labeled)
- Campground; picnic area

Cemetery: small; large	Cem

Roads and related features

Roads on Provisional edition maps are not classified as primary, secondary, or light duty. They are all symbolized as light duty roads.

- Primary highway
- Secondary highway
- Light duty road
- Unimproved road
- Trail
- Dual highway
- Dual highway with median strip

Railroads and related features

- Standard gauge single track; station
- Standard gauge multiple track
- Abandoned

Transmission lines and pipelines

Power transmission line; pole; tower	
Telephone line	Telephone
Aboveground oil or gas pipeline	
Underground oil or gas pipeline	Pipeline

Figure 7 Standardized topographic map symbols used on USGS maps.
English units still prevail, although a few USGS maps are in metric. [From USGS, Topographic Maps, 1969.]

Appendix B

The Canadian System of Soil Classification (CSSC)

Canadian efforts at soil classification began in 1914 with the partial mapping of Ontario's soils by A. J. Galbraith. Efforts to develop a taxonomic system spread countrywide, anchored by universities in each province. Regional differences in soil classification emerged, further confused by a lack of specific soil details. By 1936, only 1.7% of Canadian soil had been surveyed (15 million hectares).

Canadian scientists needed a taxonomic system based on observable and measurable properties in soils specific to Canada. This meant a departure from Marbut's 1938 U.S. classification. Canada's first taxonomic system was introduced in 1955, splitting away from the soil classification effort in the United States and the Fourth Approximation stage. Classification work progressed through the Canada Soil Survey Committee after 1970 and was replaced by the Expert Committee on Soil Survey in 1978, all under Agriculture Canada.

The **Canadian System of Soil Classification (CSSC)** provides taxa for all soils presently recognized in Canada and is adapted to Canada's expanses of forest, tundra, prairie, frozen ground, and colder climates. As in the U.S. Soil Taxonomy system, the CSSC classifications are based on observable and measurable properties found in real soils rather than idealized soils that may result from the interactions of genetic processes. The system is flexible in that its framework can accept new findings and information in step with progressive developments in the soil sciences.

Categories of Classification in the CSSC

Categorical levels are at the heart of a taxonomic system. These categories are based on soil profile properties organized at five levels, nested in a hierarchical pattern to permit generalization at several levels of detail. Each level is referred to as a category of classification. The levels in the CSSC are briefly described here, as adapted from *The Canadian Soil Classification System*, 2nd ed., Publication 1646 (Ottawa: Supply and Services Canada, 1987, p. 16).

- **Order:** Each of nine soil orders has pedon properties that reflect the soil environment and effects of active soil-forming processes.
- **Great Group:** Subdivisions of each order reflect differences in the dominant processes or other major contributing processes. As an example, in Luvic Gleysols (great group name followed by order) the dominant process is gleying—reduction of iron and other minerals—resulting from poor drainage under either grass or forest cover with Aeg and Btg horizons (see Table 1).
- **Subgroup:** Subgroups are differentiated by the content and arrangement of horizons that indicate the relation of the soil to a great group or order or the subtle transition toward soils of another order.
- **Family:** This is a subdivision of a subgroup. Parent material characteristics such as texture and mineralogy, soil climatic factors, and soil reactions are important.
- **Series:** Detailed features of the pedon differentiate subdivisions of the family—the essential soil-sampling unit. Pedon horizons fall within a narrow range of color, texture, structure, consistence, porosity, moisture, chemical reaction, thickness, and composition.

Soil Horizons in the CSSC

Soil horizons are named and standardized as diagnostic in the classification process. Several mineral and organic horizons and layers are used in the CSSC. Three mineral horizons are recognized by capital letter designation, followed by lowercase suffixes for further description. Principal soil-mineral horizons and suffixes are presented in Table 1.

Four organic horizons are identified in the Canadian classification system. *O* is further defined through subhorizon designations. Note that for organic soils, such layers are identified as *tiers*. These organic horizons are detailed in Table 2.

The Nine Soil Orders of the CSSC

The nine orders of the CSSC, and related great groups, are summarized in Table 3 with a general description of properties, related great groups, an estimated percentage of land area for the soil order, a fertility assessment, and any applicable Soil Taxonomy equivalent.

Figure 1 is a generalized map of the distribution of principal soil orders in relation to physiographical regions in Canada. This grouping allows you to easily compare soils across Canada. A summary of the nine soil orders appears in Table 3. Please consult the *National Atlas of Canada*, 5th edition, for a detailed map of Canadian soils (**http://atlas.gc.ca/**).

The *Soil Landscapes of Canada* (SLC) site at **http://sis.agr.gc.ca/cansis/nsdb/slc/intro.html** is most useful! Here you will find a wonderful assortment of landscape and soil profile photographs, arranged geographically across Canada from east to west. The site also has an interactive GIS on-line mapping application. Version 2.2 SLC Component Mapping (December 1996) is operational and involves the CSSC and the Canadian Land

Table 1 Three Mineral Horizons and Mineral Horizon Suffixes Used in the CSSC

Symbol	Mineral Horizon Description
A	Forms at or near the surface; experiences *eluviation*, or leaching, of finer particles or minerals. Several subdivisions are identified, with the surface usually darker and richer in organic content than lower horizons (***Ah***); or a paler, lighter zone below that reflects removal of organic matter with clays and oxides of aluminum and iron leached (removed) to lower horizons (***Ae***).
B	Experiences *illuviation*, a depositional process, as demonstrated by accumulations of clays (***Bt***), sesquioxides of aluminum or iron, and possibly an enrichment of organic debris (***Bh***), and the development of soil structure. Coloration is important in denoting whether hydrolysis, reduction, or oxidation processes are operational for the assignment of a descriptive suffix.
C	Exhibits little effect from pedogenic processes operating in the ***A*** and ***B*** horizons, except the process of gleysation associated with poor drainage and the reduction of iron, denoted (***Cg***), and the accumulation of calcium and magnesium carbonates (***Cca***) and more soluble salts (***Cs***) and (***Csa***).

Symbol	Horizon Suffix Description
b	A buried soil horizon.
c	Irreversible cementation of a pedogenic horizon, e.g., cemented by $CaCO_3$.
ca	Lime accumulation of at least 10 cm thickness that exceeds in concentration that of the unenriched parent material by at least 5%.
cc	Irreversible cemented concretions, typically in pellet form.
e	Used with ***A*** mineral horizons (***Ae***) to denote eluviation of clay, Fe, Al, or organic matter.
f	Enriched principally with illuvial iron and aluminum combined with organic matter, reddish in upper portions and yellowish at depth, determined through specific criteria. Used with ***B*** horizons alone.
g	Gray to blue colors, prominent mottling, or both, produced by intense chemical reduction. Various applications to ***A***, ***B***, and ***C*** horizons.
h	Enriched with organic matter: accumulation in place or biological mixing (***Ah***) or subsurface enrichment through illuviation (***Bh***).
j	A modifier suffix for ***e***, ***f***, ***g***, ***n***, and ***t*** to denote limited change or failure to meet specified criteria denoted by that letter.
k	Presence of carbonates as indicated by visible effervescence with dilute hydrocholoric acid (HCl).
m	Used with ***B*** horizons slightly altered by hydrolysis, oxidation, or solution, or all three to denote a change in color or structure.
n	Accumulation of exchangeable calcium (Ca) in ratio to exchangeable sodium (Na) that is 10 or less, with the following characteristics: prismatic or columnar structure, dark coatings on ped surfaces, and hard consistence when dry. Used with ***B*** horizons alone.
p	***A*** or ***O*** horizons disturbed by cultivation, logging, and habitation. May be used when plowing intrudes on previous ***B*** horizons.
s	Presence of salts, including gypsum, visible as crystals or veins or surface crusts of salt crystals, and by lowered crop yields. Usually with ***C*** but may appear with any horizon and lowercase suffixes.
sa	A secondary enrichment of salts more soluble than Ca or Mg carbonates, exceeding unenriched parent material, in a horizon at least 10 cm thick.
t	Illuvial enrichment of the ***B*** horizon with silicate clay that must exceed in overlying ***Ae*** horizon by 3 to 20% depending on the clay content of the ***Ae*** horizon.
u	Markedly disrupted by physical or faunal processes other than cryoturbation.
x	Fragipan formation—a loamy subsurface horizon of high bulk density and very low organic content. When dry it has a hard consistence and seems to be cemented.
y	Affected by cryoturbation (frost action) with disrupted and broken horizons and incorporation of materials from other horizons. Application to ***A***, ***B***, and ***C*** horizons and in combination with other suffixes.
z	A frozen layer.

Resource Network (CLRN). The component mapping involves a GIS model consisting of layers that include the major characteristics of soil and land for all of Canada. You can select a spatial area and a variety of attributes to display on the map (drainage class, soil type, rooting depth, local surface form, slope, and vegetation cover, among others).

Table 2 Four Organic Horizons Used in the CSSC

Symbol	Description
O	Organic materials, mainly mosses, rushes, and woody materials
L	Mainly discernible leaves, twigs, and woody materials
F	Partially decomposed, somewhat recognizable **L** materials
H	Indiscernible organic materials
O is further defined through subhorizon designations:	
Of	Readily identifiable fibric materials
Om	Mesic materials of intermediate decomposition
Oh	Humic material at an advanced stage of decomposition—low fiber, high bulk density

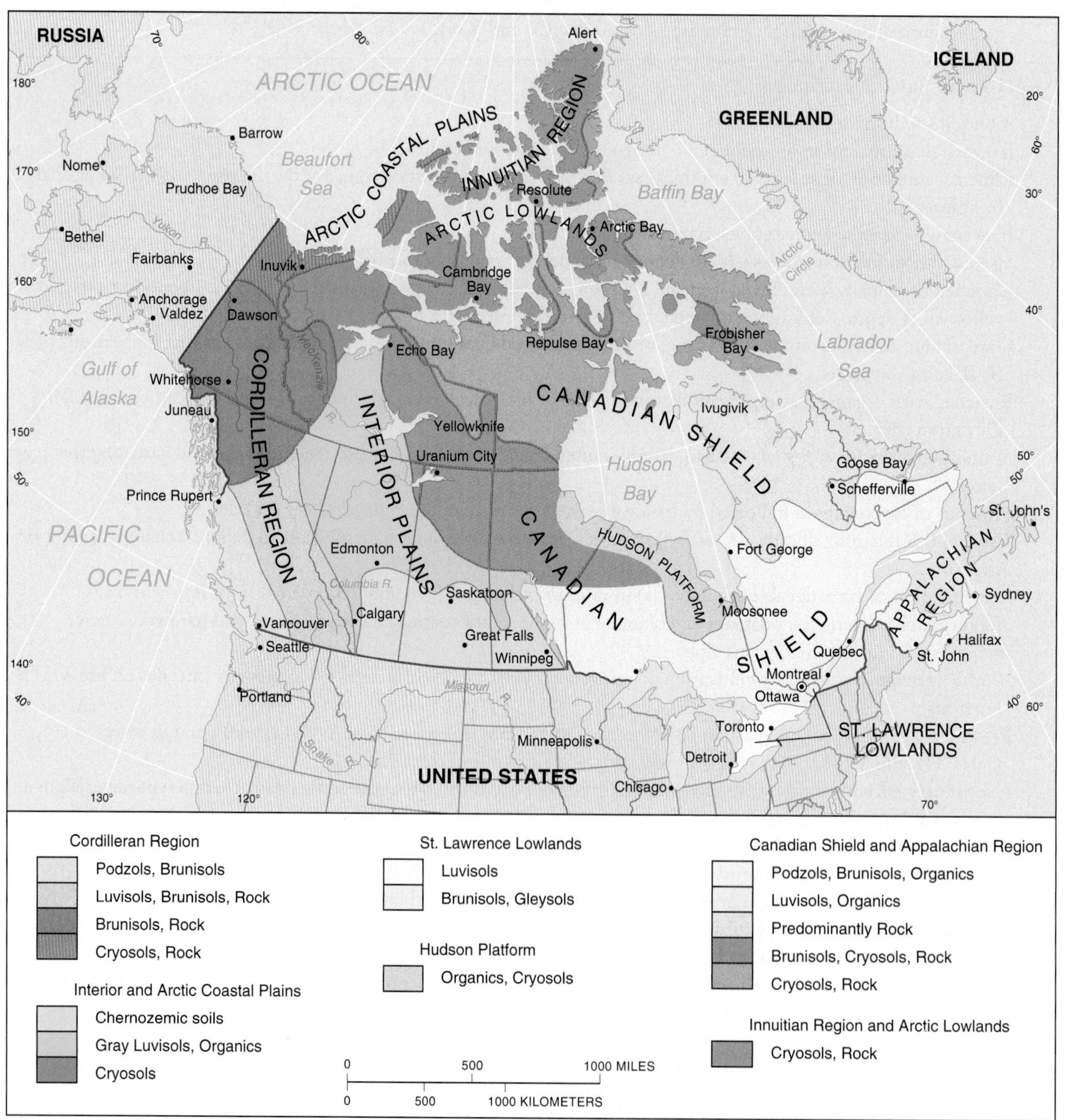

Figure 1 Soil orders of Canada.
Principal soil regions of the Canadian System of Soil Classification (CSSC) as related to major physiographic regions. [After maps prepared by the Land Resources Research Institute, Geological Survey of Canada, and the Canadian Soil Survey Committee.]

Table 3 Nine Orders of the Canadian System of Soil Classification

Order Great Group	Characteristics[a]	Fertility
Chernozemic (Russian, *chernozem*) Brown (more moist) Dark brown Black Dark gray (less moist) (38 subgroups)	Well to imperfectly drained soils of the steppe-grassland-forest transition, Southern Alberta, Saskatchewan, Manitoba, Okanagan Valley, BC, Palouse Prairie, BC. Accumulation of organic matter in surface horizons. Most frozen during some winter months with soil-moisture deficits in the summer. A diagnostic **Ah** is typical (although **Ahe, Ap** are present) at least 10 cm thick or 15 cm if disturbed by cultivation. Mean annual temperature >0°C and usually <5.5°C. (5.1%, 470,000 km^2; Soil Taxon. = Mollisols.)	High; wheat-growing
Solonetzic (Russian, *solonetz*) Solonetz Solodized Solonetz Solod (27 subgroups)	Solonetz denotes saline or alkaine soils. Well to imperfectly drained mineral soils developed under grasses in semiarid to subhumid climates. Limited areas of central and north-central Alberta. Noted for a **B** horizon that is very hard when dry but swells to a sticky, low-permeability mass when wet. A saline **C** horizon reflects nature of parent materials. (0.8%, 73,700 km^2; Soil Taxon. = Natric horizon of Mollisols and Alfisols.)	Variable (medium) about 50% cultivated, remainder in pasture
Luvisolic Gray brown Luviso Gray Luvisol (18 subgroups)	Eluviation-illuviation processes produce a light-colored **Ae** horizon and a diagnostic **Bt** horizon. Soils of mixed deciduous-coniferous forests. Major occurrence is the St. Lawrence lowland. Luvisols do not have a solonetzic **B** horizon, evidence of Gleysolic order and gleying, or organics less than in the Organic order. Permafrost within 1 m of surface and 2 m if soils are cryoturbated. (10.3%, 950,000 km^2; Soil Taxon. = Boralfs, Udalfs—suborders of Alfisols.)	High
Podzolic (Russian, *podzol*) Humic Ferro-humic Podzol Humo-ferric Podzol (25 subgroups)	Soils of coniferous forests and sometimes heath, leaching of overlying horizons occurs in moist, cool to cold climates. Iron, aluminum and organic matter from **L, F,** and **H** horizons are redeposited in podzolic **B** horizon. A diagnostic **Bh, Bhf,** or **Bf** is present depending on great group. Dominant in western British Columbia, Ontario, and Québec. (22.6%, 2,083,000 km^2; Soil Taxon. = Spodosols, some Inceptisols.)	Low to medium depending on acidity
Brunisolic (French, "brown") Melanic Brunisol Entric Brunisol Sombric Brunisol Dystric Brunisol (18 subgroups)	Sufficiently developed to distinguish from Regosolic order. Soils under forest cover with brownish **Bm** horizons, although various colors are possible. Also, can be with mixed forest, shrubs, and grass. Diagnostic **Bm, Bfj,** thin **Bf,** or **Btj** horizons differentiate from soils of other orders. Well to imperfectly drained. Lack the podzolic **B** horizon of podzols although surrounded by them in St. Lawrence lowland. (8.8%, 811,000 km^2; Soil Taxon. = Inceptisols, some Psamments [Aquents in Entisols.])	Medium (variable)
Regosolic (Greek, *rhegos*) Regosol Humic Regosol (8 subgroups)	Weakly developed limited soils, the result of any number of factors: young materials; fresh alluvial deposits; material instability; mass-wasted slopes; or dry, cold climatic conditions. Lack solonetzic, illuvial, or podzolic **B** horizons. Lack permafrost within 1 m of surface, or 2 m if cryoturbated. May have **L, F, H,** or **O** horizons, or an **Ah** horizon if less than 10 cm thick. Buried horizons possible. Dominant in Northwest Territories and northern Yukon, now designated as Cryosols under CSSC. (1.3%, 120,000 km^2; Soil Taxon. = Entisols.)	Low (variable)
Gleysolic (Russian, *glei*) Luvic Gleysol Humic Gleysol Gleysol (13 subgroups)	Defined on the basis of color and mottling that results from chronic reducing conditions inherent in poorly drained mineral soils under wet conditions. High water table and long periods of water saturation. Rather than continuous they appear spotty within other soil orders and occasionally may dominate an area. A diagnostic **Bg** horizon is present. (1.9%, 175,000 km^2; Soil Taxon. = Various aquic suborders, a reducing moisture regime.)	High to medium
Organic Fibrisol Mesisol Humisol Folisol (31 subgroups)	Peat, bog, and muck soils, largely composed of organic material. Most water-saturated for prolonged periods. Are widespread in association with poorly to very poorly drained depressions, although Folisols are found under upland forest environments. Exceed 17% organic carbon and 30% organic matter overall. (4.2%, 387,000 km^2; Soil Taxon. = Histolsols.)	High to medium given drainage, available nutrients
Crysolic (Greek, *kyros*) Turbic Cryosol Static Cryosol Organic Cryosol (15 groups)	Dominate the northern third of Canada, with permafrost closer to the surface and composed of mineral and organic soil deposits. Generally found north of the treeline, or in fine-textured soils in subarctic forest, or in some organic soils in boreal forests. **Ah** horizon lacking or thin. Cryoturbation (frost action) common, often denoted by patterned ground circles, polygons, and stripes. Subgroups based on degree of cryoturbation and the nature of mineral or organic soil material. (45%, 4,150,000 km^2; Soil Taxon. = Gelisols, Cryoquepts, Inceptisols, and pergelic temperature regime in several suborders.)	Not applicable

*Estimated percentage and square kilometers of Canada's land area and Soil Taxonomy equivalent are given in parentheses.

Review Questions

1. Why did Canada adopt its own system of soil classification? Describe a brief history of events that led up to the modern CSSC system.
2. Which soil order is associated with the development of a bog? Explain its use as a low-grade fuel.
3. Describe the podzolization process occurring in northern coniferous forests. What are the surface horizons like? What management strategies might enhance productivity in these soils? Name the soil order for these areas.
4. Compare and contrast Interior Plains soils with those of the southeastern Canadian Shield.
5. What processes inhibit soil development in the extreme north? Explain.
6. Briefly describe what you found on the *Soil Landscapes of Canada* (SLC) Web site. Please try the SLC Component Mapping feature for your specific area or region (**http://sis.agr.gc.ca/cansis/nsdb/slc/intro.html**).
7. Which of the nine soil orders is characteristic of the area where you live, or where you attend college? How did you determine the answer?

Appendix C

The Köppen Climate Classification System

The Köppen climate classification system was designed by Wladimir Köppen (1846–1940), a German climatologist and botanist, and is widely used for its ease of comprehension. The basis of any empirical classification system is the choice of criteria used to draw lines on a map to designate different climates. **Köppen-Geiger climate classification** uses *average monthly temperatures*, *average monthly precipitation*, and *total annual precipitation* to devise its spatial categories and boundaries. But we must remember that boundaries really are transition zones of gradual change. The trends and overall patterns of boundary lines are more important than their precise placement, especially with the small scales generally used on world maps.

Take a few minutes and examine the Köppen system, the criteria, and the considerations for each of the principal climate categories. The modified Köppen-Geiger system has its drawbacks, however. It does not consider winds, temperature extremes, precipitation intensity, quantity of sunshine, cloud cover, or net radiation. Yet the system is important because its *correlations with the actual world are reasonable* and the *input data are standardized and readily available.*

Köppen's Climatic Designations

Figure 6.4 in Chapter 6 shows the distribution of each of Köppen's six climate classifications on the land. This generalized map shows the spatial pattern of climate. The Köppen system uses capital letters (A, B, C, D, E, H) to designate climatic categories from the equator to the poles. The guidelines for each of these categories are in the margin of Figure 1.

Five of the climate classifications are based on thermal criteria:

- A Tropical (equatorial regions)
- C Mesothermal (Mediterranean, humid subtropical, marine west coast regions)
- D Microthermal (humid continental, subarctic regions)
- E Polar (polar regions)
- H Highland (compared to lowlands at the same latitude, highlands have lower temperatures—recall the normal lapse rate—and more efficient precipitation due to lower moisture demand)

Only one climate classification is based on moisture as well:

- B Dry (deserts and steppes)

Within each climate classification additional lowercase letters are used to signify temperature and moisture conditions. For example, in a *tropical rain forest Af* climate, the *A* tells us that the average coolest month is above 18°C (64.4°F, average for the month), and the *f* indicates that the weather is constantly wet, with the driest month receiving at least 6 cm (2.4 in.) of precipitation. (The designation *f* is from the German *feucht*, for moist.) As you can see on the climate map, the *tropical rain forest Af* climate dominates along the equator and equatorial rain forest.

In a Dfa climate, the *D* means that the average warmest month is above 10°C (50°F), with at least one month falling below 0°C (32°F); the *f* says that at least 3 cm (1.2 in.) of precipitation falls during every month; and the *a* indicates a warmest summer month averaging above 22°C (71.6°F). Thus, a Dfa climate is a *humid-continental, hot-summer* climate in the microthermal category.

What's in a Boundary?

Originally, Köppen proposed that the isotherm boundary between mesothermal C and microthermal D climates be a coldest month of −3°C (26.6°F) or lower. That might be an accurate criterion for Europe, but for conditions in North America, the 0°C isotherm is considered more appropriate. The difference between the 0°C and −3°C isotherms for January covers an area about the width of the state of Ohio. Remember, these isotherm lines are really transition zones and do not mean abrupt change from one temperature to another.

A line denoting at least one month below freezing runs from New York City roughly along the Ohio River, trending westward until it meets the dry climates in the southeastern corner of Colorado. The line marking −3°C as the coldest month runs farther north along Lake Erie and the southern tip of Lake Michigan. From year to year, the position of the 0°C isotherm for January can shift back and forth several hundred kilometers as weather conditions vary. The map in Figure 1 uses the 0°C isotherm for the C-D climate boundary.

Köppen Guidelines

The Köppen guidelines and map portrayal are in Figure 1. First, check the guidelines for a climate type, then the color legend for the subdivisions of the type, then check out the distribution of that climate on the map. You may want to compare this with the climate map in Figure 6.5 that presents causal elements that produce these climates.

Figure 1 World climates and their guidelines according to the Köppen classification system.

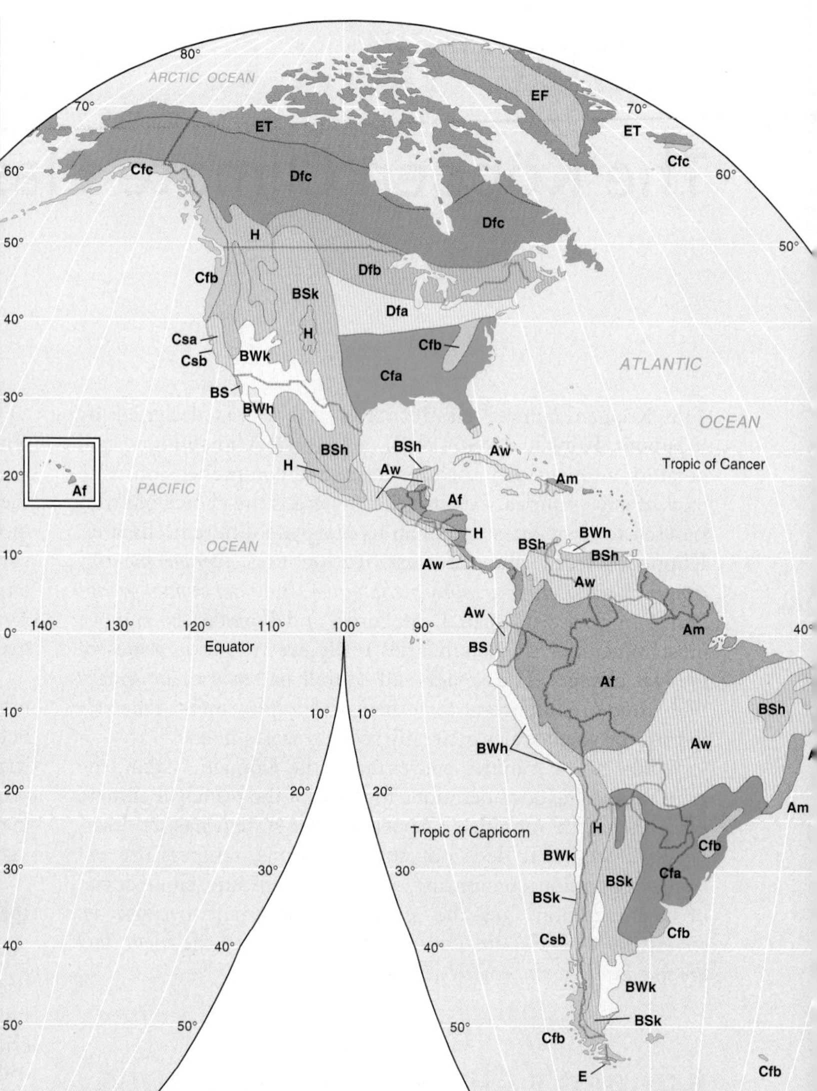

Köppen Guidelines
Tropical Climates — A

Consistently warm with all months averaging above 18°C (64.4°F); annual water supply exceeds water demand.

Af — Tropical rain forest:
f = All months receive precipitation in excess of 6 cm (2.4 in.).

Am — Tropical monsoon:
m = A marked short dry season with 1 or more months receiving less than 6 cm (2.4 in.) precipitation; an otherwise excessively wet rainy season. ITCZ 6–12 months dominant.

Aw — Tropical savanna:
w = Summer wet season, winter dry season; ITCZ dominant 6 months or less, winter water-balance deficits.

Mesothermal Climates — C

Warmest month above 10°C (50°F); coldest month above 0°C (32°F) but below 18°C (64.4°F); seasonal climates.

Cfa, Cwa — Humid subtropical:
a = Hot summer; warmest month above 22°C (71.6°F).
f = Year-round precipitation.
w = Winter drought, summer wettest month 10 times more precipitation than driest winter month.

Cfb, Cfc — Marine west coast, mild-to-cool summer:
f = Receives year-round precipitation.
b = Warmest month below 22°C (71.6°F) with 4 months above 10°C.
c = 1–3 months above 10°C.

Csa, Csb — Mediterranean summer dry:
s = Pronounced summer drought with 70% of precipitation in winter.
a = Hot summer with warmest month above 22°C (71.6°F).
b = Mild summer; warmest month below 22°C.

Microthermal Climates — D

Warmest month above 10°C (50°F); coldest month below 0°C (32°F); cool temperate-to-cold conditions; snow climates. In Southern Hemisphere, occurs only in highland climates.

Dfa, Dwa — Humid continental:
a = Hot summer; warmest month above 22°C (71.6°F).
f = Year-round precipitation.
w = Winter drought.

Dfb, Dwb — Humid continental:
b = Mild summer; warmest month below 22°C (71.6°F).
f = Year-round precipitation.
w = Winter drought.

Dfc, Dwc, Dwd — Subarctic:
Cool summers, cold winters.
f = Year-round precipitation.
w = Winter drought.
c = 1–4 months above 10°C.
d = Coldest month below −38°C (−36.4°F), in Siberia only.

Dry Arid and Semiarid Climates — B

Potential evapotranspiration* (natural moisture demand) exceeds precipitation (natural moisture supply) in all B climates. Subdivisions based on precipitation timing and amount and mean annual temperature.

Earth's arid climates.

BWh — Hot low-latitude desert

BWk — Cold midlatitude desert
BW = Precipitation less than 1/2 natural moisture demand.
h = Mean annual temperature >18°C (64.4°F).
k = Mean annual temperature <18°C.

Earth's semiarid climates.

BSh — Hot low-latitude steppe

BSk — Cold midlatitude steppe
BS = Precipitation more than 1/2 natural moisture demand but not equal to it.
h = Mean annual temperature >18°C.
k = Mean annual temperature <18°C.

Polar Climates — E

Warmest month below 10°C (50°F); always cold; ice climates.

ET — Tundra:
Warmest month 0–10°C (32–50°F); precipitation exceeds small potential evapotranspiration demand*; snow cover 8–10 months.

EF — Ice cap:
Warmest month below 0°C (32°F); precipitation exceeds a very small potential evapotranspiration demand; the polar regions.

EM — Polar marine:
All months above −7°C (20°F), warmest month above 0°C; annual temperature range <17 C° (30 F°).

*Potential evapotranspiration = the amount of water that would evaporate or transpire if it were available — the natural moisture demand in an environment; see Chapter 7.

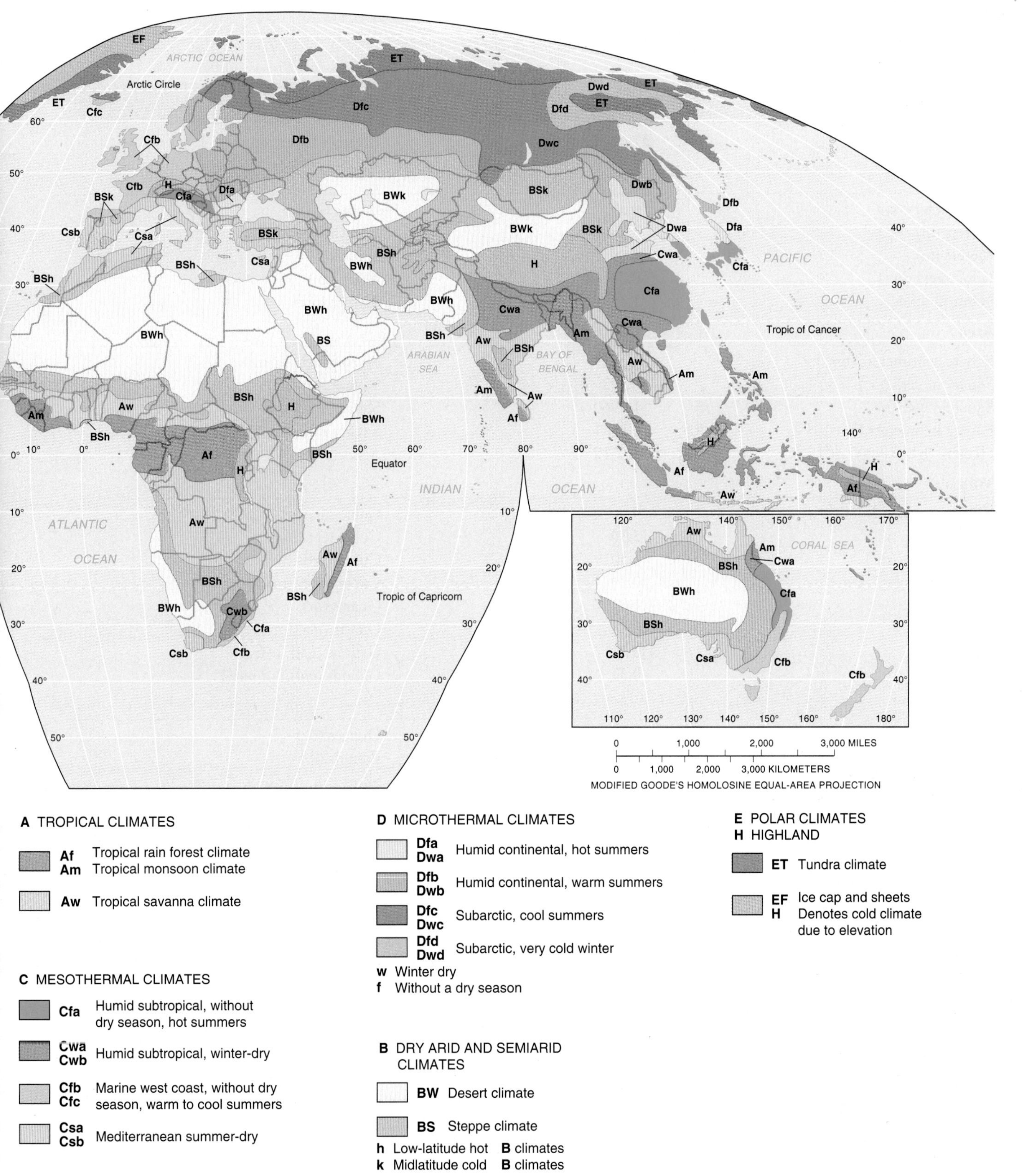
ARCTIC OCEAN
Arctic Circle
PACIFIC OCEAN
Tropic of Cancer
ARABIAN SEA
BAY OF BENGAL
Equator
INDIAN OCEAN
ATLANTIC OCEAN
Tropic of Capricorn
CORAL SEA
0 1,000 2,000 3,000 MILES
0 1,000 2,000 3,000 KILOMETERS
MODIFIED GOODE'S HOMOLOSINE EQUAL-AREA PROJECTION
A TROPICAL CLIMATES
Af Tropical rain forest climate
Am Tropical monsoon climate
Aw Tropical savanna climate
C MESOTHERMAL CLIMATES
Cfa Humid subtropical, without dry season, hot summers
Cwa Cwb Humid subtropical, winter-dry
Cfb Cfc Marine west coast, without dry season, warm to cool summers
Csa Csb Mediterranean summer-dry
D MICROTHERMAL CLIMATES
Dfa Dwa Humid continental, hot summers
Dfb Dwb Humid continental, warm summers
Dfc Dwc Subarctic, cool summers
Dfd Dwd Subarctic, very cold winter
w Winter dry
f Without a dry season
B DRY ARID AND SEMIARID CLIMATES
BW Desert climate
BS Steppe climate
h Low-latitude hot B climates
k Midlatitude cold B climates
E POLAR CLIMATES
H HIGHLAND
ET Tundra climate
EF Ice cap and sheets
H Denotes cold climate due to elevation

Appendix D

Metric to English

Metric Measure	Multiply by	English Equivalent
Length		
Centimeters (cm)	0.3937	Inches (in.)
Meters (m)	3.2808	Feet (ft)
Meters (m)	1.0936	Yards (yd)
Kilometers (km)	0.6214	Miles (mi)
Nautical mile	1.15	Statute mile
Area		
Square centimeters (cm^2)	0.155	Square inches ($in.^2$)
Square meters (m^2)	10.7639	Square feet (ft^2)
Square meter (m^2)	1.1960	Square yards (yd^2)
Square kilometers (km^2)	0.3831	Square miles (mi^2)
Hectare (ha) (10,000 m^2)	2.4710	Acres (a)
Volume		
Cubic centimeters (cm^3)	0.06	Cubic inches ($in.^3$)
Cubic meters (m^3)	35.30	Cubic feet (ft^3)
Cubic meters (m^3)	1.3079	Cubic yards (yd^3)
Cubic kilometers (km^3)	0.24	Cubic miles (mi^3)
Liters (L)	1.0567	Quarts (qt), U.S.
Liters (L)	0.88	Quarts (qt), Imperial
Liters (L)	0.26	Gallons (gal), U.S.
Liters (L)	0.22	Gallons (gal), Imperial
Mass		
Grams (g)	0.03527	Ounces (oz)
Kilograms (kg)	2.2046	Pounds (lb)
Metric ton (tonne) (t)	1.10	Short ton (tn), U.S.
Velocity		
Meters/second (mps)	2.24	Miles/hour (mph)
Kilometers/hour (kmph)	0.62	Miles/hour (mph)
Knots (kn) (nautical mph)	1.15	Miles/hour (mph)
Temperature		
Degrees Celsius (°C)	1.80 (then add 32)	Degrees Fahrenheit (°F)
Celsius degree (C°)	1.80	Fahrenheit degree (F°)
Additional water measurements:		
Gallon (Imperial)	1.201	Gallon (U.S.)
Gallons (gal)	0.000003	Acre-feet

1 cubic foot per second per day = 86,400 cubic feet = 1.98 acre − feet

ADDITIONAL ENERGY AND POWER MEASUREMENTS

1 watt (W) = 1 joule/s	1 W/m^2 = 2.064 $cal/cm^2 day^{-1}$
1 joule = 0.239 calorie	1 W/m^2 = 61.91 $cal/cm^2 month^{-1}$
1 calorie = 4.186 joules	1 W/m^2 = 753.4 $cal/cm^2 year^{-1}$
1 W/m^2 = 0.001433 cal/min	100 W/m^2 = 75 $kcal/cm^2 year^{-1}$
697.8 W/m^2 = 1 $cal/cm^2 min^{-1}$	

Solar constant:
1372 W/m^2
2 $cal/cm^2 min^{-1}$

English to Metric

English Measure	Multiply by	Metric Equivalent
Length		
Inches (in.)	2.54	Centimeters (cm)
Feet (ft)	0.3048	Meters (m)
Yards (yd)	0.9144	Meters (m)
Miles (mi)	1.6094	Kilometers (km)
Statute mile	0.8684	Nautical mile
Area		
Square inches (in.^{2})	6.45	Square centimeters (cm^2)
Square feet (ft^2)	0.0929	Square meters (m^2)
Square yards (yd^2)	0.8361	Square meters (m^2)
Square miles (mi^2)	2.5900	Square kilometers (km^2)
Acres (a)	0.4047	Hectare (ha)
Volume		
Cubic inches (in.^{3})	16.39	Cubic centimeters (cm^3)
Cubic feet (ft^3)	0.028	Cubic meters (m^3)
Cubic yards (yd^3)	0.765	Cubic meters (m^3)
Cubic miles (mi^3)	4.17	Cubic kilometers (km^3)
Quarts (qt), U.S.	0.9463	Liters (L)
Quarts (qt), Imperial	1.14	Liters (L)
Gallons (gal), U.S.	3.8	Liters (L)
Gallons (gal), Imperial	4.55	Liters (L)
Mass		
Ounces (oz)	28.3495	Grams (g)
Pounds (lb)	0.4536	Kilograms (kg)
Short ton (tn), U.S.	0.91	Metric ton (tonne) (t)
Velocity		
Miles/hour (mph)	0.448	Meters/second (mps)
Miles/hour (mph)	1.6094	Kilometers/hour (kmph)
Miles/hour (mph)	0.8684	Knots (kn) (nautical mph)
Temperature		
Degrees Fahrenheit (°F)	0.556 (after subtracting 32)	Degrees Celsius (°C)
Fahrenheit degree (F°)	0.556	Celsius degree (C°)
Additional water measurements:		
Gallon (U.S.)	0.833	Gallons (Imperial)
Acre-feet	325,872	Gallons (gal)

Additional Notation

Multiples	Prefixes	
1,000,000,000 = 10^9	giga	G
1,000,000 = 10^6	mega	M
1,000 = 10^3	kilo	k
100 = 10^2	hecto	h
10 = 10^1	deka	da
1 = 10^0		
0.1 = 10^{-1}	deci	d
0.01 = 10^{-2}	centi	c
0.001 = 10^{-3}	milli	m
0.000001 = 10^{-6}	micro	μ

Glossary

The chapter in which each term appears **boldfaced** in parentheses is followed by a specific definition relevant to the key term's usage in the chapter. Cross-reference terms appear in *italics*.

Abiotic (1) Nonliving; Earth's nonliving systems of energy and materials.

Ablation (14) Loss of glacial ice through melting, sublimation, wind removal by deflation, or the calving off of blocks of ice.

Abrasion (11, 12, 14) Mechanical wearing and erosion of bedrock accomplished by the rolling and grinding of particles and rocks carried in a stream, moved by wind in a "sandblasting" action, or imbedded in glacial ice.

Absorption (3) Assimilation and conversion of radiation from one form to another in a medium. In the process the temperature of the absorbing surface is raised, thereby affecting the rate and quality of radiation from that surface.

Active layer (14) A zone of seasonally frozen ground that exists between the subsurface permafrost layer and the ground surface. The active layer is subject to consistent daily and seasonal freeze-thaw cycles (see *permafrost*, *periglacial*).

Actual evapotranspiration (7) Actual amount of evaporation and transpiration that occurs (ACTET); derived in the water-balance equation by subtracting the deficit (DEFIC) from potential evapotranspiration (POTET).

Adiabatic (5) Pertaining to the heating and cooling of a descending or ascending parcel of air through compression and expansion, without any exchange of heat between the parcel and the surrounding environment.

Adsorption (15) The process whereby cations become attached to soil colloids, or the adhesion of gas molecules and ions to solid surfaces with which they come into contact.

Advection (3) Horizontal movement of air or water from one place to another (compare with *convection*).

Advection fog (5) Active condensation formed when warm, moist air moves laterally over cooler water or land surfaces, causing the lower layers of the overlying air to be chilled to the dew-point temperature.

Aggradation (11) The general building up of land surface because of deposition of material; opposite of *degradation*. When the sediment load of a stream exceeds the stream's capacity, the stream channel is filled through this process.

Air mass (5) A distinctive, homogeneous body of air in terms of temperature and humidity that takes on the moisture and temperature characteristics of its source region.

Air pressure (2) Pressure produced by the motion, size, and number of gas molecules and exerted on surfaces in contact with the air. Normal sea level pressure, as measured by the height of a column of mercury (Hg), is expressed as 1013.2 millibars, 760 mm of Hg, or 29.92 inches of Hg. Air pressure can be measured with mercury or aneroid barometers (see listing for both).

Albedo (3) The reflective quality of a surface, expressed as the percentage of reflected insolation to incoming insolation; a function of surface color, angle of incidence, and surface texture.

Alfisols (15) A soil order in the Soil Taxonomy. Moderately weathered forest soils that are moist versions of Mollisols, with productivity dependent on specific patterns of moisture and temperature; rich in organics; most wide-ranging of the soil orders in the Soil Taxonomy classification.

Alluvial fan (12) Fan-shaped fluvial landform at the mouth of a canyon, generally occurs in arid landscapes where streams are intermittent.

Alluvial terraces (11) Level areas that appear as topographic steps above the stream, created by a stream as it scours with renewed downcutting into its floodplain; composed of unconsolidated alluvium (see *alluvium*).

Alluvium (11) General descriptive term for clay, silt, and sand, transported by running water and deposited in sorted or semisorted sediment on a floodplain, delta, or streambed.

Alpine glacier (14) A glacier confined in a mountain valley or walled basin, consisting of three subtypes: valley glacier (within a valley), piedmont glacier (coalesced at the base of a mountain, spreading freely over nearby lowlands), and outlet glacier (flowing outward from a continental glacier).

Altitude (2) The angular distance between the horizon (a horizontal plane) and the Sun (or any point in the sky).

Altocumulus (5) Middle level, puffy clouds that occur in several forms: patchy rows, wave patterns, a "mackerel sky," or lens-shaped "lenticular" clouds.

Andisols (15) A soil order in the Soil Taxonomy. Derived from volcanic parent materials in areas of volcanic activity. A new order created in 1990 of soils previously considered under Inceptisols and Entisols.

Anemometer (4) A device to measure wind velocity.

Aneroid barometer (4) A device to measure air pressure using a partially emptied, sealed cell (see *air pressure*).

Antarctic high (4) A consistent high-pressure region centered over Antarctica; source region for an intense polar air mass that is dry and associated with the lowest temperatures on Earth.

Anthropogenic atmosphere (2) Earth's next atmosphere, so named because humans appear to be the principal causative agent.

Anticline (9) Upfolded strata in which layers slope away from the axis of the fold, or central ridge (compare *syncline*).

Anticyclone (4) A dynamically or thermally caused area of high atmospheric pressure with descending and diverging air flows that rotate clockwise in the Northern Hemisphere and counterclockwise in the Southern Hemisphere (compare *cyclone*).

Aphelion (2) The most distant point in Earth's elliptical orbit about the Sun; reached on July 4 at a distance of 152,083,000 km (94.5 million mi); variable over a 100,000-year cycle (compare *perihelion*).

Aquatic ecosystem (16) An association of plants and animals and their nonliving environment in a water setting.

Aquiclude (7) A body of rock that does not conduct water in usable amounts; an impermeable layer.

Aquifer (7) Rock strata permeable to groundwater flow.

Aquifer recharge area (7) The surface area where water enters an aquifer to recharge the water-bearing strata in a groundwater system.

Arctic tundra (16) A biome in the northernmost portions of North America, Europe, and Russia, featuring low ground-level herbaceous plants as well as some woody plants.

Arête (14) A sharp ridge that divides two cirque basins. Means "fish bone" in French. Arêtes form sawtooth and serrated ridges in glaciated mountains.

Aridisols (15) A soil order in the Soil Taxonomy; largest soil order. Typical of dry (arid and semiarid) climates; low in organic matter and dominated by calcification and salinization.

Artesian water (7) Pressurized groundwater that rises in a well or a rock structure above the local water table; may flow out onto the ground.

Asthenosphere (8) Region of the upper mantle just below the lithosphere known as the plastic layer; the least rigid portion of Earth's interior; shatters if struck yet flows under extreme heat and pressure.

Atmosphere (1) The thin veil of gases surrounding Earth that forms a protective boundary between outer space and the biosphere; generally considered to extend about 480 km (300 mi) elevation from Earth's surface.

Auroras (2) A spectacular glowing light display in the ionosphere, stimulated by the interaction of the solar wind with oxygen and nitrogen gases at high latitudes; called aurora borealis (Northern Hemisphere) and aurora australis (Southern Hemisphere).

Autumnal (September) equinox (2) The time around September 22–23 when the Sun's declination crosses the equatorial parallel; all places on Earth experience days and nights of equal length. The Sun rises at the South Pole and sets at the North Pole (compare *vernal equinox*).

Available water (7) The portion of capillary water that is accessible to plant roots; usable water held in soil moisture storage.

Axial parallelism (2) Earth's axis is parallel to itself throughout the year; the North Pole points to near Polaris.

Badland (12) Rugged topography, usually of relatively low, varied relief and barren of vegetation; associated with arid and semiarid regions and rocks with low resistance to weathering.

Bajada (12) A continuous apron of coalesced alluvial fans, formed along the base of mountains in arid climates; presents a gently rolling surface from fan to fan (see *alluvial fan*).

Barrier beach (13) Narrow, long depositional feature, generally composed of sand, that forms offshore and is roughly parallel to the coast; may appear as barrier islands and long chains of barrier beaches.

Barrier island (13) Generally, a broadened barrier beach.

Barrier spit (13) A depositional form that develops when transported sand or gravel in a barrier beach or island is deposited in long ridges that are attached at one end to the mainland and partially cross the mouth of a bay.

Basalt (8) A common extrusive igneous rock; its mafic composition is fine-grained, comprising the bulk of the ocean floor crust, lava flows, and volcanic forms.

Base level (11) A hypothetical level below which a stream cannot erode its valley, and thus the lowest operative level for denudation processes; in an absolute sense represented by sea level extending back under the landscape.

Base load (11) Those coarser materials dragged along a streambed by a stream's traction (compare *suspended load*; see *traction*).

Basin and Range Province (12) A region of dry climates, few permanent streams, and interior drainage patterns in the western United States; composed of a sequence of *horsts* and *grabens*.

Batholith (8) The largest plutonic form exposed at the surface; an irregular intrusive mass (>100 km^2; >40 mi^2) that invades crustal rocks, cooling slowly so that crystals develop (see *pluton*).

Bay barrier (13) An extensive sand spit that encloses a bay, cutting it off completely from the ocean and forming a lagoon; produced by littoral drift and wave action; sometimes referred to as a baymouth bar (see *barrier spit*, *lagoon*).

Beach (13) The portion of the coastline where an accumulation of sediment is in motion.

Bed load (11) Coarse materials that are dragged along the bed of a stream by traction or by the rolling and bouncing motion of saltation; involves particles too large to remain in suspension (see *traction*, *saltation*).

Bedrock (10) The rock of Earth's crust that is below the soil and basically unweathered; such solid crust sometimes is exposed as an outcrop.

Biodiversity (16) A principle of ecology and biogeography: The more diverse the species population in an ecosystem (in number of species, quantity of members in each species, and genetic content), the more risk is spread over the entire community, which results in greater overall stability, greater productivity, and increased use of soil nutrients, as compared to a monoculture of no diversity.

Biogeochemical cycles (16) The various circuits of flowing elements and materials (carbon, oxygen, nitrogen, phosphorus, water) that combine Earth's biotic (living) and abiotic (nonliving) systems; the cycling of materials is continuous and renewed through the biosphere and the life processes.

Biogeography (16) The study of the distribution of plants and animals and related ecosystems; the geographical relationships with related environments over time.

Biomass (16) The total mass of living organisms on Earth or per unit area of a landscape; also, the weight of the living organisms in an ecosystem.

Biome (16) A large terrestrial ecosystem characterized by specific plant communities and formations; usually named after the predominant vegetation in the region (see *terrestrial ecosystem*).

Biosphere (1) That area where the atmosphere, lithosphere, and hydrosphere function together to form the context within which life exists; an intricate web that connects all organisms with their physical environment; sometimes called *ecosphere*.

Biotic (1) Living; Earth's living system of organisms.

Blowout depressions (12) Eolian erosion whereby deflation forms a basin in areas of loose sediment. Size may range up to hundreds of meters (see *deflation*).

Bolson (12) The slope and basin area between the crests of two adjacent ridges in a dry region.

Boreal forest (16) See *needleleaf forest*.

Brackish (13) Descriptive of seawater with a salinity of less than 35‰; for example, the Baltic Sea (contrast *brine*).

Braided stream (11) A stream that becomes a maze of interconnected channels laced with excess sediments. Braiding often occurs with a reduction of discharge that affects a stream's transportation ability or an increase in sediment load.

Breaker (13) The point where a wave's height exceeds its vertical stability and the wave breaks as it approaches the shore.

Brine (13) Seawater with a salinity of more than 35‰; for example, the Persian Gulf (contrast *brackish*).

Calcification (15) The illuviated accumulation of calcium carbonate or magnesium carbonate in the B and C soil horizons.

Caldera (9) An interior sunken portion of a composite volcanic crater; usually steep-sided and circular, sometimes containing a lake; also found in conjunction with shield volcanoes.

Canadian System of Soil Classification (Appendix B) A classification system for all soils presently recognized in Canada and adapted to Canada's particular forest, tundra, prairie, frozen ground, and colder climates. The CSSC is organized at five levels of generalization (order, great group, subgroup, family, and series), with each level referred to as a category of classification. The system is described in *The Canadian Soil Classification System*, 2nd edition, published by Agriculture Canada, 1987 (compare *Soil Taxonomy*; see *Gelisols*). Also illustrated at the *Soil Landscapes of Canada* (SLC) Web site: **http://sis.agr.gc.ca/cansis/nsdb/slc/intro.html**.

Capillary water (7) Soil moisture, most of which is accessible to plant roots; held in the soil by the water's surface tension and cohesive forces between water and soil (see also *available water*, *field capacity*, and *wilting point*).

Carbonation (10) A process of chemical weathering by a weak carbonic acid (water and carbon dioxide) that reacts with many minerals containing calcium, magnesium, potassium, and sodium—especially limestone—transforming them into carbonates.

Carbon monoxide (2) An odorless, colorless, tasteless combination of carbon and oxygen produced by the incomplete combustion of fossil fuels or other carbon-containing substances; toxicity to humans is due to its affinity for hemoglobin, displacing oxygen in the bloodstream; CO.

Carnivore (16) A secondary consumer that principally eats meat for sustenance. The top carnivore in a food chain is considered a tertiary consumer (compare *herbivore*).

Cartography (1) The making of maps and charts; a specialized science and art that blends aspects of geography, engineering, mathematics, graphics, computer science, and artistic specialties.

Cation-exchange capacity (CEC) (15) The ability of soil colloids to exchange cations between their surfaces and the soil solution; a measured potential that indicates soil fertility (see *soil colloid*, *soil fertility*).

Chaparral (16) Dominant shrub formations of Mediterranean dry-summer climates; characterized by (sclerophyllous) scrub and short, stunted, and tough forests; derived from the Spanish *chapparo;* specific to California (see *Mediterranean shrubland*).

Chemical weathering (10) Decomposition and decay of the constituent minerals in rock through chemical alteration of those minerals. Water is essential, with rates keyed to temperature and precipitation values. Chemical reactions are active at microsites even in dry climates. Processes include *hydrolysis*, *oxidation*, *carbonation*, and solution.

Chlorofluorocarbon compounds (CFCs) (2) A manufactured molecule (polymer) containing chlorine, fluorine, and carbon; inert, and possessing remarkable heat properties; also known as halogens. After slow transport to the stratospheric ozone layer CFCs react with ultraviolet radiation, freeing chlorine atoms that act as a catalyst to produce reactions that destroy ozone; manufacture banned by international treaties.

Chlorophyll (16) A light-sensitive pigment that resides within chloroplast (organelle) bodies in leaf cells of plants; the basis of photosynthesis.

Cinder cone (9) A landform of tephra and scoria, usually small and cone-shaped and generally not more than 450 m (1500 ft) in height; with a truncated top.

Circle of illumination (2) The division between lightness and darkness on Earth; a day-night great circle.

Cirque (14) A scooped-out, amphitheater-shaped basin at the head of an alpine glacier valley; an erosional landform.

Cirrus (5) Wispy filaments of ice-crystal clouds that occur above 6000 m (20,000 ft); appear in a variety of forms, from feathery hairlike fibers to veils of fused sheets.

Classification (6) The process of ordering or grouping data or phenomena in related classes; results in a regular distribution of information; a taxonomy.

Climate (6) The consistent, long-term behavior of weather over time, including its variability; in contrast to *weather*, which is the condition of the atmosphere at any given place and time.

Climatic regions (6) Areas of similar climate, which contain characteristic regional weather and air mass patterns.

Climatology (6) A scientific study of climate and climatic patterns and the consistent behavior of weather and weather variability and extremes over time in one place or region; including the effects of climate change on human society and culture.

Climographs (6) Graphs that plot daily, monthly, or annual temperature and precipitation values for a selected station; may also include additional weather information.

Closed system (1) A system that is shut off from the surrounding environment so that it is entirely self-contained in terms of energy and materials; Earth is a closed material system (see *open system*).

Cloud (5) An aggradation of moisture droplets and ice crystals that are suspended in air and are great enough in volume and density to be visible; basic forms include stratiform, cumuliform, and cirroform.

Cloud-albedo forcing (3) An increase in albedo (the reflectivity of a surface) caused by clouds due to their reflection of incoming insolation.

Cloud-greenhouse forcing (3) An increase in greenhouse warming caused by clouds because they can act like insulation, trapping longwave (infrared) radiation.

Col (14) Formed by two headward eroding cirques that reduce an arête (ridge crest) to form a high pass or saddlelike narrow depression.

Cold desert and semidesert (16) A type of desert biome found at higher latitudes than warm deserts. Interior location and rain shadow locations produce these cold deserts in North America.

Cold front (5) The leading edge of a cold air mass; identified on a weather map as a line marked with a series of triangular spikes pointing in the direction of frontal movement (compare *warm front*).

Community (16) A convenient biotic subdivision within an ecosystem; formed by interacting populations of animals and plants in an area.

Composite volcano (9) A volcano formed by a sequence of explosive volcanic eruptions; steep-sided, conical in shape; sometimes referred to as a stratovolcano, although composite is the preferred term (compare *shield volcano*).

Condensation nuclei (5) Necessary microscopic particles on which water vapor condenses to form moisture droplets; can be sea salts, dust, soot, or ash.

Conduction (3) The slow molecule-to-molecule transfer of heat through a medium, from warmer to cooler portions.

Cone of depression (7) The depressed shape of the water table around a well after active pumping from an aquifer. The water table adjacent to the well is drawn down during the process of water removal.

Confined aquifer (7) An aquifer that is bounded above and below by impermeable layers of rock or sediment.

Consumers (16) Organisms in an ecosystem that depend on producers (autotrophs—organisms capable of using carbon dioxide as their sole source of carbon) for their source of nutrients (compare *producer*).

Continental divide (11) A ridge or elevated area that determines the drainage pattern of drainage basins; specifically, that ridge in North America that separates drainage to the Pacific in the west from drainage to the Atlantic and Gulf in the east and to Hudson Bay and the Arctic Ocean in the north.

Continental effect (3) A qualitative designation applied to regions that lack the temperature-moderating effects of the sea and that exhibit a greater range between minimum and maximum temperatures, both daily and annually (see *marine effect, land-water heating differences*).

Continental glacier (14) A continuous mass of unconfined ice, covering at least 50,000 km^2 (19,500 mi^2); most extensive as ice sheets covering Greenland and Antarctica.

Continental platforms (9) The broadest category of landforms, including those masses of crust that reside above or near sea level and the adjoining undersea continental shelves along the coastline.

Continental shield (9) Generally old, low-elevation heartland regions of continental crust; various cratons (granitic cores) and ancient mountains exposed at the surface.

Contour lines (Appendix A) Isolines on a topographic map that connect all points at the same elevation relative to a reference elevation called the vertical datum.

Convection (3) Vertical transfer of heat from one place to another through the actual physical movement of air; involves a strong vertical motion.

Convectional lifting (5) Air passing over warm surfaces gains buoyancy and lifts, initiating adiabatic processes.

Convergent lifting (5) Air flows in conflict force lifting and displacement of air upward, initiating adiabatic processes.

Coordinated Universal Time (UTC) (1) The official reference time in all countries, formerly known as Greenwich Mean Time; now measured by six primary standard atomic clocks whose time calculations are collected in Paris, France, at the International Bureau of Weights and Measures (BIPM); the legal reference for time in all countries and broadcast worldwide.

Coral (13) A simple, cylindrical marine animal with a saclike body that secretes calcium carbonate to form a hard external skeleton, and landforms called reefs; lives symbiotically with nutrient-producing algae; presently in a worldwide state of decline due to bleaching (loss of algae).

Core (8) The deepest inner portion of Earth, representing one-third of its entire mass; differentiated into two zones—a solid iron inner core surrounded by a dense, molten, fluid metallic-iron outer core.

Coriolis force (4) The apparent deflection of moving objects on Earth from a straight path, in relationship to the differential speed of rotation at varying latitudes. Deflection is to the right in the Northern Hemisphere and to the left in the Southern Hemisphere; it produces a maximum effect at the poles and zero effect along the equator.

Crater (9) A circular surface depression formed by volcanism (accumulation, collapse, or explosion); usually located at a volcanic vent or pipe; can be at the summit or on the flank of a volcano.

Crevasses (14) Vertical cracks that develop in a glacier as a result of friction between valley walls, or tension forces of extension on convex slopes, or compression forces on concave slopes.

Crust (8) Earth's outer shell of crystalline surface rock, ranging from 5 to 60 km (3 to 38 mi) in thickness from oceanic crust to mountain ranges.

Cryosphere (14) The portion of the hydrosphere and ground that is perennially frozen, generally at high latitudes and elevations.

Cumulonimbus (5) A towering, precipitation-producing cumulus cloud that is vertically developed across altitudes associated with other clouds; frequently associated with lightning and thunder and thus sometimes termed a thunderhead.

Cumulus (5) Bright and puffy cumuliform clouds up to 2000 m in altitude (6500 ft).

Cyclone (4) A dynamically or thermally caused area of low atmospheric pressure with converging and ascending air flows. Rotates counterclockwise in the Northern Hemisphere and clockwise in the Southern Hemisphere (compare *anticyclone*; see *midlatitude cyclone, tropical cyclone*).

Daylength (2) Duration of exposure to insolation, varying during the year depending on latitude; an important aspect of seasonality.

Daylight saving time (1) Time is set ahead 1 hour in the spring and set back 1 hour in the fall in the Northern Hemisphere. In the United States and Canada time is set ahead on the first Sunday in April and set back on the last Sunday in October—only Hawai'i, Arizona, portions of Indiana, and Saskatchewan exempt themselves.

Debris avalanche (10) A mass of falling and tumbling rock, debris, and soil; can be dangerous because of the tremendous velocities achieved by the onrushing materials.

Declination (2) The latitude that receives direct overhead (perpendicular) insolation on a particular day; migrates annually through 47° of latitude between the Tropics of Cancer (23.5° N) and Capricorn (23.5° S).

Decomposer (16) Bacteria and fungi that digest organic debris outside their bodies, and absorb and release nutrients in an ecosystem (see *detritivores*).

Deficit (7) In a water balance, the amount of unmet, or unsatisfied, potential evapotranspiration (DEFIC).

Deflation (12) A process of wind erosion that removes and lifts individual particles, literally blowing away unconsolidated, dry, or noncohesive sediments (see *blowout depression*).

Delta (11) A depositional plain formed where a river enters a lake or an ocean;

named after the triangular shape of the Greek letter *delta*, Δ.

Denudation (10) A general term that refers to all processes that cause degradation of the landscape: weathering, mass movement, erosion, and transport.

Deposition (11) The process whereby weathered, wasted, and transported sediments are laid down; deposited by air, water, or ice.

Desert biome (16) Arid landscapes of uniquely adapted dry-climate plants and animals.

Desertification (12) The expansion of deserts worldwide, related principally to poor agricultural practices (overgrazing and inappropriate agricultural practices), improper soil-moisture management, erosion and salinization, deforestation, and the ongoing climatic change; an unwanted semi-permanent invasion into neighboring biomes.

Desert pavement (12) In arid landscapes, a surface formed when wind deflation and sheet flow remove smaller particles, leaving residual pebbles and gravels to concentrate at the surface; resembles a cobblestone street (see *deflation*, *sheet flow*).

Detritivores (16) Detritus feeders and decomposers that consume, digest, and destroy organic wastes and debris. *Detritus feeders*—worms, mites, termites, centipedes, snails, crabs, even vultures, among others—consume detritus and excrete nutrients and simple inorganic compounds that fuel an ecosystem (compare with *decomposers*).

Dew-point temperature (5) The temperature at which a given mass of air becomes saturated, holding all the water it can hold. Any further cooling or addition of water vapor results in active condensation.

Diagnostic subsurface horizon (15) A soil horizon that originates below the epipedon at varying depths; may be part of the A and B horizons; important in soil description as part of the Soil Taxonomy.

Differential weathering (10) The effect of different resistance in rock, coupled with variations in the intensity of physical and chemical weathering.

Diffuse radiation (3) The downward component of scattered incoming insolation from clouds and the atmosphere.

Discharge (11) The measured volume of flow in a river that passes by a given cross section of a stream in a given unit of time; expressed in cubic meters per second or cubic feet per second.

Dissolved load (11) Materials carried in chemical solution in a stream derived from minerals such as limestone and dolomite, or from soluble salts.

Downwelling current (4) An area of the sea where a convergence or accumulation of water thrusts excess water downward; occurs, for example, at the western end of the equatorial current or along the margins of Antarctica (compare *upwelling currents*).

Drainage basin (11) The basic spatial geomorphic unit of a river system; distinguished from a neighboring basin by ridges and highlands that form divides.

Drainage pattern (11) A geometric arrangement of streams in a region; determined by slope, differing rock resistance to weathering and erosion, climatic and hydrologic variability, and structural controls of the landscape.

Drawdown (7) See *cone of depression*.

Drumlin (14) A depositional landform related to glaciation that is composed of till and is streamlined in the direction of continental ice movement; blunt end upstream and tapered end downstream with a rounded summit.

Dry adiabatic rate (DAR) (5) The rate at which a parcel of air that is less than saturated cools (if ascending) or heats (if descending); a rate of 10 C° per 1000 m (5.5 F° per 1000 ft) (see also *adiabatic*).

Dunes (12) A depositional feature of sand grains deposited in transient mounds, ridges, and hills; extensive areas of sand dunes are called sand seas.

Dust dome (3) A dome of airborne pollution associated with every major city; may be blown by winds into elongated plumes downwind from the city.

Dynamic equilibrium model (10) The balancing act between tectonic uplift and reduction rates of erosion, between the resistance of crust materials and the work of denudation processes. Landscapes evidence ongoing adaptation to rock structure, climate, local relief, and elevation.

Earthquake (9) A sharp release of energy that produces shaking in Earth's crust at the moment of rupture along a fault or in association with volcanic activity. The moment magnitude scale expresses earthquake magnitude; Mercalli scale describes earthquake intensity (see *moment magnitude scale*).

Earth systems science (1) An emerging science of Earth as a complete, systematic entity. The study of an integrated set of chemical, biological, and physical systems that include the processes of a whole Earth system. A study of planetary change resulting from system operations; includes a desire for a more quantitive understanding among components, rather than qualitative description.

Ecological succession (16) The process whereby different and usually more complex assemblages of plants and animals replace older and usually simpler communities; communities are in a constant state of change as each species adapts to conditions; ecosystems do not exhibit a stable point or successional climax condition as previously thought (refer to *primary succession*, *secondary succession*).

Ecology (16) The science that studies the interrelationships among organisms and their environment and among various ecosystems.

Ecosphere (1) Another name for the *biosphere*.

Ecosystem (16) A self-regulating association of living plants, animals, and their nonliving physical and chemical environment.

Ecotone (16) A boundary transition zone between adjoining ecosystems that may vary in width and represent areas of tension as similar species of plants and animals compete for the resources (see *ecosystem*).

Effusive eruption (9) An eruption characterized by low viscosity, basaltic magma, with low-gas content readily escaping. Lava pours forth onto the surface with relatively small explosions and little tephra; tends to form shield volcanoes (see *shield volcano*, *lava*, *pyroclastic*; compare *explosive eruption*).

Elastic-rebound theory (9) A concept describing the faulting process, in which the two sides of a fault appear locked despite the motion of adjoining pieces of crust, but with accumulating strain they rupture suddenly, snapping to new positions relative to each other, generating an earthquake.

Electromagnetic spectrum (2) All the radiant energy produced by the Sun placed in an ordered range, divided according to wavelengths (see *wavelength*).

Eluviation (15) The downward removal of finer particles and minerals from the upper horizons of soil (compare *illuviation*).

Empirical classification (6) A climate classification based on weather statistics or other data; used to determine general climate categories (compare *genetic classification*).

Endogenic system (8) The system internal to Earth, driven by radioactive heat derived from sources within the planet. In response, the surface is fractured, mountain building occurs, and earthquakes and volcanoes are activated (compare *exogenic system*).

Entisols (15) A soil order in the Soil Taxonomy. Specifically lacks vertical development of horizons; usually young or undeveloped; found in active slopes, alluvial-filled floodplains, poorly drained tundra.

Environmental lapse rate (2, 5) The actual lapse rate in the lower atmosphere at any particular time under local weather conditions; may deviate above or below the average normal lapse rate of 6.4 C° per 1000 m (3.5 F° per 1000 ft); compare *normal lapse rate*.

Eolian (12) Caused by wind; refers to the erosion, transportation, and deposition of materials; spelled *aeolian* in some countries.

Epicenter (9) The surface area directly above the subsurface focus where movement along the fault plane was initiated. Shock waves radiate outward from the epicenter area.

Epipedon (15) The diagnostic soil horizon that forms at the surface; not to be confused with the A horizon; may include all or part of the illuviated B horizon.

Equal area (1) A trait of a map projection; indicates the equivalence of all areas on the surface of the map, although shape is distorted.

Equatorial and tropical rain forest (16) A lush biome of tall broadleaf evergreen trees and diverse plants and animals. The dense canopy of leaves usually is arranged in three levels.

Equatorial low-pressure trough (4) A thermally caused low-pressure area that almost girdles Earth, with air converging and ascending all along its extent; also called the intertropical convergence zone (ITCZ).

Equilibrium line (14) The area of a glacier where accumulation (gain) and ablation (loss) are balanced.

Erg desert (12) Sandy deserts, or areas where sand is so extensive that it constitutes a *sand sea*.

Erosion (11) Denudation by wind, water, and ice, which dislodges, dissolves, or removes surface material.

Esker (14) A sinuously curving, narrow deposit of coarse gravel that forms along a meltwater stream channel, developing in a tunnel beneath the glacier.

Estuary (11) The point at which the mouth of a river enters the sea and freshwater and seawater are mixed; a place where tides ebb and flow.

Eustasy (5) Refers to worldwide changes in sea level that are not related to movements of land but rather to a rise and fall in the volume of water in the oceans.

Eutrophication (16) A natural process in which lakes receive nutrients and sediment and become enriched; the gradual filling and natural aging of water bodies.

Evaporation (7) The movement of free water molecules away from a wet surface into air that is less than saturated; the phase change of water to water vapor.

Evaporation fog (5) A fog formed when cold air flows over the warm surface of a lake, ocean, or other body of water; forms as the water molecules evaporate from the water surface into the cold, overlying air; also known as a steam fog or sea smoke.

Evaporation pan (7) A weather instrument; a standardized pan from which evaporation occurs, with water automatically replaced and measured; an evaporimeter.

Evaporite (8) Chemical sediments formed from inorganic sources when water evaporates and leaves behind a residue of salts previously in solution.

Evapotranspiration (7) The merging of evaporation and transpiration water loss into one term (see *potential evapotranspiration* and *actual evapotranspiration*).

Exfoliation dome (10) A dome-shaped feature of weathering, produced by the response of granite to the overburden removal process, which relieves pressure from the rock. Layers of rock slough (sluff) off in slabs or shells in a sheeting process.

Exogenic system (8) The external surface system, powered by insolation, that energizes air, water, and ice and sets them in motion, under the influence of gravity. Includes all processes of landmass denudation (compare *endogenic system*).

Exosphere (2) An extremely rarefied outer atmospheric halo beyond the thermopause at an altitude of 480 km (300 mi); probably composed of hydrogen and helium atoms, with some oxygen atoms and nitrogen molecules present near the *thermopause*.

Explosive eruption (9) A violent and unpredictable eruption—the result of magma that is thicker, stickier (more viscous), and higher in gas content and silica than that of an effusive eruption; tends to form blockages within a volcano; produces composite volcanic landforms (see *composite volcano*; compare *effusive eruption*).

Faulting (9) The process whereby displacement and fracturing occurs between two portions of Earth's crust; usually associated with earthquake activity.

Feedback loop (1) A portion of the system output cycles back as an information input, causing changes that guide further system operations (see *negative feedback* and *positive feedback*).

Field capacity (7) Water held in the soil by hydrogen bonding against the pull of gravity, remaining after water drains from the larger pore spaces; the available water for plants (see *available water*; *capillary water*).

Fire ecology (16) Recognizes fire as a dynamic ingredient in community succession. Controlled fires secure plant reproduction and prevent the accumulation of forest litter and brush; widely regarded as a wise forest management practice.

Firn (14) Snow of a granular texture that is transitional in the slow transformation from snow to glacial ice; snow that has persisted through a summer season in the zone of accumulation.

Firn line (14) The snowline that is visible on the surface of a glacier, where winter snows survive the summer ablation season; analogous to a snowline on land (see *ablation*, *equilibrium line*).

Fjord (14) A drowned glaciated valley, or glacial trough, along a coast that is filled by the sea.

Flash flood (12) A sudden and short-lived torrent of water that exceeds the capacity of a stream channel; associated with desert and semiarid washes.

Flood (11) A high water level that overflows the natural (or artificial) banks along any portion of a stream.

Floodplain (11) A low-lying area near a stream channel, subject to recurrent flooding; alluvial deposits generally mask underlying rock.

Fluvial (11) Stream-related processes; from the Latin *fluvius* for "river" or "running water."

Fog (5) A cloud, generally stratiform, in contact with the ground, with visibility usually restricted to less than 1 km (3300 ft).

Folding (9) A process that bends and deforms beds of various rocks subjected to compressional forces.

Food chain (16) The circuit along which energy flows from producers, who manufacture their own food, to consumers; a one-directional flow of chemical energy, ending with detritivores.

Food web (16) A complex network of interconnected food chains.

Formation class (16) That portion of a biome that concerns the plant communities only, subdivided by size, shape, and structure of the dominant vegetation.

Friction force (4) The effect of drag by the wind as it moves across a surface; may be operative through 500 m (1640 ft) of altitude. Surface friction slows the velocity of the wind and therefore reduces the effectiveness of the Coriolis force.

Frost action (10) A powerful mechanical force produced as water expands 9% of its volume as it freezes and can exceed the tensional strength of rock, breaking it.

Funnel cloud (5) The visible swirl extending from the bottom side of a cloud, which may or may not develop into a tornado. A tornado is a funnel cloud that has extended all the way to the ground (see *tornado*).

Fusion (2) The process of forcibly joining, under extreme temperature and pressure, positively charged hydrogen and helium nuclei; occurs naturally in thermonuclear reactions within stars, such as our Sun.

Gelisols (15) A new soil order in the Soil Taxonomy, added in 1998, describing cold and frozen soils at high latitudes or high elevations; characteristic tundra vegetation (see *Canadian System of Soil Classification in* Appendix B for similar types).

General circulation model (GCM) (6) Complex computer-based climate models that produce generalizations of reality and predictive forecasts of future weather and climate conditions.

Genetic classification (6) A type of classification that uses causative factors to determine climatic regions; for example, an analysis of the effect of interacting air masses (compare *empirical classification*).

Geodesy (1) The science that determines Earth's shape and size through surveys, mathematical means, and remote sensing (see *geoid*).

Geographic information system (GIS) (1) A computer-based data processing tool or methodology used for gathering, manipulating, and analyzing geographic information to produce a holistic, interactive analysis.

Geography (1) The science that studies the interdependence among geographic areas, natural systems, processes, society, and cultural activities over space—a spatial science. The five themes of geographic education include: location, place, movement, regions, and human-Earth relationships.

Geoid (1) A word that describes Earth's shape, literally, "the shape of Earth is Earth-shaped." A theoretical surface at sea level that extends through the continents; deviates from a perfect sphere.

Geologic cycle (8) A general term characterizing the vast cycling (hydrologic, tectonic, and rock) in and on the lithosphere. It encompasses the hydrologic cycle, tectonic cycle, and rock cycle.

Geologic time scale (8) A listing of eras, periods, and epochs that span Earth's history; reflects the relative relationship of various layers of rock strata and the absolute dates as determined by scientific methods such as radioactive isotopic dating.

Geomagnetic reversal (8) With an uneven regularity the magnetic field fades to zero, then phases back to full strength, but with the magnetic poles reversed. Reversals have occurred nine times during the past 4 million years.

Geomorphic threshold (10) The threshold up to which landforms change before lurching to a new set of relationships, with rapid realignments of landscape materials and slopes.

Geomorphology (10) The science that analyzes and describes the origin, evolution, form, classification, and spatial distribution of landforms.

Geostrophic wind (4) Winds moving between pressure areas along paths that are parallel to the isobars. In the upper troposphere the pressure gradient force equals the Coriolis force so that the amount of deflection is proportional to air movement (see *isobar, pressure gradient force, Coriolis force*).

Geothermal energy (8) The energy potential of boiling steam produced by subsurface magma in near contact with groundwater. Active examples include Iceland, New Zealand, Italy, and northern California.

Glacial drift (14) The general term for all glacial deposits, both unsorted (*till*) and sorted (*stratified drift*).

Glacial ice (14) A hardened form of ice, very dense in comparison to normal snow or *firn*; under pressure within a glacier is capable of downhill movement.

Glacier (14) A large mass of perennial ice resting on land or floating shelflike in the sea adjacent to the land; formed from the accumulation and recrystallization of snow, which then flows slowly under the pressure of its own weight and the pull of gravity.

Glacial surge (14) The rapid, lurching movement of a glacier.

Gleization (15) A process of humus and clay accumulation in cold, wet climates with poor drainage.

Global positioning system (GPS) (1) Latitude, longitude, and elevation are accurately calibrated using this handheld instrument that calibrates radio signals from satellites.

Goode's homolosine projection (Appendix A) An equal-area projection developed in 1925 by Dr. Paul Goode; combining two oval projections: the holographic and the sinusoidal; presented interrupted in the text.

Graben (9) Pairs or groups of faults that produce downward faulted blocks; characteristic of the basins of the interior western United States (see *horst* and *Basin and Range Province*).

Graded stream (11) A condition in a stream of mutual adjustment between the load carried by a stream and the related landscape through which the stream flows, forming a state of dynamic equilibrium among erosion, transported load, deposition, and the stream's capacity.

Gradient (11) The drop in elevation from a stream's headwaters to its mouth, ideally forming a concave slope.

Granite (8) A coarse-grained (slow-cooling) intrusive igneous rock of 25% quartz and more than 50% potassium and sodium feldspars; characteristic of the continental crust (compare *basalt*).

Gravitational water (7) That portion of surplus water that percolates downward from the capillary zone, pulled by gravity to the groundwater zone.

Gravity (2) The mutual force exerted by the masses of objects that are attracted one to another; produced in an amount proportional to each object's mass.

Great circle (1) Any circle of circumference drawn on a globe with its center coinciding with the center of the globe. An infinite number of great circles can be drawn, but only one parallel is a great circle—the equator (compare *small circle*).

Greenhouse effect (3) The process whereby radiatively active gases (carbon dioxide, water vapor, methane, and CFCs) absorb insolation and reradiate the energy at longer wavelengths, which are retained longer, delaying the loss of infrared to space. Thus, the lower troposphere is warmed through the radiation and reradiation of infrared wavelengths. The approximate similarity between this process and that of a greenhouse explains the name.

Greenwich Mean Time (GMT) (1) Former world standard time, now known as Coordinated Universal Time (UTC) (see *Coordinated Universal Time*).

Ground ice (14) Subsurface water that is frozen in regions of permafrost. The moisture content of areas with ground ice may vary from nearly absent in regions of drier permafrost to almost 100% in saturated soils.

Groundwater (7) Water beneath the surface that is beyond the soil-root zone; a major source of potable water.

Groundwater mining (7) Pumping an aquifer beyond its capacity to flow and recharge; an overuse of the groundwater resource.

Gulf Stream (3) A strong northward-moving warm current off the east coast of North America, which carries its water far into the North Atlantic.

Habitat (16) That physical location in which an organism is biologically suited to live. Most species have specific habitat parameters, or limits.

Hail (5) A type of precipitation formed when a raindrop is repeatedly circulated above and below the freezing level in a cloud, with each cycle adding more ice to the hailstone until it becomes too heavy to stay aloft.

Herbivore (16) The primary consumer in a food chain, which eats plant material formed by a producer that has synthesized organic molecules (compare *carnivore*).

Heterosphere (2) A zone of the atmosphere above the mesopause, 80 km (50 mi) in altitude; composed of rarified layers of oxygen atoms and nitrogen molecules; includes the ionosphere.

Histosols (15) A soil order in the Soil Taxonomy. Formed from thick accumulations of organic matter, such as beds of former lakes, bogs, and layers of peat.

Homosphere (2) A zone of the atmosphere from the surface up to 80 km (50 mi), composed of an even mixture of gases, including nitrogen, oxygen, argon, carbon dioxide, and trace gases.

Horn (14) A pyramidal, sharp-pointed peak that results when several cirque glaciers gouge an individual mountain summit from all sides.

Horst (9) Upward-faulted blocks produced by pairs or groups of faults; characterized by the mountain ranges of the interior of the western United States (see *graben* and *Basin and Range Province*).

Hot spots (8) An individual point of upwelling material originating in the asthenosphere; tend to remain fixed relative to migrating plates; some 100 are identified worldwide; exemplified by Yellowstone National Park, Hawai'i, and Iceland.

Human-Earth relationships (1) One of the oldest themes of geography (the human-land tradition); includes the spatial analysis of settlement patterns, resource utilization and exploitation, hazard perception and planning, and the impact of environmental modification and artificial landscape creation.

Humidity (5) Water vapor content of the air. The capacity of the air to hold water vapor is mostly a function of the water vapor temperature and air temperature.

Humus (15) A mixture of organic debris in soil, worked by consumers and decomposers in the humification process; characteristically formed from plant and animal litter laid down at the surface.

Hurricane (5) A tropical cyclone that is fully organized and intensified in inward-spiraling rainbands; ranges from 160 to 960 km (100 to 600 mi) in diameter, with wind speeds in excess of 119 kmph (65 knots, or 74 mph); a name used specifically in the Atlantic and eastern Pacific (compare *typhoon*).

Hydration (10) A physical weathering process involving water, although not involving any chemical change; water is added to a mineral, which initiates swelling and stress within the rock, mechanically forcing grains apart as the constituents expand (contrast to *hydrolysis*).

Hydraulic action (11) The erosive work accomplished by the turbulence of water; causes a squeezing and releasing action in joints in bedrock; capable of prying and lifting rocks.

Hydrograph (11) A graph of stream discharge over a period of time (minutes, hours, days, years) at a specific place on a stream. The relationship between stream discharge and precipitation input is illustrated on the graph.

Hydrologic cycle (7) A simplified model of the flow of water and water vapor from place to place as energy powers system operations. Water flows through the atmosphere, across the land, where it is also stored as ice, and within groundwater.

Hydrology (11) is the science of water, its global circulation, distribution, and properties, specifically water at and below Earth's surface.

Hydrolysis (10) A chemical weathering process in which minerals chemically combine with water; a decomposition process that causes silicate minerals in rocks to break down and become altered (contrast with *hydration*).

Hydrosphere (1) An abiotic open system that includes all of Earth's water.

Ice age (14) A cold episode, with accompanying alpine and continental ice accumulations, that has repeated roughly every 200 to 300 million years since the late Precambrian era (1.25 billion years ago); includes the most recent episode during the Pleistocene Ice Age, which began 1.65 million years ago.

Iceberg (14) Floating ice created by calving ice (a large piece breaking off) or large pieces breaking off and floating adrift; a hazard to shipping because the ice is 91% submerged.

Ice cap (14) A dome-shaped glacier, less extensive than an ice sheet ($<$50,000 km^2, 19,300 mi^2), although it buries mountain peaks and the local landscape.

Ice field (14) An extensive form of land ice, with mountain ridges and peaks visible above the ice; less than an ice cap or ice sheet.

Ice sheet (14) An enormous continuous continental glacier. The bulk of glacial ice on Earth covers Antarctica and Greenland in two ice sheets.

Igneous rock (8) One of the basic rock types; it has solidified and crystallized from a hot molten state (either magma or lava). (Compare *metamorphic rock, sedimentary rock.*)

Illuviation (15) The downward movement and deposition of finer particles and minerals from the upper horizon of the soil; a depositional process. Deposition usually is in the B horizon, where accumulations of clays, aluminum, carbonates, iron, and some humus occur (compare *eluviation;* see *calcification*).

Impermeability (7) Subsurface structure that obstructs water flows in the groundwater system.

Inceptisols (15) An order in the Soil Taxonomy. Weakly developed soils that are inherently infertile; usually young soils that are weakly developed, although they are more developed than Entisols.

Industrial smog (2) Air pollution associated with coal-burning industries; may contain sulfur oxides, particulates, carbon dioxide, and exotics.

Infiltration (7) Water access to subsurface regions of soil moisture storage through penetration of the soil surface.

Insolation (2) Incoming solar radiation that is intercepted by Earth.

International Date Line (1) The 180° meridian; an important corollary to the prime meridian at 0°, Greenwich, London, England; established by the treaty of 1884.

Intertropical convergence zone (ITCZ) (4) See *equatorial low-pressure trough.*

Ionosphere (2) A layer in the atmosphere above 80 km (50 mi) where gamma, X-ray, and some ultraviolet radiation is absorbed and converted into infrared, or heat energy, and where the solar wind stimulates the auroras.

Isobar (4) An isoline connecting all points of equal pressure.

Isostasy (8) A state of equilibrium formed by the interplay between portions of the lithosphere and the asthenosphere; the crust depresses with weight and recovers with the melting of the ice or removal of the load, in isostatic rebound.

Isotherm (3) An isoline connecting all points of equal atmospheric temperature.

Jet stream (4) The most prominent movement in upper-level westerly wind flows; irregular, concentrated, sinuous bands of *geostrophic wind,* traveling at 300 kmph (190 mph).

Joints (10) Fractures or separations in rock without displacement of the sides; increases the surface area of rock exposed to weathering processes.

Kame (14) A depositional feature of glaciation; a small hill of poorly sorted sand and gravel that accumulates in crevasses or in ice-caused indentations in the surface.

Karst topography (10) Distinctive topography formed in a region of chemically weathered limestone with poorly developed surface drainage and solution features that appear pitted and bumpy; originally named after the Kr̆s Plateau of Yugoslavia.

Katabatic wind (4) Air drainage from elevated regions, flowing as gravity winds. Layers of air at the surface cool, become denser, and flow downslope; known worldwide by many local names.

Kettle (14) Forms when an isolated block of ice persists in a ground moraine, an outwash plain, or a valley floor after a glacier retreats; as the block finally melts, it leaves behind a steep-sided hole that frequently fills with water.

Kinetic energy (2) The energy of motion in a body; derived from the vibration of the body's own movement and stated as temperature.

Köppen-Geiger climate classification (Appendix C) An empirical climate classification system based on average monthly temperature, average monthly precipitation, and total annual precipitation. Capital letters A, B, C, D, E, and H designate climate categories.

Lacustrine deposit (14) Deposit associated with lake-level fluctuations; for example, benches or terraces marking former shorelines.

Lagoon (13) A portion of coastal seawater that is virtually cut off from the ocean by a bay barrier or barrier beach; also, the water surrounded and enclosed by an atoll.

Land-sea breeze (4) Wind along coastlines and adjoining interior areas created by different heating characteristics of land and water surfaces—onshore (landward) breeze in the afternoon and offshore (seaward) breeze at night.

Landslide (10) A sudden, rapid, downslope movement of a cohesive mass of regolith and/or bedrock in a variety of mass-movement forms under the influence of gravity—form of *mass movement.*

Land-water heating differences (3) The differences in the way land and water heat, as a result of contrasts in transmission, evaporation, mixing, and specific heat capacities. Land surfaces heat and cool faster than water and are characterized as having aspects of continentality, whereas water is regarded as producing a marine influence.

Latent heat (5) Heat energy that is absorbed or released in the phase change of water and is stored in one of the three states—ice, water, or water vapor; includes the latent heat of melting, freezing, vaporization, evaporation, and condensation.

Latent heat of condensation (5) The heat energy released in a phase change from water vapor to liquid; 585 calories are released from 1 gram of water vapor that condenses at 20°C (68°F).

Latent heat of evaporation (5) The heat energy required to change phase from liquid to water vapor; under normal sea-level pressure, 585 calories must be added to 1 gram of water (at 20°C) to achieve a phase change to water vapor.

Latent heat of vaporization (5) The heat energy absorbed from the environment in a phase change from liquid to water vapor at the boiling point; under normal sea-level pressure, 540 calories must be added to each gram of boiling water to achieve a phase change to water vapor.

Lateral moraine (14) Debris transported by a glacier that accumulates along the sides of the glacier and is deposited along these margins.

Laterization (15) A pedogenic process operating in well-drained soils that are found in warm and humid regions; typical of Oxisols. High precipitation values leach soluble minerals and soil particles; soils are usually reddish or yellowish colors.

Latitude (1) The angular distance measured north or south of the equator from a point at the center of Earth. A line connecting all points of the same latitudinal angle is called a parallel (compare *longitude*).

Lava (8, 9) Magma that issues from volcanic activity onto the surface; the extrusive rock that results when magma solidifies (see *magma*).

Life zone (16) An altitudinal zonation of plants and animals that form distinctive communities. Each life zone possesses its own temperature and precipitation relations.

Lightning (5) Flashes of light, caused by tens of millions of volts of electrical charge igniting the air to temperatures of 15,000°C to 30,000°C (*see thunder*).

Limestone (8) The most common chemical sedimentary rock (nonclastic); lithified calcium carbonate ($CaCO_3$), which is very susceptible to chemical weathering by acids in the environment, including carbonic acid in rainfall.

Limiting factor (16) The physical or chemical factor that most inhibits (either through lack or excess) biotic processes.

Lithification (8) The compaction, cementation, and hardening of sediments into sedimentary rock.

Lithosphere (1) Earth's crust, and that portion of the uppermost mantle directly below the crust, that extends down to 70 km (45 mi). Some use this term to refer to the entire Earth.

Littoral zone (13) A specific coastal environment; that region between the high water line during a storm and a depth at which storm waves are unable to move sea-floor sediments.

Loam (15) A mixture of sand, silt, and clay in almost equal proportions, with no one texture dominant; an ideal agricultural soil.

Location (1) A basic theme of geography dealing with the absolute and relative position of people, places, and things on Earth's surface.

Loess (12) Large quantities of fine-grained clays and silts left as glacial outwash deposits; subsequently blown by the wind great distances and redeposited as a generally unstratified, homogeneous blanket of material covering existing landscapes; in China, loess originated from desert lands.

Longitude (1) The angular distance measured east or west of a prime meridian from a point at the center of Earth. A line connecting all points of the same longitude is called a meridian.

Longshore current (13) A current that forms parallel to a beach as waves arrive at an angle to the shore; generated in the surf zone by wave action, transporting large amounts of sand and sediment.

Lysimeter (7) A weather instrument; device for measuring potential and actual evapotranspiration; isolates a portion of a field so that the moisture moving through the plot is measured.

Magma (8) Molten rock from beneath the surface of Earth; fluid, gaseous, under tremendous pressure, and either intruded into country rock or extruded onto the surface as lava (see *lava*).

Magnetosphere (2) Earth's magnetic force field, which is generated by dynamolike motions within the planet's outer core; deflects the solar wind toward each pole.

Mangrove swamp (13) Wetland ecosystems between 30° N or S and the equator; tend to form a distinctive community of mangrove plants (compare *salt marsh*).

Mantle (8) An area within the planet representing about 80% of Earth's total volume, with densities increasing with depth; occurs above the core and below the crust; is rich in iron and magnesium oxides and silicates.

Map (1, Appendix A) A generalized view of an area, usually some portion of Earth's surface, as seen from above at a greatly reduced size (see *scale* and *map projection*).

Map projection (1, Appendix A) The reduction of a spherical globe onto a flat surface in some orderly and systematic realignment of the latitude and longitude grid.

Marine effect (3) Regions that are dominated by the moderating effects of the ocean and that exhibit a smaller minimum and maximum temperature range than continental stations (see *land-water heating differences*).

Mass movement (10) All unit movements of materials propelled by gravity; can range from dry to wet, slow to fast, small to large, and free-falling to gradual or intermittent; sometimes used interchangeably with *mass wasting*.

Meandering stream (11) The sinuous, curving pattern common to graded streams, with the outer portion of each curve subjected to the greatest erosive action and the inner portion receiving sediment deposits (see *graded stream*).

Mean sea level (MSL) (13) The average of tidal levels recorded hourly at a given site over a long period of time, which must be at least a full lunar tidal cycle.

Medial moraine (14) Debris transported by a glacier that accumulates down the middle of the glacier when two glaciers merge and their lateral moraines combine; forms a depositional feature following glacial retreat.

Mediterranean shrubland (16) A major biome dominated by Mediterranean dry summer climates and characterized by scrub and short, stunted, tough (sclerophyllous) forests (see *chaparral*).

Mercator projection (1) A cylindrical, true-shape projection developed by Gerardus Mercator in 1569. All straight lines represent true constant directions, or rhumb lines. It presents false notions of the size (area) of midlatitude and poleward landmasses but presents true compass direction (see *rhumb line*).

Mercury barometer (4) A device that measures air pressure with a column of mercury in a tube that is inserted in a vessel of mercury.

Meridian (1) See *longitude*; a line designating an angle of longitude.

Mesocyclone (5) A large rotating circulation initiated within a parent cumulonimbus cloud at the mid-troposphere level; generally produces heavy rain, large hail, blustery winds, and lightning; may lead to tornado activity.

Mesosphere (2) The upper region of the homosphere from 50 to 80 km (30 to 50 mi) above the ground; designated by temperature criteria and very low pressures.

Metamorphic rock (8) Existing rock, both igneous and sedimentary, that goes through profound physical and chemical changes under increased pressure and temperature. Constituent mineral structures may exhibit foliated or nonfoliated textures.

Meteorology (5) The scientific study of the atmosphere that includes a study of the atmosphere's physical characteristics and motions, related chemical, physical, and geological processes, the complex linkages of atmospheric systems, and weather forecasting.

Methane (6) A radiatively active gas that participates in the greenhouse effect; derived from the organic processes of burning, digesting, and rotting in the presence of oxygen; CH_4.

Microclimatology (3) The study of climates at or near Earth's surface.

Midlatitude broadleaf and mixed forest (16) A biome in moist continental climates in areas of warm-to-hot summers and cool-to-cold winters; relatively lush stands of broadleaf forests trend northward into needleleaf evergreen stands.

Midlatitude cyclone (5) An organized area of low pressure, with converging and ascending air flow producing an interaction of air masses; migrates along storm tracks. Such lows or depressions form the dominant weather pattern in the middle and higher latitudes of both hemispheres (see *wave cyclone*).

Midlatitude grassland (16) The major biome most modified by human activity; so named because of the predominance of grass-like plants, although deciduous broadleafs appear along streams and other limited sites; location of the world's breadbaskets of grain and livestock production.

Mid-ocean ridge (8) Submarine mountain ranges that extend more than 65,000 km (40,000 mi) worldwide and average more than 1000 km (620 mi) in width; centered along sea-floor spreading centers (see *sea-floor spreading*).

Milky Way Galaxy (2) A flattened, disk-shaped mass estimated to contain up to 400 billion stars; includes our Solar System.

Miller cylindrical projection (Appendix A) A compromise map projection that avoids the severe distortion of the Mercator projection (see *map projection*).

Mineral (8) An element or combination of elements that form an inorganic natural compound; described by a specific formula and qualities of a specific nature.

Model (1) A simplified version of a system, representing an idealized part of the real world.

Mohorovičić discontinuity (8) The boundary between the crust and the rest of the lithospheric upper mantle; named for the Yugoslavian seismologist Mohorovičić; a zone of sharp material and density contrasts, also called the Moho.

Moist adiabatic rate (MAR) (5) The rate at which a parcel of saturated air cools in ascent; a rate of 6 C° per 1000 m (3.3 F° per 1000 ft). This rate may vary with moisture content and temperature, from 4 C° to 10 C° per 1000 m (2 F° to 6 F° per 1000 ft) (see *adiabatic*; compare *dry adiabatic rate*).

Moisture droplet (5) Initial composition of clouds. Each droplet measures approximately 0.002 cm (0.0008 in.) in diameter and is invisible to the human eye.

Mollisols (15) A soil order in the Soil Taxonomy classification that has a humus-rich organic content high in alkalinity; some of the world's most significant agricultural soils.

Moment magnitude scale (9) An earthquake magnitude scale. Considers the amount of fault slippage, the size of the area that ruptured, and the nature of the materials that faulted in estimating the magnitude of an earthquake. Replaces the Richter scale (amplitude magnitude), especially valuable in assessing larger magnitude events.

Monsoon (4) From the Arabic word *mausim*, meaning "season"; refers to an annual cycle of dryness and wetness, with seasonally shifting winds produced by changing atmospheric pressure systems; affects India, Southeast Asia, Indonesia, northern Australia, and portions of Africa.

Moraine (14) Marginal glacial deposits (lateral, medial, terminal, ground) of unsorted and unstratified material.

Mountain-valley breeze (4) A light wind produced as cooler mountain air flows downslope at night, and as warmer valley air flows upslope during the day.

Movement (1) A major theme in geography involving migration, communication, and the interaction of people and processes across space.

Natural levees (11) Long, low ridges that occur on either side of a river in a developed floodplain; depositional byproducts (coarse gravels and sand) of river-flooding episodes.

Needleleaf forest (16) Forests of pine, spruce, fir, and larch, stretching from the east coast of Canada westward to Alaska and continuing from Siberia westward across the entire extent of Russia to the European Plain; called the taiga (a Russian word) or the boreal forest; principally in the humid continental and subarctic climates. Includes montane forests that may be at lower latitudes at elevation.

Negative feedback (1) A feedback loop that tends to slow or dampen response in a system; promotes self-regulation in a system; far more common than positive feedback in living systems (see *feedback loop, positive feedback*).

Net primary productivity (16) The net photosynthesis (photosynthesis minus respiration) for a given community; considers all growth and all reduction factors that affect the amount of useful chemical energy (*biomass*) fixed (chemically bound) in an *ecosystem*.

Net radiation (NET R) (3) The net all-wave radiation available; the final outcome of the radiation balance process between incoming and outgoing shortwave and longwave energy.

Niche (16) The basic function, or occupation, of a lifeform within a given community; the way an organism obtains its food, air, and water (compare *habitat*).

Nickpoint (knickpoint) (11) The point at which the longitudinal profile of a stream is abruptly broken by a change in gradient; for example, a waterfall, rapids, or cascade.

Nimbostratus (5) A rain-producing, dark, grayish stratiform cloud characterized by gentle drizzles.

Nitrogen dioxide (2) A reddish-brown choking gas produced in high-temperature combustion engines; can be damaging to human respiratory tracts and to plants; participates in photochemical reactions and acid deposition.

Normal fault (9) A type of geologic fault in rocks. Tension produces strain that breaks a rock with one side moving vertically relative to the other side along an inclined fault plane (compare *reverse fault*).

Normal lapse rate (2, 5) The average rate of temperature decrease with increasing altitude in the lower atmosphere; an average value of 6.4 C° per km, or 1000 m (3.5 F° per 1000 ft).

Occluded front (5) In a cyclonic circulation, the overrunning of a surface warm front by a cold front and the subsequent lifting of the warm air wedge off the ground; initial precipitation is moderate to heavy.

Ocean basins (9) The physical containers (a depression in the lithosphere) for Earth's oceans.

Oceanic trench (8) The deepest single features of Earth's crust; associated with *subduction* zones. The deepest is the Mariana Trench near Guam, which descends to 11,033 m (36,198 ft).

Omnivore (16) A consumer that feeds on both producers (plants) and consumers (meat)—a role occupied by humans, among other animals compare *consumer, producer*).

Open system (1) A system with inputs and outputs crossing back and forth between the system and the surrounding environment. Earth is an open system in terms of energy (compare *closed system*).

Orogenesis (9) The process of mountain building that occurs when large-scale compression leads to deformation and uplift of the crust; literally the birth of mountains.

Orographic lifting (5) The uplift of migrating air masses in response to the physical presence of a mountain, a topographic barrier. The lifted air cools adiabatically as it moves upslope; may form clouds and produce increased precipitation (see *rain shadow*).

Outgassing (5) The release of trapped gases from rocks, forced out through cracks, fissures, and volcanoes from within Earth; the terrestrial source of Earth's water.

Outwash plain (14) Glaciofluvial deposits of stratified drift from meltwater-fed, braided, and overloaded streams; beyond a glacier's morainal deposits.

Oxbow lake (11) A lake that was formerly part of the channel of a meandering stream; isolated when a stream eroded its outer bank, forming a cutoff through the neck of a looping meander (see *meandering stream*).

Oxidation (10) A chemical weathering process whereby oxygen oxidizes (combines with) certain metallic elements to form oxides; most familiar as the "rusting" of iron in a rock or soil (Ultisols, Oxisols), that produces a reddish-brown stain of iron oxide (Fe_2O_3).

Oxisols (15) A soil order in the Soil Taxonomy. Tropical soils that are old, deeply developed, and lacking in horizons wherever well-drained; heavily weathered, low in cation exchange capacity, and low in fertility.

Ozone layer (2) See *ozonosphere*.

Ozonosphere (2) A layer of ozone (O_3) occupying the full extent of the stratosphere 20 to 50 km (12 to 30 mi) above the surface; the region of the atmosphere where ultraviolet wavelengths are principally absorbed and converted into heat. Its thinning and depletion relates to human use of CFCs and other compounds. International treaties strive to facilitate recovery and foster protection for the ozonosphere.

Paleoclimatology (6, 14) The science that studies the climates of past ages.

Paleolake (14) An ancient lake, such as Lake Bonneville or Lake Lahonton, associated with former wet periods when the lake basins were filled to higher levels than today.

PAN (2) See *peroxyacetyl nitrates*.

Pangaea (8) The supercontinent formed by the collision of all continental masses approximately 225 million years ago; named by Wegener (1912) in his continental drift theory (see *plate tectonics*).

Parallel (1) See *latitude*; a line, parallel to the equator, that designates an angle of latitude.

Parent material (10) The unconsolidated material, from both organic and mineral sources, that is the basis of soil development.

Particulate matter (PM) (2) Dust, dirt, soot, salt, sulfate aerosols, fugitive natural particles, or other material particles suspended in air.

Paternoster lake (14) One of a series of small, circular, stair-stepped lakes, formed in individual rock basins aligned down the course of a glaciated valley; named because they look like a string of rosary (religious) beads.

Pedon (15) A soil profile extending from the surface to the lowest extent of plant roots or to the depth where regolith or bedrock is encountered; imagined as a hexagonal column; the basic soil sampling unit.

Percolation (7) The process by which water permeates through to the subsurface environment; vertical water movement through soil or porous rock.

Periglacial (14) Cold-climate processes, landforms, and topographic features along the margins of glaciers, past and present, that occupy more than 20% of Earth's land surface; including *permafrost*, *frost action*, and *ground ice*.

Perihelion (2) That point in Earth's elliptical orbit about the Sun where Earth is closest to the Sun; occurs on January 3 at 147,255,000 km (91,500,000 mi); variable over a 100,000-year cycle (compare *aphelion*).

Permafrost (14) Forms when soil or rock temperatures remain below 0°C (32°F) for at least 2 years in areas considered periglacial; criterion is based on temperature and not on whether water is present (see *periglacial*).

Permeablity (7) The ability of water to flow through soil or rock; a function of the texture and structure of the medium.

Peroxyacetyl nitrates (PAN) (2) A pollutant formed from photochemical reactions involving nitric oxide (NO) and volatile organic compounds (VOCs). PAN produces no known human health effects but is particularly damaging to plants.

Phase change (5) The change in phase, or state, between ice, water, and water vapor; involves the absorption or release of latent heat.

Photochemical smog (2) Air pollution produced by the interaction of ultraviolet light, nitrogen dioxide, and hydrocarbons; produces ozone and PAN through a series of complex photochemical reactions. Automobiles are the major source of the contributive gases.

Photosynthesis (16) The joining of carbon dioxide and hydrogen in plants, under the influence of certain wavelengths of visible light; releases oxygen and produces energy-rich organic material (sugars and starches).

Physical geography (1) The science concerned with the spatial aspects and interactions of the physical elements and processes that make up the environment: energy, air, water, weather, climate, landforms, soils, animals, plants, microorganisms, and Earth.

Physical weathering (10) The breaking and disintegrating of rock without any chemical alteration; sometimes referred to as mechanical or fragmentation weathering.

Pioneer community (16) The initial plant community in an area; usually found on new surfaces or those that have been stripped of life—for example, surfaces created by mass movements of land or land disturbed by human activities.

Place (1) A major theme in geography focused on the tangible and intangible characteristics that make each location unique.

Plane of the ecliptic (2) A plane intersecting all the points of Earth's orbit.

Planetesimal hypothesis (2) Early protoplanets formed from the condensing masses of a nebular cloud of dust, gas, and icy comets; a formation process now being observed in other parts of the galaxy.

Planimetric map (Appendix A) A basic map showing the horizontal position of boundaries, land-use activities, and political, economic, and social outlines.

Plate tectonics (8) The conceptual model that encompasses continental drift, sea-floor spreading, and related aspects of crustal movement; accepted as the foundation of crustal tectonic processes.

Plateau basalts (9) An accumulation of horizontal flows formed when lava spreads out from elongated fissures onto the surface in extensive sheets; associated with effusive eruptions; also known as flood basalts (see *basalt*).

Playa (12) An area of salt crust left behind by evaporation on a desert floor, usually in the middle of a bolson or valley; intermittently wet and dry.

Plinthite (15) An ironstone hardpan structure formed in Oxisol subsurfaces or surface horizons that are exposed to repeated wetting and drying sequences; quarried in some areas for building materials.

Pluton (8) A mass of intrusive igneous rock that has cooled slowly in the crust; forms in any size or shape. The largest partially exposed pluton is a *batholith*.

Podzolization (15) A pedogenic process in cool, moist climates; forms a highly leached soil with strong surface acidity because of humus from acid-rich trees.

Point bar (11) The inner portion of a stream meander that receives sediment fill.

Polar easterlies (4) Variable weak, cold, and dry winds moving away from the polar region; an anticyclonic circulation.

Polar front (4) A significant zone of contrast between cold and warm air masses; roughly situated between 50° and 60° N and S latitude.

Polar high-pressure cells (4) A weak, anticyclonic, thermally produced pressure system positioned roughly above each pole.

Polypedon (15) The identifiable soil in an area, with distinctive characteristics differentiating it from surrounding polypedons forming the basic soil mapping unit; composed of many pedons.

Porosity (7) The total volume of available pore space in soil; a result of the texture and structure of the soil.

Positive feedback (1) Feedback that amplifies or encourages responses in a system (see *negative feedback*, *feedback loop*).

Potential evapotranspiration (7) POTET, or PE; the amount of moisture that would evaporate and transpire if adequate moisture were available; the amount lost under optimum moisture conditions—the moisture demand.

Precipitation (7) Rain, snow, sleet, and hail—the moisture supply (PRECIP).

Pressure gradient force (4) Causes air to move from areas of higher barometric pressure to areas of lower barometric pressure due to pressure differences.

Primary succession (19) Succession that occurs among plant species in an area of new surfaces created by mass movement of land, areas exposed by a retreating glacier, cooled lava flows and volcanic eruption landscapes, surface mining and clear-cut logging scars, or an area of sand dunes, with no trace of a former community (see *primary community*).

Prime meridian (1) An arbitrary meridian designated as 0° longitude; the point from which longitude is measured east or west; agreed to in an 1884 treaty.

Process (1) A set of actions and changes that occur in some special order; analysis of processes is central to modern geographic study.

Producers (16) Organisms that are capable of using carbon dioxide as their sole source of carbon, which they chemically fix through photosynthesis to provide their own nourishment; also called an autotroph.

Pyroclastic (9) An explosively ejected rock fragment launched by a volcanic eruption; sometimes described by the more general term tephra.

Radiation fog (5) Formed by radiative cooling of the surface, especially on clear nights in areas of moist ground; occurs when the air layer directly above the surface is chilled to the dew-point temperature, thereby producing saturated conditions.

Rain gauge (7) A standardized device that catches and measures rainfall.

Rain shadow (5) The area on the leeward slopes of a mountain range; in the shadow of the mountains, where precipitation receipt is greatly reduced compared to windward slopes (see *orographic lifting*).

Reflection (3) The portion of arriving energy that returns directly back to space without being converted into heat or performing any work (see *albedo*).

Refraction (3) The bending effect that occurs when insolation enters the atmosphere or another medium; the same process by which a crystal, or prism, disperses the component colors of the light passing through it.

Region (1) A geographic theme that focuses on areas that display unity and internal homogeneity of traits; includes the study of how a region forms, evolves, and interrelates with other regions.

Regolith (10) Partially weathered rock overlying bedrock, whether residual or transported.

Relative humidity (5) The ratio of water vapor actually in the air (content) compared to the maximum water vapor the air is able to hold (capacity) at that temperature; expressed as a percentage (compare *vapor pressure*, *specific humidity*).

Relief (9) Elevation differences in a local landscape; an expression of local height difference of landforms.

Remote sensing (1) Information acquired from a distance, without physical contact with the subject; for example, photography, orbital imagery, or radar.

Respiration (16) The process by which plants derive energy for their operations; essentially, the reverse of the photosynthetic process; releases carbon dioxide, water, and heat into the environment (compare *photosynthesis*).

Reverse fault (9) Or compression fault, one side of the fault moves upward vertically in comparison to the other side (compare *normal fault*).

Revolution (2) The annual orbital movement of Earth about the Sun; determines year length and the length of seasons.

Rhumb line (1) A line of constant compass direction, or constant bearing, that crosses all meridians at the same angle. A portion of a *great circle*.

Robinson projection (Appendix A) An oval projection developed by A. Robinson in 1963.

Roche moutonnée (14) A glacial erosion feature; an asymmetrical hill of exposed bedrock; displays a gently sloping upstream side that has been smoothed and polished by a glacier and an abrupt, steep downstream side.

Rock (8) An assemblage of minerals bound together (like granite), or may be a mass of a single mineral (like rock salt).

Rock cycle (8) A model representing the interrelationships among the three rock-forming processes—igneous, sedimentary, and metamorphic—shows how each can be transformed into another rock type.

Rockfall (10) Free-falling movement of debris from a cliff or steep slope, generally falling straight down or bounding downslope.

Rossby waves (4) Undulating horizontal motions in the upper-air westerly circulation at middle and high latitudes.

Rotation (2) The turning of Earth on its axis; averages 24 hours in duration; determines day-night relation; counterclockwise when viewed from above the North Pole; from above the equator, west to east or eastward.

Salinity (13) The concentration of natural elements and compounds dissolved in solution, as solutes; measured by weight in parts per thousand (‰) in seawater.

Salinization (15) A pedogenic regime that results from high potential evapotranspiration rates in deserts and semiarid regions. Soil water is drawn to surface horizons, and dissolved salts are deposited as the water evaporates.

Saltation (11, 12) The transport of sand grains (usually larger than 0.2 mm, or 0.008 in.) by stream or wind, bouncing the grains along the ground in asymmetrical paths.

Salt marsh (13) Wetland ecosystem characteristic of latitudes poleward of the 30th parallel (compare *mangrove swamp*).

Sand sea (12) An extensive area of sand and dunes; characteristic of Earth's erg deserts.

Saturated (5) Air that is holding all the water vapor that it can hold at a given temperature.

Scale (1) The ratio of the distance on a map to that in the real world; expressed as a representative fraction, graphic scale, or written scale.

Scarification (10) Human-induced mass movements of Earth materials, such as large-scale open-pit mining and strip mining.

Scattering (3) Deflection and redirection of insolation by atmospheric gases, dust, ice, and water vapor; the shorter the wavelength, the greater the scattering, thus skies in the lower atmosphere are blue.

Scientific method (1) An approach that uses applied common sense in an organized and objective manner; based on observation, generalization, formulation of a hypothesis, and ultimately the development of a theory.

Sea-floor spreading (8) As proposed by Hess and Dietz, the mechanism driving the movement of the continents; associated with upwelling flows of magma along the worldwide system of mid-ocean ridges (see *mid-ocean ridge*).

Secondary succession (19) Succession that occurs among plant species in an area where vestiges of a previously functioning community are present; an area where the natural community has been destroyed or disturbed, but where the underlying soil remains intact.

Sediment (10) Fine-grained mineral matter that is transported and deposited by air, water, or ice.

Sedimentary rock (8) One of three basic rock types; formed from the compaction, cementation, and hardening of sediments derived from former rocks (compare *igneous rock*, *metamorphic rock*).

Seismic waves (8) The shock waves sent through the planet by an earthquake or underground nuclear test. Transmission varies according to temperature and the density of various layers within the planet; provide indirect diagnostic evidence of Earth's internal structure.

Seismograph (9) A device that measures seismic waves of energy transmitted throughout Earth's interior or along the coast.

Sensible heat (2) Heat measured with a thermometer; a measure of the concentration of kinetic energy from molecular motion.

Sheet flow (11) Water that moves downslope in a thin film as overland flow, not concentrated in channels larger than rills.

Sheeting (10) A form of weathering associated with fracturing or fragmentation of rock by pressure release; often related to exfoliation processes (see *exfoliation dome*).

Shield volcano (9) A symmetrical mountain landform built from effusive eruptions; gently sloped, gradually rising from the surrounding landscape to a summit crater; typical of the Hawaiian Islands (see *effusive eruption*; compare *composite volcano*).

Sinkhole (10) Nearly circular depressions created by the weathering of karst landscapes; also known as a doline; may collapse through the roof of an underground space (see *karst topography*).

Slopes (10) Curved, inclined surfaces that bound landforms.

Small circle (1) Circles on a globe's surface that do not share Earth's center; for example, all parallels other than the equator (compare *great circle*).

Soil (15) A dynamic natural body made up of fine materials covering Earth's surface in which plants grow; composed of both mineral and organic matter.

Soil colloid (15) Tiny clay and organic particles in soil; provide chemically active sites for mineral ion adsorption (see *cation-exchange capacity*).

Soil creep (10) A persistent mass movement of surface soil where individual soil particles are lifted and disturbed by the expansion of soil moisture as it freezes, or even by grazing livestock or digging animals.

Soil fertility (15) The ability of soil to support plant productivity when it contains organic substances and clay minerals that absorb water and certain elemental ions needed by plants (see *cation-exchange capacity*).

Soil horizon (15) The various layers exposed in a pedon; roughly parallel to the surface and identified as O, A, E, B, and C, residing above bedrock designated R.

Soil-moisture recharge (7) Water entering available soil storage.

Soil-moisture storage (7) The retention of moisture within soil; represents a savings account that can accept deposits (soil moisture recharge) or experience withdrawals (soil moisture utilization) as conditions change (acronym Δ STRGE).

Soil-moisture utilization (7) Plants extract moisture for their needs; increasing inefficiency of withdrawal as the soil storage is reduced.

Soil science (15) Interdisciplinary science of soils. Pedology concerns the origin, classification, distribution, and description of soil. Edaphology focuses on soil as a medium for sustaining higher plants.

Soil Taxonomy (15) A soil classification system based on observable soil properties actually seen in the field; published in 1975 by the U.S. Soil Conservation Service, revised in 1990 and 1998 by the Natural Resources Conservation Service to include 12 soil orders (compare *Canadian System of Soil Classification*, Appendix B).

Soil-water budget (7) An accounting system for soil moisture using inputs of precipitation and outputs of evapotranspiration and gravitational water.

Solar constant (2) The amount of insolation intercepted by Earth on a surface perpendicular to the Sun's rays when Earth is at its average distance from the Sun; a value of 1370 watts per m^2, or 1.968 calories/cm^2 per minute; averaged over the entire globe at the thermopause.

Solar wind (2) Clouds of ionized (charged) gases emitted by the Sun and traveling in all directions from the Sun's surface. Effects on Earth include auroras, disturbance of radio signals, and possible influences on weather.

Solum (15) A true soil profile in the *pedon*; ideally, a combination of A and B horizons.

Spatial (1) The nature or character of physical space, as in an area; occupying or operating within a space. Geography is a spatial science; spatial analysis its essential approach.

Spatial analysis (1) The examination of spatial interactions, patterns, and variations over area and/or space; a key integrative approach of geography.

Specific heat (3) The increase of temperature in a material when energy is absorbed; water is said to have a higher specific heat than a comparable volume of soil or rock.

Specific humidity (5) The mass of water vapor (in grams) per unit mass of air (in kilograms) at any specified temperature. The maximum mass of water vapor that a kilogram of air can hold at any specified temperature is termed its maximum specific humidity (compare *vapor pressure*, *relative humidity*).

Speed of light (2) Specifically, 299,792 km per second (186,282 mi per second), covering more than 9.4 trillion km per year (5.9 trillion mi per year)—a distance known as a light-year; at light speed Earth is 8 minutes and 20 seconds from the Sun.

Spheroidal weathering (10) A chemical weathering process in which the sharp edges and corners of boulders and rocks are weathered in thin plates, creating a rounded, spheroidal form.

Spodosols (15) A soil order in the Soil Taxonomy classification that occurs in northern coniferous forests; best developed in cold, moist, forested climates in humid continental or subarctic regions; lacks humus and clay in the A horizons, with high acidity associated with podzolization processes.

Stability (5) The condition of a parcel, whether it remains where it is or changes its initial position. The parcel is stable if it resists displacement upward; it is unstable if it continues to rise.

Steady-state equilibrium (1) The condition that occurs in a system when rates of input and output are equal and the amounts of energy and stored matter are nearly constant around a stable average.

Stomata (16) Small openings on the undersides of leaves, through which water and gases pass.

Storm surge (5) Large quantities of seawater pushed inland by the strong winds associated with a tropical cyclone.

Stratified drift (14) Sediments deposited by glacial meltwater that appear sorted; a specific form of *glacial drift* (compare *till*).

Stratigraphy (8) An analysis of the sequence, spacing, and spatial distribution of rock strata.

Stratocumulus (5) A lumpy, grayish, low-level cloud, patchy with sky visible, sometimes present at the end of the day.

Stratosphere (2) That portion of the *homosphere* that ranges from 20 to 50 km (12.5 to 30 mi) above Earth's surface; temperatures range from −57°C (−70°F) at the tropopause to 0°C (32°F) at the stratopause. The functional *ozonosphere* is within the stratosphere.

Stratus (5) A stratiform (flat, horizontal) cloud generally below 2000 m (6500 ft).

Strike-slip fault (9) Horizontal movement along a faultline, that is, movement in the same direction as the fault; also known as a transcurrent fault. Such movement is described as right-lateral or left-lateral, depending on the relative motion observed.

Subduction zone (8) An area where two plates of crust collide and the denser oceanic crust dives beneath the less dense continental plate, forming deep oceanic trenches and seismically active regions (see *oceanic trench*).

Sublimation (5) A process in which ice evaporates directly to water vapor or water vapor freezes to ice (deposition).

Subpolar low-pressure cell (4) A region of low pressure centered approximately at 60° latitude in the North Atlantic near Iceland and in the North Pacific near the Aleutians, as well as in the Southern Hemisphere. Air flow is cyclonic; it weakens in summer and strengthens in winter.

Subsolar point (2) The only point receiving perpendicular insolation at a given moment—the Sun directly overhead (see *declination*).

Subtropical high-pressure cells (4) Dynamic high-pressure areas covering roughly the region from 20° to 35° N and S latitudes; responsible for the hot, dry areas of Earth's arid and semiarid deserts.

Sulfate aerosols (2) Sulfur compounds in the atmosphere, principally sulfuric acid; principal sources relate to fossil fuel combustion; scatters and reflects insolation.

Sulfur dioxide (2) A colorless gas detected by its pungent odor; produced by the combustion of fossil fuels that contain sulfur as an impurity (forming SO_2); can react in the atmosphere to form sulfuric acid (H_2SO_4), a component of acid deposition.

Summer (June) solstice (2) The time when the Sun's declination is at the Tropic of Cancer, at 23.5° N latitude; June 20–21 each year (compare to *winter solstice*).

Sunspots (2) Magnetic disturbances on the surface of the Sun; occurring in an average 11-year cycle; related flares, prominences, and outbreaks produce surges in solar wind.

Surface creep (12) A form of *eolian* transport that involves particles too large for *saltation*; a process whereby individual grains are impacted by moving grains and slide and roll.

Surplus (7) SURPL; the amount of moisture that exceeds potential evapotranspiration; moisture oversupply when soil moisture storage is at field capacity.

Suspended load (11) Fine particles held in suspension in a stream. The finest particles are not deposited until the stream velocity nears zero.

Swell (13) Regular patterns of smooth, rounded waves in open water; can range from small ripples to very large waves.

Syncline (9) A trough in folded strata, with beds that slope toward the axis of the downfold.

System (1) Any ordered, interrelated set of materials or items existing separate from the environment, or within a boundary; energy transformations and energy and matter storage and retrieval occur within a system.

Taiga (16) See *needleleaf forest*.

Tarn (14) A small mountain lake, especially one that collects in a cirque basin behind risers of rock material.

Tectonic processes (8) Driven by internal energy from within Earth; refers to large-scale movement and deformation of the crust.

Temperate rain forest (16) A major biome of lush forests at middle and high latitudes; occurs along narrow margins of the Pacific Northwest in North America, among other locations; includes the tallest trees in the world.

Temperature (3) A measure of sensible heat energy in the atmosphere and other media, indicates the average kinetic energy of individual molecules within the atmosphere.

Temperature inversion (2) A reversal of the normal decrease of temperature with increasing altitude; can occur anywhere from ground level up to several thousand meters; functions to block atmospheric convection and thereby trap pollutants.

Terranes (9) A migrating crustal piece, dragged about by processes of mantle convection and plate tectonics. Displaced terranes are distinct in their history, composition, and structure from the continents that accept them.

Terrestrial ecosystem (16) A self-regulating association characterized by specific plant formations; usually named for the predominant vegetation and known as a biome when large and stable (see *biome*).

Thermal equator (3) An isoline on an isothermal map that connects all points of highest mean temperature.

Thermopause (2) A zone approximately 480 km (300 mi) in altitude that serves conceptually as the top of the atmosphere; an altitude used for the determination of the solar constant.

Thermosphere (2) A region of the heterosphere extending from 80 to 480 km (50 to 300 mi) in altitude; contains the functional ionosphere layer.

Thrust fault (9) A low-angle fault plane, relative to the horizontal, where an overlying block shifts (thrusts) over an underlying block.

Thunder (5) The violent expansion of suddenly heated air, created by lightning discharges, sending out shock waves as an audible sonic bang.

Tide (13) Patterns of daily oscillations in sea level produced by astronomical relationships among the Sun, the Moon, and Earth; experienced in varying degrees around the world.

Till (14) Direct ice deposits that appear unstratified and unsorted; a specific form of *glacial drift* (compare *stratified drift*).

Till plains (14) Large, relatively flat plains composed of unsorted glacial deposits behind a terminal or end moraine. Low-rolling relief and unclear drainage patterns are characteristic.

Tombolo (13) A landform created when coastal sand deposits connect the shoreline with an offshore island outcrop or sea stack.

Topographic map (Appendix A) A map that portrays physical relief through the use of elevation contour lines that connect all points at the same elevation above or below a vertical datum, such as mean sea level.

Topography (9) The undulations and configurations that give Earth's surface its texture; the heights and depths of local relief, including both natural and human-made features.

Tornado (5) An intense, destructive cyclonic rotation, developed in response to extremely low pressure; associated with mesocyclone formation.

Total runoff (7) Surplus water that flows across a surface toward stream channels; formed by sheet flow, combined with precipitation and subsurface flows into those channels.

Traction (11) A type of sediment transport that drags coarser materials along the bed of a stream (see *bed load*).

Trade winds (4) Northeast and southeast winds that converge in the equatorial low pressure trough, forming the intertropical convergence zone.

Transmission (3) The passage of shortwave and longwave energy through space, the atmosphere, or water.

Transparency (3) The quality of a medium (air, water) that allows light to shine through it.

Transpiration (7) The movement of water vapor out through the pores in leaves; the water is drawn by plant roots from the soil moisture storage.

Transport (11) The actual movement of weathered and eroded materials by air, water, and ice.

Tropical cyclone (5) A cyclonic circulation originating in the tropics, with winds between 30 and 64 knots (39 to 73 mph); characterized by closed isobars, circular organization, and heavy rains (see *hurricane* and *typhoon*).

Tropical savanna (16) A major biome containing large expanses of grassland interrupted by trees and shrubs; a transitional area between the humid rain forests and tropical seasonal forests and the drier, semiarid tropical steppes and deserts.

Tropical seasonal forest and scrub (16) A variable biome on the margins of the rain forests, occupying regions of lesser and more erratic rainfall; the site of transitional communities between the rain forests and tropical grasslands.

Tropic of Cancer (2) The northernmost point of the Sun's declination during the year; 23.5° N latitude.

Tropic of Capricorn (2) The southernmost point of the Sun's declination during the year; 23.5° S latitude.

Troposphere (2) The home of the biosphere; the lowest layer of the homosphere, containing approximately 90% of the total mass of the atmosphere; extends up to the *tropopause*, defined by a temperature of −57°C (−70°F); occurring at an altitude of 18 km (11 mi) at the equator, 13 km (8 mi) in the middle latitudes, and at lower altitudes near the poles.

True shape (1) A map property showing the correct configuration of coastlines; a useful trait of conformality for navigational and aeronautical maps, although areal relationships are distorted (see *map projection*; compare *equal area*).

Tsunami (13) A seismic sea wave, traveling at high speeds across the ocean, formed by sudden and sharp motions in the seafloor, such as sea-floor earthquakes, submarine landslides, or eruptions from undersea volcanoes.

Typhoon (5) A tropical cyclone in excess of 65 knots (74 mph) that occurs in the western Pacific; same as a hurricane except for location.

Ultisols (15) A soil order in the Soil Taxonomy. Features highly weathered forest soils, principally in the humid subtropical climatic classification. Increased weathering and exposure can degenerate an Alfisol into the reddish color and texture of these more humid, tropical soils. Fertility is quickly exhausted when Ultisols are cultivated.

Unconfined aquifer (7) The zone of saturation in water-bearing rock strata, with no impermeable overburden, and recharge generally accomplished by water percolating down from above.

Undercut bank (11) A steep bank formed along the outer portion of a meandering stream; produced by lateral, erosive, undercutting action of a stream (compare *point bar*).

Uniformitarianism (8) An assumption that physical processes active in the environment today are operating at the same pace and intensity that has characterized them throughout geologic time; proposed by Hutton and Lyell.

Upslope fog (5) Forms when moist air is forced to higher elevations along a hill or mountain and is thus cooled (compare *valley fog*).

Upwelling current (4) Ocean currents that cause surface waters to be swept away from the coast by surface divergence or offshore winds. Cool, deep waters, which are generally nutrient rich, rise to replace the vacating water (compare *downwelling current*).

Urban heat island (3) Urban microclimates, which are warmer on the average than areas in the surrounding countryside because of various surface characteristics.

Valley fog (5) The settling of cooler, more dense air in low-lying areas; produces saturated conditions and fog.

Vapor pressure (5) That portion of total air pressure that results from water vapor molecules; expressed in millibars (mb). At a given temperature the maximum capacity of the air is termed its saturation vapor pressure.

Ventifacts (12) The individual pieces and pebbles etched and smoothed by eolian erosion—abrasion by wind-blown particles.

Vernal (March) equinox (2) The time when the Sun's declination crosses the equatorial parallel and all places on Earth experience days and nights of equal length; around March 20–21 each year. The Sun rises at the North Pole and sets at the South Pole (compare *autumnal equinox*).

Vertisols (15) A soil order in the Soil Taxonomy. Features expandable clay soils; composed of more than 30% swelling clays; occurs in regions that experience highly variable soil moisture balances through the seasons.

Volcano (9) A landform at the end of a conduit or pipe that rises from below the crust and vents to the surface. Magma rises and collects in a magma chamber deep below, resulting in eruptions that are effusive or explosive, forming the mountain landform.

Warm front (5) The leading edge of an advancing warm air mass, that is unable to push cooler, passive air out of the way; tends to push the cooler, underlying air into a wedge shape; noted on weather maps with a series of rounded knobs placed along the front in the direction of the frontal movement (compare *cold front*).

Wash (12) An intermittently dry streambed that fills with torrents of water after rare precipitation events in arid lands.

Watershed (11) The catchment area of a drainage basin; delimited by divides (see *drainage basin*).

Waterspout (5) An elongated, funnel-shaped circulation formed when a tornado takes place over water.

Water table (7) The upper limit of groundwater; that contact point between zones of saturation and aeration in an unconfined aquifer (see *zone of aeration*, *zone of saturation*).

Wave (13) Undulations of ocean water produced by the conversion of solar energy to wave energy; produced in a generating region or a stormy area of the sea.

Wave-cut platform (13) A flat, or gently sloping, table-like bedrock surface that develops in the tidal zone, where wave action cuts a bench that extends from the cliff base out into the sea.

Wave cyclone (5) An organized area of low pressure, with converging and ascending air flow producing an interaction of air masses; migrates along storm tracks. Such lows or depressions form the dominant weather pattern in the middle and higher latitudes of both hemispheres (see *midlatitude cyclone*).

Wavelength (2) Measurement of waveform; the actual distance between the crests of successive waves. The number of waves passing a fixed point in one second is called the frequency of the wavelength (see *electromagnetic spectrum*).

Wave refraction (13) A process that concentrates energy on headlands and disperses it in coves and bays; a coastal straightening process.

Weather (5) The short-term condition of the atmosphere, as compared to climate, which reflects long-term atmospheric conditions and extremes. Temperature, air pressure, relative humidity, wind speed and direction, daylength, and Sun angle are important measurable elements that contribute to the weather.

Weathering (10) The related processes by which surface and subsurface rock disintegrate, or dissolve, or are otherwise broken down. Rocks at or near Earth's surface are exposed to physical, organic, and chemical weathering processes.

Westerlies (4) The predominant wind flow pattern from the subtropics to high latitudes in both hemispheres.

Wetland (13) A narrow, vegetated strip occupying many coastal areas and estuaries worldwide; highly productive ecosystems with an ability to trap organic matter, nutrients, and sediment.

Wilting point (7) That point in the soil moisture balance when only hygroscopic water and some bound capillary water remains. Plants wilt and eventually die after prolonged stress from a lack of available water.

Wind (4) The horizontal movement of air relative to Earth's surface; produced essentially by air pressure differences from place to place; also influenced by the Coriolis force and surface friction. Turbulence adds wind updrafts and downdrafts and a vertical component to this definition.

Wind vane (4) A weather instrument used to determine wind direction; winds are named for the direction from which they originate.

Winter (December) solstice (2) That time when the Sun's declination is at the Tropic of Capricorn, at 23.5° S latitude, December 21–22 each year. The day is 24 hours long south of the Antarctic Circle. The night is 24 hours long north of the Arctic Circle (compare to Summer [June] solstice).

Zone of aeration (7) A groundwater zone above the water table, which may or may not hold water in pore spaces.

Zone of saturation (7) A groundwater zone below the water table, in which all pore spaces are filled with water.

Index